METHODS IN MOLECULAR BIOPHYSICS

Second Edition

Current techniques for studying biological macromolecules and their interactions are based on the application of physical methods, ranging from classical thermodynamics to more recently developed techniques for the detection and manipulation of single molecules. Reflecting the advances made in biophysics research over the past decade, and now including a new section on medical imaging, this new edition describes the physical methods used in modern biology.

All the key techniques are covered, including mass spectrometry, hydrodynamics, microscopy and imaging, diffraction and spectroscopy, electron microscopy, molecular dynamics simulations, and nuclear magnetic resonance. Each method is explained in detail using examples of real-world applications. Short asides are provided throughout to ensure that explanations are accessible to life scientists, physicists, and those with medical backgrounds.

The book remains an unparalleled and comprehensive resource for graduate students of biophysics and medical physics in science and medical schools, as well as for research scientists looking for an introduction to techniques from across this interdisciplinary field.

Nathan R. Zaccai is a Research Associate at the Cambridge Institute for Medical Research, University of Cambridge. His current research focuses on the molecular and thermodynamic basis of the transport and presentation at cell surfaces of proteins involved in pathogen evasion and host immunity.

Igor N. Serdyuk (1939–2012) was Professor of Molecular Biology and Head of the Laboratory of Nucleoprotein Physics at the Institute of Protein Research, Pushchino, Russia.

Joseph Zaccai is Directeur de Recherche Emeritus at the Centre Nationale de la Recherche Scientifique and Visiting Scientist at the Institut Laue-Langevin and Institut de Biologie Structurale, Grenoble. His current research interests include the exploration of the role of dynamics and physical chemical limits for life. He has many years of experience in teaching biophysics to biologists, medical students, and physicists.

REVIEWS FROM THE FIRST EDITION

I first asked what methods in molecular biophysics I would expect to use as a biochemist and structural biologist. This text book provides an introduction to the physics of each of [the techniques used by my own group] as well as a review of the applications. . . . [It] will be in demand by third year undergraduates in the many courses run by physicists to introduce them to biological themes. It would also be used by the many post-graduate students doing . . . research degrees as well as post-doctorals in chemical biology, biochemistry, cell biology and structural biology research groups. . . . In summary, this is a valuable contribution to the field. . . . this is an area which has advanced tremendously and the major texts in biophysical methods are now simply out of date. The text covers the methods that young researchers and some undergraduates will wish to learn. I am sure that it will find itself on the shelves of many laboratories throughout the world. There is nothing quite like it at the moment.

Sir Tom Blundell FRS, FMedSci, Professor and Head, Department of Biochemistry, University of Cambridge

Thank you very much for giving me the opportunity to preview this wonderful text book. It has outstanding breadth while maintaining sufficient depth to follow modern experiments or initiate a deeper understanding of a new subject area. I love the 'Physicist's' and 'Biologist's Boxes' to address specific subjects for researchers with different backgrounds. This is one of the most comprehensive and highly relevant texts on biophysics that I have encountered in the last 10 years, clearly written and up-to-date. It is a must-have for biophysicists working in all lines of research, and certainly for me.

Nikolaus Grigorieff, Professor of Biochemistry, Brandeis University

[This is] a wonderful up-to-date treatise on the many and diverse methods used . . . in the fields of molecular biophysics, physical biochemistry, molecular biology, biological physics and the new and emerging field of quantum nanobiology. The wide range of methods available . . . in these multidisciplinary fields has been overwhelming for most researchers, students and scientists [who fail] to fully appreciate the utility and usefulness of the methods [other than their own]. [In many cases, this has] created disagreements and . . . controversy. The only way to understand and appreciate fully the problems in quantum nanobiology and their complexity is to utilize and fully understand the many diverse methods covered by the authors in this very fine treatise . . . [It] should be in the library of any serious researcher in the many diverse multidisciplinary fields working on problems in quantum nanobiology. . . . They will be greatly rewarded by an ability to see and view the problems and their complexity through different perspectives, aspects and points of view, . . .

Karl J. Jalkanen, Associate Professor of Biophysics, Quantum Protein Centre, Technological University of Denmark

This most welcome text provides an up-to-date introduction to the vast field of biophysical methods. Written in an accessible style with an eye to a broad audience, it will appeal to biologists who wish to understand how to determine how macromolecules function and to scientists with a physics or physical chemistry background who wish to know how measurement of the physical world can impact our understanding of biological problems. The book succeeds in unifying disparate approaches under the aegis of developing an understanding of how macromolecules work.

Importantly, the text also provides the relevant historical background, an invaluable guide that will aid in the appreciation of what has gone before and should serve to orient them towards the future and what may be possible. It is a valuable resource for novice and seasoned biophysicists alike.

Dan Minor, California Institute for Quantitative Biomedical Research University of California, San Francisco

Methods in Molecular Biophysics is now the book I consult first when faced with an unfamiliar experimental technique. Both classic analytical techniques and the latest single-molecule methods appear in this single comprehensive reference.

Philip Nelson, Department of Physics, University of Pennsylvania, and author of Biological Physics

The authors provide an overview of many of the major recent accomplishments in the use of physical tools to investigate biological structure. There are interesting historical and biographical comments that lead the reader into understanding contemporary concepts and results. The book will be valuable both for students and research scientists.

Michael G. Rossmann, Hanley Professor of Biological Sciences, Purdue University

The melding of physics, chemistry and biology in modern science has changed our view of the natural world and opened avenues for detailed understanding of the origin of biological regulation. *Methods in Molecular Biophysics* provides an up-to-date view of classical biophysics, theory and practice of modern chemical biology and represents an essential text for the interdisciplinary scientist of the 21st Century. A great achievement and presentation awaits the student who reads this book, along with an excellent reference for the seasoned practitioner of biophysical chemistry.

Milton H Werner, Laboratory of Molecular Biophysics, The Rockefeller University

The methods, concepts, and discoveries of molecular biophysics have penetrated deeply into the fabric of modern biology. Physical methods that were once seemingly arcane are now commonplace in modern cell biology laboratories. This well written, thorough, and elegantly illustrated book provides the connections between molecular biophysics and biology that every aspiring young biologist needs. At the same time, it will serve physical scientists as a guide to the key ideas of modern biology.

Stephen H. White, Professor, Department of Physiology and Biophysics, University of California at Irvine

Methods in Molecular Biophysics offers a well-written, modern and comprehensive coverage of the properties of biological macromolecules and the techniques used to elucidate these properties. The authors have done a great service to the biophysics community in providing a long-needed update and expansion of previous texts on analysis of biological macromolecules. The choice and organization of material is especially well done. This book will be of considerable value not only to students, but also, due to the scope and breadth of coverage, to experienced researchers. I enthusiastically recommend *Methods in Molecular Biophysics* to anyone who wishes to know more about the techniques by which the properties of biological macromolecules are determined.

David Worcester, Department of Biological Sciences, University of Missouri – Columbia

METHODS IN MOLECULAR BIOPHYSICS

Structure, Dynamics, Function for Biology and Medicine
Second Edition

NATHAN R. ZACCAI
University of Cambridge

IGOR N. SERDYUK
Formerly of the Institute of Protein Research, Pushchino, Moscow Region

JOSEPH ZACCAI
Institut Laue-Langevin

CAMBRIDGE
UNIVERSITY PRESS

University Printing House, Cambridge CB2 8BS, United Kingdom

One Liberty Plaza, 20th Floor, New York, NY 10006, USA

477 Williamstown Road, Port Melbourne, VIC 3207, Australia

4843/24, 2nd Floor, Ansari Road, Daryaganj, Delhi – 110002, India

79 Anson Road, #06-04/06, Singapore 079906

Cambridge University Press is part of the University of Cambridge.

It furthers the University's mission by disseminating knowledge in the pursuit of
education, learning, and research at the highest international levels of excellence.

www.cambridge.org
Information on this title: www.cambridge.org/9781107056374
10.1017/9781107297227

© Nathan R. Zaccai, Joseph Zaccai, and Igor N. Serdyuk 2017

First published 2017

Printed in the United Kingdom by Clays, St Ives plc in 2017

A catalog record for this publication is available from the British Library.

Library of Congress Cataloging-in-Publication Data
Names: Zaccai, Nathan R., author. | Serdyuk, Igor N., author. | Zaccai, G. (Giuseppe), author.
Title: Methods in molecular biophysics : structure, dynamics, function for Biology and
Medicine / Nathan R. Zaccai, Igor N. Serdyuk, Joseph Zaccai.
Description: Second edition. | Cambridge, United Kingdom ; New York, NY, USA :
Cambridge University Press, 2017. | Igor N. Serdyuk's name appears first in the
previous edition. | Includes bibliographical references and index.
Identifiers: LCCN 2016046859 | ISBN 9781107056374 (hardback : alk. paper)
Subjects: | MESH: Biophysical Phenomena | Chemistry Techniques, Analytical–methods |
Diagnostic Imaging | Macromolecular Substances–chemistry
Classification: LCC QH505 | NLM QT 34 | DDC 571.4–dc23 LC record available at https://lccn.loc.gov/2016046859

ISBN 978-1-107-05637-4 Hardback

To Ol'ga, Brinda, Missy

CONTENTS IN BRIEF

CONTENTS

Contents

PART K MEDICAL IMAGING 627

PREFACE TO THE FIRST EDITION

André Guinier, whose fundamental discoveries contributed to the X-ray diffraction methods that are the basis of modern structural molecular biology, died in Paris at the beginning of July 2000, only a few weeks after it was announced in the press that a human genome had been sequenced. The sad coincidence serves as a reminder of the intimate connection between physical methods and progress in biology. Not long after, Max Perutz, Francis Crick and then David Blow, the youngest of the early protein crystallographers, passed away. The period marked the gradual closing of the era in which molecular biology was born and the opening of a new era. In what has been called the post-genome sequencing era, physical methods are now increasingly being called upon to play an essential role for the understanding of biological function at the molecular and cellular levels.

Molecular biophysics classical text books published in the previous decades have been overtaken not only by significant developments in existing methods such as those brought about by the advent of synchrotrons for X-ray crystallography or higher magnetic fields in NMR, but also by totally new methods with respect to biological applications, such as mass spectrometry and single molecule detection and manipulation. Our ambition in attempting this book was to be as up-to-date and exhaustive as possible. In their respective parts, we covered classical and advanced methods based on mass spectrometry, thermodynamics, hydrodynamics, spectroscopy, microscopy, radiation scattering, electron microscopy, molecular dynamics and NMR. But rapid progress in the field (we couldn't very well ask the biophysics community to stop working during the few years it takes to write and prepare a book!), and the requirement to keep the book to a manageable size meant that certain methods are either omitted or not perfectly up-to-date.

The key-word in molecular biophysics is *complementarity*. The Indian story of the six blind men and the elephant (see Frontispiece) is an appropriate metaphor for the field. Each of the blind men touched a different part of the elephant, and concluded on its nature: a big snake said the man who touched the trunk, the tusks were spears, its side a great wall, the tail a paint brush, the ears huge fans, the legs were tree trunks. We could add a seventh very short-sighted man to the story who can see the whole elephant but as a blurred grey cloud to illustrate diffraction methods. As we wrote in the Introduction, the ideal molecular biophysics method does not exist. It would be capable of observing not only the positions of atoms in molecules *in vivo*, but also the atomic motions and conformational changes that occur as the molecules are involved in the chemical and physical reactions associated with their biological function, regardless of the timescale involved. No single experimental technique is capable of yielding this information. Each provides us with a partial field of view with its clear regions and areas in deep shadow. In the 21st century, physical methods have to cope with very complicated biological problems, whose solution will depend on the ability to transfer structural and functional knowledge from the operation of a single molecule to the cellular level, and then to the whole organism. The splendor and complexity of the task is humbling, but the challenge will be met.

We are deeply obliged to Professor Don Engelman of Yale University, USA, and Professor Pierre Joliot of the Institut de Biologie Physico Chimique, France, who agreed to write forewords for the book. Outstanding scientists and teachers, each is both major actor and observer in biophysical research and the development of modern biology. We are very grateful to Brinda Muthusamy who painted the frontispiece. Grateful thanks also to expert colleagues for critical discussions on the different methods: Martin Blackledge and the members of the NMR laboratory, Christine Ebel, Dick Wade, Hugues Lortat-Jacob, Patricia Amara, the members of the laboratory of mass spectrometry, all of the Institut de Biologie Structurale, France, Regine Willumeit of the GKSS, Forschungszentrum Geesthacht, Germany, Victor Aksenov of the Joint Institute of Nuclear Research, Russia, Lesley Greene, Christina Redfield, Guillaume Stewart-Jones, Yvonne Jones and David Stuart of the University of Oxford, UK, Jonathan Ruprecht and Richard Henderson of the Laboratory of Molecular Biology, UK, Simon Hanslip and Robert Falconer of the University of Cambridge, UK. We gratefully acknowledge support from the Radulf Oberthuer Foundation, Germany, the Institut de Biologie Structurale and the Institut Laue Langevin, France, and the Cyril Serdyuk Company, Ukraine. We are indebted to Gennadiy Yenin of the Institute of Protein Research, Russia for drawing figures and scientific illustrations, and to Aleksandr Timchenko, Margarita. Shelestova, Margarita Ivanova, Tatyana Kuvshinkina, and Albina Ovchinnikova (Institute of Protein Research, Russia) for technical assistance. And finally, we would like to thank all our colleagues, friends and families, and the staff of Cambridge University Press, who supported us with much patience, understanding and encouragement.

Igor N. Serdyuk, Nathan R. Zaccai, Joseph Zaccai
August 2005

PREFACE TO THE SECOND EDITION

As we wrote in the preface to the first edition, our ambition in attempting *Methods in Molecular Biophysics* was to be as up-to-date and exhaustive as possible, considering the rapid progress in the field. Judging by broad readers' responses, the book usefully fulfilled its mission. The historical introduction to each method and "physicist" and "biologist" boxes were especially appreciated. Criticism focused on the inclusion of methods which even if once important are no longer topical, and relative inattention to emerging methods that were subsequently proven to be very powerful. Scientific predictions are, of course, particularly difficult to make, especially as progress may come from difficult to foreknow technical breakthroughs. The development of new detector systems, which now permit approaching atomic resolution in cryo-electron microscopy, comes to mind. The unwieldy size and weight of the first edition also invited justified criticism (it is interesting to note that the Russian edition is in two tomes). In this second edition, we have chosen a different book format that we hope will be easier and more pleasant to handle. We have carefully gone through the text to reorganize, bring up-to-date, and prune each of the chapters. We have added a new section on medical imaging so that the book now includes the range of topics covered in most medical school biophysics courses.

To the list of colleagues gratefully acknowledged in the first edition preface, we would like to add Frank Gabel, Institut de Biologie Structurale, Grenoble, for his critical reading of the first edition to suggest corrections and improvements, and expert colleagues who checked the updates, revisions, and additions in the second edition: Elisabetta Boeri Erba, Martin Blackledge, Dimitrios Skoufieas of the Institut de Biologie Structurale, Grenoble; Harriet Crawley-Snowdon, James Edgar, Antoni Wrobel, of the Cambridge Institute for Medical Research, University of Cambridge; Antony Fitzpatrick, Laboratory of Molecular Biology, Cambridge; Massimo Antognozzi, School of Physics, University of Bristol; Lotte Stubkjaer Fog, Medical Physicist, Section for Radiotherapy, Oncology Clinic, Rigshospitalet, Copenhagen; Alberto Bravin, Bio-medical Beam Line, European Synchrotron Radiation Facility, Grenoble; Jeremy Smith, Governor's Chair and Director, University of Tennessee/Oak Ridge National Laboratory Center for Molecular Biophysics.

Many thanks also to our friends and families, and the staff of Cambridge University Press, who supported the project with much patience, understanding, and encouragement.

It is with sadness that we recall the memory of Igor Serdyuk, our co-author, who died suddenly in Spring 2012.

Nathan R. Zaccai, Joseph Zaccai
June 2016

INTRODUCTION

Molecular Biophysics at the Beginning of the Twenty-First Century: From Ensemble Measurements to Single-Molecule Detection

The ideal biophysical method would be capable of measuring atomic positions in molecules *in vivo*. It would also permit visualization of the structures that form throughout the course of conformational changes or chemical reactions, regardless of the timescale involved. At present there is no single experimental technique that can yield this information.

1 A BRIEF HISTORY AND PERSPECTIVES

Molecular biology was born with the double-helix model for DNA, which provided a superbly elegant explanation for the storage and transmission mechanisms of genetic information (Fig. 1). The model by **J.D. Watson** and **F.H.C. Crick** and supporting fiber diffraction studies by **M.H.F. Wilkins, A.R. Stokes**, and **H.R Wilson**, and **R. Franklin** and **R.G. Gosling**, were published in a series of papers in the **April 25, 1953** issue of *Nature*, and marked a major triumph of the physical approach to biology.

The Watson and Crick model was based only in part on data from X-ray fiber diffraction diagrams. The patterns, which demonstrated the presence of a helical structure of constant pitch and diameter, could not provide unequivocal proof for a more precise structural model. One of the "genius" aspects of the discovery was the realization that A–T and G–C base pairs have identical dimensions; as the rungs of the double-helix ladder, they give rise to a constant diameter and pitch. From a purely "diffraction physics" point of view, a variety of helical models was compatible with the fiber diffraction diagram, and other authors proposed an alternative model for DNA, the so-called "side-by-side model," coupling two single DNA helices. This shows that if molecular biology were ever to be established, it was important to obtain the structure of biological molecules in more detail than was possible from fiber diffraction. Considering the dimensions involved, about $1\,\text{Å}\,(0.1\,\text{nm})$ for the distance between atoms, X-ray crystallography appeared to be the only suitable method. Major obstacles remained to be overcome, such as obtaining suitable crystals, coping with the large quantity of data required to describe the positions of all the atoms in a macromolecule, and solving the phase problem.

Protein crystals had already been obtained in the 1930s. It was not until 1957, however, that **Max Perutz** and **John Kendrew** found a way to solve the crystallographic phase problem by isomorphous replacement using heavy-atom derivatives. This permitted the structure of myoglobin to be solved in sufficient detail to describe how the molecule was folded. The difficulties encountered with protein crystallization, and the labor-intensive nature of the crystallographic study itself (this was before powerful computers, and long calculations were essentially performed by "post-doctoral hands") appeared to doom protein crystallography to providing rare, unique information on the three-dimensional structure of a very few biological macromolecules. Structural molecular biologists, therefore, continued the development and improvement of methods that could not provide atomic resolution but have complementary advantages for the study of macromolecular structures. These methods, at the boundary between thermodynamics and structure, had already played crucial roles in the century before the discovery of the double helix. The discovery of biological macromolecules is itself tightly interwoven with the application of physical concepts and methods to biology (biophysics).

The application of physics to tackle problems in biology is certainly older than its definition as biophysics. The *Encyclopædia Britannica* suggests that the study of bioluminescence by **Athanasius Kircher** in the **seventeenth century** might be considered as one of the first biophysical investigations. Kircher showed that an extract made from fireflies could not be used to light houses. The relation between biology and what would become known as electricity has preoccupied physicists for centuries. **Isaac Newton**, in the concluding paragraph of his *Principia* **(1687)**, reflected that

> *all sensation is excited, and the members of animal bodies move at the command of the will, namely, by the vibrations of this Spirit, mutually propagated along the solid filaments of the nerves, from the outer organs of sense to the brain, and from the brain into the muscles. But these are things that cannot be explained in few words, nor are we furnished with that sufficiency of experiments which is required to an accurate determination and demonstration of the laws by which this electric and elastic Spirit operates.*

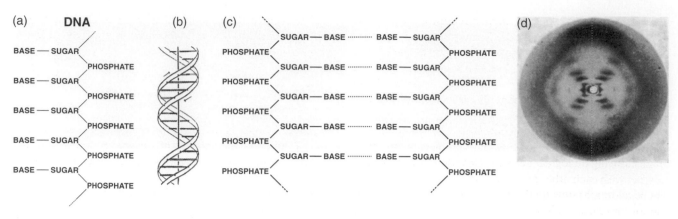

Fig. 1 (a) Chemical organization of a single chain of DNA. (b) This figure is purely diagrammatic. The two ribbons symbolize the two phosphate–sugar chains, and the horizontal rods the pairs of bases holding the chains together. The vertical line marks the helix axis. (c) Chemical organization of a pair of DNA chains. The hydrogen bonding is symbolized by dotted lines. (d) X-ray fiber diffraction of the B-form of DNA. The figures are facsimiles from the original papers of Watson and Crick (1953) and Franklin and Gosling (1953).

One hundred years later, **Luigi Galvani** and **Alessandro Volta** performed experiments on frogs' legs that would lead to the invention of the electric battery. They also laid the foundations of the science of electrophysiology, even though, because of the excitement caused by the electric battery, it was well into the **nineteenth century** before the study of animal electricity was developed further, notably by **Emil Heinrich Du Bois-Reymond**. Another nineteenth-century branch of biophysics, however, that dealing with diffusion and osmotic pressure in solutions, would later overlap with physical chemistry, and is more directly relevant to the discovery and study of biological macromolecules. The first papers published in *Zeitschrift für Physikalische Chemie* (**1887**) were concerned with reactions in solution, because biological processes essentially take place in the aqueous environment inside living cells.

The thermal motion of particles in solution ("Brownian" motion) was discovered by **Robert Brown (1827)**. **Abbé Nollet**, a professor of experimental physics, first described osmotic pressure in the early **nineteenth century** from experiments using animal bladder membranes to separate alcohol and water. The further study and naming of the phenomenon is credited to the medical doctor and physiologist **René J. H. Dutrochet (1828)**, who recognized the important implication of osmotic phenomena in living systems and firmly believed that basic biological processes could be explained in terms of physics and chemistry. The theory of osmotic pressure was developed by **J. Van't Hoff (1880)**. **George Gabriel Stokes** (middle of the **nineteenth century**) is best known for his fundamental contributions to the understanding of the laws governing particle motion in a viscous medium, but he also named and worked on the phenomenon of fluorescence. The laws of diffusion under concentration gradients were written down by **Adolf Fick (1856)**, by analogy with the laws governing heat flow.

Macromolecules, although large as molecules, are still much smaller than the wavelength of light. They could not be seen through direct observation by using microscopes, which had already shown the existence of cells in biological tissue and of structures within the cells such as the chromosomes (from the Greek, meaning "colored bodies"). From the knowledge gained from experiments on solutions it gradually became apparent that the biochemical activity of proteins, studied by **Emil Fischer (1882)**, is due to discrete macromolecules. In **1908, Jean Perrin** applied a theory proposed by **Albert Einstein (1905)** to determine Avogadro's number from Brownian motion. The theory of macromolecules is due to **Werner Kuhn (1930)**, after **Hermann Staudinger (1920)** proposed the concept of macromolecules as discrete entities, rather than colloidal structures made up of smaller molecules. The discovery of X-rays by **Wilhelm Conrad Röntgen (1895)**, and its application to atomic crystallography in the **1910s** through the work of **Peter Ewald, Max von Laue**, and **William H**. and **W. Laurence Bragg** laid the groundwork for the observation of atomic structural organization within macromolecules almost half a century later. **Theodor Svedberg (1925)** made the first direct "observation" of a protein as a macromolecule of well-defined molar mass by using the analytical centrifuge he had invented. In parallel, the atomic theory of matter became accepted as fact. There was rapid progress in X-ray diffraction and crystallography, electron microscopy and atomic spectroscopy. The novel experimental tools, provided by the new understanding of the interactions between radiation and matter, were carefully honed to meet the challenge of biological structure at the molecular and atomic levels. Physicists, encouraged by the example of **Max Delbrück**, who chose to study the genetics of a bacteriophage (a bacterial virus) as a tractable model in the **1940s**, and **Erwin Schrödinger's** influential book *What is Life?* (**1944**), which discussed whether or not biological processes could be accounted for by the known laws of physics, turned to biological problems in a strongly active way.

At the beginning of the **twenty-first century**, biophysics is dominated by two methods, X-ray crystallography and nuclear magnetic resonance (NMR), which play the key role in determining three-dimensional structures of biological macromolecules to high resolution. But even if all the protein structures in different genomes were solved, crucial questions would still remain. What is the structure and dynamics of each macromolecule in the crowded environment of a living cell? How does macromolecular structure change during biological activity? How do macromolecules interact with each other in space and time? These questions can be addressed only by the combined and complementary use of practically all biophysical methods. Mass spectrometry can determine macromolecular masses with astonishing accuracy. Highly sensitive scanning and titration microcalorimetry are applied to determine the thermodynamics of macromolecular folding and stability, and are joined by biosensor techniques in the study of binding interactions. There has been a rebirth of analytical ultracentrifugation, with the advent of new, highly precise and automated instrumentation, and it has joined small-angle X-ray and neutron scattering in the study of macromolecular structure and interactions in solution and the role of hydration. A femtosecond time resolution has been achieved for the probing of fast kinetics by optical spectroscopy. Light microscopy combined with fluorescence probes can locate single molecules inside cells. Scanning force microscopy is determining the profile of macromolecular surfaces and their time-resolved changes. Electron microscopy is approaching atomic resolution and is most likely to bridge the gap between single-macromolecule and cellular studies. Neutron spectroscopy is providing information on functional dynamics of proteins within living cells. Synchrotron radiation circular dichroïsm can access a wider wavelength range in the vacuum ultraviolet for the study of electronic transitions in the polypeptide backbone.

Up to the **late 1970s**, biophysics and biochemistry had only dealt with large molecular ensembles for which the laws of thermodynamics are readily applicable. One hundred microliters of a 1 mg ml^{-1} solution of hemoglobin, for example, contains 10^{15} protein molecules; a typical protein crystal contains of the order of 10^{15} macromolecules. In their natural environment, however, far fewer molecules are involved in any interaction and exciting new methods have been developed that allow the study of single molecules. Single molecules can now be detected and manipulated with hypersensitive spectroscopic and even mechanical probes such as atomic force microscopy, with which a single macromolecule can even be stretched into novel conformations. Conformational dynamics can be measured by single-molecule fluorescence spectroscopy. Fluorescence resonance energy transfer can measure distances between donor and acceptor pairs in single molecules, in vitro or in living cells. Near-field scanning optical microscopy can identify and provide dynamics information on single molecules in the condensed phase.

The historical development of each of the biophysical methods outlined above is discussed in more detail in the corresponding section of this book.

2 LANGUAGES AND TOOLS

Physike in Greek is the feminine of *physikos* meaning *natural*. Physics is the science of observing and describing Nature. When one of the authors (J. Z.) was a student at Edinburgh University, physics was taught in the department of Natural Philosophy. The word *philosophy*, love of wisdom, conveyed quite accurately how the wisdom of the observer is brought to bear in science. The observer plays his role through the tools he uses in his experiments and the language he uses to describe his results. Modern science covers so many diverse areas that it is impossible to master an understanding of all the tools and languages involved. Biophysics students are familiar with the language difficulties of trying to communicate with "pure" physicists, on the one hand, and "pure" biologists, on the other, despite decades of interdisciplinary teaching and research in universities. Rather than bemoaning this fact, we should recognize that it reflects the richness and depth of each discipline, expressed in its own sophisticated language, and developed in its own set of observational tools. Clearly, physics and biology have different languages, but it is important to appreciate that within each discipline also there are different languages. Language influences tool development, which in turn contributes to refining the concepts described by language. Biophysicists have to be fluent in the various languages of physics and biology and be able to translate between them accurately. This is a very difficult and sometimes impossible task, as any good language interpreter can testify, each language having its own specificity and viewpoint.

Biophysics deals, to a large extent, with the structure, dynamics, and interactions of biological macromolecules. What are biological macromolecules? Their biological activity is described in the language of biochemistry and molecular spectroscopy; they were discovered through their hydrodynamic and thermodynamic properties; they are visualized by their radiation scattering properties, and their pictures are drawn in beautiful color as physical particles. To each language there corresponds a set of tools, the instruments and methods of experimental observation. Progress in probing and understanding biological macromolecules has undoubtedly been based on advances in the methods used. Physical tools capable of ever-increasing accuracy and precision require a parallel development in biochemical tools (often themselves of physical basis, like electrophoresis or chromatography, for example) to provide meaningful samples for study. The word *meaningful* is a key word in the previous sentence. It refers to the relevance of the study with respect to *biology* (from the Greek *bios*, life, and *logos*, word or reason), i.e., biophysics has the goal of increasing our understanding of life processes. It should be distinguished from *biological physics*, which deals with the properties of biological matter, for example to design nanomachines based on DNA.

3 LENGTH AND TIMESCALES IN BIOLOGY

Biological events occur on a wide range of length and timescales – from the distance between atoms on the ångström scale to the size of the Earth as an ecosystem, from the femtosecond of electronic rearrangements when retinal absorbs a photon in the first step of vision to the 10^9 years of evolution. Observation tools have been developed that are adapted to the different parts of the length and timescales. The cell represents a central threshold for biological studies (Fig. 2). With a usual size of the order of 1–10 μm, cells can be seen under the light microscope. Also, the durations of cellular processes, which are of the order of seconds to minutes, can be observed and measured with relative ease. If we imagined diving into a eukaryotic cell through its plasma membrane, we would see other

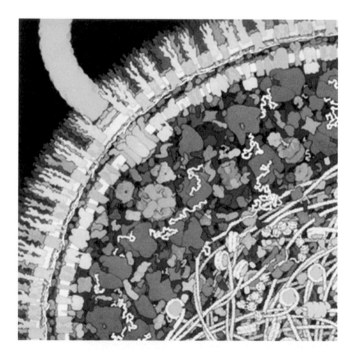

Fig. 2 A "realistic" drawing of the bacterium *Escherichia coli*, based on available experimental data. A flagellum, the double cell membrane and its associated proteins and glycoproteins are shown in hues of green; ribosomes and other protein and nucleic acid cytoplasmic components are in violet and blue; nascent polypeptide chains are in white; DNA and its associated proteins are in yellow and orange. The scale is given by the size of the bacterium of about 1 μm, or the double membrane thickness of about 10 nm. (www.scripps.edu/mb/goodsell)

membrane structures that separate distinct compartments like the nucleus or mitochondria, large macromolecular assemblies such as chromatin, ribosomes, chaperone molecular machines or multienzyme complexes. Looking for progressively smaller structures we would find RNA and protein molecules, then peptides and other small molecules, water molecules and ions, and finally the atoms that make them up (Fig. 3).

The smaller the length, the shorter the time, the heavier is the implication of sophisticated physical instrumentation and methods for their experimental observation.

The femtosecond (10^{-15} s) is the shortest time of interest in molecular biology; it corresponds to the time taken by electronic reorganization in the light-sensitive molecule, retinal, upon absorption of a photon, in the first step in vision. Time intervals of this order can be measured by laser spectroscopy (the distance covered by light in 1 fs is 3×10^{-7} m, or 300 nm, about one half the wavelength of visible light). Thermal fluctuations are in the picosecond (10^{-12} s) range; DNA unfolds in microseconds; enzyme catalysis rates are of the order of 1000 reactions per second; protein synthesis takes place in seconds, etc. The longest time of interest in molecular biology is, in fact, geological time, corresponding to the more than 1000 million years of molecular evolution (Fig. 4).

4 THE STRUCTURE–FUNCTION HYPOTHESIS

This book describes the application of classical and advanced physical methods to observe biological structure, dynamics, and interactions at the molecular level. Intensive research since the 1950s has emphasized the fundamental importance of biological activity at this level. The

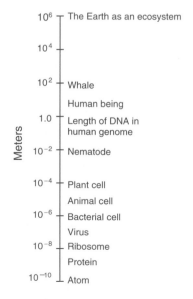

Fig. 3 Length scales in biology.

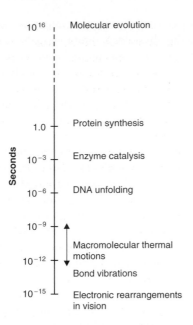

Fig. 4 Timescales in biology.

structure–function hypothesis is the foundation of molecular biology. One of its implications is that if a protein exists today in an organism, it is because it fulfills a certain biological function and its "structure" has been selected by evolution. The discovery and study of nucleic acids and proteins as macromolecules with well-defined structures has allowed an unprecedented understanding of processes such as the storage and transmission of genetic information, the regulation of gene expression, enzyme catalysis, immune response or signal transduction. In parallel, it became apparent that we could act on biological processes by acting on macromolecular structures, and powerful tools were developed not only to further fundamental scientific understanding but also to apply this knowledge in biotechnology or in drug design pharmacology.

The concept of "structure" should be understood in the broadest sense. The three-dimensional organization of a protein is not rigid but can adapt to its ligands according to the hypothesis of "configurational adaptivity" or "induced fit." Also, many proteins have been found that display a highly flexible random-coil conformation under physiological conditions. An intrinsically disordered protein could adopt a permanent structure through binding, but there are cases of proteins with intrinsic disorder that are biologically active while remaining disordered. A large proportion of gene sequences appear to code for long amino acid stretches that are likely either to be unfolded in solution or to adopt non-globular structures of unknown conformation.

Events taking place on the ångström and picosecond scales have profound consequences for life processes over the entire range of length and timescales – from the length and time associated with a cell, via those associated with an organism to those associated with the relation between an organism and its environment. The development of high-throughput techniques for whole-genome sequencing, for the analysis of genomic information (bioinformatics), for the identification of all the proteins present in a cell (functional proteomics), for determining how this population responds to external conditions (dynamic proteomics), and for protein structure determination (structural genomics) has opened up a new era in molecular biology whose revolutionary impact still remains to be assessed.

Biological macromolecule structures usually appear in pictures as static structures. A more precise definition would be "ensemble and time-averaged" structures. The atoms in a macromolecular structure are maintained at their average positions by a balance of forces. Under the influence of thermal energy, the atoms move about these positions. Dynamics, from the Greek *dynamis* meaning strength, pertains to forces. Structure *and* motions result from forces. It is common usage in biophysics, however, to separate *structure* from *dynamics*, considering the first as referring to the length scale (i.e., to the time-averaged configuration) and the second as referring to the timescale (i.e., to energy and fluctuations). The separation into two separate concepts is validated by the fact that the methods used to study structure and dynamics are usually quite separate and specialized. Modern experiments, however, often address both an average structure and how it changes with time.

5 COMPLEMENTARITY OF PHYSICAL METHODS

We know of the existence of macromolecules only through the methods with which they are observed. No single method, however, provides all the information required on a macromolecule and its interactions. Each method gives a different view of the system in space and time: the methods are *complementary*.

Biological macromolecules take up their active structures only in a suitable solvent environment. The forces that stabilize them are weak forces (of the order of kT, where k is Boltzmann's constant and T is absolute temperature), which arise in part from interactions with the solvent. The study of biological macromolecules, therefore, cannot be separated from the study of their aqueous solutions. The macromolecules are usually studied in dilute or concentrated solutions, in the lipid environment of membranes, or in crystals. Protein molecules or nucleic acid molecules in the unit cell of a crystal are themselves surrounded by an appreciable number of solvent molecules, and there are aqueous layers on either side of membranes. According to the experimental method used, we shall consider biological macromolecules in solution as "physical particles" (mass spectrometry, single-molecule detection ...), "thermodynamics particles" (osmotic pressure measurements,

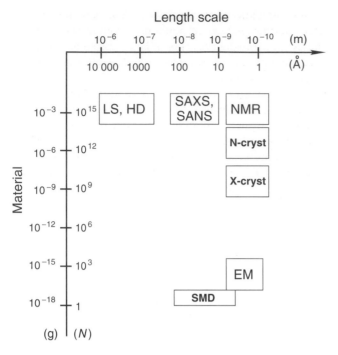

Fig. 5 Length resolution achieved and amount of material required for the sample for experiments using different physical methods to determine structure. Abbreviations: g, grams; N, number of molecules (assuming a molecular weight of the order of 100 000); LS, light scattering; HD, hydrodynamics; SAXS, SANS, X-ray and neutron small-angle scattering, respectively; NMR, nuclear magnetic resonance in solution; N-cryst, neutron crystallography; X-cryst, X-ray crystallography; EM, electron microscopy; SMD, single-molecule detection methods.

calorimetry...), "hydrodynamics particles" (viscosity, diffusion, sedimentation...) or "radiation interaction particles" (spectroscopy, diffraction and microscopy).

The length resolution scale achieved, the techniques involved, and the sample mass required for some biophysical methods are illustrated in Fig. 5.

6 THERMODYNAMICS

It is a result of classical thermodynamics that many properties of solutions, such as an increase in boiling point, freezing point depression, and osmotic pressure, depend on the number concentration of the solute. At constant mass concentration, therefore, these thermodynamics parameters vary sensitively with the molecular mass of the solute. Thus, for example, macromolecular masses and interactions have been determined from osmotic pressure measurements.

Macromolecular folding itself and the stabilization of active biological structures follow strict thermodynamics rules in which interactions with solvent play a determinant role. Sensitive calorimetric measurements of heat capacity as a function of temperature showed very clearly that stabilization free energy presents a maximum at a temperature close to the physiological temperature, the stability of the folded particle decreasing at lower as well as higher temperatures. The interpretation is the following. The behavior of the chain surrounded by solvent is much more complex than if it were in a vacuum. Enthalpy may rise, decrease, or even not change upon folding, because bonds can be made equally well within the macromolecule and between the chain and solvent components. Similarly for entropy, the loss of chain configuration freedom upon folding may be more than compensated for by a loss of degrees of freedom for the solvent molecules around the unfolded chain, for example through the exposure of apolar groups to water molecules. A water molecule in bulk has the freedom to form hydrogen bonds with partners in all directions. Apolar groups cannot form hydrogen bonds, so water molecules in their vicinity lose some of their bonding possibilities; their entropy is decreased.

In a protein solution, the heat capacity is strongly dominated by the water, and that of the macromolecules represents a very small part of the measured total. High-precision microcalorimeters were built to allow experiments on protein solutions to be performed. Nevertheless, early calorimetric studies on biological macromolecules concentrated on relatively large effects such as sharp transitions as a function of temperature. They led to a fundamental understanding of the energetics of protein folding. There are now important modern developments in the field. Very sensitive nanocalorimeters have been developed as well as analysis programs to treat the thermodynamics information and relate it to structural data. The energetics of intramolecular conformational changes, of complex formation and of interactions between partner molecules can now be explored in detail for proteins and nucleic acids. We should recall, however, that calorimetry (like all thermodynamics-based methods) provides measurements of an ensemble average over a very large number of particles (typically of the order of 10^{15}), even if results are usually illustrated in a simple way in terms of changes occurring in one particle.

7 HYDRODYNAMICS

The first hints of the existence of biological macromolecules as discrete particles came from observations of their hydrodynamic behavior. The language of macromolecular hydrodynamics is the language of fluid dynamics in the special regime of low Reynolds numbers. The Reynolds number in hydrodynamics is a dimensionless parameter that expresses the relative magnitudes of inertial and viscous forces on a body moving through a fluid. Bodies with the same Reynolds number display the same hydrodynamics behavior. Because of this, it is possible, for example, to determine the behavior of an airplane wing from wind-tunnel studies on a small-scale model. The Reynolds numbers of a small fish and a whale are 10^5 and 10^9, respectively.

Reynolds numbers in aqueous solutions for biological macromolecules and their complexes, from small proteins to large virus particles and even bacteria, are very small. For example, it is 10^{-5} for a bacterium swimming with a velocity of about $10^{-3}\,\mathrm{cm\,s^{-1}}$. Inertial forces are negligible under such conditions, so that the motion of a particle through the fluid depends only on the forces acting upon it at the given instant; it has no inertial memory. Particle diffusion through a fluid under the effects of thermal or electrical energy, and sedimentation behavior in a centrifugal field can be predicted by relatively simple equations in terms of macromolecular mass and frictional coefficients that depend on shape. The *resolution* defines the detail with which a particle structure is described. Hydrodynamics provides a *low-resolution* view of a biological macromolecule, for example as a two- or three-axis ellipsoid, but it is also very sensitive to particle flexibility and particle–particle interactions. Modern hydrodynamics includes a number of novel experimental methods. In addition to the classical approaches of analytical ultracentrifugation to measure sedimentation coefficients and dynamic light scattering to measure diffusion coefficients, we now have free electrophoresis to measure transport properties in solution, fluorescence photobleaching recovery to monitor the mobility of individual molecules within living cells, time-dependent fluorescence polarization anisotropy and electric birefringence to calculate rotational diffusion coefficients, fluorescent correlation spectroscopy and localized dynamic light scattering to measure macromolecular dynamics.

8 RADIATION SCATTERING

We see the world around us because it scatters light, which is detected by our eyes and analyzed by our brains. In a diffraction experiment, waves of radiation scattered by different objects interfere to give rise to an observable pattern, from which the relative arrangement (or structure) of the objects can be deduced. The interference pattern arises when the wavelength of the radiation is similar to or smaller than the distances separating the objects. In some cases, the waves forming the pattern can be recombined by a lens to provide a direct image of the object. Atomic bond lengths are close to one ångström unit ($10^{-10}\,\mathrm{m}$ or $0.1\,\mathrm{nm}$), and three types of radiation are used, in practice, to probe the atomic structure of macromolecules by diffraction experiments: X-rays of wavelength about $1\,\text{Å}$, electrons of wavelength about $0.01\,\text{Å}$, and neutrons of wavelength about 0.5–$10\,\text{Å}$. Visible light scattering, with wavelengths in the 400–$800\,\mathrm{nm}$ range, provides information on large macromolecular assemblies and their dynamics. X-rays, however, because they permit studies to atomic detail, provided the foundation on which structural biology has been built and is developing.

Neutron diffraction studies of biological membranes, fibers, and macromolecules and their complexes in crystals and in solution became possible in the 1970s with the development of methods that make full use of the special properties of the neutron.

Following the limitations of staining techniques, cryo-electron microscopy was developed to visualize subcellular and macromolecular structures to increasing resolution.

In the last decade of the twentieth century, the availability of intense synchrotron sources caused a revolution in macromolecular crystallography by greatly increasing the rate at which structures could be solved. Efficient protein modification, crystallization, data collection, and analysis approaches were developed for macromolecular crystallography. Extremely fast data-collection times made it possible to use time-resolved crystallography to study kinetic intermediates in enzymes. In parallel, field emission gun electron microscopes were applied and new methods developed to solve single-particle structures. Spallation sources for neutron scattering promise highly improved data-collection rates.

Light, X-rays, and neutrons are scattered weakly by matter and require samples containing very large numbers of particles in order to obtain good signal-to-noise ratios. Experiments provide ensemble-averaged structures. Modern electron microscopy methods, on the other hand, allow single macromolecular particles to be visualized.

9 SPECTROSCOPY

In spectroscopy, the radiation has exchanged part of its energy with the sample, through absorption effects or excitations due to particle internal or global dynamics, resulting in a change in the wavelength (frequency or color) of the outgoing beam with respect to the incident beam. Since absorption depends on the location of an atom in a structure, certain types of spectroscopic experiment may also be used to study structure. Nuclear magnetic resonance spectroscopy is sensitive to close to atomic resolution. The frequency of absorbed radiation can be measured as a function of time with an accuracy better than one part in a million. The precise nature of the signal depends on the chemical environment of the nucleus; hence structural information is obtained. In magnetic resonance imaging (MRI), millimeter resolution is obtained with meter wavelength probes by placing the body to be observed in magnetic field gradients and by focusing on nuclei in a given chemical environment; an absorption resonance then corresponds to a given field value and therefore to a precise location. As with diffraction, for which the wavelength matches the structural resolution required, the beam energy in spectroscopy is chosen so that differences due to sample excitations or absorption can be measured readily. In general, therefore, radiation of different wavelengths is used for diffraction and for spectroscopic experiments.

Electromagnetic radiation							
λ (m)	10^{-15}	10^{-12}	10^{-9}	10^{-6}	10^{-3}	1	10^3
ν (s^{-1})	10^{24}	10^{21}	10^{18}	10^{15}	10^{12}	10^9	10^6

Cosmic rays Gamma rays X-rays Ultraviolet Visible Infrared Radio

E (eV)	10^9	10^6	10^3	1	10^{-3}	10^{-6}	10^{-9}
E/k (K)	10^{13}	10^{10}	10^7	10^4	10	10^{-2}	10^{-5}

Neutrons

	10^{-10}	10^{-9}
λ (m)	10^{-10}	10^{-9}
v (m s^{-1})	4000	400
E (eV)	3×10^{-2}	3×10^{-4}
E/k (K)	400	4

Fig. 6 Wavelength, energy, and frequency for electromagnetic and neutron radiation. The scales in the figure give approximate orders of magnitude. The precise values for the constants are obtained from: $\nu\lambda = c$ where $\nu\lambda$ are the frequency and wavelength, respectively, of electromagnetic radiation and c is the speed of light (3×10^8 m s^{-1}); $E = h\nu$ (where E is energy and h is Planck's constant ($6.626 = 10^{-34}$ J s $= 4.136 \times 10^{-15}$ eV s); the temperature equivalent of energy, 1 eV/$k = 11604.5$ K, where k is Boltzmann's constant. In the neutron case, $\lambda = h/mv$ (where v m s^{-1} is neutron speed), and $E = \frac{1}{2}mv^2$, where m is neutron mass (1.6726×10^{-27} kg).

Coherent spectroscopy, in which radiation fields of well-defined phase are used, created unprecedented opportunities to study dynamics and time-evolving structures. The "spin echo" method, applied to NMR and neutron spectroscopy, was extended by the "photon echo" method when coherent lasers became available. Two-dimensional spectroscopy, first developed for NMR, measures the coupling within networks of vibrational modes. It has been applied to the infrared region to determine the structure of small molecules. The most exciting aspect of two-dimensional infrared spectroscopy is the combination of its sensitivity to structure and time resolution down to the femtosecond.

Taking electromagnetic radiation as an example, atomic diffraction requires X-ray wavelengths, while intramolecular vibrations correspond to infrared energies (Fig. 6). In NMR spectroscopy, the probing electromagnetic radiation is in the radio-frequency range, corresponding to meter wavelengths. Note that with neutron radiation, wavelengths of about 1Å (corresponding to interatomic distances and fluctuation amplitudes) have associated energies of about 1 meV (corresponding to the energies of atomic

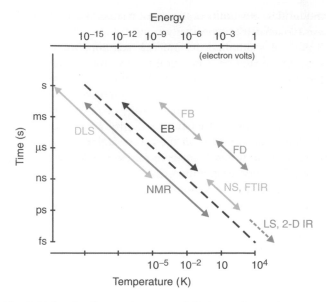

Fig. 7 Molecular timescales, associated energies and temperatures of various biophysical methods. The range follows the dashed black diagonal but the arrows have been displaced horizontally for clarity. Abbreviations: DLS, dynamic light scattering; NMR, nuclear magnetic resonance; EB, electric birefringence; NS, neutron spectroscopy; FTIR, Fourier transform infrared spectroscopy; LS, laser spectroscopy; 2-D IR, two-dimensional infrared spectroscopy; FD, fluorescence depolarization.

fluctuations), so that diffraction and spectroscopy experiments can be performed simultaneously to measure atomic amplitudes and frequencies of motion in macromolecules. Molecular timescales and corresponding energies and temperatures are shown in Fig. 7 for different biophysical methods.

10 SINGLE-MOLECULE DETECTION

Until the 1980s, biochemical and biophysical studies of biological macromolecules suffered the fundamental disadvantage of always having to deal with very large numbers of particles, whereas under in-vivo conditions they function as single particles in a dynamic heterogeneous environment. Structures, dynamics, and interactions were (and predominantly still are) observed and measured as ensemble averages. Furthermore, enzymatic, binding, or signaling reactions are in general stochastic, so that the kinetics of a protein activity measurement, for example, is also hidden in an ensemble average when measured in a large molecular population, even if the reaction is triggered contemporaneously for the entire sample.

Single macromolecules had been visualized by electron microscopy, but only in the last decade have methods become available to observe them while they were active. The development of single-molecule detection (SMD) techniques now allows the observation as well as the manipulation of single macromolecules in action. SMD is based on the two key technologies of *single-molecule imaging under active conditions* and *nanomanipulation*. Single-molecule signals that are detectable with good signal-to-noise ratios are given by fluorescent labels, which are observed using fluorescent optical microscopy. Applying total reflection and evanescent field techniques, the resolution of the method is several-fold better than the diffraction limit given by the wavelength of light. Single-molecule nanomanipulation techniques include capturing biomolecules using a glass needle or beads trapped by the force exerted by a focused laser beam (optical tweezers), and probing molecular forces with atomic force or scanning probe near-field microscopy. The forces involved are in the piconewton range, comparable to the thermal forces stabilizing the active macromolecular structures.

Erwin Schrödinger wrote in 1952 that we would never be able to perform experiments on just one electron, one atom, or one molecule. In the early 1980s, however, scanning tunneling microscopy was invented by G. Binning and H. Rohrer and radically changed the ways scientists view matter. Mechanical experiments to measure the piconewton forces that structure a single macromolecule became possible (Comment 1).

In optical tweezers instruments (Fig. 8a) one or two laser beams are focused on a small spot, creating an optical trap for polystyrene beads. One end of a single molecule (DNA, for example) is attached to a bead, while the other end is attached to a moveable surface, which, in this example, is another bead on a glass micropipette. The opposing force is measured as the molecule is stretched by moving the micropipette.

In magnetic tweezers instruments (Fig. 8b), one end of the single molecule is attached to a glass fiber, while the other end is attached to a magnetic bead. A magnetic field exerts a constant force on the bead. The extension and rotation of the molecule as a function of the applied force is then measured.

In an atomic force microscopy experiment (Fig. 8c), one end of the molecule is attached to a surface and the other to a cantilever. As the surface is pulled away, the deflection of the cantilever is monitored from the position of a reflected laser beam.

(a)

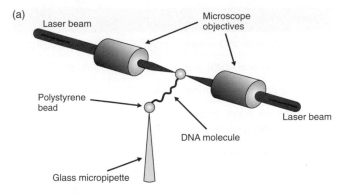

(b)

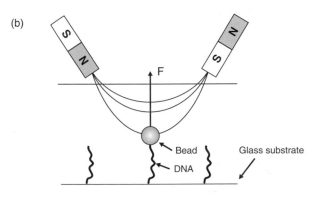

(c)

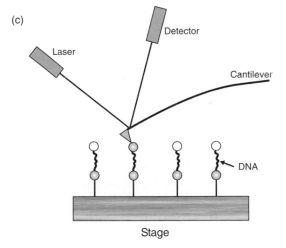

Fig. 8 A schematic view of three main techniques used in single-molecule force studies: (a) optical tweezers, (b) magnetic tweezers, and (c) atomic force microscopy (Carrion-Vazquez et al., 2000).

COMMENT 1 ENTROPIC FORCE

The typical energy scale for a macromolecule is thermal energy: $k_BT = 4 \times 10^{-21}$ J. Since the length scale of biomacromolecules is of order of 1 nm, the force scale is on the order of the piconewton (10^{-12} N). Therefore an entropic force can be calculated as $k_BT/(1\,nm)$, which is equal to 4 pN at 300 K.

The experiments allow a new structural parameter to be accessed within a single molecule: *force* (Table 1). The upper boundary for force measurements in micromanipulation experiments is the tensile strength of a covalent bond (in the eV/Å range or about 1000–2000 pN). The smallest measurable force limit is set by the Langevin force (about 1 fN), which is responsible for the Brownian motion of the sensor (size of the order of 1 μm).

Note that the total range of forces in Table 1 covers only three orders of magnitude. Until single-molecule

TABLE 1 THE RANGE OF FORCES IN MACROMOLECULES	
Tensile strength of a covalent bond	1000–2000 pN
Deformation of a sugar ring	700 pN
Breaking of double-stranded DNA	400–580 pN
Unfolding the β-fold immunoglobulin domain of the muscle protein titin	180–320 pN
Adhesive force between avidin and biotin	140–180 pN
Structural transition of uncoiling double-stranded DNA upon stretching	60–80 pN
Structural transition of double-stranded DNA upon torsional stress	~20 pN
Individual nucleosome disruption	20–40 pN
Unfolding triple helical coiled-coil repeating units in spectrin	25–35 pN
RNA–polymerase motor	14–27 pN
Structural transition of RNA hairpin in ribozyme under stretching (folding–unfolding)	~14 pN
Separation of complementary DNA strands (room temperature, 150 mM NaCl, sequence-specific)	10–15 pN
Stall force of the myosin motor	3–6 pN
Force generated by protein polymerization in growing microtubules	3–4 pN

techniques became available, information on protein stability could only be obtained by measuring the loss of structure under denaturing conditions (by using temperature, chemical agents, or pH) from which folding free energy could be calculated for an ensemble average of molecules. Free energy, however, does not provide direct information on mechanical stability. For mechanical stability, it is important to know how the total energy varies as a function of spatial coordinates. Several proteins were studied to measure the force required to unfold a single molecule. These studies revealed very large differences in magnitude (which can reach the order of a factor of 10) between the unfolding forces for different protein domains whose melting temperatures are very similar. These results demonstrated that the mechanical stability of a protein fold is not directly correlated with its thermodynamic stability. We expect the analysis of the mechanical properties of macromolecules to set the foundation of a new field of study, *mechanochemical biochemistry*.

11 BIOPHYSICS AND MEDICINE

The unprecedented insights into biological function gained from biophysics have created opportunities to rationally manipulate biological responses. In the most optimistic and perhaps simplistic case, once characterized, the molecular basis of a disease can be addressed. For example, evolving mutations on viral replication have led to the continuous structure-guided development of anti-HIV drugs.

This aspect of biophysics is complemented by the direct application of biophysical methods to the diagnosis and even treatment of disease. The same physical principles that underlie methods to observe molecules in vitro can also be applied in vivo, in the context of an organism. The practice of medicine has consequently been transformed by the advent of sonograms, MRI, CAT, and PET scans, which provide astonishing information about the workings of the human body, as well as the progression of disease. Fetal health can be monitored; cardiovascular disease and the spread of cancerous cells, for example, can be detected and measured prior to the appearance of symptoms.

BIOLOGICAL MACROMOLECULES AND PHYSICAL TOOLS

A

MACROMOLECULES IN THEIR ENVIRONMENT

A1

A1.1 HISTORICAL REVIEW

The discovery of biological macromolecules is tightly interwoven with the history of physical chemistry, which formally emerged as a discipline in **1887**, when the journal founded by **Jacobus Van't Hoff** and **Wilhelm Ostwald**, *Zeitschrift für Physikalische Chemie*, was first published. Interestingly, the first papers were concerned with reactions in solution, because biological processes essentially take place in the aqueous environment inside living cells.

The **nineteenth-century** discoveries of solution properties that led to our knowledge about biological macromolecules are described briefly in the Introduction. We must also mention the Grenoble chemist **François-Marie Raoult (1886)**, who formulated the freezing-point depression law that made it possible to determine the molecular weight of dissolved substances, and **Hans Hofmeister (1895)**, a medical doctor and physiologist who was interested in the diuretic and laxative effects of salts, and classified them according to how they modified the solubility of protein in aqueous solutions. The *Hofmeister series* was later established as a ranking order of the "salting-out," or precipitating, efficiency of ions. **Gilbert Newton Lewis** introduced the concepts of *activity* in **1908**, and of *ionic strength*, with **Merle Randall** in **1921**. In **1911, Frederick George Donnan** published a paper on the membrane potential developed during dialysis of a non-permeating electrolyte. **Peter Debye** and **Erich Hückel (1923)** proposed a theory for electrolyte solutions. In recent decades, modern methods, such as dynamic light scattering and small-angle neutron scattering, developed for the characterization of polymers, and especially polyelectrolytes, have contributed significantly to our current understanding of biological macromolecules in solution.

There is now a growing interest in the behavior of proteins in non-aqueous solvents and even in vacuum, mainly in the context of biotechnology, but also with respect to whether or not water is essential to life. Proteins, as active biological particles, have evolved in the presence of water, however, and, to a large extent, cannot be considered separately from their aqueous environment. We recall that even crystals of biological macromolecules contain an appreciable amount of solvent and should be considered as organized macromolecular solutions.

A1.2 MACROMOLECULAR SOLUTIONS

A solution is a homogeneous mixture at the molecular level of two or more components. The majority component is the *solvent*; the others are the *solutes*. We shall deal mainly with macromolecular aqueous solutions, in which the solvent is water, and the solutes are macromolecules and other small molecules, such as simple salts.

A1.2.1 Concentration

The solute concentration can be defined in various ways.

- The *weight or mass fraction* is the mass of solute per unit weight of solution (or per 100 weight units of solution if it is expressed as a *weight or mass percentage*).
- The usual unit of molecular mass in biochemistry is the dalton (Comment A1.1).
- The *molarity* is the number of moles of solute per liter of solution. Expressing a concentration in molar terms has the advantage of being more relevant than using weight fraction with respect to the *colligative* properties of the solution (i.e., properties that depend only on the number of solute particles rather than on the mass of solute or its specific properties; see also Section A1.2.3).
- The *molality* is the number of moles of solute per kilogram of solvent. Expressing solute concentration in

---

COMMENT A1.1 BIOLOGIST'S BOX: MOLECULAR MASS UNITS

Molar mass is in $g\,mol^{-1}$ or SI units of $kg\,mol^{-1}$.

Relative molecular mass or *molecular weight* is a dimensionless quantity, defined as the ratio of the mass of a molecule relative to 1/12 the mass of the carbon isotope ^{12}C. The molar mass of ^{12}C is very close to $12\,g\,mol^{-1}$. Molar mass (in $kg\,mol^{-1}$) can, therefore, be converted to molecular weight by dividing by $10^{-3}\,g\,mol^{-1}$ (the equivalent of multiplying by 1000 and cancelling units).

Biochemists use *molecular mass* expressed in *daltons* (Da) (1 Da = 1 atomic mass unit = 1/12 of the mass of one atom of ^{12}C).

these terms has the advantage that molality is obtained by weighing masses of solute and solvent, whereas molarity depends also on measuring a solution volume. Masses are invariant, while the volume of the solution is a function of temperature and pressure.

- The *mole fraction* and *volume fraction* definitions are similar to that of *weight fraction* but refer to moles or volume of solute per total moles or total volume, respectively.

The usual ways of measuring concentrations in protein and nucleic acid solutions are discussed in Comment A1.2.

A1.2.2 Partial Volume

The partial volume of a solute is equal to the volume change of the solution upon addition of the solute, under given conditions. The partial volume is not simply the volume occupied by the added solute, because its presence may lead to a volume change in the solvent. The partial volumes of charged molecules in aqueous solution are an interesting illustration of solvent effects. The water molecule can be represented by a small electric dipole. Liquid water has a rather loose hydrogen bonded structure (see Section A1.3.4); when in the presence of ions, water molecules orient around the charges, effectively taking up a smaller volume than in the bulk liquid. The effect is called *electrostriction*. The partial volume of a charged ionic solute may be negative, therefore, as is the case for the Na^+ and Mg^{2+} ions, for example. The decrease in volume due to electrostriction for these ions is greater than the volume they effectively occupy in the solution. The partial volume of the K^+ ion is slightly positive; electrostriction does not quite compensate for the volume occupied by the ion. Note that ions cannot be added separately to a solution, so that partial ionic volumes were obtained by interpolation from data on different neutral salts. It is also a consequence of electrostriction that ionic partial volume values are solvent composition-dependent; in general they increase with salt concentration, for example, because the water has already suffered some electrostriction.

The *partial specific volume* of a solute is the volume change of the solution per gram of added solute.

The *partial molal volume* of a solute is the volume change of the solution per mole of added solute.

The partial volumes of ions and biological macromolecules, in usual units, are given in Comment A1.3. Nucleic acids and to some extent carbohydrates, depending on their specific chemical natures, are polyelectrolytes and their partial volume values are strongly solvent salt-concentration-dependent. Interestingly, the partial volumes of proteins correspond well to the sum of their amino acid component volumes and they are not salt-concentration-dependent despite the fact that protein surfaces may show significant charge. This is because there are compensating effects on the volume from electrostrictive protein–solvent

The concentration of protein or nucleic acid solutions cannot be measured simply by weighing an amount of material into the solvent. Protein and nucleic acid powders obtained from lyophilization or precipitation always contain an unknown quantity of hydration water and salt ions, which are necessary for the maintenance of an active conformation. Furthermore, extremely low protein and nucleic acid concentrations are sufficient for many experimental methods, and it is not possible to weigh micrograms or less of material with precision. Protein concentrations are measured by colorimetric assays (e.g., the Bradford assay), in which an indicator interacts chemically with the polypeptide, or by spectrophotometry, in which the absorption of light at a given wavelength is proportional to the amount of material present. The absorbance at 280 nm is particularly sensitive to the presence of tryptophan, tyrosine, and cysteine amino acid residues, and for most proteins it is of the order of 1 for a 1 mg ml^{-1} solution and a path length of 1.0 cm. The exact value, however, varies with the number of sensitive residues in the protein. Nucleic acids show a strong absorbance at 260 nm (1 absorbance unit corresponds to about 40 micrograms per milliliter), which is used to measure their concentration in solution. The colorimetric and spectrophotometric measurements yield *relative* values with respect to a calibration series. When absolute concentration values are required, e.g., as they are for the interpretation of small-angle scattering data (Chapter G2), the colorimetric or spectrophotometric values have to be calibrated on an absolute scale for the specific macromolecule, e.g., by precise amino acid analysis of the sample.

interactions on the one hand, and a looser surface packing of amino acid residues on the other (see also Chapter D2).

A1.2.3 Colligative Properties

Colligative properties of solutions (from the Latin *ligare*, to bind) are properties that depend only on the number of solute molecules per volume and not on the mass or the nature of the molecules. The discovery of the colligative properties of solutions played an essential role in early physical chemistry, by allowing accurate measurements of molecular weight, which in turn provided evidence for the very existence of atoms and molecules. *Raoult's law* states that in an *ideal solution* at constant temperature the partial pressure of a component in a liquid mixture is proportional to its mole fraction. Colligative properties related to Raoult's law and applicable to dilute solutions of non-volatile molecules are the rise in boiling point and

Ion	Partial molal volume (ml mol^{-1})
Na$^+$ in water	−5.7
Na$^+$ in sea-water (~0.725 molal NaCl)	−4.4
K$^+$ in water	4.5
K$^+$ in sea-water (~0.725 molal NaCl)	5.9
Mg^{2+} in water	−30.1
Mg^{2+} in sea-water (~0.725 molal NaCl)	−27.0
Cl$^-$ in water	22.3
Cl$^-$ in sea-water (~0.725 molal NaCl)	23.3

From Millero (1969).

Macromolecule	Partial specific volume (ml g^{-1})
Proteins	0.73 (0.70–0.75)
Carbohydrates	0.61 (0.59–0.65)
RNA	0.54 (0.47–0.55)
DNA	0.57 (0.55–0.59)

Note that the spread in values for proteins corresponds to different
protein molecules, while that for the other macromolecules also takes
into account variations with solvent salt concentration. For example, a
given RNA molecule has a partial specific volume of 0.50 ml g^{-1} in water
and 0.55 ml g^{-1} in high salt concentration.

decrease in freezing point that result when a solute is
dissolved in a solvent. The temperature differences
between the values for the ideal solution and those for the
pure solvent are proportional in each case to the number of
solute molecules present. These laws can still be applied to
non-ideal solutions by applying the concepts of *chemical
potential* and *activity*.

A1.2.4 Chemical Potential and Activity

Consider the box shown in Fig. A1.1, with a barrier separating solutions of different molar concentration, C_A, C_B,
on either side, respectively. If we open a breach in the
barrier, there is a net flow of solute molecules from the

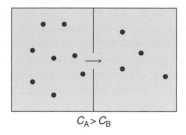

$$C_A > C_B$$

Fig. A1.1 Chemical potential (see text).

high-concentration to the low-concentration side, similar
to water flowing down a gravity potential gradient. A potential can, therefore, be associated with the solute concentration. The chemical potential, μ, of a solute is the *free
energy* gain upon addition of one mole into the solution
(see also Chapter C1).

$$\mu = \Delta G/\Delta C$$

In solution thermodynamics, the free energy difference
between two solutions of concentrations C_A, C_B, (in
number of moles per volume of solution), respectively, is
given by:

$$G_A - G_B = \Delta G = -RT \ln \frac{C_A}{C_B} \tag{A1.1}$$

where R and T are the gas constant and absolute temperature, respectively. The expression results from integration
of the Boltzmann equation, which describes how molecules in a perfect gas at constant temperature distribute
according to their free energy:

$$\frac{p_A}{p_B} = \exp\left(-\frac{G_A - G_B}{RT}\right) \tag{A1.2}$$

where p_A, p_B are pressures at constant volume (proportional to molar concentration) associated with states of
free energy G_A, G_B, respectively. In a trivial rewriting
Eq. (A1.1) becomes

$$\Delta G = \mu(C_A - C_B) = -RT \ln \frac{C_A}{C_B} \tag{A1.3}$$

Expressing the relation in terms of free energy or chemical potential *differences*, rather than absolute values,
avoids having to define a standard free energy or chemical
potential (e.g., that associated with an ideal solution at a
given concentration, temperature, and pressure). The
equations apply to ideal solutions, i.e., solutions in which
the solute molecules behave as point particles and do not
interact in any way with each other (or with the vessel).

G. Lewis introduced the concept of *activity* in 1908 to
account for deviations from ideal behavior in solutions.
Writing Eq. (A1.3) in terms of activity, rather than
concentration:

$$\Delta G = -RT \ln \frac{a_A}{a_B} \tag{A1.4}$$

The activity of solute A is given by $a_A = \gamma_A C_A$ and γ_A is an activity coefficient in the appropriate units and is equal to 1 for an ideal solution. Activity coefficients are obtained experimentally, for example, from deviations from Raoult's law. By replacing *concentration* with *activity*, equations that were derived in the ideal case could be applied in practice to *real* solutions.

A1.2.5 Temperature

The rise in boiling point and decrease in freezing point of solutions due to the presence of solute are not very useful properties when dealing with biological macromolecules such as proteins or nucleic acids – first, because the mole fraction of macromolecule in even highly concentrated solutions is very low (Comment A1.4); second, because biological macromolecules are usually not stable in pure water solvent and the molar concentration of buffer solutes and ions strongly dominates the effect; and, finally, because proteins and nucleic acids are usually not stable at the boiling and freezing points of water. Proteins and nucleic acids have evolved to be able to fold into their stable and active conformations in limited solvent conditions, and in limited ranges of the thermodynamic parameters of temperature and pressure (see Part C). It is interesting to note, however, that there exist organisms, called *extremophiles* (lovers of the extreme), which have adapted to various extreme environmental conditions, including temperature (Comment A1.5).

A1.2.6 Osmotic Pressure

Osmotic pressure is a colligative property that applies equally usefully to small molecule solutes and macromolecules (Fig. A1.2). Its importance for biological processes has been appreciated since its discovery. Named from the Greek *osmos*, impulse, osmosis is the phenomenon that occurs when two solutions of different concentrations are separated by a semipermeable membrane. A semipermeable membrane is a membrane that allows solvent molecules (water) to cross it unhindered, but not the solute. Osmotic pressure is *a hydrostatic pressure* due to

the solvent that develops on the more concentrated side of the membrane; it is due to water trying to move from the dilute to the concentrated side of the membrane in an attempt to equalize its chemical potential. The pressure builds up on the high concentration side, because the compartment volumes are constant. The osmotic pressure Π is given by

$$\Pi V = NRT \qquad (A1.5)$$

where R and T are the gas constant and absolute temperature, respectively, and N is the number of moles of solute in the volume V. The mass concentration C and the molar mass M of solute are related to N and V by

$$\frac{N}{V} = \frac{C}{M}$$

The difference between the osmotic pressure in the two compartments in Fig. A1.2 is given by $\rho g h$, where ρ is the density of the solvent and g is the acceleration due to gravity:

$$(\Pi_A - \Pi_B) = \frac{C_A - C_B}{M} RT \qquad (A1.5a)$$

When only one compartment contains the solute (i.e., C_B is zero), Eq. (A1.5a) provides an especially sensitive measure of molar mass, which has been used extensively in the field of polymers.

Note that the apparent mathematical identity of Eq. (A1.5) with the perfect gas equation (expressed in molar units) has no physical basis. As we pointed out above, osmotic pressure, Π, is a hydrostatic pressure due to the solvent and does not arise from solute molecules "pushing" like gas molecules against the membrane on the high-concentration side.

When there is more than one solute in the solution, Eq. (A1.5) becomes

$$\Pi V = \sum_j N_j RT \qquad (A1.6)$$

where the sum is over all non-diffusible solutes, i.e., solutes for which the membrane is impermeable. The effective osmotic pressure across a membrane, therefore, depends also on the quality of the membrane with respect to the different solutes. For example, in order to calculate the osmotic pressure across a dialysis membrane with a "cut-off" of 10 kD, we apply Eq. (A1.5), but we only count molecules with a higher molecular mass than 10 kD.

A1.2.7 Virial Coefficients

Equations (A1.5) and (A1.6) assume we are dealing with ideal solutions, in which the solute particles do not interact with each other. They can be considered as good approximations when the solutions are very dilute. Although interparticle interactions can be taken into account by a

COMMENT A1.5 EXTREMOPHILES

Microbial life has adapted to various extreme environments, including very high or very low pH (alkalophiles and acidophiles, respectively), high salt concentration (halophiles, see Section A1.3), high pressure in deep ocean trenches (barophiles), temperatures close to 0 °C in polar or glacier waters (psychrophiles, from the Greek *psychro* meaning cool; *cryo* meaning cold is usually applied to subzero temperatures), temperatures of about 70 °C in thermal springs, shown in Fig. A1.5.1 (thermophiles) and even temperatures as high as 113 °C (the highest known for a living organism) for *Methanocaldococcus jannaschi*, which lives in deep marine hydrothermal vents (hyperthermophiles). The proteins of extremophile organisms have very similar folds to those from the *mesophiles* (defined in an anthropomorphic way as living at around 37 °C). Their shifted stability and activity optima with respect to environmental conditions arise from amino acid substitutions that modify the internal forces appropriately (see Chapter C2).

Figure A1.5.1.

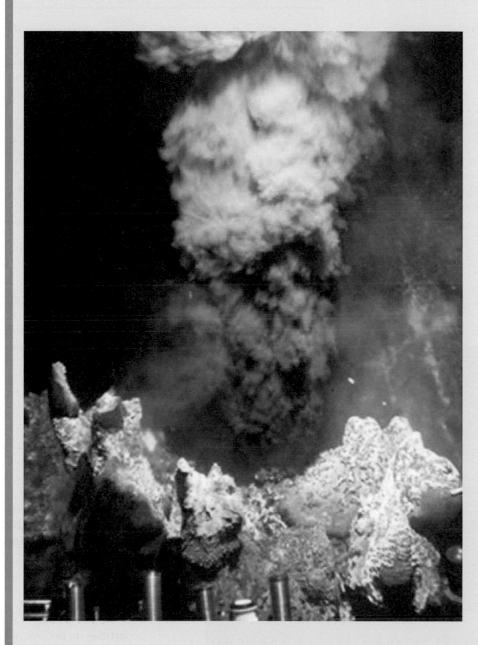

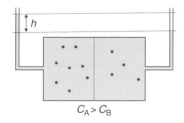

Fig. A1.2 Definition of osmotic pressure (see text).

chemical potential analysis, using *activity coefficients*, an approach in terms of *virial coefficients* (from the Latin *vires*, forces) has been more widely used for macromolecules in solution.

We now rewrite the single solute equation (Eq. (A1.5) with $N/V = C/M$) for the non-ideal solution; the osmotic pressure of a non-ideal solution is given by an expansion in powers of concentration

$$\frac{\Pi}{CRT} = \frac{1}{M} + A_2 C + A_3 C^2 + \cdots \tag{A1.7}$$

where $A_2, A_3 \ldots$ are the second, third $\ldots$ virial coefficients, respectively. Virial coefficients are determined from experimental measurements and provide valuable quantitative data on particle–particle interactions. The coefficients may be positive (corresponding to repulsive interactions) or negative (corresponding to attractive interactions).

A1.3 MACROMOLECULES, WATER, AND SALT

Salt solutions have important specific and non-specific effects on the conformational stability of macromolecules – so much so that proteins from the extreme halophiles (Comment A1.6) have evolved specific adaptation

---

COMMENT A1.6 EXTREME HALOPHILES

Extreme halophiles are archaeal and bacterial organisms (see Chapter A2) that can live only in high-salt environments, such as salt lakes like the Dead Sea, the Great Salt Lake, or the Lac Rose in Senegal (shown in Fig. A1.6.1), and in natural or artificial salt flats. They contain carotenoids and are responsible for the pink color of these environments. The carotenoids enter the food chain and give their color to flamingos and salmon flesh, for example. Extreme halophiles are distinguished from moderate halophiles and halotolerant organisms by the fact that they compensate for the high osmotic pressure of the environment, due to close-to-saturated NaCl, by a correspondingly high concentration of KCl in their cytoplasm, instead of so-called compatible solutes like glycerol. All their biochemical machinery, therefore, functions in an environment that is usually deleterious to protein stability and solubility. Halophilic proteins have adapted to this environment in interesting ways. Instead of "protecting" their structure from the salt, as might have been expected, for example by surrounding themselves with a strongly bound water shell, they actually incorporate large numbers of salt ions and water molecules to stabilize their fold and maintain their solubility.

Figure A1.6.1.

mechanisms in order to be stable, soluble, and active in the high salt concentrations found in the cytoplasm of these organisms. Specific salt effects on macromolecules are due to ion association or binding. They depend on both salt and macromolecule type. A small solvent concentration of sodium or potassium and magnesium ions, for example, is necessary for transfer RNA (tRNA) to achieve its folded conformation in solution. At solvent concentrations of about $0.1\,M$ NaCl and $1\,mM$ $MgCl_2$ the Na^+ and Mg^{2+} counter-ions neutralize the repulsion between the phosphate groups in the main chain of the nucleic acid so that it can take up a compact conformation. The effect is not purely electrostatic but contains a steric component. If the salt is $N(CH_3)_4Cl$, the $N(CH_3)_4^+$ counter-ion does not allow the tRNA to fold correctly, presumably because the $N(CH_3)_4^+$ is too large. Specific ion binding also plays an essential role in stabilization and switching mechanisms in certain proteins, such as calmodulin, the amylases, the parvalbumins, for example, which bind Ca^{2+}. Salt effects that are similar for all macromolecules have been named non-specific, even though they could depend strongly on ion type. They include ionic strength effects at low salt concentrations. Salts also have an important action on macromolecules through modifications of the water structure. Such an effect is usually evident at high salt concentrations (in the molar range). Ammonium sulfate is a salt that is frequently used at high concentrations in biochemistry to precipitate or to crystallize proteins. The sulfate ion acts on the water structure to reduce the solubility of apolar (hydrophobic) groups. In fact, in the early days of biochemistry, proteins were classified as globulins or albumins according to the ammonium sulfate concentration required for their precipitation.

A1.3.1 Ionic Strength and Debye–Hückel Theory

The discovery of the dissociation of strong electrolytes into separate ions in aqueous solution caused great excitement in physical chemistry. Ionic solutions are very far from *ideal* even when highly dilute, because of the electrostatic interaction between charges. The concept of *activity* (see above) was introduced for strong electrolytes. Lewis and Randall, in order to take into account the effects of ions of different valency, suggested that the mean activity of a completely dissociated electrolyte in dilute solution depends only on the ionic strength, i, of the solution,

$$i = \frac{1}{2}\sum_j C_j z_j^2 \tag{A1.8}$$

where C_j and z_j are, respectively, the molar concentration and charge of ion j. The value of i calculated for a $1\,mM$ solution of NaCl is $1\,mM$. The value of i calculated for a $1\,mM$ solution of $MgCl_2$ is $2.5\,mM$. Of course, the expression for the ionic strength can be calculated for solutions of any concentration, and unfortunately workers in the field

---

COMMENT A1.7 BIOLOGIST'S BOX: POISSON'S EQUATION

Poisson's equation is a linear partial differential equation of the second order named after the nineteenth-century physicist Siméon-Denis Poisson, which arises in the treatment of electrostatics.

---

have written of ionic strength values for solution concentrations of $1\,M$ or even higher. We wrote "unfortunately" because it should not be forgotten that the concept of ionic strength is applicable only to (very) dilute solutions, in which the "nature" of the ions and their interactions with water are neglected so that they can be considered to act as point charges. In a strict sense, ionic strength considerations should be limited to concentrations in the millimolar range.

The Debye–Hückel theory was essential for the understanding of the behavior of dilute electrolyte solutions. The theory combines Poisson's equation, the general form of Coulomb's law of electrostatics (Comment A1.7), with statistical mechanics to calculate electrostatic potential at a point in the solution, in terms of the concentration and distribution of ionic charges and the dielectric constant of the solvent. The theory established a relation between the ionic strength and the activity coefficient of an electrolyte solution for low-ionic-strength values (for which the concept is valid). It predicts that the activity coefficient decreases with rising ionic strength, in agreement with observations. At higher ionic strength values (in which the concept is no longer valid), the activity coefficient rises and in some cases may become greater than 1. Attempts to extend an approach similar to the Debye–Hückel theory to the high-concentration domain take into account specific ionic properties such as *solvation* (specific solute–water interactions).

A1.3.2 Polyelectrolytes and the Donnan Effect

Polyelectrolytes are macromolecular ions. DNA, for example, exists as a neutral salt in powder form; the negatively charged phosphate groups that are covalently bound in the macromolecule are neutralized by positive counter-ions, for example Na^+. The counter-ions dissociate from the macromolecule in solution, giving negatively charged polyions (or macroions) and "free" Na^+ ions. Interestingly, the distribution of counter-ions in the electrostatic field of the polyion can be calculated in low-ionic-strength conditions from Debye–Hückel theory.

Consider the dialysis experiment in Fig. A1.3. The figure shows a dilute solution of NaCl in a vessel separated into two compartments by a membrane that is perfectly permeable to water and to the Na^+ and Cl^- ions. A 70-nucleotide $Na \cdot tRNA$ salt is dissolved in the left-hand compartment of the vessel (the red ellipsoid); the macromolecule dissociates into a tRNA polyion with 70 negative charges and

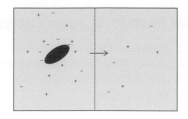

Fig. A1.3 The Donnan effect (see text).

Fig. A1.4 Density increments. The solvent made up of components 1 (water) and 3 (salt) is light blue. The macromolecule (component 2) is shown as a red ellipsoid. Its solvation shell made up of bound water and salt is shown in darker blue.

70 Na$^+$ ions (red in the figure). The polyions cannot cross the membrane. Three electroneutral components were used to make up the solution: component 1 is water; component 2 is the Na·tRNA salt macromolecule; and component 3 is NaCl. The negative charges on component 2 are restricted to the left-hand compartment, and, in order to maintain electroneutrality on either side of the membrane, there will be more small positive ions (Na$^+$) than unbound negative ions (Cl$^-$) in that compartment. Since there are also 70 positive counter-ions added per macromolecule, the result is an outflow of component 3 from the left-hand compartment to the right-hand compartment. The phenomenon is called the Donnan effect. The distribution of ions across the membrane to establish chemical potential equilibrium between the left- and right-hand compartments is termed the Donnan distribution.

A1.3.3 Macromolecule–Solvent Interactions

Again we consider two solutions separated by a dialysis membrane (Fig. A1.4). On the left-hand side, the solution is made up of the three components: (1) water, (2) macromolecule (which is too large to diffuse across the dialysis membrane), and (3) a small solute, e.g., salt (which can diffuse across the membrane). The solution on the right-hand side does not contain the macromolecule. The water and diffusible solute move between the compartments to equalize their chemical potentials, μ_1, μ_3. In general, the macromolecule interacts specifically with water and the small solute – through hydration, ion binding, or the Donnan effect, for example. How do these interactions affect the distribution of components 1 and 3 across the membrane?

Equation (A1.9) defines the *density increment* of the solution on the left-hand side due to the presence of a concentration c_2 of the macromolecule:

$$\left(\frac{\partial \rho}{\partial c_2}\right)_{\mu_1,\mu_3} = \frac{\rho' - \rho}{c_2} \tag{A1.9}$$

where ρ and ρ' are the mass densities of the solutions in the left- and right-hand compartments, respectively.

The density increment is a bulk property of the solution. It can be measured precisely by weighing a given volume of solution at constant temperature in a densimeter. The presence of the macromolecule perturbs the solvent around it. If, for example, it binds salt and water in a ratio that is different from their ratio in the solvent, then this results in salt or water flowing in or out of the dialysis compartment to compensate and maintain constant chemical potential of diffusible components across the membrane. The mass density increment expresses the increase in mass density of the solution per unit macromolecular concentration. Not only does it account for the presence of the macromolecule itself, but also for its interactions with diffusible solvent components. It can be written

$$(\partial \rho / \partial c_2)_{\mu} = (1 + \xi_1) - \rho\,({}^{\circ}\bar{v}_2 + \xi_1 \bar{v}_1) \tag{A1.10}$$

where the subscript μ is shorthand for constant μ_1, μ_3, and ξ_1 is an interaction parameter in grams of water per gram of macromolecule, and $\bar{v}_x$ is the partial specific volume (in milliliters per gram) of component x.

The parameter ξ_1 does not represent water "bound" to the macromolecules; it represents the water that flows into a dialysis bag to compensate for the change in solvent composition caused by the association (or repulsion) of both water and small solute components with the macromolecule. The mass density increment is the difference in mass between the volume of 1 g of macromolecules and ξ_1 g of water, on the one hand, and the same volume of bulk solvent, on the other.

An equation similar to Eq. (A1.10) can be derived in terms of ξ_3, representing grams of the small solute per gram of macromolecule; ξ_1 and ξ_3 are related by $\xi_1 = -\xi_3/w_3$, where w_3 is the molality of the solvent in grams of component 3 per gram of water. The *preferential interaction parameters* ξ_1 and ξ_3 are thermodynamic representations of the solvent interactions of the macromolecule in the given solution conditions.

A1.3.4 Water, Salt and the Hydrophobic Effect

Whereas at very low salt concentrations (neglecting specific effects related to the properties of different ions) the ionic strength concept appears to be justified, it is certainly not the case at high salt concentrations. The structure of liquid water and its specific interactions with solutes play an essential role in defining the behavior of macromolecules in solution.

(a)

(b)

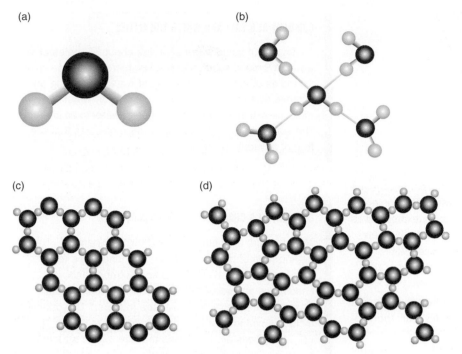

Fig. A1.5 Water structures: (a) a single molecule, the large dark atom is oxygen, the two light atoms are hydrogen; (b) directional tetrahedral hydrogen bonding around one molecule; (c) a two-dimensional model of an ordered water or ice lattice; (d) a two-dimensional model of a disordered water lattice. (The figures were kindly provided by Professor John Finney.)

(c)

(d)

Despite an intensive effort in molecular modeling to account for the extensive thermodynamics data on pure water and solutions, water structure remains incompletely understood at the molecular level. We describe its main features in qualitative terms; bearing in mind that they are based on carefully measured experimental thermodynamic quantities. Liquid water is a highly dynamic system of directional hydrogen bonds. Each molecule can participate in four hydrogen bonds and we can consider it as a small body, the oxygen atom that can "accept" two bonds, with two hydrogen atoms that can act as donors, stretching out at fixed angles like two arms (Fig. A1.5).

The water molecules in the liquid phase are constantly rearranging in order to form hydrogen bonds with different partners, like a person in the middle of a crowd briefly touching the bodies of different pairs of the surrounding people, while being touched by others (Fig. A1.6). In thermodynamics terms, forming a hydrogen bond leads to a decrease in enthalpy (negative ΔH) while the configurational freedom of having different partners with which to form bonds leads to an increase in entropy (positive ΔS). Both features contribute to a decrease in free energy ($\Delta G = \Delta H - T\Delta S$) and are, therefore, favored thermodynamically (see Part C).

What happens to the liquid water picture in the presence of solute? A solute molecule, in general, perturbs the dynamics of water molecules with which it is in contact. If the solute is polar (i.e., ionically charged, or neutral but capable of forming hydrogen bonds) it may orient the surrounding water molecules in a preferential way compared to when they are in the bulk liquid. We saw in Section A1.2.2 how a small charged ion (Li$^+$ or Na$^+$, for

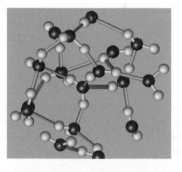

Fig. A1.6 Three-dimensional model of water. The structure is highly dynamic with hydrogen bonds forming, breaking, and reforming in different patterns. In bulk water, each molecule has the configurational freedom of forming such bonds in all directions around itself. This is not the case in the presence of a solute. (The figure was kindly provided by Professor John Finney.)

example) "pulls" in the water molecules toward itself and causes electrostriction. If the solute is apolar (i.e., non-ionic and incapable of forming hydrogen bonds), it still perturbs the water structure because water molecules in its vicinity have fewer hydrogen bonding possibilities. They lose the configurational freedom of being able to make hydrogen bonds in any direction and their entropy decreases. Apolar solutes, therefore, are poorly soluble in water. Furthermore, in a temperature range close to room temperature their solubility curve is anomalous: The solubility decreases with rising temperature (Fig. A1.7). The temperature range concerned is where the effects of the solute on the entropy of the solvent (inducing a reduction

21

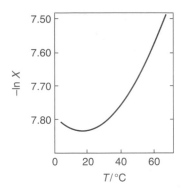

Fig. A1.7 Solubility of an apolar solute in water. The curve shown is that for benzene in water. *X* is solubility expressed as the mole fraction. The Gibbs free energy of transfer is given by $-RT \ln X$ (Franks *et al.*, 1963). (Figure reproduced with permission.)

in conformational freedom or in the number of hydrogen bonding possibilities available) dominate the thermodynamics of the system.

A phenomenon named the *hydrophobic effect* results from the low solubility of apolar solutes in water. Because contact between the solute and water is entropically unfavorable (see Part C), apolar solutes tend to aggregate to minimize their surface of interaction with the surrounding water molecules leading to an apparent hydrophobic (water fearing) force bringing them together.

Salt solutes perturb the water structure around them according to specific properties of the ions, especially their charge and volume. At high salt concentrations, a large portion of the water solvent is affected (Comments A1.8, A1.9), which, in turn, affects the solubility of other solutes. So-called "salting-out" ions reduce the solubility of apolar solutes in water even further; adding such ions to a solution savors precipitation ("salting out") of apolar solutes. Biological macromolecules (proteins and nucleic acids) display heterogeneous surfaces, with various charged as well as apolar patches, leading to complex hydration interactions, folding, and solubility behavior (see also Part C). The *Hofmeister series* is a classification of different ions according to how, at high solvent concentrations, they stabilize the native fold of and reduce the solubility of proteins in aqueous solutions (Table A1.1). The series has been established from phenomenological observations and appears to be valid for a wide variety of processes, from protein solubility to helix coil transitions in certain polymers and the stabilization of the folded or unfolded form of macromolecules in general. *Salting-in* is the process in which the solubility of the macromolecule is increased by adding salt. Strong salting-in ions also destabilize the folded structure at high concentration, presumably because the unfolded chain offers a larger number of binding sites. The exact order of ions in the Hofmeister series is not strict and may vary for different proteins, for example. The classification of anions in terms of stabilizing

TABLE A1.1 THE HOFMEISTER SERIES EXPRESSED SEPARATELY FOR CATIONS AND ANIONS. OF THE ANIONS, PHOSPHATE IS THE MOST SALTING-OUT, WHILE CHLORIDE IS NEUTRAL AND THIOCYANIDE THE MOST SALTING-IN. OF THE CATIONS, AMMONIUM, POTASSIUM AND SODIUM ARE NEUTRAL IN THEIR SALTING BEHAVIOR, WHILE CALCIUM AND MAGNESIUM ARE MORE SALTING-IN THAN LITHIUM
Cations:
phosphate > sulfate > acetate > chloride < bromide < chlorate < thiocyanide
Anions:
ammonium, potassium, sodium < lithium < calcium, magnesium

(salting-out) and destabilizing (salting-in) ions, nevertheless, appears to have general validity (Comment A1.9).

Specific salt effects on biological macromolecules are used extensively in biochemistry and biophysics as an important part of the battery of methods for fractionation, solubilization, crystallization, etc. The essential role of the solvent in maintaining the integrity of an active biological macromolecule cannot be overestimated and in biochemistry it is vital to work in well-defined, buffered solvent conditions (Comment A1.10).

A1.4 CHECKLIST OF KEY IDEAS

- Biological macromolecules as active biological particles cannot be considered separately from their aqueous environment.
- The *partial volume* of a solute is the volume change of the solution due to the presence of the solute.
- The solute concentration in a solution can be expressed in terms of mass per volume or a value that depends on the number concentration of molecules per volume (*molarity* or *molality*).
- *Colligative* properties of solutions depend on the number concentration of solute molecules.
- *Raoult's law* states that in an ideal solution at constant temperature the partial pressure of a component in a liquid mixture is proportional to its mole fraction.
- A consequence of Raoult's law is that the freezing point depression (or rise in boiling point) of a solution with respect to the pure solvent is proportional to the mole concentration of solute.
- The *chemical potential* of a solute is the free energy created by its presence in the solution.
- The concept of *activity* was introduced to replace *concentration* and account for deviations from ideal behavior in solutions.
- Laws derived for ideal solutions in terms of concentration remain valid in non-ideal circumstances if concentration is replaced by activity.
- The activity of a solute is equal to its concentration multiplied by an *activity coefficient* that is equal to 1 for an ideal solution.
- Biological macromolecules are folded and active in limited ranges of solvent conditions, temperature, and pressure.

- *Extremophiles* are organisms that live under extreme conditions of solvent, temperature, or pressure.
- Macromolecules from extremophiles have adapted to be stable and active in the solvent, temperature, and pressure conditions of their environment.
- A *semipermeable membrane* is permeable to water but not to solutes.
- *Osmotic pressure* is a hydrostatic pressure that develops on the high-concentration side of a semipermeable membrane separating solutions of different molar concentration.
- Osmotic pressure is a colligative property; at constant temperature, it is proportional to the difference in molar concentration of solute on either side of the membrane.
- In non-ideal solutions, osmotic pressure can be expressed as a concentration power series; the coefficients of the different terms are called the *virial coefficients*.
- Salts have important effects on the stability and solubility of macromolecules in solution, according to chemical type and concentration.
- Salt effects act through salt–water interactions or direct binding to the macromolecules.
- *Ionic strength* is a concept that is valid only at very low (millimolar) salt concentrations, in which only the concentration and valency of ions are taken into account, and not their specific character.
- *Debye–Hückel theory* provides a means of calculating the electrostatic potential at a point in an electrolyte solution in terms of the concentration and distribution of ionic charges and the dielectric constant.
- Debye–Hückel theory is applicable in low ionic concentrations, in which the concept of ionic strength is valid.
- The distribution of ions around a polyelectrolyte, such as DNA, can be calculated, at low ionic strength, by Debye–Hückel theory.
- The conventional numbering of components in a macromolecular solution containing a further small solute, such as salt, is as follows: component 1 is water, component 2 is the macromolecule, and component 3 is salt.
- Macromolecule–water–small solute interactions can be calculated from *density increment* measurements under dialysis conditions, in terms of *preferential interaction parameters*.
- The structure of liquid water and its interactions with small solutes play an essential role in defining the behavior of macromolecules in solution.
- Liquid water is a highly dynamic system of directional hydrogen bonds.
- The *hydrophobic effect* results from the low solubility of apolar groups in water because they cannot form hydrogen bonds.
- The hydrophobic effect results in an apparent *hydrophobic force* bringing together apolar groups in aqueous solution.

- Dissolved salts modify the hydrophobic effect according to their type and concentration.
- *Kosmotropic* ions, "water structure makers," are small, high-charge-density ions that increase the solubility of apolar groups in water.

- *Chaotropic* ions, "water structure breakers," favor the hydrophobic effect.
- The *Hofmeister series* is a phenomenological classification of ions according to their ability to stabilize the native fold and precipitate proteins and nucleic acids in solution.

Suggestions for Further Reading

Von Hippel, P. and Schleich, T. (1969). The effects of neutral salts on the structure and conformational stability of macromolecules in solution. In *Structure of Biological Macromolecules*, eds., S. N. Timasheff and G. D. Fasman. New York: Marcel Dekker Inc., pp. 417–575. This chapter is still an excellent reference on the effects of salts on macromolecules, even though it is more than 40 years old.

Collins, K. D. (1997). Charge density-dependent strength of hydration and biological structure. *Biophys. J.*, **72**, 65–76.

Madern, D., Ebel, C., and Zaccai, G. (2000). Halophilic adaptation of enzyme. *J. Extremophiles*, **4**, 95–98.

Price, P. B. (2000). A habitat for psycrophiles in deep Antarctic ice. *Proc. Natl. Acad Sci. USA*, **97**, 1247–1251.

Jaenicke, R. (2000). Do ultrastable proteins from hyperthermophiles have high or low conformational rigidity? *Proc. Natl. Acad Sci. USA*, **97**, 2912–2940.

Papers presented at a Royal Society (UK) meeting on water and life are published in (2004). *Phil. Trans. R. Soc. Lond. B.*, **359**.

Zaccai, G. (2011). Life in high salt (a molecular perspective). In *Origins of Life: An Astrobiological Perspective*, eds., P. Garcia-Lopez, M Gargaud, and H Martin. Cambridge: Cambridge University Press.

Seckbach, J. and Oren, A. (2013). *Polyextremophiles: Life Under Multiple Forms of Stress (Cellular Origin, Life in Extreme Habitats and Astrobiology)*. New York: Springer.

Marty, V., Jasnin, M., Fabiani, E., *et al.* (2013). Neutron scattering: a tool to detect in vivo thermal stress effects at the molecular dynamics level in micro-orgranisms. *J. R. Soc. Interface*, **10**, 20130003

MACROMOLECULES AS PHYSICAL PARTICLES

A2

A2.1 HISTORICAL REVIEW AND BIOLOGICAL APPLICATIONS

Late 1600s

R. Boyle questioned the extremely practically oriented chemical theory of his day and taught that the proper task of chemistry was to determine the composition of substances.

1750

A.-L. Lavoisier studied oxidation, and correctly understood the process. He demonstrated the quantitative similarity between chemical oxidation and respiration in animals. He is considered the father of modern chemistry.

Late 1700s

The work of **J. Priestley**, **J. Ingenhousz**, and **J. Senebier** established that photosynthesis is essentially the reverse of respiration.

1800s

The development of organic chemistry, despite the strong opposition of *vitalists* (who believed that transformations of substances in living organisms did not obey the rules of chemistry or physics but those of a *vital force*), led to the birth of biochemistry. In **1828**, **F. Wöhler** performed the first laboratory synthesis of an organic molecule, urea. During the **1840s**, **J. V. Liebig** established a firm basis for the study of organic chemistry and described the great chemical cycles in Nature. In **1869** a substance isolated from the nuclei of pus cells was named nucleic acid. **O. Avery**'s experiments of **1944** on the transformation of *pneumococcus* strongly suggested that nucleic acid was the support of genetic information.

1860s

L. Pasteur is considered the father of bacteriology. He proved that microorganisms caused fermentation, putrefaction, and infectious disease, and developed chemical methods for their study. In **1877** Pasteur's "ferments" were named *enzymes* (from the Greek *en* "in" and *zyme* "leaven or yeast.") In **1897 E. Buchner** showed that fermentation could occur in a yeast preparation devoid of living cells.

1882

E. Fischer showed that proteins are very large molecules built of amino acid units; he also discovered the phenomenon of *stereoisomerism* in carbohydrates.

1926

Urease, the first enzyme to be crystallized in a pure form, was shown to be a protein by **J. B. Sumner**. In the **1960s**, **M. Perutz** and **J. Kendrew** solved the molecular structures of hemoglobin and myoglobin, respectively, by crystallography, thus proving that proteins had well-defined structures. They were awarded the Nobel Prize.

1940

F. A. Lipmann proposed that adenosine triphosphate (ATP), which had been isolated from muscle in **1929**, is the energy exchange molecule in many cell types.

1953

J. D. Watson and **F. Crick** published the double-helix structure of DNA, based on chemical intuition and the fiber diffraction data of **R. Franklin** and **M. Wilkins.**

1955

F. Sanger determined the amino acid sequence of insulin, thus proving proteins are well-defined linear polymers of amino acids, an achievement for which he was awarded the Nobel Prize. He went on to develop powerful methods for the determination of nucleotide sequences in DNA and RNA, which earned him a share (with **P. Berg** and **W. Gilbert**) in a second Nobel Prize.

1970s

C. Woese classified living organisms in three kingdoms, *Bacteria*, *Archaea*, and *Eukarya*, following sequence analysis of ribosomal RNA (Fig. A2.1). The classification was confirmed and refined when whole genome sequences became available.

1980s

S. Altman isolated ribonuclease P, an enzyme that contains both protein and RNA active components, and opened the way for the discovery of many other types

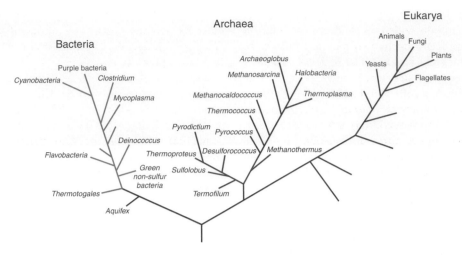

Fig. A2.1 An "unrooted" philogenetic tree of life showing the three kingdoms resulting from ribosomal RNA sequence analysis. Further work showed that the likely root of the tree lies between the Bacteria and the Archaea. Bacteria and Archaea are *prokaryotes*, unicelluar organisms, whose genetic material is not contained in a nucleus. Archaea were so named because they were thought to be closer to the root of early life forms. They have specific characteristics as well as characteristics in common with either Bacteria or Eukarya. Most of the known extremophile organisms (see Chapter A1) are found in the Archaea. Some of them, like the anaerobic methanogens found in cow gut and marshland, are extremely common.

of catalytic RNA. He was awarded the Nobel Prize. In **1986**, **W. Gilbert** coined the phrase *RNA World* to denote a hypothetical stage in the evolution of life, already suggested to have existed by **C. Woese, F. H. C. Crick**, and **L. E. Orgel**, in which RNA combined the role of a genetic information storage molecule with the catalytic properties necessary for a form of self-replication. The RNA World hypothesis led to a strong revival of interest in origin-of-life studies.

1990s and since 2000

Lee *et al.* (1993) and **Reinhart *et al.* (2000)** discovered microRNAs in the nematode *Caenorhabditis elegans*. The following decade witnessed intense activity in the RNA field with the discovery of different micro RNAs and of their implication in regulating gene expression. **Medina *et al.* (2010)** discovered that in mice overexpression of a single miRNA was sufficient to cause cancer. **Salmena *et al.* (2011)** proposed the hypothesis that competing endogenous RNAs (ceRNA) communicate with and regulate other RNA transcripts by competing for shared miRNAs, suggesting it may be the Rosetta Stone of a hidden RNA language.

The spectacular development of molecular biology was accompanied by equally spectacular progress in methods for solving the structures of biological macromolecules and for studying their interactions. These will not be reviewed here as they constitute the subject of the book and are treated in the other chapters. In this chapter we describe the structural components and organization of biological macromolecules.

A2.2 BIOLOGICAL MOLECULES AND THE FLOW OF GENETIC INFORMATION

A gene is represented by a well-defined nucleotide sequence in the cellular DNA. There are four different nucleotides. The genetic code relates the 64 possible nucleotide triplets (*codons*) to the 20 amino acids found in proteins. Several triplets correspond to the same amino acid, allowing for stability at the protein level with respect to mutation at the DNA level, and there are codons corresponding to a "stop" signaling the end of the chain. In protein synthesis, the gene is transcribed into messenger RNA and subsequently translated into the corresponding protein on the ribosome, according to the fundamental dogma of molecular biology:

$$DNA \leftrightarrow DNA \leftrightarrow RNA \rightarrow protein$$

The first double arrow represents the replication of a DNA mother molecule into identical daughter molecules; the second double arrow represents DNA to RNA transcription and RNA to DNA reverse transcription (observed in *retroviruses*); and the last arrow stands for translation.

The scheme represents a flow of information and not a sequence with respect to the evolution of macromolecules, since the final product, protein, is required in each and every step. The processes are catalyzed and highly regulated by protein transcription factors, protein enzymes such as the polymerases, protein translation factors, and ribosomes, themselves made up of protein and RNA. The end result is a protein with an amino acid sequence (*primary structure*) that has a one-to-one correspondence with the gene that initially coded for it. The protein has to undergo a folding process through solvent and intrachain interactions before it gains biological activity. Its *secondary structure* represents local structures, such as helices or extended chain conformations stabilized by intrachain or interchain hydrogen bonding; its *tertiary structure* is the structure achieved when the secondary structure elements pack to form a compact three-dimensional fold; its *quaternary structure* is the structure obtained when different proteins or copies of the same protein molecule associate to form a complex.

DNA has often been discussed using polymer terminology. It is composed of a linear chain of repeating structural units, the nucleotides. Each nucleotide is made up of a constant phosphate–ribose group, to which is bound a variable base taken from a group of four: adenine (A), guanine (G), cytosine (C), and thymine (T). DNA differs from usual synthetic polymers, however, in that it forms specific structures, according to its environment and interactions with proteins, which are intimately related to its biological function.

It is often useful to define a physical object by writing what it is not. A protein, perhaps even more so than DNA, is not a *classical* polymer (Fig. A2.2). This statement might appear to be contradictory, since a protein is often described as a polymer of amino acid residues. Its *biological* activity, however, conveys special properties to a protein molecule – properties that are different from those of the classical polymers, whose study led to the laws of polymer science. (Chemists have moved, however, toward the synthesis of polymers with some of the properties that are specific to proteins.)

First, with respect to composition and molar mass, a classical polymer chain is composed of a small number of different units repeated a large number of times. A sample is made up of chains of different length, and the molar mass is defined as an average over a statistical distribution of chain lengths. A protein chain has a well-defined beginning (the N-terminal) and a well-defined end (the C-terminal); its primary structure (the amino acid sequence) and molar mass are perfectly defined (e.g., all the 10^{15} molecular chains in a tenth of a milligram of a given protein have an *identical* mass, to a precision restricted only by the natural isotopic variability); these properties result strictly from the structure of the gene that codes for the protein, and have been confirmed experimentally by mass spectrometry.

Second, with respect to solution conformation, the classical polymer chain takes up a solvent-dependent conformation in solution, which again can be described by an average over a distribution of different conformations. Water-soluble and membrane protein chains fold in the solvent environment in which they are active to form well-defined compact structures. These structures are *essentially* identical for all the molecules in a pure protein sample. Protein crystallography and NMR studies of proteins (also on samples containing about 10^{15} molecules) yield structures where atomic positions may be defined with a precision better than 1 Å.

The polypeptide chain that makes a protein has a precise length defined by its gene, with an N-terminal beginning and a C-terminal end. The chain is composed of a number of amino acid residues. There are 20 naturally occurring amino acids. In physiological conditions it takes up well-defined tertiary and quaternary structures which

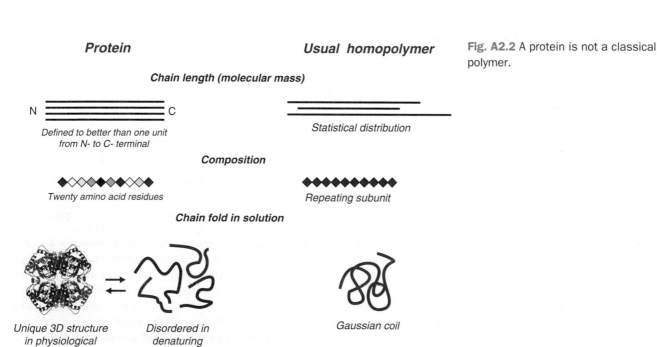

Protein   **Usual homopolymer**

Chain length (molecular mass)

N ——— C
Defined to better than one unit from N- to C- terminal

Statistical distribution

Composition

Twenty amino acid residues

Repeating subunit

Chain fold in solution

Unique 3D structure in physiological conditions   *Disordered in denaturing conditions*

Gaussian coil

Fig. A2.2 A protein is not a classical polymer.

unfold under denaturing conditions (heat or in certain solvents such as urea). Denatured protein attains conformations that are similar to homopolymers (Gaussian coils). Disordered protein structures, however, are not necessarily biologically inactive. The complete genome sequences now available have shown that a large proportion of gene sequences appear to code long amino acid stretches that are unfolded in physiological conditions.

In addition to DNA, RNA, and protein, a cell contains two other major classes of molecule – carbohydrates and lipids – which play essential structural and functional roles. Carbohydrates and lipids are not coded directly in the genome but are the substrates and products of protein enzymes in complex metabolic pathways.

A2.3 PROTEINS

Essentially all the molecules in a living organism are either proteins or products of protein action. Except for the ribozymes, which constitute a small class of RNA molecules, all enzymes (the biochemical catalysts of metabolic reactions) are proteins. Several hormones (molecules with regulatory functions through their interaction with protein receptors) are themselves proteins, while others are protein components such as single amino acids or small peptides (the only other major hormone classes are steroids and amines). Proteins execute transport functions (e.g., hemoglobin, the oxygen carrier in red blood cells). They constitute the basis of muscle, of the conjunctive tissue that maintains the body structure in animals, and of skin and hair. Proteins are the components of elaborate recognition and signaling pathways (e.g., in the immune system as well as in a variety of cellular responses to external stimuli that can be chemical or physical, as in the case of photons in the processes of vision and photosynthesis). Proteins constitute the pumps and channels that establish the electrochemical gradients across membranes, which are essential to cellular bioenergetics, and play important roles in processes such as nerve transmission. And this is certainly not an exhaustive list of the biological functions accomplished by proteins.

Proteins are mainly made up of 20 amino acid types, linked in a linear chain by peptide bonds. Their biological activity results from the higher levels of organization achieved by the chain in its physiological solvent environment. The *protein* is defined as the object with biological activity, while *polypeptide* refers to its chemical composition. Non-amino acid *prosthetic groups* may be associated with a protein and be essential for its activity. Examples are the oxygen-binding heme groups in hemoglobin, myoglobin and certain respiratory chain proteins; retinal, which provides the light sensitivity in rhodopsin, the protein associated with vision; and chlorophyll in plant proteins associated with photosynthesis. In fact, since the natural amino acids are colorless (their main absorption is in the

UV), the color of a protein is always associated with a prosthetic group: red for the heme, purple for retinal, green for chlorophyll. Such light-absorbing groups allow particularly useful spectroscopic approaches to the study of their associated proteins (see Part E).

A2.3.1 Chemical Composition and Primary Structure

The general structure of an amino acid and its modification when it forms peptide bonds and enters a polypeptide chain are shown in Fig. A2.3. The *primary structure* of a protein is the sequence of amino acids in its polypeptide. The 20 main naturally occurring amino acid side-chains are shown with their properties in Fig. A2.4.

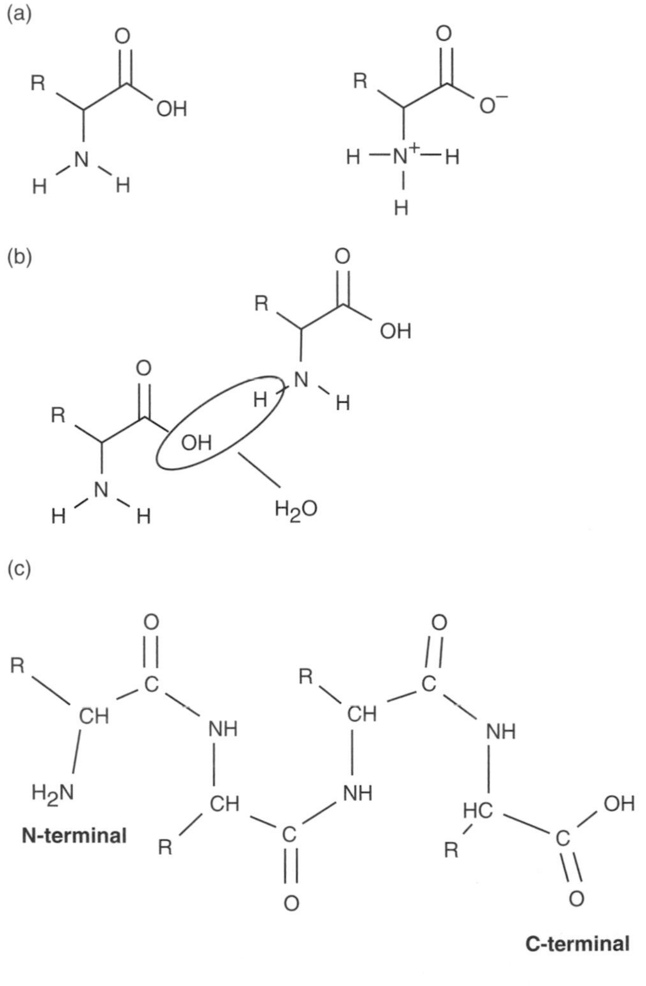

Fig. A2.3 (a) An amino acid in its neutral form, and in the doubly charged ion (zwitterion) form it can have in solution. R is a variable side-chain. (b) Amino acids polymerize to form a polypeptide chain, with the loss of a water molecule for each peptide bond formed. (c) The chain has an N-terminal end and a C-terminal end. The N–C-terminal direction corresponds to the direction in which the gene is translated and the chain is synthesized on the ribosome.

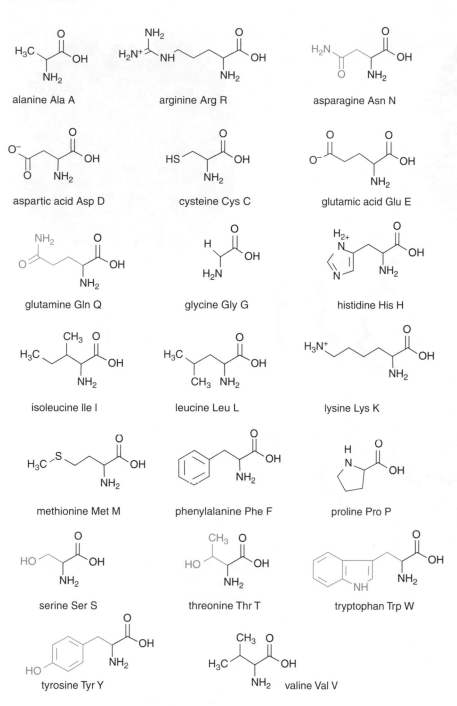

Fig. A2.4 The 20 main natural amino acids. The three-letter and one-letter codes are given under each. The main chain is in crimson, apolar side-chains are in black, neutral polar side-chains in green, positively charged (basic) chains in blue, and negatively charged (acidic) chains in red. Histidine is shown in its charged form (below pH 6.0).

alanine Ala A

arginine Arg R

asparagine Asn N

aspartic acid Asp D

cysteine Cys C

glutamic acid Glu E

glutamine Gln Q

glycine Gly G

histidine His H

isoleucine Ile I

leucine Leu L

lysine Lys K

methionine Met M

phenylalanine Phe F

proline Pro P

serine Ser S

threonine Thr T

tryptophan Trp W

tyrosine Tyr Y

valine Val V

Except for glycine, amino acids can exist in two *enantiomeric* forms depending on the side of the α-carbon to which the side-chain binds. Natural amino acids are in the L (*laevo*, or left-handed) form (according to the direction of rotation of the plane of polarized light).

A2.3.2 Structures of Higher Order

A protein structure can be depicted in different ways. The pictures in Fig. A2.5 were drawn from the same structural model, resulting from the analysis of crystallographic data. Each emphasizes a different aspect. The protein in our

example is the metabolic enzyme, malate dehydrogenase, from the "halophilic" (which lives in a high-salt environment) organism *Haloarcula marismortui*.

The secondary, tertiary and quaternary structures of the protein are all seen in the top part of Fig. A2.5a, which traces the polypeptide fold in a *ribbon representation*. *Secondary structures* are favored local chain conformations arising from chemical and steric constraints. The most common secondary structures found in proteins are α-helices and β-strands (see below); they are depicted, respectively, as ribbon helices and arrows. The protein is a tetramer. Its *tertiary structure* is the three-dimensional

(a)

(b)

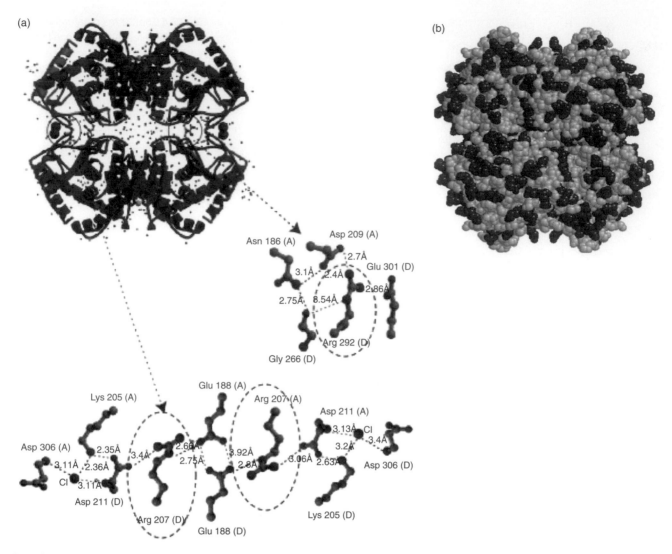

Fig. A2.5 Different ways of drawing the same protein structure in order to emphasize different features. **(a)** Top: a *ribbon diagram* showing how the secondary structure elements (ribbon helices for α-helices and arrows for β-sheets) are organized to form the tertiary structure of each subunit and how the four subunits are organized in the tetrameric quaternary structure; bottom: *ball-and-stick* atomic "zooms" into the interdimer interfaces to show the complex ion-binding salt bridges in these regions of the structure. **(b)** An atomic *space-filling* representation showing the protein surface; negatively charged atoms are red, positively charged atoms are blue.

conformation of the subunit (given by the coordinates of the constituent atoms). Its *quaternary structure* is the organization of subunits in the tetramer. A high-resolution zoom into a section of the structure is shown in *ball and stick* representation in the bottom part of Fig. A2.5a. It illustrates the detailed relationship between the different chemical groups. Such an illustration is often used to show the active site interactions in an enzyme, for example. In this case, the picture is of complex salt bridges (ionic bonding between charged amino acids) and solvent salt ion binding, which are related to protein stability in high salt concentrations.

In the *space-filling model* of Fig. A2.5b, a Van der Waals sphere is drawn around each atom to provide a picture of the protein surface. Negatively and positively charged atoms are

in red and blue, respectively. The surface has a net negative charge, which appears to be a general property of proteins from organisms adapted to high salt concentrations.

Secondary Structure

The atoms in the main chain of a polypeptide (N–C_α–C–N– ...) cannot lie in a straight line or rotate freely because of the properties of the bonds between them.

In particular, the peptide bond restricts the atoms in the group to lie in the same plane so that the chain has limited flexibility (Fig. A2.6).

The chain conformation can be described by the angles ϕ and ψ, formed by the peptide plane and the C_α atoms on either side. The ϕ rotation is clockwise facing the NH from the C_α atom, and the ψ is clockwise facing the CO from the

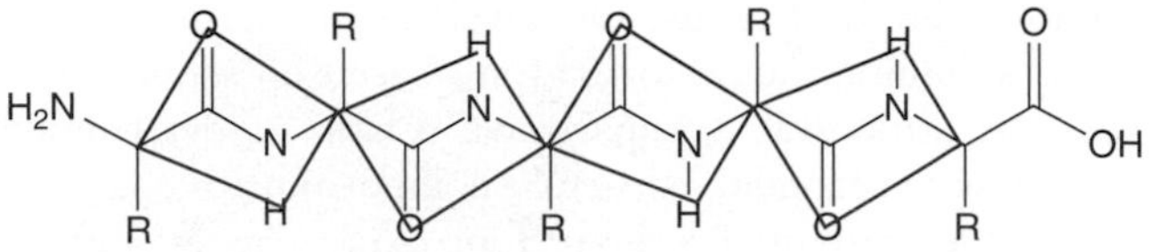

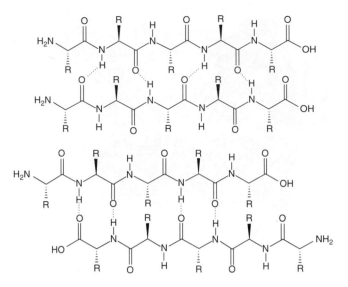

Fig. A2.6 The planar peptide bonds in a polypeptide chain.

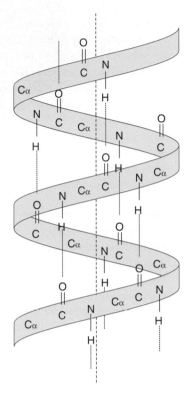

Fig. A2.7 The α-helix. The main chain carboxyl group of amino acid *j* is hydrogen bonded to the NH group of amino acid *j* + 4.

Fig. A2.8 Parallel (top) and antiparallel (bottom) β-sheets.

A β-sheet in which polar and apolar amino acids alternate, for example, displays a polar face on one side and an apolar face on the other.

Tertiary Structure

The tertiary structure results from weak (non-covalent, except for the disulfide bond between two cysteines) interactions between the amino acids in the polypeptide chain. The structure function hypothesis is based mainly on tertiary structure, and it is interesting to note how its general features for a particular protein family can be strongly conserved by evolution, even for molecules with widely varying primary structures.

Most proteins have structural similarities with other proteins. The study of these similarities is important not only in order to understand the evolutionary and functional relationships involved but also because it can assist in the analysis in structural terms of the huge amount of sequence data produced by the genome projects. The tertiary structures of protein subunits can be divided into *single domain* and *multidomain* proteins according to distinct features. For example, the metabolic enzymes called dehydrogenases share a common dinucleotide cofactor domain, called the *Rossman fold*, after the scientist who discovered it; the diphtheria toxin protein subunit is strikingly divided into three recognizable domains, each with a distinct organization associated with its function (Fig. A2.9).

Domain structures have been classified in groups or families, following the analysis of the entire solved structure database. These structures are accessible at the protein data bank web page.

The CATH protein structure classification is a hierarchical domain classification program in five levels: class (C-level), architecture (A-level), topology or fold family (T-level), homologous superfamily (H-level), and sequence

C_α atom. Constant ϕ and ψ values result in the chain assuming a regular helical conformation. A *Ramachandran plot* is a two-dimensional representation of ϕ–ψ pairs and the corresponding secondary structures (see Chapter G3). Some of these structures, like the α-helix, for example, are stabilized by internal hydrogen bonds and favored energetically (Fig. A2.7). The "straight" chain with side-chains projecting on alternate sides is called a β-strand; β-sheets are made up of β-strands joined together by hydrogen bonds. They can be parallel or antiparallel (Fig. A2.8). α-helices and β-sheets are the most common secondary structures found in the structures deposited in protein data banks, which can be accessed on the web.

Secondary structures are asymmetric, not only because of the directionality of the main chain, but also because of the asymmetric position of amino acid side-chains on the α-carbon (except for glycine). We recall that natural amino acids are in the L (left-handed) enantiomeric configuration. The side-chains therefore all point outwards in the right-handed α-helix configuration. Side-chains in β-sheets project outwards alternately above and below the sheet.

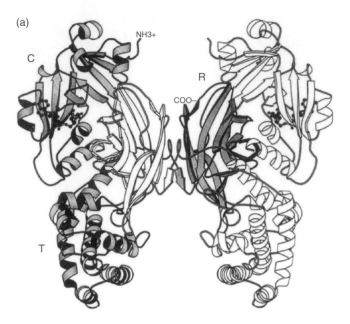

(a)

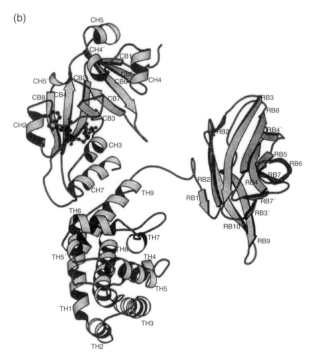

(b)

Fig. A2.9 (a) Ribbon model of the diphtheria toxin (DT) dimer, showing the three-domain structure, C, T, R. One monomer is in gray and the other in white. The dimer illustrates the phenomenon of *domain swapping*, in which monomers exchange domains R. Each domain is associated with a specific function of the toxin. The beta protein domain R (similar in tertiary structure to immunoglobulin domains) is the receptor recognition domain in the first step of the toxin's entry into a cell. In the process of endocytosis the DT is carried to an endosome within the cell. The alpha protein domain T (similar in tertiary structure to transmembrane proteins) allows the toxin to cross the endosome membrane in order to release the catalytic domain C into the cytoplasm, after cleavage of the polypeptide chain between C and T. The alpha–beta

family (S-level). The last two levels are for proteins that share a common ancestor and are based on sequence as well as structural comparisons. Class is determined according to secondary structure composition and packing within the domain. There are four major classes: mainly alpha, which comprises domains whose secondary structure is mainly α-helix; mainly beta, which includes domains organized in β-sheets; alpha–beta, which includes both alternating alpha–beta secondary structures and alpha plus beta secondary structures; the fourth class includes domains with low secondary structure content. Architecture describes the overall shape of the domain, determined by the orientations of the secondary structure elements. Topology depends on both the overall shape and connectivity of the secondary structures.

SCOP (structural classification of proteins) provides another domain organization database, which is based on all solved structures. The levels of the SCOP hierarchy are *species*, *proteins*, *families*, *folds*, and *classes*.

Quaternary Structure

The organization of protein subunits at the quaternary structure level has been selected by evolution because of functional advantages that range from simple stability considerations as in the case of the oligomerization of enzymes from hyperthermophilic organisms that have to be stable and active at temperatures close to 100 °C (see Chapter A1), via the subtle tuning of activity as in the case of *allostery* (from the Greek "other structure," Comment A2.1), to extremely coordinated complex reaction mechanisms as in the case of the large molecular machines in the cell, such as ribosomes (Fig. A2.14, see also Fig. H2.11), proteasomes, thermosomes (see Chapter G2) and chaperone complexes. Large organized structures made up of different protein

---

COMMENT A2.1 PHYSICIST'S BOX: ALLOSTERY

Allostery is a change in structure caused by an *effector*, which facilitates a certain activity, for example the binding of a substrate to an enzyme. Hemoglobin is the oxygen carrier in blood. The protein is a tetramer with four bound heme groups, one to each subunit. The subunit conformations can be in either a relaxed (R) or tense (T) form. Oxygen acts as both the allosteric effector and substrate of the protein. When it binds to one of the subunits it causes all of the subunits to go into the R form, which has a high affinity for oxygen binding.

---

domain C is an enzyme that catalyzes a reaction that is lethal to the cell. Note that the active site of C is hidden by R so that the intact protein is inactive. Similarly, the dimer is less active in cell entry because the R domains shield each other. **(b)** Ribbon drawing of an "open" monomer. (Bennet *et al.*, 1994.) (Figure reproduced with permission from Protein Science.)

molecules as found in muscle or microtubules (see Chapter H2), for example, can also be considered as quaternary structures.

A2.4 NUCLEIC ACIDS

There are two main classes of nucleic acid: deoxyribonucleic acid (DNA) and ribonucleic acid (RNA). Similarly to proteins, which have a polypeptide primary structure, nucleic acids are polynucleotides, unbranched polymers of a certain chemical type of subunit, the nucleotide. DNA is the depository of genetic information. It is found associated with proteins in the chromosomes of cells. Its structure is predominantly double-stranded, with a helical organization. RNA displays a range of biological functions (and new ones are still being discovered), and presents an even greater variety of structural organization. It is usually made up of a single strand, which nevertheless turns around and interacts with itself to form intricate secondary structures. Messenger RNA (mRNA) is transcribed from DNA and carries the genetic message to the ribosome. Transfer RNA (tRNA) is a family of small RNA molecules (of about 70 nucleotides); each type of tRNA carries a specific amino acid to the ribosome and acts as an adapter between the mRNA and the growing polypeptide chain. Ribosomes are themselves organized by a number of small and large RNA molecules, ribosomal RNA (rRNA). Since the purification of the enzyme RNase P, which contains both RNA and protein active components, many types of catalytic RNA molecules have been and are being discovered. These include the so-called ribozymes, and portions of rRNA that play important catalytic roles in translation, microRNAs like siRNA, small interfering RNA molecules involved in gene-silencing by RNA interference (RNAi) with mRNA. RNAi is currently a hot topic for study because of its potential applications in cancer and viral infection therapies. tmRNA is a remarkable molecule that combines transfer RNA and messenger RNA functions. It codes for and adds a C-terminal peptide tag to an unfinished protein on a stalled ribosome, which directs the aborted protein for proteolysis. Small nucleolar RNAs (snoRNA) are localized in the eukaryotic cell nucleolus and are involved in rRNA biogenesis by defining nucleotide modification sites. The genome of certain viruses is based on double-stranded viral RNA rather than DNA. This is the case for retro-viruses, like the HIV virus, which also provide the machinery for retro-transcription that transcribes the viral RNA into DNA so that it can incorporate the infected cell's genetic material.

A2.4.1 Chemical Composition and Primary Structure

Like the amino acids, nucleotides are made up of constant and variable parts. The constant part forms the main chain of the polynucleotide and is made up of a *nucleoside*, which is constituted of two covalently linked groups: a deoxyribose group (in DNA) or a ribose group (in RNA) and a phosphate group. The variable part is a nitrogen-containing base bound to the sugar group. There are four main naturally occurring bases in DNA: adenine, thymine, guanine, and cytosine. The same bases occur in RNA, except for thymine, which is replaced by uracil. In certain molecules the bases can be found in modified chemical forms (e.g., via *methylation* of DNA, or in more complex ways in tRNA). Chemical modification usually has important biological significance with respect to the interactions of the nucleic acid with other molecules. We recall that cells of different types within the same organism contain identical DNA molecules. The different nature of the cells is due to the fact that different sets of genes may be transcribed in each type. *DNA methylation*, i.e., the addition of methyl groups to specific cytosine residues in a DNA chain, appears to be involved with transcription inhibition in vertebrate cells, with distinct methylation patterns associated with each cell type. Nucleotides and polynucleotide chain elongation are shown in Fig. A2.10.

A2.4.2 Structures of Higher Order

DNA

DNA is predominantly double-stranded, and its secondary and tertiary structures are based on base pairing. The classical Watson–Crick pairs are A–T, G–C, leading to the rule $A + C = T + G$ obeyed by DNA composition. An important stereochemical consideration that contributed to the discovery of the double helix is that the A–T pair, joined as in Fig. A2.11 by three hydrogen bonds, has the same width as the G–C pair joined by three hydrogen bonds, so that either pair fits into a regular double-helix structure. The double-helical tertiary structure allows for sequence specificity through access to the major and minor grooves (Fig. A2.12). DNA can take up a variety of double-helical tertiary structures, depending on composition and hydration (Fig. A2.13). This structural variability of DNA is likely to be exploited in biological functions. There is, for example, clear evidence of the existence of proteins that bind to the B, A, and Z double-helical conformations. DNA can also fold into structures that involve more than two strands, such as triplexes and quadruplexes. G-quadruplex (guanine tetrad motif) structures have been found in vitro for telomeric sequences (sequences at the end of eukaryotic chromosomes). The verification of their occurrence in vitro is at present a hot topic in the field.

RNA

The genome of some RNA viruses is double-stranded. In general, however, the rule $A + C = U + G$ does not hold for RNA because the structures are predominantly single-stranded. The polynucleotide, nevertheless, bends back on

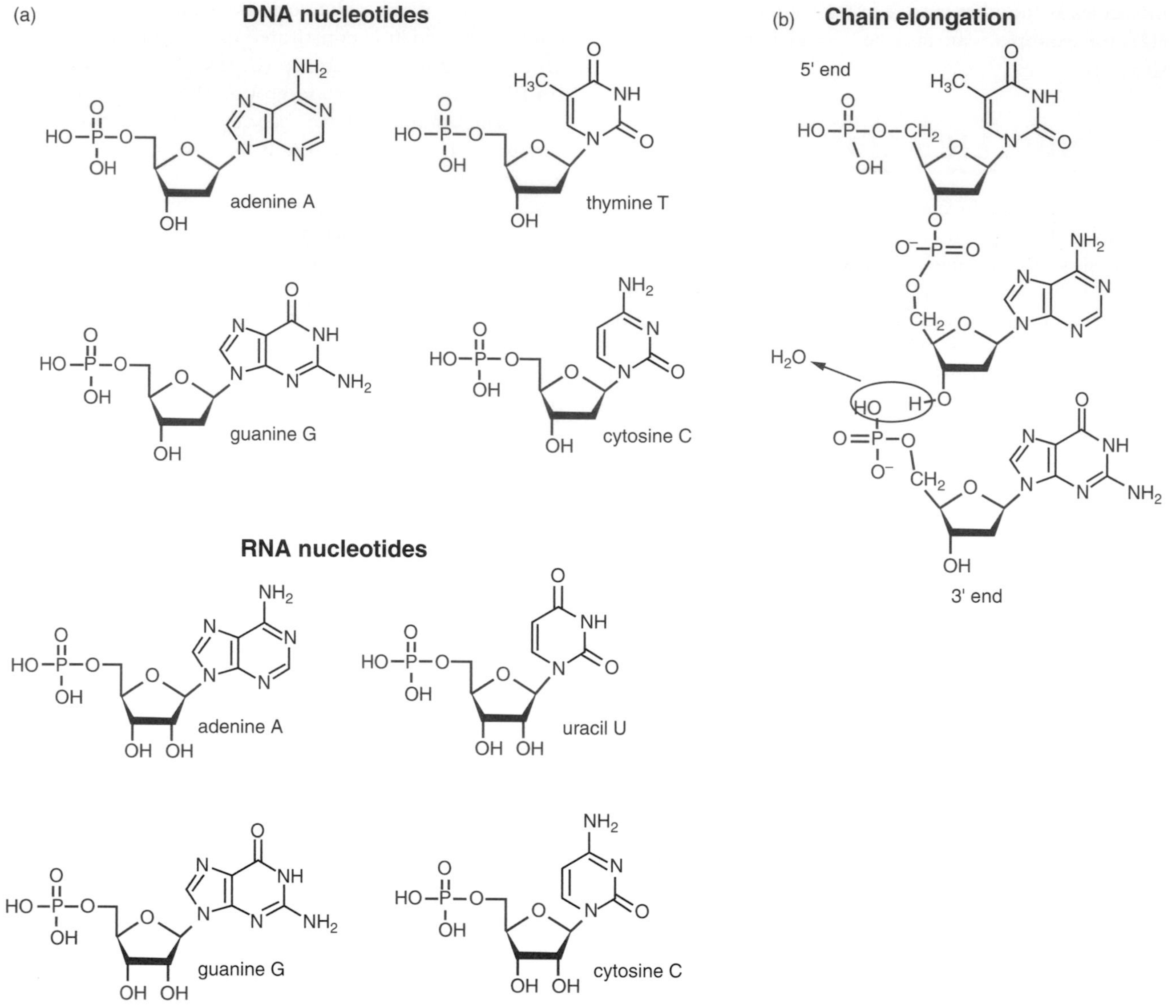

(a) **DNA nucleotides**

adenine A

thymine T

guanine G

cytosine C

RNA nucleotides

adenine A

uracil U

guanine G

cytosine C

(b) **Chain elongation**

5' end

H_2O

3' end

Fig. A2.10 (a) Chemical structures of the nucleotides, with the constant part shown in blue, made up of a deoxyribose (in DNA) and ribose (in RNA) and a phosphate group bound to the 5′ carbon of the sugar (sugar carbons are numbered clockwise 1′, 2′, etc., starting to the right of the ring oxygen atom), and the four natural bases in black. The base and sugar group together are called a nucleoside. (b) Chain elongation is via the loss of a water molecule and association of the oxygen on the 3′ carbon with the phosphate group of the next nucleotide in the chain. The polynucleotide chain displays a 5′ and a 3′ end. A gene is read in the 5′ to 3′ direction, which is also the direction of chain growth during replication and transcription.

itself in hairpin loops to form double-helical stem regions of A–U and G–C base pairs, leading to a much richer variety of secondary and tertiary structures than for DNA.

The family of tRNA molecules displays typical *clover leaf* secondary structures and *L-shaped* tertiary structures. In general, however, RNA secondary and tertiary structures can be extremely complex as is the case for 23 S ribosomal RNA, for example (Fig. A2.14). Secondary structure predictions for the larger molecules result in a number of possible patterns. RNA tertiary structures have proven difficult to study because of their relatively labile character.

A2.5 CARBOHYDRATES

Carbohydrate means "watered carbon" and members of this molecular family can be represented by the general formula $C_x(H_2O)_x$ with more or less minor modifications. Carbohydrates serve living organisms as structural components and as energy and carbon sources, as signaling molecules and as mediators of cell–cell interactions as well as of interactions between different organisms. *Saccharide* (from the Latin "*saccharum*" through the Greek "*sakcharon*" meaning "sugar") entities are the basic components of the highly complex carbohydrate molecules

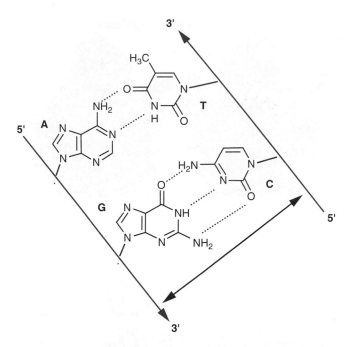

Fig. A2.11 Watson–Crick base-pairs, A–T and G–C. Note the equal widths of their structures, which permits the formation of a regular double helix.

encountered in *glycobiology* (a term coined in 1988 to recognize the combination of the traditional discipline of carbohydrate biochemistry with the modern understanding of the role of complex sugars in cellular and molecular biology).

Biological carbohydrates are divided into *monosaccharides* (single sugar subunits, e.g., *glucose*, from the Greek *"glycys,"* "sweet"; *-ose* is the generic nomenclature ending chosen for the monosaccharides), *disaccharides* (two covalently bound sugar subunits, e.g., *sucrose*), *oligosaccharides* ("a few" or several covalently bound sugar subunits), and *polysaccharides* ("many" covalently bound sugar subunits) – the division between the oligo- and polysaccharide groups being fairly loose.

Cellulose, a large linear polymer of glucose subunits, is the principal structural component in plants and the most common natural polysaccharide. *Glycogen* (from "sweet" and the root of *"gennaein,"* "to produce" or "give birth") is a complex branched polysaccharide of glucose subunits, stored as an energy source in liver and muscle cells. Linear and branched polysaccharide mixtures compose the *starch* (from the old English *"stercan"* meaning "to stiffen") granules, which represent the main energy storage mechanism in plants.

All cell types and many biological macromolecules carry complex arrays of oligosaccharide chains called *glycans*, which also occur as free standing entities. Most glycans are bound to excreted macromolecules or to protein or lipid molecules on the outer surfaces of cells, which are surrounded by a specific carbohydrate-rich "shell" called the *glycocalyx* (from the Greek *"kalyx,"* a

Fig. A2.12 The DNA double helix showing how the bases (in gray) are accessible for interaction in major and minor grooves.

covering, *not* from the Latin *"calix,"* a cup). Glycans participate in specific cell–cell and cell–macromolecule recognition events, cell matrix interactions in higher organisms, and interactions between different organisms, such as the ones between a parasite and its host. They are topics of intense study because of their implications for disease through immune response or tumor proliferation, for example. Glycans are immunogenic and are recognized and bound specifically by *antibody* molecules. *Lectins* are non-antibody proteins that bind carbohydrates without modifying them.

Polypeptides and polynucleotides are linear chains, each containing one type of bond between their constituent amino acids or nucleotides, respectively. Glycans, on the other hand, result from a very large number of subunit bonding possibilities that can create infinitely complex branched polymeric structures. There are two kinds of linkage (called α and β) between any pair of several positions to form a disaccharide from two monosaccharide subunits. We can easily imagine how the structural possibilities increase dramatically as the number of subunits increases. Fortunately, however, naturally occurring glycans contain relatively few

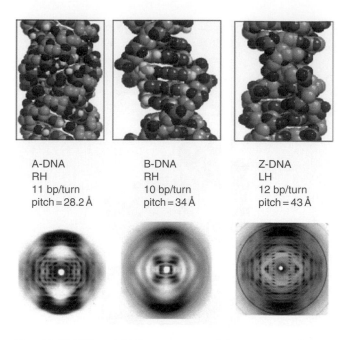

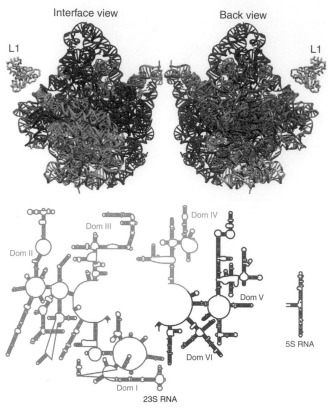

Fig. A2.13 Different DNA structures and the corresponding fiber diffraction diagrams. Right-handed A-DNA changes to right-handed B-DNA when the relative humidity is increased from 75% to 92%; Z-DNA is a *left-handed* helix observed for poly GC sequences in high salt (Fuller *et al.*, 2004). (Figure reproduced with permission from the Royal Society.)

Fig. A2.14 Secondary and tertiary structures of 23 S rRNA from the 50 S subunit (Ramakrishnan and Moore, 2001). (Figure reproduced with permission from Elsevier.)

monosaccharide subunit types in a limited number of combinations (if it were otherwise, structural studies of glycans, which are already very difficult, might be close to impossible).

A2.5.1 Chemical Composition and Primary Structure

The chemical structures of monosaccharides were first described by E. Fischer, who also discovered their stereoisomerism, at the end of the nineteenth century. Glucose, fructose, and galactose (we recall the -ose suffix is typical in carbohydrate nomenclature; five- and six-carbon carbohydrates, for example, are named *pentoses* and *hexoses*, respectively), for example, are stereoisomers; they have the same chemical composition ($C_6H_{12}O_6$) yet differ by the structural arrangement of the atoms, which gives them distinct properties and characteristics. Biological systems are very sensitive to the stereoisomer state (we have already mentioned that natural amino acids are in the L enantiomeric form), since active-site recognition by an enzyme, for example, depends on a precise atomic structural arrangement.

Monosaccharides are divided into two classes, *aldoses* and *ketoses*, according to whether they contain a functional *aldehyde* group (–CH=O) or *ketone* group (>C=O). They are then divided into subclasses according to the number of carbon atoms: aldotriose (three carbons), aldotetrose (four

carbons), ketotriose, ketotetrose, etc. The open and closed forms of glucose, mannose, and galactose are shown in Fig. A2.15. The Fisher and Haworth projections are accurate with respect to the orientation of the groups, but misleading in that the ring is not planar but can take various "chair"-like conformations, as shown.

Common monosaccharide subunits found in glycans of higher animals are given in Table A2.1. The molecules in Table A2.1 are dominant in higher animal (and of course human) glycobiology; several other types are found in the lower animals, plants, and microorganisms. Note how chemical modifications involving the monosaccharide hydroxyl, amino, and carboxyl groups further increase the variety of structures and enrich polysaccharide biological functionality.

In solution, most sugar molecules form cyclic five- or six-member ring structures that do not display free aldehyde or ketone groups. Ring closure creates a further asymmetric center at the original carbonyl carbon, termed the *anomeric* carbon, so that D-glucose, for example, can exist in two forms called α and β. Glycans are often found associated with a non-carbohydrate moiety, usually protein or lipid, to form a *glycoconjugate*, and are defined according to the linkage between them. When a protein or a lipid forms a glycoconjugate it is said to be *glycosylated*.

Microheterogeneity renders the study of protein glycosylation particularly difficult. It describes the observation

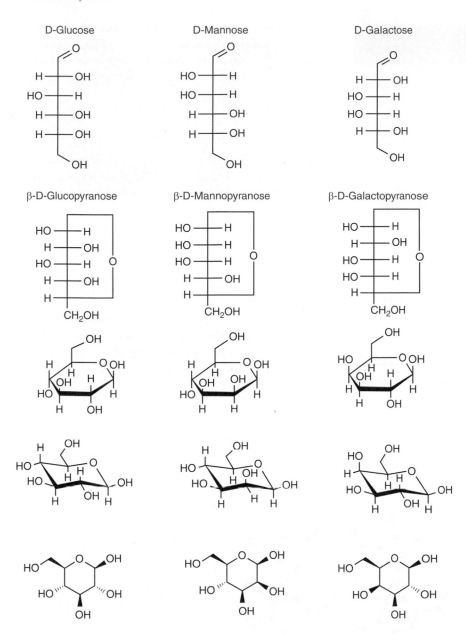

Fig. A2.15 Row 1: the open form of three aldohexoses. The closed forms are shown in different projections in the following rows: row 2, Fisher projection; row 3, Haworth projection; row 4, the "chair" projection; row 5, the stereo projection.

that a range of different glycans are found on the same glycosylation site of a given protein.

As carbohydrate primary structures are solved they are deposited in a database that can be consulted by computer. The existence of branching significantly increases the complexity of these primary structures and special software has been developed to access and analyze them.

A2.5.2 Higher-Order Structures

Cellulose is a homopolymer of glucose subunits, organized in a well-defined structure that has been studied by electron microscopy and diffraction (Fig. A2.16). Homo- and hetero-polysaccharides associated with protein or lipid also form ordered structures that fulfill structural roles in connective tissue or cell walls.

High-resolution structures of protein glycoconjugates, showing the nature of the linkage and three-dimensional atomic organization, have been solved by X-ray crystallography and NMR. Solution NMR is the preferred technique because even though mono- and disaccharide subunits are relatively rigid molecules, oligosaccharides usually show a degree of flexibility that makes crystallization difficult (see also Section J3.2.3).

Parasite, bacterial, and viral infections often involve specific binding between adhesion microbial proteins and glycans presented on the host cell surface. Cholera toxin is an adhesion protein that binds to a cellular oligosaccharide. The structure of the cholera toxin adhesion protein with a bound sugar has been solved by X-ray crystallography to high resolution, and provides a good picture of interactions between carbohydrate and protein atoms (Fig. A2.17).

TABLE A2.1. MONOSACCHARIDE SUBUNITS IN GLYCANS OF HIGHER ANIMALS
Pentose: five carbon sugar xylose (Xyl)
Hexoses: neutral sugars including glucose (Glc), galactose (Gal), and mannose (Man) (Fig. A2.15)
Hexosamines: hexose with either a free or N-acetylated amino group on carbon 2, N-acetylglucosamine (GlcNAc), N-acetylgalactosamine (GalNAc)
Deoxyhexoses: hexose without the hydroxyl group on carbon 6, fucose (Fuc)
Uronic acids: hexose with a carboxylate on carbon 6, glucuronic acid (GlcA), iduronic acid (IdA)
Sialic acids (Sia): nine-carbon acidic sugars, mainly N-acetyl neuraminic acid (Neu5Ac, NeuNAc, NeuAc, or NANA)

A2.6 LIPIDS

Lipids may be defined loosely as a class of molecules whose properties are dominated by a water-insoluble moiety. They constitute a diverse group of compounds involved in many aspects of cellular function. Lipid biochemistry is correspondingly rich; lipid molecules serve as sources of energy, they participate in complex pathways such as blood-clot formation, and are necessary for the normal function of the nervous system. With respect to macromolecular structure and dynamics, however, we are concerned mainly with the structural roles of lipids in biological membranes and their interaction with membrane-associated proteins.

A2.6.1 Chemical Composition

The main lipid components of biological membranes can be presented schematically as being made up of a water-soluble charged or polar headgroup and water-insoluble apolar hydrocarbon chains. The general composition of phosphoglycerides and phospholipids is given in Fig. A2.18. The chains are composed of fatty acids (black) that can be *saturated* (containing only single carbon–carbon bonds) or *unsaturated* (containing one or more double bonds). They are associated by ester linkages to the glycerol group (red). The headgroups are shown in blue.

Biological membranes contain many different types of lipid. In addition to phosphatidyl ethanolamine and phosphatidyl choline, higher animal membranes, for example, contain significant amounts of sphingolipids and cholesterol (Fig. A2.19). Cholesterol is a *steroid*, which fits between the fatty acid chains and changes the fluidity of the membrane structure. The apolar chains in lipids are not necessarily fatty acid derivatives. Lipid hydrocarbon chains in the membranes of Archaea, for example, are isoprenoid derivatives, similar to the *phytol* chains attached to chlorophyll in plants, containing alternate

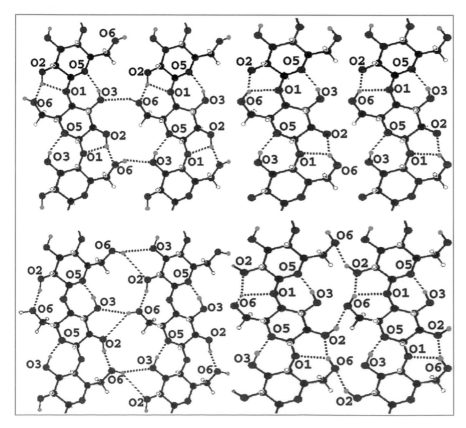

Fig. A2.16 Cellulose structure and hydrogen bonding pattern. Carbon, oxygen, hydrogen, and deuterium atoms are colored black, red, white, and green, respectively. Hydrogen bonds are represented by dotted lines. Four alternative hydrogen bonding patterns have been identified, and are shown on the four panels (Nishiyama *et al.*, 2002) (Figure reproduced with permission from the American Chemical Society.)

(a)

(b)

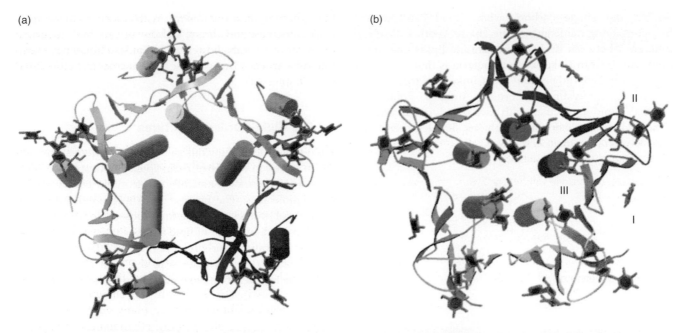

Fig. A2.17 A comparison of the receptor-binding sites of the cholera toxin CT and Shiga toxin SHT families (bound sugars are shown in stick representation) **(a)** CT B pentamer showing five binding sites in the ganglioside oligosaccharide. **(b)** SHT family pentamer showing 15 binding sites for the ganglioside (three per monomer labeled I,II,III). (Fan *et al.*, 2000). (Figure reproduced with permission from Elsevier.)

(a)

(b)

(c)

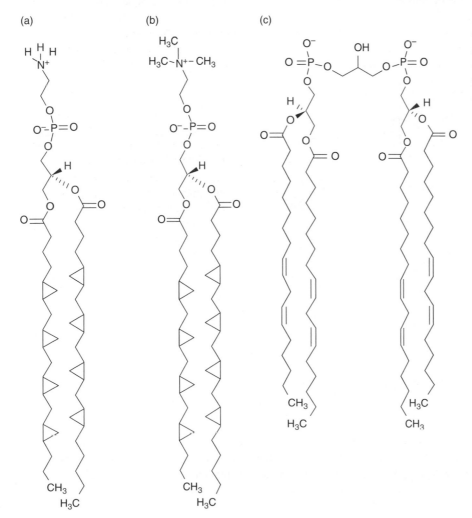

Fig. A2.18 (a) Phosphatidyl ethanolamine; **(b)** phosphatidyl choline; and **(c)** cardiolipin. For **(a)** and **(b)** the name describes the headgroup and the fatty acid chains may have different compositions (see text).

double and single carbon–carbon bonds and methyl branches along the length of the chain. Another chemical difference between Archaeal membrane lipids and membrane lipids from Eukarya and Bacteria is that the chains are ether-linked rather than ester-linked to the glycerol backbone. Interestingly, certain Archaea contain a "double" lipid molecule, with two headgroups one on either side of the hydrocarbon chains. These glycerol dialcyl glycerol tetraethers were first discovered in

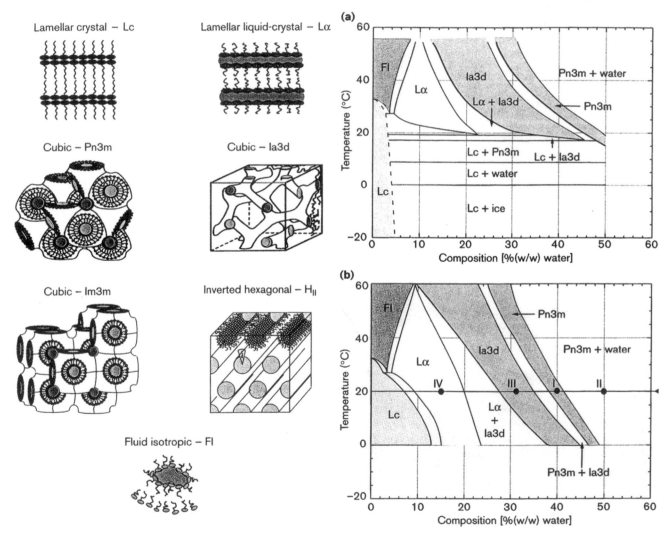

Fig. A2.19 Cholesterol.

hyperthermophilic organisms, in which they presumably help to increase membrane stability at very high temperatures. More recently, however, similar structures have been found in Archaea living in cool environments, below 20 °C (see Chapter A1).

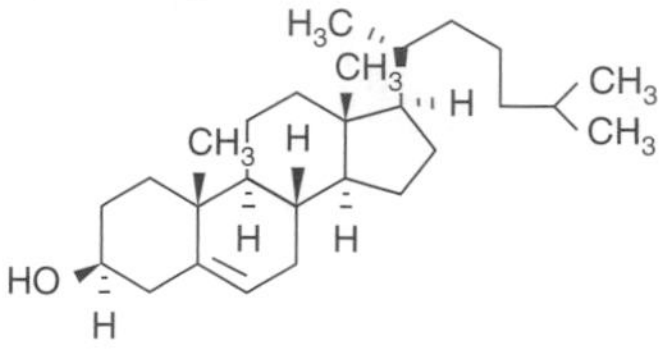

A2.6.2 Higher-Order Structures

Lipid molecules in aqueous solution spontaneously associate in order to sequester their apolar, hydrophobic moieties from thermodynamically unfavorable contacts with water molecules. The process leads to the existence of a number of interesting structural phases for lipid–water mixtures, as a function of lipid composition, water-to-lipid ratio, temperature, and pressure (Fig. A2.20).

Lamellar bilayer structures in the fluid phase (L_α), in which the hydrocarbon chains are disordered, are the most

Fig. A2.20 Structural phases (presented schematically on the left) and the phase diagram for monoolein–water mixtures. Monoolein is a single hydrocarbon lipid that is not found in natural membranes but is commonly used in membrane protein crystallization. The fluid lamellar bilayer phase (L_α) corresponds to the state of most natural membranes (Caffrey, 2000). (Figure reproduced with permission from Elsevier.)

likely phases in biological membranes, providing an extremely efficient, essentially planar, permeability barrier, which still allows functional flexibility and lateral diffusion motions of associated membrane proteins. With natural membrane lipids, the L_α phase is favored by lipid mixtures containing unsaturated hydrocarbon chains, higher temperatures, and water content. It is interesting to note that in organisms exposed to low temperatures, the membrane lipid composition adapts in order to maintain a fluid bilayer structure. In hyperthermophilic Archaea, a "bilayer" structure is obtained with a monolayer of the double-polar head lipid found in these organisms. The cubic phase has been suggested but has not been proven to occur in certain natural membranes.

A2.6.3 Lipids and Membrane Proteins

Membrane proteins are intimately associated with lipids. They have a hydrophobic character, which makes their study by the usual biochemical methods very difficult. Membrane protein studies have, therefore, lagged far behind those of water-soluble proteins. The development of detergent methods allowed active membrane proteins to be purified out of their lipid environment and in some cases to be crystallized. A Nobel Prize was awarded for the first high-resolution membrane protein structure obtained by X-ray crystallography from crystals containing detergent. Obtaining membrane protein crystals, however, remains a formidable task. The phase diagram in Fig. A2.20 was studied in the context of membrane protein biophysics. The cubic phase of monoolein, in particular, was found to favor membrane protein crystallization for high-resolution structural studies (see Part G).

A2.7 CHECKLIST OF KEY IDEAS

- Living organisms are classified, by their genome sequences into three kingdoms: *Bacteria, Archaea*, and *Eukarya*.
- The flow of genetic information in all known organisms is from DNA to RNA to protein.
- The scheme is a product of a high degree of evolution since sophisticated regulated catalytic mechanisms by the final product, protein, are required for all the other steps: from DNA replication, RNA transcription and reverse transcription to genetic translation to protein biosynthesis itself.
- Biological macromolecules, although they are made up of a concatenation of subunits, have evolved to fulfill specific functions and have specific properties that are very different from those of classical polymers.
- All the molecules in a living organism are either proteins or can be considered as products of protein action.
- Proteins are made up of properly folded polypeptides of amino acid residues, and may include *prosthetic*

groups with specific properties (such as the heme group, which binds oxygen).
- Color in proteins (e.g., the red in hemoglobin, or green in chlorophyll binding proteins) is always due to a prosthetic group, because amino acids absorb in the UV region.
- There are 20 main amino acids in natural proteins, with a variety of chemical characteristics: acid and base, polar and non-polar, aliphatic and aromatic.
- *Primary structure* is the subunit sequence in the macromolecule; *secondary structures* are favored local chain conformations arising from chemical and steric constraints; *tertiary structure* is the three-dimensional conformation of the macromolecular chain (given by the coordinates of the constituent atoms); *quaternary structure* is the organization of different or similar chains in a macromolecular complex.
- The secondary structures of proteins can be expressed on a *Ramachandran plot* in terms of angles of rotation of the peptide planes in the chain around the so-called alpha-carbons, to which the amino acid side-chains are bound.
- α-helices and β-sheets are the main secondary structures found in proteins.
- The tertiary structure results from weak (non-covalent, except for the disulfide bond between two cysteines) interactions between the amino acids in the polypeptide chain.
- Protein domains with distinct features have been identified in the solved tertiary structures.
- Protein domains have been classified in groups or families, in levels, according to their architecture, topology, homology, and sequence.
- The organization of proteins in quaternary structures has been selected by evolution because of functional advantages with respect to stability, allosteric control, and the coordination of complex reaction mechanisms by large molecular machines in the cell, such as ribosomes, proteasomes, thermosomes, and chaperone complexes.
- DNA is the repository of genetic information; its structure is predominantly double-stranded, with a helical organization depending on its environment.
- RNA displays a range of biological functions and presents a great variety of structural organization.
- RNA is usually made up of a single strand that, nevertheless, turns around and interacts with itself to form intricate secondary structures.
- Ribozymes are RNA molecules with catalytic activity; their discovery strengthened the *RNA World* hypothesis of primitive life forms based on RNA, both for genetic storage and catalysis of biological reactions.
- Since the purification of the enzyme RNase P, which contains both RNA and protein active components, many types of catalytic RNA molecules have been and are being discovered. These ribozymes, and portions of rRNA play important catalytic roles in translation, microRNAs like siRNA, small interfering RNA molecules involved in gene-silencing by RNA interference (RNAi) with mRNA.

These are intensely studied because of their implications in cancer and viral infection and potential therapeutic applications.

- DNA is composed of a polynucleotide chain of four main nucleotides, denoted by their bases: adenine, thymine, guanine, and cytosine.

- In RNA, thymine is replaced by uracil.

- Bases may present chemical modification relevant to biological activity such as methylation in DNA or the more complex modifications observed for tRNA.

- Carbohydrates serve living organisms as structural components, energy, and carbon sources, signaling molecules and mediators of cell–cell interactions as well as of interactions between different organisms.

- Polysaccharides can be simple linear chains or more complex branched chains composed of the same or different covalently bound monosaccharide units.

- All cell types and many biological macromolecules are *glycosylated* by binding complex arrays of oligosaccharide chains called *glycans*, to form glycoconjugates.

- Glycans are immunogenic and are recognized and bound specifically by *antibody* molecules.

- *Lectins* are non-antibody proteins that bind glycans without modifying them.

- Monosaccharides can exist in different structural forms called *stereoisomers*.

- *Microheterogeneity* renders the study of protein glycosylation particularly difficult. It describes the observation that a range of different glycans are found on the same glycosylation site of a given protein.

- Parasite, bacterial, and viral infections often involve specific binding between adhesion microbial proteins and glycans presented on the host cell surface.

- Lipids are the main components leading to the passive permeability barrier in biological membranes.

- The main lipid components of biological membranes can be presented schematically as being made up of a water-soluble charged or polar headgroup, and water-insoluble apolar hydrocarbon chains.

- Lipid molecules in aqueous solution spontaneously associate in order to sequester their apolar, hydrophobic moieties from thermodynamically unfavorable contacts with water molecules.

- Lipid–water mixtures display various interesting structural phases as a function of lipid composition, water-to-lipid ratio, temperature, and pressure.

- The so-called L_α phase, in which the lipid hydrocarbon chains are in a fluid, liquid crystalline state, is the most relevant for natural biological membranes.

- The so-called cubic phases obtained under artificial conditions have been useful for the crystallization of membrane proteins.

Suggestions for Further Reading

Woese, C. (1967). *The Genetic Code*. New York: Harper and Row.

Crick, F. H. C. (1968). The origin of the genetic code. *J. Mol. Biol.*, **38**, 367–379.

Orgel, L. E. (1968). Evolution of the genetic apparatus. *J. Mol. Biol.*, **38**, 381–393.

Dickerson, R., and Geiss, I. (1969). *The Structure and Action of Proteins*. Menlo Park, CA: Benjamin Cummings.

Gilbert, W. (1986). The RNA World. *Nature*, **319**, 618.

Lee, R. C., Feinbaum, R. L., and Ambros, V. (1993) The *C. elegans* heterochronic gene lin-4 encodes small RNAs with antisense complementarity to lin-14. *Cell*, **75**, 843–854.

Merrit, E. A., Sarfaty, S., van den Akker, F., *et al.* (1994). Crystal structure of cholera toxin B-pentamer bound to receptor G_{M1} pentasaccharide. *Prot. Sci.*, **3**, 166–175.

Hirao, I., and Ellington, A. D. (1995). Re-creating the RNA World. *Curr. Biol.*, **5**, 1017–1022.

Branden, C., and Tooze, J. (1999). *Introduction to Protein Structure*. New York: Garland Science.

Varki, A., Cummings, R., Esko, J. *et al.* (eds) (1999). *Essentials of Glycobiology*. New York: CSHL Press.

Langan, P., Nishiyama, Y., and Chanzy, H. (1999). A revised structure and hydrogen bonding system in cellulose II from a neutron fibre diffraction analysis. *J. Am. Chem Soc.*, **121**, 9940–9946.

Caffrey, M. (2000). A lipid's eye view of membrane protein crystallization in mesophases. *Curr. Opin. Struct. Biol.*, **10**, 486–497.

Schouten, S., Hopmans, E. C., Pancost, R. D., and Damste, J. S. S. (2000). Widespread occurrence of structurally diverse tetraether membrane lipids: Evidence for the ubiquitous presence of low-temperature relatives of hyperthermophiles. *Proc. Natl. Acad. Sci. USA*, **97**, 14 421–14 426.

Reinhart, B. J., Slack, F. J., Basson, M., *et al.* (2000) The 21-nucleotide let-7 RNA regulates developmental timing in *Caenorhabditis elegans*. *Nature*, **403**, 901–906.

Dykxhoorn, D. M., and Novina, C. D. (2003). Killing the messenger: Short RNAs that silence gene expression. *Nat. Rev. Mol. Cell Biol.*, **4**, 457–467.

Medina, P. P., Nolde, M., and Slack, F. J. (2010) OncomiR addiction in an in vivo model of microRNA-21-induced pre-B-cell lymphoma. *Nature*, **467**, 86–90.

Salmena, L., Poliseno, L., Tay, Y., Kats, L., and Pandolfi, P. P. (2011) A ceRNA hypothesis: The Rosetta Stone of a hidden RNA language? *Cell*, **146**, 353–358.

www.glycopedia.eu. A glycobiology website with news, e.chapters, and resources such as for 3-D visualization of complex carbohydrates, polysaccharides, and glyco-conjugates.

UNDERSTANDING MACROMOLECULAR STRUCTURES

A3

A3.1 HISTORICAL REVIEW

In **1963 Richard Feynman** traced to **Pythagoras** (*c*.**500 BC**) the first example, outside geometry, of the discovery of a numerical relationship in Nature. Pythagoras' discovery, which led to the foundation of a school of thought with mystic beliefs in the power of numbers, was that two strings under the same tension but of different lengths give a pleasant sound when plucked together if the ratio of their lengths is that of two small integers. We now say that a ratio of 1:2 corresponds to an octave, a ratio of 2:3 to a fifth, and so on, which are all harmonic-sounding chords. Feynman analyzed this discovery in terms of three characteristics: its basis in experimental observation; the use of mathematics as a tool for understanding Nature; and its concern with aesthetics (the "pleasant" quality of the sound). With easy hindsight (as he readily admits!) Feynman wrote that if Pythagoras had been more impressed by the first point on the importance of experimental observation the science of physics might have had an earlier start.

We discovered the existence and seek our biophysical understanding of macromolecules through experimentation, and we use mathematical tools not only to set up experiments and analyze the results, but also for the description of the studied "objects" themselves. The basis of aesthetics may still remain as mysterious as in Pythagoras' time, but there is no doubt that the "beauty" of the DNA double helix and the satisfyingly elegant way in which it provided an explanation for how genetic information is stored and transmitted was an essential inspiration in the development of modern molecular biology. Similarly, the usual, colorful illustrations of protein structural models are undoubtedly aesthetic and certainly play a role in the acceptance and understanding of these models by biologists, who might have been put off by a purely mathematical interpretation of the data, even if it were more accurate.

Experiments on biological macromolecules are difficult to perform and interpret. Sample material is fragile and a good biophysical experiment must rest on a firm biochemical foundation. Biological macromolecules are much smaller than the wavelength of light and cannot be "seen." We gain information on their structures and dynamics by shining suitable electromagnetic or other types of radiation on a sample and analyzing what emerges. The sample may absorb or scatter the radiation. Energy differences between the incident and emergent radiation are analyzed by *spectroscopy*, whereas in *diffraction* experiments there is no energy difference and the information is obtained from the intensity of the scattered beam as a function of scattering angle. The interpretation of such experiments relies entirely on the physics and mathematics of *waves* and their interactions, be it in the classical description of crystallography in terms of interference of scattered waves, or in the quantum chemical description necessary for the interpretation of electromagnetic spectroscopy. The work of **Joseph Fourier (1768–1830)** on heat transmission in terms of waves has led to an extraordinary set of mathematical tools for the interpretation and understanding of the interactions between radiation and matter.

A useful approach to understanding macromolecular structures is based on an understanding of the forces underlying them – the forces acting to maintain the atoms in their positions in the correctly folded active conformation of the molecule. If we achieve an understanding of these forces, then we shall also understand how the atoms in the structure move about their mean locations. Expressed in the language of physics: Our aim is to reach a sufficient understanding of the *force field* or *potential energy function* around each atom in a structure, in order to simulate its molecular dynamics.

In **1975**, the first molecular dynamics simulation of the time course of atomic motions in a biological macromolecule, the small stable protein bovine pancreas trypsin inhibitor (BPTI) for which an accurate X-ray crystallographic structure was available, was published by **J. A. McCammon, B. R. Gelin**, and **M. Karplus**. Despite it being limited to a duration of less than 10 ps by the computing power available then, the BPTI study was effectively instrumental, together with the 1955 hydrogen exchange experiments of **K. Linderstrom-Lang** and his collaborators, in establishing the view that proteins are not rigid bodies but dynamic entities whose internal motions must play a role in their biological activity.

The decades following the publication of the BPTI simulation, the **1980s** and **1990s**, saw a significant extension of the calculations to larger proteins and longer durations, as higher-resolution structures and more powerful computers became available. Parallel developments in experimental approaches provided information on macromolecular energies and internal motions and were essential in order to validate the simulation methods. These included microcalorimetry (see Part C), the analysis of temperature factors in crystallography (see Part G), fluorescence depolarization (see Part D), NMR (see Part J), inelastic neutron scattering (see Part I) and Mössbauer spectroscopy, Fourier transform infrared spectroscopy (see Part E), and various fast kinetics measurements using laser flashes to trigger reactions. Synergistic relationships were established between molecular dynamics calculations and NMR and X-ray crystallography, starting with the incorporation of energy minimization in structural refinement. The dynamical transition in proteins was discovered by Mössbauer spectroscopy (**F. Parak** and collaborators) and inelastic neutron scattering (**W. Doster** and collaborators). **H. Frauenfelder** and collaborators published the conformational substate (CS) model for proteins, providing a physical framework for the understanding of macromolecular structure and dynamics.

Molecular dynamics simulations have now been applied on the supramolecular and even cellular scale. Experimental techniques are being refined continuously and new ones are being developed, such as real-time crystallography and kinetic cryocrystallography, in which intermediate structures in an enzyme catalytic cycle are "frozen" and examined separately.

A3.2 BASIC PHYSICS AND MATHEMATICAL TOOLS

Practically all the experiments designed to analyze biological macromolecules (with the exception of calorimetry and classical solution physical chemistry methods such as the ones used to determine osmotic pressure, viscosity, etc.) rely completely or in part on the observation of their interaction with radiation. This is patently obvious for diffraction- and spectroscopy-based methods, like crystallography or NMR, but it is also true for hydrodynamics-based methods such as analytical ultracentrifugation or dynamic light scattering, in which the presence of the macromolecule in the solution is detected via its absorption or scattering of radiation. For most purposes, the sample is a *black box* into which we shine radiation of known properties and analyze what comes out (Fig. A3.1). In order to be able to deal with the interactions between radiation and matter, we require some knowledge of the mathematical tools that describe waves, and also of quantum mechanics, in which it is useful to consider the particle-like properties of radiation beams and radiation-like properties of moving particles.

A3.2.1 Waves

Sines and Cosines
The sine or cosine function is the simplest way to describe the periodic rise and fall of a wave (Fig. A3.2). A wave is

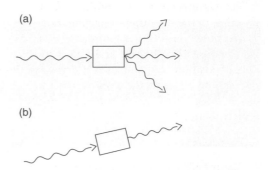

Fig. A3.1 The sample is a "black box" into which radiation of known wavelength is shone: **(a)** diffraction experiment – the scattering intensity is measured as a function of angle; **(b)** spectroscopy experiment – the energy of the emerging beam is analyzed.

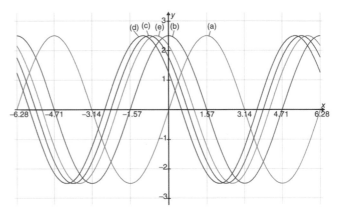

Fig. A3.2 Waves represented by functions:
(a) $Y = A \sin X$; (b) $Y = A \cos X$;
(c) $Y = (A/\sqrt{2})(\cos X - \sin X)$;
(d) $Y = (A/2)(\cos X - \sqrt{3} \sin X)$;
(e) $Y = (A/2)(\sqrt{3} \cos X - \sin X)$ (see text).

characterized by various parameters. Its *amplitude*, A, represents the maximum rise above the average level; the *intensity* of the wave is equal to the square of its amplitude; its *wavelength*, λ, is the repeat distance of a cycle in space; its *frequency*, f, is the number of cycles in unit time. The propagation velocity of the wave, v (the distance covered by a point in the cycle per unit time) is, therefore, given simply by its frequency multiplied by its wavelength:

$$v = f\lambda \tag{A3.1}$$

The $\sin X$ and $\cos X$ functions cycle with a period of 2π (Fig. A3.2). The waves with which we are concerned propagate in space or in time, or in both.

We first consider X in space, as in the case of a standing wave or a propagating wave observed at a specified time. X can then be written in terms of the wavelength as $kx = (2\pi/\lambda) x$, where x is distance, so that the wave is described by

$$Y = \cos kx \text{ or } Y = \sin kx \tag{A3.2}$$

where we have assumed unit amplitude for simplicity.

The *wavevector*, **k**, of magnitude $k = (2\pi/\lambda)$, is defined as pointing in the direction of propagation.

We now consider the parameter X as describing time; we are observing the rise and fall of the wave at one particular position in space, as a function of time. As above, taking into account the cycling with a period of 2π, we write $X = \omega t = 2\pi f t$, where t is time, and (again assuming unit amplitude)

$$Y = \cos \omega t \text{ or } Y = \sin \omega t \quad \text{(A3.3)}$$

$\omega = 2\pi f$ is called the angular frequency of the wave.

The phase difference, δ, describes the extent to which two waves of the same wavelength and frequency are out of step with each other (Fig. A3.2). If one wave is described by $\cos \omega t$, for example, then the other will be described by $\cos (\omega t + \delta)$. Waves with different phases are described by the functions $\cos(\omega t + \delta_1)$, $\cos(\omega t + \delta_2)$, $\cos(\omega t + \delta_3)$, etc. We note that the cosine and sine functions are related by a phase difference $\delta = \pi/2$ (Fig. A3.2a, b).

We recall from the mathematics of sines and cosines that

$$\cos (\omega t + \delta_1) = \cos \delta_1 \cos \omega t - \sin \delta_1 \sin \omega t \quad \text{(A3.4)}$$

Since $\cos \delta_1$ and $\sin \delta_1$ are constants, we can write

$$\cos (\omega t + \delta_1) = a_1 \cos \omega t + b_1 \sin \omega t \quad \text{(A3.5)}$$

In other words, a wave of given frequency and phase can be represented by the sum of a cosine and a sine with appropriate coefficients (a_1, b_1, respectively, in Eq. (A3.5)).

The wave described by the function $(\cos X - \sin X)$ is out of phase by $\delta = \pi/4$ with the wave described by $\cos X$ (Fig. A3.2b, c). (This can also be calculated from Eq. (A3.4); $\sin \delta = \cos \delta = 1/\sqrt{2}$ for $\delta = \pi/4$ (45°).) Similarly, waves that are out of phase with $\cos X$ by $\pi/3$ and $\pi/6$ are described by $(\cos X - \sqrt{3} \sin X)$ and $(\sqrt{3} \cos X - \sin X)$, respectively (Fig. A3.2d, e).

A sum of waves of the same wavelength and frequency but with different amplitudes and different phases can be expressed mathematically, therefore, as a sum of sine and cosine functions with appropriate coefficients. Two extreme cases are illustrated in Fig. A3.3: the sum of two waves with a phase difference of $n\pi$, where n is zero or an even integer, leading to *constructive interference*; and the sum of two waves with a phase difference of $n'\pi$, where n is an odd integer, leading to *destructive interference*.

The interference of waves of different frequency leads to more complicated pictures. Consider two waves of equal amplitude and angular frequencies, ω_1, ω_2, respectively. Applying the sine and cosine rules, their sum is given by

$$\cos \omega_1 t + \cos \omega_2 t = 2 \cos \tfrac{1}{2}(\omega_1 + \omega_2)t \, \cos \tfrac{1}{2}(\omega_1 - \omega_2)t \quad \text{(A3.6)}$$

The sum of waves of different frequency and different amplitude, A_1, A_2, respectively, is given by

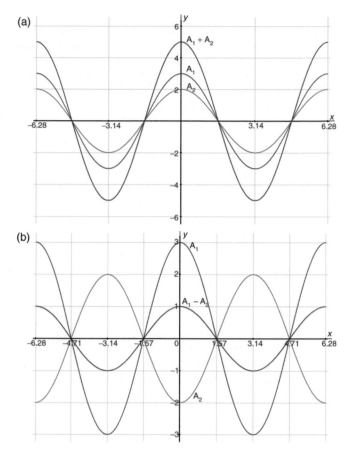

Fig. A3.3 **(a)** Constructive interference of two waves *in phase* (or out of phase by 2π, 4π, 6π, etc.)
$A_1 \cos X + A_2 \cos (X + n\pi) = (A_1 + A_2) \cos X$
where n is zero or an even integer. **(b)** Destructive interference of two waves, which are perfectly out of phase (phase difference $n'\pi$, where n' is an odd integer)
$A_1 \cos X + A_2 \cos (X + n'\pi) = (A_1 - A_2) \cos$.

$$A_1 \cos \omega_1 t + A_2 \cos \omega_2 t$$
$$= \cos \tfrac{1}{2}(\omega_1 + \omega_2)t \left[A_1 \cos \tfrac{1}{2} (\omega_1 - \omega_2)t + A_2 \cos \tfrac{1}{2}(\omega_1 - \omega_2)t \right] \quad \text{(A3.7)}$$

The wave sum for an equal-amplitude case is illustrated in Fig. A3.4. To a very good approximation, the resulting wave can be considered as having an average frequency $1/2$ $(\omega_1 + \omega_2)$, and an amplitude that oscillates with a lower frequency, $1/2(\omega_1 - \omega_2)$. The resulting wave is said to be *modulated* by the lower-frequency oscillation.

The phenomenon of *beats* occurs when ω_1 and ω_2 are close to each other ($\sim\omega$), as when we listen to the vibration of two strings of slightly different tension during the tuning of a guitar. The frequency of the *note* corresponds to $\sim\omega$, but its intensity *pulsates* or beats with a much lower frequency ($\omega_1 - \omega_2$); this frequency is twice the frequency of the amplitude oscillation, because the intensity is equal to the square of the amplitude (Fig. A3.4). The phenomenon of beats is observed in dynamic light-scattering experiments,

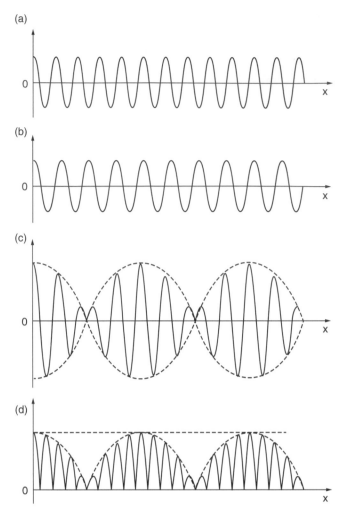

Fig. A3.4 The sum part **(c)** of two waves of slightly different frequency **(a)** and **(b)**. **(d)** The variation in intensity (the square of the amplitude) of the resultant wave on a relative scale.

in which light waves of slightly different frequency interfere after being scattered by molecules moving with different velocities (Doppler effect, see Chapter D4).

The velocity of a wave, as defined in Eq. (A3.1), corresponds to the velocity of propagation of a point of constant phase (e.g., a crest in the wave); it is called the *phase velocity* of the wave. Using the definitions of wavevector and angular frequency, given above, we rewrite Eq. (A3.1) as

$$v = \omega/k \qquad (A3.8)$$

which is the usual definition of phase velocity.

When we have a superposition of waves of different ω and k values traveling in the same direction, as in Fig. A3.4, the *group velocity*, v_g, of the resulting wave is the velocity of propagation of the modulation. It can be shown that

$$v_g = \frac{d\omega}{dk} \qquad (A3.9)$$

The relation between the ω and k values of a wave is called the *dispersion relation*. The phase and group velocities are

clearly equal for waves described by a dispersion relation ω proportional to k.

In order to calculate the group velocity of the resultant wave in Fig. A3.4, i.e., the velocity of the modulation, we first add to Eq. (A3.7) the oscillation along the length dimension, x:

$$A_1 \cos(\omega_1 t - k_1 x) + A_2 \cos(\omega_2 t - k_2 x)$$
$$= \cos\frac{1}{2}[(\omega_1 + \omega_2)t - (k_1 + k_2)x]$$
$$\times \left\{ A_1 \cos\frac{1}{2}[(\omega_1 - \omega_2)t - (k_1 - k_2)x] \right.$$
$$\left. + A_2 \cos\frac{1}{2}[(\omega_1 - \omega_2)t - (k_1 - k_2)x] \right\} \qquad (A3.10)$$

The time and length terms have opposite signs because, as the wave moves forward, a crest at a certain point, for example, results from the arrival of a crest that was one wavelength *behind* in the *previous* time period of the wave. The modulation arises from the cosine terms in the curly brackets on the right-hand side of Eq. (A3.10), and its speed is equal to $v_{mod} = (\omega_1 - \omega_2)/(k_1 - k_2)$. In the limit of very small frequency and wave vector differences v_{mod} is equal to the group velocity defined in Eq. (A3.9).

Complex Exponentials

Complex numbers were invented in order to solve equations such as $x^2 = -1$, but because of their properties they became powerful mathematical tools for a large range of problems in physics.

A complex number a is written $x + iy$, where i is the square root of -1; x is called the "real" part of the complex number and y the "imaginary" part. Note that in mathematical terms there is nothing imaginary about that part!

Sums and products of complex numbers are themselves complex numbers:

$$\begin{aligned}(x + iy) + (x' + iy') &= (x + x') + i(y + y') \\ (x + iy)(x' + iy') &= (xx' - yy') + i(xy' + x'y) \\ a^* &= x - iy \\ aa^* &= x^2 + y^2 \end{aligned} \qquad (A3.11)$$

where a^* is called the complex conjugate of a.

The complex number a can be represented on an xy plot as a radial line of length A (called the amplitude) making an angle ϕ (called the phase) with the x-axis, the "real" axis; the y-axis is the "imaginary" axis (Fig. A3.5). Such a plot is called an *Argand diagram*. We see from the figure that $A = \sqrt{(x^2 + y^2)}$, and the complex number can now be written as $a = A(\cos\phi + i\sin\phi)$. We know from mathematics, however, that the *complex exponential*

$$\exp i\phi = \cos\phi + i\sin\phi \qquad (A3.12)$$

so that $a = A \exp i\phi$.

Feynman called Eq. (A3.12) "the most remarkable formula in mathematics ... our jewel"! The formula embodies the connection between algebra (complex numbers) and geometry (sines and cosines defined as ratios of sides in

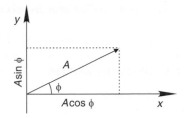

Fig. A3.5 The complex number $a = x + iy$ expressed geometrically on an xy plane. It can also be written $a = A \exp i\phi$, where A is called the amplitude and ϕ, the phase.

$x = A \cos \phi$

$y = A \sin \phi$

$A = \sqrt{(x^2 + y^2)}$

right-angle triangles), and it is in fact amazing how its properties greatly simplify the mathematical operations in which it is involved. The main one of these properties is

$$\exp[i(\phi_1 + \phi_2)] = \exp(i\phi_1) \exp(i\phi_2) \qquad (A3.13)$$

multiplying the exponential by its complex conjugate, we have

$$\exp(i\phi) \exp(-i\phi) = \exp 0 = 1$$

A complex number can be written, therefore, in terms of real and imaginary parts as $x + iy$ or in terms of an amplitude value and a phase angle in a complex exponential as $A \exp i\phi$. Note that, similarly to Eq. (A3.11), the square of the amplitude of the complex exponential is not obtained by squaring the exponential but by multiplying it by its complex conjugate

$$aa^* = A \exp(i\phi A) \ \exp(-i\phi) = A^2 \qquad (A3.14)$$

There are considerable advantages in describing waves in terms of complex exponentials, because they are mathematically easier to work with than sums of sines and cosines (Comment A3.1).

COMMENT A3.1 SUMMING COMPLEX EXPONENTIALS

We can derive Eq. (A3.7) in just a few lines by using complex exponentials.

$A_1 \exp i\phi_1 + A_2 \exp i\phi_2$

$= A_1 \exp i\frac{1}{2}(\phi_1 + \phi_2 + \phi_1 - \phi_2)$

$\quad + A_2 \exp i\frac{1}{2}(\phi_1 + \phi_2 - \phi_1 + \phi_2)$

$= \left[\exp i\frac{1}{2}(\phi_1 + \phi_2) \right]$

$\left[A_1 \exp i\frac{1}{2}(\phi_1 - \phi_2) + A_2 \exp - i\frac{1}{2}(\phi_1 - \phi_2) \right]$

Equation (A3.7) is the "real" part of the line above. The derivation using sine and cosine rules is much more complicated.

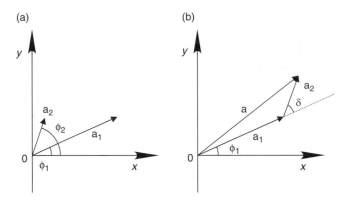

Fig. A3.6 (a) Two waves represented as complex numbers, a_1, a_2, on an Argand diagram and **(b)** their sum, a, expressed as a vector sum of a_1 and a_2. Note that a etc. denote the complex number and not the length of the line, which is A etc. as in Fig. A3.5. The phases of a_1, a_2 are ϕ_1, ϕ_2, respectively, and $\phi_2 - \phi_1 = \delta$.

A wave in space of amplitude A and wavevector magnitude k is written $A \exp ikx$. A wave in time of amplitude A and angular frequency ω is written $A \exp i\omega t$. A wave oscillating in both space and time is represented by $A \exp[i(kx - \omega t)]$. As in the previous section, the time term has a negative sign because, as the wave propagates, a crest at a certain point, for example, results from the arrival of a crest that was one wavelength *behind* in the *previous* time period of the wave.

Two waves of amplitudes A_1, A_2, respectively, and phases, ϕ_1, ϕ_2, respectively, are represented on an Argand diagram by $a_1 = A_1 \exp i\phi_1$ and $a_2 = A_2 \exp i\phi_2$ and it is then straightforward to calculate their sum, $A \exp i\phi$, from either algebra or geometry, by treating a_1 and a_2 as vectors in the xy plane (Fig. A3.6, and Eq. (A3.14)):

$$A \exp i\phi = A_1 \exp i\phi_1 + A_2 \exp i\phi_2$$

Equating the real parts we have

$$A \cos \phi = A_1 \cos \phi_1 + A_2 \cos \phi_2 \qquad (A3.15)$$

and equating the imaginary parts we have

$$A \sin \phi = A_1 \sin \phi_1 + A_2 \sin \phi_2$$

The properties of the resulting wave are derived following an analysis as in Comment A3.1.

We now write the two waves in terms of the phase difference between them, $\delta = \phi_2 - \phi_1$, and calculate the amplitude of the resulting wave

$$a = a_1 + a_2 = A_1 \exp i\phi_1 + A_2 \exp[i(\phi_1 + \delta)]$$

$$aa^* = \{[A_1 \exp i\phi_1 + A_2 \exp i(\phi_1 + \delta)]\}$$

$$\quad \{[A_1 \exp - i\phi_1 + A_2 \exp - [i(\phi_1 + \delta)]\}$$

$$= \exp(i\phi_1) \exp(-i\phi_1)[A_1 + A_2 \exp i\delta][A_1 + A_2 \exp(-i\delta)]$$

$$= A_1^2 + A_2^2 + A_1 A_2 [\exp i\delta + \exp(-i\delta)]$$

$$= A_1^2 + A_2^2 + 2A_1 A_2 \cos \delta \qquad (A3.16)$$

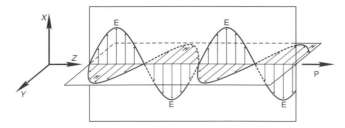

Fig. A3.7 Propagation of a plane polarized electromagnetic wave. The electric field oscillates along the x-axis and the magnetic field oscillates along the y-axis. The wave propagates along the z-axis.

The amplitude of the resulting wave is equal to $(A_1^2 + A_2^2 + 2A_1 A_2 \cos \delta)$. And, as another illustration of the power of using complex exponentials, we note that the last line of Eq. (A3.16) is simply the *cosine rule* applied to the triangle with sides formed by $a_1, a_2, a,$ in Fig. A3.6.

Polarization

The electric and magnetic fields in an electromagnetic wave oscillate along directions perpendicular to the propagation direction of the wave (Fig. A3.7). The light illustrated is said to be *linearly or plane polarized*, because the electric and magnetic fields oscillate along straight lines (the x- and y-axes, respectively). However, this need not be the case. The electric field (or magnetic field) that describes the light is described by an oscillating vector, which can lie in any direction provided this is perpendicular to the propagation direction. In other words, if the propagation direction is along the z-axis, the electric field vector, **E**, may have components, E_x, E_y, along the x- and y-axes, respectively.

The light is linearly or plane polarized when E_x and E_y oscillate *in phase*. Cases for different amplitude values for each of the components are illustrated in Fig. A3.8.

When E_x and E_y oscillate with a phase difference, δ, the light is said to be *elliptically polarized* (*circularly polarized* in the special cases of $\delta = +\pi/2$ or $-\pi/2$), because the tip of the electric field vector traces an ellipse (or a circle) in time (Fig. A3.9). Using the cosine or complex exponential notation introduced above, and assuming unit amplitude,

$$E_x = \cos \omega t \quad \text{or} \quad 1$$

$$E_y = \cos(\omega t + \delta) \quad \text{or} \quad \exp\ i\delta$$

The tip of the electric field vector turns in a clockwise direction for phase angles $0 < \delta < \pi$, and anticlockwise

for $\pi < \delta < 2\pi$. The light is linearly or plane polarized for $\delta = 0$ or π.

A3.2.2 Simple Harmonic Motion

A Simple Harmonic Oscillator

A swinging pendulum or a mass on a spring moving periodically up and down are mechanical systems that provide us with good examples of *simple harmonic motion*. We consider the motion of a mass, m, attached to a spring (Fig. A3.10).

The mass undergoes simple harmonic motion in one dimension, described by an equation of motion given by

$$F = -ky = m \frac{\mathrm{d}^2 y}{\mathrm{d}t^2} \tag{A3.17}$$

where F is the force exerted on the particle (e.g., by the stretched spring), y is the mass displacement from an equilibrium position, and k is a force constant, which depends upon the stiffness of the spring. The negative sign indicates that F is a *restoring force*. The force is proportional to the displacement and acts in the opposite direction. Thus, it tends to restore the mass to its original position. Equation (A3.17) is an expression of Hooke's law, which states that in an elastic body the strain (deformation) is proportional to the stress (force per area). The solution of Eq. (A3.17) is in the form of a wave

$$y = A \cos \omega_m t \tag{A3.18}$$

which describes a periodic motion, where ω_m is the natural vibrational angular frequency of the mass, and A is the maximum displacement from equilibrium, the *amplitude* of the motion. Of course, we could also have written a wave solution in terms of complex exponentials.

In order to find the relation between ω_m and the force constant k, we calculate the second time derivative of y (the acceleration) in Eq. (A3.18), substitute in the equation of motion (Eq. (A3.17)) and rearrange

$$\omega_m = \sqrt{k/m} \tag{A3.19}$$

We arbitrarily set the potential energy, E, of the system as equal to zero when the mass is in its equilibrium position. As the spring is compressed or stretched, E increases by an amount equal to the work required to displace the mass:

$$\mathrm{d}E = -F \mathrm{d}y \tag{A3.20}$$

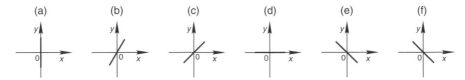

Fig. A3.8 Linearly polarized light: electric field components along x- and y-axes of different amplitude oscillating in phase: **(a)** $E_y = 1$; $E_x = 0$; **(b)** $E_y = 1$; $E_x = 1/2$; **(c)** $E_y = 1$; $E_x = 1$; **(d)** $E_y = 0$; $E_x = 1$; **(e)** $E_y = 1$; $E_x = -1$; **(f)** $E_y = -1$; $E_x = 1$.

(a) (b) (c) (d) (e)

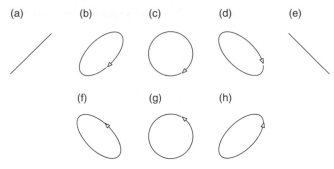

(f) (g) (h)

Fig. A3.9 Linearly, elliptically, or circularly polarized light: E_x and E_y are assumed to have the same amplitude, and to oscillate with phase ωt and $\omega t + \delta$, respectively: **(a)** $\delta = 0$; **(b)** $\delta = \pi/4$; **(c)** $\delta = \pi/2$; **(d)** $\delta = 3\pi/4$; **(e)** $\delta = \pi$; **(f)** $\delta = 5\pi/4$; **(g)** $\delta = 3\pi/2$; **(h)** $\delta = 7\pi/4$. The light in cases **(a)** and **(e)** is linearly polarized, it is circularly polarized in **(c)** and **(g)**, and elliptically polarized in the other cases.

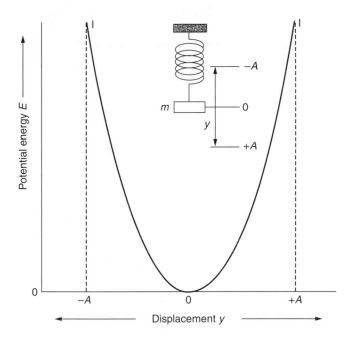

Fig. A3.10 Potential energy diagram for a simple harmonic oscillator.

Combining Eq. (A3.20) with Eq. (A3.17) and integrating, we derive the following relation for the potential energy of the oscillator as a function of displacement:

$$E = \tfrac{1}{2}ky^2 \qquad (A3.21)$$

which describes a parabola (Fig. A3.10). The potential energy is a maximum when the spring is stretched or compressed to the amplitude, A, and it decreases to zero at the equilibrium position.

The total energy of the system is equal to its potential energy plus the kinetic energy of the moving mass. Note that contrary to the potential energy dependence, the kinetic energy is a maximum as the mass passes through the

Fig. A3.11 Two masses connected by a spring of force constant, k.

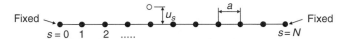

Fig. A3.12 A linear chain of $N+1$ atoms. The boundary conditions are that atoms, $s=0$ and $s=N$ are fixed; u_s is the displacement of atom s and a is the atom spacing.

equilibrium position, and decreases to zero at $y = +A$ and $y = -A$, where the mass stops before reversing the direction of its motion. The total energy of the oscillator, E_{total}, is exchanged back and forth between potential and kinetic energy, and it can be shown that it is a constant during all phases of the motion:

$$E_{total} = \tfrac{1}{2}kA^2 \qquad (A3.22)$$

The above equations may be modified to describe the behavior of a system consisting of two masses m_1 and m_2 connected by a spring (in the absence of gravity) (Fig. A3.11), simply by substituting the *reduced mass* $m_{1,2}$ for m, where

$$m_{1,2} = \frac{m_1 m_2}{m_1 + m_2} \qquad (A3.23)$$

The vibrational angular frequency of the system is then given by

$$\omega_m = \sqrt{\frac{k}{m_{1,2}}} = \sqrt{\frac{k(m_1 + m_2)}{m_1 m_2}} \qquad (A3.24)$$

Normal Modes in One Dimension
The motions of a set of masses coupled by Hooke's law springs, or simple harmonic potentials such as the one illustrated in Fig. A3.10, can be seen as resulting from a superposition of fundamental vibrations called *normal modes*.

An analysis of the normal modes in a linear chain of atoms under specific boundary conditions is illustrated in Fig. A3.12. The chosen boundary conditions are that the end atoms do not move, like the ends of a guitar string. And, similarly to a plucked guitar string, the fundamental modes correspond to a set of waves in the line, while leaving the end atoms fixed.

The longitudinal (parallel to the line) or transverse (perpendicular to the line) wave displacement of atom s is given by

$$u_s \propto \sin sQa \qquad (A3.25)$$

where Q is called a wavevector. (It is not written as a vector because our example is in one dimension; the wavelength of the atomic displacements is given by $\lambda = 2\pi/Q$.) The first boundary condition, $u_s = 0$ for $s = 0$, is automatically satisfied by Eq. (A3.5). The second boundary condition, $u_s = 0$ for $s = N$, sets limits on the values of Q. It is satisfied by choosing

$$Q = \frac{\pi}{Na}, \frac{2\pi}{Na}, \frac{3\pi}{Na}, ..., \frac{N\pi}{Na} \tag{A3.26}$$

Note that the solution for the maximum value of Q ($Q = \pi/a$) results in $u_s = 0$ for all atoms. There are, consequently, $N-1$ normal modes for the line of atoms with the given boundary conditions. Each normal mode is defined by its Q vector and corresponds to a standing wave described by the equation

$$u_s = u(0) \exp(-i\omega_Q t) \sin sQa \tag{A3.27}$$

where $u(0)$ is an initial amplitude, ω_Q is the angular frequency of the mode, and t is time.

We recall that the mathematical relation between ω_Q and Q is called a *dispersion relation*. For the above example, the dispersion relation is given by

$$\omega_Q = \begin{cases} \dfrac{Na}{2\pi}, & for -\dfrac{\pi}{a} \leq Q \leq \dfrac{\pi}{a} \\ 0, & otherwise \end{cases} \tag{A3.28}$$

Normal Modes in Three Dimensions

The equation of motion of a mass, m, undergoing simple harmonic motion in one dimension is given in Eq. (A3.17). Extending the analysis to a body of N harmonically coupled masses moving in three dimensions, the m and k terms are replaced by two $3N \times 3N$ matrices, respectively, of effective masses and force constants between all mass pairs, and y is replaced by a $3N$-dimensional coordinate vector. By analogy with Eq. (A3.18), the solutions for the three-dimensional case form a set of $3N$ periodic functions. Six of these, however, correspond to the translational and rotational degrees of freedom of the body as a whole, for which the masses do not move relative to each other, so that the system has $3N-6$ normal modes. It can be shown that if the masses are in a line (as in the example discussed below) the number of normal modes corresponds to $3N-5$.

The 3N-6 degrees of freedom of a diatomic molecule are shown in Fig A3.13; three correspond to translational displacements (along the x-, y-, and z-axes); two are rotational modes (about the y- and z-axes); and there is only one vibrational (bond stretching) mode as expected from the $(3N-5)$ relation for masses on a line.

In the harmonic approximation, the vibrations of atoms in a macromolecule result from a superposition of the normal modes. Atomic collective motions calculated using harmonic potentials for two low-frequency normal modes in the small protein BPTI are shown in Fig. A3.14. Note the

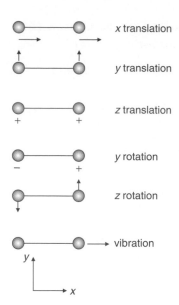

Fig. A3.13 Intrinsic degrees of freedom of a diatomic molecule. The z-axis points out of the page. Because the atoms are considered as point masses, there is no x rotation.

(a)   (b)

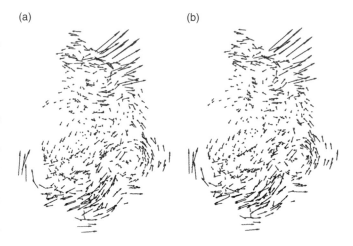

Fig. A3.14 Collective atomic displacements for two low-frequency normal modes from a calculation for BPTI. The mode frequencies are 3.56 ps^{-1}, 0.21 ps^{-1}, for **(a)**, **(b)**, respectively (Gō *et al.*, 1983).

relatively large displacements of some of the atoms. In each normal mode, all atoms move with the same phase, i.e., they achieve maximum and minimum displacements and pass through their equilibrium positions simultaneously.

A3.2.3 Fourier Analysis

Periodic Functions, Fourier Series, and Fourier Transforms

We showed above that a periodic function could be expressed as a sum of sines and cosines. In fact, according to a theorem due to Fourier, *any periodic function* may be represented by a sum of cosines and sines, known as a *Fourier series* (Eq. A3.29):

$$y(t) = a_0 + a_1 \cos \omega t + b_1 \sin \omega t + a_2 \cos 2\omega t + b_2 \sin 2\omega t$$
$$+ a_3 \cos 3\omega t + b_3 \sin 3\omega t + ...$$

$$= \sum_{n=0}^{\infty} (a_n \cos n\omega t + b_n \sin n\omega t) \qquad (A3.29)$$

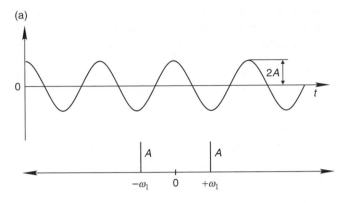

(a)

The cosine and sine terms describe waves of increasing frequency in integer multiples, $n\omega$, of a fundamental frequency, ω, with amplitudes, a_n, b_n, respectively. We can also write the Fourier series in terms of complex exponentials:

$$y(t) = \sum_{n=-\infty}^{+\infty} A_n \, \exp \, i \, n\omega t \qquad (A3.30)$$

where A_n is the amplitude of the wave of frequency $n\omega$, and the sum is over negative as well as positive integers because they are all required to establish the correspondence between Eqs. (A3.29) and (A3.30). Consider the simple cosine wave as an illustration of the relationship between the complex exponential and sine cosine series (see also Fig. A3.15):

$$y(t) = a \, \cos \omega t = \sum_{n=0}^{\infty} (a_n \cos n\omega t + b_n \sin n\omega t)$$

Clearly $a_1 = a$, and all the other a_n, b_n are zero. In terms of complex exponentials,

$$y(t) = a \cos \omega t = \sum_{n=-\infty}^{\infty} A_n \exp \, n \, i\omega t$$

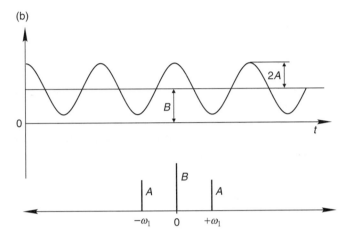

(b)

has two terms, for $n = \pm 1$ so that the sine terms making up the imaginary part cancel out,

$$y(t) = a \cos \omega t \quad = \quad A_{-1} \exp(-i \, \omega t) + A_1 \exp(+i \, \omega t)$$
$$= \quad 2A \cos \omega t$$

where $A_{-1} = A_1 = A$, and $a = 2A$.

Another way to describe the series expansion for $y(t)$ is to consider its wave components separately, each defined by an amplitude A_n and corresponding frequency value $n\omega$. This yields a set of lines called *Fourier components*; the amplitude of each component is called a *Fourier coefficient*; and the function $A_n(n\omega)$ is called the *Fourier transform* of $y(t)$.

The Fourier transform of a wave form, which is described perfectly by $A \cos \omega_1 t$, for example, has Fourier components of amplitude, A, at $n\omega = \pm\omega_1$. A wave resulting from the sum of two cosine functions as in Eq. (A3.7) has Fourier components on $n\omega = \pm\omega_1, \pm\omega_2$, of height, A_1, A_2, respectively, and so on (Fig. A3.15). Note from Fig. A3.15b that the Fourier transform of a constant function, $y(t) = B$, is a single line at the origin ($n\omega = 0$); i.e., a constant level can be described by a wave of zero frequency or infinite wavelength.

Fourier analysis of a periodic function is equivalent to *spectral analysis*, in which light is separated by a prism or diffraction grating into its component colors, or to analysis of a musical sound in terms of its harmonics. Fourier

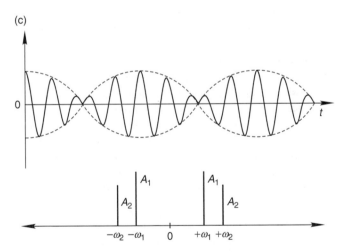

(c)

Fig. A3.15 Waves and their corresponding Fourier transforms: (a) a wave described by $2A \cos \omega t$ oscillating about zero; (b) a wave described by $B + 2A \cos \omega t$ oscillating about the line at plus B; (c) a wave described by $2A \cos \omega t + 2A_2 \cos \omega_2 t$, oscillating about zero.

analysis is also equivalent to normal mode analysis of the complex vibration pattern of a set of masses coupled by Hooke's law springs (see above).

Fourier devised a mathematical method to derive the values of the coefficients A_n in the series, and calculated that

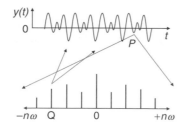

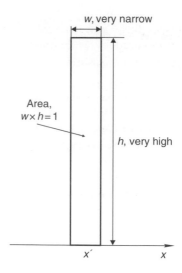

Fig. A3.16 The periodic function $y(t)$ and its Fourier transform $A_n(n\omega)$. Each point P on $y(t)$ is fully defined by the entire set of $A_n(n\omega)$ components and each component Q is completely defined by the full periodic pattern in $y(t)$.

$$A_n = \frac{2}{T}\int_0^T y(t)\exp(-i\,n\omega t)\,dt \qquad (A3.31)$$

where $T = 2\pi/\omega$ represents one *period* of the fundamental oscillation of frequency ω, i.e., the periodic repeat value of the variable t.

Since integration is equivalent to taking the sum of the function at very small intervals, the symmetry apparent in Eqs. (A3.30) and (A3.31) is quite remarkable. It essentially tells us that, on the one hand, each Fourier component is fully defined by the fundamental pattern of the function y, while, on the other hand, the value of y at each point t is, itself, defined by the entire set of Fourier components (Fig. A3.16).

We note another important property of Fourier transforms: if y is defined as a function of t, then its Fourier transform is defined as a function of frequency, i.e., a parameter that is proportional to $1/t$. The application of Fourier analysis and this reciprocal relation is quite general. In the case of a wave in space, if y is a function of x, then its Fourier transform is a function of the wave vector magnitude Q, which is proportional to $1/x$. The analogous equation to Eq. (A3.30) is written

$$y(x) = \sum_{n=-\infty}^{+\infty} A_n \exp i n\, Qx \qquad (A3.32)$$

Because of this reciprocal relation, the space of x is sometimes called *real space* and the space of Q, in which the Fourier transform is defined, is called *reciprocal space*.

The Dirac Delta Function

The lines drawn for the Fourier coefficients in Fig. A3.15 were given heights proportional to their values. Although this is quite illustrative and pleasing to the eye, it is not mathematically correct. In fact, each line corresponds mathematically to an infinitely narrow curve, which has, nevertheless, a surface area equal to the value of the Fourier coefficient. The mathematical function called the *Dirac delta function* describes such a curve of unit surface area. A Dirac delta function at position $x = x'$ in one dimension is written $\delta(x - x')$. The function has the following properties:

Fig. A3.17 The Dirac delta function.

if $x = x'$, then $\delta(x - x') = \infty$;

if $x \neq x'$, then $\delta(x - x') = 0$;

The surface area of the curve is equal to unity, so that $\int \delta(x - x')\,dx = 1$. Since the function is extremely narrow, the integral need not be from $-\infty$ to $+\infty$, but only over a small range of x around x' (Fig. A3.17). It can be shown that

$$\delta(x - x') = \int_{-\infty}^{+\infty} \exp[i\,Q(x - x')]\,dQ \qquad (A3.33)$$

where Q is a wavevector magnitude as described above. There is a striking similarity between Eqs. (A3.32) and (A3.33). Equation (A3.33) also appears to be a Fourier series. In this case, however, the terms of the series are not separated by integer increments in n but by very small dQ intervals.

The Fourier Integral and Continuous Fourier Transform

Equation (A3.33) describes a *Fourier integral*. It shows that Fourier analysis is not limited to functions of a periodic character. We recall that any periodic function can be represented by a sum of waves with frequencies increasing in integer multiples. We now generalize this statement and write that *any function $y(x)$ can be represented as a Fourier integral in terms of a sum of waves of *continuously* decreasing wavelength (or of continuously increasing frequency if x is replaced by t):

$$y(x) = \frac{1}{2\pi}\int_{-\infty}^{+\infty} F(Q)\exp(iQx)\,dQ$$

$$\qquad (A3.34)$$

$$y(t) = \frac{1}{2\pi}\int_{-\infty}^{+\infty} F(\omega)\exp(i\omega t)\,d\omega$$

where the $1/2\pi$ factor results from replacing a sum by an integral.

In the corresponding Fourier transform, the A_n coefficients are now replaced by a continuous function $F(Q)$ (or $F(\omega)$).

$$F(Q) = \int y(x) \exp(-iQx)dx$$
$$F(\omega) = \int y(t) \exp(-i\omega t)dt \qquad (A3.35)$$

Equation (A3.34) is called the *reverse Fourier transform of* $F(Q)$. Fourier transforms of useful non-periodic functions are given in Fig. A3.18.

The bell-shaped *Gaussian* and *Lorentzian* functions often occur in molecular biophysics. The statistical distribution of data measured on a large number of molecules (ensemble) follows a Gaussian curve (normal distribution). Practically all time-dependent phenomena can be described by an exponential decay in the time domain (e.g., the signal in dynamic light scattering (Chapter D4) and quasi-elastic neutron scattering (Chapter I2), related to translational diffusion; fluorescence depolarization (Chapter D3), or electric birefringence, related to rotational diffusion, etc.). The Fourier transform of an exponential function is a Lorentzian curve, so that the time-dependent phenomenon can be described by a Lorentzian in the frequency or energy domain.

In three dimensions, the Fourier transform is written in terms of vectors

$$F(\mathbf{Q}) = \int y(\mathbf{r}) \exp(-i\mathbf{Q} \cdot \mathbf{r})d\mathbf{r} \qquad (A3.36)$$

Convolution

Consider the functions in Fig. A3.19. Clearly, the function $C(x)$ in line (c) must be related mathematically to those in lines (a) and (b), $A(x)$, $B(x)$, respectively. It results, in fact, from an operation called convolution. $C(x)$ is equal to the convolution of $A(x)$ and $B(x)$.

Mathematically, the convolution of two one-dimensional functions $f(x)$, $g(x)$ is written

$$f(x) \otimes g(x) = \int_{-\infty}^{\infty} f(u)g(x-u)du \qquad (A3.37)$$

where $\otimes$ is the convolution symbol. The convolution operation is illustrated in Fig. A3.20a, b for two simple functions, centered on $x = 0$ and x', respectively. We see in Fig. A3.20c the situation of one of the terms in the integral, for $x = x'$ and $u = u'$. The function $g(x-u)$ has a finite value since it has been displaced and it is now centered on u'; we

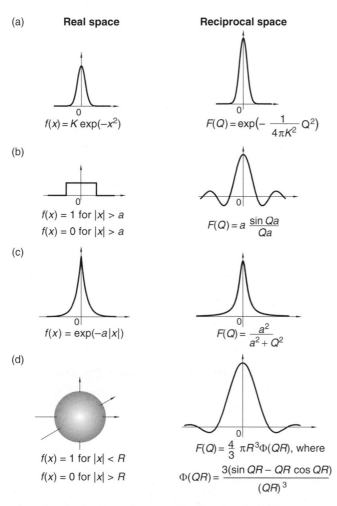

(a) **Real space**   **Reciprocal space**

(a) $f(x) = K \exp(-x^2)$ $F(Q) = \exp\left(-\dfrac{1}{4\pi K^2}Q^2\right)$

(b) $f(x) = 1$ for $|x| > a$; $f(x) = 0$ for $|x| > a$ $F(Q) = a\dfrac{\sin Qa}{Qa}$

(c) $f(x) = \exp(-a|x|)$ $F(Q) = \dfrac{a^2}{a^2 + Q^2}$

(d) $f(x) = 1$ for $|x| < R$; $f(x) = 0$ for $|x| > R$ $F(Q) = \dfrac{4}{3}\pi R^3 \Phi(QR)$, where $\Phi(QR) = \dfrac{3(\sin QR - QR \cos QR)}{(QR)^3}$

Fig. A3.18 Some non-periodic functions and their Fourier transforms. (a) The Fourier transform of a Gaussian of width $1/K$ is itself a Gaussian of width K. (b) The Fourier transform of a slit of width $2a$ is an oscillating function with a large central maximum of Q width $2\pi/a$ and subsidiary maxima that rapidly fade away. (c) The Fourier transform of an exponential decay is a bell curve with longer wings than a Gaussian called a Lorentzian function. (d) The Fourier transform of a sphere of radius R is a function with spherical symmetry called a Bessel function with a large central maximum of Q width about $2\pi/R$ and weak oscillations on either side of it.

(a)
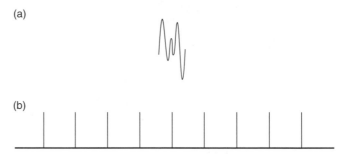

(b)

(c)

Fig. A3.19 The function on line (c) results from the *convolution* of the function in line (a) with the set of delta functions on line (b).

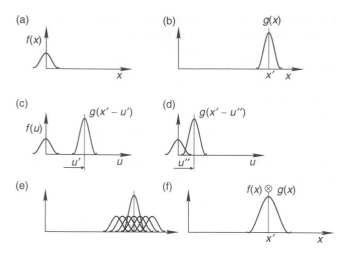

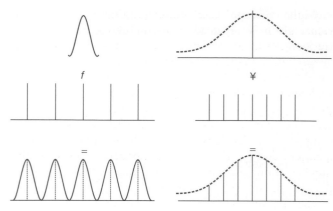

Fig. A3.20 An illustration of *convolution*: in **(a)** and **(b)** are the two functions whose convolution product is to be calculated; **(c)** and **(d)** show the terms in the integral of Eq. (A3.37) for $x = x'$ and $u = u'$ and u'', respectively; **(e)** shows how the result can be seen as each point in g being multiplied by f; **(f)** shows the result of the convolution.

Fig. A3.21 Left panel: The convolution of a function (red curve) with a lattice represented by a set of delta functions. Right panel: The Fourier transform of the repeating function is equal to the Fourier transform of the red curve multiplied by the Fourier transform of the lattice.

see, however, that $f(u')$ is equal to zero at this point, so that the term in the integral is zero. Clearly, the term is non-zero only if both $g(x-u)$ and $f(u)$ are non-zero. This only happens for x close to x', otherwise g is zero, and for values of u that shift g so that it *overlaps* with $f(u)$, i.e., u'' close to zero (Fig. A3.20d). The result of the convolution is a function in which each point in g has been multiplied by f (Fig. A3.20e). If $f(x)$ is not centered on zero but on $x+x''$, the convolution is centered on $x'+x''$. It can be shown that $f(x) \otimes g(x) = g(x) \otimes f(x)$, so that we should obtain the same result by scanning either of the functions over the other.

The convolution product plays a very important role in physical measurement. Consider a sample that absorbs only in a very narrow wavelength band of radiation in a spectroscopy experiment. It is likely that the instrument used for the measurement itself has an insensitivity to wavelength that is broader than the band to be observed. The result of the measurement is the convolution product of the band shape function and instrument *wavelength resolution* function.

The convolution product in three dimensions is written in terms of vectors

$$f(\mathbf{r}) \otimes g(\mathbf{r}) = \int_{-\infty}^{\infty} f(\mathbf{u}) g(\mathbf{r} - \mathbf{u}) du \qquad (A3.38)$$

Calculating the Fourier transform of a convolution product leads to an extremely useful result (shown in one dimension for simplicity)

$$FT[f(x) \otimes g(x)] = FT f(x) \times FT g(x) \qquad (A3.39)$$

where FT stands for the Fourier transform operation. Equation (A3.39) states that in order to obtain the Fourier

transform of the *convolution product* of two functions, we simply *multiply* together the Fourier transforms calculated separately for each. By using the terms *real* and *reciprocal space*, we can write that convolution in real space leads to multiplication in reciprocal space (and vice versa).

We illustrate the result in Eq. (A3.39) with an example taken from crystallography (Fig. A3.21). A lattice is described by a set of Dirac delta functions. Placing any function on each node of the lattice can be seen as the result of a convolution product of the lattice and the function (left-hand panel in the figure). The right-hand panel shows the corresponding Fourier transforms. The Fourier transform of the repeating function is obtained by multiplying the Fourier transform of the function (the broad red bell curve) with the Fourier transform of the lattice (a set of delta functions separated at nQ_d, where n is an integer and $Q_d = 2\pi/d$, where d is the lattice repeat in this one-dimensional example). The resulting transform is the set of red lines (delta functions of decreasing amplitude) in the bottom of the right-hand panel.

A3.2.4 Quantum Mechanics

The concepts of quantum mechanics are necessary for the description of the interaction between radiation and matter on an atomic scale. Quantum mechanics was largely developed to provide an interpretation of spectroscopic observations, in particular of the discrete lines in atomic spectra, and it is essential for many of the methods described in this book. Even a succinct description of quantum mechanics, however, would be too long to be included here, and only a few key concepts are summarized below.

Planck's Constant, Energy Quanta and Photons

In quantum mechanics, energy may be transmitted during a given time from one system to another only in discrete multiples of a universal minimum limiting value. This value, given by Planck's constant ($h = 6.626 \times 10^{-34}$ J s), can be seen as an indivisible packet of *energy multiplied by time*. In a quantum mechanical description, a wave of frequency, v, which oscillates v times a second, transmits well-defined packets of energy, hv, which are called *quanta* (the singular is *quantum*). The energy of a wave in quantum mechanics is described by two components. The first component, which is fully defined by its frequency, is its quantum energy. The second component is given by the number of quanta carried per unit time, which is proportional to the intensity of the wave.

The quantum of a wave of electromagnetic radiation is called a *photon*. The quantum energy of visible light, for example, depends on its color. Quanta of red light are of lower energy than quanta of blue light, but a more intense beam of red light may still carry more energy. The wavelength associated with a photon is inversely proportional to the frequency, $\lambda = c/v$, where c is the speed of light. The quantum energy of a beam of electromagnetic radiation is fully defined, therefore, by either the frequency (hv) or the wavelength (hc/λ).

Equations relate intensity, momentum, and energy in the wave and particle pictures for neutrons ("classical" particles), electrons ("relativistic" particles), and photons (particles of zero rest mass). The De Broglie equation relates wavelength, λ, and momentum, $\mathbf{p}$; $\mathbf{u}$ is a unit vector in the direction of wave propagation; $\mathbf{k}$ is the wave vector, $k = 2\pi/\lambda$; $\hbar$ is Planck's constant divided by 2π; m_n and m_e, are the mass of the neutron and electron, respectively. Other symbols have their usual meanings.

The Wave–Particle Duality

It is convenient for the analysis of its behavior to picture a beam of radiation either as a stream of particles or as a wave. Since the events involving the beam happen in times and over distances that are very much smaller than can be grasped by our common-sense experience, the question "Is the beam made up of particles or waves?" is meaningless. Suffice it to say that the beam can be described in either way. And, since the two descriptions must be self-consistent, there are rules linking them. Thus, the momentum and energy in terms of mass and velocity of the particle description are related by equations involving Planck's constant to the wavevector and frequency of the wave description. The relations are given in Table A3.1 for three cases: a beam of neutrons (particles of rest mass m_n moving "slowly" with velocity u); a beam of electrons (particles of rest mass m_e, moving at a velocity close to the velocity of light); and a beam of photons (particles of rest mass zero, moving at the speed of light).

De Broglie's relation states that the wavelength associated with a moving particle is inversely proportional to its momentum, with the proportionality constant equal to Planck's constant, $\lambda = h/p$. Faster particles are therefore associated with shorter wavelengths and vice versa. Note that the neutron, because of its large mass, is a particle with unique properties for the study of matter. If we express the

TABLE A3.1 PARTICLES AND WAVES

	Particle	Wave
Intensity	Rate = number of particles per unit time	Square of the amplitude $= A^2$
Momentum	Neutrons: $\mathbf{p} = m_n v$	Neutrons: $\mathbf{p} = \hbar \mathbf{k} = (h/\lambda)\mathbf{u}$
	Electrons: $\mathbf{p} = \dfrac{m_e v}{\sqrt{1 - v^2/c^2}}$	Electrons: $\mathbf{p} = \hbar \mathbf{k} = (h/\lambda)(\mathbf{u})$
	Photons: $\mathbf{p} = \hbar k = (h/\lambda)\mathbf{u}$	Photons: $\mathbf{p} = \hbar \mathbf{k} = (h/\lambda)(\mathbf{u})$
Energy	Neutrons: $E = \dfrac{1}{2}m_n v^2 = \dfrac{1}{2}p^2/m_n v^2$	Neutrons: $E = hv = \hbar\omega = \dfrac{h^2}{2\lambda^2 m_n} = \dfrac{\hbar^2 k^2}{2m_n}$
	Electrons: $E = \dfrac{1}{2}p^2/m_e v^2 = \dfrac{m_e v^2}{2(1 - v^2/c^2)}$	Electrons: $E = hv = \hbar\omega = \dfrac{h^2}{2\lambda^2 m_e} = \dfrac{\hbar^2 k^2}{2m_e}$
	Photons: $E = hv = \hbar\omega = hc/\lambda$	Photons: $E = hv = \hbar\omega = hc/\lambda$

kinetic energy of a neutron in terms of temperature, room temperature neutrons (*thermal* neutrons) have an associated wavelength between 1 and 2 Å; *cold* neutrons (about 10 K) have wavelengths close to 10 Å; while *hot* neutrons have wavelengths of a fraction of an ångström.

Heisenberg's Uncertainty Principle

The only way that we can measure the position, momentum, or energy of a particle or wave is by interfering with it in some way. The perturbation due to the measurement cannot be infinitely small but is itself limited by the minimum energy–time packet represented by Planck's constant, which of course is not negligible for measurements on an atomic scale. Heisenberg's uncertainty principle is a result of this. Expressed in two inequalities, it relates momentum–spatial coordinate and energy–time, respectively:

$$\Delta p\,\Delta x \geq h$$
$$\Delta E\,\Delta t \geq h \qquad \text{(A3.40)}$$

where x is a spatial coordinate. The uncertainty principle states that in order to increase the precision of a position measurement (i.e., make Δx as small as possible) at the atomic level, for example, we have to sacrifice knowledge of its momentum, and vice versa. In other words, we can hope to know *either* how fast a particle is moving *or* where it is, but we cannot hope to have both bits of information at once.

An illustration of the uncertainty principle is provided by the wave description of matter. We recall from the section on Fourier analysis above that a non-periodic function can be represented by a sum of waves of different wavelengths. A particle that is localized in space can, therefore, be described by such a sum of waves. Since the waves have a range of different wavelengths we can only describe its momentum with a large uncertainty. Consider now a particle whose momentum is known precisely. It is represented by a wave of well-defined wavelength (e.g., a single sine wave). But such a wave extends infinitely in space, so that we have no idea where the particle might be! Similar arguments can be developed for energy and time measurements.

The uncertainty principle is also illustrated rather nicely by the resolution condition in crystallography (see Chapter G1, and Fig. A3.22). It is a consequence of the reciprocal relation between a crystal structure and the waves it scatters (which in effect are represented by its Fourier transform) that the smallest distance, d_{min}, that can be resolved in a crystallography measurement is related to the maximum observed value of scattering vector magnitude, Q_{max}, by

$$d_{min} = \frac{2\pi}{Q_{max}}$$

where

$$\mathbf{Q}_{max} = \mathbf{k}_{max} - \mathbf{k}_0 \qquad \text{(A3.41)}$$

and $\mathbf{k}_0$ and $\mathbf{k}_{max}$ are the wavevectors of the incident and diffracted beams, respectively. We know from Table A3.1 that the momentum of a wave is expressed in terms of its

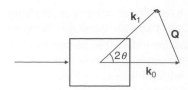

Fig. A3.22 We "see" the atomic arrangement in a crystal by the way it scatters radiation. The crystal (depicted as a rectangle) is put in a beam described by a wave of wavevector $\mathbf{k}_0$ and the distance between atomic planes is measured by observing the intensity of scattered waves as a function of wavevector $\mathbf{k}_1$. The scattering vector $\mathbf{Q}$ is defined as the difference between $\mathbf{k}_1$ and $\mathbf{k}_0$.

wavevector, so that Q_{max} can be seen as proportional to the difference in momentum between the diffracted and incident waves. The spatial resolution, d_{min}, can be seen as Δx, the minimum spacing that can be determined in the experiment:

$$\Delta p = \frac{h}{2\pi} Q_{max}$$
$$\Delta x = \frac{2\pi}{Q_{max}}$$

so that

$$\Delta p\,\Delta x = h \qquad \text{(A3.42)}$$

The Schrödinger and Dirac Approaches

It is a consequence of the uncertainty principle that the results of quantum mechanical calculations (e.g., the solution of the equations of motion of a system of atoms) are not in terms of determined values but in terms of probabilities. Thus, trajectories of varying probability are calculated for a beam of particles hitting a screen in which there is a small slit; the probability pattern found corresponds to the diffraction fringes observed experimentally. In the Schrödinger *wave mechanics* approach the probability density in the beam of particles before and after they cross the slit is depicted as the square of the amplitude of a wave function. The Dirac approach is in terms of the probabilities of different final *states*, given an initial *state*, expressed as discrete elements in a mathematical matrix.

Energy Levels

A fundamental concept of quantum mechanics is that the energy of systems such as atomic nuclei or electrons is quantized in levels that are occupied according to specific statistical laws. It is the basis of spectroscopy experiments (see Part E) that transitions to higher levels can be stimulated by the absorption of radiation and that radiation is emitted when the system relaxes to a lower level. *The energy quanta*, hv, *absorbed or emitted correspond exactly to the differences between levels*.

The energy levels of an atomic electron are characterized by *orbital* and *spin* quantum states. In a classical analogy, these may be described as discrete energy states corresponding to the electron orbiting around the nucleus and spinning on its axis, respectively. The analogy should not be taken too far, however, because it is the very essence of quantum mechanics that it was invented because events could not be described in a classical manner. The occupation of electronic states is governed by strict rules; e.g., a state of given orbital and spin quantum numbers may only be occupied by one electron. The lowest-energy state is called the *ground state*, and states of higher quantum number are called the *first excited state*, *second excited state*, etc.

Atoms that are bonded together to form a molecule vibrate and rotate like the balls connected by springs we discussed in the section on simple harmonic motion. In quantum mechanics, these vibrations and rotations also correspond to discrete energy levels characterized by quantum numbers. The "springs" of the atomic bonds comprise shared electrons whose energy levels are themselves affected by the molecular vibrational and rotational energies (Fig. A3.23). The lowest (ground) state and first excited electronic states in a molecule are each described by a potential energy well with respect to a conformational coordinate, in which the electrons may be seen as vibrating and rotating with the molecule. Contrary to the classical case, in quantum mechanics only discrete energy levels may be occupied in these potential wells. Note the large separation between electronic levels compared with the vibrational and rotational transitions, and that difference between vibrational energy levels is larger than that between rotational levels.

Atomic vibrations in a molecule may be discussed, to a first approximation, in terms of simple harmonic motion. Above, we derived the potential energy function of a classical simple harmonic oscillator. The quantum mechanical potential energy of a simple harmonic oscillator made up of two bonded atoms of reduced mass $m_{1,2}$ is of the form

$$E = \left(n + \frac{1}{2}\right)\hbar\omega$$

where

$$\omega = \sqrt{\frac{k}{m_{1,2}}} \tag{A3.43}$$

and n is a vibrational quantum number, which can take only positive integer values including zero, $\hbar$ is Planck's constant divided by 2π, k is a force constant, and $m_{1,2}$ is reduced mass (compare Eq. (A3.24)). In contrast to classical mechanics, where vibrators can assume any potential energy, quantum vibrators can take only certain discrete energies (Fig. A3.23). Transitions in energy levels can be brought about by absorption of radiation, provided the energy of the radiation exactly matches the difference in energy levels ΔE between the quantum states.

The potential energy of electronic vibrations in a bond between two atoms, however, is not perfectly described by a harmonic model. For example, as the two atoms approach one another, Coulombic repulsion between the two nuclei produces a force that acts in the same direction as the restoring force on the bond; thus the potential energy can be expected to rise more rapidly than predicted by the harmonic approximation. Qualitatively, the potential energy curves takes the anharmonic form shown in Fig. A3.24 (and in Fig. A3.23). Such curves deviate by varying degrees from harmonic behavior, depending upon the nature of the bond and the atoms involved.

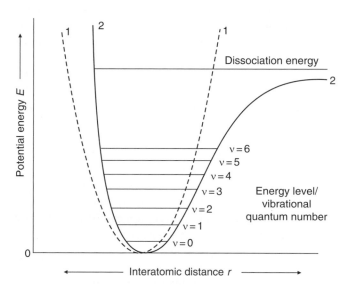

Fig. A3.24 Potential energy diagram for electronic vibrations in a bond between two atoms: curve 1, harmonic oscillator; curve 2, anharmonic oscillator. Note that the harmonic and anharmonic curves nearly coincide at low potential energies. The dissociation energy is that required to "break the spring" between the atoms.

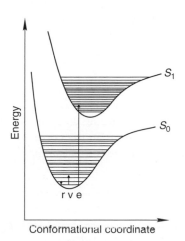

Fig. A3.23 Potential energy diagram of the lowest (S_0) and first excited (S_1) electronic states in a molecule: r, v, and e are rotational, vibrational, and electronic transitions, respectively. (After Cantor and Schimmel, 1980.)

A3.2.5 Measurement Space, Mathematical Functions, and Straight Lines

Practically all experimental measurements in biophysics are performed in a parameter space that is related to real space by a mathematical transformation. In crystallography, for example, the *measurement* space is *reciprocal* space (the space in which diffraction is measured), which is related to real space (the space of atomic coordinates) by Fourier transformation. In hydrodynamics, the variation of a parameter (translational or rotational diffusion, for example) is plotted as a function of time in the measurement space of the experiment. A model provides the means to transpose this information to real space (for example, if a sphere is assumed its radius will be determined).

It is important to emphasize that the experiment always takes place in measurement space; the information in real space is usually model dependent and will be obtained with a limited accuracy.

In order to simplify the mathematical interpretation of experimental data it is often possible to express complicated equations in linear form (see Comment A3.2).

A3.3 DYNAMICS AND STRUCTURE, KINETICS, KINEMATICS, RELAXATION

The concept of *dynamics* is often used to describe time dependence (or motions), as opposed to *structure*, which describes a static or (more accurately) time-average view. The Greek origin of the word, *dynamis*, means strength, however, and refers to *forces*, and *dynamics* (a singular noun) is the branch of *mechanics* in physics that deals with the motion of objects and the forces that act to produce such motion.

In physics, dynamics is divided into *kinetics* (from the Greek *kineein*, to move), which is concerned with the relationship between moving objects, their masses, and the forces acting upon them, and *kinematics*, which is concerned only with the motion of objects, without consideration of forces. Kinematics and the word *cinema* share the same Greek origin (*kinema*, movement). In chemistry, *kinetics* refers to the study of the *rates of chemical reactions*.

Relaxation refers to the return to equilibrium of a disturbed system. The simplest relaxation can be described by an exponential decay:

$$A(t) = A(0) \exp(-t/\tau) \tag{A3.44}$$

The parameter $A(t)$ represents the deviation of a property of the system from equilibrium at time t. It is $A(0)$ at time 0 and decays to zero at infinite time. The relaxation time, τ, is the time at which A has decayed to 1/e (about 1/2.7) of the value of $A(0)$ (see Chapter D3).

In this chapter, we discuss the present understanding of biological macromolecular structure in terms of the acting

COMMENT A3.2 BIOLOGIST'S BOX: MATHEMATICAL FUNCTIONS AND STRAIGHT LINES

Parameters measured in an experiment often occur in complicated forms in mathematical equations. Consider, for example, the angular dependence of scattered intensity from a solution of macromolecules:

$$I(Q) = I(0) \exp\left(-\frac{1}{3} R_G^2 Q^2\right)$$

where Q is related to the scattering angle. The formula is known as the Guinier approximation (see Chapter G2). Two experimental parameters can be derived from $I(Q)$: $I(0)$, which is related to the molar mass of the particle in solution; and R_G, the radius of gyration, which is a measure of the particle shape and dimensions. The easiest way to obtain these parameters is to *linearize* the equation:

$$\ln I(Q) = \ln I(0) - \frac{1}{3} R_G^2 Q^2$$

In this form a plot of $\ln I(Q)$ against Q^2 will yield a straight line of intercept $\ln I(0)$ and slope $-\frac{1}{3} R_G^2$.

The application of linearization procedures in order to express a mathematical function as a straight line is generally useful in biophysical studies. Another example of the application is in hydrodynamics. The translational friction suffered by particles of *the same shape* in solution is proportional to the cube root of the molar mass (linear dimension)

$$f = k \; M^{1/3}$$

where k is a constant. We can linearize the equation in the following way:

$$\log \; f = \log \; k + \tfrac{1}{3}\log \; M$$

and plot the *f versus M* values on a double logarithmic scale. If the particles are indeed of the same shape, the equation is obeyed and the data fall on a straight line with slope 1/3 (see Chapter D2).

forces. We, therefore, concentrate on the internal dynamics that define the macromolecule as a physical particle (i.e., the dynamics of the atoms in the correctly folded macromolecule) rather than on the dynamics of the particle moving as a rigid body, which is treated in Part D on hydrodynamics.

A3.3.1 Macromolecular Stabilization Forces

The forces that maintain atoms in position in a correctly folded, biologically active macromolecular structure are known (see Section C2.4.9). They arise from van der Waals interactions, hydrogen bonds, electrostatic interactions,

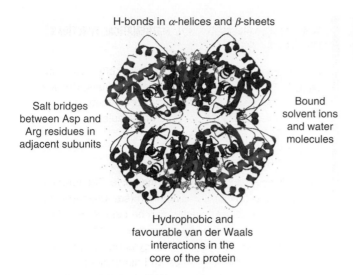

H-bonds in α-helices and β-sheets

Salt bridges between Asp and Arg residues in adjacent subunits

Bound solvent ions and water molecules

Hydrophobic and favourable van der Waals interactions in the core of the protein

TABLE A3.2 LENGTH AND TIMESCALES IN MACROMOLECULAR DYNAMICS		
Type	**Amplitude (Å)**	**Time (s)**
Electronic changes	0.01	10^{-15}
Atomic bond fluctuations	0.01–0.1	10^{-14}–10^{-13}
Side-chain motions	1	10^{-12}–10^{-9}
Side-chain rotations in protein interior domain motions	5	10^{-4}–1
	1–5	10^{-9}–1
Protein folding, complex formation, conformational changes . . .	1 to >5	10^{-6}–10^{3}

Fig. A3.25 Interactions stabilizing the folded form of the malate dehydrogenase tetramer from the halophilic archaeon *Halobacterium marismortui*. The protein does not contain cysteine residues. Its physiological environment is close to saturated in KCl and it requires high salt concentration to be folded and stable. In fact, weak electrostatic interactions lead to specific binding of solvent ions that participate in the stabilization. The orange balls between the subunits in the picture are chloride ions (Richard *et al.*, 2000).

hydrophobic interactions, and S–S bonds between cysteine residues (Fig. A3.25).

The *stabilization energy* of a macromolecular structure is the difference in free energy between the folded and unfolded forms. Covalent bonds, which are very strong with respect to temperature, are not usually broken when a macromolecule unfolds, so that (except for S–S bonds) they do not contribute to the stabilization energy of macromolecular tertiary or quaternary structure. Biological macromolecules are *soft* because the energies associated with the stabilization forces are *weak* in the sense that they are of the order of thermal energy at physiological temperatures. This softness may at first appear surprising because proteins, for example, are known from crystallography and calorimetry to form compact structures in the core of which the atoms are tightly packed; also, the electrostatic (Coulombic) interaction between two charges is long-range and very intense, and S–S bonds are essentially covalent bonds. Stabilization forces, however, are all strongly environment-dependent. The electrostatic interaction in biological macromolecules is shielded by the dielectric properties of water and by solvent counter-ions. S–S bonds are easily broken in a reducing environment to form two separate SH groups. Hydrogen bonds of different strengths are formed between donor and acceptor groups in the macromolecule and between the macromolecule and solvent ions. The hydrophobic interaction is environment-dependent by definition, since it arises from entropic effects due to the low solubility in water of certain

chemical groups (see also Sections A1.3.4, C2.4.9). Even the van der Waals interaction, which arises from the close packing of atoms, is environment-dependent, because certain atoms pack closely better than others, and it has been suggested as an important driving force of protein folding.

Molecular dynamics simulations represent our theoretical understanding of macromolecules as *physical particles*. Despite the fact that the types of macromolecular stabilization forces are well known, our quantitative knowledge of the determinants involved is very far from complete, mainly because of complex environment effects. Such calculations, therefore, still have to remain firmly anchored by experimental results.

A3.3.2 Length and Timescales in Macromolecular Dynamics

The amplitudes of atomic motions in macromolecules at ambient temperature (300 K) range from 0.01 Å to >5 Å for time periods from 10^{-15} s to 10^3 s (a femtosecond for electronic rearrangements to about 20 min for protein folding or local denaturation) (Table A3.2). Different biophysical methods are adapted to each time and length scale (see Introduction). Laser triggered optical spectroscopy can now reach subfemtosecond resolution and spans the entire 18 orders of magnitude of the timescale to kinetic measurements taking minutes. NMR is sensitive to dynamics from the picosecond upwards. Neutron scattering can resolve frequencies in the 10^9–10^{12} s^{-1} range and amplitudes in the 1–5 Å range.

A3.3.3 A Physical Model for Protein Dynamics

Our current understanding of protein dynamics (see Chapter I1) is based on experimental results from a wide range of biophysical methods, in particular, time-resolved optical and neutron spectroscopies, NMR, and crystallography.

A protein is a densely packed object with the atoms occupying definite average positions corresponding to the

COMMENT A3.3 FLASH PHOTOLYSIS EXPERIMENTS ON MYOGLOBIN AT LOW TEMPERATURES (SEE ALSO SECTION E1.3.3)

Myoglobin is a small monomeric oxygen-binding protein (molecular weight about 16 K) found in muscle. It was one of the first (with hemoglobin, the related oxygen carrier protein in red blood cells, which can be seen as a tetramer of myoglobin-like subunits) whose structure was solved by X-ray crystallography. The Nobel Prize was awarded to Max Perutz and John Kendrew for these studies. Hemoglobin and myoglobin have been studied extensively by crystallography and functionally important structural differences have been elucidated between the unbound (deoxy-) states and various ligand-bound states. Myoglobin contains a prosthetic heme group with an iron atom, which binds the oxygen molecule. Myoglobin has a higher affinity for carbon monoxide than for oxygen, which is the reason why CO is a poison (it blocks the oxygen-binding site). Its CO-binding properties, however, have provided important tools for the study of the protein. The binding of CO to the heme group in myoglobin can be followed readily by its effect on the protein's absorption spectrum for visible light. In a flash photolysis experiment, rebinding kinetics is observed after the CO bond to the heme iron is broken by a laser flash. At low temperatures (below 200 K), the CO does not diffuse out of the protein. In position A when it is bound, it released by flash photolysis to position B in the heme pocket of the protein.

Contrary to expectations for a simple relaxation model (Eq. (A3.34)), it has been observed that rebinding kinetics did not fit a single exponential in time, but is constituted of a mixture of fast and slow processes. This can be explained in one of two ways: Either the proteins make up a homogeneous population with several B sites in each (e.g., a "fast" and a "slow" rebinding site) or the proteins make up a heterogeneous population with different molecules displaying different sites.

It was possible to distinguish unambiguously between the two models by *multiple flash* experiments, using different flash rates. When the flash rate is intermediate between the fast and slow rates, very different behavior is expected for the homogeneous and inhomogeneous protein models. It has been established that the inhomogeneous protein model is the correct one. The *pure protein* population at low temperatures, therefore, is made up of molecules with slightly different structures, represented by the *conformational substates*. Direct rebinding from the heme pocket becomes exponential at higher temperatures, with a homogeneous protein population fluctuating between the conformational substates (Frauenfelder *et al.*, 1988.) (Figures reproduced with permission from Annual Reviews.)

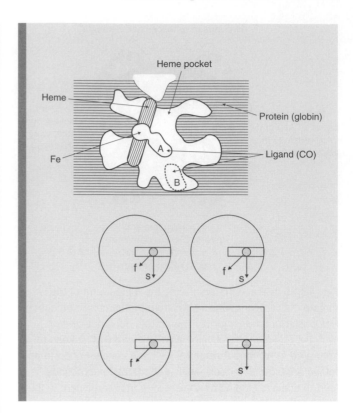

native structure. Substantial restoring forces act on the atoms so that protein molecular dynamics resembles that of an amorphous solid. Only to a first approximation can the protein be seen as forming a continuous elastic medium, with local side-chain motions and global domain motions displaying simple Hooke's law character (see Section A3.2.2).

Conformational Substates in the Energy Landscape and Protein-Specific Motions

The *conformational substate* (CS) model has been put forward for proteins, following experiments on carbon monoxide binding to the heme group in myoglobin (Comment A3.3). An analysis of the rebinding kinetics indicates a heterogeneous protein population at low temperature (below about 200 K). At higher temperatures, the ligand overcomes a number of energy barriers on its path to rebinding in a way that suggests the protein fluctuates between subtly different conformations. A schematic energy landscape characteristic of the CS model is shown in Fig. A3.26.

Below about 200 K, the protein is trapped in one of a number of energy minima – one of the CS (bottom panel in Fig. A3.26).

The structural difference between two CS may be very small, e.g., due to a different orientation of just one amino acid side-chain. With increasing temperature there comes a point where the macromolecule has sufficient thermal energy to "jump" the barrier from one CS to another. The resulting motions have been called *protein-specific motions*. In other words, the CS represent a "static" disorder in

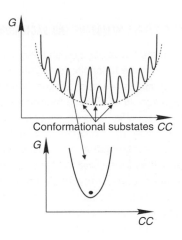

Fig. A3.26 One-dimensional schematic diagram of the CS model and potential free energy (G) landscape of a protein. CC is a conformational coordinate (from Frauenfelder *et al.*, 1988). (Figure reproduced with permission from *Annual Reviews*.)

protein structure at low temperature – each molecule in a population having a slightly different structure (e.g., a side-chain in conformation A in one molecule and in conformation B in another); and they represent a "dynamic" disorder at higher temperature (e.g., the side-chain moving, within the same molecule, between conformations A and B). In fact, careful analysis of the CO rebinding showed that the *energy landscape* of the protein presents a complex hierarchy, with each CS potential well itself divided into further CS. The CS model has received support from various other experimental methods, including analysis of the temperature dependence of Debye–Waller factors in X-ray crystallography, which clearly demonstrated the static disorder in the crystal population.

The Dynamical Transition and Mean Effective Force Constants

Neutron scattering experiments showed that the mean square fluctuation, $\langle u^2 \rangle$, of atoms in myoglobin has a temperature dependence as shown in Fig. A3.27. Note that the magnitude of the fluctuations is of the order of an ångström. The break in slope at 180 K was called a *dynamical transition*. Since the atomic fluctuations in the protein are due to thermal energy, the temperature axis actually represents the energy of the system. Below about 180 K (–93 °C), $\langle u^2 \rangle$ increases linearly with energy, with a backward extrapolation to zero at zero absolute temperature (Comment A3.4). The straight-line dependence can be accounted for by a model in which the atoms move in simple harmonic potentials, for which the restoring force is proportional to the displacement from the mean position. The energy of a set of harmonic oscillators varies with the mean square displacement with a proportionality constant, $\langle k \rangle$, equal to the mean elastic force constant or "spring" constant (Section A3.2.2). The value of $\langle k \rangle$ can, therefore, be calculated from the inverse of the slope of $\langle x^2 \rangle$ versus T

(Fig. A3.27). Converting T to energy units by multiplying by Boltzmann's constant, the mean force constant maintaining the atoms in the myoglobin structure at low temperature was calculated to be $2\,\mathrm{N\,m^{-1}}$, corresponding to the macroscopic force constant for a fairly stiff elastic band (Comment A3.5). Trehalose is a natural disaccharide, which is synthesized by desert and other organisms that have to survive extreme drought. It coats and protects their cellular structures while their metabolism is arrested and they are waiting for better, wetter, times. Myoglobin encased in a trehalose glass (circle data in Fig. A3.27) is stiffer than the protein in water with a $\langle k \rangle$ value of $3\,\mathrm{N\,m^{-1}}$.

In order to avoid terms like rigidity or stiffness, which are often used qualitatively, the mean effective force constant derived from the temperature dependence of the mean square fluctuation was called the *resilience* of the structure. Note that there is no dynamical transition in trehalose, leading to the suggestion that the protective

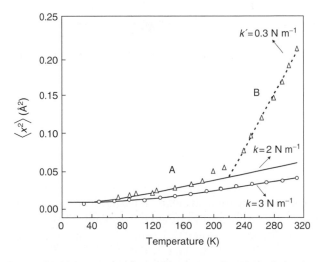

Fig. A3.27 Mean square fluctuations as a function of absolute temperature, measured by neutron scattering (see Chapter I2) in myoglobin surrounded by water (Δ) and in a trehalose glass (o). The break in slope at about 180 K is called a dynamical transition. Effective force constants k and k' were calculated for different parts of the curves as explained in the text.

COMMENT A3.5 FORCE CONSTANTS ON ATOMS AND IN ELASTIC BANDS

(a) An effective force constant of about $2\,N\,m^{-1}$ holds an atom in the myoglobin structure in place at low temperature. (b) An elastic band of force constant $2\,N\,m^{-1}$ stretches by 1 cm under a weight of 2 g. The force exerted by the weight is $0.002 \times 10\,N$.

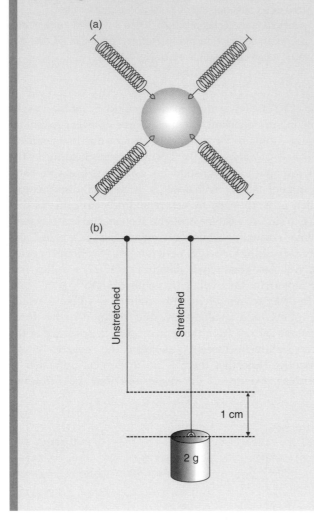

COMMENT A3.6 PURPLE MEMBRANE AND BACTERIORHODOPSIN

Purple membrane forms patches on the cell membrane of the archaeal extreme halophile (see Chapter A1) *Halobacterium salinarum*. It is an extraordinary membrane, made up of a single membrane protein type, organized with specific lipids on a highly ordered two-dimensional lattice. The existence of the lattice permitted high-resolution crystallographic studies by electron microscopy of the natural membrane (see Part G), so that we know a lot about its structure. The protein was called *bacteriorhodopsin* (BR) because it binds a molecule of the chromophore, retinal, which gives the purple color to the membrane. Before this discovery, retinal had been found only in the rhodopsins, the vision proteins in animals. In *H. salinarum*, BR functions as a light-activated proton pump, transferring one proton out of the cell (against its potential gradient) for each photon absorbed by the retinal. A millisecond photocycle of color changes, associated with the proton pump activity, provide a powerful tool for the study of structure function relations in purple membrane by spectroscopic methods (see Section E1.3.3).

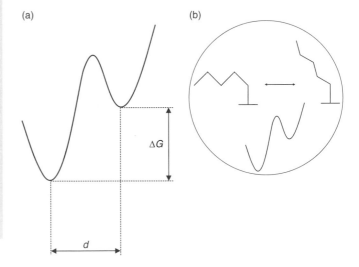

effect of the sugar is due to its trapping the macromolecule in a stiff harmonic state to relatively high temperatures, which precludes the motions that would lead to unfolding.

In terms of the CS model, the protein atoms, below the dynamical transition, oscillate in harmonic potential wells. The dynamical transition then corresponds to their being able to fluctuate between different CS wells in the potential energy landscape. The data in Fig. A3.27 have been modeled successfully by using a two-well potential model (two CS) for all the atoms in the protein surrounded by water (Fig. A3.28). This is clearly an oversimplification because different protein atoms move in different potentials, but it does provide an illustration of the link between the dynamical transition and the CS model. Even though the motions

Fig. A3.28 (a) A double-well free energy (G) potential for protein atoms. Such a potential would be illustrated in practice by side-chains being able to sample two different local energy minima in the structure. (b) By using a quasi-harmonic approximation, an approximate effective force constant can be calculated as $\langle k' \rangle \sim \delta G/2d^2$.

are not simple harmonic above the dynamical transition, it is still possible to consider a quasi-harmonic approximation in order to calculate an effective phenomenological force constant from the inverse slope of $\langle u^2 \rangle$ versus T. The value for myoglobin is $0.3\,N\,m^{-1}$, about ten times "softer" than for the protein at low temperature.

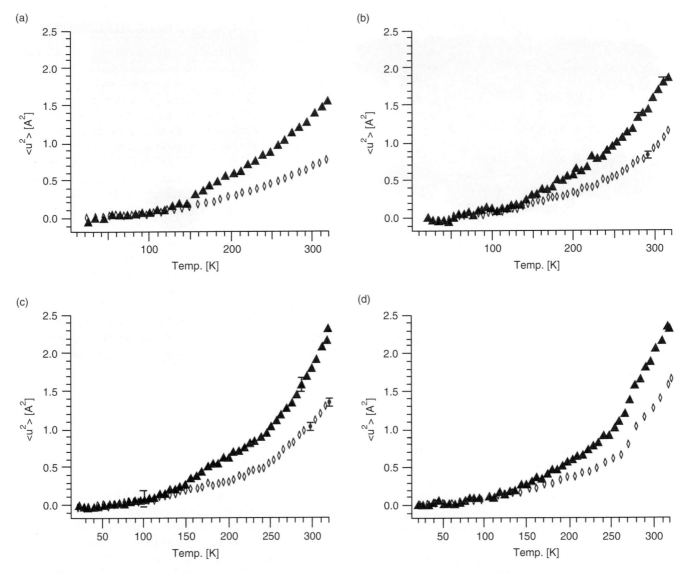

Fig. A3.29 BR dynamics measured by neutron scattering (see Part I). Mean square fluctuations are plotted versus temperature for the native membrane (▲) and a labeled membrane (◊) in which the data correspond to the motions of specific amino acids and retinal in the core of the protein. Parts **(a)**–**(d)** correspond to progressively higher hydration; 0%, 75%, 86%, 93% relative humidity, respectively (Lehnert, 2002).

Dynamical transitions have been observed in various other protein samples, as well as in myoglobin, and they may well represent a general feature of protein molecular dynamics. The values of the mean square fluctuations and effective force constants appear to be specific to each case and related to protein stability function and activity. It has been shown in cryocrystallography experiments (see Part G), for example, that the enzyme ribonuclease A cannot bind its substrate at temperatures below the dynamical transition. If the substrate is bound at a higher temperature and the protein cooled to below the transition, then it cannot be released. The additional dynamical flexibility of the protein, permitting it to sample different CS, appears therefore to be necessary for this enzyme's activity. Correlations have also been established between the dynamic transition and activity for bacteriorhodopsin (BR), a

membrane protein that functions as a light-driven proton pump (Comment A3.6; Fig. A3.29).

There are similarities between the dynamical transition in proteins and the glass transition in amorphous materials, but they cannot be considered as completely analogous. Proteins are complex, heterogeneous structures whose dynamics has evolved to fulfill specific functions. Figure A3.29 shows that in BR, for example, the mean square fluctuations and effective force constants are different for different amino acid groups within the protein structure. The active core of the proton pump mechanism of the protein lies around the retinal, which remains bound to a lysine residue, via a so-called Schiff base linkage, about half way down the membrane thickness. During the BR photocycle (see Comment A3.6), the Schiff base proton changes its orientation from being accessible to the

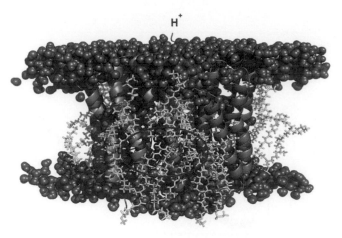

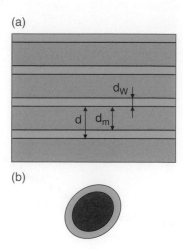

Fig. A3.30 A schematic diagram of purple membrane, showing a BR molecule as a ribbon diagram and lipids in blue. Two lipid molecules and the labeled groups discussed in Fig. A3.29 are shown in ball-and-stick representation. Converting light energy to chemical energy, BR pumps a proton out of the cell for each photon absorbed. Experiments demonstrated that the core and extracellular half of BR (within the red line), which act as the valve of the pump, are more resilient (stiffer) than the membrane as a whole. (Zaccai, 2000.)

Fig. A3.31 (a) Purple membrane samples for neutron scattering are made up of a regular stack of membrane and water layers, of periodicity d. The membrane thickness, d_m, is constant and equal to about 50 Å; the water layer thickness, d_w, varies with RH from the thickness of one layer of water (about 3 Å) at 75% RH to the thickness of three layers at 93% RH. **(b)** Schematic representation of the hydration shell (light blue) around a lysozyme (red). The shell thickness corresponds to approximately one molecular layer and its density is about 10% higher than bulk water (Weik *et al.*, 2004; Svergun *et al.*, 1998; Merzel and Smith, 2003). (See also Section D1.3.)

COMMENT A3.7 SALT AND RELATIVE HUMIDITY

The partial pressure of water above a saturated salt solution is a constant for a particular salt type at a given temperature and pressure. Relative humidity (RH) can, therefore, be regulated under experimental conditions by putting the sample in a closed temperature-controlled vessel, in the presence of an appropriate saturated salt solution. At 25 °C and normal pressure, for example, the RH above a saturated LiCl solution is 15%; for the same conditions it is 75% for NaCl, 86% for KCl, and 93% for KNO_3. An advantage of these salts is that the corresponding RH values are not very sensitive to small temperature variations. Zero percent ("dry") and 100% ("wet") relative humidity values are more difficult to define precisely. A "dry" atmosphere may be obtained with silica gel, but "even drier" conditions are achieved by pumping a vacuum over P_2O_5. One hundred percent RH is obtained, in principle, above pure water, but even small temperature gradients in the walls lead to condensation and fluctuating RH values within the vessel.

(Fig. A3.30), as expected for a stiffer environment required to regulate the valve function of the retinal binding site.

Solvent Effects, Membrane, and Protein Hydration

From a dynamics standpoint, a protein structure cannot be considered separately from its environment. The effect of trehalose on myoglobin dynamics is striking (Fig A3.27). Similarly, we note in Fig. A3.29 that the mean square fluctuations in BR are strongly hydration-dependent above a transition temperature of about 150 K. Hydration was defined, in the experiments, by putting the sample in a controlled relative humidity environment (Comment A3.7). Above 150 K, both the labeled and native sample have higher fluctuations and become softer with increasing relative humidity, illustrating the effects of hydration on the membrane and protein dynamics. It is interesting to recall that purple membrane is obtained from an extreme halophile that lives in a close to saturated salt environment (see Chapter A1). Saturated NaCl and KCl solutions correspond, respectively, to relative humidity values of 75%, and 86%. Parts (b) and (c) of Fig. A3.29 might, therefore, be the closest representations of purple membrane physiological conditions. Purple membrane samples in the neutron scattering experiments were made up of stacks of alternating membrane and water layers (Fig. A3.31). The proton pump photocycle of purple membrane (also measured in stack samples) is inhibited below about 75% RH, a condition in which there is one layer of water between the membranes in the stack – a hydration level that appears to be essential

extracellular side of the membrane to being accessible to the intracellular side. The retinal linkage to the protein, therefore, acts as the valve of the pump, allowing the unidirectional transmission of the proton. Experiments have been designed in this context to examine the dynamics of different parts of the BR structure. The active core of the protein presents lower mean square fluctuations and higher effective force constants than the protein average

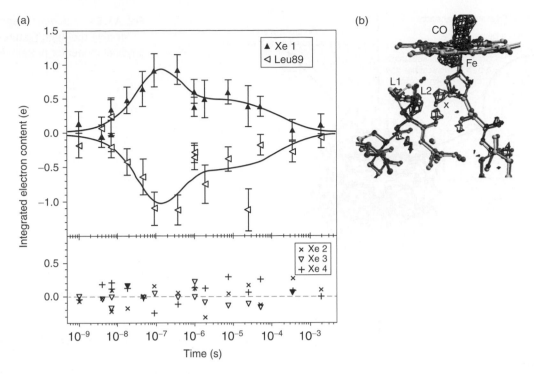

Fig. A3.32 (a) The time courses of the integrated electron content of the positive difference density at the Xe 1 binding site and of the integrated electron content of the negative Leu89 feature. The time courses of the other three Xe binding sites (Xe 2, Xe 3, and Xe 4) are shown for comparison; they indicate no occupation by CO on the timescale. The solid line represents a fit of the time course of the Xe 1 density by two exponential phases and a bimolecular phase, fixed to that of ligand rebinding. **(b)** Difference electron density map of the Xe 1 region at 362 ns. Positive density at the Xe 1 site is labeled X, while positive and negative densities indicating rearrangement of the Leu89 side-chain are labeled L1 and L2, respectively (corresponding to two conformational substates). The CO- and deoxymyoglobin structures are in red and blue ball-and-stick representation, respectively. Positive and negative difference electron density appears as blue and red "nets," respectively. The heme with the iron atom (Fe) at its center and the position of the bound CO are seen at the top of the structure (Srajer *et al.*, 2001). (Figure reproduced with permission from the American Chemical Society.)

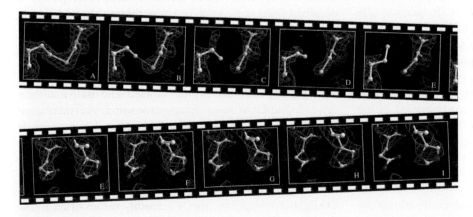

Fig. A3.33 Movie of the radiation damage to a glutamic acid side-chain (bottom film band) and a cysteine–cysteine bond (top film band) in acetylcholinesterase at 100 K (from Weik *et al.*, 2000). (Figure reproduced with permission from *Proceedings of the National Academy of Sciences* (USA).)

to ensure the minimum dynamic level for protein activity in the membrane. Careful measurements of activity as a function of hydration in lysozyme and other soluble enzymes have produced a similar conclusion. It is now generally accepted that at least one hydration layer around globular soluble proteins is required for activity. Depending on the size of the protein, this corresponds to a minimum hydration of between about 0.2 and 0.4 g of water per gram of

protein. It has also been observed experimentally and successfully simulated in a molecular dynamics calculation that the density of this first hydration shell is about 10% higher than that of bulk liquid water (Fig. A3.31).

Motion Pictures of Intermediate Structures

Kinetic crystallography provides the closest approximation we currently have to motion picture (cinema) images of

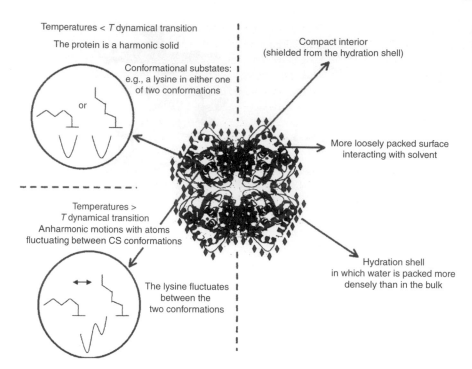

Temperatures < *T* dynamical transition

The protein is a harmonic solid

Conformational substates:
e.g., a lysine in either one
of two conformations

or

Temperatures >
T dynamical transition
Anharmonic motions with atoms
fluctuating between CS conformations

The lysine fluctuates
between the
two conformations

Compact interior
(shielded from the hydration shell)

More loosely packed surface
interacting with solvent

Hydration shell
in which water is packed more
densely than in the bulk

Fig. A3.34 Schematic diagram illustrating the main features of the physical model for protein dynamics.

macromolecules. (The method should really be called kinematic crystallography.) Macromolecular crystallography produces beautiful pictures of structures to atomic detail, and it was a logical consequence to explore its possibilities for obtaining images of a structure as a function of time. The extremely high intensity of broad wavelength range X-ray beams produced in synchrotrons and the possibility of obtaining them in periodic pulses as short as a few picoseconds made time-resolved crystallography appear to be an attainable goal. But a crystal contains about 10^{15} molecules, and the information on atomic positions is averaged not only over time, but also over all the molecules in the crystal. In other words, in order to observe structural changes with time, all the molecules should be in phase, undergoing the same changes at the same time. Photoactivation is extremely fast and reaction synchrony has been achieved by using a laser flash to trigger the conformational changes simultaneously in the entire crystal, either if the protein itself naturally binds a light-sensitive molecule (e.g., the heme in hemoglobin and myoglobin (Comment A3.3) or the retinal in rhodopsin and BR (Comment A3.6)) or by the use of *caged compounds*. Binding a *blocking* chemical group to a substrate or cofactor can effectively cage it in and greatly reduce its affinity for the enzyme, or, if it binds to the enzyme, effectively stop the reaction from proceeding. By choosing a blocking group bond that is photolabile (i.e., sensitive to light), after diffusion into the crystal, the caged compound can be released by a laser flash of appropriate wavelength very rapidly and homogeneously to initiate the reaction.

Real-time crystallography provides motion pictures of working photosensitive proteins down to nanosecond resolution. For various technical reasons, however, in the middle of 2004 most published real-time crystallography studies of enzyme reactions using caged compounds related to second or even minute time resolution, and there was only one study reporting millisecond resolution. A complementary and perhaps more promising approach to obtaining motion pictures of proteins than real-time crystallography is based on trapping transient structural intermediates at *cryotemperatures* (from the Greek *cryo*, very cold). This approach has been very fruitful in studies of naturally photosensitive proteins and in studies using caged compounds. The effective time resolution is obtained from the choice of temperature. The reaction is stopped if it reaches an energy barrier that cannot be crossed because of insufficient thermal energy.

Various authors have investigated the structural intermediates formed when CO bound to myoglobin dissociates and migrates out of the protein molecule by using real-time crystallography at room temperature or cryocrystallography. The results of these experiments were described in terms of motion pictures. A careful examination of these movies is particularly instructive about some of the general features of protein dynamics involved in biological function. The first frame of the movie corresponds to CO-ligated myoglobin, the last frame to the structure of deoxymyoglobin (Comment A3.3). An X-ray crystallography study provides information on the electron density distribution in a structure. Structural changes in real time appear as positive electron density peaks, indicating the new presence of atoms at these locations, or negative peaks, indicating that atoms have moved away from these sites. A short laser pulse triggers CO dissociation. The very first changes, a movement of the iron atom and "doming" of the heme plane, take place too rapidly to be resolved in the time frames of the movie, which follow changes in the crystal structure occurring between 1 ns and 1 ms. On this timescale, structural fluctuations in the protein structure open up a diffusion path for the CO, which proceeds outwards from its binding pocket by pausing at a number of specific "docking" sites, where it interacts favorably with protein side-chains. Four xenon-binding sites, labeled Xe 1 to Xe 4, have been identified as potential such docking sites. Meanwhile, the protein structure changes toward the conformation of the unligated deoxymyoglobin. In particular, there is a rearrangement of the side-chain of Leu89 to accommodate CO in the Xe 1 site. The time courses of the integrated electron content of the positive difference density at the Xe 1 binding site and of the integrated electron content of the negative Leu89 feature are shown in Fig. A3.32a. The map showing the corresponding difference electron density peaks at 362 ns is in Fig. A3.32b.

We recall, the structure observed in a crystallographic experiment corresponds to that adopted by an appreciable number of molecules in the crystal. The frames in the movie, therefore, illustrate specific intermediates (or energetically favored CS) in the relaxation between CO-ligated and deoxymyoglobin structures. The crystallographic movies do not show continuous changes in atomic positions between one intermediate and the next, indicating that individual molecules in the crystal take different paths across the energy barriers. The picture fits well with the hypothesis that a large number of fluctuation-like motions on the nanosecond timescale result in the much slower conformational changes between intermediates and the opening of a way out for the CO molecules.

Figure A3.33 illustrates another type of crystallographic motion picture. It shows the time course under irradiation of two locations in the structure of an enzyme called acetylcholinesterase, which is involved in nerve transmission. The amino acid side-chains in the initial structure are shown as ball-and-stick, and the observed electron density is shown as a blue "net." The bottom panel shows the progressive decarboxylation of a glutamic acid side-chain due to radiation damage. The top panel shows the breaking of an S–S bond between two cysteines. The nine data sets, A–I, were collected at 100 K. At 150 K the damage affected more residues including those in the active site, which indicates that the side-chains can sample different conformations at the higher temperature.

The main features of the current physical model for protein dynamics are summarized and illustrated in Fig. A3.34.

A3.4 CHECKLIST OF KEY IDEAS

- Molecular biophysics is a predominantly experimental science, in which even *theoretical* approaches such as *molecular dynamics simulations* have to be based firmly on observation.

- Practically, all the experiments in biophysics (with the exception of calorimetry and classical solution physical chemistry methods such as the ones used to determine osmotic pressure, viscosity, etc.) rely completely or in part on the observation of the interaction between macromolecules and radiation.

- The mathematical tools required include those dealing with *the general properties of waves, complex exponentials, simple harmonic motion and normal modes*, and *Fourier analysis*.

- Experiments are performed in a *measurement space* related to real space by a mathematical transformation, such as *reciprocal space* in crystallography, which is related to real space by Fourier transformation.

- The concepts of quantum mechanics are essential for the description of the interaction between radiation and matter on an atomic scale.

- *Dynamics* (a singular noun) is the branch of *mechanics* in physics that deals with the motion of objects and the forces that act to produce such motion. Dynamics is divided into *kinetics*, which is concerned with the relationship between moving objects, their masses, and the forces acting upon them, and *kinematics*, which is concerned only with the motion of objects, without consideration of forces.

- In chemistry, *kinetics* refers to the study of the *rates of chemical reactions*.

- *Relaxation* refers to the return to equilibrium of a disturbed system.

- The folded native structures of biological macromolecules are maintained by forces arising from hydrogen bonds, salt bridges, or screened electrostatic interactions, so-called hydrophobic interactions, and van der Waals interactions.

- The amplitudes of atomic motions in macromolecules at ambient temperature range from 0.01 Å to >5 Å for time periods from 10^{-15} s to 10^3 s (a femtosecond for electronic rearrangements to about 20 min for protein folding or local denaturation).

- Our understanding of protein dynamics is based on results from various spectroscopic experiments, and crystallographic studies of time-averaged structures and transient intermediates.

- At low temperature, below about 200 K, a protein structure can be represented by one of many slightly different conformational substates, CS; above about 200 K, the protein can sample several CS by fluctuating between them.

- The change between the low-temperature harmonic vibration dynamics regime, where the protein is trapped in one CS, and the higher temperature regime, where it fluctuates between CS, is called a *dynamical transition*.

- Below the dynamical transition, mean square atomic fluctuations measured by neutron scattering are found to increase linearly with temperature, as expected for a set of simple harmonic oscillators, and a mean force constant can be calculated from the inverse slope of the dependence.

- Above the dynamical transition, an effective force constant can be calculated from neutron scattering, mean square fluctuation, and temperature dependence data, by applying a quasi-harmonic approximation.

- Harmonic force constants below the dynamical transition for proteins are of the order of newtons per meter; above the dynamical transition, effective force constants are ten times softer.

- Protein dynamics is sensitively solvent-dependent, and measured mean square fluctuations and force constants indicate greater flexibility and lower *resilience* with increasing relative humidity (water partial pressure) of the macromolecular environment.

- One hydration layer appears to be necessary for a protein to achieve the dynamic level required for biological activity.

- A trehalose glass maintains proteins in a high-resilience harmonic state to high temperatures, protecting them from unfolding.

Suggestions for Further Reading

General university course books on mathematics for biophysics, physical chemistry and chemical thermodynamics.

Zaccai, G. (2013) The ecology of protein dynamics. *Curr. Phys. Chem.*, special issue on quantum nanobiology and biophysical chemistry, **3**, 9–16.

MASS SPECTROMETRY

B

MASS AND CHARGE

B1

B1.1 HISTORICAL REVIEW

1897

J. J. Thomson made the first measurement of the mass-to-charge ratio of elementary particle "corpuscles," which later became known as electrons. This can fairly be considered as the birth of mass spectrometry.

1918–1919

A. Dempster and **F. Aston** developed the first mass spectrographs. Photographic plate was used as the array detector. The instruments were used for isotopic relative abundance measurements.

1951

W. Pauli and **H. Steinwedel** described the development of a quadrupole mass spectrometer. The application of superimposed radio-frequency and constant potentials between four parallel rods acted as a mass separator in which only ions within a particular mass range perform oscillations of constant amplitude and are collected at the far end of the analyzer.

1959

K. Biemann was the first to apply electron ionization mass spectrometry to the analysis of peptides. Later it was shown that for sequence determination, peptides had to be derivatized prior to analysis by a direct probe.

1968–1970

M. Dole was the first to bring synthetic and natural polymers into the gas phase at atmospheric pressure. This was done by spraying a sample solution from a small tube into a strong electric field in the presence of a flow of warm nitrogen, to assist desolvation. First experiments on lysozyme demonstrated the phenomenon of multiple charging.

1974

D. Torgerson introduced plasma desorption mass spectrometry. This technique uses ^{252}Cf fission fragments to desorb large molecules from a target. It was the first of the particle-induced desorption methods to demonstrate that gas-phase molecular ions of proteins could be produced from a solid matrix.

1974

B. Mamyrin made the most important contribution to the development of time-of-flight (TOF) mass spectrometry. He constructed the so-called reflectron device, which had been proposed by **S. Alikanov** in **1957**. The reflectron essentially improves mass resolution in the TOF mass spectrometer.

1978

N. Commisarow and **A. Marshall** adapted Fourier transform methods to ion cyclotron resonance spectrometry and built the first Fourier transform mass instrument. Since that time, interest in this technique has increased exponentially, as has the number of instruments.

1981

M. Barber discovered fast atom bombardment (FAB), a new ion source for mass spectrometry. The mass spectrum of an underivatized undecapeptide, Met-Lys-bradykinin of $M = 1318$ was obtained by bombarding a small drop of glycerol containing a few micrograms of the peptide with a beam of argon atoms of a few kiloelectron-volts. The technique revolutionized mass spectrometry and opened it to the biologist.

1984

R. Willoughby and, independently, **M. Aleksandrov** proposed the coupling of liquid chromatography and mass spectrometry for analyzing high-molecular-weight substances delivered by a liquid phase.

1988

J. Fenn and, independently, **M. Yamashita** were able to bring biological macromolecules into the gas phase at atmospheric pressure. They proposed a new type of ionization technique called electrospray ionization (ESI) to generate intact biological molecular ions by spraying a very dilute solution from the tip of a needle across an electrostatic field gradient of a few kilovolts. **M. Karas** and **F. Hillencamp** and, independently, **K. Tanaka** developed a new ionization technique called matrix-assisted laser desorption–ionization (MALDI). It was shown that proteins up to a molecular weight of 60 000 could be ionized if embedded in a large molar excess of a UV-absorbing matrix and irradiated with a laser beam. Taking advantage of high resolution, mass measurement accuracy, and ion-trapping capabilities, MALDI provides not only molecular mass information but also structural information for various peptides and oligonucleotides. **K. Tanaka** received the 2002 Nobel Prize in Chemistry for his contribution to mass spectrometry.

1992–1999

The molecular specificity and sensitivity of MALDI-MS gave rise to a new technology for direct mapping and imaging of biological macromolecule distributions present in a single cell or in mammalian tissue. By rastering the ion beam across a sample, and collecting a mass spectrum for each point from which ions are desorbed, it is possible to create mass-resolved images of molecular species across a cell surface or in a piece of tissue.

2000–2014

Mass spectrometry has developed into an important analytical tool in the life sciences. Soft-ionization techniques, such as ESI and MALDI, allow routine mass measurements of proteins and nucleic acids with high resolution and accuracy. Mass spectrometry has become one of the most powerful experimental tools for the direct observation of gas-phase biological complexes, their assembly, and their disassembly in real time. Mass spectrometry combined with isotopic labeling, affinity labeling, and genomic information has become a powerful tool in biophysics, and its applications in structural biology, medicine, and therapeutics are still growing rapidly. With the rapid development of new instruments, such as the Orbitrap, computational tools and analytical strategies, mass spectrometry has become an essential tool to study molecular and cellular processes in living cells and organisms. MS-based proteomics has contributed importantly to the development of systems biology.

B1.2 INTRODUCTION TO BIOLOGICAL APPLICATIONS

Since the 1980s, mass spectrometry (Comment B1.1) has become an important analytical tool in structural biology. This is a result of the ability to produce intact, high-molecular-mass gas-phase ions of various biological macromolecules. Several ionization techniques such as MALDI and ESI revolutionized mass spectrometry, opened it up to biology. New methods for ultrasensitive protein characterization such as the Orbitrap analyzer have been developed, providing a detection limit of approximately 30 zmol (30×10^{-21} mole) for proteins with molecular <100 kDa. Individual ions from polyethylene

COMMENT B1.1 THE TERM "MASS SPECTROSCOPY"

We would like to warn the reader against the term "mass spectroscopy." The term "mass spectroscopy" is not correct because it bears no relation to real spectroscopic techniques described in Parts E, I, and J. The mass spectrum depends mainly on the stability of ions produced and collected during the experiment. The stability of ions strongly depends on experimental conditions and therefore prediction of a mass spectrum is practically impossible.

COMMENT B1.2 ABSOLUTE AND RELATIVE MASSES

A mass spectrometer measures mass to charge ratio (m/z) where m is a relative mass value. The instrument needs to be calibrated with standard compounds, whose *mass* values are known very accurately. The carbon scale is used most frequently with $^{12}C = 12.000000$.

COMMENT B1.3 MOLECULAR MASS AND MOLECULAR WEIGHT

Some confusion may arise when M_r is used to denote relative molecular mass. M_r is a relative measure and has no units. However, M_r is equivalent in magnitude to M and the latter does have units and for high-mass biological macromolecules the dalton is usually used. Note that molecular weight (which is a force and not a mass) is an incorrect term in this case.

glycol to DNA, with masses in excess of 10^8 Da, can be isolated (Comments B1.2 and B1.3).

Mass spectrometry has now become the method of choice for a number of important aspects of protein structure: precise protein and nucleic acid mass determination in a very wide mass range, peptide and nucleotide sequencing, identification of protein post-translational modifications, protein structural changes, folding and dynamics, identification of subpicomole quantities of proteins, identification of isotope labeling. Mass spectrometry has been developed powerfully for proteomics and contributed significantly to the development of systems biology. Mass spectrometry-based techniques also allowed the characterization of fully functional biological subcellular complexes and organelles as well as intact bacteria, opening a large perspective for applications in functional proteomics, bacterial taxonomy, and medicine. All of the above are developed more fully in Chapter B2.

B1.3 IONS IN ELECTRIC AND MAGNETIC FIELDS

An ion that is accelerated out of a source acquires kinetic energy

$$E_{\text{kin}} = zV_{\text{acc}} = \frac{mv^2}{2} \tag{B1.1}$$

where z is the charge of the ion, V_{acc} represents the potential difference that defines the acceleration region, m is the ion mass, and v is the ion velocity. When entering a homogeneous magnetic field **B** perpendicular to its trajectory, the ion experiences the Lorentz force **F** (Comment B1.4), which is perpendicular to both **B** and **v** (Fig. B1.1). The resulting trajectory of the ion in a magnetic field is a circle with radius r, because the Lorentz force just balances the centrifugal force

---

$$F = zvB = \frac{mv^2}{r} \tag{B1.2}$$

The mass to charge ratio m/z is given by

$$\frac{m}{z} = \frac{B^2 r^2}{2V_{\text{acc}}} \tag{B1.3}$$

As seen from Eq. (B1.3) the lightest ions have the smallest radius of curvature. The radius increases as the mass of the ions and the strength of the electric field grow.

B1.4 MASS RESOLUTION AND MASS ACCURACY

B1.4.1 Mass Resolution

The ability to separate mass signals is affected by the resolving power of the mass spectrometer. The resolution R in mass spectrometry is defined as $R = m/\Delta m$, where Δm is the mass difference of two neighboring masses, m and $m + \Delta m$, of equal intensity, with signal overlap of 10%. A resolution of 100 000 makes it possible to distinguish an ion of mass 100 000 Da from one of mass 100 001, i.e., to ten parts per million.

Because peaks in a mass spectrum have width and shape, it is necessary to define the extent of overlap between adjacent peaks when determining the resolution. There are two definitions in widespread use, and it is essential to know which is being used when resolution figures are quoted (Fig. B1.2). The first one is the so-called 10% valley definition in which the two adjacent peaks each contribute 5% to the valley in between them. The second one is the "full-width, half-maximum" (FWHM) definition. The resolution of a peak using this definition is the mass of the peak (in daltons) divided by the width (in daltons) measured at the half-height of the peak. A useful rule of thumb is that the value for the resolution determined using the FWHM definition is approximately twice that obtained using the 10% valley definition. A resolution of 1000 using the 10% valley definition is approximately equivalent to a resolution of 2000 using the FWHM definition. Note that, for proteins, resolving the isotopes in the protonated molecular ion envelope is possible in the case

---

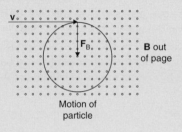

COMMENT B1.4 LORENTZ FORCE

The Lorentz force (**F**) is experienced by an ion of charge (z) moving in an electromagnetic field. $\mathbf{F} = z(\mathbf{E} + \mathbf{v} \times \mathbf{B})$, where **E** represents the electric field strength and $\mathbf{v} \times \mathbf{B}$ is the vector product of the magnetic field strength, **B**, and the ion's velocity, **v**.

Fig. B1.4.1

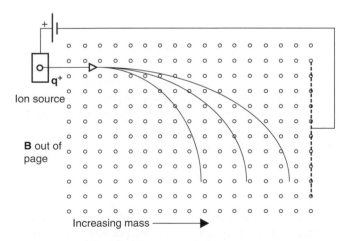

Fig. B1.1 Charged particles in electric and magnetic fields.

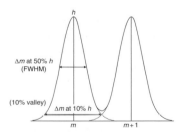

Fig. B1.2 Two definitions of mass resolution in mass spectrometry (see text for details). (After Carr and Burlingame, 1996.)

of very high resolution using FTICR mass spectrometers (Section B1.6.6).

B1.4.2 Molecular Mass Accuracy

The molecular mass accuracy of a measurement is defined as the difference between the measured and calculated masses for a certain ion. The accuracy is stated as a percentage of the measured mass (e.g., molecular mass = $10\,000 \pm 0.01\%$) or as parts per million (ppm) (e.g., molecular mass = $10\,000 \pm 100$ ppm). As the mass considered increases, the absolute mass error corresponding to the percentage or ppm error also increases proportionally.

Usually, accurate mass measurements do not require highest mass resolution, provided that for the observed signal there is only one species at that mass. The main factor limiting accurate molecular mass determination for high-mass biological macromolecules is peak overlap. For MALDI the peaks correspond to $[M+H]^+$, $[M+Na]^+$, and $[M+matrix]^+$.

High mass resolution is usually deemed to be a requirement for accurate mass measurements, but under appropriate circumstances (sample ion completely separated from background ions), measurements with comparable accuracy may be made at low resolution (see Comments B1.5–B1.7).

B1.5 IONIZATION TECHNIQUE

B1.5.1 From Ions in Solution to Ions in the Gas Phase

The transfer of ions from the gas phase to solution is a natural process. In the presence of solvent molecules such as H_2O, naked gas-phase ions such as Na^+ spontaneously form ion–solvent molecule clusters, $Na^+(H_2O)_n$. If the pressure of the solvent vapor is somewhat above the saturation vapor pressure, these clusters grow to small droplets.

The transfer of ions from solution to the gas phase is a desolvation process, which requires energy (it is endoergic) and hence does not occur spontaneously. For example, the free energy required when Na^+ ions are transferred from aqueous solution to the gas phase is very large, about 98 kcal/mol. The energy required for ion transfer to the gas phase in most analytical mass spectrometry methods is very high and is supplied by complex high-energy collision cascades or highly localized heating.

Ions may be produced in three different ways. First, by removing an electron from the molecule to produce a positively charged cation, which can be accelerated in either an increasing negative gradient field or decreasing positive gradient field. Second, by adding an electron to form an anion. In this case the accelerating fields are exactly the opposite to what they were for cations. Third, by removal or addition of protons. In this case, the mass of the

---

COMMENT B1.5 MONOISOTOPIC MASS

Most chemical elements have a variety of naturally occurring isotopes, each with a unique mass and natural abundance. The monoisotopic mass of an element refers specifically to the lightest stable isotope of the element. For example, there are two principal isotopes of carbon, ^{12}C and ^{13}C, with masses of 12.000 000 and 13.003 355 and natural abundances of 98.9% and 1.1%, respectively. Similarly, there are two naturally occurring isotopes for nitrogen, ^{14}N and ^{15}N, with masses of 14.003 074 (monoisotopic mass) and 15.000 109, and natural abundance of 99.6% and 0.4%, respectively. A monoisotopic peak means that all the carbon atoms in the molecule are ^{12}C, all the nitrogen atoms are ^{14}N, all the oxygen atoms are ^{16}O, etc. The monoisotopic mass of the molecule is thus obtained by summing the monoisotopic masses of each element present.

COMMENT B1.6 BIOLOGIST'S BOX: MEASURED MASS

Measurements are made on a large, statistical ensemble of molecules and consist not only of species having just the lightest isotopes of the element present, but also of some percentage of species having one or more atoms of one of the heavier isotopes. The contribution of these heavier isotope peaks in the molecular ion cluster depends on the abundance-weighted sum of each element present. The theoretical probability of occurrence of these isotope clusters may be precisely calculated by solving the polynomial expression shown below:

$$(a + b)^m$$

where a is the percentage natural abundance of the light isotope, b is the percentage natural abundance of the heavy isotope, and m is the number of atoms of the element concerned in the molecule.

Calculations show that for small molecules such as n-butane (C_4H_{10}) there is a small but significant probability (~4%) that natural n-butane will have a molecule containing a ^{13}C atom. The probability of there being two or three ^{13}C atoms is negligible. For biological macromolecules containing several hundred carbon and nitrogen atoms the isotopic distribution pattern becomes extremely complicated. It can be calculated, however, with commercially available programs.

---

resulting ion differs by ± 1 from the mass of the original neutral one. The molecule analyzed influences the choice of ionization technique. For peptides, proteins, and protein complexes, ESI and MALDI are mainly used.

COMMENT B1.7 AVERAGE MASS

The chemical average mass of an element is simply the sum of the abundance-weighted masses of all of its stable isotopes (e.g., 98.9% for ^{12}C and 1.1% for ^{13}C, to give the isotope weighted average mass of 12.011 for carbon). The average mass of the molecule is then the sum of the chemical average masses of the elements present.

The relationship between the monoisotopic mass, average mass, and peak top mass is shown in Figs. B1.7.1 and B1.7.2 for a protein of mass 2.5 kDa and a protein with mass of 25 kDa respectively. The important consequence of the contribution of the heavier isotope peaks is that for peptides with masses greater than 2000 Da, the peak corresponding to the monoisotopic mass is no longer the most abundant in the isotopic cluster (Fig. B1.7.2). With increasing molecular mass, the peak top mass continues to shift upward relative to the monoisotopic mass. Above masses of 8000 Da, the monoisotopic mass has an insignificant contribution to the isotopic envelope (Fig. B1.7.2). Whether the monoisotopic mass or the average mass should be used when measuring and reporting molecular masses depends on the mass of the substance and the resolving power of the mass spectrometer. Another very interesting point is that at very high resolution, the satellite peaks become visible.

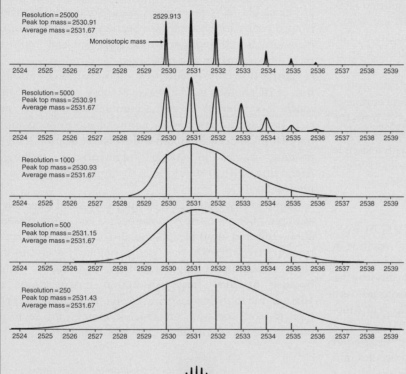

Fig. B1.7.1 The molecular ion cluster for the oxidized β-chain of insulin (formula $C_{97}H_{151}N_{25}O_{46}S_4$) is shown at various resolutions. The asymmetry of the cluster becomes less apparent as resolution is decreased and the peak top mass and the average mass become almost identical (Carr and Burlingame, 1996).

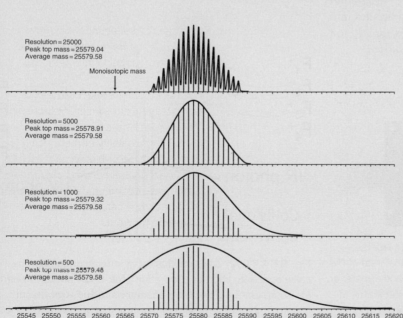

Fig. B1.7.2 The molecular ion cluster for the protein HIV-p24 (formula $C_{1129}H_{1802}N_{316}S_{13}$) is shown at various resolutions. The position of the monoisotopic mass is indicated by the arrow (Carr and Burlingame, 1996).

B1.5.2 Laser Desorption, Matrix-Assisted Laser Desorption Ionization, and Photodissociation MS

In laser desorption ionization (LDI), laser radiation is focused onto a small spot with a very high power density that gives an extremely high rate of heating. This leads to the formation of a localized laser "plume" of evaporated molecular species, either from adsorbed material or from the solid substrate itself. Direct LDI of intact biological molecules without using the matrix is limited to molecular masses of about 1 kDa. The mass range limitation gave rise to the development of MALDI.

The MALDI process differs from direct laser desorption because it utilizes a specific matrix material mixed with the sample. From this point of view, MALDI is similar to FAB; the latter using liquid matrices to provide soft ionization. However, MALDI provides much softer ionization than FAB, which allows the analysis of large molecules up to 1000 kDa with minimum fragmentation.

The details of energy conversion and sample desorption and ionization are still not fully known. A general outline of the mechanism is presented in Fig. B1.3. Energy from the laser beam is absorbed by the chromophore matrix, which rapidly expands into the gas phase, carrying with it sample molecules. Ionization occurs by proton transfer between excited matrix molecules and sample molecules, presumably in the solid phase, and also by collisions in the expanding plume.

The matrix is the key component in the MALDI technique. The matrix functions as an energy "sink" resulting in longer sample life. The material to be analyzed is mixed with an excess of matrix, which preferentially absorbs the laser radiation. Commonly used matrix materials are aromatic compounds that contain carboxylic acid functional groups. The aromatic ring of the matrix acts as a chromophore for the absorption of laser irradiation leading to desorption of matrix and sample molecules into the gas phase. The matrix not only increases sample ion yield, but also prevents its extensive fragmentation.

The energization processes afforded by collisional activation and photoactivation by infrared (IR) or ultraviolet (UV) photons are illustrated in Fig. B1.4. The IR laser efficiently couples with molecular vibrational modes, while the UV laser can excite electronic modes in aromatic molecules. The irradiation period may extend from a few nanoseconds to hundreds of milliseconds, depending on the photon flux of the laser and the energy deposition per photon. Because most laser sources are pulsed, TOF and Fourier transform ion cyclotron resonance (FTR-ICR) mass spectrometers have been most widely used with MALDI (Section B1.6). A mass accuracy of $\pm 0.01\%$ (± 1 Da at a molecular mass of 10 kDa) can be achieved under favorable conditions. If high-resolution conditions are available, it is possible to resolve individual carbon isotope peaks, for example (see Section B2.1). The MALDI technique is still under active development and improvements are occurring at a rapid rate.

Similarly to collisional activation, activation by absorption of IR photons is a step-wise process – however, with much less internal energy accumulation (~0.1 eV) per step. Because of the step-wise nature of energy deposition, both IR photoactivation and collisional activation tend to promote access to energy fragmentation pathways depending on the rate of ion activation (collision rate or photon flux).

In photodissociation MS to study biological molecules, multiple photons must be absorbed to cause ion dissociation. Infrared multiple photon dissociation mass spectrometry (IRMPD-MS) has been particularly useful in the characterization of oligonucleotides and nucleic acids because of their large absorption cross-sections at 10.6 μm due to their phosphate backbone, resulting in very high photodissociation efficiencies upon exposure to as little as 1–2 ms of continuous wave IR radiation.

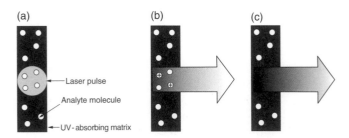

Fig. B1.3 Schematic mechanism for MALDI using lasers: **(a)** absorption of radiation by the matrix; **(b)** dissociation of the matrix, phase change to super-compressed gas, and transfer of charges to sample molecules; **(c)** expansion of the matrix at supersonic velocity, entrainment of sample molecules in expanding matrix plume, and transfer of charge to molecule.

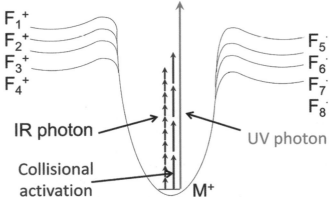

Fig. B1.4 Energy diagram illustrating energy deposition by collisional activation or absorption of IR or UV photons where M+ represents a selected precursor ion and F+ represent various fragment ions with different activation energies. (After Brodbelt, 2014.)

B1.5.3 Electrospray Ionization (ESI)

ESI produces intact ions from sample molecules directly from solutions at atmospheric pressure. Ions are formed by applying a 1–5 kV voltage to a sample solution emerging from a capillary tube, at a low flow rate (1–20 nl min^{-1}). The high electric potential, which is applied between the tip of the capillary tube and a counter-electrode located a short distance away, causes the liquid at the tip of the tube to be dispersed into a fine spray of charged droplets (Fig. B1.5). The production of positive or negative ions is determined by the polarity of the voltage applied to the capillary.

An attractive feature of the electrospray process is the formation of multiply-charged molecular species, if the sample molecule can accept more than one charge (Comment B1.8). Because of multiple charging, high-mass ions can be detected within a low m/z range. The shifted scale makes the high-resolution detection of large mass ions possible because the mass-resolving power is inversely proportional to m/z. The attainable mass accuracy for measuring molecular masses with the ESI technique in conjunction with Fourier transform mass spectrometry is about 0.001–0.005%. If high-resolution conditions are available, the individual carbon isotope peaks can be resolved (Comment B1.7).

Finally, it should be pointed out that ESI is one of the more gentle ionization methods available, yielding no molecular fragmentation in practice.

B1.6 INSTRUMENTATION AND INNOVATIVE TECHNIQUES

The first mass spectrometer was built in 1913, when J. J. Thomson proposed using fixed magnetic and electric fields to separate two different isotopes of the noble gas neon (Ne), by making use of the different behavior of charged particles according to momentum and energy in an electromagnetic field. The essential requirements to obtain a mass spectrum are to produce ions in the gas phase, to accelerate them to a specific velocity using electric fields, to introduce them into a suitable mass analyzer for separation, and finally to detect each charged entity of a particular mass sequentially in time (Fig. B1.6).

Sample introduction systems consist of controlled leak devices, through which sample vapor is introduced from a reservoir, various direct insertion probes for the injection of low-volatility liquids, and combinations with various chromatographic techniques. The ions that are produced in a number of ways in the ionization chamber (Section B1.5) are analyzed according to their mass-to-charge ratio in the mass analyzer.

Analyzers belong to quadrupole, magnetic sector, ion trap, time-of-flight (TOF), or Fourier transform (FT) generic types, depending on the physics of mass analysis. They could be further combined in various ways. The types of

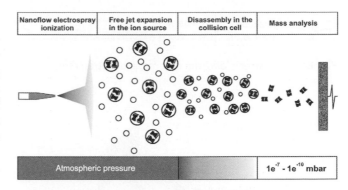

Fig. B1.5 Schematic representation of the passage of ions from the nanoflow electrospray needle to the detector of the mass spectrometer. Protein solution, typically 1–2 µl of 5 µM concentration, is placed in a fine-drawn capillary of internal diameter approximately 10 µm. A voltage of several kilovolts is applied to the gold-plated needle, causing an electrospray of fine droplets. The positively charged droplets are electrostatically attracted, dissolvated, and focused in the mass spectrometer for detection. (After Rostom, 1999.)

COMMENT B1.8 NUMBER OF ATTACHED PROTONS

In general, the maximum number of protons that attach to a peptide or protein under ESI conditions correlates well with the total number of basic amino acids (Arg, Lys, His) plus the N-terminal amino group, unless it is acylated. However, the accessibility of these basic sites is an important factor. The distribution of charge states thus depends on pH, temperature, and any denaturing agent present in the solution. This information can be used to probe conformational changes in the protein.

For example, for bovine cytochrome c, the most abundant ion has ten positive charges when electrospraying a solution at pH 5.2, but 16 charges at pH 2.6. A similar effect is observed upon reduction of disulfide bonds. Hen egg white lysozyme with four disulfide bonds shows a charge distribution centered at 12^+, but upon reduction with DTT (dithiothreitol), a new cluster appears centered around 15^+.

---

mass analyzer most used for proteins, peptides, and protein complexes are: quadrupole mass filter, TOF, and ion traps (quadrupole, Orbitrap, and FT cyclotron resonance devices). Ion detection after mass analysis can be performed by destructive or non-destructive techniques (see below). Modern mass spectrometers have almost total computer control over the various parts of a spectrometer, with advanced software available for data handling and interpretation. In fact, so many spectra are usually collected and analyzed during an experiment that they are rarely all shown.

Fig. B1.6 Schematic diagram of a mass spectrometer.

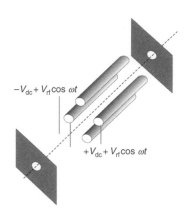

Fig. B1.7 Quadrupole mass filter. In the quadrupole mass filter instrument, one pair of rods has a negative dc voltage, $-V_{dc}$, applied and the other pair a positive dc voltage, $+V_{dc}$. There is also a superimposed radio-frequency (rf) voltage, $V_{rf}\cos\omega t$, which is 180° out of phase between rod pairs. In an ideal situation, rods with hyperbolic cross-section would be used. In order to scan between $m/z = 1$ and 500, the dc voltage is varied between 0 and 300 V and the ac voltage between 0 and 1500 V. The ac frequency is in the megahertz range. (After Gordon, 2000.)

B1.6.1 Quadrupole Mass Filter

Mass separation in a quadrupole mass filter is based on achieving a stable trajectory for ions of specific m/z values in a rapidly changing electric field. An idealized quadrupole mass filter consists of four parallel cylindrical rods of circular cross-section, as shown in Fig. B1.7. To one pair of diagonally opposite rods a negative direct current (−dc) voltage and an alternating radio-frequency (rf) voltage are applied. To the other pair of rods, a positive dc voltage of opposite polarity and the inverse (180° out of phase) rf voltage are applied. Mass filtering occurs as these voltages are scanned, but the ratio of dc to rf voltage is kept constant. For a given set of field conditions, only certain trajectories are stable, allowing ions of specific mass to be transmitted in the direction of the detector. Ions that have unstable trajectories come in contact with the rods and are not transmitted, hence the term filter.

One of the advantages of a quadrupole mass filter over an ESA instrument is the low voltage applied to the ion source (5–20 V) compared to several kiloelectron-volts for ESA instruments. The low voltage makes interfacing to liquid chromatography easier.

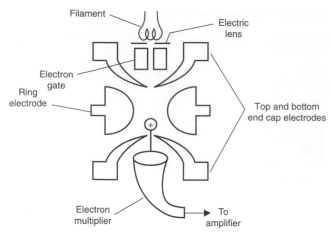

Fig. B1.8 A longitudinal cross-section of a quadrupole ion trap. Ions are created within the trap by radial injection of a pulse of electrons through holes in the ring electrode or axial injection through an end cap. For given values of m/z the ions are held in stable orbits, provided the correct amplitudes and frequency of dc and rf potentials are applied between the end caps and the ring. (After Gordon, 2000.)

B1.6.2 Quadrupole Ion Trap

The quadrople ion trap was originally developed by physicists who were interested in increasing the observation time available for spectroscopic measurements on elementary particles. The quadrupole ion trap is based on the same principle as the quadrupole mass filter, except that the quadrupole field is generated within a three-dimensional device consisting of a ring electrode and two end caps, as shown in Fig. B1.8. The ring electrode is a hyperboloid of one sheet. It is similar to a torus except that the cross-section of the ring is hyperbolic.

Ions that are produced in the trap itself or in an external ion source are stored in the trap. By raising the rf potential, the trajectories of ions of successive m/z values are made unstable and these ions are ejected out of the trap where they are detected by means of an electron multiplier. In contrast to the quadrupole filter, where the ions with stable trajectories are detected, ions with unstable trajectories are detected in the ion trap.

A schematic view of an atmospheric pressure MALDI ion-trap mass spectrometer is shown in Fig. B1.9.

B1.6.3 Ion Cyclotron Resonance Mass Spectrometry (ICR-MS)

As an ion-trapping technique, ICR-MS differs substantially from mass spectrometry that uses ion transmission to separate masses (Comment B1.9). In ICR, ions trapped in magnetic and dc electric fields are detected when the frequency of an applied rf field comes into resonance with the cyclotron frequency (Comment B1.10). The resonance

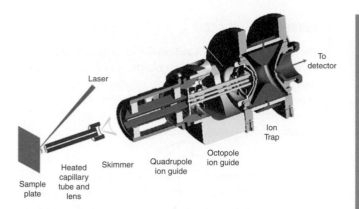

Fig. B1.9 Schematic view of a MALDI mass spectrometer based on an ion trap.

COMMENT B1.9 ION CYCLOTRON PRINCIPLE

In 1932, E. Lawrence and S. Livingstone demonstrated that a charged particle moving perpendicular to a uniform magnetic field is constrained to a circular orbit in which the angular frequency of its motion is independent of the particle's orbital radius and is given by the cyclotron equation (Eq. (B1.4)). Lawrence showed that cyclotron motion of a particle could be excited to a larger orbital radius by applying a transverse alternating electric field whose frequency matched the cyclotron frequency of the particle. The significance of Lawrence's discovery was that a particle could be excited to very large kinetic energy by use of only modest electric field strength. An alternating voltage of 1 kV would, after 1000 cyclotron cycles, accelerate the particle to a kinetic energy of 1 MeV.

COMMENT B1.10 ION CYCLOTRON FREQUENCIES

It follows from Eq. (B1.4) that ions of different m/z have unique cyclotron frequencies. At a magnetic field strength of 6 T, an ion of $m/z = 36$ has a cyclotron frequency of 2.6 MHz, whereas an ion of $m/z = 3600$ has a cyclotron frequency of 26 kHz. Equation (B1.5) also shows that increasing the magnetic field linearly increases the cyclotron frequencies of the ions, making high-mass ions easier to detect over the environmental noise in the low-kilohertz region. Additional benefits of increasing the magnetic field include an improvement in mass-resolving power and the extension of the upper mass limit.

It should be noted that Eq. (B1.4) does not account for the presence of the electric field produced by two trapping plates and can be considered as a first approximation.

Tesla (T)

The standard unit of magnetic flux density in the SI system:

$$1\,T = 1\,kg\,s^{-2}A^{-1}$$

Torr

A unit of pressure, being that necessary to support a column of mercury 1 mm high at 0 °C at standard gravity:

$$1\,Torr = 133.322\,Pa$$

Pascal (Pa)

The standard unit of pressure in the SI system:

$$1\,Pa = 1\,kg\,m^{-1}\,s^{-2}$$

frequency ω_c is directly proportional to the strength of the magnetic field (typically 3–7 T) and inversely proportional to the mass-to-charge ratio, m/z, of the ions:

$$\omega_c = \frac{Bz}{m} \qquad (B1.4)$$

Once formed, ions in the ICR-MS analyzer cell are constrained to move in circular orbits of radius r

$$r = \frac{mv}{zB} \qquad (B1.5)$$

with the motion confined perpendicular to the magnetic field (xy plane) but not restricted parallel to the magnetic field (z-axis) (Fig. B1.10). Ion trapping along the z-axis is accomplished by applying an electrostatic potential to the two plates on the ends of the cell. The trapped ions can be in the cell for up to several hours, provided that a high vacuum (10^{-8}–10^{-9} Torr) is maintained to reduce the number of destabilizing collisions between the ions and residual neutral molecules.

After formation by an ionization event, trapped ions of a given m/z have the same cyclotron frequency but a random position in the cell. The net motion of the ions under these conditions does not generate a signal on the receiver plates of the ICR-MS cell because of their random location. To detect cyclotron motion, an excitation pulse must be applied to the ICR-MS cell so that the ions spatially "bunch" together into a coherently orbiting ion packet. As a result, the net coherent ion motion produces a time-dependent signal on the receiver plates.

Fourier transform ion cyclotron resonance mass spectrometry (FTICR-MS) is a further development of the ICR technique. Time-domain signals are digitized and subjected to Fourier transformation to generate an ICR frequency-domain signal which can subsequently be converted into a mass spectrum (Section B1.6.6).

The essential advantages of FTICR-MS are: (1) an extremely high mass resolution; (2) a wide range of m/z values detected simultaneously; (3) the ability to study ion–molecule reactions at low pressure.

Finally, we would like to point out the similarities between ICR and NMR (Comment B1.11).

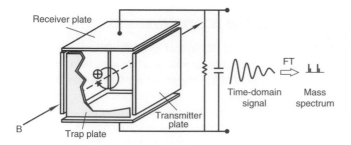

Fig. B1.10 FTICR cell, identifying the trapping plates, the transmitter plates, the receiver plates and direction of the magnetic field (**B**). (After Buchanan and Hettich, 1993.)

B1.6.4 Orbitrap Analyzer

The Orbitrap mass analyzer consists of three electrodes, with two cup-shaped outer electrodes facing each other and a central electrode that holds the trap together and aligns it. When voltage is applied between the outer and the central electrodes, the resulting electric field is strictly linear along the axis. Oscillations along this direction will therefore be purely harmonic. The radial component of the field bends the ion trajectory toward the central electrode while tangential velocity creates an opposing centrifugal force. With correctly chosen parameters, the ions remain on a nearly circular spiral inside the trap, while the axial electric field caused by the conical shape of the electrodes pushes ions toward the widest part of the

trap, initiating harmonic axial oscillations. Outer electrodes are then used as receiver plates for image current detection of these axial oscillations. The digitized image current in the time domain is Fourier-transformed into the frequency domain and converted into a mass spectrum. Orbitrap mass spectrometers calculate the mass of an ion based on the axial frequency measured as it rotates around the central electrode inside the trap itself. (This is the same basic concept underlying FTICR instruments, which differ from Orbitraps in the manner in which they induce that orbital rotation). The Orbitrap is considered as a potential alternative for FTICR because of its high resolution and good mass accuracy combined with a lower price.

The development of pulsed injection from an external ion storage device of the C-trap type (Comment B1.12) effectively allowed the decoupling of the Orbitrap analyzer from any preceding ion source, ion transmission device, or analyzer. Any device capable of selecting or transmitting precursor ions as well as any fragmentation technique can, therefore, be interfaced to the Orbitrap. A hybrid linear ion-trap/Orbitrap mass spectrometer, incorporating a C-trap, is shown in Fig. B1.11.

B1.6.5 TOF Mass Spectrometer

Mass analysis in a TOF mass spectrometer is based on the principle that ions of different m/z values have the same energy but different velocities, after acceleration out of the ion source. It follows that the time required for each ion to pass the drift tube is different for different ions: low-mass ions are quicker to reach the detector than high-mass ions. From Eq. (B1.1) we derive the expressions for the velocity u of an ion of mass m and charge z

$$u = \left(\frac{2zV_{\text{acc}}}{m}\right)^{1/2} \tag{B1.6}$$

and for the time t spent to cover a length L

$$t = \left(\frac{m}{2zV_{\text{acc}}}\right)^{1/2} L \tag{B1.7}$$

Equation (B1.7) shows that with an accelerating voltage of 20 kV and L of 1 m, a singly charged ion of mass 1 kDa has a velocity of about 6×10^4 m s^{-1} and the time spent traversing the drift tube is 1.4×10^{-5} s.

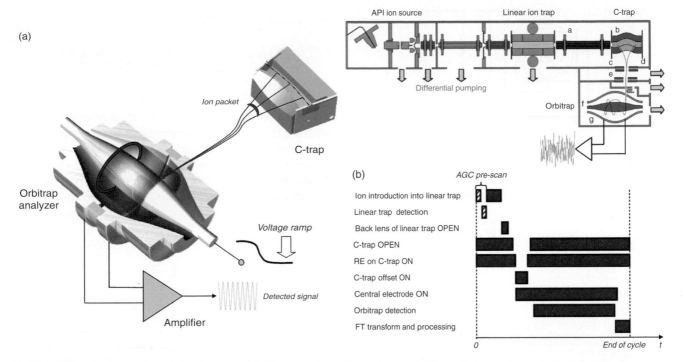

Fig. B1.11 (a) Cross-section of the C-trap and Orbitrap analyzer (ion optics and differential pumping not shown). Ions are stored in the rf-only bent quadrupole of the C-trap, then the rf is ramped down and a high-voltage pulse is applied across the trap, each *m/z* being ejected in a short packet. The packets from the C-trap enter the analyzer during the voltage ramp and spread into oscillating rings that induce current detected by the differential amplifier. (Zubarev and Makarov, 2013). See the figure for a description of **(b)**. (From Makarov *et al.* 2006).

It is evident that for a TOF mass analyzer the suitable ionization techniques are those by which ions are generated in a pulsed regime: using [252]Cf fission particles, a laser pulse, and introduction of ions from continuous ionization sources (EI, ES, and so on) with pulsed deflection of an ion beam or pulsed extraction from an ion source. The pulse gives the start signal for data acquisition.

The TOF method can be advantageous compared with scanning technologies because of its "unlimited" mass range, high transmission (most of the ions injected into the analyzer are detected), high speed (the experiment involves nearly simultaneous detection of the mass spectrum on the microsecond timescale), and the potential for high duty factors (percentage of ions formed that are detected). A major drawback is the low mass resolving power. From Eq. (B1.7) it follows that *m/z* is proportional to t^2, which leads to the formula for resolution:

$$R = \frac{m}{\Delta m} = \frac{1}{2}\frac{t}{\Delta t} \tag{B1.8}$$

Standard linear TOF instruments typically have a resolution no greater than 1000.

A significant improvement of the resolution in the TOF method can be obtained by using an electrostatic mirror or "reflectron" and the orthogonal TOF mass spectrometer (o-TOF-MS).

A reflectron TOF mass spectrometer is based on the fact that high-energy ions penetrate deeper into the reflection electric field and, therefore, spend more time there than low-energy ions. Because they must traverse a greater distance, the more energetic ions arrive at the detector at the same time as the less energetic ones. With the reflectron, the resolution of the TOF mass spectrometer increases up to 6000.

The main feature of o-TOF-MS is its use of orthogonal dimensions, *x* and *y*, respectively, for the continuous ion beam and distance over which the TOF is measured. Ions are sampled from a nearly parallel ion beam from a continuous ion source. The electric fields are designed to apply a force that is strictly and exclusively at right angles to the axis of the ion beam. The decoupling of the ion beam velocity spread from the TOF axis leads to the resolving power advantage of orthogonal acceleration. The resolving power of such instruments is about 4000. o-TOF-MS is highly compatible with the reflectron geometry.

B1.6.6 Fourier Transform Mass Spectrometry (FT-MS)

In most mass spectrometers, ions are detected by electrical current when they hit the surface of a device such as an

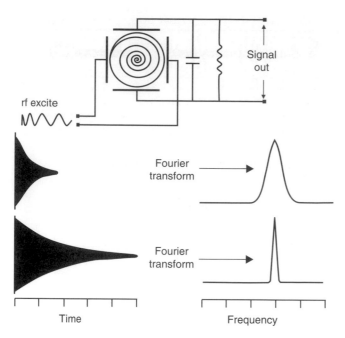

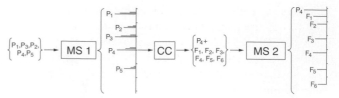

Fig. B1.13 Principle of tandem mass spectrometry. MS 1 and MS 2 are the first and the second mass spectrometer, respectively. CC is the collision cell. A mixture of five peptides is scanned to produce the spectrum of the five $(M + H)^+$ ions $(P_1–P_5)$. After the scan only one selected ion (P_4) passes into the collision cell. The fragments $(F_1–F_6)$ produced upon collision-induced decomposition of the precursor ion (part of which remains intact) are then mass analyzed by scanning MS 2 to record the product ion spectrum. (After Biemann, 1992.)

Fig. B1.12 General scheme of the cyclotron motion of excited ions in the FTICR cell. The resulting time-domain signal is then Fourier transformed to the frequency domain, from which the mass spectrum is obtained. (Carr and Burlingame, 1996.)

electron multiplier (destructive method). Although this method is widely used and very sensitive, it has the disadvantage that the ions must be destroyed in order to be detected. In other words, an ion signal can be measured only once. In FT-MS the detection method is fundamentally different: A strong magnetic field traps ions inside an analyzer cell and electrical signals produced by their cyclotron motion are detected by a pair of metal electrodes connected to a high-impedance amplifier (Fig. B1.12).

The image current detection method "senses" the number of ions without removing them from the analyzer cell and without destroying them (non-destructive method). A signal from the same ion can be measured repeatedly. The detection problem is similar to that in NMR (Comment B1.9, and Part J). With FT-MS, detection of just a few hundred ions produces a signal that contains complete information on the frequencies and abundances of all the ions trapped in the cell. Because frequency can be measured precisely, the mass of an ion can be determined to one part in 10^9 or better. It should be noted that resolution in FTICR-MS is mass-dependent; ultra-high resolution can be obtained at low mass. The sensitivity of FTICR is so high that the method has been successfully applied to study individual multiply-charged macro-ions.

B1.6.7 Tandem Mass Spectrometry (MS-MS)

To obtain structural information by mass spectrometry the molecule must undergo fragmentation of one or more

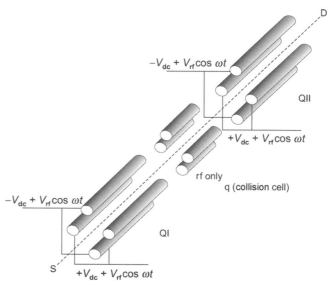

Fig. B1.14 Layout of the triple quadrupole system. QI and QII are the first and second quadrupole systems, respectively. The third quadrupole q, is used as the collision cell. S, source; D, detector; rf, radio-frequency. Such a geometry is named Q_1qQ_2. (After Gordon, 2000.)

bonds in such a manner that ions are formed, the m/z ratio of which can be related to the structure. We recall that "soft" ionization methods, such as FAB, MALDI, and ESI, generate single molecular ions that contain insufficient excess energy to fragment. However, by converting the kinetic energy of the ion into vibrational energy, fragmentation can be achieved. This can be done in MS-MS using a special collision cell.

The most common MS-MS experiment is the product ion scan. In the experiment, ions of a given m/z value are selected with the first mass spectrometer (MS 1, Fig. B1.13). The selected ions are passed into the collision cell (CC), typically filled with helium, argon, or xenon. The ions are activated by collision, and induced to fragment. The product ions are then analyzed with the second mass spectrometer (MS 2), which is set to scan over an appropriate mass range. Since it takes only 1–2 min to record the spectrum, one can then set MS 1 for the next precursor ion and obtain its collision spectrum, and so on.

There are two main types of instrument that allow MS-MS experiments. The first is made of two mass spectrometers assembled in tandem. Two mass analyzing quadrupoles, or two magnetic analyzer instruments or hybrids containing one magnetic and one quadrupole spectrometer are representative cases. The second type of MS-MS instrument consists of analyzers capable of storing ions: the ICR (Section B1.6.3) and the quadrupole ion trap (Section B1.6.2) mass spectrometers. These devices allow the selection of particular ions by ejection of all others from the trap. The selected ions are then excited and caused to fragment during a selected time period, and the ion fragments can be observed with a mass spectrometer. The process may be repeated to observe fragments of fragments, over several generations. The instruments exploit a sequence of events in time.

An alternative approach is to use the triple quadrupole design (Fig. B1.14; Section B1.5.2), which, although much cheaper, suffers from poor sensitivity and mass limitation. The first quadrupole, QI, is used as a mass spectrometer, a selected peak being injected into the collision cell (CC), and the decomposition products are analyzed in the second quadrupole, QII.

Finally, there are also "hybrid" instruments, which are so-named because they combine the use of magnetic sectors, quadrupoles and TOF instruments in linear and orthogonal projections.

B1.7 CHECKLIST OF KEY IDEAS

- A mass spectrometer measures mass/charge where mass is not in absolute units. The instrument needs to be calibrated with standard compounds, whose mass values are known very accurately.

- The ESI technique produces intact ions from samples directly from solutions at atmospheric pressure by spraying a very dilute solution from the tip of a needle across an electrostatic field gradient of a few kilovolts.
- A unique feature of ESI is the formation of multiply-charged molecular species. ESI is the most gentle ionization method, yielding no molecular fragmentation in practice.
- The MALDI technique produces intact ions from the sample mixed with specific matrix material. Proteins up to a molecular weight of 60 000 could be ionized if embedded in a large molar excess of a UV-absorbing matrix and irradiated with a laser beam. Taking advantage of high resolution, mass measurement accuracy, and ion-trapping capabilities, MALDI provides not only molecular mass information but also structural information for various peptides and oligonucleotides.
- In photodissociation MS, an IR laser efficiently couples with molecular vibrational modes, while a UV laser can excite electronic modes in aromatic molecules.
- Ions produced in a number of ways in the ionization chamber of a mass spectrometer are analyzed according to their mass-to-charge ratio in the mass analyzer.
- Analyzers belong to quadrupole, magnetic sector, ion trap, time-of-flight (TOF), or Fourier transform (FT) generic types, depending on the physics of the device. They could be further combined in various ways.
- The types of analyzer most used for proteins, peptides, and protein complexes are: quadrupole mass filter, TOF, and ion traps (quadrupole, Orbitrap, and Fourier transform cyclotron resonance devices).
- Ion detection after mass analysis can be performed by destructive or non-destructive techniques.
- Modern mass spectrometers have almost total computer control over their various parts, with a spectrometer with advanced software available for data handling and interpretation. In fact, so many spectra are usually collected and analyzed during an experiment that they are rarely all shown.

Suggestions for Further Reading

Historical and General Reviews
Griffiths, I. W. (1997). J. J. Thomson: The centenary of his discovery of the electron and his invention of mass spectrometry. *Rapid Commun. Mass Spectr.*, **11**, 2–16.
Konermann, L., Vahidi, S., Sowole, M. A. (2014) Mass spectrometry methods for studying structure and dynamics of biological macromolecules. *Anal. Chem.*, **86**, 213–232.

Ionization Techniques
Smith, D. R., Loo, J. A., Loo, R. R. O., Busman, M., and Udseth, H. R. (1991). Principles and practice of electrospray ionization: Mass spectrometry for large polypeptides and proteins. *Mass Spectr. Rev.*, **10**, 359–451.
Muddiman, D. C., Gusev, A. I., and Hercules, D. M. (1995). Application of secondary ion and matrix-assisted laser desorption–ionization time-of-flight mass spectrometry for

the quantitative analysis of biological molecules. *Mass Spectr. Rev.*, **14**, 383–429.

Gordon, D. B. (2000). Mass spectrometric techniques. In *Principles and Techniques of Practical Biochemistry*, eds., K. Wilson and J. Walker. Cambridge: Cambridge University Press, chapter 11.

Instrumentation and Innovative Techniques

Caprioli, R. M., and Suter, M. J.-F. (1995) Mass spectrometry. In *Introduction to Biophysical Methods for Protein and Nucleic Research*, eds., J. A. Glasel and M. P. Deutscher. New York: Academic Press, chapter 4.

Amster, I. J. (1996). Fourier transform mass spectrometry. *J. Mass Spectr.*, **31**, 1325–1337.

Hofmann, E. (1996). Tandem mass spectrometry: A primer. *J. Mass Spectr.*, **31**, 129–137.

Dienes, T., Pastor, J. S., Schürch, S., *et al.* (1996). Fourier transform mass spectrometry: Advancing years (1992–mid 1996). *Mass Spectr. Rev.*, **15**, 163–211.

Guilhaus, M., Mlynski, V., and Selbi, D. (1997). Perfect timing: Time-of-flight mass spectrometry. *Rapid Commun. Mass Spectr.*, **11**, 951–962.

Belov, M. E., Gorshkov, M. V., Udeseth, H. R., Anderson, G. A., and Smith, R. D. (2000). Zeptomole-sensititivity electrospray ionization: Fourier transform ion cyclotron resonance mass spectrometry proteins. *Anal. Chem.*, **72**, 2271–2279.

Makarov, A., Denisov, A., Kholomeev, A., *et al.* (2006). Performance evaluation of a hybrid linear ion trap/Orbitrap mass spectrometer. *Anal. Chem.*, **78**, 2113–2120.

Zubarev, R. A., and Makarov, A. (2013). Orbitrap mass spectrometry. *Anal. Chem.*, **85**, 5288–5296.

B2

B2.1 PROTEINS

ESI and MALDI are increasingly useful for the analysis of protein, protein complexes, and interactions. These two ionization techniques have been exploited, for example, to study protein folding, to characterize non-covalent, native, protein complexes, and map protein interactions. A view of a mass spectrometry laboratory is shown in Comment B2.1.

B2.1.1 Mass Determination

With the development of ESI and MALDI, the mass analysis of proteins became a routine procedure. For many reasons, peptides and proteins are particularly suited to these ionization techniques.

Figure B2.1 shows portions of the mass spectra of horse cytochrome c and horse myoglobin obtained by ESI-FTICR mass spectrometry. Sample concentrations were 0.4 nM (Comment B2.2). The total consumed amount for each protein was 135 zmol (about 80 000 molecules).

Figure B2.2 depicts a high-resolution MALDI mass spectrum of [Arg^8]-vasopressin. The base peak at m/z 1084.446 is the "monoisotopic" peak for the intact protonated peptide. The three peaks at higher mass, each separated by 1 Da, result from the incorporation of one or more of the less abundant carbon isotopes into the molecule, with ^{13}C at a natural abundance of 1.108% being the main contribution. For an example of the computation of protein molecular mass, the reader is referred to Comment B2.3.

B2.1.2 Proteomics

The proteome is defined as the full complement of proteins in a cell, representing the products of expression of the genome together with the influence of post-translational modifications. In contrast to the genome, the proteome is dynamic and highly dependent not only on the type but also on the state of the cell. The term "proteomics" was proposed in the 1990s from "protein" and "genomics." The aim of proteomics is to identify and quantify all the proteins of a proteome, including expression, cellular localization, interactions, post-translational modifications, and, importantly, turnover as a function of time. Considering, for example, that there is of the order of 100 000 protein forms encoded by the more than 20 000 genes of the

On the left side of Fig. B2.1.1, the nano-ESI-Q-TOF instrument has been modified to perform native MS investigation of protein complexes. Using conditions optimized for preserving non-covalent interactions, the measured mass of an intact complex (or a subcomplex) reveals the stoichiometry of the subunits. By dissociating the complex under controlled conditions (i.e., non-covalent bonds are broken), the stoichiometry is confirmed and also the subunits located within the core are distinguished from the peripheral proteins.

In the center of the laboratory a high-performance liquid chromatography system coupled with an ESI-TOF is used to determine the exact mass of proteins. This allows the evaluation of the sample features such as purity and homogeneity, which are key requirements for successful structural projects.

On the right side a MALDI-TOF is used to determine the mass of proteins and peptides. MALDI-MS is an attractive alternative to ESI-TOF due to its salt tolerance and the simplicity of data acquisition and interpretation.

Fig. B2.1.1 The mass spectrometry (MS) laboratory of the Institute of Structural Biology (Grenoble, France) (photo © CEA/www.denis-morel.com).

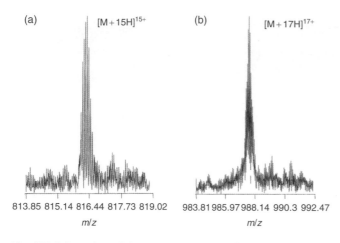

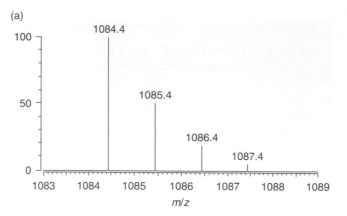

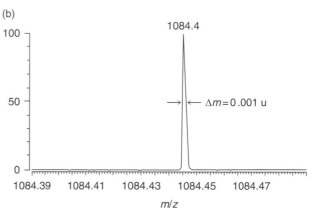

Fig. B2.1 A portion of the mass spectrum of (a) horse cytochrome *c* and (b) horse myoglobin obtained by ESI-FTICR mass spectrometry. (Belov *et al.*, 2000.)

COMMENT B2.2 UNIT PREFIXES

10^{-9}	nano	n
10^{-12}	pico	p
10^{-15}	femto	f
10^{-18}	atto	a
10^{-21}	zepto	z

Fig. B2.2 High-resolution MALDI mass spectrum for [Arg[8]]-vasopressin: (a) narrow-band acquisition from *m/z* 1080 to 1090; (b) an expanded mass axis to show a mass resolution of 1 100 000. (Li *et al.*, 1994.)

human genome, proteomics poses a formidable challenge for modern biophysics.

There have been rapid advances in the resolution, mass accuracy, sensitivity, and scan rate of mass spectrometers (see Chapter B1). Whereas mass spectrometry has been rather limited in the study of genomes, it plays a cardinal role in the characterization of proteomes. This follows the development of new technologies for peptide/protein separation, the mass spectrometry procedure itself, isotope labeling and computational data analysis. In October 2014, a special feature in *Nature Methods* (volume 10(11)) listed mass spectrometry-based proteomics in the ten areas of methods development with the most impact on biological research over the last decade.

Successes of MS-based proteomics include the study of protein–protein interactions on a proteome-wide scale, the mapping of organelles, the concurrent description of the malaria parasite genome and proteome, the generation of diverse protein profiles from various cell species leading, for example, to rapid identification of bacteria involved in infection.

MS-based high-throughput proteomics approaches fall in one of two general strategies, *bottom-up* and *top-down*, with *middle-down* as a hybrid of the two (Comment B2.4).

In "bottom-up" MS-based proteomics, a protein mixture is first digested into short peptides, and subsequently analyzed. In the "top-down" approach, intact proteins are analyzed without prior proteolytic digestion (thus maintaining the native structure, including post-translational modifications). A variant method ("middle-down") analyzes larger peptide fragments (>3 kDa). It combines benefits of both bottom-up and top-down approaches (for example, by generating peptides that contain post-translational modifications (PTMs)).

The bottom-up sequencing approach performed on a mixture of proteins, in the context of discovering and identifying what is there, is called *shotgun proteomics* by analogy to shotgun genomic sequencing. The method provides an indirect measurement through peptides derived from proteolytic digestion of intact proteins. The resulting peptide mixture is fractionated and subjected to liquid chromatography (LC) and tandem MS analysis (MS-MS, see Section B1.6.7). As shown in the last line of Figure B2.4.1, peptide identification is achieved by comparing the experimental mass spectra with theoretical ones from virtual digestion of a protein database. Similarly, peptide sequences are then assigned to proteins. The identified

COMMENT B2.3 BIOLOGIST'S BOX: COMPUTATION OF PROTEIN MOLECULAR MASS

A mass spectrum is a plot of the intensity as a function of mass-to-charge ratio. The peak in the spectrum with highest intensity is called the base peak. Generally, the spectrum is normalized to the intensity of the base peak, resulting in relative intensities.

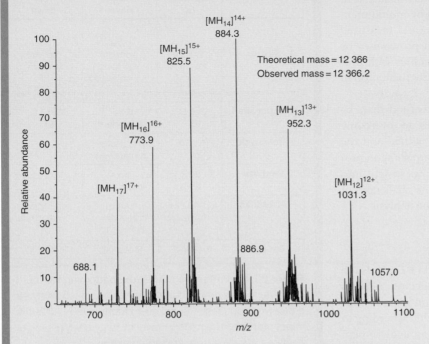

Fig. B2.3.1 Electrospray mass spectrum of multiply-charged cytochrome *c* (16 951.5 Da) at low resolution from 12+ to 18+. (After Gordon, 2000.)

The figure shows the electrospray mass spectrum of multiply-charged cytochrome *c*. The molecular mass of the protein can be calculated easily according to the mathematical formalism presented below, remembering that *z* values are integers. It is assumed that the ions are adducts of neutral molecules and protons.

The molecular mass, *M*, of the neutral molecule can be found from recorded masses m_1 and m_2 (equivalent to the *m/z* values) and the number of charges or protons added n_1 and n_2. Such that

$$M = n_2(m_2 - 1) \tag{B2.1}$$

where

$$n_1 = n_2 + 1 \text{ and } n_2 = \frac{m_1 - 1}{m_2 - m_1}$$

By taking peaks in pairs, n_2, and hence *M*, can be calculated from the recorded masses.

Applying Eq. (B2.1) to calculate the molecular mass of cytochrome *c* (see Fig. B2.3.1) using two peaks, $m_1 = 952.3$ and $m_2 = 1031.3$,

$$n_2 = \frac{m_1 - 1}{m_2 - m_1} = \frac{951.3}{1031.3 - 952.3} \approx 12.04$$

or $Z = 12$. This means that 12 positive charges are associated with a relative mass 1031.3. The molecular mass calculated from this peak is given by $M = n_2(m_2 - 1)$ is 12 363.6.

Taking the next two peaks with relative masses $m_1 = 884.3$ and $m_2 = 952.3$, we have:

$$n_2 = \frac{m_1 - 1}{m_2 - m_1} = \frac{883.3}{952.3 - 884.3} \approx 12.989$$

or $Z = 13$, i.e., positive charges associated with a relative mass 952.3. The molecular mass calculated from this peak is 12 366.9.

Continuing this procedure for other pairs of peaks, we have:

- from the two peaks with relative masses $m_1 = 825.5$ and $m_2 = 884.3$, a molecular mass of 12 366.2;
- from the two peaks with relative masses $m_1 = 773.9$ and $m_2 = 825.5$, a molecular mass of 12 367.5;
- from the two peaks with relative masses $m_1 = 825.5$ and $m_2 = 884.3$, molecular mass of 12 366.4.

So we can conclude that the observed molecular mass average, calculated from the five peaks, is 12 366.2. The theoretical mass of cytochrome *c* is 12 366.

proteins could be further scored and grouped based on their peptides, since the assignment may not be unique.

Top-down proteomics is used to characterize native proteins and has obvious advantages for the determination of PTMs and protein isoforms. The method has managed to determine more than 1000 native proteins by multidimensional separations from complex samples, and measured molecular mass up to 200 kDa. Top-down proteomics by MS has limitations, however, due to difficulties with protein fractionation, protein ionization, and fragmentation in the gas phase.

Middle-down proteomics is a hybrid method that, by analyzing larger peptide fragments, tries to overcome some of the limitations of the top-down method and the problem of peptide redundancy in bottom-up shotgun proteomics. Middle-down proteomics still allows insight into post-translational modifications, without the challenges of maintaining the native structures during the analysis.

B2.1.3 Protein Sequencing

MS-MS, tandem mass spectrometry (Section B1.6.7), is the most suitable technique for protein sequencing. MS-MS can sequence not only the 20 common amino acids, but also known or unknown modified amino acids according to their mass. MS-MS is fast, sensitive, and can analyze peptide mixtures directly.

A number of mass spectrometric approaches have been devised for sequencing large peptides and proteins. These include: (1) MS-MS approaches combined with enzymatic or chemical degradation to form oligopeptides (<3 kDa) as a first step, with MS of the resulting products as a second step in a bottom-up approach; (2) using ESI-FTMS as the first step of degradation, in a top-down MS approach (Comment B2.4); (3) combining MS-MS and classical Edman degradation (Comment B2.5).

Computation of Protein Sequence

To obtain structural information by mass spectrometry, the molecule to be studied must undergo fragmentation of one or more bonds in such a manner that ions are formed. Protein-derived peptides are linear molecules whose structures have two constraints: The linear backbone consists of only repeating α-amino acid–amide bonds, and the substituent groups of each α-carbon are side-chains made of one of the 20 naturally occurring amino acids. "Soft" ionization methods generate singly- or poly-protonated molecular ions that contain insufficient excess energy to fragment. Fragmentation can be achieved by using a special collision cell in tandem mass spectrometry (Section B1.6.7).

Although different types of fragmentation may occur, there is a predominance of peptide bond cleavages. The loss of one amino acid residue at a time gives rise to peaks in the spectrum, which differ sequentially by the mass of the amino acid minus H_2O (the so-called residue mass, Table B2.1).

COMMENT B2.4 PROTEOMIC STRATEGIES: BOTTOM-UP VS. MIDDLE-DOWN VS. TOP-DOWN VS. MIDDLE-DOWN

	Bottom-up	Middle-down	Top-down
Mixture	Lysate, cellular fractions, coIP…	Lysate, cellular fractions, coIP…	Lysate, cellular fractions, coIP…
Fractionation	Gel, LC, IEF…	Size-dependent	Gel, LC, IEF…
Proteolysis	Trypsin…	Restricted proteolysis	
Peptides	SCX, IEF…	Size-dependent	
LC-MS/MS	Peptide fragments	Large peptide fragments	Protein fragments
Database	Protein ID	Protein ID	Protein ID

Figure B2.4.1 Proteomic strategies: bottom-up vs. top-down vs. middle-down. The bottom-up approach analyzes proteolytic peptides. The top-down method measures the intact proteins. The middle-down strategy analyzes larger peptides resulting from limited digestion or more selective proteases. One or more protein or peptide fractionation techniques can be applied prior to MS analysis and database searching. From Zhang *et al.* (2013).

Abbreviations:
LC: liquid chromatography
IEF: isoelectric focusing
SCX: strong cation exchange chromatography (for phosphoprotomics to identify phosphorylation in PTM, post-translational modifications)
CoIP: collagenase treatment

The sequential mass differences represent exactly the primary structure of the peptide.

The fragmentation of an idealized peptide is shown in Fig. B2.3. The bond cleavages indicated at the top of the figure lead to fragments of types a_n, b_n, and c_n, if the charge is retained on the N-terminal fragment. The exception is that for a c_n ion two daltons have to be added because it retains the original protonating hydrogen and picks up one more from the other side of the peptide bond. The x_n-, y_n-, and z_n-type ions are formed when the charge is retained on the C-terminal, with the y_n ions having added two hydrogen atoms analogous to the c_n ions. From a single, complete ion series we can deduce the amino acid sequence of a peptide, although leucine and isoleucine, and lysine and glutamine, cannot be differentiated in

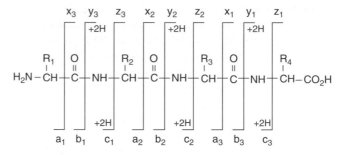

COMMENT B2.5 EDMAN DEGRADATION

In 1950, P. Edman proposed a chemical method for the step-wise removal of amino acid residues from the N-terminus of a polypeptide or protein. The series of reactions has come to be known as the Edman degradation, and the method remains the most effective chemical means for polypeptide sequencing. The Edman method requires a free amino group at the N-terminus. The peptides in which the terminal amino group is blocked (e.g., by a formyl, acetyl, or acyl group) have to be cleaved by either chemical or enzymatic degradation. Edman degradation is carried out in an automated analyzer. Sequence data can now be obtained from as little as 10–100 ng of protein.

Fig. B2.3 Fragmentation of an idealized protonated peptide. (After Biemann, 1992). The nomenclature is the following: An ion is classified as either a, b, or c type if this charge is on the N-terminal fragment; if the charge is on the C-terminal fragment, the ion type is either x, y, or z type. A subscript indicates the number of residues in the fragment. In addition to the proton(s) carrying the charge, c ions and y ions abstract an additional proton from the precursor peptide. In addition to the proton(s) carrying the charge, c ions and y ions abstract an additional proton from the precursor peptide.

TABLE B2.1 RESIDUE MASS [–NH–CHR–CO–] OF AMINO ACIDS

| Amino acid | Letter code | | Mass |
	Three	Single	
Glycine	Gly	G	57
Alanine	Ala	A	71
Serine	Ser	S	87
Proline	Pro	P	97
Valine	Val	V	99
Threonine	Thr	T	101
Cysteine	Cys	C	103
Isoleucine	Ile	I	113
Leucine	Leu	L	113
Aspargine	Asn	N	114
Aspartic acid	Asp	D	115
Glutamine	Gln	Q	128
Lysine	Lys	K	128
Glutamic acid	Glu	E	129
Methionine	Met	M	131
Histidine	His	H	137
Phenylalanine	Phe	F	147
Arginine	Arg	R	156
Tyrosine	Tyr	Y	163
Tryptophan	Trp	W	186

this manner. A single, complete ion series is not generally observed, and overlapping N- and C-terminal ions are used to determine the full sequence (Fig. B2.3).

The fragmentation processes of protonated peptides are now so well understood and reproducible that a set of empirical rules has been devised.

(1) The highest mass peak in the FAB (see below) spectrum represents $(M + H)^+$.
(2) The mass difference (Δ) between sequence ions represents the amino acid (aa) residual mass.
(3) The sum of the amino acid residues should give the total molecular mass.

In fast atom bombardment mass spectrometry (FAB-MS), the ions are obtained by bombarding a small drop of glycerol containing a few micrograms of the peptide with a beam of argon atoms of a few kilo-electronvolts of energy. The application of the fragmentation rules in the case of FAB-MS of an idealized protonated peptide from the following m/z data 128, 185, 299, 396, 528 is presented below. The highest mass peak in the FAB spectrum is m/z 528, hence $M = 527$. The set of m/z data can be converted to sequence of peptides in the following manner:

m/z	128	185	299	396	527
Δ		57	114	97	131
aa		Gly	Asn	Pro	Met

The lowest mass sequence ion, m/z 128 is Lys, hence the sequence from the N-terminal end is Met–Pro–Asn–Gly–Lys.

The sum of the aa residues is 527 Da. This agrees with the mass spectrometry value for M.

"End-Sequencing"

A high-sensitivity and high-throughput system based on orthogonal MALDI MS-MS (Section B1.6.7) provides automated recognition of fragments corresponding to the N- and C-terminal amino acid residues. The pulsed feature of the instruments enhances the low-mass region of the spectra by approximately one order of magnitude. The low-mass range of the spectrum is very simple to interpret, given the fact that only a few amino acid combinations lead to the observed low fragment masses. Typically, the sum of the masses of the second N-terminal ion (the b_2 ion), the third N-terminal ion (the b_3 ion), and the two C-terminal fragments of the peptide (y_1 and y_2) can be determined. Given the mass accuracy in the low ppm range, peptide end sequencing on one or two tryptic peptides is sufficient to uniquely identify a protein from gel samples in the low silver-stained range (a few nanograms of protein per band). Figure B2.4 shows the "end sequencing" principle. The combination of N- and C-termini sequencing unambiguously identified the peptide as VHLVGIDIFTGK (Val–His–Leu–Val–Gly–Ile–Phe–Thr–Gly–Lys) and the protein as a eukaryotic translation initiation factor protein 5A.

B2.1.4 Protein Folding and Dynamics

The speed, accuracy, and sensitivity of ESI and MALDI have been exploited in the development of several different mass-spectrometry-based approaches for studying protein

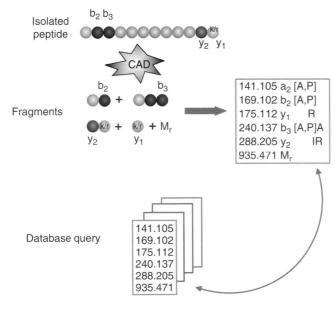

Fig. B2.4 The "end sequencing" principle. Isolated peptides are fragmented. The masses of the low-mass fragments such as y_1, y_2, b_2, and b_3 fragment ions plus the accurate molecular mass of the intact peptide are used to "end sequence" the peptide. The pattern and the molecular mass uniquely identify the peptide in the database. (After Nielsen *et al.*, 2002.)

folding and dynamics and mapping protein function. In many cases, mass spectrometry has provided data on the functional properties of a protein complementary to data obtained from traditional techniques such as circular dichroïsm, NMR, and fluorescence spectroscopy. Also, the relative speed and ease with which ESI and MALDI can be used to acquire very accurate molecular mass information on limited amounts of sample has made possible the acquisition of data that are not readily obtainable by other techniques.

Ion-Mobility Mass Spectrometry

Ion-mobility MS (IM-MS) is an excellent technique for the rapid analysis of molecular conformations. It is routinely used by airport security, for example, for the detection of explosives and illegal substances. The principles of an ion-mobility experiment are the following.

A pulse of ions is injected into a chamber (drift cell) filled with a known gas at a known pressure. An electric field is applied across the chamber and the time taken for the ions to pass through is measured. Upon injection, the ions experience an electrostatic force, moving them along the chamber – a force that is countered by collisions with gas molecules. At low electric field strength and high gas molecule number density (high gas pressure), the ions are said to move in the *low field limit*. The low field mobility, K is defined by:

$$v_d = KE \tag{B2.2}$$

where v_d is the drift velocity, E is the electric field, and K is a constant defined as the *low field mobility*. The mobility constant K depends on ion shape, charge, and the gas pressure in the chamber. Charge and gas pressure effects at a given temperature are described together by a rotationally averaged collision cross-section (Ω) for each ion. The MS experiment consists in determining drift velocity accurately under known electric field strength, gas pressure, and temperature, in order to determine Ω and K.

Folded and Unfolded States

In the early 1990s it was shown that folded and unfolded proteins produced different distributions of charge states in their ESI spectra. Proteins electrosprayed from solution conditions that preserve their native conformation tend to have a narrow distribution with a low net charge, whereas proteins electrosprayed from denaturing solutions produce a broad distribution centered on a much higher charge (Fig. B2.5). The difference in the distribution of charge states is believed to be related to the accessibility of ionizable groups. It is likely, for example, that in an unfolded state the basic amino acids (Arg, Lys, His residues) are more accessible to accumulating charge than when they are in the native state (Fig. B2.5).

The folding state of a protein in solution can be monitored by the charge state distribution produced during ESI. Thus, in the case of acid-induced unfolding of

Folded protein

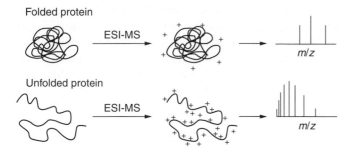

Unfolded protein

Fig. B2.5 The multiple-charging characteristics of folded and unfolded proteins in ESI. (Winston and Fitzgerald, 1997.)

cytochrome c (M = 12 360 Da) the observed changes clearly indicate a highly cooperative unfolding behavior (see Fig. B2.6a). In contrast, the unfolding of horse heart apo-myoglobin (myoglobin after loss of the heme group M = 16 951.5 Da) is accompanied by gradual shifts in the maximum of the observed charge state distribution (Fig. B2.6b). The observations suggested that ESI-MS can be considered as a general experimental method for assessing the cooperativity of protein folding transitions.

It is interesting to note that processes resembling folding and unfolding of equine cytochrome c ions in vacuo were observed by ESI-MS, raising the question of the role of water in protein folding.

Protein Folding Intermediates

Taking advantage of the fact that charge distribution depends on the folded state of a protein, a novel approach for studying protein folding using "time-resolved ESI" has been proposed. With a time resolution of 0.1 s, ESI has been used to monitor folding processes for cytochrome c and myoglobin. In the case of the first protein, no conformational intermediates between folded and unfolded states were detected. In contrast, a similar experiment with myoglobin revealed the presence of intermediates during its acid-induced denaturation (Fig. B2.7). The initial experiments produced only qualitative information. The extraction of quantitative information from ESI mass spectra became possible after the procedure of deconvolution of the charge-state distribution was introduced.

The ESI mass spectrum of any protein can be represented by a linear combination of charge-state distributions, called "basis functions," which can be approximated

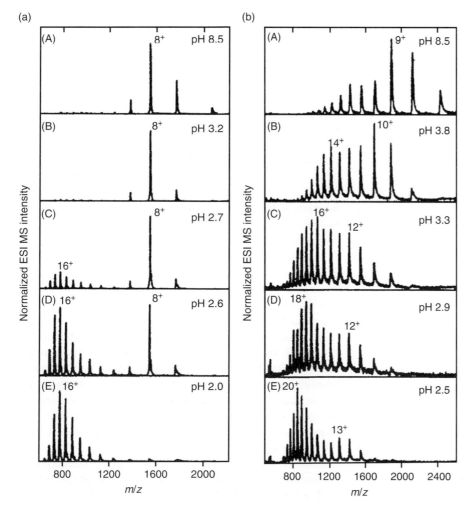

(a)

(b)

Fig. B2.6 (a) ESI mass spectra of cytochrome c recorded at different pH: **(A)** pH 8.5, **(B)** pH 3.2, **(C)** pH 2.7, **(D)** pH 2.6, and **(E)** pH 2.0. The pH was adjusted by addition of ammonium hydroxide and/or acetic acid. **(b)** ESI mass spectra of apo-myoglobin recorded at different pH: **(A)** pH 8.5, **(B)** pH 3.8, **(C)** pH 3.3, **(D)** pH 2.9, and **(E)** pH 2.5. The pH was adjusted by addition of ammonium hydroxide and/or acetic acid. (Konermann and Douglas, 1998a.)

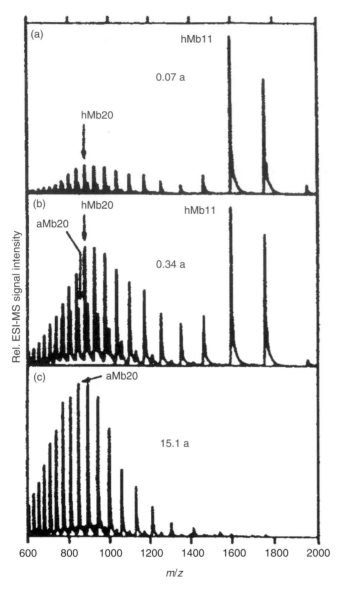

Fig. B2.7 Illustration of "time-resolved ESI" experiments following the acid denaturation of myoglobin: (a) hMb11 – the folded heme–myoglobin intermediate, (b) hMb20 – a partially unfolded heme–myoglobin intermediate, (c) aMb20 – unfolded myoglobin. The decay of the peak intensity of the intermediate was of the order of 0.4 s, and correlated well with the lifetime obtained in a solution phase. (Konerman and Douglas, 1998a.)

by a Gaussian distribution. The intensity changes are represented by a weighting factor, which accounts for the relative contribution to the overall charge-state distribution. In this way, an observed ESI mass spectrum can be considered as a sum of the contributions from each protein conformation (conformer). Figure B2.8 shows ESI mass spectra of holo-myoglobin (hMb) and apo-myoglobin (aMb) over a wide pH range (2.5–7.4). The hMb spectra exhibit a very narrow charge-state distribution at pH 4.5 and above. The spectrum contains only two peaks, for charge states +8 and +9, respectively. Further decrease of

solution pH (i.e., to pH 4) results in large-scale conformational changes, as manifested by the appearance of the highly protonated (low m/z) protein ions and partial dissociation of the heme group from the protein (Fig. B2.8c). Further decrease of solution pH down to 2.5 leads to a continuous increase of the average charge state of protein ions and disappearance of the protein–heme complex ions.

Unlike hMb, the aMb spectrum exhibits a multimodal character even at neutral pH (Fig. B2.8f). In addition to +9 and +8 ions, a wide distribution of less abundant ion peaks is seen in the spectrum at charge states ranging from +10 to +23 (Fig. B2.8). At pH 2.5, the aMb spectrum is indistinguishable from that of hMb, fully consistent with the expectation that any interaction between the heme group and the acid-destabilized form of the protein would be minimal.

The results of deconvolution of some of the charge-state distributions using basis functions are shown in Fig. B2.9a–f. Only one basis function is required to obtain a satisfactory fit to the spectra at pH 4.5 and above for hMb, while three basis functions are needed for data fitting in the pH range 2.5–4.5. Finally, a fourth basis function has to be added to the set in order to fit the data at low pH levels both for hMb and the aMb. The four basis functions were assigned to four different conformational states of the protein (in order of decreased folding): native (N), so-called "pH 4 intermediate" (I), extended conformation (E), and unfolded state (U). All the ESI spectrum features, interpreted as a linear combination of ionic contributions from N, I, E, and U, are fully consistent with the existing picture of the acid unfolding of the protein using a wide variety of other experimental techniques.

The experiments give an excellent illustration of the unique ability of ESI to monitor distinct populations of folding intermediates.

Intrinsically Disordered Proteins (IDPs)

See also Chapter J3 for an introduction to IDPs and complementary NMR studies.

In a study of SIC-1 protein (Fig. B2.10), using native ESI-MS in combination with limited proteolysis, gel filtration, and circular dichroïsm (CD), it has been shown that the SIC-1 protein is in a highly disordered state. Its C-terminal fragment, however, which includes an inhibitory domain, was resistant to proteolysis, indicating a degree of secondary or tertiary structure. The authors then used IM-MS to probe the structure of the SIC1 Δ214 fragment. The degree of ESI charging under non-denaturing conditions and ion-mobility derived cross-sections informed on the structural compactness of the fragment. It was deduced from the drift time plot that the monomer and a low level of dimer each possess two distinct conformations. The lower charge states of the monomer displayed a higher mobility, indicating a more compact conformation.

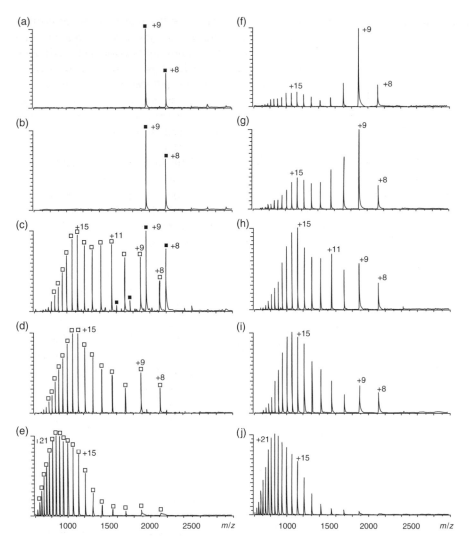

Fig. B2.8 Positive ion ESI mass spectra of **(a)–(e)** holo-myoglobin (*h*Mb) and **(f)–(j)** apo-myoglobin (*a*Mb) acquired at pH 7.4 (**(a)**, **(f)**), 4.5 (**(b)**, **(g)**) 4.0 (**(c)**, **(h)**), 3.5 (**(d)**, **(i)**) and 2.5 (**(e)**, **(j)**). Intact *h*Mb ion peaks are indicated with filled squares. (Dobo and Kaltashov, 2001.)

B2.2 NON-COVALENT COMPLEXES AND NATIVE (TOP-DOWN) MS

B2.2.1 Protein Complexes

The physiologically active, native forms of many proteins are multimeric, with active sites often at subunit interfaces. The strength of non-covalent interactions generally arises from a multitude of relatively weak bonds and can vary widely. It is reflected in the dissociation constants (K_D) typically determined for a specific set of solution conditions. Clearly, the detection of weakly bound, thermally sensitive complexes requires gentle ESI interface conditions (Comment B2.6). Several non-covalent complexes have been reported to remain intact in ESI-MS experiments. They include metals, heme groups, and peptides bound to protein as well as multimeric protein complexes, oligonucleotides, enzyme–substrate and receptor–ligand complexes, and large RNA–protein complexes.

One of the most impressive illustrations of ESI-MS was the observation of very tight complexes between proteins and other molecules. The ESI spectrum of human cytoplasmic receptor for cyclosporin exhibits an abundant $(M + 7H)^{7+}$ ion at m/z 1688.7 (Fig. B2.11a). Upon addition of the immunosuppressive drug with molecular mass $M = 804$ Da a new peak appears at m/z 1803.1, corresponding to the FKBP–FK506 complex (1:1) in the 7^+ charge state (Fig. B2.11b). The same effect is observed with rapamycin, which has $M = 913$ Da, and it is even possible to estimate the relative ratio of their binding constants from the relative peak height when adding a 1:1 mixture of the rapamycin (Fig. B2.11c). Nevertheless, a substantial amount of free FKBP was also detected in the mass spectra in both cases. Using this methodology it was possible to monitor the hydrolysis of hexa-N-acetylglucosamine by hen egg white lysozyme.

The MS study of native, non-covalent interactions requires not only the use of appropriate ionization conditions, but also instruments that are specifically modified to transmit and detect such large assemblies. The characteristics of MS instruments used to study non-covalent complexes are summarized in Table B2.2. MS spectra of

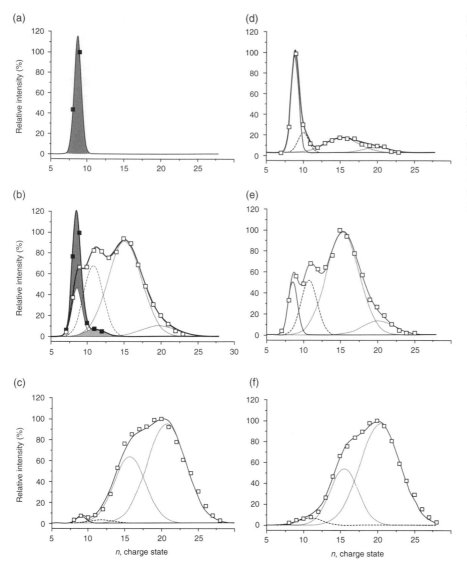

Fig. B2.9 Curve fitting of positive ion charge-state distributions in ESI mass spectra of **(a)**–**(c)** *h*Mb and **(d)**–**(f)** *a*Mb acquired at pH 7.4 (**(a)**, **(d)**), 4.0 (**(b)**, **(e)**), and 2.5 (**(c)**, **(f)**). Experimental data points are shown with squares (■ for intact *h*Mb ions and □ for *a*Mb ions). The Gaussian curves represent the weighted basis functions used for curve fitting (shaded for *h*Mb). The thick solid lines represent the summation of weighted basis functions. (Dobo and Kaltashov, 2001.)

intact protein complexes from an Orbitrap mass spectrometer with masses ranging from 150 kDa to 800 kDa are shown in Fig. B2.12.

Acidic and organic solvents are avoided in top-down or native MS, and volatile buffers such as ammonium acetate are preferred. Fragmentation is accomplished by electron capture dissociation (ECD) and electron transfer dissociation (ETD) (Comment B2.7).

Native, top-down MS can investigate non-covalent complexes of proteins with small ligands as well as large intact protein assemblies. Figure B2.13 illustrates how the interaction network of subunits within a protein complex can be generated through a multistep process using native MS.

B2.2.2 Ribosomes, Ribosomal Subunits and Ribosomal Proteins

Ribosomes 70 S *E. coli* in a buffer containing 5 mM Mg^{2+} were projected into the gas phase of a mass spectrometer

by means of nanoflow ESI techniques (Fig. B2.14a). By lowering the Mg^{2+} concentration in the solution, the particles were found to dissociate into 30 S and 50 S subunits (Fig. B2.14b). The resolution of the charge states in the spectrum of the 30 S subunit enabled its mass to be determined as $852\,187 \pm 3918$ Da, a value within 0.6% of that calculated from its constituent proteins and 16 S RNA. Further dissociation into smaller macromolecular complexes and then individual proteins can be induced by subjecting the particles to increasingly energetic gas-phase collisions. Of the 56 proteins in this complex, 55 were observed as complete proteins (without fragmentation) and a wide variety of PTMs was observed. It is interesting to note that the ribosomal protein of highest molecular mass (S1, 61 kDa) was not detected.

A similar study using another approach has been carried out on the yeast ribosome. All the proteins of this complex were isolated, denatured, subjected to trypsin digestion, and desalted. The peptide mixture was then separated by chromatography and was observed by MALDI-TOF mass

(a)

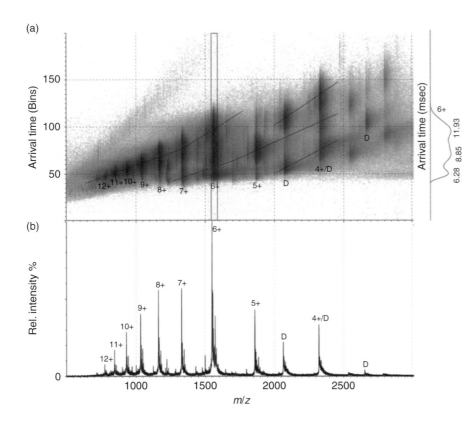

(b)

Fig. B2.10 Ion mobility-mass spectrum of the SIC Δ214 fragment. **(a)** Drift plot and arrival time distribution (top-right corner; for the 6+ charge state) indicate how multiple conformations coexist which are coincident in the mass spectrum in **(b)**. There is also a low level of dimer present (peaks labeled "D" in **(b)**; the 4+ monomer charge state and the 8+ dimer coincide in the mass spectrum, but not in the drift plot. (Brocca *et al.*, 2011; adapted with permission in the review by Konijnsberg *et al.*, 2013).

spectrometry. Of the predicted 78 proteins in the ribosomal complex, only three were not identified.

B2.3 NUCLEIC ACIDS

The analysis of nucleic acids by MS lags behind that of proteins. The main reason for this is the high affinity of nucleic acids for sodium ions, which can greatly reduce the ionization efficiency. Moreover, the generation of intact molecular ions of oligomers of more than two nucleotides turned out to be difficult when using classical ionization techniques such as electron impact and chemical ionization, due to the high polarity of nucleic acids and a high tendency for their molecular ions to fragment. In 1982, the capability of the FAB technique to generate intact molecular ions of nucleotides was demonstrated. Ultrafast sequencing of oligonucleotides by FAB mass spectrometry was proposed in the following year. The introduction of ESI and MALDI, and techniques such as infrared multiple photon dissociation (IRMPD) (see Section B1.5.2) made way for a significant improvement of the accessible mass range for the analysis of nucleic acids. Notable advances have followed in oligonucleotide sequencing, mixture analysis, studies of non-covalent complexes, and microscale sample handling.

Computation of Nucleotide Sequence

Figure B2.15 shows the shorthand structure of the oligodeoxyribonucleotide d(A-C-T-C-G-A-T-G), with bonds

COMMENT B2.6 DETECTION OF WEAKLY BOUND COMPLEXES

In most cases, the solution conditions that are needed to maintain an intact complex are not optimal for normal ESI operation. Thus, for maximum sensitivity, solutions of pH 2–4 for positive ionization and pH 8–10 for negative ionization are typical for polypeptide analysis. Many protein complexes are denatured in solution at pH values outside the pH 6–8 range. It is clear therefore that protein solutions for analysis in ESI-MS should be maintained close to physiological conditions of neutral pH and ambient temperature such that the protein remains close to its native state. However, ESI-MS with neutral pH solutions of proteins generally demands a more extended *m/z* mass spectrometer than typically required when using conventional conditions.

It is not yet fully understood which weakly bound complexes known to exist in solution are observable by ESI-MS, or what minimum binding strength may be required for an ESI-MS observation. Evidence from a growing body of literature suggests that the ESI-MS observation for these weakly bound systems reflects, to some extent, the nature of the interaction found in the condensed phase. However, the results of all ESI-MS experiments show that each biomolecular system has its inherent experimental features and experience obtained from studying one protein complex may not be the proper preparation for investigating the properties of another one.

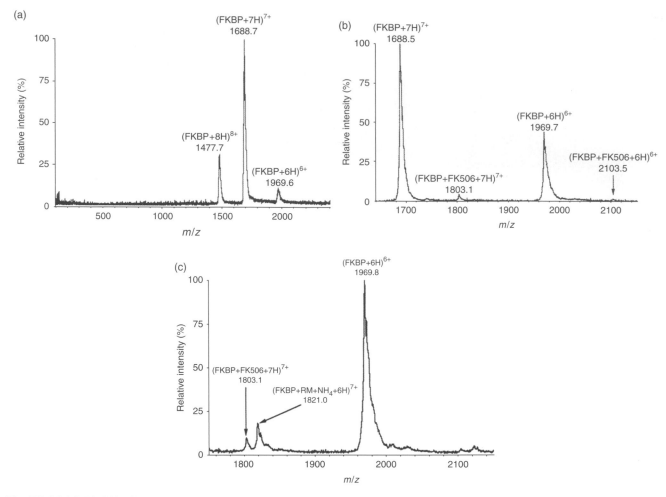

Fig. B2.11 **(a)** ESI-MS of human cytoplasmic receptor binding protein (FKBP) for cyclosporin at pH 7.5. The molecular mass, M, of FKBP is 11 812 Da. The envelope of multiply-charged ions ranges from the $(M+6H)^{6+}$ to the $(M+8H)^{8+}$ charge state of FKBP. **(b)** ESI of FKBP with small substance FK506 ($M = 804$ Da). **(c)** Competitive binding of FKBP with FK506 and rapamycin (RM). The molecular mass of RM is 912 Da. (Ganem *et al.*, 1991.)

TABLE B2.2 FWHM IS FULL WIDTH AT HALF MAXIMUM; "DISSOCIATION" INDICATES THAT NON-COVALENT INTERACTIONS ARE BROKEN; "FRAGMENTATION" INDICATES THAT COVALENT BONDS ARE CLEAVED. SEE SECTION B1.6, FOR THE INSTRUMENTS. RESOLUTION IS DEFINED IN SECTION B1.4 (AFTER BOERI ERBA, 2014).

	Quadruple-TOF	**FTICR MS**	**Orbitrap**
m/z range	~30 000	15 000	24 000
Resolution	~5000 FWHM	High, depending on magnet strength	25 000 FWHM at *m/z* 5000 16 000 FWHM at *m/z* 10 000
Dissociation	Parent ion selection	No parent ion selection	Parent ion and monomer selection
Fragmentation	Very limited	Highly informative	Highly informative

marked that on breakage give rise to the main fragment ions. In the same way as for a peptide sequence, a set of empirical rules has been devised for a nucleotide sequence. The four possible phosphodiester bound cleavages are labeled a, b, c, and d for the fragments containing the 5′-OH group and w, x, y, and z for fragments containing the 3′-OH group. The subscript indicates the position of the cleavage, counting the number of bases from the appropriate terminal group. An example of the application of these rules in the case of FAB-MS is presented in Fig. B2.15.

(a)

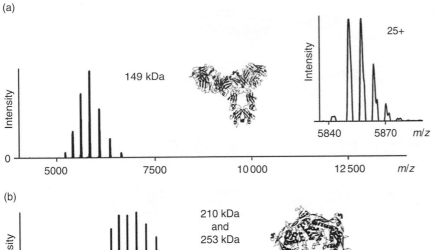

149 kDa

25+

5840   5870   *m/z*

5000   7500   10 000   12 500   *m/z*

Fig. B2.12 Mass spectra of intact macromolecules acquired using a modified Orbitrap. Spectra of **(a)** an IgG antibody (a zoom of 25+ charge state of IgG1 is also shown); **(b)** oligomers of bacteriophage HK97 capsid; **(c)** yeast 20S proteasome; and **(d)** *E. coli* GroEL. Crystal structures are also presented. (Reproduced with permission from Rose *et al.* copyright (2012) Nature Publishing Group) (Boeri Erba, 2014.)

(b)

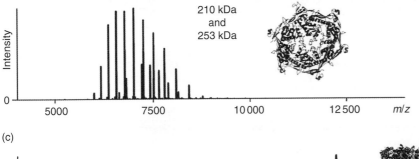

210 kDa
and
253 kDa

5000   7500   10 000   12 500   *m/z*

(c)

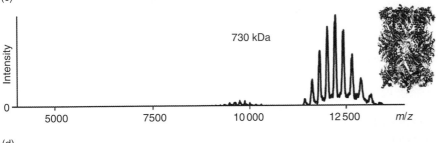

730 kDa

5000   7500   10 000   12 500   *m/z*

(d)

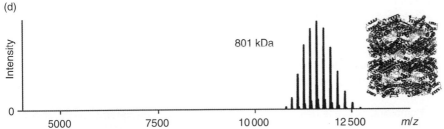

801 kDa

5000   7500   10 000   12 500   *m/z*

COMMENT B2.7 FRAGMENTATION BY ECD AND ETD

Fragment ions observed in an MS-MS spectrum depend on primary sequence, energy considerations, charge state, etc. Fragments need to carry at least one charge in order to be detected.

In ECD, gas-phase ions are fragmented through trapped multiply protonated molecules capturing low-energy electrons. Odd-electron ions are formed and subsequently fragmented. ECD appears to be very bond-specific. It favors peptide backbone cleavages, while preserving weak interactions (e.g., involving PTM or non-covalent ligand bindings).

ETD is based on the capture of electrons originating from radical anions, and induces peptide backbone fragmentation, preserving labile PTMs.

ECD and ETD produce primarily c-type and z-type ions. Sequence coverage is comparable for both techniques, and approaches that of the more commonly used collisionally activated dissociation (CAD) technique.

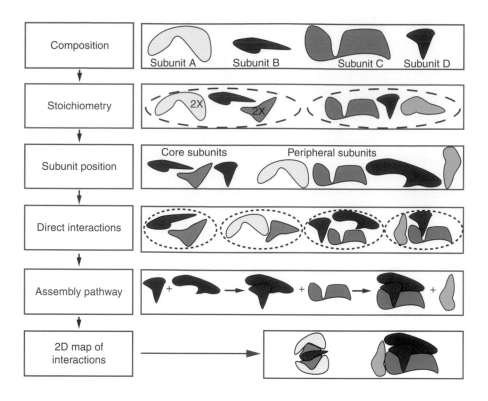

Fig. B2.13 A two-dimensional interaction network of subunits within a protein complex can be generated through a multistep process using native MS.

Step 1 (composition): Under denaturing conditions, the subunits are chromatographically separated and their masses are determined.

Step 2 (stoichiometry): Using MS conditions optimized for preserving non-covalent interactions (e.g., using ammonium acetate buffer), the measured mass of an intact complex (or a subcomplex) reveals the stoichiometry of the subunits.

Step 3 (subunit position): A series of MS-MS spectra indicates whether the subunits are located within the core or at the periphery of the assembly.

Step 4 (direct interactions): By adding organic solvent, overlapping subcomplexes (e.g., dimers, trimers) are generated. The composition of the different subcomplexes reveals the direct interactions between the subunits.

Step 5 (assembly pathway): Individual subunits can be mixed in solution and a mass shift can be detected if a subcomplex is formed.

Step 6 (two-dimensional map of interactions): Combining all these data allows one to draw an accurate interaction network of a protein complex. (Boeri Erba, 2014)

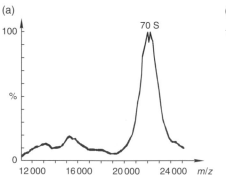

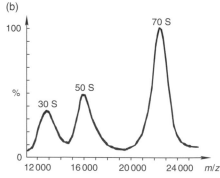

Fig. B2.14 Nanoflow ESI mass spectra of 70 S ribosome particles: **(a)** in the presence of 5 mM Mg^{2+} and **(b)** after a three-fold dilution of the solution. (Rostom *et al.*, 2000.)

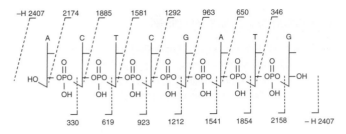

Fig. B2.15 The shorthand structure of the octatomic oligodeoxyribonucleotide d(A–C–T–C–G–A–T–G) with bonds marked that on breakage give rise to the main fragment ions. The dashed lines show the main fragmentations with attached mass, those above corresponding to the ions with 5′-phosphate ends (5′-P) and those below to the ions with 3′-phosphate ends (3′–P). (Grotjahn *et al.*, 1982.)

The sequence of an oligodeoxyribonucleotide can be determined from the following set of *m/z* data

m/z (5′–P):	2407, 2174, 1885, 1581, 1292, 963, 650, 346
m/z (3′–P):	330, 619, 923, 1212, 1541, 1854, 2158, 2407

The necessary rules are the following:

(1) The mass differences (Δ) in *m/z* values between sequence ions represent the nucleotide residual mass. This means that the set of *m/z* data can be converted to sequence of oligonucleotides in the following manner:

(5′-P) Δ		233*	289	304	289	329	313	304
m/z 2407		2174	1885	1581	1292	963	650	346*
Nucleotide	A	C	T	C	G	A	T	G

The first nucleotide is A (molecular mass is $233 + 96$ $(PO_4H) - 17(OH) + 1(H) = 313$B2). The last nucleotide is G (molecular mass is $346 - 17(OH) = 329$). So, the sequence is

d(A-C-T-C-G-A-T-G)

The reader can check the sequence by applying the same rules to a set of *m/z* values for the (3′-P) end.

(2) The highest mass peak in the FAB spectrum is *m/z* 2407, which represents $(M - H)^-$. Hence $M = 2408 + 96$ $(PO_4H) - 34(2OH) = 2470$.

The summation of the nucleotide residues = 2470 Da. This agrees with the mass spectrometry value for *M*.

B2.3.1 Oligonucleotide Mixture Analysis

The direct combination of liquid chromatography and electrospray ionization techniques in mass spectrometry should have been highly beneficial in the analysis of a nucleotide mixture. However, developments in this field have been relatively limited owing to the incompatible demands of the two

methods. Anion exchange and reversed-phase liquid chromatography utilize a salt-containing mobile phase for efficient separation, which reduces the quality of mass spectra. This problem has been solved by employing a new mobile phase additive (hexafluro-2-propanol). The broad potential of this new method was demonstrated for synthetic homopolymers of thymidine up to 75 bases, fragments based on the pBR322 plasmid sequence, and phosphorothioate ester antisense oligonucleotides. This approach is particularly useful for the characterization of DNA probes, polymerase chain reaction (PCR) primers, as well as for materials used in clinical trials on a level below 10 pmol.

B2.3.2 Non-Covalent Complexes

Due to the gentle nature of the ESI process, complementary oligonucleotide duplexes can be transferred with fidelity to the gas phase. Highly accurate mass measurements of the complexes formed provide strict evidence for the identity and stoichiometry of the constituent subunits. A number of successful measurements for protein–oligonucleotide complexes including a small protein called gene V protein (molecular mass 9688 Da) and bovine serum albumin (molecular mass 66 497 Da) have been reported. Such experiments are highly dependent on experimental conditions, however, and the question remains as to whether or not the measurements reflect real solution structures.

B2.3.3 Large and Very Large Nucleic Acids

Attempts to analyze large nucleic acids by MALDI-MS with lasers emitting in the UV region have been only moderately successful. The state-of-the-art mass upper limit is ~90 kDa for DNA and ~150 kDa for RNA, and is restricted mainly by ion fragmentation.

Using an IR laser and a glycerol matrix was found to be the most gentle combination for the intact desorption and ionization of nucleic acids in a broad range of the mass, from small oligonucleotides to molecules of more than 2000 nt. Thus, for a synthetic 21-nucleotide DNA (molecular mass 6398 Da), a mass resolution of 1200 has been obtained, similar to that for proteins of comparable mass.

Detection of very large DNA molecules was accomplished in the original FTICR experiment when a single pass of the ions permitted remeasurement on the same ions more than 450 times. This greatly improved the signal-to-noise ratio and the resulting precision of mass and charge measurement for very large ions. Figure B2.16 shows the mass spectrum of 2.88 MDa ions of DNA carrying more than 250 charges obtained with low mass resolution (about 25).

At first sight, ESI appears to be a superior method to MALDI for the intact desorption of large nucleic acids. However, mass assignment is usually very poor (with an uncertainty of around 10%). The largest nucleic acids whose masses have been accurately determined by ESI-MS have a mass of about 40 kDa.

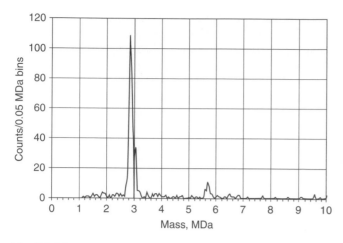

Fig. B2.16 Histogram of the average mass obtained from a DNA sample. The dominant peak at 2.88 MDa corresponds to the sodium adducted bacterial plasmid (pBR322) and the much smaller peak at 5.85 MDa corresponds to pBR322 dimer ions. (After Benner, 1997.)

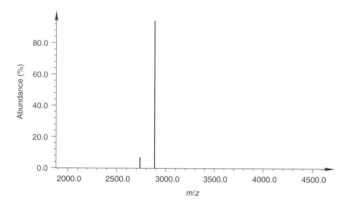

Fig. B2.17 FTICR spectrum of the ion at m/z 2883 obtained for Coliphage T4 DNA. The calculated molecular mass is about 90.9 ($\pm$9.1) 10^6 Da. (Smith *et al.*, 1996.)

Coupling of ESI techniques with FTICR-MS allows the analysis of individual multiply-charged macroions. Figure B2.17 shows FTICR spectra obtained for Coliphage T4 DNA (expected molecular mass 111.5 MDa). The signal from individual Coliphage T4 DNA ions at a pressure of $\leq 1.0 \times 10^{-9}$ Torr was detected more than 1.5 h after the initial excitation.

B2.3.4 DNA Sequencing

For mass spectrometry to be a useful method for sequencing DNA, the ionization method must be capable of generating ions of DNA containing several hundred bases with sufficient mass resolution and mass accuracy to identify peaks corresponding to a difference of one base or better. From the methodological point of view,

mass spectroscopy approaches can be subdivided into two groups as sequence information can be derived either directly from gas-phase fragmentation, or by mass measurements of chain cleavage products formed in the condensed phase (mass analysis of Sanger sequencing reaction products, or ladder sequencing).

The most straightforward approach to using mass spectrometry for DNA sequencing is to replace the polyacrylamide gel separation of the Sanger sequencing (Comment B2.8) products by mass analysis. Instead of having one lane for each of the four dideoxy termination reactions, four mass spectra are obtained, one for each of the A, C, T, G bases. Each sequencing solution contains a mixture of DNA strands several hundred bases in length. The mass spectrometer must be able to generate ions from this mixture at a resolution sufficient to distinguish between two bases in a strand up to 500 bases in length. This corresponds to a mass resolution of 500 for a mass range in excess of 100 kDa.

Mass spectrometric ladder sequencing involves cleaving successive nucleotides from the strand with an enzyme and measuring the resulting change in mass. The mass resolution and mass measurement accuracy requirements are more stringent for ladder sequencing than for Sanger sequencing. A mass difference of 9 Da must be measured to distinguish A from T, whereas for Sanger sequencing the ability to determine a mass difference of ~300 Da is sufficient to define the sequence.

In the gas-phase fragmentation methods, all of the sequencing process is done in the mass spectrometer by identification of fragment ions resulting from collision-induced dissociation. Sequencing by gas-phase fragmentation has the potential of being much faster than methods that rely on solution-phase chemistry.

Each approach has advantages and drawbacks. Thus, direct sequencing by gas-phase fragmentation is simple from the experimental standpoint, but is limited to chain lengths of ~80 nucleotides. This is because the mass resolution of MALDI degrades rapidly and it is difficult to separate two oligonucleotides that differ by one base if the DNA size is larger than 80 bases. On the other hand, indirect sequencing methods, although requiring pre-mass-spectrometry sample treatment of varying complexity, are capable of reaching chain lengths of ~100 nucleotides. However, in some cases the quality of the results is improved by lack of enzyme cleavage peculiarity.

At the beginning of the Human Genome Project (HGP), mass spectrometrists surmised that mass spectrometry would eventually allow DNA sequencing. When the HGP was completed, a total of about three billion bases had been identified directly from electrophoretic gel sequencing (see Comment B2.9) without using mass spectrometry. The real advantage of electrophoretic gel sequencing was that DNA with up to 1200 nucleotides could be analyzed. Such lengths were inaccessible to mass spectrometry, which is limited to <100 bases. Mass spectrometry is best suited for many genotyping applications in which fast and accurate mass measurements of short DNA lengths are required. Many mass spectrometrists continue to share the opinion that the potential advantages of mass spectrometry DNA sequencing are quite substantial, and that in the long run mass spectrometry will replace electrophoretic sequencing.

B2.3.5 Mass Spectrometry of RNA

RNA displays a rich diversity of properties, and the nucleic acid field is flourishing with the discovery and characterization of more and more vital functions involving RNA molecules (Comment B2.10).

COMMENT B2.9 HUMAN GENOME PROJECT (HGP)

In the HGP, DNA sequencing was performed by electrophoretic separation of fragments generated by the Sanger method. The four reaction product mixtures were loaded into lanes and separated according to their electrophoretic mobility, an intrinsic property that corresponds roughly to the length of the DNA fragment. Depending on the length of the gel, the separation can take several hours. The extended period was compensated for by performing several analyses in parallel. Automated sequencers running 96 lanes in parallel worked day and night and a total of about three billion bases were identified directly from the gels.

COMMENT B2.10 RNA TYPES

Further to the first discovered three RNA types involved in protein synthesis, messenger (mRNA), transfer (tRNA), and ribosomal (rRNA), ~30 different types (in 2014) of RNA with a wide range of functions have been characterized. They include signal recognition particle (SRP RNA, involved in membrane integration), small nuclear and small nucleolar (snRNA and snoRNA, involved in splicing and nucleotide modification, respectively), micro (miRNA involved in gene regulation), and of course information-carrying RNA in virus genomes and self-propagation small satellite RNA and retro-transposons.

Post-synthesis modifications, which play essential roles in DNA and RNA biology, are well characterized by MS (reviewed by Fabris, 2011; Giessing and Kirpekar, 2012).

Comprehensive lists of RNA modifications can be found in the RNA Modification Database (http://rna-mdb.cas.albany.edu/RNAmods) and Modomics, a database of RNA modification pathways (http://modomics.genesilico.pl).

The RNA World hypothesis of origin of life studies proposes a period that preceded our present DNA/protein world, in which information carrying and catalysis were both driven by properties inherent within RNA molecules.

Since 2010, effective developments have permitted mass spectrometry to answer important questions in DNA and especially RNA biology, through the detailed characterization of post-synthesis modifications, the enzymes involved, and nucleic acid interactions with proteins and small molecules.

RNA is a poly-anion in interaction with its counterions. In solution it will be distributed on numerous species (e.g., $RNA.Na^+$, $RNA.K^+$, $RNA.2\ Na^+...$). This will result in reduced signal-to-noise during mass spectrometry. The problem is usually solved by replacing the counterion with "volatile" cations, such as ammonium- or nitrogen-containing bases (e.g., protonated amines). The complex in solution will dissociate to release ammonia when the analyte is transferred into the gas phase for MS analysis.

The extremely polar nature of RNA also makes it difficult to bring into the gas phase and MS analysis of nucleotide oligomers was difficult before the advent of MALDI and ESI. Surprisingly, there is little difference in the MALDI-MS sensitivity in positive and negative ion mode for nucleic acids, whereas the negative ion mode sensitivity of ESI is around an order of magnitude better.

Because of the particular chemical nature of nucleic acids, in-source fragmentation is more prevalent than for peptides of the same mass. Similar nomenclature to that of DNA nucleotides (Fig. B2.15) is applied for RNA. In RNA fragmentation, c and y ions are generally the dominating backbone fragments in collision induced dissociation (CID), for both MALDI and ESI ionization (Comment B2.11).

101

COMMENT B2.11 MASS SPECTROMETRY OF TRANSFER RNA

Transfer RNA is the most heavily modified of RNA species. In a top-down MS-MS study, multiply-charged anions from the major fragmentation components were isolated in the gas phase and subjected to ion trap CID without subsequent ion/ion reactions (Huang *et al.*, 2010).

B2.4 COMPLEX CARBOHYDRATES

Biological carbohydrates constitute a diverse group of polymers playing various roles in cellular processes, ranging from energy storage (glycogen) and structural support (cellulose, chitin) to signaling, adhesion, and protein modifications. Biological carbohydrates range from simple monosaccharides to megadalton homopolymers, from precisely structured signaling oligosaccharides to complicated mixtures of carbohydrate-modified proteins and lipids.

The main difficulty for the structural identification of carbohydrate oligomers is the large number of isomers arising both from variation in linkage positions between monomer residues and from the possibility of multiple linkages to a single residue (branching). Hence, for a straight-chain oligomer, structural identification requires the sequence of glycosyl linkage types, i.e., the linkage position and anomeric configuration, as well as the sequence of monomer residues.

Many of the structurally interesting biological carbohydrates are glucoconjugates, e.g., glycoproteins, glycolipids, and glycolipoproteins. Conjugation also presents an additional dimension of structural isomers in terms of the specific location of the carbohydrate on a glycoprotein. The combination of conjugation and isomeric structural complexity has prevented the development of standard sequencing methodologies for carbohydrates, which have long been available for polypeptide or nucleic acid analysis.

Compared with the methodologies currently in use for protein and nucleic acid sequencing, the methods used for glycan analysis are far more diverse and are still being developed.

Below we give two simple examples of studies of carbohydrates by mass spectrometry. Recent reviews of the field are given in Comment B2.12.

B2.4.1 Oligosaccharides

Compared with peptides, native oligosaccharide substrates are difficult to ionize. However, various derivatization strategies have been reported, including permethylation. Permethylated oligosaccharides undergo well-characterized cross-ring fragmentation, which enables the assignment of particular linkage types to the carbohydrate structure. Cleavage of glycosidic bonds produces fragments that are easily distinguished by differences in the mass arising from methylation.

COMMENT B2.12 OLIGOSACCHARIDE MS AND GLYCOMICS

Kailemia *et al.* (2013) have reviewed recent developments in oligosaccharide MS.

Sulfated glycosaminoglycans (GAGs) are important components of animal physiology. They are difficult to characterize, however, because of the heterogeneous nature of GAG chains and their dynamic spatial and temporal expression patterns. Mass spectrometry approaches are currently being developed to contribute to the high throughput "glycomics" of GAGs. The main areas addressed include disaccharide analysis, oligosaccharide profiling and tandem mass spectrometric sequencing (Aia, 2013).

Glcβ1—4Glcβ1—4Glcβ1—4Glcβ1—4Glcβ1—4Glcβ1—4Glc

Fig. B2.18 Structure of maltoheptaose, Examples of monosaccharide abbreviations are given in Table A2.1.

Linear oligosaccharides and N-linked protein oligosaccharide mixtures (Fig. B2.18) have been studied successfully using ESI-quadrupole ion trap mass spectrometry. MS-MS experiments with orders higher than two offer a number of ways to enhance MS-MS spectra and to derive information not present in spectra from mass-spectrometer and MS-MS experiments (Fig. B2.19). Collisional activation of permethylated oligosaccharide molecular ions (MS-MS (or MS^2), as illustrated with maltoheptaose (Fig. B2.19), produces abundant fragments from glycosidic bond cleavages. This indicates composition and sequence, and weak cross-ring cleavage products, which denote specific linkages within the oligosaccharide. Through the trapping and further dissociation of these fragments (MS^n), cross-ring cleavage products can be confirmed and their relative abundances increased to facilitate interpretation.

Methylation also enables one to establish the branching motif of a given sugar; since only unoccupied linkage sites are derivatized, the extent of methylation is inversely related to the degree of branching at a given monosaccharide.

B2.4.2 Glycopeptides

Numerous nuclear and cytoplasmic proteins are modified by single N-acetylglucosamine residues attaching to the hydroxyl side-chain of serine or threonine (O-GlcNAc). This PTM is believed to be a regulatory modification similar to phosphorylation since it is dynamic, rapidly changing in response to external cellular stimuli, and fluctuating in a cell-cycle-dependent manner. The current method of site mapping, which involves galactosyltransferase labeling, generation of glycopeptides by proteolysis, purification by

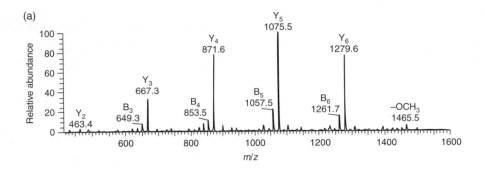

Fig. B2.19 **(a)** MS^2 spectrum of permethylated maltoheptaose, [M + Na] = m/z 1497.8. Scheme of the ion fragments produced upon **(b)** MS^2, **(c)** MS^3, and **(d)** MS^4. The nomenclature system used is based upon that introduced by Domon and Castello in 1987. (After Weiskopf *et al.*, 1997.)

several rounds of HPLC and gas-phase, and manual Edman sequencing, is very tedious and requires about 10 pmol of pure, labeled glycopeptide.

A very sensitive ESI-MS method was developed for selective detection of glycopeptides and identification of the exact sites of glycosylation at the low picomole level.

Synthetic glycopeptides have been generated (Fig. B2.20) and used to demonstrate that O-GlcNAc-modified peptides can be rapidly identified in complex mixtures by HPLC-coupled ESI-MS due to partial loss of linked O-glycan ($M = 204$ Da) at modest orifice potential. The resulting mass spectra are presented in Fig. B2.21. The site of the

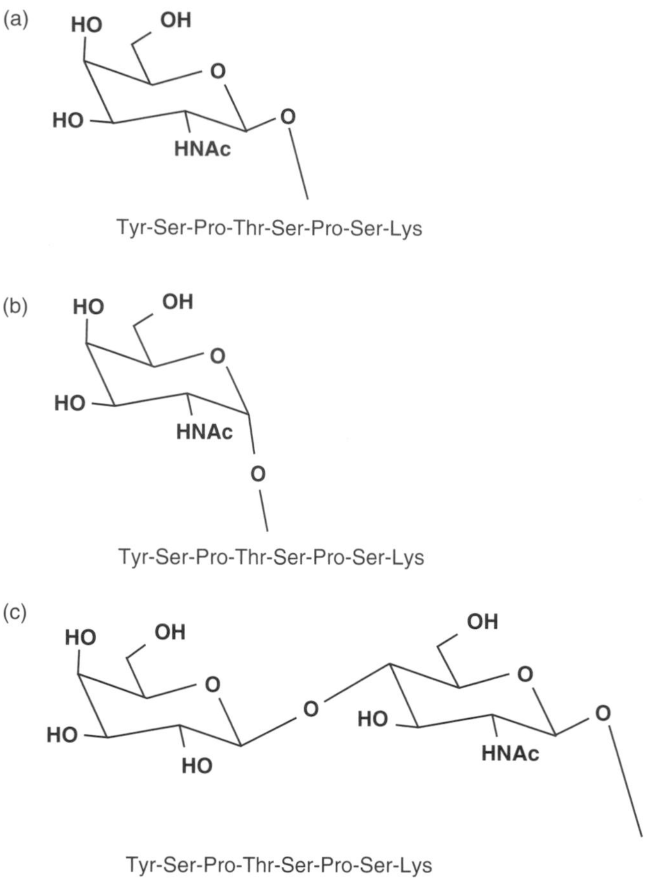

(a)

Tyr-Ser-Pro-Thr-Ser-Pro-Ser-Lys

(b)

Tyr-Ser-Pro-Thr-Ser-Pro-Ser-Lys

(c)

Tyr-Ser-Pro-Thr-Ser-Pro-Ser-Lys

Fig. B2.20 Structure of three synthetic glycopeptides: **(a)** O-GlcNAc peptide; **(b)** O-GalNAc peptide; **(c)** LacNAc peptide. (Greis *et al.*, 1996.)

glycosylation was directly identified by collision-induced dissociation (CID, Section B1.6.7) of the glycopeptide after removal of O-GlcNAc by alkaline β-elimination.

The conversion of glycosylserine to 2-aminopropeonic acid (see the structures in Fig. B2.20) by β-elimination both decreased the mass of the glycopeptide by 222 and resulted in a CID fragmentation representing the loss of 69 units of molecular mass instead of 87 (Ser) at the position of the glycosylserine.

The results show that this method may be a powerful tool for determining the sites of O-GlcNAc modification on proteins of low abundance such as transcription factors and oncogenes (see Comment B2.13).

B2.5 LIPIDOMICS AND MEMBRANE PROTEIN INTERACTIONS

Lipidomics is a branch of metabolomics that aims at full analysis of lipid species, their biological roles, and involvement in diseases (Comment B2.14). MS-based methods (mainly ESI and MALDI), together with chromatographic and spectroscopic approaches, are the main analytical tools of lipidomics. In direct-injection analysis, MS characterizes a sample without previous separation of the different lipids. The method is rapid, accurate, reproducible, and highly sensitive. Furthermore, MS imaging (see below) can provide visualization and distribution information, which is especially useful for the investigation of many biological processes.

Membrane proteins serve vital biological roles and are prime drug targets. A complete survey of more than 1000 putative integral, peripheral, and lipid-anchored membrane proteins from brewers' yeast *Saccharomyces cerevisiae* were affinity purified in the presence of non-denaturing detergents and the identities of the co-purifying proteins were determined by tandem mass spectrometry. The results were then used to derive a high-confidence physical interaction map (interactome) of membrane protein–protein interactions and putative heteromeric complexes (Fig. B2.22).

B2.6 MASS SPECTROMETRY IN MEDICINE

A variety of problems in clinical chemistry involving nucleic acid and protein analysis can be addressed by mass spectrometry. It has already replaced electrophoresis for analyzing products from routine molecular biological procedures. It allows the sizing of DNA amplified from the polymerase chain reaction (Comment B2.15), the sequencing of short oligonucleotides, the detection of genetic changes and mutations, and many other applications. Technology platforms for full comparative DNA analysis based on the generation of short diagnostic products and precise measurements of their masses by MALDI-TOF have been developed. They are the first steps in preclinical and clinical drug trials moving from a statistics-based approach to a non-statistics-based approach, enabling individual patient-specific response profiles for drug candidates.

Perspectives in clinical applications of MS have been reviewed by Strathman and Hoofnagle (2011). Current toxicology screening methods are mainly immunoassay-based. These assays are qualitative, however, and LC-MS/MS can provide highly sensitive and specific quantitative confirmatory data to establish the true identity of a compound. The quantitative sensitivity of MS-based methods has also had important applications in endocrinology. LC-MS/MS, for example, is currently unique in being able to separate vitamin D_2 from vitamin D_3 as well as provide information on the epimeric form of vitamin D. The low specificity of immunoassays determination of steroid hormones has also led a growing demand for LC-MS/MS. The application of mass spectrometry to steroid analysis, however, requires a high degree of technical competence, skill, and experience to provide the needed improvement for measures of endocrine function.

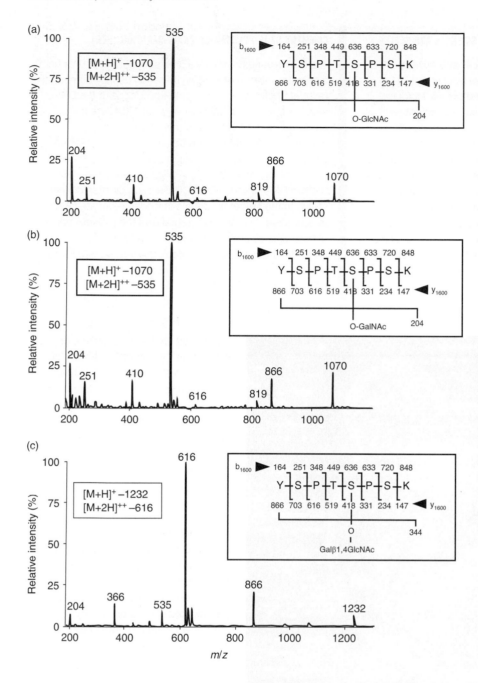

Fig. B2.21 ESI mass spectra of the HPLC-purified glycopeptides obtained by triple quadrupole MS: **(a)** O-GlcNAc peptide; **(b)** O-GalNAc peptide; **(c)** LacNAc peptide. The insets represent the predicted b and y ion mass fragments for each glycopeptide. The nomenclature system used is the same as for proteins. The calculated average masses for the glycopeptides are 1069 for the O-GlcNAc and O-GalNAc glycopeptides, and 1231 for O-LacNAc peptide. The mass spectra of both the O-GlcNAc (Fig. B2.20a) and O-GalNAc (Fig. B2.20b) glycopeptides were consistent with the expected mass as demonstrated by the major peaks at mass charge ratios (m/z) of 535 and 1070. These peaks correspond to the doubly protonated and singly protonated forms, respectively, of the glycopeptides. Other prominent peaks at 866 and 204 (corresponding to the deglycosylated peptides and O-GlcNAc or O-GalNAc) as well as those at 819, 616, and 251 were generated by fragmentation of the glycosidic and peptide bonds (see inset for predicted fragmentation patterns). The mass spectrum of the O-LacNAc peptide displays the same characteristics. (Greis et al., 1996.)

COMMENT B2.13 ALKALINE-INDUCED β-ELIMINATION

Alkaline-induced β-elimination is one of the most common methods used to liberate O-linked glycans from proteins and peptides. The method converts glycosylated serine and threonine residues to alanine and 2-aminobutyric acid, respectively.

A SWATH-MS combined with selected reaction monitoring (SRM) to identify clinical biomarkers in a proteomics approach is illustrated in Fig. B2.23. SWATH is a recently developed data acquisition independent (DIA)

method, which converts molecules in a physical sample into perpetually re-usable digital maps.

Clearly, the significant developments in MS applications in clinical protein analysis and metabolic measurements to compare healthy and diseased states in cells will have a powerful impact on medicine.

MALDI-MS is also being applied to characterize microbial pathogenesis in diagnostic microbiology. MALDI-MS can rapidly identify bacteria based on the molecular profiles of small cell populations. Systematic molecular profiling across tissue sections represents an imaging modality (2D and 3D MALDI-MS, see Section B2.8), enabling region-specific molecular measurements to be made *in situ*. Furthermore, MALDI-MS provides a means to

COMMENT B2.14 LIPIDOMICS FOR THE LIFE SCIENCES AND HEALTH.

Lipidomics and its application in cancer and Alzheimer's disease is discussed in a recent review by Li *et al.* (2014).

Lipidomic methodologies highlighted the physiological importance of individual lipid molecular species rather than changes in an overall lipid class. Results identified lipid changes in cancer cells, raising hope of identifying biomarkers of the disease (Zhang and Wakelam, 2014).

Lipids play a dominant role in the central and peripheral nervous systems. Lipid–protein complexes surround neurons and provide electrical insulation for signal transmission. Lipids are involved in vesicle formation and fusion in synapses. They provide means of rapid signaling and cell motility. MS-based lipidomics and its application to neurosciences is reviewed by Enriquez-Algeciras and Bhattacharya (2013).

COMMENT B2.15 POLYMERASE CHAIN REACTION (PCR)

The PCR is used to amplify a precise fragment of DNA from a complex mixture of starting material, usually termed the template DNA. It requires some knowledge of the DNA sequence, which flanks the fragment of DNA to be amplified (target DNA). From this information two oligonucleotide primers can be chemically synthesized, each complementary to a single stretch of DNA to the 3'-side of the target DNA, one oligonucleotide for each of the two DNA strands. The PCR consists of three defined steps. In the first step, double-stranded template DNA is denatured. The second step allows the hybridization of two oligonucleotide primers to bind to their complementary sites, which flank the target DNA. In the third step, DNA synthesis is carried out by DNA polymerase.

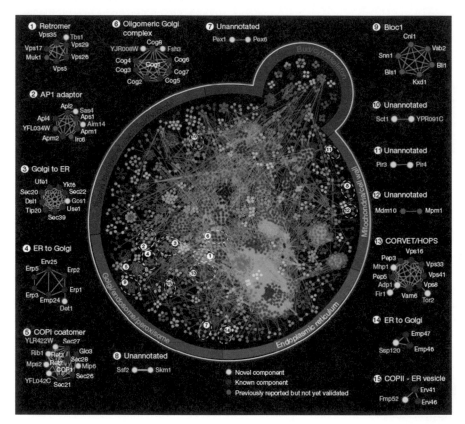

Fig. B2.22 Global organization of yeast membrane protein (MP) complexes. Predicted MP clusters (subunits shown as similarly colored nodes) inferred from the integrated network of high-confidence PPI (edges), demarcated according to primary compartment annotations. Representative complexes at the periphery highlight some of the findings of our study, including novel complexes and known complexes with new components. Our purifications were most successful for MPs localized to the Golgi and endoplasmic reticulum, a bias reflected in the highlighted examples. For each complex, previously reported components (red nodes), novel subunits (yellow nodes), and previously reported but not yet validated interactors (pink nodes) are displayed. (From Babu *et al.*, 2012).

study pathogen–host interaction and to discover potential markers of infection by analyzing molecules in serum and tissue samples (review by Moore *et al.*, 2014).

B2.7 BACTERIA AND BACTERIAL TAXONOMY

The taxonomic identification of bacteria based on constituent proteins represents a logical extension of the methods

developed for bacterial proteomic studies (Comment B2.16). Ideally, this approach should be rapid and yet be based on a sufficiently large group of proteins so that a unique mass spectrum fingerprint for the organism or strain can be obtained. Figure B2.24 shows spectra obtained from four strains of *E. coli*. They are typical of "whole cell" or "intact cell" spectra that show a large number of peaks between *m/z* 3000 and *m/z* 10 000. The different *E. coli* strains share many peaks in common, but also display unique features. It is important to note that such spectra

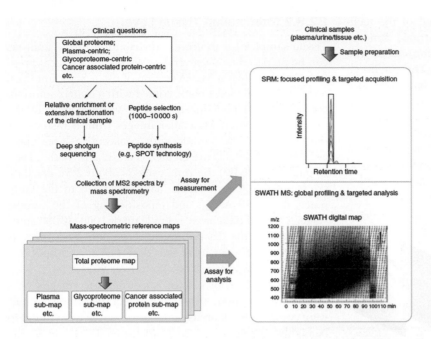

Fig. B2.23 The generation and targeted navigation of mass spectrometric reference maps associated with specific clinical questions. MS2 spectra are obtained either via the deep-sequencing analysis of the real sample or by the shotgun identification of the synthetic peptides representing the proteins of interest. The MS2 information is collected as assays, yielding the mass spectrometric reference maps. The targeted navigation can then be achieved by either SRM-based targeted profiling or SWATH MS-based global profiling. Note that the MS assays from the reference maps are important for both the targeted measurement by SRM and the data extraction step in SWATH analysis. (From Liu *et al.*, 2013).

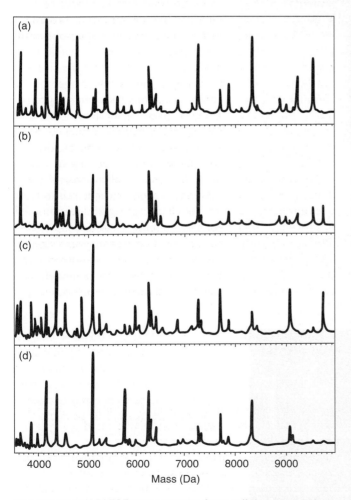

Fig. B2.24 MALDI-TOF mass spectra from cellular suspensions of four strains of *E. coli*: **(a)** BLR, **(b)** XL1 blue, **(c)** RZ1032, and **(d)** CSH23, that show the typical "whole cell" pattern of peaks in the interval from *m/z* 3000 to 10 000. (Lay, 2001.)

COMMENT B2.16 TAXONOMIC IDENTIFICATION OF BACTERIA BY MALDI-TOF

The Biomerieux company proposes an automated system for bacterial and fungal identification, VITEK MS, incorporating a MALDI-TOF mass spectrometer and a database that provides rapid, robust, and accurate identifications. (www.biomerieux-industry.com).

depend on the growth stage of the cells. Bacteria respond rapidly to environmental changes, and the production of stress proteins when cells are stored, handled, or cultured over different time periods results in non-identical samples. Such considerations must be taken into account when MALDI-TOF-MS is applied to identify bacteria.

B2.8 IMAGING MASS SPECTROMETRY

While conventional methods such as transmission and scanning electron microscopy are routinely used to create images of biological samples, they are generally limited to providing mainly morphological information (see Part H). The molecular specificity and sensitivity of MALDI-MS is being employed in the development of a new technology for direct mapping and imaging of biomacromolecules present in single-cell or tissue sections.

B2.8.1 Single-Cell Level

Frozen-hydrated freeze-fractured biological samples provide the best targets for direct molecular characterization by imaging MALDI-TOF-MS. The potential of the technique has been demonstrated on concentrated *Paramecium* cells that were freeze-fractured and introduced into a TOF secondary ion mass spectrometer. The single-cell organisms were ideal for study because their large size (~180–310 μm long) allowed the unambiguous positioning of individual cells directly in the path of the primary ion beam. Images of the distribution of molecular species across a cell surface with a submicrometer ion probe beam were obtained. The distribution of dimethylsulfoxide and cocaine added to the *Paramecium* cells was also imaged.

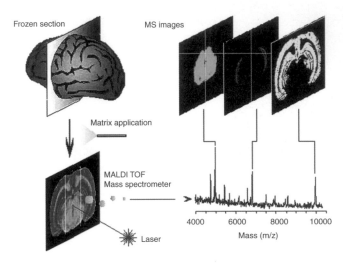

Fig. B2.25 Schematic demonstration of methodology developed for the spatial analysis of tissue by MALDI-MS. Frozen sections are mounted on a metal plate, coated with a UV-absorbing matrix and placed in the mass spectrometer. A pulsed UV laser desorbes and ionizes molecules in the tissue and the *m/z* values are determined using a TOF analyzer. From a raster over the tissue and measurement of the peak intensities over thousands of spots, mass spectrometric images are generated of specific molecular mass value distributions. (Stoeckli *et al.*, 2001.)

B2.8.2 Mammalian Tissue Level

Tissue samples for molecular distribution image analysis can be prepared using several protocols. In a typical preparation procedure, a frozen section of tissue is mounted on a stainless steel target plate, coated with a matrix solution, then dried and introduced into the vacuum inlet of the mass spectrometer. Molecular images from a raster over the surface of the sample with consecutive laser spots (~25 μm in diameter) are created (Fig. B2.25). With a laser frequency of 20 Hz, the time cycle is about 2.5 s per data point, including acquisition, data download to the computer, data processing, and repositioning of the sample stage. A typical data array is made up of 1000–30 000 spots depending on the desired image resolution. The image is made up of the intensity of ions desorbed at each spot in a molecular mass range of 500 Da to over 80 kDa.

Figure B2.26 shows mass spectrometric images of a mouse brain section. Although many of the protein signals were common to all areas of the brain, some were found to be highly specific to a given brain region. For example, the protein at *m/z* 8258 (Fig. B2.26b) was present in the regions of the cerebral cortex and the hippocampus; the protein at *m/z* 6716 was localized in the regions of the substantia nigra and medial geniculate nucleus; and the peptide at *m/z* 2564 was in the midbrain. Identification of the proteins can be done through extraction, high-performance liquid chromatography fractionation, proteolysis, mass spectrometric sequencing of one or more of the fragments, and protein database searching. The potential of the new technology has been illustrated in cancer research by identification of specific tumor markers in proliferating tissue.

The above examples illustrate applications of two-dimensional imaging MS. However, multiple sections of the same sample can also be characterized and recombined to provide a three-dimensional analysis of tissues. Three-dimensional IMS has been applied to microbes growing on agar plates to image secretions in three spatial dimensions (Fig. B2.27). The three-dimensional IMS approach has been used to analyze in-vitro interactions of *Pseudomonas*

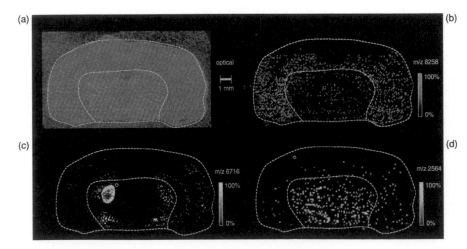

Fig. B2.26 Mass spectrometric images of a mouse brain section: **(a)** optical image of a frozen section mounted on a gold-coated plate; **(b)** *m/z* 8258 in the regions of the cerebral cortex and the hippocampus; **(c)** *m/z* 6716 in the regions of the substantia nigra and medial geniculate nucleus; **(d)** *m/z* 2564 in the midbrain. (Stoeckli *et al.*, 2001).

Optical   *m/z* 412

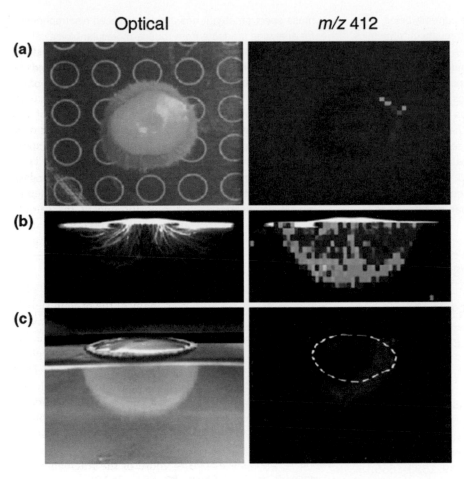

Fig. B2.27 IMS of the ion at *m/z* 412 from a colony of *C. albicans* grown on agar. **(a)** An overhead two-dimensional IMS analysis of *m/z* 412 reveals that the ion is not very abundant on the agar surface. **(b)** A two-dimensional IMS analysis of the transected colony reveals the ion at *m/z* 412 is abundantly secreted into the agar. **(c)** Reconstructed three-dimensional analysis of *m/z* 412 showcases the power of three-dimensional analysis. Not only is the ion readily detected in the agar, but it also reveals that the distribution is not uniform around the colony. (Adapted by permission from Watrous *et al.*, 2013; Macmillan Publishers Ltd: The ISME Journal. Copyright 2013).

aeruginosa and *Candida albicans*, two opportunistic pathogens that have previously been co-isolated from the lungs of cystic fibrosis patients (review by Moore *et al., 2014*).

B2.9 CHECKLIST OF KEY IDEAS

- The speed, accuracy, and sensitivity of ESI and MALDI have been exploited in the development of several different mass-spectrometry-based approaches for studying protein function, structure, and dynamics.
- Mass spectrometry plays a cardinal role in proteomics. In "bottom-up" MS-based proteomics, a protein mixture is first digested into short peptides, and subsequently analyzed. In the "top-down" approach, intact proteins are analyzed without prior proteolytic digestion (thus maintaining the native structure, including PTMs). A variant method ("middle-down") analyzes larger peptide fragments. It combines benefits of both bottom-up and top-down approaches (for example, by generating peptides that contain PTMs).
- MS-MS, tandem mass spectrometry is the most suitable technique for protein sequencing.

- Ion-mobility MS (IM-MS) is an excellent technique for the rapid analysis of molecular conformations.
- The folding state of a protein in solution can be monitored by the charge-state distribution produced during ESI. It has provided information, for example, on folding intermediates and intrinsically disordered proteins.
- In native or top-down MS, the study of non-covalent interactions led to the development of instruments such as the Orbitrap mass spectrometer, which has characterized intact protein complexes with masses ranging from 150 kDa to 800 kDa.
- A two-dimensional interaction network of subunits within a protein complex can be generated through a multistep process using native MS.
- Mass spectrometry may be best suited to many genotyping applications for which fast and accurate mass spectrometric measurements of short DNA lengths are required.
- Effective developments have permitted mass spectrometry to answer important questions in DNA and especially RNA biology, through the detailed characterization of post-synthesis modifications, the enzymes involved, and nucleic acid interactions with proteins and small molecules.

- Mass spectrometry approaches are currently being developed to contribute to the high throughput "glycomics" of biological glycans. The main areas addressed include disaccharide analysis, oligosaccharide profiling, and tandem mass spectrometric sequencing.
- A very sensitive ESI-MS method was developed for selective detection of glycopeptides and identification of the exact sites of glycosylation at the low picomole level.
- Lipids play a dominant role in the central and peripheral nervous systems and MS-based lipidomics has important applications in the neurosciences.
- A variety of problems in clinical chemistry involving nucleic acid and protein analysis can be addressed by

mass spectrometry. It has already replaced electrophoresis for analyzing products from routine molecular biological procedures. **MS applications in clinical protein analysis and metabolic measurements to compare healthy and diseased states in cells will have a powerful impact on medicine.**

- MALDI-MS can rapidly identify bacteria based on the molecular profiles of small cell populations. The method is also being applied in diagnostic microbiology to characterize microbial pathogenesis.
- MALDI-MS provides direct mapping and imaging of biological macromolecule distributions present in the single cell or in mammalian tissues.

Suggestions for Further Reading

General Review
Konermann, L., Vahidi, S., and Sowole, M. A. (2014). Mass spectrometry methods for studying structure and dynamics of biological macromolecules. *Anal. Chem.*, **86**, 213–232.

Proteins and MS-Based Proteomics
Levy, E. D., Boeri Erba, E., Robinson, C. V., and Teichmann, S. A. (2008). Assembly reflects evolution of protein complexes. *Nature*, **453**, 1262–1265.

Uetrecht, C., Rose, R. J., van Duijn, E., Lorenzen, K., and Heck, A. J. (2010). Ion mobility mass spectrometry of proteins and protein assemblies. *Chem. Soc. Rev.*, **39**, 1633–1655.

Cui, W., Rohrs, H. W., and Gross, M. L. (2011) Top-down mass spectrometry: Recent developments, applications and perspectives. *Analyst*, **136**, 3854–3864.

Sabido, E., Selevsek, N., and Aebersold, R. (2012). Mass spectrometry-based proteomics for systems biology. *Curr. Op. Biotechnol.*, **23**, 591–597.

Zhang, Y., Fonslow, B. R., Shan, B., *et al.* (2013). Protein analysis by shotgun/bottom-up proteomics. *Chem. Rev.*, **113**, 2343–2394.

Sharon, M. (2013). Biochemistry: Structural MS pulls its weight. *Science*, **340**, 1059–1060.

Boeri Erba, E. (2014). Investigating macromolecular complexes using top-down mass spectrometry. *Proteomics*, **14**, 1259–1270.

Ion Mobility for Structured and Disordered Conformations
Jurneczko, E., and Barran, P. E. (2011). How useful is ion mobility mass spectrometry for structural biology? The relationship between protein crystal structures and their collision cross sections in the gas phase. *Analyst*, **136**, 20–28.

Brocca, S., Testa, L., Sobott, F., *et al.* (2011). Compaction properties of an intrinsically disordered protein: Sic1 and its kinase-inhibitor domain. *Biophys. J.*, **100**, 2243–2252.

Konijnenberg, A., Butterer, A., and Sobott, F. (2013). Native ion mobility-mass spectrometry and related methods in structural biology. *Biochim. et Biophys. Acta*, **1834**, 1239–1256.

RNA
Huang, T.-Y., Liu, J., and McLuckey, S. A. (2010). Top-down tandem mass spectrometry of a tRNA via ion trap collision-induced dissociation. *J. Am. Soc. Mass Spectrom.*, **21**, 890–898.

Fabris, D. (2011). MS analysis of nucleic acids in the post-genomic era. *Anal. Chem.*, **83**, 5810–5816.

Giessing, A. M., and Kirpekar, F. (2012). Mass spectrometry in the biology of RNA and its modifications. *J. Proteomics*, **75**, 3434–3449.

Glycan and Glycopeptide Analysis
Leymarie, N., and Zaia, J. (2012). Effective use of mass spectrometry for glycan and glycopeptide structural analysis. *Anal. Chem.*, **84**, 3040–3048.

Han, L., and Costello, C. E. (2013). Mass spectrometry of glycans. *Biochem. Biokhimiia*, **78**, 710–720.

Kailemia, M. J., Ruhaak, L. R., Lebrilla, C. B., and Amster, I. J. (2013). Oligosaccharide analysis by mass spectrometry: A review of recent developments. *Anal. Chem.*, **86**, 196–212.

Aia, J. (2013). Glycosaminoglycan glycomics using mass spectrometry. *Mol. Cell Proteomics*, **12**, 885–892.

Lipidomics
Enriquez-Algeciras, M., and Bhattacharya, S. K. (2013). Lipidomic mass spectrometry and its application in neuroscience. *World J. Biol. Chem.*, **4**(4), 102–110.

Li, M., Yang, L., Bai, Y., and Liu, H. (2014). Analytical methods in lipidomics and their applications. *Anal. Chem.*, **86**, 161–175.

Zhang, Q., and Wakelam, M. J. O. (2014). Lipidomics in the analysis of malignancy. *Adv. Biol. Regul.*, **54**, 93–98.

Membrane Proteins

Babu, M., Vlasblom, J., Pu, S., *et al.* (2012). Interaction landscape of membrane–protein complexes in *Saccharomyces cerevisiae*. *Nature*, **489**, 585–598.

MS in Medicine and Bacteria

Liu, Y., Huttenhain, R., Collins, B., and Aebersold, R. (2013). Mass spectrometric protein maps for biomarker discovery and clinical research. *Expert Rev. Mol. Diag.*, **13**, 811–825.

Moore, J. L., Caprioli, R. M., and Skaar, E. P. (2014). Advanced mass spectrometry technologies for the study of microbial pathogenesis. *Curr. Op. Microbiol.*, **19**, 45–51.

Imaging MS

Gessel, M. M., Norris, J. L., and Caprioli, R. M. (2014). MALDI imaging mass spectrometry: Spatial molecular analysis to enable a new age of discovery. *J. Proteomics*, **107**, 71–82.

THERMODYNAMICS

C

THERMODYNAMIC STABILITY AND INTERACTIONS

C1

C1.1 HISTORICAL OVERVIEW AND BIOLOGICAL APPLICATIONS

The term thermodynamics is derived from the Greek *therme*, meaning heat, and *dynamis*, meaning strength or force. Thermodynamics, as a science, has its beginnings in the **nineteenth century**, with the first experiments exploring the relationship between heat and work, the definitions of the concepts of temperature and energy and the clear enunciation of the first and second laws of thermodynamics. Classical thermodynamics uses a phenomenological approach based upon these laws, in contrast to statistical thermodynamics, which tries to establish a more fundamental understanding of heat in terms of the kinetics of large assemblies of atoms or molecules. The concept of work was generalized beyond mechanical work to include all forms of energy such as electric, magnetic, chemical, and radiation energy. The validity of thermodynamics was established for all types of system, from a volume of perfect gas at a given pressure to thermonuclear plasma, encompassing magnetic systems, liquid–vapor systems, macromolecular solutions, chemical reactions, etc. In this chapter we are concerned, in particular, with the applications of both classical and statistical equilibrium thermodynamics to biological macromolecules, their solutions and interactions.

Early in the **seventeenth century, Galileo Galilei** and his followers constructed the first thermometers, allowing reproducible measurements of temperature. **Joseph Black (1728–1799)** built the first calorimeter; he is considered the founder of calorimetry. His measurements led to the theory of *caloric*, a conserved elastic fluid that had weight and was associated with temperature. **Antoine-Laurent Lavoisier (1780)** performed the first calorimetric measurements in biology. He studied energy exchange in animals (even though the concept of energy had not yet been enunciated) by placing a guinea pig in a more elaborate ice calorimeter (developed with **P. S. Laplace**) than that of **Black**. The heat output was measured by the rate at which the ice melted; heat input was calculated by subtracting measurements on feces and urine from measurements on food intake. **Thomas Young (1807)** introduced the term *energy* (from the Greek *en* meaning "in" and *ergon* meaning "work") but the distinction between force and energy was

not made clearly at the time. **Sadi Carnot**, a military engineer, introduced the concept of the heat engine cycle, published in his memoir "*Sur la puissance motrice du feu*" **(1824)** ("*On the motor power of fire*") and interpreted his results in terms of caloric theory. **Julius Robert von Mayer (1842)**, a medical doctor, announced the equivalence of heat and mechanical energy from physiological considerations. In **1845**, he published a numerical value for the mechanical equivalent of heat. **James Prescott Joule**, a Manchester brewer and amateur scientist, provided convincing experimental evidence that heat and work could be converted one into the other **(1840s)**. **Lord Kelvin (William Thomson) (1849)** pointed out the incompatibility between Joule's results and the caloric theory used by Carnot to interpret his experiments. The problem was solved by **Rudolf Julius Clausius (1850)**, who explicitly stated the first and second laws of thermodynamics; his paper marked the birth of the new science of thermodynamics. Clausius defined the property *entropy* (from the Greek *en*, meaning "in," and *tropos*, meaning "turn," and used in the sense of transformation, so that entropy is transformation content). The property of *energy* was defined by Kelvin, who also discussed various temperature scales. **James Clerk Maxwell** derived mathematical relations between thermodynamics parameters that now bear his name. **J. Willard Gibbs (1873, 1874)** published three definitive papers discussing the conditions of equilibrium in mathematical terms so elegant that they became the foundations of classical thermodynamics and physical chemistry. In **1918, Walther Nernst** stated the third law, essentially completing the exposition of classical thermodynamics.

Mechanical theories of heat received great impetus from Joule's experiments. The science of statistical thermodynamics is based on the relation between heat and the motions of atoms and molecules. It developed in parallel with classical thermodynamics, mainly through the work of **Hermann Ludwig Ferdinand von Hemholtz, Clausius, Maxwell, Ludwig Boltzmann,** and **Gibbs**.

Calorimetry was developed in the **twentieth century** for the careful study of the physical chemistry of solutions. **John T. Edsall (1930s)** measured thermodynamics parameters for the dissolution of amino acids and other organic compounds in water. With respect to the application of thermodynamics ideas to protein folding, **Christian Anfinsen** proposed the "thermodynamic hypothesis" in the **1950s**, following experiments on ribonuclease. The hypothesis states that all the information required for a protein to fold into its active structure under physiological conditions is contained in its sequence. The first highly sensitive calorimeter for protein solution studies was built in the **1960s** by **Peter Privalov** and his collaborators, and in the following decades differential scanning calorimetry and, later, isothermal titration calorimetry were used to study protein folding and binding interactions, respectively. Experiments allowing the measurement of thermodynamic parameters with great precision became possible with the development of ever more sensitive calorimeters – microcalorimeters and, with the approach of the year **2000**, nanocalorimeters (the micro and nano prefixes referring to the precision with which heat can be measured in these devices). Important microcalorimetric work on proteins was undertaken by **Privalov**, **Julian Sturtevant**, **Ernesto Freire**, and their collaborators.

Sensitive biosensor devices that measure the binding between molecules in solution and immobilized ligands became available in the **1990s**, and entered routine operation in many biochemistry laboratories.

Non-equilibrium thermodynamics is beyond the scope of this book, but it should be mentioned in this brief historical survey because of its important applications in many disciplines, including experimental physics, reaction chemistry, biology, ecology, and even sociology. The field of irreversible and non-equilibrium thermodynamics has known a period of strong development since the **late 1950s**, mainly through the work of **Ilya Prigogine** and his collaborators. They introduced the term *dissipative structures* to define open systems that could be maintained in far-from-equilibrium states by a flow of energy or matter. It is remarkable that spectacular examples of self-organization in biology, such as cell division, differentiation and morphogenesis, can be modeled in terms of dissipative structures.

There is great scope in biophysics for applying modern thermodynamics, based on high-resolution structural information (see Part G) and the tools of molecular biology. The stability of a given protein under different conditions, for example, can be understood and tuned to a fine degree, by studying the folding–unfolding transition of various specially designed mutants. For thermodynamics as for other biophysical methods, a high-resolution structure acts as a starting point for understanding rather than as an end in itself.

C1.2 THE LAWS OF THERMODYNAMICS

The language of classical and statistical thermodynamics is a difficult but extremely useful one (in fact it cannot be avoided). It is essential for the organization and understanding of experimental data on the energetics of defined systems, from very large assemblies of particles to single molecules. In writing this chapter we have not attempted more than a brief description of the principles governing the language. There are many good books on the subject and the reader is strongly encouraged to learn the language as thoroughly as possible, because thermodynamics is fundamental to the development and applications of biophysical methods.

Classical thermodynamics is a phenomenological science, concerned with precise observations on defined systems (e.g., a macromolecular solution). Statistical thermodynamics attempts an explanation of the observations in terms of the motions of the particle components of the system (e.g., the solute and solvent molecules that make up a solution).

Only a brief overview of the laws of classical thermodynamics is given in this chapter. We discuss the most useful parameters and the mathematical relations between them required mainly for the understanding of calorimetric and binding experiments on the folding and interactions of biological macromolecules (Comment C1.1). The basic principles presented are also valid and useful for the understanding of macromolecules in solution (Part A), hydrodynamics experiments (Part D), scattering experiments (Part G), spectroscopy (Part E), and single-molecule studies (Part F). The treatment is limited to equilibrium thermodynamics, which provides an adequate basis for the understanding of biological macromolecules and their (mainly in-vitro) interactions. Non-equilibrium thermodynamics, which is essential for the analysis of higher orders of biological complexity, is beyond the scope of this book.

COMMENT C1.1 BIOLOGIST'S BOX: THERMODYNAMICS AND MATHEMATICS

It would be gravely misleading to attempt an explanation of *thermodynamics* without using mathematics. The triumph and usefulness of thermodynamics are due to its formulation of a few laws in terms of mathematical equations, which establish precise relations between measurable quantities. *Partial differential equations*, in particular, which themselves follow definite mathematical rules, provide a simple and elegant way of expressing the changes expected in an experimental parameter (e.g., the pressure of a gas), when another one is changed (e.g., the temperature), while a third parameter is kept constant (e.g., the volume). These equations not only tell us that the pressure will rise, as we might expect intuitively, but also by exactly how much for a given temperature change. This example is a simple illustration of the application of thermodynamics, but similar strict predictions can be made for much more complex situations, such as chemical reactions or protein folding.

C1.2.1 Fundamental Definitions and the Zeroth Law

Adiabatic, Isothermal Systems and the Concept of Equilibrium

We consider a system to be a collection of matter contained by walls with which it does not react chemically. An *adiabatic* (from the Greek for "does not allow passage") wall isolates the system perfectly from external effects. Walls that are not adiabatic are said to be *diathermal* (from the Greek *dia*, meaning "through," and *therme*, meaning "heat"). Two systems separated by a diathermal wall are *isothermal*. The condition of the system is called its *state*. The properties of a given state can be determined by experimental measurements. Two states are identical when all the values measured for their properties are identical. A system enclosed in an adiabatic vessel will, in time, reach a state in which its properties no longer evolve; the system is then said to be in *equilibrium*. Note that equilibrium denotes a static (unchanging) condition only with respect to macroscopic properties. A gas or solution at equilibrium, for example, would be seen to be a highly dynamic system if we observed the Brownian motion of the molecules or solute particles, respectively.

The Zeroth Law

The zeroth law of thermodynamics was stated after the first and second laws. It essentially defines the concept of temperature. As this concept is useful for the understanding of the other laws, it was called the zeroth law.

The zeroth law states that if two bodies, A and B, are separately in equilibrium with a third body, C, then each of A and B is in equilibrium with the other.

A consequence of the law is that there exists an empirical function of the state of a system, the temperature, which takes the same value for different systems in equilibrium.

Functions of State and Conjugate Variables

An observable parameter, whose value depends only on the condition of a system and not on the means by which the condition was attained, is called a function of state. Temperature is a function of state. Internal energy, entropy, pressure, and volume are other examples of functions of state. A parameter that depends on the size of the system is called an extensive variable, whereas a parameter that depends on the condition but not on the size of the system is called an intensive variable. Volume, internal energy, and entropy are all extensive variables; pressure, temperature, and the pH value of a solution, for example, are intensive variables.

In the thermodynamic equations, which describe relations between different functions of state, the terms appear in units of energy. When an energy term contains two parameters, the pair is made up of so-called *conjugate variables*. Pressure and volume are conjugate variables, as are entropy and temperature (see Eq. (C1.5), for example).

C1.2.2 The First Law and Energy

The first law of thermodynamics has been worded in various ways. In essence, it states that

The amount of work needed to change the state of a system from an initial to a final state depends solely on the change accomplished, and not on the means by which the work is performed, nor on the intermediate states by which the system may pass.

The first law defines *internal energy* (U) as a property of the initial and final states whose value changes by the amount of work done on the system.

Energy Conservation

Conservation of energy (with no distinction between whether it is mechanical, chemical, heat, or any other type of energy) is a consequence of the first law of thermodynamics. It is interesting to note, however, that the first law cannot be derived from a statement of conservation of energy.

According to the first law, the change, ΔU, in the internal energy of a system is given by

$$\Delta U = Q + W \qquad \text{(C1.1)}$$

where W and Q are, respectively, the amount of mechanical work (or chemical work, or any work other than in the form of heat) done on the system and the heat given to the system (Comment C1.2).

Chemical Potential

In the case of a solution, solute molecules are brought together into a volume of solvent. The chemical potential, μ, of a solute is the energy required to add one molecule (or one mole of molecules, according to definition) into the solution. We can split the work, W, in Eq. (C1.1) into a chemical potential term and a mechanical work term

$$\Delta U = Q + \mu \Delta n - P \Delta V \qquad \text{(C1.2)}$$

COMMENT C1.2 DEFINITION OF HEAT

Equation (C1.1) defines heat. For a system enclosed by adiabatic walls, the change in internal energy is given by $\Delta U = W$, so that the value of the quantity Q is a measure of the extent to which the walls are not adiabatic and *heat* can flow into the system from the outside. It can be shown that Q has the properties we usually associate with heat:

- The addition of heat to a system changes its state.
- Heat is conveyed from one system to another by conduction, convection, or radiation.
- In a calorimetric experiment on an adiabatically enclosed system, heat is conserved.

where Δn is the change in number of solute molecules between the initial and final states (depending on the scale used, this can also be expressed in terms of moles). We have expressed the mechanical work in terms of pressure, P, and volume change, ΔV. Chemical potential is an intensive variable. Concentration is the corresponding, conjugate, extensive variable. For a change at constant pressure, the chemical potential term is related to a function called the Gibbs free energy (Section C1.3.1).

C1.2.3 The Second Law and Entropy

A direct implication of the concept of equilibrium is that many types of change that would satisfy the first law simply do not occur in practice. The observed fact that an isolated system tends toward equilibrium (for example, solute molecules moving to fill a vessel homogeneously after the removal of a wall that confined them to part of the vessel) excludes the occurrence of the reverse path (the molecules moving back to one corner of the vessel, in the example) from taking place, even though it would not violate the first law.

The Second Law

We can express the directionality of certain thermodynamic changes in terms of the subjective notions of hotter and colder. There is a clear tendency for heat to flow from a hotter to a colder body. Clausius' formulation of the second law includes this concept:

It is impossible to devise an engine which, working in a cycle, shall produce no effect other than the transfer of heat from a colder to a hotter body.

Reversibility

Carnot demonstrated, following his experiments with heat engines, that the closest we can get to converting heat into work is by using an engine in which every process is reversible. Heat flow in such an ideal engine would be between bodies of essentially the same temperature (Fig. C1.1). If we cool one body just a little, heat will flow to it from the other; if now we heat the same body just a little, heat will flow the other way. The process is reversible. Similarly, we can imagine exerting work in a reversible manner on a volume of gas that is isothermal with its surroundings at a temperature T by moving a frictionless piston very slowly in one direction or the other so that heat Q flows in or out of the gas, while its temperature stays very close to T (Fig. C1.2).

Entropy

Clausius expressed the tendency of an isolated system (one which cannot exchange energy or matter with its surroundings) to reach equilibrium by defining a function of state, the entropy S, which increases with time until it reaches a maximum at equilibrium

$$\Delta S \geq 0 \tag{C1.3}$$

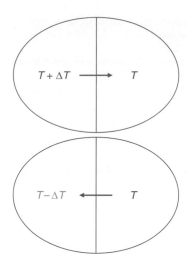

Fig. C1.1 Reversible heat transfer between two bodies.

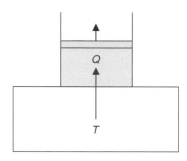

Fig. C1.2 Exerting work in *a reversible* manner on a volume of gas.

The equality (zero change in entropy) is valid in the limit of a reversible change, during which, in any case, the system is considered as being very close to equilibrium.

Exploring the consequences of the second law, it was found that entropy is related to the heat term in Eq. (C1.1) by

$$\Delta S = Q/T \tag{C1.4}$$

Equation (C1.4) defines the units of entropy in terms of energy and temperature. We can now rewrite Eq. (C1.2) as

$$\Delta U = T\Delta S + \mu\Delta n - P\Delta V \tag{C1.5}$$

Entropy and Disorder

The example we gave above of molecules moving to fill the entire volume available to them as the system evolves toward equilibrium is a representation of an intuitive concept that the equilibrium state is a disorganized state. Its components sample all possible configurations available to them. Boltzmann, in his development of statistical thermodynamics, expressed this concept quantitatively and related entropy, S, to the number of configurations, p, available to the system:

$$S = k_B \ln p \tag{C1.6}$$

where k_B is Boltzmann's universal constant (Comment C1.3).

Equation (C1.6) combined with the second law of thermodynamics (Eq. (C1.3)) indicates that at equilibrium the number of configurations sampled by the system is a maximum. Reconsidering Eq. (C1.4) in the light of Eq. (C1.6), we conclude that TS is the heat energy invested by the system in order to sample a number p of configurations at temperature T (in kelvins). We shall see in Section C1.3.1 that this energy is not "free energy," i.e., it is not available to perform work.

C1.2.4 The Third Law and Absolute Zero

The Nernst heat theorem, also called the third law of thermodynamics, states that

The entropy of any body is zero at the zero of the absolute temperature scale.

The classical formulation did not have a zero point for entropy (or temperature) and only *differences* in these quantities between one state and another were defined. The third law provided the zero point of an absolute entropy scale.

Since it has now been shown that there is no process in which entropy can become zero, alternative statements of the third law are:

By no finite series of processes is the absolute zero attainable,

and

As the temperature tends to zero, the magnitude of the entropy change in any reversible process tends to zero.

Absolute Temperature

The law defines the absolute zero of the temperature scale. Absolute temperature is now expressed in kelvins. It is zero at absolute zero and rises in degree steps, where the degree is defined as one-hundredth of the temperature difference between the freezing and boiling points of water at atmospheric pressure.

The third law plays a valuable role in physical chemistry. By setting the zero point for entropy and an absolute temperature scale, it allowed the calculation of *equilibrium constants* from the thermal properties of reactants (Section C1.3.2), and of *total internal energy* in calorimetry (Section C1.3.3).

C1.3 USEFUL CONCEPTS AND EQUATIONS

Classical thermodynamics is built on a set of mathematically simple equations, which establish how different variables are related to each other. The equations can also be written in terms of variable changes by applying the rules of differential geometry. Maxwell's relations, for example, relate changes in the pressure P, volume V, entropy S, and temperature T of a fluid system (Comment C1.4). In what follows, we shall concentrate on a few concepts from chemical applications of classical and statistical thermodynamics that are useful for the understanding of biological macromolecules and their interactions.

C1.3.1 Free Energy and Allied Concepts

Enthalpy

The enthalpy H of a system is defined as

$$H = U + PV \tag{C1.7}$$

The term enthalpy is from the Greek for "warmth within." Under constant pressure (a usual condition for solutions of biological molecules) the enthalpy of a system at temperature T represents the total energy required to heat the system from absolute zero to T.

The change in enthalpy, ΔH, during a chemical reaction is the heat absorbed or released in the breaking and formation of bonds. ΔH represents the experimentally measured quantity in calorimetry.

Gibbs Free Energy

The Gibbs free energy, G, is defined as

$$G = U + PV - TS = H - TS \tag{C1.8}$$

G represents the part of the energy in a system that is not used for the population of the different entropy configurations. The next step in the reasoning is as follows: Because the Gibbs energy is not involved in maintaining thermal agitation, it can be transformed into useful work, hence its name *free energy*.

A reversible chemical reaction (Comment C1.5) can be written as

$$A \rightleftharpoons B \tag{C1.9}$$

At constant temperature (the system is maintained in isothermal contact with a large constant temperature bath, as is the case for molecules in dilute solution for example) the free energy released by the reaction is given by

$$\Delta G = \Delta H - T\Delta S \tag{C1.10}$$

The second law postulates that an isolated system evolves in order to maximize its entropy. Analogously, a system that exchanges energy with its surroundings evolves in order to minimize its free energy. Then, the spontaneous sense of the reaction in Eq. (C1.8) is in the direction for which $\Delta G < 0$. When two states of a system are at equilibrium, their free energy difference is zero, $\Delta G = 0$.

The Gibbs free energy is a very useful parameter in the analysis of binding events. It is dependent on the changes in enthalpy (ΔH) and entropy (ΔS) of the ligands upon binding. The enthalpy term reflects internal energy changes associated with bond formation and breakage; the entropy term reflects the number of configurations available to the system.

The Free Energy Landscape and the Boltzmann Distribution

When there are states of different energies available to a system, they are populated to different extents as a function of temperature (Fig. C1.3).

The probability that a state of free energy ΔG_j at temperature T is populated is given by Boltzmann statistics:

$$C_j = \exp\left(-\Delta G_j / RT\right) \tag{C1.11a}$$

so that

$$\Delta G_j = -RT \ln C_j \tag{C1.11b}$$

The probability of a state of energy RT is $\exp(-1)$ or about 0.36, and, as expected, lower-energy states have a greater

---

COMMENT C1.5 CHEMICAL REACTIONS INVOLVING BIOLOGICAL MACROMOLECULES

A chemical reaction involves bonds of various types being made or broken between atoms. Folding and ligand binding interactions in biological macromolecules are chemical reactions in this general sense. Covalent bonds have associated energies of the order of $10^4\, k_B T$. Van der Waals bonds, hydrogen bonds, electrostatic bonds in aqueous solution, so-called hydration and hydrophobic bonds are all called weak bonds, because their associated energies are of the order of $k_B T$ for T close to room temperature, 300 K. Folding (a) and ligand (b) interactions may involve the formation and breakage of both covalent and weak bonds:

$$P + L \leftrightarrow PL$$

where P is protein and L is ligand.

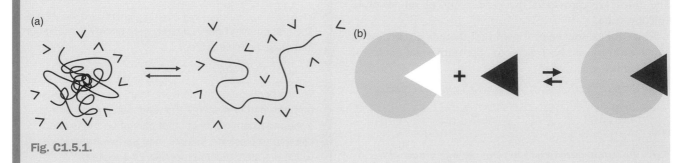

Fig. C1.5.1.

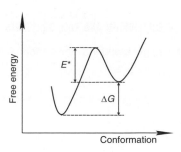

Fig. C1.3 A free energy diagram of a system as a function of its conformation. Two states separated by a free energy difference, ΔG, are shown in the example. The excess free energy barrier, E^*, is called the activation energy between the two states.

likelihood to be occupied than high-energy ones (e.g., the probabilities for free energies, $RT/2$ and $2RT$, are about 0.60 and 0.13, respectively).

The Partition Function

The partition function Q is defined as the sum of the statistical weights of all the states accessible to the system:

$$Q = \sum_{j=0}^{N} C_j = \sum_{j=0}^{N} \exp\left(-\Delta G_j/RT\right) \tag{C1.12}$$

All the thermodynamics variables of a system can be written in terms of Q. Equation (C1.12) defines the *canonical ensemble partition function* for a system containing a constant number of particles and a constant average energy.

Defining state $j = 0$ as the reference state with respect to which the thermodynamics variables are expressed as *excess* quantities, Eq. (C1.12) can be rewritten as

$$Q = 1 + \sum_{j=1}^{N} \exp\left(-\Delta G_j/RT\right) \tag{C1.12a}$$

The population of state j, P_j, is given by the ratio of the statistical weight of that state over the sum of the statistical weights of all states accessible to the system

$$P_j = \exp\left(-\Delta G_j/RT\right)/Q \tag{C1.12b}$$

The average free energy of the system, $\langle\Delta G\rangle$, is written

$$\langle\Delta G\rangle = \sum_{j=1}^{N} P_j \Delta G_j = -RT \ln Q \tag{C1.12c}$$

The average excess enthalpy is

$$\langle\Delta H\rangle = \sum_{j=1}^{N} P_j \Delta H_j = RT^2 \left(\partial \ln Q/\partial T\right) \tag{C1.12d}$$

and the average excess entropy is

$$\langle\Delta S\rangle = \sum_{j=1}^{N} P_j \Delta S_j = RT \left(\partial \ln Q/\partial T\right) + R \ln Q \tag{C1.12e}$$

Equations similar to Eqs. (C1.12c)–(C1.12e) apply to any property of the system. In general, the experimental average value of a property α of the system given by

$$\langle\alpha\rangle = \sum_{j=1}^{N} P_j \alpha_j \tag{C1.12f}$$

where α_j is the value of α associated with state j.

Equation (C1.12f) provides the basis for theoretical or model calculations of $\langle\alpha\rangle$ to compare with experimental observations.

Chemical Potential

We recall from the discussion of Eq. (C1.2) that the chemical potential, μ, of a solute is the *energy* required to add one mole of molecules to the solution. This *energy* is in fact *free energy*, so that chemical potential can be equated to the free energy change resulting from the addition of one mole of solute to the solution. The chemical potential or change in free energy of a solution due to the dissolution of a solute to a concentration N_j (moles per unit volume) is given by

$$N_j \mu_j = \Delta G_j = -RT \ln N_j \tag{C1.13}$$

Note that by comparing Eqs. (C1.11b) and (C1.13) we can equate N_j and C_j; the solution defined by the concentration N_j represents a state C_j of the system.

For a solution containing several different solutes:

$$\Delta G = -RT \sum \ln N_j = -RT \ln N_1 N_2 N_3 \ldots = -RT \ln \Pi N_j \tag{C1.13a}$$

The Biological Standard State

Since we are dealing with differences in potentials or between energy levels, it is useful to define *standard states* with respect to which these differences are measured. The temperature T is taken as 298.15 K (25 °C) unless otherwise defined. Gases are at 1 atm partial pressure. For aqueous solutions, solutes have 1 M concentrations. When protons are involved in a reaction, the *biological standard state* is defined as pH 7.

Chemical Reactions

We rewrite the reaction in Eq. (C1.9)

$$A_1 + A_2 + A_3 + \ldots \rightleftharpoons B_1 + B_2 + B_3 + \ldots$$

where A_j are reactants and B_j are products. By using Eq. (C1.13a) we calculate the free energy change in the reaction as it proceeds from left to right, in terms of reactant and product concentrations.

$$\begin{aligned} \Delta G &= -RT \ln \Pi[B] - RT \ln \Pi[A] \\ &= -RT \ln \left(\Pi[B]/\Pi[A] = -RT\% \ln L \right. \end{aligned}$$
$$\text{(where } K = [B_1][B_2][B_3]\ldots/[A_1][A_2][A_3]\ldots). \tag{C1.14}$$

It is a consequence of the definition that $K = 1$ for the standard conditions. The standard free energy $\Delta G°$ is defined as the free energy change when a reacting system starts at $K = 1$ and ends at equilibrium, $K = K_{eq}$, where there is no difference between the free energy of the products and reactants:

$$\Delta G° = -RT \ln K_{eq} \qquad (C1.15)$$

The free energy change for the reaction starting from any value of K is now written

$$\Delta G = \Delta G° - (-RT \ln K) = \Delta G° + RT \ln K \qquad (C1.16a)$$

or

$$\Delta G = -RT \ln K_{eq}/K \qquad (C1.16b)$$

Note that $\Delta G°$ is equal to zero only if $K_{eq} = 1$ (i.e., when concentrations are such that $[B_1][B_2][B_3] \cdots = [A_1][A_2][A_3]\cdots$).

If $K > 1$, ΔG is negative and the reaction occurs spontaneously in standard conditions (25 °C etc.). If $K < 1$, ΔG is positive and the reaction requires energy input in order to occur under standard conditions.

Van't Hoff Analysis

Equation (C1.15) provides the basis of a van't Hoff analysis for a two-state system ($A \rightleftharpoons B$). In such an analysis, standard free energy, and consequently enthalpy and entropy, are calculated from the measured equilibrium concentrations of the two states.

Standard free energy, enthalpy, and entropy are related by Eq. (C1.10)

$$\Delta G° = \Delta H° - T\Delta S° \qquad (C1.17)$$

Combining with Eq. (C1.15), we obtain $\ln K_{eq} = -\Delta H°/RT + \Delta S°/R$, from which we derive the van't Hoff equation

$$\Delta H_{v'tH} = RT^2 \partial \ln K_{eq}/\partial T \qquad (C1.18)$$

C1.3.2 Binding Studies

Binding Constants and Affinity

If molecule A binds to molecule B to form the complex AB, the thermodynamic equilibrium dissociation constant, K_d, is defined in terms of the different species' concentrations, [A], [B], and [AB], at equilibrium:

$$K_d = \frac{[A][B]}{[AB]} \qquad (C1.19)$$

Here, we have followed the binding experiment convention and written the equilibrium constant as the dissociation constant. Referring to Eq. (C1.14) and following $K_d = 1/K_{eq}$, *where the association constant, K_{eq}, is a measure of the binding "strength" or affinity of the interaction*. The constant K_d is related to the chemical on-rate constant k_{on} for the association of A and B, and the chemical off-rate constant k_{off} for the dissociation of the complex AB by (Comment C1.6):

$$K_d = k_{off}/k_{on} \qquad (C1.20)$$

COMMENT C1.6 EQUILIBRIUM AND RATE CONSTANTS

Consider the single-step binding reaction

$$A + B \rightleftharpoons AB$$

The rate of the forward reaction, which converts A and B to AB is

$$k_{on} \times [A][B]$$

The rate of the reaction converting AB back to A and B is

$$k_{off} \times [AB]$$

Equilibrium is reached when these two rates are equal:

$$k_{on} \times [A][B] = k_{off} \times [AB]$$

Rearranging gives

$$k_{off}/k_{on} = [A][B]/[AB] = K_d$$

Binding to Many Sites

If A has more than one site for the binding of B, different conditions may apply. The binding is said to be *cooperative* if the occupation of one site favors the occupation of another. The binding is *anticooperative* if, on the contrary, the occupation of a site causes a decrease in the binding affinity of another. These types of complex behavior can be identified and analyzed in binding experiments. In the case in which the sites are *identical and independent*, the dissociation constants for each site are identical and do not change as a function of whether or not other sites are occupied. If molecule A has n identical and independent binding sites for a *ligand* B, the *average number*, $\bar{n}$, of ligands actually bound per molecule A is given by

$$\bar{n} = \frac{n[B]}{K_d + [B]} \qquad (C1.21)$$

where [B] is the free ligand equilibrium concentration. *Scatchard* analysis recasts this equation as

$$\frac{\bar{n}}{[B]} = \frac{n}{K_d} - \frac{\bar{n}}{K_d} \qquad (C1.22)$$

Both the average number bound, $\bar{n}$, and the free ligand concentration, [B], are obtained from a binding experiment at equilibrium. Plotting $\bar{n}/[B]$ versus $\bar{n}$ then yields a straight line of slope $-1/K_d$ and the intercept on the x-axis is n, the number of sites. A Scatchard plot is shown in Chapter C4.

Binding Free Energy

Knowledge of K_d enables the calculation of the energetics involved in the interaction between a ligand and its receptor. The difference in Gibbs free energy between bound and unbound states can be attributed to various chemical and physical factors. We recall Eq. (C1.10) ($\Delta G = H - T\Delta S$) and that the reaction is spontaneous in the direction of decreasing free energy, i.e., $\Delta G < 0$. The enthalpy term, ΔH, reflects internal energy changes associated with the inter-molecular bonding interactions. The entropy term, ΔS, provides a description of the degrees of freedom of the components involved in the binding reaction. Since whether or not the interaction takes place depends on ΔG and not on the other terms separately, large negative (unfavorable) entropy changes can occur on binding if there are compensating enthalpy differences. These may reflect the formation of favorable atomic interactions, like salt bridges and hydrogen bonds.

Calculating Standard Enthalpy and Entropy Changes Upon Binding

We recall Eq. (C1.15) for the standard free energy of the reaction:

$$\Delta G^\circ = -RT \ln K_{eq}$$

in terms of $k_{d,}$ is

$$\Delta G^\circ = RT \ln K_d \qquad (C1.23)$$

Using Eq. (C1.10) and rearranging, we obtain

$$\ln K_d = -\frac{\Delta H^\circ}{RT} + \frac{\Delta S^\circ}{R} \qquad (C1.24)$$

Measuring affinities as a function of temperature we obtain

$$\frac{\partial}{\partial T}(\ln K_d) = -\frac{\partial}{\partial T}\left(\frac{\Delta H^\circ}{RT}\right) + \frac{\partial}{\partial T}\left(\frac{\Delta S^\circ}{R}\right) \qquad (C1.25)$$

Since ΔH° and ΔS° are temperature-independent, Eq. (C1.25) reduces to

$$\frac{\partial}{\partial T}(\ln K_d) = \frac{\Delta H^\circ}{R} \times \frac{1}{T^2} \qquad (C1.26)$$

which can be rewritten as

$$\frac{\partial(\ln K_d)}{\partial(1/T)} = -\frac{\Delta H^\circ}{R} \qquad (C1.26a)$$

A plot of $\ln K_d$ against $1/T$ (an Arrhenius plot) therefore gives a straight line with a slope proportional to the change in standard enthalpy upon binding. The associated change in standard entropy can then be determined from Eqs. (C1.23) and (C1.10).

Avidity

When the binding event is due to a network of smaller events occurring at the interface between the molecules, the overall binding strength is called the *avidity* and, analogously to affinity, is measured through the overall equilibrium constant. The avidity then results from a set of affinities for the different sites involved. The avidity, how-ever, is larger than the sum of affinities (Comment C1.7).

An important consequence of the relationship between the standard ΔG and avidity is that a small change in one of the sites in the network can lead to a very large change in the overall equilibrium constant.

COMMENT C1.7 AVIDITY

The standard free energy of binding of a network of weak independent sites is given by

$$\Delta G = \sum_{n=1}^{k} \Delta G_n = \Delta G_1 + \Delta G_2 + \Delta G_3 + \dots$$

and

$$\Delta G = \sum_{n=1}^{k} (-RT \ln K_{d,n})$$
$$= -RT \ln K_{d,1} - RT \ln K_{d,2} - RT \ln K_{d,3} - \dots$$

where $K_{d,j}$ is the dissociation equilibrium constant of binding event j.

The overall K_d is called the avidity of the reaction. It is given by the product (and not the sum) of the individual affinities:

$$K_d = K_{d,1} \times K_{d,2} \times K_{d,3} \times \dots$$

C1.3.3 Calorimetry and Binding

The complete set of thermodynamics equations describing a system, in a given temperature range, can be calculated from the functional dependence of enthalpy on tempera-ture, which can be obtained, in principle, from careful calorimetric measurements.

Calorimetry measures heat exchange during tempera-ture-induced changes in a system. Temperature (T) and enthalpy (H) are conjugate extensive and intensive vari-ables (Section C1.2.1). When a change in the state of a system (e.g., protein unfolding or ligand binding) results from a change in temperature, there is a corresponding change in enthalpy. Heat is absorbed ($\Delta H > 0$) if the change results from an increase in temperature or heat is released ($\Delta H < 0$) if the change results from a decrease in temperature.

Heat Capacity

The quantity measured in a calorimetry experiment on a macromolecular solution is the heat capacity at constant pressure, C_p:

$$C_p = (\partial H/\partial T)_P \qquad (C1.27)$$

The corresponding enthalpy and entropy functions, $H(T)$ and $S(T)$, are obtained by integrating C_p in the temperature range:

$$\Delta H = H(T_0) + \int_{T_0}^{T} \Delta C_p(T) dT \qquad (C1.28)$$

$$\Delta S = S(T_0) + \int_{T_0}^{T} (\Delta C_p(T)/T) dT \qquad (C1.28a)$$

We recall from Section C1.3.1 that the reference temperature, T_0, is equal to absolute zero in physics, so that under constant pressure the enthalpy of a system at temperature T represents the total energy required to heat the system from absolute zero to T. In biophysics, however, it is usual to take the reference temperature T_0 to be equal to room temperature (298.15K).

In the case of a reaction, the H and S terms in the equations above are replaced by ΔH and ΔS, corresponding to the respective *changes* during the reaction. The heat capacity change during a reaction at temperature T is given by

$$\Delta C_T = \Delta C_{p,T_0} + \int_{T_0}^{T} (d\Delta C_{P,T}/dT)dT \qquad (C1.29)$$

where the subscripts define temperature and pressure conditions.

The Effects of Temperature on Binding

Subsuming the equations above and Eq. (C1.15), which relates the dissociation constant to standard free energy change, the standard free energy change can be written

$$\Delta G_T^{\circ} = \Delta H_{T_0}^{\circ} + T\Delta S_{T_0}^{\circ} + T\Delta C_{T_0}^{\circ}\left[1 - \frac{T_0}{T} - \ln\left(\frac{T_0}{T}\right)\right]$$
$$+ \frac{1}{2}TT_0\left(\frac{d\Delta C_{T_0}^{\circ}}{dT}\right)\left[\frac{T_0}{T} - \frac{T}{T_0} + 2\ln\left(\frac{T}{T_0}\right)\right]$$
$$(C1.30)$$

If the heat capacity is the constant in the temperature range, the equation simplifies to

$$\Delta G_T^{\circ} = \Delta H_{T_0}^{\circ} + T\Delta S_{T_0}^{\circ} + T\Delta C_{T_0}^{\circ}\left[1 - \frac{T_0}{T} - \ln\left(\frac{T_0}{T}\right)\right] \qquad (C1.31)$$

which is a form of the *van't Hoff equation*. $\Delta H_{T_0}^{\circ}$, $\Delta S_{T_0}^{\circ}$ and $\Delta C_{T_0}^{\circ}$ are estimated from measurements of a non-linear fit of ΔG_T° measured values at different temperatures.

C1.3.4 Activation Thermodynamics

The thermodynamic path of a reaction or conformational change often includes an energy barrier that must be overcome. Whereas the reaction or change will occur if the equilibrium ΔG is favorable ($\Delta G < 0$), the rate at which it occurs depends on the height of the intervening energy barrier (Fig. C1.4). The time constants of a reaction or change observed at different temperatures provide a measure of the thermodynamics parameters involved.

The temperature dependence of the rate constant, k, is given by the Arrhenius equation

$$k = A\exp\left(\frac{-E^*}{RT}\right) \qquad (C1.32)$$

where A is a constant. In *statistical thermodynamics*, A includes the probability of the change taking place (e.g., the probability of collision between two interacting partners), and E^* is activation energy. In *transition state*

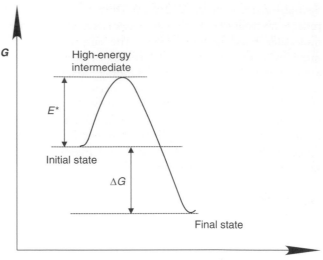

Fig. C1.4 Free energy profile of a reaction or conformational change. E^* is the activation energy or activation enthalpy depending on the theory used (see text).

theory, A depends on the *activation entropy* and E^* is the *activation enthalpy*. A is given in the quantum mechanical *Eyring transition state model* as

$$A = \frac{k_B T}{h}\exp\left(\frac{\Delta S^*}{R}\right) \qquad (C1.33)$$

where k_B is the Boltzmann constant, h is Planck's constant, and ΔS^* is activation entropy. Combining Eqs. (C1.32) and (C1.33), and putting $E^* = \Delta H^*$, the activation enthalpy, yields

$$\ln\left(\frac{k}{T}\right) = \ln\left(\frac{k_B}{h}\right) + \frac{\Delta S^*}{R} - \frac{\Delta H^*}{RT} \qquad (C1.34)$$

Activation enthalpy and activation entropy can be obtained directly from the slope and intercept of the line $\ln(k/T)$ versus $1/T$ (called an *Eyring plot*).

C1.4 CHECKLIST OF KEY IDEAS

- Classical thermodynamics is a phenomenological science, concerned with precise observations on defined systems.
- Macromolecular solutions can be considered as thermodynamic systems.
- An *adiabatic* system is perfectly isolated from external effects; an *isothermal* system is in perfect thermal contact with its surroundings.
- Two systems are in identical *states* when all the values measured for their properties are identical.
- A system enclosed in an adiabatic vessel will, in time, reach a state in which its properties no longer evolve; it is then said to be in *equilibrium*.
- The zeroth law of thermodynamics states that if two bodies, A and B, are separately in equilibrium with a third body, C, then each of A and B is in equilibrium with the other.

- The first law of thermodynamics states that the amount of work needed to change the state of a system from an initial to a final state depends solely on the change accomplished, and not on the means by which the work is performed, nor on the intermediate states by which the system may pass.
- The second law of thermodynamics states that it is impossible to devise an engine which, working in a cycle, produces no effect other than the transfer of heat from a colder to a hotter body.
- The *entropy* of a system is a function of state, which increases with time until it reaches a maximum at equilibrium.
- The entropy is a direct function of the number of different configurations (disorder) available to the system.
- The heat energy invested by a system in order to sample all the configurations available to it at a given temperature is equal to its entropy multiplied by the temperature.
- The third law of thermodynamics states that the entropy of any body is zero at the zero of the absolute temperature scale.
- The third law, by setting the zero point for entropy and an absolute temperature scale (the kelvin scale), allows the calculation of *equilibrium constants* from the thermal properties of reactants, and of *total internal energy* in *calorimetry*.
- The *enthalpy* of a system at temperature T represents the total energy required to heat the system from absolute zero to T.
- The change in enthalpy during a chemical reaction is the heat absorbed or released in the breaking and formation of bonds.
- The *Gibbs free energy* represents the part of the energy in a system that is not used for the population of the different entropy configurations.
- A system that exchanges energy with its surroundings evolves in order to minimize its free energy.
- When two states of a system are in equilibrium, their free energy difference is zero.
- All the thermodynamic variables of a system can be written in terms of its partition function, which is defined as the sum of the statistical weights of all the states accessible to the system.
- The *chemical potential* of a solute is the *free* energy required to add one mole of molecules to the solution.
- The *biological standard state* is defined at a temperature of 298.15 K (25 °C), 1 atm partial pressure for gases, 1 M concentration for solutes in aqueous solution and pH 7.

- The equilibrium constant of a chemical reaction is equal to the mathematical product of reaction product concentrations divided by the mathematical product of reactant concentrations, when the free energies of products and reactants are equal.
- The van't Hoff equation relates the enthalpy change to the equilibrium constant of a chemical reaction.
- The equilibrium constant of an association reaction is a measure of the binding strength or *affinity* of the interaction.
- Binding is said to be *cooperative* if the occupation of one site favors the occupation of another; binding is *anticooperative* if, on the contrary, the occupation of a site causes a decrease in the binding affinity of another; sites are *identical and independent* if the dissociation constants for each are identical and do not change as a function of whether or not other sites are occupied.
- A Scatchard plot relates the number of binding sites to the dissociation constant and free ligand concentration.
- Equilibrium thermodynamic parameters can be obtained from the Arrhenius plot relating dissociation constants to temperature.
- For a binding event due to a network of smaller events occurring at the interface between the molecules, the overall binding strength is called the *avidity* and, analogously to affinity, is measured through the overall equilibrium constant.
- A calorimetry experiment on a macromolecular solution measures the heat capacity at constant pressure.
- The corresponding enthalpy and entropy functions in calorimetry are obtained by integrating the heat capacity and the heat capacity divided by the temperature, respectively, in the temperature range of the experiment.
- The rate of a reaction or conformational change depends upon the height of the activation energy barrier in the path between the higher and lower free energy states; activation thermodynamics parameters can be obtained from the Eyring plot relating rate constants to temperature.

Suggestions for Further Reading

Feynman, R. P., Leighton, R. B., and Sands, M. (1963). *The Feynman Lectures on Physic*, **vol. 1**. Reading, MA: Addison Wesley.

Price, N. C., Dwek, R. A., Ratcliffe, R. G., and Wormald, M. R. (2001). *Principles and Problems in Physical Chemistry for Biochemists*. 3rd edn. Oxford: Oxford University Press.

DIFFERENTIAL SCANNING CALORIMETRY

C2.1 HISTORICAL OVERVIEW

The application of differential scanning calorimetry (DSC) to protein studies dates from the **mid-1960s**, when the first instruments with sufficient sensitivity were developed independently by **P. L. Privalov**, **S. J. Gill**, and **J. M. Sturtevant**. Effort was put into this development because a number of problems in biophysics had arisen that could not be solved without measuring small heat effects in relatively dilute samples. In the following decades, experiments allowing the measurement of thermodynamics variables with progressively greater precision became possible with the development of ever more sensitive calorimeters. Microcalorimeters and nanocalorimeters have now become available (the micro and nano prefixes referring to the precision with which heat can be measured in these devices).

C2.2 BASIC THEORY

A full quantitative thermodynamics description of a macromolecule in solution for a given temperature range may be obtained from the DSC measurement of its partial specific heat capacity at constant pressure, C_p, as a function of temperature. We recall from Chapter C1

$$\Delta H = \int \Delta C_p(T) \mathrm{d}T \tag{C2.1}$$

$$\Delta S = \int \frac{\Delta C_p(T)}{T} \mathrm{d}T \tag{C2.1a}$$

where ΔC_p is the heat capacity change due to a given macromolecular process (e.g., protein unfolding) and ΔH and ΔS are the associated enthalpy and entropy changes, respectively.

C2.3 EXPERIMENTAL CONSIDERATIONS

C2.3.1 Instrument Specifications

The general specifications of a DSC instrument for studies of biological macromolecules can be summarized in three points:

(1) very high sensitivity and precision;
(2) an operational range from negative to positive temperatures;
(3) the possibility of scanning down as well as up in temperature.

Modern instruments of different makes are available commercially. Typical specifications include macromolecule requirements of less than $50\,\mu g$ per test, a temperature range from $-10\,°C$ to $130\,°C$, $\sim 10\,\mu J\,deg^{-1}$ sensitivity, and, of course, powerful and user-friendly computer control to pilot the experiments and for online analysis.

The reversibility of processes should be checked by scanning down as well as up in temperature, with precautions taken to avoid irreversibility due to secondary effects such as aggregation or chemical degradation at high temperatures.

Calorimetric measurements down to $-10\,°C$ (in supercooled solutions) are not trivial but they are of special interest because they allow certain processes, such as cold denaturation, the folding of an α-helix or of a single-stranded oligonucleotide, to be followed to completion. Measurements below $-10\,°C$ are not possible in practice because of the high probability of spontaneous freezing in the aqueous solutions at these temperatures.

Measurements above $100\,°C$ are equally difficult to perform; they require pressure cells and special precautions in sample preparation and experimental design. The heating rate could be increased, for example, to minimize exposure time to high temperatures. High-temperature measurements, up to about $130\,°C$ (above this temperature the chemical degradation of biological molecules is too fast), are important in order to study the proteins and complexes of hyperthermophile organisms, which are stable and active beyond $100\,°C$ (see Chapter A1).

C2.3.2 Sensitivity of Heat Capacity Measurements

The variable, $C_p(T)_{obs}$, measured in DSC is *the heat capacity difference at constant pressure as a function of temperature between a macromolecular solution and the same volume of solvent alone*. The heat capacity of the solvent being greater than that of an equal volume of solution, the observed heat capacity difference is negative. It is related to the partial specific heat capacity of the macromolecule by

$$C_p(T)_{obs} = C_p(T)_{macrom.} m(T)_{macrom.} - C_p(T)_{solv.} \Delta m(T)_{solv.}$$
(C2.2)

where $C_p(T)_{macrom.}$ is the *partial specific heat capacity* (i.e., the heat capacity per mass of substance) of the macromolecular solute at temperature T, $m(T)_{macrom.}$ is the mass of macromolecule in the measuring cell at temperature T, $C_p(T)_{solv.}$ is the partial heat capacity of the solvent at temperature T, and $\Delta m(T)_{solv.}$ is the mass of solvent displaced by the presence of mass m of macromolecules. The mass of displaced solvent can be calculated from the partial specific volumes of the macromolecule and solvent, $\bar{v}(T)_{macrom.}$, $\bar{v}(T)_{solv.}$, respectively:

$$\Delta m(T)_{solv.} = m(T)\bar{v}(T)_{macrom.}/\bar{v}(T)_{solv.}$$
(C2.3)

By combining Eqs. (C2.2) and (C2.3), and putting in typical values for the parameters, we see that the observed difference in heat capacity is very small compared with the heat capacity measured for the solution (Comment C2.1).

C2.3.3 Sample Requirements

Macromolecular sample solutions for calorimetry experiments should be well-characterized, very pure, and homogeneous. It has been estimated that no more than 3% of contaminants can be tolerated for quantitative studies. Similarly, the absolute concentration should be known with a precision of a few percent. The solvent should be

chosen such that the macromolecules remain soluble in the temperature range of the study, the macromolecular states to be studied are attainable, and transitions between these states are reversible. The solution should be sufficiently dilute to reduce the risk of aggregation.

In the decades when the fundamental thermodynamics of protein folding was being established, DSC sample cells had volumes of the order of 1 ml; sample macromolecular concentrations were of the order of 0.1–1 mg ml^{-1}, depending on the sensitivity required. Since then, instruments have attained greatly improved characteristics by integrating hardware and software innovations, and measurements on samples of as little as 50 μg of protein have become possible.

C2.4 THE HEAT CAPACITY OF PROTEINS

C2.4.1 The Heat Capacity Versus Temperature Curve

A molar heat capacity as a function of temperature curve, obtained for a protein in a DSC experiment, is illustrated in Fig. C2.1. The native state is assigned as the reference relative to which the thermodynamic parameters of the system are defined (see Chapter C1), so that the excess heat capacity, ΔC_p, becomes the significant parameter measured from the curve in Fig. C2.1. Contributions to ΔC_p can be split into two terms

$$\Delta C_p(T) = \Delta C_p(T)_{tr.} + \Delta C_p(T)_{bl.}$$
(C2.4)

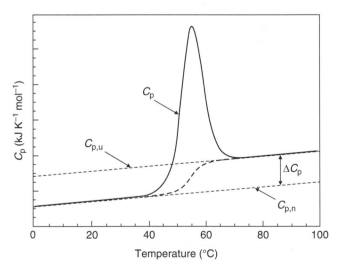

Fig. C2.1 Molar heat capacity, C_p, of a protein observed as a function of temperature by DSC $C_{p,u}$ and $C_{p,n}$ are the molar heat capacities of the unfolded and native states, respectively. ΔC_p is the heat capacity change associated with the transition. When there are many states above the main transition temperature, the excess heat capacity is expressed as an average over all these states, $\langle \Delta C_p \rangle$.

The first term refers to the transition peaks in the scan, while the second describes the baseline shift associated with the transition (the sigmoid shift between the two dashed lines in Fig. C2.1).

C2.4.2 Partition Function Analysis of the Heat Capacity Curve

In the analytical approach described in Chapter C1, the value of a thermodynamic variable, $\langle a \rangle$, is expressed as a sum of contributions from the different states of the system, which are themselves expressed in terms of a partition function (Eqs. (C1.12f), (C1.12b))

$$\langle a \rangle = \sum_{j=0}^{N} P_j a_j$$
$$P_j = \exp\left(-\Delta G_j/RT\right)/Q$$

where Q is the partition function and P_j and a_j are, respectively, the values of the relative population and thermodynamic variable associated with state j. If state $j=0$ is taken as the reference native state, then the sum is from $j=1,\ldots,N$; and $\langle a \rangle = \langle \Delta C_p \rangle$ represents the excess heat capacity associated with the population of states other than the native state. Note that in a rigorous analysis the $\langle \rangle$ brackets should have been included for the terms in Fig. C2.1 and Eq. (C2.4), since in general there could be an ensemble of denatured states for the protein.

The excess heat capacity is given by the gradient of the excess enthalpy versus temperature curve (the inverse of Eq. (C2.1a)):

$$\langle \Delta C_p \rangle = (\partial \langle \Delta H \rangle / \partial T)_p \tag{C2.5}$$

Splitting the right-hand side of Eq. (C2.5), in a similar way to Eq. (C2.4) and expressing the terms as sums over the different states, we obtain

$$\langle \Delta C_p \rangle = \sum_{j=1}^{N} \Delta H_j (\partial P_j/\partial T)_P + \sum_{j=1}^{N} P_j \Delta C_{pj} \tag{C2.6}$$

The first term (by analogy with the transition term, subscript tr, in Eq. (C2.4) is the heat capacity associated with the transitions from the native state $j=0$ to the states, $j=1,\ldots,N$. In a modeling approach (see below), this term may correspond to a calculated value, which would be compared with the experimentally measured $C_{p,tr.}$. The second term (by analogy with the baseline term, subscript bl, in Eq. (C2.4)) is the sum of the heat capacities of the states other than the native state, each weighted by its population, P_j.

The partition function analysis of the curve is quite general and applies to any monomeric (so that the number of moles present in the solution does not change upon unfolding) system at equilibrium, independently of transition mechanisms and of the number of intermediate states involved.

C2.4.3 Two-State Transition: Calorimetric and Van't Hoff Enthalpies are Equal

The first DSC studies of small single-domain proteins established that their temperature unfolding could be modeled adequately as a cooperative two-state transition. The native structure unfolds very quickly so that only native and unfolded states coexist in the temperature range of the transition (see Fig. C1.5.1). Any transient intermediates involved would be too short-lived to be observed through their contribution to the heat capacity curve.

The modeling approach is based on a comparison of the calorimetric transition enthalpy, integrated from the DSC curve, with a value calculated using the van't Hoff equation (Chapter C1), assuming a two-state transition. If the two values are found to be the same, then the transition can be considered to be between two states with negligible intermediates.

The experimental calorimetric transition enthalpy is given by

$$\Delta H_{tr.,\,exp.} = \int \Delta C_{p,tr.,\,exp.}(T)\,dT \tag{C2.7}$$

where $\Delta C_{p,tr,exp}(T)$ (Eq. (C2.4) and Fig. C2.1) is the value in the transition peak above the $\Delta C_{p,bl,exp}(T)$ experimental baseline value.

The van't Hoff equation relates enthalpy and the temperature-dependent equilibrium constant between the states (Comment C2.2):

$$\Delta H_{v'tH} = RT^2 \partial \ln K_{eq}/\partial T \tag{C2.8}$$

which is the slope of $\ln K$ versus $1/RT$. We recall from Chapter C1 that for a two-state transition

$$A \rightleftharpoons B$$

the equilibrium constant at temperature T is given by

$$K(T) = \frac{[B(T)]}{[A(T)]} \tag{C2.9}$$

COMMENT C2.2 EQUILIBRIUM CONSTANTS AND THERMODYNAMICS VARIABLES

We recall from Section C1.3.1 that

$$\Delta G = -RT \ln K$$

Combining with the fundamental relation

$$\Delta G = \Delta H - T\Delta S$$

we obtain

$$\ln K = -\Delta H/RT + \Delta S/R$$

from which we derive the van't Hoff equation

$$\Delta H_{v'tH} = RT^2 \partial \ln K/\partial T$$

(ΔH is the slope of $\ln K$ versus $1/RT$).

where [B(T)], [A(T)] are equilibrium concentrations. The equilibrium constant, $K(T)$, can therefore be measured from whichever experiment will give the ratio [B(T)]/[A(T)], as a function of temperature. For example, $K(T)$ can be calculated from circular dichroïsm spectroscopy, if the signals from states A and B are significantly different as is the case for the unfolding transition in proteins (see Chapter E4).

It is interesting to note that within the two-state assumption the population ratio between the states can be obtained from the shape of the DSC curve itself (Comment C2.3).

COMMENT C2.3 VAN'T HOFF ENTHALPY FROM THE DSC CURVE

Figure C2.3.1 represents a schematic heat capacity curve after baseline subtraction of the pure native and unfolded state contributions. The heat capacity represents an equilibrium measurement, i.e., the different states contributing to each point on the curve are at equilibrium with each other. The population ratio of states N and U, P_N/P_U, can be calculated for each temperature point on the curve. At the left-hand baseline level P_N and P_U are 1 and 0, respectively; at the right-hand baseline level they are 0 and 1, respectively. At the peak, B, of the curve, $P_N = P_U = 0.5$. At point A, for example, the ratio can be calculated from the ratio of the areas $a/2b$, where a is the area under the curve up to the blue dashed line, and $2b$ is the total area. By applying these geometrical considerations to the curve, it was shown that the van't Hoff enthalpy is given by

$$\Delta H_{v'tH} = 4RT^2 \frac{\Delta C_p(T_{tr})}{\Delta H_{tr}}$$

where $\Delta C_p(T_{tr})$ is the heat capacity peak value at the transition temperature T_{tr} and ΔH_{tr} is the total heat absorbed in the transition peak.

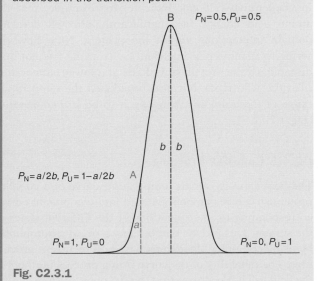

Fig. C2.3.1

If the calorimetric and van't Hoff transition enthalpies calculated from Eqs. (C2.7) and (C2.8), respectively, are equal, then the transition can be considered to be cooperative between two states. In practice, this means that any intermediate state would be too short-lived to be detected.

C2.4.4 Calorimetric and Van't Hoff Enthalpies are Not Equal: Cooperative Domains

Useful information on the transition is also obtained when the van't Hoff and calorimetric enthalpies are not equal. The calorimetric transition enthalpy measured by DSC represents the actual heat change associated with the transition per mole of the system. The validity of the van't Hoff analysis, on the other hand, depends on the calculated populations of states corresponding in fact to the true populations. The van't Hoff enthalpy reflects the heat associated with the transformation of one mole of cooperative unit within the system.

When the cooperative domain extends over the whole molecule, the van't Hoff and calorimetric enthalpies are equal. When the cooperative domain is smaller than the molecule, as would be the case for a multidomain protein, the van't Hoff enthalpy is smaller than the calorimetric enthalpy.

A van't Hoff enthalpy that is larger than the calorimetric enthalpy is an indication that the cooperative domain is larger than the molecule, i.e., it provides evidence of intermolecular interactions such as oligomerization.

In all cases where the van't Hoff and calorimetric enthalpies are not equal, the transition proceeds via intermediate states (Comment C2.4).

C2.4.5 Folding Intermediates and Effects of Mutations

Multidomain Proteins

The partition function equations developed in Section C2.4.2 have been used as a basis to calculate computer-simulated heat capacity functions for transitions that involve more than two states.

COMMENT C2.4 CALORIMETRIC AND VAN'T HOFF ENTHALPIES FOR MONODOMAIN, TWO-DOMAIN MACROMOLECULES, AND DIMERS

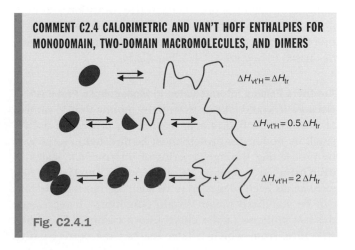

Fig. C2.4.1

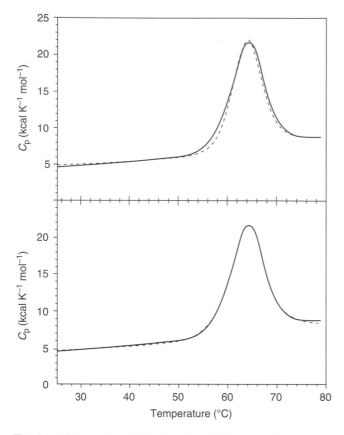

Fig. C2.2 Experimental and simulated DSC scans of hen egg white lysozyme. The solid curve is the experimental scan at pH 2.5, in both panels. The dashed lines are computer-simulated curves assuming a two-state transition (top) and a three-state transition (bottom). (Privalov *et al.*, 1995.) (Figure reproduced with permission from Elsevier.)

Fitting simulated functions to experimental curves provides an efficient approach to the thermodynamics analysis of DSC data. The precision of DSC nanocalorimeters is such that very small deviations between simulated and experimental curves can be observed and interpreted (Fig. C2.2). Hen egg white lysozyme is a two-domain protein and the data in Fig. C2.2 establish that its temperature-induced unfolding is not a cooperative transition over both domains. The independent unfolding of the two domains is more pronounced in the related protein equine lysozyme, discussed in Section C2.4.7 (Fig. C2.5).

Are there Folding Intermediates in Monodomain Proteins?

Because it would take an essentially infinite time to sample all possible configurations, the folding process of even a small monodomain protein must be directed in some way by proceeding through a series of intermediate states of decreasing free energy (Comment C2.5). The *molten globule* has been suggested to be such an intermediate that can be stabilized under certain conditions. It represents the fast collapse of the chain into a compact liquid-like structure, from which it later reaches the correct native

structure (see Chapter E4). The molten globule state would solve the Levinthal paradox by greatly reducing the configurational space search. The properties of the molten globule and its very existence, however, are controversial. Various authors have defined it in different ways, especially with respect to its dynamics and degree of structure. P. L. Privalov reviewed calorimetric data in the search for protein folding intermediates. He concluded that observed intermediate states were either misfolded forms obtained under conditions inappropriate for folding or partially unfolded forms that retained a correctly folded subpart of the molecule. These subparts have all the characteristics of cooperative folding domains. Small monodomain proteins were found to fold very rapidly, with any transient intermediates being too unstable to be observed.

Mutant Stability

Changes in protein structure resulting from the mutation of as few as one or two amino acid residues may result in significantly different experimental DSC curves (Fig. C2.3). It is tempting to assign a change in protein stability due to a mutation to enthalpic or entropic contributions of the amino acid replacement itself. Such an assignment should be done with great care, however, because the stability change may very well be due to a change in cooperativity of the unfolding process. Unlike other measures of protein stability, the DSC curve is extremely sensitive to changes in cooperativity. It can indicate, for example, that the amino acid change has significantly affected domain–domain interactions within the protein. Note how the serine to lysine replacement in T_4 lysozyme has not only destabilized the protein (it unfolds at a lower temperature than the wild type), but also broadened the temperature range of the unfolding, showing it to be less cooperative (Fig. C2.3).

C2.4.6 Complex Proteins

The heat capacity versus temperature curve of a complex protein structure can be separated into component curves corresponding to the transitions of the different states of the system. This is done without any a-priori assumptions, by using computer algorithms based on recurrent procedures. The reliability of the curve fitting process (authors in the field call it "deconvolution" but it is not a deconvolution in the strict sense) depends mainly on the precision of the experimental heat capacity curve, and, therefore, on the microcalorimetry instrumentation. Ideally, each component curve corresponds to a two-state transition, for which it

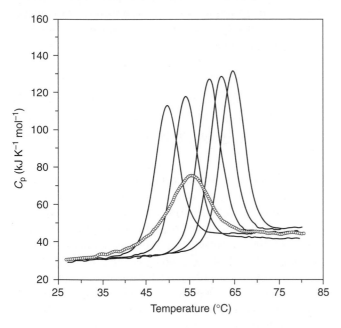

Fig. C2.3 DSC scans of T_4 lysozyme. The solid lines are the data for pH 2.8, 3.0, 3.3, 3.5, 3.7 (left to right). The circles are the data points for the S44A mutant of the protein at pH 4.0. (Carra *et al.*, 1996.) (Figure reproduced with permission from *Biophysical Journal*.)

defines a transition temperature and transition enthalpy. Component curves may be assigned to structural units in the complex from model compound information, or by studying fragments, for example. Observing whether or not component transition temperatures and enthalpies vary independently with conditions such as pH provides information on the relationships between the structural units. The method has been applied successfully to multidomain proteins, protein–protein and protein–nucleic acid complexes, and fibrous protein structures such as the muscle proteins and collagens. The calorimetric analysis of the melting process of the coil–coil structure of the myosin rod is shown as an example in Fig. C2.4. The complexity of the melting profile increases with increasing fragment size. All the profiles could be fitted by different combinations of the same six component peaks, each corresponding to a two-state transition. Each component was then assigned to a discrete cooperative unit in the structure by estimating the approximate size of the cooperative block from the transition enthalpy (assuming a constant specific melting enthalpy along the rod). The analysis yielded very interesting results that could not be obtained by other methods. The least stable cooperative block (component 1) was close to the middle of the rod, precluding the rod melting by an "unzippering" process. It is also noteworthy that the block corresponding to component 4 was apart from the others and located between the easily proteolized flexible regions (hatched in Fig. C2.4), which do not take part in the melting process.

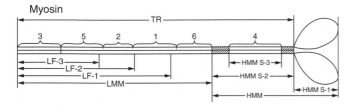

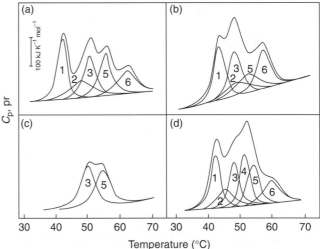

Fig. C2.4 Calorimetric analysis of myosin. Left-hand panel: Schematic view of myosin. The hatched parts are susceptible to complete proteolysis. LMM and HMM (light and heavy meromyosin, respectively) are the main fragments obtained by trypsin or pepsin treatment. The HMM fragment is further split by papain as shown. LMM is further split by trypsin into the fragments shown. The numbered curly brackets denote the cooperative blocks revealed by the calorimetric analysis shown in the right-hand panel. Right-hand panel: The partial heat capacity functions of myosin fragments in 0.5 M KCl, 25 mM potassium phosphate buffer, pH 6.5: **(a)** trypsin fragment LMM; **(b)** pepsin fragment LMM; **(c)** LF-3; **(d)** TR, the total rod. The two-state transition fitted peaks have been numbered according to increasing stability and assigned to the structure as shown in the left-hand panel. (Privalov, 1982.) (Figure reproduced with permission from Elsevier.)

C2.4.7 Solvent Effects on the Transition and the Absolute Partial Heat Capacity Difference Between Folded and Unfolded States of a Macromolecule

The solvent plays an important role in the stabilization of both the folded and native states of a protein (see Chapter C1), and in defining the nature of the transition between them. Protein–solvent interactions may be specific, as in the case of ion binding to specific sites, or more general through, for example, pH, ionic, or small solute effects on water structure affecting the hydration and solubility of protein polar and apolar groups.

Ion-Binding Stabilization

Equine lysozyme is made up of two domains that unfold independently, giving two peaks in the DSC scan. One of the domains is stabilized significantly by calcium ion binding. The peak positions in temperature, their relative heights and widths all depend significantly on the $CaCl_2$ concentration in the solvent (Fig. C2.5).

Effects of pH

The unfolding temperature of barnase varies from below 25 °C at pH 1.8 to above 50 °C at pH 5.5 (Fig. C2.6). We recall that knowledge of $\langle \Delta C_p(T) \rangle$, the heat capacity

difference between the native and unfolded states, provides a full characterization of the thermodynamics of the transition. The pure native and unfolded states, however, are usually realized at different temperatures, so that $\langle \Delta C_p(T) \rangle$ can only be estimated by extrapolation or model calculation (see, for example, the dashed and dash–dot lines in Figs. C2.5 and C2.6). Solvent effects, if they do not affect the heat capacity of the protein, could be very useful in this context. The pH variation illustrated in Fig. C2.6, by shifting the unfolding transition, allows the measurement of $\langle \Delta C_p(T) \rangle$ values at the same temperature.

C2.4.8 Heat Capacity Calculations from Structural Data

The heat capacity of macromolecules is due mainly to internal vibrational and rotational modes and solvent interactions. Reliable free energy predictions are an essential component of approaches to protein and ligand design. It is not possible, however, to predict the heat capacity of a complex system from *ab initio* considerations. Nevertheless, methods to calculate heat capacities of proteins in their various states have been developed, based on DSC information and data from model compounds. In turn, these methods have greatly enriched the analysis of DSC data, through the interplay between simulated and experimental curves.

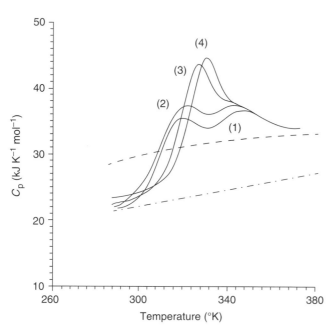

Fig. C2.5 Ion binding to equine lysozyme: DSC temperature scans of equine lysozyme at pH 4.5, in the presence of various concentrations of $CaCl_2$: curve (1) 0.00 mM; (2) 0.10 mM; (3) 0.75 mM; (4) 1.50 mM. The dashed line is the heat capacity function calculated for the unfolded peptide, and the dot–dash line is that of the native state obtained by linear extrapolation, using the slope specific for native globular proteins (Section C2.4.8). (Griko *et al.*, 1995) (Figure reproduced with permission from Elsevier.)

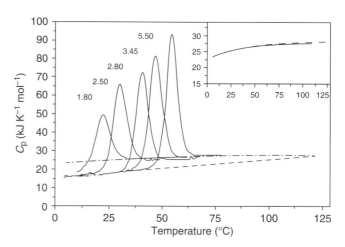

Fig. C2.6 Effects of pH: DSC temperature scans of barnase in solutions of different pH (indicated by each curve). The dashed and dot–dash lines, respectively, are the partial molar heat capacities of the native protein and unfolded proteins.
Inset: the solid line is the heat capacity versus temperature of irreversibly denatured barnase (obtained by heating the protein to 100 °C); the dashed line is the calculation for the heat capacity of the unfolded peptide obtained by summing the heat capacities of individual amino acid residues. (Privalov and Makhatadze, 1992.) (Figure reproduced with permission from Elsevier.)

Calculated Heat Capacities of Native and Unfolded Proteins

The heat capacity at constant pressure of an unfolded peptide can be calculated accurately and reliably by summing the heat capacities of the individual amino acids (e.g., see the dashed line in Fig. C2.5 and the inset in Fig. C2.6). We would expect a more complicated situation, however, with respect to the native state.

The heat capacity of a native or unfolded protein in aqueous solution can be considered as being made up of a set of contributions each referring either to the intrinsic chemical properties of the macromolecule or to solvent effects. Extensive work on model organic compounds has confirmed that these contributions to the heat capacity are additive. The heat capacity of a protein can be written

$$C_p = C_{p,a} + C_{p,b} + C_{p,c} + C_{p,other} \qquad (C2.10)$$

$C_{p,a}$ is called the primary heat capacity and contains the atomic and covalent bond contributions. It depends only on the amino acid composition of the macromolecule. $C_{p,b}$ contains the contributions from non-covalent bonds in the folded protein. $C_{p,c}$ contains the contributions due to hydration or protein–solvent interactions, in general. $C_{p,other}$ contains other contributions – for example, those due to the protonation state of groups with ionizable side-chains such as histidine, aspartic acid, glutamic acid, arginine, and lysine. It is usually small compared with the other terms in Eq. (C2.10), or can be corrected for quite easily, so we shall neglect it in the following discussion.

Polypeptide Chain Energies: The Primary Heat Capacity

The primary heat capacity, $C_{p,a}$, is the heat capacity of the anhydrous unfolded polypeptide. It can be calculated accurately by summing contributions from individual amino acids (Table C2.1) and from peptide bond formation (Comment C2.6).

Heat capacities of amino acid residues in solution are tabulated in Makhatadze and Privalov (1990), Privalov and Makhatadze (1990), and Makhatadze and Privalov (1996).

Internal Protein Interactions: The Heat Capacity of an Anhydrous Folded Protein

The heat capacity of an anhydrous folded protein is equal to $C_{p,a} + C_{p,b}$. When expressed in units per mass, the heat capacities of different soluble proteins turn out to be quite similar, with the following values: $1.17\,kJ\,K^{-1}\,g^{-1}$ for the heat capacity calculated from the primary structure; $1.21\,kJ\,K^{-1}\,g^{-1}$ for the experimental value for the anhydrous state at 25 °C; $1.46\,kJ\,K^{-1}\,g^{-1}$ for the folded state in solution; and $1.96\,kJ\,K^{-1}\,g^{-1}$ for the unfolded state in solution. Note also that the measured heat capacities of anhydrous proteins are very close to the calculated primary heat capacity values, showing that most of the absolute heat capacity originates in the covalent structure. In the case of the anhydrous protein, $C_{p,a}$ accounts for 97.5% of the experimental heat capacity (see also Comment C2.6).

TABLE C2.1 EXPERIMENTAL HEAT CAPACITIES OF AMINO ACIDS

Amino acid	$C_p{}^a$ (J K mol^{-1})
Ala	122.7
Arg	234.4
Asn	160.9
Asp	155.8
Cys	163.0
Gln	184.9
Glu	175.7
Gly	99.5
His	216.2
Ile	189.0
Le	201.7
Lys	205.4
Met	290.6
Phe	203.7
Ser	136.1
Thr	147.8
Trp	239.0
Tyr	217.1
Val	169.5

^{a} Values for anhydrous amino acids at 25 °C are from Hutchens (1970).

COMMENT C2.6 HEAT CAPACITY OF PEPTIDE BOND FORMATION

The average heat capacity contribution of peptide bond formation with the release of a water molecule was calculated from the difference between experimental heat capacities of dipeptides and the corresponding individual amino acids, all in the anhydrous state. It is equal to $-39.2 \pm 0.74\,J\,K^{-1}\,mol^{-1}$.

The heat capacity of anhydrous proteins has been shown to increase linearly with temperature, with a slope, $4.10 \times 10^{-3}\,J\,K^{-2}\,g^{-1}$, which is close to constant for all the proteins studied.

Heat Capacity of a Native Protein in Solution

According to Eq. (C2.10), the heat capacity of a native protein in solution is equal to $C_{p,a} + C_{p,b} + C_{p,c,native}$. It is larger than that of the anhydrous protein, revealing mainly the positive magnitude of the hydration term, $C_{p,c}$. At 25 °C, hydration accounts for about 15% of the total heat capacity of a native protein in solution. It is also responsible for the larger (linear) temperature dependence observed for that heat capacity ($6.80 \times 10^{-3}\,\mathrm{cal\,K^{-2}\,g^{-1}}$), compared to that of the anhydrous folded state ($4.1 \times 10^{-3}\,\mathrm{J\,K^{-2}\,g^{-1}}$).

The hydration term varies according to the protein. It reflects the role of the macromolecular surface composition in the solvent interactions, especially with respect to the accessible surface area of polar and apolar groups. The hydration contribution of each group has been established from model compound studies to be proportional to its solvent accessible area. The heat capacity of polar group hydration is negative, while that of apolar group hydration is positive. Qualitatively, therefore, the larger heat capacity of native proteins in solution compared to the anhydrous folded state is as expected from the fact that, on average, 55% of the surface of a soluble protein is made up of apolar groups.

The Heat Capacity of an Unfolded Peptide in Solution

The heat capacity of an unfolded peptide in solution is equal to $C_{p,a} + C_{p,c,\mathrm{unfolded}}$, where the hydration term $C_{p,c,\mathrm{unfolded}}$ is the sum of the appropriate values for the individual amino acids in the protein sequence, N- and C-terminal groups. The partial heat capacities of amino acid residues in solution have been measured accurately and are well known over the temperature range 5–125 °C (see Table C2.1 for references).

The heat capacity of the unfolded state in solution is about $0.4\,\mathrm{J\,K^{-1}\,g^{-1}}$ larger than that of the native state (Comment C2.7). It is dominated by the hydration terms arising from the higher solvent exposure of apolar groups, rather than the lack of non-covalent interactions within the chain, in the unfolded state. Unlike that of the native state in solution, however, the heat capacity of the unfolded state in solution does not vary linearly with temperature. In the range 0–100 °C, it is approximated well by a second-order temperature polynomial.

Global Fit of the Heat Capacity by a Single Mathematical Function

A single mathematical function, based on Eq. (C2.10), has been derived for the heat capacity as a function of temperature of proteins in their various states (Comment C2.8). We recall Eq. (C2.10), neglecting the last term,

$$C_p = C_{p,a} + C_{p,b} + C_{p,c} \tag{C2.11}$$

and express each of the heat capacities on the right-hand side in terms of a number of parameters (v, w, p, q, a, b) and protein structural properties (M_W, molecular weight; A_{buried}, the total buried area inaccessible to the solvent; A_{apolar} and A_{polar}, the apolar and polar solvent-accessible areas, respectively):

$$C_{p,a} = [v + w(T - 25)]M_W \tag{C2.12}$$
$$C_{p,b} = [p + q(T - 25)]A_{\mathrm{buried}} \tag{C2.13}$$
$$C_{p,c} = a(T)A_{\mathrm{apolar}} + b(T)_{\mathrm{Apolar}} \tag{C2.14}$$

where

$$a(T) = a_1 + a_2(T - 25) + a_3(T - 25)^2 \tag{C2.14a}$$
$$b(T) = b_1 + b_2(T - 25) + b_3(T - 25)^2 \tag{C2.14b}$$

The equations above were fitted to the entire protein thermodynamics database available. The best values obtained for the parameters are given in Table C2.2.

The resulting unique set of parameters gives remarkably good predictions, in most cases, not only of the measured heat capacity of a given protein state, but also of its temperature dependence – in particular, the almost linear and quadratic temperature curves of the native state and unfolded states in solution, respectively.

The heat capacity of proteins in different states is expressed as a single mathematical function in terms of parameters that could be fitted from experimental data (Eqs. (C2.12)–(C2.14)) (Gomez et al., 1995).

Although the average parameter values given in Table C2.2 appear to be good approximations in most cases, it should be pointed out that hydration heat capacities do

---

COMMENT C2.7 HEAT CAPACITY VALUES

Expressed in units per gram, the average experimental heat capacity of the anhydrous folded proteins is $1.22\,\mathrm{J\,K^{-1}\,g^{-1}}$; their average primary heat capacity is equal to $1.19\,\mathrm{J\,K^{-1}\,g^{-1}}$, so that the non-covalent interaction term is estimated to contribute only about $0.03\,\mathrm{J\,K^{-1}\,g^{-1}}$.

The heat capacities of the native and unfolded states in solution are larger than the values corresponding to the anhydrous states. Because of the positive contribution of apolar capacity, values vary from protein to protein according to surface composition (for the native states) and amino acid composition (for the unfolded states). The difference between the heat capacities of the unfolded and native states in solution ranges between 0.4 and $0.7\,\mathrm{J\,K^{-1}\,g^{-1}}$.

COMMENT C2.8 PROTEIN HEAT CAPACITY CALCULATIONS

Heat capacity calculations for macromolecules based on composition, hydration values, and structural parameters such as the solvent accessible surfaces of different groups have been developed essentially by the groups of Privalov and collaborators and Freire and collaborators (reviewed in Privalov and Privalov (2000) and Freire (1995)).

TABLE C2.2 PARAMETERS REPRESENTING THE BEST FIT TO THE HEAT CAPACITIES OF THE PROTEIN DATABASE

Parameter	Value (units)
V	$1.18\ (\mathrm{J\,K^{-1}\,g^{-1}})$
W	$41.0 \times 10^{-4}\ (\mathrm{J\,K^{-2}\,g^{-1}})$
P	$36.5 \times 10^{-3}\ (\mathrm{J\,K^{-1}\,mol^{-1}\,\text{Å}^{-2}})$
q	$27.0 \times 10^{-4}\ (\mathrm{J\,K^{-1}\,mol^{-1}\,\text{Å}^{-2}})$
a_1	$1.89\ (\mathrm{J\,K^{-1}\,mol^{-1}\,\text{Å}^{-2}})$
a_2	$11.0 \times 10^{-4}\ (\mathrm{J\,K^{-2}\,mol^{-1}\,\text{Å}^{-2}})$
a_3	$-17.6 \times 10^{-5}\ (\mathrm{J\,K^{-3}\,mol^{-1}\,\text{Å}^{-2}})$
b_1	$-1.11\ (\mathrm{J\,K^{-1}\,mol^{-1}\,\text{Å}^{-2}})$
b_2	$12.0 \times 10^{-4}\ (\mathrm{J\,K^{-2}\,mol^{-1}\,\text{Å}^{-2}})$
b_3	$18.1 \times 10^{-5}\ (\mathrm{J\,K^{-2}\,mol^{-1}\,\text{Å}^{-2}})$

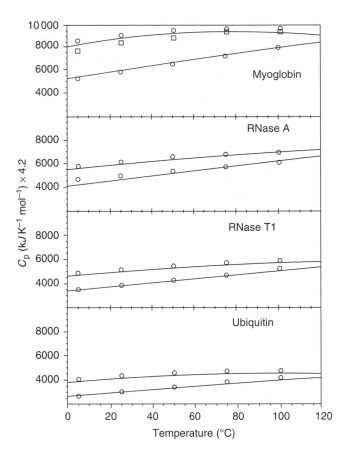

Fig. C2.7 Calculated (lines) and measured heat capacities (small circles) of the unfolded (top curve in each panel) and native states (lower curve) of different soluble proteins. The calculations were performed by using the parameters in Table C2.2. Data points for the native state above 50 °C were extrapolated. In the case of unfolded myoglobin the squares are calculated values taken from Privalov and Makhatadze (1990). (Freire, 1995) (Figure reproduced with permission from Elsevier.)

vary considerably according to the chemical group. A better approximation is obtained by dividing apolar groups further into aliphatic and aromatic groups; the latter, in fact, have a slightly polar character. The corresponding average parameters at 25 °C are $2.14\ \mathrm{J\,K^{-1}\,mol^{-1}\,\text{Å}^{-2}}$ for aliphatic groups and $1.55\ \mathrm{J\,K^{-1}\,mol^{-1}\,\text{Å}^{-2}}$ for aromatic groups, so that the parameter a_1 (given above as an average for aliphatic and aromatic groups) can itself be divided accordingly into two terms.

Heat Capacity Changes upon Protein Folding

The heat capacity difference between the unfolded and native states in solution is dominated by the difference in hydration heat capacity:

$$\Delta C_{\mathrm{p,unfolded-native}} = \Delta C_{\mathrm{p,b,unfolded-native}} + \Delta C_{\mathrm{p,c,unfolded-native}}$$
(C2.15)

where $\Delta C_{\mathrm{p,unfolded-native}}$ is the difference between the heat capacities of the unfolded and native states in solution.

The $\Delta C_{\mathrm{p,c}}$ term accounts for more than 90% of ΔC_{p} at 25 °C. On average, the contribution of apolar hydration decreases by half as the temperature increased from 25 °C to 100 °C, from $1.89\ \mathrm{J\,K^{-1}\,mol^{-1}\,\text{Å}^{-2}}$ at 25 °C to about $0.96\ \mathrm{J\,K^{-1}\,mol^{-1}\,\text{Å}^{-2}}$ at 100 °C, while that of polar hydration goes from $-1.11\ \mathrm{J\,K^{-1}\,mol\,\text{Å}^{-2}}$ at 25 °C to practically zero at 100 °C. The $\Delta C_{\mathrm{p,b}}$ term is negative because non-covalent interactions are disrupted in going from the native to the unfolded state. Its magnitude is smaller than the hydration term; it increases with temperature to an average value of $-0.25\ \mathrm{J\,K^{-1}\,(mol\ buried)^{-1}\,\text{Å}^{-2}}$ at 100 °C. The net effect of hydration and non-covalent interactions on the temperature dependence of the ΔC_{p} of unfolding is that its extrapolated value tends to vanish as the temperature approaches 120 °C (Fig. C2.7).

Reliability of the Calculations

Because of the importance of free energy predictions in protein and ligand design, accurate, structure-based, heat capacity calculations are extremely useful. Such calculations can be performed reliably for a completely unfolded protein in solution (so that all the amino acids are solvent accessible), based on its sequence and tabulated values for the individual amino acids. The heat capacity of a *native* protein in solution cannot be calculated from its sequence, because it depends on the macromolecular surface and dynamics of the folded structure. When the native structure is known, however, a parametrized approach, as described above, could be a very good approximation (Fig. C2.7), when the protein belongs to the same structural family (e.g., small monodomain) as the proteins in the database utilized to derive the parameters. In principle, however, such calculations should be treated with care, because the heat capacities of individual chemical groups may vary significantly from best-fit values obtained for all the proteins in a given database. The difference in the chain dynamics heat capacity is expected to be small between the native and unfolded states, so that calculations of the heat capacity *difference* upon unfolding are more reliable than calculations on the native state, because they are dominated by hydration terms (Comment C2.7). The hydration heat capacity contribution of a given group is proportional to its solvent accessible area. In the unfolded state, the group is assumed to be totally accessible to solvent, while its solvent accessible area in the native protein can be determined accurately from the structure coordinates.

C2.4.9 Protein Stabilization Forces

Protein stabilization forces arise from the competition between various protein–protein and protein–solvent interactions. They favor the folded structure in a given temperature interval. Calorimetric studies have made fundamental contributions to our understanding of these forces. As we saw above, the ΔC_p between the unfolded and folded states is accounted for primarily by hydration interactions. Figure C2.7 shows its temperature dependence for small monodomain proteins, and how it tends to very small values at the highest temperatures. What does a full thermodynamic analysis of the experimental heat capacity in terms of entropy and enthalpy terms yield about stabilization forces in protein folding?

Chain Folding in a Vacuum

A straightforward thermodynamic analysis of protein folding in a vacuum (or for conditions in which solvent effects can be neglected) is shown in Fig. C2.8. The chain is maintained in its folded configuration by internal bonds such as hydrogen bonds, van der Waals, or electrostatic interactions. Breaking these bonds leads to a temperature-independent increase in internal energy, so that, upon unfolding, ΔH is positive and constant with temperature. Now, considering entropy, the unfolded state clearly has many more degrees of freedom than the folded one, so that ΔS is also positive upon unfolding. The free energy change, $\Delta G = \Delta H - T\Delta S$, upon unfolding therefore decreases with temperature and crosses zero at T_m, as shown in Fig. C2.8b; the folded state is favored above T_m, and the unfolded state is favored below T_m, so that T_m can be seen as a chain melting temperature.

Solvent Effects in Protein Folding

It has been clearly established, however, from careful calorimetric measurements that the simple picture shown in Fig. C2.8 is not applicable to proteins. The ΔH, ΔS, and therefore ΔG upon unfolding calculated from the calorimetric heat capacity curve all vary with temperature in complex ways, establishing that solvent effects in protein folding are far from negligible.

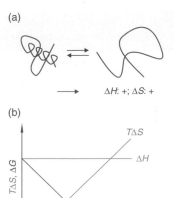

Fig. C2.8 Thermodynamics of protein folding in a vacuum. **(a)** schematic representation of the equilibrium; upon unfolding ΔH and ΔS are both positive. **(b)** The changes in enthalpy (ΔH), entropy (ΔS), and free energy (ΔG) upon unfolding. T_m is the melting temperature of the chain.

COMMENT C2.9 RELATING ΔH, ΔS, AND ΔC_p

We recall that the enthalpy and entropy terms are calculated from the heat capacity measurements according to Eq. (C2.1):

$$\Delta H = \int \Delta C_p(T)\, dT$$

$$\Delta S = \int \frac{\Delta C_p(T)}{T}\, dT$$

The converse of the equations is that ΔC_p is related to the slopes of the ΔH, and ΔS temperature dependence.

The general shapes of thermodynamic functions derived from calorimetric data for myoglobin (Fig. C2.9; Comment C2.9) are fairly typical of small globular proteins, although they may be shifted in temperature or magnitude according to the specific nature of each protein.

The native state is stabilized in a narrow temperature range by a positive unfolding free energy that peaks close to room temperature, falling off at both higher and lower temperatures. The maximum stabilization free energy is interestingly small. At about $50\,\mathrm{kJ\,mol^{-1}}$ it corresponds, for example, to the enthalpy gained by breaking two or three hydrogen bonds in the protein interior (Table C2.3), when hundreds are involved in internal protein and protein–solvent interactions (Fig. C2.10; Table C2.3). For comparison, the covalent bond has an associated energy of about $300\,\mathrm{kJ\,mol^{-1}}$. With the exception of S–S bonds that can be formed between cysteine side-chains in non-reducing conditions, covalent bonds do not intervene in

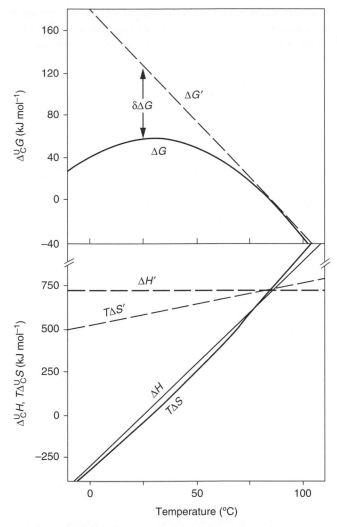

Fig. C2.9 The temperature dependence of thermodynamic functions for myoglobin unfolding. The enthalpy, entropy, and free energy differences upon unfolding are calculated from the heat capacity as a function of temperature. The dashed lines correspond to the functions that would have been obtained in the absence of solvent, so that $\delta\Delta G$ is due to protein–solvent interactions. (Privalov and Khechinashvili, 1974.) (Figure reproduced with permission from Elsevier.)

native fold stabilization. Bonds involved in stabilization of the native protein fold are weak bonds, with associated energies corresponding to a few kT.

The Hydrophobic Effect

The stabilization of the protein native state could be discussed in solubility terms. The protein folds in order to "hide" as many as possible of the "insoluble" apolar groups in the chain from solvent contact. In this sense, the observation that the unfolding free energy has a maximum close to 30 °C is reminiscent of the anomalous temperature dependence of the solubility of apolar compounds (see Fig. A1.7).

TABLE C2.3 ENERGIES ASSOCIATED WITH PROTEIN STABILIZATION FORCES

"Bond"	Energy (kJ mol^{-1})	Equivalent in RT at 300 K (~2.5 kJ mol^{-1})
Electrostatic in water at ~1 Å^a	~20	~10×
Hydrogen bond^b	~4 to 25	~2 × to 10×
Hydrophobic^c	~4	~2×
Van der Waals	~0.5	~0.1×

^{a} The Coulombic interaction between charged groups depends on the charge magnitudes, their separation and the dielectric constant of the medium; the energy given is for unit charges separated by 1 Å in water.

^{b} The hydrogen bond energy depends strongly on the environment; it is minimum in water and maximum in a vacuum or in an otherwise apolar environment.

^{c} The energy given corresponds to the approximate free energy change when a CH, CH$_2$, or CH$_3$ group is shielded from water contact. Note that the precise free energy changes are different for the three apolar group types.

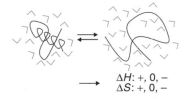

ΔH: +, 0, −
ΔS: +, 0, −

Fig. C2.10 Protein folding in a solvent. Internal protein bonds, broken upon unfolding, could reform with solvent molecules. The unfolding enthalpy, therefore, could equally well be positive, negative, or zero. The same is true for the unfolding entropy. The gain in chain configurational entropy could be more than compensated for by a loss in the configurational freedom of solvent molecules in contact with protein groups.

The solubility of apolar compounds in water is minimum close to room temperature, corresponding to a maximum transfer free energy. The dominant term at this temperature is a large negative entropy change, which has been interpreted as arising from an ordering of water molecules around the apolar groups. The structure of liquid water results in a large entropy due to the configurational freedom of each water molecule to make two hydrogen bonds with the nearest neighbors in all directions (see Chapter A2); this freedom is clearly reduced when the water molecule finds itself close to an apolar group that does not offer hydrogen bonding possibilities. This phenomenon is called the *hydrophobic effect*.

A system evolves in order to avoid the thermodynamically unfavorable exposure of apolar groups to water. The hydrophobic effect then results in an effective *hydrophobic force*, bringing together apolar molecules in aqueous solution, in order to minimize their water contact surfaces. A free energy loss of about $4\,\text{kJ}\,\text{mol}^{-1}$ results from the burying of a CH, CH_2, or CH_3 group away from contact with water (Table C2.3). Although the hydrophobic effect is dominated by the negative entropy change of apolar group hydration, an enthalpic contribution favoring the van der Waals packing of the apolar molecules in their pure liquid state is not negligible.

The hydrophobic effect has been suggested to be the main driving force behind protein stabilization. Careful analysis of calorimetric data, such as those shown in Fig. C2.9, however, introduced considerable qualifications on its role in the stabilization of the native protein fold.

The Enthalpy and Entropy of Native Protein Stabilization: Apolar and Polar Group Hydration and the Compactness of the Protein Interior

We can see from the curves in Fig. C2.9 that, close to room temperature, the respective enthalpy and entropy changes due to protein–solvent interactions compensate to a large extent the enthalpy changes due to internal protein bonds and entropy changes associated with chain configurational freedom. At the maximum value of unfolding ΔG, ΔS is zero, so that the native protein is stabilized by a small enthalpy term. At lower temperatures, ΔS is negative and the stabilization is entropy driven, while at higher temperatures enthalpy terms dominate. Note how, in Fig. C2.9, $\Delta G'$, the stabilization free energy in the absence of solvent interactions, is everywhere higher than ΔG and that the difference between them tends to vanish with increasing temperature. The hydrophobic effect and solvent interactions, in general, therefore, make the native state *less* stable with respect to the vacuum condition.

The entropy of apolar group solvation dominates toward the lower temperature end of the ΔG curve. It leads to a destabilization of the native protein fold as the temperature decreases, which has been called cold denaturation (Comment C2.10). Total cold denaturation of a protein (ΔG crossing zero) is not observed in the usual physiological solvents, because it would occur at subzero temperatures.

COMMENT C2.10 DESTABILIZATION OF PROTEIN COMPLEXES AT LOW TEMPERATURE

The mechanisms of cold denaturation also act on protein complexes. It is well known that certain protein complexes dissociate if their sample solutions are put in a cold room. F_1-ATPase, the soluble part of the membrane-associated F_1F_0-ATPase, is an example of such a cold-sensitive structure. It can be concluded that hydrophobic forces dominate the protein–protein interactions within the complex.

It has been observed in denaturing solvents, however, such as concentrated guanidinium hydrochloride solutions, which shift the thermodynamics variable curves toward higher temperature.

Further analysis has emphasized the non-negligible role of polar group hydration. We recall from Section C2.4.8 that, in contrast to the positive contribution of apolar group hydration, polar group hydration provides a negative contribution to ΔC_p, which vanishes at very high temperature. We recall (Comment C2.9) that ΔC_p is related to the slope of the entropy temperature dependence. The exposure of apolar and polar groups has an ordering effect on water molecules, be it in different ways – apolar groups, because they cannot form hydrogen bonds, and polar groups, because they can form hydrogen bonds with preferential orientations. Close to room temperature, the hydration entropies of apolar and polar groups are negative and similar. The hydration entropy of polar groups has a negative slope and decreases even further with temperature, however, while that for apolar groups has a positive slope and approaches zero above $100\,°C$. It must be emphasized that the "general" values assumed for the heat capacities of apolar and polar group hydration in the calculations of Section C2.4.8 are, at best, good approximations. The value and temperature dependence of hydration heat capacity depend on the specific nature of the group, and within the apolar family they are significantly different for aliphatic and aromatic groups (aromatic groups having a degree of polar character).

Analysis of calorimetric data after hydration effects have been accounted for has shown that the tight van der Waals packing of the protein interior plays a significant part in native fold stabilization. From a consideration of the hydrophobic effect, we would expect as many as possible of the apolar groups to gather together away from water exposure. The hydrophobic force, however, is due mainly to an undirected entropic effect, i.e., one which cannot provide a mechanism for specific interactions. In the case of hydrocarbons in water, the aggregation of apolar molecules forms a liquid phase. This is *not* the case for the protein interior, which has been shown by inspection of thousands of protein crystal structures to be more like a compact *amorphous solid*. Individual van der Waals interactions are very weak (Table C2.3). The sum of van der Waals energies from the specific packing of all the groups in the protein interior is quite significant, however, and could well be an important driving force behind the cooperativity of the native fold.

Detailed structural analyses of the macromolecular surface and internal packing density are required in order to describe the energetics of protein stabilization and to make the most efficient use of calorimetric data. Considerable effort is being devoted to the development of methods to perform such analyses on protein structures that have been solved by crystallography or NMR (Comment C2.11).

COMMENT C2.11 THERMODYNAMICS AND PROTEIN STRUCTURE

Note the compact packing of atoms in Fig. 2.11.1. The red and blue spheres on the protein surface represent negatively and positively charged side-chains, respectively. The main processes involved in the difference in stability between the native and unfolded states are:

- Surface group hydration: This is proportional to the solvent accessible area for each group, which can be calculated from the structure in the native state and assumed to be total in the unfolded state.
- Detailed packing of the native protein interior (which can be analyzed in the structure).

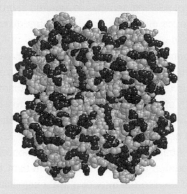

Fig. C2.11.1 Space-filling model of a protein structure solved by X-ray crystallography.

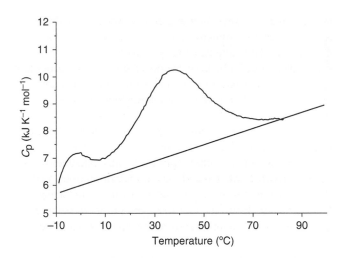

Fig. C2.11 Partial heat capacity as a function of temperature of the single strand deoxyoligonucleotide 5′-GCGAACAATCGG-3′ at a concentration of $1.88\,\mathrm{mg\,ml^{-1}}$. The total unfolding enthalpy from $-10\,°\mathrm{C}$ to $80\,°\mathrm{C}$ is equal to $77\,\mathrm{kJ\,mol^{-1}}$. (Privalov and Privalov, 2000) (Figure reproduced with permission from Elsevier.)

C2.5 NUCLEIC ACIDS AND LIPIDS

The arguments and analytical approach in DSC developed above for protein fold stabilization are readily extended to other macromolecules such as nucleic acids (Fig. C2.11), as well as to the interactions involved in complex formation between macromolecules and between macromolecules and smaller ligands.

The energetics of DNA double-helix formation has been examined through the study of oligonucleotides. The residual structure of oligonucleotides appears as the temperature decreases below $80\,°\mathrm{C}$. The structure is formed in a diffuse process with quite a low enthalpy of formation, which ends below $0\,°\mathrm{C}$, requiring supercooled conditions for its observation.

DSC is also an important tool for the study of membrane formation when lipids are mixed with water. Complex phase diagrams are obtained as a function of temperature and water/lipid ratio. The energetics of such systems is examined by measuring the temperature dependence of the heat capacity for each phase and the peaks associated with the transitions. Review references are given in the suggestions for further reading.

C2.6 CHECKLIST OF KEY IDEAS

- A full quantitative thermodynamics description of a macromolecule in solution for a given temperature range may be obtained from the measurement of its partial specific heat capacity at constant pressure, as a function of temperature, by DSC.
- Specifications for a modern DSC include macromolecule requirements of less than $50\,\mu\mathrm{g}$ per test, a temperature range from $-10\,°\mathrm{C}$ to $130\,°\mathrm{C}$, $\sim 10\,\mu\mathrm{J\,deg^{-1}}$ sensitivity, and the possibility of scanning down as well as up in temperature to check the reversibility of processes.
- In a heat capacity versus temperature scan of a macromolecule, the native state is assigned as the reference relative to which the thermodynamics parameters of the system are defined, so that the excess heat capacity, with respect to the native state becomes the significant parameter.
- A partition function analysis of the heat capacity curve is quite general and applies to any monomeric (so that the number of moles present in the solution does not change upon unfolding) system at equilibrium, independently of the transition mechanisms and of the number of intermediate states involved.
- When the cooperative domain extends over the whole molecule, the van't Hoff and calorimetric enthalpies are equal, and a two-state transition may be assumed.
- In all cases where the van't Hoff and calorimetric enthalpies are not equal the transition proceeds via intermediate states.
- When the cooperative domain is smaller than the molecule, as would be the case for a multidomain

protein, the van't Hoff enthalpy is smaller than the calorimetric enthalpy.

- A van't Hoff enthalpy that is larger than the calorimetric enthalpy is an indication that the cooperative domain is larger than the molecule, i.e., it provides evidence of intermolecular interactions such as oligomerization.
- There is no calorimetric evidence for a *molten globule* intermediate for small monodomain proteins.
- The difference in stability measured by calorimetry between a native protein and a mutant may very well be due to a change in cooperativity of the unfolding process, due to the mutation, and not to enthalpic or entropic contributions due to the amino acid replacement itself.
- The heat capacity versus temperature curve of a complex protein structure can be separated into component curves, corresponding to the transitions of the different states of the system without any a-priori assumptions by using computer algorithms based on recurrent procedures.
- The solvent plays an important role in the stabilization of both the folded and native states of a protein, and protein–solvent interactions may be specific, as in the case of ion binding to specific sites, or more general through, for example, pH, ionic, or small solute effects on water structure affecting the hydration and solubility of protein polar and apolar groups.
- Methods to calculate heat capacities of proteins in their various states have been developed based on DSC information and data from model compounds.
- The primary heat capacity, the heat capacity of an anhydrous unfolded polypeptide, can be calculated accurately by summing contributions from individual amino acids and peptide bond formation.
- The heat capacities of anhydrous folded proteins are very similar.
- The heat capacity of a native protein in solution is larger than that of the anhydrous folded protein, revealing mainly the positive magnitude of the hydration term.
- The hydration term varies according to the protein, reflecting the role of the macromolecular surface composition in the solvent interactions, especially with respect to the accessible surface area of polar and apolar groups.
- The heat capacity of an unfolded peptide in solution is equal to the primary heat capacity and a term equal to the sum of the hydration values for the individual amino acids in the protein sequence, and the N- and C-terminal groups.

- The heat capacity of the native state in solution varies linearly with temperature; the heat capacity of the unfolded state in solution is approximated well by a second-order temperature polynomial in the range 0–100 °C.
- Because of the importance of free energy predictions in protein and ligand design, accurate structures based on heat capacity calculations are extremely useful.
- The heat capacity of a *native* protein in solution cannot be calculated from its sequence, because it depends on the macromolecular surface and dynamics of the folded structure, but parametrized approaches have been developed that provide very good approximations.
- Calorimetric studies have made fundamental contributions to our understanding of protein stabilization forces.
- Detailed structural analyses of the macromolecular surface and internal packing density are required in order to describe the energetics of protein stabilization and to make the most efficient use of calorimetric data.
- The *hydrophobic effect* results in an effective *hydrophobic force*, bringing together apolar molecules in aqueous solution, in order to minimize their water contact surfaces.
- The entropy of apolar group solvation leads to a destabilization of the native protein fold as the temperature decreases, which has been called cold denaturation.
- Analysis of calorimetric data after hydration effects have been accounted for has shown that the tight van der Waals packing of the protein interior plays a significant part in native fold stabilization.
- Subzero temperatures and supercooled conditions are required to observe the full heat capacity curve of a nucleic acid corresponding to the appearance of oligonucleotide structure.

Suggestions for Further Reading

Privalov, P. L. (1980). Scanning microcalorimeters for studying macromolecules. *Pure Appl. Chem.*, **52**, 479–497.

Privalov, G. P., and Privalov, P. L. (2000). Problems and prospects in microcalorimetry of biological macromolecules in "Energetics of biological macromolecules." *Meth. Enzymol.*, **323**, 31–62.

Weber, P. C., and Salemme, F. R. (2003). Applications of calorimetric methods to drug discovery and the study of protein–protein interactions. *Curr. Opin. Struct. Biol.*, **13**, 115–121.

Tristram-Nagle, S., and Nagle, J. F. (2004). Lipid bilayers: Thermodynamics, structure, fluctuations, and interactions. *Chem. Phys. Lipids*, **127**, 3–14.

ISOTHERMAL TITRATION CALORIMETRY

C3

C3.1 HISTORICAL REVIEW

T. Wiseman, S. Williston, J. F. Brandts, and **L. N. Lin** published a paper in **1989** with the title: "Rapid measurement of binding constants and heats of binding using a new titration calorimeter." The term isothermal titration calorimetry (ITC) was introduced by **E. Freire** and colleagues in **1990**. The method is unique in providing not only the magnitude of the enthalpy change upon binding, but also, in favorable experimental conditions, values for the binding affinity and entropy changes. Because these parameters fully define the *energetics* of the binding process, ITC is playing an increasingly important role in the detailed study of protein–ligand interactions and the associated molecular design approaches, in particular with respect to drug design. In the **1990s**, a number of critical reviews of ITC results and analytical developments were published.

C3.2 EXPERIMENTAL ASPECTS AND EQUATIONS

C3.2.1 Measuring Protocol and Samples

The reaction cell in ITC has a volume close to 1 ml and contains one of the reactants. The other reactant is added to it by injection in small volumes (close to 10 μl), and stirred in. The amount of power (in millijoules per second or in watts) required to maintain a constant temperature difference between the reaction bath and a reference cell is measured by the calorimeter. The heat absorbed or released by the chemical reaction is determined from the integral of the power curve over the appropriate time (Fig. C3.1).

In the ITC experiment of Fig. C3.1, the 0.6 ml calorimeter cell was filled with protein solution (68.8 mM pancreatic RNase A), and the ligand solution (0.5 mM cyclic monophosphate) was added in precise 5 μl volumes through an injection syringe that also stirred the resulting solution to ensure proper mixing. The quantity plotted on the y-axis is the time dependence of the power needed to maintain a constant temperature difference between the measuring and a reference cell. Each peak corresponds to one injection, and its integral (after subtraction of the

background due to ligand dilution, mechanical mixing effects etc.) to the heat associated with the amount of binding (q_j in Eq. (C3.1)).

The peaks are progressively smaller with each injection because there is less free protein available for binding. When the protein is saturated, the peak height represents the background level. The integral heat effect of the binding as a function of injection number is plotted in Fig. C3.1b.

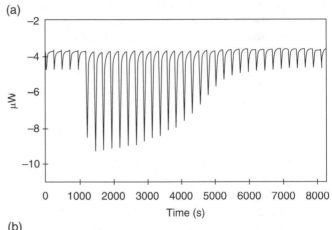

(a)

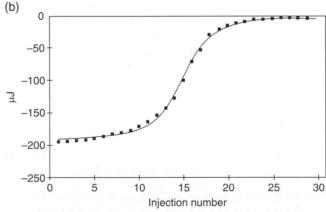

(b)

Fig. C3.1 ITC experimental curves of cyclic monophosphate binding in RNase A. **(a)** The power required to maintain a constant temperature difference between the measuring and the reference cell, as a function of time; each peak corresponds to one injection. **(b)** The integral heat effect due to the binding as a function of injection number. (Privalov and Privalov, 2000) (Figure reproduced with permission from Elsevier.)

C3.2.2 Binding Enthalpy and Heat Capacity

The amount of heat q_j absorbed or released in the injection j (Fig. C3.1) is equal to

$$q_j = v\Delta H\Delta[L_{j,\text{bound}}] \qquad (C3.1)$$

where v is the volume of the reaction cell and $\Delta[L_{j,\text{bound}}]$ is the change in concentration of bound ligand after the jth injection, so that q_j is proportional to the amount of ligand, $v\Delta[L_{j,\text{bound}}]$, that binds to the protein. The constant of proportionality is the enthalpy change per mole of bound ligand, ΔH. The change in binding heat capacity can be determined by repeating the titration at different temperatures (Eq. (C3.2) and Fig. C3.2)

$$\Delta C_P = \partial\Delta H/\partial T \qquad (C3.2)$$

If the amount of added ligand is small with respect to the affinity of the reaction, we can assume that all the added ligand is bound and hence derive the corresponding enthalpy change directly by using Eq. (C3.1). If, furthermore, the association constant, K_a, of the reaction is known from other measurements (such as fluorescence titration or surface plasmon resonance, see Chapters C4, D2), the free energy change, ΔG and hence the entropy change, ΔS, of the reaction can be calculated from ΔG and ΔH by using the classical Eq. (C3.3), and the reaction thermodynamics is fully defined:

$$\Delta G = -RT \ \ln \ K_a = \Delta H - T\Delta S \qquad (C3.3)$$

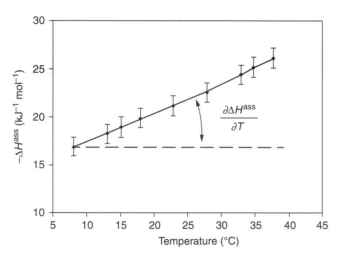

Fig. C3.2 Binding enthalpy and heat capacity. Plot of the binding or association enthalpy, ΔH^{ass}, as a function of temperature for a protein–DNA interaction (the HMG box of Sox-5 protein with a 12-base-pair DNA duplex containing the recognition sequence). The binding heat capacity is given by the initial slope of the line ($-6\,\text{kJ}\,\text{mol}^{-1}\,\text{K}^{-1}$) at the lower temperatures. The break in slope at the higher temperatures is due to temperature-induced alterations in the components of the complex. (see also Section C3.3.3)

Depending on the number of sites and other aspects of the binding model, the mathematical expression relating affinity constants to Eq. (C3.1) may be quite complicated.

C3.2.3 Affinity Constants

In the simple case of a protein, P, having one binding site for a ligand, L, we write

$$P + L \overset{K_a}{\rightleftharpoons} PL$$
$$K_a = \frac{[PL]}{[P][L]} \qquad (C3.4)$$
$$P_{\text{total}} = [P] + [PL]$$
$$L_{\text{total}} = [L] + [PL]$$

where P_{total} and L_{total} are the total concentrations of protein and ligand, respectively, and the square brackets indicate molar concentrations. Combining Eqs. (C3.1) and (C3.4)

$$q_j = v(\Delta H)P_{\text{total}}\left(\frac{K_a[L]_j}{1+K_a[L]_j} - \frac{K_a[L]_{j-1}}{1+K_a[L]_{j-1}}\right) \qquad (C3.5)$$

where $[L]_k$ is the free ligand concentration after the kth titration. Equation (C3.5) can also be written in a more complicated form in terms of total ligand concentration, by solving Eq. (C3.4).

By using the appropriate set of binding equations, the analysis can be extended to reactions with more than one binding site or involving equilibrium constants between many different states. Complementary experiments, with other methods, are necessary to fully solve such problems, however, because of the larger number of parameters.

In cases for which the reaction enthalpy is large enough to be measurable over a range of ligand concentrations with respect to the affinity constant, the titration series also provides a measure of the affinity constant itself (Fig. C3.3). A parameter, c, was defined to relate binding affinity and experimental conditions

$$c = K_a[P] \qquad (C3.6)$$

The value of c must be <1000 for the K_a of the reaction to be determined by ITC, which, in practice, sets an upper limit of 10^8–$10^9\,\text{M}^{-1}$ on the binding constant (Fig. C3.3).

Reactions with High Binding Affinities

For values of c beyond 1000, ITC curves lose their characteristic shape and no longer contain information concerning the binding constant (Fig. C3.3). This has been an important limitation in the application of ITC to drug design, because lead compounds have to be optimized to nanomolar or even higher affinities (see section on drug design, below). A protocol has been developed, however, that allows high affinity binding constants to be determined in ITC experiments, by using ligand competition. Three titrations are performed on the protein solution. First, the binding thermodynamics parameters of a weak

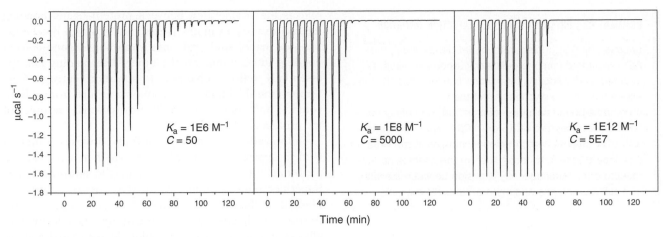

Fig. C3.3 ITC curves for affinity constants of different strength. Simulated ITC curves are presented in the three panels to illustrate the effects of increasing binding affinity. The following parameters were used for the simulation: $[p]=0.05$ mM; $[I]=0.6$ mM; $\Delta H=-42$ kJ mol^{-1}; K_a and c are shown in the figure, where $c=K_a[p]$. The binding enthalpy can always be measured accurately. The binding affinity, however, can only be determined in cases illustrated by the first panel ($c<1000$), in which the peaks are progressively smaller. If the affinity is too high with respect to experimental conditions ($c>1000$), only a lower limit can be set on its value. Microcalories per second can be converted to microwatts by multiplying by 4.2. (Leavitt and Freire, 2001) (Figure reproduced with permission from Elsevier.)

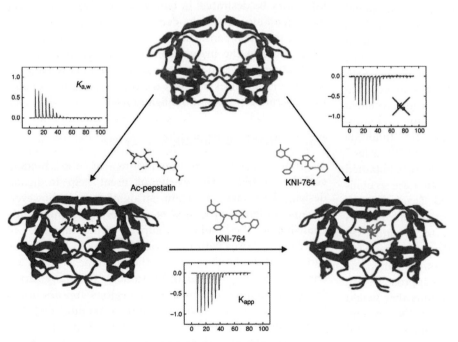

Fig. C3.4 Displacement titration to measure high-affinity binding constants (see text). The experiments on the inhibitor–HIV-1 protease interaction were performed at 25 °C in 10 mM acetate, pH 5.0, 2% DMSO. (Velazquez-Campoy et al., 2001) (Figure reproduced with permission from Elsevier.)

inhibitor are determined. Second, the protein is titrated by the high-affinity ligand in order to determine its binding enthalpy. In the third titration, *the displacement titration*, the high-affinity ligand is titrated into a solution in which the protein has been prebound to the weaker inhibitor. The method is illustrated in Fig. C3.4 by a study of inhibitors of HIV-1 protease, an important drug target in the fight against AIDS (Comment C3.1). The binding enthalpy of the inhibitor–HIV-1 protease interaction was measured accurately by ITC (Fig. C3.4, top right-hand panel, the

y-axes in these panels are in mcal M^{-1}) but not its association constant K_a, which is too high to be determined by ITC. The displacement titration (binding constant K_{app}, determined in Fig. C3.4, bottom panel) was performed in the presence of the weak inhibitor acetyl-pepstatin (binding constant, $K_{a,w}$, determined in the top left-hand panel). The selection of a weak competitive inhibitor with a binding enthalpy of opposite sign increases the sensitivity of the experiment. K_a is determined from K_{app} by using the equation:

Infection by HIV (human immunodeficiency virus), the AIDS (acquired immune deficiency syndrome) virus, is initiated by the specific interaction between gp120, the external envelope glycoprotein of the virus, and membrane-bound cellular receptor CD4, and obligatory chemokine receptors CCR5 or CXCR4. The interaction with CD4 triggers conformational changes in the virus envelope that lead to recognition of the chemokine receptors and subsequently to fusion between the viral and cell membranes and internalization of the virus.

$$K_{app} = \frac{K_a}{1 + K_{a,w}[X]}$$

where [X] is the concentration of the weak inhibitor, which can be adjusted to give a value of K_{app} in the desired range. The values found for the affinity constants were $K_{a,w} = 2.4 \times 10^6 \, M^{-1}$ and $K_a = 3.1 \times 10^{10} \, M^{-1}$.

Drug Design

An important medical application of three-dimensional structure determination of enzymes involved in disease is the design of high-affinity inhibitors that could be potential pharmaceutical agents. Affinities in the nanomolar range (dissociation constants of 10^{-9} M) are required for a drug to be useful. Briefly, the drug design approach is divided into a number of steps. After having chosen the target protein, a very large combinatorial search is effected for lead compounds, which could interact with its active site. Hundreds of thousands of such compounds are found in "libraries." Many of these compounds are natural molecules obtained from plants, for example, with structures that are so complex they could never have been invented. Taxol, an antimitotic drug, purified from yew bark, is an example of such a structure. Promising lead compound–protein complexes are characterized by different methods, including ITC, which can provide essential information on binding energetics. Chemical modeling and simulation approaches are then applied to propose structural variations that would increase the affinity of the compound for the protein. Modified inhibitors are synthesized and characterized and the process is repeated until a satisfactory compound is obtained.

C3.3 APPLICATIONS

C3.3.1 Entropic Versus Enthalpic Optimization

We see from the classic Eq. (C3.3) that the same binding constant could be achieved from different combinations of entropic and enthalpic contributions. The binding enthalpy reflects the competition between protein–ligand bonding interactions (e.g., van der Waals and hydrogen bonds), on the one hand, and protein–solvent and ligand–solvent bonding interactions, on the other. The binding entropy essentially reflects changes in solvation entropy (e.g., through the shielding of hydrophobic groups from solvent contact) and changes in the conformational entropies of the protein and ligand.

Optimization of binding affinity invokes different tactical approaches depending on whether it is achieved predominantly via the enthalpic or the entropic terms. Superhigh binding affinity requires a synergy of favorable binding enthalpy and binding entropy contributions. Entropic binding is favored by the design of conformationally constrained ligands with a high degree of hydrophobicity. HIV-1 protease inhibitors (Fig. C3.4) displaying these thermodynamics properties have been approved for clinical use. Such ligands exhibit very high specificity toward their targets. This is an important contribution to high binding affinity when dealing with an immutable target, but it also makes the ligand more susceptible to drug-resistant mutations or ineffectual against related but not identical target molecules. A certain degree of ligand flexibility may be desirable, in which case the resulting lower binding affinity due to the decrease in entropy upon binding has to be compensated by other favorable interactions. Note also that the hydrophobicity of a ligand affects its solubility. The selection of the entropy/enthalpy balance, therefore, influences not only the binding affinity but also the general properties of the ligand itself.

C3.3.2 Relating Binding Energy and Structure

Accurate structural interpretations of the binding enthalpy would obviously be of great help in ligand design. They are far from straightforward, however, because different processes of roughly equal magnitude contribute to the binding enthalpy. A structural survey of protein complexes with low-molecular-weight ligands has shown that the binding sites in the unbound proteins are characterized by low structural stability. The stabilization by ligand binding of these protein regions appears in the thermodynamics as a conformational binding enthalpy term. Despite the inherent difficulties of the problem, there are attempts to parametrize the binding enthalpy in terms of structural features, in a similar way to that developed for differential scanning calorimetry (DSC) (see Chapter C2).

C3.3.3 Combining ITC and Other Biophysical Methods

ITC, Mass Spectrometry, Analytical Ultracentrifugation (AUC) and Surface Plasmon Resonance (SPR)
The energetics of the HIV gp120-CD4 binding reaction (Comment C3.1) has been studied by a series of biophysical

experiments: MALDI-MS and AUC to characterize the masses of the proteins and complexes, ITC to characterize the binding affinities and energetics, SPR to characterize binding kinetics and circular dichroïsm (CD) to characterize secondary structure rearrangements upon binding (see also Chapters B1, D4, C4, E4).

Soluble forms of the two proteins were obtained by expression in Chinese hamster ovarian cells and *Drosophila* cell lines, respectively. gp120 was prepared as the full-length glycosylated protein and as a "core gp120," which had certain amino acid deletions and the removal of most of the sugar groups. The masses of the expressed proteins were checked by MALDI and AUC, and their assembly state in the complex was determined by AUC (Fig. C3.5).

The data shown in Fig. C3.5 (squares) are for a mixture of CD4 and core gp120. The solid line represents the best fit (the differences with the data are expressed as residuals in the top panel). The masses of CD4 and core gp120 were 45 kDa and 38 kDa, respectively. The complex was measured at 83 kDa by AUC, showing it to be a one-to-one assembly of the two proteins. The component curves of the fit are also shown in the figure: The long dashed line is for the one-to-one complex, the dotted line is for excess CD4, and the baseline is the dash–dot line.

The binding thermodynamics of the interaction was determined by ITC (Fig. C3.6). Figure C3.6 shows the power data and integrated enthalpy for the titration of full length (part (a)) and core (part (b)) gp120 by CD4. Best-fit lines are for binding enthalpies of –63 and –62 kcal (mol of CD4)$^{-1}$ for full length and core gp120, respectively. Equilibrium dissociation constants K_D were determined as 5 and 190 nM, for the full length and core gp120, respectively. Figure C3.6c shows the temperature dependence of the CD4 binding enthalpy for full length (open circles) and core (filled circles) gp120. The slopes of the best-fit lines gave binding heat capacities of –1.2 and –1.8 kcal mol^{-1} K^{-1}, respectively.

The thermodynamic parameters in Table C3.1 were measured or calculated from the ITC data in Fig. C3.6. We note that these parameters are essentially similar for the full length and core gp120, indicating common binding mechanisms. The apparently large difference in dissociation constant values (5 nM and 190 nM for the full length and core gp120, respectively) in fact corresponds to just over 2 kcal mol^{-1} in binding free energy, less than the energy of formation of one hydrogen bond.

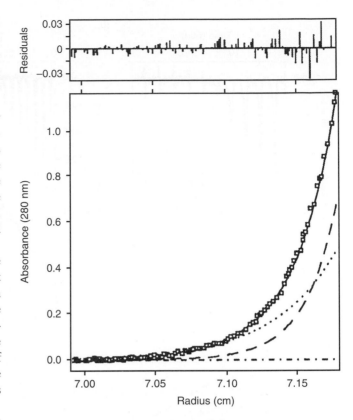

Fig. C3.5 Sedimentation equilibrium analysis of the gp120–CD4 interaction (see text). (Myszka *et al.*, 2000) (Figure reproduced with permission from *Proceedings of the National Academy of Sciences (USA)*.)

The thermodynamic parameters for CD4–gp120 binding were compared to those of other protein–protein interactions. Thirty protein–protein complexes were sampled, as well as 13 antibody–antigen interactions, and two T-cell receptor–major histocompatibility complex (MHC) peptide interactions (see also Fig. C4.7).

The thermodynamics of the CD4–gp120 interaction is qualitatively different from that of the 30 protein–protein complexes sampled, which had small positive ΔG, ΔH, and $-T\Delta S$. It is similar, however, to that of antibody–antigen and T-cell receptor–MHC peptide interactions. The favorable binding enthalpy of the CD4–gp120 interaction $(\Delta H^{\circ} \sim -63$ kcal mol$^{-1})$ is strikingly large, while the binding entropy $-T\Delta S^{\circ} \sim 52$ kcal mol$^{-1})$ is also very large but unfavorable. CD4–gp120 complex formation, therefore, involves a large number of bonding interactions, such as hydrogen bonds or van der Waals

TABLE C3.1 THERMODYNAMICS PARAMETERS OF THE CD4–GP120 INTERACTIONS AT 37 °C					
	ΔG° (kcal mol^{-1})	ΔH° (kcal mol^{-1})	$-T\Delta S^{\circ}$ (kcal mol^{-1})	ΔC° (kcal K^{-1})	K_D (nM)
Full length gp120	−11.8 ± 0.3	−63 ± 3	51.2 ± 3	−1.2 ± 0.2	5 ± 3
Core gp120	−9.5 ± 0.1	−62 ± 3	52.5 ± 3	−1.8 ± 0.4	190 ± 30

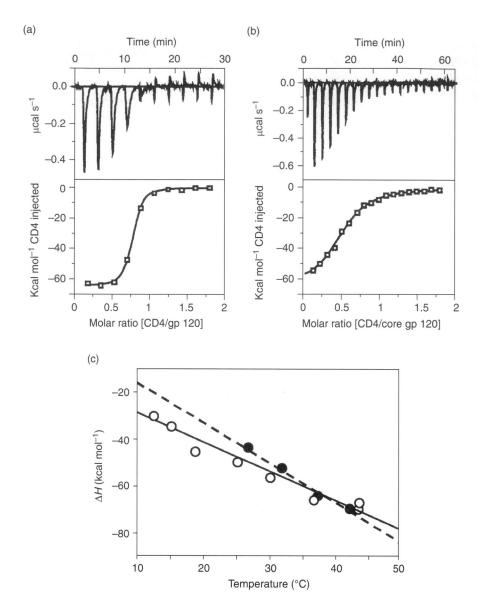

Fig. C3.6 ITC analysis of the gp120–CD4 interaction: **(a)** power data and integrated enthalpy for the titration of the full length gp120 by CD4; **(b)** power data and integrated enthalpy for the titration of the gp120 core by CD4; **(c)** temperature dependence of the CD4 binding enthalpy to full length (open circles) and core (filled circles) gp120. (Myszka *et al.*, 2000) (Figure reproduced with permission from *Proceedings of the National Academy of Sciences (USA)*.)

interactions, as well as substantial loss of degrees of freedom. Analysis of the high-resolution crystal structure suggests that the entropy decrease is not accounted for by trapped water molecules, but must involve major conformational rearrangements of the proteins. The large binding enthalpy and binding entropy values compensate to give low values for the dissociation constants and binding free energy, similar to those for other complexes.

The binding heat capacities (Table C3.1) are significantly greater than the values (-0.2 to -0.7 kcal mol^{-1} K^{-1}) usually observed for protein–protein complexes. Recalling the structural interpretation of heat capacity changes discussed above, this suggests extensive apolar surface area is buried upon CD4–gp120 complex formation. This is not observed at the binding site in the crystal structure of the complex, suggesting that this apolar surface area is buried elsewhere in the structure through major conformational

COMMENT C3.2 STRUCTURAL MODELING OF THERMODYNAMICS PARAMETERS

In an independent approach to that of Freire and his collaborators (Chapter C2), Spolar and Record (1994) have related thermodynamic parameters from calorimetry to the number of residues that reorganize during binding, by analyzing high-resolution structures.

rearrangements. In another modeling approach (Comment C3.2), binding heat capacity and entropy were interpreted in terms of a number of residues that reorganize during binding. Application of the method to the CD4–gp120 complex gave values for the number of residues that were close to 100, among the largest reported for a protein–protein interaction.

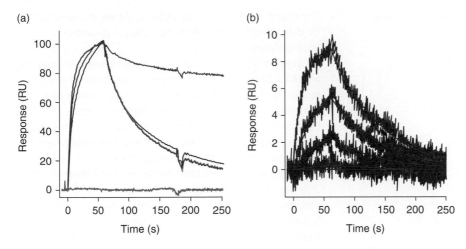

Fig. C3.7 SPR analysis of CD4–gp120 binding kinetics. **(a)** Normalized binding responses of CD4 (1 μM) over full length gp120 (black curve), glycosylated core (red curve), core gp120 (blue curve), and unmodified sensor (gray curve). Responses were normalized to 100% bound at the beginning of the dissociation phase. **(b)** Kinetic data for core gp120 binding to a CD4 surface, injected at varying concentrations from 975 (top curve) to 36 nM, the baseline is for 0 nM added. The data were globally fitted (lines) by a single site interaction model. (Myszka *et al.*, 2000) (Figure reproduced with permission from *Proceedings of the National Academy of Sciences (USA)*.)

Given the strong indications from the thermodynamic data of major conformational changes in the proteins upon binding, the study continued with a CD investigation. A substantial change in ellipticity was observed during complex formation, consistent with a reduction in random coil content. These data suggest that binding induces structural rigidity.

The kinetics of CD4–gp120 binding has been analyzed by SPR (Fig. C3.7; see also Section C4.4.3). The similar and relatively slow association rate constants measured for both gp120 ($6.72 \times 10^4 \, M^{-1} s^{-1}$ and $6.27 \times 10^4 \, M^{-1} s^{-1}$, respectively) are consistent with major structural rearrangements during binding.

The different equilibrium dissociation constants found for the full length and core gp120 are consistent with the values measured by ITC. They arise from different dissociation rate constants probably because of a loss of favorable contacts in the core gp120. Glycosylation had little effect on binding kinetics.

The flexibility of gp120 and the masking of its receptor and chemokine binding sites contribute to the fact that monomers of this protein, shed by the virus, elicit mainly non-neutralizing antibodies and are very inefficient in eliciting broadly neutralizing antibodies – a decoy strategy that may help the virus evade humoral immune response.

The gp120 protein is organized on the virus in a trimeric complex with another protein called gp41. The complex is stable and inert. Binding to CD4 and the resulting large conformational change in gp120 induces a properly timed metastable state in gp41 that triggers fusion between the viral envelope and the cell membrane.

The results from hydrodynamics, thermodynamics, and spectroscopy when combined with the crystallographic structure of the CD4–gp120 complex present an intriguing picture of the role of gp120 in the biology of HIV infection.

ITC and DSC
The gradient of the binding enthalpy temperature can be interpreted as a binding heat capacity only if the temperature change does not induce alterations in any of the reaction components. If the apparent heat capacity changes with temperature, as it does in the reaction shown in Fig. C3.2, it is an indication that such alterations with temperature do occur. In these cases, the correct approach is to combine ITC with DSC to study the heat capacity of each of the reaction components in the appropriate temperature range.

We treat here the example of the HMG box–DNA interaction. The temperature dependence of the binding enthalpy has already been shown in Fig. C3.2. The molar partial heat capacity functions of the reaction components, measured by DSC, are shown in Fig. C3.8. Clearly, the interpretation of the binding heat capacity of the interaction is considerably complicated by the unfolding of the protein (dotted line in Fig. C3.8) and partial dissociation of the complex (solid line) in the temperature range of the study. The HMG box–DNA reactions were expressed as a set of equilibrium equations, similar to Eq. (C3.4). By fitting the observed heat capacity functions of the complex and its free components, all the thermodynamic parameters of the system were evaluated.

The analysis revealed that there are two stages in the temperature-induced changes of the complex (solid line in Fig. C3.8). In the first, there is a gradual increase in heat capacity reflecting accumulation of thermal energy. In the second, there is a highly cooperative dissociation of the

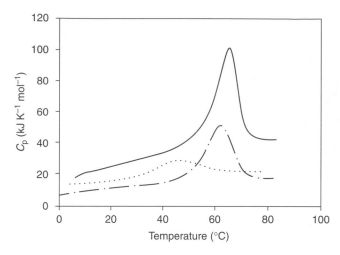

Fig. C3.8 Component partial molar heat capacities for the HMG box–DNA interaction. The dash–dot line is for 12-base-pair DNA duplex; the dotted line is for Sox-5 HMG box protein; the solid line is for their complex. (Privalov and Privalov, 2000) (Figure reproduced with permission from Elsevier.)

complex with concomitant unfolding of its protein and DNA duplex components. The second stage proceeds with a significant peak in heat capacity, which drops sharply with temperature, due to the decrease in complex population.

- The change in binding heat capacity during the reaction can be determined by repeating the titration at different temperatures.
- In cases for which the reaction enthalpy is large enough to be measurable over a range of ligand concentrations with respect to the affinity constant, the titration series also provides a measure of the affinity constant itself.
- Promising lead compound–protein complexes in *drug design* can be characterized by ITC, which provides essential information on binding energetics.
- The same binding constant can be achieved from different combinations of entropic and enthalpic contributions; binding enthalpy reflects the competition between protein–ligand bonding interactions, on the one hand, and protein–solvent and ligand–solvent bonding interactions, on the other; binding entropy essentially reflects changes in solvation entropy and changes in the conformational entropies of the protein and ligand.
- ITC has been combined with mass spectrometry, AUC, and SPR to provide as full a description as possible of the energetics and structural changes during a binding reaction.
- In reactions for which the heat capacity of the components may change with temperature, ITC is combined with DSC, which is used to measure the heat capacity of each of the reaction components in the appropriate temperature range.

C3.4 CHECKLIST OF KEY IDEAS

- In ITC, the amount of power required to maintain a constant temperature difference between a reaction bath and a reference cell during a titration is measured by the calorimeter; the heat absorbed or released by the chemical reaction is determined from the integral of the power curve over the appropriate time.

Suggestions for Further Reading

Seelig, J. (2004). Thermodynamics of lipid–peptide interactions. *Biochim. Biophys. Acta*, **1666**, 40–50.

Velazquez-Campoy, A., Leavitt, S. A., and Freire, E. (2004). Characterisation of protein–protein interactions by isothermal titration calorimetry. *Methods Mol. Biol.*, **261**, 35–54.

SURFACE PLASMON RESONANCE AND INTERFEROMETRY-BASED BIOSENSORS

C4

C4.1 HISTORICAL OVERVIEW AND INTRODUCTION TO BIOLOGICAL PROBLEMS

1801

Thomas Young observed that two slits placed in front of a light source created a pattern of intensity fringes on a screen. This phenomenon is the basis of modern interferometers, which have now been adapted to measure ligand–receptor interactions.

1902

R. W. Wood observed for the first time the phenomenon of surface plasmon resonance (SPR), which provided a simple and direct sensing technique for probing refractive index changes that occur in the very close vicinity of a thin metal film surface.

1968

E. Kretschmann and **H. Z. Raether** proposed the basic configuration of an SPR sensor, with an optical system containing a prism coupled to a reaction cell.

1982–1983

B. Liedberg and colleagues realized the potential of SPR for macromolecular binding studies since the change in refractive index is dependent on the molecules accumulating at the metal surface. They adsorbed an antibody specific to immunoglobulin G onto a gold sensing film, resulting in the selective binding and detection of the protein.

1980s–1990s

Several biosensors, based on either interferometry or SPR that allowed the simple, rapid, and non-labeled assay of various biochemical analytes such as proteins, DNA, and small compounds became commercially available.

1993

The potential to analyze weak affinities by SPR instrumentation was demonstrated by **S. Davis** and coworkers in their studies of cell surface receptor interactions.

2000–

Optical arrays were developed to analyze the interactions of thousands of different molecules simultaneously, with high selectivity and sensitivity. Biosensors are used in the study of intermolecular interactions with a broad range of affinities under a variety of chemical conditions and temperatures. The configuration of the biosensors makes them particularly advantageous in the rapid screening of putative ligands. These instruments have become essential tools for proteomics. In conjunction with other biophysical techniques, such as isothermal titration calorimetry, analytical centrifugation, and mass spectrometry, they provide very precise thermodynamic analysis, characterization, and identification of binding partners and binding events. The advantages of biosensor experiments compared with other binding studies include the absence of labeling, the small amounts of material required (in the nanogram range) and the wide variety of systems that can be studied (ranging from 200 Da molecules to unicellular organisms). In particular, a high degree of purity is not essential.

C4.2 MEASURING SURFACE BINDING

C4.2.1 Layout of a Biosensor

A number of biosensors have been developed for the study of biological macromolecule interactions. Surface binding experiments typically involve immobilizing a molecule (*the ligand*) on a surface and monitoring its interaction with a second molecule (*the analyte*) in solution. The set-up has the advantage of mimicking many biological recognition events, such as those between cell-bound receptors and soluble signaling proteins.

In the biosensor method, the build-up of bound analyte concentration leads to a change in refractive index (Comment C4.1) near the surface. Two types of instrument have been developed to measure this change with high sensitivity. One is an interferometer and the other is based on the phenomenon of SPR. The biochemical binding constants, such as the equilibrium dissociation constant and the on and off rates of a reaction (see Chapter C1) can be calculated directly from the change in refractive index of the solution over time.

A biosensor comprises a flow cell attached to an optical device. Buffer and analytes enter and exit the cell, and biochemical interactions are monitored through changes in refractive index.

C4.2.2 SPR Biosensor

Several biosensors have been designed around the phenomenon of SPR. Due in large part to its relative simplicity and high sensitivity, this method has become very popular.

Total internal reflection occurs at an interface between two non-absorbing media (see Part F). When light is shone upon the interface between two transparent media of different refractive index above a *critical angle* of incidence, most of the light coming from the side with the higher refractive index is reflected back. Some of the incident light, however, leaks into the lower refractive index medium as an electrical field intensity called an *evanescent field wave*. The amplitude of the evanescent wave decreases exponentially with distance from the interface, with a decay constant of about one light wavelength.

If the interface between the two media is coated with a thin layer of a suitable conducting material, such as a metal, the p-polarized component of the evanescent field wave penetrates the metal layer (Fig. C4.1). In fact, at a specific angle of incidence, plane polarized light excites the delocalized surface electrons (or plasmons) of the metal, which results in a larger evanescent wave. As a consequence, at this angle of resonance, the intensity of the reflected light decreases *drastically* due to the energy transferred to the plasmons.

The two media in the SPR biosensor are, respectively, a glass prism and a flow cell in which the biochemical reaction occurs (Fig. C4.2). A thin film of metal separates the two. The conditions for *surface plasmon excitation* at the interface between the metal and the biochemical solution are achieved by matching the projection of the wave vector of the incident light in the direction of the interface (k_x)

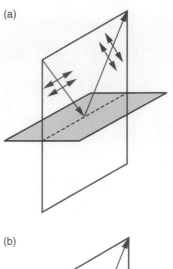

(a)

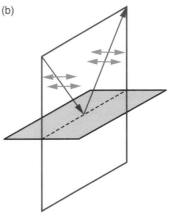

(b)

Fig. C4.1 (a) p-polarized light: The electric field vector (blue arrows) lies parallel to the plane (blue square) formed by the incident and reflected beams (red lines). **(b)** s-polarized light: the electric field vector (green arrows) lies perpendicular to the plane formed by the incident and reflected beams and is, therefore, parallel to the surface (green plane).

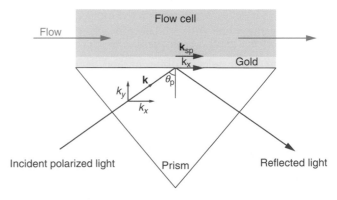

Fig. C4.2 General principle of biomolecular interaction analysis using SPR. The polarized light at incident angle θ_p excites the surface plasmons at the gold interface. $\boldsymbol{k}_{sp}$ is the surface plasmon wave vector, and $\boldsymbol{k}$ is the wave vector of the incident beam with components k_x and k_y parallel and perpendicular to the surface, respectively.

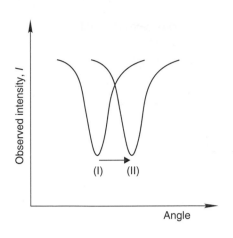

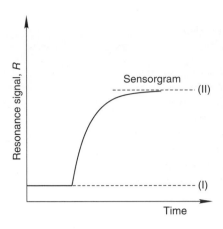

Fig. C4.3 General principles of biomolecular interaction analysis using SPR. An increased sample concentration on the sensor chip surface causes a corresponding increase in the refractive index, which alters the SPR angle. The angle change is monitored as a change in the detector position for the reflected intensity dip (from I to II).

and wave vector (k_{sp}) of the surface plasmon oscillation and are given by

$$k_x = k_{sp} \qquad (C4.1)$$

with

$$k_x = \frac{\omega}{c} \sqrt{\varepsilon_{prism}} \times \sin\theta, \qquad k_{sp} = \frac{\omega}{c} \sqrt{\frac{\varepsilon_{metal} \times \varepsilon_{cell}}{\varepsilon_{metal} + \varepsilon_{cell}}}$$

$$\sin^2\theta_p = \frac{(\varepsilon_{metal} \times \varepsilon_{cell})}{(\varepsilon_{metal} + \varepsilon_{cell})} \times \frac{1}{\varepsilon_{prism}} \qquad (C4.2)$$

where ω is the angular frequency of the incident wave, c is the speed of light, and ε is the wavelength-dependent complex dielectric permittivity. The incidence angle θ_p at which SPR conditions are satisfied therefore depends on the refractive index of the material on the non-illuminated side of the metal (the flow cell in this case). The prism enables a range of incidence angles to be observed simultaneously in a wedge of light beams. When the resonance condition described in Eq. (C4.2) is satisfied, there is a strong absorption dip within the angular dependence of the wedge of reflected light (Fig. C4.3). At optical wavelengths, the SPR condition is fulfilled by several metals, of which gold and silver are the most commonly used (Comment C4.2). The resonance is influenced by the refractive index in the evanescent wave path. Consequently, the refractive index

beyond the penetration distance of about 600 nm (the wavelength of the incident light) does not affect the experimental outcome, and, more decisively, the biosensor signal directly correlates with the amount of protein interacting near the surface. Results are plotted as a *sensorgram*, which represents changes in resonance signal as a function of time (Fig. C4.3).

C4.2.3 Interferometers as Biosensors

The speed of light c_n in a medium of refractive index n is given by

$$c_n = \frac{\omega}{n} \sqrt{\varepsilon} \qquad (C4.3)$$

where ω is the light wave's angular frequency, and ε is the dielectric permittivity of the medium.

In the dual wave-guide interferometer, a laser beam travels through two parallel optical wave-guides, A and B, one of which (A) contains the sample. The sample, because it has a different refractive index, introduces a velocity change, dephasing the beam emerging from A with respect to that emerging from B. The two waves interfere to give a classical Young's pattern of fringes related to the refractive index of the sample. Any changes in the refractive index of the sample alter the speed of propagation of the laser beam and cause a shift in the Young's fringes. The magnitude of the shift is calibrated by using known refractive index changes. The optical configuration is shown in Fig. C4.4.

In the biosensor interferometer, the sample is the flow cell in which molecular reactions take place. Receptors are attached to the cell surfaces and potential analytes flow past. In contrast to SPR biosensors, the experiment is sensitive to refractive index changes throughout the volume of the cell.

C4.2.4 Other Types of Biosensor

A variety of physical phenomena can form the basis of biosensor devices. Coupled plasmon-wave-guide resonance spectroscopy (CPWR), for example, combines features of both SPR and wave-guide instruments. Optical diffraction gratings have also been incorporated into SPR instruments.

COMMENT C4.2 SURFACE PLASMON PROPAGATION

Once light has been converted into a surface plasmon mode on a flat metal surface, it propagates but gradually attenuates owing to losses arising from absorption in the metal. This attenuation depends on the dielectric function of the metal at the oscillation frequency of the surface plasmon. Gold and silver are the best candidates for the metal film in the visible light region.

Although silver films yield a more distinct SPR spectrum than gold ones, this metal tends to be unstable chemically. Thin gold films are therefore a better choice for SPR biosensing applications.

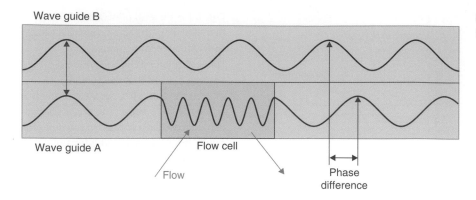

Wave guide B

Wave guide A

Flow

Flow cell

Phase
difference

Fig. C4.4 Biosensor-based interferometry.

Instruments are designed to reproduce as closely as possible natural biological environments in which to monitor intermolecular interactions.

C4.2.5 Coupling Ligands to a Surface

The preparation of the ligand surface in the flow cell poses special challenges which have been met by developments in the field of surface nanoscience. Ligands in biosensor experiments are usually proteins, but they can also be small molecules, DNA, or putative drugs tethered to a surface in the flow cell. Long-lived, highly stable associations prevent signal drift, and enable efficient measurement and control of the amount of analyte binding.

Proteins can be bound via their hydrophobic surface, or covalently coupled to the chip via their amino, carboxyl, or sulfydryl groups. With these types of bonding, the ligands are likely to adopt a random orientation on the chip surface, and the analyte-binding interface is not always accessible. This affects the maximum binding capacity but not the measured rates and dissociation constants (see Section C4.3).

Tags are often used in coupling. Chosen wisely they can orient a receptor protein as it is naturally found at the cell membrane's surface. Chips have been specifically designed to bind biotin, GST (glutathione S-transferase), polyhistidine tags, and antibodies. More realistic systems, involving lipid-embedded ligands with lateral fluidity, have also been developed.

A biosensor chip can be re-used because after each experiment the ligand surface is *regenerated* so as to remove any remaining bound analyte. A new binding analysis can then be performed.

C4.3 BINDING BETWEEN A SOLUBLE MOLECULE AND A SURFACE

C4.3.1 Thermodynamics of Surface Interactions

Recall from Chapter C1 that the thermodynamic equilibrium dissociation constant for the association of A and B is given by

$$K_d = \frac{[\text{A}] \cdot [\text{B}]}{[\text{AB}]} = \frac{k_{\text{on}}}{k_{\text{off}}}$$

In biosensor instruments, k_{on} is the rate of analyte binding to the surface and k_{off} is the rate at which the analyte is removed from the surface. Typical ranges of values are $K_d = 10^{-7}$–10^{-9} M, $k_{\text{on}} = 10^3$–10^8 M^{-1} s^{-1}, and $k_{\text{off}} = 10^0$–10^{-6} s^{-1}.

A typical sensorgram is shown in Fig. C4.5. The biosensor is calibrated such that the measured signal, R, is proportional to the concentration of bound analyte. The signal varies over time according to the molecular concentration in solution and to how fast the molecule dissociates and associates from the flow cell-bound receptor.

C4.3.2 Measurement of the Equilibrium Constant

The shape of the sensorgram reveals the kinetics of the interaction. The standard equation for the change in signal over time, for a one-to-one interaction, is

$$dR(t)/dt = k_{\text{on}}C(t)[(R_{\text{max}} - R(t))] - k_{\text{off}}R(t) \qquad \text{(C4.4)}$$

for an analyte at concentration C. Here, R_{max} is the maximal response. The equilibrium dissociation constant K_d of an interaction is determined from the equilibrium response. The equilibrium signal R_{equ} is reached when $dR(t)/dt = 0$, and the previous equation can be rewritten as,

$$R_{\text{equ}} = [1/(k_{\text{off}}/k_{\text{on}} + C)]CR_{\text{max}} \qquad \text{(C4.5a)}$$

Or, by using the equilibrium constant $K_d = k_{\text{off}}/k_{\text{on}}$,

$$R_{\text{equ}} = [1/(K_d + C)]CR_{\text{max}} \qquad \text{(C4.5b)}$$

which is similar to a Langmuir isotherm (Comment C4.3). In the steady-state affinity analysis, the equilibrium dissociation constant K_d is calculated from the dependence of the equilibrium plateau signal on the concentration of the analyte. The standard method of obtaining an estimate of K_d is to measure the resultant equilibrium response of R_{equ} for various concentrations C_k of ligand. Recasting the previous equation into linear form gives the following,

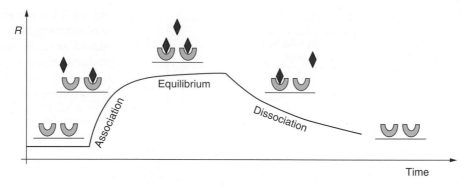

Fig. C4.5 A sensorgram is a plot of the SPR signal versus time. The increase in resonance units from the initial baseline represents the association of molecules. The plateau represents the steady-state phase of the interaction where binding is balanced by dissociation from the complex. The decrease in resonance units from the plateau represents the dissociation from the surface after removing the analyte from the running buffer.

$$(1/C_k)R_{equ}(C_k) = (-1/K_d)R_{equ}(C_k) + (1/K_d)R_{max} \tag{C4.6}$$

The *Scatchard* analysis is based on the binding responses when at equilibrium and the corresponding analyte concentrations (see Chapter C1). The slope of R_{equ}/C_k plotted against R_{equ} is $-1/K_d$ and its x-axis intercept is R_{max}/K_d (see the example plot in Fig. C4.6). Note that Eq. (C4.6) allows K_d and R_{max} to be calculated from the equilibrium responses at different concentrations without actually reaching the R_{max} plateau (the plateau corresponds to saturating the ligand binding surface with analyte). In fact, with weak affinity interactions (K_d in the micromolar range (~0.1–500 μM)), saturation of the sensor surface is seldom achieved.

During an experiment, the analyte is usually diluted serially to achieve ideally a 100-fold range of concentrations. The initial concentration value should be a magnitude higher than the dissociation constant (Fig. C4.6), and the last dilution at least a magnitude lower.

The standard free energy change upon binding, $\Delta G°$, can be calculated from the dissociation constant as outlined in Chapter C1. The standard entropy and enthalpy contributions to the binding can be calculated from the

temperature dependence of $\Delta G°$ via a non-linear fit of Eq. (C1.28).

C4.3.3 The Determination of the k_{off} and k_{on} of an Interaction

Binding interactions can have similar affinities (expressed as K_d), but can in fact result from very different kinetics.

When the analyte concentration in the flow is reduced to zero, bound analytes that detach from the ligand are not replaced and the signal progressively returns to its initial value at a rate that depends on the binding strength (Fig. C4.5). In the kinetic analysis of a one-to-one binding model, the dissociation rate k_{off} is deduced from the exponential decrease of the response (R) over time (t):

$$R(t) = R(t_0) \exp[-k_{off}(t - t_0)] \tag{C4.7}$$

The time t_0 denotes the instant the signal started decreasing, i.e., when more molecules detach from the ligand surface than bind to it.

In the association phase, the measured signal increases rapidly, depending on the concentration C of analyte. Equation (C4.4) can be reformulated as

$$\frac{dR(t)}{dt} = -(k_{on} \times C_k + k_{off})R(t) + k_{on}C_k R_{max} \tag{C4.8}$$

and the k_{on} rate determined from the linear regression plot of $dR(t)/dt$ against $R(t)$ in the association phase (Fig. C4.5). Fast association rates make this estimate quite error-prone, and k_{on} is better established from

$$k_{on} = K_d \times k_{off} \tag{C4.9}$$

Equations C4.7 and C4.8 are now regularly fit against the sensorgram data to yield the k_{on}, k_{off} and K_d of the interaction. Unlike the steady-state analysis based on the R_{max},

---

COMMENT C4.3 THE LANGMUIR ISOTHERM

The Langmuir isotherm was initially proposed for an equilibrium in which a gas is in contact with a solid:

$$X/M = abC_e/(1 + bC_e)$$

where X is the mass of gas adsorbed, M is the mass of the solid, C_e is the equilibrium concentration remaining in solution and a and b are Langmuir constants. The relation is called an *isotherm* because it is only valid at constant temperature.

(a)

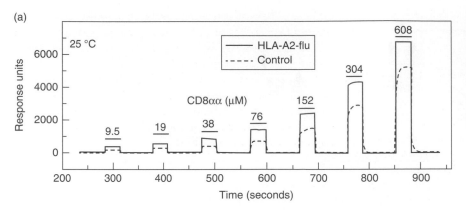

(b)

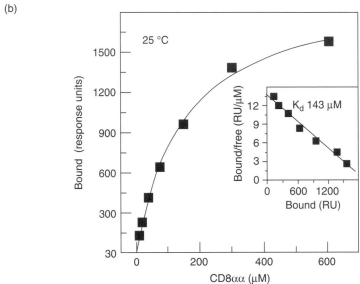

Fig. C4.6 Analysis of the binding between the ligand protein CD8αα and a HLA-A2 molecule (analyte). **(a)** Serial dilutions show a decrease in signal above the control. **(b)** The binding response and Scatchard plot (inset). (Wyer *et al.*, 1999) (Figure reproduced with permission from *Immunity*.)

it is also no longer necessary for all the analyte to dissociate from the chip's surface before addition of further analyte (Fig. C4.7).

Importantly, the correctness of a binding model must be demonstrated by plotting theoretical binding curves using the resultant rates.

Activation Processes

The rate of a reaction measured as a function of temperature provides information on the activation thermodynamics parameters. Recall Eq. (C1.34)

$$\ln\left(\frac{k}{T}\right) = \ln\left(\frac{k_B}{h}\right) + \frac{\Delta S^*}{R} - \frac{\Delta H^*}{RT}$$

In the binding experiment, the rate constant, k, in the equation is equal to k_{off} for the dissociation interaction and $C^{\circ}k_{on}$ for the association reaction, where C° is the standard concentration (1 M). The activation enthalpy and activation entropy are then obtained directly from the slope and intercept of the *Eyring plot*, $\ln(k/T)$ versus $1/T$.

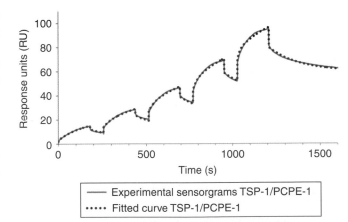

Fig. C4.7 Multiple addition of analyte results in a see-saw shaped sensorgram. Thrombospondin-1 (TSP-1) forms a high-affinity complex with extracellular matrix glycoprotein PCPE-1. Kinetic titration, referred to as single-cycle kinetics, of TSP-1 with PCPE-1. Injection of TSP-1 (12.5–200 nM) over PCPE-1 covalently immobilized on a Biacore CM5 sensor chip. Gray line, experimental sensorgram; broken black line, fitted curve. (Salza *et al.*, 2014).

C4.4 EXPERIMENTAL ANALYSIS

C4.4.1 Scope of Analytes

In early experiments, biosensors were used mainly for analysis of large molecules (>5 kDa), because the refractive index signal is mass-dependent. Biosensors were able to identify *E. coli* strains from which ligands they bound, and to measure binding affinities between viral particles. Current biosensors are sufficiently sensitive to directly detect analytes as small as 100 Da. Even smaller molecules have been characterized indirectly through *inhibition* or *competition assays*, in which the strength of binding is measured by mixing the molecule with another much bulkier molecule that also binds the ligand. The affinity of the smaller molecule for the ligand can be calculated from the assay, which determines the extent to which the bulkier analyte is prevented from binding by the addition of the smaller molecule. A *caveat* of the method is that the two analytes must bind at the same site on the ligand.

C4.4.2 Experimental Controls and Pitfalls

A number of experimental controls must be performed to assess the reliability of the analysis and avoid potential pitfalls. Non-specific binding is assessed by using either blank cells (that do not contain ligand) or cells displaying molecules that are known not to bind. The Scatchard-like plot assumes that the maximum binding capacity of the receptor to the ligand is invariant in time. This is often not the case and experimental procedures must be adapted accordingly. The ligand may lose its binding capacity through, for example, denaturation or degradation. Repeating the experiment at similar analyte concentrations is a potential way of assessing ligand decay.

Concentrations of either ligand or analyte that are too high may cause steric hindrance and prevent molecules from binding due to overcrowding at the flow cell surface. The diffusion of analytes from bulk solvent to the ligand surface may be impeded. This would lead to wrong estimates for the on and off rates. Such *mass transport* effects can be identified by varying the flow rate of the analyte, but also limited by reducing ligand density on the chip surface.

Very fast on-rates will lead to ligand competition for a limited number of analytes as well as rebinding, and to an underestimation of the dissociation rate. Difficulty in measuring high-affinity interactions can also originate from a slow dissociation rate. For example, for an analyte whose k_{off} is in the order of $10^{-6}\,s^{-1}$ it would take almost three hours ($10^4\,s$) for only 1% of bound material to dissociate from the ligand. The observed affinity should be confirmed by swapping the ligand and the analyte with each other. The new measurement series then indicates whether the binding is influenced by the coupling method to the cell, by the presence of aggregates, or by errors in determining concentrations of active analyte.

Biosensor instruments should not be used for certain types of molecular interactions. Attaching the ligand to a surface may alter its binding properties. Binding may also be precluded if large ligand conformational changes or oligomerization are involved.

C4.4.3 Cell–Cell Interactions

Biosensors have been applied successfully to the study of cell–cell interactions, since the immobilization of a normally membrane-bound ligand to a flow cell mimics to a certain extent the in-vivo situation.

Antigen recognition by T-cells is the key event controlling the adaptive immune responses and has been studied extensively with SPR biosensors. The T-cell receptor (TCR) recognizes antigens presented by major histocompatibility complex (MHC) molecules. The antigen is usually a peptide derived from proteins synthesized by the cell. The T-cell checks for the presence of unusual peptide antigens as these indicate that something is wrong. For example, T-cells recognize and destroy cells displaying viral antigens during a viral infection. The interaction between the MHC peptide complex and the TCR controls the fate of the MHC-presenting cell. The measured affinity is usually low ($K_d \sim 0.1$–$500\,\mu M$) as it is due to slow association and fast dissociation reactions. The large k_{off} (~ 0.01–$5\,s^{-1}$) corresponds to half-lives of 70–0.1 s and is highly significant because it indicates that, once formed, the TCR–peptide–MHC complex is more stable than other cell–cell recognition molecule interactions. SPR measurements of the binding at different temperatures further showed that the binding is characterized by unusually favorable enthalpic changes and highly unfavorable entropic changes (Fig. C4.8 (see also Section C3.3.3; Fig. C3.7)).

A number of different MHC–peptide complexes and TCRs have been studied using biosensors, and a broad correlation between affinity/half-life and functional effect has been observed. This suggests that the duration of binding determines the outcome of TCR–peptide–MHC interactions; the longer interaction offers the opportunity for other T-cell molecules to assemble at the contact point between the two cells, and to initiate the death of the antigen-presenting cell.

C4.4.4 SPR and Mass Spectrometry

Combining a biosensor with mass spectrometry makes it possible to link analyte binding and kinetic analysis with its identification.

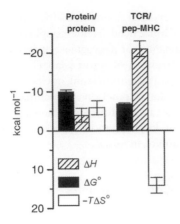

Fig. C4.8 Comparison of the average thermodynamic parameters of 30 protein–protein interactions and of the TCR–peptide–MHC interaction. (Willcox *et al.*, 1999) (Figure reproduced with permission from *Immunity*.)

The biosensor can isolate binding partners for a receptor by analyzing whole-cell lysate. Molecules of interest are separated from the crude extract as they are retained by their ligand in the flow cell. The analyte is then eluted and subsequently identified by mass spectrometry. For proteins, a proteolytic step can be included in the mass spectrometric analysis in order to identify post-translational modifications.

In another application, mass spectrometry and SPR can be combined to analyze the competition of two analytes for the same ligand. Figure C4.9 shows results on the competition between two drugs (saquinavir and indinavir) for a sensor surface containing immobilized HIV-1 protease. Mass spectrometry analysis identified the contribution of each inhibitor to the dissociation phase.

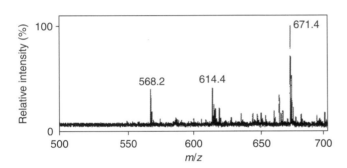

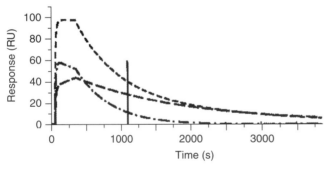

Fig. C4.9 Binding of saquinavir and indinavir to HIV-1 protease. (a) MALDI-TOF spectrum showing saquinavir (670 Da) and indinavir (613 Da) eluted from the surface. (b) Simulated sensorgram from the mass spectrometry data for the dissociation of saquinavir and indinavir to the sensor chip. The vertical line indicates the time at which the ratio between the two inhibitors should be 0.4, corresponding to the concentration ratio between species in (a). Dashed line, saquinavir plus indinavir; long dashed line, saquinavir; dot–dash line, indinavir. (Mattei *et al.*, 2004.)

C4.5 CHECKLIST OF KEY IDEAS

- The phenomenon of SPR occurs when monochromatic, p-polarized light is reflected on a metal-coated interface between two media of different refractive index.
- A biosensor binding experiment involves immobilizing a molecule on a surface (the ligand) and monitoring its interaction with a second molecule in solution (the analyte).
- A change in local macromolecular concentration is associated with a change in refractive index.
- If light is shone upon the interface between two transparent media of different refractive index, above a critical angle of incidence, total internal reflection occurs, i.e., the light coming from the side with higher refractive index reflects back.
- Refractive index changes in a biosensor experiment can be measured by SPR or interferometry.
- The binding progress curve is called a sensorgram.
- A number of tags have been developed to secure the ligand to the sensor chip.
- If the diffusion of the analytes from the bulk solvent to the chip's surface is impeded, this mass transport effect gives misleading binding data.
- Serial dilutions of analytes over a 100-fold range of concentrations are often used in biosensor experiments in order to provide a precise measure of the thermodynamics parameters.
- The dissociation rate k_{off} is deduced from the decrease of the binding response over time.
- The association constant k_{on} rate can be determined from the association phase using different concentrations of analyte.

- The correctness of a binding model must be demonstrated by plotting theoretical binding curves using the resultant rates.
- The *Eyring plot*, k/T against $1/T$, where k is k_{off} or $C^{\circ}k_{on}$, provides the values of the activation entropy and enthalpy of the reaction.
- Analytes can range in size from molecules with a mass of a few hundred daltons to whole cells.
- Cell–cell interactions have been mimicked in biosensor experiments.
- Combining SPR with mass spectrometry makes it possible to link binding and kinetic analysis with analyte identification.

Suggestions for Further Reading

Mullet, W. M., Lai, E. O. C., and Yeung, J. M. (2000). Surface plasmon resonance-based immunoassays. *Methods*, **22**, 77–91.

Schultz, D. A. (2003). Plasmon resonant particles for biological detection. *Curr. Opin. Biotech.*, **14**, 13–22.

Barnes, W. L., Dereux, A., and Ebbesen, T. W. (2003). Surface plasmon subwavelength optics. *Nature*, **424**, 824–830.

Rich, R. L. and Myszka, D. G. (2010). Grading the commercial optical biosensor literature: Class of 2008: "The Mighty Binders." *J. Mol. Recognit.*, **23**, 1–64.

HYDRODYNAMICS

D

BIOLOGICAL MACROMOLECULES AS HYDRODYNAMIC PARTICLES

D1

D1.1 HISTORY AND INTRODUCTION TO BIOLOGICAL PROBLEMS

Traditionally, hydrodynamics deals with the behavior of bodies in fluids and, in particular, with phenomena in which a force acts on a particle in a viscous solution. Very eminent scientists, such as **Isaac Newton**, **James Clerk Maxwell**, **Lord Rayleigh (J. W. Strutt)**, and **Albert Einstein**, started their careers with major contributions to the science of hydrodynamics. Note that not only are the discoveries from more than 100 years ago still highly relevant today, but also that they continue to stimulate important new developments in the field.

1731

The science of hydrodynamics arose from the book *Hydrodynamics* by **Daniel Bernoulli**, which contained the "Bernoulli law" relating pressure and velocity in an incompressible fluid.

1821

Botanist **Robert Brown** described the random, thermal motions of small plant particles suspended in water, a phenomenon that was later named Brownian motion. In **1855 Adolf E. Fick** published a phenomenological description of translational diffusion and deduced the fundamental laws governing transport phenomena in solutions. In the **1990s**, the method of video-enhanced microscopy was proposed for the direct observation of Brownian motion of labeled macromolecules in a membrane.

1845

G. Stokes showed that the translational friction for a sphere is proportional to its radius and to the viscosity of its surrounding solvent. In **1856** he demonstrated that, for small angular velocity, the rotation of the sphere is characterized by a single parameter, which is proportional to the cube of the linear dimension. The way was then clear for a direct determination of particle dimensions from hydrodynamic measurements. In **1880**, **Stokes** analyzed rotational friction and deduced

an expression to relate the rotational friction coefficient of a particle to its volume.

1850

T. Graham observed that egg albumin diffuses much more slowly than common compounds such as salt or sugar. In **1861** he used dialysis to separate mixtures of slowly and rapidly diffusing solutes, and made quantitative measurements of diffusion on many substances. On the basis of this work he classified matter in terms of *colloids* and *crystalloids*.

1879

Publication of *Hydrodynamics* by **Sir Horace Lamb** in **1879**.

1887

Osborne Reynolds pointed out that the ratio of inertial and viscous forces is a key feature of the characterization of any fluid movement. In the **1970s**, **Howard Berg** and **Edward Purcell** applied this idea to describe the movement of different objects (from molecules to animals) in solution. It was shown that the movement of particles with molecular dimensions (between 1 nm and 1 μm) was described in terms of so-called low Reynolds numbers. This means that biological macromolecules "live" in a world without inertia.

1905

Albert Einstein developed the theory of Brownian motion, which enabled him to characterize molecular motions in simple solutions and gases quantitatively. A year later he derived an equation relating the diffusion coefficient of a macromolecule in solution to its coefficient of translational friction and demonstrated that the specific viscosity of a suspension of rigid spheres is proportional to their volume fraction, and is independent of their radius. In **1940 Robert Simha** obtained the equation for the viscosity of a solution of ellipsoids of revolution and in **1981 Stephen Harding** and **Arthur Rowe** solved the viscosity equation for a three-axis ellipsoid.

1926

L. Mandelshtam recognized that the translational diffusion coefficient of macromolecules could be obtained from the spectrum of scattered light. However, the lack of spatial coherence and the non-monochromatic nature of conventional light sources rendered such experiments impossible until **1964**, when **H. Cummins, N. Knable**, and **Y. Yeh** used an optical-mixing technique to resolve spectrally the light scattered from dilute suspensions of polystyrene latex spheres. In **1967**, **S. Dubin, S.J. Lunacek**, and **G. Benedek** measured the translational diffusion coefficients of bovine serum albumin, lysozyme, tobacco mosaic virus, and DNA by dynamic light scattering.

1927

T. Svedberg established that the molecular weight of a protein can be calculated from its sedimentation behavior, partial specific volume, and diffusion coefficient. This stimulated the development of new and improved methods and apparatus for the study of the diffusion process itself.

1964

Herman Z. Cummings, following the theory of **Robert Pekora**, published the first experimental paper on dynamic light scattering and demonstrated that the diffusion coefficients of latex particles in solution can be extracted by this method. This work confirmed the theoretical predictions of **Leonid Mandelshtam**, in **1923**, concerning the modulation of scattered light intensity by Brownian motion. It marked the beginning of a new trend in structural biology for the rapid and accurate determination of macromolecular diffusion coefficients from dynamic light scattering.

1972–1974

D. Magde, **E.L. Elson**, and **W.W. Webb** published a rigorous formalism for fluorescence correlation spectroscopy (FCS), highlighting the great potential of the method for the measurement of macromolecular diffusion coefficients. In **1990**, **R. Rigler** and coworkers reached the single-molecule detection limit by combining FCS with confocal fluorescence microscopy.

1978

D. Teller made a first attempt to calculate the friction coefficient of proteins from atomic coordinates. **R. Venable** and **R. Pastor** calculated the frictional properties of proteins using a detailed picture of the distribution of different amino acids in proteins. **J. Garcia de la Torre** extended the calculations to nucleic acids. In **1995–1999** several approaches were proposed for constructing hydrodynamic models of the biological macromolecules on the basis of their atomic coordinates. Special approaches were put forward to try to model the frictional properties of short DNA fragments and closed circular DNA.

1994–2000

The spectacular progress in solving protein and nucleic acid structures to high resolution by X-ray crystallography and NMR stimulated the development of novel approaches to calculate hydrodynamic parameters from atomic-level structural details. It was shown that the frictional parameters of a protein can be calculated with an accuracy of about 1–3% from its atomic structure by including a hydration shell.

Since 2000

Modern hydrodynamics is undergoing a renaissance, and is one of the recognized approaches for determining the size, shape, flexibility, and dynamics of biological macromolecules. Modern hydrodynamics includes many novel experimental physical methods: fluorescence photobleaching recovery to monitor the mobility of individual molecules within living cells; time-dependent fluorescence polarization anisotropy to calculate Brownian rotational diffusion coefficients for macromolecules; and fluorescent correlation spectroscopy and localized dynamic light scattering to study the dynamical properties of macromolecules. But in spite of all these achievements we must remember that *hydrodynamics is a low-resolution method* (Comment D1.1). It operates on a few parameters only. The highest level of data interpretation that can be achieved by using direct methods is to define a particle as a three-axis body (Section D2.5.3).

COMMENT D1.1 BIOLOGIST'S BOX: THE UNITS OF FORCE AND VISCOSITY IN HYDRODYNAMICS

Because hydrodynamics is a technique developed decades ago, calculations have traditionally been performed in cgs units.

The dyne is the unit of force in the cgs system. One dyne is the force necessary to accelerate a one-gram mass by one centimeter per second per second:

$$dyne = g\,cm\,s^{-2}$$

All fluids possess a definite resistance to change of shape. This property, a sort of inertial friction, is called viscosity. The unit of viscosity, defined as the tangential force per unit area ($dyne\,cm^{-2}$) required to maintain unit difference in velocity ($1\,cm\,s^{-1}$) between two parallel planes separated by $1\,cm$ of fluid, is the poise:

$$1\,poise = 1\,dyne\,s\,cm^{-2} = g\,cm\,s^{-1}$$

Kinematic viscosity is the ratio of viscosity to density. The cgs unit of kinematic viscosity is the stoke:

$$1\,stoke = 1\,cm^2\,s^{-1}$$

D1.2 HYDRODYNAMICS AT A LOW REYNOLDS NUMBER

In order to construct reasonable physical models for flow systems involving biological particles, it is necessary to make a number of simplifications. In this section it is assumed that the flow is laminar and, further, that it is sufficiently "slow" that inertial effects need not be considered in the equations of motion, which describe the movement of particles relative to fluid (solvent). The approximation is justified, since biological systems of interest consist of very small particles, and even though the particles move rapidly with respect to the container wall, in, for example, the viscosity and flow birefringent methods, they still move slowly with respect to the fluid surrounding them.

D1.2.1 Reynolds Number

We consider an object moving with some velocity through a fluid of specific density and viscosity. The Reynolds number, R, is a dimensionless parameter, which determines the relative importance of inertial and viscous effects

$$\text{Reynolds number} = \frac{\text{fluid density} \times \text{speed} \times \text{particle size}}{\text{viscosity}}$$

$$= R = \frac{\rho u l}{\eta} \qquad (D1.1)$$

The ratio was proposed as a significant intrinsic number to characterize a system more than 100 years ago by Reynolds. When the Reynolds number is low, viscous forces dominate. If it is high, inertial forces dominate.

D1.2.2 Movement at Low Reynolds Number

We calculate the Reynolds number for a virus $500\,\text{Å}$ ($5 \times 10^{-6}\,\text{cm}$) in diameter moving in water with a speed of

order $10^{-3}\,\text{cm}\,\text{s}^{-1}$. Taking $\rho = 1\,\text{g}\,\text{cm}^{-3}$ and $\eta = 10^{-2}\,\text{g}\,\text{cm}^{-1}\,\text{s}^{-1}$, we obtain a Reynolds number of 5×10^{-7}, i.e., the Reynolds number for the virus is negligibly small. A small Reynolds number means that the virus molecule will stop moving immediately when the force acting on it disappears. Of course, the virus is still subject to Brownian motion, so in reality it does not stop.

Calculations show that for large biological complexes including bacteria in water, the Reynolds number is also very small (Comment D1.2). So all biological macromolecules from small proteins to bacteria live in a world without inertia, where viscous forces predominate (Comment D1.3).

D1.3 HYDRATION

In hydrodynamic experiments, a biological macromolecule moves with a certain amount of bound solvent, thus defining the concept of a hydrated particle as a core of particle material and an envelope of bound water (see Section A3.3.3). Figure D1.1 shows that hydration is manifested as increased size or volume of the core particle. The hydrated volume, V_{hyd}, is larger than the "dry" volume, V_{anh}, which can be obtained from molecular mass, M, and the partial specific volume, $\bar{v}$, of the protein:

$$V_{anh} = M\bar{v}/N_A \qquad (D1.2)$$

Protein hydration, δ, which is denoted in $g\,g^{-1}$, expresses the ratio of the mass of the bound water to that of the protein

$$\delta = \frac{\text{grams (water)}}{\text{grams (protein)}} \qquad (D1.3)$$

If ρ is the density of the solvent, then we have

$$\delta = (V_{hyd}/V_{anh} - 1)\rho\bar{v} \qquad (D1.4)$$

There are two interpretations of the δ value. The first is based on the uniform expansion hypothesis (Fig. D1.2). It originates in the classical representation of globular proteins as ellipsoidal particles. For a particle of arbitrary shape, uniform expansion assumes that the linear dimension, l, of the particle is expanded by constant factor, h,

---

COMMENT D1.2 DIFFERENT OBJECTS IN WATER

E. Purcell was the first to calculate the Reynolds numbers for bacteria and fish (Purcell, 1977). He considered that a bacterium is $10^{-4}\,\text{cm}$ in diameter and swims with a velocity of the order of $2 \times 10^{-3}\,\text{cm}\,\text{s}^{-1}$. Taking $\rho = 1\,\text{g}\,\text{cm}^{-3}$ and $\eta_0 = 10^{-2}\,\text{g}\,\text{cm}^{-1}\,\text{s}^{-1}$, he obtained a Reynolds number of 10^{-5}, i.e., very small. The bacterium therefore lives in a world *without inertia*.

The same calculation for a fish of length $l = 10\,\text{cm}$, moving with velocity $\sim 100\,\text{cm}\,\text{s}^{-1}$ in water yields a Reynolds number of about 10^5. This is an example of hydrodynamics at a high Reynolds number. The fish lives in a water medium *with inertia*.

For a whale $l = 10\,\text{m}$ ($1000\,\text{cm}$) moving with velocity $36\,\text{km}\,\text{h}^{-1}$ ($1000\,\text{cm}\,\text{s}^{-1}$) in water the Reynolds number is about 10^8. The whale swims in a water medium with *very large inertia*.

COMMENT D1.3 DEFINITION OF MOVEMENT AT VERY LOW REYNOLDS NUMBER

The best definition of movement at very low Reynolds number is by E. Purcell: "What You are doing at the moment is entirely determined by the forces that are exerted on You *at the moment*, and by *nothing in the past*." (Purcell, 1977).

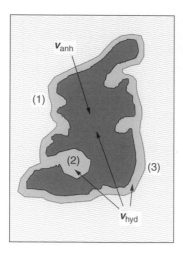

Fig. D1.1 Schematic presentation of a rigid hydrated particle. Hydration influences the overall shape of the protein in the sense of smoothing out some structural details such as pockets or cavities. (After Garcia de La Torre, 2001.)

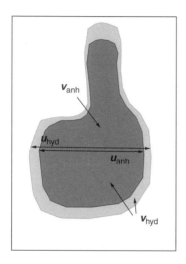

Fig. D1.2 Illustration of the δ value as a uniform expansion in globular proteins. (After Garcia de La Torre, 2001.)

$$h = u_{\text{hyd}}/u_{\text{anh}} \qquad \text{(D1.5)}$$

such that

$$h^3 = V_{\text{hyd}}/V_{\text{anh}} \qquad \text{(D1.6)}$$

It follows that in this representation h is related to δ by

$$h = \left(1 + \frac{\delta}{\bar{v}\rho}\right)^{1/3} \qquad \text{(D1.7)}$$

The uniform expansion is applicable for compact particles, but is not realistic for very elongated or rod-like particles (Comment D1.4).

The second interpretation of the δ value is based on the assumption that the anhydrous core is coated by a bound water shell which has a constant thickness t_{h}, measured in

COMMENT D1.4 UNIFORM EXPANSION FOR COMPACT AND ELONGATED PARTICLES

For a typical globular protein in water if $\rho = 1$ cm^3 g^{-1}, $\bar{v} = 0.73$ g cm^{-3} and $\delta = 0.3$ g g^{-1}, then $h = 0.12$, which corresponds to 12% in linear dimensions and 41% in volume.

For a very elongated particle that is 20 Å in diameter and 200 Å in length, hydration leads to an increase in diameter to approximately 22 Å. The same hydration applied to the particle length leads to an increase to 280 Å, i.e., 40 Å at each end. Evidently this result is not realistic because it leads to abnormal hydrodynamic solution properties.

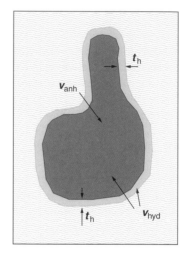

Fig. D1.3 Illustration of the interpretation of the δ value as a thick uniform hydration shell; t_{h} is the average thickness of the hydration shell. (After Garcia de La Torre, 2001.)

the direction normal to the protein surface (Fig. D1.3). The hydration shell is considered as an intrinsic property of all proteins (Chapter D3).

Early interpretations of hydrodynamic data led to hydration levels that varied widely from protein to protein. For example, hydration values deduced from diffusion coefficients and intrinsic viscosity are in a broad range, from 0.1 to more than 1 g of water per gram of protein. These results were obtained by modeling proteins as spheres or ellipsoids, for which the translation, friction, and intrinsic viscosity are known analytically.

The use of models obtained from detailed protein structures for the calculation of translation, rotational friction, and intrinsic viscosity leads to a much smaller hydration range, from 0.3–0.4 g of water per gram of protein, corresponding to less than a single molecular layer in the hydration shell. The development of this type of calculation (Chapter D3) allows hydrodynamic measurements to join

The estimation of hydration from hydrodynamic properties of a protein is sensitive to several types of error because the extent of hydration is determined as "a small difference of two large values" for the hydrated and dry volumes (Eq. (D1.4)). Its main source of uncertainty is found in the experimental errors in the data of hydrodynamic and other solution properties. Many of the tabulated data for common proteins are up to 60 years old, and it is evident that for a quantitative, more accurate evaluation of hydration more precise data are required.

We note here the link developed in the 1980s between hydrodynamics and small-angle scattering for the study and determination of macromolecular hydration, described in more detail In Chapter G2.

Mathematical definitions of stick and slip boundary conditions are particularly complex. Interested readers can find them in the specialized literature (Hu and Zwanzig, 1974).

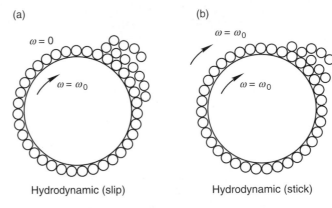

Fig. D1.4 Hydrodynamic slip (a) and stick (b) boundary conditions; ω is the frequency of rotation.

other techniques such as NMR, infrared spectroscopy, calorimetry, and small-angle X-ray and neutron scattering to provide a unified picture of protein hydration.

The following picture of protein hydration has emerged (Fig. D1.3). A protein is hydrated at a definite level corresponding to a 1.2 Å thick hydration shell on average. The local density of water in the hydration shell is about 10% higher than that of bulk water. If in a hydrodynamic experiment a hydration value that differs greatly from the usual levels were required to fit the data, this is an indication that the hydrodynamic equivalent model used is probably wrong (Comment D1.5).

Because of their net negative charge in solution, more water is associated with RNA or DNA molecules, leading to salt-dependent hydration values of about $0.6 \pm 0.2\,\mathrm{g\,g^{-1}}$. For glycosylated proteins and for carbohydrates, hydration values are also larger $(0.5\,\mathrm{g\,g^{-1}})$ owing to the generally higher affinity for water of these glycopolymers.

D1.4 FRICTION

D1.4.1 "Stick" and "Slip" Boundary Conditions

In classical hydrodynamic theory two extreme cases of solvent–particle interaction called "slip" and "stick" boundary conditions are usually considered (Comment D1.6). In the "slip" approximation, there is no interaction between solvent and particle and the solvent slips over the particle surface (Fig. D1.4a). The other extreme is represented by the "stick" approximation, in which the first solvent layer sticks to the particle surface and moves with it (Fig. D1.4b).

It is important to note that the value of the coefficient in the equation connecting measured and calculated hydrodynamic values depends on the boundary conditions (see Eqs. (D1.11), (D1.12), (D1.14), and (D1.15) and (D1.16)). It is generally accepted that for small protein molecules

(<2000 Da) "slip" conditions are more applicable, whereas for the large Brownian particles (>5000 Da) "stick" boundary conditions hold.

D1.4.2 Hydrodynamic Quantities

The hydrodynamic methods presented in Table D1.1, below, allow the *translational* and *rotational friction coefficients*, and the *intrinsic viscosity* of biological macromolecules to be determined. These, in turn, depend on the viscosity of the solvent and also on particle properties, which are of significant interest in the characterization of macromolecular structures and interactions.

Translational Friction Coefficient

The translational friction influences translational diffusion, high-speed sedimentation and electrophoretic mobility. In each case, a force **F** acts on the particles and causes them to accelerate. The movement of a particle of mass m due to this force is described by the fundamental relation in mechanics: Force is equal to mass times acceleration

$$\mathbf{F} = m\,\mathrm{d}\mathbf{u}/\mathrm{d}t \tag{D1.8}$$

where **u** is the particle velocity and t is time.

In a viscous solution, the motion is opposed by solvent drag. The force, $\mathbf{F}_{\mathrm{frict}}$, due to this friction is proportional to the velocity of the particle and is in the opposite direction (Fig. D1.5). The proportionality constant is defined as the friction coefficient f:

TABLE D1.1 METHODS CURRENTLY USED TO DETERMINE DIFFUSION COEFFICIENTS OF BIOLOGICAL MACROMOLECULES

Approach	Experimental method	Diffusion coefficient
Observations of the average microscopic motion of particles in an ensemble	Fluorescence photobleaching recovery (FPR Section D1.5.5)	Tracer diffusion coefficient, D_{tracer}
Direct observation of the diffusional motion of single fluorescent particles	Confocal fluorescent microscopy (Section F1.3.1)	Tracer diffusion coefficient, D_{tracer}
Stochastic appearance and disappearance of molecules in a small volume	Number fluctuation in dynamic light scattering (Chapter D4)	Tracer diffusion coefficient, D_{tracer}
Stochastic appearance and disappearance of fluorescent molecules in a very small volume	Fluorescence correlation spectroscopy (Section D4.6)	Tracer diffusion coefficient, D_{tracer}
Mutual motion of particles in an assembly	Dynamic light scattering (Chapter D4)	Mutual diffusion coefficient, D_{mutual}
Macroscopic change in concentration	Spreading boundary technique (Section D1.5.5)	Translational diffusion coefficient, D_{transl}

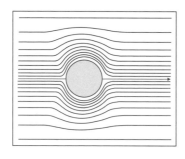

Fig. D1.5 Streamlines for flow around a sphere. The sphere is moving to the right at a constant velocity in a stream of viscous liquid. Solvent drag creates a force on the particle opposite to the velocity direction.

COMMENT D1.7 UNITS IN COMMONLY USED CGS UNITS

The translational frictional coefficient is in "grams per second" ($g\,s^{-1}$).

The unit of the rotational frictional coefficient is "per second" (s^{-1}).

Specific viscosity is dimensionless.

$$\mathbf{F}_{frict} = -f\mathbf{u} \qquad (D1.9)$$

The negative sign denotes that the force is in the direction opposite to the velocity (see Comment D1.7).

When the two opposing forces are equal in magnitude, acceleration is nil, and the particle, therefore, moves with a constant velocity, $\mathbf{u}$, given by

$$\mathbf{u} = \mathbf{F}/f \qquad (D1.10)$$

where $\mathbf{F}$ is the magnitude of each force, when $\mathbf{F} = -\mathbf{F}_{frict}$.

Equation (D1.10) relates this constant velocity with the frictional coefficient and the magnitude of the two

COMMENT D1.8 TIME REQUIRED TO REACH CONSTANT VELOCITY FOR MACROMOLECULES

We can estimate from Eq. (D1.10) that the time required to reach constant velocity in a macromolecular solution is very small. The molar mass of a 300 amino-acid residue protein is about $33\,000\,g\,mol^{-1}$; its f value is $5 \times 10^{-8}\,g\,s^{-1}$, and m is about $5 \times 10^{-20}\,g$. The final velocity is achieved in no longer than $10^{-12}\,s$ (1 ps). This is very fast, and is close to the relaxation time of thermal vibrations in the molecule.

opposing forces. It provides the basis for the determination of the translational friction coefficient by the experimental methods in Table D1.1.

It follows from Eq. (D1.9) that the greater the particle friction, the greater the force that needs to be applied to make it move with constant velocity. If the ratio of the applied force to the coefficient of friction ($\mathbf{F}/f$) is large, the stable velocity may be too high to be reached in the experiment. However, if the ratio $\mathbf{F}/f$ is sufficiently small, the final velocity is achieved almost instantaneously after application of the force (Comment D1.8).

Stokes derived the relation for a spherical particle in two extreme solvent interaction cases (Comment D1.9). In the "stick" approximation (Stokes' law), the relation between the translational friction coefficient f and solvent viscosity η_0 for a sphere of radius R_0 is

$$f = 6\pi\eta_0 R_0 \qquad (D1.11)$$

The Stokes relation for "slip" conditions is

$$f = 4\pi\eta_0 R_0 \qquad (D1.12)$$

Thus the friction is reduced by one-third when the boundary conditions are changed.

COMMENT D1.9 ON THE SIMILARITY OF EQUATIONS DESCRIBING THE PROPERTIES OF A SPHERE IN REACTION KINETICS, ELECTROSTATICS, AND HYDRODYNAMICS

It is interesting to note that mathematical equations describing the properties of a sphere (with radius R_0) in reaction kinetics, electrostatics, and hydrodynamics are similar. Indeed, the diffusion-controlled reaction rate of particles (with diffusion constant, D) wandering toward an absorbing sphere is given by

$$k = 4\pi D R_0$$

In electrostatics, the capacitance, C, of a conducting sphere is

$$C = R_0$$

The translational friction coefficient, f of a sphere in a solvent with viscosity η_0 under "stick" boundary condition is given by Stokes' law (Eq. (D1.11))

$$f = 6\pi\eta_0 R_0$$

All three quantities are proportional to the radius of a sphere (they scale as the radius of the sphere).

Rotational Friction Coefficient

Rotational friction influences rotational diffusion, the Kerr effect, the Maxwell effect, and fluorescence polarization (see Section D1.7.2). In analogy with translational motion we may define a rotation frictional coefficient. If a constant torque $\mathbf{F}_{rot}$ is placed on a particle in a fluid, the particle will reach a constant angular velocity, ω, after a transient period (Fig. D1.6). The parameter relating the angular velocity to torque is the rotational friction coefficient θ:

$$\omega = \mathbf{F}_{rot}/\theta \tag{D1.13}$$

This equation is the analogue of Eq. (D1.4). It provides a basis for the determination of the rotational friction coefficient by many of the experimental methods presented in Table D1.1.

According to Stokes, in the "stick" approximation the relation between the rotational friction coefficient and the solvent viscosity for a sphere of volume V is given by:

$$\theta = 8\pi\eta_0 V \tag{D1.14}$$

In the case of "slip" boundary conditions, the rotational friction coefficient is zero.

Intrinsic Viscosity

The viscosity of a pure solvent is denoted by η_0. Addition of macromolecules should raise the viscosity to a new value η. This occurs because the large macromolecules, which extend across the streaming lines, greatly enhance resistance to flow. As a result, the viscosity of a macromolecular solution is always greater than that of pure solvent.

In viscosity measurements we determine the local energy dissipation produced when large particles are introduced into a solvent (Fig. D1.7). Let η_0 denote the solvent viscosity in the absence of particles and η the viscosity when the number concentration of particles is C. It is assumed that the solution contains a monodisperse suspension of spherical particles. Einstein's relation for specific viscosity, in the case of "stick" conditions, is

$$(\eta - \eta_0)/\eta_0 = 2.5\Omega \tag{D1.15}$$

where Ω is the volume fraction of the solution occupied by spheres. Under "slip" conditions the relation for the specific viscosity is

$$(\eta - \eta_0)/\eta_0 = 1\Omega \tag{D1.16}$$

It should be noted that radius of the spheres does not enter into Eqs. (D1.15)–(D1.16).

Hydrodynamic Equivalent Bodies

Equations (D1.11), (D1.14), and (D1.15) are very useful for work on large biological macromolecules (more than several kilodaltons). They open the way to express a measured value for the friction coefficients and viscosity in terms of the radius, R_0, of a hydrodynamic equivalent sphere. R_0 has been named the Stokes radius. The concept can be extended to other hydrodynamic equivalent shapes (Fig. D1.27).

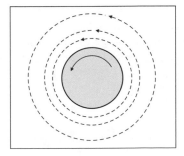

Fig. D1.6 Streamlines for flow around a sphere. The sphere rotates about an axis normal to the plane of the page with constant angular velocity, ω.

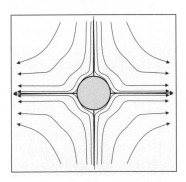

Fig. D1.7 Streamlines for flow around a sphere in a liquid.

It should be clearly understood that the description of the particle in terms of a hydrodynamic equivalent does not mean that it has this shape, only that it has the same hydrodynamic properties. In spite of the fact that this appears to be a rough approximation, in many cases the hydrodynamic equivalent body describes the behavior of the particle in a real experiment quite reliably.

In the case of a rigid sphere, only one parameter (the sphere volume or its radius) is sufficient to describe it fully. To describe an ellipsoid of revolution (the shape of a rugby football), it is necessary to determine two parameters, characterizing its volume and axial ratio. To describe a three-axis ellipsoid of rotation, three parameters are required, characterizing its volume and two axial ratios.

The groups of methods presented in Table D1.1 differ in their comparative sensitivity relative to these parameters. Thus, all methods based on translational friction are sensitive to the linear dimensions of the particle, whereas methods based on rotational friction and viscosity are sensitive to its volume (the cube of linear dimensions).

D1.5 DIFFUSION

D1.5.1 Translational Diffusion Coefficients

The translational diffusion coefficient, D_{transl}, can be determined by different types of measurement. Historically, D_{transl} was defined via macroscopic fluctuations in the local concentration (Fig. D1.8) and determined experimentally by the spreading boundary technique (Section D1.5.5).

The tracer diffusion coefficient, D_{tracer}, is obtained from a *microscopic* approach, in which the average motion of individual particles is considered (Fig. D1.9). Experimentally, D_{tracer} is determined by FPR photobleaching recovery (Section D1.5.5), nanovid microscopy (Section D1.5.5), and number-fluctuation techniques in dynamic light scattering under non-Gaussian statistics (Section D4.5) or fluorescence correlation spectroscopy (Section D4.6).

The mutual diffusion coefficient, D_{mutual}, is also a *microscopic* diffusion coefficient. It describes the mutual motions of particles in an assembly (Fig. D1.10). Experimentally, D_{mutual} can be determined by dynamic light scattering under Gaussian statistics (Section D4.4).

In the case of non-interacting particles, the three types of diffusion measurement lead to the same result, the D_{transl} diffusion coefficient. Table D1.1 presents a summary of hydrodynamics methods currently in use for the determination of the various diffusion coefficients. The methods in Table D1.1 are listed according to the approaches defined above, the experimental techniques used and the type of diffusion coefficient. The chapters and sections in

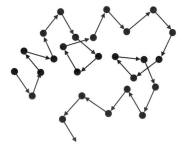

Fig. D1.9 "Tracer" diffusion as chaotic motion of a single particle.

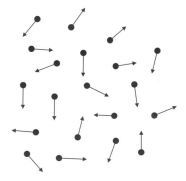

Fig. D1.10 "Mutual" diffusion as a mutual motion of particles in an assembly.

Fig. D1.8 Diffusion as a macroscopic change in concentration.

which each experimental technique is treated in detail and applications are discussed are also given.

As we have already noted, for non-interacting Brownian particles the three diffusion coefficients provide equivalent results, i.e., D_{tracer}, D_{mutual}, and D_{transl} are identical.

D1.5.2 Microscopic Theory of Diffusion

The diffusion of macromolecules in solution can be considered as a phenomenon connected with Brownian motion. Owing to thermal energy, macromolecules in solution are in permanent chaotic motion (Comment D1.10).

It was shown by Einstein that for one-dimensional diffusion the mean square displacement $\langle x^2 \rangle$ is proportional to time t:

$$\langle x^2 \rangle = 2D_1 t \tag{D1.17}$$

and

$$\langle x^2 \rangle^{1/2} = (2D_1 t)^{1/2} \tag{D1.18}$$

Equations (D1.17) and (D1.18) show that a knowledge of the mean-square displacement $\langle x^2 \rangle$ is sufficient to determine D_1 and vice versa.

Consider that motions in the x-, y-, and z-directions are independent. If $\langle x^2 \rangle = 2D_1 t$, then $\langle y^2 \rangle = 2D_2 t \langle z^2 \rangle = 2D_3 t$. In two dimensions, the square of the distance from the origin to the point (x, y) is $r^2 = x^2 + y^2$; therefore

$$\langle r^2 \rangle = 4 Dt \tag{D1.19}$$

where D is the average of the diffusion coefficients D_1 and D_2 in a two-dimensional random walk. In three dimensions, $r^2 = x^2 + y^2 + z^2$, and

$$\langle r^2 \rangle = 6 Dt \tag{D1.20}$$

where D is average of the D_1, D_2, and D_3 diffusion coefficients in a three-dimensional random walk. Again, knowledge of the $\langle r^2 \rangle$ is sufficient to determine D and vice versa.

COMMENT D1.10 MEAN-SQUARE VELOCITY OF A PARTICLE IN BROWNIAN MOTION

A particle at absolute temperature T has on average a kinetic energy associated with movement along each axis of $kT/2$, where k is Boltzmann's constant. A particle of mass m and velocity v_x on the x-axis has a kinetic energy $mv_x^2/2$. On average $\langle mv_x^2/2 \rangle = kT/2$, where $\langle\rangle$ denotes an average over time or over an ensemble of similar particles. From this relationship we can compute the mean-square velocity as

$$\langle v_x^2 \rangle = kT/m \tag{A}$$

and the root-mean-square velocity as

$$\langle v_x^2 \rangle^{1/2} = (kT/m)^{1/2} \tag{B}$$

Diffusion coefficients determined using Eqs. (D1.17), (D1.19), and (D1.20) should be considered as *tracer* diffusion coefficients, D_{tracer}. Four experimental methods currently used for determining D_{tracer} are presented in Table D1.1. Examples of calculations of the diffusion–velocity and diffusion–time relationships are given in Comments D1.11 and D1.12.

D1.5.3 Macroscopic Theory of Diffusion and Fick's Equations

Fick's First Equation

Fick's first equation states that the net flux J_x is always proportional to the first power of the solute concentration gradient dC/dx with $-D$ as the constant proportionality:

$$J_x = -D[dC/dx] \tag{D1.21}$$

If the particles are uniformly distributed, the slope is 0, i.e., $dC/dx = 0$ and the system is at equilibrium. If the slope is constant, i.e., $dC/dx = $ constant, J_x is also constant. This occurs when C is a linear function of x, as shown in Fig. D1.11.

Fick's Second Equation

Fick's second equation follows from the first, provided that the total number of the particles is conserved, i.e., that particles are neither created nor destroyed. Consider the box of volume $A\delta$ shown in Fig. D1.12. In the period

---

COMMENT D1.11 BIOLOGIST'S BOX: DIFFUSION AND VELOCITY

From Eq. (A) in Comment D1.10 we can calculate the instantaneous velocity of a small particle. Consider two examples.

Sucrose in a Vacuum

Sucrose has a molecular mass of 342 Da. This is the mass in grams of one mole or 6.02×10^{23} molecules; the mass of one molecule is $m = 5.7 \times 10^{-23}$ g. The value of kT at room temperature, 293 K, is 4.04×10^{-14} g cm^2 s^{-2}. Therefore, $\langle u_x^2 \rangle^{1/2} = 8.3 \times 10^3$ cm s^{-1}. If collisions were absent the sucrose molecule would cross a typical swimming pool in about 1 s. According to Eq. (B) in Comment D1.10 the velocity of a small particle is inversely proportional to the square-root of its molecular weight.

Biological Macromolecules

For proteins with a typical molecular mass of 20 kDa the velocity is 1.09×10^3 cm s^{-1}. For the DNA with a molecular mass of 900 000 kDa the velocity is 33 cm s^{-1}.

From Eq. (D1.18) we can calculate the instantaneous velocity of a small particle. Consider a few examples.

Urea in Water

The diffusion coefficient of urea in water is $118\,cm^2\,s^{-1}$. A particle with such a diffusion coefficient diffuses a distance $x = 10^{-4}\,cm$ in a time $t = x^2/2D = 5 \times 10^{-4}\,s$, or about 0.5 ms. It diffuses a distance $x = 1\,cm$ in a time $t = x^2/2D = 5 \times 10^5\,s$, or about 14 h. It is clear that diffusive transport takes a long time when distances are large.

Proteins in Water

A particle with a diffusion coefficient D of the order $10^{-6}\,cm^2\,s^{-2}$ (a small globular protein) diffuses a distance $x = 10^{-4}\,cm$ in a time $t = x^2/2D = 5 \times 10^{-3}\,s$, or about 5 ms. It diffuses a distance $x = 1\,cm$ in a time $t = x^2/2D = 5 \times 10^{-5}\,s$, or about 138 h.

Tobacco Mosaic Virus in Water

The tobacco mosaic virus has a small diffusion coefficient ($0.44 \times 10^{-7}\,cm^2\,s^{-2}$) and diffuses very slowly: It travels a distance $x = 0.5\,cm$ in a time $t = x^2/2D = 2.5 \times 10^6\,s$, or about 700 h. This explains why the moving boundary method in which the minimal required diffusing distance is about a few millimeters is never used for measurements of diffusion coefficients of such large molecules.

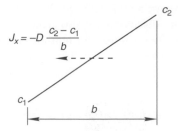

Fig. D1.11 The flux due to a concentration gradient. Molecules move from right to left only because there are more particles on the right than on the left. (After Berg, 1983.)

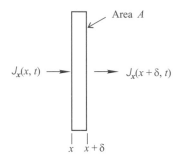

Fig. D1.12 Fluxes through the faces of a thin box extending from position x to position $x+\delta$. The area of each face is A.

of time τ, $J_x(x)A\tau$ particles will enter from the left and $J_x(x+\delta)A\tau$ will leave from the right. The number of the particles per unit volume in the box therefore increases at the rate

$$\frac{1}{\tau}[C(t+\tau) - C(\tau)] = \frac{1}{\tau}[J_x(x+\delta) - J_x(x)]\frac{A\tau}{A\delta}$$
$$= -\frac{1}{\delta}[J_x(x+\delta) - J_x(x)]$$

In the limit $\delta \to 0$ and $\tau \to 0$ this means that

$$dC/dt = -dJ_x/dx \qquad (D1.22)$$

or using Fick's first law (Eq. (D1.21)), that

$$dC/dt = D d^2C/dx^2 \qquad (D1.23)$$

This equation states that the time rate of change in concentration, dC/dt, is proportional to the second concentration derivative of the solute, d^2C/dx^2, where D is the proportionality constant.

In three dimensions the concentration changes in time as

$$dC/dt = D\nabla^2 C \qquad (D1.24)$$

where ∇^2 is the three-dimensional Laplacian, $d^2/dx^2 + d^2/dy^2 + d^2/dz^2$.

If the problem is spherically symmetric, the flux is radial

$$J_r = -DdC/dr, \qquad (D1.25)$$

and

$$dC/dt = D(1/r^2)d(r^2 dC/dr)/dr \qquad (D1.26)$$

D1.5.4 Solutions to Fick's Equations

Time-Dependent Solutions of Fick's Equations

Consider that we have diffusion from a column of liquid initially containing particles at concentration C_0 into a column of liquid that initially does not contain any particles (Fig. D1.13). In this case, the initial conditions are $C = C_0$ for $x > 0$ and $C = 0$ for $x < 0$ and Eq. (D1.24) has the solution

$$C(x,t) = \frac{C_0}{2}\left[1 - 2/\pi^{1/2}\right]\int_0^{(x/2)(Dt)^{1/2}} \exp\left(-y^2\right) dy \qquad (D1.27)$$

The integral in Eq. (D1.27) is known as the probability integral; it is a function of $(x/2)(Dt)^{1/2}$ and varies in value from 0 to 1/2 as $(x/2)(Dt)^{1/2}$ varies from 0 to ∞. A graphical representation of the Eq. (D1.27) is shown in Fig. D1.14a. By taking derivatives of $C(x,t)$ with respect to x or t, we obtain

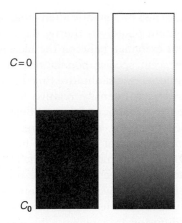

Fig. D1.13 Diffusion from a column of liquid initially containing particles at a concentration C_0 (bottom left) into a column of liquid initially containing no particles (top left). Spreading of the boundary after a certain time (right).

$$dC/dx = C_0/(4\pi Dt)^{1/2} \exp\left(-x^2/4Dt\right) \tag{D1.28}$$

The behavior of this equation is shown in Fig. D1.14b, while Eq. (D1.30) shows that the maximum value of dC/dx will occur at $x = 0$, and its value is $C_0/(4\pi Dt)^{1/2}$.

Steady-State Solutions of Fick's Equations

In the steady-state limit $dC/dt = 0$, and Eq. (D1.24) reduces to

$$\nabla^2 C = 0 \tag{D1.29}$$

For the case of spherical symmetry

$$(1/r)d\left(r^2 dC/dr\right)dr = 0 \tag{D1.30}$$

One other important case is diffusion to a spherical absorber. Consider a spherical absorber of radius R_0 in an infinite volume, as shown in Fig. D1.15. Each particle reaching the surface of the sphere is captured, so that the concentration on the surface at $r = R_0$ is 0. At $r = \infty$ the

concentration is C_0. With these boundary conditions Eq. (D1.29) has a solution

$$C(r) = C_0(1 - R_0/r) \tag{D1.31}$$

and the flux, Eq. (D1.22), is

$$J_r(r) = DC_0 R_0/r^2 \tag{D1.32}$$

If we define the diffusion current I as $I = 4\pi r^2 J_r(r)$, then

$$I = 4\pi DR_0 C_0 \tag{D1.33}$$

if C_0 is expressed in particles per cubic centimeter and I is in particles per second. This current is proportional not to the area of the sphere but to its radius. It comes from the fact that as the radius R_0 increases, the area increases as R_0^2, but the concentration gradient to which flux is proportional decreases as $1/R_0$.

It is important to note that Eq. (D1.29) is analogous to Laplace's equation for the electrostatic potential in charge-free space. This implies that the diffusion current to an isolated absorber of any size and shape can be written as

$$I = 4\pi DCC_0 \tag{D1.34}$$

where C is the electrical capacitance (in cgs units C is in centimeters, see Comment D1.9) of an isolated conductor of that size and shape. The resemblance between Eqs. (D1.33) and (D1.34) is evident. Since the electrical capacitance of bodies with different shapes is known, now we can use Eq. (D1.34) in many practical cases to calculate the frictional properties of molecules.

D1.5.5 Experimental Methods for Directly Determining Diffusion Coefficients

Spreading Boundary Technique

In this technique, a sharp boundary is initially set up between a solution of uniform concentration C_0 and pure solvent. The concentration, C, as a function of distance, x, from the boundary at a given time is then measured, for

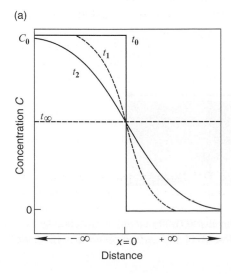

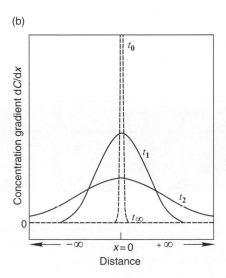

Fig. D1.14 (a) Concentration as a function of position at times t_0, t_1, t_2, and t_∞. The initial concentration of particles at time t_0 is C_0. At infinite time, the concentration is uniform throughout the column and equals $C_0/2$. **(b)** Derivatives of concentration with respect to x as a function of position at times t_0, t_1, t_2, and t_∞ for the same process as described in **(a)**.

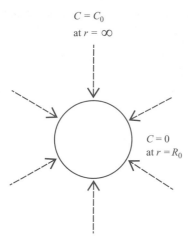

$C = C_0$
at $r = \infty$

$C = 0$
at $r = R_0$

Fig. D1.15 A spherical absorber of radius R_0 in an infinite medium containing particles at initial concentration C_0. The dashed arrows are lines of flux.

example, by monitoring the light absorption peak of the diffusing solute (Eq. (D1.27)). A value of D at each time t is calculated from $C(x)$.

An alternative analysis of the experimental concentration dependence is based on Eq. (D1.28), which describes the dependence dC/dx on x (Fig. D1.14b). The maximum height of the curve, H_{max}, is equal to $kC_0/(4\pi Dt)^{1/2}$ and the area, A, under the curve is equal to kC. The diffusion coefficient is calculated from

$$(A/H_{max})^2 = 4\pi Dt \qquad (D1.35)$$

Fluorescence Photobleaching Recovery

Fluorescence photobleaching recovery (FPR), also called *fluorescence recovery after photobleaching* (FRAP), is a tracer technique that measures the diffusion of a labeled solute. If the solution contains differently labeled species, their transport coefficients can be obtained separately from the mobility of each tracer.

The method is simple both in concept and in application. A small volume containing mobile fluorescent molecules is exposed to a brief, intense pulse of light, which causes irreversible photochemical bleaching of the fluorophore in that region (Fig. D1.16). Fluorescence in the bleached

region is then excited by a greatly attenuated beam in order to avoid significant photolysis during the recovery phase. The subsequent exchange between the bleached and non-bleached fluorescent species populations in the sampled volume is monitored by quantitative time-lapse microscopy and the fluorescence intensity relative to the prebleach period is plotted as a function of time. Diffusion coefficients are determined from the rate of fluorescence recovery, resulting from transport of fluorophore into the bleached region (Fig. D1.17). The time profile of the FPR pattern can also be used to monitor the nature of the transport, and to distinguish between diffusive and flow motions (Fig. D1.18).

FPR may well be the technique most widely used to study lateral diffusion in a plane – the geometry corresponding to extended regions of cell plasma membranes, reconstituted membranes, and thin layers of solution or cytoplasm. It has been shown by FPR that several classes of cell surface proteins and glycoproteins undergo lateral movements within the membrane plane, which have considerable functional significance. Direct experimental measurements have been made of the apparent diffusion coefficients of different membrane proteins *in situ*. These have yielded estimates ranging from 5×10^{-9} cm^2 s^{-1} for rhodopsin in photoreceptor membranes to less than 10^{-12} cm^2 s^{-1} for fluorescence in labeled surface proteins in the human erythrocyte. Values of 10^{-11} cm^2 s^{-1} were recorded for fluoresce in labeled concanavalin A receptor complexes on the plasma membrane of cultured rat myoblasts. Finally, we note that the diffusion coefficients of macromolecules determined by the FPR method often have *no direct relation* to the shape of the molecule nor usually to the character of the mobility of the molecules under specific conditions. The same can be written for the method of nanovid microscopy (Comment D1.13)

D1.6 TRANSLATIONAL FRICTION AND DIFFUSION COEFFICIENTS

D1.6.1 Einstein–Smoluchowski Relation

The kinetic theory of diffusion presented above allows us to estimate the value of the diffusion coefficients by

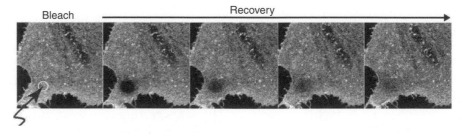

Bleach   Recovery

Fig. D1.16 Fluorescence photobleaching recovery: The fluorescent species in the small volume (white circle) is bleached by a short, high-intensity laser pulse; the recovery of fluorescence in the volume due to the diffusion of unbleached particles is observed as a function of time. (After Bastiaens and Pepperkok, 2000.)

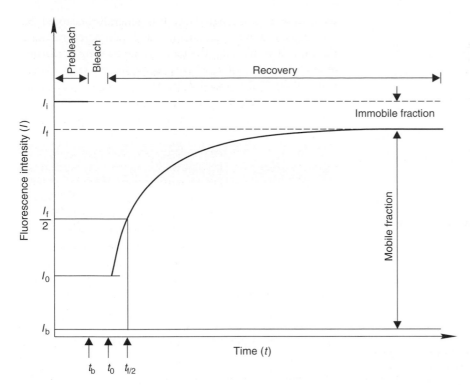

Fig. D1.17 An idealized plot of fluorescence intensity (I) as a function of time shows the parameters of a quantitative FPR experiment. The bleached region is monitored during a prebleach period to determine the initial intensity I_i. The region is then bleached using high-intensity illumination from time t_b to t_0, and recovery is monitored starting at t_0 until I reaches a final I_f, when no further increase can be detected. Some methods calculate the effective diffusion coefficient, D_{eff}, directly from the time ($t_{1/2}$) required to reach half the final intensity ($I_{f/2}$) or from a fit of all the curves (solid line). (Adapted from Bastiaens and Pepperkok, 2000.)

computing the velocity at which a particle drifts through the medium when exposed to an externally applied force. In practice, the velocity at which the particle moves in response to such a force is infinitesimal in comparison with the instantaneous root-mean-square velocity given by Eq. (A) of Comment D1.10. It means that particles diffuse much as they would in the absence of the field, but with a small persistent directional bias, as shown in Fig. D1.19.

Now we deduce the equation connecting the diffusion coefficient of a particle D with its frictional coefficient f using Berg's "random walk with drift" approach.

Consider a particle of mass m at position x on which acts an externally applied force in the direction $+x$. According to definition, the coefficient of translational friction is a coefficient of proportionality in the equation:

$$F_x = fu_x \tag{D1.36}$$

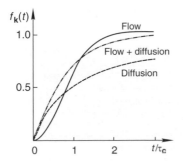

Fig. D1.18 Theoretical fluorescence recovery curves expressed as the fractional recovery $f_k(t)$ for pure diffusion, flow, and a combination of the two. (After Koppel, 1979.)

where F_x is the active force in the x direction and u_x is the average velocity of the particle. From the other side, in accordance with Newton's second law, the force causes the particle to accelerate uniformly to the right with acceleration $a = F_x/m$. According to the random walk approach a particle steps once every τ seconds to the right with an initial velocity $+v_x$ or to the left with initial velocity $-v_x$. A particle starting at position x with an initial velocity $+u_x$ moves in time τ a distance $\delta_+ = u_x + a\tau^2/2$, while a particle starting at position x with an initial velocity $-u_x$ moves in time τ a distance $\delta_- = -u_x + a\tau^2/2$. Since a step to the right and one to the left are equally probable, the average displacement in time τ is $a\tau^2/2$, and the particle drifts to the right with an average velocity

COMMENT D1.13 "NANOVID MICROSCOPY"

In 1988 M. Sheets devised nanovid microscopy, which enables small colloidal gold particles to be located with a light microscope with nanometer precision. One year later, he applied the method to track the motion of 40 nm diameter gold particles attached to the lectin, concanavalin A, on the surface of living cells.

Diffusion coefficients of macromolecules determined by this experimental method as a rule have *no direct relation* to the shape of molecule and usually demonstrate the character of the mobility of the molecules of interest under specific conditions (Sheetz *et al.*, 1989).

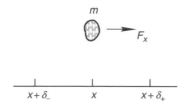

Fig. D1.19 A particle of a mass m subjected to an externally applied force F_x while undergoing a one-dimensional random walk. (After Berg, 1983.)

$$u_d = \tfrac{1}{2}a\tau = \tfrac{1}{2}F_x\tau/m \qquad (D1.37)$$

In the model random walk, $f = 2m/\tau$. Multiplying both the numerator and the denominator of Eq. (D1.37) in the form $u_d = F_x/f$ by (δ/τ^2) and remembering that $u_x = \delta/\tau$ and $D = \delta^2/2$, we find that $f = mu_x^2/D$. But by Eq. (A) of Comment D1.10, $mu_x^2 = kT$; therefore $f = kT/D$, or

$$D = kT/f \qquad (D1.38)$$

The result is known as the Einstein–Smoluchowski relation (Comment D1.14), and is very general. It does not depend on any assumptions made about the structure of the particle or the details of its motion. The particle always moves through the medium with a velocity proportional to the externally applied force. In the case of diffusion, the constant of proportionality is D/kT. In the case of sedimentation, this constant is $s/m(1 - \bar{v}\rho_0)$ (Chapter D2). In the case of free electrophoresis this constant is μ/Q, where μ is the electrophoretic mobility and Q is the charge of the particle.

The Einstein–Smoluchowski relation shows that if we can measure D, we can calculate f, the coefficient of friction, which gives us information about the dimension and shape of the molecules.

D1.6.2 Diffusion Coefficients of Biological Macromolecules

The diffusion coefficient of a macromolecule is a function of the solvent viscosity and the temperature at which the

---

measurement was carried out. It is generally accepted that any measured diffusion coefficient should be corrected to the value it would be if the measurements were performed at 20 °C in pure water, where η_{20w}, is the viscosity of pure water at 20 °C (Tables D1.2, D1.3):

$$D_{20,\mathrm{w}} = D^o(293/T)(\eta_{\mathrm{soln},T}/\eta_{\mathrm{w},20}) \qquad (D1.39)$$

TABLE D1.2 DIFFUSION COEFFICIENTS OF SOME GLOBULAR PROTEINS IN AQUEOUS SOLUTION

Protein	Molecular mass (Da)	Diffusion coefficient $D^0_{20,\mathrm{w}}$ (Fick units)
Somatostatin	1636	24.5
Gramicidin (dimer form)	2500	18.4
Lipase (milk)	6669	14.5
Ribonuclease A (bovine pancreas)	12 640	13.1
Cytochrome c (bovine heart)	13 370	11.4
Lysozyme (chicken egg white)	13 930	11.2
Profilin	14 800	10.6
Myoglobin	17 190	10.3
Cellulase	22 000	9.8
Chymotrypsinogen A (bovine pancreas)	23 240	9.5
Insulin	34 400	7.9
Carboxypeptidase B	34 280	8.2
Ovalbumin	45 000	7.8
Albumin	68 600	6.4
Hemoglobin (pig heart)	68 000	6.9
Citrate synthase	97 938	5.8
Lactic dehydrogenase (beef heart)	133 000	5.1
Aldolase (rabbit muscle)	156 000	4.8
Nitrogenase (bovine liver)	220 000	4.5
Catalase	250 000	4.1
Apoferritin (horse spleen)	467 000	3.6
Urease (Jack bean)	482 700	3.5
Glutamate dehydrogenase	1 015 000	2.5

1 Fick unit = $10^{-7}\,\mathrm{cm}^{-2}\mathrm{s}^{-1}$.

Diffusion coefficients are taken from: Zipper and Durchschlag (2000); Banachowicz et al. (2000); Zhou (2001); Hellweg et al. (1997); Smith (1970); Garcia de la Torre (2001); Byron (1997); Brune and Kim (1993).

COMMENT D1.14 DERIVATION OF EQ. (D1.38)

As pointed out in Berg's book in the strict sense, such a derivation of Eq. (D1.38) has no good physical basis. Of course, in reality particles do not step at a fixed interval or start each step at a fixed velocity. There are distributions of step intervals, directions, velocities owing to exchange of energy between molecules. But the end result is the same. The essential point is that a particle is accelerated by the externally applied force, but this force is so small (see Section D1.2.2) that it forgets about this acceleration when it exchanges energy with molecules of the fluid. The cycle repeats again and again. This is the reality for molecules at low Reynolds number (Berg, 1983).

TABLE D1.3 DIFFUSION COEFFICIENTS OF HOMOGENEOUS DOUBLE-STRAND DNA

Particle	Molecular mass (Da)	$D^0_{20,w}$ (Fick units)
8 bp	5304	15.26
12 bp	7956	13.41
20 bp	13 260	10.86
89 bp	59 007	4.27
104 bp	68 952	3.88
124 bp	82 212	3.41
2311 bp	1 532 193	0.46
DNA-pUC19	1 829 880	0.35
DNA-pDS1	2 538 637	0.29

Values are taken from: Eimer and Pecora (1991); Tirado et al. (1984); Sorlie and Pecora (1988); Seils and Dorfmuller (1991); Jolly and Eisenberg (1976).

Here, $D_{20,w}$ is the diffusion constant in pure water at 20 °C, T is the absolute temperature, and $\eta_{soln,T}$ is the viscosity of the real solution used at temperature T.

Equation (D1.39) has no solid theoretical background and is based mainly on phenomenological consideration. Note that $D_{20,w}$ is a quantity that can be defined even if the species does not exist in water at 20 °C. Diffusion coefficients of biological macromolecules are normally obtained at finite concentration and should be extrapolated to zero concentration (Comment D1.15).

Table D1.2 lists some representative measured values of $D_{20,w}$ and M of some proteins, starting from small synthetic peptides (somatostatin) and finishing with large oligomeric protein complexes (glutamate dehydrogenase). Table D1.3 gives diffusion coefficients of homogeneous double-stranded DNA.

D1.6.3 Dependence of the Diffusion Coefficient on the Molecular Mass of Globular Proteins

Assume that diffusing molecules are spherical particles. In this case a combination of the Einstein–Smoluchowski relation (Eq. (D1.38)) and Stokes' law (Eq. D1.11) opens the way for the calculation of the radius, R_0, of a sphere from the experimentally measured diffusion coefficient D and the solvent viscosity η_0:

$$R_0 = kT/6\pi\eta_0 D \qquad (D1.40)$$

Thus, from D and η_0 we can calculate R_0, and if the density of a molecule is ρ, mass is given by

$$m = (4/3)\pi R_0^3 \rho \qquad (D1.41)$$

and the molecular mass M is given by

$$M = (4/3)\pi R_0^3 \rho N_A \qquad (D1.42)$$

The important point brought out of this sample case is that diffusion is inversely proportional to the cube root of the molecular mass of spherical particles

$$D \sim M^{-1/3} \qquad (D1.43)$$

Figure D1.20 shows the dependence of the experimental values of the diffusion coefficients on the molecular mass for globular proteins starting with the tetradecapeptide somatostatin and ending with the large proteins (the oligomeric protein, glutamate dehydrogenase from bovine liver) using the data in Table D1.2. The plot of log D against log M correlates very well with a straight line of slope –0.336. In the limit of experimental error the slope coincides with

COMMENT D1.15 CONCENTRATION DEPENDENCE OF DIFFUSION COEFFICIENTS

The concentration dependence of translational diffusion can be characterized by a coefficient k_D in the equation

$$D_c = D_0(1 - k_D C) \qquad (A)$$

The coefficient k_D is specified at low solute concentration by the algebraic sum of the effects of the "excluded volume" effect and hydrodynamic friction retardation and is analogous to coefficients k_s, which describes concentration dependence in sedimentation coefficients (Chapter D2). In sedimentation these two effects are of similar magnitude but opposite sign. It follows, as has long been appreciated, that k_D is small in magnitude and of variable sign.

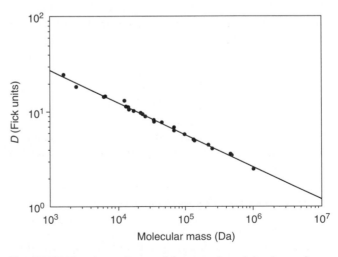

Fig. D1.20 The dependence of the experimental values of diffusion coefficients on the molecular mass for globular proteins on a double logarithmic scale.

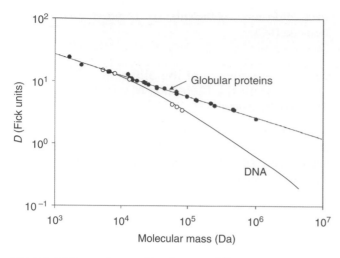

Fig. D1.21 Dependence of log $D_{20,w} \times \bar{v}^{1/3}$ on the logarithm of the molecular mass, for globular proteins (•) and for the DNA molecules (○) in range of molecular mass 8 bp–6 kbp. Data are taken from Tables D1.2 and D1.3, assuming partial specific volumes of DNA and protein of $0.56 \, cm^3 \, g^{-1}$ and $0.73 \, cm^3 \, g^{-1}$, respectively.

the expected theoretical value $-1/3$ for spherical particles (Eq. (D1.43)).

D1.6.4 Dependence of the Diffusion Coefficient on the Molecular Mass of DNA

Figure D1.21 demonstrates the dependence of diffusion coefficients on molecular mass for double-stranded DNA using the data of Table D1.3. For DNA molecules ranging from 70 bp (molecular weight ~46 kDa) to 6 kbp (molecular weight ~4 MDa), the dependence is described by one straight line:

$$D = 276M^{-0.72} \qquad \text{(D1.44)}$$

It is interesting to note that the frictional properties of short DNA fragments (12–20 bp) are similar to those of globular proteins of corresponding molecular weight (Fig. D1.21). At the same time, these properties are very different for large molecular weights. This fact reflects the principal difference between these two homologous series. In the globular protein series, asymmetry remains constant with increasing molecular weight, whereas in the DNA fragment series, asymmetry grows with molecular weight. The reader can find the same phenomenon in Section D2.10.3 in which we discuss the sedimentation behavior of DNA.

D1.6.5 The Limits to Stokes' Law

Figure D1.22 shows the dependence of the experimental values of diffusion coefficients on the molecular mass for a wide range of masses (from gases to large macromolecules) in a double logarithmic plot. As seen from the figure, the dependence can be subdivided into three different regions.

These correspond to three different modes of diffusion in liquids. Region I relates to the "flow" mode of the "large" molecules, like proteins, in liquids. In the "flow" mode the "large" molecules experience the solvent as a continuum. The molecules correspond to particles moving at very low Reynolds number ($<10^{-4}$). In this case the movement of the molecules obeys Stokes' law with "stick" boundary conditions (Eq. (D1.11)). It is possible to assume that the lower limit for the application of Stokes' law in "stick" boundary conditions is about 1000 Da (Fig. D1.22). Thus we may define "large" molecules as molecules which have a mass of 1000 Da or more.

Region III relates to the "lattice" mode of "small" molecules in liquids. In the "lattice" mode "small" molecules move in a medium composed of particles of a similar size to the molecules (i.e., in the self-diffusion of a liquid). These molecules, including ions and very small molecules like glycerol (92 Da), move through the solvent by a "jump and wait" mechanism for a significant fraction of their net movement, taking advantage of cavities in the liquid. Consequently the molecules do not experience the solvent as a continuum. The diffusional coefficient calculated from the Einstein–Smoluchowski relation is generally greater than the value predicted by Stokes' law. We define "small" molecules as molecules, which have masses of no more than 100 Da.

Region II is the transition region between two modes of diffusion in liquids. Molecules move through the solvent by mixed mechanisms. Region I gradually turns into region II, whereas there is a jump between regions II and III (Fig. D1.22). We define the molecules in this interval of masses (100–1000 Da) as "medium-sized" molecules. For "medium-sized" molecules "slip" conditions are more applicable; however, Eq. (D1.40) is not used in practice.

D1.7 HYDRODYNAMIC EXPERIMENTS

Experiments in hydrodynamics can be divided into four groups.

In the first, we find experiments that measure the *equilibrium velocity* of the particles. Translational diffusion is observed when the effective force arises from particle concentration gradients in the solution. When the acting force is gravitational (either under natural gravity or through ultracentrifugation), the phenomenon is called sedimentation. If the acting force is electrical in nature, the phenomenon is called electrophoresis.

The second group includes experiments in which the *rate of particle rotation* under the action of a pair of forces (a torque) is determined. If a velocity gradient in the solvent plays the role of an orienting force, the phenomenon is known as the Maxwell effect or flow birefringence. If the force is of an electrical nature, the phenomenon is called the Kerr effect or electric birefringence.

TABLE D1.4. METHODS CURRENTLY USED TO DETERMINE TRANSLATIONAL FRICTION COEFFICIENTS, F, OF BIOLOGICAL MACROMOLECULES

Experimental method	Measured value	The range of measured values	The applicability range
Translational diffusion (this chapter)	Translational diffusion coefficient D	$5 \times 10^{-5} - 5 \times 10^{-12} \, (\mathrm{cm^2 s^{-1}})$	Applicable to biological macromolecules of different shapes
Sedimentation velocity (Chapter D2)	Sedimentation coefficient s (in Svedberg units)	$0.5 - 200000 \, (10^{-13} \mathrm{s})$	Not applicable to very small or very large molecules
Electrophoretic mobility	Electrophoretic mobility μ	$1 \times 10^{-4} - 5 \times 10^{-4} \, (\mathrm{cm^2 \, V^{-1} s^{-1}})$	Applicable mainly to DNA molecules
Fluorescence correlation spectroscopy (Chapter D4)	Translational diffusion coefficient D	$5 \times 10^{-5} - 5 \times 10^{-12} \, (\mathrm{cm^2 \ s^{-1}})$	Applicable to fluorescent labeled biological macromolecules of different shapes
Recovery of fluorophore after photobleaching (this chapter)	Translational diffusion coefficient D	$5 \times 10^{-5} - 5 \times 10^{-12} \, (\mathrm{cm^2 s^{-1}})$	Applicable to fluorescent labeled biological macromolecules of different shapes

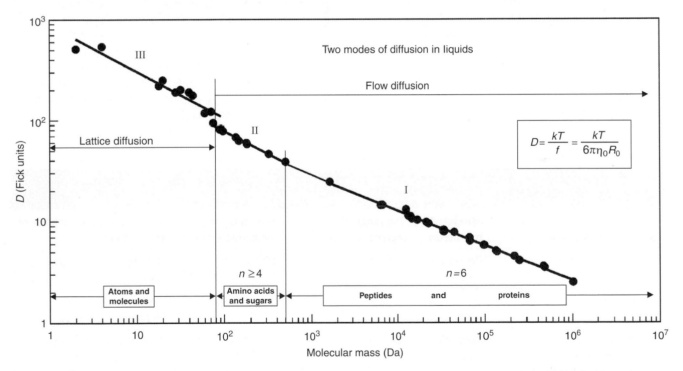

Fig. D1.22 Dependence of diffusion coefficients on the molecular mass M for different substances on a double logarithmic scale. Region I shows the data for globular proteins. The data are taken from Table D1.2. The equation of the straight line is

$$D = 276M^{-0.336} \tag{I}$$

Region II shows the data for small molecules, amino acids and sugars. The equation of the straight line is

$$D = kM^{-0.47} \tag{II}$$

Region III shows the data for atoms and gases. The equation of the straight line is

$$D = kM^{-0.42} \tag{III}$$

In Eqs. (I)–(III) M is in daltons and D in Fick units (1 Fick $= 1 \times 10^{-7} \, \mathrm{cm^2 \, s^{-1}}$).

The third group is represented by viscosity experiments that measure *energy loss* due to friction of the molecule in the solution.

In the fourth group are experiments in which no external force acts on the particles and their displacements and rotations occur only under the action of *thermal agitation* or *Brownian motion*. The behavior of a particle in fluorescence and dynamic light scattering experiments is of this type.

Table D1.4 present a summary of hydrodynamic methods currently in use. Note, that in spite of the variety of experimental methods presented, ultimately, only three parameters can be obtained: translational and rotational friction coefficients and intrinsic viscosity.

D1.7.1 Measurement of Translational Frictional Coefficients

Table D1.4 shows the main experimental methods currently used to determine translational friction coefficients, f, of biological macromolecules, their measured values, and the range of measured values. It also includes the range of applicability (in dimensions and shape) of each of the methods and gives the chapters where each experimental technique is treated in detail and where applications and the applicability range of each method are discussed.

Actual translational frictions of rigid microscopic particles can be determined by observing the settling rates of their macroscopic models in a high-viscosity fluid at low Reynolds number (Comment D1.16).

COMMENT D1.16 DIRECT DETERMINATION OF THE TRANSLATIONAL FRICTION COEFFICIENT

The standard apparatus consists of a glass cylinder (usually ~10 cm in internal diameter, 100 cm high), immersed in a water bath at constant temperature. Two circular marks approximately 7–10 cm from each end of the cylinder indicate the distance over which the settling particle is timed. Rigid models are constructed from balls using different materials. The diameters of the balls range from 0.25 cm to 1 cm. Multi-sphere particles are constructed by gluing matched balls together with a minimum amount of epoxy cement.

Settling rates for non-spherical particles (cylinders, dimers, and tetramers) are determined in two orientations for different particle sizes and different fluid viscosities, covering Reynolds number in the range 0.0001–0.01. For each multi-sphere particle, the quantity of interest is the ratio of the settling rate of a single sphere to that of the multi-sphere particle. The ratio should be free from any influence of the experimental set-up, especially wall effects, and should not be dependent on particle size.

Such simple experiments on macroscopic models allow the determination of the translational friction properties of microscopic particles of arbitrary shape. Theoretical values coincide with experimental ones in the limit of experimental error (0.1%) in all cases for which analytical results are known.

TABLE D1.5 METHODS CURRENTLY USED FOR THE DETERMINATION OF ROTATIONAL FRICTION COEFFICIENTS, Θ, OF BIOLOGICAL MACROMOLECULES[a]

Experimental method	Measured parameter	The range of measured Θ in s^{-1}	The range of measured τ values	The range of applicability of the method
Electric birefringence	Relaxation time τ	$1.7 \times 10^6 - 0.3$	100 ns–500 ms	Not applicable to very small spherical molecules
Flow birefringence	Orientation angle χ	$5 \times 10^4 - 1.7$	3 µs–100 ms	Applicable to highly elongated molecules
Fluorescence and phosphorescence depolarization (Chapter D3)	Relaxation time τ	$50 \times 10^6 - 1.7 \times 10^6$	3–100 ns	Applicable to small spherical molecules
Dynamic light scattering (Chapter D4)	Decay of correlation function	$50 \times 10^6 - 0.17$	1 ns–1 s	Applicable to biological macromolecules of different shapes
Nuclear magnetic resonance (Chapter J3)	Relaxation time τ	$50 \times 10^6 - 1.7 \times 10^6$	3–100 ns	Applicable to small spherical or quasi-spherical molecules

[a] It should be noted that workers in different experimental methods frequently use different definitions of the relaxation time. Workers in electric and flow birefringence usually use the term *rotational relaxation time*, denoted as ρ, whereas workers in magnetic resonance and fluorescence polarization spectroscopy customarily use *rotational correlation time*, denoted as ϕ. They are related by $\phi - \rho/3$, $\phi - 1/6D$, $\rho = 1/2D$. In this book τ_r means rotational relaxation time, ι_c means rotational correlation time.

D1.7.2 Measurement of Rotational Frictional Coefficients

Experimental methods for the measurement of rotational frictional coefficients can be divided into two main groups

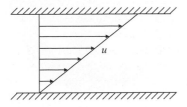

Fig. D1.23 Velocity and gradient distribution in a cylinder (infinitely narrow gap).

(see also Section D1.4.2). In the first, we find experiments in which the rate of particle rotation under the *action of a pair* of forces (a torque) is determined. If a velocity gradient in the solvent takes the role of an orienting force, the phenomenon is known as birefringence in flow or flow birefringence (the Maxwell effect). If the force is electrical in nature, the phenomenon is called electric birefringence (the Kerr effect). In the second group, we place phenomena in which no external force acts on the particles and their rotations occur only *under Brownian motion*. The behavior of a particle in fluorescence and dynamic light scattering experiments is of this type.

Table D1.5 contains a summary of five methods currently used to determine rotational friction coefficients.

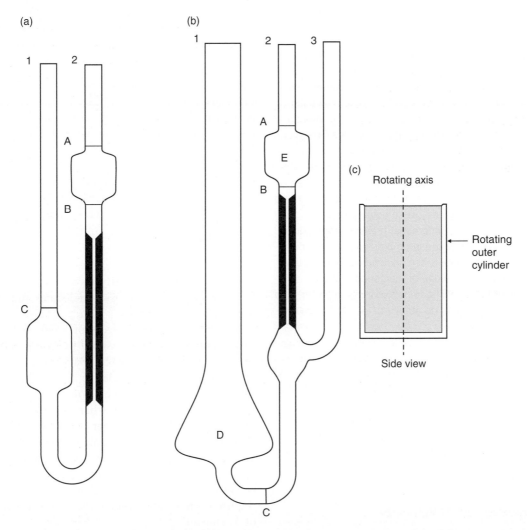

Fig. D1.24 Capillary viscometers: **(a)** Ostwald, **(b)** Ubbelohde, and **(c)** Couette viscometers. The Ostwald instrument is used in the following way. The solution is added at opening 1 until the liquid level at rest is at scratch C. Suction is then applied at opening 2 until the liquid level is above scratch A. The suction is removed and the liquid falls owing to the difference in height between the two arms. The time t required for the meniscus to move between scratches A and B is measured. Because of the change in relative liquid heights, the flow rate is not constant. The Ubbelohde instrument is used in the following way. The solution is added at opening 1 until the liquid level at rest is at scratch C. Opening 3 is closed and suction is applied at opening 2, until the liquid has been drawn above A. Opening 3 is then opened with opening 2 closed, which allows bulb D to drain. Opening 2 is then opened and the times for the meniscus to pass scratches A and B are determined. To vary the concentration, the liquid in bulb E can be diluted because the amount of liquid in X does not determine the volume of liquid contained between A and the bottom of the capillary.

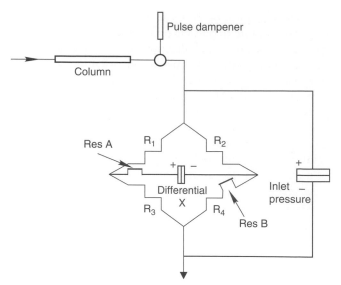

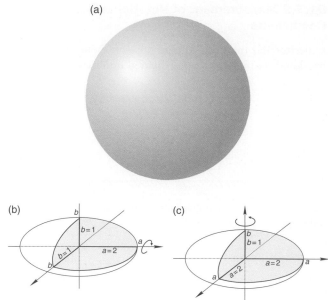

Fig. D1.25 Principles of the pressure imbalance differential viscometer. At "baseline conditions" (i.e., solvent only) the differential pressure across the bridge is zero because there is solvent in all four capillaries, R_1–R_4. When the molecular solution enters the bridge from the column it fills capillaries R_1, R_2, and R_3, while solvent from a delay reservoir (Res B) remains in capillary R_4 and prevents the entry of the macromolecular solution. The transducer X measures the differential pressure ΔP and Y measures the inlet pressure P_i. The reservoir Res A located out of the flow stream acts to compensate the volume so that any flow rate fluctuations cause equal pressure changes on each side of the differential pressure transducer Y. The difference in viscosity causes a pressure imbalance in the bridge that is proportional to the specific viscosity of the sample solution. From knowledge of the concentration (measured using an off- or online refractive index detector), the reduced specific viscosity can be obtained. (Adapted from Dutta et al., 1991.)

Fig. D1.26 Sphere (a) and two ellipsoids of revolution ((b), (c)) with equal volumes. A prolate ellipsoid (b) is a rod-like shape, generated by rotating an ellipse around its long semiaxis a; the two short semiaxes, b, are identical. The oblate ellipsoid (c) is a disc shape, generated by rotating an ellipse around its short semiaxis b; the two long semiaxes, a, are identical. For either kind of ellipsoid, the axial ratio (p) is defined as a/b, the ratio of the long to the short semiaxes. The near right octant of each ellipsoid is cut away to show the long (a) and short (b) axes.

The table also lists the range of experimental values of rotational coefficients and the relaxation times accessible to each method and its range of applicability (in dimensions and shape).

Note that electric and flow birefringence will not be treated in depth as these techniques can be used in only very specific cases.

D1.7.3 Measurement of Viscosity

The internal friction or viscosity of a liquid appears during flow when a non-zero velocity gradient is set up. The simplest example is a laminar flow with a constant velocity gradient $G = du_x/dy$ at a right angle to the direction of flow (Fig. D1.23).

Intrinsic viscosity is one of the oldest molecular parameters. Einstein considered the intrinsic viscosity of a suspension of spherical particles in 1906. It was understood

later that intrinsic viscosity is very sensitive to the conformation of biological macromolecules and can be measured with high precision.

Traditionally, the viscosity of a solution of biological macromolecules was measured by timing either the flow of the liquid through a capillary tube or the rotation speed of one cylinder relative to another (Fig. D1.24). Although the instrumentation for both of these approaches has become either automatic or semiautomatic with online computer data treatment, the basic design has remained the same since Yang's and Zimm's articles were published in the early 1960s. Neither approach is attractive in protein biochemistry, however, because of the relatively large quantities of material required – normally greater than 1 ml at concentrations of 5–6 mg ml^{-1}.

An important development was the appearance of an instrument based on a quite different principle – the so-called "differential" viscometer (Fig. D1.25). With this instrument, measurements can be performed at a lower concentration (about 1 mg cm^{-3}) and on smaller volumes, making the intrinsic viscosity attractive for use in protein biochemistry once more. However, in spite of this innovation, viscosity is not now popular as

(a)

(b)

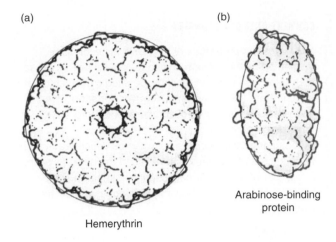

Arabinose-binding
protein

Hemerythrin

Fig. D1.27 (a) A sphere as the hydrodynamic equivalent body for hemerythrin; **(b)** an ellipsoid of revolution with axial ratio 2:1 as the hydrodynamic equivalent body for arabinose-binding protein.

a method for studying the structure of biological macromolecules.

D1.7.4 Prediction of Hydrodynamic Properties

Modern hydrodynamics allows the prediction of the frictional properties of biological macromolecules of any shape. The computing procedure used essentially depends on the particle shape. Examples of sphere and ellipsoid models are shown in Fig. D1.26.

Particles of "Round" Shape
Biological macromolecules of a "round" shape can be approximated reasonably well by a sphere or a two-axis ellipsoid of revolution as hydrodynamic equivalents (Figs. D1.26, D1.27). Hydrodynamic properties for such particles can be calculated analytically.

Particles with a "Broken" Shape
As a first approximation, many biological macromolecules can be modeled as rigid rods. These include small, rigid, rod-like monomers, oligomers, duplex oligonucleotides, α-helical polypeptides, and rod-like proteins and viruses. The difficulty for all hydrodynamic theories that attempt to model the translation and rotation of rod-like molecules arises from the effects of the sharp ends of the rod ("end correction"). The friction of a *cube* which has sharp edges and corners is one of the major unsolved problems of hydrodynamics (Fig. D1.28).

At present there are several different approaches to calculating the frictional properties of particles with a "broken" shape: modeling the *entire* particle with a set of spheres; modeling the *surface* of a particle with a set of small equal spheres; modeling the *surface* of a particle with a set of panel elements; and using their electrostatic counterparts. These approaches allow only an approximation of the friction properties of particles with a "broken" shape.

Particles of Arbitrary Shape
The detailed description of a particle with a more complex shape (Fig. D1.29) might require the determination of a far greater number of parameters than can be obtained from hydrodynamic measurements alone. The approaches above are used to calculate hydrodynamic characteristics of such biological macromolecules with good accuracy.

Particles with a Known Three-Dimensional Structure
A sphere or an ellipsoid of revolution can quite satisfactorily describe the hydrodynamic features of several biological molecules. A three-axis spheroid, in particular, describes the shape of most globular proteins. However, many biological macromolecules, e.g., tRNA, immunoglobulins, and phosphoglyceratemutase, whose shapes are far from being spherical or ellipsoidal (Fig. D1.30) require more complex models for the calculation of their hydrodynamic properties.

(a)

(b)

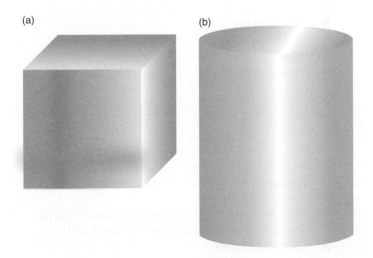

Fig. D1.28 Particles with a "broken" shape have sharp edges. They include **(a)** the cube and **(b)** the circular cylinder. Their frictional properties can be calculated only as an approximation.

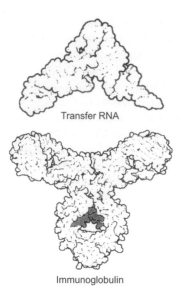

Fig. D1.29 Examples of particles of arbitrary shape: tRNA and immunoglobulin. Their frictional properties can be calculated only as an approximation.

Fig. D1.30 The structure of the lysozyme molecule with atomic resolution. The frictional properties of the molecules can be predicted with accuracy of a few percent.

The impressive progress in solving protein and nucleic acid structures to high resolution by X-ray crystallography and NMR stimulated the development of novel approaches to calculating hydrodynamic parameters from atomic-level structural details. Several algorithms have been proposed to calculate its hydrodynamic properties, including: modeling the particle with a set of beads; modeling the protein with a set of beads in C-α carbon positions; and modeling the surface of the protein with a set of panel elements. Comparison with experiment shows that the algorithms predicted diffusion coefficients with an accuracy of about 1% and intrinsic viscosity with an accuracy of about 2–3%. Special approaches have been proposed for DNA molecules (Comment D1.17).

COMMENT D1.17 DNA GEOMETRY

Most dynamic properties of long DNA molecules are dominated by their length and bending flexibility, whereas their thickness and, particularly, detailed cross-sectional structure are less important. However, for short DNA fragments, cross-sectional structures are more important.

The geometry of a single helix can be described by the following set of quantities: A, radius of the helix; P, pitch; n, number of turns (not necessarily an integer). If the helical axis is z, then the parametric equations of the helix can be written as

$$X = A\cos(t + \varphi)$$
$$Y = A\sin(t + \varphi)$$
$$Z = Pt/2\pi$$

where the value of the phase angle, φ, can be arbitrary and t is a continuous parameter that goes from $t = 0$ to $t = 2\pi n_t$. The two strands of a double helix can also be described by an equation using different values for the phase angle of each strand. In the bead model of the helix, beads of radius R_0 are placed along the contour of the helical line and are chosen such that the beads touch.

D1.8 CHECKLIST OF KEY IDEAS

- A dimensionless parameter, the Reynolds number, characterizes the ratio of inertial and viscous forces acting on a particle in a viscous medium.
- The movement of biological particles with molecular dimensions (10–10 000 Å) is described in terms of low Reynolds number hydrodynamics, where viscous forces dominate.
- Biological macromolecules move with a certain amount of bound solvent, thus defining the concept of a hydrated particle as a core of particle material and an envelope of bound water.
- In classical hydrodynamic theory, two extreme cases of solvent–particle interaction called "slip" and "stick" boundary conditions are typically considered.
- The results from all hydrodynamics methods can be described by only three hydrodynamic parameters (translational and rotation frictional coefficients and intrinsic viscosity).
- The description of a particle in terms of a hydrodynamic equivalent does not mean that it has this shape, but that it has the same hydrodynamic properties.
- The diffusion of macromolecules in solution can be considered as a phenomenon connected with Brownian motion.

- FPR, also called *fluorescence recovery after photobleaching* (FRAP), is a tracer technique that measures the diffusion of a labeled solute.
- The Einstein–Smoluchowski relation shows that if we can measure *the diffusion coefficient*, we can calculate the *coefficient of friction*, which gives us information about the molecules' dimension and shape.
- The coefficient of translational friction of a sphere is proportional to its radius and to the viscosity of the solvent within which the particle moves. Consequently, diffusion is inversely proportional to the cube root of the molecular mass of spherical particles.
- Hydrodynamic properties for globular proteins whose detailed atomic structures are known can be calculated by using one of several algorithms with an accuracy of about 1–3%, assuming a 1 Å hydration shell (for all proteins).

Suggestions for Further Reading

Hydrodynamics at Low Reynolds Number

Purcell, E. M. (1977). Life at low Reynolds number. *Am. J. Phys.*, **45**, 3–11.

Hydration

Kuntz, I. D., and Kauzmann, W. (1974). Hydration of proteins and polypeptides. *Adv. Prot. Chem.*, **28**, 239–345.

Finny, J. L. (1996). Overview lecture: Hydration processes in biological and macromolecular systems. *Faraday Discuss. Chem. Soc.*, **103**, 1–395.

Wuthrich, K., Billeter, M., Guntert, P., *et al.* (1996). NMR studies of the hydration of biological macromolecules. *Faraday Discuss. Chem. Soc.*, **103**, 245–253.

Garcia de la Torre, J. (2001). Hydration from hydrodynamics: General consideration and applications to bead modelling to globular proteins. *Biophys. Chem.*, **93**, 159–170.

Zhou, H.-X. (2001). A unified picture of protein hydration: Prediction of hydrodynamic properties from known structures. *Biophys. Chem.*, **93**, 171–179.

Perkins, S. J. (2001). X-ray and neutron scattering analyses of hydration shells: A molecular interpretation based on sequence predictions and modelling fits. *Biophys. Chem.*, **93**, 129–139.

Engelsen, S. B., Monteiro, C., Herve de Penhoat, C., and Perez, S. (2001). The dilute aqueous solvation of carbohydrates as inferred from molecular dynamics simulations and NMR spectroscopy. *Biophys. Chem.*, **93**, 103–127.

Determination of Particle Friction Properties

Happel, J., and Brenner, H. (1973). *Low Reynolds Number Hydrodynamics*. 2nd edition. Groningen: Noordhoff Int.

Hu, C.-M., and Zwanzig, R. (1974). Rotational friction coefficients for spheroids with the slipping boundary conditions. *J. Chem. Phys.*, **60**, 4354–4357.

Byron, O. (1997). Construction of hydrodynamic bead models from high resolution x-ray crystallographic or nuclear magnetic resonance data. *Bioph. J.*, **72**, 406–415.

Harding, S. E. (1998). The intrinsic viscosity of biological macromolecules: Progress in measurement, interpretation and application to structures in dilute solution. *Prog. Biophys. Mol. Biol.*, **68**, 207–262.

Zipper, P., and Durchschlag, H. (2000). Prediction of hydrodynamic and small angle scattering parameters from crystal and electron microscopic structure. *J. Appl. Cryst.*, **33**, 788–792.

Prediction of Particle Friction Properties

Brune, D., and Kim, S. (1994). Predicting protein diffusion coefficients. *Proc. Natl. Acad. Sci. USA*, **90**, 3835–3839.

Zhou, H.-X. (1995). Calculation of translational friction and intrinsic viscosity: II. Application to globular proteins. *Biophys. J.*, **69**, 2298–2303.

Garcia de la Torre, J., Huertas, M. L., and Carrasco, B. (2000). Calculation of hydrodynamic properties of globular proteins from their atomic-level structure. *Biophys. J.*, **78**, 719–730.

Allison, S. A. (2001). Boundary element modelling of biomolecular transport. *Biophys. Chem.*, **93**, 197–213.

ANALYTICAL ULTRACENTRIFUGATION

D2

D2.1 HISTORICAL REVIEW

1913

A. Dumansky proposed the use of ultracentrifugation to determine the dimensions of colloidal particles.

1923

T. Svedberg and **J. B. Nichols** constructed the first centrifuge with an optical system to follow particle behavior in a centrifugal field. One year later, **Svedberg** noted the decrease in absorbance at the top of the cell during centrifugation of a hemoglobin solution.

1926

Svedberg made the first measurements of protein molecular weights (hemoglobin and ovalbumin) by sedimentation equilibrium and in **1927** he determined the molecular weight of hemoglobin by using a combination of sedimentation and diffusion data. These pioneering studies led to the undeniable conclusion that proteins are truly macromolecules, made up of a large number of atoms linked by covalent bonds (Comment D2.1).

1929

O. Lamm deduced a general equation describing the behavior of the moving boundary in the ultracentrifuge field. The exact solution of the equation is an infinite series of integrals, which can only be computed by numerical integration. In later work, the Lamm equation was solved analytically for specific limiting cases (**H. Faxen, W. J. Archibald, H. Fujita**).

1930s

Schlieren optical systems were designed by **J. St. L. Philpot** and **H. Svenson**, and independently by **L. G. Longsworth**; these allowed a representation of the concentration gradient (or, more precisely, the refractive index increment) as a function of distance in the centrifuge sample cell. Physicists started to use **Perrin**'s hydrodynamics theories for ellipsoids of revolution to interpret the frictional coefficients deduced from sedimentation experiments in terms of the shape and hydration of macromolecules.

Important achievements include demonstrating that proteins are individual molecules of a definite molecular mass, as well as, even before the tobacco mosaic virus was visualized by electron microscopy, the prediction of its rod-like shape and dimensions.

1940s

The Spinco Model E analytical centrifuge, an extremely reliable electrically driven instrument, was constructed and became commercially available. In **1942**, **W. J. Archibald** proposed new methods for determining molecular mass from analytical ultracentrifugation (AUC) data. In **1943**, **E. J. Cohn** and **J. T. Edsall** showed that protein partial specific volumes could be successfully calculated from the atomic composition and partial specific volumes of component residues. The first textbook devoted to analytical centrifugation, *The Ultracentrifuge* by **T. Svedberg** and **K. O. Pedersen**, appeared in **1942**. This monograph became known as the bible of ultracentrifugation.

1950s

This decade marked the beginning of the widespread use of the sedimentation method. Using density gradient ultracentrifugation and isotope labeling, **M. Meselson** and **F. W. Stahl** proved the semiconservative mechanism of DNA replication. The experiments of **A. Tissier** and **J. D. Watson (1958)**, **F.-C. Chao** and **H. K. Schachman (1959)** led to the discovery of ribosomes. Another classic textbook, *Ultracentrifugation in Biochemistry*, by **H. K. Schachman**, was published in **1959**.

1960s

The first scanning photoelectric absorption optical system was developed. In this period there was widespread use of the ultracentrifuge to study proteins, ribosomes, DNA, and viruses. Essential contributions to the theory

---

COMMENT D2.1 THEODOR SVEDBERG

Theodor Svedberg was awarded the Nobel Prize for his work on colloidal systems and not for inventing the analytical centrifuge.

and practice of AUC analysis were made by **J. W. Williams, K. E. van Holde, H. K. Schachman, D. A. Yphantis**, and **H. Fujita**.

1970s

These were the golden years of AUC. By 1980, there were about 2000 analytical ultracentrifuges routinely operating in the world. The incorporation of a monochromator in the absorption optics allowed extremely dilute solutions to be studied. Rayleigh interference optics yielded highly accurate data for non-absorbing solutes, and was used mainly for the determination of macromolecular mass by sedimentation equilibrium. In **1971** the introduction of the differential sedimentation method opened the way for the observation of small differences between sedimentation coefficients (**M. W. Kirschner** and **H. K Schachman**). In **1978** the development of new data analysis methods, which removed the contribution of diffusion from sedimentation velocity boundaries, yielded integral distributions of sedimentation coefficients (**K. E. van Holde** and **W. O. Weischet**). A third landmark textbook on ultracentrifugation, *Foundation of Ultracentrifugation Analysis*, by **H. Fujita** appeared in **1975**.

1980s

In the early 1980s the sedimentation method started to lose popularity among biochemists. The reasons for this were two-fold: first, because of the development of gel electrophoresis and gel chromatography, which require very small quantities of material; second, in spite of the fact that the ultracentrifuge was linked to a computer, the possibilities for fast data treatment were very limited. Data analysis was a tedious process and still based on pattern photography (Comment D2.2).

1990s

The situation changed dramatically with the appearance of a new generation of instruments, which were highly automated for data collection and analysis. The Beckman XL ultracentrifuge includes two different optical detection systems: a UV absorption system that makes it possible to study proteins or nucleic acids at a very low concentration ($\sim$2–3 µg ml^{-1}); and a Rayleigh interference optical system that can detect macromolecules with low or no light absorbance. A Schlieren optical system became unnecessary, owing to the fact that precise derivative patterns can be obtained from integral curves by computer calculation. New data treatment programs are able to provide numerical solutions of the Lamm equation in the case of complex systems.

Since 2000

AUC is currently undergoing a renaissance. It can be applied to molecular weights from several hundreds to tens of millions of daltons. It is now the recognized

---

COMMENT D2.2 "ANALYTICAL ULTRACENTRIFUGATION REBORN"

H. Schachman, in 1989 Nature article titled "Analytical ultracentrifugation reborn," wrote:

> But by 1980 – despite the almost frenetic activity devoted to the design of new cells, the incorporation of different optical systems, the development of additional treatments and the application of ultracentrifugation to a diverse host of important biological problems – the use of the instruments came to a rather abrupt end. What happened? Were better, more reliable, and more versatile techniques developed that could be used instead of ultracentrifugal methods? Or were the questions asked by protein chemist and molecular biologist in the 1980s so different that sedimentation techniques had become obsolete?.... But molecular biologists in the early 1980s who were interested in molecular interactions between different proteins, or between proteins and nucleic acids, were content with "yes or no" answers rather than equilibrium constant and free energy changes. For them, filter binding assays, Sephadex columns and polyacrylamide gels provided the desired information. Now, however, with literally hundreds of proteins produced by the technique of site-directed mutagenesis awaiting characterization, precise techniques for physical chemical investigations are needed.

(Schachman, 1989)

---

method for accurate determination of the purity, mass, shape, self-association, and other binding properties of macromolecules in solution, and is rapidly becoming, once again, a necessary technique in most biological laboratories.

D2.2 INSTRUMENTATION AND INNOVATIVE TECHNIQUE

Below, we describe briefly the main characteristic features of modern analytical ultracentrifuges, paying special attention to the Beckman Instruments Optima XL (Comment D2.3). The introduction in the 1990s of this series of instruments has brought about substantial improvements in the accuracy, precision, and range of data that could be obtained. This was made possible by digital data acquisition, microprocessor-controlled experimental parameters, high-quality optical components, and by the introduction of a Rayleigh interference system that permits the acquisition of data from macromolecular solutions of low optical absorbance.

The main parts of an analytical ultracentrifuge are a rotor contained in a refrigerated and evacuated protective chamber, an optical system to observe the concentration distribution in the sample during centrifugation, and a data acquisition and analysis system.

D2.2.1 Rotors and Cells

Rotors must be capable of withstanding large gravitational stresses. At 60 000 rev min^{-1}, a typical ultracentrifugation rotor generates a centrifugal field in the cell of about 300 000 g (see Comment D2.4). Under these conditions, a mass of 1 g has an apparent weight of 300 kg (Comment D2.4).

A schematic diagram of a rotor and sedimentation cell is shown in Fig. D2.1. The rotor is solid, with holes to hold the sample cells. The simplest type of rotor incorporates two cells: the analytical cell and the counterpoise cell, which acts as a counterbalance. Cells have upper and lower plane windows of optical-grade quartz or synthetic sapphire. A variety of analytical cells are available with volume capacities between 0.02 and 1.0 cm^3.

A sector-shaped cell is essential in velocity experiments since the sedimenting particles move along radial lines. In a sample compartment with parallel sides, sedimenting molecules at the periphery would collide with the walls and cause convection. Sectors that diverge more widely than radial lines would also cause convection.

Double-sector cells permit us to account for absorbing components in the solvent, and to correct for the redistribution of solvent components (Fig. D2.2a). A sample of the solution is placed in one sector, and a sample of solvent alone acts as a reference in the second compartment. Boundary-forming cells (Fig. D2.2b) allow the layering of the solvent over a sample of the solution while the cell is spinning at a moderately low speed. These cells are useful for setting up a sharp boundary in diffusion coefficient determinations, using the Spreading Boundary

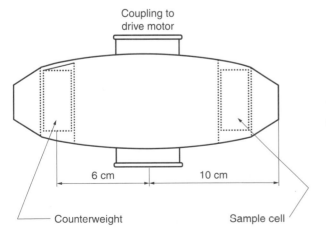

Coupling to drive motor

6 cm   10 cm

Counterweight   Sample cell

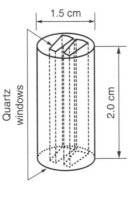

1.5 cm

2.0 cm

Quartz windows

Fig. D2.1 A typical analytical centrifuge rotor and sample double-sector-shaped cell.

(a)

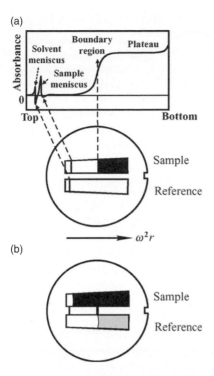

(b)

Fig. D2.2 Two types of cells used in AUC: **(a)** double-sector-shaped cell; **(b)** boundary-forming cell. Note the two small connecting channels between compartments. Liquid flow across the lower channel occurs only in the centrifugal field. The upper channel permits the return flow of air.

(a)

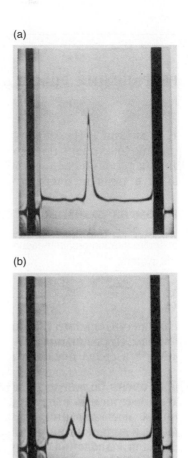

(b)

Fig. D2.3 Sedimentation profile of **(a)** 70 S ribosome *E. coli* and **(b)** 30 S and 50 S ribosomal subunits, obtained using the Schlieren optical system.

Technique (Section D1.5.5), or for the examination of small molecules, for which the rate of sedimentation is insufficient to produce a sharp boundary that clears the meniscus.

D2.2.2 Optical Detection Systems

Depending on the type of optical system used, sedimentation in an ultracentrifuge has traditionally been observed as either the concentration or the concentration gradient as a function of radius. For example, the Schlieren optical system displays the boundary in terms of the refractive index gradient, while in the Rayleigh optical system it is in terms of the refractive index, and in the absorption optical system, in terms of optical density.

The Schlieren optical system dominated the field for 70 years and has since passed into history (Comment D2.5). Many brilliant achievements in molecular biology have been accomplished using these optics, including the discovery of ribosomes (Fig. D2.3). The Beckman XL ultracentrifuge contains only two optical detection systems: absorption optics in the UV and Rayleigh interference optics. Schlieren-type sedimentation profiles can be obtained by direct differentiation of the

---

COMMENT D2.5 SCHLIEREN OPTICAL SYSTEM

In the Schlieren optical system (named for the German word meaning "streaks"), light is deflected by passing through a region in the cell where the concentration and hence the refractive index is changing. The optical system converts the radial deviation of the light into a vertical displacement of an image in the camera. The displacement is proportional to the refractive gradient. Thus the Schlieren image provides a measure of the refractive gradient, dn/dr, as a function of radial distance, r.

---

centrifugation absorption and interference pattern, using computer analysis.

The absorption optics system is based on the proportional relation between the concentration of the solute in a solution and its optical density. Most biological macromolecules absorb light in the near UV region. Nucleic acids

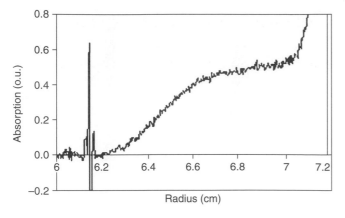

Fig. D2.4 Typical sedimentation profile measured by an absorbance optics system.

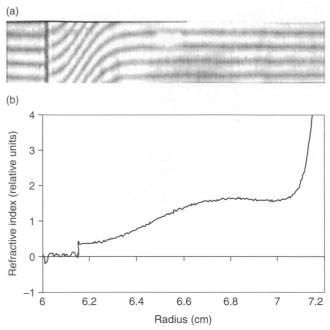

Fig. D2.5 A typical sedimentation profile measured by interference optics. (a) The fringe pattern produced by a boundary in Rayleigh interference optics. The refractive index difference (Δn) between two points of the cell can be calculated from the number of fringes N crossed in going between these two points from the classical relationship $N = a(\Delta n)/\lambda$, where a is the cell thickness and λ is the wavelength of light used. Since the refractive index is proportional to the concentration, the number of fringes can be used to monitor concentration as a function of radius in the cell. (b) A graphical representation of fringe displacement as a function of radius obtained by the software of the XL-I centrifuge.

COMMENT D2.6 THE RAYLEIGH INTERFEROMETER

An interference pattern is produced by splitting a beam of coherent light and passing it through paired sectors of a cell in a spinning rotor. When the two beams are merged after passage through the sectors, the waves form an interference pattern. If both sectors contain identical solutions the pattern is of straight interference fringes. However, when one sector contains the solvent and the other contains a macromolecular solution, the fringe pattern is shifted to an extent that corresponds to the concentration difference between the sample and solvent cells.

---

have a very strong absorption in the region 258–260 nm, while proteins are characterized by an absorption peak close to 280 nm (Section E1.3.2). Figure D2.4 shows a typical sedimentation profile measured using an absorbance optics system.

The most commonly used interference optics is based on the Rayleigh interferometer (Comment D2.6). The system can be applied to macromolecules that do not absorb significantly in the UV–visible range. Figure D2.5a shows a fringe pattern produced by a boundary in Rayleigh interference optics and Fig. D2.5b illustrates fringe displacement as a function of radius calculated by the software of the XL-I ultracentrifuge.

Significantly higher sensitivity would be expected from an analytical ultracentrifuge with fluorescence detection, because much lower concentrations would be detectable than with UV absorption optics. Furthermore, molecules with different fluorophores could be recorded selectively, even if they were present as a minor component in a

mixture. Sedimentation profiles of bovine serum albumin (BSA), recorded for two different concentrations with a fluorescence detection system, are shown in Fig. D2.6. The lowest concentration (6.5 ng ml^{-1}) is 4–5 orders of magnitude lower than would be detectable with UV absorption optics for proteins (Comment D2.7). Finally, we point out that turbidity has also been measured and interpreted by AUC (Comment D2.8).

D2.2.3 Data Acquisition

With the introduction of the Beckman Instruments Optima XL analytical ultracentrifuge, data acquisition in sedimentation experiments has now become largely automated. Computer programs for the analysis of sedimentation data commonly rely on the graphical transformation of experimental data to obtain integral distributions of the sedimentation velocity boundary. Because of the improved data quality, it has now become feasible to extract

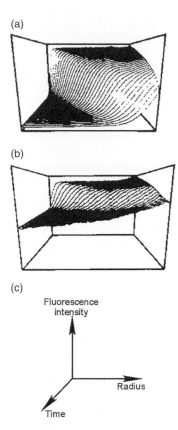

(c)

Fig. D2.6 Two sedimentation profiles of BSA recorded with a fluorescence detecting system. BSA was labeled by FITC. (a) The highest concentration, 65 ng ml^{-1}; (b) the lowest concentration, 6.5 ng ml^{-1}; (c) axis directions of the fluorescence: intensity (Z), radius (X) and time (Y). A fluorescence detecting system was developed for the Spinco Model E analytical ultracentrifuge and was installed in place of the standard Schlieren optics (Schmidt and Reisner, 1992).

COMMENT D2.7 PROTOTYPE FLUORESCENCE DETECTOR FOR THE XL-I ANALYTICAL ULTRACENTRIFUGE

A prototype fluorescence detector for the XL-I analytical ultracentrifuge has been developed for sedimentation velocity and equilibrium experiments. It is capable of detecting concentrations as low as 300 pM for fluorescein-like labels. The radial resolution of the detector is comparable to that of the absorbance system (MacGregor et al., 2004).

additional information from the sedimentation velocity boundary, such as diffusion coefficients and partial concentrations, and in some cases even to determine association constants (Comment D2.9).

COMMENT D2.8 GRAVITATIONAL-SWEEP SEDIMENTATION

In 1972 a light-scattering/turbidity detector was installed in an ultracentrifuge to monitor the concentration changes of polymer molecules in the cell during centrifugation (Scholtan and Lange, 1972). This method was further developed for high-resolution, submicron, particle size distribution analysis of polymers (Machtle, 1999). The main technical feature of this method is gravitational-sweep sedimentation in which the rotor speed is increased reproducibly and exponentially from 0 to 40 000 rpm over a one-hour period. This innovation allowed extremely broadly distributed polymer dispersions (from 10 to 3000 nm) to be studied without fractionation (see Section D2.5.3 for further discussion).

COMMENT D2.9 HOMEPAGES FOR AUC

Currently, there are three homepages on the web where we can find online information that is useful to both the novice and the expert in the ultracentrifugation field.

The Beckman homepage (www.beckman.com), besides containing information about their commercial products, also has an extensive list of background scientific papers and information about product development and data analysis for the XL-A and XL-I instruments.

The RASMB (reversible association in structural and molecular biology) site was created as a mail server to allow communication between research groups (www.rasmb.org).

Bo Demeler's Analytical Ultracentrifugation page is hyperlinked via RASMB (www.ultrascan.uthscsa.edu). The site has downloadable software but in addition presents a mailing list with past communications, collaborations, and fee-for-service information.

D2.3 THE LAMM EQUATION

The Lamm equation describes the transport process in the ultracentrifuge. It was derived by introducing drift into Fick's diffusion equation. As discussed in Section D1.5.3, the flux is proportional to the concentration gradient with the constant of proportionality equal to minus the diffusion coefficient, $-D$ (Eq. (D1.21)). Assuming that all the particles in the cell drift in the direction $+x$ with a speed u, then the flux at point x should increase by an amount $uC(x)$, where $C(x)$ is the local concentration. Fick's first equation then becomes

$$J_x = -D[dC/dx] + uC(x) \tag{D2.1}$$

Recalling that $u = s\omega^2 x$ we obtain

$$J_x = -D[\mathrm{d}C/\mathrm{d}x] + s\omega^2 x C(x) \tag{D2.2}$$

The first and second terms on the right-hand side of Eq. (D2.2) correspond to transport by diffusion and sedimentation, respectively.

Equation (D2.2) was derived for a solution in an ideal infinite cell without walls and is not a strictly correct description of the sedimentation–diffusion process under real sector cell experimental conditions. The cross-section of a sector cell is proportional to r. The appropriate continuity equation is

$$\left(\frac{\mathrm{d}C}{\mathrm{d}t}\right) = -\frac{1}{r}\left(\frac{\mathrm{d}(Jr)}{\mathrm{d}r}\right) \tag{D2.3}$$

Combining Eqs. (D2.2) and (D2.3) we obtain the following partial differential equation, known as the Lamm equation:

$$\left(\frac{\mathrm{d}C}{\mathrm{d}t}\right)r = -\frac{1}{r}\left\{\frac{\mathrm{d}}{\mathrm{d}r}\left[\omega^2 r^2 sC - Dr\left(\frac{\mathrm{d}c}{\mathrm{d}r}\right)t\right]\right\}t \tag{D2.4}$$

The Lamm equation describes diffusion with drift in an AUC sector cell, under real experimental conditions.

D2.4 SOLUTIONS OF THE LAMM EQUATION FOR DIFFERENT BOUNDARY CONDITIONS

D2.4.1 Exact Solutions

The general analytical solution of Eq. (D2.4) is an infinite series of integrals that can be calculated only by numerical integration. However, exact solutions of Eq. (D2.4) exist in two limiting cases: "no diffusion" and "no sedimentation."

No Diffusion

In the absence of diffusion, the Lamm equation has an exact solution for a homogeneous macromolecular solution:

$$C_2(x,t) = \begin{cases} 0 & \text{if } x_m < x < \bar{x} \\ C_0 \exp\left(-2s\omega^2 t\right) & \text{if } \bar{x} < x < x_b \end{cases} \tag{D2.5}$$

where C_0 is the initial concentration, and $\bar{x} = x_m \exp\left(s\omega^2 t\right)$. A graphical presentation of Eq. (D2.5) is shown in Fig. D2.7. The molecule concentration changes sharply from zero to a value that is independent of x at any time. The step function defines the boundary. The region $x > \bar{x}$ is called the plateau. The plateau level decreases with time owing to sectorial dilution (Fig. D2.7). The position of the step function also changes with time.

No Sedimentation

In the limiting case of no sedimentation occurring, the Lamm equation (D2.4) is given by

$$\left(\frac{\mathrm{d}C}{\mathrm{d}t}\right)r = -D\left(\frac{\mathrm{d}^2 C}{\mathrm{d}t^2}\right)t \tag{D2.6}$$

The concentration gradient becomes

$$\left(\frac{\mathrm{d}C}{\mathrm{d}t}\right)r = -C_0(\pi Dt)^{1/2}\exp\left(\frac{-x^2}{4Dt}\right) \tag{D2.7}$$

The solution of this equation was discussed in Section D1.5.4. The diffusion coefficient is determined by measuring the standard deviation of the Gaussian fit to the boundary function, which is equal to $(2Dt)^{1/2}$. In practice, the concentration gradient curves are measured as a function of time, according to the procedure described for pure diffusion (Fig. D1.14). This approach is valid if the term describing sedimentation flow $\omega^2 r^2 sC$ is essentially smaller than the diffusion term. In practice, this condition is fulfilled for small globular proteins (s value about 2 S) at low speed (2000–6000 rpm) with a synthetic boundary cell. The time-dependent spreading of the boundary is measured after overlaying the macromolecular solution with buffer.

D2.4.2 Analytical Solutions

Analytical solutions of the Lamm equation date from before 1930, but were limited to very specific boundary conditions (Comment D2.10). More realistic experimental

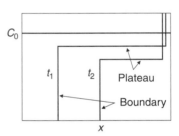

Fig. D2.7 Graphical presentation of Eq. (D2.5). In the absence of diffusion, a sharp boundary is formed and moves down the cell in time. There is a flat plateau below the boundary. The concentration in the plateau decreases with time owing to sectorial dilution.

COMMENT D2.10 ANALYTICAL SOLUTIONS OF THE LAMM EQUATION WITH VARIOUS SPECIFIC BOUNDARY CONDITIONS

Faxen-type solutions: The centrifugation cell is considered as an infinite sector, diffusion is small ($D/\omega^2 sx \ll 1$)), and only early sedimentation times are considered ($\omega^2 st \ll 1$).

Archibald solutions: s and D are considered to be constant.

Fujita-type solutions: An extended Faxen solution for the case when D is constant, but s depends on concentration as $s = s^0(1 - k_s c)$.

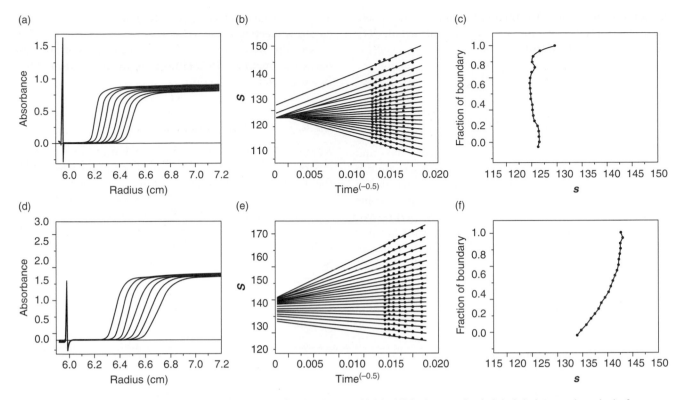

Fig. D2.8 Analysis of sedimentation velocity boundaries by the van Holde–Weischet method: **(a)–(c)** data and analysis for bacteriophage T7 MLD capsid II; **(d)–(f)** data and analysis for bacteriophage T7 MLD capsid II, in which the capsids contained measurable amounts of subgenomic, packaged DNA. The speed of the runs was 10 000 rpm. (Hansen *et al.*, 1994.)

situations can now be considered. The van Holde–Weischet and the Stafford methods have been developed, each of which enables the simultaneous determination of sedimentation and diffusion coefficients and, as a consequence, of molecular mass. Which method is applicable depends on the relation between the respective contributions of diffusion and sedimentation to the moving boundary.

The van Holde–Weischet Method

The van Holde–Weischet method is based on the fact that sedimentation is proportional to the *first power of time*, while diffusion is proportional to *the square root of time* (compare Eqs. (D2.18) and (D2.28)). It follows that extrapolation to infinite time must eliminate the contribution of diffusion to the boundary shape. In a first step, the distance between the baseline and the plateau for each scan is divided into N (usually a few tens) horizontal divisions. N apparent sedimentation coefficients s^* are calculated as

$$s^* = \ln\left[r/r_{\mathrm{m}}\right]/\omega^2 t \tag{D2.8}$$

where r and r_{m} are the radius of the division and the meniscus radius, respectively. The s^* values are then plotted versus the inverse square root of the time of each scan. The y-intercept, which corresponds to infinite time, is equal to the diffusion-corrected s value.

> **COMMENT D2.11 METHOD OF VAN HOLDE–WEISCHET**
>
> In practical applications, the method requires a well-defined plateau region and simulations which can be used to demonstrate the validity of the approach.

For a homogeneous sample, all lines should converge to the same limit, s. If the sample is not homogeneous, the average of the y-intercepts is used to calculate a weight-average sedimentation coefficient, s_{w}. The results are then plotted as a fraction of total sedimenting material versus s. Such a plot permits the characterization a priori of sample quality. Figure D2.8 shows sedimentation velocity boundaries (parts (a) and (d)) and an analysis by the van Holde–Weischet method of a homogeneous (parts (a)–(c)) and a heterogeneous (parts (d)–(f)) sample (Comment D2.11).

The van Holde–Weischet method has been introduced by B. Demeler into the software platform ULTRASCAN (www.cauma.uthscsa.edu).

The Stafford Method

The second approach is the Stafford method, in which the sedimentation coefficient distribution is computed from the time derivative of the sedimentation velocity concentration profile. An apparent sedimentation coefficient distribution (i.e., uncorrected for diffusion), $g(s^*)_t$, can be calculated from $(dC/dt)_r$ as

$$g(s^*)_t = (dC/dt)_{\text{corr}}(1/C_0)\left[\omega^2 t^2 / \ln(r_m/r)(r/r_m)^2\right] \quad \text{(D2.9)}$$

where s is the sedimentation coefficient, ω is the angular velocity of the rotor, C_0 is the initial concentration, r and r_m are the radius and radius of the meniscus, respectively. With this approach, velocity data are obtained in the standard way (see Fig. D2.9a). Closely spaced data sets are subtracted, giving time derivative profiles, as shown in Fig. D2.9b. Each boundary yields a Gaussian-shaped profile. A plot of $g(s^*)$ has essentially the same properties as a Schlieren gradients plot (dC/dr plot). The curves are then normalized for time. As a result, the separate Gaussian distributions are superimposed. The y-axis is renormalized so that the areas under the curves are equal to the concentration. The resulting peak is a Gaussian whose position and width give the sedimentation and diffusion coefficients, respectively (Fig. D2.9c).

An improved commercial version of the Stafford time derivative method (DCDTPLUS) is available from Philo (www.jphilo.mailway.com).

D2.4.3 Numerical Solutions

Numerical solutions of the Lamm equation for a continuous particle size distribution can be computed by stating the problem as an integral equation. Regularization is then used for its numerical inversion (Comment D2.12). The approach requires prior knowledge of the partial specific volumes and shapes of the molecules (in terms of frictional ratios, f/f_0), and of the density and viscosity of the solvent. Sedimentation coefficient or molar mass distributions can be obtained at relatively high resolution from the analysis. An example application is given in Section D2.5.2.

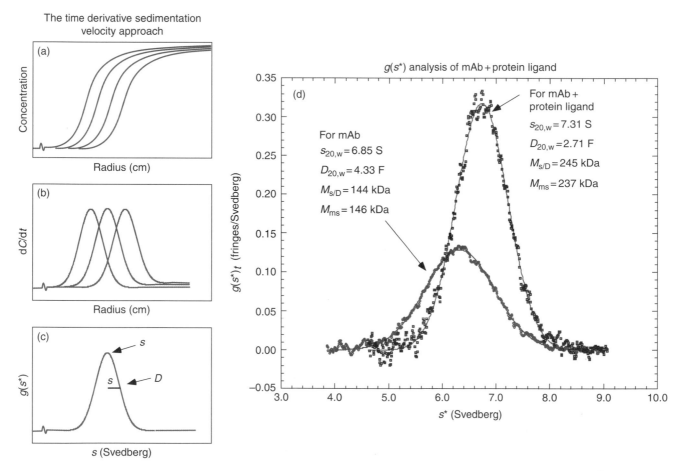

Fig. D2.9 Illustration of the Stafford method ((a)–(c), see text) Typical results are given in (d). Here we see the $g(s^*)$ results for a monoclonal antibody (mAb) and a mAb–protein complex. The molecular mass determined for mAb from s and D was 144 kDa and that from MALDI-MS was 146 kDA, indicating an error of 1.4% for the AUC value. For the mAb–protein complex, a molecular mass of 245 kDa was calculated from s and D, indicating an error of 3.8% when compared to the MALDI-MS of 237 kDA. The very good precision comes from the fact that all the data from 40 boundaries (~1000 points per curve) were used in the analysis. (Hensley, 1996.)

COMMENT D2.12 CONTINUOUS PARTICLE SIZE DISTRIBUTION

In the absence of interactions between particles, the experimentally observed sedimentation profiles of a continuous size distribution can be described as a superposition of subpopulations, $C(M)$, of particles of mass between M and $M + dM$. If $L(M, r, t)$ denotes the sedimentation profile of a monodisperse species of mass M and radius r at time t, the problem is described by a Fredholm integral equation of the first kind,

$$s(r,t) = \int C(M)L(M,r,t)\,dM + \varepsilon \qquad \text{(A)}$$

where $s(r, t)$ denotes the experimentally observed signal with an error ε.

The equation is encountered in similar form in particle characterization problems occurring in many other techniques, such as dynamic light scattering (Chapter D4), for example. In the case of sedimentation velocity AUC on a dilute solution, the kernel $L(M, r, t)$ of Eq. (A) is the solution of the Lamm equation.

The continuous size distribution analysis program SEDIT of P. Schuck is available from www.analyticalultra centrifugation.com.

D2.5 SEDIMENTATION VELOCITY

Sedimentation velocity and sedimentation equilibrium are the two main complementary methods in AUC. They are based on different experimental protocols but use the same instrumentation (Table D2.1). Sedimentation velocity, in which a high rotor speed and a long solution column are used to maximize the resolution, provides information about size, shape, and molecular interactions. Sedimentation equilibrium, in which low rotor speeds and a short solution column are used to minimize the time needed to reach equilibrium, provides information about molecular mass, association constants and stoichiometry. Figure D2.10 shows typical data from a sedimentation velocity experiment (a) and from a sedimentation equilibrium experiment (b).

D2.5.1 Macromolecules in a Strong Gravitational Field

The usual way of determining masses is to observe the movement of particles under the action of known forces, since the operational definition of mass depends on the relative acceleration of an unknown and a standard mass when acted upon by the same force. The most convenient force to use on a macroscopic scale is the force of gravity. For molecules, however, this force is so small that it is far outweighed by forces generated in collisions from the random motion of surrounding molecules.

TABLE D2.1 THE TWO TYPES OF EXPERIMENT IN AUC

	Sedimentation velocity	Sedimentation equilibrium
Angular velocity	Large (chosen according to the particles' sedimentation properties)	Small
Analysis	As a function of time (ex: 3–6 hours)	At equilibrium (ex: after 24 hours)
Measurement	Forming a boundary (sedimentation profile as function of time)	The particle distribution in the cell
Calculated parameters	Shape, mass composition	Mass composition

A simple calculation shows that, at the molecular level, gravitational potential energy is smaller than thermal energy by a factor of about a few hundred (Comment D2.13). It follows that the experimental detection of macromolecular movement due to natural gravity is impossible. A way to avoid the difficulty is to increase the effective gravitational potential energy to a value greater than kT, by putting the particles in a cell rotating at high speed. Let us consider a particle of mass m and density ρ in a sector-shaped cell that is in a rotor spinning about the z-axis with angular velocity ω moving through a solvent of density ρ_0. The molecule is acted on by centrifugal, F_c, buoyant, F_b, and frictional drag forces, F_d (Comment D2.14, Fig. D2.11).

The centrifugal force is proportional to the product of the particle mass m and the linear acceleration $\omega^2 r$

$$F_c = m\omega^2 r \qquad \text{(D2.10)}$$

where r is the distance from the center of rotation. Under the influence of this force the particle moves through the surrounding medium radially out from the axis of rotation.

The buoyant force (Comment D2.15) is equal in magnitude to the weight of the fluid displaced by the particle

$$F_b = -m_0\omega^2 r \qquad \text{(D2.11)}$$

The resulting net "gravitational" force is

$$F_c + F_b = (m - m_0)\omega^2 r = m\omega^2 r(1 - \rho_0/\rho) \qquad \text{(D2.12)}$$

Substituting the density ρ by its reciprocal, the partial specific volume $\bar{v}$ (see Section D2.8) we obtain

$$F_c + F_b = m\omega^2 r(1 - \bar{v}\rho_0) \qquad \text{(D2.13)}$$

The frictional force experienced by a molecule depends on its velocity u and frictional coefficient f (Eq. (D1.9))

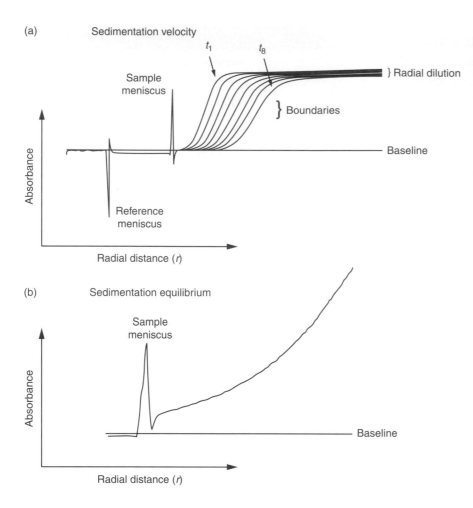

(a) Sedimentation velocity

(b) Sedimentation equilibrium

Fig. D2.10 (a) Boundaries in a sedimentation velocity experiment. Eight successive scans on a sample of reconstituted oligonucleosomes in a low salt buffer are shown. The speed of this run was 20 000 rpm. The direction of sedimentation is from left to right. The scans were collected at 8 min intervals (the earliest recorded scan is labeled "t_1," the latest recorded scan is labeled "t_8"). The baseline is the absorbance of the solvent alone. The centrifuge cell is composed of two sector-shaped compartments. The radial position of the air–solvent interface in the sample sector is labeled "sample meniscus." **(b)** Sedimentation equilibrium data on the same oligonucleosome sample, after centrifugation at 1600 rpm for 72 hours. The lower rotor speed resulted in the formation of an equilibrium concentration gradient rather than the formation of a moving boundary as in **(a)**. Equilibrium was achieved as judged by the fact that scans measured after 76–90 hours were identical with the one after 72 hours shown in the figure. (Hansen *et al.*, 1994.)

COMMENT D2.13 BIOLOGIST'S BOX: COMPARISON OF GRAVITATIONAL POTENTIAL AND THERMAL ENERGIES

We shall calculate the difference in gravitational potential energy between two points separated by a vertical distance L for a molecule of molecular mass M. Take $M = 10^5$ Da, and $L = 1$ cm. Then $M = 10^5/(6.03 \times 10^{26}) = 1.6 \times 10^{-22}$ kg (by using the number of molecules in a kilogram mole). The acceleration due to gravity is 10 m s^{-2}, the potential energy difference, E_p, is $E_p = mgh = 1.6 \times 10^{-22} \times 10 \times 10^{-2} = 1.6 \times 10^{-23}$ J. The thermal agitation energy is $\sim kT$, and at room temperature $kT = 1.38 \times 10^{-23}$ J K^{-1}, $293 \times K = 400 \times 10^{-23}$ J. We conclude that the gravitational potential energy is smaller than the thermal energy by a factor of 250, and therefore experimental detection of particle sedimentation due to gravity is impossible. An analogous calculation for $M = 10^9$ Da and $L = 10$ cm shows that in this case gravitational potential energy is larger than the thermal energy by a factor of 400 and experimental detection of the particle sedimentation due to gravity is possible.

COMMENT D2.14 MECHANICAL AND THERMODYNAMIC APPROACHES

The discussion in the text is in terms of a simple mechanical model for sedimentation processes. Some of the ambiguities that arise from this type of treatment can be avoided by using a thermodynamic approach.

$$F_d = -fu$$

A constant velocity is reached when the total force on the particle is zero, i.e.,

$$F_c + F_b + F_d = 0 \tag{D2.14}$$

$$m\omega^2 r(1 - \bar{v}\rho_0) = fu \tag{D2.15}$$

In order to put things on a mole basis we multiply Eq. (D2.14) by Avogadro's number. Placing experimentally measured parameters on the left-hand side of the equation and molecular parameters on the right-hand side we obtain:

$$s \equiv u/\omega^2 r = M(1 - \bar{v}\rho_0)/N_A f \tag{D2.16}$$

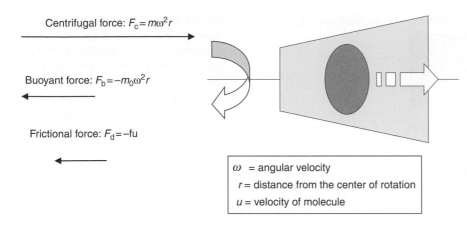

Centrifugal force: $F_c = m\omega^2 r$

Buoyant force: $F_b = -m_0\omega^2 r$

Frictional force: $F_d = -fu$

Fig. D2.11 Forces on the particle in a centrifuge cell (see text).

ω = angular velocity
r = distance from the center of rotation
u = velocity of molecule

COMMENT D2.15 ARCHIMEDES' FORCE: MACROSCOPIC AND MICROSCOPIC APPROACHES

Archimedes' principle states that "the buoyancy of a body in a liquid equals the weight of the liquid in a volume equal to the volume of the body." It is well known that the law was formulated for macroscopic bodies under hydrodynamics conditions and, therefore, it would seem that it is not relevant to individual molecules. In this connection the question arises: Can a molecule in a liquid be considered as a body immersed in the same liquid? In contrast to the liquid molecule, the macromolecule can be considered as a physical body possessing finite volume from which liquid molecules are excluded. The buoyancy force arises from the liquid pressure gradient across the volume. (See also Section G1.7.)

The ratio of the measured velocity of the particle to its centrifugal acceleration is called the *sedimentation coefficient*, s. Its dimension is time in seconds. The sedimentation coefficients of most biological particles are very small, and for convenience the basic unit is taken as 10^{-13} s, a unit that was called 1 svedberg (1 S), in recognition of Teo Svedberg's fundamental pioneering work in AUC.

D2.5.2 Determination of Sedimentation and Diffusion Coefficients from the Moving Boundary

Routine Mode AUC: *s* and *D* are Well Defined

AUC in routine mode is used to analyze dilute solutions of one type of macromolecule (a monodisperse solution) or of a mixture of a few different types for which s and D are well defined. The aqueous solvent typically contains a pH buffer and about 100 mM salt to screen electrostatic interactions between the macromolecules.

In a sedimentation velocity experiment, the centrifuge cell contains an initially homogeneous solution (the concentration is constant over the length of the cell). In the process of centrifugation, the particles move along gravitational force lines, the meniscus is cleared of macromolecules, and a moving boundary forms between solvent depleted of macromolecules and the solution of finite concentration (Fig. D2.12). The velocity of the moving boundary is given by the rate of change in its radial coordinate, dr_b/dt (Fig. D2.12). The velocity is expressed as a scalar quantity u, since the direction is well defined in the radial direction. We obtain, from Eq. (D2.16)

$$u = dr_b/dt = \omega^2 rs$$

After integrating,

$$\ln r_b(t)/r_b(t_0) = \omega^2 s(t - t_0) \qquad (D2.17)$$

The slope of $\ln r_b(t)/r_b(t_0)$ versus $(t - t_0)$ gives $\omega^2 s$ and hence s (Fig. D2.12). The method is applicable *only* to "ideal" solutions of a single type of macromolecule with no interparticle interactions. For slightly asymmetric or polymodal boundaries, the second moment method is a useful approach (Comment D2.16).

Complex Mode AUC: Analysis of the Boundary in Terms of *s* and *D*

Sedimentation velocity analysis of complex systems requires a more elaborate approach than the routine mode described above. In the case of very large particles, for which separation is achieved during the time of the experiment, a high-resolution analysis can be performed based on the spatial derivative of the sedimentation profiles, dc/dr, or by the related method of observing the time course of sedimentation at a single radial position. For smaller particles, however, diffusion broadening of the sedimentation boundary makes it more difficult to resolve subpopulations of the distribution. The Stafford or van Holde–Weischet methods (Section D2.4.2) are very useful in this regime. Both have the virtue of *model-free* analysis.

If a model is available for the sedimentation behavior of the macromolecules, size-distribution analysis can provide superior information in terms of precision in the derived

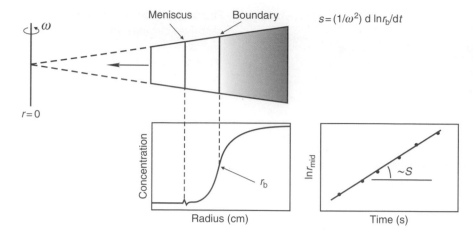

$s = (1/\omega^2)\, d \ln r_b / dt$

COMMENT D2.16 THE SECOND MOMENT METHOD

The second moment method provides a very useful approach for obtaining accurate estimates of the sedimentation coefficient in cases that exhibit slightly asymmetric or polymodal boundaries. The method permits the calculation of the position of a hypothetical boundary that would have been observed in the absence of diffusion and polydispersity (Fig. D2.13). The rate of movement of the equivalent boundary position gives the weight average sedimentation coefficient.

The second moment approach has generally been applied to data obtained with the Schlieren optical system, from which the equivalent boundary position was calculated from the second moment of the refractive index gradient curve:

$$\langle r^2 \rangle = \frac{\int\limits_{r_m}^{r_p} \frac{dc}{dr} r^2 dr}{\int\limits_{r_m}^{r_p} \frac{dc}{dr} dr}$$

However, when the data are obtained as concentration versus radius instead of concentration gradient versus radius, it is convenient to compute the equivalent position from the following relation:

$$\langle r^2 \rangle = \frac{\int\limits_{r_m}^{r_p} r^2 dc}{\int\limits_{r_m}^{r_p} dc}$$

parameters, although this is computationally more difficult than the Stafford or van Holde–Weischet methods. Figure D2.13 shows an example depicting the sequential steps in the analysis procedure for experimental sedimentation profiles.

Numerical solutions of the Lamm equations with applications for direct fitting of ultracentrifuge data are available (for example in Peter Schuck's program SEDFIT).

If ρ, η, $\bar{v}$, and f/f_0 are inadequately chosen, the $C(s)$ distribution is obtained with decreased resolution, while the derived $C(M)$ distribution is definitely wrong (s is an experimental parameter, M is calculated from the s values and shape and $\bar{v}$ estimates). For heterogeneous systems, and particularly for broad or complex species distributions, a regularization procedure has to be applied. As a result, the description of the sedimentation profiles in terms of a continuous distribution permits the determination of the level of non-homogeneity in the sample. Furthermore, evidence of interactions between the macromolecules can be provided from the changes in the apparent distribution of sedimentation coefficients with sample concentration.

D2.5.3 Highly Heterogeneous Systems

It is usual, in routine sedimentation velocity experiments, for a single rotor speed to be chosen for the entire run. Consequently, the range of observable sedimentation coefficients can be severely limited. For example, at 50 000 rpm and a usual acceleration rate of 400 rpm s^{-1}, the largest particle half way down the cell has a sedimentation coefficient of about 280 S. The limitation can be removed and the observable s range expanded by approximately three orders of magnitude, by varying the rotor speed during the run, starting with a relatively low speed so that the largest particles can be easily observed. The speed is gradually increased and the run continued at full speed until the smallest species of interest have cleared the solution. The data from each speed are combined into a single continuous distribution function.

A speed protocol is suggested in Table D2.2 that provides an observable s range from over 200 000 S to 2.0 S in a single experiment. The centrifuge is run for 600 seconds at each speed until the maximum speed is reached. The

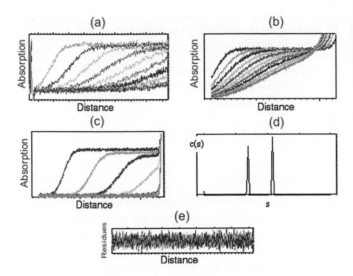

Fig. D2.13 Obtaining a discrete distribution of sedimentation coefficients, $C(s)$, by the SEDFIT program. (**a**) Experimental files for a solution comprising a 14 kDa fragment of heparin and a protein of 34 kDa (Interferon-γ) in the same proportion. Ten profiles obtained at 42 000 rpm and 10 °C are shown. (**b, c**) Simulated profiles with $\rho = 1.0105$ g ml^{-1}, $\eta = 1.429$ cp, $\bar{v} = 0.64$ mlg^{-1}, $f/f_0 = 1.25$. The values of ρ and η are those of the solvent, those of $\bar{v}$ and f/f_0 are reasonable assumptions corresponding to globular, usually hydrated complexes with about 1 g heparin per 1 g protein. (**b**) corresponds to the smallest considered particle with $s = 1$ S and a corresponding calculated D value of 10×10^{-7} cm^2 s^{-1}. (**c**) corresponds to the largest considered particle with $s = 5$ S and a corresponding calculated D value of 4.5×10^{-7} cm^2 s^{-1} (see the text). Comparison of (**b**) and (**c**) shows clearly how broadening of the sedimentation profiles is taken into account by the program. SEDFIT was asked to simulate 100 sets of sedimentation profiles corresponding to s values between $s = 1$ S and $s = 5$ S. The best combination of these profiles for the description of the experimental profiles corresponds to the continuous distribution $C(s)$ presented in (**d**). (**e**) shows the superposition of the differences between the profiles calculated with the distribution of particles of (**d**) and the experimental profiles. The $C(s)$ distribution on (**c**) suggests there are two types of species in solution. (Courtesy Chr. Ebel.)

Speed (rpm)	Time at each speed (s)	s^* (at 6.5 cm) (Svedberg units)
0–6000	15	~500.000
6000	600	4220
9000	600	1300
13 000	600	550
18 000	600	250
25 000	600	125
35 000	600	62
50 000	3600	31
50 000	3600	7.6
50 000	3600	4.4
50 000	3600	3.0
50 000	3600	2.3

TABLE D2.2 TYPICAL SPEED PROTOCOL; RADIUS OF THE MENISCUS = 5.9 CM (STAFFORD AND BRASWELL, 2004)

s^* (6.5 cm) = ln (6.5/5.9)/$j(\omega^2 dt)$.

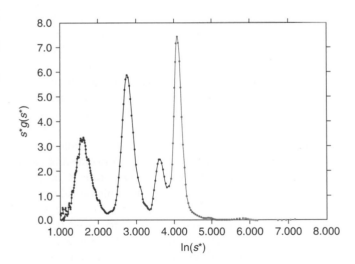

Fig. D2.14 Plot of $s*g(s*)$ verses ln $(s*)$ for a *Limulus polyphemus* hemeocyanin solution (2 mg ml^{-1}). The rotor velocity was ramped through speeds of 12, 18, 25, 35, and 50 krpm, and between 50 and 90 scans were made at each speed (~20 min), except at the highest speed. The run was continued at 50 krpm until it was obvious that the solution in the lower half of the cell was clear of protein. In this case the run was held at 50 krpm for three hours in order to bring most of the 5 S material to the lower quarter of the cell. The points acquired at each speed are shown in a different color. The areas under the peaks give the concentration for the boundary corresponding to each peak distribution. (Stafford and Braswell, 2004.)

rotor is then maintained at 50 000 rpm until the smallest material has sedimented more than half way to the bottom of the cell. Figure D2.14 shows the application of the method (Comment D2.17) to a hemeocyanin solution. The data were analyzed according to the time derivative method of Stafford (Section D2.4.2), and plotted as $s^*g(s^*)$ versus ln (s^*) because of the large range covered in s^*. The hemeocyanin sample resolved into four peaks (s values of ~5, 16, 40 and 60 S), corresponding to monomers, hexamers, 24-mers, and 36-mers, demonstrating how an extremely wide range of sedimentation coefficients can be accommodated in a single multi-speed run.

D2.5.4 Sedimentation Coefficients of Biological Macromolecules

Sedimentation coefficients of biological macromolecules are normally obtained in buffered solutions whose viscosity and density may differ from those of water. Sedimentation coefficients can also be measured at different temperatures. It is customary to correct sedimentation data to the "standard values" that would be observed if the measurements were performed in pure water at 20 °C

$$s_{20,w} = s_{exp} \frac{1 - \bar{v}\rho_{20,w}}{1 - \bar{v}\rho_{exp}} \frac{\eta_{exp}}{\eta_{20,w}} \qquad (D2.18)$$

where $s_{20,w}$ is the sedimentation coefficient of the molecule in water at 20 °C, s_{exp} is the experimentally measured sedimentation coefficient of the molecule, η_{exp} is the viscosity of the solvent at the experimental temperature T (°C), $\eta_{20,w}$ is the viscosity of water at 20 °C, $\rho_{20,w}$ is the density of water at 20 °C, ρ_{exp} is the density of the solvent at given temperature T (°C), and $\bar{v}$ is the partial specific volume of the molecule. Note that $s_{20,w}$ can be defined even if the species does not exist in water at 20 °C.

Sedimentation coefficients of biological macromolecules are normally obtained at finite concentration and should be extrapolated to zero concentration (Comment D2.18). The sedimentation coefficients of biological macromolecules cover a very large range of values. For small peptides and globular proteins they range from 0.4 S to 100 S; for ribosomal RNA, from 4 S to 30 S; for ribosomes and polysomes, from 20 S to 100 S; for viruses, from 40 S to 1000 S; for lysosomes, from 4000 S; for membranes, from 100 000 S; for mitochondria, from 20 000 S to 70 000 S; and for nuclei, from 4 000 000 S to 40 000 000 S. Generally, the larger the particle, the larger is its s value. Very large molecules may sediment in the Earth's gravitational field.

Table D2.3 lists some representative measured values of $s_{20,w}$, $\bar{v}$, and M of proteins, starting from small synthetic peptides (peptide I, molecular mass 2 kDa) and finishing with large oligomeric protein complexes (hemeocyanin, molecular mass about 9 MDa). Table D2.4 presents measured values of $s_{20,w}$ for native DNA and DNA fragments.

D2.5.5 Differential Sedimentation for Measuring Small Changes in Sedimentation Coefficients

Very small changes in sedimentation coefficient can be measured by the technique of differential sedimentation. The uncertainties due to the rotor speed and temperature are the principal errors in the determination of sedimentation coefficients. They can be eliminated, essentially, by placing the sample solution and solvent in two double-sector cells in the same rotor. Schlieren optics were used in the first experiments with two double-sector cells. One of the cells was assembled using a wedge window, so that its image was displaced on the photographic plate, which then revealed two separate Schlieren patterns.

The technique proposed by Kirschner and Schachman in 1971 is analogous to difference spectroscopy (Chapter E1). Each compartment in one double-sector cell is filled with a solution. If the two solutions are identical, the Rayleigh interference pattern consists of a series of straight, parallel, interference fringes. If the boundary in one solution moves more slowly than in the other, the difference in their positions produces a pattern of curved fringes, similar to that observed with a Schlieren optical system.

The method was tested by measuring differences in sedimentation properties between a solution of bushy stunt virus (BSV) in H_2O buffer and a solution of BSV in an

TABLE D2.3 SEDIMENTATION COEFFICIENTS, PARTIAL SPECIFIC VOLUMES OF PROTEINS IN AQUEOUS SOLVENT. BOLD TYPE INDICATES MOLECULES THAT HAVE NO GLOBULAR CONFORMATION.

Peptides and proteins	Molecular mass (Da)	Sedimentation coefficient $s^0_{20,w}$ (Svedberg units)	Partial specific volume $\bar{v}$ (cm³g⁻¹)
Small synthetic peptide I	2049	0.37	0.706
Small synthetic peptide II	2777	0.45	0.715
Small synthetic peptide III	3505	0.52	0.720
Lipase[a]	6669	1.14	0.7137
Insulin (dimer)	11 466	1.60	0.744
Cytochrome C	12 400	1.80	0707
Ribonuclease A (bovine pancreas)	13 683	1.78	0.696
Lysozyme (chicken egg white)	14 305	1.91	0702
α-lactalbumin (bovine milk)	14 180	1.92	0.704
Myoglobin	17 836	1.98	0.741
Papain	23 350	2.42	0.723
α-chymotrypsin	25 000	2.40	0.736
Chymotrypsinogen A (bovine pancreas)	25 767	2.58	0.736
Elastase	25 900	2.60	0.73
Subtilisin BPN	27 537	2.77	0.731
Carbonic anhydrase	30 000	3.0	0.735
Riboflavin-binding protein	32 500	2.76	0.720
Carboxypeptidase	34 472	3.2	0.725
Pepsin	34 160	2.88	0.725
β-lactoglobulin A, dimer (bovine milk)	36 730	2.87	0.751
β-lactoglobulin	35 000	3.08	0.75
Kinesin motor domain construct K366	41 404	3.25	0.733
Albumin ovum	44 000	3.6	0.74
Bovine serum albumin	66 300	4.50	0.735
Hemoglobin (human)	64 500	4.60	0.749
Anthax protective antigen	85 000	5.01	0.762
Tropomyosin	**93 000**	**2.6**	**0.71**
Lactate dehydrogenase (dogfish)	138 320	7.54	0.741
β-lactoglobulin A, octamer (bovine milk)	146 940	7.38	0.751
GPD, apo (bakers' yeast)	142 870	7.60	0.737
Aldolase	156 000	7.40	0.742
Malate synthetase	170 000	8.25	0.735
Catalase (bovine liver)	248 000	11.3	0.730
Glutamate dehydrogenase (bovine liver)	312 000	11.4	0.749

TABLE D2.3 (CONT.)

Peptides and proteins	Molecular mass (Da)	Sedimentation coefficient $s^0_{20,w}$ (Svedberg units)	Partial specific volume $\bar{v}(cm^3g^{-1})$
Fibrinogen	**330 000**	**7.6**	**0.706**
Apoferritin	467 000	17.6	0.750e
Apoferritin (horse spleen)	502 000	18.3	0.728
Urease	482 700	18.6	0.742
Myosin	**570 000**	**6.43**	**0.728**
Glutamate dehydrogenase	1 015 000	26.6	0.73
Hemoglobin (snail)	3 500 000	58.9	0.747
Hemocyanin	8 950 000	105.8	0.728

[a] Data are taken from different literature sources

H$_2$O–D$_2$O mixture. The results indicated that both the precision and the accuracy of the difference sedimentation technique are within 5% even for values of $\Delta s/s$ as low as 0.002.

The technique has been applied to a variety of proteins in the molecular mass range (30 000–300 000 Da) and it has been shown that changes in sedimentation coefficients of as low as 0.01 S can be measured with an accuracy better than ±5%.

D2.6 MOLECULAR MASS FROM SEDIMENTATION AND DIFFUSION DATA

Substituting the Einstein relation for the frictional coefficient in terms of the diffusion coefficient (Eq. (D2.15)) into Eq. (D2.7) we have

$$\frac{s}{D} = \frac{M(1 - \bar{v}\rho_0)}{RT} \qquad (D2.19)$$

Rewriting, we obtain the first Svedberg equation for the determination of macromolecular mass (Comment D2.19):

$$M = sRT/D(1 - \bar{v}\rho_0) \qquad (D2.20)$$

The values of $\bar{v}$ and ρ_0 need to be determined for the same conditions as for s and D (Comment D2.20).

In deriving Eq. (D2.20) we assumed that the frictional coefficients affecting diffusion, f_d, and sedimentation, f_s, are identical. The assumption is strictly correct only from a theoretical point of view. In reality, f_d and f_s are measured under different experimental conditions: f_d, under the action of zero or weak forces, and f_s, in a high gravitational field. Strong gravitational forces may deform flexible molecules or orient asymmetric ones. Extrapolation of s to zero angular velocity eliminates this effect and we can consider the value of f_s obtained in this way as identical to f_d.

COMMENT D2.19 MOLAR MASS

The use of SI units in the Svedberg equation (D2.20) yields a value of M technically referred to as the *molar mass*, with units of kilogram per mol.

We recall (see Chapter B1) that the *relative molecular mass* or *molecular weight ratio* is a dimensionless relative quantity, defined as the ratio of the mass of a molecule relative to 1/12 of the mass of the carbon isotope ^{12}C (very close to 1 g mol^{-1}).

Molar mass (in kilograms per mol) can be converted to a molecular weight by dividing by 10^{-3} g mol^{-1} (the equivalent of multiplying by 1000 and cancelling units).

Biochemists continue to use *molecular mass expressed in daltons* (Da) or atomic mass units (1 Da = 1 atomic mass unit, which equals one-twelfth of the mass of one atom of ^{12}C).

COMMENT D2.20 BIOLOGIST'S BOX: CALCULATION OF MOLECULAR MASS

A macromolecule with $\bar{v} = 0.74$ sediments in H$_2$O at 20 °C; its sedimentation coefficient $s^0_{20,w}$ is 14.2 S and its diffusion coefficient is $D^0_{20} = 5.82 \times 10^{-7} cm^2$.

According to Eq. (D2.20)

$$M = \frac{s^0_{20w}}{D^0_{20}} \frac{RT}{1 - \bar{v}\rho_0}$$

$$= \frac{14.2 \times 10^{-13}(8.31 \times 10^7)293}{5.82 \times 10^{-7}(1 - 0.74 \times 0.998)} = 227 \text{ kDa}$$

TABLE D2.4 SEDIMENTATION COEFFICIENTS OF DOUBLE-STRANDED DNA

Particle	Molecular mass (bp)	Sedimentation coefficient $s_{20,w}^0$
Fragments of DNA, prepared by digestion of PM2 phage with the restriction endo-nuclease Hae III (Kovacic and van Holde, 1972)	12	1.86[a]
	20	2.54[a]
	50	3.51
	94	4.58
	117	4.92
	145	5.20
	160	5.41
	263	6.24
	288	6.37
	322	6.70
	498	7.62
	592	7.89
	642	8.03
	794	8.59
	854	8.72
	1310	9.99
	1606	10.67
	1735	10.78
Native DNA:		
Ø×174RF	5386	14.3
T7 DNA	39.937	32.0
λ DNA	48.502	34.4
T4 DNA	168.903	62.1
T2 DNA	172.000	64.5
Highly fractionated samples by gel chromatography: ratio M_w/M_n is about 1.1	368[a]	7.2
	289[a]	6.3
	172[a]	5.7
	136[a]	5.4
	133[a]	5.4
	111[a]	4.7
	68[a]	3.9

[a] Sedimentation coefficients of 12 and 20 bp DNA fragments are calculated from diffusion coefficients according to the Svedberg equation (Eq. (D2.20)).

D2.7 SEDIMENTATION EQUILIBRIUM

D2.7.1 Molecular Mass

What happens to a solution of macromolecules if it is centrifuged for a very long time? Large molecular mass solutes form a pellet at the bottom of the centrifugation cell, but macromolecules of medium or small molecular mass are distributed near the bottom of the cell in a way that does not change in time, because of compensation between gravitational and diffusion forces.

In a sedimentation equilibrium experiment, a volume of an initially uniform solution of macromolecules is centrifuged at a lower angular velocity than is required for sedimentation velocity experiments (Comment D2.21). As macromolecules begin to sediment, the process of diffusion opposes the gravitational force. After an appropriate period of time, an equilibrium is attained in which the macromolecular concentration increases exponentially toward the bottom of the cell, in the same way that gases in the atmosphere distribute exponentially in the Earth's gravitational field (Fig. D2.15). The molecular mass of the sedimenting macromolecules can be calculated from the concentration values at different levels in the cell.

An equation describing the equilibrium sedimentation distribution is obtained by setting the total flux equal to zero in Eq. (D2.13), since, at equilibrium, there are no changes in concentration with time:

$$\left(\frac{dC}{dt}\right)r = -\frac{1}{r}\left\{\frac{d}{dr}\left[\omega^2 r^2 sC - Dr\left(\frac{dc}{dr}\right)t\right]\right\}t = 0 \qquad (D2.21)$$

Consequently,

$$D(dC/dr)_t - \omega^2 rCs = 0$$

By rearranging and substituting D from the Svedberg equation (Eq. (D2.20)) we obtain

$$\frac{d\ln(C)}{d(r^2/2)} = \frac{1}{r}C\frac{dC}{dr}\frac{\omega^2 s}{D} = \frac{M(1-\bar{v}\rho)\omega^2}{RT} \qquad (D2.22)$$

The macromolecular concentration distribution between the meniscus at a and point r obeys an exponential law, called the second Svedberg equation:

$$C(r) = C(a)\exp\left[\omega^2 M(1-\bar{v}\rho)(r^2 - a^2)/2RT\right] \qquad (D2.23)$$

A plot of ln [$C(r)$] vs $r^2/2$ is presented in Fig. D2.16a. In the case of a monodisperse solution, the slope of the straight line is proportional to the macromolecular mass. In a polydisperse solution, containing macromolecules of different molecular mass, each will be distributed according to Eq. (D2.23). The higher molecular masses concentrate selectively toward the bottom of the cell, while the lower molecular masses dominate the distribution at the top, and the plot of ln [$C(r)$] vs $r^2/2$ shows an upward curvature (Fig. D2.16b). The tangential slope at each point on the curve yields the average molecular mass at the corresponding point in the cell (Comment D2.22).

Since the derivation of Eq. (D2.23) did not rely on the molecular shape, the molecular masses obtained in this fashion are independent of this parameter. This is the power of the method. Macromolecular shape affects only the rate at which equilibrium is reached, not the final distribution.

Equation (D2.23) can be used to estimate the rotation speed needed for an equilibrium measurement. With a typical optical system, an accurate measurement can be made when the concentration falls by a factor of 2 across a 1 mm sample height. If the molecular mass of the particle is 50 000 Da, a = 6 cm, and $\bar{v}$ = 0.75 cm^{3}g^{-1}, then ω = 1000 rad s^{-1} or 10 400 rpm.

Figure D2.17 shows optimum speeds for an equilibrium run if either the molecular weight or the sedimentation coefficient of the sample can be estimated.

A set of at least three speeds is chosen to yield significantly varied data for diagnostics.

D2.7.2 Binding Constants

Measuring the concentration dependence of an effective average molecular mass (for example, to determine the

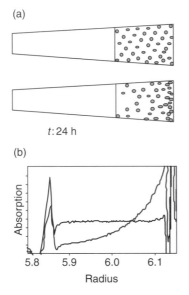

Fig. D2.15 (a) Schematic representation of sedimentation equilibrium. The flow of macromolecules due to sedimentation increases with radial distance. This process is balanced at equilibrium by the reverse flow from diffusion, which increases with concentration gradient. (b) At equilibrium, the resulting concentration distribution is exponential with the square of the radial distribution. The molecular weight of the macromolecule is directly related to the concentration gradient at equilibrium according to Eq. (D2.23).

COMMENT D2.22 THE AVERAGE MOLECULAR MASS

The average molecular mass calculated for a polydisperse solution is based on the proportion (by number) of the various components present. This type of average is known as a *number-average* molecular mass, M:

$$M_n = \sum N_i m_i / \sum N_i == \sum n_i m_i$$

where n_i is the number fraction of molecules with mass m_i.

The *weight-average* molecular mass, M_w, is defined as

$$M_w = \sum N_i m_i^2 / \sum N_i m_i = \sum w_i m_i$$

where w_i is the mass fraction of molecules with masses m_i.

For example, for an equimolar mixture of just two molecules with masses of 10 000 Da and 100 000 Da, respectively,

$$M_n = 55\,000\,\text{Da}$$
$$M_w = 91\,818\,\text{Da}$$

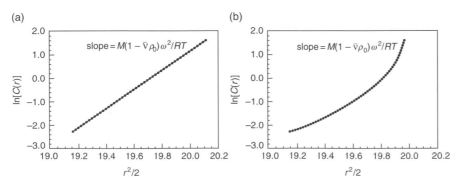

Fig. D2.16 (a) Plot of the logarithm of the equilibrium concentration versus the radius squared for a monodisperse macromolecule solution. The slope of this plot is proportional to the molecular weight of the macromolecule. (b) Plot of the logarithm of the equilibrium concentration versus the radius squared for a polydisperse macromolecule solution.

amount of monomers and oligomers) is a highly useful way to describe different kinds of association phenomenon (see also Chapter C1).

For a heterogeneous associating system of the form $A + B \rightleftarrows AB$ the equilibrium sedimentation distribution is described by three exponentials, one for each molecular mass species (the exponentials are written in condensed form as σ_i):

$$C(r) = C_A(r)\sigma_A + C_B(r)\sigma_B + C_{AB}(r)\sigma_{AB} \quad (D2.24)$$

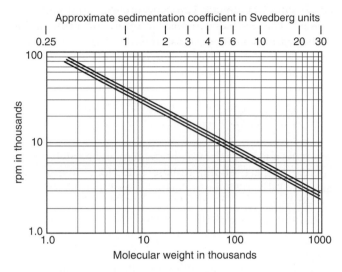

Approximate sedimentation coefficient in Svedberg units

Fig. D2.17 Optimum speeds for equilibrium runs if either the molecular weight or the sedimentation coefficient can be estimated. (After Chervenka, 1969.)

The pre-exponential term for the AB association can be recast in terms of the concentrations of A, B, and the dissociation equilibrium constant, K_{AB}

$$C(r) = C_A(r)\sigma_A + C_B(r)\sigma_B + \frac{C_A(r)C_B(r)}{K_{AB}}\sigma_{AB} \quad (D2.25)$$

Recall from the law of mass action that

$$K_{AB} = \frac{C_A(r)C_B(r)}{C_{AB}(r)} \quad \text{and} \quad C_{AB}(r) = \frac{C_A(r)C_B(r)}{K_{AB}} \quad (D2.26)$$

The dissociation equilibrium constant may therefore be determined directly from the equilibrium sedimentation data by applying Eq. (D2.26).

As an example, Fig. D2.18 shows the data and analysis for two associating proteins forming a monomer $\leftrightarrows$ dimer $\leftrightarrows$ tetramer reversibly associating system. The values determined for $K_{1,2}$ and $K_{2,4}$ were 8.3×10^{-6} M and 2.0×10^{-6} M, respectively. Figure D2.18b shows the fit as a sum of three exponentials and Fig. D2.18a shows the distribution of residuals. The computed relative concentrations of monomer, dimer, and tetramer are plotted as a function of total monomer concentration in Fig. D2.18c. Note the wide concentration range (10^{-3} to 10^{-8} M) that can be analyzed usefully in a sedimentation equilibrium experiment.

In a second example, Fig. D2.19 shows the concentration dependence of molecular weight, M_w, for carbonic anhydrase in a range covering about three orders of magnitude. The solid line describes a trimer–hexamer equilibrium, and the dashed line a dimer–trimer equilibrium. It is clearly seen that the data preclude a dimer–tetramer equilibrium (dashed line).

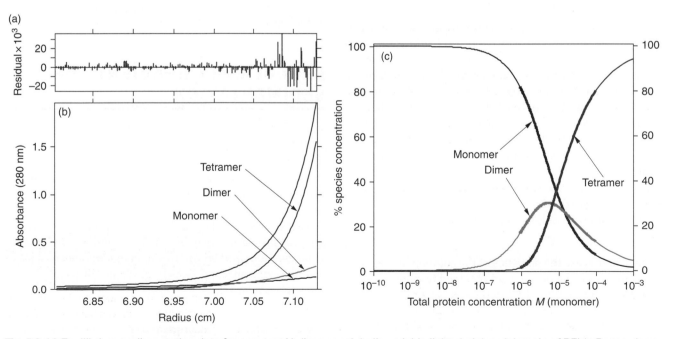

Fig. D2.18 Equilibrium sedimentation data for a mutant V_L (immunoglobulin variable light chain) and domain of REI (a Bence–Jones protein) analyzed in terms of a monomer $\leftrightarrow$ dimer $\leftrightarrow$ tetramer model (see text). The thickened portions of the curves in (c) indicate the concentration range where the analysis was performed. The thin portions of the curves were obtained by extrapolation (After Hensley, 1996.)

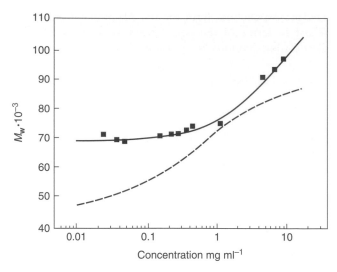

Fig. D2.19 Concentration dependence of molecule weight, M_w, for carbonic anhydrase from *Methanosarcina thermophila*. The solid line describes a trimer–hexamer equilibrium. The data are in agreement with crystallographic data demonstrating that in the crystal two trimers dimerize to form a hexamer by association of the N-terminals of six monomers into a six-stranded β-structure with a hydrophobic core. The dashed line was calculated for a dimer–tetramer equilibrium. (After Belhke and Ristau, 1997.)

D2.8 THE PARTIAL SPECIFIC VOLUME

Determination of the molecular mass of a macromolecule by using sedimentation equilibrium (Section D2.7) or by combining sedimentation and diffusion data (Section D2.6) requires an accurate knowledge of its partial specific volume, $\bar{v}$. For a simple two-component system containing one type of solute dissolved in one type of solvent (in our case buffer), $\bar{v}$ is the thermodynamically defined partial specific volume of the solute, which is related to the volume increase of the solution when adding a given quantity of dry macromolecules (see Chapter D1).

The partial specific volume is usually determined from a series of precise density measurements on solutions containing a different weight concentration, w, of solute:

$$\rho = \rho_0 + w(1 - \rho_0 \bar{v}) \tag{D2.27}$$

where ρ and ρ_0 are the densities of the solution and solvent, respectively. The values of $\bar{v}$ are obtained from the slope of ρ plotted as a function of w. In the case of a dilute macromolecular solution, $\bar{v}$ is independent of the macromolecular concentration and the plot is a straight line. Partial specific volume, $\bar{v}$, is usually expressed in units of cubic centimeter per gram.

Inspection of Eq. (D2.27) shows that extremely precise density measurements are needed to obtain accurate values of $\bar{v}$. In order to obtain 1% accuracy in $\bar{v}$ (required for 3–4% accuracy in M in the case of protein), density values must be more precise than 1 part in 10^5 at a solute concentration of 1 mg ml^{-1}. It is possible but difficult to obtain the necessary precision with most of the standard techniques, which include pycnometry, magnetic float devices, electrobalance with plummet instruments, vibrating tube densitometers, and AUC in solvents with different isotope compositions. We must keep in mind, however, that such measurements are very tedious and difficult. Fortunately, there is a way to overcome the difficulty, at least for proteins and carbohydrates. The partial specific volumes of proteins or carbohydrates can be calculated with adequate precision from the amino acid or monosaccharide composition, respectively, on the assumption of additivity of residue volumes,

$$\bar{v} = \sum_i \bar{v}_i W_i \Big/ \sum_i W_i \tag{D2.28}$$

where W_i and $\bar{v}_i$ are the weight fraction and partial specific volume, respectively, of residue i. Residue volumes are given in Table D2.5 for amino acids and in Table D2.6 for carbohydrates.

Careful studies of the problem have shown that the initial approach gives the closest agreement with experiment, while later attempts to refine the method, for example by attributing different volumes to amino acid residues according to whether they are found in the core or on the surface of the molecule only made matters worse (Comment D2.23). Typical $\bar{v}$ values for the main types of biological macromolecule are listed in Table D2.7.

D2.9 DENSITY GRADIENT SEDIMENTATION

Density gradient sedimentation in AUC has a long history in the study of nucleic acids and proteins, and is still applied in a variety of studies, with topics ranging from the composition of genomes to the characterization of protein–detergent, protein–nucleic acids, and protein–lipid complexes.

Below, we describe the classical approaches based on *analytical zonal sedimentation velocity* and *density gradient sedimentation equilibrium*, in which macromolecules move through concentration density gradients of smaller cosolutes (Comment D2.24).

D2.9.1 Velocity Zonal Method

A sample solution containing a mixture of particles is layered on a preformed gradient of increasing density (Fig. D2.20a). Upon centrifugation, the particles sediment

TABLE D2.5 PARTIAL SPECIFIC VOLUMES AT 25 °C, $\bar{v}$, AND MOLECULAR MASS, M, OF AMINO ACID RESIDUES[a]

Amino acid	M	$\bar{v}$ (cm³g⁻¹)
Glycine	57	0.64
Alanine	71	0.74
Serine	87	0.63
Threonine	101	0.70
Valine	99	0.86
Leucine	113	0.90
Isoleucine	113	0.90
Proline	97	0.76
Methionine	131	0.75
Phenylalanine	147	0.77
Cystine	103	0.61
Tryptophan	186	0.74
Tyrosine	163	0.71
Histidine	137	0.67
Arginine	156	0.70
Lysine	128	0.82
Aspartic acid	115	0.60
Glutamic acid	129	0.66
Glutamine	128	0.67
Asparagine	115	0.60

[a] Data are taken from Durchschlag (1965).

TABLE D2.6 PARTIAL SPECIFIC VOLUMES AT 20 °C, $\bar{v}$, AND MOLECULAR MASS, M, OF CARBOHYDRATES[a]

Carbohydrate	M	$\bar{v}$ (cm³g⁻¹)
Fructose	180	0.614
Fucose	164	0.671
Galactose	180	0.622
Glucose	180	0.622
Hexose	180	0.613
Sucrose (0.05 M)	342	0.613
Sucrose (0.15 M)	342	0.616
Sucrose (1 M)	342	0.620
Raffinose (25 °C)	486	0.608

[a] Data are taken from Chervenka (1969).

TABLE D2.7 APPROXIMATE VALUES OF PARTIAL SPECIFIC VOLUME $\bar{v}$ FOR THE MAIN TYPES OF BIOLOGICAL MACROMOLECULE

Macromolecules	$\bar{v}$ (cm³ g⁻¹)
Proteins	0.70–0.75
Carbohydrates	0.59–0.65
RNA	0.47–0.55
DNA	0.55–0.59

COMMENT D2.23 PREDICTION OF $\bar{v}$

The partial specific volume is now routinely calculated from the chemical composition of proteins with the program SEDNTEP (Harpaz et al., 1994).

COMMENT D2.24 DYNAMIC DENSITY GRADIENT

A new *dynamic density gradient from sedimenting cosolute* method was developed by Schuck and colleagues. It is based on a numerical solution of the Lamm equation with spatial and temporal variations of the local solvent density and viscosity (Schuck, 2004).

through the gradient to separate zones based on their sedimentation velocity. The centrifugation is terminated before particles in the separate zones reach the bottom of the tube (Fig. D2.20b). The maximum density of the gradient should be chosen not to exceed that of the densest particle to be separated. The density and viscosity of the gradient medium and the velocity of centrifugation also affect the separation. Sucrose and glycerol solutions are the most frequently used as a non-ionic gradient material for velocity zonal centrifugation. Linear 5–20% sucrose gradients are a traditional choice for the determination of the sedimentation coefficients of protein and nucleoprotein particles. Velocity zonal centrifugation separates molecules in the mixture according to their sedimentation coefficients, which are determined by their size, shape, and buoyant density. The method also allows the estimation of relative molecular masses (Comment D2.25).

To illustrate the method, Fig. D2.21 shows the separation of 22 S RNP particles by centrifugation in a sucrose concentration gradient. Peak I corresponds to the 22 S RNP.

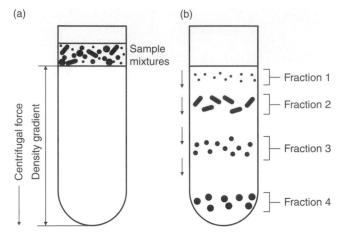

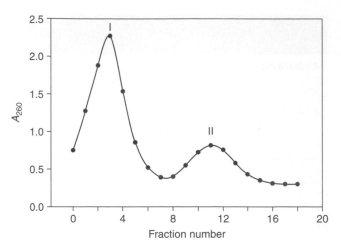

Fig. D2.20 Velocity zonal separation of a mixture of particles using a density gradient. **(a)** A sample containing mixtures of particles of varying size, shape, and density is added on the top of a preformed density gradient. The gradient is higher in density toward the bottom of the tube. **(b)** Centrifugation results in separation of the particles depending on their size, shape, and buoyant density. For globular proteins, those with larger molecular mass are likely to sediment to positions toward the bottom of the gradient, compared to those with smaller molecular mass that remain near the top of the gradient. Fractions of defined volume are collected from the gradient.

Fig. D2.21 Isolation of the 22 S RNP particles by centrifugation in a 5–20% sucrose gradient at 50 000 rpm for 3 h at 20 °C in an SW 55 rotor. (After Ulitin et al., 1997.)

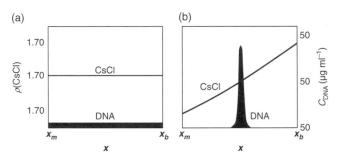

Fig. D2.22 Equilibrium ultracentrifugation of DNA in a CsCl gradient: **(a)** the density gradient distribution at the beginning of the experiment; **(b)** the density gradient at equilibrium. (After Cantor and Schimmel, 1980.)

COMMENT D2.25 SEDIMENTATION COEFFICIENTS FROM VELOCITY ZONAL CENTRIFUGATION

Sedimentation coefficients obtained by velocity zonal centrifugation in a sucrose gradient may be erroneous for any protein undergoing reversible self-association or interaction with other macromolecules. The source of error comes from the molecular crowding effect due to high sucrose concentration that shifts an equilibrium to favor polymeric states of the macromolecule. In this case the use of classical AUC methods is strongly recommended for the determination of sedimentation coefficients.

D2.9.2 Equilibrium Sedimentation in a Density Gradient

The large forces that can be generated in AUC have been used to great advantage in a sedimentation–equilibrium method in which the macromolecules are spun through a small-solute density gradient, which is itself formed by the centrifugal field. Such a gradient may be obtained by spinning a solution of a heavy salt, such as CsCl or RbCl, until equilibrium is reached.

Consider a three-component molecular solution: solvent molecules (component 1); the macromolecules under study (component 2); and small solutes (component 3).

According to Eq. (D2.23) the small solute is distributed in the cell in just the same way as a large molecule. Because of the small molecular mass of component 3, we can expand the exponential in Eq. (D2.23) and only keep the first terms

$$C_3(r)/C_3(a) = 1 + M_3(1 - \bar{v}\rho)\omega^2(r^2 - a^2)/2RT \qquad (D2.29)$$

Equation (D2.29) shows that a parabolic concentration distribution of component 3 is produced in the sample cell. The density of the solvent is an approximately linear function of the concentration of component 3 so that, at equilibrium, a parabolic density gradient is established (Fig. D2.22). If the density, ρ, of component 2 falls between the two extreme density values of the solvent, the macromolecules tend to gather in the region of the cell where their buoyant density is zero, i.e., $\rho = \rho_0$. The width of the component 2 band is a function of the macromolecular mass and of the steepness of the local density gradient. The bands may be very narrow for large masses because of their small diffusion coefficients. Native DNA, for example, produces such a narrow band that DNA containing ^{15}N ("heavy nitrogen") is clearly separated from the natural abundance, ^{14}N ("light nitrogen"), macromolecule by equilibrium centrifugation in a caesium chloride density gradient (Comment D2.26). Based on this fact one of the *gold medal experiments* in structural molecular biology was performed on the mechanism of DNA replication by Messelson and Stahl in 1957.

The most striking feature of the Watson–Crick double-helix model was its postulate that the two DNA strands are complementary, and that replication is based on new complementary strands forming on each strand. The result is two daughter duplex DNA molecules, each of which contains one strand from the parental DNA. The process is called *semi-conservative* replication and is shown in Fig. D2.23.

Messelson and Stahl grew *E. coli* cells in a medium in which the sole nitrogen source was ^{15}N-labeled ammonium chloride. The ^{15}N-containing *E.coli* cell culture was then transferred to a light ^{14}N medium and allowed to continue growing. Samples were harvested at regular intervals. The DNA was extracted and its buoyant density determined by centrifugation in CsCl density gradients. The

---

COMMENT D2.26 ^{15}N ("HEAVY") AND ^{14}N ("LIGHT") ISOTOPES

At sedimentation equilibrium in CsCl, the difference in density of DNA molecules containing the ^{15}N isotope and the ^{14}N isotope is equal to 0.014 g cm^{-3}. As a result, the difference between bands is equal to 0.5 mm in a standard ultracentrifuge cell at 40 000 rpm.

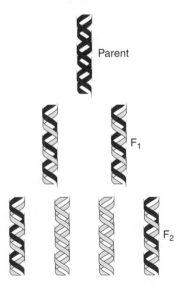

Fig. D2.23 The semi-conservative mechanism of DNA replication. Each F_1 duplex contains one parent strand. The F_2 generation consists of two hybrid DNAs and two totally new DNAs. Newly replicated strands are shown in gray; parental strands in black.

isolated DNA showed a single band in the density gradient, midway between the light ^{14}N-DNA and the heavy ^{15}N-DNA bands (Fig. D2.24c). After two generations in the ^{14}N medium the isolated DNA exhibited two bands, one with a density equal to light DNA and the other with a density equal to that of the hybrid DNA observed after one generation (Fig. D2.24d). After three generations in the ^{14}N medium the DNA still has two bands, similar to those observed after two generations (Fig. D2.24e). The results were exactly those expected from the semi-conservative replication hypothesis.

D2.10 MOLECULAR SHAPE FROM SEDIMENTATION DATA

We recall that Eq. (D2.19) relates the sedimentation coefficient, s, to the molecular properties, M, D, and $\bar{v}$:

$$s = \frac{MD(1 - \bar{v}\rho_0)}{RT} = \frac{MD(1 - \bar{v}\rho_0)}{s} \tag{D2.30}$$

The sedimentation coefficient (itself related to the translational friction coefficient) depends on particle volume and shape. It is usually expressed in terms of the radius of an equivalent sphere (the Stokes radius) but can equally well be accounted for by different volume–shape combinations – for example, an elongated particle of smaller volume than the Stokes radius sphere (Section D2.2.1). It follows from Eq. (D2.2) that the direct determination of molecular shape from the sedimentation coefficient is not possible. The sedimentation coefficient can be explained, for example, in terms of either a spherical particle of definite $\bar{v}$ and total volume, or an elongated particle of the same $\bar{v}$ and smaller total volume.

Molecules of the same shape but different molecular mass form a so-called homologous series. In this case, the relation between M and s is a follows:

$$s = K_s M^a \tag{D2.31}$$

where K_s is a coefficient that depends on the partial specific volume. The value of the exponent, a, depends on the *shape* of the molecules in the series (Section D2.6). The usefulness of such an analysis is illustrated below for different homologous series (see also Comment D2.27).

D2.10.1 Homologous Series of Quasi-Spherical Particles: Globular Proteins in Water

We recall (Section D2.2.1) that the frictional coefficient of a sphere of radius R_0 under slip boundary conditions in a solvent of viscosity η_0 is given by Stokes' law (Eq. D2.1):

$$f_0 = 6\pi\eta_0 R_0$$

The radius of the sphere is related to its mass via its volume, V_0

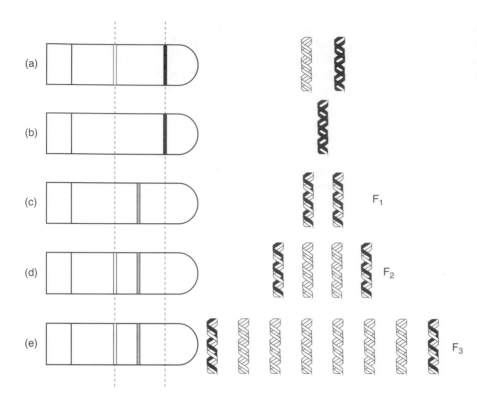

Fig. D2.24 Results of the Messelson–Stahl experiment: **(a)** control experiment, a mixture of heavy (^{15}N) and light (^{14}N) DNAs; **(b)** parent DNA is the heavy DNA; **(c)** DNA after one division in the ^{14}N medium; **(d)** DNA after two divisions in ^{14}N; **(e)** DNA after three divisions in ^{14}N.

$$\frac{4}{3}\pi R_0^3 = V_0 = M\bar{v}/N_A \tag{D2.32}$$

and we can write

$$f_0 = 6\pi\eta_0(3M\bar{v}/4\pi N_A)^{1/3} \tag{D2.33}$$

Inserting this result into Eq. (D2.19), we obtain, after rearrangement,

$$s^* = s_{20,w}^0 \bar{v}^{1/3}/(1 - \bar{v}\rho) = M^{2/3}/6\pi\eta_0 N_A^{2/3}(3/4\pi)^{1/3} \tag{D2.34}$$

The log–log dependence of s^* on M is a straight line of slope 2/3. When $s_{20,w}^0$ is expressed in Svedberg units and all other quantities in cgs units, Eq. (D2.34) becomes

$$s^* = 12.0 \times 10^{-3}M^{2/3} \tag{D2.35}$$

We now explore to what extent globular proteins behave as a homologous series of spherical particles. A log–log plot of s^* as a function of M is shown in Fig. D2.25 for globular proteins with a wide range of molecular masses using the data presented in Table D2.3. A good straight-line fit of log s^* versus log M plot is obtained according to a single equation:

$$s^* = 9.1 \times 10^{-3}M^{0.65} \tag{D2.36}$$

which establishes that globular proteins do in fact form a homologous series. Comparison of Eqs. (D2.35) and (D2.36) shows that though the exponents in these equations coincide, the coefficients are different. Small deviations

---

COMMENT D2.27 KUHN–MARK–HOUWINK RELATIONS

For a homologous series of particles the following *KMX* relations are valid:

$$[\eta] = K_{[\eta]}M^a \tag{A}$$

$$s = K_sM^a \tag{B}$$

$$D = K_D M^{-b} \tag{C}$$

The relations between constants α, a, and b are:

$$b + a = 1 \tag{D}$$

$$b = (1 + \alpha)/3 \tag{E}$$

Thus, for a homologous series of rigid rod-like particles:

$$\alpha = 1.7, \quad a = 0.15, \quad b = 0.85$$

for a homologous series of spherical particles:

$$\alpha = 0, \quad a = 2/3, \quad b = 1/3$$

for a homologous series of Gaussian coils in an "ideal" solvent:

$$\alpha = 0.5, \quad a = 0.5, \quad b = 0.5$$

and for homologous series of Gaussian coils in a "good" solvent:

$$\alpha = 0.8, \quad a = 0.4, \quad b = 0.6$$

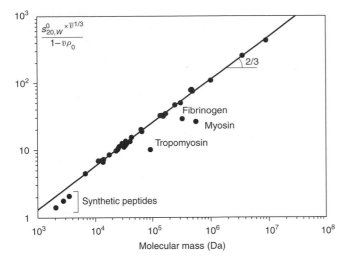

Fig. D2.25 The log–log plot of the dependence $s_{20,w}^0 \bar{v}^{1/3}/(1 - \bar{v}\rho_0)$ on M for proteins in a wide mass interval using the data presented in Table D2.3. For a globular proteins plot the slope of this is 2/3, which is typical for quasi-spherical particles. Exceptions to the rule are: synthetic peptides (unfolded particles), tropomyosin, fibrinogen, and myosin (rod-like particles).

from perfect spherical shapes and the existence of hydration shells modify the relation between the Stokes radius and the partial specific volume assumed in Eq. (D2.1), without changing the power law. We conclude from the fact that their sedimentation behavior can be described by a single equation, Eq. (D2.36), that globular proteins are very close to spherical in shape, and hydrated to about the same degree.

Table D2.3 includes three proteins (tropomyosin, fibrinogen, myosin) that do not obey Eq. (D2.27). Their sedimentation coefficients are smaller than that of a globular protein of the same molecular mass (Fig. D2.25). There are two possible causes for such behavior: Either the molecule has an elongated (rod-like) shape or it is unfolded. The shape of a non-globular protein can be derived *directly* from sedimentation measurements if the molecule can be broken up into fragments that themselves form a homologous series. Collagen is a good example of such an analysis (Comment D2.28).

D2.10.2 Homologous Series of Random Coils: Proteins in Guanidine Hydrochloride

Most proteins are dissociated into single polypeptide chains and unfolded in 6 M guanidine hydrochloride (GuHCl) or 8 M urea containing 0.1 M mercaptoethanol (EtSH). Figure D2.26 shows the dependence of $s_{20,w}^0/(1 - \Phi'\rho)$ on n, where n is the number of amino acid residues in a polypeptide chain and Φ' is the apparent partial specific volume for proteins in 6 M GuHCl. The plot is a good straight line that follows the equation

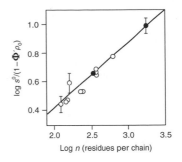

Fig. D2.26 A log–log plot showing the dependence of $s_{20,w}^0/(1 - \Phi'\rho)$ on n, where n is the number of amino acid residues and Φ' is the apparent partial specific volume for proteins in 6 M guanidine hydrochloride. (After Tanford et al., 1967.)

$$s^0/(1 - \bar{v}\rho) = 0.286 n^{0.47} \tag{D2.37}$$

The sedimentation coefficient varies with molecular mass in the manner expected for random coil polymers in a solvent close to "ideal" (Comment D2.29).

D2.10.3 From Slightly Flexible Rod to Nearly Perfect Random Coil: DNA

The study of DNA molecules is a good example of how to extract a molecular shape from sedimentation data. It is well known from early studies that the log–log dependence

---

COMMENT D2.28 COLLAGEN AND ITS FRAGMENTS

Sonic irradiation of soluble calf-skin collagen causes fragmentation of the macromolecules into shorter pieces that retain the three-stranded, helical structure. The dependence of the sedimentation coefficient on molecular mass is small and can be written as

$$s_{20,w}^0 = 0.232 M^{0.20} \tag{A}$$

Referring to the small value of the exponent for elongated structures in Table D2.3, the 0.20 exponent in Eq. (A) is a solid argument in favor of the fact that the collagen and its fractions form a homologous series of solid rod-like molecules.

COMMENT D2.29 "IDEAL" AND "GOOD" SOLVENTS IN POLYMER CHEMISTRY

The term "ideal" or "theta" solvent in polymer chemistry means a solvent in which the second virial coefficient is equal to zero. The term "good" solvent means a solvent in which the second virial coefficient is positive.

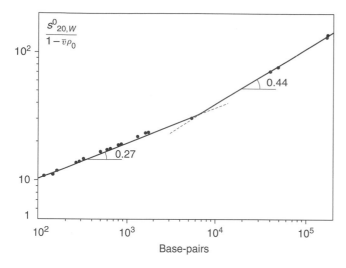

Fig. D2.27 Dependence $s^0_{20,w}$ on molecular mass (in base-pairs) corrected for the buoyancy factor $(1 - \bar{v}\rho_0)$ for DNA fragments. Data are taken from Table D2.4.

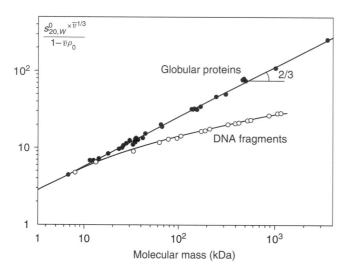

Fig. D2.28 Dependence of log $s^0_{20,w}$ on the logarithm of molecular mass, corrected for the buoyancy factor $(1 - \bar{v}\rho_0)$ for globular proteins (•) and for the PM2 *Hae* III fragments of DNA (○) in the small molecular mass range. Data are taken from Tables D2.2 and D2.3. Sedimentation coefficients for 12 bp and 20 bp DNA fragments are calculated from experimental values for their diffusion coefficients: 13.4 and 10.9 Fick units, respectively, assuming a partial specific volume $\bar{v} = 0.56$ cm³g⁻¹.

of sedimentation data on molecular weight for DNA cannot be described by *one* straight line in the full molecular weight interval.

Now that homogeneous fragments of double-stranded DNA can be prepared easily and their precise molecular weight calculated from the sequence, we can reconsider in greater detail the sedimentation coefficient dependence on molecular weight (Fig. D2.27). The new set of

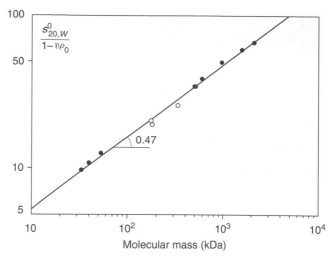

Fig. D2.29 Dependence of the logarithm of the sedimentation coefficient corrected for the buoyancy factor, $(1 - \bar{v}\rho_0)$ versus the logarithm of the molecular mass for different ribosomal RNAs (•) and their fragments (○) using the data presented in Table D2.8.

data obtained confirms that the log–log dependence is not a straight line, but has an inflection near a molecular weight of 3.5 MDa (Fig. D2.27). For fragments ranging from 170 kbp (molecular weight ~112 MDa) to nearly 5.4 kbp (molecular weight ~3.5 MDa) the dependence is described by

$$s = K_s M^{0.440} \tag{D2.38}$$

while in the range 200–5400 bp, it is described by

$$s = K_s M^{0.273} \tag{D2.39}$$

Comparing the new exponent values (Eqs. (D2.38) and (D2.39)) with the early ones reveals a clearer definition of the random coil properties of large molecular weight DNA (a increases from 0.40_5 to 0.44_0), and a more pronounced degree of chain stiffness in the low-molecular-weight range (a decreases from 0.32_5 to 0.27_3).

It is interesting to note that the frictional properties of short DNA fragments (12–20 bp) are similar to those of globular proteins of corresponding molecular weight (Fig. D2.28). At the same time these properties are very different for large molecular weight. This fact reflects the principal difference between these two homologous series. In the globular protein series, asymmetry remains constant with increasing molecular weight, whereas in the DNA fragment series, asymmetry grows.

D2.10.4 Ribosomal RNAs, Ribosomal Particles, and RNP Complexes

Figure D2.29 shows log–log plot dependence of $s^0_{20,w}/(1 - \bar{v}\rho)$ for ribosomal RNAs and their fragments

TABLE D2.8 RIBOSOMAL RNAS AND THEIR FRAGMENTS IN BUFFER CONTAINING 100 MM KCL WITHOUT MG IONS ($\bar{v} = 0.54\,cm^3g^{-1}$)

Particle	Molecular mass (in bases)	Sedimentation coefficient $s^0_{20,w}$ (Svedberg units)	References
5 S RNA of chloroplast	100	4.5	1
5 S RNA of E. coli	120	5.0	1
5 S RNA of rat	160	5.8	1
3'-fragment (1000–1542) of 16 S RNA E. coli	542	9.0	2
5'-fragment (1–530) of 16 S RNA E. coli	530	9.5	2
3'-fragment (535–1542) of 16 S RNA E.coli	1007	12.0	2
5'-fragment (1–996) of 16 S RNA of E.coli	996	12.0	2
16 S RNA of E.coli	1542	16.0	1
16 S RNA of chloroplast	1490	16.0	1
16 S RNA of T. thermophilus	1515	16.0	1
18 S of yeast	1789	18.0	1
18 S RNA of yeast	1825	18.0	1
23 S RNA of E.coli	2904	23.0	1
28 S RNA of rat	4750	28.0	1
RNA of TMV	6363	31.0	1

Reference 1: data from Spirin (1999).
Reference 2: data from Serdyuk (unpublished).

TABLE D2.9 RIBOSOMAL PARTICLES, SUBPARTICLES, AND RNP COMPLEXES

Particle	Molecular mass (kDa)	Sedimentation coefficient $s^0_{20,w}$ (Svedberg units)	Partial specific volume $\bar{v}$ (cm^3g^{-1})	$\dfrac{s^0_{20,w}}{(1-\bar{v}\rho)}$	References
T4 RNP complex of E.coli	70	5.8	0.63	15.7	1
T3 RNP complex of E. coli	93	6.1	0.64	16.9	1
Central domain of 30 S T. th. (547–895)	195	12.0	0.61	30.8	2
5'-domain of T. th. (1–539)	242	13.0	0.58	31.0	3
3' domain of T. th. (890–1373)	280	15.4	0.61	39.5	4
"Beheaded" 30 S of T. th.	427	22S	0.59	53.7	5
30 S E. coli	910	30.8	0.612	79.4	5
50 S E. coli	1550	50.0	0.597	124	5
70 S E. coli	2650	70.0	0.610	179	5
80 S from yeast	3700	80.0	0.630	216	5
80 S from Artemia salina	3800	81.0	0.630	219	6
80 S from Eugena	4400	86.0	0.630	232	6

Data are taken from:
1 Agalarov and Williamson (2000).
2 Agalarov et al. (1999).
3 Agalarov et al. (1998).
4 Selivanova et al. (unpublished result).
5 Ulitin et al. (1997).
6 Serdyuk et al. (1999).

(b)

(a)

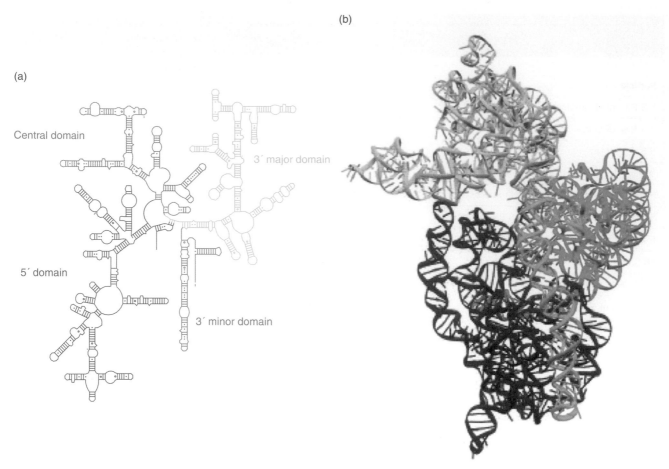

Central domain

3′ major domain

5′ domain

3′ minor domain

Fig. D2.30 (a) Three major domains of 16 S RNA of the small ribosomal subunit. The 5′ domain (nucleotides 1–563) corresponds to the central body of the 30 S subunit, the central domain (nucleotides 564–915) is involved in the formation of the side bulge or platform, whereas the 3′ major domain (nucleotides 919–1396) is localized in the head region of the subunit; there is also an additional 3′ minor domain (residues 1397). (Ramakrishnan and Moore, 2001.) (b) The "beheaded" 30 S ribosomal subunit corresponds to the 5′ end and central domains of the 16 S RNA from *T. thermophilus*. The procedure is based on oligodeoxyribonucleotide-directed cleavage of protein-deficient ribonucleoprotein derivatives of the 30 S ribosomal subunits with ribonuclease H (Serdyuk, *et al.*, 1999.)

(Comment D2.30) in buffer without Mg^{2+} ions using the data presented in Table D2.8. The straight line follows the equation

$$s_{20,w}^0/(1 - \bar{v}\rho) = K'' M^{0.47} \tag{D2.40}$$

with a slope equal to 0.47, which is close to the slope of the straight line for proteins in 6 M GuHCl (Eq. (D2.37)). Such a slope is typical for Gaussian coils in solvent close to "ideal" (Section D2.6).

Figure D2.29 also shows data for ribosomal particles and their subparticles using the data presented in Table D2.9. The straight line follows the equation

$$s_{20,w}^0/(1 - \bar{v}\rho) = K' M^{0.66} \tag{D2.41}$$

with a slope equal to 0.66, which is typical for compact quasispherical particles (Section D2.6). Using the same approach as above (Section D2.9.1), we note that the exponents in Eqs. (D2.40) and (D2.41) coincide, but the coefficients are different. We conclude that globular proteins

COMMENT D2.30 FRAGMENTS OF 16 S rRNA

Fragments of the 16 S rRNA of *T. thermophilus* representing the 3′ domain (nucleotides 890–1515), 5′ domain (nucleotides 1–539), and the central domain (nucleotides 547–895) were prepared by transcription in vitro. Incubation of those fragments with a total of 30 S ribosomal proteins of *T. thermophilus* resulted in the formation of the specific ribonucleoprotein complexes (Agalarov et al., 1999).

and ribosomal particles have similar compact shapes but different hydration.

A comparison of the sedimentation characteristics of RNP complexes, modeling the three major domains (Fig. D2.30), with those of the original ribosomal particles indicates that the complexes have the same compactness as the

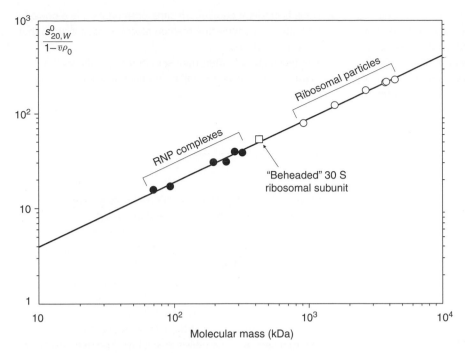

Fig. D2.31 Dependence of the logarithm of the sedimentation coefficient corrected for the buoyancy factor $(1 - \bar{v}\rho_0)$ versus the logarithm of the molecular mass for different ribosomal particles and their subparticles (○), and RNP particles (●) using the data presented in Table D2.9.

original particles. This follows from the fact that all RNP complexes lie on the continuation of the straight line formed by ribosomal units and subunits in the log–log plot of sedimentation versus molecular weight (Fig. D2.31). The result demonstrates clearly that each of the three major parts (domains) of the 30 S subunit from *T. thermophilus* are capable of self-assembly in vitro relatively independently of each other and that the 22 S RNP complex can be considered as a compact "beheaded" derivative of the 30 S ribosomal subunit.

D2.11 CHECKLIST OF KEY IDEAS

- The introduction in the 1990s of the new Beckman Instruments Optima XL ultracentrifuge opened a new epoch in AUC.
- Two complementary methods can be applied in an analytical ultracentrifuge: sedimentation velocity and sedimentation equilibrium.
- Sedimentation velocity provides information about the size, shape, and interactions between molecules; a high rotor speed and a long solution column are used to maximize the resolution.
- Sedimentation equilibrium provides information about the molecular mass, association constants, and stoichiometries; low rotor speeds and short solution columns are used to minimize the time needed to reach equilibrium.
- The ratio of the measured velocity of the particle to its centrifugal acceleration is called the sedimentation coefficient, s. (1 S (Svedberg unit) = 10^{-13} s.)

- The Lamm equation describes the behavior of a macromolecular solution in the ultracentrifuge field.
- The exact solution of the Lamm equation is an infinite series in terms of integrals, which can be computed only by numerical integration.
- The application of modern computer facilities enables numerical solutions for the Lamm equation and approximate analytical solutions to be proposed under given boundary conditions.
- The partial specific volume of a protein can be calculated very accurately from its amino acid composition, according to the Cohn and Edsall method.
- The molecular mass of a macromolecule can be determined by a combination of sedimentation and diffusion coefficients, according to Svedberg's equation, or directly by the sedimentation equilibrium method.
- The slope of the dependence of the sedimentation constant on molecular mass, on a double logarithmic scale, is related to shape for a homologous series of macromolecules.
- Globular proteins in aqueous solvent form a homologous series of quasi-spherical particles with exponent a equal to 2/3 in the dependence of s on M.
- Proteins in GuHCl form a homologous series of random coils with exponent a equal to 0.47.
- Collagen molecules and products of their sonic fragmentation form a homologous series of rod-like particles with exponent a equal to 0.2.
- The dependence of s on M for DNA molecules is described by two exponents: $a_1 = 0.273$ in the molecular weight range $1.3 \times 10^4 < M < 3.5 \times 10^6$, and $a_2 = 0.440$ for large molecular weights (beyond 3.5×10^6).

- Ribosome particles and subparticles, the "beheaded" 30 S subunit, and RNP complexes, modeling the three major domains of 30 S subunits, form a homologous series of quasi-spherical particles with exponent *a* equal to 2/3.
- Ribosomal RNAs and their fragments in buffer without Mg ions form a homologous series of random coils with exponent *a* equal to 0.47.

- Density-gradient equilibrium ultracentrifugation is a powerful technique for separating macromolecules according to their buoyant densities.
- The technique of differential sedimentation allows the measurement of very small differences in sedimentation coefficients.

Suggestions for Further Reading

Historical Review
Schachman, H. K. (1989). Analytical ultracentrifugation reborn. *Nature*, **341**, 259–260.
Schachman, H. K. (1992). Is there a future for the ultracentrifuge? In *Analytical Ultracentrifugation in Biochemistry and Polymer Science*, eds., S. E. Harding, A. J. Rowe, and J. C. Horton. Cambridge: Royal Society of Chemistry.

Instrumentation and Innovative Techniques
Schachman, H. K. (1959). *Ultracentrifugation in Biochemistry*. New York: Academic Press.
Geibeler, R. (1992). The Optima XL-A: A new analytical ultracentrifuge with a novel precision absorption optical system. In *Analytical Ultracentrifugation in Biochemistry and Polymer Science*, eds., S. E. Harding, A. J. Rowe, and J. C. Horton. Cambridge: Royal Society of Chemistry,
Ralston, G. (1993). *Introduction to Analytical Ultracentrifugation.* Fullerton, CA: Beckman Instruments.
Van Holde, K. E., Johnson, W. C., and Ho, S. P. (1998). *Principles of Physical Biochemistry*. Upper Saddle River, NJ: Prentice Hall.

Velocity Sedimentation
Van Holde, K. E. (1975). Sedimentation analysis of proteins. In *The Proteins*, eds. H. Neurath and R. I. Hill, **vol. VI**. 3rd edition. New York: Academic Press, pp. 225–291.
Perkins, S. J. (1986). Protein volumes and hydration effects: The calculation of partial specific volumes, neutron scattering matchpoints and 280 nm absorption coefficients for proteins and glycoproteins from amino acid sequences. *Eur. J. Biochem.*, **157**, 169–180.
Harding, S. E., Rowe, A. J., and Horton, J. C. (eds.) (1992). *Analytical Ultracentrifugation in Biochemistry and Polymer Science*. Cambridge: The Royal Society of Chemistry.
Stafford, W. F., III (1994). Boundary analysis in sedimentation velocity experiments. *Methods Enzymol.*, **240**, 478–501.

Harding, S. E. (1994). Determination of macromolecular homogeneity, shape, and interactions using sedimentation velocity analytical centrifugation. In *Methods in Molecular Biology*, eds. C. Jones, B. Mulloy, and S. Thomas, **vol. 22**. Totowa, NJ: Humana Press.
Hansen, J. C., Lebowitz, J., and Demeler, B. (1994). Analytical utracentrifugation of complex macromolecular systems. *Biochemistry*, **33**, 13 155–13 163.
Hensley, P. (1996). Defining the structure and stability of macromolecular assemblies in solution: The re-emergence of analytical ultracentrifugation as a practical tool. *Structure*, **4**, 367–373.
Laue, T. M., and Stafford, W. F., III (1999). Modern application of analytical ultracentrifugation. *Annu. Rev. Biophys. Biomol. Struct.*, **28**, 75–100.
Carruthers, L. M., Schirf, V. R., Demeler, B., and Hansen, J. C. (2000). Sedimentation velocity analysis of macromolecular assemblies. *Methods Enzymol.*, **321**, 66–80.
Shuck, P. (2000). Size distribution analysis of macromolecules by sedimentation velocity ultracentrifugation and Lamm equation modelling. *Biophys. J.*, **78**, 1606–1619.

Sedimentation Equilibrium
Cantor, C., and Schimmel, P. (1980). *Biophysical Chemistry. Part II. Technique for the Study of Biological Structure and Function*. San Francisco, CA: W. H. Freeman and Company.
Oberfelder, R. W., Consler, T. G., and Lee, J. C. (1985). Measurement of changes of hydrodynamic properties by sedimentation. *Meth. Enzymol.*, **117**, 27–40.
Schuster, T. M., and Laue, T. M. (1994). *Modern Analytical Ultracentrifugation*. Boston, MA: Birkhauser.
Laue, T. M. (1995). Sedimentation equilibrium as thermodynamic tool. In *Methods in Enzymology*, eds. G. K. Ackers and M. L. Jonson. New York: Academic Press, pp. 427–452.

FLUORESCENCE DEPOLARIZATION

D3

D3.1 HISTORICAL REVIEW

1926
E. Gaviola created the first instrument for fluorescence measurement, which he called a fluorometer.

1934
F. Perrin formulated the relation between fluorescence depolarization and Brownian rotary motion for hydrodynamic spherical molecules. In 1936 he extended his treatment to symmetric top molecules.

1952
G. Weber first applied Perrin theory to biological macromolecules. Numerous fluorescence polarization measurements were carried out under steady-state conditions, utilizing constant illumination. Most of them were performed on dye-conjugated macromolecules to determine the harmonic mean rotational relaxation time. Weber's classical review of the application of polarized fluorescence, particularly to proteins, appeared in the 1950s.

1961
A. Jublonski proposed following the rotational relaxation process directly by measuring the decay of the fluorescence polarization as a function of time. He stressed that time-dependent anisotropy can be interpreted more directly and definitely than its time-average value observed with constant illumination.

1966
P. Wahl, and two years later **L. Stryer**, experimentally measured the time decay of fluorescence polarization on dye-conjugated globular proteins.

1969
T. Tao derived relations between the time-dependent fluorescence polarization anisotropy and the Brownian rotational diffusion coefficients of macromolecules. It was shown that in the most general case of a completely asymmetric body, five exponentials appear in anisotropy. In **1972** analogous results were obtained by other groups (**G. G. Belford, R. L. Belford**, and **G. Weber**; **T. J. Chuang** and **K. B. Eisenthal**).

Mid-1990s
A few groups reported the synthesis and spectral properties of long-lifetime, highly luminescent, metal–ligand complexes (**B. A. De Graff** and **J. N. Demas, J. R. Lakowicz**'s group). These very photostable probes with microsecond decay times allowed measurements of rotational correlation times as long as 2 μs and detection of high-molecular-mass analytes in fluorescence polarization immunoassays.

1980–1995
J. R. Lakowicz's group made an essential contribution to the theory and practice of depolarized fluorescence spectroscopy. In 1987 the textbook *Principles of Fluorescence Spectroscopy*, by **J. R. Lakowicz**, appeared.

2000 to present
Fluorescence depolarized spectroscopy is very useful for investigating the rotational motion of macromolecules in aqueous solution in the nanosecond and microsecond time ranges. It has opened the new possibility of measuring inter-domain protein movement, lateral diffusion in membranes, large protein and protein–nucleic acid complexes.

D3.2 INTRODUCTION TO BIOLOGICAL PROBLEMS

The theoretical foundations of fluorescence spectroscopy were established in the first half of the twentieth century by pioneers including E. Gaviola, Jean and Francis Perrin (father and son), P. Pringsheim, S. Vavilov, T. Förster, and, more recently, G. Weber. In 1961 A. Jublonski proposed following the rotational relaxation process directly by measuring the decay of the fluorescence polarization as a function of time. He stressed that time-dependent anisotropy can be interpreted more directly and definitely than its time-average value observed with constant illumination.

For many years, the most common measurements were of fluorescent intensity only, but since the 1980s access to instrumentation for measuring fluorescence decays has improved markedly, largely as a result of the more widespread availability of lasers and improvements in electronics. An investigator may now choose to use either time-correlated single-photon counting or multi-frequency phase fluorimetry and be confident of getting the same answers.

Since the 1980s, essential experimental and theoretical contributions to the time-resolved fluorescence spectroscopy were made by J. Lakowicz, D. Millar, and R. Cherry and their groups. Time-resolved measurements can now be used in two ways. First, measurement of the decay of the total fluorescence intensity following pulsed excitation may be used as a signal of an *event* (protein conformational changes, ligand binding, protein–nucleic acids interactions, and so on). Second, the values of fluorescent parameters may be used to understand details of *the structure and dynamics of the fluorophore and its environment*. However, our incomplete understanding of the details of photophysics of most fluorophores, such as tryptophan and tyrosine, is a principal reason why interpretation of fluorescence in terms of protein structure and dynamics is still so difficult.

The other type of time-resolved fluorescence experiment involves measurement of the polarization anisotropy decay to characterize the *rotation* of an entire macromolecule or its separate parts. For many years, the most common measurements were done in the nanosecond timescale. Now fluorescence is capable of detecting microsecond dynamics. This change in timescale was made possible by the development of new metal–ligand complexes, which display decay times ranging from 10 ns to 10 μs. It has opened the possibilities of measuring protein domain movement, lateral diffusion in membranes, and microsecond rotational correlation times.

This chapter focuses on the use of fluorescence anisotropy decay methods, which provide hydrodynamic data describing the rotational diffusion of biological macromolecules. In Section F3.5 we will discuss the energy transfer Förster effect.

D3.3 THEORY OF FLUORESCENCE DEPOLARIZATION

In this section we briefly discuss the theory of fluorescence spectroscopy as applied to the study of biological macromolecules in order to understand how rotational friction coefficients are determined from depolarized fluorescent measurements.

D3.3.1 Fluorescence as a Physical Phenomenon

When an electron is excited to an orbital with a high electronic energy, the excess energy is stored. In most cases this energy is transferred as heat. This is called *non-radiative* energy transfer. In some molecules the electronic energy states are changed so that the energy can be reradiated as the electron drops from the excited state back to

the ground state. Non-radiative energy transfer is illustrated in Fig. D3.1a. Because of the relative configurations of the excited and ground states, the electron can easily make a transition from a vibrational state in the excited system to a nearly degenerate level in the ground state, and it can continue to undergo radiationless energy loss until it reaches the ground state. *Radiative* energy transfer occurs when the electronic configurations of the ground end excited states are arranged as shown in Fig. D3.1b. Again, absorption of a photon leads to the transition of the electron to an excited state with transfer of some energy to the solvent as heat. However, the lowest vibrational level of the excited state is too far above any vibrational level of the ground state for energy to be relinquished except with radiation of a photon. The absorption and emission of light in radiative energy transfer is illustrated by the energy-level diagram suggested by Jablonski (Fig. D3.2).

The absorption is quite rapid, taking about 10^{-15} s. Following light absorption, several processes usually occur. A fluorophore is usually excited to some higher vibrational singlet level S_2. Usually, molecules in condensed phases rapidly relax to the lowest vibrational singlet level of S_1 (Comment D3.1). This process is called internal conversion and generally occurs in 10^{-12} s. Since fluorescence lifetimes are typically near 10^{-8} s, internal conversion is generally complete prior to emission. Hence, fluorescence emission generally results from the thermodynamically equilibrated excited state.

Molecules in the S_1 state can also undergo conversion to the first triplet state T_1. Emission from T_1 is called *phosphorescence*, and generally is shifted to longer wavelengths (low energy). Conversion of S_1 to T_1 is called intersystem crossing. Transition from T_1 to the ground state is forbidden according to quantum mechanical rules, and as a result the rate constant for such an emission is

(a)　　　　　　　　　　(b)

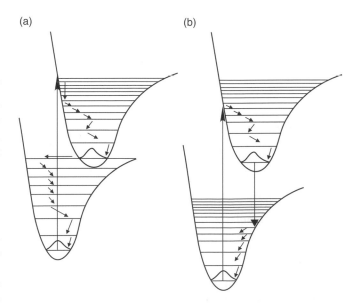

Fig. D3.1 (a) Non-radiative and **(b)** radiative energy transfer.

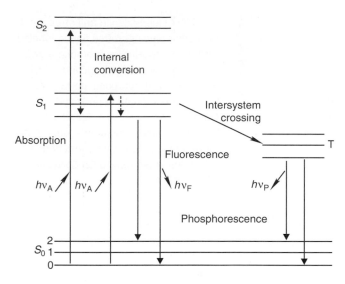

Fig. D3.2 Diagram of the absorption and emission of light (the Jablonski diagram). The ground and first and second electronic states are depicted by S_0, S_1, and S_2, respectively. At each of these electronic energy levels the fluorophores can exist in a number of vibrational energy levels, depicted by 0, 1, 2, etc. (Lakowicz, 1999.)

COMMENT D3.1 SINGLET AND TRIPLET STATES

The Pauli exclusion principle states that two electrons in an atom cannot have the same set of four quantum numbers. This restriction requires that no more than two electrons can occupy an orbital and, furthermore, those two must have opposed spin states. The spins are then said to be paired. Because of spin pairing, most molecules exhibit no net magnetic field and are thus said to be diamagnetic – i.e., they are neither attracted nor repelled by static magnetic fields. A molecular electronic state in which all electron spins are paired is called a *singlet* state, and no splitting of electronic energy levels occurs. When one of a pair of electrons in a molecule is excited to a higher energy level, either a singlet or a *triplet* state is formed. In the excited singlet state, the spin of the promoted electron is still paired with the ground-state electron. In the triplet state, however, the spins of two electrons have become unpaired and are thus parallel.

---

several orders of magnitude smaller than those of fluorescence. Different molecular states are shown in Fig. D3.3, where the arrows represent the direction of spin.

D3.3.2 Lifetime of Fluorophore and Rotational Correlation Time

The time interval between excitation and emission for many small molecules in water solution is of the order of 10^{-8} s,

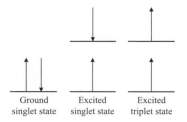

Fig. D3.3 Three different states of electron. Note that the excited triplet state is less energetic than the corresponding excited singlet state.

while the rotational relaxation time is considerably shorter. It may be expected that by the time the emission takes place, random orientation of the excited molecules will have been regained and the fluorescence will be unpolarized. A large molecule such as a protein, however, will have undergone on the average only a relatively small rotation due to its Brownian movement during a time interval of the order of 10^{-8} s. Consequently, the fluorescent light that it emits is partially polarized (Fig. D3.4).

In practice, very few proteins are naturally fluorescent. Most proteins must be converted into fluorescent derivatives by attaching to them suitable fluorescent groups, which have a lifetime of approximately a few nanoseconds. Comment D3.2 gives some typical intrinsic (native) probes, their synthetic analogues, and extrinsic fluorescent probes.

D3.3.3 Steady-State Fluorescence Depolarization

The rotational properties of fluorescent molecules may be determined by a method due to Perrin. Suppose that the solution is illuminated with a beam of plane-polarized light (Fig. D3.5). The exciting light travels along the x-direction, polarized with the electric vector along the z-axis. Two components of the emitted light are measured: $I_\parallel$ is polarized along the z-axis, and $I_\perp$ is polarized along the x-axis; μ is the transition dipole of the fluorescent molecule. The anisotropy or polarization, denoted by A and P, respectively, are then calculated as the difference between these values divided by a sum:

$$A = \frac{I_\parallel - I_\perp}{I_\parallel + 2I_\perp} \tag{D3.1}$$

$$P = \frac{I_\parallel - I_\perp}{I_\parallel + I_\perp} \tag{D3.2}$$

The denominator in Eq. (D3.1) is the total light that would be observed if no polarization were used. Referring to Fig. D3.6 and considering the polarization, we can see that $I_\parallel$ and $I_\perp$ are emitted along the x-axis as well as the y-axis, but $2I_\perp$ is emitted along the z-axis. The total intensity I is proportional to the sum of the emitted light along the three mutually orthogonal Cartesian axes, or to $I = I_\parallel + 2I_\perp$.

To calculate fluorescence polarization, we need to calculate the probability of exciting molecules with particular

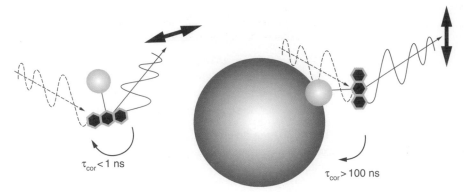

Fig. D3.4 Fluorescence anisotropy and molecular binding. A free ligand labeled with a fluorescein-like fluorophore (left) rotates on the subnanosecond timescale (with rotational correlation time τ_{cor}) and emits depolarized fluorescence. The same ligand when bound to a large macromolecule (right) rotates much more slowly and hence maintains anisotropy. (After Sportsman (2003), Fig. 1.)

$\tau_{cor} < 1$ ns

$\tau_{cor} > 100$ ns

COMMENT D3.2 FLUORESCENCE PROBES

Fluorescence probes can be broadly placed into three categories: intrinsic probes, analogues of intrinsic probes, and extrinsic probes.

Intrinsic probes are any naturally occurring molecules that exhibit sufficient fluorescence to be of use in practice. Examples include the aromatic amino acids, especially tryptophan and tyrosine, NADH, FAD, some porphyrins, some modified nucleic acids (such as the Y base in some tRNAs), and chlorophylls.

In analogues of intrinsic probes, residues in an intrinsic probe are replaced with analogues that have unique spectroscopic properties; examples include 5-hydroxyltryptophan and 7-azatryptophan.

Extrinsic probes are probes that can be introduced into the target system to form a complex, either covalent or non-covalent. Thousands of probes are now available. Examples of common non-covalent probes are dimethylaminonaphthalene sulfonyl chloride (dansyl chloride), 8-anilino-1-naphthalene sulfonate (ANS), 1,6-diphenyl-1,3,5-hexatriene (DPH), etidium bromide, sulfonyl chloride, isothiocyanate, succinimidyl ester, iodacetamide, and maleimide. These and many others are available to target a wide variety of biological macromolecules.

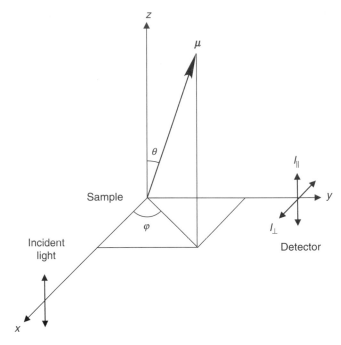

Fig. D3.5 The experimental set-up for measuring the depolarization of fluorescent light. The angles φ and θ define the orientation of μ with respect to the fixed space axes. The polarization is independent of the angle φ and varies with θ, becoming zero for radiation emitted along the z-direction.

orientations, and then to compute the probability that these molecules will emit light polarized in certain directions. First, consider a rigid, isotropic molecule. The molecule must be large enough that no appreciable molecular rotations occur on the fluorescence timescale (typically <100 ns). Calculation shows that if the emitting and absorbing transition dipoles are parallel (i.e., the same electronic transition both absorbs and emits), then $I_\| = 3/5$ and $I_\perp = 1/5$. For a totally rigid molecule $P = 1/2$ and $A = 2/5$, as one can see by substituting the results of the calculations into Eqs. (D3.1) and (D3.2). When the emission transition dipole of a molecule is perpendicular to the absorption

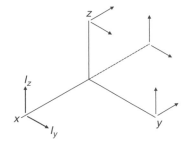

Fig. D3.6 Definition of anisotropy A: Non-polarized light incident along x can be resolved into y- and z-polarized components. The total emission can be found by adding the emission along the three Cartesian axes.

transition dipole, $P = 1/2$ and $A = 2/5$. In general, the polarization and anisotropy for rigid molecule are given by

$$P_0 = \frac{3\cos^2\zeta - 1}{\cos^2\zeta + 3} \tag{D3.3}$$

and

$$A_0 = \frac{3\cos^2\zeta - 1}{5} \tag{D3.4}$$

where ζ is the angle between the absorption and emission transition dipoles.

In the other extreme, the dipole can rotate fast enough to randomize its orientation during the lifetime of the excited state. In this case, by the time emission occurs, all memory of the original photoselection is lost. $I_\parallel = I_\perp$, and the polarization and anisotropy are both zero. Most macromolecules of biological interest fall between the extreme cases just described: Their rotational motions are not negligible on the fluorescence timescale, but they cannot tumble fast enough to achieve random orientation.

D3.3.4 Time-Resolved Fluorescence Depolarization

There are two types of experiments that can be performed using time-resolved fluorescence techniques. The first involves measurement of the time for the *decay of the emission* following a brief excitation pulse. This is used to determine the fluorescence lifetime of the fluorophore, which reflects the average time that the molecule remains in the excited singlet state. The fluorescence lifetime is strongly dependent on the molecular properties of the fluorophore environment and can vary from a few picoseconds to tens of nanoseconds.

The other type of time-resolved fluorescence experiment is based on measurement of the fluorescence *anisotropy decay*. The anisotropy decay monitors reorientation of the emission dipole during the excited-state lifetime and gives information on local fluorophore motion, segmental motion, and the overall rotational diffusion of macromolecules.

Suppose that a fluorescent probe is rigidly attached to the macromolecules (Fig. D3.4). It is evident that the observed polarization is some intermediate value between Eq. (D3.3) and zero. To compute this value, we must analyze the relative rates of emission and macromolecular rotational motion. The observed polarization measurements are a function of both the lifetime and the relaxation time of the protein. This link between the fluorescence lifetime and the observable motion can be seen from the Perrin equation:

$$A_0 = \frac{A_F}{1 + \tau_F/\tau_{cor}} \tag{D3.5}$$

In this expression A_0 is the steady-state anisotropy and A_F is the anisotropy of the fluorophore in the absence of

rotational motion. If the lifetime τ_F is much longer than the rotational correlation time τ_{cor}, the anisotropy A_0 can be taken to be zero. If the lifetime is much shorter than the correlation time, the anisotropy is A_F. The anisotropy A_0 is sensitive to the rate of rotational diffusion only if τ_{cor} is comparable to the lifetime of fluorophore τ_F; to which extent molecular parameters such as rotational diffusion tensor components and orientation of electronic transition moments parameters can be obtained in a deterministic way from a time-resolved fluorescence anisotropy experiment has been analyzed carefully (Comment D3.3).

D3.4 INSTRUMENTATION

The two most popular methods used to record time-resolved fluorescence data are the harmonic-response and impulse-response methods. In the harmonic-response method, time-resolved fluorescence parameters are deduced by analyzing the response of the emission to sinusoidally modulated excitation. The light modulators and signal processing circuitry required for these measurements can be incorporated into the design of standard spectrofluorimeters.

The impulse-response method involves direct observation of the time decay of emission following a short excitation pulse. The excitation pulses are typically derived from mode-locked dye lasers or titanium–sapphire lasers, which produce picosecond or subpicosecond pulses and are widely tunable in the visible and UV regions of the spectrum. Emission decay profiles are recorded by the time-correlated, single-photon-counting technique using fast

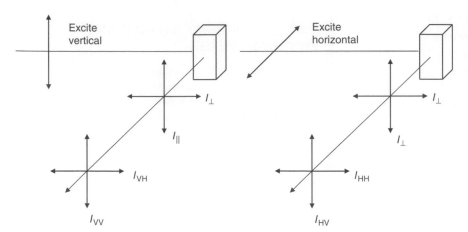

Fig. D3.7 Schematic diagram for the L-format method for the measurements of fluorescence anisotropy: vertical and horizontal excitation.

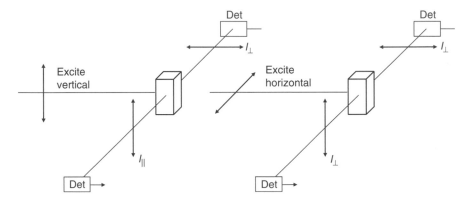

Fig. D3.8 Schematic diagram for the T-format method for the measurements of fluorescence anisotropy: vertical and horizontal excitation.

photomultiplier tubes. The use of such a device allows the recording of the fluorescence decay profiles with a high degree of precision and a time resolution of the order of a picosecond or less. Fluorescence anisotropy is technically a rather simple measurement. Two methods are commonly used. These are the L-format and T-format methods. In the L-format method, a single emission channel is used (Fig. D3.7). Fluorescence emitted from a sample is detected at right angles to the excitation beam, and these two beams define the scattering plane. For rapid fluorescence anisotropy measurements, two photomultiplier tubes are usually employed in a T-shaped configuration to measure simultaneously the emitted fluorescence light under the two respective relative polarizations, parallel and perpendicular to the polarization of the incoming beam (Fig. D3.8). In practice, a correction factor G is introduced into Eqs. (D3.1) and (D3.2) to take into account the differences in sensitivity of the detection system in the two polarizing directions (Comment D3.4).

A simplified detection system uses only one photomultiplier tube without a polarizer. Here, the polarization of the excitation light is altered between vertical (V) and horizontal (H) by means of a photoelastic modulator (Comment D3.5). As shown in Fig. D3.9a, when the excitation light is vertically polarized, the sum of the emitted intensities I_{VH} and I_{VV} associated with the μ_x and μ_z components, respectively, is detected. Hence, this measurement yields

---

COMMENT D3.4 CORRECTION FACTOR G

In real measurements a correction factor G is introduced into Eqs. (D3.1) and (D3.2):

$$A = \frac{I_\parallel - GI_\perp}{I_\parallel + 2GI_\perp}$$

$$P = \frac{I_\parallel - GI_\perp}{I_\parallel + GI_\perp}$$

The alignment of the optical system can be checked and adjusted using scattering from dilute suspensions of a non-fluorescent substance (e.g., glycogen or colloidal silica in water). The scattered light is 100% polarized, hence the measurements should yield an anisotropy of $A = 1.00$. The alignment is adequate when the measured value is 0.98 or larger (Lakowicz, 1999).

$$A_v = I_{VH} + I_{VV} \tag{D3.6}$$

The situation for horizontally polarized excitation is depicted in Fig. D3.9b. In this case the signal detected for horizontally polarized excitation yields:

$$A_H = 2 \times I_{VH} \tag{D3.7}$$

and the anisotropy is calculated from I_{VH} and I_{VV} using the resulting expression:

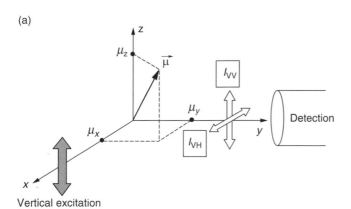

Vertical excitation

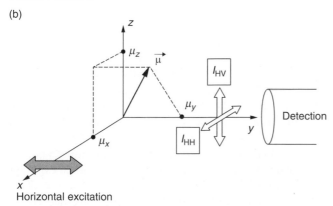

Horizontal excitation

Fig. D3.9 Schematic view of the experimental procedure for measuring fluorescence anisotropy: **(a)** the excitation polarization is vertical, and detection without a polarizer gives $I_{VH} + I_{VV}$; **(b)** the excitation polarization is horizontal, and detection gives $I_{HH} + T_{HV}$. The latter is equal to $2 \times I_{VH}$ if it is assumed that the sample is randomly oriented. Without any additional correction A was found to be 0.95. (After Canet *et al.*, 2001.)

$$A = \frac{A_V - A_{II}}{A_V + 0.5 \times A_H} \tag{D3.8}$$

This method does not require the G factor correction (see Comment D3.4).

D3.5 DEPOLARIZED FLUORESCENCE AND BROWNIAN MOTION

D3.5.1 Steady-State or Static Polarization

If constant illumination is employed, the measurement is called steady-state or static polarization. In this case the expressions for average static polarization and anisotropy are:

$$\bar{P} = \frac{3}{1 + 10(1 + \tau_F/\tau_{cor})(3\cos^2\zeta - 1)^{-1}} \tag{D3.9}$$

$$\bar{A} = \frac{3\cos^2\zeta - 1}{5(1 + \tau_F/\tau_{cor})} \tag{D3.10}$$

If measurements are made at such high values of η/T that the macromolecules are unable to rotate during the lifetime of the excited state, then $\tau_F/\tau_{cor} = 0$, and the polarization and anisotropy become

$$\bar{P}_0 = \frac{3\cos^2\zeta - 1}{\cos^2\zeta + 3} \tag{D3.11}$$

$$\bar{A}_0 = \frac{3\cos^2\zeta - 1}{5} \tag{D3.12}$$

These values, called the limiting polarization and anisotropy, are identical to those derived for a rigid system (Eqs. (D3.3) and (3.4)).

Using Eqs. (D3.11) and (D3.12) for $\bar{P}_0$ and $\bar{A}_0$, we can rewrite Eqs. (D3.9) and (D3.10) in a convenient form, called the Perrin equations:

$$\frac{1}{\bar{P}} - \frac{1}{3} = \left(\frac{1}{\bar{P}_0} - \frac{1}{3}\right)\left(1 + \frac{\tau_F}{\tau_{cor}}\right) = \left(\frac{1}{\bar{P}_0} - \frac{1}{3}\right)\left(1 + \frac{\tau_F kT}{V_h \eta}\right) \tag{D3.13}$$

$$\frac{1}{\bar{A}} = \frac{1}{\bar{A}_0}\left(1 + \frac{\tau_F}{\tau_{cor}}\right) = \frac{1}{\bar{A}_0}\left(1 + \frac{\tau_F kT}{V_h \eta}\right) \tag{D3.14}$$

On the right-hand side of the Perrin equations, the rotational correlation time τ_{cor} has been expressed using Eq. (D2.13).

Equations (D3.13) and (D3.14) can be tested for consistency by plotting $1/\bar{P}_0$ or $1/\bar{A}_0$ against T/η, when a straight line should be obtained. The slope yields the hydrodynamic volume (V_h), provided that the fluorescence decay time (τ_F) is known. The intercept at $T/\eta \rightarrow 0$ yields the limiting anisotropy or polarization. Figure D3.10 shows an example of such a Perrin plot for protein bovine serum albumin.

D3.5.2 Fluorescence Anisotropy Decay Time

Consider what happens if the sample is excited by a pulse of polarized light, and the time dependence of $I_{\parallel}$ and $I_{\perp}$ is measured. If the emission and absorption dipoles are parallel, the earliest photons to be emitted are highly likely to be z-polarized because the molecules have not had time to reorient. The last protons to be emitted should have

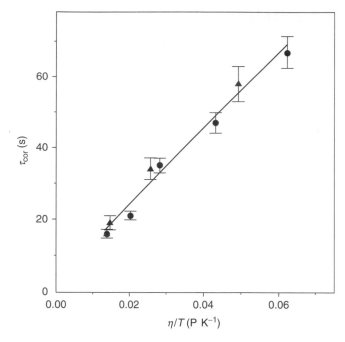

Fig. D3.10 Rotational correlation time τ_{cor} of the erythrosine–BSA complex in different mixtures of glycerol/buffer as a function of η/T, at 273 K constant temperature (•) or at 92% w/w glycerol constant composition (▲). (Ferrer et al., 2001.) The value of τ_{cor} extrapolated to 1 cP, 273 K is 40 ± 2 ns.

random polarization because by then the system has experienced considerable rotational motion. Therefore, if the polarization or anisotropy is measured as a function of the time of fluorescence emission, it decays from the initial values of P_0 or A_0 to final values of zero. It is clear that the rate of decay is a measure of the rate of rotational motion.

The rotational behavior of a sphere can be described by one parameter, the rotational correlation time τ_{cor}. If we define the rotational correlation time τ_{cor} as

$$\tau_{cor} = \frac{1}{6\Theta} = \frac{V_h \eta_0}{kT} \tag{D3.15}$$

then

$$A(t) = (2/5)e^{-t/\tau_{cor}} \frac{3\cos^2\zeta - 1}{2} \tag{D3.16}$$

Thus, the decay of the fluorescence anisotropy of a spherical molecule is a single exponential.

For a rigid ellipsoid the results become more complicated. Explicit equations have been derived only for rigid ellipsoids of revolution, for which $A(t)$ decays as the sum of three exponential terms. The decay of anisotropy of the fluorescence is given by

$$\frac{A(t)}{A_0} = \alpha_1 \exp\left(-\frac{t}{\tau_{1cor}}\right) + \alpha_2 \exp\left(-\frac{t}{\tau_{2cor}}\right) + \alpha_3 \exp\left(-\frac{t}{\tau_{3cor}}\right) \tag{D3.17}$$

where the pre-exponentials α_i depend only on the relative orientation of the electronic transition moments of the

emitting label (dye). If a protein is randomly labeled with an emitting label:

$$[\alpha_1 = \alpha_2 = \alpha_3 = 2/5] \tag{D3.18}$$

the rotational correlation times τ_{cor} are related to the rotational diffusion coefficients of the ellipsoid $D_\parallel$ and $D_\perp$ by.

$$\tau_{cor} = \left(4D_\parallel + 2D_\perp\right)^{-1}, \tau_{2cor} = \left(D_\parallel + 5D_\perp\right)^{-1},$$
$$\tau_{3cor} = \left(6D_{\perp r}\right)^{-1} \tag{D3.19}$$

The dependence of the diffusion coefficients on molecular shape, temperature and solvent viscosity can be expressed in compact form by the Stokes–Einstein–Debye equation:

$$D_i = \frac{kT}{6\eta V_h g_i}(i \rightarrow \parallel, \perp) \tag{D3.20}$$

where V_h is the hydrodynamic volume and η is the solvent viscosity. The Perrin factors g_i depend only on the axial ratio of the ellipsoid $p = a/b$ (where a is the symmetry axis) and are given by

$$g_\parallel = \frac{2(p^2 - 1)}{3p(p - S)} \tag{D3.21}$$

$$g_\perp = \frac{2(p^4 - 1)}{3p(2p^2 - 1)S - p} \tag{D3.22}$$

$$S_{prol} = (p^2 - 1)^{-1/2} \ln\left[p + (p^2 - 1)\right]^{-1/2} \tag{D3.23}$$

$$S_{obl} = (1 - p^2)^{-1/2} \arctan\left[p^{-1}(1 - p^2)\right]^{-1/2} \tag{D3.24}$$

In practice, it is rare that the accuracy of the data justifies fitting to more than two exponentials. In this case, the best one can do is to construct various plausible models for the structure and see how they fit the data. One conclusion is evident: An asymmetric molecule has a larger τ_{cor} than a sphere.

For rigid prolate ellipsoids with an axial ratio ≤ 5 and for any oblate ellipsoids $A(t)$ is very close to a single exponential function. Under these conditions the observed decay can be an approximate function with an average correlation time

$$\bar{\tau}_{cor} = \sum_1^3 \alpha_i \tau_{icor} \tag{D3.25}$$

This average $\bar{\tau}_{cor}$ is different from the harmonic mean value that is obtained from the slope of the $A(t)$ function at $t = 0$.

D3.5.3 Rotational Correlation Time of Globular Proteins

Most proteins studied using polarization techniques exhibit a single rotational correlation time. Figure D3.11 shows an example of anisotropy decay data for anthraniloyl chymotrypsin. The measured rotational correlation

TABLE D3.1 ROTATIONAL CORRELATION TIMES $\bar{\tau}_{cor}$ (20 °C) OF PROTEINS DETERMINED BY DIFFERENT METHODS

Protein	Molecular mass (kDa)	Rotational correlation time $\bar{\tau}_{cor}$ (20 °C) (ns)	Method	Reference
Gramicidin	2.5	2.8	DLS	f
Xfin-Zinc finger	2.93	2.4	NMR	a
BPTI	6.16	4.4	NMR	a
Eglin c	8.15	6.2	NMR	a
Calbindin-D9k apo	8.43	5.1	NMR	a
Calbindin-D9k apo	8.43	4.9	NMR	a
Ubiquitin	8.54	5.4	NMR	a
Cytochrome b_5	9.61	6.1	NMR	a
Barstar c/40/82A	10.14	7.4	NMR	a
Trp-repressor	11.89	23.1	NMR	a
Ribonuclease	13.68	8.3	NMR	a
Azurin	13.95	4.8	ns DPF	b
Lysozyme	14.32	8.3	NMR	a
Lysozyme	14.30	10	DLS	c
Staphylococcal nuclease	15.51	13.3	NMR	a
Interleukin-1β	17.4	12.4	NMR	a
β-lactoglobulin (monomer)	18.40	9.4	ns DPF	h
β-lactoglobulin (monomer)	18.40	8.6	μs DPF	k
Leukemia inh. factor	19.1	14.9	NMR	a
S. aurease nuclease B	20.00	9.9	trypt. DPF	h
p21^{ras}	21.00	16	ns DPF	e
HIV-1 protease	21.58	13.2	NMR	a
Trypsin	25.00	14.3	ns DPF	d
Chymotrypsin	25.00	3.7	ns DPF	d
Savinase	26.70	12.4	NMR	a
Carbonic anhydrase	30.0	12.4	ns DPF	d
β-lactoglobulin (dimer)	36.00	22.5	ns DPF	d
Apoperoxidase	40.00	28.0	ns DPF	d
Ovalbumin	43.5	30.2	μs DPF	h
F_{AB} fragment Ig G	47.50	33	ns DPF	d
Serum albumin	66.50	40	μs DPF	g
β-lactoglobulin	18.40	8.6	μs DPF	h
Sucrase-isomaltase	221.0	112	μs DPF	h

Abbreviations: DLS, dynamic light scattering; NMR, nuclear magnetic resonance; ns DPF, depolarized fluorescence in the nanosecond range; μs DPF, depolarized fluorescence in the microsecond range; trypt. DPF, depolarized fluorescence using tryptophan residue.

Data are taken from:

a Garcia de la Torre (2001).

b Kroes et al. (1998).

c Dubin et al. (1971).

d Yguerabide et al. (1970).

e Hazlett et al. (1993).

f Michielsen and Pecora (1981).

g Ferrer et al. (2001).

h Cherry and Schneider (1976).

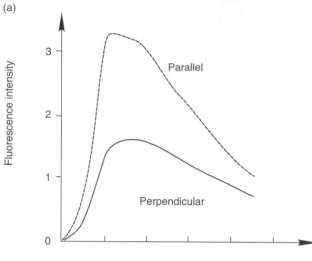

(a)

(b)

Fig. D3.11 (a) Fluorescence decay data of anthraniloyol-Ser[193]-α-chymotrypsin for individual polarized components. **(b)** Anisotropy on a logarithmic scale after removing the effects of the exciting pulse (*G* correction). (Stryer, 1968.)

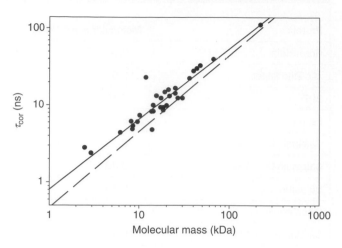

Fig. D3.12 Dependence of the rotational correlation times of 32 different proteins on molecular mass. Data are taken from Table D3.1. The dashed straight line corresponds to data predicted by Eq. (D3.26).

COMMENT D3.6 ROTATIONAL RELAXATION τ_{RELAX} AND CORRELATION τ_{COR} TIMES

Please note that workers in magnetic resonance and electric birefringence sometimes use the rotational relaxation time τ_{relax} and rotational correlation time τ_{cor}. The rotational correlation time τ_{cor} and rotational relaxation time τ_{relax} are related by $\tau_{cor} = \tau_{relax}/3$, $\tau_{cor} = (6D)^{-1}$ and $\tau_{relax} = (2D)^{-1}$. The simplest case is that of a rigid sphere, for which $A(t)$ decays exponentially, $A(t) = A_0 e^{-6\Theta t} = A_0 e^{-t/\tau_{cor}}$, molecular weight where Θ is the rotational diffusion coefficient.

---

time is about 15 ns. A measurement of the decay constant permits the hydrodynamic volume V_h to be calculated if the viscosity of the solvent is known. Equation (D3.15) shows that the larger the molecule, the slower decay of the fluorescence anisotropy.

Simple calculations show that τ_{cor} can be estimated from Eq. (D3.15):

$$\tau_{cor} = \frac{M}{2.55} \times 10^{-12} s \qquad (D3.26)$$

Roughly, τ_{cor} is 1 ns for each 2500 Da of protein molecular mass, if the protein has an approximately spherical shape with a hydration level of 0.3–0.4 g g^{-1} (Chapter D1).

Rotational correlation times $\bar{\tau}_{cor}$ of proteins determined using depolarized fluorescence are presented in Table D3.1. The table also contains protein correlation times obtained by NMR (Chapter J1) and depolarized dynamic light scattering (Chapter D4).

The dependence of rotational correlation times on molecular mass of 32 different proteins is plotted in Fig. D3.12 (Comment D3.6). The dashed straight line corresponds to the data predicted by Eq. (D3.26). Figure D3.12 shows that practically all proteins (with the exception of carboanhydrase) have a rotational correlation time sufficiently large to correspond to the motion of spherical particles. The difference between predicted and experimental rates arises because the protein is not spherical. Molecules with elongated shapes rotate more slowly. The slope of the dashed straight line, drawn through the experimental points, shows that the rotational correlation time in the limit of experimental error is proportional to molecular mass in the agreement with Eq. (D3.26).

Most of the fluorescent data presented in Fig. D3.12 were obtained in the nanosecond range. Such conditions put a serious limitation on the range of molecular mass of biological macromolecules available for such measurements. From Eq. (D3.7) it follows that macromolecules

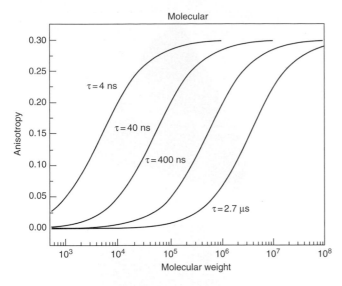

Fig. D3.13 Dependence of the anisotropy on molecular mass for probes with decay times from 4 to 2700 ns. For these calculations Eq. (D3.5) was used. The volume V of a protein is calculated from its molecular weight with the assumption of 20% hydration. (After Lakowicz *et al.*, 2000.)

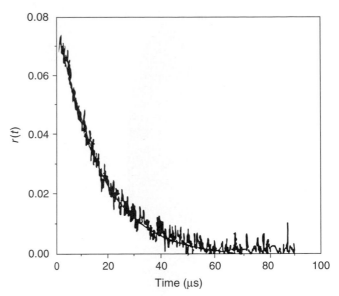

Fig. D3.14 The decay of the phosphorescence anisotropy of the erythrosine–BSA complex and the fitted monoexponential function with rotational correlation time $\tau = 16 \pm 1.5$ μs (92% w/w glycerol/buffer, 18 °C). (Ferrer *et al.*, 2001.) Erythrosin is 2′,4′,5′,7′-tetraiodofluorescein.

with molecular masses in the range 10–50 kDa have relaxation times in the range 4–20 ns, a range that corresponds to the singlet lifetimes of most organic dyes. Figure D3.13 shows the dependence of anisotropy on the molecular mass according to Eq. (D3.5). The figure demonstrates that for a protein with a molecular mass of 100 kDa labeled with a 4 ns decay time probe the emission is expected to be completely polarized. This explains why for a protein the upper molecular mass limit for measuring rotational diffusion with a nanosecond lifetime probe is near 40–50 kDa.

It is now possible to use fluorescence in the microsecond timescale. This change in timescale has been made possible by the development of metal–ligand complexes that display decay time ranging from s to more than 10 μs, a range that corresponds to triplet lifetimes. Figure D3.14 shows the decay of the phosphorescence anisotropy of the erythrosin–BSA complex on the microsecond timescale in a buffer containing glycerol. These measurements are consistent with the absence of independent motion of large protein segments in solution, in the time range from nanoseconds to fractions of a millisecond, and give a single rotational correlation time τ_{cor} for BSA (1 cP, 20 °C) = 40 ± 2 ns (see also Fig. D3.10).

Combining the values of the rotational correlation time (40 ± 2 ns) and intrinsic viscosity (4.1 cm^3 g^{-1}) allows a choice to be made between the heart-shaped structure of human serum albumin observed in single crystals and the standard, textbook hydrodynamic model of BSA – a cigar-shaped ellipsoid with dimensions of 140 Å × 40 Å (Fig. D3.15). A detailed analysis of the BSA hydrodynamics

was based on two bead modeling methods. In the first, BSA was modeled as a triangular prismatic shell with optimized dimensions of 84 Å × 84 Å × 84 Å × 31.5 Å, whereas in the second the atomic-level structure of human serum albumin obtained from crystallographic data was used to build a shell model of dimensions 80Å × 80Å × 80Å × 30Å (Fig. D3.16). In both cases, the predicted and experimental rotational diffusion coefficients were in good agreement. These data show that the elongated ellipsoid model used for decades to interpret the protein solution hydrodynamics at neutral pH is inappropriate.

D3.6 DEPOLARIZED FLUORESCENCE AND MOLECULAR INTERACTIONS

Depolarized fluorescence can be used to analyze most molecular interactions, including protein–protein, DNA–protein, and antigen–antibody binding. Depolarized fluorescence is unique among the methods used to analyze molecular binding because it gives a direct, nearly instantaneous measure of a tracer's bound/free ratio, even in the presence of free tracer. Most such measurements can be taken in real time, allowing the kinetic analysis of association and dissociation reactions. Below we will give a few examples of the application of depolarized fluorescence to studies of molecular protein–protein interactions.

Fluorescence polarization equilibrium binding studies usually involve titrating a small amount of fluorescently

(a)

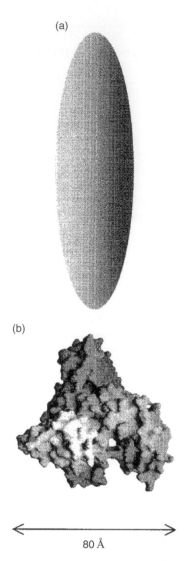

(b)

80 Å

Fig. D3.15 (a) The elongated ellipsoid (140 Å × 40 Å) used as the standard model for conformation in a solution of human and bovine serum albumin in all textbooks. **(b)** The heart-shaped structure of human serum albumin at the same scale, adapted from the Brookhaven Protein Data Bank X-ray coordinates, file 1bm0. (After Ferrer *et al.*, 2001.)

(a)

(b)

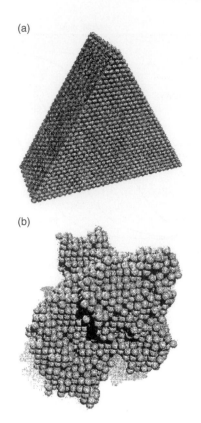

Fig. D3.16 (a) Rough shell triangular model of BSA with dimensions of 80Å × 80Å × 80Å × 30Å; **(b)** rough shell model derived from the atomic-level structure of human serum albumin. (After Ferrer *et al.*, 2001.)

labeled protein with a range of concentrations of a second protein. As the fraction of bound tracer increases, the polarization increases. Equilibrium binding constants are calculated from the resulting binding curve (i.e., polarization versus unlabeled protein concentration). Figure D3.17 shows the effect of thrombin activation on factor XIII A and B subunit association (Comment D3.7). Thrombin cleavage of rA_2 to rA'_2 results in a dramatic decrease in the affinity of rA_2 for labeled B2 (294 nM for rA'_2 versus 19 nM for rA_2), thereby explaining A_2B_2 complex dissociation.

Another example concerns the action of degradation enzymes, such as proteases, DNases, and RNase. The action of such enzymes usually results in a change in

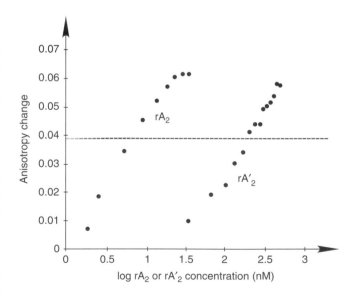

Fig. D3.17 Effect of thrombin activation on factor XIII A and B subunit association. Aliquots of rA_2 or thrombin-activated rA'_2 were added to a starting solution of B_2. The horizontal dashed line shows the half maximum anisotropy change. (After Checovich *et al.*, 1995.)

Blood coagulation factor XIII, or the fibrin stabilizing
factor, is a plasma zymogen of the enzyme responsible
for strengthening the clot network and for endowing it
with a higher resistance to lysis. A zymogen or proenzyme
is an inactive enzyme precursor that requires a
biochemical change, such as hydrolysis, in order to
become active in catalysis. Factor XIII is made up of two
different subunits in an AB protomeric structure thought
to associate into an A_2B_2 tetramer. Only the A subunits
are modified by thrombin, realizing an N-terminal
activation peptide. The hydrolytically cleaved A subunit is
denoted A′.

The implementation of conventional multiplexed assays
for high-throughput studies of protein–ligand interactions
is expensive due to the large amounts of reagents
required and the need for automation. A homogeneous
fluorescence anisotropy-based binding assay was
implemented in an automated microfluidic chip to
simultaneously interrogate more than 2000 pairwise
interactions. The platform was used to determine the
binding affinities between chromatin-regulatory proteins
and different post-translationally modified histone
peptides.

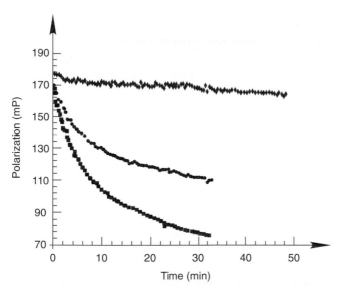

Fig. **D3.18** Time course of fluorescein-labeled casein digested
with trypsin. The reaction was started by the addition of 5 pmol
fluorescein-labeled casein with: 200 ng trypsin (bottom plot);
100 ng trypsin (middle plot): no trypsin (top plot). mP,
millipolarization units. (After Checovich *et al.*, 1995.)

molecular size, and can therefore be followed using
depolarized fluorescence. Figure D3.18 shows a time
course of trypsin activity using fluorescein-labeled casein.
Almost any substrate can be fluorescently labeled and used
to measure the activity of a specific enzyme. The standard
curve for the casein-based assay is linear over at least a
100-fold concentration range, and the detection limit is ten
times more sensitive than is possible with the more
common acid-precipitation assays.

The sensitivity of fluorescence makes it extremely useful
for high-throughput experiments (Comment D3.8).

D3.7 CHECKLIST OF KEY IDEAS

- Luminescence is the phenomenon of light
  emission from an electronically excited state of a
  molecule.
- Emission from a *singlet* state to the ground state is
  termed *fluorescence*; in fluorescence the interval
  between excitation and emission is of the order
  of 10^{-8} s.
- Emission from a *triplet* state to the ground state is
  termed *phosphorescence*; in phosphorescence, the
  interval between excitation and emission ranges from
  milliseconds to seconds.
- In practice, very few proteins are naturally fluorescent;
  most proteins and practically all nucleic acids must be
  converted into fluorescent derivatives by attaching to
  them suitable fluorescent probes.
- Fluorescent probes can be placed into three
  categories: intrinsic probes (aromatic groups in
  proteins), chemical analogues of intrinsic probes
  (e.g., 5-hydroxyltryptophan), and extrinsic probes for
  proteins and nucleic acids, of which there are
  thousands.
- Most biological macromolecules have rotational motion
  on a timescale which is longer than the fluorescence
  timescale, so they cannot tumble fast enough to achieve
  random orientation.
- A rotational correlation time τ_{cor} can be measured from
  the decay of the fluorescence or phosphorescence
  polarization as a function of time only, if the rate of
  macromolecular rotational diffusion is comparable to the
  fluorophore lifetime.
- Globular proteins with molecular masses in the range
  10–50 kDa have relaxation times in the range 4–20 ns, a
  range that corresponds to the singlet lifetimes of most
  organic dyes.

- The rotational correlation times of larger proteins can be measured with microsecond probes. For a phosphorescence decay time of 2.7 µs it is possible to measure the correlation time of macromolecules with molecular masses of a few million daltons.
- Fluorescence anisotropy of a spherical molecule exhibits single-exponential decay; roughly, τ_{cor} is 1 ns for each 2500 Da of protein molecular mass if the protein is approximately spherical.

- For rigid non-spherical molecules, the experimental rotational correlation time is always greater than that predicted for a sphere, because elongated molecules rotate more slowly.
- The decay of the fluorescence anisotropy of a general rigid body follows a multiexponential function (up to five exponentials); however, in practice it is rare that the accuracy of the experimental data justifies fitting to more than two exponentials.

Suggestions for Further Reading

Historical Review and Introduction to Biological Problems

Tao, T. (1969). Time dependent fluorescence depolarization and Brownian rotational diffusion coefficients of macromolecules. *Biopolymers*, **8**, 609–632.

Prendergast, F. G. (1991). Time-resolved fluorescence techniques: Methods and application in biology. *Curr. Opin. Struct. Biol.*, **1**, 1054–1059.

Checovich, W. J., Bolger, R. E., and Burke, T. (1995). Fluorescence polarization: A new tool for cell and molecular biology. *Nature*, **375**, 254–256.

Theory of Fluorescence Depolarization

Weber, G. (1953). Rotational Brownian motion and polarization of the fluorescence of solutions. *Adv. Protein Chem.*, **8**, 415–459.

Small, E. W., and Isenberg, I. (1977). Hydrodynamic properties of a rigid molecule: Rotational and linear diffusion and fluorescence anisotropy. *Biopolymers*, **16**, 1907–1928.

Szubiakowski, J. P. (2014) Identifiability analysis of rotational diffusion tensor and electronic transition moments measured in time-resolved fluorescence depolarization experiment. *J. Chem. Phys.*, **140**, http://dx.doi.org/10.1063/1.4881257.

Instrumentation

Cherry, R. J., and Schneider, G. (1976). A spectroscopic technique for measuring slow rotational diffusion of macromolecules. 2: Determination of rotational correlation times of proteins in solution. *Biochemistry*, **24**, 3657–3661.

Kroes, S. J., Canters, G. W., Giardi, G., Van Hoek, A., and Visser, A. J. W. G. (1998). Time-resolved fluorescence study of azurin variants: Conformational heterogeneity and tryptophan mobility. *Biophys. J.*, **75**, 2441–2450.

Lakowicz, J. R., Gryczynski, I., Piszczek, G., et al. (2000). Microsecond dynamics of biological macromolecules. *Meth. Enzymol.*, **323**, 473–509.

Sportsman, J. R. (2003). Fluorescence anisotropy in pharmacologic screening. *Meth. Enzymol.*, **361**, 505–529.

Depolarized Fluorescence and Brownian Motion

Cherry, R. J., and Schneider, G. (1976). A spectroscopic technique for measuring slow rotational diffusion of macromolecules. 2: Determination of rotational correlation times of protein in solution. *Biochemistry*, **15**, 3657–3661.

Lakovicz, J. R. (ed.). (1999) *Principles of Fluorescence Spectroscopy*. 2nd edition. New York: Kluwer Academic/Plenum.

Ferrer, M. L., Duchowicz, R., Carrasco, B., Garcia de la Torre, J., and Acuna, A. U. (2001). The conformation of serum albumin in solution: A combined phosphorescence depolarization-hydrodynamic modeling study. *Biophysical J.*, **80**, 2422–2430.

Depolarized Fluorescence and Molecular Interactions

Hazlett, T. L., Moore, K. J. M., Lowe, P. N., Jameson, D. M., and Eccleston, J. F. (1993). Solution of p21[ras] proteins bound with fluorescent nucleotides: A time-resolved fluorescence study. *Biochemistry*, **32**, 13 575–13 583.

Millar, D. P. (2000). Time-resolved fluorescence methods for analysis of DNA–protein interactions. *Meth. Enzymol.*, **323**, 442–459.

Canet, D., Doering, K., Dobson, C. M., and Dupont, Y. (2001). High-sensitivity fluorescence anisotropy detection of protein-folding events: Application to α-lactalbumin. *Biophys. J.*, **80**, 1996–2003.

Cheow, L. F., Viswanathan, R., Chin, C. S., et al. (2014) Multiplexed analysis of protein–ligand interactions by fluorescence anisotropy in a microfluidic platform. *Anal Chem.*, **86**, 9901–9908

DYNAMIC LIGHT SCATTERING AND FLUORESCENCE CORRELATION SPECTROSCOPY

D4

D4.1 HISTORICAL REVIEW

1869

J. Tyndall performed the first experimental studies on light scattering from aerosols. He explained the blue color of the sky by the presence of dust in the atmosphere.

1871

Lord Rayleigh presented the theory of scattering from assemblies of non-interacting particles that were sufficiently small compared with the wavelength of light. According to Rayleigh, scattering by a gas occurs because of the fluctuation of the molecules around a position of equilibrium. Rayleigh obtained the formulae that explained the blue color of the sky as being due to the molecules in the atmosphere preferentially scattering blue light in comparison to red light.

1906

L. Mandelshtam raised the question on the nature of scattered light once again. He pointed out that Rayleigh's arguments do not fully explain the scattering phenomenon. According to Mandelshtam, light scattering is also due to the random fluctuations of molecules near the position of equilibrium. As a result of random fluctuations of order λ^3, where λ is the scattering wavelength, the number of macromolecules will vary with time. It follows from the Mandelshtam theory that the translational and rotational diffusion coefficients of macromolecules could be obtained from the spectrum of the scattered light. In **1924**, **L. Mandelshtam** described theoretically and experimentally what he termed *combination* light scattering, which later was renamed Raman scattering. As early as **1926** he recognized that the translational diffusion coefficient of macromolecules can be obtained from the spectrum of the light they scatter. However, at that time, lack of spatial coherence and monochromaticity in conventional light sources rendered such experiments impossible.

1914

L. Brillouin predicted a doublet in the frequency distribution of scattered light due to scattering from thermal sound waves in a solid. This doublet was later named after him. **E. Gross** performed a series of light-scattering experiments on liquids in 1932. He observed not only the Brillouin doublet but also a central peak, the position of which was unshifted. In **1933 L. Landau** and **G. Placzek** gave a theoretical explanation of the central line from a thermodynamic point of view and calculated the ratio of the integrated intensity of the central line to that of the doublet.

1916

M. von Smoluchowski gave the first theoretical description of the amplitude and the temporal decay of number fluctuations in a diffusion system.

1947

L. Gorelik conceived the idea of *optical mixing spectroscopy*. He called this method demodulated analysis of scattered light. In **1955 A. Ferrester**, **R. Gudmunsen**, and **P. Johnson** realized this idea in practice. The technique was called optical-beating and was used as an alternative to conventional high-resolution spectroscopy. The signal-to-noise ratio was very poor and the technique was not widely adopted for conventional spectroscopy at that time.

1964

R. Pecora showed that the frequency profile of scattered light was broadened by translational diffusion of macromolecules. He postulated that the half-width at half-height of the central peak is a direct measure of translational diffusion coefficient. In **1964 H. Z. Cammins**, **N. Knable**, and **Y. Yeh** used an optical-mixing technique to resolve light scattering from diluted suspensions of polystyrene latex spheres. They used a He–Ne laser as the light source and what is now commonly called the "heterodyne" optical-beating technique. In **1967 S. B. Dubin, S. J. Lunacek**, and **G. B. Benedek** reported Rayleigh line width measurements for a series of biological macromolecules. These pioneering theoretical and experimental works have led to the current rich and diverse field. In **1976 B. J. Berne** and **R. Pecora** published their book *Dynamic Light Scattering with Application to Chemistry, Biology and Physics*.

1972

B. R. Ware and **W. H. Flygare** reported dynamic light-scattering studies of bovine serum albumin in the presence of a static electric field. They showed that the effect of the electric field is to superimpose a drift velocity that is proportional to the electrophoretic mobility of the species on the random Brownian motion of the charged particles. This technique is referred to as electrophoretic light scattering (ELS).

1972–1974

D. Magde, E. L. Elson, and **W. W. Webb** published a rigorous formalism of fluorescence correlation spectroscopy (FCS) with its various modes of possible applications, highlighting its potential.

1976

S. B. Dubin, N. A. Clark, and **G. B. Benedek** measured the rotational diffusion coefficient of lysozyme using a high-resolution spherical Fabry–Perot interferometer. Combining this result with the translational diffusion coefficient showed that lysozyme in solution has very nearly the same molecular dimensions as crystalline lysozyme. In **1991** Pecora extended this approach to short oligonucleotides.

1982

Z. Kam and **R. Rigler** used cross-correlation for the separation of translational diffusion from the other contributions to the dynamic fluctuations of laser light-scattering. Rotational diffusion of asymmetric particles, conformational relaxation of random coils, and association–dissociation dynamics were determined by cross-correlating scattering intensity fluctuations at different angles.

1990

R. Rigler and coworkers made the final breakthrough for the FCS method. They reached the single-molecule detection limit by combining FCS with a confocal set-up, thus increasing the signal-to-noise ratio dramatically. By tightly focusing a laser beam and inserting a pinhole into the image plane, maximum lateral and axial resolution were achieved.

1994

M. Eigen and **R. Rigler** triggered an important further development by proposing the application of dual-color cross-correlation for diagnostic purposes. The underlying idea was to separate single-labeled reaction educts from dual-labeled reaction products to discriminate against an excess of free single-labeled species and thus enhance the specificity of detection. In **1997**, **P. Schwille** and coworkers successfully monitored a hybridization reaction of two differently labeled oligonucleotides by the dual-color cross-correlation technique. In **2000**,

K. G. Heinze and coworkers reported the application of dual-color two-photon cross-correlation to determine enzymatic cleavage of a DNA substrate by endonuclease.

1997

R. Bar-Ziv and collaborators proposed the technique of localized dynamic light scattering (LDLS) for the study of dynamical properties of a single object at the nanometer level. This technique was applied to the measurement of biomolecular force constants and to probe the viscoelastic properties of complex media.

Since 2000

Dynamic light scattering is routinely used for fast (a few minutes) and accurate (1–2% error) measurements of translational and rotational diffusion coefficients of macromolecules, particularly small globular proteins and oligonucleotides. In parallel, FCS has evolved into a whole family of related methods sharing the basic principle of fluorescence fluctuation analysis.

D4.2 INTRODUCTION TO BIOLOGICAL PROBLEMS

In the mid-1960s molecular biology benefited from the development of a new optical method that has had several names and which, like R. Pecora, we'll call *dynamic light scattering* (DLS). Successive developments over 15 years have shown that the method adds the *time dimension* to the classic probe of incoherent, total-intensity scattering of light (Comment D4.1). Intensive work has been devoted to the analysis of the dynamics of different processes that contribute to fluctuations in intensity of the scattered light. Experiments established the effect of diffusion on broadening and the effect of directed motion on the Doppler shift of the spectrum of light scattered from a wide range of particles of different sizes, shapes, and flexibility.

Dynamic light scattering is able to provide very accurate measurements of translational diffusion coefficients of macromolecules in solution. There are moreover many variations of the original light-scattering technique (first introduced in 1964), which include:

Dynamic light scattering: This is a generic term encompassing all of the light-scattering methods that provide information on molecular dynamics.

Quasi-elastic light scattering: This name draws attention to the fact that the process of scattering gives a new wave, the scattered wave, whose central frequency is that of the incident light beam, but the amplitude and phase of this scattered wave are frequency modulated.

Photon correlation spectroscopy: This is an experimental technique that employs photon counting in the computation of the autocorrelation function, as opposed to using an analogue signal as in the carrier correlators.

Light is a non-perturbative probe that can be used to obtain information about the structure and dynamics of molecules. Maxwell's equations form the basis of the description of all electromagnetic phenomena. These equations identify light as a transverse electromagnetic wave that oscillates in both space and time, i.e., the direction of oscillation is perpendicular to its direction of propagation.

The mode of interaction of light with matter depends upon the electronic structure of the material as determined by its quantum mechanical properties. If the energy of the incident photon (hv, where h is Planck's constant and v is the frequency of light) is equal to the difference in energy between two states in the system, then the photon may be adsorbed by the system. As an oscillating electric field, light also distorts the distribution of the charges in the system and consequently the accelerated charges also emit radiation in the form of scattered light. If there is no exchange of energy between the photon and the system, then the frequency of the scattered light, v_s, is equal to v_0, and the process is referred to as *elastic light scattering*. If energy is exchanged between the photon and the system, then v_s differs from v_0, and the process is referred to as *inelastic light scattering*.

The response of a system to an external electric field is called the *polarization* of the system. The magnitude of the polarization depends on the amplitude of the applied electric field and the ability of the charge distribution to be "deformed" by the external stimulus. The capacity of the system to be distorted is referred to as the *polarizability* of the system. In general, the polarizability of a system is not homogeneous throughout and the *scattering of light is a result of fluctuations in the polarizability of the medium*.

If light is polarized in the direction perpendicular to the plane defined by the source, the scattered cell, and the detector, then the beam is said to have vertical polarization. If the polarization of both the incident and the scattered light is in the vertical direction, then the process is referred to as *polarized light scattering*, denoted by I_{VV}. *Depolarized light scattering*, in which the horizontal component of the scattered light is monitored, is denoted by I_{VH}. *Unpolarized light* can be thought of as a combination of several polarized beams with random orientation about the axis of propagation.

Optical mixing spectroscopy: This term is used to underline that a technique of very high resolution might be used for the detection of dynamic processes that describe the mobility of biological macromolecules in solution. There are many other variations (intensity fluctuation spectroscopy, Rayleigh broadening, etc.). We will consider all these name variants as synonyms.

In this chapter we describe two types of DLS experiments. In the first, intensity fluctuates as the *molecule distribution* (orientation and position) in a microscopic volume ($\sim\lambda^3$) changes with time. We will call this type of experiment DLS under Gaussian statistics (i.e., for a large number of particles). In the second, intensity fluctuates due to fluctuations in the number of molecules in the microscopic volume. We will call this type of experiment DLS under non-Gaussian statistics (i.e., for a small number of particles).

This chapter will conclude with FCS, a logical development of DLS. However, unlike DLS, which is a *coherent* phenomenon, FCS is an *incoherent* phenomenon. Its emergence is mainly connected with the shift from studying collective properties of macromolecules in a large volume to measuring their individual properties in a very small volume. In fact, FCS cannot be performed on an ensemble of molecules, but is ideally applied to the study of just a single molecule in an observed volume.

D4.3 DYNAMIC LIGHT SCATTERING AS A SPECTROSCOPY OF VERY HIGH RESOLUTION

It is fundamental to statistical physics that a system of many particles never exhibits perfect uniformity. The thermodynamic variables describing the system constantly fluctuate around their most probable values. Since these fluctuations occur very slowly compared with the frequency of a light wave, a very high-resolution spectrometer is required. The resolution needed is so high that conventional optical spectroscopy is inadequate and we need to use the technique of optical mixing spectroscopy.

The laser beam incident on the scattering medium can be considered as a *"carrier wave"* oscillating at an optical frequency of about 5×10^{14} Hz. The process of scattering gives a new wave, the scattered wave, whose central frequency is that of the incident light beam but whose amplitude and phase are modulated in synchrony with fluctuations of the medium. To obtain the information contained in the modulation, one must demodulate the scattered field. The demodulation of *fast* processes is conventionally accomplished with a spectrometer such as a diffraction grating for processes with a timescale of 10^{-11}–10^{-10} s or a Fabry–Perot interferometer for processes with a timescale of 10^{-10}–10^{-7} s. Such a device operates as an optical filter and therefore cannot resolve spectral components narrower than ~10 MHz.

Dynamic processes describing the mobility of biological macromolecules in solution are referred to as *slow* processes and can produce spectral line widths as narrow as 10 Hz, six orders of magnitude narrower than that detectable optically. Hence, very high-resolution data are required. This technique is called *optical mixing spectroscopy* (Comment D4.2).

In order to see how to resolve such narrow lines we must remember that similar problems arose with radio-frequency. In that case, the resolution problem was solved by the development of "homodyne" and "heterodyne" receivers. These devices first shift the central frequency of the carrier wave down to a lower frequency by a process of non-linear mixing. Then, at the lower frequency, where narrow band filters are available, the analysis is performed by sweeping the narrow filter across the spectrum, which is now centered at a low frequency. Thus, whereas the optical spectrometer filters at the optical frequency, the radio-frequency is filtered only after the carrier wave has been shifted down, by means of a mixing element, to a convenient low frequency. Optical mixing spectroscopy is the application of the radio-frequency technique to the optical frequency region, where the required mixing element is a photomultiplier.

D4.3.1 Fluctuations and Time-Correlation Functions

In order to obtain dynamic information from a fluctuating system we can adopt one of two different but entirely equivalent approaches. We may analyze the fluctuating system in terms of various frequency components of differing magnitude, thus yielding a *frequency spectrum* for fluctuations of property A, $\langle \Delta A^2(v) \rangle$, where v is the frequency. However, an often more precise method can be used in which one measures the *time correlation function* of the fluctuations $\langle A(t), A(t+\tau) \rangle$ directly. Experiments involving analysis by the time correlation technique are currently more common and it is these we shall discuss.

Construction of Time-Dependent Correlation Functions
In the laboratory, the transformation of the fluctuating signal into a correlation function is performed using a device called a correlator. The signal must usually be presented to the correlator in a digital form and the experiments designed accordingly. In light scattering, a light signal is detected as a photon count rate and the input to the correlator is a series of pulses.

A schematic diagram of a correlator is shown in Fig. D4.1. Note that commercially available digital correlator cards in a personal computer can now perform autocorrelation with submicrosecond time resolution in virtually real time.

Field and Intensity Correlation Functions
Since light scattering is an interference phenomenon, the scattered field at the detector depends on the local

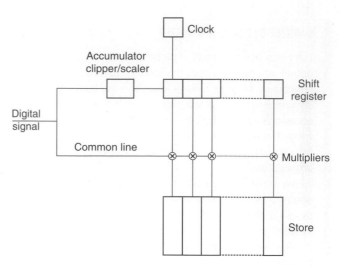

Fig. D4.1 Schematic diagram of a correlator for the analysis of fluctuating processes. The digitized noise signal is fed to two channels: an accumulator and a common line in the correlator. The accumulator counts the pulses in the train for a length of time Δt governed by a crystal clock. At the end of this time the contents of the accumulator are moved to the first stage of a shift register. Simultaneously, the accumulator is reset to zero and begins to count pulses for a second identical time interval. This process continues throughout the experiment so that the shift register contains a continuously updated past history of the pulse train extending back in time by $n \times \Delta t$, where n is the number of stages in the shift register and is called the sampling time which can be chosen typically in the range 2 ns to 1 s. A plot of counter contents against time interval is proportional to the noise correlation function.

inhomogeneity in the system. A completely homogeneous system gives no net scattered field.

For a single particle, the induced dipole **p**, which is the source of the scattering, is determined by the incident field **E** and by the particle polarizability α

$$\mathbf{p} = \alpha \cdot \mathbf{E} \tag{D4.1}$$

The light radiated by all the individual induced dipoles is collected at the detector where the net electric field depends on the relative positions and orientations of all particles in the scattering volume. A typical light-scattering geometry is shown in Fig. D4.2. The correlation function of the scattered electric field, $G^{(1)}(\tau)$, sometimes called the first-order correlation function which carries the undistorted information about this motion, can be defined by

$$G^{(1)}(\tau) = \langle \mathbf{E}(t)\mathbf{E}(t+\tau) \rangle \tag{D4.2}$$

In practice, of course, it is the intensity fluctuations rather than the field fluctuations that are measured. For the intensity $(\mathbf{I} \sim \langle \mathbf{E}^2 \rangle)$ a normalized correlation function, $G^{(2)}(\tau)$, sometimes called the second-order correlation function, is given by

$$G^{(2)}(\tau) = \langle \mathbf{I}(t)\mathbf{I}(t+\tau) \rangle \tag{D4.3}$$

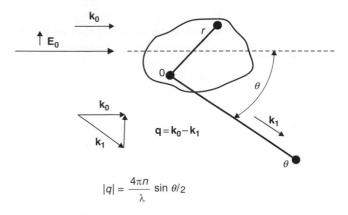

$$|q| = \frac{4\pi n}{\lambda} \sin \theta/2$$

Fig. D4.2 A typical light-scattering geometry. $\mathbf{E}_0$ is the incident laser field; $\mathbf{k}_0$ and $\mathbf{k}_1$ are the incident and scattered wave vectors; $\mathbf{q} = \mathbf{k}_0 - \mathbf{k}_1$ is the wave vector of the scattering density fluctuation as defined by the scattering angle θ and the wavelength of incident light λ; n is the refractive index of the medium.

For systems containing a large number of independent scatterers, the Gaussian approximation holds and

$$G^{(2)}(\tau) = 1 + |G^{(1)}(\tau)|^2 \tag{D4.4}$$

Correlation Functions and Frequency Spectra of Fluctuations

As mentioned previously, the frequency spectrum and the correlation function are entirely equivalent descriptions of the dynamical behavior of a property of an equilibrium system, being related by the Fourier transformation:

$$I(\nu) = (2\pi)^{-1} \int G(\tau) \exp(-2\pi i \nu t) d\tau \tag{D4.5}$$

where ν is the Hertzian frequency. Thus, if the intensity $I(\nu)$ has a Lorentzian form (shaped as a singly peaked function) around zero frequency, then the correlation function $G(\tau)$ shows a simple exponential decay with a time constant equal to the reciprocal of the Lorentzian half-width. This simple form occurs for all elementary stochastic processes and accounts for the behavior of most of the systems that we shall discuss. In order to show the link between correlation functions and frequency spectra, some simple examples are shown in Fig. D4.3.

D4.3.2 Measurements of the Dynamic Part of Scattered Light

Figure D4.4 schematically illustrates various techniques used in light experiments: (a) filter; (b) homodyne; and (c) heterodyne. These techniques are briefly described below.

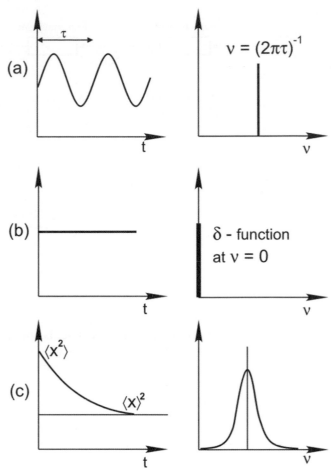

Fig. D4.3 Correlation functions and frequency spectra for various models of fluctuation processes: **(a)** sinusoidal time dependence; **(b)** invariant $A(t)$; **(c)** simplest correlated noise (see also Chapter A3).

Filter Technique

The demodulation of *fast* processes is conventionally accomplished with a spectrometer, such as a diffraction grating for processes with a timescale of 10^{-11}–10^{-10} s or a Fabry–Perot interferometer for processes with a timescale 10^{-10}–10^{-7} s (Comment D4.3). Such devices operate as optical filters and therefore cannot resolve spectral components narrower than ~ 10 MHz. Because the rotational periods of small globular proteins lie between 10 MHz and 1 MHz (see Table D2.2), a high-resolution spherical Fabry–Perot interferometer can be successfully used to measure them (see below).

Photon Correlation Spectroscopy

The apparatus necessary to study the self-beat spectrum (see below) is not complicated, although some care in its design is required if optimum signal-to-noise ratio is to be obtained. A diagram of a typical experimental set-up is

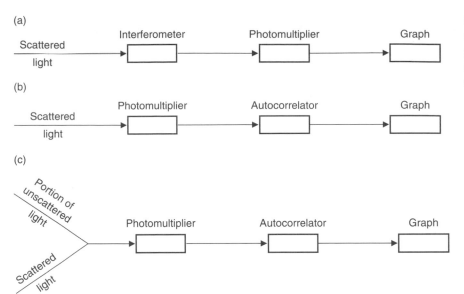

Fig. D4.4 Schematic illustration of the various techniques in light experiments: **(a)** filter method; **(b)** homodyne technique; and **(c)** heterodyne technique.

shown in Fig. D4.5. The lens L_1 focuses the laser beam in the center of the scattering cell. The aperture lens L_2, a pinhole, and the photomultiplier, which is arranged on an arm rotating about an axis through the center of the cell, establish a scattering angle of 2θ. The position of the elements on the arm is adjusted so that an image of the scattering volume is focused onto the pinhole. The electronic signal correlator is used to process the signal.

Figure D4.6 shows the general features of the $G^{(2)}(\tau)$ correlation function. The scattered intensity always consists of time-dependent and time-independent parts. The time-independent term, as measured by a dc voltmeter, is the Rayleigh scattering commonly used in the study of macromolecules, and is proportional to the number of particles in the scattering volume.

Note that in photon correlation spectroscopy no "filter" is inserted between the scattering medium and the photomultiplier.

COMMENT D4.3 FABRY–PEROT INTERFEROMETER

A Fabry–Perot interferometer consists of two plane dielectric mirrors held parallel to each other. The inner surfaces of the mirrors are highly reflecting (99%). In light-scattering experiments, light usually enters the interferometer normal to the mirrors and is then reflected back and forth in the interferometer cavity. The condition for survival in the cavity of a wave of wavelength λ is that an integral number of half wavelengths must fit in the cavity, i.e., $m(\lambda/2)=d$, where m is an integer and d is the spacing between the mirrors. At this wavelength there is a maximum in the intensity of the transmitted light. If $d=0.1$ cm and $\lambda=5000$ Å, then $m=2\times10^3$. The difference between successive wavelengths, $\Delta\lambda$, that can pass through the interferometer for this value of d is $\Delta\lambda \sim 2.5$ Å. The corresponding frequency range is $\Delta\nu \sim 3\times10^{11}$ Hz. Thus a proper setting of d in the interferometer selects only one frequency component (wavelength) of the light-scattering spectrum. Some workers who want especially high resolution use spherical mirrors rather than the flat mirrors described above. Spherical mirrors increase the light-gathering power and also allow easier scanning of a spectrum.

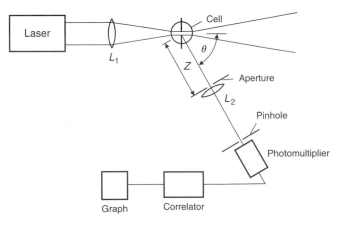

Fig. D4.5 Basic layout of the photon correlation spectrometer used to observe self-beat spectra. The optimal signal is obtained when the pinhole has the same radius as the image of the laser beam. A typical value for the distance, L_1, between scattering molecules and slit is about 1 m and the dimension d_1 of the slit in front of PM is about 0.1–0.2 mm. To preserve the coherency of scattering it is necessary to make the dimensions of the illuminating area (slit d_2) small, while still putting in as much incident power as possible.

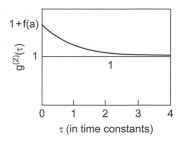

Fig. D4.6 The $G^{(2)}(\tau)$ correlation function expected in an experiment.

Homodyne and Heterodyne Techniques

In most of the applications discussed below, a "self-beating" technique (or homodyne) is employed in which the scattered light correlates with itself. Another experimental variant of correlation spectroscopy uses the heterodyne technique. In this the scattered light beats against a reference signal, usually the incident laser beam. This gives a somewhat simpler correlation function but the experiment is rather more difficult to do. In some experiments, such as electrophoretic light scattering and Doppler velocimetry, only the heterodyne technique is used (see below).

D4.3.3 Diffusion Coefficients from DLS

Mutual Diffusion Coefficient D_m

There are many different experimental techniques for determining the diffusion coefficients of macromolecules (Chapter D1), among which is DLS. The best results for DLS experiments are obtained with dilute and semi-dilute solutions of macromolecules. We call the diffusion coefficient determined by DLS D_m to underline that DLS is fundamentally a coherence phenomenon. D_m cannot be determined from DLS experiments on a single level.

Concentration Dependence of D_m

As the concentration of the solute molecules is increased we expect to find changes in the observed correlation function reflecting two separate effects. First, the diffusion constant changes due to the thermodynamic and hydrodynamic influence responsible for virial coefficients and modified frictional coefficients, respectively. Perhaps of more interest in terms of the information to be gained about the macromolecules are the changes due to the association of the solute molecules. In this case the correlation function is governed by the diffusion constants of the monomers, dimers, and larger aggregates. It is no longer a single exponential, but rather a sum of exponentials, each term representing a different species of molecule.

Multiple Decay Analysis of the Correlation Function

The distribution of internal relaxation times in the correlation function leads to the problem of solving Fredholm integral equations of the first kind in the form

$$G^{(1)}(q,t) = \int_0^\infty G(\Gamma) \exp(-Gt)\mathrm{d}\Gamma \qquad (D4.6)$$

where $G^{(1)}(q,t)$ is the normalized time and scattering vector-dependent first-order function (also called the electric-field autocorrelation function) and $G(\Gamma)$ is the distribution function (or frequencies) determined from $G^{(1)}(q,t)$ (Comment D4.4).

The analysis of data for a single time constant is quite straightforward. Any of the weighted least-squares fitting programs commonly available is satisfactory and may be expected to converge rapidly. The analysis of the multi-exponential signals is, however, of greater difficulty and even at times impossible, especially if the time constants differ by a factor of 2 or less. Nonetheless, the combination of DLS with multiple angle laser light scattering can provide sufficient data to solve multiple-decay analysis problems (Comment D4.5; see also Section G2.6).

Continued

COMMENT D4.5 (CONT.)

exclusion chromatography (HPSEC, also called gel permeation chromatography) and field flow fractionation (flow FFF), which sort a complex sample solution into fractions of similar molecular weight and size particles, and "user-friendly" analysis software taking advantage of the progress in the power of personal computers, these instruments provided a readily accessible means for the determination of macromolecular mass and size distributions on an absolute scale.

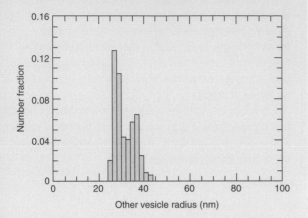

Fig D4.5.1 Number-weighted size distribution of the mixture of vesicles with two original samples of diameters 27 nm and 36 nm. (Korgel *et al.*, 1998.)

To illustrate the power of this technique, a mixture of vesicles of different average size was analyzed by flow FFF/MALLS. The peaks of the two models of the distribution occur at 27 nm and 36 nm, very close to the peaks in the histogram of the two original vesicle samples. The resolution of this bimodal size distribution illustrates the possible superiority of flow FFF/MALLS for resolving a heterogeneous population. In comparison, DLS alone cannot resolve peaks in such a bimodal distribution. If DLS could be coupled effectively to a flow FFF device, improved resolution might be achievable. However, the explicit time dependence of DLS makes it more amenable to batch as opposed to flow measurements.

D4.4 DYNAMIC LIGHT SCATTERING UNDER GAUSSIAN STATISTICS

In the interpretation of $G^{(1)}(t)$ or $G^{(2)}(t)$, a long list of assumptions is normally made. Rotations and internal movements (if present) are assumed to be independent of both each other and translational motion. Furthermore, for most solutions it is supposed that the solutions are sufficiently dilute that the translational motions of different scatterers are independent. Finally, all the relaxations are usually assumed to have exponential decays (Brownian translational and rotational motions). In these simplifying circumstances, for a single solute

$$G^{(1)}(q,t) = A \exp\left(-D_t q^2 t\right)\left[1 + \sum_j B_j \exp\left(-t/\tau_j\right)\right]$$

$$[1 + C \exp\left(-t/\tau_c\right)] \qquad (D4.7)$$

The first exponential term describes the translational diffusion of molecules with diffusion coefficient D_t, the second is concerned with rotation about the jth molecular axis (B_j are the relevant polarizability anisotropy components). The final exponential describes decay of the correlation function due to the internal motion of the solute species with a characteristic relaxation time τ_c. A, B, and C are constants giving the magnitude of the respective fluctuations.

A fundamental problem in analyzing DLS data from complex systems is to *decompose* Eq. (D4.7) into the individual exponential decay components, given a measured data set $G^{(1)}(t)$ or $G^{(2)}(t)$. This problem will be discussed below.

D4.4.1 Particles that are Small Compared to the Wavelength of the Incoming Light

As a first example, we shall discuss the application of light scattering to particles that are much smaller than the wavelength of the incoming light. In this case the coefficients B_j and C in Eq. (D4.7) are zero and only the first term remains, so that

$$G^{(1)}(\tau) = a_1 \exp\left(-D_t q^2 \tau\right) \qquad (D4.8)$$

and

$$G^{(2)}(\tau) = 1 + a_1 \exp\left(-2D_t q^2 \tau\right) \qquad (D4.9)$$

From the single exponential decay of the measured $G^{(2)}(t)$ autocorrelation function, one obtains the translational diffusion coefficients under study (Comment D4.6).

When the system becomes more complex – for example, when the sample contains a mixture of N species with diffusion coefficients D_i, $G^{(1)}(t)$ and $G^{(2)}(t)$ become sums of single exponential decays with rate constants $q^2 D_i$:

$$G^{(1)}(\tau) = \sum_{i=1}^{N} a_i \exp\left(-D_i q^2 \tau\right) \qquad (D4.10)$$

$$G^{(2)}(\tau) = \text{constant} + \left[\sum_{i=1}^{N} a_i \exp\left(-D_i q^2 \tau\right)\right]^2 \qquad (D4.11)$$

D4.4.2 Rigid Particles of Dimension Comparable to the Wavelength of Light

If the intensity of light scattered by a molecule depends upon its orientation with respect to the laboratory, it is possible to obtain information concerning the rotational diffusion coefficient D_R of the molecule from DLS. Two separate experimental approaches lead to the required orientation dependence. The first approach is based on using polarized light, whereas the second one operates with depolarized light.

Polarized Component of Scattered Light

If a molecule is non-spherical in shape and has one dimension comparable to the wavelength of light or larger, the scattering varies with orientation because of interference effects between light scattered from different positions on the molecule. For example, if we consider light scattered in the backward direction from a long rod-shaped molecule of length L, we find that when the molecule is perpendicular to the incident light beam, it scatters more light than when it is parallel, because in the former case all scattered waves are in phase, whereas in the latter there is a certain amount of destructive interference.

The theory shows that the result for the simpler case when the rotational and translational motions are uncoupled is

$$G^{(1)}(\tau) = \sum_{l\,\mathrm{even}} B_l \exp\left(-D_t q^2 + l(l+1)D_R\right)\tau \qquad (\text{D4.12})$$

where l takes on the values $0, 2, 4 \ldots$ and the B_l depends upon qL as shown in Fig. D4.7. It is seen that for small values of qL, as would be obtained even for light scattered from large molecules if a small scattering angle is used, B_0 is the only important coefficient and only the translational diffusion constant may be obtained. For large qL, however, B_2 becomes important and a second exponential time constant enters $G^{(1)}(\tau)$ from which D_R may be determined. The

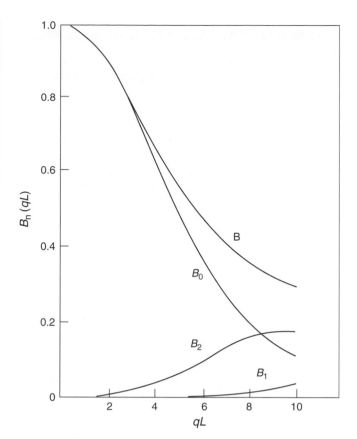

Fig. D4.7 The relative amplitudes of the components of the correlation function given in Eq. (D4.12) that result when light is scattered from a rod-shaped molecule of length L. The curve labeled B gives the sum of all B_n and hence represents the total scattered light. (Pecora, 1968.)

usual procedure is to extrapolate the small-angle single exponential to large angles. The difference is analyzed in terms of an exponential with time constant $D_t q^2 + 6D_R$.

Depolarized Component of Scattered Light

The second approach arises when the polarizability of the scattering molecule depends on the relative orientations of the electric field of the incident light and the axis of the molecule. In this case the polarizability is said to be anisotropic and it is found that the scattered light is no longer polarized as it would be for a molecule with isotropic polarizability but has a component (usually small) polarized in a direction perpendicular to that of the normal scattering. If the self-beat spectrum of the depolarized light is obtained, it can be shown that $G^{(1)}(\tau)$ contains a single exponential provided the molecule has cylindrical symmetry. This is a remarkable result as it allows D_R to be determined from the analysis of a single exponential measuring scattering rather than the more complicated analysis required by Eq. (D4.12).

$$G^{(1)}(\tau) = \exp(-6D_R)\tau \qquad (\text{D4.13})$$

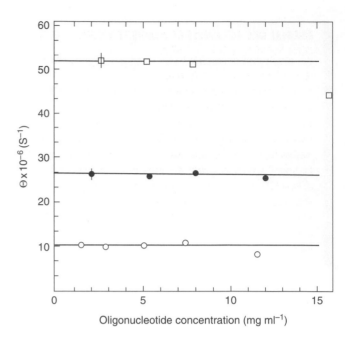

Fig. D4.8 Rotational diffusion coefficients versus concentration corrected to 20 °C for the three oligonucleotides. Plot symbols: ○ = 20-mer, ● = 12-mer, and □ = 8-mer. (Eimer and Pecora, 1991.)

Unfortunately, depolarized scattering is usually very weak (Comment D4.7). In spite of that, the rotational diffusion coefficient of the tobacco mosaic virus (TMV) has been obtained using this technique and found to be $350\,\text{s}^{-1}$, in good agreement with results obtained by other techniques. In the case of lysozyme, results for D_R and D_t, combined with Perrin's expression for the rotational and translational diffusion coefficients of ellipsoids, showed that lysozyme in solution has very nearly the same molecular dimensions as crystalline lysozyme. Note that these results were obtained in the late 1960s using a simple wave analyzer to give a spectrum of scattered light.

In the early 1990s, R. Pecora and colleagues applied this technique to measurements of the rotational diffusion coefficients of a homologous series of small oligonucleotides. The scattered light at 90° in the *VH* geometry, where *V* and *H* correspond to directions that are vertical and horizontal with respect to the scattering plane, was frequency analyzed by passing it through a Fabry–Perot interferometer. The piezoelectrically driven instrument was equipped with a set of 750 MHz free spectral range confocal mirrors to allow the analysis of slow relaxation processes of the order of nanoseconds. Figure D4.8 shows the concentration dependence of three B-duplex oligonucleotides 8, 12, and 20 base pairs in length.

The combination of depolarized and polarized DLS provides a powerful method for obtaining the hydrodynamic dimensions and asymmetry of short rod-like particles (Eq. (D2.34)).

D4.4.3 Flexible Macromolecules: DNA

In addition to the rigid body motions already discussed, when the molecule under observation can flex on a scale comparable to the wavelength of light, the relaxation times of the flexing modes can be obtained by careful analysis of the scattered light. Once again the phenomenon responsible for an additional contribution to the field correlation

function is the change in light scattering intensity as the molecule assumes a new conformation due to interference effects between the light waves scattered from different segments of the same molecule.

Many efforts have been made during the last 40 years to extract from DLS measurements internal relaxation times for flexible molecules, like DNA. Three strategies have generally been followed to obtain this information. In the first, a double-exponential fitting was used, assuming that for sufficiently long molecules (more than 200 nm) the correlation function at long times is dominated by the single time constant

$$G^{(1)}(\tau) = D_t q^2 + 2/\tau_1 \qquad (D4.14)$$

where τ_1 is the relaxation time of the first normal mode of oscillation of the molecule. But this is only correct for a very limited range of q that is not known in advance (Comment D4.8).

In the second strategy, D_t is determined in the low-q range. Using the measured D_t value, the q dependence of the internal modes has been determined. Extrapolation of the internal relaxation times for $q \to 0$ yields the longest internal relaxation time τ_1.

In the third strategy, one can obtain an apparent diffusion coefficient D_{app} by approximation of the autocorrelation function through a forced single-exponential fit. In this case, an increase of D_{app} with increasing q is observed (Fig. D4.9). The low-q part of the curve approaches the

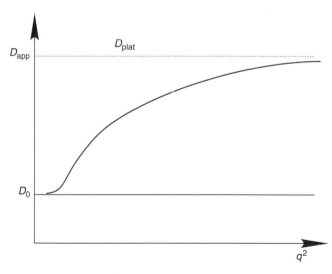

Fig. D4.9 Behavior of the apparent diffusion coefficient of a flexible polymer in a DLS experiment as a function of the square of the scattering vector q^2. (Adapted from Langowski *et al.*, 1992.)

D4.4.4 Macromolecules in Uniform Motion: Electrophoretic Light Scattering

Doppler Shift and Correlation Function

The light scattered from moving particles is Doppler shifted in frequency:

$$\Delta v = \frac{1}{2\pi} \mathbf{q} \cdot \mathbf{u} \tag{D4.15}$$

By measuring Δv one can measure the velocity of the particles, because the scattering vector $\mathbf{q}$ is known from the angle of scattering (Comment D4.9). ELS is a technique for the rapid measurements of electrophoretic drift velocities via the Doppler shifts of scattered light (Comment D4.10). The principal advantage of the technique is its ability to perform the electrophoretic characterization of many particles simultaneously.

When a solution of charged molecules is subject to an electric field, the particles drift toward the electrode of opposite polarity with a characteristic velocity. The effect of the electric field is that an additional term must be

translational diffusion coefficient D_t, the middle-q part corresponds to translational and rotational contributions, while at the high-q end a plateau value is reached which corresponds to the motion of the smallest independent subunits.

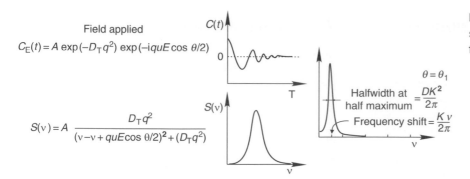

Field applied

$$C_E(t) = A \exp(-D_T q^2) \exp(-iquE \cos \theta/2)$$

$$S(v) = A \frac{D_T q^2}{(v-v+quE\cos \theta/2)^2+(D_T q^2)}$$

$\theta = \theta_1$

Halfwidth at half maximum $= \dfrac{DK^2}{2\pi}$

Frequency shift $= \dfrac{K v}{2\pi}$

Fig. D4.10 Schematic diagram of spectral shapes and correlation functions observed in ELS.

included in the correlation function, assuming the absence of reorientations and chemical reactions:

$$G(\tau) = A \exp\left(-D_t q^2 \tau\right) \exp\left[iquE \cos\left(\theta/2\right)\right] \qquad (D4.16)$$

where E is the magnitude of the applied electrical field and u is the electrophoretic mobility. Equation (D4.16) shows that the peak width caused by diffusion exhibits an angular dependence proportional to the square of the amplitude of the scattering vector $\mathbf{q}$, whereas the Doppler-induced peak shift is only linearly proportional to the absolute value of $\mathbf{q}$. Thus the analytical resolution of the technique may be improved, when contribution diffusion to the line width is decreased, by working at a low scattering angle.

Figure D4.10 shows the theoretical shapes of the correlation function and the resulting spectrum. In an electric field the autocorrelation function is expected to be a damped cosine whose frequency is related to the electrophoretic drift velocity of the particles and whose damping constant is related to their diffusion coefficient. In a zero field the spectral function is Lorentzian (see Chapter A3); in the presence of the electric field the spectral function is a well-shifted asymmetrical Lorentzian of width similar to the zero-frequency-centered heterodyne peak.

Experimental Geometry

ELS can be considered a special embodiment of both laser Doppler velocimetry (Comment D4.11) and DLS. However,

the application of an external electric field and the relatively slow particle velocities encountered in ELS necessitate the use of special light-scattering configurations and sample chambers distinct from those normally employed in DLS experiments (Fig. D4.11). Light from the main laser beam scattered by particles through some angle θ falls on a photodetector simultaneously with unscattered light from a reference beam derived from the same laser source. The scattered light is shifted in frequency relative to the reference beam because of the Doppler effect by an amount given by

$$\Delta v = \frac{1}{2\pi}\mathbf{q}\cdot\mathbf{u} = \frac{2nU}{\lambda_0} \sin\left(\theta/2\right) \qquad (D4.17)$$

It should be noted that when $\mathbf{u}$ is perpendicular to the scattering plane there is no frequency shift (Doppler shift), but when $\mathbf{u}$ is parallel to $\mathbf{q}$ there is a maximum Doppler shift.

The range of frequency difference typically encountered in ELS is only about 100 Hz. This shift is so small that no optical filters of sufficient resolution are available. However, it can be easily measured because the detector is sensitive to the frequency difference between the scattered light and the reference beam (heterodyne detection).

Applications

In principle ELS can be applied to the study of the electrokinetic properties of any charged species in solution, but in

---

COMMENT D4.11 LASER DOPLER VELOCIMETRY

In laser Doppler velocimetry, laser light illuminates the flow, and light scattered from particles in the flow is collected and processed. In practice, a single laser beam is split into two equal-intensity beams that are focused at a common point in the flow field. An interference pattern is formed at the point where the beams intersect, defining the measuring volume. Particles moving through the measuring volume scatter light of varying intensity, some of which is collected by a photodetector. The resulting frequency of the photodetector output is related directly to particle velocity.

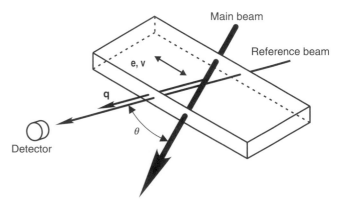

Fig. D4.11 Scattering geometry in an ELS instrument. **e** and **v** are the electric field and particle velocity, respectively. **q** is the scattering vector. (Adapted from Langley, 1992.)

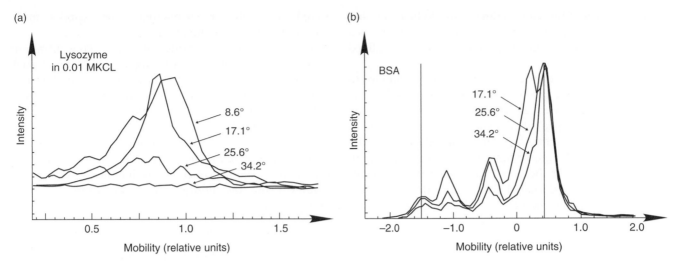

Fig. D4.12 Frequency spectra in ELS. **(a)** Free lysozyme: diffusion broadening smears out the peak except at the two smaller scattering angles. **(b)** Multimers of bovine serum albumin: three frequency spectra were obtained simultaneously at three different scattering angles in one experimental run, and the results scaled are plotted on the same mobility axis. (Adapted from Langley, 1992.)

practice the application of the technique, and particularly its efficacy *vis-à-vis* other electrophoretic techniques, must be considered separately for different particle-size domains. ELS is very difficult to apply to small globular proteins, such as lysozyme, largely because of the substantial contribution of diffusion to the ELS line width (Fig. D4.12a). The resolution of ELS essentially increases with increasing molecular mass. The electrophoretic mobility spectrum of a preparation of BSA is shown in Fig. D4.12b. At least five species of different mobility are resolved.

Note that using a gravitational field is a possible alternative to using an electric field. Both sedimentation and diffusion coefficients can be measured simultaneously from the spectrum of scattered light, and from these the molecular weight can be calculated using the Svedberg equation (Eq. (D2.20)).

D4.5 DLS UNDER NON-GAUSSIAN STATISTICS

D4.5.1 Scattering of a Small Number of Particles (Number Fluctuations)

Let us imagine that we are performing an experiment to monitor the concentration of a solute in a large volume of particles. Generally speaking, in this case the larger the number of particles being studied, the more easily are the fluctuations detected by DLS. The total intensity of scattered light received by a detector is the sum of contributions from scattering events from many particles. Changes in intensity due to *number fluctuations* are usually very small (Comment D4.12).

Now let us imagine that we are performing an experiment to monitor the concentration of a solute in a small

volume within a large sample of solution and that this defined volume has permeable walls. Experiments of this type were first carried out at the beginning of the nineteenth century using light microscopes. The object of such studies was to determine the mean number $\langle N \rangle$ of particles (in this case colloidal) per unit volume. If, however, we were to obtain a number of instantaneous "snapshots"

COMMENT D4.12 WHY DO WE NOT READILY NOTICE NUMBER FLUCTUATIONS IN A MACROSCOPIC SAMPLE?

Straightforward application of thermodynamic fluctuation theory gives the magnitude of these number fluctuations in terms of the mean square deviation $\langle \Delta N^2 \rangle$ from the average value $\langle N \rangle$. Angle brackets will be used henceforth to indicate an equilibrium ensemble time average. The Poisson distribution is a discrete probability distribution for the counts of events that occur randomly in a given interval of time or volume. $\langle N \rangle$ represents the mean number of events per interval. (Note that the Gaussian distribution is derived from the limit of the Poisson distribution for large values of $\langle N \rangle$).

The Poisson distribution theorem states that

$$\langle \Delta N^2 \rangle = \langle N \rangle \tag{A}$$

or

$$\langle \Delta N^2 \rangle^{1/2} / \langle \Delta N \rangle = \langle \Delta N \rangle^{1/2} \tag{B}$$

so that the relative magnitude of the fluctuation is inversely proportional to the square root of the number of particles. This is why we do not readily notice the fluctuations in a macroscopic sample where $\langle N \rangle = 10^{20}$.

of the volume as a function of time, it would be found that the number of particles counted would vary with time as a result of particle diffusion in and out of the defined volume.

For the case to be considered here, two contributions to the time dependence are important. One is caused by the time fluctuations due to particle diffusion in and out of the scattering volume. The other is caused by the diffusion of particles within the scattering volume over distances of the order of the wavelength of light, which thus causes temporal fluctuations. If the two time fluctuating terms are assumed to be uncorrelated with each other, then, for homodyne detection, the normalized correlation function is given by

$$G^{(2)}(\tau) = A\left[N_0 \exp\left(-\gamma t\right) + B N_0^2 \exp\left(-2D_t q^2 t\right)\right] \quad \text{(D4.18)}$$

The first term on the right-hand side results from fluctuations in the total number of particles within the scattering volume and the second from a diffusional interference term. N_0 is the average number of particles in the scattering volume; γ is the characteristic time required for a particle to diffuse across the scattering volume and is of the order $L^2/24N_0 D_t$ (where L is a characteristic dimension of the scattering volume); A and B are the constants.

At high concentrations, the second term dominates and the experimentally observed decay constant yields the diffusion constant. At sufficiently low concentrations, however, N_0 is comparable to $B N_0^2$, and both components contribute to the observed correlation function. Figure D4.13 shows results for an avian myelblastosis virus

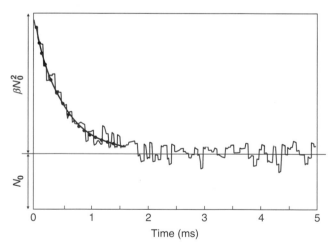

Fig. D4.13 Time correlation function of the scattered light intensity from a solution of avian myelblastosis virus ($M = 4 \times 10^8$, $s = 680\,S$, $D_t = 0.31 \times 10^7\,cm^2\,s^{-1}$) of concentration sufficiently low (2×10^7 particles per ml) to see the number fluctuations. The scattering volume is $10^{-3}\,ml$. The presence of the approximately flat baseline, displaced from the correlator zero, to which the exponential decays is indicative of the number fluctuations expected at low concentrations. (Salmeen et al., 1975.)

sample in which number fluctuation noticeably contributes to the correlation function.

However, the scattering at such concentrations ($\sim$2000 particles in the scattering volume) and at such a molecular mass ($= 4 \times 10^8$) is so weak that the virus diffusion coefficient cannot be determined precisely from either the interference fluctuation (the second term in Eq. (D4.18)) or the number fluctuations (the first term in Eq. (D4.11)). The effect of the first term can be fully suppressed if the process being studied is *incoherent* (as in fluorescence). This has been exploited in studies using the fluorescence correlation technique (Section D4.6).

D4.5.2 Cross-Correlation (Method of Two Detectors)

Number fluctuations can be observed from DLS, provided the coherent part of light scattering usually prevailing can be essentially reduced. This can be achieved by violation of spatial coherency (Eq. (D4.7)) with appropriate optical arrangements as well as by cross-correlating the intensity fluctuation of the same volume using two detectors. Figure D4.14a shows the cross-correlation function due to rotational motion of the TMV around its short axis of symmetry at various scattering angles. It is evident that the cross-correlation function does not depend on scattering angle. Unlike the correlation due to rotational diffusion relaxation of rigid molecules, the measured cross-correlation function for *E. coli* plasmid DNA due to internal degrees of freedom depends on the scattering vector (Fig. D4.14b).

D4.5.3 Scattering of Single Particles

The simplest illustrative case of non-Gaussian statistics is the single particle. In this case, phase correlations are *absent*, diffusion broadening is completely eliminated, and studies of pure form-factor fluctuations become possible. In practice, however, probing single-particle dynamics of submicron dimensions is impossible because of the small signal-to-noise ratio. For larger micron-dimensions, such studies become possible by using the method of localized dynamic light scattering (LDLS). The basic idea is to perform a DLS experiment with a single scatterer placed in a highly inhomogeneous field of light. Light from the microscope is collected and detected at different scattering angles by optical fiber probe coupled photomultiplier. The signal is then fed into a digital correlator that calculates the intensity autocorrelation function. The computed dynamic spectrum of a fluctuating object ranges from several decades in time down to the nanosecond regime. The applicability of this approach has been demonstrated by measuring the force constants of a single bead and a pair of beads trapped by laser tweezers (see Chapter F2).

(a)

(b)

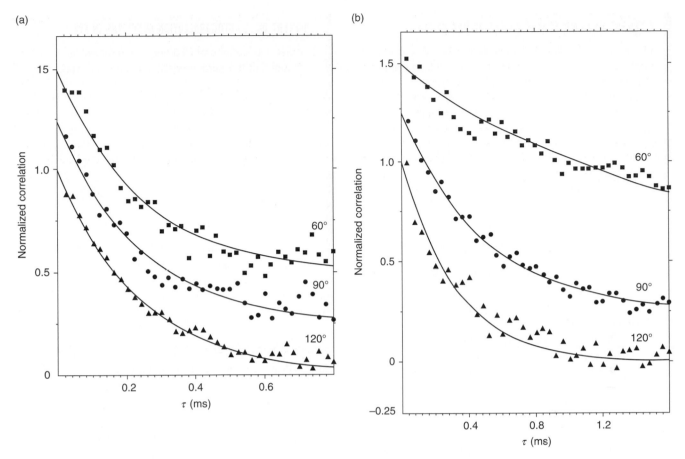

Fig. D4.14 (a) Cross-correlation measured from TMV scattering. The fitted correlation time is $240 \pm \mu s$, microseconds, independently of scattering angle. The correlations at 90° and 60° are shifted along the ordinate by 0.25 and 0.5 units, respectively. **(b)** Cross-correlation of fluctuation scattering from *E. coli* plasmid DNA molecules. The fitted correlation times are 400, 120, and 80 μs at scattering angles of 60°, 90°, and 120°, respectively. The correlations at 90° and 60° are shifted along the ordinate by 0.25 and 0.5 units, respectively (After Kam and Rigler, 1982.)

D4.6 FLUORESCENCE CORRELATION SPECTROSCOPY

D4.6.1 General Principles of FCS

The analytical strategy of FCS was introduced in the 1970s to measure chemical kinetics and the associated modulation of molecular diffusion by analysis of concentration fluctuations of a small ensemble (~1000) of particles. The power of the technique was demonstrated by measuring diffusion and the binding kinetics of the small fluorescent drug ethidium to DNA, as well as by the molecular weight determination of giant phage DNA (Comment D4.13). This early research formulated the still-used mathematical framework of the general FCS problem in terms of the eigenvalues and eigenvectors of the coupled diffusion and chemical kinetic rate equations.

FCS can be considered as a branch of DLS under non-Gaussian conditions because fluorescence is fundamentally an *incoherent* phenomenon. FCS is a method in which

the fluorescence intensity arising from a very small volume containing fluorescent molecules is analyzed (correlated) to obtain information about the processes that give rise to fluctuations in the fluorescence. One important process is Brownian motion, which leads to the stochastic appearance and disappearance of fluorescent molecules in the small observation volume. The relative fluctuation amplitude of a signal is inversely proportional to the number of molecules simultaneously measured (Eq. (A) of Comment D4.13). Thus, the presence of a large number of molecules, as typically encountered in a macroscopic medium, suppresses the effect of fluctuations and only the ensemble average is observed. This simple argument illustrates that FCS measurements require signals from a single molecule or a very small number of molecules in order not to mask the signal fluctuation.

Experimentally, FCS can be performed on small sample volumes, often of less than a femtoliter. An FCS instrument, in which illumination is focused on a zone less than 1 μm in diameter, plus the selection of the fluorescence

COMMENT D4.13 MOLECULAR WEIGHT OF DNA BY FCS

In the mid-1970s the FCS technique was successfully applied to the determination of the molecular weight of DNA molecules from different sources. These experiments were done in solutions containing about 1000 molecules to provide a reasonable signal-to-noise ratio. Recalling Eq. (B) of Comment D4.12, we can write

$$\left\langle \left(\frac{\Delta N}{\langle N \rangle}\right)^2 \right\rangle = \left\langle \left(\frac{\Delta C}{\langle C \rangle}\right)^2 \right\rangle = \frac{1}{N} \qquad (A)$$

where C is the (wt/vol) concentration of the molecules. The fewer the number of molecules, the larger are the fractional fluctuations. Therefore, for given concentration $\langle C \rangle$, the size of the fluctuations increases with molecular weight. By measuring these fluctuations (via any parameter that is sensitive to concentration), one can determine from Eq. (A) the number of molecules N in a given volume V within the measured fluctuations. Then knowing the average concentration $\langle C \rangle$, the molecular weight is determined from

$$M = \langle C \rangle \left(\frac{\Delta C}{\langle C \rangle}\right)^2 V N_A \qquad (B)$$

The molecular weights of DNA molecules from different sources were: for T2 phage DNA, 1.14×10^8; and replicating E. coli DNA, 3.9×10^9. The molecular weight of nuclei and individual chromosomal DNA molecules of Drosophila melanogaster were obtained as 3×10^{11} and 4×10^{10}, respectively.

(Adapted from Weissman et al., 1976.)

COMMENT D4.14 AUTOCORRELATION FUNCTIONS IN FCS

In most applications of FCS one of the following definitions of the autocorrelation function is used:

$$G^{\delta F}(\tau) = \frac{\langle (\delta F(t) \delta F(t+\tau) \rangle}{\langle F(t) \rangle^2}$$

or

$$G^F(\tau) = \frac{\langle F(t) F(t+\tau) \rangle}{\langle F(t) \rangle^2}$$

where $\langle \rangle$ denotes the time average,

$$\delta F(t) = F(t) - \langle F(t) \rangle$$

denotes the fluctuation around the mean intensity and for a long time average of F (no bleaching),

$$G^{\delta F} = G^F - 1$$

emitted only from this region using a pinhole in the image plane, allows a concentration of 1 nM to be analyzed; this corresponds to an average of *less than one molecule* in the studied volume. Thus, the FCS technique is particularly attractive for researchers striving for a quantitative assessment of interactions of a small molecular quantity in biologically relevant systems.

D4.6.2 Basics and Applications

The general principles of FCS are illustrated in Fig. D4.15. The fluctuations in the fluorescence signal are recorded (horizontal arrow) from which the autocorrelation function is computed (vertical arrow). The fluctuations are

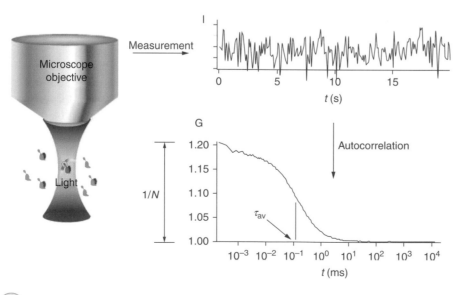

Fig. D4.15 General principles of FCS (see text). (After Bastiaens and Pepperkok, 2000.)

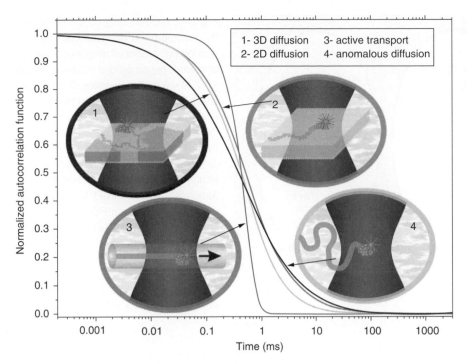

Fig. D4.16 Model autocorrelation curves for different particle mobility: (**1**) three- and (**2**) two-dimensional diffusion; (**3**) active transport and (**4**) anomalous diffusion (see text). (Adapted from Hausten and Schwille, 2003.)

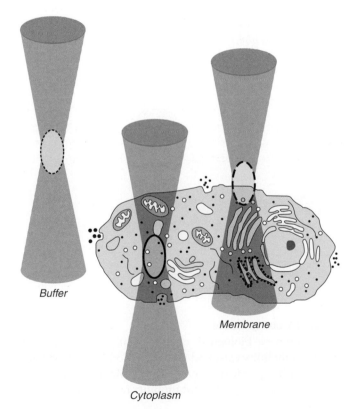

Fig. D4.17 Autocorrelation measurements assess a wide range of dynamic behavior in vitro and in vivo. The FCS focal volume, which is approximately 0.2 fl, can be positioned in, e.g., buffer (outside the cell), the cytoplasm, or the plasma membrane of a cell.

caused by diffusion of fluorescent molecules through the cavity or by changes in fluorescence over time caused by chemical kinetics. From the autocorrelation function the averaged dwell time (τ_{av}) and number of molecules in the cavity (N) can be derived (see below and Comment D4.14). Together with the known size of the light cavity, this can be used to determine the diffusion coefficient D of the fluorescence probe. D can then be used to evaluate the formation of molecular complexes as the diffusion of molecular complexes (red and green structures in Fig. D4.15) is slower than that of free protein molecules (green structure), resulting in a smaller value of D.

Three- and Two-Dimensional Diffusion

When a laser beam with a Gaussian beam profile is used for excitation and the fluorescence is collected through a confocal pinhole, the probe volume generated can be described to a good approximation by a three-dimensional Gaussian function. If r and l are the characteristic radial and axial dimensions of the volume (i.e., where the Gaussian function drops to $1/e^2$ of its volume at maximum), the relation between the autocorrelation function $G(\tau)$ and the diffusion time τ_{av} is:

$$G(\tau) = \frac{1}{N}\left(\frac{1}{1+\frac{\tau}{\tau_{av}}}\right)\left(\frac{1}{1+\left(\frac{r}{l}\right)^2\frac{\tau}{\tau_{av}}}\right)^{1/2} \qquad \text{(D4.19)}$$

The diffusion time τ_{av} characterizes the average time it takes for a molecule to diffuse through the radial part of the observation volume of the microscope.

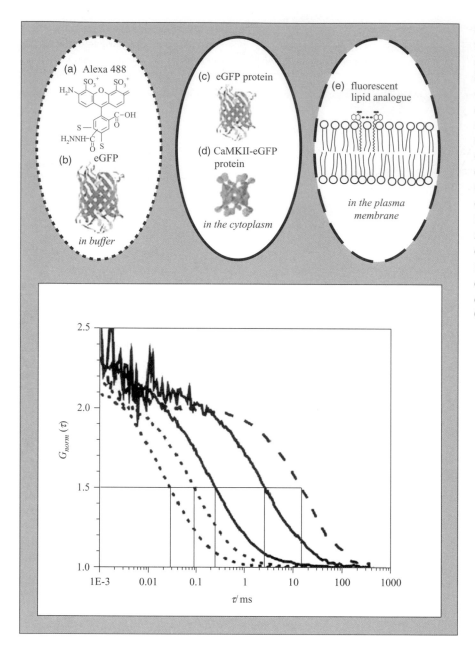

Fig. D4.18 Depending on the probe and the environment, diffusion coefficients of different orders of magnitude are obtained: **(a)** small dye molecule (Alexa 488, M ~ 0.6 kDa) in water, $D = 3 \times 10^{-6}$ $cm^2\,s^{-1}$; **(b)** green fluorescent protein (eGFP) in aqueous solution (M ~ 27 kDa); **(c)** the same protein in the cytoplasm of a HEL cell, $D = 3 \times 10^{-7}$ $cm^2\,s^{-1}$; **(d)** large protein complex (multimeric complex of calmodulin-dependent kinase II-eGFP fusion protein) in the cytoplasm of a HEK cell, $D = 3 \times 10^{-8}\,cm^2\,s^{-1}$; **(e)** two-dimensional diffusion of a fluorescent lipid analogue (long-chain carbocyanine dye "dif C_{18}") in the plasma membrane of a HEK cell, $D = 6 \times 10^{-9}\,cm\,s^{-1}$. (Adapted from Bacia and Schwille, 2003.)

However, the diffusion time is not a constant and changes with the size of the observation volume, which depends on the laser wavelength and the optics of the instrument. The relationship between the diffusion time and the diffusion coefficient is

$$\tau_{av} = \frac{r^2}{4D} \tag{D4.20}$$

Full three-dimensional sampling of diffusion can be performed by creating defined light cavities through two-photon excitation (Section F3.2.2). In this case the diffusion time is reduced by a factor of 2: $r^2/8D$.

For large structural parameters r/l or two-dimensional diffusion, as expected in planar membranes, Eq. (D4.19) simplifies to

$$G(\tau) = \frac{1}{N}\left(\frac{1}{1 + \tau/\tau_{av}}\right)^{1/2} \tag{D4.21}$$

Different Mobility Modes

In a complex biological environment, many different modes of mobility exist and no conventional diffusion constant can be specified (Section D2.3). Model autocorrelation curves showing different mobility modes are shown in Fig. D4.16.

The figure shows that the autocorrelation curve for active transport displays the sharpest decay, whereas the one for anomalous diffusion decreases rather slowly. If one treats the autocorrelation curve as a kind of distribution function of residence times within the focal volume,

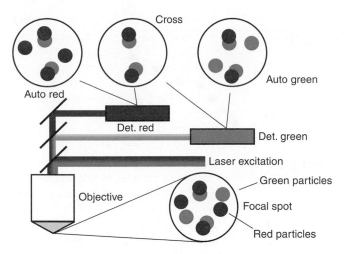

Fig. D4.19 Schematic of the measurement principle: green and red lasers illuminate the sample, which contains diffusing species R, G, and GR. Photodetector Det. Red detects R and GR; photodetector Det. Green detects G and GR.

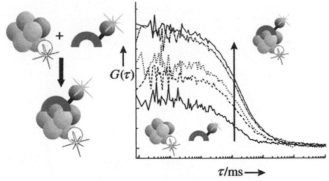

Fig. D4.20 The amplitude of the autocorrelation curve is a direct measure of the number of the doubly labeled particles. Only coordinated movements of the two different chromophores result in non-zero cross-correlation amplitudes. For an enzymatic cleavage reaction of a doubly labeled DNA strand, the maximum amplitude is measured before starting the reaction. The cleavage reaction can be monitored online via the gradually decreasing cross-correlation amplitude that directly reflects the decreasing number of doubly labeled particles. (Adapted from Haustein and Schwille, 2003.)

this behavior becomes clearer. Particles diffusing within a biological membrane may exhibit a large variety of different mobilities, depending on the time-dependent local environment of each individual chromophore. This results in curves that are typically spread out across several orders of magnitude in time. Exactly the opposite situation is physically realized in plug flow experiments. Here, all the particles are expected to have a velocity. Only the ellipsoidal shape of the focal volume causes a small variation in the residence times, so that the resulting decay of the autocorrelation curve is very abrupt. The same principles may be applied to two- and three-dimensional diffusion.

Dynamic Time Range

FCS commonly accesses dynamics in the time range 1 µs to 100 ms. For convenient plotting of this very wide dynamic range, a logarithmic timescale is generally used (Figs. D4.17 and D4.18). Depending on the surrounding medium, the molecular mobility changes by several orders of magnitude. The corresponding diffusion coefficients range from 3×10^{-6} cm^2 s^{-1} for free dye in an aqueous buffer solution to 10^{-10} cm^2 s^{-1} for a large receptor in the cell membrane.

Since the FCS temporal resolution is below 100 ns, limitations in FCS for in-vivo applications appear in the slow time range. When analyzing slowly moving molecules, for example, those interacting with the cytoskeleton, nuclear DNA, or membranes, one has to ensure that the molecules under study are not photobleached during their transit time through the focal volume.

Multicomponent Systems

In a multicomponent system, the correlation curve is evaluated, assuming a model of two or more diffusing components, by a standard Marquardt non-linear least-

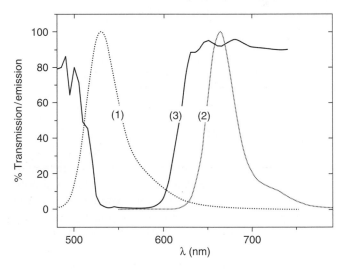

Fig. D4.21 Emission spectra of rhodamine green (**1**) and CY-5 (**2**), as well as the transmission characteristic of a second dichroic mirror (**3**). (After Schwille *et al.*, 1997.)

squares fitting routine. Although this has been successfully carried out in some cases, extensive calibration measurements of all species must be carefully made to fix, for example, the diffusion coefficients or related parameters. The dual-color cross-correlation scheme described below eliminates these preliminary calibration steps by introducing two spectroscopically separable fluorescence labels that allow simultaneous measurements of two reaction partners and their products.

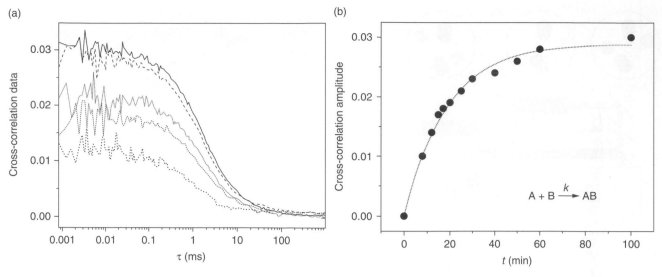

Fig. D4.22 (a) Time course of the cross-correlation function in the renaturation (hybridization) reaction. Light dots: 8 min incubation; dots: 15 min; short dashes: 25 min; dashed line: 60 min; solid line: 120 min. The fraction of GR molecules represented by the amplitude increases with time. The diffusion time (curve decay time) remains the same. **(b)** Plot of the cross-correlation amplitude during the reaction (points). By fitting the irreversible second-order reaction (short dashes), we get $k = 8 \times 10^5 \, M^{-1} \, s^{-1}$. (After Schwille *et al.*, 1997.)

D4.6.3 Dual-Color Fluorescence Cross-Correlation Spectroscopy

Dual-color fluorescence cross-correlation spectroscopy uses separate lasers to excite two different color probe molecules, and independent red and green detector channels are used for fluorescence detection (Fig. D4.19). The correlation between the two channels is then measured, as opposed to the autocorrelation of a single channel in standard FCS measurements.

The dye system is designed to have a green species (G) and a red species (R), as well as an increasing fraction of green-and-red substance (GR) due to the reaction of both partners. Whereas one detector each should record pure G and R, GR is detected in both of them (Fig. D4.19). Cross-correlation of the detector signals is therefore a means of measuring the reaction product GR independently of fluorescent educts. In contrast to a single-color autocorrelation analysis, the method in principle yields a *yes-or-no* decision about the presence of a doubly labeled reaction product, without the necessity of mathematical evaluation of the correlation curve. The most prominent example of cross-correlation is the dual-color mode schematically depicted in Fig. D4.20. Whereas autocorrelation measurements are sensitive only to signal variations within one channel, cross-correlation analysis is used to compare the signals arising from two differently labeled fluorescent species. Only temporally coordinated fluctuations in both channels, i.e., colors, give rise to cross-correlation.

A number of experiments applying this technique to different kinds of reactions have been carried out to probe interactions between different molecular species. One of them was the monitoring of a hybridization reaction of two differently labeled DNA molecules. In this experiment,

two complementary ssDNA molecules were each fluorescently tagged at one end, one strand with rhodamine green and the other with Cy-5 (Fig. D4.21). The hybridization reaction kinetics was followed using cross-correlation FCS analysis. The main advantage of the dual-color FCS approach is that there is no correlation between the two channels unless individual molecules produce both red and green fluorescence signals. In other words, only DNA molecules that have associated with their complementary strand produce a correlated signal. The amplitude of the cross-correlation signal is thus directly proportional to the concentration of hybridized DNA, and the hybridization kinetics can be interpreted directly as changes in the time zero amplitude of the measured cross-correlation signal. This is in contrast to one-color FCS measurement in which the ssDNA and dsDNA molecules both contribute to the amplitude of the correlation signal, requiring some analysis to discern the equivalent information. Figure D4.22a shows the time course of the cross-correlation function over 120 min. Figure D4.22b shows a plot of the cross-correlation amplitude values versus time. The kinetic rate parameter may be evaluated directly from these data.

D4.7 CHECKLIST OF KEY IDEAS

- DLS adds the *time dimension* to the classic probe of incoherent, total-intensity scattering of light.
- In a solution of macromolecules, the local concentration continuously fluctuates due to Brownian motion of the particles; since these fluctuations are *very slow* compared with the frequency of the light wave, a very high-resolution spectrometer is required.

- The resolution needed to determine the translational diffusion coefficients of macromolecules is so *high* that conventional optical spectroscopy is inadequate and the technique of optical mixing spectroscopy is required; however, this resolution is good enough to determine the rotational diffusion coefficients of small globular proteins.

- Time-dependent correlation functions provide a concise method of expressing the degree to which two dynamical properties are *correlated* over a period of time; in the laboratory, the transformation of the fluctuating signal into a correlation function is performed using a device called a correlator.

- XThe correlation function and the frequency spectrum are entirely equivalent descriptions of the dynamical behavior of a property of an equilibrium system, being related by Fourier transformation.

- Two different experimental techniques are employed in DLS; in the "self-beating" (or homodyne) technique the scattered light correlates with itself, whereas in the heterodyne technique the scattered light beats against a reference signal, usually the incident laser beam.

- For small molecules, the translational diffusion contribution is by far the largest component in the fluctuation of scattering intensity.

- For molecules of a size approaching the wavelength of light, rotational diffusion and internal flexibility (if present) contribute a sizable amplitude to the fluctuation.

- A fundamental problem in analyzing DLS data from complex systems is the *decomposition* of the measured correlation function into the individual exponential decay components.

- Cross-correlating measurements at different angles are used to separate translational diffusion from the other contributions such as rotational diffusion of asymmetric particles and conformational relaxation of random coils.

- In standard DLS experiments, the fluctuation of molecules in a large (*macroscopic*) volume is analyzed; in this case the larger the number of particles being studied the more easily are the temporal fluctuations detected.

- ELS is a technique for rapid measurement of electrophoretic drift velocities via the Doppler shift of scattered light.

- If the scattering volume contains a relatively small number of particles (*about a few thousand*) the fluctuating intensity possesses *two* commensurable contributions: The first results from fluctuations in the total number of particles within the scattering volume and the second results from a diffusional interference term.

- If the scattering volume contains just a few particles, the fluctuating intensity has only a single contribution which results from fluctuations in the total *number* of particles within the scattering volume.

- FCS is a method in which the stochastic appearance and disappearance of fluorescent molecules in a small (*microscopic*) volume is analyzed.

- FCS is an *incoherent phenomenon*, and cannot be observed in an ensemble of molecules. FCS is best performed in the presence of a single molecule in the observed volume.

- Dual-color fluorescence cross-correlation spectroscopy uses separate lasers to excite two different color probemolecules, and independent red and green detector channels are used for fluorescence detection.

Suggestions for Further Reading

Historical Review and Introduction to Biological Problems

Elson, E. L., and Magde, D. (1974). Fluorescence correlation spectroscopy. I: Conceptual basis and theory. *Biopolymers*, **13**, 1–27.

Berne, B. J., and Pecora, R. (1976). *Dynamic Light Scattering*. New York: Wiley.

Bastianes, P. I. H., and Pepperkok, R. (2000). Observing proteins in their natural habitat: The living cell. *TIBS*, **25**, 631–637.

Haustein, E., and Schwille, P. (2003). Ultrasensitive investigations of biological systems by fluorescence correlation spectroscopy. *Methods*, **29**, 153–166.

DLS as a Spectroscopy of Very High Resolution

Dubin, S. B., Lunacek, J. H., and Benedek, G. B. (1967). Observation of the spectrum of light scattered by solutions of biological macromolecules. *Proc. Natl. Acad. Sci. USA*, **57**, 1164–1171.

Benedek, G. B. (1969). Optical mixing spectroscopy in physics, chemistry, biology and engineering. In *Polarization: Matière and Rayonnement*. Paris: Les Presses Universitaires de France, pp. 4–84.

Doherty, J. V., and Clarke, H. R. (1980). Noisy solutions: A source of valuable kinetic information, *Sci. Prog. Oxf.*, **66**, 385–419.

DLS under Gaussian Statistics

Wada, A., Suda, N., Tsuda, T., and Soda, K. (1968). Rotary-diffusing broadening of Rayleigh lines scattered from optically anisotropic macromolecules in solution. *J. Chem. Phys.*, **50**, 31–35.

Dubin, S. B., Clark, N. A., and Benedek, G. B. (1971). Measurement of the rotational diffusion coefficient of lysozyme by depolarized light scattering: Configuration of lysozyme in solution. *J. Chem. Phys.*, **54**, 5158–5164.

Ware, B. R., and Haas, D. D. (1983). Electrophoretic light scattering. In *Fast Methods in Physical Biochemistry and Cell Biology*, eds., R. I. Sha'afi and S. M. Fernandez. New York: Elsevier.

Seils, J., and Dorfmuller, T. H. (1991). Internal dynamics of linear and superhelical DNA as studied by photon correlation spectroscopy. *Biopolymers*, **31**, 813–825.

Eimer, W., and Pecora, R. (1991). Rotational and translational diffusion of short rodlike molecule in solution: Oligonucleotides. *J. Chem Phys.*, **94**, 2324–2329.

Chu, B. (1991). *Laser Light Scattering: Basic Principles and Practice*. San Diego, CA: Academic Press.

Langowski, J., Kremer, W., and Kapp, U. (1992). Dynamic light scattering for study of solution conformation and dynamics of superhelical DNA. *Meth. Enzymol.*, **211**, 431–448.

Langley, K. H., (1992). Developments in electrophoretic laser light scattering and some biochemical application. In *Laser Scattering in Biochemistry*, eds., S. E. Harding, D. B. Sattelle, and V. A. Bloomfield. Cambridge: Royal Society of Chemistry.

DLS under Non-Gaussian Statistics

Weissman, M., Schindler, H., and Feher, G. (1976). Determination of molecular weights by fluctuation spectroscopy: Application to DNA. *Proc. Natl. Acad Sci. USA*, **73**, 2776–2780.

Kam, Z., and Rigler, R. (1982). Cross-correlation laser scattering. *Biophys. J.*, **39**, 7–13.

Bar-Ziv, R., Meller, A., Tlusty, T., *et al.* (1997). Localized dynamic light scattering: Probing single particle dynamics at the nanoscale. *Phys. Rev. Lett.*, **78**, 154–157.

Meller, A., Bar-Ziv, R., Tlusty, T., *et al.* (1998). Localized dynamic light scattering: A new approach to dynamic measurements in optical microscopy. *Biophys. J.*, **74**, 1541–1548.

General Principles of FCS

Magde, D., Elson, E. L., and Webb, W. W. (1972). Thermodynamic fluctuations in a reacting system: Measurement by fluorescence correlation spectroscopy. *Phys. Rev. Lett.*, **29**, 705–708.

Eigen, M., and Rigler, R. (1994). Sorting single molecules: Application to diagnostics and evolutionary biotechnology. *Proc. Natl. Acad Sci. USA*, **91**, 5740–5747.

Schwille, P., and Kettling, U. (2001). Analyzing single protein molecules using optical methods. *Curr. Opin. Biotech.*, **12**, 382–386.

Bacia, K., and Schwille, P. (2003). A dynamic view of cellular processes by in vivo fluorescence auto- and cross-correlation spectroscopy. *Methods*, **29**, 74–85.

Dual-Color Fluorescence Cross-Correlation Spectroscopy

Schwille, P., Meyer-Almes, F. J., and Rigler, R. (1997). Dual-color fluorescence cross-correlation spectroscopy for multicomponent diffusional analysis in solution. *Biophys. J.*, **72**, 1878–1886.

Koltermann, A., Kettling, U., Stephan, J., Winkler, T., and Eigen, M. (2001). Dual-color confocal fluorescence spectroscopy and its application in biotechnology. In *Fluorescence Correlation Spectroscopy: Theory and Application*, eds., R. Rigler and E. Elson. Heidelberg: Springer-Verlag.

Stephan, J., Dorre, K., Brakmann, S., *et al.* (2001). Towards a general procedure for sequencing single DNA molecules. *J. Biotechnol.*, **86**, 255–267.

OPTICAL SPECTROSCOPY

E

VISIBLE AND IR ABSORPTION SPECTROSCOPY

E1

E1.1 BRIEF HISTORICAL REVIEW AND BIOLOGICAL APPLICATIONS

1704

In *Opticks* **Isaac Newton** dealt with the formation of a *spectrum* by a prism, and the composition of white light and its dispersion. The Latin word, *spectrum*, means an appearance; a spectrum is obtained when radiation is broken up into its color or wavelength distribution.

1800

The astronomer **William Herschel** discovered infrared (IR) radiation.

1801

The physicist **Johann Wilhelm Ritter** discovered ultraviolet (UV) radiation.

1814

Joseph von Frauenhofer showed that the Sun's spectrum contained dark lines (later named Frauenhofer lines), indicating that light of the corresponding color was missing because of absorption.

1850–1900

August Beer stated the empirical law, which was named after him, that there is an exponential dependence between the transmission of light through a substance, the concentration of the substance, and the path length of the beam through it. The law is also known as the Beer–Lambert law or the Beer–Lambert–Bouguer law, in recognition of the work of **Pierre Bouguer (1729)** and **Johann Heinrich Lambert (1760)**. **Gustav Kirchhoff**'s discovery that each pure substance has a characteristic spectrum provided the basis for analytical spectroscopy. **Gustav Kirchhoff** and **Robert Bunsen** identified the chemical elements in the Sun by analyzing its spectrum. **Johann Jacob Balmer** identified a numerical series in the spectrum of hydrogen. **Joseph John Thomson** discovered the electron. **Max Planck** introduced the concept of quanta in the treatment of heat radiation and laid the foundation of quantum theory. He was awarded the Nobel Prize in **1918**.

1911–1913

Ernest Rutherford proved that atoms comprise positively charged nuclei surrounded by electrons. **Niels Bohr** made the first theoretical calculations of the discrete energy states in the hydrogen atom and the related wavelengths of the emitted radiation lines.

Quantum and wave mechanics, which was developed in the **1920s** by **Erwin Schroedinger**, **Werner Karl Heisenberg**, **Wolfgang Pauli**, and **Paul Adrien Maurice Dirac**, now provides a firm basis for the theoretical understanding and application of spectroscopic methods to study matter.

The first mid-IR spectrometer was constructed less than 35 years after the discovery of IR radiation, and within 90 years IR spectroscopy was finding applications in astronomy and organic and atmospheric chemistry.

1949

E. K. Blout and **R. C. Mellors** demonstrated that IR spectra could provide information concerning the molecular structure of human and animal tissue. This avenue of exploration was not vigorously pursued, however, because of instrumentation limitations and the complexities of the systems under investigation and of the spectra produced.

1950

A. Elliot and **E. J. Ambrose** demonstrated that the IR absorption spectrum of a protein is sensitive to conformation. Since then fundamental theoretical calculations on the subject have been extensively developed by **T. Miyazawa** and coworkers and **S. Krimm** and his colleagues. Nine characteristic amide vibrational bands or group frequencies were identified in the **1980s**. Among them, the amide I band, which is due almost entirely to the C–O stretch vibration, turned out to be the most useful probe for determining protein secondary structure in solution. However, the low sensitivity of conventional dispersive IR spectrometers and the strong water (H_2O) absorption at wavelengths in the amide I region severely limited early experiments to the solid state or heavy water (D_2O) solutions.

1961

H. T. Miles investigated the tautomeric forms of the DNA bases, making use of the power of IR spectroscopy to identify chemical groups. He discovered that the correct tautomer for cytosine is the amine-carbonyl form that is taken for granted today.

1994

After the introduction of Fourier transform IR (FTIR) spectroscopy in **1980**, **B. R. Singh** and coworkers demonstrated the potential usefulness of amide III spectra for the determination of the type of secondary structure of proteins in aqueous solution.

... and now

Proteins and nucleic acids have specific absorption signatures in the UV and IR spectral ranges. Biological macromolecules are themselves colorless, but they are often associated with prosthetic groups, such as retinal in the vision proteins or the heme group in myoglobin, hemoglobin, and the cytochromes, which absorb visible light. The absorption spectra of these chromophores (from the Greek *chroma*, color, and *phoros*, bearing) depend on their local environment and can be used to probe conformational changes and dynamics in the macromolecule during its biological active cycle. Enzyme activity can be measured with great sensitivity through the observation of the specific optical absorption patterns of substrates, products, and cofactors.

Spectrophotometers in the UV–visible spectral range have become indispensable tools in biochemistry and biophysics laboratories for the systematic analysis and characterization of macromolecular solutions. The determination of the absorption of UV and visible light might well be the most-performed measurement on biological macromolecules. Experiments using UV–visible absorption range from being aimed at the initial detection of material and concentration measurements to sophisticated temperature-dependence approaches used to measure macromolecular dynamics.

Linear IR spectroscopy is one of the classical tools for the study of the structures and interactions of small molecules. At first it appeared to be too ambitious to apply this technique to biological macromolecules because of their enormous number of vibrational modes. It was thought that the resulting overlap in absorption bands would not allow the extraction of detailed structural information from the IR spectra. Biological macromolecules, however, exhibit an intrinsic order of repeating units: the peptide bond in the protein backbone, the phosphate ester bond between the 3'-hydroxyl group on the sugar residue of one nucleotide and the 5'-phosphate group of the next nucleotide in nucleic acids, and the two-dimensional array of a limited class of molecules in lipid membranes. IR spectra of biological macromolecules are simpler than at first expected, therefore, and detailed band analysis provides useful information on macromolecular structure and interactions. In the case of proteins, linear IR spectroscopy

provides insights into secondary structure, complementary to the information obtained from circular dichroïsm (see Chapter E4). For nucleic acids, information can be obtained on the overall structure and interactions with small molecules such as intercalating drugs or metal ions.

IR spectroscopy has powerful applications in elucidating the molecular mechanisms of protein activity, by providing information on the interaction of participating molecules. The IR *difference* spectra formed between different states of an enzyme, for example, only contain bands of those groups that undergo changes during the transition from one state to the other. In parallel experiments, time-resolved IR spectroscopy provides information on the reaction kinetics. The difference spectroscopy approach, in fact, has greatly increased the power of the FTIR method by allowing the assignment of peaks in often formidably complex spectra, by comparing spectra of the same protein containing amino acids that have been modified by mutation or chemical modification.

Multidimensional IR spectroscopy has been proposed as a means to determine the structure and dynamics of biological assemblies. It provides a wealth of new information beyond that obtained from linear spectroscopy (see Chapter E2).

E1.2 BRIEF THEORETICAL OUTLINE

The full quantum chemistry theory of UV–visible–IR absorption spectroscopy is in terms of the interactions between the electromagnetic wave and the electrons of the bound atoms in sample molecules. Such a treatment is beyond the scope of this book and here we give only a *qualitative* outline of the theory. We emphasize, however, that the usefulness of absorption spectroscopy in the laboratory is based on the *quantitative* interpretation of the absorption curves of biological macromolecules in terms of accurately determined coefficients. Band assignments have been made reliably from the study of model compounds. For example, the absorption of the peptide bond at 190 nm was established from studies of simple amides, amino acids, and polypeptides in solid films and aqueous solutions.

In an absorption spectroscopy experiment, electromagnetic radiation of a given wavelength is observed after it passes through a sample to see how much of it has been absorbed. The power of the method is based on the fact that the molecules in the sample and the light beam exchange only specific amounts of energy, depending on their particular molecular states and dynamics (see Chapter A3). As described by quantum mechanics, radiation energy is absorbed in order to allow transitions between these well-defined molecular states. In a classical mechanics description, the interaction of an electromagnetic wave

TABLE E1.1 ABSORPTION SPECTROSCOPIES

λ (m)	v (Hz)	E (kJ mol^{-1})	Method
~3	~10^8	~10^{-4}	Radio-frequency NMR
~0.1–0.01	~10^9–10^{10}	~10^{-1}–10^{-2}	Microwave rotational and EPR spectroscopy
~10^{-4}–10^{-5}	~10^{12}–10^{13}	~1–10	IR vibrational spectroscopy
~8×10^{-7}–4×10^{-7}	~5×10^{14}–10^{15}	~200	Visible electronic spectroscopy
~10^{-7}	~4×10^{15}	~10^3	UV electronic spectroscopy
~10^{-10}	~10^{18}	~10^6	X-ray absorption electronic spectroscopy
~10^{-13}	~10^{21}	~10^9	γ-ray Mössbauer spectroscopy

with matter is treated in terms of induced dipoles. An electromagnetic wave is constituted of mutually perpendicular oscillating electric and magnetic fields (see Fig. A3.7). The fields are *transverse* with respect to the wave; i.e., they are also perpendicular to the direction of propagation. As the wave passes through a molecule, depending on its frequency and on how strongly the charges are maintained in place (the *polarizability*), it causes positive and negative charges to oscillate with opposite phase in the electric field plane. This results in the induction of dipoles with an oscillating *dipole moment* (charge multiplied by separation distance). The electromagnetic wave is represented by discrete energy quanta, hv, where h is Planck's constant and v is the frequency of the wave (see Chapter A3). Similarly, the electronic charges in the molecule can only form oscillating dipoles of certain defined energies and in certain defined directions with respect to the beam, because of the properties of the atoms and bonding patterns involved.

Absorption spectroscopy has been applied to biological macromolecules in the full range of the electromagnetic spectrum (Table E1.1). In this chapter we discuss the absorption of near-UV, visible, and IR radiation. Near-UV, with wavelengths of about 300 nm and energies of about 4 eV or 400 kJ mol^{-1}, and visible radiation, with wavelengths between 400 and 800 nm and energies of about 2 eV or 200 kJ mol^{-1}, correspond to transitions between electronic orbital states, and can be used to probe the different types of bonds in molecules. The spectral range of IR radiation, with wavelengths between 10 μm and 100 μm and energies between about 100 and 10 meV (10 kJ mol^{-1} to 1 kJ mol^{-1}) includes thermal energy at ambient temperature (about 25 meV) and can be used to study vibrational states and molecular dynamics.

E1.2.1 The Extinction Coefficient and Absorbance

Consider the simple set-up in Fig. E1.1. The intensity of light absorbed, $-\mathrm{d}I$, by the molecules in the thin sample slab of thickness, $\mathrm{d}l$, and unit area is proportional to the number of moles in the slab, $C\mathrm{d}l$, and to the incident intensity, I:

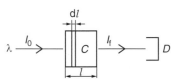

Fig. E1.1 Schematic of an experiment to measure absorbance. A beam of wavelength, λ, and intensity, I_0, is shone on a sample cell of path length, l, containing a solution of concentration, C (mol liter^{-1}). The final transmitted intensity, I_f, is measured by the detector, D.

$$-\mathrm{d}I = C\varepsilon'(\lambda)I\mathrm{d}l \qquad (E1.1)$$

where ε', the proportionality constant, depends on the wavelength and on the particular molecular type. It is called the *molar extinction coefficient* and has usual units of (mol liter^{-1})$^{-1}$ cm^{-1}, or M^{-1} cm^{-1}.

Integrating Eq. (E1.1) over the sample path length, l, we obtain

$$\ln(I_0/I_f) = C\varepsilon'(\lambda)l \qquad (E1.2)$$

where I_f is the final intensity transmitted by the sample (Fig. E1.1).

Equation (E1.2) is usually expressed in terms of the base 10 logarithm:

$$\log(I_0/I_f) = C\varepsilon(\lambda)l \equiv A(\lambda) \qquad (E1.3)$$

Equation (E1.3) is the Beer–Lambert law. The dimensionless (measured) quantity, $A(\lambda)$, is called the *absorbance* or *optical density* (OD) of the sample. OD can be measured accurately by modern spectrophotometers in the range of about 0.1 to ≤3. Note that for a value of 3, a fraction of only 1/1000 of the intensity is transmitted. The molar extinction coefficients, ε' and ε, are related by $\varepsilon = \varepsilon'/e$, where $e = 2.303$. Published values are usually given for ε.

E1.3 THE UV–VISIBLE SPECTRAL RANGE

The UV–visible spectral range corresponds to transition energies of the order of 100–1000 kJ mol^{-1} between states,

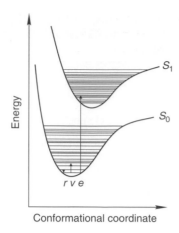

Fig. E1.2 Potential energy diagram of the lowest (S_0) and first excited (S_1) electronic states in a molecule. *r*, *v*, and *e* are rotational, vibrational, and electronic transitions, respectively. (After Cantor and Schimmel, 1980.)

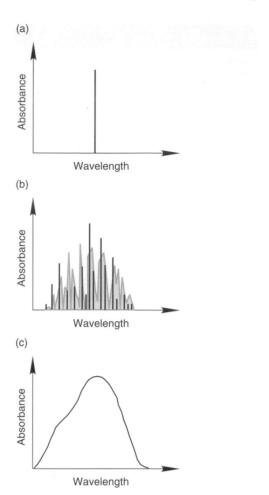

Fig. E1.3 Absorption spectrum of a molecule with one type of electronic transition: **(a)** theoretical, at very low temperature; **(b)** theoretical at ambient temperature; **(c)** observed. (After Cantor and Schimmel, 1980.)

which matches the difference between the ground (lowest-energy) and first excited electronic states (S_0 and S_1, respectively) in small molecules. These energies are much greater than thermal energy at ambient temperature ($2.5 \, \text{kJ mol}^{-1}$ at 300 K).

Each electronic state is itself split into finer energy levels arising from rotational and vibrational energies of the molecule itself (Fig. E1.2). For small molecules, the energy differences between vibrational states are of the order of $40 \, \text{kJ mol}^{-1}$, also larger than ambient thermal energy, and it results from statistical quantum mechanics that essentially all the molecules can be considered as occupying the lowest vibrational energy levels in the electronic ground state.

Energy states corresponding to molecular rotations are separated by less than $5 \, \text{kJ mol}^{-1}$, so that many rotational energy levels could be occupied at ambient temperature.

Electronic transitions, such as the one depicted in Fig. E1.2, become possible when light of the appropriate wavelength is shone on the molecule. If there were no rotational or vibrational states, the absorption spectrum would be characterized by a sharp line for each type of transition in the molecule (Fig. E1.3a). Transitions are possible, however, between the lowest S_0 level and the different vibrational and rotational levels in S_1, resulting in a set of closely spaced lines in the absorption spectrum (Fig. E1.3b), whose intensity (*absorbance*) is related to the probability of the transition. In practice, the lines are not only very close together but also each is broadened by various effects, e.g., due to an inhomogeneous local molecular environment or coupling with solvent if the molecule is in solution, so that only a smooth envelope is observed (Fig. E1.3c).

For most applications, UV–visible absorption spectra are modeled in terms of single electronic bands, without taking line shape into account. But the line broadening and shape contain information on vibrational and rotational states and an approach has been developed to study protein molecular dynamics from line behavior as a function of temperature (Comment E1.1).

Biological macromolecules are studied mainly in aqueous solution. Measurements have to be limited to above 170 nm, because water itself absorbs so strongly below that wavelength that even a micron-thick layer appears opaque.

Electronic bands in the macromolecule are also perturbed through coupling with the polar water molecules. These interactions depend on distance and mutual

COMMENT E1.1 MOLECULAR DYNAMICS FROM BAND SHAPES

Lorenzo Cordone and his collaborators have developed a method to characterize the molecular dynamics of myoglobin from a detailed shape analysis of the experimental visible absorption spectrum of the heme group measured as a function of temperature (Leone *et al.*, 1994).

orientation and, since the macromolecular solution is essentially disordered, they are different for each individual molecule, resulting in further broadening of the absorption line.

E1.3.1 UV–Visible Spectrophotometers and Measurement Strategies

A *spectrophotometer* is an instrument that measures the intensity of a light beam as a function of its *color*, or wavelength. In absorption experiments, a spectrophotometer provides an accurate quantitative measurement of the fraction of light that passes through a defined path length of solution. Common laboratory spectrophotometers operate in the UV–visible spectral range and sometimes also in the near-IR spectral region. Instruments operating in the middle-IR spectral region have to satisfy different technical requirements, in particular with respect to the photosensor device (the light detector in the instrument) and sensitivity, since all substances emit thermal radiation as IR light (see Section E1.4). Absorption spectrophotometers are either double-beam or single-beam instruments. A double-beam spectrophotometer measures the ratio of light intensity on two different optical paths (e.g., through a sample solution cell and through a cell containing solvent alone). The light intensity with respect to a fixed reference is measured on single-beam instruments. In a single-beam instrument, sample solution and solvent absorbance are measured sequentially; the solvent $A(\lambda)$ curve is stored on the instrument's computer and automatically subtracted from the sample solution curve. The optical path in a type of single-beam spectrophotometer is shown in Fig. E1.4.

Modern spectrophotometers are computer-driven and incorporate diverse software packages for specific applications such as time-based kinetics or temperature scans, and for quite a refined analysis of the measured absorbance curves.

Kinetics

For a two-state reaction or conversion, in which each state has a specific absorption spectrum, the spectral change, as the reaction proceeds, displays an *isobestic* point (the point at which the absorbances of the two states are equal). An isobestic point is also observed in cases where the equilibrium between two absorption spectral states is shifted by changing solvent conditions, temperature or pressure, for example (Fig. E1.5). Note, however, that an isobestic point is a necessary but not sufficient condition for a two-state transition, since an *intermediate* state may well show isobestic points with the initial and final states at the same wavelength.

Difference Spectra

Difference spectra are very useful in the study of processes characterized by spectral changes. Figure E1.6 shows difference absorption spectra as a function of time after photoexcitation for retinal in bacteriorhodopsin (see Comment A3.6). Note how the isobestic point (where the difference spectrum crosses zero) varies for the shorter delay times (Fig. E1.6a). This indicates the progression of the reaction through *intermediates* with different spectral characteristics. An apparent single isobestic point after longer times (Fig. E1.6b) suggests the later reaction steps can be interpreted in terms of a two-state transition.

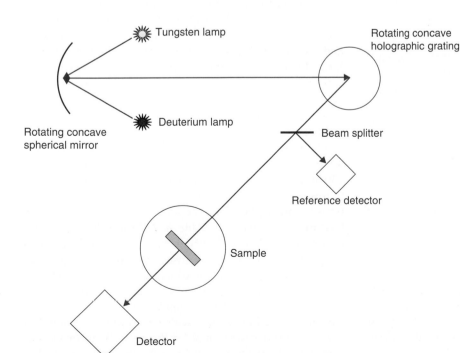

Fig. E1.4 The optical path in a single/split beam spectrophotometer. The tungsten and deuterium lamps provide light in the visible and UV ranges, respectively. The rotating concave holographic grating geometrically spreads out the beam into its wavelength components, so that the entire wavelength range is measured simultaneously by a photodiode array (detector). A beam splitter diverts part of the beam to a reference detector for continuous calibration.

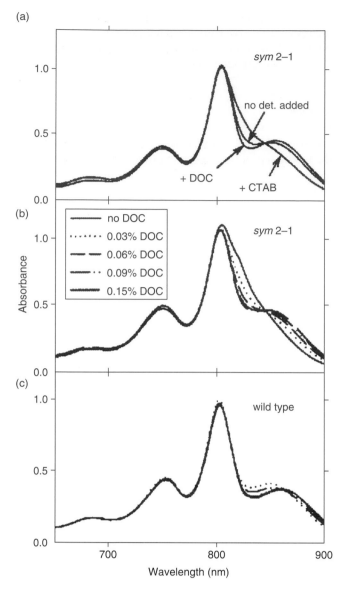

Fig. E1.5 Absorption spectra of the photosynthetic reaction center (the solar energy converter) of the purple bacterium *Rhodobacter capsulatus*, and a mutant (*sym2-1*): **(a)** addition to *sym2-1* of 0.3% negatively charged detergent deoxycholate (DOC) or positively charged detergent cetyltrimethylammonium bromide (CTAB); **(b)** *sym2-1* dialyzed in standard buffer after addition of various amounts of DOC; **(c)** same as **(b)** for the wild-type protein. The detergent induces conversion between two spectral forms with an *isobestic* point close to 860 nm. (Eastman *et al.*, 2000.) (Figure reproduced with permission from the American Chemical Society.)

Derivative Spectroscopy

The detection sensitivity of small spectral changes can be amplified by *derivative spectroscopy*. Small shoulders in an absorbance curve, for example, yield large oscillatory features when expressed as higher derivatives. Good examples of the effect are given in Section E1.4.

The derivative approach is used mainly to identify and separate small effects in the aromatic region of protein spectra, close to 280 nm (see below).

E1.3.2 UV Absorption Spectra of Proteins

The absorption of UV light by proteins has been analyzed in detail and proposed as a structural probe from the very early days of molecular biology (Comment E1.2). The absorption of proteins in this spectral range arises mainly from electronic bands in the peptide group (at 170–220 nm), aromatic amino acid side-chains close to 280 nm, and prosthetic groups, cofactors, enzyme substrates, or inhibitors (in the full UV–visible range depending on the group). Color in a protein is always due to a prosthetic group that absorbs in the visible region of the spectrum. The visible absorption spectra of protein-associated molecular groups are treated separately below.

Spectra of amide groups in crystals show electronic coupling, suggesting that the peptide ordering in secondary structures is reflected in the absorption spectra.

The molar extinction coefficient of a lysine polypeptide, expressed per amino acid residue, is shown Fig. E1.7 as a function of wavelength in the region of peptide group absorption.

The conformation dependence of the spectra in Fig. E1.7 is obvious, and attempts have been made to estimate the α-helix content in proteins from measured molar extinction values at 190 nm by interpolating between the random coil and helix values. Such measurements are not entirely satisfactory, however, because of the low reliability of the assumptions involved with respect to the effects of other than helix and random coil conformations, as well as with respect to corrections for amino acid side-chain contributions (Fig. E1.8). Secondary structure estimates in proteins are more reliably performed by IR absorption and circular dichroïsm (CD) measurements (see Chapter E4). Histidine, tryptophan, tyrosine, and phenylalanine side-chains present extinction coefficients greater than 5000 in the peptide absorption region (190–210 nm). The chemical formulae of the amino acid side-chains are given in Chapter A2.

The near-UV is more easily accessible than the far-UV for accurate measurement of absorption profiles. Above 230 nm, the strongest individual amino acid contribution to absorption in proteins is from tryptophan, followed by tyrosine, then phenylalanine and cystine. Trp absorption has an ε maximum of 5600 at 279.8 nm; Tyr has an ε maximum of 1429 at 274.6 nm – both measured in aqueous solution at pH 6 (Fig. E1.9). Also shown in the figure is the absorption curve of the phenylalanine side-chain with an ε maximum of 197 at 257.4 nm, comparable to the ε values for cystine of 360 at 250 nm and 280 at 260 nm (curve not shown). These four amino acids have to be considered in practice in order to account for the near-UV absorption of a protein. Absorption contributions are additive, so that

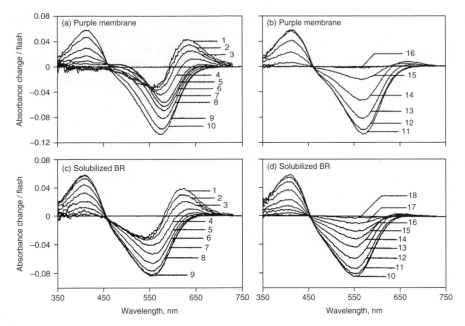

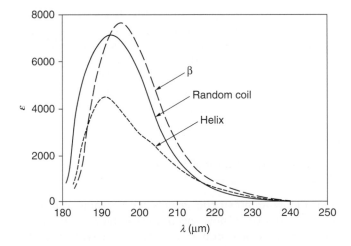

Fig. E1.6 Difference absorption spectra of bacteriorhodopsin in purple membrane as a function of delay time after photoexcitation. Numbers between 1 and 16 correspond to delay times between 100 ns and 600 ms, in non-equal intervals (parts **(a)** and **(b)** are for times shorter and longer than 1 ms, respectively). See also Section E1.3.3. (Váró and Lanyi, 1991.) (Figure reproduced with permission from the American Chemical Society.)

each amino acid type contributes to the observed absorbance in proportion to how many units of that type there are in the protein. The contribution of Phe and Cys absorption can be quite significant, therefore, in a protein that contains very few Trp and Tyr residues.

The molar absorption coefficients in Fig. E1.9 are plotted on a log scale, which emphasizes qualitative similarities. Fine structure can be seen in the curve maxima (it is especially clear in the spectrum for Phe) resulting from electronic transitions to different vibrational states. These features become sharper at low temperatures, at which broadening due to coupling with rotational modes is reduced.

The UV absorption curve of Tyr depends on its various ionization states and is especially sensitive to pH. Between pH 6 and pH 13, the peak maximum in the near-UV, at 274.6 nm, shifts to higher wavelengths by about 20 nm and increases in intensity by almost a factor of 2 (Fig. E1.10). This sensitivity has been used in the accurate titration of Tyr residues in proteins as well as in the separate determination of Tyr and Trp contributions to an observed absorption spectrum (the absorbance of Trp is much less sensitive to pH than that of Tyr).

Protein Concentration Measurements

Protein concentration measurements are usually performed by measuring the absorbance of the solution at wavelengths close to 280 nm. Such a determination provides only a *relative* concentration value, however, unless the molar extinction coefficient of the specific protein has been determined independently. We recall that biophysical methods involving determinations of molar mass or molar scattering power require accurate and *absolute* determinations of concentration (see Comment A1.2). Molar extinction coefficients at 280 nm, calculated from amino acid compositions (in particular, Tyr, Trp, and Phe contents), have been found to agree reasonably well with experimental absorbance measurements on unfolded proteins in guanidine hydrochloride, but values for native proteins may differ considerably because of the environmental effects on

Fig. E1.7 UV absorption spectra of poly-L-lysine hydrochloride in aqueous solution: random coil at pH 6.0, 25 °C; helix at pH 10.8, 25 °C; β-form at pH 10.8, 52 °C. The x-axis is the wavelength in nanometers (the "milli-micron" is an old fashioned unit, now replaced by "nanometre"). (From Rosenheck and Doty, 1961, reviewed by Wetlaufer, 1962.)

COMMENT E1.2 UV SPECTRA OF PROTEINS AND AMINO ACIDS

The treatment given in this chapter is based on a review by D. B. Wetlaufer published in *Advances in Protein Chemistry* in 1962 (Wetlaufer, 1962).

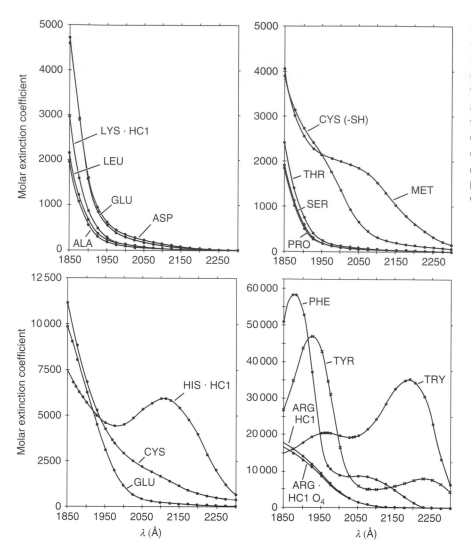

Fig. E1.8 The far-UV absorption spectra of amino acids in aqueous solution at pH 5, except for cystine (pH 3.0). The chloride ion contribution was subtracted from the basic amino acids measured as hydrochlorides. TRY is an old-fashioned abbreviation for tryptophan. CYS (-SH) is cysteine. CYS is cystine. Other abbreviations are the usual ones. (From Holzwarth *et al.*, 1962, reviewed by Wetlaufer, 1962.) (Figure reproduced with permission from Elsevier.)

the absorbance of the amino acids discussed above. In early work, molar absorption coefficients were found by determining concentration on an absolute scale via a dry weight measurement (at 105 °C in vacuo, corrected for ash) or a nitrogen determination. The modern approach is based on quantitative amino acid analysis. The molar extinction coefficients of a few different proteins are given in Comment E1.3.

E1.3.3 Visible Absorption Spectra of Protein-Associated Groups

Chromophore prosthetic groups with strong absorption bands in the UV–visible spectral range have greatly facilitated spectroscopic studies of protein activity. Usually, the chromophores are themselves part of the active site, their electronic configuration, reflected in the observed absorption bands and extinction coefficients, providing a sensitive measure of the progression of a reaction. The light absorption properties of a protein-associated chromophore may be of direct biological relevance as is obviously the case for

retinal in vision, or chlorophyll in photosynthesis. Chromophores can be as simple as a single metal ion that changes its color according to its oxidation state, such as copper in the blue protein azurin with absorption bands at 781 and 625 nm, or an organic molecule, such as retinal in rhodopsin (see below), whose main absorption band shifts in a wide range between 400 and 600 nm, depending on its local environment. Protein-linked chromophores can also be quite complex, such as the metal clusters in ceruloplasmin (absorption band at 793 nm), iron–sulfur clusters in ferredoxins, porphyrin associated with iron in the heme groups of myoglobin, hemoglobin and the cytochromes (different bands between 300 and 800 nm giving the red color to the proteins), porphyrin associated with magnesium in chlorophyll (three sets of bands, respectively, at 360–400, 580–590, close to 780 nm, giving a green color of course), or the flavin (vitamin B2) and flavin derivative groups that are involved in electron transfer reactions in many enzymes (bands close to 450 nm).

Proteins that have to ensure complex, highly regulated, electron transfer reactions, such as photoreaction centers

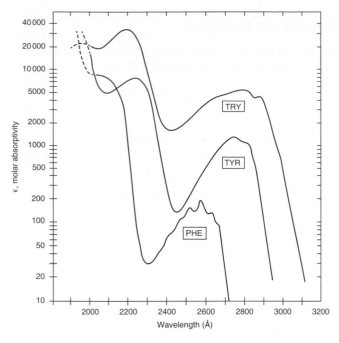

Fig. E1.9 The near-UV absorption spectra of the aromatic amino acids in aqueous solution at pH 6. TRY is TRP. The *x*-axis is wavelength in ångströms. (Reviewed by Wetlaufer, 1962.) (Figure reproduced with permission from Elsevier.)

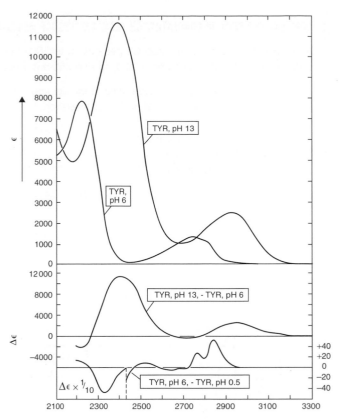

Fig. E1.10 Absorption spectra and difference spectra of Tyr measured for various pH values. The *x*-axis is wavelength in ångströms. The *y*-axis scale on the right-hand side of the bottom panel refers to the range above 2430 Å. (Reviewed by Wetlaufer, 1962.)

in photosynthesis or different cytochromes in respiration chains, may include many different chromophore prosthetic groups within the same structure (Fig. E1.11).

Reactions and associated changes in absorption properties of a chromophore can be initiated and followed very rapidly by using laser flash triggers. Below we briefly describe two examples of such studies that have had significant impacts on molecular biophysics: the rebinding of CO to the heme group in myoglobin following flash photolysis, and the light-activated proton pump photocycle in bacteriorhodopsin.

Flash Photolysis of CO–Myoglobin

Results from flash photolysis experiments on CO–myoglobin constituted one of the bases for the development of current ideas on protein dynamics and its relation to biological function, especially with respect to the conformational substates and dynamical transition models (see Chapter A3). The absorption spectra of hemoglobin and oxygen-bound hemoglobin are in Fig. E1.12. The absorption spectrum of the protein-bound heme contains three main groups of bands, called the Soret bands, between 325 and 500 nm; the a, b bands between 500 and 600 nm; and the "charge transfer" band or band III, in the near-IR, above 650 nm. CO binds much more strongly than oxygen to the iron atom of the heme group (this is why CO is a poisonous gas). A picture of CO bound to myoglobin is given in Fig. A3.32b. CO binding to myoglobin leads to a shift in the Soret band, a merging of the a, b bands, and a

disappearance of band III. In flash photolysis experiments, the Fe–CO bond is broken by a laser flash (photodissociation) and the rebinding kinetics is then followed, with nanosecond time resolution, through the changes in the absorption spectrum (Fig. E1.13).

The time dependence kinetics of the Soret band and band III shifts are very similar (Fig. E1.14). Information was derived from the non-exponential kinetics on the important role played by protein dynamics in the competition between ligand rebinding (geminate recombination, meaning "repeated," from the Latin *geminus*, twin) and ligand escape from the heme pocket through the protein structure.

A relation between protein relaxation and ligand rebinding was first formulated for NO geminate recombination, because they occur on similar timescales. In contrast, the recombination rate for CO is several orders of magnitude slower than protein relaxation rates under normal temperatures and solvent conditions. Experiments to examine the effects of protein relaxation on CO rebinding kinetics (see Comment A3.3) have been performed in viscous, glycerol-containing, solvents, and at low temperature to slow down protein relaxation rates.

COMMENT E1.3 PROTEIN CONCENTRATION FROM ABSORBANCE MEASUREMENTS (SEE ALSO COMMENT A1.2)

In order to measure the absorbance peak close to 280 nm correctly, it is essential to display the full background-corrected absorbance curve between about 320 nm and 250 nm. Such a curve is shown schematically in Fig. E1.3.1.

(a) The figure illustrates the absorbance of the ribosomal protein S13, for which there is significant light scattering from the solution, perhaps because of protein aggregation. In such a case, because the background is sloping 'the correct reading is h and not h'!

(b) The reading at 280 nm can be used for the concentration measurement of the ribosomal protein S6, because the background above about 320 nm is flat and at zero. Furthermore, the low absorbance at 260 nm indicates there is no contamination by ribosomal RNA.

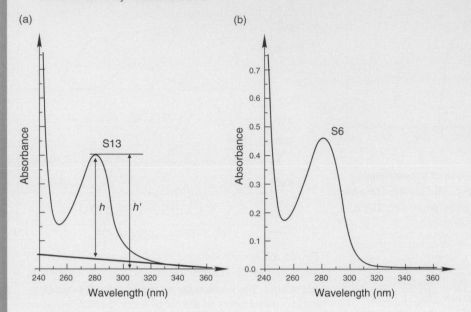

Figure E1.3.1

Examples of Extinction Coefficients for Proteins

The wavelength (λ, nm), observed ε (M^{-1} cm^{-1}) and molecular weight of ribonuclease (A), carboxypeptidase A (B), and bovine serum albumin (C) are:

Protein	λ	ε	M
A	277.5	9 800	13 680
B	278	66 700	34 400
C	280	43 600	66 000

The corresponding absorbance values for 1 mg ml^{-1} solutions in 1 cm optical path length cells are 0.72 for ribonuclease, 1.94 for carboxypeptidase A, and 0.66 for bovine serum albumin. Note that absorbance values for 1 mg ml^{-1} solutions, 1 cm optical path are of the order of 1. The higher value for carboxypeptidase A is because this protein has six Trp residues. Ribonuclease has no Trp residues and bovine serum albumin has 2.

(From the review by Wetlaufer (1962) which gives the original references.)

The absorption band changes result from the influence of the porphyrin and protein environment on the electronic orbitals of the iron atom. Interpretations of the spectral changes in structural terms have been attempted from correlations with crystallography data on different protein states (see Chapter A3) and from studies of different mutants (Fig. E1.15). Two important histidines are associated with the heme group in myoglobin: H93, which is bound by a covalent bond to the iron, is called the *proximal histidine*; H64, the *distal histidine*, is on the other side of the heme plane, pointing toward the ligand–Fe binding site. The figure shows the large differences in protein relaxation

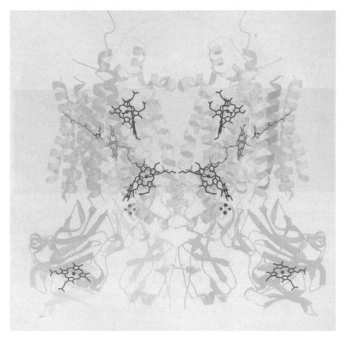

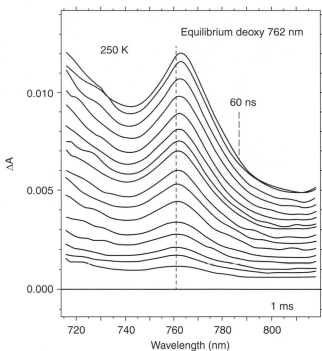

Fig. E1.11 The dimer of the cytochrome b6f complex from *Chlamydomonas reinhardtii*. The stromal side is on the top, the luminal side is on the bottom. The protein subunits are shown as light color ribbons. The light orange rectangle indicates the location of the membrane. The complex binds a number of chromophore prosthetic groups: heme (red), beta carotene (orange), chlorophyll (green); the iron atoms of iron–sulfur groups are shown as red dots. The blue molecule is tridecylstigmatellin, an analogue of quinone, which is a substrate of the complex. We are grateful to Daniel Picot for the figure. The cytochrome b6f complex structure is described in Stroebel *et al.* (2003).

Fig. E1.13 Time-resolved band III spectra of the recovery of CO–myoglobin after flash photolysis. The equilibrium unbound, deoxy-myoglobin spectrum maximum is given by the dashed line. The first measurement was after 60 ns (top), the last curve (bottom) was taken after 1 ms. The experiment was performed in 75% glycerol-containing solvent at 250 K. The transient spectra are offset vertically for clarity. (After Franzen and Boxer, 1997.)

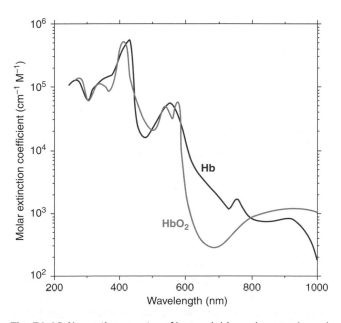

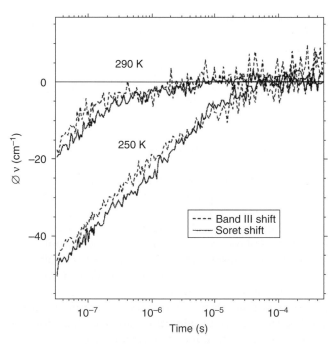

Fig. E1.12 Absorption spectra of hemoglobin and oxygen-bound hemoglobin.

Fig. E1.14 The time dependence kinetics of the Soret band and band III shifts are very similar. Experimental conditions were as for Fig. E1.13. (Franzen and Boxer, 1997.)

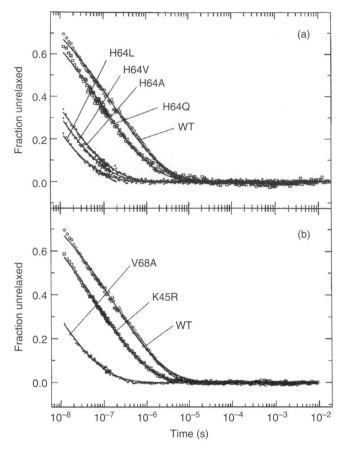

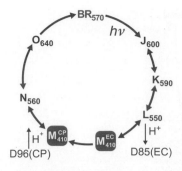

Fig. E1.16 The photocycle of bacteriorhodopsin (BR). The capital letters denote each spectral state of the retinal. The subscripts are the maximum absorption wavelengths in nanometers. Between L and M the Schiff base is accessible to the extracellular (EC) side. A proton is released from amino acid residue D85 to the EC side of the membrane. Within M there is an irreversible transition of the Schiff base that becomes accessible to the cytoplasmic (CP) side. Between M and N, a proton is bound to D96 on the CP side. See also Comment E1.4. (Haupts *et al.*, 1997.)

Fig. E1.15 Time dependence of Soret band shifts following photodissociation of CO–myoglobin for the wild-type protein (WT) and different distal histidine mutants. Experimental conditions were as for Fig. E1.13. Note the six orders of magnitude timescale on the x-axis. (After Franzen and Boxer, 1997.)

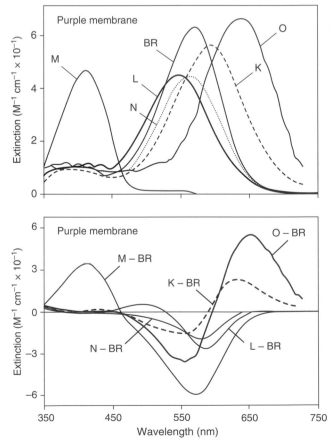

rates and CO rebinding, as monitored by the Soret band shift following CO flash dissociation, for different distal histidine mutants.

The Photocycle of Bacteriorhodopsin

Bacteriorhodopsin, the light-activated proton pump in the purple membrane of *Halobacterium salinarum*, was described in Comment A3.6. Absorption of green light, with a maximum wavelength of 569 nm, initiates a photocycle of observable color changes in the retinal (Fig. E1.16). The photocycle was established by careful measurements of absorbance change kinetics following photoactivation in wild-type bacteriorhodopsin and various bacteriorhodopsin mutants (e.g., see Fig. E1.6). By using a small number of reasonable assumptions, it was possible to calculate the absorption spectra of the different states making up the photocycle (Fig. E1.17).

The analysis led to the discovery of two "M" states in the photocycle with similar absorbance curves but different structures. Interpretation models were proposed in which

Fig. E1.17 Absorbance (top) and absorbance differences (bottom) of the different states in the bacteriorhodopsin photocycle. (Váró and Lanyi, 1991.) (Figure reproduced with permission from the American Chemical Society.)

COMMENT E1.4 LIGHT AND BACTERIORHODOPSIN (SEE ALSO COMMENT A3.6)

In absence of light, bacteriorhodopsin is in the *dark-adapted* state, and the retinal configuration is a mixture of all-*trans* and 13-*cis*. Upon illumination, all-*trans* retinal accumulates in the *light-adapted* state, and the photocycle is activated from that state. Retinal remains bound to bacteriorhodopsin via a Schiff base linkage during the entire photocycle. The orientation and protonation state of the Schiff base vary, however, in the different intermediates. Different models persist for the primary reaction and the electronic origin of the J_{600} intermediate formed about 500 fs after absorption of a photon by the retinal. The K_{590} intermediate, which can be trapped below 150 K and in which retinal is in the 13-*cis* configuration, is formed 5 ps later. After a lifetime of 1 μs, K leads to L_{550}, which decays to M_{410} after 100 μs. During the L to M transition D85 becomes protonated from the Schiff base and a proton is released on the EC side.

The Schiff base is deprotonated only in the blue-shifted M_{410} state of the photocycle, which is also the longest-lived intermediate with a lifetime in the order of milliseconds. The switch or valve function of the proton pump (see Chapter A3) occurs during the M state, in which the accessibility of the Schiff base changes from the EC to the CP side (Fig. E1.16), associated with a large protein conformational change. The Schiff base is reprotonated between M and N from D96, which is itself reprotonated from the CP side. The retinal isomerizes back to all-*trans* between N_{560} and O_{640}, which decays in milliseconds to the ground state with the reprotonation of D85. (Haupts *et al.*, 1997, 1999.)

the M states constitute the "switch" (Comment E1.4), by relating absorbance in the visible with other results mainly from IR spectroscopy and X-ray crystallography on different states trapped at low temperatures.

E1.3.4 UV Absorption Spectra of Nucleic Acids

The electronic transitions of the sugar and phosphate moieties of a nucleotide are well beyond 200 nm in the UV and the aromatic bases dominate nucleic acid absorption at more easily observable wavelengths. Chemical derivatives of purine and pyrimidine bases exhibit strong absorption in the 260–290 nm spectral range, through the electronic states of their conjugated double-bond structures (see Chapter A2 for nucleotide chemical structures). The purine and pyrimidine bases in aqueous solution have distinct absorption spectra, which vary characteristically with pH (Fig. E1.18, Table E1.2).

Despite the differences between the spectra of the individual bases, the maxima all occur close to 260 nm, and the

TABLE E1.2 PH DEPENDENCE OF PURINE AND PYRIMIDINE ABSORBANCE MAXIMA[a]

Base	pH	λ_{max} (nm)
Adenine	1	262.5
Adenine	7	260.5
Adenine	12	269
Cytosine	1	276
Cytosine	7	267
Cytosine	14	282
Guanine	1	248, 276
Guanine	7	246, 276
Guanine	11	274
Thymine	4	264.5
Thymine	7	264.5
Thymine	12	291
Uracil	4	259.5
Uracil	7	259.5
Uracil	12	284

[a] Data taken from Sober (1997).

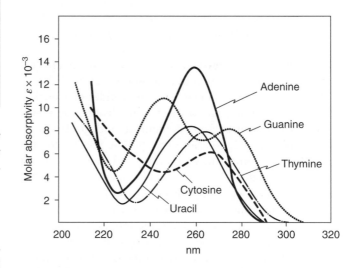

Fig. E1.18 UV absorption curves of purine and pyrimidine bases at pH 7. (Davidson and Secrest, 1972.) (Figure reproduced with permission from Elsevier.)

minima close to 230 nm. This leads to UV absorbance curves of nucleic acids that are similarly smooth, with a peak at 260 nm and a minimum close to 230 nm, from which it is not possible to extract base composition information.

The molar extinction coefficients of nucleotides at 260 nm are of the order of $10^4 \, M^{-1} \, cm^{-1}$. The UV absorption spectrum of nucleic acids can be measured with

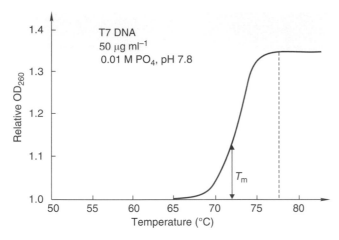

Fig. E1.19 Increase in the absorbance at 260 nm of T7 DNA as it melts with temperature. T_m is the melting temperature. (After Cantor and Schimmel, 1980.)

concentrations as low as a few micrograms per milliliter. It provides a sensitive microdetection method for a variety of applications, ranging from chromatography to AUC (see Chapter D4). Nucleic acid extinction values are much larger than those of amino acid side-chains close to 280 nm, except for Trp ($5600\,M^{-1}\,cm^{-1}$) and Tyr ($1460\,M^{-1}\,cm^{-1}$) (see above). There are relatively few aromatic residues in a protein, however, while in any nucleic acid all the nucleotides display strong UV absorption, so that the UV absorption spectrum of a nucleic acid–protein complex is strongly dominated by the nucleic acid component. The absorbance ratio at 280 nm/260 nm in UV spectrophotometry is, therefore, a sensitive test for nucleic acid or nucleotide (e.g., many enzyme cofactors are nucleotide derivatives) contamination in a protein solution.

The extinction coefficient (per nucleotide) of a nucleic acid is smaller than that of the free nucleotides so that it increases significantly upon denaturation, degradation, or hydrolysis (Fig. E1.19). Note the similar spectra for the denatured (with no tertiary structure) and enzymatic digest (essentially made up of mononucleotides) samples. The decrease in extinction coefficient (*hypochromicity*, from the Greek meaning under-color) upon tertiary structure formation is due to strong electronic interactions between the stacked bases. The increase in extinction coefficient is called the *hyperchromic effect* (from the Greek, over-color). Hypo- and hyperchromicity are useful properties that allow nucleic acid stability to be easily assessed by measuring melting temperature as a function of temperature by UV spectrophotometry (see Chapter C2).

E1.4 IR ABSORPTION SPECTROSCOPY

The energy of IR radiation corresponds to the small energy differences between vibrational and rotational states in molecules (see Fig. E1.2). As we discussed above, radiation absorption by a molecule is related to electric dipole changes, so that in the case of IR radiation these changes must be associated with vibrational or rotational motions.

E1.4.1 IR Spectrometers

The IR region of the electromagnetic spectrum encompasses the wavelength range from 0.78 to 1000 μm. In IR spectroscopy, the frequency of a band is usually expressed in terms of wave numbers in units of cm^{-1} (wave number is the inverse of wavelength; recall that velocity = frequency × wavelength, so that wave number = frequency/velocity). The wave number range of IR radiation is from about 12 800 to 10 cm^{-1}. From considerations of the applications as well as the instrumentation, the IR spectrum is conveniently divided into near-, mid-, and far-IR regions; the rough limits of each are shown in Table E1.3. The applications and methodology associated with each of the three IR spectral regions are considerably different. Measurements in the *near*-IR region are made with a spectrophotometer similar in design and components to the instruments described for UV–visible spectrometry (see above).

Until the early 1980s, instruments for the *mid*-IR region were largely of the dispersive type, based on diffraction gratings. Since that time, however, there has been a dramatic change in mid-IR instrumentation. Current instruments are of a Fourier transform (see Chapter A3) type and the method is called Fourier transform IR (FTIR) spectroscopy.

In the past, the *far*-IR region of the spectrum has had limited use because of significant experimental difficulties. The few available sources of this type of radiation are notoriously weak and are further attenuated by the need for order-sorting filters that must be used to prevent radiation of higher grating orders from reaching the detector. FTIR spectrometers, with their much higher throughput, have largely overcome this problem and made the far-IR spectral region much more accessible to chemists and biologists.

FTIR Spectrometers

All IR spectrometers include a broad-band source (usually an incandescent ceramic material), optics for collimating the IR light and directing it through a sample, an IR detector, and some means for analyzing the wavelength dependence of transmitted light. In conventional *dispersive* spectrometers, this last function is accomplished by spatially spreading the light with a diffraction grating, and selecting individual spectral elements with a narrow slit. Only a narrow wavelength range reaches the sample and detector at any time. The collection of a spectrum covering a broad wavelength range requires sequential scanning over different spectral elements.

TABLE E1.3 IR SPECTRAL REGIONS

Region	Wavelength range (μm)	Wave number range (cm^{-1})	Frequency range (Hz)
Near	0.78–2.5	12 800–4000	$3.8 \times 10^{14} - 1.2 \times 10^{12}$
Middle	2.5–50	4000–200	$1.2 \times 10^{14} - 6.0 \times 10^{12}$
Far	50–1000	200–10	$6.0 \times 10^{12} - 3.0 \times 10^{11}$
Most used	2.5–15	4000–670	$1.2 \times 10^{14} - 2.0 \times 10^{13}$

Attenuated Total Reflection

IR transmission measurements of biological molecules are often impeded by the intense absorption of water near 1650 and 3300 cm^{-1} bands. For this reason most transmission measurements must be made on very thin samples (10 μm) in D_2O, or under partially dehydrated conditions. Attenuated total reflection (ATR) avoids many problems associated with transmission measurements by limiting the effective sample thickness to a thin layer near the surface of the internal reflection element (IRE). At each point of internal reflection there is an evanescent electromagnetic wave that penetrates through the surface into the medium of lower refractive index (see Section F3.2.3 for details). In the case of a 3 mm thick by 50 mm long IRE crystal, it is routine to obtain 10–20 reflections. The actual penetration depth is described by Comment F3.1. For the case of crystalline germanium surrounded by an aqueous medium, this equation gives a penetration depth of ~5 μm at 1000 cm^{-1}. Thus, ATR measurements solve the problem of thickness of the sample and create appropriate conditions for the transmission measurement of biological samples. These conditions are very suitable for biomembrane studies: A multilayer membrane sample can be kept at the interface between the internal reflection element and well-defined buffer solution without the bulk water contributing significantly to the IR absorption spectrum.

A more efficient approach is to detect the signals from different wavelengths simultaneously. In the case of visible and near-UV dispersive spectrometers, this is accomplished by using multichannel area detectors. Such detectors, however, are not available for the IR spectral region (400–4000 cm^{-1}). The alternative approach employed in FTIR spectrometers is to pass the broadband IR beam through an interferometer before sending it through the sample and then to a single very sensitive detector (Fig. E1.20). In this case, the intensity of each wavelength component varies as a cosine function of the optical path length difference, ΔL, from a beam splitter to fixed and moving mirrors; the frequency of each cosine function is the reciprocal of the associated IR wavelength (i.e., the wave number $\bar{v}$). The single-beam spectrum (the detected IR intensity as a function of $\bar{v}$) is obtained from a Fourier transform of the *interferogram*, which is the detected intensity as a function of ΔL. Most modern IR spectrometers are Fourier transform instruments.

E1.4.2 Molecular Vibrations

The relative positions of atoms in a molecule fluctuate continuously as a consequence of a multitude of different types of vibrations and rotations about their bonds (see Section A3.2.2). For a simple diatomic or triatomic molecule, it is easy to define the number and nature of such excitations and to relate these to the energies of absorbed radiation. Such an analysis becomes difficult if not impossible for molecules made up of more atoms. Not only do they contain a large number of vibrating centers, but also interactions can occur among them and must be taken into account.

Vibrations fall into one of two basic categories: *stretching* and *bending vibrations*. A stretching vibration involves a continuous change in interatomic distance along the axis of the bond between two atoms. Bending vibrations are characterized by a change in the angle between two bonds and are of four types: *scissoring*, *rocking*, *wagging*, and *twisting* (see Fig. E1.21). All of the vibration types shown in Fig. E1.21 are possible in a molecule containing more than two atoms. In addition, interaction or coupling between vibrations can occur if they involve a bond to a single central atom. Coupling results in a change in the characteristics of the vibrations involved.

Vibration and Polarized Light

The various types of vibration of a hypothetical nonlinear chemical group made up of three nuclei are illustrated in Fig. E1.22. In-plane stretching can be either *in-phase* or *out-of-phase*. There are also in-phase and out-of-phase in-plane bending modes. Finally, out-of-plane bending can be in-phase or out-of-phase. We recall that electromagnetic radiation interacts with vibrating dipoles in the molecule. In order for the oscillations of the IR light to excite the in-phase stretching or in-plane bending modes, the beam needs to be *polarized along* the symmetry axis of the group. For the out-of-phase stretching the light needs to be polarized in-plane and across the symmetry axis. For the in-phase out-of-plane bending mode, the light needs to be polarized perpendicularly to the plane of the group. Out-of-phase and out-of-plane bending modes do not have dipole motions that can be driven by radiation absorption (see Section E1.4.3), and thus will not be excited.

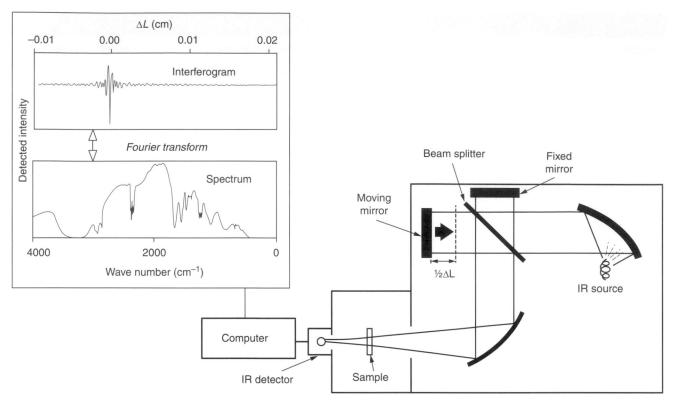

Fig. E1.20 Simplified layout of an FTIR spectrometer. Steady illumination from a broad-band IR source is modulated by a Michelson interferometer, i.e., a beam splitter that divides the collimated light along different paths to two mirrors. The light reflected from the fixed and moving mirrors is recombined at the beam splitter and is directed through the sample and into the IR detector. The path length difference is measured very precisely with a He–Ne laser directed into the Michelson interferometer (not shown). A computer stores the digitized interferogram and converts it via Fourier transformation to an IR spectrum. (Braiman and Rothschild, 1988.)

E1.4.3 IR-Active and IR-Inactive Modes

A permanent dipole moment is associated with a *heteronuclear* diatomic molecule (or its chemical bond), because the *electronegative* values of the bonded atoms are different so that there is a separation between the centers of negative and positive charge. If the bond vibrates with the normal mode frequency, ν_m (see Chapter A3), then the dipole moment oscillates at the same frequency. According to classical electromagnetic theory, an oscillating dipole absorbs electromagnetic radiation of the same frequency as that of its oscillation. A normal mode of vibration that gives rise to an oscillating dipole in the IR spectral range is said to be IR-*active*. Conversely, a normal mode of vibration that does not give rise to an oscillating dipole moment (e.g., the stretching of a *homonuclear* diatomic molecule) cannot lead to IR absorption and is said to be IR-*inactive*.

As an example, we consider the carbon dioxide molecule (O=C=O), which has four normal modes of vibration (see Section E1.4.5). Symmetry considerations dictate that two of these are bond-stretching modes, and two are valence angle-bending modes. One stretching mode is symmetrical, involving the simultaneous extension or the simultaneous contraction of each oxygen–carbon double bond. This mode is IR-inactive, because no molecular dipole moment change accompanies the in-phase displacements of the two oxygen atoms with respect to the central carbon. The other stretching mode is antisymmetrical; the carbon–oxygen bonds vibrate out of phase, which is essentially equivalent to linearly displacing the central carbon atom between two fixed oxygen atoms. This mode is IR-active, because it is associated with a large dipole moment fluctuation.

E1.4.4 Quantum Mechanical Treatment of Vibrations

The Harmonic Approximation

Molecular vibrations can be described in terms of quantum mechanical simple harmonic oscillators and normal modes (see Chapter A3). We recall that in contrast to classical mechanics, in which vibrators can assume any potential energy, quantum vibrators can take only certain discrete energy values, E. In the case of two masses, m_1 and m_2, connected by a Hooke's law spring, E is given by Eq. (E1.4) (see also Eq. (A3.43)):

(a)

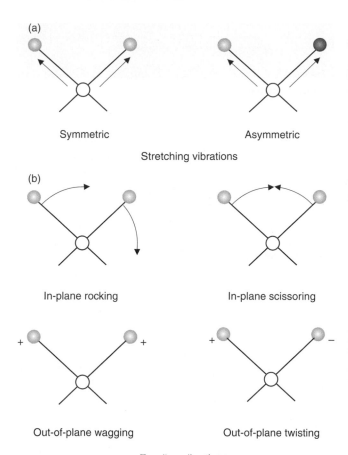

Symmetric   Asymmetric

Stretching vibrations

(b)

In-plane rocking   In-plane scissoring

Out-of-plane wagging   Out-of-plane twisting

Bending vibrations

Fig. E1.21 Types of molecular vibration. Note that + indicates motion out of the page toward the reader; − indicates motion away from the reader.

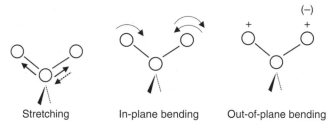

Stretching   In-plane bending   Out-of-plane bending

Fig. E1.22 In-phase and out-of-phase vibrations for three non-linear atoms.

$$E = \left(n + \frac{1}{2}\right) \frac{h}{2\pi} \sqrt{\frac{k}{m_{1,2}}} \tag{E1.4}$$

where n is the vibrational quantum number, which can take only positive integer values including zero, h is Planck's constant, $m_{1,2} = m_1 \, m_2/(m_1 + m_2)$, and k is the spring force constant (Comment E1.5).

Transitions between vibrational energy levels can be brought about by the absorption of radiation only if the energy of the radiation exactly matches the difference in energy between levels and certain *selection rules* are obeyed

---

COMMENT E1.5 THE FORCE CONSTANTS OF CHEMICAL BONDS

IR absorption measurements allow the evaluation of the force constants for various types of chemical bond by using Eq. (E1.4). Generally, k has been found to lie in the range between 3×10^2 and $8 \times 10^2 \, \text{N m}^{-1}$ for most single bonds, with $5 \times 10^2 \, \text{N m}^{-1}$ serving as a reasonable average value. Double and triple bonds have associated force constants of about two and three times this value (1×10^3 and $1.5 \times 10^3 \, \text{N m}^{-1}$, respectively). By inserting these average experimental values, Eq. (E1.6) can be used to estimate the wave number of the fundamental absorption peak.

COMMENT E1.6 SELECTION RULES

Quantum theory indicates that transitions can take place only between levels for which the vibrational quantum number differs by unity; i.e., there is a *selection rule* that $\Delta n = \pm 1$. Since the vibrational levels are equally spaced (Eq. (E1.4)), only a single absorption peak is observed for a given molecular vibration.

(Comment E1.6). The energy difference, ΔE, between the vibrational quantum states is given by

$$\Delta E = h v_m \frac{h}{2\pi} \sqrt{\frac{k}{m_{1,2}}} \tag{E1.5}$$

where v_m can be seen as a fundamental frequency associated with the set of vibrational modes. In IR spectroscopy, the frequency of a band is usually expressed in terms of wave numbers

$$\bar{v} = \frac{1}{2\pi c} \sqrt{\frac{k}{m_{1,2}}} = 5.3 \times 10^{-12} \sqrt{\frac{k}{m_{1,2}}} \tag{E1.6}$$

where $\bar{v}$ is the wave number of an absorption peak in cm^{-1}, k is the force constant for the bond in newtons per meter, c is velocity of light in centimeters per second, and the reduced mass, $m_{1,2}$, has units of kilogram (Comment E1.5).

Anharmonicity

Even from a qualitative standpoint, the description of a bond between two atoms in terms of a harmonic spring cannot be fully satisfactory. As the two atoms approach one another, for example, Coulombic repulsion between the two nuclei produces a force that acts in the same direction as the bond restoring force; thus the potential energy can be expected to rise more rapidly than predicted by the harmonic approximation. The corresponding potential energy curve has a shape similar to the one in Fig. A3.24. Such curves depart from harmonic behavior to varying degrees, depending upon the nature of the bond and the atoms involved. Anharmonicity leads to ΔE not being constant but becoming smaller at higher energies (curve 2 in Fig. A3.24). The selection rule is no longer rigorously followed; as a result, transitions of $\Delta n = \pm 2$ or $\Delta n = \pm 3$ are observed. Such transitions are responsible for the appearance of *overtone lines*, frequencies approximately two or three times that of the fundamental line. Overtone absorption lines are usually of low intensity, however, and may not be observed.

E1.4.5 Vibrational Modes of Polyatomic Molecules

Molecular normal modes of vibration are discussed in Chapter A3. It has been shown rigorously that only the normal modes of vibration of a molecule are capable of the interaction with electromagnetic radiation that gives rise to IR absorption. In the case of biological macromolecules, most of the $3N-6$ normal modes are highly localized – meaning that, to a good approximation, each involves only a small group of atoms moving as if they were isolated from the vibrations of other molecular groups. For example, if the N atoms of the protein are distributed among K sets of identical groups each of n atoms, and if each such group exhibits the same or nearly the same conformation and environment in the macromolecule, then the collection of $3nk-6$ hypothetical spectral bands is reduced to $3n-6$ (see also Section A3.2.2). Consider the case of the amide I mode of an α-helical polypeptide chain of k residues. To a first approximation, each peptide–CONH group in the α-helix will have the same vibrational frequency, resulting in the appearance of a single amide I band, rather than k distinct bands.

The vibrational frequencies of polyatomic molecules are determined, in principle, by the masses and geometrical arrangement of their constituent atoms and the interatomic forces resulting from the distortion of the equilibrium configuration. The frequencies and atomic displacements can be calculated by using empirical, classical mechanics, potential energy functions (see Chapter A3) incorporating simplifying assumptions according to the complexity of the molecule and its environment. Fairly rigorous analysis is now possible for molecules containing 10–20 atoms. Normal mode calculations (by definition in the harmonic approximation) of much larger biological macromolecules are also possible (see Chapter A3).

Amide Bands of Polypeptides and Proteins

The IR spectra of polypeptides and proteins exhibit several relatively strong absorption bands, which are approximately constant in frequency and intensity from one sample to another. These bands are associated with the vibrations of the CONH group, the structural unit common to these molecules.

Characteristic bands of the CONH group of polypeptide chains are similar to absorption bands exhibited by

TABLE E1.4 FREQUENCY RANGE FOR CHARACTERISTIC ABSORPTION BANDS OF SECONDARY AMIDES IN THE CRYSTALLINE STATE (BASED ON MODEL COMPOUNDS)	
In-plane modes	
Amide A	~3300 cm^{-1}
Amide B	~3100 cm^{-1}
Amide I	1597–1672 cm^{-1}
Amide II	1480–1575 cm^{-1}
Amide III	1229–1301 cm^{-1}
Amide IV	625–767 cm^{-1}
Out-of-plane modes	
Amide V	640–800 cm^{-1}
Amide VI	537–606 cm^{-1}
Amide VII	~200 cm^{-1}

secondary amides, in general. This has led to the commonly accepted procedure of naming the prominent absorption bands in proteins and polypeptides the "amide I band," "amide II band," etc. There is a total of nine such bands, usually called amide A, amide B, and amide I–VII, in order of decreasing frequency. An isolated planar CONH group would give rise to five in-plane (C=O stretching, C–H stretching, N–H stretching, OCN bending, CNH bending) modes and one out-of-plane (C–N torsion) normal mode. The frequency ranges of the different amide bands in model compounds are given in Table E1.4. Such bands are shown in Fig. E1.23 for N-methylacetamide, a model for the trans peptide group in proteins, which has 30 normal modes of vibration (Fig. E1.24). Neglecting the six methyl hydrogen atoms, there are 12 normal modes arising from the remaining six skeletal atoms (CCONHC).

There are four main characteristic IR bands from secondary amides such as N-methylacetamide (Fig. E1.25):

(1) In dilute solution (no hydrogen bonding) a sharp band observed at 3400–3460 cm^{-1} (amide A) is easily assigned to an essentially pure N–H stretching motion by comparison with related molecules and by shifts caused by deuterium labeling of the H atom. Upon hydrogen bonding the band shifts to 3120–3320 cm^{-1} and another, usually weaker, band appears close to 3100 cm^{-1} (amide B). The relatively high frequency of the N–H stretching vibration (3100–3400 cm^{-1} range) isolates this type of motion from other molecular vibrations in N-methylacetamide (Fig. E1.23).

(2) All secondary amides, polypeptides, and proteins show a strong band in the 1600–1700 cm^{-1} region (amide I), which shifts to higher frequencies in dilute solution. By simple qualitative considerations, based on comparison with related molecules, and by the frequency shift upon

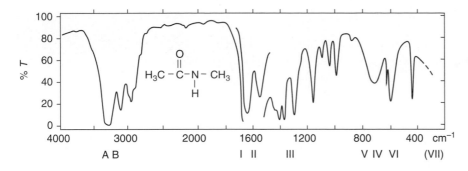

Fig. E1.23 Characteristic amide bands as exhibited by a capillary film of N-methylacetamide. The amide VII band is not shown.

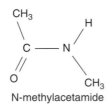

N-methylacetamide

Fig. E1.24 N-methylacetamide is an analogue for the *trans* peptide group in proteins.

dilution (related to the breaking of hydrogen bonds), the band has been assigned to a mode essentially involving the stretching vibration of the C=O bond.

(3) The amide II band absorbs between 1510 and 1570 cm^{-1}. Unlike the bands associated primarily with C=O stretching and N–H stretching, the amide II band is not easily interpreted by qualitative considerations.

(4) Bands in the amide III spectral region (1350–1570 cm^{-1}) are predominantly due to the in-phase combination of N–H in-plane bending and C–N stretching vibrations and are highly sensitive to the secondary structure.

E1.4.6 Resolution Enhanced FTIR Spectra

The analysis of IR spectra in terms of protein structure is not straightforward and presents serious conceptual and practical problems, despite the well-recognized conformational sensitivity of the IR-active bonds. Bands in both the amide I and amide III regions are broad, not resolved into individual components corresponding to different secondary structure elements.

Resolution enhancement or band-narrowing methods are applied to resolve broad overlapped bands into individual bands. FTIR spectroscopy presents several advantages over conventional dispersive techniques for this type of analysis through the application of *second derivative* and *Fourier self-deconvolution* techniques, including higher resolution, sensitivity, signal-to-noise ratio, and frequency accuracy.

Second Derivative Spectra
Because FTIR spectra are digitally encoded with one data point n every $\Delta k/(2^m) = \Delta W$ frequency units, where Δk is the

nominal instrumental resolution (in cm^{-1}) and n is the number of times the interferogram is zero-filled prior to Fourier transformation, the second derivative of the spectrum may be calculated by a straightforward analytical method. In particular, at data point n the value of the second derivative A_n'' (in units of absorbance /(wave number)^2) is given by

$$A_n'' = (A_{n+1} - 2A_n + A_{n-1})/(\Delta W)^2 \tag{E1.7}$$

where A_n is the absorbance at data point n in the measured spectrum. (Note that this function gives the second derivative spectrum without any smoothing.)

The intrinsic shape of a single IR absorption line of an isolated molecule may be approximated by a Lorentzian function (the Fourier transform of an exponential decay function, see Chapter A3)

$$A = \frac{\sigma}{\pi\left(\sigma^2 + k'^2\right)} \tag{E1.8}$$

$$= \frac{\sigma}{\sigma\pi\left(1 + Bk'^2\right)} \tag{E1.8a}$$

where A is the absorbance, 2σ is the width at half height, $k' = k - k_0$ is the frequency referred to the band center at k_0, and $B = 1/\sigma^2$. The second derivative of Eq. (E1.8a) is

$$A'' = \frac{-2B\left(1 - 3Bk'^2\right)}{\sigma\pi\left(1 + Bk'^2\right)^3} \tag{E1.9}$$

The peak frequency for the second derivative is identical to the frequency of the original band center, k_0. The half-width of the second derivative, σ^{II}, is related to the half-width of the original line by

$$\sigma^{II} = \sigma/2.7 \tag{E1.10}$$

and the peak intensity of the second derivative is

$$A_o'' = -2A_o/\sigma^2 \tag{E1.11}$$

Thus, the peak height of the second derivative is proportional to the original peak height (with opposite sign) and inversely proportional to the square of the original half-width.

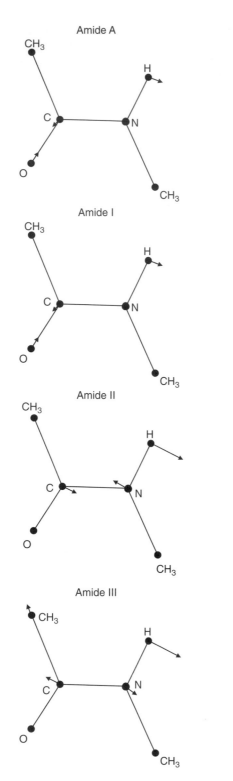

Fig. E1.25 Atomic displacement vectors for the amide A, I, II, and III modes as calculated for N-methylacetamide.

The main advantage of using second derivative spectra lies in the ease with which the peak frequencies of unresolved components can be identified. Measured and second derivative spectra of bovine α-chymotrypsin in D_2O solution are shown in Fig. E1.26. The strong amide I band,

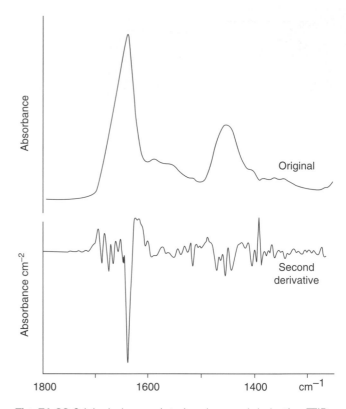

Fig. E1.26 Original, deconvoluted and second derivative FTIR spectrum of bovine α-chymotrypsin in D_2O solution from 1250 to 1800 cm^{-1}. (Susi and Byler, 1986.)

which is unresolved in the original spectrum, is resolved into seven components by the second derivative technique. The interpretation of second derivative spectra is more complex in the case of overlapping bands that deviate from a Lorentzian shape. The above formulae, nevertheless, still provide a reasonable approximation in such cases.

Fitting Deconvoluted Spectra

Observed IR spectra result from the *convolution* (see Chapter A3) of the sample absorption spectrum with an instrumental resolution function (Comment E1.7). An important analytical approach in IR spectroscopy involves curve fitting of deconvoluted amide bands by iterative adjustment of component band positions, heights, and relative weights (fraction of the total peak area). Component bands are assigned to different types of secondary structure (α-helices, β-sheets, turns, and irregular loops), and their fractional areas to the fractions of the corresponding secondary structure in the sample protein.

Figure E1.27 demonstrates that the amide I spectra of α-chymotrypsin can be fitted by components with easily measurable areas. Authors call the process deconvolution, but in actual fact it is a decomposition into component spectra (see Section C2.4.6). The bands close to 1627, 1637, and 1674 cm^{-1} are associated with β-strands, the band near 1653 cm^{-1} with the α-helix, and the bands close to 1687, 1681, and 1665 cm^{-1} with turns.

The usefulness of FTIR spectroscopy as a tool for protein denaturation studies is illustrated in Fig. E1.28b. Treatment of α-chymotrypsin with guanidine hydrochloride induced dramatic changes in the spectrum. The bands above 1300 cm^{-1} (most likely corresponding to α-helix) moved to lower wave numbers and their intensities decreased, whereas the intensity of bands between 1248 and 1270 cm^{-1} (most likely corresponding to random coils) increased sharply. The band at 1248–1249 cm^{-1} increased in strength from 20% to 38%. This indicates that this band represented a borderline assignment between a coil and a β-sheet. The native protein bands at 1290 and 1279 cm^{-1} decreased from 20% to 11% after treatment with 6 M guanidine hydrochloride and the 1279 band shifted to 1282 cm^{-1}. The spectral range was, therefore, tentatively assigned to a β-turn.

The amide III spectra of the same protein is shown in Fig. E1.28a. The spectral analysis revealed that the 1323, 1311, and 1304 cm^{-1} bands are associated with the α-helix, the bands near 1249, 1234, and 1222 cm^{-1} with β-sheets, and the 1289, 1278, and 1260 cm^{-1} bands with other structures in native α-chymotrypsin.

E1.4.7 From Amide Bands to Protein Secondary Structure

Amide I, Amide II, and Amide III

The amide I mode (1600–1700 cm^{-1}) of the peptide group is the most widely used band in studies of protein secondary

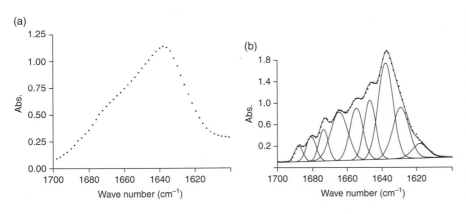

(a)

(b)

Fig. E1.27 Amide I bands of α-chymotrypsin measured at 5% w/v in D$_2$O, path length 0.0075 cm: **(a)** digitized original FTIR spectrum (dots); **(b)** deconvoluted spectrum (dots), and individual Gaussian components (solid lines) (deconvolution constants: σ = 6.5 cm^{-1}, K = 2.4). The solid overall curve is obtained by summation of the components (Susi and Byler, 1986.)

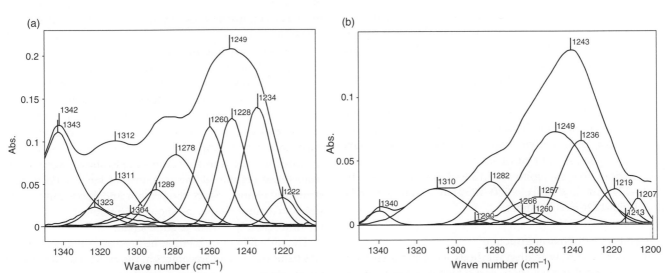

Fig. E1.28 Curve fitting analysis of FTIR spectra of α-chymotrypsin in the amide III region: **(a)** native state, concentration 1 mg ml^{-1} in H$_2$O buffer; **(b)** treated with 6 M guanidine hydrochloride. Deconvolution constants: σ = 20 cm^{-1}, K = 2.0. (After Cai and Singh, 1999.)

COMMENT E1.8 WATER ABSORPTION

Unfortunately, water absorbs IR light just where the most interesting groups found in biological macromolecules also absorb. To some extent, this difficulty can be overcome by using dried films of the macromolecule or by working in D_2O solution, in which solvent absorption is shifted to less interesting regions around 1200 and 2500 cm^{-1}. The absorption of water can also be brought within tolerable limits with the high accuracy of modern FTIR instruments by using cells of 6 μm path length.

COMMENT E1.9 EVALUATING THE AMIDE I BAND

Readers can find a critical review of the methods evaluating the amide I band in the literature (Surewicz et al., 1993).

structure. Due to the strong absorption of water (H_2O) between 1640–1650 cm^{-1}, most structure determinations by amide I are performed in D_2O solution (Comment E1.8). However, uncertainty in the NH/ND exchange process may cause a certain degree of ambiguity. Also the serious overlap of the random coil and the α-helix band in the amide I region makes it difficult to accurately predict the α-helix contents in a protein (Comment E1.9).

Even though the intensity in the amide II region (1510 and 1570 cm^{-1}) is relatively strong, it is not very sensitive to secondary structure changes in proteins. Furthermore, the amide II bands strongly overlap with bands originating from amino acid side-chain vibrations.

Bands in the amide III spectral region (1350–1200 cm^{-1}) are predominantly due to the in-phase combination of N–H in-plane bending and C–N stretching vibrations and are highly sensitive to the secondary structure. In contrast to amide I band, there is no H_2O interference in this region. And even though the signals of these bands are ~(5–10)-fold weaker than those of amide I, they are easily resolved and better defined, making the amide III spectral region quite suitable for quantitative analysis of protein secondary structure. The assignment of spectral bands in the amide III region is as follows: 1330–1295 cm^{-1}, α-helix; 1295–1270 cm^{-1}, β-turns; 1270–1250 cm^{-1}, random coils; and 1250–1220 cm^{-1}, β-sheets.

Comparison with Known Structure

A number of closely related methods were proposed that avoid spectral deconvolution, and which are based on the use of a calibration matrix of IR spectra of proteins of known (i.e., determined by X-ray crystallography or NMR) secondary structure. This type of data treatment is conceptually very similar to that used in the analysis of CD spectra of proteins (Chapter E4).

Table E1.5 compares secondary structure estimates from FTIR spectroscopy using the amide I and amide III bands with the analysis from CD spectra and from X-ray crystallography. The agreement between the FTIR results and the *structural* values derived from X-ray crystallography is quite good (Comment E1.10).

Finally we emphasize that the methodologies currently used for quantitative estimation of protein secondary structure content from IR spectra are not without serious shortcomings. Limitations in the quantitative assessment of protein secondary structure appear to be common to all low-resolution spectroscopic techniques, including CD. Since the potential sources of error in the CD and spectroscopic methods are different, these two techniques are highly complementary and should be used in conjunction.

E1.4.8 IR Difference Spectroscopy

IR spectra are very sensitive to structural alterations as small as those involving a single atomic coordinate. A change in hydrogen bonding distance of only 0.002 Å, for example, shifts the frequency of the amide A (N–H stretch) vibration of a peptide group by approximately 1 cm^{-1}, well within the detection limits of most FTIR spectrometers. In contrast, X-ray crystallography studies of proteins rarely have a resolution below 1 Å. However, it is very difficult, in practice, to detect small, localized structural changes in a biological macromolecule by IR spectroscopy. All groups in the molecule essentially have IR-active vibrations and, therefore, tend to give rise to a multitude of overlapping spectral bands. For example, bacteriorhodopsin, a 248-amino-acid (26 kDa) membrane protein containing an all-*trans*-retinal prosthetic group (see Chapter A3), is made up of about 1900 heavy (non-hydrogen) atoms that display about 5700 different skeletal vibrational modes.

A simple way around this problem is based on the principle of difference spectroscopy. Figure E1.29 illustrates bacteriorhodopsin undergoing a structural change from state A to state B. The IR spectrum of the protein in state A (top) consists of the sum of contributions from all of the skeletal vibrations. In contrast, the absorbance difference spectrum, computed by subtracting the spectrum recorded in state A from that in state B, shows signals only from those regions of the protein that have undergone a structural change. Since the protein undergoes only a very small structural change as shown, the intensity of peaks in the difference spectrum is quite small.

The retinal binding membrane proteins, Channelrhodopsins (ChRs), are photoreceptors in the eyespot of unicellular green algae. They initiate phototactic and photophobic responses, which induce the cells to swim toward optimum light conditions. When heterologously expressed in animal cells, ChRs act as light-gated cation channels and are hence

TABLE E1.5 COMPARISON OF SECONDARY STRUCTURE DETERMINATIONS (%) BY FTIR, USING THE AMIDE I AND AMIDE III SPECTRAL REGIONS, FROM THE ANALYSIS OF CD SPECTRA AND FROM X-RAY CRYSTALLOGRAPHY DATA

Protein	α-helix	β-sheet	β-turn	Random coil	Method
α-chymotrypsin	15	47	20	18	Amide III
	8	50	27	15	X-ray
	9	47	30	14	Amide I
	8–15	10–53	2–12	38–70	CD
Concanavalin A	18	49	25	8	Amide III
	3	60	22	15	X-ray
	8	58	26	8	Amide I
	3–25	41–49	15–27	9–36	CD
Cytochrome c	43	15	26	15	Amide III
	48	10	17	25	X-ray
	42	21	25	12	Amide I
	27–46	0–9	15–28	28–41	CD
Hemoglobin	65	4	11	20	Amide III
	87	0	7	6	X-ray
	78	12	10	–	Amide I
	68–75	1–4	15–20	9–16	CD
IgG	10	60	17	13	Amide III
	3	67	18	12	X-ray
	3	64	28	5	Amide I
Lysozyme	43	16	18	23	Amide III
	45	19	23	13	X-ray
	40	19	27	14	Amide I
Myoglobin	61	11	7	21	Amide III
	85	0	8	7	X-ray
	85	0	8	7	Amide I
	67–86	0–13	0–6	11–30	CD
Ribonuclease A	21	41	25	12	Amide III
	23	46	21	10	X-ray
	15	40	36	9	Amide I
	13–30	21–44	11–22	19–50	CD
Trypsin	14	49	23	13	Amide III
	9	56	24	11	X-ray
	9	44	38	9	Amide I

Partially from Bandecar (1992).

being developed as optogenetic tools. Difference FTIR contributes powerfully to the characterization of light-driven transitions in ChRs (Comment E1.11).

Genetic engineering is a powerful approach to the assignment of IR bands from a protein. Mutants should show simple difference spectra with peaks corresponding to the changed residues. As always with such studies, however, the interpretation should be done carefully because mutations can have structural effects beyond the target amino acid residue.

E1.4.9 Time-Resolved IR Spectroscopy

Vibrational spectroscopy experiments have an extremely fast intrinsic time resolution, and, in general, changes in the state of chemical bonds can be detected in a period as short as their characteristic vibrational periods, i.e., 10–100 fs.

Photobiological Systems

Photobiological systems such as chromoproteins (from the Greek *chroma*, color) constitute excellent samples for FTIR difference spectroscopy studies, because light is an ideal reaction trigger. In the simplest approach, difference spectra are formed between an initial state before illumination and a final state after illumination. In most cases, however, the photoreaction passes through a number of intermediates.

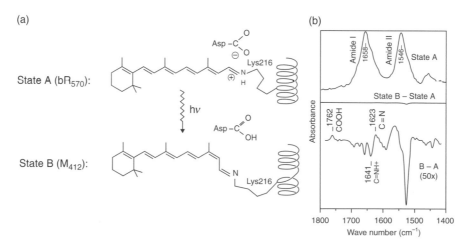

Fig. E1.29 The basic principles of FTIR difference spectroscopy applied to membrane proteins. **(a)** Simplified structure of the active site of a membrane protein (in this case, bacteriorhodopsin), in two different states involved in its physiological reaction (in this case, light-induced proton pumping. **(b)** The spectrum of the IR absorbance in state A (top), the spectrum of the difference in IR absorbance between states Λ and B (middle), and the difference spectrum on an enlarged scale (bottom). (Braiman and Rothschild, 1988.)

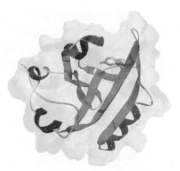

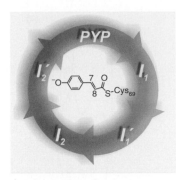

Fig. E1.30 Overall structure of the photoactive yellow protein (PYP). PYP is a relatively small protein (14 kDA) containing *para*-hydroxycinnamate (pCA) as a prosthetic group bound to Cys 69 via a thioester linkage. The secondary structure is represented as a ribbon (α-helices in red and β-sheets in green) and the molecular surface of PYP is in the background. The chromophore is shown in orange. (Adapted from Heberle and Gensch, 2001.)

Fig. E1.31 Photocycle of PYP. The chemical structure of the chromophore is depicted in the center. (After Heberle and Gensch, 2001.)

Studies of the yellow-light photoreceptor PYP (Fig. E1.30) provide excellent examples of how the IR technique can contribute to understanding the processes involved in phototransduction. Upon absorption of a photon by its covalently attached chromophore, PYP undergoes a cyclic sequence of reactions (Fig. E1.31). The ground state P (λ_{max} of 446 nm) is converted into intermediate I_1 (λ_{max} of 465 nm) in ~3 ns. I_1 is then converted into the long-lived blue-shifted intermediate I_2 (λ_{max} of 350 nm), which returns to the ground state. Global fit analysis of the

absorbance changes yields four apparent time constants, $\tau_1 = 113$ μs, $\tau_2 = 1.5$ ms, $\tau_1 = 189$ ms, $\tau_4 = 583$ ms.

A three-dimensional representation of the light-induced IR absorbance changes during the photocycle of PYP, measured simultaneously from 1900 to 1000 cm^{-1}, is shown in Fig. E1.32. The 30 ns time resolution permitted the monitoring of the light cycle reaction starting from intermediate I_1 and proceeding via the intermediate I_2 back to the ground state P.

E1.4.10 DNA Conformation

Nucleic acid vibrations arise from different parts of the macromolecule, resulting in IR absorption bands in several

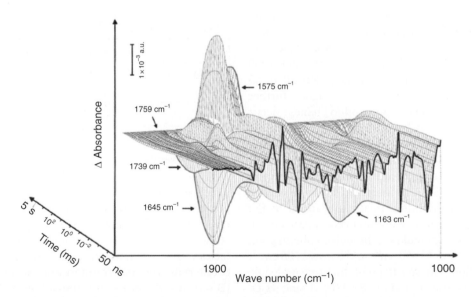

Fig. E1.32 Three-dimensional representation of the IR absorbance changes in PYP. The absorbance changes are shown as a function of wave number (1900–1000 cm^{-1}) and time on a logarithmic scale (50 ns–5 s). The first difference spectrum, recorded 50 ns after the laser flash excitation, is highlighted in black and represents an I_1–P difference spectrum. The subsequent difference spectra show the reaction from I_1 via I_2 back to the ground state P. Vibrations that characterize changes of specific functional groups during the photocycle were selected for kinetic analysis. Thus, the C=O stretching vibrations of Glu 46 were followed at 1759 cm^{-1} and 1739 cm^{-1}. The coupled C–C/C=C stretching vibration at 1575 cm^{-1} monitors the protonated *cis*-chromophore, etc. (Adapted from Brudler *et al.*, 2001.)

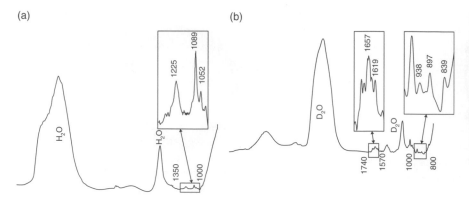

COMMENT E1.12 INFRARED SPECTRUM – STRUCTURE CORRELATIONS IN NUCLEIC ACIDS

For detailed discussions of nucleic acid spectra–structure correlations, established on the basis of known three-dimensional X-ray structures of oligonucleotide single crystals, interested readers are referred to the review of Thomas and Tsuboi (1993). Another type of information which can be obtained from the FTIR spectra concerns recognition of DNA sequences by a wide variety of molecules, such as oligonucleotides (triple-helical structures), drugs, and proteins (Taillandier and Liquier, 1992).

TABLE E1.6 MAIN IR MARKER BANDS (IN CM^{-1}) OF A-, B-, AND Z-DNA[a]

Conformation			Assignment
A	**B**	**Z**	**In-plane base double**
1705	**1715**	**1695**	**stretching**
		1433	A–T bases
1418	1425	1408	Deoxyribose
1375	1375		dA · dG anti
		1355	dA · dG syn
1335	1344		dT
1335	1328		dA
		1320	dG
1275	1281		T
		1265	G
1240	1225	1215	Antisymmetric phosphate stretching
1188			Deoxyribose
		1065	Deoxyribose

[a] Data from Taillandier and Liquier (1992).

spectral regions. The main vibrations are observed between 1800 and 700 cm^{-1}. Four domains can be distinguished in this region: absorption bands due to the stretching vibrations of double bonds in the base planes are between 1800 and 1500 cm^{-1}; between 1500 and 1250 cm^{-1} we find the bands that are particularly dependent on the glycosidic torsion angle in the base sugars; further strong sugar and phosphate absorption bands occur between 1250 and 1000 cm^{-1}; below 1000 cm^{-1}, there are bands due to the vibrations of the phosphodiester main chain coupled to vibrations of the sugar groups.

A DNA FTIR spectrum recorded in H_2O solution is presented in Fig. E1.33a. The enlarged area shows the spectral domain corresponding to the vibrations of the phosphate groups. We observe an important contribution of the solvent H_2O, which makes it difficult to obtain good data in the region around 1600 cm^{-1} and below 1000 cm^{-1}. To avoid this problem, the spectrum of the same molecule is recorded in D_2O solution (Fig. E1.33b). Using H_2O/ D_2O solvent combinations makes it possible to examine the complete spectral range necessary for nucleic acid studies.

Different types of information on DNA structure can be obtained by IR spectroscopy (Comment E1.12). For example, it is possible to characterize the molecular geometry by using what is usually called "IR marker bands,"

which are conformation sensitive. Table E1.6 lists the most prominent IR marker bands of DNA double-helical structures (A, B, and Z conformations). These bands demonstrate the existence of a conformational transition when factors such as temperature, hydration, concentration, counterion type and counterion concentration are varied (B → A, B → Z, helix → coil transitions, etc.). In the case of conformation mixtures, computer simulations based on the observed marker bands and "pure form" spectra are used to evaluate quantitatively the amount of each geometric form in the sample. FTIR spectra also contain useful information in the recognition of DNA sequences by a wide variety of molecules, such as oligonucleotides (triple-helical structures), drugs, and protein samples.

E1.5 CHECKLIST OF KEY IDEAS

- In an absorption spectroscopy experiment, a beam of electromagnetic radiation is observed after it passes through a sample to see how much of it has been absorbed at each wavelength.
- The measurement is made with a spectrophotometer.
- Radiation energy is absorbed in order to allow transitions between quantum mechanical electronic states of well-defined energies in the sample molecules. These can be seen as oscillating electric dipoles, induced by the oscillating electric field of the wave.
- Near-UV and visible radiation energies correspond to transitions between electronic orbital states, and can be used to probe the different types of bonds in molecules.
- The range of lower energies of IR radiation includes thermal energy at ambient temperature and can be used to study vibrational states and molecular dynamics.
- Absorption as a function of wavelength is characterized by a dimensionless *absorbance* or *optical density* value, or by a *molar extinction coefficient* in units of (mol liter^{-1})$^{-1}$ cm^{-1}, or M^{-1} cm^{-1}.
- UV–visible absorption spectra are usually modeled in terms of single electronic bands, without taking line shape into account. But there is significant line broadening due to lower-energy transitions between vibrational and rotational states, an inhomogeneous environment and coupling with solvent molecules.
- For a two-state reaction in which each state has a specific absorption spectrum, the kinetic plot of the absorption spectra displays an *isobestic* point, at which the different spectra cross.
- Difference spectra are very useful in the study of processes characterized by spectral changes.
- The detection sensitivity of small spectral changes can be amplified by *derivative spectroscopy*.
- The absorption of proteins in the UV spectrum arises mainly from electronic bands in the peptide group (at 170–220 nm), aromatic amino acid side-chains close to 280 nm, and prosthetic group cofactors, enzyme substrates, or inhibitors.
- Amino acid residues do not absorb in the visible spectrum, and color in a protein is always due to a prosthetic group such as the heme, which gives hemoglobin its red color.
- Protein concentration measurements are usually performed by measuring the absorbance of the solution at wavelengths close to 280 nm. Such a determination provides only a *relative* concentration value, however, unless the molar extinction coefficient of the specific protein has been determined independently.

- Protein-associated chromophores can be as simple as a single metal ion that changes its color according to its oxidation state, or a small organic molecule, such as retinal in rhodopsin, or quite complex, such as metal clusters, iron–sulfur clusters, porphyrin associated to iron in heme, or to magnesium in chlorophyll, or flavin and flavin derivative groups.
- In flash photolysis experiments on myoglobin, the Fe–CO bond is photodissociated by a laser flash and the rebinding kinetics is then followed by observing absorption changes with nanosecond time resolution.
- Upon absorption of a photon, the retinal binding membrane protein bacteriorhodopsin displays a millisecond photocycle of absorption changes associated with proton pump activity. One proton is pumped out of the cell for each photon absorbed.
- Nucleic acids can be identified by their strong absorption close to 260 nm with molar extinction coefficients of the order of 10^4 M^{-1} cm^{-1}.
- The relative positions of atoms in a molecule fluctuate continuously as a consequence of a multitude of different types of *vibrations* and *rotations* about their bonds.
- A *stretching* vibration involves changes in the length of an interatomic bond along its axis. *Bending* vibrations are characterized by a change in the angle between two bonds.
- For a simple diatomic or triatomic molecule, it is easy to define the number and nature of stretching and bending vibrations and to relate these to absorption energies; such an analysis becomes prohibitively difficult for larger molecules.
- A stretching vibration deviates from harmonic behavior by varying degrees, depending upon the nature of the bond and the atoms involved.
- In order to absorb IR radiation, a molecule must undergo a net change in dipole moment as a consequence of its vibrational or rotational motion.
- The IR spectra of polypeptides and proteins exhibit several relatively strong absorption bands associated with the CONH (amide) group, which change little in frequency and intensity from one molecule to another.
- Of all the amide modes of the peptide group, the single most widely used one in studies of protein secondary structure is the *amide I* mode (1600–1700 cm^{-1}, stretching vibration of the C=O bond); however, due to the strong absorption of water between 1640–1650 cm^{-1}, most structure determinations based on the amide I mode are performed in D_2O solution.
- The intensity of bands in the *amide II* region (1480–1575 cm^{-1}) is relatively strong, but it is not very sensitive to protein secondary structure changes; furthermore, the amide II bands are strongly overlapped by bands originating from amino acid side-chain vibrations.
- The *amide III* bands (1200–1350 cm^{-1}) are predominantly due to the in-phase combination of N–H in-plane bending and C–N stretching vibrations and highly sensitive to protein secondary structure; in contrast to the amide I band, there is no H_2O interference in this spectral region.

- Despite a well-recognized conformational sensitivity of protein IR bands, the analysis of the spectra in terms of protein structure is not straightforward and presents serious conceptual and practical problems.
- The estimation of secondary structural elements using amide I and amide III modes correlates quite well with secondary structure estimations from X-ray crystallography.
- Methods currently used to extract information on protein secondary structure from IR spectra are based on empirical correlation between the frequencies of certain vibrational modes and types of

secondary structure of polypeptide chains such as α-helix, β-sheet, β-turn, and random coil.
- IR difference spectra of different enzyme states contain bands only of those groups which undergo changes during the transition from one state to the other. Difference spectra are, therefore, significantly simpler to interpret than spectra corresponding to the enzyme states. IR difference spectra of protein mutants can be used to assign bands to particular amino acid residues.
- Time-resolved IR spectroscopy is the principal technique for the study of reaction dynamics.

Suggestions for Further Reading

Historical Review

Seibert, F. B. (1995). Infrared spectroscopy applied to biochemical and biological problems. *Meth. Enzymol.*, **246**, 501–526.

Tinoco, I., Jr. (1995). Optical spectroscopy: General principles and overview. *Meth. Enzymol.*, **246**, 13–18.

Jakson, M., Sowa, M. G., and Mantsch, H. H. (1997). Infrared spectroscopy: A new frontier in medicine. *Biophys. Chem.*, **68**, 109–125.

Chalmers, J. M., and Griffiths, P. R. (eds.) (2002). *Handbook of Vibrational Spectroscopy*, 5 volumes. Chichester: John Wiley and Sons.

Gauglitz, G., and Vo-Dinh, T. (eds.) (2003). *Handbook of Spectroscopy*, 2 volumes. Weinheim: Wiley-VCH Verlag GmbH KgaA.

Brief Theoretical Outline

Taillandier, E., Firon, M., and Liquier, J. (1991). In *Spectroscopy of Biological Macromolecules*, eds. R. E. Hester and R. B. Girling. Cambridge: Royal Society of Chemistry.

Mathies, R. (1995). Biomolecular vibrational spectroscopy. *Meth. Enzymol.*, **246**, 377–389.

Sauer, K. (1995). Why spectroscopy? Which spectroscopy? *Meth. Enzymol.*, **246**, 1–10.

Tinoco, I. Jr, Sauer, K., and Wang, J. C. (1998). Molecular structure and interactions: Physical chemistry. In *Principles and Applications in Biological Science.* Upper Saddle River, NJ: Prentice Hall.

Applications in the UV–Visible–IR Spectral Range

Miyazawa, T., and Blout, E. R. (1961). The infrared spectra of polypeptides in various conformations: Amide I and II bands. *JACS*, **83**, 712–719.

Susi, H., and Byler, D. M. (1986). Resolution-enhanced Fourier transform infrared spectroscopy of enzymes. *Meth. Enzymol.*, **130**, 290–311.

Braiman, M. S., and Rotschild, K. J. (1988). Fourier transform infrared techniques for probing membrane protein structure. *Ann. Rev. Biophys. Chem.*, **17**, 541–570.

Bandecar, J. (1992). Amide modes and protein conformation. *Biochim. Biophys. Acta*, **1120**, 123–143.

Martin, J.-L., and Vos, M. H. (1992). Femtosecond biology. *Annu. Rev. Biophys. Biomol. Struct.*, **21**, 199–222.

Taillandier, E., and Liquier, J. (1992). Infrared spectroscopy of DNA. *Meth. Enzymol.*, **211**, 307–352.

Surevicz, W. K., Mantsch, H. H., and Chapman, D. (1993). Determination of protein secondary structure by Fourier transform infrared spectroscopy: A critical assessment. *Biochemistry*, **32**, 389–394.

Arrondo, J. L. R., Muga, A., Castresana, J., and Goni, F. M. (1993). Quantitative studies of the structure of proteins in solution by Fourier-transform infrared spectroscopy. *Progr. Biophys. Mol. Biol.*, **59**, 23–56.

Mantele, W. (1993). Reaction-induced infrared difference spectroscopy for the study of protein function and reaction mechanisms. *TIBS*, **18**, 197–202.

Gevert, K. (1993). Molecular reaction mechanisms of proteins as monitored by time-resolved FTIR spectroscopy. *Curr. Opin. Struct. Biol.*, **3**, 769–773.

Miura, T., and Thomas, G. J. Jr, (1995). Optical and vibrational spectroscopic methods. In *Introduction to Biophysical Methods for Protein and Nucleic Acid Research*, eds. J. A. Glasel and M. P. Deutcher. London: Academic Press.

Heberle, J., and Gensch, T. (2001). When FT-IR spectroscopy meets X-ray crystallography. *Nature Struct. Biol.*, **8**, 195–197.

Muders, V., Kerruth, S., Lorenz-Fonfria, V. A., *et al.* (2014) Resonance Raman and FTIR spectroscopic characterization of the closed and open states of channelrhodopsin-1. *FEBS Lett.*, **588**, 2301–2306.

TWO-DIMENSIONAL IR SPECTROSCOPY

E2

The present chapter is included in Part E with the other optical spectroscopy methods; however, the development of two-dimensional infrared (2D-IR) spectroscopy is strongly based on two-dimensional NMR, and it is easier to understand after reading the relevant sections in Part J, which the reader is strongly encouraged to do first.

E2.1 HISTORICAL REVIEW AND INTRODUCTION TO BIOLOGICAL PROBLEMS

1950

O. Hann proposed coherent spectroscopy – the use of radiation fields with well-defined phase properties – to extract information about atoms and molecules. The "spin echo" experiment in nuclear magnetic resonance was the first demonstration of the possibilities of coherent spectroscopy.

1957

R. P. Feynman, **F. L. Vernon Jr.**, and **R. W. Hellwarth** published a landmark paper pointing out that if coherent light fields were ever created, it would be possible to use these same methods on optical transitions. The invention of the laser in **1960** was followed quickly by a demonstration of the "photon echo" – the optical version of the "spin echo."

1998

R. M. Hochstrasser and collaborators proposed 2D-IR spectroscopy, in analogy with two-dimensional NMR, for the determination of time-evolving structures. The spins associated with the different nuclei in NMR are replaced in the IR experiments by a network of vibrational modes whose coupling can be used to determine molecular structure and dynamics. Structures of dipeptides, tripeptides, and pentapeptides were determined by 2D-IR spectroscopy. The most exciting aspect of 2D-IR spectroscopy, however, is the combination of its sensitivity to structure with time resolution.

Conventional FTIR spectroscopy can monitor protein secondary structure features on the nanosecond–second timescale. Pulsed IR methods can probe vibrations on timescales down to the femtosecond.

E2.2 LINEAR AND MULTIDIMENSIONAL SPECTROSCOPY

Linear spectroscopy, such as visible or IR absorption, Raman scattering, and one-dimensional NMR, produces one-dimensional projections of electronic and nuclear interactions onto a single frequency (or time) axis. For simple molecules, direct information can be obtained about energy levels and absorption cross-sections. The situation is very different for complex molecules, with strongly overlapping levels. Here, the microscopic information is highly averaged, and is often totally buried under broad, featureless line shapes, whose precise interpretation often remains a mystery. Multiple pulse techniques have the capacity to induce electronic and vibrational degrees of freedom into non-equilibrium states, and to monitor their subsequent evolution, yielding ultrafast snapshots of dynamical events such as energy or charge transfer pathways, photoisomerization, and structural fluctuations.

E2.3 PRINCIPLES OF 2D-IR SPECTROSCOPY

E2.3.1 Pump Probe Experiments

Heterodyne two-pulse IR correlation spectroscopy (IR COSY) has the same measurement principles as NMR COSY (Chapter J2). In both techniques there is a preparation and a coherence transfer step (Fig. E2.1). In 2D-IR spectroscopy, the first pulse (pump) creates a vibrational coherence analogous to the spin coherence created by the first radio-frequency pulse in NMR, but unlike the $\pi/2$ pulse of NMR, this interaction is weak. After some delay, a pair of weak IR pulses (probe pulses) interrupts the initial coherence evolution to generate a free induction decay (FID) from the transferred coherence. The FID electric field is the measured signal in both techniques. When the third

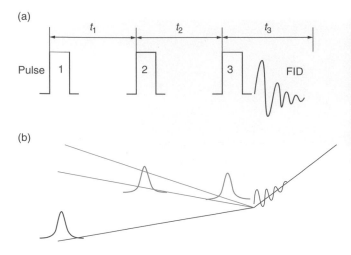

(a)

Pulse

(b)

Fig. E2.1 Typical pulse sequences used in **(a)** NMR spectroscopy and **(b)** heterodyned 2D-IR spectroscopy. The three pulses in both techniques are separated by time delays t_1 and t_2, and the FID is measured versus t_3. In heterodyned 2D-IR, the three IR laser beams (in color) and the emitted FID signal beam all travel in different directions, and thus are spatially separated from one another. The electric field of the signal beam cannot be measured directly, as it is in NMR, so it is overlapped with a fourth local-oscillator beam that heterodynes the signal (not shown). When $t_2 = 0$, the 2D-IR spectrum is analogous to a COSY experiment, otherwise it is similar to NOESY. (After Zanni and Hochstrasser, 2001.)

COMMENT E2.1 PULSED LASER TECHNIQUES

Pulsed laser techniques have made giant strides forward in recent years with the result that complete control of optical and IR electric fields is now possible: The shapes, frequencies, timing, and phases of laser pulses can now be manipulated in ways that were thought to be possible only for radio waves. The feasibility of impulsive excitations, which require laser pulses shorter than the corresponding nuclear motion periods, has opened up new possibilities for conducting multiple-pulse coherent measurements to disentangle complex electronic and nuclear motions.

IR pulse follows some time after, the pulse sequence is analogous to NMR NOESY or a stimulated spin echo. Like R. Hochstrasser, we refer to this technique as THIRSTY (three IR pulse-stimulated echo spectroscopy). The variable delay between the second and third pulses allows time for vibrational coherence and population transfer to occur, analogous to spin transfer in NMR. The signals in the 2D-IR experiments are created by the responses of the network of vibrational modes excited by the pulses (Comment E2.1). They can be spatially separated from the transmitted fields and independently measured by a reasonable choice of input beam directions (Fig. E2.1). Such spatial selection of phase is not possible in NMR because of the long wavelengths of the radio-frequency fields.

The vibrational FID is measured as a function of the three time intervals and yields a three-dimensional time grid whose Fourier transforms give the IR COSY and THIRSTY spectra. The diagonal peaks in these spectra correspond to the vibrational frequencies in the structure distribution, but the cross-peaks show up only if the modes are coupled. To obtain well-resolved 2D-IR spectra, the FID periods need only to be measured for a few tens of picoseconds, thus the method can be used to monitor the structures of intermediates in kinetic experiments.

The first 2D-IR experiments concerned the amide I mode of peptides, which are mainly C=O vibrators

(see Chapter E1). In such cases, all of the relevant frequencies of an interacting ensemble of modes can be bracketed by the spectral bandwidth of 120 fs IR laser pulses. The two-dimensional correlation spectra yield the relevant structural and dynamical information (see below).

Two-dimensional IR spectroscopy experiments have been developed incorporating two pulses of different frequency (*dual-frequency 2D-IR spectroscopy*), one in the C=O bond (amide I) and the other in the N–H (amide II bond for peptide) region. They have been applied to measure the coupling and angular relation between C=O and N–H modes (Fig. E2.2).

In a dual-pulse experiment on N-methylacetatamide, for example, a *difference frequency generator* produces the first IR pulse of frequency $\omega_1 = 1675$ cm^{-1} propagating toward the sample in direction $\mathbf{k}_1$, and the *local oscillator* pulse, along $\mathbf{k}_{LO}$, required for the heterodyne technique (see Chapter D4). The output of the second difference frequency generator is split into two equal parts to generate two beams of frequency $\omega_2 = 1540$ cm^{-1} and vector directions $\mathbf{k}_2$ and $\mathbf{k}_3$ (Figs. E2.2 and E2.3).

E2.3.2 Selection Rules for Two-Dimensional Spectroscopy

A vibrational motion must cause a change in the permanent dipole moment for it to be IR active and revealed in a linear spectrum. This condition forms a selection rule for linear spectroscopy. In order to observe a signal in a two-dimensional spectrum, a further consideration should be applied. It is well understood that the non-linear optical response from a set of harmonic oscillators vanishes due to destructive interference involving adjacent vibrational levels. Thus observing a non-linear signal requires a deviation from harmonic behavior. The following conditions must be fulfilled: (1) anharmonicity in the ground state potential; (2) non-linear dependence of the transition dipole on vibrational coordinates; and (3) energy-level-dependent dynamics. These conditions form the set of selection rules for 2D-IR spectroscopy, and affect the

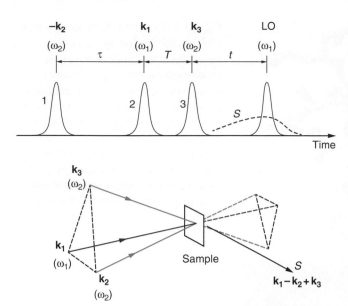

Fig. E2.2 The top panel shows the time sequence in a dual-frequency three-pulse photon-echo experiment. S (dashed line) is a typical signal envelope. LO denotes the local oscillator pulse. Three phase-locked IR pulses, labeled 1, 2, and 3, with wave vectors $\mathbf{k}_1$, $\mathbf{k}_2$, and $\mathbf{k}_3$, respectively, are incident on the sample at time intervals τ, between 1 and 2, and T, between 2 and 3. The bottom panel shows the spatial arrangement of the three beams interacting with the sample. The signal envelope is observed in the $\mathbf{k}_1 - \mathbf{k}_2 + \mathbf{k}_3$ direction. The IR beams were tuned such that one is centered on the amide I band ($\mathbf{k}_1$, ω_1), and the other is close to the amide II band ($\mathbf{k}_2$, and $\mathbf{k}_3$, ω_2). (After Rubtsov et al., 2003.)

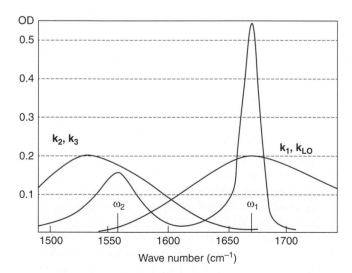

Fig. E2.3 The linear spectrum on an optical density (OD) scale of NMA in DMSO. The laser spectra centered near ω_1 and ω_2 are shown also. LO is the local oscillator pulse. (After Rubtsov et al., 2003.)

positions, amplitudes, and shapes of the various peaks in a two-dimensional spectrum.

E2.3.3 NMR and 2D-IR Spectroscopy: Similarity and Difference

One of the major achievements of NMR is its capacity to disentangle hopelessly complicated spectra by spreading them over several frequencies or time axes. Carefully designed sequences of radio-frequency pulses induce coherence transfers that make it possible to expose coupling between the spins on different nuclei or to eliminate dipolar interactions which dominate the line widths in solid-state NMR (Part J). Magnetic resonance is the elder brother of coherent laser spectroscopy, and the two share many basic concepts. The ability to shape and control radio-frequency pulses pre-dated similar advances in short pulse laser techniques by many decades, thus many of the optical and IR non-linear techniques have their analogues in early NMR developments.

The time-window and information content of electronic and vibrational multidimensional techniques are very different from NMR. For example, a sequence of IR pulses transfers coherence among the components of a network of vibrators. This network has much faster and different dynamical properties than the networks of spins that are interrogated by a sequence of radio-frequency pulses in NMR. The assumptions and theoretical techniques necessary to convert optical data into the relevant structural and dynamical information are fundamentally different from those invoked in NMR. The new peaks representing coupling between excitations, the distributions of structures in the equilibrium ensemble, and the intensity of the signals and their line shapes present direct signatures of molecular structure through distance and angular relations between chromophores or vibrations. They also expose the dynamics of electronic or nuclear excitations through the spectral density of their local environments. Among systems already studied are aggregates of chromophores, hydrogen-bonded complexes and liquids, the secondary and tertiary structure of polypeptides, and protein conformational dynamics. As with NMR, isotope labeling very usefully provides an extra dimension for peak assignment in the 2D-IR spectra.

E2.4 FROM AMIDE BANDS TO PROTEIN TERTIARY STRUCTURE

We recall that IR spectra of the amide transition provide information about secondary structural motifs in proteins and peptides (see Chapter E1). The amide I band, which consists mainly of the stretching motion of the peptide

(a)

(b)

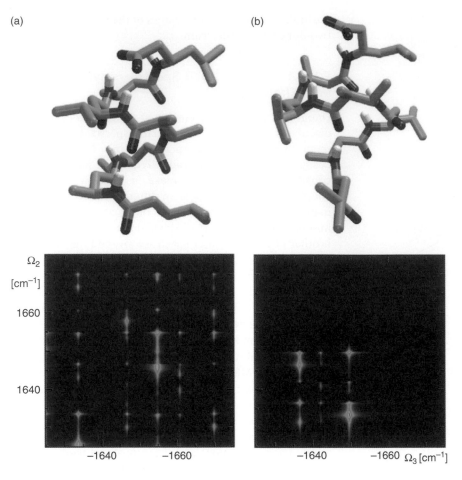

Fig. E2.4 Simulated 2D-IR spectra of a β-heptapeptide. The spectra were calculated from two instantaneous conformations during a molecular dynamic simulation when the peptide is: **(a)** folded and **(b)** partially unfolded. The simulations treated the peptide as a coupled set of vibrations and vibrational dynamics were deliberately omitted in order to produce narrow lines. There are clear signatures in the spectra for the structural changes between the folded and unfolded states. (Adapted from Zanni and Hochstrasser, 2001.)

backbone C=O groups, is a strong IR absorber that is spectrally isolated from other vibrational modes, such as those of the amino acid side groups. Although the group spectrum varies slightly for different positions in a protein structure, the observed spectra are not resolved for proteins and the peaks together constitute a single amide I band.

E2.4.1 Simulations of 2D-IR Spectroscopy

The sensitivity of 2D-IR spectroscopy to peptide structure is readily demonstrated by simulations, such as those shown in Fig. E2.4, which examine the coupling between the amide I vibration modes of a β-heptapeptide. The folded β-heptapeptide exhibits a characteristic 2D-IR spectrum of diagonal peaks and cross-peaks, whereas the spectrum of the partially unfolded peptide has a different distribution of couplings and frequencies (Comment E2.2).

E2.4.2 Determination of Peptide Structures

The first detailed experimental application of IR-COSY involved the acetylproline–NH_2 dipeptide and led to the determination of its structure. The peaks generated from

the FIDs are shown in Fig. E2.5 for the peptide in chloroform. Each vibrational mode produced a pair of peaks along the diagonal (A, B, C) that are separated by the degree of anharmonicity. Each pair of modes results in

---

COMMENT E2.2 2D-IR SPECTRA AND POLYPEPTIDE STRUCTURE

The distribution of polypeptide structures and their nuclear motions are reflected in the shapes and frequencies of the diagonal peaks and cross-peaks in a 2D-IR spectrum. The 2D-IR spectrum of a polypeptide depends on the interpeptide potential energy function and, in the simulation shown in Fig. E2.4, the potential was chosen as a dipole–dipole interaction. When a complete, more realistic distribution of structures is used, the sharp peaks in Fig. E2.4 become spread along the diagonal to form the elliptical profiles seen in the spectra. The cross-peaks are also broadened into shapes that signal an interaction between the peptide units. The minor axes of these ellipses do not depend on the structure distribution, however, but are determined by the dephasing of the vibrational modes.

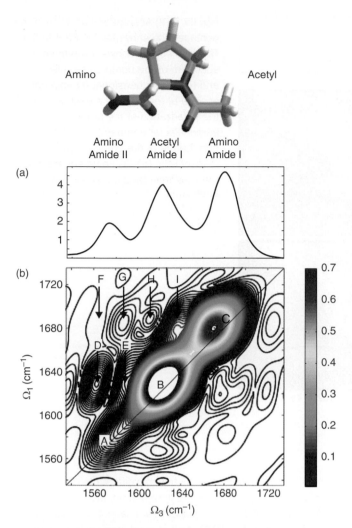

Fig. E2.5 The FTIR and heterodyne 2D-IR (IR-COSY) spectra of acetylproline–NH₂ (top) in chloroform. The three peaks in the linear spectrum **(a)** correspond to the three diagonal features A, B, C in the absolute magnitude spectra **(b)**. Peaks A and C are from amide I and II vibrational modes that are located on the amino end of the molecule, and peak B is from the amide I mode on the acetyl end. The cross-peaks are labeled D–I. (After Zanni and Hochstrasser, 2001.)

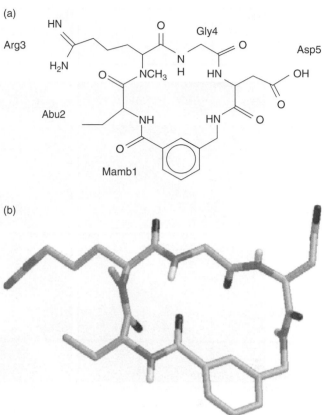

Fig. E2.6 **(a)** The chemical formula and **(b)** the three-dimensional structure of the cyclic pentapeptide (*cyclo*-Mamb–Abu–Arg–Gly–Asp). The peptide, which is stabilized by a hydrogen bond between the Mamb-1–Abu-2 peptide bond and the Arg-3–Gly-4 peptide bond was specifically designed to form a single, well-defined conformation in solution with an almost ideal type II β-turn centered at the Abu–Arg residues. (Adapted from Hamm *et al.*, 1999.)

two pairs of cross-peaks. There are two cross-peaks between each of the acetyl bands and the amino band (cross-peaks F and G between peaks A and C, and cross-peaks H and I between peaks B and C), which indicates that acetylproline–NH₂ must adopt two conformations in chloroform.

A *de novo* designed peptide was the first system whose 2D-IR spectrum was understood quantitatively. The peptide is a cyclic pentapeptide (*cyclo*-Mamb–Abu–Arg–Gly–Asp) for which both the NMR and X-ray structures are known. Its chemical formula and three-dimensional structure are shown in Fig. E2.6. The structure was specifically

designed to form a single, well-defined conformation in solution with an almost ideal type II β-turn. The linear absorption spectrum of the peptide in the region between 1570 cm⁻¹ and 1680 cm⁻¹ exhibits five partially resolved bands (Fig. E2.7a). A multiline Lorentzian fit (see Chapter A3) yields peak positions at 1673 cm⁻¹, 1648 cm⁻¹, 1620 cm⁻¹, 1610 cm⁻¹, and 1584 cm⁻¹. There are no other strong IR bands between 1500 cm⁻¹ and 2000 cm⁻¹. The 2D-IR spectra of the peptide are shown in Fig. E2.7b, c. The time delay between pulses was set such that the broad frequency bandwidth probe pulse (second laser pulse) followed *c.*800 fs after the fast-rising edge of the pump pulse (first laser pulse). Further information is obtained from the use of laser pulses with different polarization. The spectra for perpendicular and parallel pump and probe polarizations are shown in Fig. E2.7b and c, respectively. The fits to the experimental data are shown in Fig. E2.7d–f.

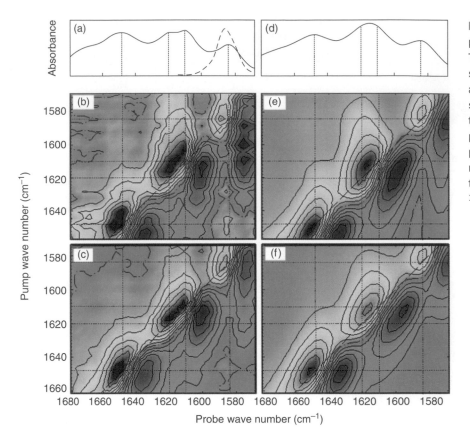

Fig. E2.7 (a) Absorption spectrum of the pentapeptide shown in Fig. E2.6 in D$_2$O. The dashed line shows a representative spectrum of the pump pulse used. **(b)** and **(c)** Two-dimensional pump-probe spectra of the peptide measured with the polarization of the probe pulse perpendicular and parallel to the polarization of the pump pulse, respectively. **(d)–(f)** Least-square fits of the experimental data. (After Hamm *et al.*, 1999.)

E2.5 CHECKLIST OF KEY IDEAS

- Two-pulse IR COSY has the same measurement principles as NMR COSY; however, the time-window and information content of electronic and vibrational multidimensional techniques are very different from NMR.
- The signals in the 2D-IR experiments are created by the responses of the network of vibrational modes excited by the two pulses and can be spatially separated from the transmitted fields and independently measured by a reasonable choice of input beam directions.
- In 2D-IR spectroscopy, the first pulse creates a vibrational coherence analogous to the spin coherence created by the first radio-frequency pulse in NMR, but unlike the $\pi/2$ pulse of NMR, this interaction is weak; after some delay, a pair of weak IR pulses (probe pulses) interrupts the initial coherence evolution to generate an FID from the transferred coherence.
- The three pulses in the 2D-IR and 2D-NMR techniques are separated by time delays t_1 and t_2, and the FID is measured versus t_3; in heterodyned 2D-IR spectroscopy, the three IR laser beams and the

emitted FID signal beam all travel in different directions, and thus are spatially separated from one another.
- The electric field of the signal beam cannot be measured directly, as it is in NMR, so it is overlapped with a fourth local-oscillator beam that heterodynes the signal; when $t_2 = 0$, the 2D-IR spectrum is analogous to a COSY experiment, otherwise it is similar to NOESY.
- Multiple-pulse techniques have the capacity to induce electronic and vibrational degrees of freedom into non-equilibrium states, and to monitor their subsequent evolution, yielding ultrafast snapshots of dynamical events such as energy or charge transfer pathways, photoisomerization, and structural fluctuations.
- To date, structures of dipeptides, tripeptides, and pentapeptides have been determined by 2D-IR spectroscopy; the most exciting aspect of 2D-IR spectroscopy, however, is the combination of its sensitivity to structure with time resolution.
- If conventional FTIR spectroscopy can monitor protein secondary structure features on the nanosecond to second timescale, then pulsed IR methods can probe vibrations on timescales down to the femtosecond.
- As with NMR, isotope labeling very usefully provides an extra dimension for peak assignment in the 2D-IR spectra.

Suggestions for Further Reading

Historical Review and Introduction to Biological Problems

Zanni, M. T., and Hochstrasser, R. M. (2001). Two-dimensional infrared spectroscopy: A promising new method for the time resolution of structure. *Curr. Opin. Str. Biol.*, **11**, 516–522.

Linear and Multidimensional Spectroscopy

Mukalmel, S. (1995). *Principles of Nonlinear Spectroscopy*. New York: Oxford University Press.

Multidimensional spectroscopies. *Chem. Phys*, (2001), **266**, 137–351. The complete issue of this journal was devoted to advances being made in optical and IR multidimensional spectroscopy.

Principles of 2D-IR Spectroscopy

Asplund, M. C., Zanni, M. T., and Hochstrasser, R. M. (2000). Two-dimensional infrared spectroscopy of peptides by phase controlled femtosecond vibrational photon echoes. *Proc. Natl. Acad. Sci. USA*, **97**, 8219–8224.

From Amide Bands to Protein Tertiary Structure

Hamm, P., Lim, M., DeGrado, W. F., and Hochstrasser, R. M. (1999). The two-dimensional IR nonlinear spectroscopy of a cyclic penta-peptide in relation of its three-dimensional structure. *Proc. Natl. Acad. Sci. USA*, **96**, 2036–2041.

Zanni, M. T., Gnakaran, S., Stenger, J., and Hochstrasser, R. M. (2001). Two dimensional infrared spectroscopy of solvent dependent conformations of acetyleproline-NH_2. *J. Phys. Chem.*, **105**, 6520–6535.

Rubtsov, I. V., Wang, J., and Hochstrasser, R. M. (2003). Dual-frequency 2D-IR spectroscopy heterodyned photon echo of the peptide bond. *Proc. Natl. Acad. Sci. USA*, **100**, 5601–5606.

NMR and 2D-IR Spectroscopy: Similarity and Difference

Keusters, D., Tan, H. S., and Warren, W. S. (1999). Role of pulse phase and direction in two dimensional optical spectroscopy. *J. Phys. Chem. A*, **103**, 10369–10380.

RAMAN SCATTERING SPECTROSCOPY

E3

E3.1 HISTORICAL REVIEW AND INTRODUCTION TO BIOLOGICAL PROBLEMS

1928

L. Mandelshtam and **C. V. Raman** independently discovered that a wavelength shift in a small fraction of scattered visible radiation depends on the chemical structure of the molecule responsible for the scattering. **Raman** was awarded the 1931 Nobel Prize in physics for the discovery of the phenomenon, which became known as Raman scattering or the Raman effect (Comment E3.1). Raman scattering is *inelastic* light scattering that results from the same type of quantized vibrational transitions as those associated with IR absorption and occurs in the same spectral region. The Raman scattering and IR absorption spectra for a given molecule are often similar. Differences between the properties that make a chemical group IR or Raman active, however, make the techniques strongly complementary rather than competitive. Raman spectroscopy has the advantages of minimal or no damage to the sample, and relatively little interference from the water signal in aqueous solution or in vivo. The possibility of observing Raman spectra from crystals as well as from solutions in vitro or in vivo provides a very useful approach to relating molecular structures solved by X-ray crystallography and conformations in solution or in living cells.

Raman spectroscopy was not widely used by molecular biophysicists until lasers became available in the **1960s**. Interference from fluorescence excited in the molecule studied or in sample impurities was also a deterrent to the general use of the method. This problem has now been largely overcome by the use of near-IR laser sources.

E3.2 CLASSICAL RAMAN SPECTROSCOPY

Classical or *non-resonance* Raman spectroscopy probes radiation scattering by vibrational modes, which have energies in the IR spectral region. The method has proven very useful in molecular biophysics for the determination of protein, nucleic acid, and membrane conformational states.

Resonance Raman spectroscopy is discussed in Section E3.3. The resonance effect is between the scattered radiation and *electronic transitions* in the sample molecules. Since these transitions occur in the visible spectral region, which is also readily accessible by the conventional lasers commonly used for Raman spectroscopy, the method has provided considerable information on visual pigments, heme, and other metalloproteins. The most studied chromophores are non-fluorescent so that the relatively weak Raman signals can be observed without contamination from broad fluorescence emission.

E3.2.1 Raman Spectra

Figure E3.1 shows a portion of a Raman spectrum obtained by irradiating a sample of carbon tetrachloride with an intense beam from an argon ion laser (488.0 nm wavelength, corresponding to 20 492 cm^{-1} wave numbers). The emitted radiation is of three types: Stokes scattering, anti-Stokes scattering, and Rayleigh scattering. The last, which corresponds to *elastic* scattering of wavelength exactly equal to that of the excitation source, is significantly more intense. The Stokes and anti-Stokes lines are *inelastic* scattering corresponding to quantum energy given to or taken away from the sample by the light beam, respectively. The x-axis in Fig. E3.1 is the wave number shift $\Delta\bar{\nu}$, defined as the difference in wave numbers (cm^{-1}) between the observed and excitation radiation. Note the three Raman peaks in the spectrum are found on both sides of the Rayleigh peak, with identical shift magnitudes. Stokes lines are at 218, 314, and 459 cm^{-1}, wave numbers below the Rayleigh peak, while anti-Stokes lines occur at 218, 314, and 459 cm^{-1} above it. Quite generally, anti-Stokes lines are appreciably less intense than the corresponding Stokes lines, indicating a higher probability for

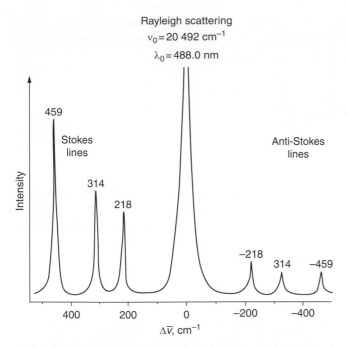

Fig. E3.1 Raman spectra for CCl_4 excited by laser radiation of $\lambda_0 = 488$ nm and $\nu_0 = 20\,492$ cm^{-1}. The number above the peaks is the Raman shift, $\Delta\bar{\nu}$ cm^{-1}. Raman shift magnitudes are independent of the excitation wavelength. Identical shift patterns would be observed for CCl_4 regardless of whether excitation was carried out with a krypton ion laser (488.0 nm), a helium–neon laser (632.8 nm), or a Nd:YAG laser (1064 nm). (Adapted from Skoog *et al.*, 1995.)

the sample to be excited by the beam rather than the other way around. In practice, only the more intense, Stokes portion of a spectrum is analyzed.

The appearance of Raman spectral lines of lower energy (longer wavelengths) than the excitation beam is analogous to the Stokes shifts found in a fluorescence experiment (Chapter D3); for this reason, Raman shifts to longer wavelengths were called *Stokes shifts*. We see in Section E3.2.2, however, that Raman and fluorescence spectra arise from fundamentally different processes, so the application of the same terminology to both may be misleading.

E3.2.2 Frequency, Intensity, and Polarization

Raman spectroscopy is a form of vibrational spectroscopy. Like IR spectroscopy, it is related to transitions between vibrational energy levels in a molecule. It differs from IR spectroscopy in that information is derived from *light scattering*, whereas in IR spectroscopy information is obtained from *absorption* processes.

Let us assume a light beam of frequency ν_{exp} is incident upon a solution of molecules. The electric field E associated with the radiation is described by the wave equation (see Chapter E1)

$$E = E_0 \cos\left(2\pi\nu_{exp}t\right) \tag{E3.1}$$

where E_0 is the amplitude of the wave. As the field propagates through the sample the electron charges respond to it. The dipole moment, m, induced in the electron cloud of a molecular bond by the oscillating electric field is given by

$$m = \alpha E = \alpha E_0 \cos\left(2\pi\nu_0 t\right) \tag{E3.2}$$

where α is the polarizability of the bond.

In order to be Raman active (i.e., to be active in inelastic light scattering), the vibration of the atoms joined by the bond must be reflected favorably by polarizability changes in the bond electrons,

$$\alpha = \alpha_0 + \left(r - r_{eq}\right)\left(\frac{d\alpha}{dr}\right) \tag{E3.3}$$

where α_0 is the polarizability of the bond at the equilibrium internuclear distance r_{eq} and r is the internuclear separation at any instant. The change in internuclear separation varies with the frequency of the vibration ν_v as given by

$$r - r_{eq} = r_m \cos\left(2\pi\nu_v t\right) \tag{E3.4}$$

where r_m is the maximum internuclear separation relative to the equilibrium position. Substituting Eq. (E3.4) into Eq. (E3.3) gives

$$\alpha = \alpha_0 + \left(\frac{d\alpha}{dr}\right)r_m \cos\left(2\pi\nu_v t\right) \tag{E3.5}$$

We can then obtain an expression for the induced dipole moment m by substituting Eq. (E3.5) into Eq. (E3.2):

$$m = \alpha_0 E_0 \cos\left(2\pi\nu_0 t\right) + E_0 r_m \left(\frac{d\alpha}{dr}\right)\cos\left(2\pi\nu_v t\right)\cos\left(2\pi\nu_0 t\right) \tag{E3.6}$$

Recall from trigonometry that

$$2\cos x \cos y = \left[\cos\left(x+y\right) + \cos\left(x-y\right)\right]$$

Applying this identity to Eq. (E3.6) gives

$$m = \alpha_0 E_0 \cos\left(2\pi\nu_0 t\right) + \frac{E_0}{2}r_m\left(\frac{d\alpha}{dr}\right)\cos\left[2\pi(\nu_0 - \nu_v)t\right]$$
$$+ \frac{E_0}{2}r_m\left(\frac{d\alpha}{dr}\right)\cos\left[2\pi(\nu_0 + \nu_v)t\right] \tag{E3.7}$$

The first term on the right-hand side of Eq. (E3.7) represents Rayleigh scattering at the excitation frequency ν_0. The second and third terms correspond, respectively, to the Stokes and anti-Stokes frequencies of $\nu_0 - \nu_v$ and $\nu_0 + \nu_v$. Here, the excitation frequency has been modulated by the vibrational frequency of the bond. It is important to note that Raman scattering requires that the polarizability of the bond varies as a function of distance: i.e., $d\alpha/dr$ in Eq. (E3.7) must be greater than zero for Raman lines to appear.

The energy shifts observed for a given bond in Raman experiments should be identical to the energies of its IR absorption band, provided that the vibrational modes involved are active toward both IR absorption and Raman scattering. Figure E3.2 illustrates the similarity of the two types of spectra. Several peaks with identical $\bar{\nu}_0$ and $\Delta\bar{\nu}$

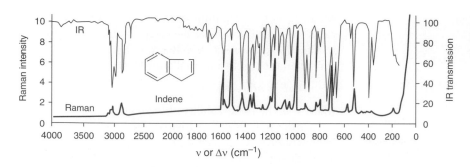

Fig. E3.2 Comparison of Raman and IR spectra for indene. (Adapted from Skoog *et al.*, 1995.)

values exist for indene. However, the relative sizes of the corresponding peaks are frequently different, and certain peaks that occur in one spectrum are absent in the other.

The differences between a Raman and an IR spectrum are not surprising when it is considered that the basic mechanisms, although dependent upon the same vibrational modes, arise from processes that are mechanistically different. IR absorption is observed for vibrational modes which change the *dipole moment* of the molecule, while Raman scattering is associated with modes that produce a change in the *polarizability* of the molecule. Because of this fundamental difference in mechanism, the Raman activity of a given vibrational mode may differ markedly from its IR activity. For example, a homonuclear molecule such as nitrogen has no dipole moment either in its equilibrium position or when a stretching vibration causes a change in the distance between the two nuclei. Thus, absorption of radiation corresponding to the vibration frequency cannot occur. On the other hand, the polarizability of the bond between the two atoms varies periodically in phase with the stretching vibrations, reaching a maximum at the greatest separation and a minimum at the closest approach. A Raman shift is therefore observed at the frequency of the vibrational mode.

It is interesting to compare the IR and Raman activities of coupled vibrational modes such as those described for the carbon dioxide molecule. In the symmetric mode, no change in the dipole moment occurs as the two oxygen atoms move away from or toward the central carbon atom; thus, this mode is IR-inactive (Fig. E3.3). The polarizability, however, fluctuates in phase with the vibration since distortion of the bonds becomes easier as they lengthen and more difficult as they shorten. Raman activity is associated with this mode. In contrast, the dipole moment of carbon dioxide fluctuates in phase with the antisymmetric vibrational mode. Thus, an IR absorption peak arises from this mode. On the other hand, as the polarizability of one of the bonds increases as it lengthens, the polarizability of the other decreases, resulting in no net change in the polarizability. Thus, the asymmetric stretching vibration is Raman-inactive.

As in the example above, parts of Raman and IR spectra can be complementary, each associated with a different set of vibrational modes within a molecule. Other vibrational modes may be both Raman- and IR-active. For example, all

of the vibrational modes of sulfur dioxide yield both Raman and IR peaks. Peak heights differ, however, because the probability for the transitions is different for the two mechanisms.

E3.2.3 Raman Spectrometers and Raman Microscopes

In order to obtain the Raman spectrum of a protein or nucleic acid, one places the sample at the focal point of a focused laser beam. The scattered light is collected by a lens system and directed through a suitable monochromator to a photon detector. The Raman spectrum consists of a plot of the scattered intensity as a function of its frequency difference with the incident laser frequency. A band at a particular frequency in the Raman spectrum corresponds to a vibrational mode of that frequency. It is a fundamental

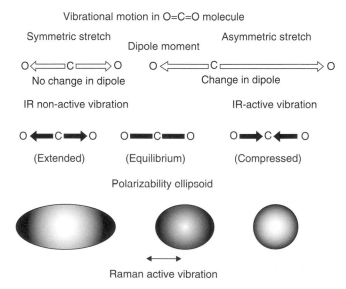

Fig. E3.3 Selection rules for IR and Raman spectra using the carbon dioxide molecule as an example. For vibration to be IR-active, the dipole moment of the molecule must change. Therefore, the symmetric stretch in carbon dioxide is not IR-active because there is no change in the dipole moment. The asymmetric stretch is IR-active due to a change in dipole moment. For vibration to be Raman active, the polarizability of the molecule must change with the vibration motion (Altose *et al.*, 2001.)

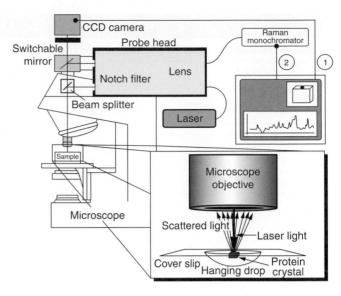

Fig. E3.4 Schematic diagram (not to scale) of a Raman microscope. Both video **(1)** and spectrographic data sets **(2)** can be displayed on the computer screen. Inset: The arrangement for observing Raman data from a single protein crystal in a hanging drop. (Adapted from Altose *et al.*, 2001.)

principle of the Raman effect that the spectrum is independent of the incident laser frequency.

Three types of Raman spectrometer are in general use for the characterization of protein or nucleic acid structure: (1) the classic laser Raman spectrometer; (2) the laser Raman microscope; and (3) the UV resonance spectrometer (Comment E3.2). The classic spectrometer uses a wavelength in the visible spectral region and has the advantage that it puts no constraints on the sample, which may be a protein or nucleic acid powder, crystal, fiber, or solution. Because these samples are transparent to visible light, there is little or no heating of the sample by the laser beam. Classical Raman spectra of protein or nucleic acid solutions have been obtained from 5–50 μl samples in a glass capillary mounted in a circulating water temperature controlled copper block, with "tunnels" for the incident and scattered beams. The scattered light is collected at right angles to the incident beam. The concentration of the solution is typically 2.0–20 mg ml^{-1}, or about 20–100 mM in nucleotide or peptide groups.

A powerful advantage of Raman microscopy is that it allows a single protein crystal to be studied *in situ* in a hanging drop (Fig. E3.4). The Raman spectrum recorded from the crystal is caused only by the focal volume of the laser beam within the crystal, which is typically ~20 μm in diameter and 50 μm in depth. In this volume, the protein concentration is about 50 mM, thus providing very high-quality data. Any Raman scattering collected from the surrounding mother liquor makes a negligible contribution because protein concentrations there are less than 1 mM. Thus, Raman microscopy can be used to screen the properties of protein crystals in a hanging drop.

In general, Raman spectroscopy and Raman microscopy provide direct comparison of the properties of a molecule in solution and in a single crystal.

E3.2.4 Protein Secondary Structure from Raman Spectra

The different conformations of polypeptide chains have remarkable differences in their Raman spectra. Table E3.1 presents a summary of some of the Raman active modes that are useful for obtaining information about the structure and dynamics of proteins. Of the modes listed in Table E3.1, one of those whose Raman spectrum is sensitive to conformation is the amide I mode. Thus, the amide I vibration changes from 1655 ± 5 cm^{-1} in the α-helical forms in H_2O to 1670 ± 3 cm^{-1} in the β-sheet form in H_2O, indicating that the carbonyl group is more strongly polarized, i.e., has a higher dipole moment in the α-helix than in the β-sheet. Hydrogen–deuterium exchange causes the frequency to decrease because of the increased mass of the deuteron. A frequency of 1632 ± 5 cm^{-1} was found for the α-helical form in D_2O, whereas a frequency of 1658 ± 3 cm^{-1} was found for the β-sheet form in D_2O.

In the estimation of protein secondary structure content in terms of percentage helix, β-strand, and turn, the amide I band is analyzed as a linear combination of the spectra of the reference proteins whose structures are known. The Raman spectra of globular proteins in the crystal and in solution are almost identical, reflecting the compact nature of the macromolecules. Thus one may use the fraction of each type of secondary structure determined in the crystalline state by the X-ray of diffraction studies to study protein in solution.

In addition to the structural information on proteins that is obtained from the vibrations of the polypeptide backbone, the vibrations of the side-chains can also give valuable information on the environment of the side-chain and sometimes on the accessibility to protons and deuterons.

The state of tyrosine in proteins is often of great importance. The ratio of the intensity of the Raman bands at 850–830 cm^{-1} is sensitive to the environment of the

TABLE E3.1 MAIN MARKER BANDS OF PROTEINS[a]

Origin and frequency ($\bar{v}\,cm^{-1}$)	Assignment	Structural information
Amide bands		
Amide I	Amide C=O stretch coupled to N–H wagging	Strong band; hydrogen bonding lowers amide I frequencies
1655 ± 5	–	α-helix (H$_2$O)
1632 ± 5	–	α-helix (D$_2$O)
1670 ± 3	–	Antiparallel β-related sheet
1661 ± 3	–	α-helix (D$_2$O)
1655 ± 3	–	Disordered structure (H$_2$O)
1658 ± 2	–	Disordered structure (D$_2$O)
Amide III	N–H in-plane bend, C–N stretch	Strong hydrogen bonding raises amide III frequencies
>1275	–	α-helix
1235 ± 5	–	Antiparallel β-related sheet
983 ± 3	–	Disordered structure (D$_2$O)
1245 ± 4	–	Disordered structure
Amino acid chains		
Tyrosine doublet 850/830	Resonance between ring fundamental and overtone	State of tyrosine
Tryptophan 880/1361	Indole ring	Ring environment
Phenylalanine 1006	Ring breathing	Conformation-insensitive
Histidine 1409	N-deuteroimidazole	Possible probe of ionization state
S–S		
510	S–S stretch	Indication of conformational heterogeneity among disulfides
525	S–S stretch	*Gauche–gauche–trans*
540	S–S stretch	*Trans–gauche–trans*
C–S		
630–670		*Gauche*
700–745		*Trans*
S–H		
2560–2580	S–H stretch	Environment, deuteration rate

[a] Table adapted from Peticolas (1995).

phenolic –OH. This doublet is due to resonance between a ring-breathing vibration and an overtone of an out-of-plane ring-breathing vibration. A strong hydrogen bond from the phenolic hydrogen to a negative acceptor results in a low value of I_{850}/I_{830} (about 3:10), whereas hydrogen bonding to the phenolic oxygen from an acidic proton donor yields a higher value (about 10:30) for this ratio.

An example of the use of this technique was in the investigation of the buried tyrosines in fd filamentous phage coat protein (Fig. E3.5, Comment E3.3). The two tyrosines at positions 21 and 24 of the fd coat protein provide a potentially useful probe of the phage structure. Several peaks arising from the tyrosines are clearly evident in the Raman spectrum of the virus. The measured

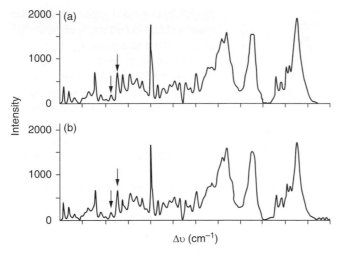

Fig. E3.6 T_6 crystal structure of insulin (Protein Data Bank (PDB) identifier 4INS).

Fig. E3.5 Laser Raman spectra of the fd phage pellets at **(a)** pH 8.1 and **(b)** pH 12.0. The positions of the two tyrosine peaks are shown by arrows. Lack of a change in the I_{850}/I_{830} ratio is evident from comparison of the peak intensities (see text). (Dunker et al., 1979.)

COMMENT E3.3 PHYSICIST'S BOX: FD FILAMENTOUS PHAGE

The filamentous phage fd is a long, thin structure of about 895 nm by about 6 nm. The phage particle contains about 2700 copies of a largely α-helical coat protein. This phage is structurally similar to Pf1 phage that is discussed in Comment J2.5.

---

intensity ratio I_{850}/I_{830} is about 10:3. According to the above interpretation, such a ratio indicates that the tyrosine OH groups hydrogen bond with a very strong acidic proton donor. Further titration experiments showed that the tyrosines and their acidic donors in the filamentous phage fd are inaccessible to the solvent. One possible explanation of this fact is that the tyrosine OH groups are the recipients of hydrogen-bonded protons arising from fairly acidic donors, yet these acidic donors do not become titrated over the range pH 7–12.

E3.2.5 Protein Conformational Dynamics in Solution and in Crystals

Insulin provides an attractive model for protein conformational dynamics because of its wealth of prior structural characterization. Insulin is a small globular protein containing two chains, A (21 residues) and B (30 residues). Stored in the pancreatic β cell as Zn^{2+}-stabilized hexamers, the hormone functions as a Zn^{2+}-free monomer. Classical crystallographic studies have focused on Zn^{2+}-stabilized hexamers (Fig. E3.6). Use of engineered insulin monomers and dimers enables direct comparison of the structure and dynamics of a protein in different states of self-assembly.

Solution Raman spectra (1550–1750 cm^{-1} region) of the T_6 hexamer, the engineered dimer designated T_2, and the engineered monomer, herein designated T_1, are shown in Fig. E3.7. Strikingly, amide I profiles broaden and change in relative peak heights on disassembly of the protein. Line widths of Raman bands associated with the polypeptide backbone exhibit progressive narrowing with successive self-assembly. A major increase in width at half-height (WHH) occurs upon dissociation of the aqueous T_6 hexamer (42.9 cm^{-1}) to the T_2 dimer (48.7 cm^{-1}); a further increase is seen in the T_1 monomer (55.1 cm^{-1}).

A further narrowing of line widths of Raman bands associated with the polypeptide backbone occurs under fibril formation. Figure E3.8 shows the amide profile of the insulin fibrils. The peak at 1673 cm^{-1} has a narrow WHH of 19.7 cm^{-1}. These results provide physical evidence for the flexibility of the isolated monomer and damping of fluctuations in native and pathological assemblies.

Additional information on conformational dynamics for insulin can be obtained from studies in a single crystal *in situ* in a hanging drop. The partial Raman spectrum of a single insulin crystal in its native conformation is shown in Fig. E3.9a, c. Raman data for a single T_6 insulin crystal in the amide I region (Fig. E3.9a) are very close to the data in solution (Fig. E3.7a).

Dramatic changes occur when the reducing agent tris(2-carboxyethyl)phosphine is added to the mother liquor containing the crystal. The α-helical 1657 cm^{-1} band (Fig. E3.9a) disappears and is replaced by a single β-sheet marker band at 1669 cm^{-1} (Fig. E3.9b). The WHH of this band of 24 cm^{-1} compares quite closely to that found for the fibrils. These findings show that native proteins in the crystalline phase can also undergo large-scale conversion to β-sheets as for solution state described above (Comment E3.4).

E3.2.6 Conformation of DNA

Some of the vibrational normal modes of DNA are conformationally sensitive. This leads to changes in frequencies

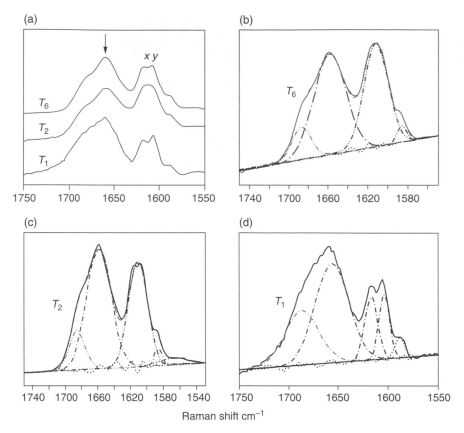

Fig. E3.7 (a) Amide I region of the Raman spectra (1550–1750 cm⁻¹) of insulin in solution in the hexameric T_6, dimeric T_2, and monomeric T_1 states. The arrow indicates the amide I maximum. Spectra are color-coded as follows: continuous line, experimental data (black); broken line, fitted sum (light purple); dash–dot line, individual components (Gaussian lines) at 1657 cm⁻¹ (red), near sum at 1657 cm⁻¹. (b)– (d) Amide I profile deconvolution 1611 cm⁻¹ (blue), at 1687 cm⁻¹ (green), and at 1585 cm⁻¹ (brown); and dotted line, residual (black). The peak at 1657 cm⁻¹ in the amide I profiles is due to contributions from α-helices, whereas the shoulder near 1680 cm⁻¹ is due to β-sheet and β-strands. Residues in disordered regions are expected to contribute weakly at 1648 cm⁻¹ and 1678 cm⁻¹. The unresolved doublet near 1610 cm⁻¹ (labeled x and y) and the shoulder at 1585 cm⁻¹ are due to ring modes of phenylalanine and tyrosine. (Adapted from Dong et al., 2003.)

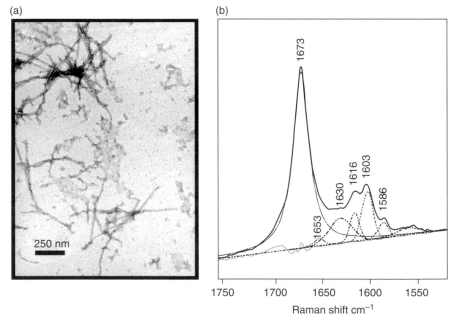

Fig. E3.8 Structure of insulin fibrils formed by gentle agitation of the protein in 60% ethanol and 0.25 M NaCl at 35 °C: (a) electron micrograph of insulin fibrils; (b) Raman spectrum of insulin fibrils obtained using a Raman microscope. The amide I region can be fitted by a line centered at 1673 cm⁻¹ and to features centered at 1653 cm⁻¹ and 1630 cm⁻¹, respectively. The peak at 1673 cm⁻¹ is due to β-sheet formation. (Adapted from Dong et al., 2003.)

and/or intensities of the Raman bands with changes in conformation. From a comparison of the Raman spectra taken from different crystals of DNA of known structure, a correlation between the frequencies and intensities of Raman bands and the conformation of the DNA has been established.

Figure E3.10 shows the Raman spectra of hexameric oligonucleotides containing only cytidine (C) and guanine (G). The spectra are of r(CGCGCG) in the A-form (top spectrum), d(CGCGCG) in the B-form, and d(CGCGCG) in the Z-form (bottom spectrum). The bands that change with conformation are highlighted in black. These are the marker bands

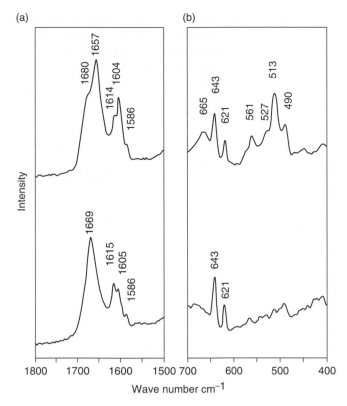

Fig. E3.9a and b Partial Raman spectra of an insulin single crystal obtained by a Raman microscope. Top: native crystal; bottom: β-transformed crystal after the reducing agent tris(2-carboxyethyl)phosphine has been added to the mother liquor. The peaks at 1615 and 643 cm^{-1} are due to tyrosine modes, and the peaks at 1604 and 621 cm^{-1} are due to phenylalanine ring modes. The ring modes are expected to only undergo small changes in intensity on secondary structure change and, thus, act as integral intensity standards. (After Zheng *et al.*, 2004.)

COMMENT E3.4 AMYLOIDOGENIC DISEASES

Protein unfolding and misfolding events that eventually lead to extensive β-sheet structure are central to amyloidogenic diseases. Thus, native soluble proteins partially unfold and aggregate, possibly using a residual "β-sheet" domain as a nucleation motif. This leads to amyloid fibrilogenesis and the formation of highly insoluble fibrils (Booth *et al.*, 1997).

---

for guanine, cytosine, and the sugar phosphate backbone as the DNA goes from the A- to the B- to the Z-form.

E3.3 RESONANCE RAMAN SPECTROSCOPY (RRS)

The central advantage of RRS is that when the wavelength of the exciting laser line is adjusted to coincide with that of an allowed electronic transition in a molecule, the intensities of certain Raman bands are enhanced relative to the off-resonance values. This technique is very selective, because only those vibrational modes that are related to the chromophoric part of the scattering molecule can be enhanced at resonance. This means that the vibrations of the chromophoric group can be monitored in biological macromolecules. If the chromophore is itself a site of functional activity, then RRS can provide structural information associated with reactivity.

This is illustrated in Fig. E3.11, which shows a dramatic increase in the intensity of the Cu–S(Cys) vibrational modes near 400 cm^{-1} as the excitation wavelength (647.1 nm) approaches that of the S → Cu charge transfer electronic transition (~600 nm) in the blue copper protein azurin.

Thus the correct identification of the vibrational modes showing RRS enhancement aids in the assignment of the resonant electronic transition and vice versa. Because vibrational frequencies are sensitive to the bond strength, the number of atoms, the geometry, and the coordination environment, this technique provides information which is complementary to that obtained by X-ray crystallography and X-ray absorption spectroscopy. See also Comment E1.11 for an RRS study of channel rhodopsins (ChRs), the retinal binding membrane proteins. Channel rhodopsins are being developed for optogenetics, which uses light to control response in neurons that have been genetically sensitized.

E3.4 SURFACE ENHANCED RAMAN SPECTROSCOPY (SERS)

Raman scattering is a very weak effect, with a cross-section between 10^{-30} cm^2 and 10^{-25} cm^2 per molecule. Such small Raman cross-sections require a large number of molecules to achieve adequate conversion rates from exciting laser photons to Raman photons.

SERS involves obtaining Raman spectra in the usual way on samples that are adsorbed on the surface of colloidal metal particles. The same phenomenon can be observed in the visible region (Comment E3.5). For reasons that are not fully understood, the Raman lines on the adsorbed molecules are often enhanced by a factor of 10^3–10^6. When surface enhancement is combined with the resonance enhancement technique discussed in the previous section, the net increase in signal intensity is roughly the product of the intensity produced by each of the techniques. In some cases enhancement factors are much larger than the ensemble-averaged values derived from conventional measurements, in the order of 10^{14}–10^{15}. Such large enhancement factors provide in single-molecule SERS effective Raman cross-sections which are of the order of 10^{-16} cm^2 per

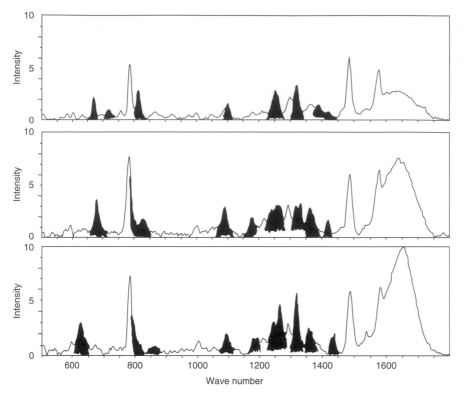

Fig. E3.10 Raman spectra of aqueous solutions of r(CGCGCG) · r(CGCGCG) in the A-form (top), d(CGCGCG) · d(CGCGCG) in the B-form (middle), and d(CGCGCG) · d(CGCGCG) in the Z-form (bottom). The top and middle spectra were of samples dissolved in 0.5 M NaCl solution, whereas the bottom spectrum is from a sample dissolved in 6 M NaCl. The marker bands for the A-, B-, and Z-forms are shaded black. All of the spectra were taken with the argon laser line at 514.5 nm. (Adapted from Peticolas and Evertsz, 1992.)

molecule. These cross-sections are comparable to the effective fluorescence cross-sections of common laser dyes, and as we will see in Chapter F3 are sufficient for single-molecule detection. A highly sensitive Raman imaging system with submicron spatial resolution can be constructed by combining atomic force microscopy with near-field optics (Chapter F2).

E3.5 VIBRATIONAL RAMAN OPTICAL ACTIVITY

Phenomena that are sensitive to molecular chirality include optical rotation of the plane polarization of a linearly polarized light beam on passing through a solution of chiral molecules (with equal and opposite optical rotation angles for mirror image enantiomers) and UV circular dichroïsm (UV CD) where left and right circularly polarized UV light is absorbed slightly differently. Such "chiroptical" techniques have a special sensitivity to the three-dimensional structure of chiral molecules and were discussed in Chapter E2. The importance of the newer vibrational optical activity methods is that they are sensitive to chirality associated with all the $3N - 6$ fundamental molecular vibrational transitions, where N is the number of atoms, and therefore have the potential to provide more stereochemical information than UV CD, which measures optical activity associated with electronic transitions and so requires an appropriate chromophore.

E3.5.1 Vibrational Circular Dichroïsm (VCD)

Proteins with different folds give characteristic VCD shapes, which vary most for the amide I mode (C=O stretch), but are also easily detectable for the broader amide II and III (very weak) modes (N–H deformation and C–N stretch). Examples are shown in Fig. E3.12.

Surveys of protein spectra show that secondary structure determines the dominant contributions to the VCD shape. Analyses of amide I (proteins in D_2O, N–D exchanged), plus amide III and III VCD and ECD (circular dichroïsm of electronic transition in UV) or FTIR data all yield secondary structure at some level. The important aspects were that VCD sensed sheet and other structural elements (including turns) differently than did ECD, which in turn was superior for helix determination. Combining them gave better determination of all components, especially minimizing the impact of outliers on the prediction scheme.

E3.5.2 Raman Optical Activity (ROA)

In the ROA technique a small difference in the intensity of Raman scattering is measured using right and left circularly polarized incident light. Interference between the waves scattered via the polarizability and optical activity tensors of the molecule yields a dependence of the scattered intensity on the degree of circular polarization of the incident light and on a circular component in the scattered light.

ROA is described by the dimensionless circular intensity difference (CID),

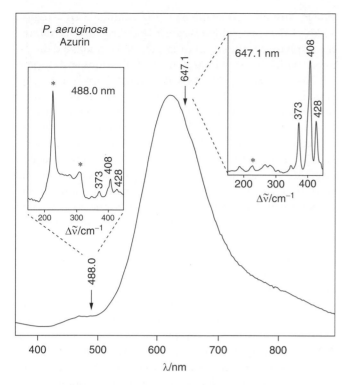

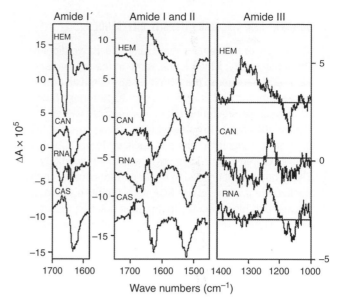

Fig. E3.11 Illustration of the selective enhancement of vibrational modes in a resonance Raman spectrum. The UV–visible absorption spectrum of a *P. aeruginosa* azurin is shown together with two different Raman spectra that derive from laser excitation within the S(Cys) → Cu(II) charge transfer absorption band at 615 nm (647.1 nm, right) and away from the absorption (488.0 nm, left). (After Spiro and Czernuszewicz, 1995.)

Fig. E3.12 Comparison of typical amide I′ (N-deuterated, left), amide I and II (middle) and amide III (right) VCD spectra for proteins in solution that have different dominant secondary structures. Spectra are for hemoglobin (HEM, top, highly helical); concanavalin A (CAN, highly β-sheet, no helix); ribonuclease A (RNA, second from bottom, sheet and helix mixed); and casein (CAS, bottom, a "random coil" protein with no extensive secondary structure). (After Keiderling, 2002.)

COMMENT E3.5 PLASMON RESONANT PARTICLES (PRP) FOR BIOLOGICAL DETECTION

The light scattering by nanometer-sized colloidal metal particles is dominated by the collective oscillation of the conduction electrons induced by the incident light. The specific color (i.e., frequency band) of the scattered light is a function of the size, shape, and material properties of the particles. Silver or gold particles with diameters of 30–120 nm efficiently scatter light in the visible spectrum (Fig. 3.5.1a). The peak scattering wavelength of a particle is generally referred to as the PR peak (Fig. 3.5.1b). Under the same excitation conditions a PRP with diameter of 100 nm can have a brightness that is equivalent to >100 000 fluorescent molecules, and does not undergo photobleaching.

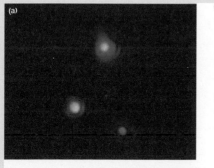
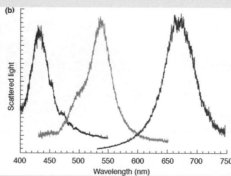

Fig. E3.5.1 (a) A color photograph of three PRPs illuminated with white light. The particles were chosen so that their PR peak wavelength would be in the red, green, and blue, respectively. (b) Spectral curves for the three particles shown in (a). (Shultz, 2003).

The theory of ROA shows that unlike the conventional Raman intensity, which is the same in the forward and backward directions, the ROA intensity is maximized in backscattering and is zero in forward scattering. The CID for right-angle scattering takes intermediate values.

A given ROA signal-to-noise ratio can be achieved at least an order of magnitude faster in backscattering than in the (easier to implement) right-angle scattering geometry. These considerations led to the important conclusion that backscattering boosts the ROA signal relative to the background biomolecules in aqueous solution (Barron *et al.*, 2000).

$$\Delta = (I^R - I^L)/(I^R + I^L) \tag{E3.8}$$

as an appropriate experimental quantity, where I^R and I^L are the scattered intensities in right and left circularly polarized incident light (Comment E3.6).

Polypeptides and Proteins

Vibrations of the backbone in polypeptides and proteins are associated with three main regions of the Raman spectrum. These are: the backbone skeletal stretch region ~870–1150 cm^{-1} originating in mainly C$_\alpha$–C, C$_\alpha$–C$_\beta$ and

C$_\alpha$–N stretch coordinates; the amide III region ~1230–1310 cm^{-1}, which is often thought to involve mainly the in-phase combination of largely N–H in-plane deformation with the C$_\alpha$–N stretch; and the amide I region ~1630–1700 cm^{-1}, which arises mostly from the C=O stretch. The extended amide III region is particularly important for ROA studies because the coupling between N–H and C$_\alpha$–H deformations is very sensitive to geometry and generates a rich and informative ROA band structure. Bands in the amide II region ~1510–1570 cm^{-1}, which originate in the out-of-phase combination of largely the N–H in-plane deformation with a small amount of the C$_\alpha$–N stretch, are not usually observed in the conventional Raman spectra of polypeptides and proteins, but sometimes make weak contributions to the ROA spectra in H$_2$O solution that are generally enhanced in D$_2$O.

Poly(L-lysine) at alkaline pH and poly(L-glutamic acid) at acid pH adopt well-defined conformations under certain conditions and are able to support an α-helical conformation stabilized by both internal hydrogen bonds and hydrogen bonds to the solvent. Backscattered Raman and ROA spectra of these samples are shown in Fig. E3.13. There are many similarities between the ROA spectra of the α-helical conformations of the two polypeptides.

A good example of the ROA spectrum of a highly α-helical protein is that of human serum albumin shown in Fig. E3.14. The amide I ROA couplet centered at

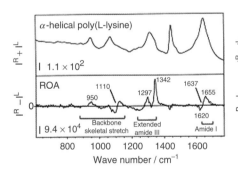
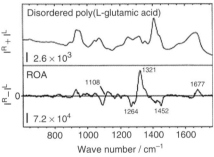

Fig. E3.13 The backscattered Raman ($I^R + I^L$) and ROA ($I^R - I^L$) spectra of poly(L-lysine) and poly(L-glutamic acid) in α-helical conformation in aqueous solution. (Adapted from Barron *et al.*, 2000.)

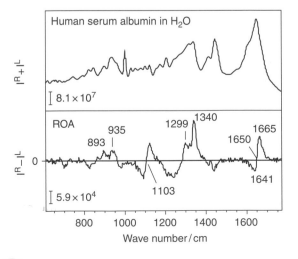
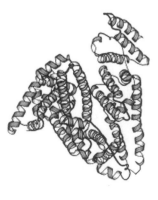

Fig. E3.14 The backscattered Raman and ROA spectra of the α-helical protein human serum albumin in H$_2$O acetate buffer solutions at pH 5.4, together with a MOLSCRIPT diagram of the X-ray crystal structure (PDB code 1ao6). The X-ray crystal structure of the protein specifies ~70% α-helix, with the rest made up of turns and long loops. (Adapted from Barron *et al.*, 2000.)

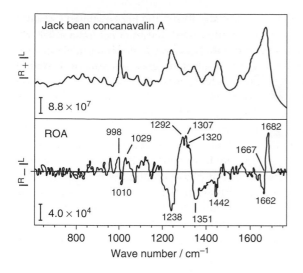

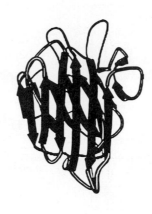

Fig. E3.15 The backscattered Raman and ROA spectra of the β-barrel protein jack bean concanavalin A in H₂O acetate buffer solutions at pH 5.4, together with a MOLSCRIPT diagram of the crystal structure (PDB code 2cna). The X-ray crystal structure of the protein specifies ~40–50% antiparallel β-sheet. (Adapted from Barron *et al.*, 2000.)

~1650 cm⁻¹, which is negative at low wave number and positive at high, appears to be a good signature of α-helix in proteins and agrees with the wave number range of ~1645–1655 cm⁻¹ for α-helix bands in conventional Raman spectra.

The two polypeptides in α-helical conformations show similar amide I ROA couplets (Fig. E3.13), but shifted to lower wave number, which may be due to the particular side-chains or the absence of disordered structure which gives an amide I ROA couplet at high wave number. A positive ROA intensity in the range ~870–950 cm⁻¹ also appears to be a signature of an α-helix. Again the two α-helical polypeptides show positive ROA bands in this region. The conservative ROA couplet centered at ~1103 cm⁻¹, which is negative at low wave number and positive at high, may also be due to an α-helix since a similar feature appears in the ROA spectra of the two α-helical polypeptides. In addition to a positive ROA band

near ~1340–1345 cm⁻¹, an α-helix also gives rise to positive ROA bands in the range ~1297–1312 cm⁻¹.

As well as providing the signature of an α-helix, the ROA spectra of the two polypeptides also provide signatures of β-structure. For example, jack bean concanavalin A in H₂O solution (Fig. E3.15) shows a negative ROA band at ~1238 cm⁻¹ that is almost unchanged in D₂O, which strongly suggests that it originates in the β-strands in the parallel β-sheet within the hydrophobic core.

Carbohydrates

Carbohydrates in aqueous solution are highly favorable samples for ROA studies, giving a rich and informative band structure over a wide range of the vibrational spectrum. The cyclodextrins are particularly interesting samples for ROA studies because they exhibit an enormous glycosidic ROA couplet centered at ~918 cm⁻¹ as compared to that observed in the corresponding α(1–4)-linked disaccharide D-maltose. An example the ROA spectrum of α-cyclodextrin is shown in Fig. E3.16.

Nucleic Acids

Although ROA studies of nucleic acids are still at an early stage, the results obtained so far are most encouraging, with ROA being able to probe three distinct sources of chirality: the chiral arrangement of adjacent intrinsically achiral base rings; the chiral disposition of the base and sugar rings with respect to the C–N glycosidic link; and the inherent chirality associated with the asymmetric centers of the sugar rings.

The ROA spectra of polyribonucleotides are generally more intense and informative than those of the corresponding nucleosides since there are fewer degrees of conformational freedom in the polymers, which reduces the tendency for cancellation of ROA signals from different conformers. The polyribonucleotide ROA spectra can be conveniently subdivided into three distinct spectral regions, each of which encompasses a well-defined

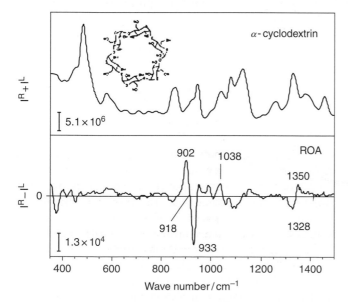

Fig. E3.16 The backscattered Raman and ROA spectra of the α-cyclodextrin in H₂O. (Adapted from Barron *et al.*, 2000.)

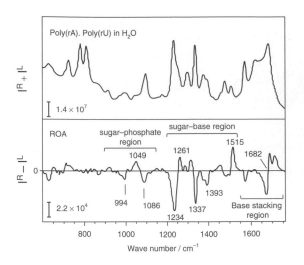

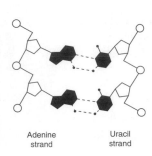

Fig. E3.17 The backscattered Raman and ROA spectra of poly(rA)·poly(rU) in H$_2$O phosphate buffer solutions at pH 7.0. (Adapted from Barron *et al.*, 2000.)

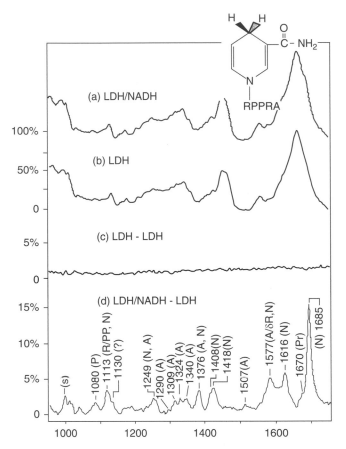

Fig. E3.18 Raman spectrum of **(a)** lactate dehydrogenase (LDH) with bound NADH and **(b)** LDH, at 4 °C in 0.1 M phosphate buffer, pH 7.2. **(c)** The result of the control measurement obtained by calculating the difference between **(a)** and **(b)** showing the real signal-to-noise ratio. **(d)** The result obtained by calculating the difference between **(a)** and **(b)**. The scale in **(c)** and **(d)** is five times that in **(a)** and **(b)**. The molecular structure of the nicotinamide group of NADH is given in spectrum **(a)** for references. Assignments of the peaks in **(d)**: A, adenin; N, nicotinamide; P, phosphate; Pr, protein; PP, pyrophosphate; R(δR), ribose; S, sulfate (After Callender and Deng, 1994.)

structural subunit. The extent and location of these regions are highlighted in the ROA spectrum of poly(rA)·poly(rU), which adopts an A-type double helix, presented in Fig. E3.17. The base stacking region ~1550–1750 cm^{-1} contains ROA patterns from base modes which are characteristic of the particular bases involved as well as being sensitive to the base stacking arrangement. The sugar–base region ~1200–1500 cm^{-1} is dominated by normal modes in which the vibrational coordinates of the sugar and base rings are mixed. The ROA band patterns in this region appear to reflect the mutual orientation of the two rings perhaps together with sugar ring conformation. ROA bands in the sugar–phosphate region ~900–1150 cm^{-1} originate in vibrations of the sugar rings and phosphate backbone. Since the C3′-endo furanose sugar pucker predominates in A-type double helices, the negative–positive–negative ROA triplet (~994, 1049, 1086 cm^{-1}) observed in this region provides a clear marker for this particular conformation. Unfortunately the ROA spectra of naturally occurring DNAs and RNAs are more complicated than those of the polyribonucleotides discussed above.

E3.6 DIFFERENTIAL RAMAN SPECTROSCOPY

In general, the protein structure and conformation should change upon ligand binding. These alterations are evident in the Raman difference spectrum, and signals from the protein should be present in the protein-ligand-minus-protein difference spectrum. Figure E3.18 shows the Raman spectra of lactate dehydrogenase (LDH) with bound NADH (i.e., reduced nicotinamide adenine dinucleotide) (Fig. E3.18a) and LDH itself (Fig. E3.18b). Figure E3.18d demonstrates the difference spectrum between the LDH/NADH binary complex spectrum and that of LDH. The Raman bands in the difference spectrum are mostly from bound NADH, although some protein bands show up owing

COMMENT E3.7 ISOTOPE EDITING

In many cases, the signal from a bound ligand is similar in size to, or smaller than, that arising from changes in protein structure. Moreover, some proteins are not stable without bound ligands. The strategy of isotope editing of the difference spectrum obviates these problems; here, isotopically labeled ligands are prepared so that the frequencies of those modes that involve motions of the isotope are shifted. Thus, a subtraction of the labeled ligand from the unlabeled one yields a difference spectrum of positive and negative peaks of just those modes, and all others cancel (Manor *et al.*, 1991).

COMMENT E3.8 HEME PROTEINS

Heme proteins are a class of biomolecules which possess a heme group embedded in a single polypeptide chain. The reactivity of the heme group varies greatly from protein to protein depending on the nature of the active site environment provided by the polypeptide side-chains, and this variability gives rise to the remarkable functional diversity that characterizes this class of proteins. For example, myoglobin (Mb) and hemoglobin (Hb) bind dioxygen reversibly and thus serve as O_2 storage (Mb) and transport (Hb) proteins. Various peroxidases and catalases have different active site environments that impart a different reactivity to the heme group, namely the ability to react to H_2O_2 to form highly oxidizing intermediates in which the oxidation equivalents are localized on the heme (and in some cases on surrounding protein residues) and which are capable of oxidizing various substrates. Still other heme proteins such as cytochrome *c* and cytochrome *b* have active site environments which render the heme unreactive to exogenous ligands, and these serve only as electron transfer proteins. Finally, one of the most important and frequently studied heme proteins is cytochrome *c* oxidase, whose function is to catalyze the four-electron reduction of dioxygen to water.

to the effect of NADH on the protein upon binding (see Fig. E3.18 caption for assignments). A completely dominant protein-ligand-minus-protein difference spectrum can be observed by the isotope editing technique (Comment E3.7).

As a rule of thumb, the Raman spectrum of a protein minus that of the perturbed protein (perturbed by bound ligand or mutated residue) is about 1% that of the protein itself. It requires a protein Raman signal with a signal-to-noise ratio in excess of 300:1. This requirement is consistent with the difference FTIR spectroscopic studies of proteins (Section E1.4.8).

E3.7 TIME-RESOLVED RESONANCE RAMAN SPECTROSCOPY

The inherent scattering processes (10^{-14} s) carries with it the implication that it is theoretically feasible to monitor events on the subpicosecond timescale. Advances in laser technology (ultra-short pulses) and development of highly sensitive detectors have now made it possible to approach these theoretical limits. Time-resolved resonance Raman methods provide the capability to probe the precise structure of fleeting intermediates which evolve and decay, even on subpicosecond timescales, throughout the course of a given biological process.

In describing the time-resolved Raman methods, we organize the treatment according to two distinctly different approaches for initiating the reaction. The first and more common approach applies to biological systems that are naturally responsive to light. In these cases, the reaction or process of interest can be initiated by a short pulse of light (called the pump pulse). Other (obviously most) biochemical systems are unresponsive to light pulses, and the reaction must be initiated by rapid mixing of the reactants.

E3.7.1 Light-Initiated Methods

Molecular systems that are susceptible to the light-initiated approach are those that undergo a sequence of structural

changes on exposure to photolysis. There are a number of ways that the photolysis step can be accomplished. By far the most common method is to use pulsed lasers; however, several methods have been devised using continuous wave lasers, and these too are useful if it is not necessary to probe at very short times. Although these latter methods provide time resolution on the microsecond timescale, the approach using a pulsed laser permits investigation at the nanosecond, picosecond, and even subpicosecond levels, the temporal resolution being determined by the widths of the laser pulses.

Heme Proteins

One of the most important and impressive applications of transient Raman studies deals with the investigation of the structural dynamics and mechanism of action of heme proteins (Comment E3.8) and, in first place, hemoglobin (Hb) and myoglobin (Mb) molecules (Comment E3.8). Hb and Mb, two of the first proteins to be structurally determined by X-ray crystallography, both contain five-coordinate, ferrous hemes at the active sites as shown in Fig. E3.19. In both cases the heme is attached to the protein through a coordinate linkage to an imidazole side-chain of a histidine residue (the so-called proximal histidine). The distal side of the heme pocket, at which exogenous ligands bind, contains non-polar amino acid residues as well as an imidazole fragment of a histidine

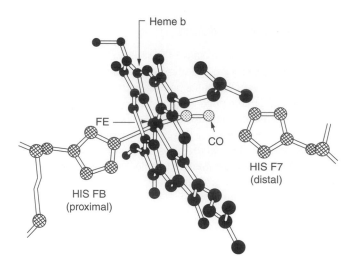

Fig. E3.19 The biological active center in hemoglobin and myoglobin molecules. The filled black circles represent the heme skeletal atoms, and the black circles with white dots represent the peripheral groups of the heme. The cross-hatched circles represent the atoms of the proximal and distal histidine and the white circles with black dots a bound CO ligand. (Kincaid, 1995.)

COMMENT E3.9 COOPERATIVITY OF O₂ BINDING

Whereas the oxygen storage protein, Mb, consists of a single polypeptide chain possessing a proteome at its active site, the oxygen transport protein, Hb, is actually a tetramer of four (two α and two β) subunits, each of which contains a heme group at its active site. The tetrameric nature of Hb permits interactions between the subunits which give rise to the phenomenon known as cooperativity. Thus the binding curve assumes a sigmoidal shape, leading to more efficient oxygen transport in the sense that it binds O₂ less readily at low O₂ levels (in tissues), but has a much higher affinity at high O₂ levels (in the lungs). This behavior is consistent with the two-state theory first proposed by Monod in which cooperative behavior is the result of the existence of a low affinity (the so-called T) state and a higher affinity (R) state; ligand binding to the T state induces a protein structural transition to the more reactive R state. The subsequent crystallographic confirmation of two distinct structures for the ligated (R) and non-ligated (T) forms provided a structural basis for the validity of the thermodynamic arguments. Though the existence of several other states is required to explain the precise details of the binding curves under some conditions, the two-state model remains a valid good approximation under most conditions of ligand binding, indicating that T–R transition is the most important factor for regulating Hb activity (see also Comment A2.1).

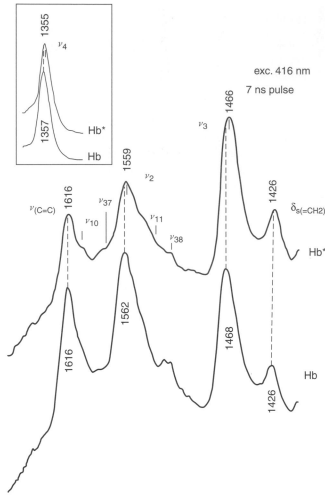

Fig. E3.20 Soret band excited time-resolved Raman spectra of deoxy Hb and Hb* (photoproduct) with 416 nm, 7 ns pulses. Concentration: 0.1 mM; laser energy, 0.5 mJ per pulse; spectral slit width, 8 cm⁻¹; collection time 10 min; and repetition rate 10 Hz. (Kincaid, 1995). Inset: made v_4 at ten times reduced scale. (Dasgupta and Spiro, 1986.)

(the so-called distal histidine), which is known to be capable of forming a hydrogen bond with certain bound ligands, including dioxygen (Comment E3.9).

Figures E3.20 and E3.21 show the spectra of the deoxyHb photoproduct on the 7 ns timescale acquired with an excitation line of 416 nm and 532 nm, respectively. As was explained in Chapter E1 under Soret band excitation (i.e., 416 nm), the totally symmetric lines (v_2, v_3, v_4) are strongly enhanced, whereas excitation at 532 nm (Q-band) provides strong enhancement of non-totally symmetric modes (v_{10}, v_{11}, v_{19}). In Figs. E3.20 and E3.21 the spectrum of the equilibrium deoxy species is presented along with that of the 7 ns photoproduct. It is obvious that all of the features associated with the heme macrocycle appear at low frequencies in the photoproduct spectra relative to their positions in the spectra of deoxyHb at equilibrium. Careful comparison

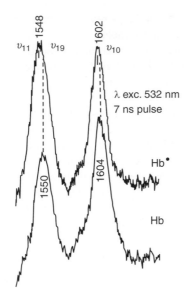

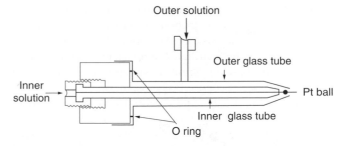

Fig. E3.22 Platinum sphere mixing device. (Adapted from Kincaid, 1995.)

Fig. E3.21 Q-band excited resonance Raman spectra of deoxyHb and Hb• with 532 nm, 7 ns pulses. Concentration, 0.5 mM; laser energy, 1 mJ per pulse; spectral slit width, 5 cm⁻¹; collection time, 30 min; and repetition rate 10 Hz. (Adapted from Kincaid, 1995.)

COMMENT E3.10 VARIOUS RESONANCE RAMAN MODES

Various resonance Raman active modes provide information about different fragments of the heme group and its immediate environment. Two high-frequency modes near 1470 cm^{-1} and 1610 cm^{-1} are sensitive to the electron density in the heme π orbitals. Other high-frequency modes (near 1470 cm^{-1}, 1610 cm^{-1}, 1540 cm^{-1}, and 1555 cm^{-1}) are called core-size markers and respond to changes in the size of porphyrin core (the distances between the pyrrole nitrogen and the center of the porphyrin core).

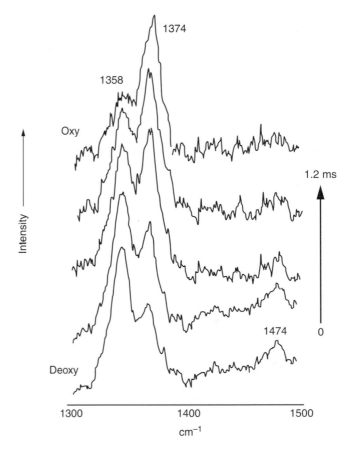

Fig. E3.23 Time-resolved Raman spectra of the reaction of deoxyhemoglobin (0.25 mM in heme) with an oxygen-saturated buffer, pH 7.0, obtained using the mixing device shown in Fig. E3.22. The excitation line was 457 mm. (Adapted from Kincaid, 1995.)

of the frequencies of these core-size markers (Comment E3.10) with a series of structurally well-defined model compounds shows that the photoproduct has a slightly expanded core relative to the deoxy structure and implies that the probable cause for this expansion is restriction of the out-of-plane displacement of the iron brought about by protein restraint on the proximal histidine. Although the preceding examples of time-resolved Raman spectroscopy studies of Hb structural dynamics provide an introduction to these issues, the detailed structural interpretation and precise temporal behavior of the various modes are quite complex and obviously beyond the scope of this book.

E3.7.2 Rapid Mixing Methods

The great majority of biological systems of interest cannot be photo-initiated. For these cases it is necessary to rely on rapid mixing methods in which the temporal resolution is dictated by the speed with which the reaction mixture can become homogeneous with respect to reactant concentrations. This mixing time is referred to as the "dead time" of the mixing apparatus, and various designs have been developed to maximize the mixing efficiency of such devices by minimizing the dead time.

All the rapid mixing time-resolved Raman spectroscopy studies so far reported have employed conventional mixing devices that were developed for use with optical (absorption or fluorescence) detection, in which the observation chamber must necessarily be optically homogeneous. However, given the fact that Raman spectroscopy is based on a scattering phenomenon, novel approaches for rapid mixing have been proposed. The use of one such device is depicted in Fig. E3.22. The essential task to be faced in rapidly and efficiently mixing two solutions is to force them to combine turbulently in a *very small* volume. In the device shown in Fig. E3.22 the two reactant solutions are fed through concentric tubes to a tip which has a platinum sphere positioned very close (10 μm) to the (~100 μm) orifice of the outer tube. In this way both solutions are forced to flow through the 10 μm circular passage formed by the sphere and the outer tube orifice. The mixing of the two solutions is extremely efficient (dead times as low as 20–40 μs). The progress of the reaction can be conveniently monitored merely by varying the distance between the tip of the outer tube and the probe laser. This approach is very useful in the study of heme proteins. For example, as shown in Fig. E3.23, the monitoring of the reaction of O_2 with deoxyhemoglobin is followed over the first 1.2 ms by monitoring modes which appear at 1358 cm^{-1} (Hb) and 1374 cm^{-1} (Hb*O_2).

E3.8 CHECKLIST OF KEY IDEAS

- Raman spectroscopy is a form of vibrational spectroscopy and is related to transitions between the different vibrational energy levels in a biological macromolecule; in this respect Raman spectroscopy is like IR spectroscopy.
- Raman spectroscopy differs from IR spectroscopy in that information is derived from *light scattering*, whereas in IR spectroscopy information is obtained from an *absorption* process.
- Raman scattering is associated with normal modes that produce a change in the *polarizability* of the molecule, while IR absorptions are observed for vibrational modes that change the *dipole moment* of the molecule.
- Raman spectroscopy of biological macromolecules is conventionally divided into two types, *resonance and non-resonance*, on the basis of the proximity of the

excitation radiation frequency to electronic transitions of the components of the sample.
- Non-resonance or classical Raman spectroscopy is very useful in the determination of protein, nucleic acid, and membrane conformational states.
- Resonance Raman spectroscopy has provided considerable information on heme proteins and other metalloproteins and visual pigments.
- The central advantage of resonance Raman spectroscopy is that when the wavelength of the exciting laser line is adjusted to coincide with that of an allowed electronic transition in a molecule, the intensities of certain Raman bands are enhanced relative to the off-resonance values.
- When a molecule is placed near the surface of a particle of metal, such as silver or gold, the Raman signal intensity is enormously enhanced; this effect is called surface-enhanced Raman scattering.
- Raman spectroscopy and Raman microscopy provide direct comparison of the properties of a molecule in solution and in a single crystal.
- In the estimation of protein secondary structure content in terms of percentage helix, β-strand, and turn, the amide I band is analyzed as a linear combination of the spectra of reference proteins whose structures are known.
- From a comparison of the Raman spectra taken from crystals of DNA of known structure, a correlation between the frequencies and intensities of Raman bands and the conformation of the DNA has been established.
- Proteins with different folds give characteristic VCD shapes, which vary most for the amide I mode (C=O stretch) but are also easily detectable for the broader amide II and III (very weak) modes (N–H deformation and C–N stretch).
- In the ROA technique a small difference in the intensity of Raman scattering is measured using right and left circularly polarized incident light.
- As well as providing a signature of the α-helix, some of the ROA spectra also provide signatures of β-structure.
- The inherent scattering processes (10^{-14} s) carries with it the implication that it is theoretically feasible to monitor events on the subpicosecond timescale; advances in laser technology (ultra-short pulses) and development of highly sensitive detectors have now made it possible to approach these theoretical limits.

Suggestions for Further Reading

Historical Review and Introduction to Biological Problems
Cary, P. R. (1982). *Biochemical Applications of Raman and Resonance Raman Spectroscopy*. London: Academic Press.

Asher, S. A. (1993). UV resonance Raman spectroscopy for analytical, physical, and biophysical chemistry: Part I. *Analyt. Chem.*, **65**, 59A–66A.

Asher, S. A. (1993). UV resonance Raman spectroscopy for analytical, physical, and biophysical chemistry: Part II. *Analyt. Chem.*, **65**, 201A–210A.

Non-Resonance Raman Spectroscopy

Williams, R. W. (1986). Protein secondary structure analysis using Raman Amide I and Amide III spectra. *Meth. Enzymol.*, **130**, 311–331.

Callender, R., and Deng, H. (1994). Nonresonance Raman difference spectroscopy: A general probe of protein structure, ligand binding, enzymatic catalysis, and the structures of other biomacromolecules. *Annu. Rev. Biophys. Biomol. Struct.*, **23**, 215–245.

Skoog, D. A., Holler, F. J., and Nieman, T. A. (1995). *Principle of Instrumental Analysis*. Philadelphia; PA: Saunders College Publishing.

Peticolas, W. L. (1995). Raman spectroscopy of DNA and proteins. *Meth. Enzymol.*, **246**, 389–415.

Tomas, G. J., Jr (1999). Raman spectroscopy of protein and nucleic acid assemblies. *Annu. Rev. Biophys, Biomol. Struct.*, **28**, 1–27.

Resonance Raman Spectroscopy

Hudson, B., and Mayne, L. (1986). Ultraviolet resonance Raman spectroscopy in biopolymers. *Meth. Enzymol.*, **130**, 331–350.

Spiro, T. G., and Chernuszevich, R. S. (1995). Resonance Raman spectroscopy of metalloprotein. *Meth. Enzymol.*, **246**, 416–459.

Vibrational Raman Optical Activity

Barron, L. D., Hecht, L., Blanch, E. W., and Bell, A. F. (2000). Solution structure and dynamics of molecules from Raman optical activity. *Prog. Biophys. Mol. Biol.*, **73**, 1–49.

Differential Raman Spectroscopy

Zheng, R., Zheng, X., Dong, J., and Carey, P. R. (2004). Proteins can convert to β-sheet in single crystals. *Protein Sci.*, **13**, 1288–1294.

Time-Resolved Resonance Raman Spectroscopy

Kincaid, J. (1995). Structure and dynamics of transient species using time-resolved resonance Raman spectroscopy. *Meth. Enzymol.*, **246**, 461–501.

OPTICAL ACTIVITY AND CIRCULAR DICHROÏSM

E4

E4.1 HISTORICAL REVIEW AND INTRODUCTION TO BIOLOGICAL PROBLEMS

Optical activity is the property of certain materials to change the angle of polarization of a beam of light that is shone through them as a function of its wavelength (*optical rotation dispersion*, ORD), or to absorb light differently according to its wavelength and polarization (*circular dichroïsm*, CD, from the Greek meaning related to two colors).

1812–1838

Jean-Baptiste Biot and **Auguste Fresnel** independently made careful studies of *optical activity*. **Biot**'s *polarimeter* (an optical instrument used to measure the angle of rotation of polarized light) used sunlight, plane-polarized by reflection off glass, as a light source. **Biot** defined [*α*] the angle of *specific rotation* of the axis of plane-polarized light by an optically active solution in terms of the concentration in grams per cubic centimeter and the optical path in *decimeters*, a definition still in use. In **1824**, **Fresnel** predicted that helical structures would be optically active.

1848

Louis Pasteur observed that *optically inactive* crystals of sodium ammonium tartrate were in fact mixtures of two classes. When separated, both classes turned out to be optically active with respective specific rotation angles of equal value but opposite in sign. The molecules making up the crystals were themselves in one of two asymmetric forms, *enantiomorphs* (from the Greek for opposite shapes), each a mirror image of the other. Since a two-dimensional molecule and its mirror image can be made to coincide and be considered to be identical, the implication was that molecules are three-dimensional (Comment E4.1). The form that turns the axis of polarized light to the right when viewed toward the light source in a *polarimeter* is called *dextro-rotatory* or the *D-form*; the form that turns the axis to the left is called the *L-form* or *laevo-rotatory* (from the Latin for right and left). The same terms apply to *circularly*

polarized light (see Chapter A3); the electric vector of right-circularly polarized light turns in the clockwise direction when viewed toward the light source, the vector of left-circularly polarized light turns anticlockwise. Above a certain temperature (26 °C for sodium ammonium tartrate) the D- and L-forms mix to make up homogeneous crystals. They are said to form a *racemate* or *racemic mixture* (from the Latin, *racemes*, a bunch of grapes (optically inactive tartaric acid was obtained from grape juice)).

1870s

Jacobus Henricus van't Hoff and **Joseph-Achille Le Bel** independently proposed the three-dimensional tetrahedral model for the carbon atom. **Van't Hoff** coined the phrase "chemistry in space."

1895

Aimé Cotton accounted for CD and anomalous ORD (in which [*α*] does not simply increase with decreasing wavelength, but may show sign changes or curve maxima or minima) as due to optically active absorption bands. CD and its associated ORD are termed the *Cotton effect*.

COMMENT E4.1 LEFT HAND, RIGHT HAND

Look at your hands. One is a mirror image of the other. If they were two-dimensional, i.e., paper-thin without the palm on one side and the back on the other, then they would superpose exactly and could be considered to be identical. The reason they do not superpose is because they are three-dimensional and asymmetrical. You could make the palms point in the same direction, for example, but then the thumb of one hand would superpose with the little finger of the other.

The word *chiral* (from the Greek, meaning handed) was introduced in 1904 by Lord Kelvin, Professor of Natural Philosophy at the University of Glasgow, with the following definition: "I call any geometrical figure, or group of points, *chiral*, and say it has *chirality*, if its image in a plane mirror, ideally realized, cannot be brought to coincide with itself."

1935

The book *Optical Rotatory Power* by **Thomas Martin Lowry** was published.

1951

Linus Pauling and **L. B. Corey** discovered α-helices and β-sheets and suggested they were components of protein structure. These are optically active structures and the hypothesis strongly stimulated the renaissance of ORD and later the development of CD to identify them in synthetic polypeptides and proteins in solution, and to study the helix–coil transition in polymers.

1954

Jen Tsi Yang and **Joseph F. Foster**, who studied the optical rotation of proteins, showed bovine serum albumin expanded reversibly in acid solution with a marked increase in laevo-rotation at the sodium D-line (negative $[\alpha]_D$) (Comment E4.2).

1955

Carolyn Cohen postulated that right-handed helices would be dextro-rotatory even if made up of left-handed amino acid constituents. In **1957**, **Yang** and **Paul Doty** published the anomalous ORD of polypeptides in helix-promoting solvents, and in the **early 1960s** the presence of right-handed α-helices in proteins was confirmed by X-ray crystallography.

1956

William Moffit, **J. T. Yang**, **D. D. Fitts**, and **John G. Kirkwood**, first independently, then together, developed the theory of ORD of the α-helix.

1960s and the following decades

High-performance commercial instruments became available that allowed the direct and precise experimental observation of CD bands in the UV spectral region, and CD, the more powerful technique, essentially replaced ORD as a sensitive measure of protein conformation in solution. The method was applied to nucleic acids and protein–nucleic acid interactions. Automation permitted the development of kinetics measurements and stopped-flow equipment was adapted to the spectropolarimeters. The vacuum-UV wavelength limit was pushed down to approach 170 nm.

1980s and the following decades

Synchrotron radiation sources became available, opening the way for an extension of CD applications from the vacuum-UV to the extreme-UV and X-ray ranges (wavelengths ≤ 100 nm). At the other end of the spectrum, ROA studies (see Section E3.5.2) are based on the observation of inelastic scattering of left- and right-handed polarized light, and VCD (see Section E3.5.1) is based on CD in the IR spectral region.

COMMENT E4.2 OPTICAL ROTATION AND PROTEIN DENATURATION

Secondary structures in a native protein, such as right-handed α-helices, display right-handed optical rotation. Upon loss of secondary structure, a denatured polypeptide shows an increase in left-handed optical rotation because of its L-amino acid constituents.

E4.2 BRIEF THEORETICAL OUTLINE

E4.2.1 Plane, Circularly, and Elliptically Polarized Light

The relationships between plane, circularly, and elliptically polarized light have been discussed in Chapter A3 (also see Figs. E4.1 and E4.2). Recall that plane-polarized light can be resolved into equal left and right circularly polarized components. Elliptically polarized light is formed when the right

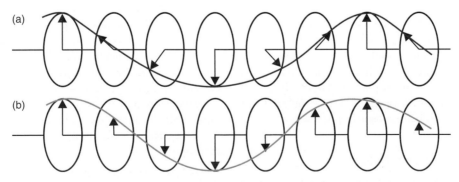

Fig. E4.1 (a) The path traced out by the tip of the electric vector (the arrow) during the propagation of right circularly polarized light traces a right-handed helix. The vector in left circularly polarized light traces a left-handed helix. Note that in both cases the length of the vector remains constant. **(b)** In linearly polarized light the electric vector stays in the same plane, but its length is modulated by the wavelength. (Johnson, 1985.) (Figure reproduced with permission from John Wiley & Sons.)

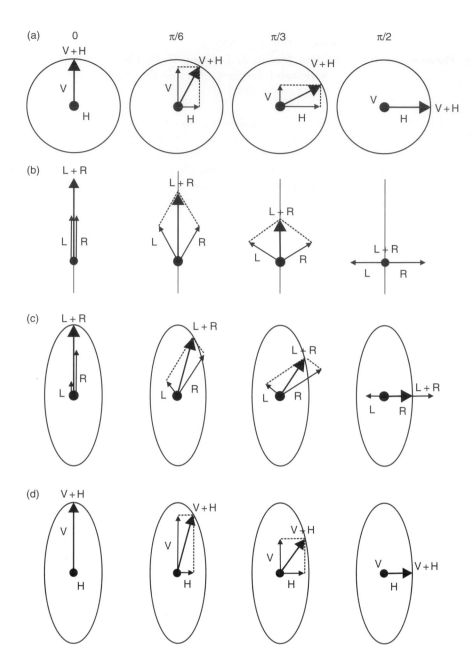

(a)

(b)

(c)

(d)

Fig. E4.2 The projection of the electric vector during light propagation as it would be seen by an observer looking into a polarimeter: **(a)** right circularly polarized light shown as made up of two plane-polarized components, V and H, of equal magnitude out of phase by 90°; **(b)** plane-polarized light shown as made up of left (L) and right (R) circularly polarized light of equal magnitude; **(c)** elliptically polarized light shown as made up of left (L) and right (R) circularly polarized light of unequal magnitude; **(d)** elliptically polarized light shown as made up of two plane-polarized components, V and H, of unequal magnitude out of phase by 90°. (Johnson, 1985) (Figure produced with permission from John Wiley & Sons.)

and left components are not equal; in CD this happens when one of the components is absorbed more than the other. The *ellipticity* is defined as the angle θ, which is usually small so that it is given simply in radians as the value of its tangent, the ratio of the minor to major axes; when the major axis of the ellipse (or axis of the original plane-polarized beam) is rotated by going through an optically active material, the *optical rotation* is defined by the angle α (Fig. E4.3).

E4.2.2 CD, Ellipticity, and ORD

The absorbance of light of a certain polarization, p, by a material is expressed in terms of an extinction coefficient, εp:

$$A_p = \log_{10}\left(I_p^o / I_p\right) = \varepsilon_p C I \qquad (E4.1)$$

where I_p^o, and Ip are, respectively, the incident and transmitted light intensities of polarization, pC is the molar concentration, and l is the optical path length in the sample. The parameter ε_p is called a *decadic* molar extinction coefficient, because it is defined with respect to the base ten logarithm (Eq. (E4.1)).

CD is defined as the difference, ΔA, between the absorption of left circularly and right circularly polarized light, A_l, A_r, respectively, for a *chiral* (from the Greek *cheir*, hand) material:

$$\Delta A = A_l - A_r = \Delta\varepsilon C l \qquad (E4.2)$$

where $\Delta\varepsilon$ is called the decadic molar CD.

CD can also be defined in terms of the *molar ellipticity*. We recall that in cases where there is a difference between A_l and A_r, the right and left circularly polarized

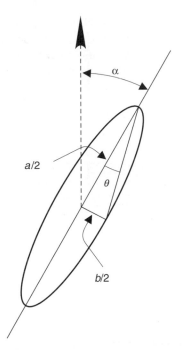

Fig. E4.3 The ellipse, with major axis a and minor axis b, formed by the electric vector projection in elliptically polarized light, showing the optical rotation angle α and the ellipticity θ. (Snatzke, 1994) (Figure reproduced with permission from John Wiley & Sons.)

components of the transmitted beam will not be equal, resulting in an elliptically polarized transmitted beam (Fig. E4.3). It can be shown that the ellipticity, θ, and CD, ΔA, are related:

$$\theta = 180(\ln 10)\Delta A/4\pi = 32.98\Delta A$$
$$[\theta] = 100\theta/Cl \qquad (E4.3)$$
$$[\theta] = 3298\Delta\varepsilon$$

where θ is now in *degrees* and the ln 10 term occurs in the equation because the CD was defined with respect to the base ten logarithm. $[\theta]$ is called the *molar ellipticity*. CD spectra are often expressed in terms of $[\theta]$ rather than $\Delta\varepsilon$, but in either case care must be exercised to specify the basis used to define the molar concentration, which could be, for example, residue concentration, protein concentration or even subunit concentration in the case of oligomeric structures (heme molarity, equivalent to subunit molarity, is usually taken for tetrameric hemoglobin).

Since CD has essentially completely supplanted ORD in molecular biophysics, we shall not discuss ORD spectra in detail. ORD and CD are related to each other by a mathematical transformation called the Kronig–Kramers transformation (Comment E4.3). The ORD can be calculated when the CD spectrum is known and vice versa. Similarly to a Fourier transformation (see Chapter A3), a CD or optical rotation value at a given wavelength is defined by the entire ORD or CD spectrum, respectively. Because of

COMMENT E4.3 CD AND ORD

The decadic molar CD, $\Delta\varepsilon$, and molar rotation, $[\phi] = 100\alpha/Cl$ (where α is the optical rotation in degrees) are related by the Kronig–Kramers transformation:

$$[\phi] = (9000 \ln 10/\pi^2)\mathbf{p}\int_0^\infty \Delta\varepsilon(\lambda)\lambda'd\lambda'/\left(\lambda^2 - \lambda'^2\right)$$

$$\Delta\varepsilon = -(2250 \ln 10\lambda)^{-1}\mathbf{p}\int_0^\infty [\phi(\lambda)]\lambda'^2 d\lambda'/\left(\lambda^2 - \lambda'^2\right)$$

where **p** indicates the "principal value" of the integral is to be taken, avoiding the point $\lambda = \lambda'$ by limiting processes (Moscowitz, 1962).

this interrelationship and the greater wavelength resolving power of CD, interest in ORD decreased significantly when CD instruments became available.

E4.2.3 Electronic Transitions, Dipole, and Rotational Strengths

The bands observed in absorption and CD spectra of macromolecules in solution correspond mainly to electronic transitions between the lowest vibrational levels in the ground-state potential energy well and various vibrational levels in the first excited-state potential energy well (see Chapter E1).

The integrated area under an absorption band for a transition from state a to state b is proportional to effective values called the *dipole strength, D_{ab}*, and *oscillator strength, f_{ab}* (Comment E4.4):

$$D_{ab} = 9.181 \times 10^{-3} \int (\varepsilon/v)dv \, (\text{debye})^2$$
$$f_{ab} = 4.315 \times 10^{-9} \int (\varepsilon/v)dv \, (\text{dimensionless}) \qquad (E4.4)$$

Note that the integration of the extinction coefficient (in units of M^{-1} cm^{-1}) in the band is performed as a function of frequency, v. A more complex equation is obtained as a function of wavelength. The numerical constant in the expression for D contains Planck's constant, the speed of light, and Avogadro's number; in the expression for f it contains Planck's constant and the mass and charge of the electron.

In an analogous way to the definitions of dipole and oscillator strength for an absorption band, the integrated intensity of the CD band associated with the transition from state a to state b is defined in terms of the *rotational strength, R_{ab}*.

$$R_{ab} = 2.295 \times 10^{-39} \int (\Delta\varepsilon/\lambda)d\lambda \, (\text{erg}\,\text{cm}^3) \qquad (E4.5)$$

COMMENT E4.4 DIPOLE STRENGTH, TRANSITION DIPOLE MOMENT, DEBYE UNITS AND OSCILLATOR STRENGTH

The dipole strength is usually given in cgs units (erg cm^3) or in units of (debye)^2. The debye (10^{-18} esu cm, where esu is the electrostatic unit of charge), is a measure of dipole moment. The dipole strength is equal to the square of the transition dipole moment. The transition dipole associated with an intense absorption band has a magnitude in the order of debyes.

The oscillator strength is proportional to the dipole strength. It is dimensionless and in a classical physics interpretation f has the meaning of the number of electrons undergoing the transition as three-dimensional harmonic oscillators (see Chapter A3).

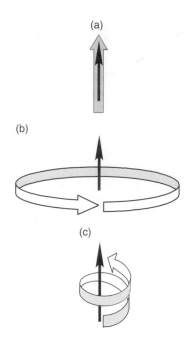

Fig. E4.4 (a) Charge displacement (in gray) due to the transition electric dipole moment vector (black arrow). **(b)** Charge displacement due to the transition magnetic moment vector. **(c)** Charge displacement due to both. (After Cantor and Schimmel, 1980.)

The dipole strength is theoretically equal to the square of the transition dipole moment ($\mu_{ab}\cdot\mu_{ab}$), the scalar product of the transition dipole moment vector with itself. The rotational strength depends on both the electric dipole and magnetic dipole transition moments:

$$D_{ab} = \mathbf{\mu_{ab}} \cdot \mathbf{\mu_{ab}}$$
$$R_{ab} = \text{Im}(\mathbf{\mu_{ab}} \cdot \mathbf{m}ba) \tag{E4.6}$$

The rotational strength is theoretically equal to the imaginary part of the scalar product of the electric and magnetic transition dipole moment vectors. This means that for a molecule to be optically active, the transition magnetic moment must have a component parallel to the transition electric dipole moment (Fig. E4.4).

Recall that a light wave consists of perpendicular oscillating and magnetic fields. Pure absorption is dominated by electric effects, which are much stronger than magnetic effects. The electric dipole transition moment can be pictured as a *linear displacement* of charge, induced by the absorption of a photon matching the energy difference of the transition from the ground state a to state b. The magnetic transition moment can be pictured as due to the *circular motion* of charge induced in the transition. The combination of the two results in a helical displacement of charge around the common dipole axis (Fig. E4.4). The implication is that an optically active molecule must be chiral (asymmetric), otherwise it would not display a preferred helical direction. Also, the helical circulation of charge is facilitated in macromolecules with an inherent helical structural organization making them strongly optically active.

E4.2.4 Rotational Strength and Structural Organization

The rotational strengths of chiral molecules and structures can be calculated, in principle, by using quantum chemistry in terms of electric and magnetic transitions. The systems studied are extremely complex, however, and the calculations difficult and practically impossible to perform analytically, so that empirical approaches are often used. In this section we give only a simplified qualitative description of the relations linking rotational strength and structural organization in macromolecules.

A chromophore is optically active if it is itself chiral; if it is perturbed by a chiral field due to the structural environment, or both. In macromolecular structure determination studies, it is the optical activity induced by the chromophore *organization* that is mainly of interest. *Electric–magnetic* and *exciton* (or *coupled-oscillator*) coupling (Comment E4.5) between the electronic states of a chromophore pair constitute structural contributions to rotational strength, which add to the individual chromophore (*one-electron*) contributions. One-electron, electric–magnetic, and exciton contributions to rotational strength can be computed analytically by quantum mechanics for a pair of identical chromophores (a chromophore dimer), each with a single absorption band. The calculation becomes prohibitively complex, however, in the case of a large heterogeneous assembly. The calculation may be simplified by using symmetry arguments and periodicity, in the case of regular repetitive structures such as a polypeptide α-helix (note, however, that an α-helix in a protein structure is not perfectly periodic along its length due to the diversity of the amino acid side-chains and the local environment).

The coupling of an electrically allowed transition dipole (linear movement of charge) on one chromophore with a magnetically allowed transition dipole (circular motion of charge) on the other is called *electric–magnetic* or μ–m coupling. The contribution to rotational strength of μ–m coupling is usually small.

The rotational strength of a chromophore dimer is dominated by a term called the *coupled oscillator* or *exciton* term, which depends on the geometry of the molecular arrangement. This term may still be large even in the case of optically inactive individual chromophores. It arises in the dynamic quantum mechanical treatment of the system as a pair of coupled oscillators, in which electrically "allowed" and magnetically "forbidden" transition dipoles are coupled by electrostatic interactions. A magnetically forbidden transition, for example, corresponds to a linear movement of charge with respect to an axis through the chromophore. Note, however, that the same motion corresponds to tangential motion along a circle centered on an external axis, which would create a magnetic moment (Fig. E4.5.1).

The third term contributing to the dimer rotational strength is called the *one-electron* term. It arises from the individual chromophore rotational strengths as they are modified by the static field due to their organization. This term also is usually weak compared to the coupled oscillator term.

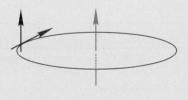

Fig. E4.5.1.

E4.3 INSTRUMENTS

The CD signal is the difference in absorption of left and right circularly polarized light, which is of the order of 10^{-3} for usual samples, so that highly sensitive instrumentation is required for its accurate measurement. Since automatic recording CD instrumentation became commercially available in the 1960s, several generations of instrument have been marketed.

E4.4 CD OF PROTEINS

If one takes into account the amount of sample required and the ease of the experimental procedure as well as measurement accuracy, CD is the most appropriate method for assessing the secondary structure of a protein in solution. The signal is strong in the far-UV, where the peptide bond absorbs, and fairly dilute solutions can be used, of the order of one-tenth of a milligram per milliliter. Circular dichroïsm measurements are extremely powerful for monitoring structural differences and conformational changes. Because of the complexity of the protein structure and the sensitivity of the spectrum to precise local conditions within the molecule, however, the interpretation in terms of structure is necessarily based on empirical criteria (see Section E4.2.4). It is known from analysis of the protein structure database that the most abundant secondary structures are the α-helix and β-sheets, and proteins have been classified according to these and their topological organization (see Chapter A2). Circular dichroïsm due to aromatic groups, such as tryptophan and tyrosine, in the near-UV (where these residues absorb, see Chapter E1) is sensitive to tertiary structure. CD also plays an important role in protein folding studies, because the CD signal of an unstructured polypeptide is distinctly different from the signals arising from secondary structures.

E4.4.1 Circular Dichroïsm of Protein Secondary Structures

Circular dichroïsm spectra observed for α proteins, β proteins, α + β proteins, α/β proteins, and disordered polypeptides are shown in Fig. E4.5. The molar ellipticity is expressed in units per decimole of amino acid residue (see also Table E1.4).

The spectra for α proteins (Fig. E4.5a) display negative bands between about 203 and 240 nm with minima at about 209 and 222 nm; the positive bands below 203 nm have a strong maximum close to 192 nm. These bands are characteristic of the α-helix, and are readily observed in model polypeptides in helix-forming solving solvents, such as poly-lysine above pH 10. The differences in band intensity between the various protein spectra reflect their different α-helical content (Comment E4.6).

The CD spectra of β proteins (Fig. E4.5b) show a greater variety and less intense bands than those for α proteins, and β proteins have been classified in different groups according to their spectrum and β structure type (see below). Regular β proteins have a spectrum similar to that of the β-sheet form of poly-lysine, with a minimum between 210 and 225 nm and a maximum between 190 and 200 nm.

The CD spectra of α + β proteins and α/β proteins can be considered as mixtures of α-helix and β-structure spectra, in which the stronger α-helix band intensities dominate (Fig. E4.5c, d).

The disordered polypeptides include denatured proteins and oligopeptides and polypeptides with disulfide bonds or associated to prosthetic groups (Fig. E4.5e). A strong negative band near 200 nm and weak positive or negative bands above 210 nm are characteristic of disordered structures.

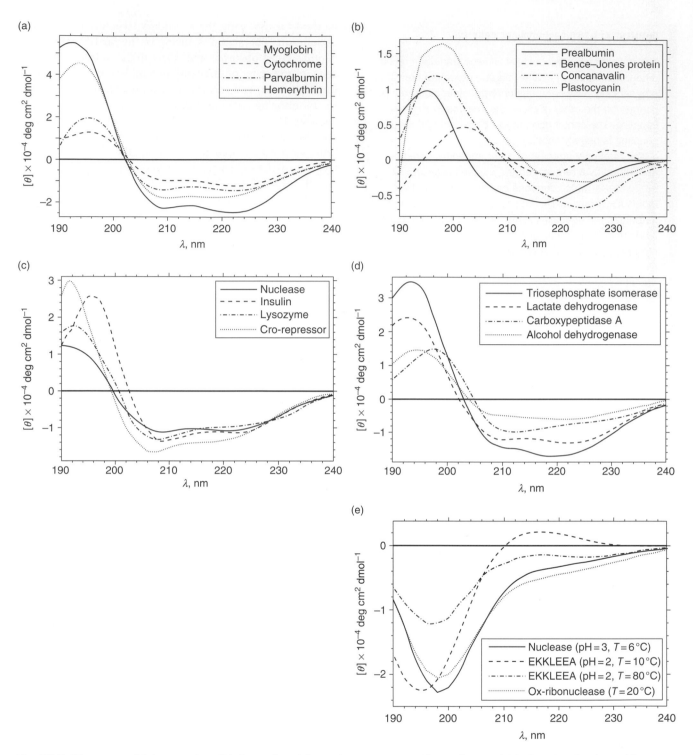

Fig. E4.5 CD spectra of **(a)** α proteins, **(b)** β proteins, **(c)** α + β proteins, **(d)** α/β proteins, and **(e)** disordered polypeptides (Venyaminov and Vassilenko, 1994). (Figure reproduced with permission from Elsevier.)

Fitting Spectra with Basic Sets

If CD coupling between different secondary structures in a protein were negligible to a first approximation, the spectrum could be considered as a simple sum of component spectral bands corresponding to specific secondary structures. The application of such an empirical approach, however, is not straightforward. CD band intensities and wavelengths due to a "perfect," long α-helix may be modeled well on an absolute per residue scale by the spectrum of α-helical polypeptides. But α-helices in a protein structure

COMMENT E4.6 α-HELIX CD

The CD spectrum of α-helical poly-lysine in solution has negative band ellipticity minima of about $-40\,000$ deg cm^2 dmole^{-1}, and a positive band maximum greater than $70\,000$ in the same units. The relative band intensities of the proteins in Fig. E4.5a reflect their relative α-helix content. Thus, myoglobin is more helical than the other proteins, with more than 50% of its residues in α-helices.

COMMENT E4.7 β STRUCTURE CD

β-rich proteins are classified according to their CD spectra as β_I or β_{II}. The β_I spectrum is similar to those of model polypeptide β-sheets (Fig. E4.5b). The β_{II} spectrum resembles that of a disordered polypeptide. It has been shown that it is compatible with the presence of a conformation called P_2 in protein structures. P_2 is the type II conformation identified for poly-proline, poly(Pro)II. It is a left-handed helix with three residues per turn (Sreerama and Woody, 2003).

COMMENT E4.8 THE OPTIMIZATION OF PROTEIN SECONDARY STRUCTURE DETERMINATION BY FTIR AND CD

The CD and FTIR spectra of a specially designed set of 50 proteins were treated by multivariate statistical analysis in order to optimize secondary structure determination in solution. The proteins in the database were selected for their structural diversity. The main two conclusions of the study were:

(1) It was not satisfactory to use subsets of the database so that a large database of proteins selected with stringent criteria must be used.
(2) The independent analysis of either FTIR or CD alone identified "wrong" secondary structure estimates, so that both complementary methods are required.

A further conclusion of the study was that the amide II band in FTIR has high information content and can be used alone (rather than amide I) for secondary structure prediction (Oberg et al., 2004).

the absence of a known structure (Comment E4.8; Fig. E4.6; Table E1.4).

Membrane Proteins

Various methods have been proposed for the interpretation of membrane protein CD spectra in terms of secondary structures, based on empirical as well as theoretical approaches (Comment E4.9). The CD study of proteins *in situ* in biological membrane fragments presents a number of difficulties not encountered in soluble protein studies. The membranes, in fact, contribute to differential light absorption and light-scattering effects resulting in optical artifact distortions in the shapes, intensities, and/or positions of the CD bands. The example of bacteriorhodopsin in purple membranes (see Chapter A3) is shown in Fig. E4.7. Light scattering by the membrane patches led to a distortion of the minima at 208 and 222 nm, as well as to a generally more intense spectrum.

Detergent solubilized membrane protein preparations are relatively free of such artifacts (Fig. E4.7). Membrane protein CD experiments should still be designed carefully, however, because different detergents may lead to subtle conformational differences, which will be seen in the spectra.

Typical spectra of membrane proteins made up predominantly of α-helices are shown in Fig. E4.7. The approaches discussed in Comment E4.9 are capable of separating peripheral and transmembrane α-helix components in membrane protein CD spectra. The CD spectra of porins, which have distinctive, transmembrane β-barrel structures, are shown in Fig. E4.8. The differences between the spectra in Fig. E4.8 reflect different minor proportions of α-helix content.

deviate from "perfection," and, in any case, they are certainly not infinitely long, and structural distortions at their edges affect the spectrum. We have seen above that β secondary structures are quite varied depending on the structure type and on the protein. In fact, the CD spectra of β-rich proteins could resemble those of either model polypeptide β-sheets or disordered polypeptides (Comment E4.7).

When coordinates from X-ray crystallography and NMR for a large number of structures started to become available in protein data banks, the opportunity arose not only to confirm secondary structure predictions made from CD spectra, but also to set up databases of CD spectra corresponding to different protein secondary structure families, which could be used as basis sets for the fitting of spectra of "unknown" proteins. The problems to be solved included careful analysis of the structures themselves in order to identify and define limits for the different secondary structures and to interpret how to deal with distortions from "perfection" within each type of secondary structure. Various databases are proposed on the web for the interpretation of CD spectra.

CD was considered a more convenient experimental approach than FTIR (see Chapter E1) for the determination of protein secondary structures. It appears from a careful analysis, however, that only the comparison of independent CD and FTIR information is able to identify "wrong" secondary structure estimates in

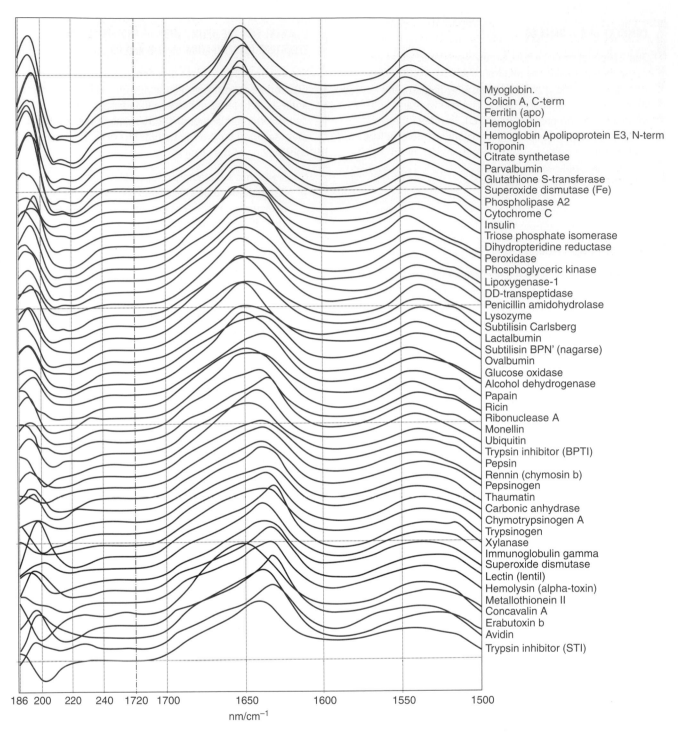

Myoglobin.
Colicin A, C-term
Ferritin (apo)
Hemoglobin
Hemoglobin Apolipoprotein E3, N-term
Troponin
Citrate synthetase
Parvalbumin
Glutathione S-transferase
Superoxide dismutase (Fe)
Phospholipase A2
Cytochrome C
Insulin
Triose phosphate isomerase
Dihydropteridine reductase
Peroxidase
Phosphoglyceric kinase
Lipoxygenase-1
DD-transpeptidase
Penicillin amidohydrolase
Lysozyme
Subtilisin Carlsberg
Lactalbumin
Subtilisin BPN' (nagarse)
Ovalbumin
Glucose oxidase
Alcohol dehydrogenase
Papain
Ricin
Ribonuclease A
Monellin
Ubiquitin
Trypsin inhibitor (BPTI)
Pepsin
Rennin (chymosin b)
Pepsinogen
Thaumatin
Carbonic anhydrase
Chymotrypsinogen A
Trypsinogen
Xylanase
Immunoglobulin gamma
Superoxide dismutase
Lectin (lentil)
Hemolysin (alpha-toxin)
Metallothionein II
Concavalin A
Erabutoxin b
Avidin
Trypsin inhibitor (STI)

186 200   220   240 1720 1700   1650   1600   1550   1500

nm/cm^{-1}

Fig. E4.6 Concatenated CD spectra (186–260 nm) and FTIR spectra (1720–1500 cm^{-1}, showing the amide I bands at around 1650 cm^{-1} and amide II bands at around 1550 cm^{-1}) of 50 reference proteins, chosen as a database by Oberg *et al.* (2004). Note that the wavelength range is not continuous over the two methods (1720 cm^{-1}, for example, corresponds to a wavelength of 5800 nm). The spectra have been rescaled and offset for better readability. The proteins are listed according to their α-helix content. Note the typical α-helix CD spectrum of myoglobin at the top. (Oberg *et al.*, 2004.) (Figure reproduced with permission from Blackwell.)

E4.4.2 Near-UV CD and Protein Tertiary Structure

We saw in Chapter E1 how the aromatic amino acids absorb strongly in the near-UV close to 280 nm. The bands also display a strong CD signal (Fig. E4.9). Bovine ribonuclease A (RNase A) does not contain Trp; it has six Tyr, three Phe and four disulfides. Analysis of the absorption spectrum close to 280 nm indicated that the Tyr residues reside in different local environments with respect to solvent exposure. We see from Fig. E4.9 that the near-UV CD spectral components of two classes of Tyr residue are distinctive; the 288 nm band corresponds to the buried residues and the 283 nm band to the solvent-exposed residues.

The near-UV CD spectrum of a protein is a sensitive measure of tertiary structure (Fig. E4.10, and see the unfolding studies discussed below).

Upon binding to sensitive bacterial cells, the antibiotic colicin A is cleaved to yield a fragment that depolarizes the cell membrane (i.e., makes it permeable to ions). In-vitro studies indicate that the fragment inserts into lipid vesicles at low pH. The analysis of the far- and near-UV CD data (Fig. E4.10) showed how, while maintaining a predominantly α-helical conformation, the peptide loses its tertiary structure at acid pH to facilitate membrane insertion.

E4.4.3 Protein Folding

CD spectra are good empirical indicators of the folding state of a protein. The aromatic region of a spectrum in the near-UV (close to 280 nm) contains information on tertiary conformation, while, as we have seen in some detail above, the peptide bond region in the far-UV CD (close to 222 nm) is sensitive to the presence of α-helix, β-sheet, and random chain conformations. Combined with stopped-flow devices, kinetic CD experiments have been used to follow protein unfolding and refolding processes (Fig. E4.11). The time axis is plotted on a logarithmic scale in Fig. E4.11c, which clearly shows that secondary structure recovery (the 222 nm signal) occurs significantly faster than tertiary recovery (the 289 nm signal). The curves in Fig. E4.11a and b were fitted by two-exponential functions, indicating the existence of intermediates in the folding pathway.

COMMENT E4.9 MEMBRANE PROTEIN CD

Fasman (1996) compares the convex constraint analysis (CCA) deconvolution algorithm, developed in his laboratory, with the fixed reference method of Chang et al. (1978), and the variable reference methods of Manavalan and Johnson (1987). The CCA algorithm extracts P common components from a database, where P is defined as an input parameter. The approach offers considerable flexibility to deal with different databases. The value of P is chosen to match the number of significant components. For example, for a database of proteins containing α-helix and random conformations but no β secondary structure, a value of $P = 3$ would be the most appropriate (for transmembrane, peripheral α-helix, and random structures, respectively). Larger values of P would force the algorithm to extract additional poorly defined components (Fasman, 1996).

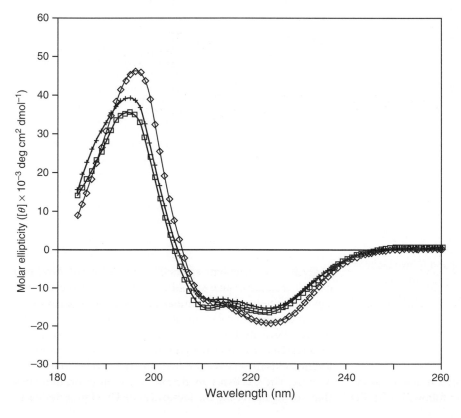

Fig. E4.7 CD spectra of bacteriorhodopsin and rhodopsin: bacteriorhodopsin in purple membranes (diamonds); bovine retina rhodopsin solubilized in the detergent laurylmaltoside (squares); light-adapted bovine rhodopsin in the same detergent (crosses). (Fasman, 1996.)

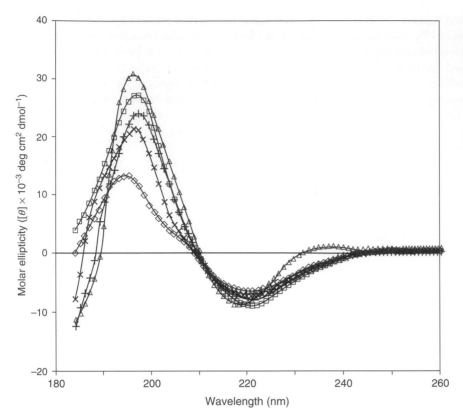

Fig. E4.8 CD spectra of porins (predominantly β-conformation proteins) solubilized in various detergents: maltoporin (triangles); *E. coli* porin (squares); *E. coli* porin in slightly different solvent conditions (crosses); phosphoporin (X); *Rhodobacter capsulatus* porin (diamonds). (Fasman, 1996.)

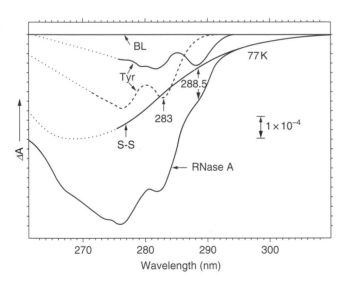

Fig. E4.9 CD spectra of bovine ribonuclease A (RNase A) in the near-UV. The dark lines are observed spectra; the lighter lines and dashed lines are Gaussian components, BL is the background level. The two Tyr lines correspond to two classes for these residues, a solvent exposed and a buried class, respectively. (Horwitz *et al.*, 1970) (Figure reproduced with permission from the American Chemical Society.)

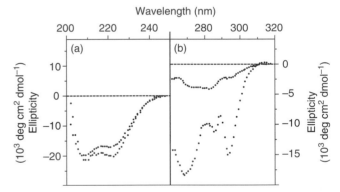

Fig. E4.10 (a) Far- and (b) near-UV CD spectra of the membrane inserting fragment of the peptide colicin A. In each panel the more intense curve corresponds to the acid conditions in which the fragment apparently loses tertiary structure (but maintains secondary structure) in order to insert into the membrane. (van der Groot *et al.*, 1991.) (Figure reproduced with permission from Macmillan Publishers.)

Molten Globules

The molten globule hypothesis has been formulated for the refolding pathway of globular proteins following the CD analysis of intermediate states (Comment E4.10). After first experiments on α-lactoglobulin, which led to the hypothesis, intermediates with similar properties were also found for other globular proteins. Recall from Chapter C2 that a native, correctly folded, globular protein is characterized by a tertiary structure with a tightly packed core, in which amino acids fit very favorably through van der Waals interactions. The CD experiments on α-lactoglobulin indicated that during unfolding the far- and near-UV signals followed

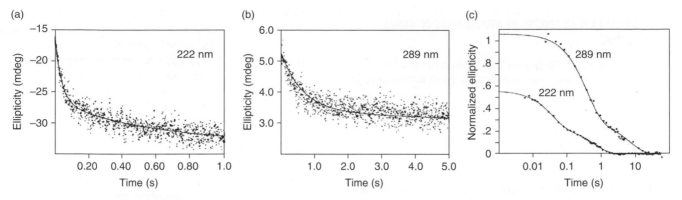

Fig. E4.11 CD stopped-flow refolding kinetics for oxidized cytochrome *c*. The reaction was triggered by a guanidine hydrochloride concentration jump from 4.3 M to 0.7 M. The time courses in the far- and near-UV are in **(a)** and **(b)**, respectively. The two folding kinetics are compared in **(c)**, after normalization. The curve fits are for a two-exponential function. Note the slower recovery of the folded state condition for the near-UV compared to the far-UV data. (Elöve *et al.*, 1992.) (Figure reproduced with permission from the American Chemical Society.)

COMMENT E4.10 THE MOLTEN GLOBULE HYPOTHESIS

Ptitsyn and his coworkers analyzed the denaturation of *α*-lactoglobulin and noted the existence of a partially unfolded intermediate that maintained the far-UV secondary structure CD signature of the folded state, but which had lost the near-UV tertiary structure signature. The observation formed the basis of the molten globule hypothesis (Dolgikh *et al.*, 1981).

A degree of controversy still remains regarding the existence of the molten globule as a folding intermediate for small globular monodomain proteins, however, since no sign of it has been found in calorimetry experiments (see Chapter C2).

---

different time courses (similar to the data shown in Fig. E4.11). An intermediate state appeared in which the secondary structure of the native state was maintained but the tertiary structure was lost. The radius of gyration measured by small-angle X-ray scattering (see Chapter G2) and other parameters established that the intermediate had a compact structure, resembling an expanded form of the tertiary structure, in which the core had lost its tight packing; this was named a molten globule.

Molten globule structures are defined by their far- and near-UV CD signatures (secondary structure similar to the native state, loss of tertiary structure), their radius of gyration (only slightly larger than the native value), and their effect on ANS (8-anilino-1-naphthalene sulfonate) fluorescence, which is sensitive to the local environment of the probe; they can penetrate the loosely packed hydrophobic core of a molten globule, but not the tightly packed core of a native, folded, globular protein.

E4.5 NUCLEIC ACIDS AND PROTEIN–NUCLEIC ACID INTERACTIONS

We recall that the nucleotide, the fundamental unit in nucleic acids, consists of phosphate, a sugar ring (ribose in RNA and deoxyribose in DNA), and an aromatic base (one of adenine, A, guanine, G, cytosine, C, and either thymine, T, for DNA or uracil, U, for RNA). The bases are planar so they do not have any intrinsic CD (see Comment E4.1). The CD signal of the bases in nucleic acid comes to a small extent from the asymmetry of the sugar, but is dominated by the hydrophobic stacking of the base-pairs in the helical structures. This technique, which is particularly sensitive to the secondary structure induced by different solvent environments and base compositions, became the basis for the investigation of nucleic acids of unknown conformation by comparing their CD to established spectra for nucleic acids of known structure (Comment E4.11).

E4.5.1 RNA

We show the CD of RNA PK5 as an example in Fig. E4.12. The molecule is a 26-nucleotide strand that displays various structures as a function of rising temperature. Circular dichroïsm and normal UV absorption spectra were measured for different temperatures and compared to the basis sets from 58 RNA molecules of known structure. The analysis yielded a pseudoknot structure at the lowest temperature, and a hairpin structure at 30 °C, with the PK5 RNA in the A-form (RNA secondary structures are discussed in Chapter A2). The strand was unfolded (melted) at 70 °C. Note how the spectra are sensitive to the different types of stacking.

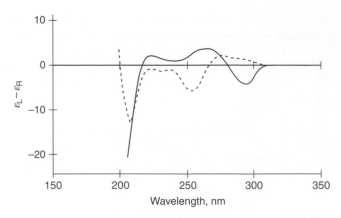

Fig. E4.13 CD spectra of poly d(GC) · poly d(GC): in the B-form in 0.1 M NaCl (full line); in the Z-form in 4 M NaCl or in 60% ethanol (dashed line). (Pohl and Jovin, 1972.) (Figure reproduced with permission from Elsevier.)

E4.5.2 DNA

It was well known that double-helical DNA can take up a variety of right-handed structures depending on the solvent conditions and hydration level (see Chapter A2). Then CD observations led to the striking discovery of Z-DNA, a left-handed, double-helical structure that is adopted by poly d(GC) · poly d(GC) in high salt concentrations or in the presence of alcohol (Fig. E4.13). The CD spectra of B- and Z-DNA are characteristically different. The B-form has a positive band close to 280 nm and a negative band close to 255 nm, while the Z-form displays a negative band at 290 nm and a positive band at 265 nm. Extending the spectra to shorter wavelengths (not shown in Fig. E4.13) revealed a further positive band close to 190 nm for B-DNA and a negative band close to 200 nm, and a positive band close to 180 nm for Z-DNA. Its CD signature guided the search for the Z-form in natural DNA (Fig. E4.14). The experiment was based on the complex supercoiling properties (when the double helix twists around itself) of circular DNA in plasmids. The CD of the

relaxed plasmid corresponds to the B-form. When the two complementary strands in the plasmid were made to associate under a certain topological constraint, however, the CD spectrum showed features characteristic of the Z-form (Fig. E4.14), indicating that left-handed regions were used to respond to the imposed supercoiling constraint. The spectrum corresponded to a Z-form content of about 40%. It is interesting to note that CD experiments established also that RNA strands with alternating purine–pyrimidine bases, such as poly r (GC) · poly r(GC) could assume the Z-form if solvent conditions were favorable.

E4.5.3 Protein–Nucleic Acid Interactions

The CD spectrum of a protein–nucleic acid complex above 250 nm is dominated by the secondary structure of the

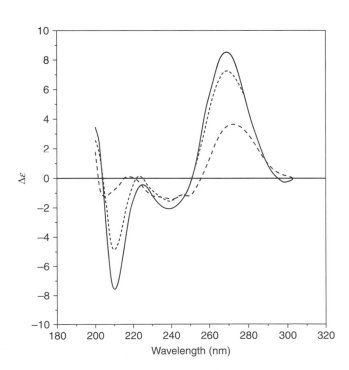

Fig. E4.12 CD spectra of RNA PK5 at different temperatures: 0 °C (line); 30 °C (dots); 70 °C (dashes) (Johnson and Gray, 1992.)

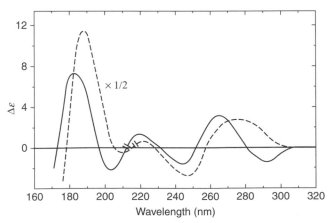

Fig. E4.14 CD spectra of a relaxed (dashed line) plasmid and the same plasmid under a certain topological constraint. The dashed line is similar to a B-form spectrum, while the full line has several Z-form features. (Johnson, 1985). (Figure reproduced with permission from John Wiley & Sons.)

nucleic acid component, and subtle changes in base stacking, for example, can be examined in this spectral region. The interpretation of measurements below 250 nm, where the protein secondary structure contribution becomes dominant, is more difficult, and should be approached with the necessary caution with respect to the spectral width examined, corrections for overlapping nucleic acid bands, and the protein database chosen. Different types of complex may lend themselves more readily to a CD analysis. For example, it is easier to correct for the CD spectra of RNA, which are less variable than those of DNA structures. The CD spectra relating to the interaction between protein and RNA in the capsid of a bacteriophage are shown in Fig. E4.15.

Bacteriophage ϕ6 contains three double-stranded (ds) RNA molecules associated with protein within the same nucleocapsid. In the intact virus, the capsid is itself surrounded by a lipid bilayer. Analysis of the isolated ds RNA and nucleocapsid spectra shows clearly how the RNA component dominates above 240 nm (Fig. E4.15). In this spectral region the encapsidated RNA displays a structure intermediate between an A-form and a dehydrated A-form, showing the effect of condensation. A subtraction of the ds RNA spectrum from the nucleocapsid spectrum above

240 nm in order to examine the protein secondary structure contribution was considered to be reliable from inspection of the spectra. The analysis yielded 50% α-helix content in the protein.

E4.6 CARBOHYDRATES

Circular dichroïsm analysis of carbohydrates can contribute to elucidating their structures, which are extremely complex in composition as well as in conformation (see Chapter A3). As with other polymeric structures, CD distinguishes between random coil and helix organization, between rigid and flexible structures, and, in favorable cases, the type of linkage between the monomer subunits (which, unlike proteins and nucleic acids, can be extremely varied). Experiments are difficult, however, because the relevant information from saccharides occurs at wavelengths below 200 nm (unless there are substituted aromatic groups), where solvent contributions dominate. The wavelength range is accessible with good laboratory instrumentation and synchrotron sources (see below), and solvent contributions can be minimized by using desolvated *film* samples.

The linkage contribution to a CD spectrum can be estimated empirically from the difference between the spectra of the oligosaccharide and the monomer subunits and modeled by using NMR, X-ray crystallography results, and molecular dynamics calculations. The example of trehalose, a dimer of specifically linked glucose subunits (see Chapter A3) is shown in Fig. E4.16. The spectra indicate a positive linkage contribution at 160 nm, which compensates for the negative peak in curve (a) and leads to a weak trehalose CD spectrum. The effect of chain length is seen in Fig. E4.17, in which are shown the spectra of a glucose monomer, maltose (a dimer), and amylose

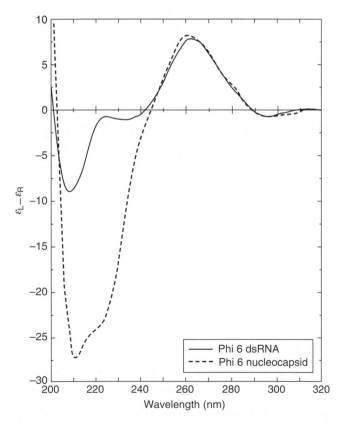

Fig. E4.15 CD spectra, expressed per mole of nucleotide of the double-stranded RNA (full line) and the protein–nucleic acid capsid (dashed line) of bacteriophage ϕ6 (quoted in Gray, Fig. 9 in Fasman (ed.), 1996). (After Steely *et al.*, 1986.)

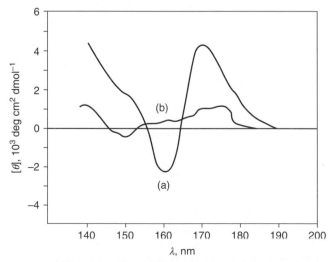

Fig. E4.16 CD spectra of (**a**) methyl α-D-glucopyranoside and (**b**) α,α-trehalose (quoted in Stevens, in Fasman, 1996).

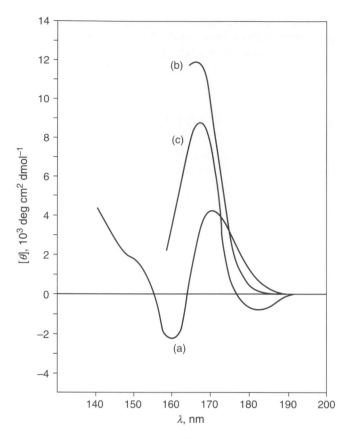

Fig. E4.17 CD spectra of: **(a)** methyl α-D-glucopyranoside; **(b)** maltose; and **(c)** amylose (quoted in Stevens, in Fasman, 1996).

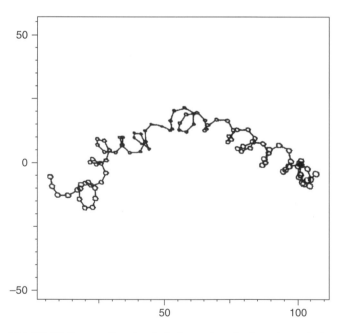

Fig. E4.18 Perspective drawing of a segment of amylose chain chosen as representative in a Monte Carlo molecular dynamics calculation. The axes represent the conformational coordinates of the chain. The scales are in ångström units. Circles represent the glycosidic oxygens and lines are virtual bonds spanning the subunits (quoted in Stevens in Fasman, 1996).

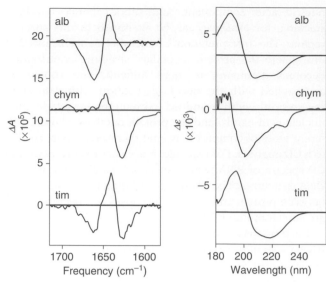

Fig. E4.19 VCD of the amide I band in D_2O solvent and electronic CD for the α protein albumin (alb), the β protein chymotrypsin (chym) and the αβ protein triosephosphateisomerase (tim) (quoted in Keiderling in Fasman, 1996).

(a homopolymer with the same linkage as maltose). There is a large positive linkage contribution in maltose below 175 nm. The difference between the maltose and amylose curves is due to the effects of "kinks" or "bends" in the longer polymer (Fig. E4.18). The long helical polymer is, in fact, made up of finite lengths of straight parts, as shown in the figure. The *persistence length* of the polymer is a measure of the straight stretches.

E4.7 CD FROM IR RADIATION TO X-RAYS

Modern instrumentation and powerful computer-based analysis techniques have extended the experimental usefulness of CD measurements to the vibrational spectral region in the IR, on the one hand, and to the far-UV, also called the *vacuum-UV* (by using intense synchrotron radiation), on the other. An example of the sensitivity of IR CD (VCD) to protein secondary structure is given in Fig. E4.19.

References on developments in VCD and synchrotron radiation CD are given for the interested reader in Fasman (1996).

E4.8 CHECKLIST OF KEY IDEAS

- Plane-polarized light can be resolved into equal left and right circularly polarized components.
- Elliptically polarized light is formed if the right and left components are not equal – in CD this happens when one of the components is absorbed more than the other.

- In an optically active material, the major axis of the ellipse is rotated by an angle called the *optical rotation* angle with respect to the direction of the incident plane-polarized beam.
- The bands observed in absorption and CD spectra of macromolecules in solution correspond mainly to electronic transitions between the lowest vibrational level in the ground-state potential energy well and various vibrational levels in the first excited state potential energy well.
- The integrated intensity of the CD band associated with the transition is defined in terms of the *rotational strength*.
- A chromophore is optically active if it is itself chiral, if it is perturbed by a chiral field due to the structural environment, or both. It is the optical activity induced by chromophore *organization* that is mainly of interest in conformational studies of macromolecules by CD.
- α-helical proteins, β-sheet proteins, and unfolded polypeptides display characteristic CD bands. β proteins show a greater CD spectral variety and less intense bands than α proteins, and have been classified in different groups according to their spectrum and β structure type. The CD spectra of α/β proteins can be considered as mixtures of α-helix and β-structure spectra, in which the stronger α-helix band intensities dominate.
- Databases of empirical CD spectra corresponding to "known" (from crystallography or NMR) protein secondary structure families (including membrane proteins), with which the spectra of "unknown" proteins could be fitted, have been set up.
- The near-UV CD spectrum of a protein is dominated by aromatic residues and is a sensitive measure of tertiary structure.
- CD spectra are good empirical indicators of the folding state of a protein; the far-UV spectral region informs on the presence or absence of secondary structure, while the near-UV informs on tertiary structure.
- Combined with stopped-flow devices, kinetic CD experiments have been used to follow protein unfolding and refolding processes.
- The molten globule hypothesis has been formulated for the refolding pathway of globular proteins following the CD analysis of intermediate states.
- The CD signal of nucleic acids is dominated by the hydrophobic stacking of the base-pairs in the helical structures, and is, therefore, sensitive to structure. CD observations led to the striking discovery of Z-DNA, a left-handed double-helical structure adopted by poly d(GC) · poly d(GC) in high salt concentrations or in the presence of alcohol.
- The CD spectrum of a protein–nucleic acid complex above 250 nm is dominated by the secondary structure of the nucleic acid component, and subtle changes in base stacking, for example, can be examined in this spectral region. The interpretation of measurements below 250 nm, where the protein secondary structure contribution becomes dominant, is more difficult.
- CD analysis of carbohydrates distinguishes between random coil and helix organization, between rigid and flexible structures, and, in favorable cases, the type of linkage between the monomer subunits (which, unlike in proteins and nucleic acids, can be extremely varied). Experiments are difficult, however, because the relevant information from saccharides occurs at wavelengths below 200 nm (unless there are substituted aromatic groups), where solvent contributions dominate. Solvent contributions have been minimized, however, by using desolvated film samples.

Suggestion for Further Reading

Fasman, G. (ed.) (1996). *Circular Dichroïsm*. New York: Plenum Press.

OPTICAL MICROSCOPY

LIGHT MICROSCOPY

F1

F1.1 HISTORICAL REVIEW

Mid-fifteenth century

The simple one-lens microscope was available at this time, with low-power magnifiers being used for the examination of insects. From these early beginnings, glass lenses of increased power were developed. **A. Leeuwenhoek** (1632–1723) produced remarkable microscopes. One of them had a resolving power of 1.35 μm, which was enough for basic cytology. Around the same time, the basic idea for the compound microscope, in which two or more lenses are arranged in such a way to form an enlarged image of an object, occurred independently to several people in the Netherlands (**H. Jansen**, his son **Zacharias**, and **H. Lippershey**). Subsequently, compound microscopes became widely available.

1830

The first achromatic microscope lenses appeared in the Netherlands at the end of the eighteenth century. In 1830, **J. Lister** published the first work in which the construction and design of achromatic lenses was placed within a theoretical framework. He showed how two separate achromatic lenses could be combined to act as a single lens so that the image rendered would be completely free from chromatic aberration and from distortion caused by curvature of the lenses. This discovery opened the way to the construction of high-power microscope lenses within the space of 50 years.

1834

Henry Fox Talbot built the first *polarization microscope* by equipping his light microscope with polarizers. It became an essential tool for mineralogists studying rock crystals.

Latter part of the nineteenth century

E. Abbe was the outstanding figure in development of the microscope. He placed the theory of microscopic image formation on a sound basis by emphasizing the importance of the light diffracted by the object in resolving its fine details. He introduced the concept of numerical aperture and developed oil-immersion lenses. Very soon this type of lens, with its much improved resolving power and high-quality images, was an essential part of the microscopist's equipment. Abbe showed that an increase of resolving power of the microscope was possible only by increasing the numerical aperture of the optical system or by using a shorter wavelength. Abbe proposed the diffraction limit of resolution power of a microscope was about 1/2 of the illuminating light's wavelength.

Early twentieth century

Developments of the microscope have included the introduction of larger stands with built-in illumination systems and higher standards of accuracy of construction. Many different types of microscopes, such as the dark-field microscope, the polarization microscope, fluorescence microscopy, and wide-field epifluorescence microscopy have been proposed. Especially notable was the introduction in **1934** of phase-contrast microscopy, which made it possible to examine living cells and tissue specimens without staining.

1928

E. Synge was one of the first to bring to the physics community the idea of the near-field scanning optical microscope (NSOM). As originally conceived, an illuminated subwavelength-diameter aperture acts as a light source of subwavelength dimensions that can be scanned close to the object to generate a superresolution image.

1934

Fritz Zernike described a phase-contrast microscope that made it possible to examine living cells and tissue specimens without staining.

1972

E. Ash and **G. Nicholls** used the idea of near-field scanning for imaging purposes. Using 3 cm microwave radiation, they achieved a resolution of $\lambda/60$ in one dimension and $\lambda/20$ for two-dimensional objects. In the **early 1980s A. Lewis'** group demonstrated that subwavelength apertures down to 15 nm gave readily detectable throughputs of transmitted or fluorescent light. Independently, **D. Pohl's** group transmitted light through a subwavelength aperture at the tip of an etched quartz rod. In **1951 J. Young** and **F. Roberts** were the first to examine the idea of confocal scanning optical microscopy. In the

1950s M. Minsky described in detail the principal features of this type of imaging. He built a microscope that enabled him to view deeper layers in a specimen with astonishing clarity, without first having to cut the specimen into thin sections.

1985

M. Petran and **M. Hadravsky** proposed a different type of confocal scanning optical microscope, the so-called tandem scanning reflected-light microscope, which operated in real time and produced an image that could be directly observed with the naked eye. They used a Nipkow disc, which contains many thousands of pinholes distributed in a spiral pattern (see Fig. F1.10). In the first experiment the microscope was transported to the top of a high mountain to obtain adequate illumination. In 1988, stimulated by Petran's work, **G. Kino** developed a modern version of the tandem scanning microscope that was called the real-time scanning optical microscope.

Late 1980s

T. Wilson and **S. J. Hewlett** proposed using a laser beam reflected off vibrating mirrors to generate a three-dimensional image. In laser scanning confocal microscopy, a cathode ray tube recreates the image. The detector signal modulates the cathode ray tube intensity and the detector scan rate is synchronized to the laser scan rate. A computer stores all confocal images. By collecting additional scans after moving the plane of focus up and down, subsequent image processing can reconstruct a three-dimensional image.

1995

Lewis' group demonstrated a new design for the NSOM probe based on a bent optical fiber. Using this optical element, a unique near-field optical head has been designed that can convert any far-field optical microscope into a multifunctional near-field optical, far-field optical, and scanned probe imaging system.

Since 2000

Near-field optics produce the highest optical resolution that has ever been achieved in light microscopy and are crashing through the barriers that have been traditionally accepted.

F1.2 LIGHT MICROSCOPY INSIDE THE CLASSICAL LIMIT

One of the fundamental properties of light is that a beam of light can convey information from one place to another. This information concerns both the source of light and also any object that has partly absorbed, reflected, or refracted the light before it reaches the observer. More information reaches the human brain through the eyes than through any other sense organ. Even so, the visual system extracts only a minute fraction of the information that is imprinted on the light that enters the eye. The main goal of an optical instrument is to extract much more information from visual surroundings.

The light microscope is an instrument for producing enlarged images of objects that are too small to be seen unaided. Such images may be viewed directly with a viewing screen or photographic apparatus or special electronic device. The value of a microscope lies in its ability not only to magnify objects, but also to make their fine details visible (Comment F1.1). These two features depend for the most part on the quality of the lens that forms the primary image. This lens is the critical element in every microscope. The objective lens determines magnification, field of view, and resolution; its quality determines light transmission, the contrast, and aberrations of the image.

F1.2.1 The Standard Light Microscope

One-Lens Microscope

Let us start with an ideal simple lens having two convex curvatures, each with a radius of curvature and focal point (called F and F'). An image is formed by a point-by-point translation of the object, which is said to be in the object plane, into an image that exists in the image plane. Figure F1.1 illustrates this translation. Two light rays coming from the point of the arrow are represented so that one travels parallel to the lens axis and the other passes through the focal point. Both pass into the lens. After passing through the lens, the light rays emerge, and the first is refracted through the focal point F, while the second now continues parallel to the lens axis. The point of intersection of the two rays is at the point of the arrow in the image plane. The geometry of the light rays obeys the relation $aa' = ff'$, with

> **COMMENT F1.1 GEOMETRICAL AND PHYSICAL APPROACHES IN OPTICS**
>
> Under certain conditions the wave nature of light is dominant, and light can be treated as a ray that travels in a straight path. These rays can be bent at the interface between two transparent objects. This forms the basis of geometric optics. Alternatively, we can emphasize the electromagnetic nature of light, in which diffraction and interference are recognized. This underlines the ideas of physical optics. From a practical point of view, these approaches are simply two sides of the same coin: the geometric approach is useful in understanding the ideas of image focus and aberration, while the physical approach leads more easily to an explanation of image contrast and the limits of resolution.

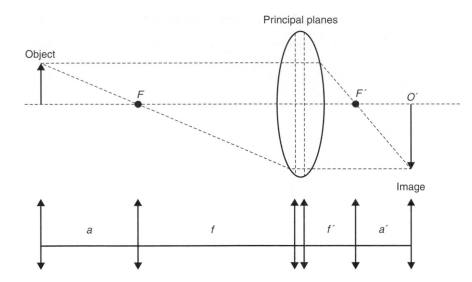

Principal planes

Object

Fig. F1.1 Optical geometry of a simple lens.

F   F'   O'

Image

a   f   f'   a'

the magnification given by $-f/a$. The meaning of the negative sign is that the image is inverted in the image plane with respect to the object plane.

An ideal lens brings all light rays from a single point in the object plane to a sharp focus in the image plane. However, real lenses fail to focus all the points in a single image plane and are said to have aberration (Comment F1.2).

Compound Microscope

In practice, a single-lens microscope requires a lens with a very short focal length, and consequently, the observer's eye has to be placed extremely close to the lens (Comment F1.3). To overcome this practical restriction, a compound microscope is used: This consists of an objective lens and a secondary-lens system, the eyepiece (Fig. F1.2). This arrangement allows an object to be placed just beyond the focal point of the objective. The objective-magnified image is further magnified by the eyepiece and then focused onto the retina of the eye.

Diffraction Limit of Resolution

Because of diffraction effects, the image of a point source of light can never be perfect, even assuming that all the imperfections in the lens are corrected.

In practice, a bright spot of light in the object plane appears as a circle of light surrounded by a series of concentric rings – the result of interference. This pattern is called the Airy disc (Fig. F1.3). The radius of the first dark ring, r, encircling the central bright spot is

$$r = \frac{0.6\lambda}{n \sin \alpha} \tag{F1.1}$$

where α is the angle made by drawing a line from the central axis of the lens to the edge of the lens, λ is the illuminating wavelength of light, and n is the refractive index of the object side of the lens. If two object points are in proximity, the ability of an observer to resolve them into

COMMENT F1.2 ABERRATIONS

Aberration commonly comes in several forms: chromatic, point-imaging, and astigmatic. Chromatic aberration occurs because the index of refraction is dependent on wavelength. Light of different wavelengths (white light) coming from a single point in the image plane is focused into separate image planes. Correctly engineered, the refractive index of the overall system can be made independent of wavelength. This lens is termed *anachromatic*.

Point-imaging aberrations occur because monochromatic rays from a single point in the object plane do not necessarily pass through the same image plane. Spherical aberration occurs when this effect is caused by the refraction of a single object point by different parts of the lens, with the consequence that the point in the image plane is not in focus. Lenses constructed to eliminate point-imaging aberrations are called *aplanatic*.

Astigmatism is another symmetry-destroying aberration in which the arrangement of object points in Cartesian space is warped so that the image does not maintain a linear point-to-point relationship with the object. Lenses that are corrected for astigmatism are called *anastigmatic*.

two clearly distinct points (i.e., the resolution) depends not on the separation of the two points but on the dimensions of their Airy discs. Equation (F1.1) shows that resolution can be improved by shortening the wavelength of the illuminating light, increasing the index of refraction on the objective lens side (for example, placing the object under oil), and increasing α. The angle α can be increased either by shortening the distance between the lens and the object, or by increasing the diameter of the lens.

Because the state of each point in the object state space is given by the complete set of these many orders of the diffraction, the only way to obtain complete information about the object's state space (image) is to collect all of the observable data carried in the totality of the diffracted orders. This ideal requires the largest possible collection angle between the object and the lens aperture (i.e., the largest α). This is called the angular aperture. However, the detail resolved is proportional not to the angular aperture but to the numerical aperture, NA. In the case of a dry lens, i.e., one in which the cone of light rays from the object passes through air, the numerical aperture is simply $\sin \alpha$. In an immersion system, in which oil (or sometimes water or glycerin) is placed between the front lens of the objective and the specimen, the numerical aperture is increased and the value of $\sin \alpha$ has then to be multiplied by the value of the refractive index n of the immersion medium. Therefore,

$$NA = n . \sin \alpha \qquad (F1.2)$$

Thus, the numerical aperture of a lens is a direct measure of the resolving power of the lens.

In any imaging system using lenses the point spread function (PSF) describes how diffraction spreads the image of a point (see Comment F1.4).

F1.2.2 The Problem of Contrast

The ability to see the details of the system under microscopic observation requires contrast, i.e., an essential difference in optical properties between point elements in the object's state space. Biological macromolecules are generally transparent to light, and their direct visualization under bright-field illumination is often impossible. Numerous contrast-enhancing techniques have been developed that increase the contrast of cells, including dark-field, phase-contrast, polarization, and interference microscopy. All these techniques are briefly discussed below.

Dark-Field Microscopy

In bright-field studies, light coming from the condenser system illuminates the object plane. In the absence of a specimen in the object plane, light uniformly enters the objective lens, giving an evenly illuminated field of view. If an opaque disc is inserted into the condenser system to exclude the unscattered beam, an annular or hollow cone of light will fall on the objective plane. The circle of light can be adjusted so that the light illuminates the object in a dark background (hence dark-field microscopy).

Though the field is dark, some light reaches this region, and any object placed in the dark field diffracts some of this light. In the resultant image, dramatically enhanced contrast is observed as the (diffracting) object appears as a silhouette, surrounded by a black (non-diffracting) field. The resolution of the object is, however, quite poor because the information carried in the low orders of diffraction is lost (Fig. F1.4a, b).

Phase-Contrast Microscopy

A common technique for visualizing cells is to convert the phase difference produced when light passes through cells

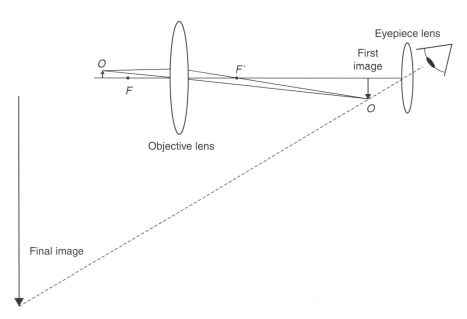

Fig. F1.2 Schematic illustration of a compound microscope.

(a)

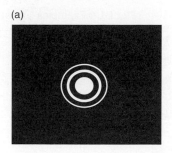

(b)

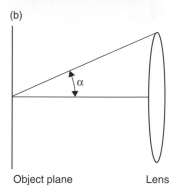

Object plane   Lens

Fig. F1.3 (a) The Airy disc is the diffraction image of each point in the image field. Resolution depends on being able to separate these diffraction patterns from one another. **(b)** The geometry used for determining the resolution of two points by a lens. A commonly used criterion for determining the resolution of an imaging system is the Rayleigh distance, the minimum distance by which two points must be separated to be distinguished when the illumination is incoherent. This Rayleigh resolution limit is theoretically half the wavelength of the imaging of light. Although this suggests that, for a wavelength of 500 nm, 250 nm resolution is possible, conventional microscopy never achieves such definition.

of different refractive indices into intensity difference. The light passing through each point of a specimen in the object plane is diffracted twice, once by the specimen and once by the objective lens. Each point of the viewed image is the result of the combined interference of each point of the object and each point of objective lens. Though the image contains the information related to the varying diffractive indices of the specimen, the phase differences are generally too small to generate sufficient interference to be perceived as intensity differences.

If the lens system is built to alter the phase of light sufficiently prior to recombination in the image plane, enough destructive interference can be generated to create contrast enhancement. In the phase-contrast microscope (Fig. F1.5), an annular light source is used just as in dark-field microscopy. The interference pattern is seen by the observer as intensity differences and contrast is increased. The phase-contrast method allows cells to be observed in substantial detail.

COMMENT F1.4 PSF IN CONFOCAL MICROSCOPY

In any lens system, be it electron, acoustic, or optical, the diffraction effect tends to spread the image of a point object. The intensity in the resulting image varies with distance from the central point. The PSF, which is simply the response of the imaging system to a point function, describes this variation. The imaging characteristics of a microscope are derived by convolving its PSF with a function characterizing the sample.

Polarization Microscopy

A polarization microscope contains two polarizers, one in the condenser and one between the objective and the eyepiece, called the analyzer (Fig. F1.6). Certain objects possess the property of form birefringence and pass only plane-polarized light when the light is parallel to the long axis of the particles in the object.

When polarized light is passed through such an object at an angle of 45°, the light is resolved into parallel and perpendicular components. If two polarizers are turned at 90° to each other, no light passes and a dark field is observed. If an object is placed at 45° to both of the polarizers, it passes some light and appears bright against the dark field. An object composed of stacked discs or long molecules in a parallel array appears bright at 45° because both of these structures are birefringent. In the process, use of a compensator provides the information necessary to indicate the positive or negative birefringence of material under investigation.

F1.3 SUBWAVELENGTH RESOLUTION WITHIN THE RESTRICTIONS OF GEOMETRICAL OPTICS

F1.3.1 Confocal Microscopy

In conventional wide-view microscopic imaging, not only is a sharp image generated from an in-focus area, signals above and below the focal plane are also acquired as out-of-focus blurs that distort the contrast and sharpness of the final image (Fig. F1.7a). Confocal microscopy is a technique for increasing the contrast of microscope images, particularly in thick samples. By restricting the observed volume, the technique keeps nearby scatterers from contributing to the detected signal (Fig. F1.7b).

Figure F1.8 shows how it is possible to view only the focal volume. A point light source is imaged at the object plane, so that the illuminated point and the source are confocal. Then the observation optics forms an image of the illuminated point on a pinhole. Thus there are three points all mutually confocal – hence the name. By moving

(a)   (b)

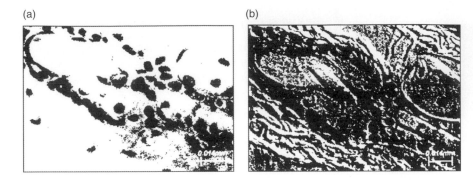

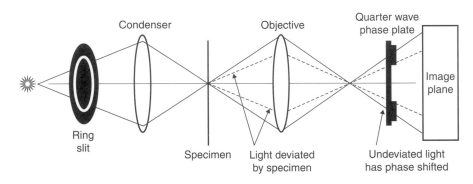

Fig. F1.5 Schematic view of the optical train in a phase-contrast microscope. The hollow cone of light passes through the object plane and falls as a ring of light on the objective lens. The image of the annular ring is thus formed in the back focal plane of the objective lens (eyepiece side). In order to induce the necessary phase shift, a quarter wave phase plate is introduced at the back focal plane. This plate is an annular disc of material, which slows the light so that the light traveling directly from the condenser annulus is retarded one-quarter wavelength. Any light that passes through a specimen placed in the objective plane and is diffracted (deviated) is not focused by the objective onto the phase plate and does not have its phase altered. The deviated and the phase-enhanced-undeviated light recombine at the image plane where substantial interference now occurs. (Adapted from Bergethon, 1995.)

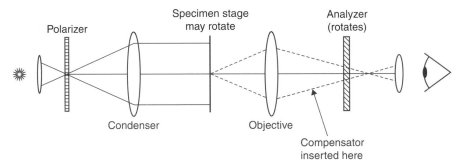

Fig. F1.6 Schematic view of the optical train of a polarization microscope. A polarized light source can be produced by placing a polarizer in the condensing system of the microscope. (Adapted from Bergethon 1995.)

the laser spot across the specimen with scanning mirrors, a sharp image from the confocal plane can be generated with a sensitive photodetector.

Three-Dimensional Image and Resolution

In biological microscopy, it is usually more important to image the interior of cells rather than the cell surface. In contrast to the regular light microscope, the confocal microscope employs a point-like illumination and detection arrangement (Fig. F1.8). Therefore, the PSF of the confocal fluorescence microscope is determined by the product of two almost identical intensity PSFs (Comment F1.4). Hence, as a good approximation the effective PSF of a confocal microscope is given by the square of the focal illumination intensity distribution in the objective lens, so that, in a simplistic photon picture, only photons from the closest vicinity of the diffraction-limited spot contribute to the signal. In a confocal microscope, the effective focus acts as a three-dimensional probe that can be scanned through a transparent specimen.

(a)

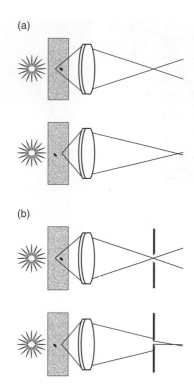

(b)

Fig. F1.7 (a) Demonstration of how a lens forms an image of two points of a thick sample, one at the focal point and one away from that point. **(b)** A pinhole conjugate to the focal point passes all the light from the focal point, and very little of the light from the out-of-focus point.

Therefore, confocal fluorescence microscopy readily produces three-dimensional images of transparent specimens such as cells.

The improvement of resolution and clarity provided by confocal microscopy is readily apparent when a photomicrograph obtained with a conventional microscope is compared with the corresponding confocal image. Figure F1.9 illustrates cultured rat hepatocytes stained with rhodamine 123, a cationic fluorophore that accumulates electrophoretically into mitochondria in response to the mitochondrial membrane potential. Figure F1.9a is an image obtained using conventional epi-illumination fluorescence microscopy. Individual mitochondria are easily identified at the periphery where the hepatocyte is thinnest, but in the thicker central portion of the cell, superimposition and out-of-focus fluorescence obscure mitochondrial outlines. Figure F1.9b is a confocal image of a hepatocyte labeled under similar conditions. In the confocal image, mitochondria are sharply defined spheres and filaments.

Scanning Process

The price for restricting the observed volume is that in the confocal microscope we can observe only one point at a time. In the scanning regime the image is built up, pixel by pixel, like a television picture. One of three approaches is

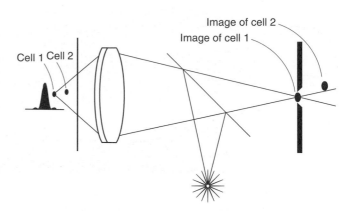

Fig. F1.8 Light from a point source is imaged at a single point of the object and that, in turn, is imaged on a small pinhole, making all these points confocal. (Webb, 1996.)

(a)　　　　　　　　　　(b)

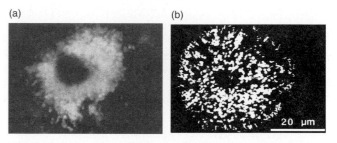

Fig. F1.9 Comparison of rhodamine-123-loaded hepatocytes viewed by conventional and confocal fluorescence microscopy: **(a)** conventional micrograph of rhodamine-123-labeled mitochondria in a cultured hepatocyte; the image was collected with a video camera using a 1.25 oil-immersion lens; **(b)** a confocal image of rhodamine-123-labeled hepatocyte loaded under similar conditions. The magnification of the two images is approximately the same. (Lemasters *et al.*, 1993.)

generally used to scan the image. The sample or the optics can be mechanically scanned. Another, somewhat faster, approach is to scan the beam using a galvanometer mirror or Bragg cells. The third approach is to use a Nipkow disc, which is a disc in which a large number of pinholes have been drilled to allow the entire image to be scanned (Fig. F1.10). This method is capable of producing real-time images at hundreds of frames per second.

F1.4 LENSLESS MICROSCOPY

The optical microscope has one serious drawback – its resolution – which results from the fundamental physics of the lens. This limit of resolution is theoretically half the wavelength of the imaging radiation (Section F1.2.1). This fact was one of the driving forces leading cell biologists to look for new high-resolution imaging tools to probe cellular structure and function. Near-field scanning optical

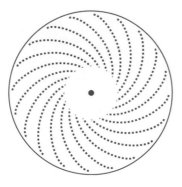

Fig. F1.10 The Nipkow disc is a disc with many holes. The holes cover about 1% of the image-plane space at any instant, and rotation of the disc maps out all of the required pixels.

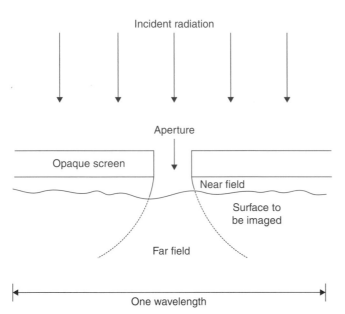

Fig. F1.11 Concept of near-field scanning optical microscopy or "optical stethoscopy." An illuminated aperture acts as a light source of subwavelength dimensions that can be scanned close to an object to generate a superresolution image.

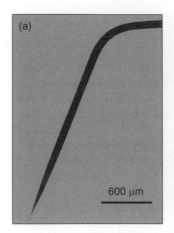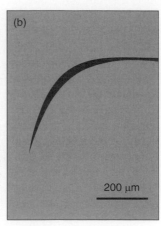

Fig. F1.12 Magnified view of: **(a)** a standard cantilevered fiber-optic probe; and **(b)** a probe that has been etched in hydrofluoric acid buffered with ammonium fluoride. The standard NSOM probe has a diameter of 125 μm and a measured spring constant of ~400–600 N m^{-1}. For the probe shown in **(b)**, the diameter has been etched down to 10–15 μm, reducing the spring constant to less than 5 N m^{-1}. (Shiku and Dunn, 1999.)

microscopy (NSOM, pronounced "ensom") is an example of the application of high-resolution imaging in cell biology. It carries optical imaging to resolutions normally associated with electron microscopy. It links the world of conventional optical microscopy with scanned probe imaging techniques, the most common of which is atomic force microscopy.

Near-Field Scanning Microscopy

The near-field optical interaction between a sharp probe and a sample of interest can be exploited to image surfaces. The easiest such probe to conceptualize consists of a subwavelength-diameter aperture in a optically opaque screen (Fig. F1.11). Light incident upon one side of the screen is transmitted through the aperture and, if the sample is within the near field, illuminates only one small region at any one time. The technique of imaging is based on the use of an "optical stethoscope" that places a small near-field light source a very short distance from the sample (less than the wavelength of light). Images are obtained by scanning the light source over the sample, and the resolution is limited only by the diameter of the light source. All mechanisms used to obtain contrast in conventional microscopy can be applied to NSOM.

Main Component of the Near-Field Microscope

The single most important element in any form of scanning probe microscopy is the probe itself. The ideal aperture is a transparent hole in a thin, perfectly conducting metal film. Real metals, however, are fairly poor conductors at optical frequencies. Therefore, the screen around the hole must have a thickness at least an order of magnitude larger than the skin depth to prevent excessive stray radiation. This requirement puts a lower limit on the aperture diameter, which cannot be substantially smaller than the film thickness.

A typical probe is a fiber-optic element, typically 10 nm in diameter, that yields a resolution down to 12 nm (<1/40) with visible light. Figure F1.12 shows examples of such optical elements. Another probe is the submicroscopic aperture at the tip of a glass pipette (Fig. F1.13). Such a bent glass structure can be easily combined with any of the available designs for force microscopy. The result is a microscope with exceptional qualities for both near-field optical imaging and normal force imaging. This microscope allows gold colloids with 80 and 40 nm diameters

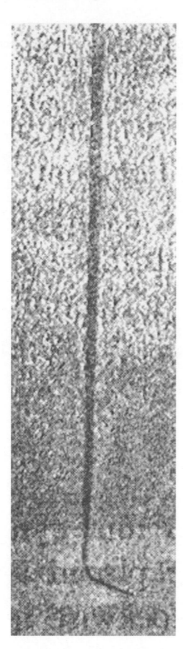

Fig. F1.13 Bent glass tip. Such a structure can be used in any of the available designs for force microscopy. (Lewis *et al.*, 1995.)

to be distinguished. This result is unique and cannot be achieved by any far-field optical system.

NSOM in Liquids

One major problem in the application of NSOM is the ability to image in liquids. Control of the distance between the sample and the near-field probe is more difficult because of damping of oscillations due both to the presence of the liquid and, in buffer, to electrostatic interactions between the ions on the surface of the near-field probe and ions in solution. One of the approaches to solving this problem is combining a near-field optical microscope and an atomic force microscope. We will discuss this approach in Chapter F2, which is devoted to atomic force microscopy.

F1.5 CHECKLIST OF KEY IDEAS

- The light microscope is an instrument for producing enlarged images of objects that are too small to be seen unaided; such images may be viewed directly with a viewing screen or photographic apparatus or special electronic device.
- Because biological macromolecules are generally transparent to light, numerous contrast-enhancing techniques have been developed including dark-field, phase-contrast, polarization, and interference microscopy.
- The diffraction limit of resolution power of a microscope is about half of the wavelength of the illuminating light; an increase of the resolving power of the microscope is possible only by increasing the numerical aperture of the optical system or by using a shorter wavelength.
- In contrast to the ordinary light microscope, the confocal microscope employs a point-like illumination and detection arrangement; by restricting the observed volume, the technique keeps nearby scatterers from contributing to the detected signal.
- The NSOM technique of imaging is based on the use of an "optical stethoscope," which places a small near-field light source a very short distance from the sample (less than the wavelength of light).
- Images in NSOM are obtained by scanning the light source over the sample, and the resolution is limited only by the diameter of the light source.
- NSOM links the world of conventional optical microscopy with scanned probe imaging techniques, the most common of which is atomic force microscopy.

Suggestions for Further Reading

Historical Review

Kino, G. S., and Corle, T. R. (1989). Confocal scanning optical microscopy. *Phys. Today*, **42**, 55–62.

Lichtman, J. W. (1994). Confocal microscopy. *Sci. Am.*, **271**, 40–45.

Light Microscopy Inside the Classical Limit

Bergethon, P. R. (1995). *The Physical Basis of Biochemistry: The Foundation of Molecular Biophysics*. New York: Springer.

Subwavelength Resolution within the Restriction of Geometrical Optics

Lemasters, J. J., Chacon, E., Zahrebelski, G., Reece, J. M., and Nieminen, A.-L. (1993). Laser scanning confocal microscopy of living cells. In *Optical Microscopy: Emerging Methods and Application*, eds., B. Herman and J. J., Lemasters. San Diego, CA: Academic Press.

Webb, R. H. (1996). Confocal optical microscopy. *Rep. Prog. Phys.*, **59**, 427–471.

Webb, R. H. (1999). Theoretical bases of confocal microscopy. *Meth. Enzymol.*, **307**, 3–20.

Lensless Microscopy

Betzig, E., and Trautman, J. K. (1992). Near-field optics: Microscopy, spectroscopy, and surface modification beyond the diffraction limit. *Science*, **257**, 189–195.

Shiku, H., and Dunn, R. C. (1999). Near field scanning optical microscopy. *Analytical Chem.*, **71**, 23A–29A.

SINGLE MOLECULE MANIPULATION AND ATOMIC FORCE MICROSCOPY

F2

F2.1 HISTORICAL REVIEW

Early 1980s

G. Binning and **H. Rohrer** proposed the scanning tunneling microscope (STM)(for which they were awarded the Nobel Prize). This invention has initiated an exciting series of novel experiments to image the surface of conducting as well as insulating solids with atomic resolution. The first attempts at imaging biological molecules using a STM date back to 1983. In 1987, individual molecules of phthalocyanine, lipid bilayers, and ascorbic acid were reported. One year later, **H. Ohtani** and collaborators imaged benzene, the molecule of Kekulé's blue dream, as three-lobed rings. In **1989** a spectacular view of the double-stranded Z-DNA molecule, the first biological macromolecule studied using an STM, was presented.

1986

A. Ashkin and coworkers proposed the idea of an optical trap (tweezers). An optical trap can be produced with a highly focused laser light, and can be used to grab, move, and apply measurable forces on micrometer-sized objects, such as dielectric microspheres. A microsphere that is chemically coupled to a molecule of interest provides a means of measuring the molecule's position and the force that it exerts.

G. Binning, C. F. Quate and **C. Gerber** invented the scanning force microscope (SFM). In this microscope a sensor tip carried by a flexible cantilever is used to touch and characterize a surface. This was a significant breakthrough; it allows biological macromolecules to be scanned in aqueous solution and gives reproducible imaging of DNA and of membrane protein crystals.

1990s

K. Bustamante and coworkers pioneered direct mechanical measurements of the elasticity of single DNA molecules using magnetic trapping. In **1994, S. Chu** and coworkers studied the relaxation properties of single DNA molecules using optical trapping, and **G. Li** and colleagues made a direct measurement of the force between complementary strands of DNA with atomic force microscopy. Mechanical properties of individual strands and double-stranded DNA have been determined.

In **1997, G. Shivashankar** and **A. Libchaber** developed a new technique for single DNA molecule grafting and manipulation using combined atomic force microscopy and an optical tweezer. These studies opened up new possibilities in biosensors and bioelectronic devices. **K. Svoboda** and coworkers applied optical trapping nanometry (optical tweezers in conjunction with nanometer-precision position detection schemes) to study a single kinesin molecule. Later, a further demonstration of the efficiency of trapping nanometry came from the observation of the forces and displacement produced by a single myosin molecule (**J. Spudlish**, **T. Yanagida**, and coworkers, **1994**), direct visualization of the rotation of the γ-subunit of F_1-ATPase (**H. Noji** and coworkers, **1997**), and unfolding individual nucleosomes by stretching single chromatin fibers (**M. Bennik** and coworkers, **2001**). After these experiments it became clear that optical trapping nanometry had fantastic potential for probing molecular mechanisms of functioning biological macromolecules, including molecular linear and rotary protein motors. A few groups demonstrated the possibility of using atomic force microscopy in the force-measuring mode, in which a single molecule is stretched between the microscopic silicon nitride tip of a flexible cantilever and a flat substrate that is mounted on a highly accurate piezoelectric positioner. Direct observation of enzyme activity (**M. Radmacher** and coworkers), the unfolding and folding of a single protein (**M. Rief** and coworkers, and **M. Kellermayer** and his colleagues, **1997**), nucleosomes (**B. D. Brower-Toland** and coworkers, **2002**), and RNA molecules (**J. Liphardt's** group, **2001**) have been demonstrated.

D. J. Keller, **Q. Zhong** and **C. A. J. Putman** independently proposed the so-called tapping mode of the atomic force microscope (AFM), in which the cantilever is oscillated vertically while it is scanned over the sample. In this mode the image is formed by displaying the reduction of the oscillation amplitude at every point of the sample. It was demonstrated that reliable height data for various biological macromolecule assemblies in solution can be obtained using the tapping mode AFM.

Since 2000

Advances in ultrasensitive instrumentation (optical and magnetic tweezers, atomic force microscopy, etc.)

provide extreme precision for position detection (such as 0.1–10 nm) and in the force regime (0.1–10 000 pN). The mechanical manipulation of single molecules has transformed many fields of biology, such as molecular motor mechanics, protein and RNA folding–unfolding, and receptor–ligand interactions. With the aid of single-molecule manipulation techniques such as atomic force microscopy, and optical and magnetic tweezers, over-stretching and supercoiling of single DNA molecules have been studied, force and displacements generated during single molecular motor reactions have been observed, and proteins have been mechanically unfolded. **C. Bustamante**, **D. J. Keller**, **P. K. Hansma**, **H. G. Hansma**, **A. Engel**, and **D. J. Müller** made essential contributions to the theory and applications of atomic force microscopy for the study of biological macromolecules. **T. Ando** and **P. K. Hansma**, and **H. G. Hansma**, independently developed fast-response devices for high speed AFM to visualize dynamic processes. **M. Miles** and his group developed ultrafast AFM using a tuning fork with resonant frequencies 10–100 kHz and fast digital processing systems.

F2.2 NANOSCALE MANIPULATION IN BIOLOGY

There have been significant advances in the characterization and manipulation of individual molecules. Optical and magnetic traps, the cantilevers of atomic force microscopy, and glass microneedles are widely used for nanoscale manipulation. Below we give a brief description of such devices.

For many years researchers have struggled to combine the high-resolution advantages of the electron microscope with the in-water operating capabilities of the optical microscope. The invention of the STM by G. Binning and H. Rohrer in 1981 opened up a new approach to achieving this goal. The STM was the first member of a new class of instruments called scanning probe microscopes, which are all based on similar principles. Atomic-resolution surface topographies recorded with the STM in vacuum fostered the expectation that the STM would become a tool for sequencing DNA molecules and for imaging protein surfaces at atomic resolution. But it turned out that the data were very difficult to reproduce and such dreams have not become reality. A significant breakthrough came in 1986 when G. Binning, C. Quate, and C. Gerber proposed AFM. In AFM, the sample is scanned with a cantilever whose deflections are recorded. An AFM could be compared to a blind person scanning the environment with a stick to explore the path ahead. AFM, in particular, has attracted the attention of biologists since it theoretically combines the two most important aspects for studying structure–function relationships of biological objects:

high-resolution imaging with high signal-to-noise ratio in the molecular/submolecular range and the ability to operate in *aqueous* environments, allowing the observation of dynamic molecular events in real time and physiological conditions.

F2.2.1 Optical Traps (Laser Tweezers)

"Optical tweezers" is the name given by A. Ashkin and coworkers to a device they invented that uses light pressure to manipulate tiny objects. By focusing a laser beam through a microscope objective, they found that particles with a high index of refraction, such as glass or plastic particles, or oil droplets, were attracted to an intense region in the beam and could be held permanently at a focal point (Fig. F2.1). Although the theory behind optical tweezers is still being developed, the basic principles are straightforward for objects either much smaller than the wavelength of light or much larger. When a small dielectric

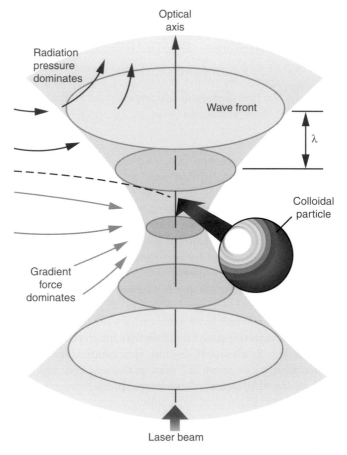

Fig. F2.1 Optical tweezers. A bead with a high index of refraction is attracted to the intense region in the laser beam. Intensity gradients in the converging beam draw small objects toward the focus, whereas the radiation pressure of the beam tends to blow them down the optical axis. Under conditions in which the gradient force dominates, a particle can be trapped near the focal point. (Adapted from Grier, 2003.)

object, such as a lipid droplet or a polystyrene or silica bead, is placed in the beam, it feels a force pulling it toward the focus of the beam. The magnitude of the exerted force is a function of number parameters: the size and shape of the object, the difference between the index of refraction of the object and the surrounding medium, and gradient in the intensity of the laser beam (Comments F2.1 and F2.2). Larger objects act as lenses, refracting the rays of light and redirecting the momentum of their photons. The resulting recoil draws them toward the higher flux of photons near the focus. This recoil is all but imperceptible for a

macroscopic lens but can have a substantial influence on mesoscopic objects.

Optical gradient forces compete with radiation pressure resulting from the momentum absorbed or otherwise transferred from the photons in the beam, which acts like a fire hose to blow particles down the optical axis. Stable trapping requires the axial gradient force to dominate, and is achieved when the beam diverges rapidly enough from the focal point. For this reason, optical tweezers are usually constructed around a microscope objective.

The basic optical trap is quite simple: A single-mode laser with Gaussian beam profile (TM_{00}) is passed through a beam expander (so that the beam fills the back aperture

COMMENT F2.1 OPTICAL TRAP AND THE DIMENSIONS OF A BEAD

Theoretical calculations for objects whose size is small relative to the wavelength of the trap light (Rayleigh particles) suggest a cubic dependence ($F_{max} \sim d^3$), where d is the size of the object. However, for trapped particles of the same approximate size as the light's wavelength (e.g., 500 nm vesicles in an 830 nm trap), this dependence is somewhat weaker, in general somewhere between linear and quadratic. Empirically, for trapped latex beads, the maximum lateral force applied on a 1.02 µm bead is 7 times that on a 0.3 µm bead, and the maximum force on a 2.97 µm bead is 2.47 times more than on the 1.02 µm bead.

COMMENT F2.2 OPTICAL TRAP AND REFRACTIVE INDEX OF A BEAD

The applied force is a function of n, the relative index of refraction, defined as $n = n_1/n_2$, where n_1 is the index of refraction of the trapped object and n_2 is the index of the surrounding medium. For lateral forces, the larger n, the larger the applied force. Relatively small corrections are needed to enable calibrations to be done in vitro, and actual measurements to be done in vivo. Thus, the applied force changes by approximately 20% when an object (e.g., a lipid droplet, $n_1 = 1.52$) is in water ($n_2 = 1.33$) versus when it is in cytoplasm ($n_2 = 1.39$). This is also true for axial forces.

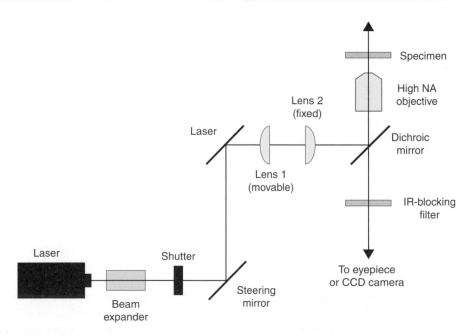

Fig. F2.2 Schematic diagram showing the path of the laser light beam from its origin at the laser to the formation of an optical trap in the specimen. The beam expander is chosen so that the laser beam just fills the back aperture of the microscope objective: if the back aperture is overfilled, laser light is lost; and if underfilled, less of the lens is used, resulting in less of a gradient in a focused beam. The shutter is used to block the laser to minimize the sample's exposure to laser light when an object is not being actively trapped. The steering mirrors are used to control the orientation of the beam, to bring it into the objective. Lenses 1 and 2 allow steering of the trap. After passing through the two lenses, the laser is brought into the microscope's optical path by reflecting it off of a dichroic mirror mounted in line with the microscope objective. (After Gross, 2003.)

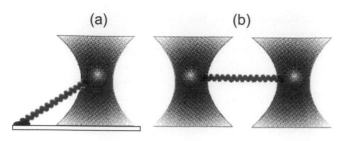 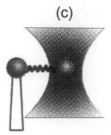

(a) **(b)** **(c)**

Fig. F2.3 Different experimental geometries using optical tweezers in which one side of a molecule is attached to the bead. The other side of the molecule is attached to: **(a)** a coverslip; **(b)** another bead; **(c)** a microneedle.

of the microscope objective, in order to use the entire lens, which generates the largest gradient in the focused beam), and then steered with a combination of mirrors and lenses into the microscope objective (Fig. F2.2). The force, F, that the trap exerts on an object is a linear function of the object's displacement, x, from the trap center, i.e., $F = -kx$, where k is a trap stiffness. Deviations from this linear relationship are observed at the very edges of the beam, typically about 200 nm from the center of the beam. The applied force increases linearly with laser power. In principle, any laser can be used to make an optical trap; in practice, IR lasers are employed because they cause the least optical damage.

For a trap based on a TM_{00} laser beam, the axial force (along the beam path) is weaker than the lateral force (pulling the object in toward the beam center). However, for both axial and lateral forces, at a given displacement from the trap center, the force that a trapped object feels is a function of its size: Up to about 4 μm, the larger the object, the more force it feels.

Biological macromolecules can be bound to polystyrene or silica beads, which are usually ~1–3 μm in diameter. A trap can be moved to steer a bead into a desired experimental geometry, e.g., to interact with a partner molecule attached to a coverslip (Fig. F2.3a). Upon binding between the two molecules, the forces and movements involved can be measured, and the interaction can be perturbed by moving the trap. Similar experiments can be performed with glass microneedles or AFMs, although such probes are typically less compliant than optical ones.

Significant technical difficulties nevertheless exist in measuring movements of single molecules. The motions occur on length scales of ångströms to nanometers and on timescales of milliseconds or less. To obtain the high spatial and temporal sensitivity required, optical tweezers are combined with a dual-beam interferometer to produce an "optical trapping interferometer." This instrument provides position detection and trapping functions simultaneously, and can provide controlled, calibratable forces in the piconewton range.

Optical tweezers use light to manipulate small particles, but only one at a time. This is because the manipulating light beam diverges around the first trapped particle, and then cannot be brought back to focus within the short distance to the next particle. So experiments must be done

Fig. F2.4 A Bessel beam. Light passing through a particular scheme of lenses can form a beam of light with a bright central spot and concentric rings of decreasing intensity; this is known as a Bessel beam, as its varying intensity is described by the zero-order Bessel function. A Bessel beam is used to create optical tweezers capable of manipulating many samples at once. (Adapted from Garces-Chavez *et al.*, 2002.)

in series, and this is a very time-consuming procedure. The first experiments in which small particles (micrometer-sized) were manipulated in *multiple planes* were demonstrated using a special kind of optical focusing to create a *"Bessel beam"* of light (Fig. F2.4).

F2.2.2 Magnetic Traps (Magnetic Tweezers)

Figure F2.5a shows an experimental setup that uses a magnetic bead to manipulate biological macromolecules. Single molecules are bound at multiple sites with one extremity attached to a treated glass coverslip and the other to a magnetic bead. Supermagnetic beads (typical diameter about a few micrometers) are made of many small ferromagnetic domains. Their magnetic susceptibility is anisotropic, and therefore their response to magnetic fields is similar to that of a compass needle. An external magnetic field can be used to rotate the beads and thus to coil and pull macromolecules. Magnetic tweezers are a useful tool, particularly between a few femtonewtons and 100 pN, with a relative accuracy of ~10% (Comment F2.3).

A further advantage of the magnetic trap technique is that measurements on DNA at constant force are trivial (the position of the magnets is just kept fixed). With cantilevers

(a)

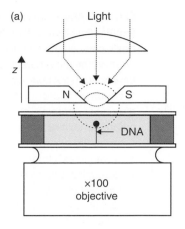

(b)

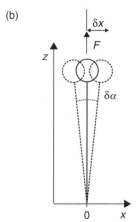

Fig. F2.5 (a) Magnetic tweezers. One end of the macromolecule is bound to a glass coverslip and the other end to a paramagnetic bead. Magnetic tweezers use the gradient of the horizontal magnetic field generated by small magnets to pull vertically on a magnetic bead anchored to a surface by a DNA molecule. A force is applied on the bead with Co–Sm magnets equipped with a polar piece to focus the field in a 2 mm gap located just above the sample. The magnetic device can spin about the optical axis, causing the paramagnetic beads to rotate. (b) Schematic of a bead undergoing transverse and longitudinal fluctuations. (After Strick et al., 1996.)

or optical tweezers, to work at constant force requires an appropriate feedback to ensure that the displacement of the sensor is kept constant. At the same time, because its stiffness depends on the force, for weak forces (<1 pN) the magnetic trap technique has a lower spatial resolution of about 10 nm than the other manipulation methods.

F2.2.3 Cantilever in the Force-Measuring Mode of the AFM

In the force-measuring mode of the AFM (see below), single molecules or pairs of interacting molecules are stretched between the silicon nitride tip of a microscopic cantilever and a flat, gold-covered substrate whose position

COMMENT F2.3 CALIBRATION OF STRETCHING FORCE

The stretching force can be determined by analysis of the Brownian fluctuations of the bead. In contrast with other techniques, the stretching force measurement is absolute and does not require a calibration of the sensor. It is based on the analysis of the Brownian fluctuation of the tethered bead, which is completely equivalent to a damped pendulum of length $l = \langle z \rangle$ pulled by a magnetic force F (along the z-axis). Its longitudinal $\left(\delta z^2 = \langle z^2 \rangle - \langle z \rangle^2 \right.$ and transverse δx^2 fluctuations are characterized by effective rigidities $k_\parallel = \partial_z F$ and $k_\perp = F/l$. By the equipartition theorem they satisfy

$$\delta z^2 = \frac{k_B T}{k_\parallel} = \frac{k_B T}{\partial_z F}$$
$$\delta x^2 = \frac{k_B T}{k_\perp} = \frac{k_B T l}{F}$$

Thus from the bead's transverse fluctuations one can extract the force pulling on the molecule (the smaller the fluctuations, the greater F) and from its longitudinal one obtains its first derivative.

This method of measurement can be used with magnetic (but not optical) traps because the variation of the trapping gradients occurs on a scale (about 1 mm) that is much larger than the scale on which the elasticity of the molecule changes (about 0.1 μm). In other words, the stiffness of the optical trap is very large compared with F/l.

is controlled by a high-precision piezoelectric positioner (Fig. F2.6). This system allows the suspended molecule(s) to be stretched with subnanometer precision. The forces acting on the molecule as it is extended are transmitted to the cantilever, causing it to bend. The amount of bending is measured by a laser beam reflected off the cantilever tip onto a photodetection system, such that the vertical deflection of the beam registers the angle at which the cantilever is bent. By calibrating the responsiveness of the cantilever, the degree of bending can be translated into the applied force with a precision at the level of a few piconewtons.

In a typical experiment, the cantilever tip is pressed into a layer of purified protein adsorbed onto the gold substrate (Fig. F2.7a). Protein molecules from the adsorbed layer affix to the cantilever by an unknown mechanism (which could involve the high pressure achieved as the cantilever tip is pressed against the gold substrate) and are stretched as the gold substrate is withdrawn. Suspended molecules resist extension and therefore cause deflection of the cantilever (Comment F2.4). The tension on the cantilever is released either when a force-induced rearrangement increases the distance between the ends of the suspended protein, or when the suspended molecule becomes detached from the cantilever or gold substrate. The resultant data may be expressed as a force extension curve (Fig. F2.7b).

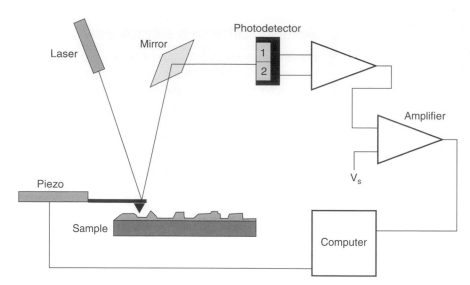

Fig. F2.6 AFM instrument design in force-regime mode. An AFM has four basic components: a microcantilever for sensing forces; a means of detecting cantilever deflection; a piezoelectric system for moving the tip in *x*, *y*, and *z* directions at atomic resolution; and a computer control system. In a typical commercial system, the cantilever is microfabricated from silicon nitride or silicon using a semiconductor fabrication method. The photodiode allows the bending motion of the cantilever to be detected to better than 1 Å. (Adapted from Wang *et al.*, 2001.)

COMMENT F2.4 CANTILEVER AND HOOKE'S LAW

The cantilever behaves as a Hookian spring with a force constant, *k*, of the order of 0.01–1 N m^{-1}. The force on the cantilever may be determined directly from the deflection of the cantilever and the force constant of the cantilever using Hooke's law, $F = kx$. The force constant *k* can be estimated from the dimensions of the cantilever using the Young's modulus of the material, but experimental determination of the force constant is preferable.

To investigate single molecules, the physics and chemistry of the tip should be well defined and tailored for its specific application. The coating of tips with a catalyst to induce reactions locally or with a metal to deposit metallic clusters has been demonstrated. C_{60} molecules attached to tips are geometrically and structurally appealing probes for AFM imaging because they combine molecular dimensions with extreme ruggedness. Chemical functionalization of carbon nanotubules by coupling basic, hydrophobic, or biochemical terminations of the carboxyl groups to the open tip ends has been reported. Specific chemical binding of pairs of molecules, one sited on the tip, the other on the surface, such as biotin–avidin or conjugated DNA strands, is being commercially developed in combination with magnetic bead technology used for immunoassays.

F2.2.4 Glass Microneedles

Using a commercial pipette puller, tapered glass microneedles (typical dimensions: shank diameter 1 mm, tip diameter 1 μm, and tapered length 1 cm) can be prepared (Fig. F2.8). The stiffness varies within a wide range from needle to needle.

Figure F2.8a shows a typical video image of a magnetic bead and microneedle under the microscope. The experiments were performed in a liquid buffer, which was retained by a ring glued with paraffin on top of a microscope slide. This assembly formed a "well" and was placed on an inverted video microscope. The face of the microscope slide in contact with the buffer was coated with an antibody against digoxigenin. The digoxigenin functionalized extremity of the construction was attached to this face, and the biotinated extremity was anchored to a microscopic bead coated with streptavidin. Figure F2.8b demonstrates the principle of force measurement using a microneedle. The bead was attached to a glass microneedle (treated with biotin), which was introduced through the free meniscus and was positioned by an *xyz* micromanipulator. The microneedle served as a force sensor, i.e., the deflection of the tip of the microneedle was monitored under a microscope as a function of the lateral displacement of the surface.

A comparison of the major features of the techniques described above is presented in Table F2.1.

F2.3 GENERAL PRINCIPLES OF AFM

Atomic force microscopy (also scanning force microscopy) *does not use lenses to form an image*, but instead uses a sharply pointed sensor, or tip, at the end of a flexible cantilever to scan and sense the topography of a sample. In the simplest mode of operation, the contact mode (see below), a scanning force microscope (SFM) resembles a phonograph; as the sample is scanned underneath the tip, the cantilever deflects up or down, tracking the surface (Fig. F2.9).

Deflections as small as 0.1 nm can be detected. A variety of methods have been devised to detect the cantilever

TABLE F2.1 COMPARISON OF ATOMIC FORCE MICROSCOPY, THE LASER TRAP, THE MAGNETIC TRAP, AND THE GLASS MICRONEEDLE

Method	Atomic force microscopy	Laser trap	Magnetic trap	Glass microneedle
Force generation	Bending of cantilever obeys Hooke's law	Intensity gradient from photons from laser source	Supermagnetic bead in magnetic field	Bending of microneedle obeys Hooke's law
Dynamic range	20 pN–20 nN	0.1–400 pN	A few fN to 100 pN	1 pN and as large as desired
Force constant	0.05 N m^{-1} (typically)	0.0002 N m^{-1}	~10^{-7} N m^{-1}	2×10^{-6} N m^{-1} (typically)

Table adapted from Wang et al. (2001).

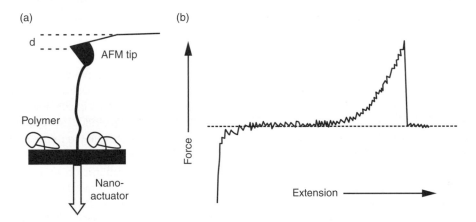

Fig. F2.7 A typical experiment in which a single molecule is stretched by an AFM tip. (a) The tip is brought into contact with the sample, which is covered by a layer of protein. Ideally, the attachment is at a known position on the protein, which could be the ends. As is often the case, the ideal situation is not the easiest to attain, and it is possible to study the extension of polymers without knowing the attachment position. (b) If a molecule is bound to the tip it can be stretched and the force is measured via the deflection of the cantilever spring as a function of the extension. When the maximum binding force is exceeded, the molecule breaks away from the tip and the tip is free again.

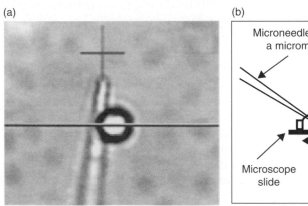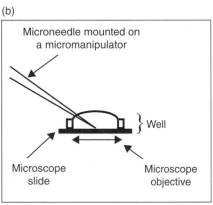

Fig. F2.8 (a) Typical video image of a magnetic bead and the microneedle under the microscope. (b) Principle of force measurement using a microneedle. (Adapted from Essevaz-Roulet et al., 1997.)

deflection. The most common approach, called the optical lever, is to reflect a laser beam off the back of the cantilever into a four-segment photodetector. The optical lever is essentially a motion amplifier. The detection of the laser spot at the photodetector is proportional to the deflection of the cantilever with a gain factor, which is typically 500–1000. This means that a deflection of 0.1 nm at the cantilever becomes a displacement of 50–100 nm at the detector, which is large enough to generate a measurable signal.

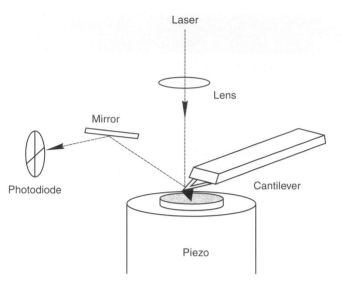

Fig. F2.9 Schematic diagram of a commercial AFM.

Fig. F2.10 The cantilever is the main element of an AFM. For biological applications, cantilevers are made of silicon nitride. The example shown is 200 μm thick and has a force constant of approximately 0.1 N m^{-1}. Thus a measurable deflection is induced by forces of a few piconewtons. (Engel *et al.*, 1999.)

Because it is relatively simple to operate, and imaging does not require an external means of contrast, the AFM has a vast range of imaging applications from single molecules to living cells. In addition, the AFM is a powerful method for single-molecule manipulation, as cantilevers are being used to exert forces on individual molecules to probe their mechanical stability and elasticity (see above).

F2.3.1 The Tip: A Key Element of Scanning Force Microscopy

The AFM is a small instrument; current designs weigh approximately 3 kg, excluding the accompanying computer and controller. The heart of the AFM is a pyramid-shaped stylus mounted on a flexible cantilever (Fig. F2.10). The shape of any tip can be described by two parameters: the end radius of curvature, R_c, and opening angle, Φ (Fig. F2.11). Although there is no generally agreed definition of resolution for an AFM, intuitively the smallest discernible features cannot be much smaller than R_c. Similarly, Φ affects how well the AFM visualizes steep features, especially around the edges of molecules.

Consider the ideal AFM tip. It should have a high aspect ratio with a cone angle $\Phi = 0°$, a radius that is as small as possible with well-defined and reproducible molecular

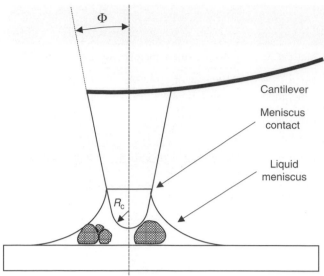

Fig. F2.11 Schematic diagram showing a tip in contact with a molecule deposited on a substrate. The size and shape of the tip are defined by its radius of curvature, R_c, and the opening angle, Φ. Standard microfabricated pyramidal tips (silicon nitride, Si$_3$N$_4$) have an R_c in the range 30–70 nm.

structure, and be mechanically and chemically robust such that its structure is not altered while imaging in fluid environments. Carbon nanotubules are the only known material that can satisfy all these critical criteria, and thus have the potential to create ideal probes for AFM imaging.

Carbon nanotubes consist of a honeycomb sp^2 hybridized carbon network (called a graphene sheet) that is rolled up into a seamless cylinder (Fig. F2.12a), which can be microns in length. Figure F2.12b shows a single needle-like carbon microtip grown on a standard pyramidal tip. Because of their flexibility, the tips are resistant to damage from tip crashes, while their slenderness permits imaging of sharp recesses in surface topography.

One of the most exciting results from AFM was the discovery that atomic resolution can be achieved when the tip is in contact with the sample while scanning. Conceptually, this "contact mode" of imaging is like using a stylus *profilometer* to measure the topography of surface atoms. At first, this idea may seem implausible, especially when one considers the macroscopic size of typical AFM tips. The AFM achieves such high resolution by using a very small loading force on the tip – typically 10^{-7}–10^{-11} N – which makes the area of contact between the tip and sample exceedingly small. This small and well-controlled loading force is the essential difference between the contact-mode AFM and earlier stylus profilometers, which typically used loading forces of the order of 10^{-4} N (Comment F2.5).

F2.3.2 Imaging Modes

The most common imaging mode is the constant force mode, sometimes referred to as the height mode, which

(a)　(b)

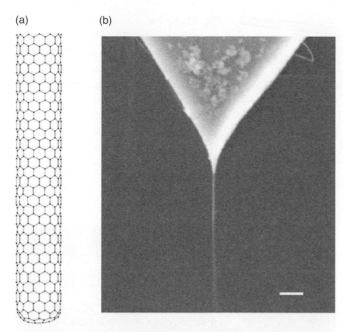

Fig. F2.12 (a) Fragment of a stick model of a single-walled carbon nanotube. Each vertex in the honeycomb mesh corresponds to an sp^2-bonded carbon atom. Radii of nanotubes range from 0.35 to 2.5 nm. **(b)** Scanning electron image of a single-walled nanotube. The scale bar is 200 nm. (Hafner *et al.*, 2001.)

COMMENT F2.5 SPRING CONSTANT OF CANTILEVER

It is perhaps surprising that it is easy to make a cantilever with a smaller spring constant than that of the equivalent spring constant between atoms. For example, the vibrational frequencies ω of atoms bound in a molecule or in a crystalline solid are typically 10^{13} Hz or higher. Combining this with the mass of the atoms, of order 10^{-25} kg, gives interatomic spring constants $k = \omega^2 m$, of the order of 10 N m^{-1}. For comparison, the spring constant of a piece of household aluminum foil that is 4 mm long and 1 mm wide is about 1 N m^{-1}.

---

displays the changes in sample height during scanning. This mode gives calibrated height information about the sample surface. Alternatively, one can monitor the small changes in cantilever deflection that signal the feedback loop to make the height adjustments; this imaging mode is called the deflection mode. In addition to raster scanning, the cantilever can oscillate at high frequency (~300 kHz). When the tip encounters elevation changes on the surface, the amplitude of the oscillation changes. The feedback system then responds to keep the amplitude of cantilever oscillation at a constant level. This mode of operation reduces the lateral forces that can push the sample around. Two types of imaging of this kind are known: non-contact

(sometimes called the attractive mode) and tapping, in which the amplitude of the oscillation is larger. In the non-contact mode, forces between the tip and the sample probably affect the oscillation, while in the tapping mode the tip actually touches the surface at the bottom of the oscillation. Below we briefly describe the main imaging modes.

Contact (Constant Force) Mode

In the constant force mode of AFM, the contact force between the tip and the sample, which is preset at a minimum value, is held constant by an electronic feedback loop. Whenever the cantilever begins to deflect upward (or downward), indicating that the tip is encountering an obstacle (or a trough) on the surface, the sample is retracted (or raised) to cancel the deflection of the cantilever, thereby keeping the tip–sample force constant (Fig. F2.13a). The surface topography is obtained by displaying the advance and withdrawal of the sample required to prevent cantilever deflection. This operation usually yields stable images, but the compression and shear force generated between the tip and surface may damage the sample. Typical operation forces in the contact mode are 1–10 nN.

Oscillation (Tapping) Mode

In the so-called tapping mode (also known as AC or intermittent contact mode), the cantilever (driven by a piezoelectric actuator) vibrates at its resonance frequency. Upon approaching the sample, the tip briefly touches the surface at the bottom of each swing, resulting in a decrease in oscillating amplitude (Fig. F13.6b). The feedback loop keeps this decrease at a preset value and a topographic image of the sample surface can be obtained. Because the restoring cantilever force should be higher than the adhesion force due to the water film present on samples, relatively stiff cantilevers with force constants ranging from 10 to 100 N m^{-1} and oscillation amplitudes between 50 and 100 nm are used. In a tapping mode AFM, the destructive influence of the lateral forces, due to the movement of the tip with respect to the sample, is virtually eliminated because the duration of tip–sample contact is short.

The oscillation of a cantilever in a liquid shows important differences compared with oscillation in air or ultrahigh vacuum. First, the cantilever motion drags the surrounding liquid, leading to an increase of the effective mass by a factor of 10–40 and a corresponding decrease of the resonant frequency. Second, the cantilever oscillation is anharmonic and asymmetric when the quality factor is low, in contrast with a tapping mode AFM in air, where the cantilever oscillation is approximately sinusoidal and symmetric. The tapping mode has been harnessed for the development of high-speed AFM (see Section F2.6).

(a)

(b)

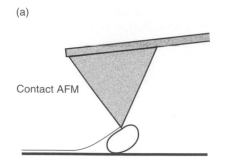

Contact AFM

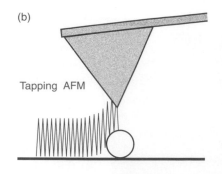

Tapping AFM

Fig. F2.13 Two basic forms of AFM. A contact-mode AFM operates with the tip actually touching the sample surface **(a)**, whereas a non-contact (tapping) AFM operates with the tip hovering a few nanometers above the surface **(b)**. In the tapping mode the cantilever oscillates vertically at its resonance frequency. The oscillation amplitude is reduced upon contact between the tip and the sample.

COMMENT F2.6 RESOLUTION IN AFM

When we ask what is the best resolution that might be achieved in experiments, one needs to look no further than the structure of the tip used for imaging, since an AFM image is a convolution of the structure of the macromolecule or macromolecular assembly under investigation and the tip structure. A well-characterized tip is essential for accurately interpreting an image, and moreover, the size of the tip defines the resolution of the image. The situation is somewhat analogous to the need for well-collimated and monoenergetic electron beams in cryoEM and well-defined X-ray sources in X-ray diffraction.

Fig. F2.14 AFM image of plasmid DNA with a ring structure on the ultrasmooth sapphire substrate wet-treated with Na_3PO_4 aqueous solution (image size: 670 nm × 450 nm). (After Yoshida et al., 1998.)

F2.4 IMAGING BIOLOGICAL STRUCTURES

The mechanism of image formation in the AFM is very different from those of optical and electron microscopes. The spatial resolution of optical and electron microscopes is an inherent property of the instrument and depends ultimately on the design and principles of operation of the microscope. In contrast, the spatial resolution of the AFM depends as much on the characteristics of the sample as on the inherent properties of the instrument. Thus, although theories of image formation for AFM exist, there is no general definition of resolution in force microscopy (Comment F2.6).

Many factors contribute to an AFM image of biological structures in addition to the topography of the sample surface. These include the size and shape of the tip, the properties of the feedback loop, and the mechanical and chemical properties of the sample and imaging environment. Thus, the finite size of the tip limits the lateral resolution, and the shape of the tip contributes significantly to many images. The part of the tip that interacts with structures on the sample surface should not be larger than the structures themselves. Hard and atomically flat surfaces can be imaged to atomic resolution using even relatively blunt tips (i.e., large R_c), because the interaction

is very local and is mediated only by the most apical atoms on the tip. On the other hand, soft samples tend to conform to the harder tip surface and tip–sample interactions take place over a larger area. One of the serious problems in AFM imaging of biological macromolecules is control of the tip–sample interaction. The interaction should ideally be highly localized and involve a very low force.

F2.4.1 Imaging DNA

One of the primary interests in AFM in biology is high-resolution imaging of DNA structures. However, even with significant instrumental developments, such as the tapping mode AFM and specially fabricated supersharp microtips, the resolution was not sufficient to resolve even the pitch of the double helix of DNA in aqueous media (Fig. F2.14; Comment F2.7). Spectacular progress was achieved in high-resolution AFM of DNA using cationic lipid bilayers on a freshly cleaved mica surface as the substrate to anchor the DNA with high affinity. Figure F2.15a shows an image of an *E.coli* plasmid. The measured width of this DNA is close to 2 nm, and a periodic modulation is reproducibly detected by the AFM. The measured period is 3.4 ± 0.4 nm, in excellent agreement with the known pitch of the double helix in B-form.

The AFM allows direct visualization of kinked DNA *in situ* (Fig. F2.16). Small DNA circles change from smoothly bent to abruptly kinked shapes. This effect is dependent on the presence of specific divalent cations. The DNA is bent smoothly in the presence of Mg^{2+}, but consists of nearly straight segments connected by kinks in the presence of Zn^{2+}.

F2.4.2 Imaging Proteins

Individual Molecules

Extraction of submolecular information from AFM topography for isolated globular proteins is a very difficult task. On soft biological samples it is not possible to achieve atomic resolution as with solid-state flat materials or two-dimensional crystals (see below). Single globular proteins or protein complexes can be described (or identified) only in a quasi-spherical approximation. The advantages of nanotube tips are demonstrated in Fig. F2.17a for isolated Immunoglobulin G (IgG). The Y-shaped structure was easily resolved. This IgG image is compared (Fig. F2.17b) with the crystal structure to emphasize the very small amount of tip-induced broadening in observed images obtained with nanotube tips.

Two-Dimensional Crystals

The discussion above shows that on soft samples, deformability leads to a pronounced indentation of the sample by the tip. This effect has led to the prediction that the lateral resolution of such a sample would be limited to several nanometers. However, studies of organic crystals or ordered molecular films show that they may be imaged at molecular

---

COMMENT F2.7 IMAGING BY AFM

Successful imaging by AFM requires that the binding of the sample to the surface be stronger than the contact interaction developed between the tip and the sample during scanning. This may be accomplished either by increasing the strength of attachment to the substrate or by reducing tip–sample forces. Experience has shown that both approaches are necessary. Mica has been used as a deposition substrate since the earliest attempts to image DNA samples. Early attempts at imaging DNA on freshly cleaved mica showed, however, that the molecules were not stably attached to the mica surface, and the scanning tip induced molecular displacement, which made the images blurred and difficult to reproduce.

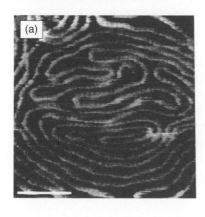

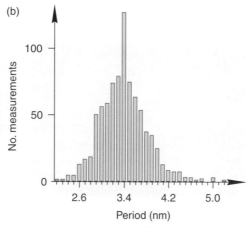

Fig. F2.15 (a) An AFM image of *E.coli* plasmid pBR322 (4.36 kb). Periodic modulations are resolved and were reproduced at different scales and with different specimens. The width of the DNA varied somewhat from AFM tip to tip, but most measurements gave a value below 3 nm. Scale bar: 40 nm. (b) A histogram of measurements of modulation periodicity. The mean value is 3.4 nm with a standard deviation of 0.4 nm, indicating that these modulations are indeed due to the pitch of the DNA double helix. (Adapted from Mou *et al.*, 1995.)

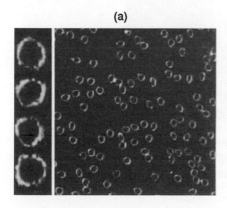

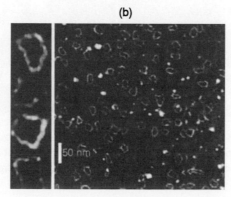

Fig. F2.16 Images of 168 bp DNA minicircles in (a) 1 mM $MgCl_2$ and (b) 1 mM $ZnBr_2$, showing a four-fold increase in kink density in Zn^{2+}. Selected molecules are displayed magnified by a factor of 4 to the left of each image. The tip was oscillated at 25 kHz with an amplitude of 5 nm and the image was acquired in five minutes. (Adapted from Han *et al.*, 1997.)

resolution. Below we give two examples of the application of AFM to studies of biological macromolecules assembled into a two-dimensional lattice (Comment F2.8).

The first example is a molecular motor of F_0F_1-adenosin triphosphate synthase. It was postulated that rotor of F_0F_1-ATP synthases comprises 12 subunits. This conclusion was mainly based on cross-linking experiments, genetic engineering, and biochemical data that suggest that four protons are required for the synthesis of one ATP. Furthermore, it was assumed that this stoichiometry would be constant regardless of the biological species from which the ATP synthases originated.

To prove this structural model, AFM was used to image isolated ion-driven rotors from chloroplast and bacterial ATP synthase. In both investigations, the number of subunits per rotor could be directly counted in the unprocessed AFM images (Fig. F2.18). Surprisingly, it was found that the rotor from ATP syntase of *Ilyobacter tartaricus* consisted of 11 subunits (Fig. F2.18a), while that from spinach chloroplast ATP synthase exhibited 14 subunits (Fig. F2.18b).

Additionally, the X-ray analyses of yeast F_0F_1-ATP synthase yielded a decameric rotor (Section F2.3.2). Complementary analysis of a single defective bacterial and chloroplast rotor shows their circular diameter to be independent of the number of subunits missing. This led to the conclusion that the subunits themselves determine the rotor diameter and thereby constrain how many subunits would fit into the rotor.

The second example is the bacteriophage ´φ29 motor. The head-to-tail connector of the *Bacillus subtilus* bacteriophage φ29 represents a rotary motor that hydrolyzes ATP to power pack double-stranded DNA into a precursor capsid. In contrast to the rotary motor of ATP synthase, the φ29 connector converts a mechanical rotation to a translational movement of DNA. In the bacteriophage, the DNA, connector, and prohead form a complex of concentric structure with 10-, 12-, and 5-fold symmetry, respectively. These constitute a moveable central spindle, intervening ball race and a static outer assembly (stator), which powers the rotor. AFM topographs of native connectors assembled into a two-dimensional lattice clearly show substructures of the connector ends (Fig. F2.19). The 12 subunits of the wide connector end can be directly revealed from unprocessed data.

F2.4.3 Biological Macromolecules at Work: High-Speed AFM

The ability of the AFM to operate in liquid makes the method attractive for the study of time-dependent changes in biological systems.

COMMENT F2.8 IMAGING OF A TWO-DIMENSIONAL LATTICE

The crucial question in such studies is whether this resolution was obtained by a point interaction or by the coherent superposition of several signals. In other words, was the periodicity of the lattice reproduced, or were individual molecules imaged?

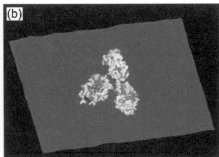

Fig. F2.17 (a) A nanotube tip image of Immunoglobulin G, which exhibits the characteristic Y-shaped structure, compared with (b) the crystal structure to emphasize the minimal tip-induced broadening. (After Hafner *et al.*, 2001.)

Fig. F2.18 Cation-driven rotors of F_0F_1-ATP synthases. The individual subunits of the cylindrical rotors represent two transmembrane α-helices connected by a polypeptide loop: (a) the sodium-driven rotor of F_0F_1-ATP synthase from *Ilyobacter tartaricus* exhibits 11 subunits; (b) the proton-driven rotor of the F_0F_1-ATP synthases from spinach chloroplast is formed by 14 subunits. (Adapted from Muller *et al.*, 2002.)

Degradation of DNA by Nuclease

Figure F2.20 shows a time-lapse sequence of the digestion of a DNA fragment by BAL 31 nuclease. The first image was taken just before the addition of the nuclease. The nuclease is not seen because it interacts only transiently with the DNA during catalysis. As the DNA is digested, it disappears from the image, leaving an increasingly large gap in the molecule.

RNA Polymerase Activity

In observing the transcription process, a paradoxical problem had to be overcome. To be able to image DNA under fluid with the AFM, not only must the DNA molecule be bound sufficiently strongly not to be distributed by the tip, it must also be bound loosely enough for the RNA polymerase (RNAP) to be able to translocate it. This compromise is achieved by adjusting the corresponding concentration of Zn^{2+} ions in the reaction chamber since certain divalent cations promote the adhesion of DNA to mica. Figure F2.21 demonstrates the position of RNAP on a 1047 bp DNA template before (Fig. F2.21a) and after (Fig. F2.21b) the addition of 0.5 µM nucleoside triphosphates (NTPs). The DNA appears to have been pulled through the RNAP and may well have been transcribed. The transcription rates were observed to be approximately 0.5–2 bases per second at ribonucleoside triphosphate concentrations of approximately 0.5 µM.

Fig. F2.19 Rotary rotors of ´φ29 bacteriophages imaged by contact mode AFM. The two-dimensional crystal exhibits a p4212 symmetry with each thick end of the connector surrounded by the four thin ends of adjacent connectors. (Adapted from Muller et al., 2002.)

Protein Conformational Changes

Outer membranes are the protecting surface barriers of Gram-negative bacteria. They are often assembled from regularly packed protein channels. A well-known example is the hexagonally packed intermediate layer of *Dienococcus radiodurans*. The pore-forming hexagonally packed intermediate units exhibit a stochastic conformational change that is best described as a switch between an open and a closed state. Figure F2.22 displays a sequence of scans over one area of the intermediate layer. Repeated imaging of the same surface at 4 min intervals showed that some pores changed their conformation from unplugged to plugged. This conformational change was fully reversible and could be observed over a long time period.

Conformational Changes at the Cell Surface

Adhesin proteins on bacterial cell surfaces have a vital role in initiating colonization and infection processes. In a study combining AFM with small-angle X-ray scattering and crystallography (see Part G), Agnew *et al.* (2011) have investigated the physical deformability of the UspA1 adhesin protein from *Moraxella catarrhalis* (Mx), a causative agent of middle ear infections in humans. UspA1 binds a range of extracellular proteins including fibronectin and CEACAM1 (epithelial cellular receptorcarcinoembryonic antigen-related cell adhesion molecule 1). Unliganded UspA1 has been shown to be densely packed, extending about 80 nm from the cell surface (Fig. F2.23).

By using a modified AFM, Agnew *et al.* have shown how the adhesive properties and thickness of the UspA1 layer vary on addition of either fibronectin or CEACAM1 (Fig. F2.24). The *in-situ* analysis was then correlated with molecular structures and conformational changes of UspA1 determined by X-ray small-angle scattering and crystallography (see Part G for the description of the methods) to provide an overall model for the adhesion. The study has therefore provided a rare direct demonstration of protein conformational change at the cell surface.

High-Speed and Ultrafast AFM

The studies above have been performed with a time resolution in the minute range. The development of high-speed atomic force microscopy (H-S AFM) permitted direct visualization of structural changes and dynamic processes

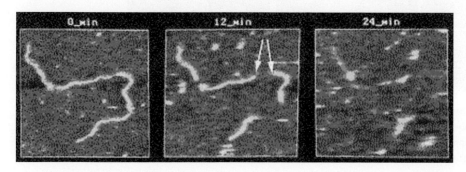

Fig. F2.20 Three sequential images (at 0, 12, and 24 minutes) show the digestion of a DNA fragment by BAl 31 restriction nuclease. The DNA fragment was deposited on mica and imaged in buffer using the tapping mode. In the frame taken at 12 min, the pair of arrows enclose the region of initial digestion of the DNA fragment by the endonuclease. (Adapted from Bustamante et al., 1994.)

in biological molecules under physiological solutions as a movie with high spatiotemporal resolution. HS-AFM has undergone major development since first experiments in the year 2000, mainly through the independent work of Ando and Hansma (reviewed by Ando, 2012). Recent results are reviewed in Eghiaian *et al.* (2014) and Ando *et al.* (2014). The small cantilevers with self-sensing and self-actuation, which had to be developed to implement HS-AFM, significantly simplified the instrument configuration.

In tapping mode, operation speed is mainly limited by the resonant frequency (f_0) of the cantilever, which depends on the following relation (for a rectangular shape of thickness t, width w, length L):

$$f_0 = \frac{1}{2\pi}\sqrt{\frac{k}{m}} \text{ and } k = Et^3 w/4L^3 \tag{F2.1}$$

where k is the spring constant and E is Young's modulus of the cantilever material.

Shorter cantilevers will have higher resonant frequencies and AFM scanning speed was increased significantly by developing cantilevers of the smallest possible size and of sufficiently soft spring constant for the high-resolution scanning of biological samples at video rate.

The H-S AFM study of myosin V (M5) action discussed in detail in the review by Ando *et al.* (2014) is an excellent

example to illustrate the power of the method. M5 is a double-headed motor protein that functions as a cargo transporter in cells (see Section F2.6 on molecular motors). M5, unlike muscle myosin, operates as a single molecule. Its structure and the actinomyosin ATPase reaction scheme are shown in Fig. F2.25. H-S AFM images at 7 fps (frames per second) of the process are shown in Fig. 2.26.

Miles' group developed ultra-fast AFM by using a tuning fork of resonant frequencies 30–100 kHz and a fast digital processing system. They obtained images of collagen fibers at a rate of 1300 fps (Picco *et al.*, 2007).

F2.4.4 The AFM Probe as a Nanoscalpel

It is possible to use the AFM tip like a mechanical "nanoscalpel." After scanning the area of interest in non-contact mode AFM, chromosome can be cut by the SFM tip with a high force (Fig. F2.27). A cross-section analysis along line AB reveals a cut width of about 100 nm. Minute amounts of material extracted at particular chromosome sites can be processed by the use of a biochemical PCR technique (see Comment B2.15).

F2.4.5 Study of Crystal Growth

A number of features of AFM make this a significant development for protein crystallographers. The kind of information found in AFM images is clearly different from that available from other microscopies, such as transmission electron microscopy. First of all, AFM images are obtained from macromolecular crystals *in situ*, i.e., as they exist in their mother liquor, and even while they are still growing, without any observable perturbation to the crystals or the growth process.

A second and particularly useful property of AFM images is that unlike transmission electron micrographs, they are not projections of the entire sample onto a single plane, but contain three-dimensional information.

A third feature is that whereas all asymmetric units (generally one macromolecule) are crystallographically

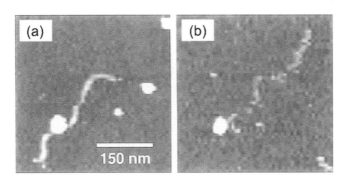

Fig. F2.21 Complex of RNAP with the 1047 bp dsDNA **(a)** before and **(b)** after the addition of 0.5 μM NTPs. These two images were acquired about 6 min apart. (After Kassas *et al.*, 1997.)

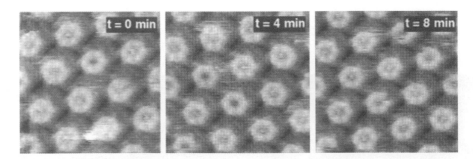

Fig. F2.22 The hexagonally packed intermediate layer of *Dienococus radiodurance* consists of hexameric units that have a central core and connecting arms. At the inner surface, the cores exhibit a central pore that occurs in "open" (unplugged) and "closed" (plugged) conformations. The defective pore at the bottom-left corner is used as a reference to align the scans. The images were recorded using the contact mode. The different shades correspond to a height range of 6 nm. (Adapted from Engel *et al.*, 1999.)

(a)

(b)

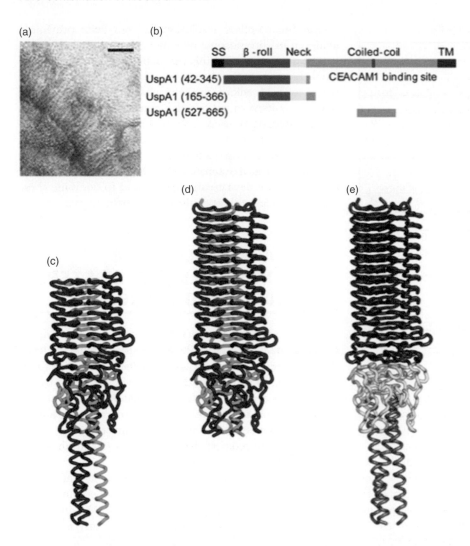

SS   β-roll   Neck   Coiled-coil   TM

UspA1 (42-345)   CEACAM1 binding site

UspA1 (165-366)

UspA1 (527-665)

(d)

(e)

(c)

Fig. F2.23 *Moraxella* cell surface, domain organization, and head region structure of UspA1. **(a)** Electron micrograph showing extended UspA1 molecules at the Mx surface (scale bar: 50 nm.). **(b)** Schematic showing full-length UspA1 comprises five regions, each colored separately and labeled. Constructs for which crystal structures are determined are also shown. TM, transmembrane. **(c–e)** Ribbon representations of the crystal structures of **(c)** UspA1 **(d)** UspA1, and **(e)** composite model of UspA1In C and D, each chain is colored separately; **(e)** is colored by domain as per **(b)**. (From Agnew *et al.*, 2011, with permission).

equivalent in the interior of a crystal, and therefore have identical chemical and physical environments, this is not true of molecules on the surface of a crystal. Thus an AFM image of a crystal surface, in general, simultaneously presents multiple, symmetrically related perspectives on a single object, the molecule, thus providing three-dimensional visual information.

Figure F2.28 shows as an example the process of two-dimensional crystallization of annexin V on supported planar lipid bilayers. It is clearly seen that AFM enables the process of two-dimensional crystal growths to be visualized in real time.

F2.5 COMBINATION OF NSOM AND AFM

The central component in NSOM is a tapered glass optical element (Section F1.4). The cantilevered nature of this optical fiber allows it to work in all the same modes as a conventional AFM probe. It is particularly suitable for the

tapping mode. Using standard AFM technology, the point of light at the tip of the cantilever can be brought to within the near-field of a surface that is to be imaged, i.e., closer to it than the dimension of the illuminating light. As the tip scans the surface of the sample, the light emanating from the aperture at the tip interacts with the surface and passes through the sample to measure the absorption or to elicit fluorescence in a region of the surface. After interaction with the surface, the light is collected by the lens of a conventional microscope, detected by a suitable detector and recorded electronically.

An example of imaging using the integrated capability of the NSOM/AFM microscope is shown in Fig. F2.29. The figure shows an NSOM image of a Giemsa-stained human chromosome and an AFM image of the topography of the same chromosome. A line scan from the NSOM image is displayed to the left of Fig. F2.25b. In this line scan, the first transition from less staining (less absorption) occurs less than 0.1 μm from the 1 μm line and is completed before the 1 μm line. The alteration from 90% to 10% transmission in such a thin sample occurs over approximately 60 nm, and this provides a measure of the effective resolution in this

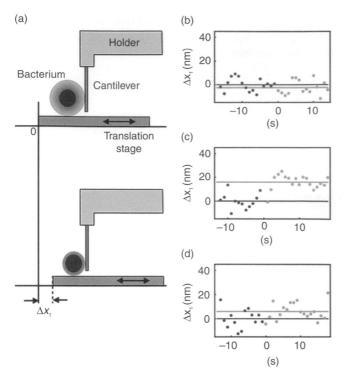

Fig. F2.24 Atomic force microscopy analysis of Mx bacteria. **(a)** A sinusoidal sideways movement is imparted to the sample stage in order to gently push a single adsorbed bacterium (blue sphere; pale-blue outer layer represents surface adhesins) against the cantilever. The change in the sample stage position ($\Delta x1$) required to achieve contact is measured upon addition of either ligand (CEACAM1 or Fn) or buffer/control protein. The graphs on the right show representative data measured for the change in the contact point position before (in red) and after (in green) the addition of **(b)** control, **(c)** CEACAM1, and **(d)** FNIII12;13;14. The red and green lines represent the mean value of the red and green points, respectively. (From Agnew *et al.*, 2011, with permission).

image. Such optical resolution has not been previously achieved in chromosome imaging.

The development of this technique has led to an instrument that can be simply added to the stage of any conventional optical microscope to enable NSOM imaging. NSOM, by its very nature, also generates a simultaneous topographic AFM image. With this advance, it became possible to integrate, with overlapping fields of view, images from a single microscope that range from low-resolution optical microscopy to high-resolution NSOM and AFM. The device can also be used to combine these imaging modes with optical-tweezer applications.

F2.6 MACROMOLECULAR MECHANICS: NANOMETER STEPS AND PICONEWTON FORCES

Macromolecular mechanics implies the monitoring of forces at the molecular scale. At the single-molecule level, the characteristic energy is given by the hydrolysis of ATP (20 kT, i.e., 80 pN nm) and the characteristic size by the diameter of a protein (a few nanometers). The resulting forces that physicists must be able to measure and to produce while studying those objects are therefore in the range of hundreds of femtonewtons to hundreds of piconewtons.

The initial applications of macromolecular mechanics involved molecules that produce active movement against a load, e.g., "molecular motors" such as myosin, which drives muscle contraction. In the 1990s many molecular motors and biological macromolecules have been observed, including: step-wise motion of a single kinesin molecule along a microtubule track, displacement of an

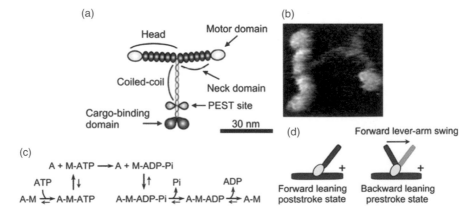

Fig. F2.25 Structure of myosin V, actinomyosin ATPase reaction scheme, and prevailing view on conformational states of myosin. **(a)** Schematic of molecular structure of myosin V. **(b)** AFM image of myosin V. **(c)** Reaction scheme of actinomyosin ATPase. **(d)** Schematic for backward-leaning prestroke and forward leaning poststroke conformations. The mark "+" indicates the plus end of an actin filament (red). The motor domain (yellow) binds to an actin filament in the same orientation, in both conformational states, while the neck (blue) is leaned in different directions. The forward lever arm swing is supposed to occur on the actin-bound head when P_i is released from myosin. The backward lever-arm swing is supposed to occur on the detached head when ATP is hydrolyzed to ADP–P_i. (From Ando *et al.*, 2014, ACS Standard ACS AuthorChoice/Editors' Choice Usage Agreement.)

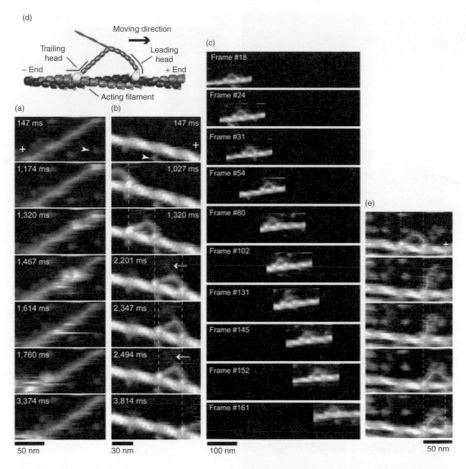

Fig. F2.26 Tail-truncated myosin V (M5-HMM) movement on actin filament captured by H-S AFM. All of the images were taken at a frame rate of 7 fps (frames per second). (a) Successive AFM images showing processive movement of M5-HMM in 1 μM ATP when positively charged lipid is absent in the planar lipid bilayer (PLB) surface. The arrowhead indicates one of the streptavidin molecules attached to the PLB surface. (b) Successive AFM images showing processive movement of M5-HMM in 1 μM ATP when positively charged lipid is present in the PLB. The arrows indicate the coiled-coil tail pointing to the minus end of actin. The arrowhead indicates one of the streptavidin molecules attached to the PLB surface. (c) Clips of successive images showing the long processive run of M5-HMM in 1 μM ATP (14 steps are recorded). (d) Schematic explaining structural features of two-headed bound M5-HMM observed in the presence of nucleotides. (e) Successive AFM images showing stepping process in 1 μM ATP. The swinging lever is highlighted with a thin white line. (From Ando *et al.*, 2014, ACS Standard ACS AuthorChoice/Editors' Choice Usage Agreement.).

actin filament by a single myosin molecule, forces and transcriptional pauses associated with RNA polymerase activity, non-linear elasticity of single polymers, reversible unfolding of single-protein domains by applied force, discrete rotations of a single F_1 subunit of the F_0F_1-adenosin triphosphate (ATP) synthase, DNA packaging in bacteriophage φ29, and chewing of one of DNA's double-helical strands by lambda exonuclease (Comment F2.9).

The parallel development of single-fluorophore detection has allowed these mechanical measurements to be combined with observations of substrate binding, protein position, and conformational change. Below we focus mainly on the use of optical trapping technology, including the rival methods that use glass microneedles and AFMs where appropriate.

F2.6.1 Linear Molecular Motors

The most intensive application of optical tweezers has been focused on linear motor proteins, such as myosin, kinesin, and dynein, which convert the energy of ATP hydrolysis into mechanical work. These proteins move along polymer substrates: myosin along actin filaments in muscle and other cells; kinesin and dinein along microtubules. Another example is RNAP, which is a highly processive molecular motor capable of moving through thousands of base pairs without detaching from the DNA template.

Do biological motors move with regular steps? What is the size of the steps? What is the speed and force of such motors? To address these questions, many instruments with the spatial and temporal sensitivity to resolve movement on a molecular scale have been used. Figure F2.30

Fig. F2.27 Topographic SFM micrograph of human chromosome 2 after DNA extraction. For nanoextraction, one line scan with approximately 50 µN contact force was performed. The arrow indicates the dissecting direction of the tip at the beginning of the cut. Next to the arrow shaft, a particle is visible. This particle can be explained as a part of the dissecting tip which broke off when the tip initially snapped onto the surface. (Thalhammer et al., 1997.)

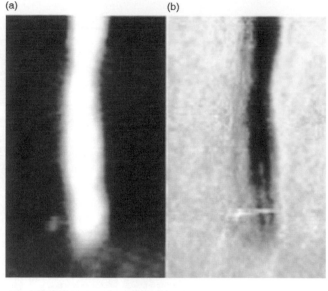

Fig. F2.29 Simultaneous AFM and NSOM imaging of a Giemsa-stained human metaphase chromosome. AFM (a) measures the topography (i.e., the height variations; brighter means higher), whereas NSOM (b) monitors the light throughput at each position; the light throughput is modulated, in this case, by the binding of the stain to the protein molecules in the chromosome (brighter means more transmission). (Adapted from Lewis et al., 1999).

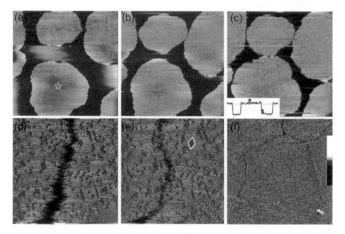

Fig. F2.28 Growth of crystalline domains of annexin V on PS-containing SPB on mica, imaged by AFM. Successive images are presented in (a)–(b). The time after injection of annexin V in the fluid cell is: (a) 19 min; (b) 23 min; (c) 26 min; (d) 29 min (e) 31 min; (f) 34 min. (a) Circular domains (one domain is labeled with a green asterisk) are homogeneously distributed over the SPB surface. These domains grow and coalesce from (a) to (e). (c). The height profile measured along the green line is shown in the inset. The three crystalline domains are of identical height and the lipid surface is flat. The height measured between the two red arrowheads is 2.6 nm. Scan size (a)–(c), 5000 nm. (d)–(e) Enlarged view of a grain boundary forming between two adjacent domains. Scan size (f), 870 nm (Adapted from Reviakine et al., 1998).

COMMENT F2.9 MOLECULAR MOTORS IN A CELL

It is likely that a given cell has 50 or more distinct molecules that operate on actin filaments and microtubules to produce the many types of movement that cells undergo. Taken together with the ATP-driven motor molecules operating along DNA, the movement of ribosomes along messenger RNA, and motors that have yet to be discovered, the number of different molecular motors in a cell is probably closer to 100.

illustrates a typical experimental configuration for manipulating linear molecular motors. The vertical arrows indicate the direction of the light. The force and displacement of the motor are detected in the trapped bead. In Fig. F2.30a, myosin, an actin filament, suspended by two separate optical traps via two beads attached to its ends, is lowered over a third, fixed bead coated with myosin S-1 fragment (see below). In Fig. F2.30b, kinesin (or dinein) coated onto a bead moves along a microtubule that is attached to the surface of a microscope coverglass (see below). In Fig. F2.30c, an RNAP bead is attached to the transcriptional downstream end of DNA so that it becomes tethered to the surface of a microscope coverglass via the RNA polymerase fixed to the coverglass (see below).

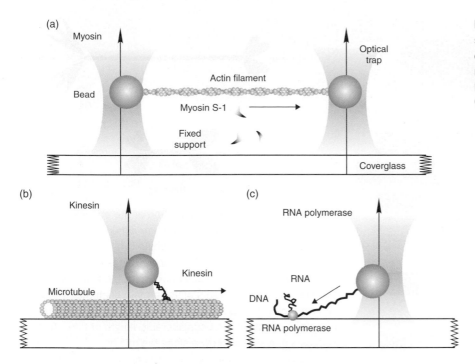

Fig. F2.30 Sketches (not drawn to scale) illustrating typical experimental configurations for manipulating linear molecular motors: **(a)** myosin; **(b)** kinesin (or dinein); **(c)** RNA polymerase. (Adapted from Wang, 1999.)

To measure the mechanical properties of single motor molecules, it is necessary to manipulate single protein molecules and allow the interaction to be monitored. A laser trap holds a bead by utilizing the attractive forces exerted by a focused laser (Fig. F2.30b, c). By moving the focal point, it is possible to move a bead that attaches to a single protein molecule. A laser trap can also measure the mechanical properties of molecular motors, because the trapped beads behave like a spring. When force is exerted on the protein, the bead becomes displaced and a restoring force is exerted by the trap toward the trap center to pull the bead back to the original position. The two forces are then balanced.

The laser trap has been used for the myosin–actin system. In contrast to processive molecular motors, conventional myosin readily dissociates from actin. To prevent the actin filament from diffusing away from the myosin during measurements, the actin filament is trapped at both ends (Fig. F2.30a).

Kinesin: Vesicle Transport along Microtubules

The most intensive application of optical tweezers has been focused on linear motor kinesin. Most kinesin proteins appear to be dimeric molecules that contain a stalk, a tail-like domain and two globular motor domains (Fig. F2.31). The tail-like domain sometimes includes non-covalently attached light chains and is presumably important for transport specificity or organelle recognition. The globular tail is referred to as the cargo domain to indicate its role in organelle movement. There are many variations in the size and arrangement of the stalks and tails relative to the motor domains.

One of the unusual features of the kinesin superfamily of molecular motors is the existence of two classes of molecule: one, as represented by kinesin, that moves toward the plus-end of the microtubules, and another, as represented by NCD, that moves toward the minus-end. The kinesin superfamily is radically different from myosin, which only moves in one direction on the actin thin filament. Interestingly, the molecular organization of kinesin and non-claret disjunctional (Ncd) differs at the motor domain residues at the N-terminus of kinesin and C-terminus of Ncd (Fig. F2.32).

Microtubule-based motors offer special opportunities for understanding how motors work. In part because of its relatively, small size, the mechanism of the action of kinesin is rapidly approaching the same level of understanding as that of myosin, and incorporates three significant new features. First, this motor works processively i.e., unlike myosin, a single molecule moves along the microtubule track for many seconds before dissociating. Second, within the kinesin family there are motors that move in one direction along a microtubule and others that move in the opposite direction. Finally, kinesin motors can be made to move slowly, permitting better time-averaging of position, and they remain attached to the substrate for a substantial fraction of the kinetic cycle, reducing the magnitude of Brownian excursions.

The atomic structures have been solved for the motor domains of two molecules belonging to the kinesin family, kinesin itself and Ncd (the product of the *Drosophila* Ncd gene). These two motors move in opposite directions on microtubule tracks, yet they have highly similar structures. It is startling that these ~45 kDa proteins are also

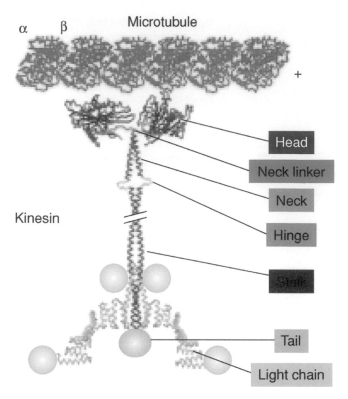

Fig. F2.31 Kinesin domain structure and associated proteins. The top structure represents a protofilament of a microtubule, the "track" for kinesin motors, containing alternating α- and β-tubulin subunits oriented so that the plus-end points to the right, toward the cell periphery for most microtubules. The blue and red structure in the middle represents a dimeric kinesin heavy chain. Each heavy chain contains a motor domain (head, red and blue) that binds to ATP and microtubules, a neck linker (cyan) whose conformation changes during the ATPase cycle, and an α-helical neck and stalk (red). The structure of the globular tail domain is not yet known and is represented as an orange sphere. The tail binds to two light chains (yellow). The light chains contain special protein-interaction motifs, which can dock onto adaptors or cargo receptors, linking kinesin to the cargo to be transported. (Adapted from Mandelkow and Mandelkow, 2002.)

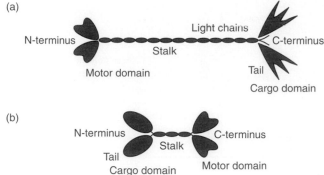

Fig. F2.32 Schematic representation of the architecture of (a) kinesin and (b) Ncd. Most kinesins are dimeric and contain two motor domains that are in close proximity and are usually held together by a section of coiled coil. For kinesin it is clear that there is an important interaction between the two heads that confers processivity on this motor protein such that a single kinesin molecule is able to remain in contact with and move along a microtubule for a considerable length of time. This suggests that kinesin moves along the microtubules in a "hand over hand" fashion: The force-generating head does not leave the microtubule until its neighbor is attached. This highly processive nature is lost in a single-headed kinesin, although a single head can still generate movement at a velocity similar to that of the intact molecule. Thus, although the connection between the heads is important, the ability to produce a significant power stroke must reside in a single head. (After Rayment, 1996.)

structurally related to the much larger (nearly double the mass) motor domains of myosin (Fig. F2.33). Significant technical difficulties nevertheless exist in measuring movements of single kinesin molecules. The motions occur on length scales of ångströms to nanometers and on timescales of milliseconds or less. To obtain the high spatial and temporal sensitivity required, optical tweezers have been combined with a dual-beam interferometer to produce an "optical trapping interferometer" (see Comment F2.6). This instrument provides position detection and trapping functions simultaneously, and can provide controlled, calibratable forces in the piconewton range.

To characterize its motion at high spatial resolution, kinesin is attached to silica beads. A trapped bead is moved

near a microtubule that is fixed on a microscope coverslip (Fig. F2.30b). Figure F2.34 shows that the kinesin advances in discrete steps of 8 nm (the tubulin repeat unit on the microtubule track), which can be clearly distinguished from Brownian motion. The steps are separated by dwell periods of variable time (Comment F2.10). The kinesin moves the bead away from the trap center, slows as the resistive force increases, and finally stalls under loads of 5–7 pN. Figure F2.35 shows the mode of operation for kinesin. Kinesin appears to "walk" along a microtubule without detaching both its feet simultaneously and probably using its feet alternately.

Dynein: Eukaryotic Flagella Motor

Bacterial flagella differ in every respect but name from eukaryotic flagella (see Comment F2.11). In the eukaryotic flagellum, dynein motors powered by ATP generate moving bends by stepping along linear microtubular tracks.

To manipulate the microtubule, optical forces are applied to streptavidin-coated latex beads that are attached as handles to biotinylated single microtubules (Fig. F2.36; see Comment F2.16). Figure F2.37 is an electron micrograph of two singlet microtubules interacting with dynein arms on a doublet microtubule. When the free end of such a microtubule is brought into contact with the

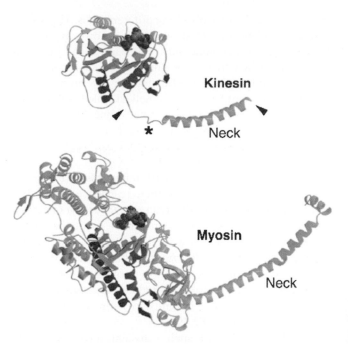

Kinesin

Neck

*

Myosin

Neck

Fig. F2.33 Comparison of the crystal structures of myosin and kinesin, looking down on the nucleotide (solid black). The structure of NCD (not shown) is almost identical to that of kinesin. The amplifier region adjacent to the motor core is the neck region of kinesin and of myosin. Note: The area between the two arrowheads in the kinesin neck region is not visible in the present crystal structure. The assignment of random coil and helix is based on secondary structure prediction and CD spectral studies; the precise boundary of the neck helix and the location of the neck region relative to the core is not known. The asterisk denotes a kinesin truncation: aa 340 in *Drosophila* kinesin and aa 332 in human kinesin. (Adapted from Vale, 1996.)

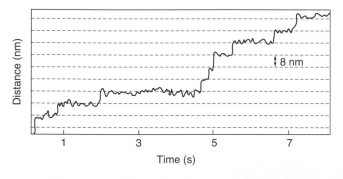

Fig. F2.34 Position record of a pulled bead. Kinesin advances in 8 nm increments, which are separated by dwell periods of variable length. (After Mehta *et al.*, 1999.)

surface of a doublet, the microtubule stops exhibiting Brownian motion and, after photolysis of caged ATP, begins to move. If the angle between the singlet microtubule and doublet is less than ~30°, the singlet microtubule gradually aligns parallel to the doublet and moves along the doublet until finally the bead escapes from the optical

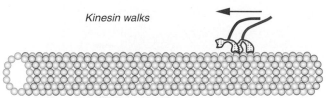

Kinesin walks

Fig. F2.35 Mode of operation for kinesin. Kinesin appears to "walk" along a microtubule without detaching both its feet simultaneously and probably using its feet alternately. (Adapted from Kinosita *et al.*, 1998.)

COMMENT F2.10 KINESIN MOVEMENT

To elucidate the kinetic scheme underlying the 8 nm advances, the distribution of dwell times between steps has been analyzed by several investigators. It was concluded that one rate-limiting process, ATP binding, precedes every 8 nm advance. Such measurements eliminate models that postulate two ATP-dependent head movements producing each 8 nm step, suggesting instead that kinesin may move through 16 nm step-by-step movements of the two heads.

COMMENT F2.11 EUKARYOTIC AND BACTERIAL FLAGELLA

In the eukaryotic flagellum, dynein motors powered by ATP generate moving bends by stepping along linear microtubular tracks. The eukaryotic flagellum undulates like a snake inside its cellular membrane. In contrast, the bacterial flagellum works like a power boat with a rotary motor turning a rigid, helical propeller.

---

trap force (at forces of >30–40 pN). However, when the microtubule is placed obliquely (at 45–90°) on the doublet the bead does not escape and force is recorded repeatedly after each flash of ultraviolet light.

Figure F2.38 shows a typical record of the force generation. Dynein arms generate a peak force of 6 pN and move the singlet microtubule in a processive manner. A remarkable feature of the dynein arm activity is the presence of oscillations (Fig. F2.38a). The force oscillates with an amplitude of ~2 pN. The maximum frequency of the oscillation at 0.75 mM ATP is ~70 Hz (Fig. F2.38b). Interestingly, the oscillation frequency is not far from the frequency of flagellar beating (~350 Hz). This discovery suggests that these oscillation forces of dynein may be the key to rhythmic beating motions of eukaryotic flagella.

Myosin: The Muscle Motor

The term myosin refers to at least 14 classes of proteins, each containing putative or demonstrated actin-based

(a)

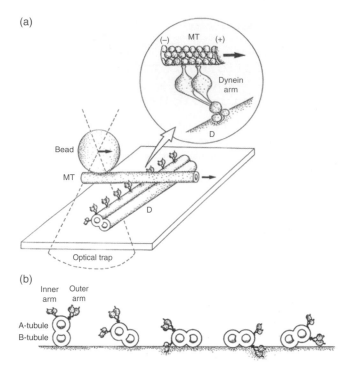

(b)

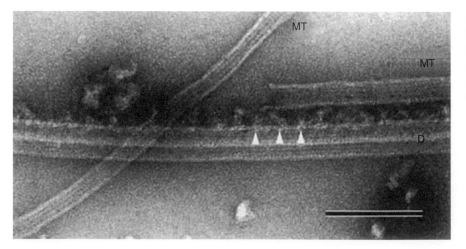

Fig. F2.36 Interaction of singlet microtubules with dynein arms presented on doublet microtubules. **(a)** Schematic diagram of the exposed dynein arms, on a doublet microtubule (D), interacting with a singlet microtubule (MT). The singlet microtubule is manipulated by means of an optically trapped streptavidine-coated bead. A two-headed arm pulls the singlet microtubule in the direction of its plus (+) end, causing the bead to move away from the center of the trap force (black arrows). **(b)** Possible orientations of a doublet microtubule on the glass surface (bottom) with outer and inner dynein arms. Dynein arms pointing upwards may be capable of interaction with a singlet microtubule (not shown). (After Shingyoji et al., 1998.)

motors and each probably filling different roles. The results described below involve skeletal muscle myosin (myosin II). Figure F2.39 schematically represents the architecture of myosin. Muscle myosin is a highly asymmetric 470 kDa protein, containing two 90 kDa globular N-terminal heads, called subfragment S1, and an α-helical coiled-coil tail that polymerizes to form the filament backbone. Myosin heads are also called cross-bridges because they can link the two sets of filaments. In S1, a "catalytic domain" (CD) contains the ATP and actin-binding sites.

Myosin II shares many structural features with kinesin. Both use ATP to move along their respective tracks, but myosin II is processive – it undergoes (at most) one catalytic cycle per diffusional encounter with its track. Although this means that a single molecule cannot move along its track for large distances, it also means that organized ensembles of molecules can move along the track at higher speeds. Myosin is thought to undergo a conformational change when it binds to actin, resulting in a "working stroke" (Fig. F2.40).

For the coupling of such transient interactions and movement to a trapped bead, a more complex geometry is needed than that described above for kinesin. Figure F2.41 shows the experimental geometry for observing a single myosin molecule binding and pulling an actin filament. The filament is attached at either end to a trapped bead. These beads are used to stretch the filament taut and move it near surface-bound silica beads that have been decorated sparsely with myosin molecules. Discrete step-wise movements averaging 11 nm were found in these experiments. The magnitudes of the single forces and displacements are consistent with the predictions of the conventional swinging cross-bridge model of muscle contraction (Comment F2.12).

The most exciting myosin experiment was performed by observing mechanical steps and fluorescent nucleotides

Fig. F2.37 Electron micrograph showing two singlet microtubules (MT) interacting with dynein arms (white arrowheads) on a doublet microtubule (D). Scale bar, 100 nm. (After Shingyoji et al., 1998.)

(a)

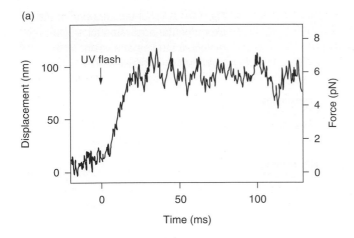

(b)

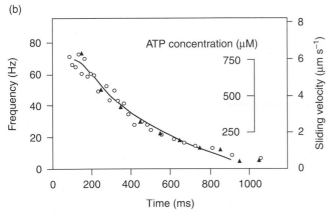

Fig. F2.38 Typical record of force generation by dynein arms. **(a)** The angle between the singlet microtubule and the doublet microtubule to which the dynein arms were attached was 89°. A UV flash was applied at time zero (arrow). Displacement refers to the displacement of the singlet microtubule over the doublet by the dynein arms. **(b)** Changes in oscillation frequency and the velocities of sliding of untrapped microtubules along doublets. The line shows changes in the estimated ATP concentration. (After Shingyoji et al., 1998.)

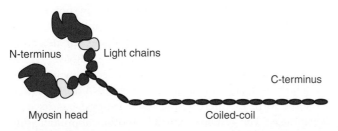

Fig. F2.39 Schematic representation of myosin architecture. (After Rayment, 1996.)

simultaneously using total internal reflection microscopy. Figure F2.41 shows the experimental apparatus. A single actin filament with beads attached to both ends is suspended in solution by optical tweezers. The suspended actin filament is brought into contact with a single

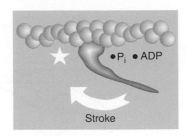

Fig. F2.40 Schematic view of a myosin molecule. For clarity, only one head (S1, the motor domain of myosin) and a short segment of the α-helical tail are shown. The small curved arrow shows the direction of the conformational change resulting in a stroke of about 10 nm. This stroke of the lever arm of S1 pulls on the myosin tail, which is attached to the myosin thick filament (not shown) and causes a relative movement of the myosin filament and the actin filament. (After Spudlish, 1994.)

one-headed myosin molecule in a myosin–rod cofilament bound to the surface of a pedestal formed on a coverslip. The displacement or the force due to actomyosin interactions is determined by measuring bead displacements with nanometer accuracy. Using total internal reflection fluorescence microscopy, individual ATPase reactions are monitored as changes in fluorescent intensity due to association–(hydrolysis–)dissociation events of a fluorescent ATP analogue, Cy3-ATP, with the myosin head. The results show that the myosin head produces a displacement of 15 nm and the force generation does not always coincide with the release of bound nucleotide. Instead the myosin head produces force several hundreds of milliseconds after a bound nucleotide is released. This finding does not support the widely accepted view that force generation is directly coupled to the release of bound ligands. It suggests that myosin has a hysteresis or memory state that stores chemical energy from ATP hydrolysis.

Many studies in crystal and solution have shown that the neck region of the myosin head undergoes conformational changes. The angle of the myosin head changes relative to the neck region. On the basis of these findings, the "swinging cross-bridge model" of force generation has been refined into the "lever-arm swinging" model, in which the neck region of the myosin head is postulated to act as a lever arm (Fig. F2.42a). The size of displacements reported varies considerably – some investigators have reported a myosin displacement of ~5 nm (per head), fully consistent with the lever-arm model, whereas others have found 10 nm and more. It is crucial to determine the myosin step size unambiguously, taking into account possible different orientations of myosin relative to the actin filament axis. The displacement varies from ~15 nm to 0 nm, with changes in the angle from 5° (near the physiologically "correct" orientation) to almost 90°. When averaged over all directions, the displacement is ~6 nm. This value is consistent with measurements in which myosin was

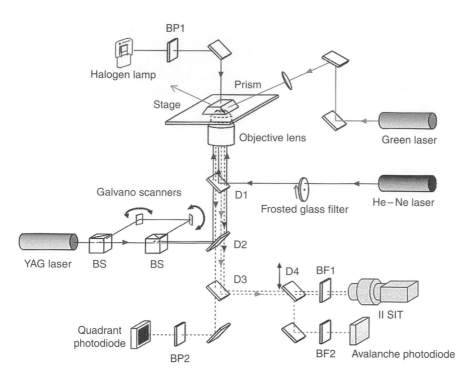

Fig. F2.41 Schematic drawing of the experimental apparatus for single-molecule studies of myosin. (After Ishijima *et al.*, 1998.)

COMMENT F2.12 HUXLEY MODEL

In 1969 H. Huxley suggested that the sliding force is developed as a result of a change in the angle of attachment in the head part of a myosin cross-bridge attached to actin. Based on a mechanical investigation of muscle, Huxley and Simmons proposed a theoretical model, called the "swinging cross-bridge model." Conformational changes in the myosin heads coupled with changes in bound ligands were observed in crystals, in solution, and in muscle fibers during contraction. The displacement expected from the observed conformational changes in the myosin head is approximately 6 nm.

---

randomly oriented on the surface. It has been concluded that the small displacements reported might be attributable to the effect of this random orientation of the myosin heads.

The rising phase of the displacements, on an expanded timescale, are shown in Fig. F2.43. The displacements do not take place abruptly, but instead develop in a step-wise fashion. The size of the substeps is ~5.5 nm. The substeps are stochastic, and the number of steps per displacement varies between one and five, which produces a total displacement of ~5 to ~30 nm.

A single myosin head moves along an actin filament independently of the load, with regular 5.5 nm steps that coincide with the distance between adjacent actin subunits in one strand of an actin filament. Multiple steps are produced during a single ATP turnover event, some of which

are backward. These results suggest that the myosin head "runs" on actin subunits using Brownian motion (Fig. F2.44). However, given that backward steps do not occur frequently (only 10% of the total number of observed steps), it seems that this Brownian motion is biased in the forward direction.

RNA Polymerase: A Processive DNA Transcription Machine

The enzyme RNAP carries out an essential step of gene expression, the synthesis of an RNA copy of the template DNA (Fig. F2.45). Each RNA molecule is synthesized in its entirety by a single molecule of RNAP moving processively along the template. Movement is powered by the free energy liberated as appropriate nucleoside NTPs are joined to the 3′- end of the nascent RNA chain, pyrophosphate (PP) is released, and the growing transcript folds.

Figure F2.46 shows different techniques for detecting DNA translocation by single molecules of RNAP. These techniques all isolate a single RNAP molecule by immobilizing it on a surface; each then observes DNA translocation by some form of microscopy. Each method has its own advantages: The tethered particle motion approach (Fig. F2.46a) is arguably the least perturbational; it neither exerts force on the DNA nor significantly alters its conformation. Since bead diffusion is more rapid than the mean velocity of DNA translocation in a polymerase catalytic cycle, attachment of the bead should not significantly affect the translocation reaction, provided the bead is attached further from the RNAP molecule than the DNA persistence length (~50 nm or 150 bp). On the other hand, the laser tweezers method (Fig. F2.46b) has a high time resolution

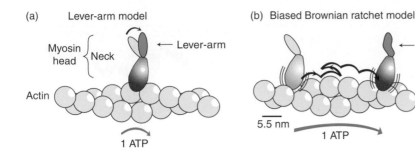

(a) Lever-arm model

(b) Biased Brownian ratchet model

Fig. F2.42 Model of an actomyosin motor: **(a)** The lever-arm model: The neck region of myosin acts as a lever arm. **(b)** The myosin head is proposed to walk on acting subunits by biased Brownian motion. Multiple steps are produced during a single ATP turnover event, some of which are backward, although backward steps do not occur frequently. (After Yanagida *et al.*, 2000.)

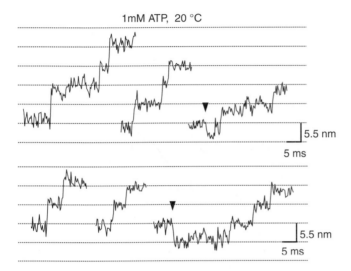

1mM ATP, 20 °C

Fig. F2.43 Myosin substeps in the rising phase of displacement recorded on an expanded timescale (After Yanagida *et al.*, 2000.)

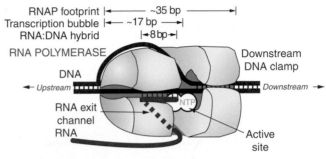

Fig. F2.45 Hypothesized structure of RNAP. Electron microscopy studies of bacterial and yeast RNAPs reveal a groove surrounded by jaw-like protrusions that may clamp around downstream duplex DNA and a tunnel that may function as a channel through which RNA exits from the active site. Chemical experiments show that RNAP protects ~17 bp of DNA on either side of the active site position, keeps the upstream half of this DNA melted, and maintains an ~8 bp RNA–DNA hybrid upstream from the active site. RNAP adds nucleotides to nascent RNA in a biparticle active site that coordinates the RNA 3'-end and the NTP. (After Gelles and Landick, 1998.)

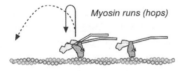

Fig. F2.44 Mode of operation for myosin. Myosin "runs" along actin, in that its two feet (the two globular parts of myosin usually referred to as "heads") are detached from actin for most of the time. In fact, myosin can run, skipping many actin monomers in a step, if other myosin molecules pull the actin filament while the first one is detached. If only one myosin molecule interacts with actin, it simply hops and will not move relative to actin while detached (except for random diffusion). (After Kinosita 1998.)

and thus far is the only method capable of measuring translocation forces as well as displacements. The surface force microscopy technique (Fig. F2.46c) is least able to observe features of the RNAP structure (e.g., the bend angle of the DNA at the polymerase) while simultaneously detecting translocation.

The most significant results from the initial round of single-molecule studies are that the enzyme is capable of translocating DNA against large opposing loads, and that the deduced efficiency of chemical-to-mechanical energy conversion is high, similar to that of the canonical cytoskeletal motor enzymes. Indeed, the single-molecule stall force is at least three times that measured for myosin and kinesin, showing that RNAP can legitimately be classed with those powerful cytoskeletal motors. The stall force against which RNAP ceased to move was determined as ~21–27 pN (Fig. F2.47a). This value exceeds the 5–7 pN that was measured for kinesin, perhaps reflecting the need for RNAP to forcefully disentangle DNA secondary structure. *E. coli* RNAP during transcript elongation progresses along DNA at speeds of about 200 nucleotides per second (Fig. F2.47b).

The three motors above are compared in Table F2.2. The lengths of genuine steps of running myosin are expected to be multiples of 5.5 nm, the distance between neighboring actin monomers. RNA polymerase is expected to step by

(a)

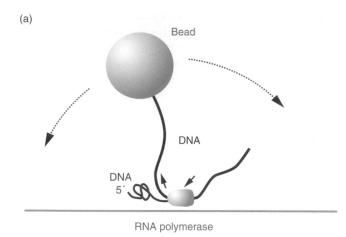

(b)

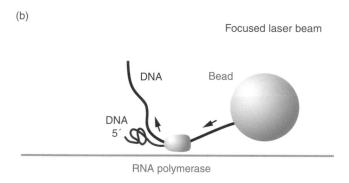

(c)

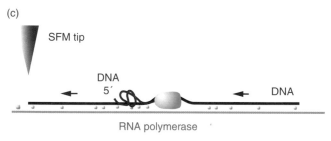

Fig. F2.46 Different techniques for studying DNA translocation by single molecules of RNAP. RNAP, bound to a glass or mica surface covered by an aqueous solution of NTPs, moves the DNA template in the indicated direction (small arrows) during transcript elongation (not to scale). (a) The tethered particle motion method. A plastic bead or colloidal gold particle undergoing rapid Brownian motion is observed by light microscopy and used to determine how the length of the DNA segment linking the bead to the polymerase changes during transcript elongation. (b) The laser tweezers method. A focused laser beam pulls on a silica or plastic bead attached to the downstream end of the DNA template, keeping the segment of DNA between the bead and the polymerase under tension. Movement of the template by the polymerase pulls the bead away from the focus of the laser beam. This displacement is detected optically. (c) The surface force microscopy method. Template DNA is loosely held against a mica surface by surface-bound cations (circles). The DNA and the position of the RNAP along the DNA contour are measured in the images. (After Gelles and Landick, 1998.)

(a)

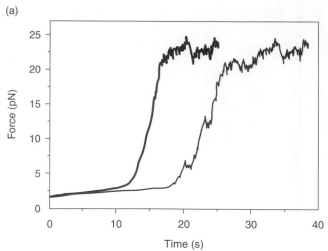

(b)

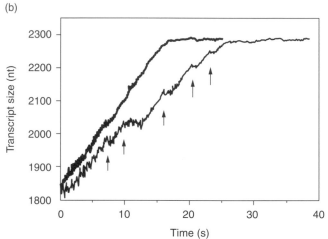

Fig. F2.47 Force and corresponding transcript length at high load. (a) The force was increased until the complex stalled at ~23 pN. (b) The bottom trace shows movement, which was interrupted occasionally by apparent transcriptional pauses (marked by arrows). The top trace shows the same movement with pauses algorithmically removed, as needed to compute a meaningful transcription velocity. (After Wang *et al.*, 1998.)

0.34 nm, the distance between base-pairs. Kinesin's 8 nm steps were measured. All these step sizes represent intervals of the structural repeats.

F2.6.2 Rotary Molecular Motors

Rotary protein motors accomplish their primary functions by rotating one group of subunits with respect to the rest. Although they constitute only a small entry in the growing catalog of cellular molecular motors, they have been studied so intensively that more is known about them than any of the others. The best characterized protein in terms of its atomic structure and biochemistry is

TABLE F2.2 COMPARISON OF NUCLEOTIDE-DRIVEN MOTORS

Motor/rail	Step size	Max. force	Max. efficiency	Processivity	Mode
Myosin/actin	Variable	3–6 pN	~20%	None–poor	Hops
Kinesin/microtubule	8 nm	5 pN	~50%	Good	Walks
RNA polymerase/DNA	0.34 nm	14–27 pN	~20%	Excellent	Crawls

ATP synthase (also called F_0F_1 ATPase), which actually consists of two rotary motors connected to a common shaft. The largest, most powerful protein motor is the bacterial flagellar motor, which is found in many species (see Comment F2.11).

F_0F_1-ATPase Motor

ATP synthase consists of two opposing rotary motors connected in series (Fig. F2.48). The soluble F_1 motor is fueled by nucleotide hydrolysis and drives the connecting γ-shaft clockwise (viewed from F_1 toward the membrane), whereas the transmembrane F_0 motor is fueled by the ion motive force and drives the γ-shaft counterclockwise (Comment F2.13). Under normal circumstances, the F_0 motor generates the larger torque and drives the F_1 motor in the synthesis direction. However, in bacteria under anaerobic conditions, when the proton motive force is low, the F_1 motor hydrolyzes ATP, driving the F_0 motor in reverse, whereupon it functions as a proton pump. The two motors illustrate two of the principal mechanisms that cells use to convert chemical energy into mechanical force. The F_0 motor is made of a transmembrane cylinder of 10–14 subunits, depending on the species (Fig. F2.49). The number of subunits may be related to the ability of the F_0 motor to also act as a proton pump in some species. The stalks coupling F_0 and F_1 are elastic, and thus F_1 does not directly see the steps of F_0. Instead, this flexible coupling transmits only the F_0 torque to F_1. Therefore, there is no fixed "stochiometry" relating the number of subunits in F_0 to the number of subunits in F_1. This elastic coupling also passes energy more efficiently between F_1 and F_0.

Rotation of F_1 had been long suspected. On the basis of extensive kinetic analysis, a model for the catalytic mechanism of F_1-ATPase was proposed by Boyer in 1993. Among other properties, this model predicted that the energy was transmitted through rotation of the γ-subunit in the center of the F_1-ATPase molecule. Boyer's model was almost entirely substantiated by experimental results, with exception of the rotation hypothesis. One year later, the X-ray crystal structure of F_1-ATPase was published. The structure is a ring formation consisting of three α-subunits alternating with three β-subunits that have catalytic sites. The rod-shaped γ-subunit spanned the center of the ring. This X-ray crystal structure provided a specific and easily understood image that permitted rotational movement of the γ-subunit.

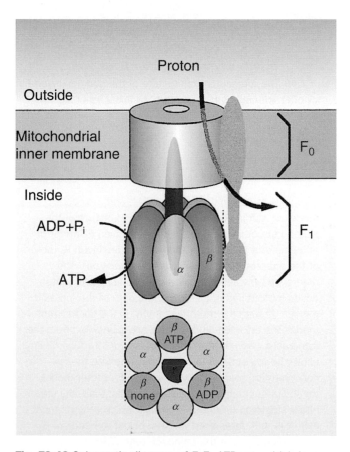

Fig. F2.48 Schematic diagram of F_0F_1-ATPase, which is a multidomain complex consisting of two units: a hydrophobic proton channel (F_0) embedded in the mitochondrial membrane and a hydrophilic catalytic unit (F_1) protruding into the mitochondria. Individual F_1-ATPase units are composed of alternating α- and β-subunits surrounding a rotating central γ-subunit. A central rotor of radius ~1 nm, formed by its γ-subunit, turns in a stator barrel of radius ~5 nm, formed by three α- and three β-subunits. When protons flow through F_0, ATP is synthesized in F_1. The synthase is fully reversible in that hydrolysis of ATP in F_1 drives a reverse flow of protons through F_0. Isolated F_1 catalyzes only the hydrolysis of ATP, and hence is called the F_1-ATP synthase. (After Kinosita et al., 1998.)

Rotation of F_1 was demonstrated only in 1998 by directly observing the motion of a fluorescent actin filament specifically bound to the rotor element (Fig. F2.50). The His-tagged F_1-ATPase was immobilized, and fluorescent

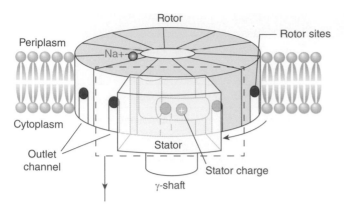

Fig. F2.49 The molecular architecture of the sodium-driven F_O motor of *Propionigenium modestum*. (Adapted from Oster and Wang, 2003.)

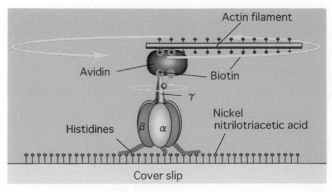

Fig. F2.50 Experimental geometry for observing the rotation of single F_1-ATP molecules. The α- and β-subunits were bound by a histidine tag to a coverslip coated with nitrilotriacetic acid, and the opposing face of the γ-subunit was attached to a fluorescent actin filament through a biotin-avidin linkage. Rotation of the filament was observed through a standard fluorescent microscope. (Adapted from Kinosita, 1998.)

COMMENT F2.13 F_O MOTOR

Extensive mutational, NMR, cryo-EM, and molecular modeling studies have provided a good picture of the general structure for both the proton-driven F_O motor of mitochondria and *E. coli*, and the sodium-driven F_O motor of the anaerobic bacterium from *P. modestum*. The difference between the models for sodium- and proton-driven motors lies in the presumed path of the ion as it passes through the motor assembly. In the sodium-motor model, the rotor ion-binding sites are accessible from the cytoplasm when outside the rotor–stator interface; this allows bound sodium ions to dissociate into the low-concentration reservoir. In the proton F_O- motor model these outlet channels are not present. Instead, the stator has a single outlet channel through which protons must exit. It is still unresolved whether or not the sodium F_O and proton F_O have the same ion path. The mechanochemical principle is the same in both models (Oster and Wang, 2003).

COMMENT F2.14 THE ABSOLUTE VALUE OF TORQUE OF THE F_1 MOTOR

Strictly speaking, the long actin filament may have altered the genuine kinetics of the F_1 motor, because the filament was subject to a large hydrodynamic friction. The orientation dependence of single-fluorophore-intensity has been exploited in order to videotape the rotation of F_1 motor in real time. These results showed that the 120° stepping is a genuine property of this molecular motor.

actin filaments were attached to the subcomplex through streptavidin for observation with an epifluorescence microscope. When ATP was added into the chamber, the actin filaments always rotated counterclockwise.

By observing this rotation at extremely low ATP concentration it was found that the rotation occurs in increments of 120°, one step per molecule of ATP hydrolyzed. In these measurements, the actin filaments attached to the γ-subunit rotated against the viscous resistance of water (Comment F2.14). The amount of rotational torque required for this movement was calculated from the length of the actin filaments and the rate of rotation to provide a mean value of approximately 40 pN nm. For a single step of 120° and, hence, a single ATP hydrolysis, this amounts to work of 80 pN nm. This value is in close

agreement with the free energy liberated by a single ATP hydrolysis under physiological conditions. From this estimate, it was concluded that F_1-ATPase can convert nearly 100% of its ATP-derived energy into mechanical work. In this respect, it appears that nature has far outperformed human engineers, making mechanistic understanding of these extraordinarily efficient motors an important goal.

The Bacterial Flagella as a Propeller Screw

Many motile bacteria are propelled by flagellar filaments, each of which is turned at its base by a rotary motor driven by a transmembrane ion gradient. These ions are either H^+ (protons), or in alkalophilic or marine bacteria, Na^+. In *E. coli*, the motor can rotate in either direction, and cells navigate toward regions rich in nutrients by controlling this direction.

The filament of *E. coli* is a long helical polymer of a protein, flagellin, with a molecular mass of about 55 kDa. The filament is connected to the hook by two junctional proteins (Fig. F2.51). Named according to its shape, the flexible hook acts as a universal joint permitting the filament and motor to rotate about different axes. The flagellar

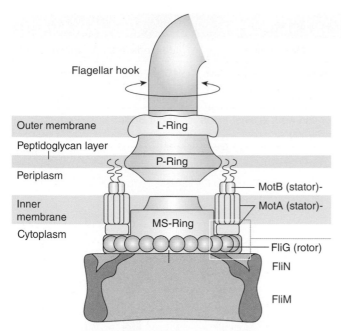

Fig. F2.51 Schematic of the bacterial flagellum as a mechanical device. The filament, hook, and motor constitute the flagellum. The remaining flagellar parts are rings. The L- and P-rings are believed to act as a bushing through which the rotating drive shaft passes. The filament or propeller lies outside the cell's inner and outer membranes and is about 10 μm long. In *E. coli*, flagella turning at speeds of 18 000 rpm push cells at 30 μm s⁻¹. It is remarkable that unlike kinesin, NCD, and myosin, flagellar motors can run in both directions. (After Oster and Wang, 2003.)

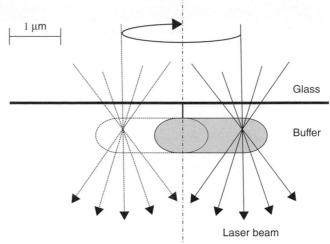

Fig. F2.52 Sketch of a bacterial cell being rotated about its tether by the optical trap (dash–dot line) aligned with the tether (short thick line). The direction of rotation is shown by the circular arrow. Two views of the cell are displayed: at $\theta = 0°$ (solid lines) and $\theta = 180°$ (dotted lines). The laser beam enters from above, through the coverslip, as shown by the crossed rays. The microscope illumination (not shown) enters from below. The tweezers can act directly on the cell or indirectly, via a trapped polystyrene bead. (After Block *et al.*, 1989.)

filament (propeller) is a 10 μm long, thin, rigid, corkscrew-shaped structure, with a helical period of about 2 μm. More so than other motors, the flagellum resembles a machine designed by a human.

Flagellar rotation has traditionally been measured by using the "tethered cell" assay, in which a single flagellum is attached to the surface, causing the cell body to counterrotate (Fig. F2.52). Whereas the flagella of swimming *E. coli* rotate at over 100 Hz and are difficult to observe directly, tethered cells rotate at ~10 Hz and are comparatively easy to monitor. The development of optical tweezers has made it possible to manipulate bacteria with forces that are sufficient to arrest actively swimming bacteria and can overcome torque generated by the flagellar motor of a bacterium tethered to a glass surface by a flagellar filament. It has been shown that a flagella motor generates ~4500 pN nm of torque. The radius of the rotor is ~20 nm, so the motor generates at its rim a force of ~200 pN, far greater than any other molecular motor (see Table F2.3). For each revolution of the motor ~1200 protons pass through it, each contributing ~6 $k_\mathrm{B}T$.

F2.6.3 The Bacteriophage φ29 DNA Packaging Motor

There are other protein motors that are thought to be rotary. One of them is the portal protein in certain viruses that package their DNA into their virial capsid.

The bacteriophage φ29 is a 19-kilobase dsDNA virus that infects *Bacillus subtilus* cells. The viral particles are formed by a prolate icosahedral capsid, or head, and tail. Between these two structures there is a connecting region, the so-called connector (Fig. F2.53). This connector, a cone-shaped dodecamer, occupies the pentagonal vertex at the base of the prohead and is the portal for DNA entry during packaging and DNA ejection during infection. The DNA packaging motor lies at a unique portal vertex of the prohead and contains: (1) the connector (the X-ray structure of which is presented in Fig. F2.54); (2) the portion of the prohead shell that surrounds the connector; (3) φ29-encoded prohead RNAs, which surround the protruding narrow end of the connector; and (4) viral ATPase which is required for DNA packing.

Bacteriophage φ29 packages its 6.6 μm long dsDNA into a 540 × 420 Å capsid by means of a portal complex that hydrolyzes ATP. This process is remarkable because the entropic, electrostatic, and bending energies of the DNA must be overcome to package the DNA to near-crystalline density.

Motor	Rotational speed	Max. torque	Max. efficiency	Processivity
$F_1\alpha\beta/F_1\gamma$	~4 rps	40 pN nm	100%	Perfect
Bacterial flagella	~300 rps[a] 1700 rps[b]	~4000 pN nm	Unknown	Perfect
Bacteriophage φ29 packaging	100 bp per second[c]	Motor can work against loads of up 57 pN	About 30%	Perfect

TABLE F2.3. COMPARISON OF THE MAIN CHARACTERISTICS OF ROTARY MOTORS

[a] Proton-driven motor.

[b] Sodium-driven motor.

[c] The packaging rate decreases as the prohead is filled.

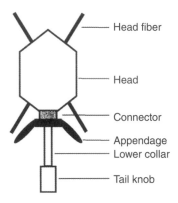

Fig. F2.53 Simplified structure of the φ29 virus. The prohead, into which the DNA is packaged, is about 540 Å long and 420 Å wide.

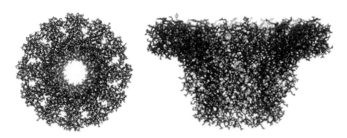

Fig. F2.54 X-ray structure of an isolated connector. The connector structure can be divided into three, approximately cylindrical, regions: the narrow end, the central part, and the wide end, having external radii of 33, 47, and 69 Å, respectively. These regions are respectively 25, 28, and 22 Å in height, making the total connector 75 Å long. The internal channel has a diameter of about 36 Å at the narrow end, increasing to 60 Å at the wide end. (After Simpson *et al.*, 2000.)

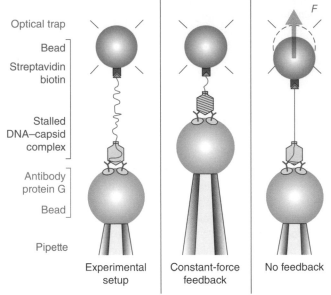

Fig. F2.55 Experimental setup in which laser tweezers are used to measure packaging activity: at the start of a measurement (left); the constant-force feedback mode (middle); and no feedback mode (right). A single φ29 packaging portal complex is tethered between two microspheres. Optical tweezers are used to trap one microsphere and measure the forces acting on it, while the other microsphere is held by a microneedle. To ensure measurement on only one complex, the density of complexes on the microsphere is adjusted so that only about one out of five to ten microspheres yields hook-ups. (Adapted from Smith *et al.*, 2001.)

Figure F2.55 shows an experimental setup in which force-measuring laser tweezers are used to follow the packaging activity of a single φ29 packaging complex in real time. Stalled, partly prepackaged complexes are attached to a polystyrene microsphere by means of the unpackaged end of DNA. This microsphere is captured in the optical trap and brought into contact with a second bead that is held by a microneedle. This bead is coated with antibodies against the phage, so a stable tether is formed between the two beads (Fig. F2.55). In the absence of ATP, the tether displays the elasticity expected for a single DNA molecule. After addition of ATP, the two microspheres move closer together, indicating packaging activity, thus demonstrating that the portal complex is a force-generating motor. In "constant-force feedback" mode (Fig. F2.55 middle), in which the microsphere distance is adjusted by feedback to maintain DNA tension at a preset value of 5 pN, packaging dynamics is monitored. Figure F2.55 shows that packaging is highly efficient: On average 5.5 min are

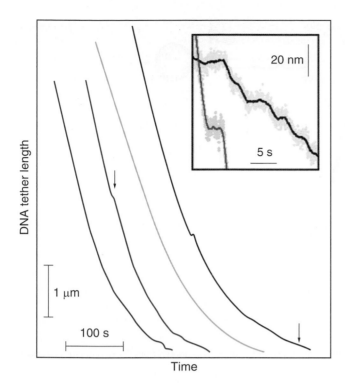

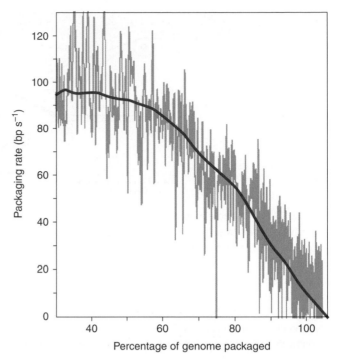

Fig. F2.56 Plots of the tether length versus time for four different DNA constructs with a constant force of ~5 pN, using a length of 1.8 times the φ29 genome. Inset: Increased detail of the regions indicated by the arrows, showing pauses (curves are shifted for clarity). (Smith *et al.*, 2001.)

Fig. F2.57 Packaging rate versus the amount of DNA packaged, relative to the original 19.3-kilobase ϕ29 genome. The red line is an average of eight measurements. Notably, the packaging rate decreases as the prohead is filled. (Smith *et al.*, 2001.)

necessary to package a length of DNA equal to the φ29 genome. Pauses in movement of variable duration (Fig. F2.56, inset) are also clearly evident.

In Fig. F2.57 the packaging rate is plotted against the percentage of genome that has been packaged. These data show a marked reduction in packaging rate from when ~50% of the genome is packaged. Initially the rate is ~100 bp s^{-1} and it gradually drops to zero as the capsid fills and the motor stalls.

The force dependence of the motor has been investigated using a "no feedback" mode in which the positions of the trap and microneedle are fixed (Fig. F2.55). In this mode the tension in the molecule increases as the motor reels in the DNA, and the bead is displaced from the center of the trap. The measurements show that the motor can work against loads of up to 57 pN on average, making it one of the strongest molecular motors reported to date.

A comparison of main characteristics of rotary motors is given in Table F2.3.

F2.6.4 Molecular Motors and Brownian Motion

The basic physical principle that governs the operation of all protein motors is quite simple: molecular motors generate mechanical forces by using intermolecular binding energy to capture "favorable" Brownian motions. They do this in two ways: (1) they may bias against unfavorable Brownian motions by a sequence of small free-energy drops, making backward steps slightly less unlikely than forward ones (Fig. F2.58a); or (2) they may rectify a long "run" of favorable thermal fluctuations by a large free-energy drop, making backward steps extremely unlikely in comparison with forward ones (Fig. F2.58b). The former is generally called a "power stroke" and the latter a "Brownian ratchet." According to the power-stroke mechanism, the ligand progresses from its initial weak binding state to its final tight binding state by biasing relatively small Brownian fluctuations of the two binding surfaces (the red curve in the free-energy landscape in Fig. F2.58a). This gradual "zipping" of intermolecular bonds is used to perform mechanical work on an external load.

One power stroke is not sufficient to make a motor, which must operate in a cycle. For this, the catalytic site must realize the tightly bound nucleotide and reset the catalytic site so that a new power stroke can be initiated.

A Brownian ratchet mechanism (Fig. F2.58b) involves a filament that polymerizes against an object whose diffusion coefficient is D. The load force F_L pushes the object to the left. If an object succeeds in diffusing to the right against the load by further than the size of a monomer, an additional monomer may polymerize onto

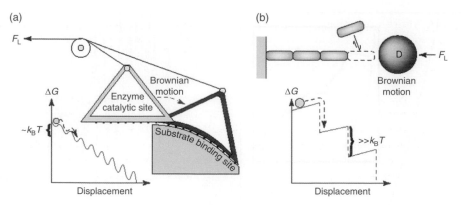

Fig. F2.58 Two extreme mechanisms for using random thermal fluctuations to drive a load. (**a**) *The power stroke.* A flexible binding site on enzyme slides over the binding surface of a fixed ligand. Its stochastic motion is driven by biased Brownian fluctuations that progressively anneal the short-range attractive forces between the enzyme and substrate "induced fit." Thus, the binding energy is converted directly into mechanical work. The red curve is the free-energy landscape seen by the system as the enzyme gradually binds to the substrate. During the binding process, the free-energy decreases encounter only small energy barriers of order k_BT. (**b**) *Brownian ratchet.* A polymer drives an object with a diffusion coefficient D by rectifying its diffusive motion. When the object diffuses against the load F_L by more than the size of a monomer, another monomer can bind to the end of the growing polymer. If binding is successful, the lengthened polymer prevents the object from being driven back to the left by the load forces. The free-energy landscape (the red curve) seen by the object is a tilted staircase. The free energy in each step is much greater than k_BT. (Adapted from Oster and Wang, 2003.)

the end of the filament. This "rectifies" the object's diffusive motion, blocking it from being pushed by the load force back to the left. Thus, the free-energy landscape resembles a tilted staircase (the red curve in Fig. F2.58b). Operation in a cycle is built into the mechanism because the binding of each new monomer reinitiates the cycle.

F2.6.5 Molecular Motors and the Second Law of Thermodynamics

The second law of thermodynamics dictates that a motor without an external free-energy supply cannot extract heat from an isothermal environment to do work. Although the notion that molecular motors run on Brownian motion appears at first glance to contravene the second law, we must remember that the second law is a statement about the average behavior of a system (e.g., the macroscopic behavior). Occasionally, and for short times, an individual molecular motor can "violate" the second law. For example, a protein-sized object can oscillate to stretch a spring but, without a free-energy supply, such individual violations are not sustainable. When averaged over many objects or over long times for one object, these violations disappear. However, proteins can capture these temporary violations by using an energy supply, such as binding free energy; such behavior does not violate the second law. This means protein motors can "extract" heat from their surroundings to do work by using a free-energy supply to bias or rectify thermal fluctuations. This cools down the motor's local environment by taking kinetic energy from the surrounding solvent molecules. Notice that the heat is

generally not paid back by the motor itself but by other metabolic processes, such as the manufacture of ATP or the pumping up of the ion gradient.

F2.6.6 DNA Mechanics

Native free large DNA (more than 4 MDa) adopts a random coil conformation, which maximizes its entropy. Pulling on the molecule reduces this entropy and costs energy. The associated entropic forces result from a reduction of the number of possible configurations of the system consisting of the molecule and its solvent, so that at full extension there is only one configuration left: a straight polymer linking both ends. To reach that configuration work has to be done against entropy; a force has to be applied.

Pioneering measurements of biopolymer elasticity employed optical tweezers and other mechanical probes to stretch DNA. Such experiments pulled macromolecules beyond their entropically determined regime in which the polymer resists an extension that constrains its range of accessible conformations and into one in which external forces disturb the structure and induce conformational changes. To produce and measure such forces on a DNA molecule, we must use single-molecule manipulation techniques. These techniques have been used to study several regimes of DNA stretching behavior in detail and have also been extended to study DNA–protein and DNA–drug interactions. High-resolution force measurements have also been used to study the stretching of a single DNA or RNA molecule. Below we give a few examples of the application of the single-molecule technique to studying the stretching of DNA and RNA molecules and their complexes with proteins.

COMMENT F2.15 ALTERNATIVE APPROACH FOR DNA IMMOBILIZATION ON A SURFACE

An interesting alternative which does not require any modification of the molecules is based on the specific adsorption of DNA by its ends on a hydrophobic surface at a pH of ~5.5. On many different hydrophobic materials (Teflon, polystyrene, silanized glass, etc.) DNA has been observed to adhere strongly and non-specifically at low pH and weakly or not at all at high pH. Between low and high pH there is a narrow range (pH = 5.5 ± 0.2) where DNA binds to the surface by its extremities only.

Molecular Yoga of DNA

The DNA structure poses some formidable mechanical problems to the cellular machinery which has to read, transcribe, and replicate the signals of the genetic code buried inside the double helix. To make the code accessible to the DNA and RNA polymerase enzymes, the molecule has to be unwound and two strands separated. The regulation of the winding and torsional stresses involved in these processes is performed by numerous enzymes known as topoisomerases. To study the function of these molecular motors one has first to understand the mechanical response of the DNA under stress.

The first step in any manipulation experiment is to anchor the DNA (preferentially via its extremities) to an appropriately treated surface (Comment F2.15). Many different methods have been developed to achieve specific DNA binding to surfaces. Most of these methods include biochemical reactions between a (possibly modified) DNA molecule and an appropriately treated surface. For example, the extremity of the molecule can be functionalized with biotin, which can interact specifically with streptavidin bound to a surface (Comment F2.16). Similarly, surfaces coated with oligonucleotides can be used to recognize the complementary extremity of DNA molecules.

Once DNA is bound to a surface by its end(s), a simple way to stretch it is to drain the solution (e.g., by pulling the surface out of it or by letting it evaporate). In Fig. F2.59 the molecule, stained with a fluorescent dye, is stretched by the hydrodynamic force. Because this force is acting locally (at the interface), DNA is stretched uniformly: ~1 μm for every 2 kilobases. Fluorescence in-situ hybridization on combed molecules allows one to obtain very accurate (~1 kbp) genomic maps. The ordering, orientation, and distance between genes and the existence of genomic rearrangements (e.g., deletions) can thus be determined.

One of the first experiments on the overstretching transition of the double helix was demonstrated on λ-phage DNA using the optical trapping technique in combination with a glass micropipette (Fig. F2.60a). When the dsDNA

COMMENT F2.16 BIOTIN (VITAMIN B7 OR H) AND STREPTAVIDIN

The interaction between biotin and streptavidin is one of the strongest known non-covalent interactions. The dissociation constant for the streptavidin–biotin complex is ~10^{-15} M. Streptavidin, from *Streptomyces avidinii*, is a tetrameric protein containing four high-affinity binding sites for the vitamin biotin. Since streptavidin is multivalent, it is able to serve as a bridge between the biotinylated DNA fragment and the biotin-containing resin. The strong interaction is extremely useful for purification of DNA-binding proteins, because DNA-affinity columns with streptavidin/biotin bridges can be washed under a wide variety of conditions (i.e., 2 M KCl and 1% SDS) without removing either the streptavidin or the biotinylated DNA fragment from the matrix.

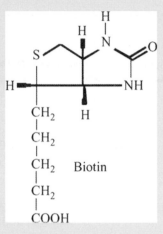

Figure F2.16.1.

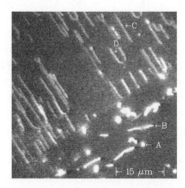

Fig. F2.59 Molecular combing of λ-phage DNA stained with YOYO1. The molecules in solution (bottom) are bound at one (A) or both (B) ends. The meniscus extends across the image from the lower left to upper right. The extended molecules left behind the meniscus are visible as straight segments (C) if bound at one end, or loops (D) if bound at both ends. In molecular combing, the anchored DNA molecules are aligned on the surface by the receding meniscus as algae on the shore are by the receding tide. The force applied on the molecule by the receding air–water interface is large enough to stretch it but not to break its bond(s). (Strick *et al.*, 2000.)

(a)

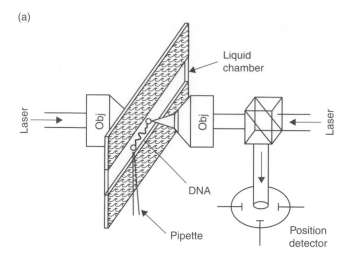

Liquid chamber

Laser · Obj · Obj · Laser

DNA

Pipette

Position detector

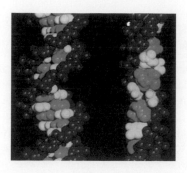

Fig. F2.61 Space-filling graphics of relaxed linear dsDNA (left) and one of the models of DNA stretched by a factor of 1.7 (right). This elongated DNA model is characterized by a strong base-pair inclination, a narrow minor groove and a diameter roughly 30% less than that of B-DNA. (After Cluzel et al., 1996.)

(b)

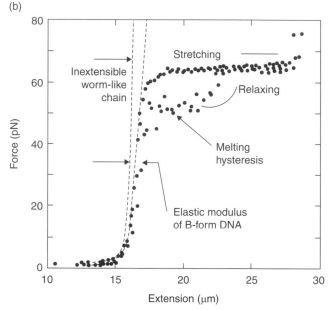

Stretching

Inextensible worm-like chain

Relaxing

Melting hysteresis

Elastic modulus of B-form DNA

Force (pN) vs Extension (μm)

Fig. F2.60 (a) Schematic view of the dual-beam laser-trap instrument. Each end of a single λ-phage dsDNA molecule (48.5 kbp, contour length about 16.4 μm) is attached to a separate microscopic latex bead. One bead is held by suction with a glass micropipette while the other bead is held in the optical trap, which also functions as a force transducer. The DNA molecule is extended by moving the pipette relative to the laser trap. Use of low numerical aperture lenses make it possible to hold beads deep inside a fluid cell, 100 μm away from the glass surface. **(b)** Stretching of λ-phage DNA in 150 mM NaCl, 10 mM Tris, 1 mM EDTA, pH 8.0. The "inextensible worm-like chain" curve corresponds to a worm-like chain with a persistence length of 53 nm and contour length of 16.4 μm. (After Smith et al., 1996.)

molecule is subject to force of 65 pN or more, it undergoes a highly cooperative transition (~2 pN) into a stable form with a 5.8 Å rise per base pair, i.e., 70% longer than the canonical B-form DNA (Fig. F2.60b). This new form of

DNA is called S-DNA (Fig. F2.61, right). When the stress is relaxed below 65 pN, the molecules rapidly and reversibly contract to their normal contour lengths. S-form DNA is stable in high salt concentrations up to forces of between approximately 150 pN (for a random sequence) and 300 pN (for poly(dG–dC)). Above these forces S-DNA exhibits the characteristic force–extension behavior of single-stranded DNA. Single-stranded DNA is more contractile than double-stranded DNA because of its high flexibility, but it can be stretched to a greater length because it no longer forms a helix.

In a magnetic tweezers setup (Fig. F2.62), by rotating the magnets placed above the sample one can twist and supercoil the anchoring DNA molecule (provided it is nick-free). At low stretching forces ($F < 0.3$ pN), the twisting of DNA results in the formation (as in a twisted tube) of plectonemes or supercoils (see below). As the torsional stress increases, it may alter the DNA structure, thus preempting its buckling. When the torque on the DNA reaches about 20 pN nm (in low salt concentrations), the overwound DNA molecule adopts a new inside-out overtwisted structure called P-DNA (Fig. F2.63). Interestingly, a P-DNA-like structure has been found in the genome of the Pf1 virus (Comment F2.17). It is very likely that this unusual structure is stabilized by the proteins of the virus's coat.

These results demonstrate that at very low forces and low degrees of supercoiling, DNA can locally undergo major structural transitions. These transitions might be relevant to the activity of RNAP, which is known to exert forces as high as 35 pN and to under(over)wind the molecule up(down)stream.

Topoisomerases and Uncoiling of DNA

Topoisomerases are enzymes that control DNA supercoiling and are responsible for disentangling molecules during replication, DNA repair, and recombination

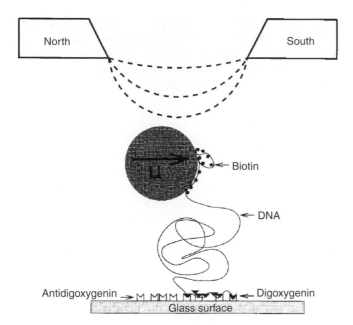

Fig. F2.62 Schematic representation of the magnetic tweezers setup. A digoxygenin (DIG)–biotin end-labeled DNA molecule (~17 000 bp) is attached at one end to a glass surface by DIG–anti-DIG bonds and at the other to a 4.5 μm magnetic bead via streptavidine–biotin links. By varying the distance between the sample and the permanent magnets, the stretching force is controlled, whereas rotating the magnets induces DNA supercoiling. (Adapted from Allemand *et al.*, 1998.)

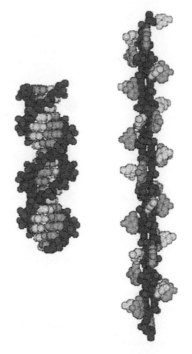

Fig. F2.63 Structure of P-DNA deduced from molecular modeling. Space-filling models of a $(dG)_{18} \cdot (dC)_{18}$ fragment into **(a)** linear B-DNA and **(b)** P-DNA conformations. In P-DNA the phosphodiester backbone winds inside the structure and its bases are exposed outside. The backbones are colored purple, and the bases are colored blue (guanine) and yellow (cytosine). The anionic oxygens of the phosphate groups are shown in red. P-DNA is a double-helical structure with a pitch of 2.6 bases per turn. (Allemand *et al.*, 1998.)

(Comment F2.18). The uncoiling and decatenation activities of these enzymes can be investigated directly by monitoring the change in extension of a coiled molecule as it is uncoiled by a single topoisomerase.

As we discussed before, an unnicked, double-stranded DNA molecule rigidly anchored at one end to a glass surface and at the other to a small magnetic bead can be stretched with a force F and twisted to a degree of supercoiling σ by translating and rotating small magnets above the sample (see Fig. F2.64). These two mechanical parameters determine the molecule extension l, which is monitored by measuring the three-dimensional position of the tethered bead (Fig. F2.64a). As one begins to rotate the magnets, the DNA's extension is unchanged and the torque increases linearly with the twist angle. When a critical torque is reached the molecule buckles (as would a twisted tube), forming a plectoneme and stabilizing the torque at its critical value (Fig. F2.65, point (1)). The appearance of these tertiary structures is well known from everyday experience, e.g., the coiling of phone cords. Thereafter it contracts regularly as overwinding generates more supercoils. Addition of topoisomerase II to the system results in an increase in l by an amount δ for every supercoil relaxed (Fig. F2.65, point (2)). In 10 mM ATP, enzyme turnover is slow and relaxation is directly observed as step-wise events of size 2δ (Fig. F2.66).

(a)

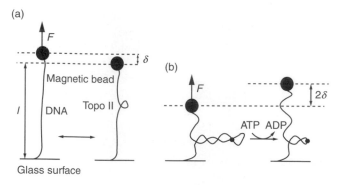

Glass surface

Fig. F2.64 (a) Sketch of a twisted DNA near the bucking instability (point (1) in Fig. F2.65) undergoing topo-II-mediated clamping in the absence of ATP. No supercoiling is relaxed, but stabilization and destabilization of a single DNA loop results in a change δ of the system's extension. **(b)** Supercoil relaxation in the plectonemic regime (point (2) in Fig. F2.65) in the presence of topo II isomerase and ATP. Each enzymatic cycle releases two supercoils, resulting in an increase 2 of δ in the system extension. (Adapted from Strick *et al.*, 2000.)

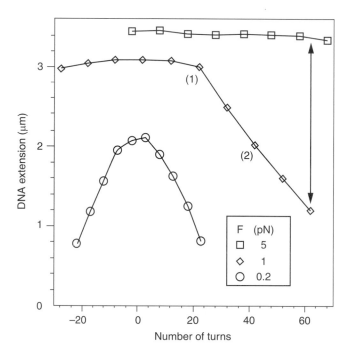

Fig. F2.65 Extension versus supercoiling behavior for stretched 11 kbp DNA. At $F = 1$ pN, the DNA needs to be overwound by about 25 turns before it reaches the buckling instability (point (1)) and forms plectonemes. Increasing the force (arrow) on DNA overwound 60 turns to 5 pN pulls out the plectonemes and causes the extension to increase. At very low forces ($F = 0.2$ pN) the buckling instability is rounded off by thermal fluctuations but the DNA does not denature when unwound. (Adapted from Strick *et al.*, 2000.)

Prokariotic type I topoisomerses, 1A, bind preferentially to single-stranded DNA. The type I enzymes relax torsion in a molecule in an ATP-independent manner by creating a break in one of the DNA strands, allowing relaxation of torsion about the intact strand.

DNA Condensation

In the presence of a critical concentration of multivalent ions, double-stranded DNA condenses and forms very compact structures. Single-molecule stretching methods are used to measure the forces that cause this DNA condensation.

Figure F2.67 shows a model for the transition in a λ-DNA molecule based on force-measuring optical tweezers. If the two ends of a DNA molecule are stretched (more than the critical extension $y_{crit} = 0.85$), they are prevented from condensation (Fig. F2.67a). At $y = y_{crit}$ thermal motion may create temporary slack, but the collapsed nuclei formed are unstable (Fig. F2.46b). At $y < y_{crit}$ the molecular slack generated coalesces into the condensed phase. The force required to prevent condensation has been shown to be constant, with a magnitude of about 2–3 pN.

Unzipping the DNA Double Helix

Unzipping the double helix was demonstrated by the mechanical separation of the complementary strands of an individual λ-phage dsDNA molecule. A specific molecular construction was designed in which one strand of the DNA to be opened is elongated by a linker arm and two specific functionalizations are introduced to obtain the desired attachments of the DNA construct to the micromechanical device.

The design of the construct used in the experiment appears in Fig. F2.68 and includes two molecules of double-stranded λ-phage DNA, each comprising 48.5 kb. The DNA to be opened is DNA-1, and the linker arm is DNA-2. Oligonucleotides (oligo-1, oligo-2, and oligo-3) are used to connect DNA-1 and DNA-2 and to introduce attachment points via a digoxigenin group and a biotin group. The other extremity of DNA-1 is capped with an oligonucleotide forming a hairpin (oligo-4), which prevents the separation of the two strands when the end of the opening process is reached. This allows repeated cycles of opening and closing. The bending of a glass microneedle is used to determine the forces required to pull apart the 3′ and 5′ extremities of the molecule (Fig. F2.69). Strand unzipping occurs abruptly at 10–15 pN and displays a reproducible "sawtooth" force variation pattern with an amplitude of ± 0.5 pN along the DNA. The following picture emerges for the characteristic forces involved for double-stranded DNA: The breaking of the double strand requires ~480 pN $\pm$ 20%; the structural transition of DNA overstretching requires 65–70 pN; and strand separation requires about 10–15 pN.

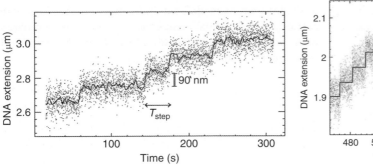

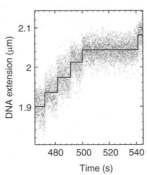

Fig. F2.66 (a) Individual time courses for the topo-II-mediated relaxation of positively supercoiled DNA stretched at $F = 0.7$ pN at 10 mM ATP. (Strick *et al.*, 2000.) **(b)** Individual time courses for the topo-I-mediated relaxation of DNA stretched at $F = 1.5$ pN in 0.25 mM MgCl$_2$. The DNA contained a symmetric mismatch of 12 nucleotides. (Adapted from Dekker *et al.*, 2002.)

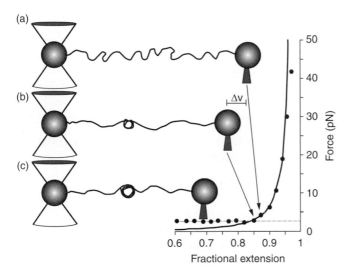

Fig. F2.67 Model for the condensing of a λ-DNA molecule constrained by force-measuring optical tweezers. **(a)** At a fractional extension y greater than the critical extension y_{crit} the free energy of the stretched DNA chain exceeds the nucleation energy for collapse. **(b)** At $y = y_{crit}$ thermal motion may create temporary slack, but the collapsed nuclei formed are unstable. **(c)** At $y < y_{crit}$ the molecular slack generated coalesces into a condensed phase. For the force–extension plot, $y_{crit} = 0.85$. (Adapted from Baumann *et al.*, 2000.)

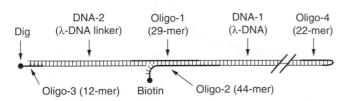

Fig. F2.68 The molecular design of the DNA construct used in the experiment. The DNA to be opened (DNA-1) and the linker arm (DNA-2) comprise double-stranded λ-phage DNA. Oligonucleotides (thick lines) are used to introduce the biotin and dig attachments and to connect DNA-11 and DNA-2 covalently. (After Essevaz-Roulet, 1997.)

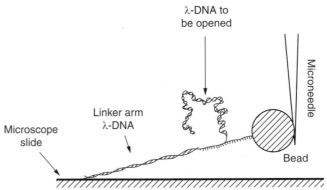

Fig. F2.69 Principle of the force measurement in which a double-stranded λ-phage DNA is forced open as the surface is displaced to the left. (After Essevaz-Roulet, 1997.)

COMMENT F2.19 NUCLEOSOMES

Nucleosomes are the fundamental organizational units of the eukaryotic genome, occurring on average every 200 bp. The foundation of the nucleosome is the nucleosome core particle, consisting of 147 bp DNA wrapped 1.65 times around an octamer of histone proteins.

Unfolding of Single Chromatin Fibers and Individual Nucleosomes

Single-molecule mechanical manipulation techniques offer a direct approach to the investigation of the forces and displacements required for enzymatic access to nucleosome-bound DNA (Comment F2.19). To study these processes, single chromatin fibers are assembled directly in the flow cell of an optical tweezers setup. A single λ-phage DNA molecule, suspended between two polystyrene beads, is exposed to a *Xenopus laevis* egg extract, leading to a chromatin assembly with concomitant apparent shortening of the DNA molecule (Fig. F2.70a–c). Assembly is force-dependent and cannot take place at a force exceeding 10 pN. The assembled single chromatin fiber is subjected to stretching by controlled movement of one of the beads, with the force generated in the molecule continuously monitored and with the second bead trapped in the optical trap. The force displays discrete, sudden drops upon fiber stretching, reflecting discrete opening events in the fiber structure (Comment F2.20).

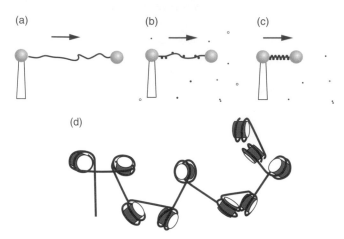

Detailed studies of the nucleosome unfolding process were done on very defined individual nucleosome core particles. Nucleosomal arrays for this study were prepared with avian core histones and a 33 684 bp DNA fragment containing 17 direct tandem repeats. The experimental setup is shown in Fig. F2.71a. An example of the force–extension curves obtained for saturated arrays (17 nucleosome) is shown in Fig. F2.71b. In the high-force range (>15 pN), these data show a sawtooth pattern composed of 17 peaks. At the end of the stretch, the chromatin curve approaches that of the full-length naked DNA (dotted red line), indicating that no histones remained attached to the DNA. The 17 peaks indicate disruption of the 17 positioned nucleosomes. At each sawtooth, DNA remains bound until a peak force is reached, leading to a sudden release of DNA and relaxation to lower tension. Uniform spacing between adjacent peaks (~27 nm) indicates that upon disruption, a relatively constant amount of DNA is released from each nucleosome core particle.

Fig. F2.70 Schematic representing chromatin assembly on a single λ-DNA molecule. (a) A single λ-DNA molecule suspended between two beads. The arrow indicates the direction of continuous buffer flow. The first bead is held by a glass micropipette using suction. The second bead is maintained downstream from the first one by the drag force. (b) X. laevis egg extract containing all core histones and numerous non-histone proteins, but lacking somatic linker histones, is introduced into the flow cell. Histone proteins bind to the single λ-DNA molecule, causing its apparent shortening. (c) Shortening of the single λ-DNA molecule continues and eventually stops. (d) Model of a chromatin fiber with 51–73 bp of linker DNA. Each cylinder represents one histone. (Adapted from Bennink et al., 2001.)

COMMENT F2.20 ON THE PERIODICITY IN THE FORCE–EXTENSION CURVES OF NUCLEOSOMES

These opening events are quantized at increments in fiber length of about ~65 nm and are attributed to unwrapping of the DNA from individual histone octamers. However, the abundance of linker histone-like proteins and other non-histone chromatin-associated proteins in oocyte nuclear extracts used for chromatin assembly in these experiments may in part explain the lack of uniformity and periodicity in the force–extension profiles.

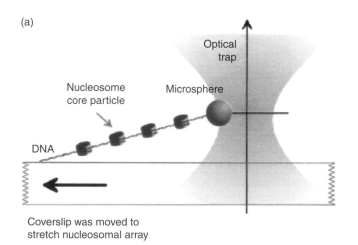

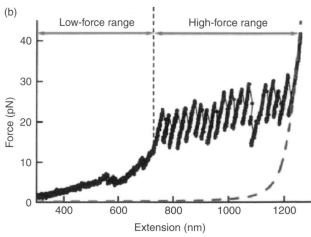

Fig. F2.71 (a) Experimental configuration of nucleosome stretching (not to scale). The DNA is labeled at one end with biotin and at the other end with digoxigenin. Before stretching, one end of each nucleosomal array is attached to the surface of an antidigoxigenin-coated microscope coverslip. A 0.48 μm diameter streptavidin-coated polystyrene microsphere is then attached to the free end of each tethered array. Once a surface-tethered microsphere is optically trapped, the coverslip is moved with a piezoelectric stage to stretch the nucleosomal DNA. **(b)** Force–extension curve of a fully saturated nucleosomal array. The force–extension characteristic of a full-length naked DNA (red dotted line) is shown for comparison. (Adapted from Brower-Toland et al., 2002.)

TABLE F2.4 FORCES IN MICROMANIPULATION EXPERIMENTS WITH DSDNA	
Breaking of the DNA double strands	400–580 pN
Structural transition of uncoiling a double strand upon stretching	60–80 pN
Structural transition of a double strand upon torsional stress	~ 20 pN
Individual nucleosome disruption	20–25 pN
Separation of complementary strands (room temperature, 150 mM NaCl, sequence-specific)	10–15 pN

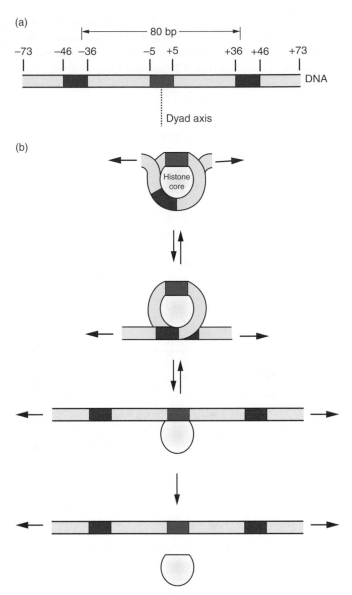

Fig. F2.72 (a) Map of critical DNA–histone interactions within an individual nucleosome core particle. **(b)** A three-stage model for the mechanical disruption of individual nucleosome. (Adapted from Brower-Toland et al., 2002.)

A suggested model for forced disassembly of each individual nucleosome is described by three stages, each involving partial unwrapping of the DNA (Fig. F2.72). The first stage of disruption releases 76 bp of the external DNA. The disruption is gradual, and only low force is required to peel DNA from the protein surface. The second and third stages of the disruption involve the sudden release of the next 80 bp of DNA. The third stage of disruption occurs at even higher loads, releases the remaining 11 bp of DNA and results in a complete dissociation of the histones from the DNA. The three-stage model of nucleosome opening suggests the way in which nucleosomes perform their dual function in the eukaryotic cell, both to maintain DNA in a condensed state and provide regulated access to the information contained therein.

The following picture emerges for the characteristic forces involved with dsDNA (Table F2.4). Breaking of double strands occurs at a force level of 400–580 pN. The structural transition of uncoiling upon stretching occurs at ~60–80 pN, whereas structural transition upon torsional stress occurs at a force level ~20 pN.

Disruption of individual nucleosomes is associated with a force of 20–25 pN. Strand separation requires 10–15 pN range and is sequence-specific (15 pN for 100% GC sequence and 10 pN for 100% AT sequence).

F2.6.7 RNA Mechanics

An RNA structure is generally separated into two levels of organization – secondary and tertiary structure. The tertiary structure is composed of secondary structural motifs that are brought together to form modules (hairpin loops, symmetric and asymmetric internal loops, junctions), domains, and the complete structure.

These structures are very complicated and bulk studies of RNA folding are often frustrated by the presence of multiple species and multiple folding pathways, whereas single-molecule studies can follow folding/unfolding trajectories of individual molecules. Furthermore, in mechanically induced unfolding, the reaction can be followed along a well-defined coordinate, the molecular end-to-end distance.

Figure F2.73a shows three types of RNA molecules representing major structural units of large RNA assemblies. P5ab is a simple RNA hairpin that typifies the basic unit of RNA structure, an A-form double helix. P5abcΔA has an additional helix and thus a three-helix junction. Finally, the P5abc domain of the *Tetrahymena thermophila* ribozyme is comparatively complex and contains an A-rich bulge, enabling P5abc to pack into a stable tertiary structure in the presence of Mg^{2+} ions. The individual RNA molecules are attached to polystyrene beads by RNA/DNA hybrid "handles" (Fig. F2.73b). One bead is held in a force-measuring optical trap, and the other bead is linked to a piezoelectric actuator through a micropipette. When the handles alone are pulled, the force increases monotonically with extension (Fig. F2.74a, red line), but when the handles with P5ab RNA are pulled, the force–extension

curve is interrupted at 14.5 pN by an ~18 nm plateau (black curve), consistent with complete unfolding of the hairpin. The force of 14.5 pN is similar to that required to unzip DNA helices.

Figure F2.74b shows stretch (blue) and relax (green) force–extension curves for P5abc domain of the *Tetrahymena thermophila* ribozyme. It is seen that the tertiary interactions in Mg^{2+} lead to substantial curve hysteresis. Forces of about 19 ± 3 pN are needed before the molecules suddenly unfold (blue curves). The blue arrow indicates the typical unfolding force when stretching; the green arrow shows a refolding transition upon relaxation. The two-step unfolding reveals two distinct kinetic barriers to mechanical unfolding of P5abc in Mg^{2+}.

Removal of Mg^{2+} removes the kinetic barriers, and folding–unfolding becomes reversible. Unfolding then begins at 7 pN (Fig. F2.74c), showing that in EDTA the A-rich bulge destabilizes P5abc. The refolding curves in Mg^{2+} and EDTA coincide, except for an offset of 1.5 pN due to charge neutralization (Fig. F2.74c, green curve). In contrast to the all-or-none behavior of P5ab, refolding of P5abc both with and without Mg^{2+} has intermediates: The force curve inflects gradually between 14 and 11 pN (Fig. F2.74c, black stars) and this inflection is followed by a fast (<10 ms) hop without intermediates at 8 pN (green arrows). The different widths of the transitions and their force separation suggest that the inflection (Fig. F2.74c, stars) marks folding of the P5b/c helices, whereas the hop (Fig. F2.74c, arrows) marks P5a helix formation.

Figure F2.75 shows a model for the unfolding of P5abc in the presence of Mg^{2+}, in which two possible unfolding paths are depicted. The blue arrow shows an unfolding path in which the molecule suddenly unfolds and increases its length to 26 nm, consistent with data indicated by the blue arrow in Fig. F2.74b. A two-step model is shown by the red arrows, in which an intermediate state 13 nm in length is indicated by the green arrow in Fig. F2.74b.

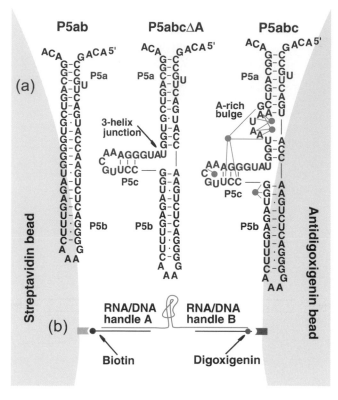

Fig. F2.73 (a) Sequence and secondary structure of P5ab, P5abcΔA, and P5abc RNAs. The five green dots represent magnesium ions that form bonds (green lines) with groups in the P5c helix and the A-rich bulge. **(b)** RNA molecules are attached between two 2 μm beads with ~500 bp RNA/DNA hybrid handles. (Liphardt *et al.*, 2001.)

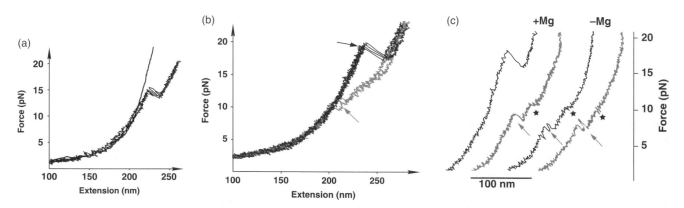

Fig. F2.74 (a) Force–extension curves for P5ab RNA when the handles alone are pulled (red line) and when the handles with P5ab RNA are pulled (black line). **(b)** Stretch (blue line) and relax (green line) force-extension curves for the P5abc domain of the *Tetrahymena thermophila* ribozyme in 10 mM Mg^{2+}. **(c)** Comparison of P5abc force–extension curves in the presence (blue curve) and the absence of Mg^{2+} (green curve). (Adapted from Liphardt *et al.*, 2001.)

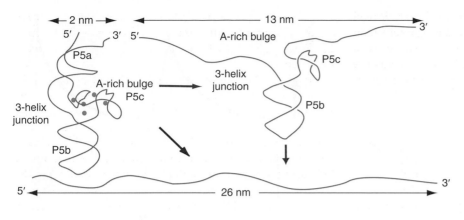

Fig. F2.75 A model for the unfolding of P5abc in the presence of Mg^{2+}, in which two possible unfolding paths are depicted. (After Liphardt et al., 2001.)

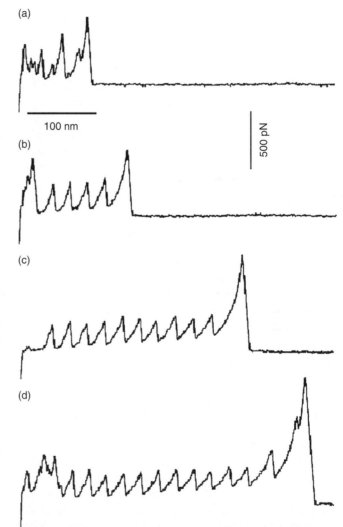

Fig. F2.76 A schematic diagram of the sequence of events during withdrawal of the gold substrate (gray box) during an AFM experiment. (Adapted from Fisher et al., 2000.)

F2.6.8 Protein Mechanics

Polyproteins

The sensitivity of AFM has allowed experiments to probe the mechanical properties of proteins. However, the heterogeneity and complexity of native proteins complicate the interpretation of AFM studies. When a protein containing multiple different domains is stretched, it is difficult to relate individual unfolding peaks in the force extension curve to specific domains and therefore to determine the mechanical properties of a specific fold. The solution to this problem was found using a special approach in molecular biology. Thus, by ligating multiple copies of the cDNA encoding a specific domain and expressing the resultant gene in bacteria, it is possible to produce a "polyprotein" consisting of multiple copies of a single protein fold. An additional benefit of using engineered polyproteins for AFM studies is that they can be constructed from domains with an altered amino acid sequence, thereby allowing dissection of the molecular determinants of mechanical stability.

Figure F2.76 shows a schematic diagram of the sequence of events during withdrawal of the gold substrate (gray box) during an AFM experiment. Prior to the experiment, a layer of proteins is allowed to adsorb onto the gold substrate. Then the AFM cantilever is pressed against the protein layer to allow adsorption onto the cantilever. Upon withdrawal of the gold substrate, the cantilever is first deflected by interactions with other molecules, such as denatured protein (in green). When these interactions break, the force on a cantilever is released.

The traces in Fig. F2.77 represent force–extension curves obtained from a sample of modular protein

Fig. F2.77 Extension of modular proteins with the AFM. A series of force–extension curves obtained from a pure sample of protein consisting of 12 identical domains. Force–extension curves can yield unfolding force peaks equal to the number of domains in the protein (as in trace (d)), but more frequently will yield fewer peaks or no regular peaks at all (as in traces (a), (b), and (c)). (After Fisher et al., 2000.)

composed of 12 identical domains. The final peak in each trace represents the detachment of the final protein molecule(s). These traces demonstrate that even when a sample of pure protein is used, spurious peaks may occur in the force–extension curve because the protein molecule may have been completely or partially denatured due to interactions with the gold substrate or entanglement with other protein molecules. Such interactions, which typically occur when the cantilever is within the gold substrate, may yield a force–extension relationship displaying a single force peak, or displaying several peaks that are irregular in amplitude and spacing (Fig. F2.77a).

The presence of the regularly spaced force peaks seen in Fig. F2.77b–d is the unmistakable fingerprint of a modular protein. These peaks correspond to the consecutive unfolding of each of the protein domains in a single protein molecule. The force–extension curves are strings of successive enthalpic and entropic portions, reflecting the unfolding of individual domains in the multidomain polypeptide chain, followed by stretching of the unfolded domain (Fig. F2.78). As such proteins are elongated as a result of the initial application of force, they undergo a typical entropic stretching at the beginning. At a certain force, one of the folded domains unfolds, adding significant length to the chain and relaxing the stress on the cantilever, which returns to its non-defected state. The denatured portion of the polypeptide chain can now undergo entropic stretching, behaving like a typical polymer chain. Further extension creates forces high enough to unfold a second domain, which is then stretched entropically, etc. The unfolding and stretching of each individual domain creates an individual peak in the force curve, leading to the characteristic *sawtooth* pattern, illustrated in Figs. F2.77 and F2.78.

Models of Elasticity

Sawtooth patterns from an engineered polyprotein may be analyzed quantitatively using models that describe the physics of polymer elasticity. Polymer chains that are free in solution exist in a coiled state since this maximizes their conformational freedom and therefore entropy. Extension of the molecule generates an opposing force due to the reduction in entropy, as the freedom of movement of the molecule is restricted. The behavior of polymers under stress may be predicted using a worm-like model of entropic polymer elasticity (the WLC model). The entropic elasticity of proteins is described by a worm-like chain equation, which expresses the relationship between force (F) and extension (x) of a protein using its persistence length (P) and its contour length (L_c)

$$F(x) = \frac{kT}{P}\left[\frac{1}{4}\left(1 - \frac{x}{L_c}\right)^2 - \frac{1}{4} + \frac{x}{L_c}\right] \qquad (F2.2)$$

The WLC model describes a polymer as a continuous string of a given total (or contour) length. Bending of the polymer at any point influences the angle of the polymer for a distance,

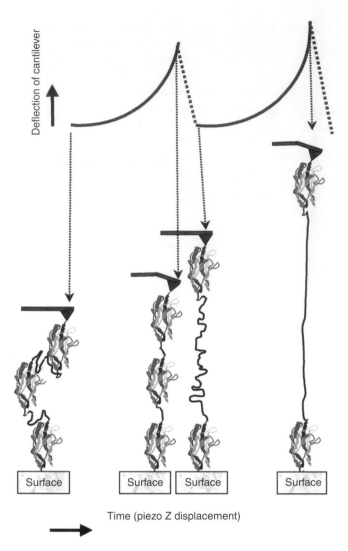

Fig. F2.78 Schematic representation of the structural transitions in multidomain proteins giving multipeak force curves (sawtooth pattern). (After Zlatanova *et al.*, 2000.)

referred to as the persistence length, that reflects the polymer flexibility. The smaller the persistence length, the greater the entropy of the polymer and the greater the resistance to extension. The persistence length and the contour length comprise the adjustable parameter of the WLC model (Comment F2.21). The WLC model with a single parameter P cannot describe the polypeptide elasticity equally well over the complete force range (0–300 pN). For force up to 50 pN, $P = 0.8$ nm describes the polypeptide elasticity well. However, a persistence length of 0.4 nm describes the polypeptide elasticity better. This is due to the fact that additional contributions from bond angle deformations become important at forces above 50 pN (Comment F2.22).

Protein Folds and Mechanical Stability

We now discuss the mechanical properties of four different proteins. Two of them are considered to have an "all beta"

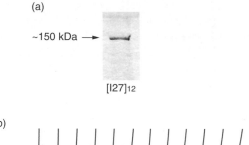

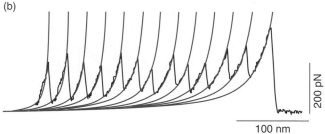

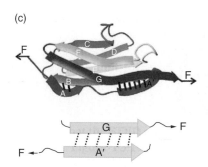

Fig. F2.79 Mechanical properties of single human cardiac titin immunoglobulin domains: **(a)** the measured molecular mass of the polyprotein (~150 kDa) in a good agreement with predicted molecular mass of a $[I27]_{12}$ concatemer; **(b)** force–extension curve for $[I27]_{12}$ with AFM; (c) top, three-dimensional structure of titin $[I27]$ domain; bottom, the dashed lines depict the topology of the critical hydrogen-bonds between β-strands under an applied force. (After Carrion-Vazquez *et al.*, 2000.)

structure and a so-called β-sandwich topology. One of these is from a typical mechanical protein (an immunoglobulin domain from titin) and the other is from a protein involved in secretion (the C2A domain from synaptotagmin I). The remaining two proteins are considered to be "all alpha" structures. One is calmodulin, a ubiquitous regulatory protein, and the other is spectrin, a major component of the membrane-associated skeleton in erythrocytes.

The $[I27]$ module of human cardiac titin, which is 89 amino acids long, has a typical IgI topology composed of seven β-strands (strands A–G), which fold into face-to-face β-sheets through backbone hydrogen bonds and hydrophobic core interactions (Fig. F2.79c). Since the hydrogen bonds in this patch are perpendicular to the direction of the applied force, unfolding requires the simultaneous rupture of this cluster of hydrogen bonds.

Stretching $[I27]_{12}$ with AFM results in a force–extension curve with peaks that vary randomly in amplitude about a value of 204 ± 26 pN (Fig. F2.79b). Fits of the WLC model to the force–extension curves of $[I27]_{12}$ give a persistence length P of about 4 nm and a variable contour length $L_c = 25$–496 nm (blue lines) with a contour length increment $\Delta L_c = 28$ nm. This persistence length is the size of a single amino acid (0.4 nm).

The mechanical function of the C2 domain protein is not known. The C2 domain is a conserved "all beta" module present in more than 40 different proteins, many of which are involved in membrane interaction and signal transduction. The first C2 domain of synaptotagmin I (C2A) is believed to be the calcium sensor that initiates membrane fusion during neurotransmitter release.

The three-dimensional structure of C2A was found to be a β-sandwich composed of 127 amino acids arranged into eight antiparallel strands with the N- and C-terminal strands pointing in the same direction (Fig. F2.80c). In contrast to the $[I27]$ domain, in which the hydrogen bonds have a "shear" topology, those in the C2A domain are in a "zipper" configuration (i.e., parallel to the direction of the applied force).

In AFM experiments, the force–extension curve of C2A polyproteins show a sawtooth pattern with force peaks of ~60 pN separated by a distance of ~38 nm.

Calmodulin is a highly conserved antiparallel "all alpha" protein that acts as a primary calcium-dependent regulator of many intracellular processes. The three-dimensional structure of calmodulin (148 residues) is dumbbell-shaped and consists of seven α-helixes distributed in a helical central region capped by two globular regions, each containing two helix–loop–helix motifs that

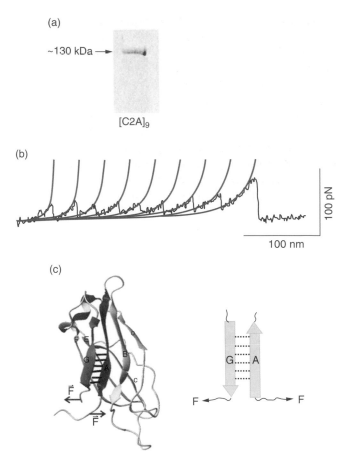

Fig. F2.80 Mechanical properties of a single C2 domain. (a) Coomassie blue staining of the purified [C2A]$_9$ protein. The measured molecular mass of the polyprotein (~130 kDa) is in a good agreement with predicted molecular mass of a [C2A]$_9$ protein. (b) Force–extension curve for [C2A]$_9$ protein with AFM. (c) left: three-dimensional structure of [C2A]$_9$ protein. Right: The dashed lines depict the topology of the critical hydrogen bonds between β-strands under an applied force. (After Carrion-Vazquez et al., 2000.)

COMMENT F2.23 MECHANICAL AND THERMODYNAMIC STABILITY OF CALMODULIN

One of the most striking features of calmodulin is its thermodynamic stability, particularly in the presence of Ca^{2+}. The protein may be exposed to 95 °C with retention of biological activity. The Ca^{2+}-free form of CaM has a melting temperature of ~55 °C, while the Ca^{2+}-bound form denatures only at temperatures exceeding 90 °C. However, a difference in mechanical stability in the presence or absence of Ca^{2+} has not been detected.

---

are responsible for Ca^{2+}-binding (Fig. F2.81c, left). In contrast to the β-sandwich topology, where there is a non-homogeneous distribution of the interstrand hydrogen bonds, the α-helix has a homogeneous distribution of

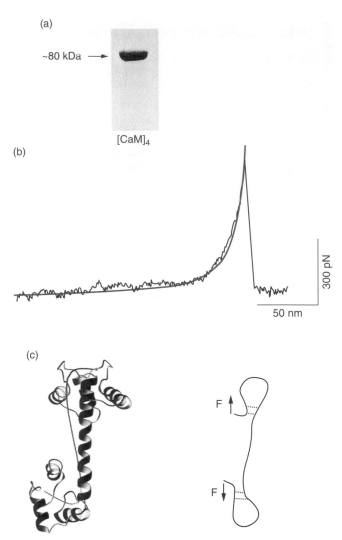

Fig. F2.81 Mechanical properties of a single polycalmodulin molecule. (a) Coomassie blue staining of the purified [CaM]$_4$ protein. The measured molecular mass of the polyprotein is ~80 kDa. (b) Stretching single calmodulin polyproteins gives force–extension curves with no evident force peaks. The force curve is well described by the WLS model (continuous lines) using a contour length of 212 nm and a persistence length of 0.32 nm. (c) Left: three-dimensional structure of rat CaM showing its α-helical structure. CaM is made of seven antiparallel α-strands and two short antiparallel β-sheet hairpins with a zipper topology that provides a structural link between the two Ca^{2+} motifs of each globular region. Right: putative mechanical topology of a calmodulin domain. (After Carrion-Vazquez et al., 2000.)

intrahelix hydrogen bonds. Figure F2.81b shows that the stretching of rat polycalmodulin (CaM$_4$) does not yield any force peaks, indicating that the unfolding forces must be below the AFM noise level in the experiments (~20 pN) (Comment F2.23). The origin of this is that two small β-hairpins of calmodulin (Fig. F2.81c, right) are in a "zipper" conformation, and therefore should offer little resistance to mechanical unfolding.

(a)

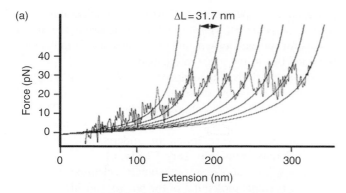

(b)

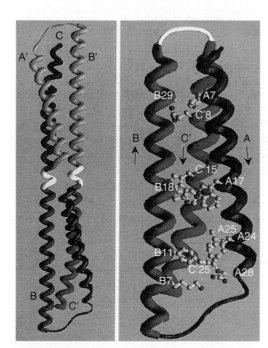

Fig. F2.82 (a) Force–extension curve. The continuous lines superimposed on the first curve are WLC fits ($p = 0.8$ Å). The gain in length upon each unfolding event is 31.7 nm, which corresponds to the 106 amino acid residues folded in each spectrin repeat. **(b)** Three-dimensional structure of spectrin. Left: Side view of the dimer. One polypeptide in the chain is shown in red hues, the other in green hues. Right: Side view of one repeat. The B, C loop (white) was inserted by model building. Some of the side-chains that pack to maintain the spacing between the α-helices are shown with carbon in yellow, nitrogen in blue, and oxygen in red. (Adapted from Rief *et al.*, 1999, and Yan *et al.*, 1993.)

The only other α-helical structure that reveals force peaks is spectrin, a major component of the membrane-associated skeleton in erythrocytes. As such, it cross-links filamentous actin and contributes to the mechanical properties of the cell. Spectrin consists of two subunits, which form laterally associated heterodimers (Fig. F2.82b). The two dimers interact and form a head-to-head dimer. The main part of both chains consists of homologous repeats. Each of these repeats (~106 amino acid residues) forms triple helical,

antiparallel coiled coils. The force required to mechanically unfold these repeats is 25–35 pN (Fig. F2.82a). The unfolding forces of the α-helical spectrin domains are 5–10 times lower than those found in domains with β-folds, like cardiac titin immunoglobulin domains. This shows that the forces stabilizing the coiled coil lead to a mechanically much weaker structure than multiply hydrogen-bonded β-sheets. On the other hand, the melting temperatures of titin domains (50–70 °C) and spectrin (53 °C) are rather similar. Melting temperatures are correlated with the free energy of activation for unfolding processes (Chapter C2). Obviously a comparison of free energies alone cannot explain the huge difference in mechanical stability.

All these experiments show that different proteins, even those with related structures, display a broad range of mechanical stability. The mechanical phenotypes of proteins may arise from differences in their topology, possibly as a result of variations in the number and position of hydrogen bonds among strands and sheets, and relative to the direction in which the force is applied.

The main results of mechanically unfolding proteins using the AFM can be summarized as follows.

(1) Folds of parallel terminal β-strands in which hydrogen bonds have a "shear" topology (for example, the I27 module of human cardiac titin) have the maximal mechanical stability (~200 pN).

(2) β-strands in which hydrogen bonds have a "zipper" configuration (e.g., the C2 domain of synaptotagmin) have folds with relatively low mechanical stability (~60 pN).

(3) The α-helical proteins in which the helices form bundles, rather than single α-helices (e.g., spectrin), are proteins with low force peaks (~30 pN). The forces stabilizing the coiled coil lead to a mechanically much weaker structure than multiply hydrogen-bonded sheets.

(4) α-helical proteins in which the α-helices have a homogeneous distribution of intrahelix hydrogen bonds (e.g., calmodulin) do not yield any force peaks.

(5) Protein mechanical stability is not correlated with thermodynamic stability: The melting temperatures of spectrin and titin Ig domains are rather similar, whereas there is a difference in their unfolding forces of up to a factor of 10.

Finally we note that the unfolding pathway depends on the pulling geometry and is associated with unfolding forces that can differ by an order of magnitude. Thus the mechanical resistance of a protein is not dictated solely by the amino acid sequence, topology, or unfolding rate constant, but depends critically on the direction of applied extension.

F2.6.9 Deformation of Polysaccharides

Many of the advances made in polysaccharide characterization have been possible because of increasingly

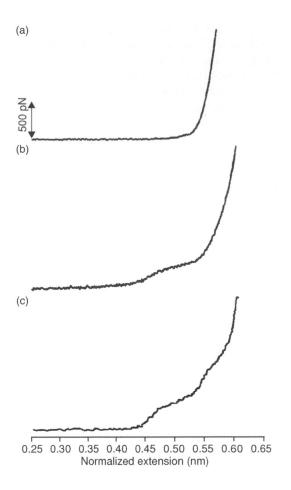

Fig. F2.83 Force–extension curves for a single polysaccharide structure: **(a)** cellulose, **(b)** amylose, and **(c)** pectin. (Brant, 1999.)

powerful mass spectrometry (Part B) and nuclear magnetic resonance (Part J). Here, we briefly describe the main results for the elasticity of single polysaccharide molecules obtained by atomic force microscopy.

Polysaccharides whose glycosidic linkages are attached equatorially to the pyranose ring (e.g., cellulose) are found to follow the FJC model of polymer elasticity (Fig. F2.83c). However, polysaccharides with axial linkage, such as amylose and pectin, were found to undergo abrupt force-induced length transitions (Fig. F2.83a, b).

Figure F2.84a shows several such force curves recorded for various single molecules of the polysaccharide dextran. All of the curves exhibit the same characteristic elastic behavior. At around 700 pN the curves deviate from a simple shape and show a kink. Using a molecular dynamics simulation it has been shown that this kink is due to a conformational transition within each dextran monomer, where the C5–C6 bond of the sugar ring flips into a new conformation, thus elongating the monomer by 0.65 Å (~10% of its length (Fig. F2.84b)).

The AFM results contradict the view that sugars are inelastic and locked into a stable conformation, raising the tantalizing possibility that force-driven sugar conformations play important roles in biological signaling as well as in the elasticity of polysaccharides.

Finally, we mention that the rupture force of single covalent bonds under an external load could be studied by AFM. For example, single polysaccharide molecules have been covalently anchored between a surface and an AFM tip and then stretched until they became detached. By using different surface chemistries for the attachment, it has been found that the silicon–carbon

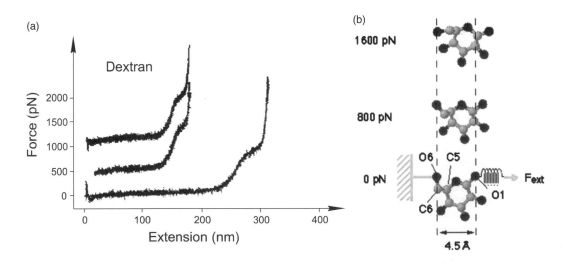

Fig. F2.84 **(a)** Characteristic shapes of the force–extension traces for dextran strands of different lengths. All curves show a kink at around 700 pN, where a bond angle within each monomer flips into a new position. The first trace shows the extension and the second trace the relaxation of the same dextran strand. No hysteresis can be observed between the cycles. Also, the force at which the transitions occur is not speed-dependent. This means that the bond flips occur on a faster timescale than the experiment, and therefore stretching is an equilibrium process. For clarity the traces are offset from each other (Rief *et al.*, 1998). **(b)** Ring conformation of dextran at different extension forces. (Adapted from Gimzewski and Joachim, 1999.)

bond ruptured at 2.0 ± 0.3 nN, whereas the sulfur–gold anchor ruptured at 1.4 ± 0.3 nN at force-loading rates of 10 nN per second.

F2.7 CHECKLIST OF KEY IDEAS

- Optical and magnetic traps, the cantilevers of atomic force microscopy and glass microneedles, are widely used for nanoscale manipulation. The force range of these facilities is from to 1 pN to 10 nN.
- Atomic force microscopy and scanning force microscopy *do not use lenses to form an image*, but instead use a sharply pointed sensor, or tip, at the end of a flexible cantilever to scan and sense the topography of a sample.
- The heart of the AFM is a tip mounted on a flexible cantilever; the ideal AFM tip should have a radius that is as small as possible, a well-defined and reproducible molecular structure, and be mechanically and chemically robust such that its structure is not altered while imaging in fluid environments.
- In the simplest mode of operation, the contact mode, the AFM resembles a phonograph. As the sample is scanned underneath the tip, the cantilever deflects up or down, tracking the surface.
- In the tapping mode of operation, the cantilever vibrates at its resonance frequency (typically a few hundred kilohertz); the cantilever could be compared to a blind person scanning the environment with a stick to explore the path ahead.
- Many factors contribute to an AFM image of biological structures in addition to the topography of the sample surface. These include the size and shape of the tip, the properties of the feedback loop, and the mechanical and chemical properties of the sample and imaging environment.
- Extraction of submolecular information from AFM topography for isolated globular proteins is a very difficult task; so far it has not been possible to achieve atomic resolution on soft biological samples.
- High-resolution imaging with high signal-to noise ratio in the molecular/submolecular range can be achieved with two-dimensional crystals of biological macromolecules.
- The ability of the AFM to operate in liquid makes the method attractive for the study of time-dependent changes in biological systems. Instrumental developments have made possible high-speed (H-S) and ultra-fast AFM that achieved video resolution of ~10 and ~1000 fps (frames per second), respectively.
- Special AFM apparatus can probe cell surface proteins.
- It is possible to use the AFM tip like a mechanical "nanoscalpel"; after scanning the area of interest in non-contact mode AFM, biological material (e.g., chromosome bands) can be cut by the AFM tip at a high force.
- AFM enables the process of two-dimensional crystal growth to be visualized in real time.
- Simultaneous AFM and NSOM imaging of a biological material links the world of optical microscopy with scanned imaging techniques: AFM measures the topography, whereas NSOM monitors at each position the light throughput.
- Kinesin (a microtubule-based motor) generates a peak force of 6 pN and appears to "walk" along a microtubule in discrete steps of 8 nm, without detaching both its feet simultaneously and probably using its feet in an alternate fashion.
- Dinein (eukaryotic flagella motor) generates a peak force of 6 pN and moves the singlet microtubule in a processive manner; a remarkable feature of the dinein arm activity is the presence of oscillations with an amplitude of ~2 pN and a maximum frequency of ~70 Hz.
- A single myosin head (part of the muscle motor myosin) moves along an actin filament in regular 5.5 nm steps that coincide with the distance between adjacent actin subunits in one strand of an actin filament: myosin molecules generate a peak force of 3–6 pN.
- RNA polymerase (DNA transcription machine) progresses along DNA at speeds of about 200 nucleotides per second and generates a peak force of 21–27 pN, perhaps reflecting the need for RNAP to forcefully disentangle the DNA secondary structure.
- In *E. coli*, the flagella motor can rotate in either direction, and cells navigate toward regions rich in nutrients by controlling this direction; the flagella motor has a rotational speed of 300–1700 rps and generates a torque of 4500 pN nm.
- F_1-adenosine triphosphatase (part of the F_0F_1-ATP synthase complex) has a rotational speed of 4 rps (under filament load) and generates a torque of ~40 pN nm; the rotation occurs in increments of $120°$.
- When the dsDNA molecule is subject to a force of 65 pN or more, it undergoes a highly cooperative transition (~2 pN) into a stable form with 5.8 Å rise per base pair; this stable form is called S-DNA.
- At low stretching forces ($F < 0.3$ pN), the twisting of dsDNA results in the formation of plectonemes or supercoils; when the torque on the dsDNA reaches about 20 pN nm the overwound DNA molecule adopts a new stable form called P-DNA.
- Breaking of the dsDNA strands occurs at a force level of about 500 pN.
- Unzipping of dsDNA occurs abruptly at 10–15 pN and displays a reproducible "sawtooth" force variation pattern with an amplitude of ± 0.5 pN along the DNA; force is sequence-dependent (~15 pN for the G–C base-pairs and ~10 pN for the A–T base-pairs).
- Opening events in a chromatin assembly are quantized at increments in fiber length of about ~27 nm and are attributed to unwrapping of the DNA from individual histone octamers; the forces measured for individual nucleosome disruptions are in the range 20–40 pN.
- The force range for ribozyme unfolding depends on the type of secondary structures inside ribozyme; a force of 14.5 pN is needed before the P5ab RNA molecule (simple hairpin) unfolds, whereas a force of 22 pN is required for the P5abc RNA molecule (an A-rich bulge and 5pc three-helix junction).

Suggestions for Further Reading

Historical Review and Introduction to Biological Problems

Rugar, D., and Hansma, P. (1990). Atomic force microscopy. *Phys. Today*, **24**, 23–30.

Engel, A. (1991). Biological applications of scanning probe microscope. *Annu. Rev. Biophys. Biophys. Chem.*, **20**, 79–108.

Bustamante, C., and Keller, D. J. (1995). Scanning force microscopy in biology. *Phys. Today*, **48**, 32–38.

Service, R. F. (1997). Chemists explore the power of one. *Science*, **276**, 1027–1029.

Mehta, A. D., Rief, M., Spudlich, J. A., Smith, D. A., and Simmons, R. M. (1999). Single-molecule biomechanics with optical methods. *Science*, **283**, 1689–1695.

Gimzewski, J. K., and Joachim, C. (1999). Nanoscale science of single molecules using local probes. *Science*, **283**, 1683–1688.

Müller, D. J., Janovjak, H., Lehto, T., Kuerschner, L., and Anderson, K. (2002). Observing structure, function and assembly of single proteins by AFM. *Prog. Biophys. Mol. Biol.*, **79**, 1–43.

General Principles of AFM

Binning, G., Quate, C. F., and Gerber, C. H. (1986). Atomic force microscope. *Phys. Rev. Lett.*, **56**, 930–933.

Bustamante, C., Erie, D. A., and Keller, D. (1994). Biochemical and structural applications of scanning force microscopy. *Curr. Opin. Struct. Biol.*, **4**, 750–760.

Dai, H., Hafner, J. H., Rinzler, A. G., Colbert, D. T., and Smalley, R. E. (1996). Nanotubes as nanoprobes in scanning probe microscopy. *Nature*, **384**, 147–150.

Smith, B. L. (2000). The importance of molecular structure and conformation: Learning with scanning probe microscopy. *Prog. Biophys. Mol. Biol.*, **74**, 93–113.

Picco, L. M., Bozec, L., Ulcinas, A., *et al.* (2007). Breaking the speed limit with atomic force microscopy. *Nanotechnology*, **18**, 44 030–44 033.

Ando, T. (2012). High-speed atomic force microscopy coming of age. *Nanotechnology*, **23**, 06200

Eghiaian, F., Rico, F., Colom, A., Casuso, I., and Scheuring, S. (2014). High-speed atomic force microscopy: Imaging and force spectroscopy. *FEBS Lett.*, **588**, 3631–3638.

Ando, T., Uchihashi, T., and Scheuring, S. (2014). Filming biomolecular processes by high-speed atomic force microscopy. *Chem. Rev.*, **114**, 3120–3188.

Imaging of Biological Structures

Mou, J., Csajkovsky, D. M., Zhang, Y., and Shao, Z. (1995). High-resolution atomic force microscopy of DNA: The pitch of the double helix. *FEBS Lett.*, **371**, 279–282.

Bustamante, C., and Rivetti, C. (1996). Vizualizing protein–nucleic acid interactions on a large scale with the scanning force microscopy. *Annu. Rev. Biophys. Biomol. Struct.*, **25**, 395–429.

Kuznetsov, Y. G., Malkin, A. J., Land, T. A., *et al.* (1997). Molecular resolution imaging of macromolecular crystals by atomic force microscopy. *Biophys. J.*, **72**, 2357–2364.

Engel, A., Lyubchenko, Y., and Muller, D. (1999). Atomic force microscopy: A powerful tool to observe biomolecules at work. *Trends Cell Biol.*, **9**, 77–80.

Hafner, J. H., Cheung, C.-L., Wooley, A. T., and Lieber, C. M. (2001). Structural and functional imaging with carbon nanotube AFM probes. *Prog. Biophys. Mol. Biol.*, **77**, 73–110.

Combination of NSOM and AFM

Lewis, A., Radko, A., Ami, N. B., Palanker, D., and Lieberman, K. (1999). Near-field scanning optical microscopy in cell biology. *TIBS*, **9**, 70–73.

Nanoscale Manipulation Techniques

Wang, M., (1999). Manipulation of single molecules in biology. *Curr. Opin. Biotechnol.*, **10**, 81–86.

Ficher, T. E., Oberhauser, A. F., Carrion-Vazquez, M., Marszalek, P. E., and Fernandez, J. M. (1999). The study of protein mechanics with the atomic force microscope. *TIBS*, **24**, 379–384.

Hegner, M. (2002). The light fantastic. *Nature*, **419**, 125.

Grier, D. (2003). A revolution in optical manipulation. *Nature*, **424**, 810–816.

Gross, S. D. (2003). Application of optical traps in vivo. *Meth. Enzymol.*, **361**, 162–174.

Macromolecular Mechanics: Nanometer Steps and Piconewton Forces

Ishijima, A., and Yanagida, T. (2001). Single molecule nanobioscience. *Trends Biochem. Sci.*, **26**, 438–444.

Molecular Motors

Finer, J. T., Simmons, R. M., and Spudlich, J. A. (1994). Single myosin molecule mechanics: Piconewton forces and nanometer steps. *Nature*, **368**, 113–119.

Spudich, J. A. (1994). How molecular motors work. *Nature*, **372**, 515–518.

Block, S. M. (1995). Nanometers and piconewtons: The macromolecular mechanics of kinesin. *Trends Cell Biol.*, **5**, 169–175.

Vale, R. D. (1996). Switches, latches and amplifiers: Common themes of G proteins and molecular motors. *J. Cell. Biol.*, **135**, 291–302.

Kitamura, K., Tokunaga, M., Iwane, A. H., and Yanagida, T. (1999). A single myosin head moves along an actin filament with regular steps of 5.3 nanometers. *Nature*, **397**, 129–134.

Adachi, K., Yasuda, R., Noji, H., *et al.* (2000). Stepping rotation of F_1-AtPase visualized through angle-resolved single-fluorophore imaging. *Proc. Natl. Acad. Sci. USA*, **97**, 7243–7247.

Oster, G., and Wang, H. (2003). Rotary protein motor. *Trends Cell Biol.*, **13**, 114–121.

DNA and RNA Mechanics

Bustamante, C., Smith, S. B., Liphardt, J., and Smith, D. (2000). Single-molecule studies of DNA mechanics. *Curr. Opin. Struct. Biol.*, **10**, 279–285.

Strick, T., Allemand, J.-F., Croquete, V., and Bensimon, D. (2000). Twisting and stretching single DNA molecules. *Prog. Biophys. Mol. Biol.*, **74**, 115–140.

Liphardt, J., Onoa, B., Smith, S. B., Tinoco, I. Jr., and Bustamante, C. (2001). Reversible unfolding of single RNA molecules by mechanical force. *Science*, **292**, 733–737.

Brower-Toland, B. R., Smith, C. L., Yeh, R. C., *et al.* (2002). Mechanical disruption of individual nucleosomes reveals a reversible multistage release of DNA. *Proc. Natl. Acad. Sci. USA*, **99**, 1960–1966.

Protein Mechanics

Carrion-Vazquez, M., Oberhauser, A. F., Fisher, T. E., *et al.* (2000). Mechanical design of proteins studied by single-molecule force spectroscopy and protein engineering. *Prog. Biophys. Mol. Biol.*, **74**, 63–91.

Fisher, T. E., Marszalek, P. E., and Fernandez, J. M. (2000). Stretching single molecules into novel conformations using the atomic force microscopy. *Nature Struct. Biol.*, **7**, 719–724.

Zlatanova, J., Lindsay, S. M., and Leuba, A. H. (2000). Single molecule force spectroscopy in biology using the atomic force microscope. *Prog. Biophys. Mol. Biol.*, **74**, 37–61.

How Strong is a Covalent bond?

Florin, E.-L., Moy, V. T., and Gaub, H. E. (1994). Adhesion forces between individual ligand–receptor pairs. *Science*, **264**, 415–417.

Rief, M., Oesterhelt, F., Heymann, B., and Gaub, H. E. (1997). Single molecule force spectroscopy on polysaccharides by atomic force microscopy. *Science*, **275**, 1295–1297.

Grandbois, M., Beyer, M., Rief, M., Clausen-Schaumann, H., and Caub, H. E. (1999). How strong is covalent bond? *Science*, **283**, 1727–1730.

Gimzewski, J. K. and Joachim, C. (1999). Nanoscale science of single molecules using molecular probes. *Science*, **283**, 1683–1688.

FLUORESCENCE MICROSCOPY

F3

F3.1 HISTORICAL REVIEW

1911–1913

Otto Heimstaedt and **Heinrich Lehmann** built the first fluorescence microscopes in Germany in 1911. These microscopes were initially used to observe auto-fluorescence in living cells, but shortly after **Stanislav Von Provazek** applied fluorescence microscopy to the study of stained biological samples.

1931

M. Goppert-Mayer gave the theoretical background of two-photon excitation in fluorescence. Thirty years later, this phenomenon was observed experimentally shortly after the invention of the laser by **W. Kaiser** and **C. Garrett.** In **1990**, **W. Denk** introduced two-photon excitation in fluorescent microscopy.

1941

Albert Coons described the fluorescent method of labeling antibodies, thus bringing forth the field of immunofluorescence.

1948

T. Forster formulated the principle of fluorescence resonance energy transfer (FRET), a phenomenon that occurs when two different chromophores (donor and acceptor) with overlapping emission/absorption spectra are separated by a suitable orientation and a distance in the range 20–100 Å. In the early 1970s, after a long period of inaction, the ground-breaking work of **L. Stryer** and **R. P Hohland** on FRET revealed the spatial proximity relationships of two fluorescence-labeled sites in biological macromolecules, thereby establishing FRET as a *spectroscopic ruler*. All of this early work used either fluorescent analogues of biomolecules or fluorescent reagents covalently or non-covalently attached to macromolecules as donors or acceptors of FRET. In the **1990s** the introduction of the green fluorescent protein (GFP) to FRET-based imaging microscopy gave new life to its use as a sensitive probe of protein–protein interactions and protein conformational changes in vivo.

1961

B. Rotman was the first to use fluorescence detection for single-molecule studies in solution. Using a fluorogenic substrate, he measured the presence of a single β-D-galactosidase molecule by detecting the fluorescent product molecules accumulated in a microdroplet through enzymatic amplification. He did not achieve single-molecule sensitivity but clearly demonstrated the great potential of fluorescence detection. In **1976**, **T. Hirschfeld** reported the use of fluorescence microscopy to detect single antibody molecules tagged with 80–100 fluorescein molecules under evanescent-wave excitation.

1964

S. Singh and **L. Bradely** predicted a three-photon absorption mechanism. In **1995** three-photon excited fluorescence spectroscopy and microscopy was demonstrated by a few groups and applied to the imaging of biological specimens and live cells.

The first lifetime measurements on single points under a microscope were carried out by a few groups in the early **1970s**.

1984

D. Axelrod introduced total internal reflection into light microscopy. In total internal reflectance fluorescence microscopy the region near an interface between two media with different refractive indices is illuminated by an evanescent wave. This wave penetrates only a short distance into the medium and selectively excites fluorescence in that region. Microscopic observation further allows one to image probes of structures fluorometrically near the interface with a spatial resolution of about half of the emission wavelength. In **2000**, **G. Cragg** and **P. So** described a modification of this technique that achieves a lateral resolution of better than one-sixth of the emission wavelength.

1985

S. Lanni's group proposed the standing wave fluorescence microscope, in which two coherent plane-wave beams from a laser cross in the specimen volume, where they interfere. In **1993** the same group described a fluorescence microscope in which the axial resolution is increased to better than 100 nm by using the principle of standing-wave excitation of fluorescence.

1989

W. Moerner's group first used a laser to see single, small organic molecules trapped inside a transparent host crystal. In **1990**, **M. Orrit**'s group showed that very high sensitivity at the single-molecule level can be achieved by probing "guest" molecules in solid hosts at cryogenic temperatures. In **1994**, **F. Guttler** and colleagues described fluorescent experiments with single pentacene molecules under a microscope at a temperature of 1.8 K. However, practical applications of single-molecule detection required measurements in solution at room temperature.

1990

R. Keller's group reported the first efficient detection of individual fluorescent molecules in solution. The observation technique involved exciting fluorescent molecules by passing them in solution through a highly focused laser beam and detecting the subsequently emitted photons. Single-molecule detection was proposed by this group as a tool for rapid base-sequencing of DNA. In **1994** the same group demonstrated a novel method of determining electrophoretic velocities by measuring the time required for individual molecules to travel a fixed distance between two laser beams.

1991

W. Whitten and colleagues solved the main problem of isolating the molecular fluorescence signal from background luminescence and Raman scattering by reducing the excitation volume to the smallest possible size with levitated micro-droplets. In **1992**, **E. Betzig** and **J. Trautman** reduced the optical excitation volume to subwavelength proportions in all three dimensions. This was accomplished using the illumination mode in near-field scanning optical microscopy.

1992

J. Lacovich's and **R. Clegg**'s groups proposed independently the principles of a new apparatus that allows lifetime imaging with simultaneous measurements at all positions in the image. In fluorescence lifetime imaging microscopes the nanosecond decay kinetics of the electronic excited-state of chromophores is mapped spatially using a detector capable of high-frequency modulation or fast gating. The spatial resolution of a microstructure on a nanosecond time-scale became possible. **S. Hell**'s group proposed so-called 4Pi-confocal microscopy, the primary goal of which is to increase the far-field resolution in the axial direction. In **1998** it was shown by the same group that the combination of two-photon excitation 4Pi-confocal fluorescence microscopy with image restoration leads to a fundamental (~six-fold) improvement in three-dimensional resolution.

1995

J. Wang and **P. Wolynes** described the mathematical formalism that can help in determinations of the microscopic origin of heterogeneity of behavior of biological systems. According to this formalism the statistics of individual activated events can exhibit intermittency and does not always obey Poisson's law and, therefore, the fluorescence decay function can exhibit a non-exponential character as a result of conformational dynamics. In **1996–1997** the non-exponential characters of two biological processes were described. Thus, **R. Rigler**'s group using confocal fluorescence microscopy monitored conformational transition for single 18-mer deoxyribonucleotide molecule linked to tetra-methylrhodamine, and **B. Cooperman**'s group using lifetime fluorescent measurements discovered two fluctuating conformational states for t-RNA^{Phe}-probe. These experiments opened a new area of molecular analysis of the biological events that can be understood only on the single-molecule level and could not be extracted from investigation of macroscopic systems.

2000 and since

Hell's group demonstrated a novel approach for superresolution imaging. It was accomplished by quenching excited organic molecules at the rim of the focal spot through stimulated emission. Along the optic axis, the spot size was reduced by up to six times. A spot volume which can be as low as 0.67 al makes this result relevant for single-molecule spectroscopy. Exciting techniques, including standing wave microscopy, 4Pi-confocal microscopy, stimulated emission, photo-activated localization microscopy (PALM), and stochastic optical reconstruction microscopy (STORM) have allowed to go beyond the standard diffraction barrier of light. Superresolution microscopy now plays a new role in biological research. Furthermore, advances in ultrasensitive instrumentation have enabled the detection, identification, and dynamical study of a single molecule in the condensed phase. These measurements have opened new avenues for scientists in many scientific regions, such as nano-chemistry, structural biology, biosensors, and molecular medicine.

The Nobel Prize in Chemistry **2014** was awarded jointly to **Eric Betzig**, **Stefan W. Hell**, and **William E. Moerner** "for the development of super-resolved fluorescence microscopy." This is an excerpt from the Nobel press release about the prize:

Two separate principles are rewarded. One enables the method stimulated emission depletion (STED) microscopy, developed by Stefan Hell in 2000. Two laser beams are utilized; one stimulates fluorescent molecules to glow, another cancels out all fluorescence except for that in a nanometer-sized volume. Scanning over the sample, nanometer for nanometer, yields an image with a resolution better than Abbe's stipulated limit. Eric Betzig and William Moerner, working separately, laid the foundation for the second method, single-molecule microscopy. The method relies upon the possibility to turn the fluorescence of individual molecules on and off. Scientists image the same area multiple times, letting just a few interspersed molecules glow each time. Superimposing these images yields a dense super-image resolved at the nanolevel.

F3.2 FLUORESCENCE MICROSCOPY INSIDE THE CLASSICAL LIMIT

F3.2.1 The Standard Wide-Field Fluorescence Microscope

Fluorescence techniques are a large part of the life science repertoire. The basic principle of conventional (*one-photon*) fluorescence is that a molecule (fluorophore) absorbs energy from an incident photon of relatively short wavelength (e.g., blue light), and then almost immediately (after a few nanoseconds) reemits a photon of lower energy (longer wavelength; e.g., green light). Because of this shift to longer wavelength, filters can be used to block out the excitation light and visualize only the fluorescence. A standard bright-field microscope can, in principle, be modified for fluorescence use simply by positioning a filter to transmit only short wavelengths (the excitation filter) with a second filter (the barrier filter) beyond the objective to block the excitation light while transmitting the emitted fluorescent light. However, this does not work well, because the excitation light is much more intense than the fluorescence light, and is difficult to block completely. Instead, modern fluorescence microscopes utilize epi-illumination in which the excitation light is introduced into the infinity space of the microscope and illuminates the specimen through the same objective lens that is used to image the fluorescence emission (Fig. F3.1). The key to this arrangement is the use of a dichroic mirror, which reflects the short-wavelength excitation light, while allowing the longer wavelength fluorescent emission to pass through toward the eyepiece. Because dichroic mirrors do not provide a perfect separation between excitation and emission wavelengths, additional excitation and barrier filters are incorporated. All three components are usually integrated within a filter "cube," and several cubes may be mounted in a turret or slider to allow ready interchange for work with different fluorophores.

F3.2.2 Two-Photon Excited Microscopy

Two-photon excitation arises from the simultaneous absorption of two photons in a single quantized event. Since the energy of a photon is inversely proportional to its wavelength, the two photons should be about twice the wavelength required for single-photon excitation

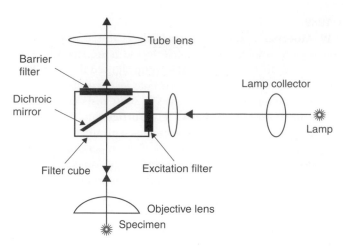

Fig. F3.1 Diagram of an epifluorescence microscope. A filter cube housing a dichroic mirror, excitation filter, and barrier (emission) filter is mounted in the infinity space behind the objective. Excitation light (from a laser or an arc lamp) is focused through the excitation filter and reflected by the dichroic mirror so that an image of the excitation source is formed at the back focal plane of the objective, resulting in even illumination of the specimen. Fluorescent light emitted by the specimen is collected through the objective, passes through the dichroic mirror, and is imaged by the usual arrangement of tube lens and eyepiece. The barrier filter serves to block any excitation light that was transmitted through the dichroic mirror.

(Fig. F3.2). For example, a fluorophore that normally absorbs UV light (~350 nm) can also be excited by two red photons (~700 nm) if they reach the fluorophore at the same time. In this case, "the same time" means within about 10^{-18} s. Because two-photon excitation depends on simultaneous absorption, the resulting fluorescence emission depends on the square of the excitation intensity.

To obtain a significant number of two-photon absorption events (where both photons interact with the fluorophore at the same time), the photon density must be approximately a million times what is required to generate the same number of one-photon absorptions. This means that extremely high laser powers are required to generate a lot of two-photon-excited fluorescence (Fig. F3.3). These powers are easily achieved using mode-locked (pulsed) lasers, where the power during the peak of the pulse is high enough to generate significant two-photon excitation, but the average laser power is fairly low. In the case of fluorescence, the resulting two-photon excited state from which emission occurs is the same singlet state that is populated during a conventional fluorescence experiment. Thus, the emission after two-photon excitation is the same as would be generated in a typical biological fluorescence experiment. Figure F3.4 shows photographs of the laser focus for one- (left) and two-photon (right) excitation. In the first case, a double cone is illuminated and spatial resolution has to be enforced by inserting a pinhole into

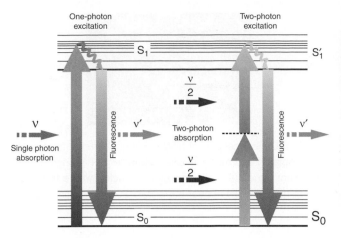

Fig. F3.2 Jablonski diagram of one-photon versus two-photon excitation. In one-photon excitation (left) the absorption of a single photon of energy v promotes the molecule from the electronic ground state into an excited state. In two-photon excitation (right) two photons of half the energy, v/2, are adsorbed simultaneously to reach the excited state. In both cases the molecule returns via emission of fluorescence to its ground state.

Fig. F3.4 Photographs of the laser focus for one-photon (left) and two-photon (right) excitation. (Adapted from Hausten and Schwille, 2003.)

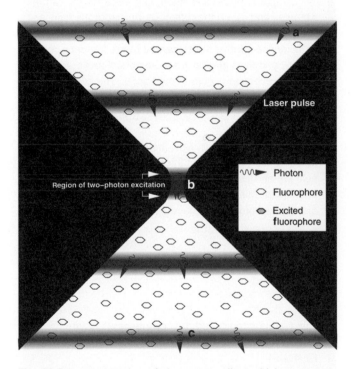

Fig. F3.3 Demonstration of photon crowding, which occurs at the focus of a microscope. As the pulsed red laser light passes through a sample containing fluorophores (blue hexagons), there is a nearly zero probability of two photons passing simultaneously within the cross-section of a single fluorophore located outside the focal point (labels "a" and "c"). This is true even during the peak of each laser pulse (denoted by the red photon symbols). However, because of the extremely high photon density at the focal point, it is possible for two of them to interact "simultaneously" with a fluorophore, which becomes excited (stippled blue hexagon, "b"). (After Piston, 1999.)

the image plane. Only molecules within a region providing sufficiently high intensities, and thus non-zero probability for the quasi-simultaneous absorption of two photons, are excited.

Three-photon excitation works in much the same way as two-photon excitation except that three photons must interact with the fluorophore at the same time. Because of the quantum-mechanical properties of fluorescence absorption, the photon density required for three-photon excitation is only about ten-fold greater than that needed for two-photon absorption. This makes three-photon excitation attractive for some experiments. For instance, an IR laser (~1050 nm) can three-photon excite a UV-absorbing fluorophore (350 nm) and simultaneously two-photon excite a green-absorbing fluorophore (~525 nm). Three-photon excitation can also be used to extend the region of useful imaging into the far-UV (i.e., use 720-nm light to excite a fluorophore that normally absorbs at 240 nm). This can be useful since UV wavelengths below ~300 nm are very problematic for standard microscopic optics.

The localization of two- or three-photon excitation to the focal point provides many advantages over confocal microscopy. In a confocal microscope, fluorescence is excited throughout the sample, but only the signal from the focal plane passes through the confocal pinhole, so background-free data can be collected (Section F1.3.1). By contrast, two-photon excitation only generates fluorescence at the focal plane, so there is no background and no pinhole is required. This dramatic difference between confocal and two-photon excitation microscopy is demonstrated by imaging the photobleaching patterns of each method (Fig. F3.5).

Two-photon excitation microscopy excels in thick multicellular systems such as brain slices, intact embryos, and other primary culture-tissue preparations. Another demonstration of the power of two-photon excitation microscopy

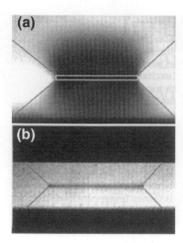

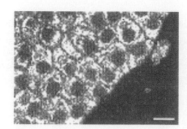

Fig. F3.6 Optical section of NAD(P)H signal arising from both the cytoplasm and mitochondria, the latter being brighter and somewhat punctate. Cell outlines and nuclei (where there is little or no NAD(P)H) appear dark. Bar, 10 μm. (Piston, 1999.)

Fig. F3.5 (a) X–Z profile of the excitation pattern in a confocal microscope formed by repeatedly scanning a single X–Y optical section in a thick film of rhodamine-formvar until fluorescence is completely bleached. The box represents the region from which data are collected by the confocal microscope for this optical section. Nearly uniform bleaching occurs both above and below the focal plane. **(b)** The same excitation pattern, but for two-photon excitation. Slanted lines denote the path taken by the excitation light to reach the focal plane, but no excitation occurs outside the focal plane. (Piston, 1999.)

is the imaging of the naturally occurring reduced pyridine nucleotides [NAD(P)H] as an indicator of cellular respiration. [NAD(P)H] has a small absorption cross-section, a low quantum yield, and absorbs in UV – thus, it is difficult to measure and has the potential to cause considerable photodamage. Figure F3.6 shows a typical image of β cell NAD(P)H autofluorescence within an intact islet, displaying a signal from both the cytoplasm and mitochondria. The outlines of single cells are visible, as are the nuclei, both of which appear dark. These imaging experiments simply cannot be performed by confocal microscopy owing to the limitations imposed by photobleaching and UV-induced photodamage.

F3.2.3 Total Internal Reflectance Fluorescence Microscopy (TIRFM)

In TIRFM, the region near an interface between two media with different refractive indices is illuminated by an evanescent wave (see Chapter C4). The evanescent wave penetrates only a short distance into the medium of lower refractive index and selectively excites fluorescence in that region (Comment F3.1). As the evanescent wave is set up in medium 2, a standing wave is set up in medium 1. The standing and evanescent waves are depicted schematically in Fig. F3.7.

Several useful experimental situations exist in which the interface is not a simple interface between two media but

---

COMMENT F3.1 "EVANESCENT WAVE"

When light propagating through a medium of high-refractive index encounters an interface with a medium of lower refractive index, it is either reflected or refracted according to Snell's law:

$$n_1 \sin \theta_1 = n_2 \sin \theta_2 \qquad (A)$$

where n_1 and n_2 are the refractive indices of the high- and low-refractive index media, and θ_1 and θ_2 are the angles of incidence relative to the normal to the interface. When $n_2 > n_1$ and θ_1 is greater than the critical angle, θ_c, total internal reflection occurs in medium 1. The critical angle is

$$\theta_c = \sin^{-1}(n_2/n_1) \qquad (B)$$

Although the incident light beam is totally internally reflected under these conditions, an electromagnetic field penetrates a small distance into medium 2. The intensity of this field decays exponentially with the distance z from the interface:

$$I(z) = I_0 \exp(-z/d_p) \qquad (C)$$

with a characteristic penetration depth

$$d_p = \frac{\lambda_0/n_1}{4\pi\sqrt{\sin^2\theta - (n_2/n_1)}} \qquad (D)$$

This field is often called the "evanescent" wave.

Note that Eq. (D) is valid for IR radiation (see Section E1.4.1).

---

rather consists of several interfaces between one or more thin layers of materials of refractive indices n_1 and n_2. These intermediate layers may be model membranes, protein layers, or membranes of whole cells. The situation that has some particularly interesting consequences for optical properties near the interface is when proteins are bound to a solid surface. Binding constants and the kinetics of the interaction of protein with different ligands was discussed in Chapter C3.

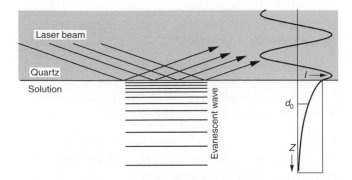

Fig. F3.7 Schematic drawing of the evanescent wave produced by a totally reflected laser beam at a quartz–buffer interface. On the right-hand side, the standing wave in a solid medium and the evanescent wave in the liquid medium are shown. When 488 nm laser light is incident at an angle of 72° on a quartz–buffer interface, the penetration depth d_p of the evanescent wave is about 97 nm. (Tamm, 1993).

F3.3 FLUORESCENCE SPECTROSCOPY OF SINGLE MOLECULES

The central problem of detecting a single absorbing molecule surrounded by quintillions of nominally transparent solvent molecules requires two deceptively simple steps: (1) establishing that only one molecule in the irradiated volume is in resonance with the laser, and (2) ensuring that the signal from the single molecule is larger than any background signal.

Optical detection of single molecules can be achieved by both frequency-modulated absorption and laser-induced fluorescence. Because of the low background and high signal-to-noise ratios, laser-induced fluorescence became the most widely used method. Now it is well established that the ultrasensitive fluorescent technique produces a higher signal-to-noise ratio when the sample background is carefully controlled.

F3.3.1 Laser-Induced Fluorescence

The fluorescence process begins with absorption of a photon, which drives the molecule from its ground molecular states to its first excited electronic state singlet state (Fig. F3.8). Absorption rates are usually limited only by the laser intensity. Emission of the fluorescence photon occurs in a few nanoseconds for most ionic dyes. The molecule has then returned to the ground electronic state and is ready for another absorption–emission cycle. The key point of the application of laser-induced fluorescence to single-molecule detection is that a single molecule can be repetitively cycled between its ground and excited electronic states with a laser beam at a wavelength resonant

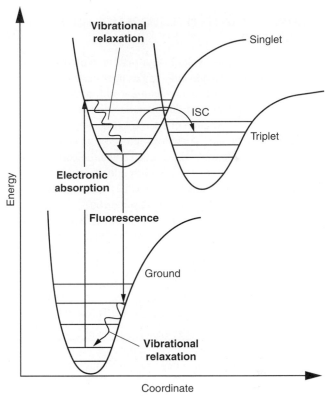

Fig. F3.8 Fluorescence cycle of a single molecule. ISC: intersystem crossing from the excited singlet state to the triplet. (Nie and Zare, 1997.)

with this transition, which yields a large number of photons (Comment F3.2).

The temporal and spectral properties of fluorescence are also advantageous to single-molecule detection. Fluorescence emission is typically red-shifted significantly with respect to the excitation frequency and is generally a much slower process, occurring in nano- to microseconds, than Rayleigh or Raman scattering, which occurs on a femtosecond scale. These properties facilitate separation of a signal from the background in both the time and frequency domains.

There are several important limitations on the magnitude of the signal that can be obtained from a single molecule. First, the fluorescent count rate, defined approximately by the absorption–emission cycle time, is limited at saturation by the finite fluorescence lifetime. Second, dyes also have a finite photochemical lifetime that limits the average number of absorption–emission cycles the molecule may undergo before photochemical destruction or "photobleaching" takes place. And finally, for some combinations of dye and solvent, singlet–triplet intersystem crossing rates (Fig. F3.8) can be significant, resulting in a substantial reduction in fluorescent count rates.

What matters most, however, is not the signal itself but the signal-to-noise ratio and the magnitude of the

By using diffraction-limited laser beams in combination with confocally imaged pinholes or fiber optics, a cylindrically shaped volume element with a radius of 200 nm and a length of 2000 nm can be intensely illuminated. The volume of such a light cavity is about 0.2 fl. One femtolitre is approximately the volume of an *E. coli* cell, and one molecule in such a volume corresponds to a concentration of 10^{-9} M. With a laser intensity of 0.5 mW, a photon density of $\sim 1 \times 10^{24}$ photons per square centimeter per second is achieved.

The ability to detect single molecules depends on the density of light quanta emitted. In the ideal case the number of quanta emitted per time interval is solely limited by the transition rate between the excited (singlet) state and the ground state. For example, for rhodamine 6G with an absorption cross-section of 1.4×10^{-16} cm^2 per molecule and a photon density of $\sim 1 \times 10^{24}$ photons cm^{-2} s^{-1}, a singlet excitation rate of 1.4×10^8 s^{-1} (per molecule) is reached. With singlet decay rate of 2.5×10^8 s^{-1} in the absence of non-radiative transition (assuming a quantum yield equal to 1) $\sim 25\%$ of the ground state is depleted. Under these conditions it is possible to detect photons at a rate $>100\,000$ per molecule per second. Taking into account the present detection efficiency and the bandwidth of detection, this value corresponds to a rate of 2×10^6 emitted quanta per second. The discrepancy between the maximal possible and the measured emission rate is partly due to existence of the triplet state and partly due to photobleaching during the excitation–deexcitation process (Eigen and Rigler, 1994).

On average at a concentration of 3.3×10^{-9} M, only one target molecule resides in the probe volume of 0.5–1.0 fl. The actual number of target molecules fluctuates over time. Assuming a Poisson distribution, the probability of finding one molecule at this concentration is 0.368, that of finding two molecules is 0.184, and that of finding more than two molecules is 0.0078. In more dilute solutions, the detection events are increasingly dominated by single-molecule events.

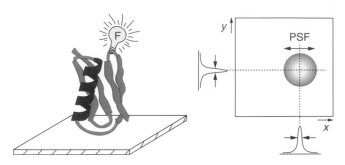

Fig. F3.9 Labeling scheme (left) and observable value (right) for the localization of a macromolecule labeled with a single fluorophore F. The point-spread function (PSF) can be localized within a few tenths of a nanometer. (Adapted from Weiss, 1999.)

single-chromophore molecules, the similarity between the photocount distributions precludes a clear determination of the number of molecules from the fluorescence burst amplitude. If the laser–molecule interaction time is long enough to measure the photobleaching kinetics, it is possible to distinguish between one or more molecules based on the time dependence of the fluorescence count rate versus time.

F3.3.2 Labeling Schemes and Observable Values

Various properties of single fluorescent probes attached to macromolecules can be exploited to provide information on molecular interaction, enzymatic activity, and conformational dynamics. "Native" fluorescence probes, such as fluorescence products, and fluorescence enzymes were successfully used to probe enzymatic turnovers of single molecules (see below). Small dye molecules that are covalently and site-specifically attached to macromolecules have also been actively used. And finally we note that another fruitful way to tag single proteins is through fusion with GFP (Section F3.6).

Several approaches to single-molecule detection can be classified by labeling schemes and observable values. Below we present some of them (Figs. F3.9–F3.14).

background signal. Source noise (from Rayleigh and Raman scattering from the solvent and fluorescence from impurities in the solvent) in these types of experiments usually dominates the background. As a rule of thumb, solvent scattering places an upper limit on the probe volume (a few picolitres).

Detection of a single molecule is usually defined by a threshold criterion. In a typical experiment, the threshold level is usually about three or more times the standard deviation in the non-fluorescent background signal. Most of the "proof" that a single molecule was detected in liquids is based on the signal-to-noise ratio, supported by small probabilities of more than one molecule occupying the probe volume in a given measurement interval (Comment F3.3).

How accurately can a distinction be made among zero, one, two, or more molecules? Just as molecular detection efficiencies are highly dependent on photocount statistics, so is the ability to clearly determine the number of molecules in a given probe volume. In the case of

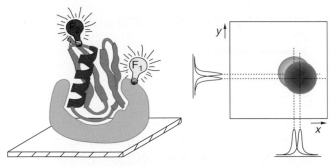

Fig. F3.10 Labeling scheme (left) and observable value (right) for the colocalization of two macromolecules labeled with two non-interacting fluorophores, F_1 and F_2. The distance between them can be measured by subtracting the center positions of the two PSFs. (Adapted from Weiss, 1999.)

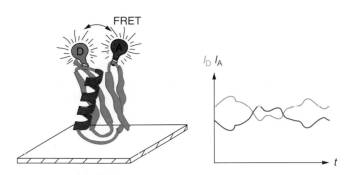

Fig. F3.11 Labeling scheme (left) and observable value (right) for intramolecular detection of conformational changes by FRET. D and A are the donor and acceptor; I_D and I_A are donor and acceptor emission intensities; t is time. (Adapted from Weiss, 1999.)

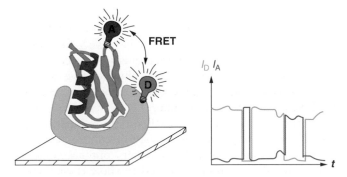

Fig. F3.12 Labeling scheme (left) and observable value (right) for dynamic colocalization and detection of association or dissociation by intermolecular spFRET. (Adapted from Weiss, 1999.)

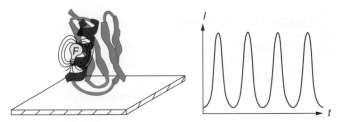

Fig. F3.13 The orientation of a single immobilized dipole can be determined by modulating the excitation polarization. The fluorescence emission follows the angle modulation. (Adapted from Weiss, 1999.)

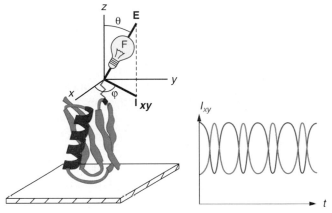

Fig. F3.14 The orientation freedom of motion of a tethered fluorophore can be measured by modulating the excitation polarization and analyzing the emission intensity. The emission dipole electric field vector **E** is defined in space by its angle ϕ in the x–y plane and angle θ, relative to the z-axis. The fluorescence emission intensity, $\mathbf{I_{xy}}$, that is detected in the x–y plane as a result of the emission dipole electric field vector **E** is defined as $\mathbf{I_{xy}} = \mathbf{I}\sin^2\phi = \mathbf{I}_x + \mathbf{I}_y$ and is also related to the fluorescence polarization angle ϕ as follows: $\cos^2\phi = \mathbf{I}x/(\mathbf{I}_x + \mathbf{I}_y)$. As defined by this spatial model, observed increases in total fluorescence intensity can be obtained if the emission dipole swings away from the z-axis (i.e., for θ approaching $90°$). (Adapted from Weiss, 1999.)

A simple, but powerful, use of single-molecule detection localizes a single fluorophore with a few tens of nanometer precision (Fig. F3.9). The dimensions of a dye molecule are much smaller than the wavelength of light it emits, and therefore it acts as a point source of light. The point spread function (PSF), the response of the optical system to this point source is a spot of light, the center of which can be localized with great accuracy (see Comment F1.4).

This positioning accuracy can be used for the colocalization of two different macromolecules (Fig. F3.10). When two macromolecules are labeled with two non-interacting fluorophores that differ in their optical properties

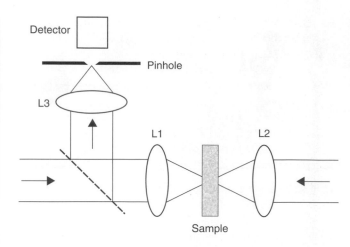

Fig. F3.15 Schematic of 4Pi-confocal microscopy. The same object point in the sample illuminates two opposing lenses (L1 and L2) of high numerical aperture. This imaging mode is technically the simplest version of 4Pi-confocal microscopy and is referred to as 4Pi-confocal microscopy of type A. (Adapted from Schrader et al., 1998.)

(absorption and emission spectra, fluorescence lifetime) they can be colocalized with nanometer accuracy and can report on association, binding, and enzymatic-turnover events.

Even higher colocalization accuracy can be obtained when the two fluorophores interact by FRET (Section F3.5). This technique, which is capable of measuring distances on a scale of 2–80 nm, relies on the distance-dependent energy transfer between a donor fluorophore and acceptor fluorophore (Fig. F3.11, see Section F3.5). This technique can also report on dynamical changes in the distance between or orientation of the two fluorophores for intramolecular and intermolecular FRET (Fig. F3.12).

The absorption and emission transition dipoles of a single fluorophore can be determined by using polarized excitation light or by analyzing the emission polarization, or both. The temporal variation in the dipole orientation of a rigidly attached (Fig. F3.13) or rotationally diffusing tethered probe (Fig. F3.14) can report on the angular motion of the macromolecule or one of its subunits.

These schemes were used to study ligand–receptor colocalization, to probe equilibrium protein structural fluctuations and enzyme–substrate interactions during catalysis, and to identify conformational states and subpopulations of individual diffusing molecules in solution.

F3.4 INCREASING THE RESOLUTION OF FLUORESCENCE MICROSCOPY

Increased resolution and "super" resolution beyond the Rayleigh limit have been achieved in fluorescence microscopy by a variety of important methodological developments.

F3.4.1 4Pi-Confocal Microscopy

The primary goal of 4Pi-confocal microscopy is an increase of the far-field resolution in the axial direction. For this purpose, the 4Pi-confocal microscope uses two opposing objective lenses of high numerical aperture with which the same object point is coherently illuminated and observed. The two spherical wave fronts add up to a single wave front so that the total aperture angle is doubled. This is accomplished either for illumination, or detection, or simultaneously for both. With two objective lenses it is not possible to obtain a complete solid angle of 4π. The acronym 4Pi, however, is a reminder of the basic idea behind the concept. Figure F3.15 sketches the 4Pi-confocal microscope for the case of coherent illumination through both objective lenses.

In standard confocal and multiphoton microscopes, a typical resolution is of the order of 150–300 nm in the lateral direction and 500–1000 nm in the axial direction. The 4Pi-confocal microscope improves the axial resolution by a factor of ~3–4 over that of a standard confocal microscope by physical means.

A further axial resolution can be achieved with standing-wave microscopy (Section F3.4.3). However, the fundamental improvement in three-dimensional resolution (about 100 nm) of specimens can be obtained by a combination of two-photon excitation 4Pi-confocal fluorescence microscopy and image restoration.

Featuring a three-dimensional resolution in the range 100–150 nm, the 4Pi-confocal (restored) images are intrinsically more detailed than their confocal counterparts. The three images in Fig. F3.16 show the confocal image, 4Pi-confocal axial image, and the 4Pi-confocal axial image after image restoration (Comment F3.4). A fundamental increase in three-dimensional resolution is evident.

F3.4.2 Stimulated Emission Depletion Microscopy

Stimulated emission depletion (STED) microscopy is based on the non-degenerate (different wavelength) pump probe scheme that excites the fluorophores with one beam and quenches them with another. A short (200 fs) visible pulse is focused to a diffraction-limited focal spot and excites all fluorophores in this volume. A second near-IR and longer (40 ps) pulse stimulates the emission from

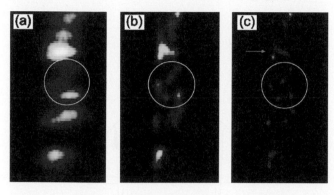

Fig. F3.16 (a) Confocal, **(b)** pointed-deconvoluted 4Pi-confocal, and **(c)** restored 4Pi-confocal *x–y* images obtained from microtubules in a mouse fibroblast cell. The images were taken consecutively at the same site in the cell. The comparison reveals a fundamental increase in the three-dimensional resolution of far-field fluorescence microscopy. (Adapted from Nagorni and Hell, 1998.)

molecules at the outer part of the focus, forcing them into the ground state immediately after they are excited. The longer duration of this second pulse provides full quenching without reabsorption.

Figure F3.17 illustrates the essentials of the technique. Figure F3.17a shows an energy-level diagram (Jablonski diagram), the absorption and emission spectra of the fluorophore, and the spectral alignment of the excitation and the STED pulses. Figure F3.17b demonstrates how the extent of the PSF is shrunk in one dimension by using this method (the illustration in one dimension is for demonstration only; shrinking is effective in all three dimensions).

Figure F3.18 shows the spot probed by a 48 nm bead stained with a fluorophore using standard confocal and STED geometries. Whereas the confocal spot features an axial full-width-at-half-maximum (FWHM) of 490 nm, that of the STED-fluorescence spot is only 97 nm. The difference in focal extent becomes obvious when comparing their *X–Z* sections shown in the insets. The STED method allows the diffraction limit to be broken and shrinks the PSF by a factor of 6 along the optical axis and by a factor of 2 in the radial direction. The spot volume decreases to 0.67 al.

It is important that the STED method is compatible with live-cell imaging. Figure F3.19 compares confocal axial images of live yeast (Fig. F3.19a) and *E. coli* (Fig. F3.19c) cells with their STED-fluorescence counterparts in Figs. F3.19b and d, respectively. In Fig. F3.19a, the axial images

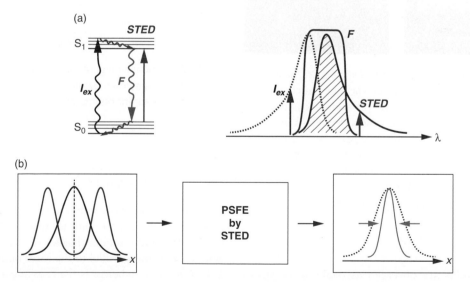

Fig. F3.17 Schematic view of STED microscopy. **(a)** Left: A Jablonski diagram showing the ground (S_0) and the first excited singlet (S_1) states of a fluorophore with the corresponding vibrational manifolds. A short pulse (I_{ex}, 200 fs) excites the fluorophore into a high vibrational state of S_1. The slow STED pulse (STED, 40 ps) stimulates emission from the lowest S_1 vibrational state into a high vibrational state of S_0. The stimulated emission depletes S_1 and quenches the fluorescence. Right: The equivalent wavelength diagram showing the absorption curve (dotted black line), the emission curve (solid black line), and the positions of I_{ex}, STED, and the filter (*F*). The filter blocks the two lasers and passes the signal (hatched green). **(b)** The combination of the excitation PSF (blue) with the STED engineered intensity distribution (red) acts as the input to PSFE by STED. The output PSF (green) is shrunk. (After Weiss, 2000.)

(a) (b)

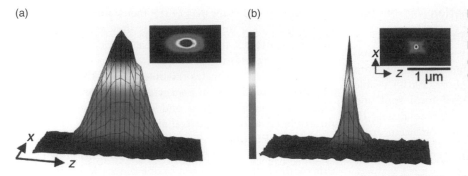

Fig. F3.18 Surface plot of the *X–Z* section of a spot probed by a 48 nm bead stained with a fluorophore in (a) confocal and (b) STED geometry. (After Klar *et al.*, 2000.)

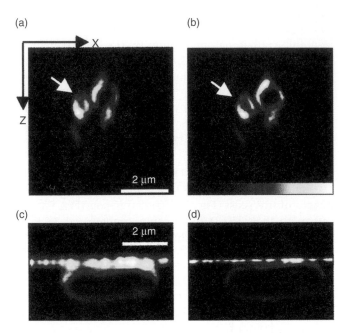

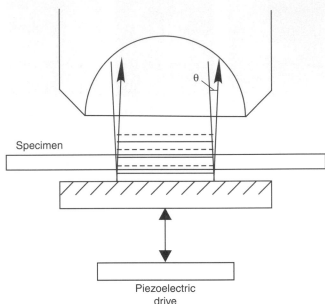

Fig. F3.19 Resolution improvement in live cells. *X–Z* images of an *S. cerevisiae* yeast cell with labeled vacuolar membrane with (a) standard confocal resolution and (b) axial resolution improved by STED. Whereas the confocal mode fails in resolving the membranes of small vacuoles, the STED microscopy better reveals their spherical structure. *X–Z* images of membrane-labeled *E. coli* show a three-fold improvement of axial resolution by STED in (d) as compared with their simultaneously recorded confocal counterparts in (c). (Klar *et al.*, 2000.)

Fig. F3.20 Optical configuration for SWFM. Fluorescence is excited in the specimen by the interference pattern of two intersecting coherent beams as described in the text. (Bailey *et al.*, 1993.)

F3.4.3 Standing-Wave Illumination Fluorescence Microscopy (SWFM)

In SWFM, two coherent plane wave beams from a laser cross in the specimen volume, where they interfere (Fig. F3.20). The nodes and antinodes of this interference field, which are planes parallel to the focal plane, have a spacing $\Delta s = \lambda/(2n \cos \theta)$, where θ is the crossing angle between two fields and n is the specimen refractive index. By controlling the angle θ, the node spacing can be varied down to a minimum value of $\lambda/2n$, when the two beams are counter-propagating along the axis of the microscope. By shifting the phase of one of the beams, the relative position of the field planes within the specimen can be adjusted, at constant node spacing. Because the standing-wave field diameter is comparable to the full field of view, images are

of the yeast vacuoles are observed as oval-shaped structures. The diffraction-limited spot in the confocal microscope overemphasizes the vertical parts of the vacuolar membrane. In contrast, because of the axially narrowed focal spot, the STED-fluorescence image in Fig. F3.19b gives a more faithful picture of the spherical shape of the vacuoles of the live yeast cell (see arrow). In particular, it reveals the spherical shape of smaller vacuoles that cannot be recognized as such by the confocal microscope.

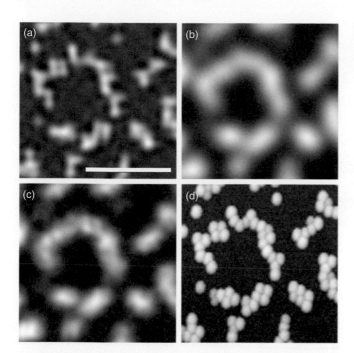

Fig. F3.21 Identical area of a sample of 100 nm diameter fluorescent polystyrene beads imaged using different techniques (bar = 1000 nm). **(a)** The image produced using a non-uniform excitation pattern with high spatial frequency components and consists of an extended two-dimensional interference field with closely spaced nodes and antinodes. **(b)** The image produced using identical lenses with standard illumination. **(c)** The confocal image. **(d)** The image recorded with AFM (Chapter F2). (After Frohn *et al.*, 2000.)

recorded directly onto an electronic camera, rather than by scanning a focal standing-wave field.

Figure F3.21 shows fluorescent beads with a diameter of 100 nm, which is far below the Rayleigh limit for standard fluorescence microscopy and also below that of confocal devices. To compare the resolving power of different techniques, the same region of water-immersed beads was imaged by using SWFM (Fig. F3.21a), standard illumination (Fig. F3.21b), and confocal scanning (Fig. F3.21c). As a reference for the actual locations of the individual beads, an atomic force microscope image is also given (Fig. F3.21d), which was acquired in air. The resolution of the standing-wave fluorescence image is clearly superior to that of the standard and also to that of the confocal image.

Table F3.1 summarizes the resolving power and probed volume of different fluorescent microscopy techniques. It is evident that the diffraction barrier responsible for a finite focal spot size and limited resolution in far-field fluorescence microscopy is fundamentally broken. The minimal spot volume is now less than 1 al, thus making new approaches relevant to single-molecule spectroscopy.

TABLE F3.1 RESOLVING CAPABILITIES AND PROBED VOLUME OF DIFFERENT FLUORESCENT MICROSCOPY METHODS.

Method	Resolving power	Probed volume
Conventional microscopy	$\lambda/2$	10 µl
Confocal microscopy	$\lambda/3$	Less than 1 fl
4Pi microscopy	$\lambda/3$	About 1 fl or less
Standing-wave illumination	$\lambda/4$–$\lambda/5$	About 1 fl or less
Stimulated emission depletion microscopy	$\lambda/6$	Less than 1 al

F3.5 FLUORESCENCE RESONANCE ENERGY TRANSFER

Fluorescence resonance energy transfer (FRET) is a technique that is more than half a century old, which has undergone a renaissance leading to applications in super-resolution microscopy. FRET relies on the distance-dependent transfer of energy from the donor fluorophore (D) to an acceptor fluorophore (A). It is one of the few tools available for measuring nanometer-scale distances and changes in distance, both in vitro and in vivo. Advances in the technique have led to qualitative and quantitative improvements, including increased spatial resolution, distance range, and sensitivity. These advances, due largely to genetically encoded dyes, such as GFP, but also to changes in optical methods and instrumentation, have opened up new biological applications.

F3.5.1 FRET as a Spectroscopic Ruler in Static and Dynamic Regimes

In FRET, a donor (D) fluorophore is excited by incident light, and if an acceptor (A) is in close proximity, the excited-state energy from the donor can be transferred. The process is non-radiative (not mediated by a photon) and is achieved through dipole–dipole interactions. The donor molecule must have an emission spectrum that overlaps the absorption spectrum of the acceptor molecule (Fig. F3.22a). FRET is useful as a means of measuring molecule interactions as the efficiency E_T at which Forster-type energy transfer occurs is steeply dependent on the distance R (nm) between the two fluorophores and is given by

$$E_T = \frac{R_0^6}{R_0^6 + R^6} \tag{F3.1}$$

R_0 (nm), the distance at which 50% energy transfer takes place (Fig. F3.22b), is expressed as

(a)

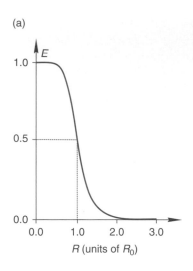

(b)

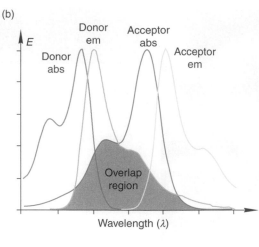

Fig. F3.22 (a) R_0 is the distance at which 50% energy transfer occurs. **(b)** Overlap of the donor emission spectrum with the acceptor excitation spectrum. (Adapted from Bastiaens and Pepperkok, 2000.)

$$R_0 = \left[k^2 \times J(\lambda) \times n^{-4} \times Q_D \right]^{1/16} \times 9.7 \times 100 \qquad \text{(F3.2)}$$

where k^2 is the relative orientation of the transition dipoles of the fluorophore, $J(\lambda)$ is the integral of the region of overlap between the donor emission and acceptor absorbance spectra, n is the refractive index of the surrounding medium, and Q_D is the quantum yield of the donor. Typically R_0 is between 20 and 60 Å, which is of the order of protein dimensions (Comment F3.5). Equation (F3.1) could be used as a *spectroscopic ruler*, i.e., by measuring E_T and knowing or calculating R_0, the distance R could be determined.

The dependence on the inverse sixth power of distance was tested in an elegant series of experiments by Stryer and Haugland in 1967. In this study, two chromophores were attached covalently to a rigid molecular framework. Then excitation transfer was measured for a series of synthetic species in which the distance between the donor and acceptor molecule differs. The transfer from a dansyl group at one end to a naphthyl group at the other end of oligoprolines with 1–12 monomer units in the rigid chain was examined. The results showed excellent agreement with the R^{-6} dependence for separations of 1.2–4.6 nm, as depicted in Fig. F3.23.

Access to long-range distance information makes FRET a valuable tool for studying nucleic acid molecules (Comment F3.6). Specifically, protein-induced DNA deformation has been characterized using this method. For example, the binding of catabolite activator protein (CAP) to straight DNA leads to a significant bending of the nucleic acid molecule (Fig. F3.17). However, only small differences in the energy transfer efficiencies can be detected for (a) in Fig. F3.24, as the dye-to-dye distance in the complex (b) in Fig. F3.24 is about 90 Å, i.e., considerably larger than R_0 (50 Å for the dye pair fluorescein-TMRh used). This illustrates a general problem. By introducing DNA structural elements such as

COMMENT F3.5 CALCULATION OF R_0

Depending upon the relative orientations of the donor and the acceptor, the k^2 factor can range from 0 to 4. For aligned and parallel transition dipoles $k^2 = 4$, and for oppositely directed and parallel dipoles $k^2 = 1$. Since the sixth root is taken to calculate the distance, variation of k^2 from 1 to 4 results in only a 26% error in R. However, if the dipoles are oriented perpendicular to one another, $k^2 = 0$, which would result in serious errors in the calculated distance. In general, the variation of k^2 does not seem to have resulted in major errors in the calculated distances. It is generally assumed that k^2 is 2/3, which is the value for donors and acceptors that randomize by rotational diffusion prior to energy transfer. This is the value that is generally assumed for calculation of R_0.

bulges, however, longer DNA molecules can be designed in which the helix ends are close to one another in space (in the range of R_0). A molecular model of a two-bulge DNA molecule with close helix ends is shown in Fig. F3.18. Double-helical B-DNA regions were used as substrates for protein binding. Molecules of 65 bp DNA, containing three dA$_5$ bulges (with the CAP-binding site in between), were constructed; the 5' end was labeled and used to monitor the binding of CAP by measuring FRET. A decrease in Förster transfer efficiencies from 21% with CAP (see (d) in Fig. F3.24) to 8% without CAP (see (c) in Fig. F3.24) was observed.

Figure F3.25 also shows the NMR structure of a DNA fragment containing a dA$_5$ bulge. Combining local (NMR) and global (FRET) structure information using molecular modeling allows a detailed model for a

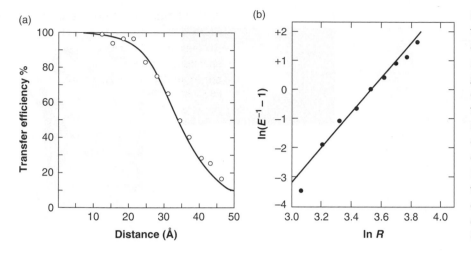

(a)

(b)

Fig. F3.23 (a) Efficiency of energy transfer as a function of a distance in dansyl-(L-prolyl)$_n$-α-naphthyl, $n = 1, \ldots, 12$. The α-naphthyl and dansyl groups were separated by a defined distance ranging from 12 to 46 Å. Energy transfer is 50% efficient at 34.6 Å. The solid line corresponds to R^{-6} dependence. **(b)** The dependence of the efficiency of energy transfer on distance is given by slope in this plot of $\ln(E^{-1} - 1)$ versus $\ln R$. The slope is 5.9, in excellent agreement with the R^{-6} dependence predicted by Förster. (After Stryer and Hougland, 1967.)

COMMENT F3.6 MEASURING OF DISTANCES IN PROTEIN STRUCTURES BY FRET

FRET is often used to measure distances in protein structures and their assemblies in solution. Proteins contain intrinsic fluorophores (tryptophan and tyrosine). In addition, they can be labeled with dyes at defined sites, e.g., cysteine residues. However, the multiple occurrence of intrinsic fluorophores and tethering sites for dyes reduces the interpretability of FRET data obtained with these chromophores.

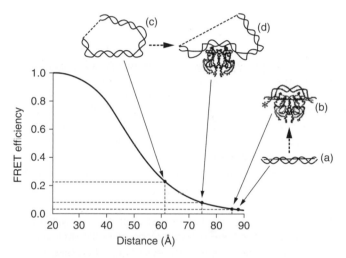

Fig. F3.24 Schematic illustration of the principle for the detection and characterization of DNA bending that utilizes DNA molecules with the helix ends close in space. The difference in the end-to-end distances of the straight 32 bp DNA **(a)** before and **(b)** after binding to CAP is small and cannot be detected by FRET. In addition, undesirable dye–protein interaction could have occurred (marked by asterisks). **(c)** The use of three-bulge DNA molecules with the helix ends close in space circumvents these problems. The dotted lines in the models symbolize FRET; the dotted arrows indicate CAP binding to DNA. (After Hillisch et al., 2001.)

large DNA molecule in solution to be deduced (see Chapter J3).

In general, FRET is better suited to detecting changes in distance rather than absolute distance because E_T depends on the orientation of the fluorophores and the attachment methods cause uncertainty in the probe position with respect to the macromolecular backbone (Comment F3.7). For an easy quantitative representation, at least one of the dyes should have rotational freedom, so that all dipole orientations occur with equal probability. In order to achieve this flexibility, the dyes are chemically attached to the DNA via alkyl chains (two to six carbon linkers).

Ensemble FRET has long been used as a very effective spectroscopic ruler for measurement of static distances in the 20–100 Å range. However, dynamic events such as the relative motion between donor and acceptor molecules cannot be detected by FRET, usually because of the lack of synchronization of these events. FRET measurements on a single pair, in contrast, allow one to look at dynamic distance changes on a millisecond timescale because only one donor–acceptor system is observed at a time; hence, time-dependent conformations can be monitored (Section F2.4.3).

F3.6 GREEN FLUORESCENT PROTEIN

Fluorescence imaging of biological reactions in living cells requires techniques to fluorescently label the macromolecules involved. These fluorescent probes need to be specific, interfere as little as possible with the reactions to be visualized, must not perturb the physiological conditions

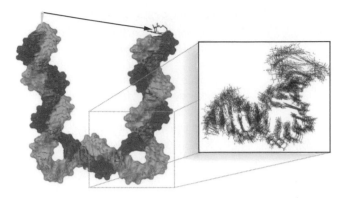

Fig. F3.25 Atomic model of a DNA molecule containing two DNA bulges. The molecule consists of three double-helical stems (16, 9 and 17 bp). The longer fluorescently labeled strand is composed of 52 bases and is colored green in the model; the shorter 42-base strand is labeled with TMRh and is depicted in orange. (Adapted from Hillisch *et al.*, 2001.)

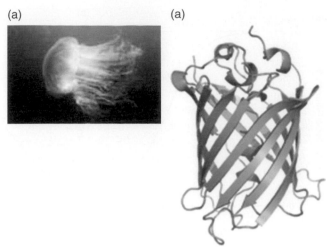

Fig. F3.26 (a) The jellyfish *Aequorea victoria*. **(b)** A schematic drawing of the backbone of GFP. The chromophore is *p*-hydroxy benzylideneimidazinone formed from the spontaneous cyclization and oxidation of the sequence – Ser^{65} (or Thr^{65})–Tyr^{66}–Gly^{67}– and is shown as a ball-and-stick model. The most distinctive feature of the fold of GFP is an 11-strand barrel wrapped around a single central helix, where each strand consists of approximately 9–13 residues. The barrel forms a nearly perfect cylinder 42 Å long and 24 Å in diameter. The top end of the cylinder is capped by three short, distorted segments, and one short, very distorted helical segment caps the bottom of cylinder (Adapted from Ormo *et al.*, 1996.)

COMMENT F3.7 MEASURING DISTANCES IN DNA STRUCTURES BY FRET

Well-defined sequences of DNA and RNA oligomers are now routinely synthesized and fluorescently labeled at the 3′ and 5′ termini, as well as within the DNA sequence. End labeling has three main advantages (when the shortest possible DNA is used): the dyes can be spatially separated from the binding protein to avoid protein–dye interactions; the DNA end sequence can be specifically designed to provide an identical chemical environment and, thus, the fluorescence properties of the dyes in a series of molecules can be derived; and the exact location of Cy3 on the DNA is known and evidence is accumulated on the position of fluorescein and rhodamine. In most studies, fluorescein and rhodamine are attached to the 5′ ends.

of the cells and must have spectroscopic properties for efficient detection. One of the most exciting advances in fluorescent *in vivo* labeling techniques was the discovery of GFP.

Wild-type GFP from the Pacific Northwest jellyfish *Aequorea victoria* is a stable, proteolysis-resistant single chain of 238 residues. The chromophore, resulting from the spontaneous cyclization and oxidation of the sequence –Ser^{65} (or Thr^{65})–Tyr^{66}–Gly^{67}–, requires the native protein to fold for both formation and fluorescence emission. The crystal structure of GFP with the sequence – Thr^{65}–Tyr^{66}-Gly^{67}– is presented in Fig. F3.26. The chromophore is completely protected from the bulk solvent and is centrally located in the molecule. Mutation of Ser^{65} to Thr (S65T) simplifies the excitation spectrum to a single peak at 488 nm.

GFP has a number of amazing properties that enable it to be used for in vivo imaging. First, GFP can be expressed in a variety of cells, where it becomes spontaneously fluorescent without the aid of a cofactor. Second, GFP can be fused to a host protein to create a fusion protein that usually retains both the fluorescence of the GFP and the biochemical function of the original host. Third, fusion proteins can be targeted to specific organelles, such as the nucleolus or endoplasmic reticulum, by adding an appropriate signaling peptide. Finally, and most importantly, mutagenesis of GFP produces many mutants with varying spectral properties that can be used as donors and acceptors of FRET (Comment F3.8).

GFP-based constructs do, however, suffer from limited sensitivity, often precluding single-cell analysis. Moreover, they are relatively large, thereby limiting spatial resolution. They should also be used with caution as it has been shown that GFP can undergo color changes upon irradiation due to photochemical changes that are independent of FRET. In addition, GFP requires hours to assemble in its final fluorescent form and is thus limited in its ability to monitor kinetic phenomena before final assembly. Nevertheless, GFP is of immense importance, and a new class of red fluorescent proteins indicate that it still has great potential.

F3.6.1 GFP as a Conformational Sensor

The introduction of GFP to FRET-based imaging microscopy enabled it to be used as a sensitive probe of protein–protein interactions and protein conformational changes in vivo. One pair of fluorescent proteins originally used for FRET is BFP as the donor and GFP as the acceptor. These proteins form FRET pairs with each other or with conventional organic dyes, and can be attached to many proteins of interest, usually at the N- or C-terminus.

Figure F3.27 shows the scheme of an experiment in which the motor protein myosin was labeled at its N- and C-termini with BFP and GFP. When BFP was excited by light at 360 nm, intramolecular FRET took place and green light was emitted at 510 nm. The orientation and distance between these probes varied as myosin underwent conformational changes associated with ATP

binding and hydrolysis. This result favors what is known as the lever-arm model of muscle contraction (Section F2.6.1).

F3.6.2 GFP as a Cellular Reporter

Genetically encoded dyes, such as GFP, revolutionized the ability to image *in vitro* and especially in living cells. In practice, the combination of two mutants, cyan-fluorescent protein (CFP) and yellow-fluorescent protein (YFP) is the most suitable for double-color imaging in living cells (Fig. F3.28).

Another approach is to use two GFPs of different colors, chosen to permit FRET from the shorter to the longer wavelength GFP. Figure F3.29 demonstrates the principal scheme in which GFP as the conformational sensor detects acceptor-sensitized emission alone.

Figure F3.30 demonstrates the imaging by FRET of an induced protein–protein interaction in individual live cells. Cyan-GFP-labeled calmodulin (CFP-CaM) and yellow-GFP-labeled calmodulin binding peptide (YFP-M13) were coexpressed. The left-hand panel shows two cells before stimulation, while the right-hand panel shows the same cells after elevation of cytosolic Ca^{2+} by 0.1 mM histamine. Low Ca^{2+} levels (left) lead to little FRET and mostly blue emission (pseudo-color green); high

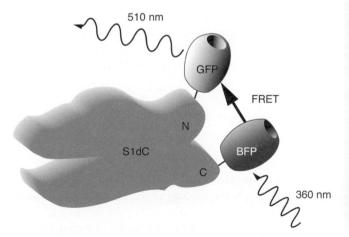

Fig. F3.27 To detect the working stroke in the myosin head in operation, fusion proteins were constructed by connecting *Aequorea victoria* GFP and BFP to the amino and carboxyl termini of the motor domain of myosin II, designated S1dC. GPF–BFP pair-based FRET was used to measure conformational changes in myosin upon ATP binding and hydrolysis. (After Suzuki *et al.*, 1998.)

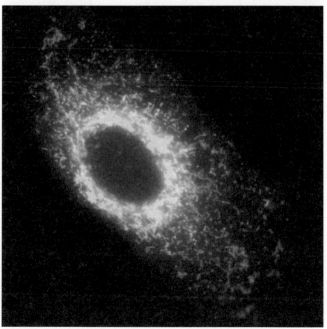

Fig. F3.28 Dual-channel fluorescence micrograph of the cell expressing two GFP fusion proteins that localize to different compartments. A fusion of the peroxisomal membrane protein 24 to CFP (pmp24-CFP) localizes to perixomes (blue) and a fusion of LCAT-like lysophospholipase to YFP (LLPL-YEP) localizes to mitochondria (yellow). (Adapted from Bastiaens and Pepperkok, 2000.)

399

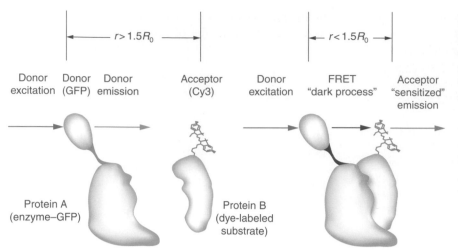

Fig. F3.29 Principal scheme in which GFP as the conformational sensor detects acceptor-sensitized emission alone. The donor GFP-labeled enzyme binds to the dye-labeled substrate. After binding, FRET between the donor and the acceptor is detected. Note that FRET is very sensitive to the distance between two phluorophores, *r*. Acceptor-sensitized emission occurs only if distance $r < R_0$. (After Bastiaens and Pepperkok, 2000.)

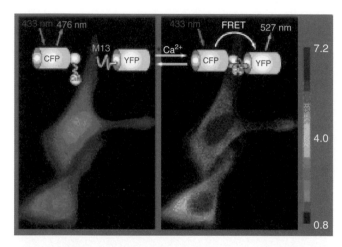

Fig. F3.30 FRET constructs for measuring intracellular calcium. CFP-labeled calmodulin (CFP-CaM) and YFP-labeled calmodulin binding peptide (YFP-M13) were coexpressed in HeLa cells. Pseudo-color hues from blue to magenta indicate increasing ratios of yellow to cyan emissions, resulting either from increased FRET (see overlaid schematic) or excess of YFP over CFP. The presence of Ca^{2+} alters the distance (or orientation) between donor and acceptor, causing the relative emission intensities at the donor and acceptor wavelengths to change. (After Tcien and Miyawaki, 1998.)

Ca^{2+} levels (right) lead to binding and FRET emission of YFP (pseudo-color red).

F3.7 FLUORESCENCE LIFETIME IMAGING MICROSCOPY (FLIM): SEEING THE MACHINERY OF LIVE CELLS

Fluorescence lifetime imaging (FLI) refers to the technique of recording a fluorescence image whereby the fluorescence lifetimes of the fluorophores in a sample are resolved at every location of the image. This can result in either the actual determination of the fluorescence lifetimes (or more often the apparent fluorescence lifetime), or the determination of a spectroscopic parameter that is possible only when the fluorescent signal is lifetime-resolved in the measurement. The spectroscopic information related to the lifetime of the fluorescence decay can then be displayed in an image format, in which every pixel of the image contains the lifetime-resolved information. Fluorescence lifetime imaging measurements are analogous to normal intensity fluorescence imaging measurements and are acquired on the same samples, except that the information related to the fluorescence lifetime is directly recorded in addition to the normal measurement of the fluorescence intensity.

Live-cell FRET measurements are more feasible using FLI because lifetimes are independent of probe concentration and light path length. By imaging donor fluorescence lifetime it is possible to calculate FRET efficiencies at each pixel E_T^i of an image:

$$E_T^i = 1 - \left(\frac{\tau_D^i A}{\tau_D^i} \right) \tag{F3.3}$$

where $\tau_D^i A$ is the lifetime of the donor in the presence of the acceptor and τ_D^i is the lifetime of the donor in the absence of acceptor.

Figure F3.31 shows the quantitative determination of protein reaction states by global analysis of FLIM data. Here the relative populations of phosphorylated ErbB1-GFP receptor were determined before (upper row) and after 60 s of epidermal growth factor (EFG) stimulation (lower row). The left-hand column shows the images of ErbB1-GFP, the middle column the images of the antiphosphotyrosine antibody (Cy3-PY72), and the right-hand

COMMENT F3.9 COUNTING SINGLE PHOTOACTIVATABLE FLUORESCENT MOLECULES BY PALM (LEE ET AL., 2012).

The paper presents an application of PALM for counting proteins within a diffraction-limited area. The intrinsic blinking of photoactivatable fluorescent proteins mEos2 and Dendra2 leads to an *overcounting* error, which constitutes a major obstacle for their use as molecular counting tags. The authors introduce a kinetic model to describe blinking to show that Dendra2 photobleaches three times faster and blinks seven times less than mEos2, making Dendra2 a better tag for molecular counting using PALM. The simultaneous activation of multiple molecules is another source of error, but it leads to molecular *undercounting* instead. A photoactivation scheme is introduced that maximally separates the activation of different molecules, thus helping to overcome undercounting. A method is also presented that quantifies the total counting error and minimizes it by balancing *over-* and *undercounting*. The method established that Dendra2 is better for counting purposes than mEos2, allowing counting *in vitro* of up to 200 molecules in a diffraction-limited spot. Finally, Lee *et al.* demonstrate that this counting method can be applied to protein quantification *in vivo* by counting the bacterial flagellar motor protein FliM fused to Dendra2 (Fig. F3.9.1).

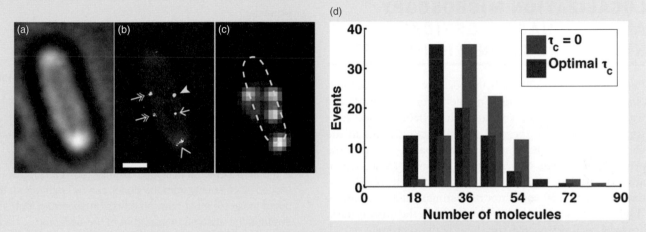

Fig. F3.9.1 Counting the number of FliM-Dendra2 molecules in the flagellar motor expressed in ΔfliM strain. (a) Bright-field and (b) FliM-Dendra2 PALM overlay image show that the motor proteins mainly localize as clusters at the cell membrane. Clusters located at the lateral periphery of the cells (arrow head) were selected for the molecular counting, with the exclusion of clusters that were elongated (chevron), located at the cell pole (chevron), or off the periphery (arrow), and surrounded by dispersed molecules (two-headed arrow). (Scale bar, 500 nm.) (c) Diffraction-limited FliM-Dendra2 image. Dashed line, contour of the cell. (d) Iterative optimal-τ_c counting method estimates 33 c1 (mean (SEM) FliM-Dendra2 molecules per cluster, which is similar to the previously quantified number of 34 molecules per flagellar motor. In contrast, the blinking non-corrected method (τ_c 1/4 0) estimates 40) 1. A total of 89 clusters were analyzed.

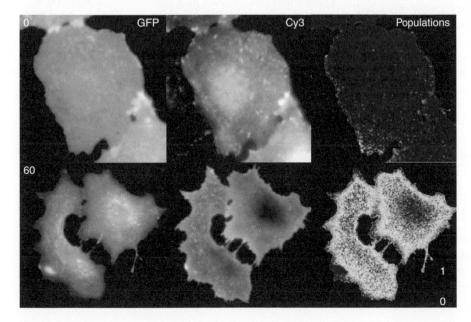

Fig. F3.31 Quantitative determination of protein reaction states by global analysis of FLIM data (see text). (After Bastiaens and Pepperkok, 2000.)

column the calculated images of the populations of phosphorylated ErbB1-GFP receptor.

By using fluorescence digital imaging microscopy, one can visualize the location of GFPs within a living cell and follow the time course of the changes in FRET corresponding to cellular events at a millisecond time resolution. The observation of such dynamic molecular events *in vivo* provides vital insights into the action of biological molecules.

F3.8 PHOTO-ACTIVATED LOCALIZATION MICROSCOPY AND STOCHASTIC OPTICAL RECONSTRUCTION MICROSCOPY

Photo-activated localization microscopy (PALM) and stochastic optical reconstruction microscopy (STORM) are wide-field microscopy methods that yield images with superresolution (i.e., beyond the diffraction limit) in the nanometer range. Both methods are based on the same principle. Consider two points in a dense cloud, which are closer to each other than the wavelength of light so that normally they could not be distinguished as separate. "Lighting" up each one individually while the other is dark, however, will permit placing them with precision.

When the sample is a volume or cloud of many points, the final resolution is not limited by diffraction but by the precision with which each point is localized. In the superresolution image, each point (e.g., molecule) in the image plane will be represented as a two-dimensional Gaussian with amplitude proportional to the number of photons collected, and width depending on precision of localization.

Although the basic principle in PALM and STORM is the same, there are differences between the two methods. PALM is usually performed using endogenously expressed fluorescent proteins, such as GFP and its mutants (see previous section). The fluorophores are photoactivated to obtain ON (active) and OFF (inactive) states. In contrast to PALM, a sample for STORM is labeled with antibodies (immunolabeling) tagged with organic fluorophore dyes, which are brighter than fluorescent proteins. Stochastic "blinking" of the dyes permits separation of neighboring molecules.

An application example is described in Comment F3.9.

F3.9 CHECKLIST OF KEY IDEAS

- Two-photon excitation arises from the simultaneous absorption of two photons in a single quantized event; in this case "simultaneous" means within about 10^{-18} s.

- In TIRFM, the region near an interface between two media with different refractive indices is illuminated by an evanescent wave that penetrates only a short distance into the medium (about 250–300 nm).
- The 4Pi-confocal microscope uses two opposing objective lenses with which the same object point is coherently illuminated and observed. The 4Pi-confocal microscope improves the axial resolution by a factor of 3–4 over that of a standard confocal microscope.
- In fluorescence microscopy, a resolution beyond the common Rayleigh limit can be achieved by different approaches, including methods based on a variation of confocal microscopy and those based on non-linear schemes.
- STED microscopy is based on the different wavelength pump probe scheme that excites the fluorophores with one beam and quenches them with another; it allows the diffraction limit to be broken and shrinks the PSF by a factor of 6 along the optical axis and by a factor of 2 in the radial direction.
- In SWFM, two coherent plane wave beams from a laser cross in the specimen volume, where they interfere.
- In FRET, a donor fluorophore is excited by incident light, and if an acceptor is in close proximity (20–100 Å), the excited-state energy from the donor can be transferred; the donor molecule must have an emission spectrum that overlaps the absorption spectrum of the acceptor molecule.
- FRET is useful as a physical phenomenon to measure molecule distances as the efficiency E at which Forster-type energy transfer occurs is steeply dependent on the distance R as R^{-6}.
- FRET is especially suited to the study of protein–nucleic acid complex structures as knowledge of the exact positions of some dyes with respect to specific DNA helix ends enables us to translate the measured distance data into precise structural information.
- Ensemble FRET is used as a very effective *spectroscopic ruler* for measurement of static distances in the 20–100 Å range; however, dynamic events such as the relative motion between donor and acceptor molecules cannot be detected by FRET.
- Single-pair FRET measurements, in contrast, allow one to look at dynamic distance changes on a millisecond timescale because only one donor–acceptor system is observed at a time; hence, time-dependent conformations can be monitored.
- FLI refers to the technique of recording a fluorescence image whereby the fluorescence lifetimes of the fluorophores in a sample are resolved at every location of the image.
- GFPs (27 kDa) enable FRET measurements to be made in living cells; these proteins can be cloned in fusion with proteins of interest and expressed in cell lines.
- PALM and STORM are wide-field microscopy methods that yield images with superresolution (i.e., beyond the diffraction limit) in the nanometer range.

Suggestions for Further Reading

Historical Review and Introduction to Biological Problems

Lakovicz, J. R. (ed.) (1999). *Principles of Fluorescence Spectroscopy*. 2nd edn. New York: Academic/Plenum.

Fluorescence Microscopy Inside the Classical Limit

Herman, B., and Jacobson, K., (eds.) (1990). *Optical Microscopy for Biology*. New York: Wiley-Liss.

Tamm, L. K. (1993). Total internal reflectance fluorescence microscopy. In *Optical Microscopy: Emerging Methods and Applications*, eds., B. Herman and J. J. Lemasters. New York: Academic Press.

Piston, D. W. (1999). Imaging living cells and tissues by two-photon excitation microscopy. *Trends Cell Biol.*, **9**, 66–69.

Fluorescence Microscopy Outside the Classical Limits

Schrader, M., Bahlmann, K., Giese, G., and Hell, S. W. (1998). 4Pi-confocal imaging in fixed biological specimens. *Biophys. J.*, **75**, 1659–1668.

Nagorni, M., and Hell, S. W. (1998). 4Pi-Confocal microscopy provides three-dimensional images of the microtubule network with 100-to-150 nm resolution. *J. Struct. Biol.*, **123**, 236–247.

Gustafsson, M. G. L. (1999). Extended resolution fluorescence microscopy. *Curr. Opin. Struct. Biol.*, **9**, 627–634.

Weiss, S. (2000). Shattering the diffraction limit of light: A revolution in fluorescence microscopy? *Proc. Nat. Acad. Sci. USA.*, **97**, 8747–8749.

Cragg, G. E., and So, P. T. C. (2000). Lateral resolution enhancement with standing evanescent waves. *Opt. Lett.*, **25**, 46–48.

Klar, T. A., Jacobs, S., Dyba, M., Egner, A., and Hell, S. W. (2000). Fluorescence microscopy with diffraction resolution barrier broken by stimulated emission. *Proc. Natl. Acad. Sci. USA*, **97**, 8206–8210.

FRET

Stryer, L. (1978). Fluorescence energy transfer as a spectroscopic ruler. *Annu. Rev. Biochem.*, **47**, 819–846.

Selvin, P. R. (2000). The renaissance of fluorescence resonance energy transfer. *Nature Str. Biol.*, **7**, 730–734.

Hillisch, A., Lorenz, M., and Diekmann, S. (2001). Recent advances in FRET: Distance determination in protein–DNA complexes. *Curr. Opin. Struct. Biol.*, **11**, 201–207.

GFP

Tcien, R. Y. (1998). The green fluorescent protein. *Annu. Rev. Biochem.*, **67**, 509–544.

Ellenberg, J., Lippincot-Schwartz, J., and Presly, J. F. (1999). Dual-colour imaging with GFP variants. *Trends Cell Biol.*, **9**, 52–60.

FLI Microscopy

Tcien, R. Y., and Miyawaki, A. (1998). Seeing the machinery of live cells. *Science*, **280**, 1954–1955.

Bastiaens, P. I. H., and Squire, A. (1999). Fluorescence lifetime imaging microscopy: Spatial resolution of biochemical processes in the cell. *Trends Cell Biol.*, **9**, 48–52.

Piston, D. W. (1999). Imaging living cells and tissues by two-photon excitation microscopy. *Trends Cell Biol.*, **9**, 66–69.

Bastiaens, P. I. H., and Pepperkok, R. (2000). Observing proteins in their natural habitat: The living cell. *TIBS*, **25**, 631–636.

Superresolution Microscopy

Clery, D. (2014) Nobel Prizes: Light loophole wins laurels. *Science*, **346**, 290–291.

X-RAY AND NEUTRON DIFFRACTION

G

THE MACROMOLECULE AS A RADIATION SCATTERING PARTICLE

G1

G1.1 HISTORICAL REVIEW AND INTRODUCTION TO BIOLOGICAL APPLICATIONS

The wave nature of light, on which diffraction phenomena are based, was first suggested by Huygens more than 300 years ago. About 100 years later, Haüy wrote an essay on the regularity of crystal forms that is considered to be the beginning of crystallography.

1690

In his *Treatise on Light*, **C. Huygens** wrote that light "spreads by spherical waves, like the movement of Sound," and explained reflection and refraction by wave constructions.

1784

R.-J. Haüy, a mineralogist, published his theory on crystal structure following observations that calcite cleaved along straight planes meeting at constant angles.

1895

J. J. Thomson discovered electrons during an investigation of cathode rays. He initially called them corpuscles.

1895

W. C. Röntgen discovered X-rays. While experimenting with electric current flow in a partially evacuated glass tube, he noted that radiation was emitted that affected photographic plates and caused a fluorescent substance across the room to emit light.

1912

P. P. Ewald's doctoral thesis on the passage of light waves through a crystal of scattering atoms led **M. von Laue** to ask what would happen if the wavelength of the light were similar to the atomic spacing, and this led to the first observations of X-ray crystal diffraction by **W. Friedrich, P. Knipping**, and **von Laue**. Because of their short wavelengths, X-rays provide a "ruler" with which to measure distances between atoms.

1912–1915

W. H. Bragg and **W. L. Bragg** interpreted diffraction in terms of reflection from crystal planes. They solved the crystal structures of NaCl and KCl and introduced Fourier analysis of the X-ray measurements.

1917

P. P. Ewald introduced the "reciprocal lattice" construction, a graphical method of expressing the geometrical conditions for crystal diffraction.

1924

W. L. Bragg and collaborators developed the use of absolute intensities in crystal analysis, leading to the solution of structures more complex than the monovalent salts.

1924

L.-V. de Broglie proposed the relation between the wave and the particle nature of matter, thus paving the way for the interpretation of scattering of particle beams (such as electrons, and later neutrons) in terms of wave diffraction.

1932

J. Chadwick discovered the neutron, one of the main constituents of atomic nuclei (neutrons and protons have about the same mass; together they make up 99.9% of an atom's mass). Neutrons are emitted spontaneously by certain radioactive nuclei and various elements undergo fission when bombarded by neutrons, emitting additional neutrons. Because they are electrically neutral, neutron beams penetrate deeply into matter. Its properties made the neutron a particularly useful probe for investigating structure and dynamics at the molecular level.

1936

W. M. Elsasser, H. v. Halban and **P. Preiswerk,** and **D. P. Mitchell** and **P. N. Powers** demonstrated diffraction of neutrons from a radium–beryllium source, and thus their wave nature according to **de Broglie**'s relation.

Late 1930s

A. Guinier developed his theory to show that X-ray scattering at small angles, around the direct beam direction, by non-crystalline solids and solutions contained information on particle size and shape.

1945 and the following years

Neutron beams from pile reactors became available for diffraction experiments and crystallography. **C. Shull** performed the first neutron diffraction experiments to investigate material structures. Diffractometers were built at Argonne National Laboratory (USA), followed by Oak Ridge National Laboratory (USA), Chalk River (Canada), and Harwell (UK). In the early **1950s**, **B. N. Brockhouse** invented the triple-axis spectrometer and measured vibrations in solids by neutron scattering. **Shull** and **Brockhouse** were awarded the Nobel Prize for Physics in 1994.

1953

F. Crick, **J. Watson**, **R. Franklin**, and **M. Wilkins** published the double-helix structure of DNA calculated from X-ray fiber diffraction and chemical model building.

Late 1960s

Research groups led by **J. C. Kendrew** and **M. Perutz** published the first ångström resolution structures of proteins (myoglobin and hemoglobin, respectively) from X-ray crystallography.

1960s and 1970s

Medium- and high-flux reactors, and later spallation sources, were built with neutron beams dedicated to the study of matter. Biophysical studies using neutrons provided information on the structure and dynamics of biological membranes and macromolecules that cannot be obtained by other methods.

1980s

The crystallization and first X-ray crystal structures of membrane proteins were obtained by using detergents. Because they are soluble only in complex solvents, the biochemical and structural study of membrane proteins lagged far behind that of water-soluble proteins.

1980s to present

Beam lines at synchrotron facilities that provide very brilliant X-ray sources for macromolecular crystallography and diffraction studies became available. Efficient protein modification, crystallization, data collection, and analysis approaches were developed for macromolecular crystallography. Extremely fast data-collection times made it possible to study kinetic intermediates in myoglobin using time-resolved crystallography. High-resolution ribosome structures were obtained from crystallography and electron microscopy.

X-rays provided the foundation on which structural biology has been built and is developing. In the **1920s**, X-ray diffraction was already being observed from complex organic crystals (long-chain carbohydrates, hexamethyl benzene, anthracene, urea), and the first polymer structural studies were performed on rubber, hair, and wool

fibers. Protein and viruses were shown to form crystals and diffraction diagrams of nerve fibers under different humidity conditions were published in the **1930s**, which led to a model for biological membrane organization.

The publication of the double-helix structure in **1953** heralded the birth of molecular biology. It was followed by the structures of myoglobin, hemoglobin, and lysozyme in the **1960s**, and of transfer RNA in the **1970s**, providing a wealth of information about structure–function relationships. Neutron radiation has the special property that it can distinguish between hydrogen and its isotope deuterium. Neutron diffraction studies, using deuterium labeling, of biological membranes, fibers, macromolecules, and their complexes in crystals and by small-angle scattering in solution contributed strongly to the understanding of biological structure.

In the last decades, the development of novel crystallographic methods plus faster computers that could implement them and the availability of intense synchrotron sources have brought about a revolution in macromolecular crystallography by greatly increasing the rate at which structures could be solved. The distinction between fundamental and applied science in current X-ray and neutron diffraction experiments on biological systems is increasingly blurred because of the use that can be made of the results in medicine, biotechnology, or food science, for example.

G1.2 RADIATION AND MATTER

In a diffraction experiment, waves of radiation scattered by different objects interfere to give rise to an observable pattern, from which the relative arrangement (or structure) of the objects can be deduced. The interference pattern arises when the wavelength of the radiation is similar to or smaller than the distances separating the objects. Radio waves with wavelengths of several meters, for example, are diffracted by buildings in a town. Atomic bond lengths are close to 1 Å unit (10^{-10} m). In practice, three types of radiation are used in diffraction experiments: X-rays of wavelength about 1 Å, electrons of wavelength about 0.01 Å, and neutrons of wavelength about 0.5–10 Å. Electron diffraction is treated with electron microscopy in Chapter H2. X-rays and neutrons are treated together in the current chapter.

A wavelength similar to or smaller than the structural scale examined is not the only criterion to define useful radiation. It should also present appropriate properties of interaction with matter; it should not be absorbed too strongly and it should be scattered with reasonable efficiency; and, of course, radiation sources of appropriate intensity should be available. X-rays and neutrons broadly satisfy these criteria. There are significant differences in the details of their interaction with matter, however, that make them strongly complementary for diffraction studies of biological molecular structure.

G1.2.1 X-ray and Neutron Scattering

We start with the neutron case, which is simpler to describe because the scattering centers can be considered as points.

Neutrons

Neutrons are scattered by atomic nuclei in a complex process. Because the neutron wavelength in diffraction experiments ($\lambda \sim 1$ Å $= 10^{-10}$ m) is so much larger than the nuclear dimensions ($\sim 10^{-15}$ m), the nuclei act as point scatterers. A point scatters a wave isotropically, i.e., equally in all directions (Comment G1.1).

Heavier elements do not dominate neutron scattering and the scattering power of different isotopes of the same element can be very different. The case of hydrogen (^{1}H) and deuterium (^{2}H or D) is of particular interest in structural biology. The neutron scattering powers of H and D are sufficiently different from each other to allow very useful labeling experiments to observe hydrogen atoms and water molecules in biological samples.

Neutrons and protons (whether they are in atomic nuclei or in beams) are quantum mechanical particles of spin one-half. The neutron scattering power of a nucleus thus also depends on the relative orientation of nuclear and neutron beam spins. Neutron and nuclear spins can be polarized (oriented) by a magnetic field. However, unpolarized beams and samples are used in most diffraction experiments, so that the same type of nucleus in a sample scatters neutrons with different power according to the spin–spin orientations. The distribution of neutron–nucleus spin–spin orientations is random in the sample, resulting in a strong *incoherent* contribution to the scattering (see Comment G1.8). The effect is largest for the proton (the hydrogen nucleus); it results in hydrogen-containing samples giving a high background signal in neutron

diffraction experiments, which does not contain structural information. The analysis of incoherent neutron scattering, however, contains information on sample molecular dynamics (see Chapter I2).

X-rays

X-rays are scattered by electrons. Atomic electron clouds are on the same length scale as the radiation wavelength (~ 1 Å) so that, in the scattering process, atoms appear as extended objects and cannot be considered as points. The scattering decreases with increasing angle. The shape of the atom "seen" by the X-ray beam is taken into account by describing the angular distribution of its scattering power in terms of a "form factor" (see Section G1.3).

The X-ray scattering power of an atom increases simply with its number of electrons. There is no isotope effect since isotopes of the same element have the same number of electrons. Heavy atoms dominate the diffraction pattern, allowing them to be used as labels in crystallography. Hydrogen atoms, with only one electron each, can only be "seen" by X-rays when they are organized with a very high degree of crystallographic order and when the high-intensity beams now available with synchrotron radiation are used.

G1.2.2 Absorption

Absorption can severely limit the penetration of radiation in a diffraction experiment to the surface layers of a sample.

X-ray absorption is due to photons exciting electrons to higher energy levels. It is, therefore, energy-dependent (and wavelength-dependent); absorption increases with increasing wavelength. In practice, X-ray absorption leads to non-negligible radiation damage in the sample that should be corrected for in diffraction experiments. Absorption is severe for wavelengths above 2.5 Å, where even air in the beam path absorbs significantly.

Neutron absorption is due to recombination with the nucleus (recall the neutron is itself a nuclear particle), resulting, for example, in nuclear fission. At the wavelengths used for diffraction studies, however, neutron absorption is very low for most nuclei, even for wavelengths of 10 Å or larger. Fortunately, isotopes of certain nuclei, such as cadmium, boron, and lithium, are notable exceptions so that they can be used for shielding or detection.

G1.2.3 Energy Momentum and Wavelength

Energy–wavelength relations for neutrons and the electromagnetic spectrum are given in Table G1.1. Neutron and X-ray photon properties are given in Comment G1.2. X-rays are electromagnetic radiation of much higher energy than visible light. Neutrons are particles of mass similar to that of a proton. Their associated wavelength is inversely proportional to their momentum by de Broglie's

---

COMMENT G1.1 SCATTERING BY A POINT ATOM

An object and the pattern of waves it scatters are related by Fourier transformation (see Chapter A3). A point scatters a wave with equal amplitude in all directions (Fig. G1.1.1a).

In other words, in the Fourier transform of a Dirac delta function (the point) is a constant amplitude independent of scattering direction (Fig. G1.1.1b).

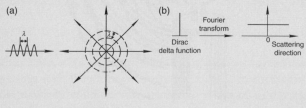

Fig. G1.1.1

TABLE G1.1 WAVELENGTHS AND ENERGIES

Wavelength λ (order of magnitude)	Electromagnetic radiation energy	Neutron λ, energy (temperature)
10 fm (0.1 Å)	Hard X-rays 124 keV	Hot neutrons 0.7 Å, 172 meV (2100 K)
100 fm (1 Å)	X-rays 12.4 keV	Thermal neutrons 1.8 Å, 24.5 meV (300 K)
1 nm (10 Å)	Soft X-rays 1.24 keV	Cold neutrons 7 Å, 1.67 meV (19 K)
10 nm (100 Å)	UV 124 eV	
100 nm (1000 Å)	Visible 12.4 eV	

relation. The velocity of neutrons of wavelengths close to 1 Å is about 4000 m s^{-1}, close to the speed of a bullet leaving the barrel of a gun (and very far from the speed of light!). One-ångström neutrons are therefore not relativistic particles; they behave like billiard balls and their momentum is simply mv, and kinetic energy is simply $1/2\ mv^2$. Neutron wavelength is, therefore, inversely proportional to the square root of the energy. The temperature of a neutron "gas" of a given wavelength is also given in Table G1.1.

G1.3 SCATTERING BY A SINGLE ATOM (THE GEOMETRIC VIEW)

Consider an atom as being constituted of a set of points that scatter radiation. When a plane wave of monochromatic radiation is incident upon the atom, each point acts as a source of spherical waves of the same wavelength, similarly to Huygens' historic construction (Comment G1.3).

Neutrons are scattered by nuclei. As we have seen above, it is a very good approximation to interpret neutron scattering by an atom in terms of scattering by a single point, and to describe it in terms of a single parameter called the *scattering amplitude*.

COMMENT G1.2 NEUTRON PROPERTIES AND CONVERSION FACTORS

As particles:

mass, $m = 1.674928 \times 10^{-27}$ kg
momentum = $\mathbf{p} = m\mathbf{v}$
energy = $E = 1/2\ mv^2$

As waves:

de Broglie associated wavelength: $\lambda = h/mv$
(h is Planck's constant: 6.6327×10^{-37} kg m^2 s^{-1})
momentum = $\mathbf{p} = h/\lambda$ (in the direction of propagation)
wave vector = $2\pi/\lambda$ (in the direction of propagation) = $2\pi\mathbf{p}/h$
energy, $E = h^2/2m\lambda^2$

$$E[\text{meV}] = 81.81/\left(\lambda^2\left[\text{Å}^2\right]\right)$$
$$= 0.0861737T[\text{K}]$$

X-ray properties and conversion factors

As particles:

Photons of energy $h\nu$ (where ν is frequency) and momentum $h\nu/c$ (in the direction of propagation, where c is the speed of light 2.99792×10^8 m s^{-1})

As waves:

Wavelength λ, energy hc/λ, momentum h/λ

$$E[\text{keV}] = 12.4/\lambda[\text{Å}]$$
$$= 4.13 \times 10^{-18}\lambda[\text{s}^{-1}]$$

COMMENT G1.3 HUYGENS' CANDLE

In *Treatise on Light*, Huygens wrote "Thus in a flame of a candle, having distinguished the points A, B, C, concentric circles described about each of these points represent the waves which come from them. And one must imagine, the same about every point of the surface and of the part within the flame" (Huygens, 1692).

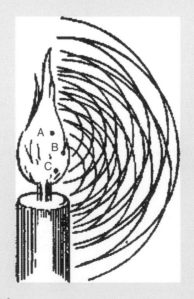

Fig. G1.3.1

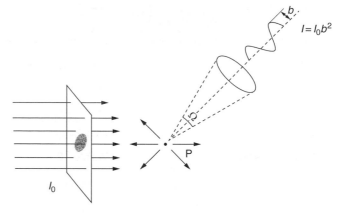

COMMENT G1.4 SCATTERING BY A SINGLE ELECTRON, THE THOMSON FACTOR, AND POLARIZATION

J. J. Thomson first derived the expression for X-ray scattering by a single electron. The scattering amplitude of an electron is called the Thomson factor:

$$f_{el} = e^2/mc^2 = 2.8 \times 10^{-15} m$$

where e and m are respectively the charge and mass of the electron and c is the speed of light; f_{el} is equal to the classical radius of an electron.

The oscillating fields of the incident electromagnetic wave set the electron to oscillate, giving off radiation of intensity I_e, at a distance r from the electron,

$$I_e = \left(I_0 \cdot f_{el}^2 \cdot 1/r^2\right) \cdot \left[1 + \left(\cos^2 2\theta\right)/2\right]$$

I_0 is the incident flux, $1/r^2$ is a measure of solid angle, and 2θ is the scattering angle.

The cosine term in the equation is the polarization factor. It takes into account the fact that the scattering process introduces partial polarization of the scattered beam even when the incident beam is unpolarized. Note that the polarization factor is close to 1 for small scattering angles.

Fig. G1.1 Scattering by a point atom. The incident flux I_0 is defined as the number of radiation particles passing through a 1 m² window per second. The point P scatters isotropically. Its scattering length, b, is defined as the amplitude of the wave scattered in unit solid angle Ω for unit incident flux. The intensity of the wave scattered in Ω per incident flux I_0 is I_0b^2. The shaded area in the incident flux window is the area of the effective cross-section (see text).

X-rays are scattered by the atomic electrons. The scattering process can be understood as the oscillating fields of the electromagnetic radiation creating oscillating dipoles in the electron cloud, which, in turn, emit radiation (see Chapter E1; Comment G1.4). In a geometrical interpretation, the atomic electron cloud behaves as an extended object in space; spherical waves scattered from each point in that object interfere to result in a scattered intensity distribution that depends on scattering angle (the *form factor*).

G1.3.1 Point Scattering: Scattering Length

First, we consider scattering by a point. In the scattering event, the point acts as the source of an isotropic spherical wave. The scattering amplitude (or length), b, of the point is defined as the amplitude of the wave observed per unit *incident flux*, in unit solid angle (in any direction, since the wave is spherical). We recall that the intensity of a wave is equal to the square of its amplitude. The scattered intensity is thus given by

$$I = I_0 b^2 \qquad (G1.1)$$

where I is the scattered intensity in unit solid angle (neutrons s⁻¹ or photons s⁻¹) and I_0 is the incident flux, in neutrons s⁻¹ m⁻² or photons s⁻¹ m⁻² (Fig. G1.1).

The total intensity scattered by the point is given by the sum over all directions (4π solid angle):

$$I_{total} = I_0 4\pi b^2 \qquad (G1.2)$$

The *scattering cross-section*, σ, of the point is then defined as $\sigma = 4\pi b^2$. It has units of [m²]. It can be understood as an effective projected area perpendicular to the incident beam (Fig. G1.1); each particle that hits the area of cross-section is scattered. The ratio of the area of cross-section to the total area in the definition of incident flux (1 m² in this case) can also be understood as the scattering *probability* associated with the point.

In general, the value of the scattering amplitude, b, is a complex number. We recall (Chapter A3) that a phase shift between two waves can be expressed in terms of a complex amplitude. A phase shift of π corresponds to a change in sign of the amplitude. By convention, a positive value of b denotes a phase shift of π between the incident and scattered waves, which is the case for X-ray and neutron scattering by most atoms (Fig. G1.2). *Anomalous scattering* refers to a phase shift other than π, resulting in a complex contribution to the scattering amplitude, usually written $f' + if''$ for X-rays (see Chapter G3).

X-ray Scattering Amplitudes

Each electron scatters X-rays with the same amplitude, known as the Thomson factor, f_{el}, for unit incident flux and in unit solid angle (see Comment G1.4). Even for a point-like electron, scattering is not isotropic because of the angle dependence of a polarization term (see Comment G1.4). The polarization term, however, is equal to 1 in the forward direction (zero scattering angle), and the waves from all the electrons in the atom interfere constructively, so that the amplitude of the wave scattered at small angles by an atom of n electrons is simply nf_{el}.

Atomic X-ray scattering amplitudes and form factors are usually tabulated simply in terms of the effective number

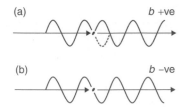

Fig. G1.2 Incident (black) and scattered (red) waves showing: **(a)** a phase difference of π, equivalent to reflection of an incident wave off a non-absorbing surface, which is the case for most atoms in X-ray and neutron scattering; **(b)** no phase difference (negative amplitude) which is the case for neutron scattering by ^{1}H. Due to the spatial extent of the electron cloud, the amplitude of X-rays scattered by the atom decreases with scattering angle, in a similar fashion to the intensity scattered by an assembly of atoms (see Section G1.5). The dependence of amplitude on scattering angle is called the *form factor* of the atom.

COMMENT G1.5 X-RAY AND NEUTRON SCATTERING AMPLITUDES

X-ray and neutron form factors, scattering amplitudes, and absorption cross-sections are tabulated on the web. *X-ray Data Booklet.* (Berkeley, CA: Lawrence Berkeley National Laboratory, University of California (up-dated versions can be found at http://xdb.lbl.gov)). *Neutron Data Booklet* (Grenoble: Institut Laue Langevin (www.ill.fr)).

COMMENT G1.6 THE BARN UNIT

The name "barn" was chosen for this very small unit of area with humorous irony since the side of a barn is usually representative of a very large area. Recall the saying for someone who does not know how to shoot that "he could not hit the side of a barn."

COMMENT G1.7 SCATTERING BY A SINGLE ATOM

What is the length of time we would have to wait to have a good probability of observing one scattered neutron from a single carbon atom?

Answer: From Eq. (G1.2):

I_{total} [neutrons s^{-1}] = $I_0 4\pi b^2$

I_0 is 10^{11} neutrons m^{-2} s^{-1}, and b for carbon is 6.65 × 10^{-15} m

I_{total} is calculated to be 5.56 × 10^{-17} neutrons s^{-1}

We should have to wait for $1/(5.56 \times 10^{-17})$ s, or almost 600 million years, to have a good probability of seeing one neutron.

How many atoms should there be in a sample to scatter 100 neutrons s^{-1}? Answer: ~10^{19} atoms or 10^{16} particles with 1000 atoms in each (e.g., a small protein).

In fact, these numbers can be achieved quite easily. For example, 10^{19} atoms of carbon represent a mass of only 0.2 mg.

---

of electrons (Comment G1.5). They are represented by real numbers, which are independent of wavelength provided the incident wavelength (energy) is far from the absorption edge of the scattering atom (see Chapter G3). Close to the absorption edge the X-ray scattering amplitude is represented by a complex number to indicate the phase shift of the scattered wave with respect to the incident wave. The effect, which is usefully exploited in protein crystallography, is called *anomalous scattering*.

Neutron Scattering Amplitude

Neutron scattering amplitudes do not increase in proportion to the size or mass of the nucleus; they are all of a similar scale between 1 and 10 fm (in scattering methods an amplitude of 1 fm has been defined as 1 Fermi unit) (Comment G1.5). This is an advantage because light elements are as "visible" as heavy ones in neutron crystallography. The isotope effect, already discussed above, is also particularly useful because of the labeling possibilities that it opens up. The hydrogen nucleus has a negative neutron scattering b value (–3.74 fm), indicating a phase shift of π for neutrons scattered by the proton relative to those scattered by, for example, the oxygen nucleus ($b = 5.85$ fm) (Fig. G1.2). Neutrons also display *anomalous*

scattering for certain nuclei. Nuclei with a large neutron absorption have an imaginary component in their b values representing a phase shift. For example, the b value of the boron isotope ^{10}B, a strong neutron absorber used for shielding, is (–0.1 – 1.066i) fm.

G1.3.2 Cross-Sections and Sample Size

X-ray and neutron atomic scattering amplitudes are of the order of 10^{-14} m (10 Fermi units); scattering cross-sections are usually given in units of 1 *barn* (10^{-24} cm^2) (Comment G1.6). The cross-section represents the effective scattering area the atom offers the incident flux (Fig. G1.1).

We can appreciate how small atomic scattering cross-sections actually are by calculating how long it would take to detect one scattered neutron from a single carbon atom placed in one of the most intense neutron beams currently available (Comment G1.7). The answer is slightly less than 600 million years. Independently of other considerations, therefore, samples for X-ray or neutron diffraction must contain a very large number of atoms.

The cross-section concept is also applicable to different forms of scattering (coherent or incoherent scattering, see below) and to absorption (Comment G1.5).

G1.4 SCATTERING VECTOR AND RESOLUTION

Consider the phase difference of a wave scattered by point P and by point O (Fig. G1.3).

A wave front W can be defined perpendicular to the wave propagation as joining points of equal phase. PM is the wave front of the incident wave when it touches P; PN is the wave front of the wave scattered in the direction 2θ. The path difference, Δ, between the waves scattered by P and those scattered by O is ON – OM. We now write this in vector notation,

$$OM - ON = \Delta = \mathbf{r} \cdot \mathbf{u}_1 - \mathbf{r} \cdot \mathbf{u}_0 = \mathbf{r} \cdot (\mathbf{u}_1 - \mathbf{u}_0) \qquad (G1.3)$$

where $\mathbf{r}$ is the vector OP and $\mathbf{u}_1$ and $\mathbf{u}_0$ are unit vectors parallel to the incident and scattered waves, respectively.

The phase difference, δ (in radians), between the two waves is given by

$$\delta = \frac{\Delta}{\lambda} 2\pi \qquad (G1.4)$$

where λ is the wavelength of the incident beam. When we place an atom of scattering amplitude f at P, the equation for the scattered wave relative to a wave of unit amplitude from O is written (see Chapter A3):

$$A = f \exp(i\delta) = f \exp\left(i\frac{2\pi}{\lambda}(\mathbf{u}_1 - \mathbf{u}_0) \cdot \mathbf{r}\right) \qquad (G1.5)$$

We introduce wave vectors $\mathbf{k}_0$ and $\mathbf{k}_1$ of magnitude $2\pi/\lambda$ in the directions of the incident and scattered waves, respectively, and a scattering vector $\mathbf{Q}$, which is the difference between them:

$$\begin{aligned} \mathbf{k}_0 &= \frac{2\pi}{\lambda}\mathbf{u}_0 \\ \mathbf{k}_1 &= \frac{2\pi}{\lambda}\mathbf{u}_1 \\ \mathbf{Q} &= \mathbf{k}_1 - \mathbf{k}_0 \end{aligned} \qquad (G1.6)$$

The wave from P can now be written simply,

$$A = f \exp(i\mathbf{Q} \cdot \mathbf{r}) \qquad (G1.7)$$

The magnitude of the scattering vector $\mathbf{Q}$ can be calculated from the geometric diagram in Fig. G1.4:

$$Q = (4\pi \sin\theta)/\lambda \qquad (G1.8)$$

The scattering vector contains in its magnitude not only the angle, 2θ, with respect to the incident wave direction in which the scattered wave is observed, but also the wavelength of the radiation, λ. In fact, expressing scattered intensity as a function of $\mathbf{Q}$ fully defines the interplay of angle and wavelength dependence so that we do not need to know either of these values separately.

The scattering vector is a very useful quantity, which essentially defines the magnification that can be achieved in the diffraction experiment. Waves from atoms separated by a vector $\mathbf{r}$ are out of phase by one cycle (2π phase angle or λ path length), leading to constructive interference when the scattering vector $\mathbf{Q}$ is parallel to $\mathbf{r}$ and for Q equal to $2\pi/r$. If the wavelength and r are on the same length scale, there is increased intensity due to the constructive interference between the two waves at an observable angle (e.g., measured in terms of centimeters on a photographic plate or detector, leading to an effective magnification of 10^8, when r and λ are in the ångström range).

The reciprocal relationship between Q and r arises from the wave construction in Fig. G1.3. It represents a fundamental property of diffraction theory. For smaller values of r, we need to go to larger values of Q to obtain the same path difference. The geometrical space in which Q is depicted is called "reciprocal space" with respect to the space of r, which is called "real" or "direct" space (see Chapter A3). In order to "resolve" shorter distances in a diffraction experiment, it is necessary to go to larger scattering vector values (achieved by increasing the observation angle and/or reducing the wavelength), in order to

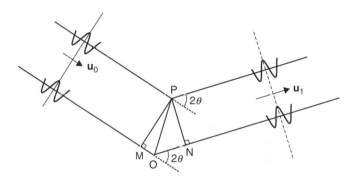

Fig. G1.3 Diffraction from two points.

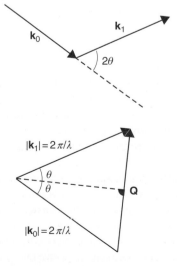

Fig. G1.4 Definition of the scattering vector.

increase the effective path difference between the scattered waves. The *resolution* of an experiment is the minimum distance between points that can be observed separately. It is given by

$$\text{resolution} \sim 2\pi/Q_{max} \qquad (G1.9)$$

where Q_{max} is the maximum value of Q for which the scattered intensity is observed.

G1.5 SCATTERING BY AN ASSEMBLY OF ATOMS

The two-atom case treated above provides the basis for building a picture of scattering by an assembly of atoms. We can consider the resulting wave in a direction defined by a given scattering vector as resulting from interference of waves from all possible atom pairs in the assembly.

G1.5.1 Coherent and Incoherent Scattering

Coherent scattering is defined as the case in which the scattered waves interfere to give a single resultant wave in a given direction, as in Fig. G1.3. The amplitudes of the waves, a_j, from each atom j are added by taking into account their phase relationships. The intensity when the waves are in phase is $\left(\sum a_j\right)^2$. If the scattering is *incoherent*, the resultant intensity is $\sum \left(a_j^2\right)$, the sum of intensities scattered individually by each of the atoms (as if the other atoms were not there!). The intermediate case, however, is the one most often observed, in which the scattered waves contain a coherent and an incoherent component.

We recall (Section G1.3.2) that we need to have very large numbers of atoms in a sample because scattering by a single atom is far too weak to observe. The same requirement holds true for a "structure" made up of two atoms separated by a given vector, for example. The sample must contain a very large number of pairs, all of which are made up of the same two atom types separated by the same vector. In other words, the sample itself presents a coherent structure. If there is disorder in the sample due to variations in either atom type or the vector separating the atoms, it contributes to incoherence.

For scattering from an assembly of atoms to be coherent, both the incident beam and sample properties must be coherent. Plane wave fronts in perfect phase must cross the entire sample during the scattering process (i.e., the *coherence length* of the source should be longer than the sample), and each equivalent atom in the sample must maintain the same scattering length (i.e., a constant amplitude and phase relationship with the incident wave) and be found in the same spatial environment (i.e., in a constant structure). If equivalent positions in a crystal structure are not all occupied by the same atom type (in X-ray scattering) or

---

COMMENT G1.8 SPIN INCOHERENCE IN NEUTRON SCATTERING

During a neutron scattering event, a nucleus of spin I combines with a neutron of spin 1/2 to form one of the two intermediate states $I + 1/2$ or $I - 1/2$, with relative weights, w_+ and w_-, respectively. Different scattering lengths, b_+ and b_-, respectively, are associated with each of these states, leading to a total scattering cross-section:

$$\sigma = S + s = 4\pi\left(w_+ b_+^2 + w_- b_-^2\right)$$

where S is the coherent and s the incoherent part:

$$S = 4\pi(w_+ b_+ + w_- b_-)^2$$
$$s = \sigma - S$$

---

by the same isotope in the same spin state (in neutron scattering), the scattered radiation has an incoherent component. Incoherent scattering in a diffraction experiment does not contain structural information and contributes to the background noise.

Compton scattering, which is inelastic, is an important source of incoherent X-ray scattering. Incoherent neutron scattering arises mainly from spin incoherence (Comment G1.8). In practice, the background in a neutron diffraction experiment on biological material is predominantly due to the strong incoherent scattering from hydrogen nuclei. In neutron inelastic scattering experiments the incoherent scattering is analyzed to provide information on sample dynamics (see Chapter I2).

G1.5.2 Elastic and Inelastic Scattering

Elastic scattering is when the scattered beam has the same energy as the incident beam. In an inelastic scattering process, the sample either loses energy to the radiation or gains energy from it (compare Stokes and anti-Stokes lines in light scattering, see Section E3.2). Standard diffraction experiments that provide information on the spatial arrangement of atoms are based on coherent elastic scattering. Inelastic scattering can be either coherent or incoherent. When the energy exchange is with sample excitations it contains information on dynamics (see Chapter I2). Energy analysis of coherent inelastic neutron scattering has, for example, shown how thermal energy propagates in waves of atomic vibration across molecules, while incoherent neutron scattering experiments provide information on individual atomic motions in protein dynamics.

Compton scattering of X-rays is inelastic and related to absorption (Comment G1.9); it is incoherent in that there is no fixed-phase relationship between the incident and scattered waves, and contributes to the background observed in an X-ray diffraction experiment.

G1.5.3 Summing Waves, Fourier Transformation, and Reciprocal Space

The two-atom example is readily extended to a particle made up of a larger assembly of atoms with a fixed spatial relationship (Fig. G1.5). Recall from Eq. (G1.7) that the scattering from an atom at a point P with respect to an arbitrary origin O:

$$A = f \exp\left(i\mathbf{Q} \cdot \mathbf{r}\right)$$

where f is the scattering amplitude of the atom, $\mathbf{Q}$ is the scattering vector, and $\mathbf{r}$ is the vector between P and O. In the case of an assembly of atoms, the scattered wave is given simply by the sum of waves individually scattered, with respect to an arbitrary origin:

$$F(\mathbf{Q}) = \sum_j f_j \exp\left(i\mathbf{Q} \cdot \mathbf{r}_j\right) \tag{G1.10}$$

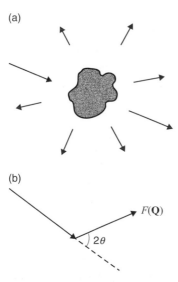

(a)

(b)

$F(\mathbf{Q})$

2θ

Fig. G1.5 (a) Scattering by an assembly of atoms in a particle. The black arrow symbolizes the incident beam, the red arrows represent the scattered waves of different amplitudes and phases in different directions, corresponding to different scattering angles and $\mathbf{Q}$ vectors. **(b)** The wave $F(\mathbf{Q})$ for a given scattering vector $\mathbf{Q}$.

where f_j and $\mathbf{r}_j$ are respectively the scattering amplitude and position vector with respect to the origin of atom j. The f_j values are Q-dependent for X-rays (they are the form factors of the atoms) but constant with Q for neutrons. The choice of origin for $\mathbf{r}_j$ is arbitrary; the phase of each scattered wave in Eq. (G1.10) is calculated relative to that of a virtual wave scattered by a point at the origin. The structural information contained in the scattered wave, however, arises from the phase differences between the scattered waves – phase differences that are independent of the origin.

Note that $F(\mathbf{Q})$ is a complex number. It defines the scattered wave for scattering vector $\mathbf{Q}$ in terms of two quantities: its amplitude and its phase. Phase information is lost when the intensity, $|F(\mathbf{Q})|^2$, of the wave is measured.

Equation (G1.10) establishes in mathematical terms that $F(\mathbf{Q})$ is the Fourier transform of the f_j, $\mathbf{r}_j$ distribution, $f(\mathbf{r})$, which in practice describes the positions of the atoms in the particle. An important property of the Fourier transform is that if A is the transform of B then B is the transform of A (see Chapter A3). The structure of the particle, $f(\mathbf{r})$, can therefore be calculated from the observed scattering amplitude, $F(\mathbf{Q})$.

Since the particle scattered amplitude is continuous as a function of $\mathbf{Q}$, it is useful to express the back Fourier transform of Eq. (G1.10) as an integral:

$$f(\mathbf{r}) = \int F(\mathbf{Q}) \exp\left(-i\mathbf{Q} \cdot \mathbf{r}_j\right) dV_{\mathbf{Q}} \tag{G1.11}$$

where $dV_{\mathbf{Q}}$ is a volume element in $\mathbf{Q}$ space (also called reciprocal space; Comment G1.10). Similarly, Eq. (G1.10) can be expressed as an integral over the particle volume:

$$F(\mathbf{Q}) = \int f(\mathbf{r}) \exp\left(i\mathbf{Q} \cdot \mathbf{r}_j\right) dV_{\mathbf{r}} \tag{G1.12}$$

There is a one-to-one relationship between the $f(\mathbf{r})$ and $F(\mathbf{Q})$ distributions. Each fully defines and is fully defined by the Fourier integral of the other (see Chapter A3).

A particle is "observed" by analyzing the waves it scatters, $F(\mathbf{Q})$, in as large a Q range as possible. Its structure (atomic arrangement) can then be calculated unequivocally with a resolution of $2\pi/Q_{\max}$ by using Eq. (G1.11). In order to do so, however, the amplitude *and* phase of each scattered wave, $F(\mathbf{Q})$, should be known.

G1.5.4 The Phase Problem

In *microscopy*, a magnified image of a particle is obtained from $F(\mathbf{Q})$ by using a lens to recombine the scattered waves while respecting the correct phase relationships. The lens essentially behaves as a Fourier transformation device.

While lenses for light and electron microscopy exist, there are none for X-rays or neutrons in the ångström resolution range. There is a phase problem, therefore, in X-ray and neutron diffraction, because only the intensity, $|F(\mathbf{Q})|^2$, of the scattered waves can be measured and phase information is lost. Solving a structure in the absence of this phase information is the main challenge faced by X-ray and neutron diffraction and crystallography.

G1.6 SOLUTIONS AND CRYSTALS

Atomic scattering cross-sections for X-rays and neutrons are extremely small, and, even with the most intense sources currently available, a very large number of macromolecular particles are required in order to observe an interpretable signal.

The level of resolution obtained from a diffraction experiment depends on how well ordered the particles are with respect to each other in the sample. Thus, for there to be intensity at a $\mathbf{Q}$ vector corresponding to 2 Å resolution, a significant number of atoms in all particles of the sample should be organized with a position accuracy of 2 Å along the vector parallel to $\mathbf{Q}$.

There are two extremes in organizational order for a sample made up of identical particles (such a sample is called *monodisperse*) (Fig. G1.6). In a dilute solution, the particles are located randomly in both position and orientation (see Chapter G2). The only order is that of the distances between atoms within each particle. These distances (not vectors since their orientations are random) represent the most information that can be derived from a diffraction

pattern of a solution sample. At the other extreme, in crystals we have particles ordered in three dimensions. Currently, the only way to obtain high-resolution structural information from diffraction experiments is by using crystallographic methods. Between the two extreme cases, we find concentrated solutions, for which interparticle diffraction can arise, and membranes and fibers, which display various degrees of one- and two-dimensional order.

G1.6.1 One-Dimensional Crystals

A crystal is a periodic arrangement in space of a repeated motif, which is called the *unit cell*. The unit cell may contain one or several macromolecules organized in a symmetrical fashion. The principles involved in calculating the diffraction from a crystal can be explained more simply for a one-dimensional periodic array.

We recall from Chapter A3 that a one-dimensional periodic array can be described as the convolution of a motif with a lattice of Dirac delta functions (Fig. G1.7). If the periodicity of the crystal is d and it contains a very large number of unit cells, crystal diffraction occurs only at the positions of the lattice diffraction with a periodicity proportional to $1/d$. In practice, this means that the waves diffracted in all other directions cancel themselves out by *destructive interference* (see Chapter A3). The peak amplitudes depend on the value of the diffraction of the single unit cell at that position in Q. It is said that the single unit cell diffraction pattern is *sampled* by the lattice diffraction pattern.

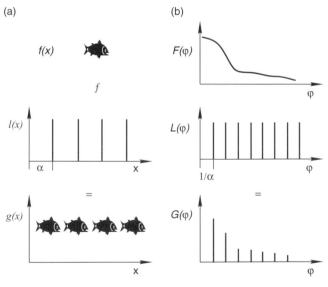

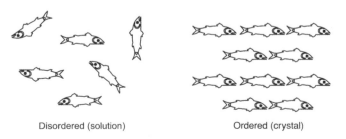

Fig. G1.6 Samples from order to disorder.

Fig. G1.7 (a) Real space: a one-dimensional crystal of N fish macromolecules seen as a convolution of a single fish with a lattice of Dirac functions of periodicity d. **(b)** The corresponding Fourier transforms in reciprocal space for the case in which N is a very large number. If N is not a large number, the Fourier transform of the lattice does not correspond to a set of delta functions but to a distribution of broader peaks.

The lattice can be described by a vector $\mathbf{R}$, of length d, displaced n times. The diffraction pattern $G(\mathbf{Q})$ produced by this arrangement is

$$G(\mathbf{Q}) = F(\mathbf{Q}) \times \sum_n \exp(in\mathbf{Q}\cdot\mathbf{R}) \qquad \text{(G1.13)}$$

where $F(\mathbf{Q})$ is the diffraction pattern of the single unit cell (Eqs. (G1.10) and (G1.12)). Each term in the sum introduces the *phase difference* (see Chapter A3) associated with successive lattice points. This is exactly equivalent to the sampling shown in Fig. G1.7.

Stacked biological membranes are good examples of one-dimensional crystals with periodicity perpendicular to the membrane plane. The detailed conformation of lipid molecules in a lipid bilayer was calculated from the one-dimensional neutron diffraction from membrane stacks (Buldt *et al.*, 1979; Zaccai *et al.*, 1979a). The myelin sheath around axons plays an important role in rapid nerve conduction in the central and peripheral nervous systems. Myelin is a lipid-rich, multilamellar assembly of membranes whose electrical insulating properties depend on the regular stacking of these plasma membranes, which practically forms a one-dimensional crystal (Denninger *et al.*, 2014).

G1.6.2 Two- and Three-Dimensional Crystals

It is straightforward to extend Eq. (G1.13) to two dimensions (as in the case of membranes with in-plane organization) and to three-dimensional crystals by using appropriate vectors $\mathbf{R}$ and $\mathbf{Q}$.

For a three-dimensional crystal, $\mathbf{R} = \mathbf{a} + \mathbf{b} + \mathbf{c}$, where $\mathbf{a}$, $\mathbf{b}$, $\mathbf{c}$ are vectors along the three axes (Fig. G1.8). By substituting for $\mathbf{R}$ in Eq. (G1.13), we write

$$G(\mathbf{Q}) = F(\mathbf{Q}) \times \sum_t \exp(it\mathbf{Q}\cdot\mathbf{a}) \times \sum_u \exp(iu\mathbf{Q}\cdot\mathbf{b})$$
$$\times \sum_v \exp(iv\mathbf{Q}\cdot\mathbf{c}) \qquad \text{(G1.14)}$$

where t, u, and v are integral numbers.

Similarly to the one-dimensional case, when the crystal contains a very large number of unit cells, all phases except the ones corresponding to the lattice diffraction positions produce destructive interference so that diffraction is observed only at

$$\mathbf{Q}\cdot\mathbf{a} = 2\pi h \qquad \mathbf{Q}\cdot\mathbf{b} = 2\pi k \qquad \mathbf{Q}\cdot\mathbf{c} = 2\pi l \qquad \text{(G1.15)}$$

It is usual in crystallography to use a scattering vector $\mathbf{S}$ instead of $\mathbf{Q}$, where $\mathbf{Q} = 2\pi\mathbf{S}$. Equation (G1.15) then becomes

$$\mathbf{S}\cdot\mathbf{a} = h \qquad \mathbf{S}\cdot\mathbf{b} = k \qquad \mathbf{S}\cdot\mathbf{c} = l \qquad \text{(G1.16)}$$

h, k, and l are integral numbers and it is conventional to describe the diffraction peak by its indices (h, k, l); these are called the *Miller indices*.

In summary, the diffraction pattern from a crystal contains two levels of information: the spatial arrangement of the unit cells can be derived from the regular spacing of the diffraction peak pattern (denoted by the *hkl* indices); the amplitudes of the diffraction pattern contain information on the arrangement of atoms within each unit cell.

G1.6.3 Disordered Systems

The construction in Fig. G1.7 can be used quite generally and not only in the case of well-ordered lattices. Figure G1.9 illustrates the example of a concentrated solution (in one dimension for simplicity but the argument also holds

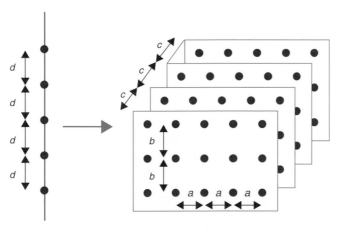

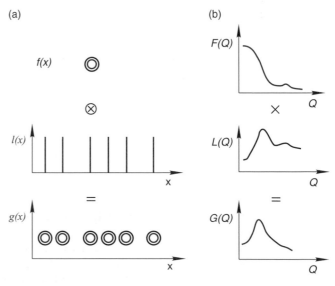

Fig. G1.8 Schematic diagram of one- and three-dimensional crystals. The red dots represent the origins of the scattered waves. In the one-dimensional lattice the origins are separated by a distance d, while in the three-dimensional lattice, the origins are displaced by distances, a, b, and c along the respective axes.

Fig. G1.9 (a) Real space: a liquid of large particles seen as a convolution of a single particle with a disordered lattice of Dirac functions. **(b)** The corresponding Fourier transforms in reciprocal space.

for three dimensions). Note that $L(Q)$, the Fourier transform of the disordered lattice, is a continuous function with a broad peak corresponding to the most frequent interparticle spacing in the liquid. $G(Q) = F(Q)L(Q)$ is also a continuous function. The peak of $G(Q)$, however, results from both the particle's internal structure and its spacing in the solution (see Chapter G2).

G1.7 RESOLUTION AND CONTRAST

Biological macromolecules and structures are generally in an aqueous environment. This is obvious for protein solutions. However, protein crystals, DNA fibers, membrane samples, etc. also contain significant volumes of solvent. A diffraction experiment on a perfectly ordered (coherent) sample should provide information on the positions of all atoms, be they in the solvent or macromolecules. In practice, however, resolution may be limited to the dimensions of a volume the size of many atoms. This is particularly true for the disordered solvent regions in a sample (Comment G1.11). In such cases, it is useful to define a *scattering density*, i.e., scattering amplitude per unit volume. The concept of *contrast* between two parts of a sample refers to the difference between their scattering densities.

Equation (G1.10) describes the resulting wave scattered by a system of atoms in a direction corresponding to a given scattering vector,

$$F(\mathbf{Q}) = \sum_j f_j \exp\left(i\mathbf{Q}\cdot\mathbf{r}_j\right)$$

Consider now the single macromolecule in solution shown schematically in Fig. G1.10. In order to calculate how the system scatters radiation, the sum in Eq. (G1.10) should be over all atoms in the solution, those within the macromolecule as well as those in the solvent. The particle is fully described by the scattering amplitudes f_j at positions $\mathbf{r}_j$, of its atoms, and it is surrounded by an infinite homogeneous solvent of scattering density $\rho°$.

We now divide the scattering system into the macromolecule, on the one hand, and the solvent, on the other (Fig. G1.11). The first part, (a), is the particle (including perturbed solvent). The second part, (b), is the homogeneous infinite solvent. Note, however, that the sum of (a) and (b) includes an extra volume of solvent, (c), when compared to the initial scattering system – the volume corresponding to that occupied by the particle. The particle in the solution scattering system, therefore, is equal to a + b – c, which is similar to Archimedes' principle for the buoyancy of a particle in a fluid (see Part D and Chapter G2). The sum of waves from each part is given in the right-hand panel of Fig. G1.11. Waves from the homogeneous solvent appear in a very narrow range close to $Q = 0$, so that (b) is not observed in practice (Comment G1.12). The scattering amplitude from the particle in solution is hence given by

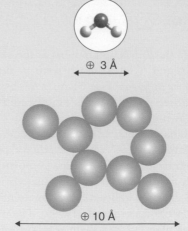
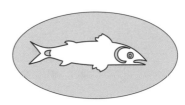

Fig. G1.10 A particle in solution. The particle is drawn as a fish to emphasize that biological macromolecules fold and attain their quaternary structures through interactions with aqueous solvent.

$$F(\mathbf{Q}) = \sum_j (f_j - \rho° v_j) \exp\left(i\mathbf{Q}\cdot\mathbf{r}_j\right) \tag{G1.17}$$

where v_j is the volume of atom j. The $f_j - \rho° v_j$ term is the *contrast amplitude* of atom j with respect to the solvent. The mean contrast amplitude of the particle is given by $\sum(f_j - \rho° v_j)$ or $\sum f_j - \rho° V$, where V is the total volume of the particle. The *mean scattering density contrast* of the particle is $\rho - \rho°$, where $\rho = \sum(f_j)/V$.

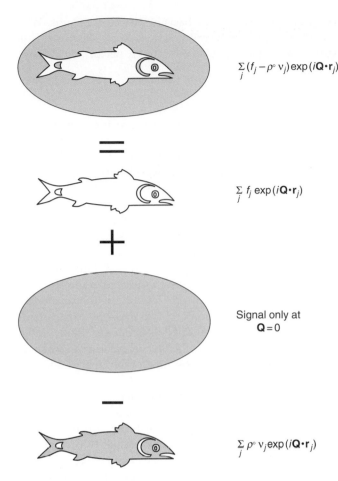

$$\sum_j (f_j - \rho^\circ v_j)\exp(i\mathbf{Q}\cdot\mathbf{r}_j)$$

$$\sum_j f_j \exp(i\mathbf{Q}\cdot\mathbf{r}_j)$$

Signal only at
$\mathbf{Q} = 0$

$$\sum_j \rho^\circ v_j\exp(i\mathbf{Q}\cdot\mathbf{r}_j)$$

Fig. G1.11 Scattering from a particle in solution.

COMMENT G1.12 SCATTERING BY A HOMOGENEOUS SOLVENT

Formally, the Fourier transform of a constant value is a delta function at the origin (see Chapter A3); recall that the larger the scattering particle the smaller the Q values in which it scatters; an infinite volume of constant scattering density can be seen as an infinitely large particle that will only scatter at $Q = 0$.

The contrast considerations are valid for small-angle scattering experiments in solution (Chapter G2) and crystallography, fiber, and membrane diffraction at low resolution only (Chapter G3).

G1.8 THE PRACTICE OF X-RAY AND NEUTRON DIFFRACTION

G1.8.1 Complementarity

The very high intensity and narrow divergence of X-ray beams currently available from synchrotron sources make

---

COMMENT G1.13 FLUX AND BRIGHTNESS

Flux is defined as the number of photons or neutrons crossing a unit area per second (e.g., the usual flux units for neutrons are neutrons $cm^{-2}\ s^{-1}$). Neutron beams have quite a large divergence, and angular resolution in a diffraction experiment can be achieved only by using slits or similar devices that reduce intensity.

With synchrotron radiation, it is useful to define an intensity unit that also takes into account the small divergence of the beam (which allows very good angular resolution while maintaining the full intensity in the beam). Brightness is defined as the number of photons crossing a unit area per second per unit solid angle of beam divergence (e.g., the usual brightness units for a synchrotron beam are photons $s^{-1}\ mm^{-2}\ mrad^{-2}$).

---

them the radiation of choice in structural biology. Neutron sources are much weaker in that respect but the special properties of the neutron, especially with respect to the scattering powers of hydrogen and deuterium, have allowed the analysis of specific problems that were difficult or impossible to address by using X-rays. Hydrogen and water molecules play essential roles in biological structures; they can be "seen" in neutron crystallography even when not highly ordered (see Chapter G3). Neutron-specific methods based on hydrogen–deuterium labeling and contrast variation have provided essential information on the organization of protein–protein and protein–nucleic acid complexes, and membranes (see Chapter G2). Neutrons also provide a uniquely suited radiation for the study of molecular dynamics. Because the energy associated with wavelengths of ~1 Å is of the order of thermal energy, they allow the simultaneous measurement of the amplitudes and frequencies of motions in a sample (see Chapter I2).

G1.8.2 Sources and Instruments

X-rays

X-ray photons are produced by an electron beam from a heated filament hitting a metal target, or in the form of synchrotron radiation from charges (electrons or positrons) accelerated by magnetic devices in a ring. X-ray sources have evolved from a sealed tube to a rotating anode to several generations of synchrotrons, providing an increase of more than ten orders of magnitude in beam brightness (Comment G1.13). This has made diffraction studies on smaller and smaller samples possible. Protein microcrystals (~10^{-5} mm^3) are routinely studied at third-generation synchrotrons. Because of the inherent difficulty in obtaining "large" crystals of biological macromolecules, the possibility of working with such minute samples is an important advantage of the method, making it applicable to a wide range of structural problems (see Chapter G3).

In the case of X-ray generators with metal anodes, the emitted radiation corresponds to electronic transitions of specific energy. The main wavelength from copper is 1.54 Å, while that from a molybdenum target is 0.71 Å. Synchrotron sources are particularly advantageous because they provide a broad spectrum from which specific wavelengths can be chosen using monochromators.

Neutrons

Neutron sources are considerably less intense than X-ray sources. Their relative weakness is compensated for to some extent by very low absorption (the absence of radiation damage compared to X-rays), associated with the possibility of using large beam cross-sections and long wavelengths, which are prohibitive for X-rays because of the high absorption. These possibilities combined with specific contrast variation techniques make neutron sources strongly competitive for low-resolution crystallography and small-angle scattering experiments, for example (Chapters G2 and G3).

A continuous flux of neutrons is produced in reactors by the fission chain reaction of ^{235}U. In spallation sources, pulsed neutron beams are produced by an accelerated proton beam impinging onto a metal target. The high-energy protons tear through the target nuclei, breaking them apart with a significant yield of fast neutrons.

The neutrons behave like a gas and are slowed down by thermal equilibration through collisions in a material called the moderator (usually light or heavy water). Neutrons of energy close to that associated with ambient temperature (300 K) are called *thermal neutrons* (energy 24.5 meV, wavelength 1.8 Å) (see Table G1.1). "Cold" and "hot" sources may also be included in a neutron source to produce neutrons at different wavelengths. The neutron spectrum from a cold source peaks at a wavelength of a few ångströms. Long-wavelength neutrons are not strongly absorbed by matter (unlike long-wavelength X-rays) and are particularly useful for small-angle scattering studies (see Chapter G2). Hot neutrons peak at a wavelength of a fraction of an ångström, and are particularly useful for high-accuracy chemical crystallography. Cold sources may also be included on a spallation source, but hot sources are unnecessary because the spallation flux already contains an appreciable fraction of energetic neutrons.

Large-Scale Facilities

Because laboratory sources provide only extremely weak beams, neutron scattering experiments are performed at large-scale facilities. Several research institutes with reactor or spallation sources dedicated to providing neutron beams and instrumentation for diffraction and spectrometry experiments in structural biology now exist in Europe, the USA, and Japan, and are open for use by research scientists on the basis of proposal systems. With the advent of synchrotron radiation, X-ray experiments in structural biology are also now largely performed at large-scale facilities.

Diffractometers

X-ray and neutron diffractometers are instruments that measure the scattered intensity from a sample as a function of incident wavelength and scattering angle. They vary in design according to the wavelength range, sample type, and angular resolution, which should be optimized for each type of experiment. Spectroscopic experimental setups are also available, in which the energy transfer between sample and radiation is measured by analyzing the wavelength (and therefore energy) of the scattered beam by the crystal reflection or time of flight (for neutrons only). The pulsed nature of the neutron beam from spallation sources allows the use of broad wavelength bands for both diffraction and spectroscopy experiments (thus considerably increasing the effective flux); under these conditions both the scattering vector and energy transfer can be analyzed by time-of-flight methods (see Chapter I2).

An X-ray or neutron diffractometer consists of three main parts: a monochromator, a sample area, and a detector (Fig. G1.12). The incident beam is usually made monochromatic by reflection off a crystal. In the neutron case only, a velocity selector device can also be used; neutron velocities are inversely proportional to wavelength (see Comment G1.2), and they are sufficiently slow for a monochromatic beam to be selected by slits in a spinning drum rotating in the 10^3 rpm range. The sample area may contain a *goniometer* (from the Greek *gonia*, angle or

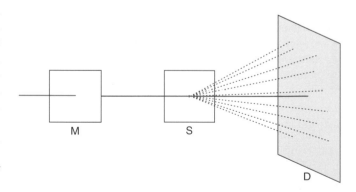

Fig. G1.12 Schematic diagram of a diffractometer. M represents the monochromator area where the wavelength (red) is selected from the source beam spectrum (blue). S is the sample area which may contain a goniometer to align the sample, temperature control devices, or other sample environment controls. The detector D converts the scattered intensity (dashed red lines) as a function of scattering angle into a signal that can be read by the instrument computer.

COMMENT G1.14 NUCLEAR REACTION FOR NEUTRON DETECTION

$n(^3He, p)^3H$	0.77 MeV
$n(^6Li, \alpha)^3H$	4.79 MeV
$n(^{10}B, \alpha)^7Li + 2.3$ MeV $+ \gamma$ (0.48 MeV)	93%
$n(^{10}B, \alpha)^7Li + 2.79$ MeV	7%
$n(^{157}Gd, Gd)e^-$	0.182 MeV

In this notation, which is used in nuclear physics, the first line, for example, is equivalent to

$$n + {}^3He \rightarrow p + {}^3H + 0.77 \text{ MeV}$$

corner) to align the sample precisely in the beam, and sample environment controls (temperature, pressure, humidity, etc.).

X-ray and neutron detectors transform the energy of the scattered radiation into a signal that can be processed computationally. They must cope with high data collection rates and accurately and rapidly record the intensity and angular position of the scattering. The oldest type of X-ray detector is photographic film. This detection process is cumbersome, however, as the film can only be used once and must subsequently be developed and scanned into a computer. Modern detectors are based on the measurement of electric currents produced directly or indirectly by the radiation in gas or solid supports.

Heavy gases, like xenon, are ionized directly by the electromagnetic properties of the X-ray photon. X-rays can also excite certain heavy metal ions to fluoresce in the visible spectral range. An image intensifier then converts the visible light into an electronic signal. *Image plate* X-ray detectors have been built based on the properties of europium ions. Eu^{2+} is excited by X-rays to Eu^{3+}, which emits violet light ($\lambda = 390$ nm) when illuminated by a red He–Ne laser ($\lambda = 693$ nm). A photomultiplier is used to detect the violet light and the plate is then "erased and reset" with white light.

Since the neutron does not carry an electric charge, its detection is based on nuclear reactions that emit either ionizing radiation or charged particles (Comment G1.14). The simplest neutron detector is the *ionization chamber*, in which a sensitive gas, usually ^3He or $^{10}BF_3$, is contained between two charged plates. Position-sensitive multiwire detectors are based on the same detection principle. Scintillation detectors are based on the light emission of ^6Li when it absorbs a neutron. Image plate detectors have also been developed for neutrons. The matrix contains Gd_2O_3, which emits an electron of sufficient energy to excite the Eu ions (as in the X-ray case) when a neutron is absorbed.

G1.9 CHECKLIST OF KEY IDEAS

- In a diffraction experiment, waves of radiation scattered by different objects interfere to give rise to an observable pattern from which the relative arrangement (or structure) of the objects can be deduced.
- Three types of radiation are used in diffraction experiments on biological macromolecules: X-rays of wavelength about 1 Å, electrons of wavelength about 0.01 Å, and neutrons of wavelength about 0.5–10 Å.
- X-rays are electromagnetic radiation of energy ~10 keV, much higher than that of visible light; neutrons are particles of mass similar to that of a proton, with associated wavelengths ~1 Å for velocities ~4km s^{-1} and energies ~10 meV.
- Neutrons are scattered by atomic nuclei, which behave as points; X-rays are scattered by atomic electron clouds, which gives rise to a *form factor* because their spatial extent is similar to the wavelength of the radiation; the X-ray scattering power of an atom increases with its electron number.
- Neutrons are sensitive to different isotopes, and the different scattering powers of hydrogen and deuterium have been especially useful in biophysical studies.
- X-ray absorption leads to non-negligible radiation damage in the sample. It increases significantly with increasing wavelength; neutron absorption is very low for most nuclei, even for wavelengths of 10 Å or more.
- The scattering *amplitude* (or *scattering length*), b, of a point atom is defined as the amplitude of the scattered wave observed per unit *incident flux* in unit solid angle; the scattering *cross-section* σ is the ratio of the total intensity scattered by the atom in unit time per unit incident flux; σ has units of area and can be understood as an effective projected area perpendicular to the incident beam; there is an analogous definition for the absorption cross-section.
- Scattering is described as a function of the *scattering vector*, a parameter that expresses the angular dependence for a given incident wavelength; in a reciprocal relation, the intensity observed at a given scattering vector magnitude Q contains information on atomic spacings down to a minimum distance equal to $2\pi/Q$.
- The *resolution* of a diffraction experiment is the minimum distance separating points that can be observed separately; it is equal to $2\pi/Q_{max}$.
- *Coherent* scattering is defined as the case in which waves scattered by a set of atoms interfere to give a single wave in a given direction, resulting from the sum of individual wave amplitudes with appropriate phases; if the scattering is *incoherent*, each atom scatters independently of the others, and the

resultant intensity is the sum of individually scattered intensities.

- Incoherent scattering does not contain structural information and contributes to the background in diffraction experiments; Compton scattering is an important source of incoherent X-ray scattering; incoherent neutron scattering arises mainly from *spin incoherence*.

- *Elastic* scattering is when the scattered beam has the same energy as the incident beam. In an *inelastic* scattering process, the sample either loses energy to the radiation or gains energy from it.

- The scattering amplitude as a function of the scattering vector (in *reciprocal space*) is calculated from the sum of waves coherently scattered by the atoms in a particle; it is related by Fourier transformation to the particle structure (as a function of the *real-space* vector).

- There is a *phase problem* in X-ray and neutron diffraction, because only the intensity and not the phase of the scattered waves can be measured.

- Atomic scattering cross-sections for X-rays and neutrons are very small, and, even with the most intense sources currently available, a very large number of macromolecular particles are required in order to observe an interpretable signal. The particles can be organized in two extreme ways: with perfect disorder as in a dilute solution or with perfect order as in crystal; the level of resolution obtained from a diffraction experiment depends on how well ordered the particles are with respect to each other in the sample.

- A crystal of particles can be represented as the *convolution* of a lattice function and the particle structure.

- At low resolution, atoms can be grouped to define the *scattering density* in a volume; the *contrast* of part of a particle with respect to another is proportional to the difference between their scattering densities.

- X-rays and neutrons have different properties that make them strongly complementary in biophysical studies; the very high intensity and narrow divergence of X-ray beams currently available from synchrotron sources make them the radiation of choice in structural biology. The different scattering amplitudes of hydrogen and deuterium have allowed the analysis of specific problems by neutron scattering that were difficult or impossible to address by using X-rays. Hydrogen and water molecules play essential roles in biological structures, and they can be "seen" in neutron crystallography even when not highly ordered. Neutron-specific methods based on hydrogen–deuterium labeling and contrast variation have provided essential information on the organization of protein–protein and protein–nucleic acid complexes and membranes. Neutrons also provide a uniquely suited radiation for the study of molecular dynamics, because the energy associated with a wavelength of ~1 Å is of the order of thermal energy, thus allowing the simultaneous measurement of the amplitudes and frequencies of motions in a sample.

SMALL-ANGLE SCATTERING AND REFLECTOMETRY

G2

G2.1 THEORY OF SMALL-ANGLE SCATTERING FROM PARTICLES IN SOLUTION

Small-angle scattering (SAS) is a very useful method in biochemistry, providing information on molecular masses, shapes, and interactions in solution (see Comment G2.1).

G2.1.1 Dilute Solutions of Identical Particles

In order to have an observable signal in X-ray or neutron SAS, a solution of the order of 100 μl containing a few milligrams per milliliter of macromolecule is required – corresponding to about 10^{15} particles for a 50 kDa protein at 1 mg ml^{-1} (we note that 1 mg ml^{-1}, the usual unit in biochemistry, is in fact equal to 1 g l^{-1}, the unit more conveniently used in equations).

We first consider solution conditions such that the particles do not influence each other, i.e., the position and orientation of each particle is totally independent of that of the others. This is the *infinite dilution* condition, for which we say there is no *interparticle interference*. In practice, it is achieved at different, low, concentrations for different macromolecules and solvents. For example, the condition may well be satisfied for a given protein at a few milligrams per milliliter, in a neutral pH buffer, but tRNA molecules, which are highly charged at pH 7 in low-salt buffer, interact with each other even at these concentrations, and it might not be possible to reach the infinite dilution condition without increasing the solvent salt concentration. In order to make sure that the infinite dilution condition is fulfilled, data should be collected as a function of concentration and extrapolated to zero concentration.

The infinite dilution condition implies that the chance of a wave being scattered by two different particles is practically nil; interference between waves scattered by atoms in different particles can be neglected and does not influence the scattering pattern.

We recall the expression for a wave scattered from a single particle as a function of scattering vector, $\mathbf{Q}$ (Eq. (G1.10)):

$$F(\mathbf{Q}) = \sum_j f_j \exp\left(i\mathbf{Q}\cdot\mathbf{r}_j\right)$$

where f_j, $\mathbf{r}_j$ are the scattering amplitude and position, respectively, of atom j. The intensity scattered by the particle is the square of the wave amplitude

$$I(\mathbf{Q}) = |\mathbf{F}(\mathbf{Q})|^2$$

If there is no interparticle interference, the scattered intensity from unit volume of solution containing N particles is given simply by summing the intensities of waves scattered by different particles (Fig. G2.1)

$$I_N(\mathbf{Q}) = \sum_{n=1}^{N} |\mathbf{F}_n(\mathbf{Q})|^2 \tag{G2.1}$$

In the identical particle (*monodisperse*) case, the particles still differ from each other with respect to their scattering properties because they take up different orientations, and we write (Fig. G2.1b)

$$I_N(\mathbf{Q}) = N\left\langle |\mathbf{F}(\mathbf{Q})|^2 \right\rangle \tag{G2.2}$$

where the $\langle \cdots \rangle$ brackets denote rotational averaging in space (the different particles take on all different orientations) and time (each particle takes on different orientations).

Note that, apart from the number N, this is equivalent to assuming that one particle moves during the experiment and takes up all orientations. Particle motion can be described fully by a translation and a rotation. Translation is equivalent to a change of origin and does not change the amplitude calculation, which depends on the positions of the atoms relative to each other and not relative to the origin. In practice, therefore, Eq. (G2.2) accounts for the N particles in the solution and for the fact that each particle is moving. Note that due to rotational averaging, Q is not a vector in $I_N(Q)$.

The rotational average in Eq. (G2.2) was calculated by Debye in 1915, and is known as the Debye formula (Comment G2.2):

$$\left\langle |\mathbf{F}(\mathbf{Q})|^2 \right\rangle = \sum_j \sum_k \left(f_j - \rho^\circ v_j\right)\left(f_k - \rho^\circ v_k\right)\frac{\sin Qr_{jk}}{Qr_{jk}} \tag{G2.3}$$

where r_{jk} is the distance between atoms j and k, and $(f_j - \rho^\circ v_j)$ are contrast amplitudes when the particle is in a solvent of scattering density ρ° (see Chapter G1).

COMMENT G2.1 BIOLOGIST'S BOX: SAS WITHOUT EQUATIONS

A good understanding of SAS analysis involves mathematical approaches that may be difficult for the biologist. Small-angle scattering is, nevertheless, a powerful technique in biochemistry. It is rapid and easy to apply on suitable samples and allows a broad characterization of macromolecular structures in solution and their interactions. Below, we summarize in non-mathematical terms the information that can be obtained from an SAS experiment, and the practical requirements for setting it up.

Questions addressed:

(1) What is the effective association state (is it a monomer, a dimer, etc.) of a protein or other macromolecule in solution or within a membrane environment?
(2) What is its shape or conformation (is it compact and globular or ellipsoidal, long and narrow, or flat and broad, star-shaped or branched ... is its structure in solution similar to its crystal structure, etc.)?
(3) Do different macromolecules in solution interact to form a complex or not?
(4) How do the answers to points 1, 2, and 3 vary as a function of solvent conditions (pH, salt, ligand, temperature, etc.)? Are there modifications in association state or conformational changes?
(5) The method of *contrast variation* in small-angle neutron scattering (SANS) allows one to render *visible* only one component within a complex structure. It is then possible to address questions 1–4 for individual components within, for example, a macromolecular machine made up of various proteins, a protein nucleic acid complex interaction, or a membrane protein in a lipid or detergent environment.

An important point is that the answers to the different questions above are obtained independently of each other. For example, SAS will inform correctly on a monomeric particle (question 1) that is very elongated (question 2), whereas the conclusion of a gel filtration experiment (which essentially measures a diffusion coefficient), for example, might be that because it has a long dimension the particle is a multimer as described by Klein *et al.* (1982).

Practical requirements for an SAS experiment are:

(1) On the order of 100 μl of solution containing a few milligrams per milliliter of macromolecule.
(2) An accurate measurement of the macromolecular concentration (in milligrams per milliliter).
(3) Access to an X-ray or neutron small-angle camera.

SAS applications are illustrated in Section G2.6 by studies of protein–nucleic acid interactions of conformational changes during the working cycle of a chaperone macromolecular machine and a study of the association state of a membrane protein in lipid vesicles.

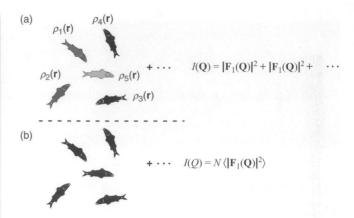

Fig. G2.1 Schematic diagram of real space structures (left) and corresponding wave intensities in reciprocal space (right). **(a)** A solution of non-interacting macromolecules described by scattering amplitude density distributions $\rho_1(\mathbf{r})$, $\rho_2(\mathbf{r})$, $\rho_3(\mathbf{r})$, etc.; **(b)** a solution of identical non-interacting macromolecules.

COMMENT G2.2 THE DEBYE FORMULA

In order to calculate the rotational average over $|\mathbf{F}(\mathbf{Q})|^2$, Debye used the classic result $\langle \exp(-i\mathbf{Q}\cdot\mathbf{r}) \rangle = \dfrac{\sin Qr}{Qr}$

COMMENT G2.3 THE TWO ASSUMPTIONS REQUIRED FOR THE INTERPRETATION OF THE SCATTERING CURVE IN TERMS OF A SINGLE PARTICLE STRUCTURE

(1) The solution is monodisperse. All the particles in the solution are identical.
(2) The solution is infinitely dilute. There is no correlation between the positions or orientations of the particles in the solution.

The scattered intensity from the solution in conditions of infinite dilution and identical particles (Comment G2.3) is given by

$$I_N(Q) = N\sum_j\sum_k m_j m_k \frac{\sin Qr_{jk}}{Qr_{jk}} \tag{G2.4}$$

where we have written $m_j = (f_j - \rho^o v_j)$.

If we write $m_j = (\rho(\mathbf{r}_j) - \rho^o)\mathrm{d}v_j$, where $\rho(\mathbf{r}_j)$ is the local scattering density of the volume $\mathrm{d}v_j$ at position $\mathbf{r}_j$ in the particle, we obtain the Debye equation in its integral form:

$$I_N(Q) = N\iint (\rho(\mathbf{r}_j) - \rho^o)(\rho(\mathbf{r}_k) - \rho^o)\frac{\sin Qr_{jk}}{Qr_{jk}}\mathrm{d}v_j\mathrm{d}v_k \tag{G2.4a}$$

The double integration is over the volume of the particle. Note that the volume of the particle should include the hydration layer or part of the solvent that is "disturbed" by the presence of the particle (see Chapter A3). *From the scattering point of view, the particle volume is defined all the way to the boundary beyond which the solvent behaves as free bulk solvent.*

It is useful to normalize $I(Q)$ by the particle concentration C in grams per liter:

$$\frac{I_N(Q)}{C} = \frac{N_A}{M} \sum_j \sum_k m_j m_k \frac{\sin Qr_{jk}}{Qr_{jk}} \qquad (G2.5)$$

where N_A (mol^{-1}) is Avogadro's number and M is the particle molar mass (grams per mole).

G2.1.2 The Scattering Curve at Small Q Values: The Guinier Approximation, the Forward Scattered Intensity, and Radius of Gyration

Expanding $(\sin Qr_{jk})/Qr_{jk}$ in the Debye equation in terms of a series in powers of Qr_{jk} (Eq. (G2.6)), we obtain (for a single particle)

$$I(Q) = \sum_{jk} \frac{m_j m_k}{Qr_{jk}} \left(Qr_{jk} - \frac{Q^3 r^3_{jk}}{3!} + \dots \right) \qquad (G2.6)$$

where we can write

$$I(Q) = \left(\sum_j m_j \right)^2 \left(1 - \frac{1}{3} R_G^2 Q^2 + \dots \right) \qquad (G2.6a)$$

and

$$I(0) = \left(\sum_{jk} m_j m_k \right) = \left(\sum_j m_j \right)^2 \text{ and}$$

$$R_G^2 = \frac{1}{2} \sum m_j m_k r_{jk}^2 \Big/ \left(\sum_j m_j \right)^2 \qquad (G2.7)$$

Note that in a mechanical analogy R_G is the radius of gyration of the m_j distribution. It can also be written

$$R_G^2 = \sum m_j r_j^2 \Big/ \sum m_j \qquad (G2.8)$$

where r_j^2 is measured from the center of mass of the m_j distribution.

Note that the radius of gyration always appears as its square (R_G^2) in the equations. This is important. Since the contrast amplitudes, m_j, may be positive or negative, we may obtain physically meaningful negative (R_G^2) values (Comment G2.4).

In the 1930s, Guinier reported that a Gaussian function was a much better approximation of the Debye formula at low Q values than just truncating at the Q^2 terms of the series in Eq. (G2.6):

$$\begin{aligned} I(Q) &= I(0) \exp\left[-(1/3)R_G^2 Q^2\right] \\ \ln I(Q) &= \ln I(0) - (1/3)R_G^2 Q^2 \end{aligned} \qquad (G2.9)$$

Equation (G2.9) is known as the Guinier approximation. It is generally valid for the scattering curve of a particle of any shape, provided QR_G is smaller than or equal to about 1.

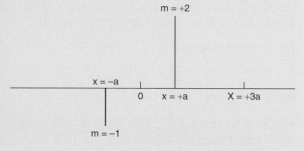
Scattering curves for a sphere and a long ellipsoid of axial ratio 1:1:5 are shown in Fig. G2.2. Also shown is the Gaussian function corresponding to the Guinier approximation. In order to observe the influence of particle shape,

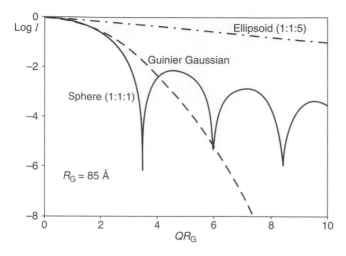

Fig. G2.2 Scattering curves of a sphere and an ellipsoid of revolution as functions of QR_G; the curve of the Guinier Gaussian is also shown. R_G is the radius of radiation in each case. Note how the curves coincide with the Guinier approximation at small QR_G.

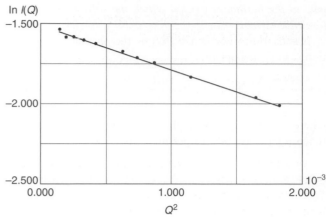

Fig. G2.3 A straight-line fit: ln $I(Q)$ versus Q^2 in the Guinier approximation. The example shows the small-angle region of neutron scattering data collected by one of the authors on a complex of the protein, seryl-tRNA synthetase, with seryl-tRNA.

independently of absolute size, the curves are drawn as a function of QR_G.

The Guinier approximation is valid to higher or lower QR_G values, depending on the shape of the particle. For ellipsoids of axial ratios about 1:1:1.7 or 1:1:0.6, the approximation is good significantly beyond the QR_G value of 1. For more compact shapes (e.g., the sphere, which is the most *compact shape*), the scattering curve deviates *below* the Guinier approximation, while for more asymmetric shapes (such as prolate or oblate ellipsoids with axial ratios 1:1: > 1.7 or 1:1: < 0.6, respectively, e.g., the 1:1:5 plot in the figure) the scattering curve deviates *above* the Guinier approximation.

During an experiment it is useful to plot the scattering curve as ln $I\langle Q \rangle$ versus Q^2 (Fig. G2.3). A straight line can then be fitted to the region where the Guinier approximation is valid. Two *experimental parameters* are obtained from the straight line: its intercept at $Q = 0$ (by extrapolation) and its slope, from which we can calculate (Eq. (G2.9)):

$$I(0) = \exp(\text{intercept})$$

and

$$R_G^2 = -3 \times \text{slope}$$

The fit is for 0.012 Å^{-1} < Q < 0.043 Å^{-1}. The radius of gyration found from the slope was 28.7 ± 0.2 Å, so that the fit corresponded to a range 0.33 < QR_G < 1.23. The $I(0)$ value calculated from the intercept of the straight line, together with the protein and tRNA concentrations, corresponded to the molar mass expected for a one-to-one complex.

When the assumptions outlined in Comment G2.3 are satisfied (i.e., the solution is effectively dilute and monodisperse), the scattering amplitude contrast of a

particle, $\sum m_j$, and its radius of gyration can be calculated from the intercept and slope, respectively, of the Guinier approximation fit to the scattering curve and the particle mass concentration. The relation between scattering amplitude contrast and molar mass is easily established from the chemical compositions of the particle and solvent (a calculation is given as an example in Comment G2.5).

The method allows one to determine independently and simultaneously the molar mass of the particle in solution (through the $I(0)$ value and concentration, in order to establish, for example, its oligomeric state in given solvent conditions), and its spatial extent (through the (R_G^2) value). This is not trivial information. Analytical centrifugation determines either molar mass (in a sedimentation equilibrium experiment) or a parameter related to molar mass and the translational frictional coefficient (the sedimentation value in a sedimentation velocity experiment) (see Part D). Gel filtration methods determine molar mass indirectly via the measurement of what is essentially a diffusion coefficient and calibration with known particles (see Part D). If the shape of the calibrating particles is different from that of the unknown particle then the measured molar mass value will be incorrect.

G2.1.3 Asymptotic Behavior of the Scattering Curve at Large Q Values: The Porod Relation

In the early 1950s, Porod introduced a particle characteristic function from which the asymptotic behavior of the scattering curve at large Q values can be calculated (Section G2.1.4). The condition to be fulfilled is that the particle has a well-defined envelope beyond which there is homogeneous solvent. Then, at high Q values, $I(Q)$ oscillates around an asymptotic line corresponding to S, the external surface area of the particle envelope. At high Q values the scattering curve decreases sharply, as Q^{-4}. Formally, a precise measurement of $I(Q)$ in this scattering

COMMENT G2.5 ABSOLUTE SCALE INTERPRETATION AND MOLECULAR MASS CALCULATIONS

For a particle of known composition we calculate its contrast amplitude per unit molar mass

$$\frac{1}{M}\sum_j m_j$$

where $\sum_j m_j = \sum_j f_j - \rho^\circ V$ and f_j are atomic scattering amplitudes, ρ° is the solvent scattering density and V is the particle volume. $V[cm^3]$ can be calculated from $V = M\bar{v}/N_A$ where M is in grams per mole and $\bar{v}$ is the partial specific volume in cubic centimeters per gram. It is usually a good approximation to assume a value between 0.73 and 0.75 for the partial specific volume of any protein (see Chapter A1).

Now M is not necessarily the molar mass of the particle in the solution. For example, if the particle is a monodisperse dimer, its molar mass in solution is $2M$. We therefore write M' for the mass in solution. Combining Eqs. (G2.5) and (G2.7) and substituting, we obtain

$$\frac{I_N(0)}{C} = \frac{N_A}{M'}\left(M'\frac{\sum_j m_j}{M}\right)^2$$

$$= N_A M'\left(\frac{\sum_j m_j}{M}\right)^2$$

so that for a monodisperse solution of non-interacting particles of known composition and partial specific volume the forward scattered intensity divided by the concentration is proportional to the molar mass.

Note that both $I_N(0)$ and C (see Comment A1.2) should be known in absolute units (e.g., cm^{-1}, gL^{-1}, respectively). In the case of SANS, absolute calibration by using the scattering of water has been shown to be effective (Jacrot and Zaccai, 1981). Absolute scale interpretation of the Guinier parameters in SANS is powerfully informative when combined with contrast variation on complex systems. The molecular weight and match point, for example, together provide a sensitive measure of composition and stoichiometry of membrane protein–lipid or protein–nucleic acid complexes.

G2.1.4 The Full Scattering Curve: The Distance Distribution Function

The scattering curve is measured in reciprocal space as a function of Q. What information does it contain on the particle structure in real space? We recall from Chapters G1 and A3 that reciprocal and real space are related mathematically by Fourier transformation.

In the SAS case, in which $I(Q)$ results from a rotational average over $|\mathbf{F}(\mathbf{Q})|^2$, the inverse Fourier transform of $I(Q)$ is a function that depends on the distance r between parts of the structure:

$$V\gamma(r) = \frac{1}{2\pi^2}\int_0^\infty Q^2 I(Q)\frac{\sin Qr}{Qr}dQ \qquad (G2.11)$$

This is similar to the Patterson function in crystallography, which is the inverse Fourier transform of *diffracted intensity*. In the equation, $I(Q)$ refers to a single particle of volume V (note that we are still within the assumptions of Comment G2.3). The function $\gamma(r)$ is a *correlation function*, defined as the average of the product of two scattering density contrast values separated by a distance r:

$$\gamma(r) = \langle(\rho(\mathbf{r}_j) - \rho^\circ)(\rho(\mathbf{r}_k) - \rho^\circ)\rangle \qquad (G2.12)$$

where $r = |\mathbf{r}_j - \mathbf{r}_k|$.

The function $V\gamma(r)$ is equivalent to a radial Patterson function. The integral from 0 to ∞ in Eq. (G2.11) is in fact equal to an integral from 0 to D_{max}, the *maximum distance* in the particle, since the scattering density contrast is zero for r greater than that value.

In the case of a particle of homogeneous scattering density ($\rho(\mathbf{r}) = \rho$), we can write

$$\gamma(r) = (\rho - \rho^\circ)^2\gamma_0(r) \qquad (G2.13)$$

where $\gamma_0(r)$ is known as Porod's characteristic function, which depends only on the particle geometry.

The same product of scattering density contrast values appears in Eq. (G2.12) and in the integral form of the Debye equation (Eq. (G2.4a)), which can now be written (as the Fourier transform of $V\gamma(\mathbf{r})$):

$$I(Q) = V\int 4\pi r^2\gamma(r)\frac{\sin Qr}{Qr}dr \qquad (G2.14)$$

or

$$I(Q) = \int p(r)\frac{\sin Qr}{Qr}dr \qquad (G2.15)$$

where

$$p(r) = 4\pi r^2 V\gamma(r) \qquad (G2.16)$$

The function $p(r)$ is called the *pair-distance distribution function* of the particle. It represents the number distribution of "distances" joining volume element pairs in the particle, weighted by the product of their scattering density contrast (Comment G2.6).

vector range allows a determination of S. In practice, however, Eq. (G2.10) is used in order to adjust the experimental background level (e.g., due to scattering from the solvent alone) so that after background subtraction from the solution scattering, $I(Q)$ follows a Q^{-4} dependence. This is because the particle scattering intensity is very weak at large Q values (2–3 orders of magnitude weaker than $I(0)$ and very sensitive to the experimental error in the background measurement).

$$\lim_{Q\to\infty}\frac{I(Q)Q^4}{2\pi I(0)} = S \qquad (G2.10)$$

COMMENT G2.6 THE DISTANCE DISTRIBUTION FUNCTION

A simple particle of unit volume is made up of two small volumes of unit scattering density contrast ($\rho = 1$) separated by unit distance ($r = 1$). The number of points a distance, r, away from each volume is proportional to the spherical surface area $4\pi r^2$.

Applying Eq. (G2.12) to derive $\gamma(r)$, the mean value of the product of ρ values separated by the distance 1 in the example in Fig. G2.6.1 is equal to $1/4\pi$. The corresponding $p(r)$ is also shown (Eq. (G2.16)). Note that $p(r)$ is in fact a *distance distribution function*. It tells us that there is one distance in the structure of value 1.

Figure G2.6.2 illustrates the same calculations as above for a particle made up of four unit scattering density contrast volumes arranged to form a square of unit side. The pair distribution function now tells us the particle contains four distances of 1 (the sides of the square) and two distances of $\sqrt{2}$ (the diagonals).

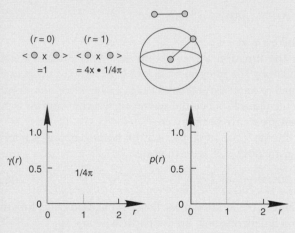

Fig. G2.6.1

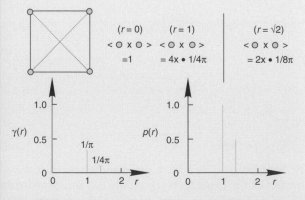

Fig. G2.6.2

The pair-distance distribution function is the inverse Fourier transform of $I(Q)$ (combining Eqs. (G2.11) and (G2.16))

$$p(r) = \frac{2}{\pi} r^2 \int_0^\infty Q^2 I(Q) \frac{\sin Qr}{Qr} dQ \qquad (G2.17)$$

Provided, therefore, that the scattering curve can be integrated from zero to infinity, it is possible to calculate the $p(r)$ function of the particle.

The measurement of $I(Q)$ to $Q = 0$ is, of course, not possible, and it is also impossible to measure the scattering curve to infinitely large Q. The evaluation of the integral can be performed if we can fill in the missing parts of the curve by extrapolation in the context of the single-particle assumptions (Comment G2.3): For example, the curve can be extrapolated to $Q = 0$ by using the Guinier approximation, and to $Q \to \infty$ at the high-Q end by using the Porod relation.

Pair-distance distribution functions for various particle shapes are drawn in Fig. G2.4.

G2.1.5 The Information Content in $p(r)$ and $I(Q)$ for a Monodisperse Solution of a Particle with a Well-Defined Envelope

The $p(r)$ (or $\gamma(r)$) and $I(Q)$ functions are different ways of expressing the same information. Ideally, each fully defines the other since they are related by Fourier transformation. Either function can be used, for example, to compare experimental results to a model calculation, although there are important practical reasons why one may be preferred instead of the other. Since $I(Q)$ is not known with infinite precision from $Q = 0$–∞, there are inherent assumptions in the calculation of $p(r)$: For example, the assumption that the data correspond to a particle with a well-defined solvent boundary, so that the Porod relation concerning the Q^{-4} asymptotic dependence can be applied at high-Q values. The information contained in $p(r)$ or $I(Q)$ sets constraints on the structure of the particle but, of course, does not unambiguously define this structure in terms of spatial coordinates. This can be understood intuitively. The structure of a particle (in real-space coordinates) can be calculated from the observation of waves (in reciprocal space) (see Chapter G1). Note that:

(1) Coordinates in real space and reciprocal space are vectors (the particle has a fixed orientation).
(2) The observed waves are defined by their amplitudes and phases (by using a complex number, for example).

In an SAS experiment information has been lost concerning these two points. Because of rotational averaging, the measurement is made as a function of Q and not of the vector $\mathbf{Q}$, and, of course, the phase information is lost, since only the intensity of the wave is observed.

The information content in the scattering curve is obtained by analyzing the properties of $p(r)$ and $I(Q)$. The

(a)

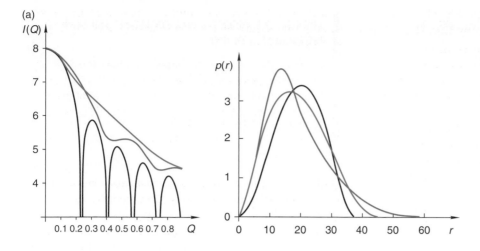

(b)

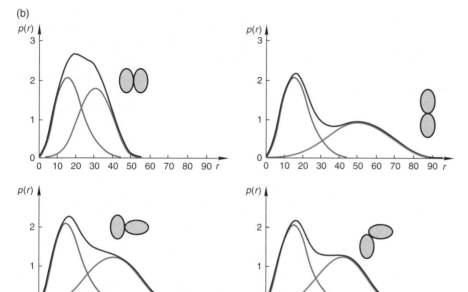

Fig. G2.4 (a) Scattering intensity and pair-distance distribution functions of a sphere (black), an oblate (flat) ellipsoid (green), and a prolate (elongated) ellipsoid (blue) of the same volume. **(b)** Pair-distance distribution functions of dimers. The blue line is the monomer function, the red line the dimer, and the green line the difference between them. (From O. Glatter.)

integral of $p(r)$ over the particle is equal to the square of its scattering amplitude contrast, which we recall from the Debye equation (Eq. (G2.4)) is equal to the forward scattered intensity $I(0)$:

$$\int_0^{D_{max}} p(r)dr = \left| \int (\rho(r) - \rho^\circ)dv \right|^2 = \left| \sum m_j \right|^2 = I(0) \qquad \text{(G2.18)}$$

COMMENT G2.7 THE RADIUS OF GYRATION

It is a useful check on the scattering curve to compare the radius of gyration calculated from the Guinier approximation at small Q values with the one obtained from the full scattering curve by using Eq. (G2.19). For example, the first may be wrong because of the presence of aggregation in the solution, the large particles affecting the scattering curve only at small Q values.

---

where D_{max} is the maximum distance (maximum value of r) in the particle.

The radius of gyration of scattering amplitude contrast is related to the second moment of the $p(r)$ distribution (Comment G2.7):

$$R_G^2 = \frac{\frac{1}{2} \int_0^{D_{max}} p(r)r^2 dr}{\int_0^{D_{max}} p(r)dr} \qquad \text{(G2.19)}$$

The second moment of the $I(Q)$ distribution is the Porod invariant:

$$C = \int_0^\infty I(Q)Q^2 dQ \qquad \text{(G2.20)}$$

Because of the integration to ∞, the Porod invariant is sensitive to the large Q tail of $I(Q)$, which usually has a large error.

The volume V and specific surface S/V of the particle (S is given in Eq. (G2.10)) are given by Eqs. (G2.21) and (G2.22), respectively:

$$V = \frac{2\pi^2 I(0)}{C} \tag{G2.21}$$

$$\frac{S}{V} = \lim_{Q\to\infty} \frac{I(Q)Q^4\pi}{C} \tag{G2.22}$$

The scattering curve, therefore, is consistent with all particle structures that present the corresponding values of $(\Sigma m)^2$, R_G^2, D_{max}, V and S.

G2.1.6 Polydisperse Solutions

Of the two assumptions discussed in Comment G2.3, we relax the assumption that the solution is monodisperse (the solution now contains particles that are not identical to each other), while maintaining the assumption about infinite dilution (there are no interactions between the particles).

Two experimentally useful cases can be distinguished. First, when the particles have similar scattering amplitude contrast radii of gyration, the scattering curve in the Guinier approximation is still fitted reasonably well by a straight line in a given Q range, where QR_G is smaller than or equal to about 1 for all particles (Fig. G2.5a). The measured $I(0)$ and R_G^2 values should now be considered as experimental parameters. They correspond to weighted averages over the scattering amplitude contrast values and radii of gyration of the particles in the solution

$$I(0) = \sum_n N_n \left(\sum_j m_j\right)_n^2 \tag{G2.23}$$

$$R_G^2 = \frac{\sum_n^n N_n \left(\sum_j m_j\right)_n^2 R_{Gn}^2}{\sum_n N\left(\sum_j m_j\right)_n^2} \tag{G2.24}$$

where there are N_n particles of scattering amplitude $(\sum_j m_j)_n$ and radius gyration R_{Gn}. The two equations present the basis for a set of very useful SAS experiments to

COMMENT G2.8 AMINOACYL tRNA SYNTHETASES AND THEIR INTERACTIONS WITH tRNA

An early SANS study of the interaction between methionine tRNA synthetase (MetRS) and tRNA^{Met} is a very good illustration of the rich potential of the application of SAS to an interacting system (Fig. G2.17, Dessen et al., 1978).

study complex formation and particle association in solution under different conditions (Comment G2.8).

We recall that the Guinier equation is a good approximation to a scattering curve at Q values for which $QR_G \sim 1$. The second experimentally useful case is when the particles have very different radii of gyration, then the Guinier approximation is still satisfied for each, but in a different Q range. This gives a broken line in the logarithm of I versus Q^2 plot if the particles fall into two sets, one of large particles and one of small ones (as when there is some aggregation in the solution), or a curved line when the particle size varies more smoothly (Fig. G2.5b, c). Small-angle scattering experiments to study each set of particles separately could be envisaged in a reasonable approximation if the size distribution is such that the Q ranges for Guinier analysis are distinct, as in Fig. G2.5b. On the other hand, practically no information can be deduced from a curved scattering curve other than that the solution is polydisperse in a given size range. An effort should then be made in the biochemical part of the experiment to reduce the polydispersity.

G2.1.7 Interactions Between the Particles

We now maintain the first assumption of Comment G2.3 concerning monodispersity, but relax the second assumption; the solution is no longer considered as infinitely dilute and each particle feels the presence of the others.

The mean distance between particles in a macromolecular solution as a function of concentration is calculated in

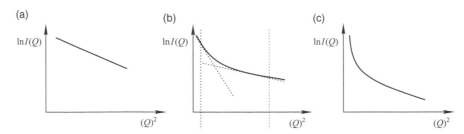

Fig. G2.5 (a) Guinier plot of a polydisperse solution of particles of about the same size. (b) Guinier plot of a solution containing small and large particles. The dotted black lines are straight-line fits to the two parts of the curve. The vertical red dotted line is where $QR_G = 1$ for the large particles. The vertical green dotted line is where $QR_G = 1$ for the smaller particles. In the example the slopes of the data lines are related by about a factor of 10 so that the radius of gyration of the smaller particle is $\sqrt{10} \approx 3$ times smaller than that of the larger particle. (c) Guinier plot of a solution containing particles of smoothly varying size.

There are N particles in a liter of solution of macromolecules of molar mass M g mol^{-1} and concentration C mg ml^{-1}:

$$N = N_A C/M$$

where N_A is Avogadro's number.

We estimate a mean distance, d, between them by assuming they are distributed on a grid in a cube of side 10 cm:

$$d = \left[(10 \times 10^8)/N\right]^{1/3} \text{Å}$$

As an example we calculate these values for hemoglobin in red blood cells. The protein concentration is about 300 g l^{-1}, a value that is similar to the concentration in cell cytoplasm in general. Hemoglobin has a molar mass of about 64 kDa. Then

$$N = 3 \times 10^{21}$$

and

$$d = 66\text{Å}$$

If approximated by a sphere, hemoglobin has a radius of about 30 Å. There is less than 10 Å or two layers of water molecules between particles in the 300 g l^{-1} solution.

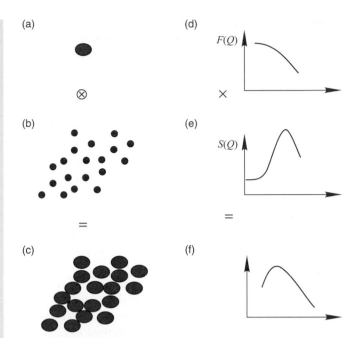

Fig. G2.6 (a) A quasi-spherical particle described by a scattering density distribution. (b) The distribution of particle centers of mass (points) in a liquid. (c) The quasi-spherical particles in a liquid described as a convolution of (a) and (b). (d) The "form" factor or scattered intensity of the particle in reciprocal space. (e) The "structure" factor of the liquid of point-like molecules. (f) The scattered intensity of the quasi-spherical liquid molecules resulting from the multiplication of (d) and (e). Recall that the Fourier transform of convolution is multiplication (see Chapter A3).

Comment G2.9. In-vivo protein concentrations are so high that in-vitro SAS experiments under similar conditions would be very far from satisfying the infinite dilution conditions. The position and orientation of each particle in a concentrated solution is strongly influenced by the others, even when there are no long-range interactions between particles, simply because there is not enough room around it for it to translate or rotate freely. For particles with approximately spherical symmetry (so that all orientations are equivalent) the influence of high concentrations on the SAS scattering curve can be calculated by using a liquid-like potential function, the interparticle interaction and packing being assimilated to that of molecules in a liquid. The problem is much more difficult to solve for asymmetric particles, unless the asymmetry is extreme (very long or very flat particles) in which case they align to form fibers or stacks.

The potential function describing molecules in a liquid, the corresponding particle distribution and the scattering factor (wave amplitude in reciprocal space as a function of Q) for point-like molecules are shown in Fig. G2.6.

We can now predict the scattering curve of a set of particles with spherical symmetry that is organized according to a given interparticle potential by applying the convolution–multiplication relation between real and reciprocal space. In the case of a repulsive potential, the scattering factor rises from Q equal to zero, whereas the form factor usually (but not always, see Section G2.3 on contrast variation) decreases. The result is a flattening of the scattering curve in the Guinier approximation region, leading to apparent radius of gyration and $I(0)$ values that are smaller than the ones corresponding to the particle in dilute solution. The scattering factor tends to 1 for Q values that are not much larger than those in the Guinier approximation region, depending on the proximity of the particles, so that the effect of interparticle interference on the scattering curve becomes negligible, and high macromolecular concentrations (where available) can be used for these "higher Q" measurements.

G2.2 MODELS AND SIMULATIONS

G2.2.1 From Structure to Scattering Curve

The X-ray or SANS curve of a known particle structure can be calculated accurately by using the Debye formula in its sum or integral form (Eqs. (G2.4 and G2.4a)).

Two Point Atoms and Two Spherical Atoms

The two-atom model is of particular interest for the label triangulation method (see below), in which a complex particle is "simplified" by contrast labeling. The scattering curve of two point atoms separated by a distance r_{12} has been calculated from the Debye formula and is given in Fig. G2.7. The scattering curve of two spherical atoms is also given in the figure, calculated by using the convolution multiplication Fourier transform relation (Chapter A3). In real space, the two-spherical-atom structure can be viewed as the convolution of two delta functions with a sphere. The scattering amplitude (in reciprocal space) of the structure is given by the product of the corresponding Fourier transforms.

Spheres and Ellipsoids

As in hydrodynamics, where friction coefficients were initially calculated for simple geometrical shapes (see Part D), a first approach to modeling scattering curves is in terms of ellipsoids. Provided the comparison is limited to very low resolution ($QR_G \leq 3$), the scattering curve of any globular particle can be simulated successfully as an ellipsoid with a given axial ratio.

The spherical particle is, of course, the extreme case. It is of particular interest because it is unique in that its internal structure expressed as the scattering density distribution as a function of radius, $\rho(r)$, directly corresponds to the rotational average. For the same reason, a sphere is also the only shape whose scattering amplitude, $F(Q)$, can be calculated directly from the square root of the scattering intensity, $I(Q)$. The scattering amplitude corresponds to the Fourier transform of the spherical distribution, which can be expressed in an analytical form by using a result first calculated by Rayleigh in 1911. The example of a homogeneous sphere is illustrated in Fig. G2.8, but the Fourier transform approach to obtaining a radial scattering density can be applied to any spherical distribution, including hollow spheres, for example, and has been used successfully to solve the internal structure of spherical viruses at low resolution (Comment G2.10).

In a range comprising larger QR_G values where the first subsidiary maximum appears, the information content is

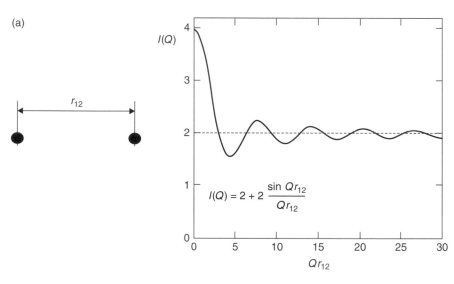

(a)

$$I(Q) = 2 + 2\,\frac{\sin Qr_{12}}{Qr_{12}}$$

(b)

Fig. G2.7 (a) Debye formula calculation (from Eq. (G2.3)) for two point atoms of unit scattering amplitude (delta functions) separated by r_{12}:

$$I(Q) = \sum_j \sum_k \frac{\sin Qr_{jk}}{Qr_{jk}}$$

$$I(Q) = 1 + 1 + 2\,\frac{\sin Qr_{12}}{Qr_{12}}$$

(b) Scattering curve of two spherical atoms of radius $a = r_{12}/4$ separated by r_{12}:

$$I(Q) = \left(2 + 2\,\frac{\sin Qr_{12}}{Qr_{12}}\right)$$

$$\left(3\,\frac{\sin Qa - Qa\cos Qa}{\langle Qa \rangle^3}\right)^2$$

(a)

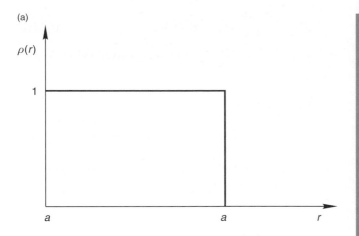

(b)

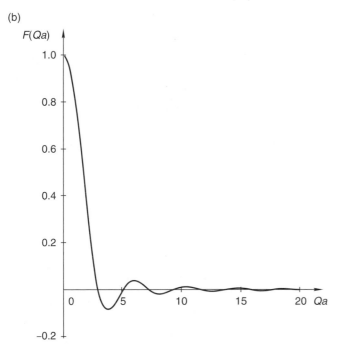

(c)

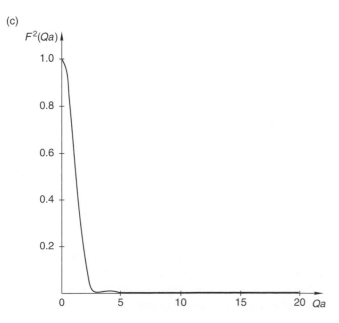

COMMENT G2.10 SPHERES AND ELLIPSOIDS

The special case of spherical particles for which (at low resolution) the spherical average structure is in fact equal to the structure has been exploited efficiently to solve the internal structure of spherical viruses by SANS and contrast variation (Jacrot et al., 1977).

Analytical integrals have been calculated for the scattering functions of ellipsoids based on the fact that for each orientation an ellipsoid presents to the beam a projection equal to that of a sphere. The intensity is then the sum of intensities from spheres of different radii. The result is a smoothing out of the spherical subsidiary maxima.

Formulae and numerical tables for the scattering intensity functions of three-axis ellipsoids are given in Mittelbach and Porod (1962).

higher and more complex shapes are required to simulate the curve successfully (see, for example, Fig. G2.4). Tabulated values of $I(Q)$ are available for oblate and prolate ellipsoids of various axial ratios as well as for a variety of other geometrical solids (Comment G2.10).

Rods and Sheets

SAS from two extreme types of asymmetric particle are of interest in structural biology, because of applications to DNA and filamentous complexes (such as certain viruses, for example), on the one hand, and membranes, on the other. These are particles where one dimension is much larger (by a factor of about 10) than the other two (rods) and particles where one dimension is much smaller (by a factor of about 10) than the other two (sheets). We can guess, intuitively, that fluctuations in the large dimensions affect the scattering curve in regions of small Q, and vice versa. When the ratio between the dimensions is large, the corresponding Q regions are well separated. We can apply this consideration to the Debye formula and derive expressions for the scattered intensity from very long or very flat

---

Fig. G2.8 (a) Scattering density contrast, $\rho(r)$; (b) scattering amplitude, $F(Qa)$; and (c) scattering intensity, $F^2(Qa)$, of a homogeneous sphere of radius a. The corresponding formulae are

(a) $\rho(r) = \begin{cases} 1 \text{ for } r \leq a \\ 0 \text{ for } r > a \end{cases}$

(b) $F(Q) = \rho V \left(3 \dfrac{\sin Qa - Qa \cos Qa}{(Qa)^3} \right)$

(c) $I(Q) = F(Q)^2$

where V is the volume of the sphere, chosen as equal to 1 in the figure.

particles that are equivalent to the Guinier approximation for globular particles.

In the case of a rod-like particle whose length $L >> r$, where r is its cross-sectional dimension (Fig. G2.9a), we define

$$I_c(Q) = I(Q)Q/\pi L$$

and a cross-sectional radius of gyration of scattering amplitude R_c:

$$R_c^2 = \frac{\sum_i m_i r_i^2}{\sum_i m_i}$$

where r_i are measured from the center of scattering mass of the cross-section. Then we have

$$I_c(Q) = \frac{(\sum m_i)^2}{L} \exp\left(-\frac{1}{2}R_c^2 Q^2\right) \qquad (G2.25)$$

which is equivalent to the Guinier approximation and valid for $QR_c \leq \sim 1$, and where L is the length corresponding to $M = \sum m_i$.

For a sheet-like particle for which $A^{1/2} >> t$, where A is the sheet area and t its thickness (Fig. G2.9b), we define

$$I_t(Q) = I(Q)Q^2/2\pi A$$

and a transverse radius of gyration of scattering amplitude R_t:

$$R_t^2 = \frac{\sum_i m_i t_i^2}{\sum_i m_i}$$

where t_i is measured from the center of scattering mass of the thickness. Then we have

$$I_t(Q) = \frac{\sum m_i}{A} \exp\left(-R_t^2 Q^2\right) \qquad (G2.26)$$

$$I_t(Q) = \frac{(\sum m_i)^2}{A} \exp\left(-R_t^2 Q^2\right) \qquad (G2.26a)$$

where $QR_t \leq \sim 1$ and A is the area corresponding to $M = \sum m_i$.

Atomic Coordinates

A scattering curve can be calculated unambiguously from the atomic coordinates of a macromolecular structure. The calculation in vacuum is a straightforward application of the Debye formula, putting in the atomic coordinates and scattering amplitudes r_j and f_j, respectively. In solution, however, a volume, v_j, has to be attributed to each atom in order to calculate the contrast amplitude of each atom, $m_j = f_j - \rho^o v_j$, and to account for the solvent contribution. Different programs are now available to perform such calculations. Note that because the neutron scattering density, ρ^o, of H_2O is small and negative (Fig. G2.10), the calculation is not very sensitive to the volume terms in ($m_j = f_j - \rho^o v_j$) and the neutron scattering curve of a macromolecule in H_2O solvent may be treated to a good approximation as if the macromolecule were in vacuum.

Programs are available to calculate X-ray (e.g., CRYSOL) or neutron (e.g., CRYSON) scattering curves from crystal structure atomic coordinates (Comment G2.11). The calculation of neutron scattering curves requires hydrogen

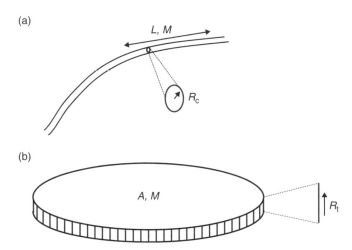

(a)

Fig. G2.9 (a) Rod-like particles of scattering mass per length M/L and cross-sectional radius of gyration R_c. **(b)** Sheet-like particles of scattering mass per area of M/A and transverse radius of gyration R_t.

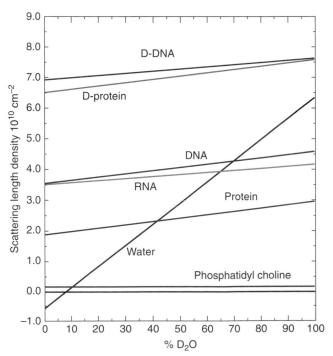

Fig. G2.10 Neutron scattering densities of various natural abundance and fully deuterated biological molecules as a function of heavy water percentage in the solvent. Neutron scattering amplitudes of amino acids and nucleotides are tabulated in Jacrot (1976), together with their partial specific volumes and exchangeable hydrogens.

COMMENT G2.11 SCATTERING CURVES FROM ATOMIC COORDINATES

The reference for the CRYSOL program is Svergun *et al.* (1995). See the web page of Dmitri Svergun for CRYSON and recent developments.

COMMENT G2.12 SPHERICAL HARMONICS

The approach is discussed in Svergun *et al.* (1996).

atom positions, which are usually not included in atomic coordinate files and have to be placed in the structure by using chemical arguments. These programs evaluate the expression

$$I(Q) = \left\langle |A_a(Q) - \rho^o A_s(Q) + \delta\rho_b A_b(Q)|^2 \right\rangle \quad (G2.27)$$

where $A_a(Q)$ is the scattering amplitude of the particle *in vacuo* calculated from the atomic positions, $\rho^o A_s(Q)$ is the scattering amplitude of the solvent excluded volume, and $\delta\rho_b A_b(Q)$ is the scattering amplitude of the hydration layer around the particle. The averaging is over all particle orientations. $A_a(Q)$ is calculated from atomic form factors and positions. The amplitude $A_s(Q)$ is evaluated by placing Gaussian spheres (dummy atoms) at the atomic positions and the particle envelope is represented by connecting the center of mass of the particle with the most distant atom in each direction. The hydration shell is modeled with a thickness and location corresponding to one layer of water (Section G2.3.1). The excluded particle volume and scattering density in the hydration shell cannot be obtained accurately from the crystal structure. If experimental data are available, these two parameters can be varied to minimize the discrepancy between the observed and calculated scattering curves.

G2.2.2 From the Scattering Curve to a Set of Structures

The *inverse scattering problem* arises because it is not possible to reconstruct a unique structure from SAS data. Because of loss of phase information and rotational averaging, there is no "inverse Debye formula," and, in principle, a very large number of structures may be compatible with the experimental scattering curve. A number of different methods, however, are being developed that successfully reduce the ambiguity and allow the calculation of sets of shapes and even low-resolution structures from measured scattering curves. The fit may be performed on the measured $I(Q)$ itself, or on its Fourier transform, the $p(r)$ function. The experimental invariants of the procedure, with which the model structure should be consistent, are $\sum m_i$, R_G^2, D_{max}, V, and S (Section G2.1.5).

Spherical Harmonics

Ab initio shape determinations from SAS data have been proposed that express the particle envelope in terms of spherical harmonics of progressively higher order

(Comment G2.12). Non-linear minimization techniques are applied to find the best fit between the calculated structure and observed scattering curve. The method works well with roughly globular shapes, which can be approximated by a low number of spherical harmonics, but particles of more complex shape with a re-entrant surface, for example, cannot be modeled successfully by this method.

Monte Carlo and Simulated Annealing Algorithms

Monte Carlo and simulated annealing methods have been proposed for *ab initio* low-resolution structure determinations from SAS data by several authors (Comment G2.13). They are based on more-or-less random searches of a pre-defined *configurational space* for structures that best fit the data, and require rapid calculations of scattering curves from large numbers of models and the definition of some goodness-of-fit parameter or "energy" function to be minimized. In one approach, the starting structure is a sphere of diameter equal to the maximum particle dimension (calculated from the scattering curve, via the $p(r)$) filled with much smaller spherical *dummy* atoms. It is straightforward to calculate the scattering curve corresponding to the model and to compare it with the experimental curve. Further constraints on the model (expressed, for example, as "penalties" in the "energy" function) can be applied, for example, with respect to the compactness of the model or the continuity of the occupied volume with no disconnected parts. The method searches for a configuration of the dummy atom model that minimizes a function expressing the discrepancy between the calculated and experimental scattering curves.

Genetic Algorithms

In an alternative approach, a *genetic algorithm* has been developed to execute the search for the "best" models to fit an experimental scattering curve (Comment G2.13). Starting models are generated with spherical beads (whose radius is chosen to be small compared with the resolution of the experiment) placed randomly on a grid. The grid represents a restricted search space, which is reasonably compatible with the available information on maximum dimension, volume, and radius of gyration. Typically the starting model set consists of a few hundred structures named "chromosomes" obtained by filling the search space with different numbers of beads. After an "evaluation and fitness" step the population is allowed to "reproduce." The part of the chromosome population that provides scattering curves that fit closest to the data is kept to become part of the next generation (usually corresponding to an

COMMENT G2.13 *MONTE CARLO* AND *GENETICS* METHODS IN SAS: REFERENCES

Svergun, D. I. (1999). Restoring low resolution structure of biological macromolecules from solution scattering using simulated annealing. *Biophys. J.*, **76**, 2879–2886.

Walther, D., Cohen, F. E., and Doniach, S. (2000). Reconstruction of low-resolution three dimensional density maps from one dimensional small-angle X-ray solution scattering data for biomolecules. *J. Appl. Crystallog.*, **33**, 350–363.

Svergun, D. I. (2000). Advanced solution scattering data analysis methods and their applications. *J. Appl. Crystallog.*, **33**, 530–534.

Svergun, D. I., Malfois, M., Koch, M. H. J., Wigneshweraraj, S. R., and Buck, M. (2000). Low-resolution structure of the sigma54 transcription factor revealed by X-ray solution scattering. *J. Biol. Chem.*, **275**, 4210–4214.

The program DAMMIN (Svergun, 2000; Svergun *et al.*, 2000), which takes a simulated annealing approach, also allows the input of information about the point symmetry of the particle.

Chacón, P., Moran, F., Díaz, J. F., Pantos, E., and Andreu, J. M. (1998). Low-resolution structures of proteins in solution retrieved from X-ray scattering with a genetic algorithm. *Biophys. J.*, **74**, 2760–2775.

Goldberg, D. E. (1989). *Genetics Algorithms in Search, Optimisation and Machine Learning*. San Mateo, CA: Addison-Wesley.

Chacón, P., Díaz, J. F., Morán, F., and Andreu, J. M. (2000). Reconstruction of protein form with X-ray solution scattering and a genetic algorithm. *J. Mol. Biol.*, **299**, 1289–1302.

The genetics approach is implemented in the program DALALGA (Chaćon *et al.*, 2000).

arbitrary fraction of 50% of the total population), while the rest of the chromosomes are eliminated to make room for new ones. Applying two genetics operators to randomly chosen chromosomes in the "best-fit" set generates a new population set. The "cross-over" operator exchanges information in two parent chromosomes according to predefined rules to yield two offspring. The "mutation" operator creates new chromosomes by copying members of the "best-fit" set with a certain (small) error rate. The "evaluation and fitness" step is then applied to the new population and the cycle is repeated until the "maximum fitness" no longer improves and convergence is obtained.

G2.3 GENERAL CONTRAST VARIATION: PARTICLES IN DIFFERENT SOLVENTS "SEEN" BY X-RAYS AND "SEEN" BY NEUTRONS

The concept of *contrast* and its application to low-resolution diffraction were discussed in Chapter G1. Contrast can be modified, in general, by changing the atomic amplitudes, f_i, and/or solvent scattering density, ρ°. Atomic amplitudes are modified by isotope labeling (for neutron scattering only) or by changing the incident radiation (since amplitudes are different for X-rays and neutrons). The solvent density can be changed by isotope labeling (for neutron scattering only), by changing the incident radiation, or by changing the composition of the solvent (the electron density of a salt or sugar solution, for example, is significantly different from that of water). The fundamental assumption of a contrast variation experiment is that *the particle structure is not modified by changing the contrast conditions*. It is important, therefore, to verify that this is the case when isotope labeling or changes in solvent composition are applied. Isotope labeling for a neutron scattering experiment, for example, does not affect electron density contrast, so that an X-ray scattering curve provides a good control that the structure has not been affected by the labeling.

Mean neutron scattering amplitude densities of biological macromolecules and water as functions of solvent water isotopic content (heavy water, D_2O, percentage) are given in Fig. G2.10. Because atomic compositions are fairly similar within each macromolecular family, it is usually quite reliable to assume the appropriate mean value for a protein or nucleic acid of unknown composition.

For comparison, the electron density of H_2O or D_2O is $0.333e/\text{Å}^3$, corresponding to 9.3×10^{10} cm^{-2} in X-ray scattering amplitude density (see Chapter G1). The mean X-ray scattering amplitude density of protein is close to 12.3×10^{10}cm^{-2}, while that of nucleic acid is slightly higher. Contrast in any solvent condition is proportional to the difference between the macromolecule and solvent scattering densities. The macromolecule scattering density contrast depends on the radiation used, as well as on the solvent composition. Recall that the scattering particle includes the hydration shell, which may itself have a non-negligible contrast with respect to the solvent (see also Chapter A3). The same particle, therefore, presents different scattering curves in SAXS and SANS from H_2O solution and in SANS from D_2O solution, to yield very useful information for solving complex structures.

G2.3.1 Two-Component Particles and the Parallel Axes Theorem

The generalized contrast variation approach was applied to ribosomes in the early 1970s. The analysis developed is valid for any two-component particle in which the scattering densities of the components are sufficiently different and for a resolution at which each component can be considered as homogeneous (see Chapter G1). Ribosomes are nucleoprotein complexes made up of several proteins and large RNA molecules. The two-component assumption and resolution condition are fulfilled for Q values in the Guinier approximation; $Q \sim 1/R_G$, where R_G is the radius of gyration of the particle.

Applying Guinier analysis to the two-component particle, the forward scattered intensity and radius of gyration squared for one particle can be written, respectively, as

$$I(0) = (m_1 + m_2)^2 \qquad (G2.28)$$

where $m_1 = f_1 - \rho^\circ V_1$, $m_2 = f_2 - \rho^\circ V_2$, and

$$R_G^2 = \frac{m_1 R_1^2 + m_2 R_2^2}{m_1 + m_2} \qquad (G2.29)$$

where the subscript 1 refers to the RNA component and the subscript 2 to the protein component, and R_1 and R_2 are the radii of gyration about the center of scattering mass of the particle.

The square root of the scattered intensity should vary linearly with solvent scattering density; this is an important check on the experiment. For example, if sample polydispersity varies with the solvent, the square root of $I(0)$ calculated from the experimental data does not correspond to that of a single particle and a plot of it versus solvent density plot is not a straight line.

Formally, the radius of gyration in SAS follows the same rules as the radius of gyration in mechanics. The parallel axes theorem states that the square of the radius of gyration of a body around an axis is equal to the square of its radius of gyration about a parallel axis through the center of mass of the body plus the square of the distance between the two axes (Fig. G2.11a).

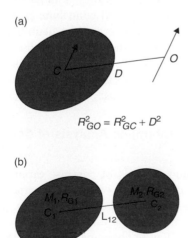

$$R_{GO}^2 = R_{GC}^2 + D^2$$

Fig. G2.11 (a) The parallel axes theorem. **(b)** The radius of gyration of a two-component body, with centers of mass at C_1, C_2, respectively, is given by Eq. (G2.31).

Applying the theorem to an n-component body, we obtain

$$(m_1 + m_2 + ...m_n)R_G^2 = m_1\left(R_{G1}^2 + D_1^2\right) + m_2\left(R_{G2}^2 + D_2^2\right)$$
$$+... + m_n\left(R_{Gn}^2 + D_n^2\right) \qquad (G2.30)$$

where R_{Gi} is the radius of gyration of compound i around its own center of scattering mass and distances D_i are between the center of scattering mass of each component and that of the complex. Combining Eq. (G2.30) and the equation defining the center of mass of the complex ($\Sigma m(\mathbf{r} - \mathbf{r}_{c\ of\ m}) = 0$), we obtain, for a two-component particle

$$R_G^2 = \frac{R_{G1}^2 m_1}{(m_1 + m_2)} + \frac{R_{G2}^2 m_2}{(m_1 + m_2)} + \frac{L_{12}^2 m_1 m_2}{(m_1 + m_2)^2} \qquad (G2.31)$$

where L_{12} is the distance between the centers of mass of 1 and 2. Equation (G2.31) has the advantage of not requiring knowledge of the center of mass coordinates (Fig. G2.11b).

By defining the respective *scattering fraction* for each component, respectively

$$x = \frac{m_1}{m_1 + m_2}, \qquad (1 - x) = \frac{m_2}{m_1 + m_2}$$

Eq. (G2.31) reduces to

$$R_G^2 = x R_{G1}^2 + (1 - x)R_{G2}^2 + x(1 - x)L_{12}^2 \qquad (G2.32)$$

The radius of gyration of each component within the complex can be calculated by extrapolation or interpolation from measurements in different solvent conditions (Fig. G2.12).

At $x = 0$ (zero contrast for component 1), the radius of gyration of component 2 is measured, and vice versa at $x = 1$. The curve is parabolic, so that at least three data points that are well spread out in the solvent scattering density range (assuming small errors!) are required to determine the parameters R_{G1}^2, R_{G2}^2, and the distance between the centers of mass of the components 1 and 2. Figure G2.12 shows the example of a ribosome study (Comment G2.14), in which data from SAS "views" of the particles by X-rays, (visible) light and neutrons in H_2O and D_2O are plotted together to yield the radii of gyration of the RNA and protein components, respectively.

Contrast Variation to See the Hydration Shell

SAS experiments resolved a controversy concerning the density of the hydration shell around proteins (see Chapter A3). X-ray and neutron data from H_2O and D_2O were analyzed together for a number of proteins, in the context of a two-component model made up in each case of the protein, which was modeled from its atomic coordinates, and the hydration shell (Fig. G2.13). As seen by X-rays, the protein component electron density is higher than that of a dense hydration shell, which is itself higher than that of the bulk solvent. For neutrons in H_2O, only the protein component is observed. The hydration shell scattering and bulk solvent

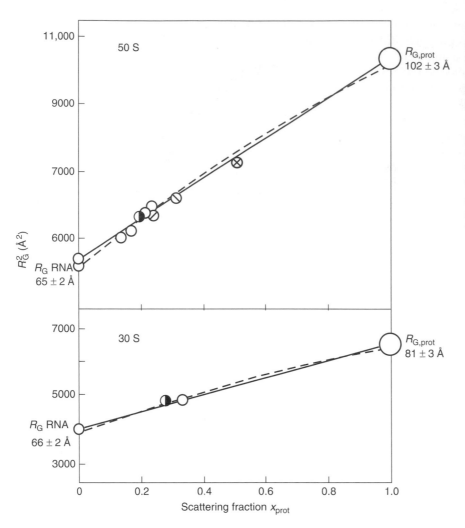

Fig. G2.12 Contrast variation radius of gyration plots of the 50 S and 30 S particles of the *E. coli* ribosome according to Eq. (G2.32). The particles are made up of two components: protein and RNA. The different data points are for measurements by X-ray, light, and neutron scattering in H_2O and D_2O buffers. The radius of gyration of the protein is obtained at the scattering fraction $x_{prot} = 1$; the radius of gyration of the RNA is obtained at $x_{prot} = 0$. (Serdyuk *et al.*, 1979.) (Figure reproduced with permission from Elsevier.)

scattering density are close to zero. For neutrons in D_2O, the hydration shell density is higher than that of the solvent (positive contrast), while that of the protein is lower (negative contrast).

COMMENT G2.14 RIBOSOMES

Ribosomes, large organized structures made up of several proteins and RNA molecules, are the main sites of protein synthesis in cells. Ribosomes were discovered by analytical centrifugation and labeled in terms of their sedimentation coefficients (see Part D). The bacterial ribosome is a 70 S particle representing the association of a 50 S "large" subunit and a 30 S "small" subunit. The crystallization of ribosome particles and the solving of their crystal structures are major triumphs of crystallography (see Chapter G3). X-ray SAS and SANS results, such as the ones in Fig. G2.12 and in Section G2.3.3, however, made important contributions to the understanding of ribosome structure and paved the way for the crystallographic success.

The result plotted for lysozyme in Fig. G2.13 established the existence of a dense hydration shell. Recall that the scattering curve of a larger particle falls more steeply with Q. The particle seen by X-rays is largest, because both protein and hydration shell contribute positive contrast; next in size is the particle seen by neutrons in H_2O, for which the hydration shell is essentially invisible; the "smallest" is the particle seen by neutrons in D_2O, for which the positive contrast of the hydration shell acts against the negative contrast of the protein part.

G2.3.2 The Stuhrmann Analysis of Contrast Variation

The Stuhrmann analysis of contrast variation interprets the contrast dependence of the full scattering curve in terms of density fluctuations in the particle. A general set of equations for contrast variation has been derived by analyzing the effect of changing solvent scattering density on the scattering curve of a particle of non-homogeneous scattering density.

The particle is described, in an integral notation, by a scattering density distribution $\rho(\mathbf{r})$, which is divided into

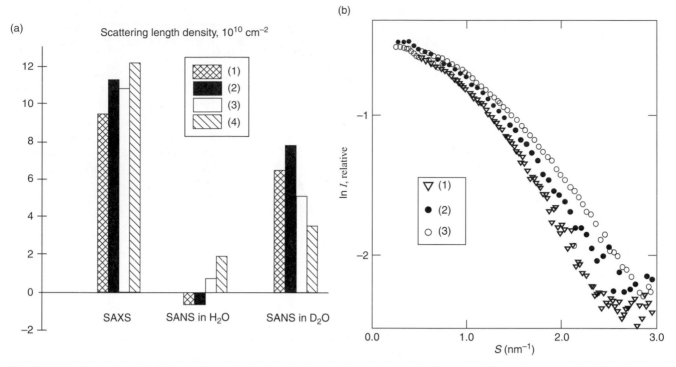

Fig. G2.13 (a) Scattering length densities of protein, hydration shell, and solvent for X-rays (SAXS) and neutrons (SANS) in H_2O and D_2O:**(1)** solvent; **(2)** a hydration shell 20% denser than bulk solvent; **(3)** disordered protein side-chains entering the hydration shell;**(4)** protein (the higher density of the protein in D_2O for neutrons is due to the exchange of labile hydrogen atoms).**(b)** Contrast variation scattering curves of lysozyme: **(1)** is from X-rays; **(2)** from neutrons in H_2O; and **(3)** from neutrons in D_2O. The curves have been normalized to the same scattering intensity at zero angle. The S on the x-axis corresponds to our parameter Q. Note that 1.0 nm^{-1} = 0.1 Å^{-1} (Svergun *et al.* 1998). (Figure reproduced with permission from the National Academy of Sciences, USA.)

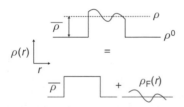

Fig. G2.14 Mean contrast and density fluctuations in Stuhrmann analysis. (Koch and Stuhrmann, 1979.)

two parts: the mean scattering density, ρ, and fluctuations about the mean, $\rho_F(\mathbf{r})$, such that (Fig. G2.14)

$$\int_V \rho_F(\mathbf{r})d\mathbf{r} = 0$$

Then the contrast at $\mathbf{r}$ in the particle can be written

$$\rho(\mathbf{r}) - \rho^\circ = \rho + \rho_F(\mathbf{r}) - \rho^\circ = \bar{\rho} + \rho_F(\mathbf{r})$$

where $\bar{\rho}$ is the mean contrast density of the particle. Putting this into the integral form of the Debye formula:

$$I(Q) = \iint (\bar{\rho} + \rho_F(\mathbf{r}_i))(\bar{\rho} + \rho_F(\mathbf{r}_j)) \frac{\sin Qr_{ij}}{Qr_{ij}} dV_i dV_j \quad \text{(G2.33)}$$

we obtain

$$I(Q) = \bar{\rho}^2 I_V(Q) + \bar{\rho} I_{VF}(Q) + I_F(Q) \quad \text{(G2.34)}$$

$I_V(Q)$, $I_{VF}(Q)$, and $I_F(Q)$ are the three characteristic functions of the particle:

$I_V(Q)$ is the scattering curve at infinite contrast; it represents the scattering of a particle of the same shape as the one under study, but of homogeneous scattering density.

$I_F(Q)$ is the scattering curve at zero contrast; it represents the scattering from density fluctuations inside the particle.

$I_{VF}(Q)$ is a cross-term.

Note that the splitting of the intensity into the three characteristic functions can still be performed in the case of non-homogeneous exchange of particle labile hydrogen atoms in a SANS H_2O/D_2O contrast variation experiment. The interpretation of the functions, however, is not as straight forward as described above.

At small values of Q, the forward scattered intensity is obtained from the Guinier approximation, and is equal to

$$I(0) = (\bar{\rho}V)^2 = \{(\rho - \rho^\circ)V\}^2 \quad \text{(G2.35)}$$

The square root of the forward scattered intensity is linear with solvent scattering density, ρ°, and it crosses zero at the particle *contrast match point*, where $\rho^\circ = \rho$ (Fig. G2.15a).

The contrast dependence of the radius of gyration is given by

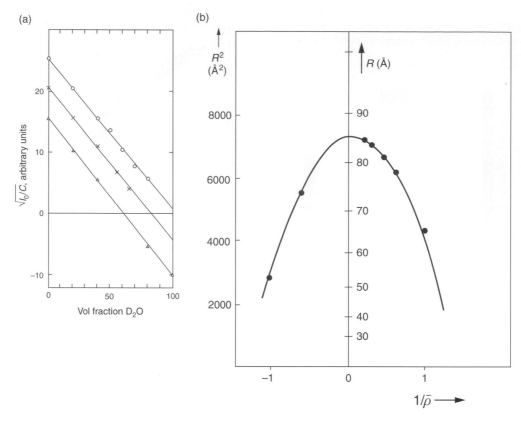

Fig. G2.15 The Stuhrmann analysis parameters:(a) square root of the forward scatter of 50 S ribosomal subunits as a function of D_2O volume fraction in the solvent. The three preparations were derived from cells grown on H_2O and different proportions of D_2O resulting in different contrast match points indicating levels of deuteration; (b) radius of gyration squared as a function of inverse contrast for 70 S ribosomes reconstituted from protonated 50 S and deuterated 30 S subunits – the strongly parabolic shape indicates the large separation between the centers of mass of the dense and less dense scattering regions (Koch and Stuhrmann, 1979). (Figure reproduced with permission from Elsevier.)

$$R_G^2 = R_V^2 + \frac{1}{\bar{\rho}V}\int \rho_F(\mathbf{r})\mathbf{r}^2 dV + \frac{1}{\bar{\rho}^2 V^2}\left\{\int \rho_F(\mathbf{r})\mathbf{r}dV\right\}^2 \quad (G2.36)$$

where we write

$$R_G^2 = R_V^2 + \alpha/\bar{\rho} + \beta/\bar{\rho}^2 \quad (G2.37)$$

R_V^2 is the radius of gyration of a particle of the same shape as the particle under study but of homogeneous density. α is the second moment of the density fluctuations in the particle. A positive value of α denotes that the higher density is toward the periphery of the particle, and vice versa. $\beta/\bar{\rho}^2$ is the separation between the center of scattering mass of the density fluctuations in the particle and the center of mass of the particle shape, when the mean contrast conditions correspond to $\bar{\rho}$. Plots of R_G^2 versus $1/\bar{\rho}$ are shown in Fig. G2.15b:

α positive: higher scattering density in the particle lies further away from its center on average than lower scattering density;

α negative: lower scattering density in the particle lies further away from its center on average than higher scattering density;

$\beta = 0$, the centers of scattering mass and density fluctuations coincide.

Recalling that three structural parameters are required to describe a two-component particle (the two radii of gyration about their respective scattering mass centers and the distance separating them), it is evident from the above equations that, in a model approach, a maximum of two particle components can be "resolved" in a contrast variation experiment from the three independent experimental parameters obtained from the scattering curve as a function of contrast at small Q values.

G2.3.3 Deuterium Labeling and Triangulation

A varied methodology has been developed to apply contrast variation concepts in SANS, by using deuterium labeling (Comment G2.15). We recall the simplicity of the scattering curve of a "two-atom" structure, and how it can be interpreted to define the distance between the atoms (Section G2.2). The principle of the label triangulation method is to deuterate a complex structure selectively in such a way that the scattering pattern is dominated by the equivalent of a

---

COMMENT G2.15 DEUTERIUM LABELING CONTRAST VARIATION

Specific labeling of biological macromolecules and complexes can be achieved by biosynthesis approaches and reconstitution. *E. coli* strains are now available that grow well in a fully deuterated medium for the expression of chosen proteins. Label triangulation using specific deuteration has been applied to the ribosome (Capel *et al.*, 1988) and the results on protein positions have been used later to solve the crystal structure (see Comment G2.14). The *triple isotopic substitution method* is based on contrast variation from different deuteration levels in the sample. It has the advantages that it can be used on concentrated solutions because it cancels out interparticle effects, and in H_2O solvents only, if D_2O has adverse effects on the sample (Serdyuk *et al.*, 1994).

COMMENT G2.16 TIME-AVERAGE FLUCTUATION THEORY

Time-averaged fluctuation theory as applied to scattering phenomena was discussed in detail by Eisenberg (1981).

The invariant particle hypothesis was developed by Luzzati and Tardieu (1980).

---

pair of atoms in order to measure the distance between them. Depending on the resolution and size of the complex, the "atoms" can be individual proteins, as in the case of the ribosome, for example. The complex structure is then "built up" from its components, pair distance by pair distance by using the surveying triangulation method.

G2.4 THE THERMODYNAMICS APPROACH IN SAS

G2.4.1 Fluctuations in Hydrodynamics and Scattering

So far in this chapter we have presented the particle approach to SAS. It assumes that radiation is scattered by particles, which have well-defined boundaries separating them from the surrounding solvent. The assumption is intrinsic, for example, in the derivation of Porod's law or the Debye equation. Historically, Rayleigh used a particle approach when he developed his theory on light scattering from the atmosphere. The particle approach, however, is not useful to account for generally more complicated scattering problems, and at the beginning of the twentieth century Smoluchowski and Einstein developed alternative theories based on the analysis of density fluctuations in the scattering system. We could look at density fluctuations as "transient particles," which scatter radiation in exactly the same way as a classical particle. The contrast in the case of density fluctuations need not arise from a different chemical composition between particle and solvent as it results from the difference in density itself. In this view, even in a pure liquid, thermal density fluctuations lead to scattering. A liquid has a finite compressibility so that its molecules cannot approach beyond certain limits; at the other extreme, the molecules cannot move too far away from each other as they would in a gas. Under the influence of thermal energy, molecules in the liquid state form dynamic

clusters by moving toward and away from each other under attractive and repulsive forces, leading to the density fluctuations.

Fluctuation theories have been developed to interpret SAS from interacting macromolecules in solution, in terms of particle–solvent and particle–particle interactions. In 1981, Eisenberg published a review discussing the application of time-averaged fluctuation theory to scattering phenomena, clearly establishing a formal link between sedimentation equilibrium (see Part D) and the forward scattering of light, X-rays, and neutrons (Comment G2.16). The thermodynamic basis of the theory is the relation between the observed scattering and directly measurable (in principle anyway) bulk parameters of the solution. For example, the number of particles in the solution is not a measurable quantity since we are unable to count the particles as we introduce them into the sample container. On the other hand, the concentration of the solute is a measurable quantity because we can weigh so many grams of material and add them to a given volume or mass of solvent.

The hydrostatic osmotic pressure (a measurable bulk property) due to a solution of N particles in a volume V is given by an equation similar to the perfect gas equation (see Chapter A1):

$$\Pi V = NRT \tag{G2.38}$$

Osmotic pressure constitutes a colligative property of the solution, because its measurement essentially "counts" the number of particles. Now, if we know how much mass of solute we have added to make up the solution, we can define a molar mass for the particles:

$$M = CV/N \tag{G2.39}$$

where C is solute concentration. If C is in grams per milliliter, M is in grams per mole. Then

$$\Pi = (C/M)RT \tag{G2.40}$$

Note that the measurement is not an absolute determination of molar mass, because it determines the ratio C/M in units of moles per milliliter, and not either of these quantities separately. The units of M are commensurate to those of C. For example, if C for a DNA sample is given in terms of "phosphate groups per milliliter," the units of M are "phosphate groups per mole."

We consider a solution made up of three components: water, a macromolecule (which is too large to diffuse across a dialysis membrane), and a small solute, e.g., salt (which diffuses across the membrane). In order to have well-defined solvent conditions, the solution is dialyzed against a large volume of solvent. It is useful to indicate the various solution components by suffixes: 1 for water, 2 for the macromolecule, 3 for the small solute.

The application of fluctuation theory to sedimentation equilibrium and forward scattering of light, X-rays, and neutrons in the solution leads to interestingly analogous equations, in terms of the derivative of osmotic pressure with respect to macromolecule concentration C_2 and $(\partial \rho / \partial C_2)_\mu$, $(\partial n / \partial C_2)_\mu$, $(\partial \rho_{el} / \partial C_2)_\mu$, and $(\partial \rho_n / \partial C_2)_\mu$, which are the mass density, refractive index, electron density, and neutron scattering density increments, respectively, at constant chemical potential μ_1 and μ_2 of diffusible solutes (see Chapter A1). The derivative of osmotic pressure with respect to C_2 can itself be written in terms of the molar mass M_2.

At vanishing C_2, the equations reduce to

$$\frac{1}{RT}\frac{d\Pi}{dC_2} = M_2^{-1} + A_2 C_2 + A_3 C_2^2 + \dots \tag{G2.41}$$

where A_i are the virial coefficients, expressing the interparticle interactions in non-dilute solutions (from the Latin *vires* meaning forces).

Sedimentation Equilibrium

$$d \ln C_2 / dr^2 = (\omega^2 / 2RT)(\partial \rho / \partial C_2)_\mu M_2 \tag{G2.42}$$

where r is distance to the center of the rotor, ω is the angular velocity and $(\partial \rho / \partial C_2)_\mu$ is the mass density increment at constant chemical potential of diffusible solvent components (see below).

Light Scattering

$$\Delta R(0) = K(\partial n / \partial C_2)_\mu^2 C_2 M_2 \tag{G2.43}$$

where $\Delta R(0)$ is the light scattering in the forward direction (zero angle) in excess of solvent scattering and K is an optical constant.

X-ray Scattering

$$I_{el}(0) = K_{el}(\partial \rho_{el} / \partial C_2)_\mu^2 C_2 M_2 / N_A \tag{G2.44}$$

where $I_{el}(0)$ is forward scattering of X-rays, K_{el} is a calibration constant, and N_A is Avogadro's number.

Neutron Scattering

$$I_n(0) = K_n(\partial \rho_n / \partial C_2)_\mu^2 C_2 M_2 / N_A \tag{G2.45}$$

where $I_n(0)$ is forward scattering of neutrons and K_n is a calibration constant.

The density increments are bulk properties of the solution. We treat the mass density increment as an example (Fig. G2.16). The presence of a macromolecule in a solution perturbs the solvent around it. If, for example, it binds salt

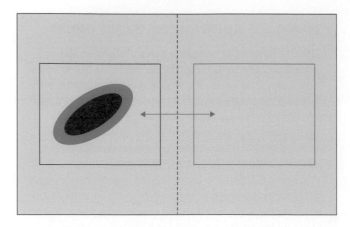

Fig. G2.16 Mass density increment. In a dialysis experiment of a particle (red ellipse) that cannot cross the membrane (dashed line), water and salt flow (double arrow) to re-establish equilibrium of diffusible components. The mass density increment at constant chemical potential of diffusible solutes is the difference between the density of a volume of the solution on the left-hand side (red rectangle) divided by particle concentration and the density of the solution on the right-hand side (green rectangle). See also Chapter A1.

and water in a ratio that is different from their ratio in the solvent, then this results in salt or water flowing in or out of the dialysis bag to compensate and maintain a constant chemical potential of diffusible components across the membrane. The mass density increment expresses the increase in mass density of the solution per unit macromolecular concentration. Not only does it account for the presence of the macromolecule itself, but also for its interactions with diffusible solvent components. It can be written

$$(\partial \rho / \partial C_2)_\mu = (1 + \xi_1) - \rho^\circ(\bar{v}_2 + \xi_1 \bar{v}_1) \tag{G2.46}$$

where ξ_1 is an interaction parameter in grams of water per gram of macromolecule, and v_x is partial specific volume (in milliliters per gram) of component x. The parameter ξ_1 does not represent water "bound" to the macromolecules. It represents the water that flows into the dialysis bag to compensate for the change in solvent composition caused by the association (or repulsion) of both water and small solute components with the macromolecule. The mass density increment is essentially equivalent to the buoyancy term arising from Archimedes' principle.

A similar equation is derived in terms of ξ_3, representing the number of grams of the small solute per gram of macromolecule; ξ_1 and ξ_3 are related by $\xi_1 = \xi_3 / w_3$, where w_3 is the molality of the solvent in grams of component 3 per gram of water. The parameter ξ_1 (or ξ_3) represents the solvent interactions of the macromolecule in the given solution conditions.

The mass density and refractive index increments are directly measurable in independent experiments on samples of the dialyzed solution, even if these measurements are not easy and require high sensitivity. Electron and scattering density increments, on the other hand, are not directly measurable in auxiliary experiments. Neutron scattering density increments are, in principle, measurable by neutron interferometry, but these also are not easy experiments.

The scattering density increments for the different experimental methods are related to solvent interactions and macromolecule properties by

$$(\partial \rho_{el}/\partial C_2)_\mu = (l_2 + \xi_1 l_1) - \rho_{el}^0 (\bar{v}_2 + \xi_1 \bar{v}_1) \qquad (G2.47)$$

$$(\partial \rho_n/\partial C_2)_\mu = (b_2 + \xi_1 b_1) - \rho_{el}^0 (\bar{v}_2 + \xi_1 \bar{v}_1) \qquad (G2.48)$$

where l_x and b_x, are electron and neutron scattering amplitudes per gram, respectively, per gram of component x. They can be calculated readily from chemical compositions.

In the same way as for the mass density increment, equations for X-ray and neutron scattering are equivalent to buoyancy terms where mass is replaced by scattering amplitudes. The equations can be used jointly to solve for unknown parameters, for example, ξ_1 and v_2. Combining the mass and X-ray equations is not very useful, because mass and electron density are close to being proportional to each other. On the other hand, atomic neutron scattering amplitudes (which can be positive or negative) are completely independent of mass or electron content, so that combining the neutron equation with mass and/or X-ray results is extremely useful.

G2.4.2 Relating the Thermodynamics and Particle Approaches

The "buoyancy" equations for the density increments are analogous to the "contrast" in the particle approach. However, the interaction parameter ξ_1 varies with solvent composition so that plotting the measured density increment, e.g., $(\partial \rho_n/\partial C_2)_\mu$, versus solvent density, e.g. ρ_n^0, does not yield a straight line. Eisenberg analyzed the "meaning" of the interaction parameter in terms of B_1 grams of water and B_3 grams of salt bound by the particle:

$$\xi_1 = B_1 - B_3/w_3 \qquad (G2.49)$$

When B_1 and B_3 are constant, ξ_1 versus $1/w_3$ is a straight line. The "bound" water and salt values may be positive or negative (when the component is excluded from the solvation shell). Applying the model to the interaction parameters, we obtain for the buoyancy equations:

$$(\partial \rho/\partial C_2)_\mu = (1 + B_1 + B_3) - \rho^0(\bar{v}_2 + B_1\bar{v}_1 + B_3\bar{v}_3)$$
$$(G2.50)$$

$$(\partial \rho_{el}/\partial C_2)_\mu = (l_2 + l_1 B_1 + l_3 B_3) - \rho^0(\bar{v}_2 + B_1\bar{v}_1 + B_3\bar{v}_3)$$
$$(G2.51)$$

$$(\partial \rho_n/\partial C_2)_\mu = (b_2 + b_1 B_1 + b_3 B_3) - \rho^0(\bar{v}_2 + B_1\bar{v}_1 + B_3\bar{v}_3)$$
$$(G2.52)$$

where the suffix 3 refers to salt. Tardieu and collaborators pointed out that whereas a straight-line dependence of ξ_1 versus $1/w_3$ is consistent with a particle that is invariant

in composition (binding constant amounts of water and salt), a straight-line dependence of density increment with solvent density is consistent with a particle in solution that is invariant in both composition and volume (see Comment G2.16).

Equations (G2.50)–(G2.52) are the contrast equations in the case of a three-component solution. Note that where the particle composition does not depend on the solvent (this is not the case for SANS experiments on solutions containing heavy water, because of the exchange of particle labile hydrogen atoms), the slope of the density increment versus solvent density is identical for the three methods and equal to the volume of the solvated particle.

G2.5 INTERACTIONS, MOLECULAR MACHINES, AND MEMBRANE PROTEINS

G2.5.1 Aminoacyl tRNA Synthetase Interactions with tRNA

Aminoacyl tRNA synthetases occupy a key position in protein synthesis (Comment G2.17). *E. coli* methionyl-tRNA synthetase (MetRS), is a dimer that binds two tRNA molecules in an *anticooperative* fashion (i.e., the binding of the first tRNA molecule to one of the subunits induces a lowering in affinity for the binding of a second tRNA to the other subunit). The forward scattered intensity and radius of gyration values have been followed as a solution of enzyme was titrated by progressive additions of tRNA, and interpreted by using Eqs. (G2.23) and (G2.24). Experiments have been performed in H_2O solvents in which the contrasts (see Fig. G2.10) of both protein and tRNA are significant, and in a 77% D_2O 23% H_2O buffer, in which tRNA contrast is close to zero so that only the protein part of the complexes was observed. The contrast variation technique permitted, in a unique way in the analysis of the measured $I(0)$ and R_G values, the separation and identification of *tRNA–protein binding stoichiometries, protein*

COMMENT G2.17 PHYSICIST'S BOX: AMINO ACYL tRNA SYNTHETASES

The ribosome is the main site of protein synthesis in the cell. It is where the genetic information encoded in messenger RNA (mRNA) is translated into a polypeptide chain (see Chapter A2). Each amino acid is encoded by a nucleotide triplet (a codon) on mRNA. Transfer RNA acts as an adaptor molecule; one of its ends binds specifically to the codon, while the corresponding amino acid, bound to the other end, is positioned for incorporation into the growing polypeptide. There is a specific tRNA molecule for each codon and amino acid. The correct binding of amino acid AA to tRNA^{AA} is catalyzed by an AA-specific enzyme: the aminoacyl tRNA synthetase.

conformational changes, and even some protein dimer dissociation associated with the binding events. The experiments established that the interaction was highly dynamic. They suggested a structural interpretation for the anticooperative binding behavior and its biological significance in favoring the alternated release of the substrate molecules after the catalytic event (Comment G2.18; Fig. G2.17).

Transfer RNA binding stoichiometries for aminoacyl tRNA synthetases were difficult to determine by standard biochemical approaches, because the charged nucleic acid often led to non-specific aggregates of protein forming around it. SANS experiments, because they could easily distinguish protein–protein from protein–nucleic acid complexes by contrast variation, played an important role in establishing these stoichiometries for different enzymes in the family and under different solvent conditions, paving the way for the successful crystallization of enzyme–tRNA complexes. One such experiment was based on changes in forward scattered intensity upon addition of tRNA^{asp} to a solution of yeast AspRS in H_2O solvent. The curve was analyzed quantitatively using Eq. (G2.23), with calculated values of m_j for protein and tRNA, respectively, according to their chemical compositions, partial specific volumes, and the solvent composition. A clear stoichiometry of two tRNA per AspRS dimer was found, not only from the break at this ratio, but also according to the absolute value of the increase in $I(0)$ as tRNA was added to the enzyme solution. Measurements at 77% D_2O, close to the contrast match of the tRNA, confirmed that there was no protein aggregation upon addition of tRNA.

G2.5.2 ATP, Solvent- and Temperature-Induced Structural Changes of the Thermosome

The thermosome is a macromolecular machine in thermophilic Archaea that has been described as a group II chaperonin (Comment G2.19). Its X-ray crystal structure has been solved and shows a rather globular conformation, which is quite different from the cylindrical structure found by cryo-electron microscopy. A SANS study of the thermosome under a variety of conditions established the relation between the two conformations and their places in the reaction cycle (Fig. G2.18).

The distance distribution function ($p(r)$) of the thermosome in solution in a standard buffer clearly indicates an "open" structure similar to the one found by electron microscopy (Fig. G2.17). Solvent (ATP or ADP binding) and temperature conditions were then found for which the experimental $p(r)$ corresponded to either the one calculated from the crystal structure (the "closed" conformation) or the one calculated from electron microscopy (the "open" conformation), leading to the model of conformational changes during the reaction cycle in Fig. G2.17b.

G2.5.3 Membrane Proteins

Membrane proteins ensure a large number of vital cellular functions. Their structural study, however, remains a major challenge because of the difficulties in maintaining their integrity during fractionation out of their complex local environment and in preparing crystals. An integral membrane protein crosses the lipid membrane; its body is surrounded by the apolar lipid environment with which it may have specific or non-specific binding interactions; at the lipid bilayer surface, the protein may interact with the lipid polar headgroups before protruding into an aqueous phase, which, in general, will have a different pH and ionic composition on each side of the membrane. The low-resolution structural organization of a membrane protein in its physiological environment is itself often difficult to solve: Is it a monomer? A dimer? The knowledge is fundamental to the understanding of functional mechanisms; SANS approaches using contrast variation have been developed specifically for the study of membrane proteins in detergent and lipid environments. Natural abundance detergent micelles can be contrast matched between 10 and 15% D_2O. Lipid vesicles, however, cannot be contrast matched simply because they are inhomogeneous over distances that are comparable to the size of the proteins to be studied. The scattering density of the polar headgroups corresponds to that of about 20% D_2O, while that of the CH, CH_2, and CH_3 groups in the chains is close to the scattering density of H_2O. In order to bypass this problem, homogeneous scattering lipid vesicles were prepared by specific deuteration in order to study the association state of the membrane protein, bacteriorhodopsin (BR) (Fig. G2.18). Bacteriorhodopsin is the protein that functions as a light-driven proton pump in the purple membrane (PM) of *Halobacterium salinarum* (see Comment A3.6). In PM, BR is organized as a trimer. What is its association state in lipid vesicles? Vesicles of dimyristoyl phosphatidyl choline (DMPC) have been prepared with two levels of deuteration (d_{63} and d_{67}) matching, respectively, in 94% D_2O (Fig. G2.19), and 99% D_2O. The Guinier curves of the protein-loaded vesicles at lipid contrast match were parallel, indicating the same radius of gyration in both cases: 16 Å, corresponding to the protein monomer, in accordance with the molar mass of 26 000 g mol^{-1} calculated from the forward scattered intensity and concentration. The use of deuterated lipid for contrast match measurements in D_2O significantly reduces the incoherent background in the measurements, which arises mainly from H nuclei (see Section G1.2.1), while maintaining high

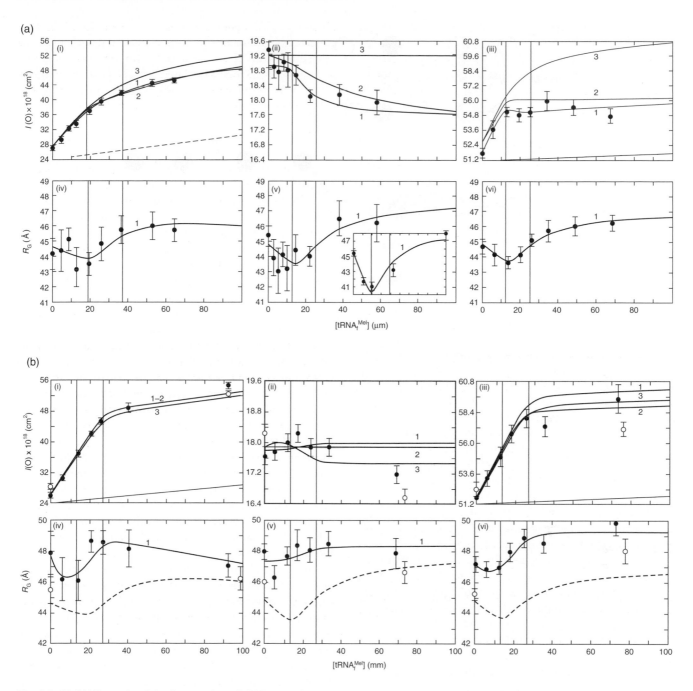

Fig. G2.17 SANS study of the interaction of tRNA specific for methionine with *E. coli* methionyl-tRNA synthetase: **(a)** Forward scattered intensity (top three panels) and radius of gyration (bottom three panels) changes while tRNA is added to a solution of the dimeric enzyme containing 10 mM $MgCl_2$. The vertical lines just below 20 and 40 μM on the x-axes correspond to tRNA: Protein stoichiometries of 1 and 2, respectively. The left-hand side panels are for the H_2O solution, in which both tRNA and protein are "visible," with the tRNA contrast being larger than the protein contrast. The middle ones are for 77% D_2O, in which the tRNA is "invisible." The right-hand panels are for D_2O, in which the tRNA contrast is smaller than the protein contrast. The data points are shown in each panel, as well as lines corresponding to different interaction models. Note the break in the intensity rise beyond a 1:1 stoichiometry in H_2O solution, indicating anticooperativity, and the fall in intensity and radius of gyration in 77% D_2O, indicating dissociation of the protein dimer. **(b)** The same as **(a)** for a solution containing 50 mM $MgCl_2$. The dashed lines in the bottom panels show the radius of gyration dependence in 10 mM $MgCl_2$. Note the almost straight line increase in intensity in the H_2O solution up to a stoichiometry of 2, indicating essentially independent binding sites under these conditions, and the larger radius of gyration (compared to the 10 mM $MgCl_2$ condition) indicating an "opening up" of the protein dimer. **(c)** The interaction model that best fitted the data. Open circles represent protein monomers, filled circles are protein monomers with bound tRNA. The dumb-bell structures are dimers, with the length of the line indicating a large or small radius of gyration. (Dessen *et al.*, 1978.) (Figure reproduced with permission from Elsevier.)

445

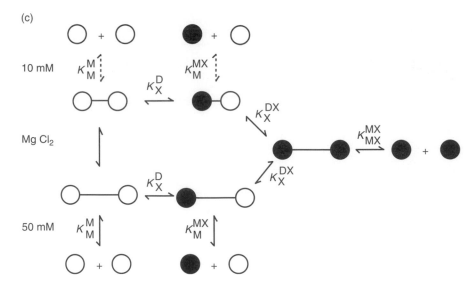

Fig. G2.17 (*cont.*)

contrast for natural abundance protein, leading to a very favorable signal-to-noise ratio for the scattered intensity.

G2.6 SAS COMBINED WITH OTHER METHODS FOR A GLOBAL STRUCTURAL STUDY

Studies combining complementary information from NMR, SAXS, and SANS are described and discussed in the NMR part of the book (Section J3.5).

In the December 2014 issue of *CELL*, authors from laboratories in Austria, Germany, Norway, Russia, Slovenia, Switzerland, and the UK described the structure and regulation of human muscle alpha-actin by combining structural results from X-ray crystallography, SAXS, and size-exclusion chromatography multiangle laser light scattering (SEC-MALLS), with ligand docking from molecular dynamics (MD) simulations, site-directed mutants, site-directed spin labeling EPR (electron spin resonance), fluorescence recovery after photobleaching (FRAP), and superresolution fluorescence microscopy with green fluorescence protein (GFP) labeling (Ribeiro *et al.*, 2014). See other chapters and sections of this book for descriptions of the various methods.

The cytoskeleton is a dynamic intracellular network. In eukaryotic cells the cytoskeleton is composed mainly of microfilaments and microtubules made up, respectively, of protein subunits actin and tubulin (Fig. G2.20).

The cytoskeleton fulfills a variety of essential functions for cell life and during cell division, when it segregates chromosomes and is involved in cytokinesis, the separation of the mother cell into two daughter cells. The cytoskeleton is also involved in motility, which is essential to all living organisms, through organelle transport within cells to movement of cells and entire organisms. The spectrin protein superfamily plays key roles in assembling the actin cytoskeleton in various cell types. Ribeiro and collaborators reported the complete high-resolution structure of the 200 kDa alpha-actinin-2 dimer from striated muscle and explored its biochemical functional implications on the cellular level by applying a battery of biophysical, biochemical, and cell biology methods. The structure provides insight into the interaction mechanisms involving the spectrin family protein and laid a foundation for studying the impact of pathogenic mutations at molecular resolution. The structure of alpha-actinin-2

(a)

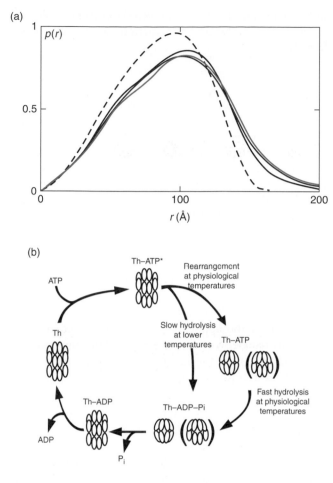

(b)

Fig. G2.18 (a) Experimental and calculated pair distribution functions of native ($\alpha\beta$) and recombinant (α) thermosomes. Red and blue lines: experimental data for ($\alpha\beta$) and (α), respectively, in a standard buffer at 25 °C; black line: calculated from the electron microscopy structure; dashed line: calculated from the crystal structure.(b) Model for open (cylindrical) and closed (globular) structures during ATPase cycling, established from the SAS data. (Gütsche *et al.*, 2001.) (Figure reproduced with permission from Elsevier.)

from X-ray crystallography and conformational changes of mutants in solution from SAXS, together with SEC-MALLS results, are shown in Fig. G2.21.

G2.7 SIZE-EXCLUSION CHROMATOGRAPHY MULTIANGLE LASER LIGHT SCATTERING

The theory of classical or static light scattering by particles in solution was developed early in the twentieth century. As described in Section G2.2.2 for small-angle X-ray scattering and SANS, the angular dependence of the absolute scattered intensity contains information on the particle mass and size. Light scattering was further and strongly

(a)

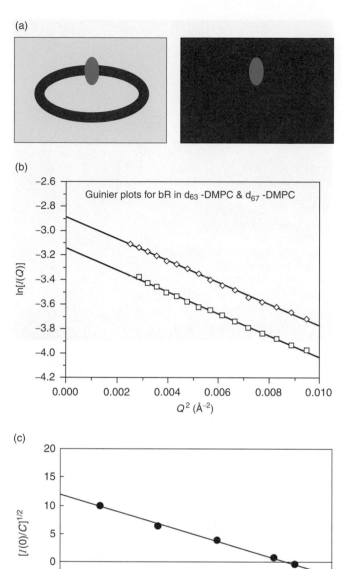

Fig. G2.19 (a) Schematic of a membrane protein (blue) in a deuterated lipid vesicle (red) in H_2O buffer (light green, left panel) and D_2O contrast match buffer for the lipid (right panel). (b) Guinier curves for bR in contrast-matched vesicles of different deuteration levels. The same radius of gyration is observed in both cases, corresponding to the protein monomer. (c) Plot of the scattered intensity divided by concentration as a function of the percentage D_2O in the solvent, revealing the contrast match point of 94% D_2O in the lipid vesicles. (Hunt *et al.*, 1997.) (Figure reproduced with permission from Elsevier.)

developed for the study of polymers, but its application to biological macromolecules suffered from a number of technical and sample characterization difficulties, and the method was used by only a few specialized biophysics

index increment due to the dissolved macromolecules, n_0, dn/dC, respectively, and the vacuum wavelength of the light λ_0:

$$K^* = 4P^2 n_o^2 (dn/dC)^2 / \lambda_o^4 N_A \qquad \text{(G2.54)}$$

where N_A is Avogadro's number. Note that for a protein, the concentration, C, can be determined in parallel, either from a measurement of the UV absorbance if the extinction coefficient is known or from a measurement of the refractive index change with concentration, since the refractive index increment of proteins is fairly constant. Coupling MALLS with high-performance size exclusion chromatography (HPSEC) and other approaches for macromolecular separation, such as flow, electrical, or thermal field-flow fractionation (FFF) techniques (see Comment D4.5), has been of particular significance in the success of the methodology for a wide range of applications.

The study described in Comment G2.20 is an excellent illustration of the application of MALLS in conjunction with small-angle X-ray scattering, NMR, and circular dichroïsm (CD) for the determination of protein association states in solution.

G2.8 REFLECTOMETRY (OR REFLECTIVITY)

G2.8.1 Background

In the last decades, neutron reflectometry (NR) and X-ray reflectometry (XRR) emerged as powerful tools to characterize the nanoscale structure of hard and soft thin films. Especially but not only in the case of soft structures in solution, such as biological membranes, neutrons have the advantage of H/D contrast variation, since, even though reflectometry and scattering are based on different phenomena, they both depend on scattering length density.

Recall (see Section G1.4) that scattering is analyzed as waves, emanating from single atoms or nuclei, interfering with each other, *without* being scattered further. Where neutrons or X-rays are reflected off a surface, however, a classical optics description applies in terms of interfaces between layers of different refractive index (Fig. G2.22). Interaction between wave and matter is not with individual atoms but with bulk properties. Note how this is similar to the behavior of sound waves at density interfaces within the body during ultrasound imaging (see Chapter K2).

Specular reflection is where the incident and reflected wave lie in the same plane and the angle of reflection is equal to the angle of incidence (θ_0), as shown in Fig. G2.22. The refractive index of a material against vacuum is given by

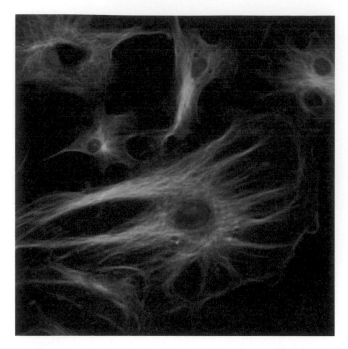

Fig. G2.20 The cytoskeleton in bovine pulmonary artery endothelial cells. Nuclei are stained blue, microtubules are marked green, and actin filaments are labeled red. (From ImageJ-Programmpaket (public domain).)

groups. Then, in the late 1980s, laser light instruments that simultaneously measured the scattering signal at several angles became available commercially. Combined with chromatographic separation techniques, which sort a complex sample solution into fractions of similar molecular weight and size particles for the MALLS experiment, and "user-friendly" analysis software taking advantage of the tremendous progress in personal computers, these instruments provided a powerful means for the absolute scale determination of macromolecular mass and size distributions.

The following equation forms the basis of the MALLS analysis:

$$\frac{K^*C}{R(\theta)} = \frac{1}{M_w P(\theta)} + 2A_2 C \qquad \text{(G2.53)}$$

where $R(\theta)$ is the light scattered by the sample per unit solid angle at an angle θ in excess of the light scattered by the solvent alone, M_W is the weight-average molecular weight and is a *form factor* (similar to the Guinier approximation, see Section G2.1.2), which contains shape information such as a weight-average radius of gyration, R_G, for example. Because of the wavelength values of visible light, only R_G values larger than about 100 Å can be measured by MALLS. A_2 is the second virial coefficient (Section G2.1.7), C is the particle concentration in mass per volume of solution, and K^* is a constant containing the specific refractive index of the solvent and the refractive

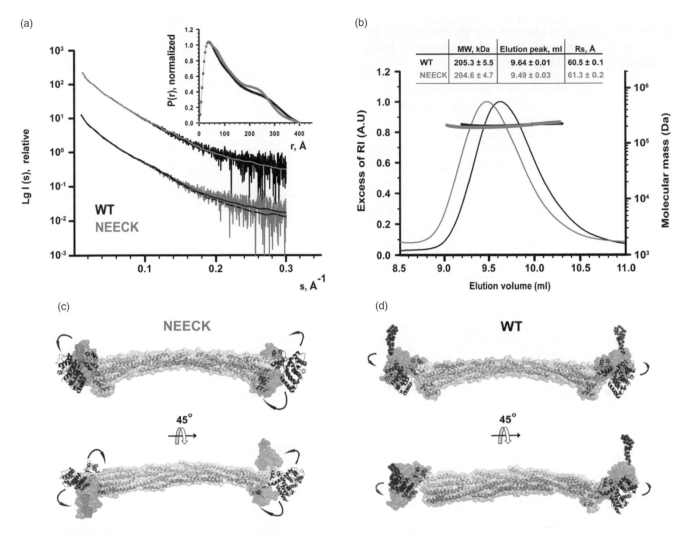

Fig. G2.21 Solution structure of alpha-actinin-2 and the NEECK mutant derived from SAXS.

(a) Experimental SAXS data of WT (black) and the NEECK mutant (green) of alpha-actinin-2. SAXS curves are computed from a rigid-body (RB) model for WT (gray) and NEECK (black). The logarithm of scattering intensity (I) is plotted as a function of the momentum transfers (s, $Å^{-1}$). Successive curves are displaced by one logarithmic unit for better visualization. Distance distribution functions (inset) P(r) for WT and NEECK assume slightly different shapes. RS modeling fits the experimental WT data with χ 1.25 (gray line) and experimental NEECK data with χ 1.14 (dashed black line). The fit discrepancy for NEECK increased to 1.32, assuming a helical neck (solid black line)·

(b) Characterization of hydrodynamic properties of alpha-actinin-2 WT and the NEECK mutant by SEC-MALLS. The lines across the protein elution volume show the molecular masses (MW) of proteins. SEC·MALLS shows that NEECK has the same MW as WT alpha-actinin-2 but a higher Stokes radius Rs (inset: data are represented as mean $\pm$ SD of three experiments), corroborating the open conformation for NEECK suggested by SAXS. AU, arbitrary units.

(c) RB model of NEECK in solvent-accessible surface representation. The neck region was modeled as a flexible linker between the rigid bodies ABD and rod, with no contact restraint. Only one RB model out of three independent BUNCH runs is shown for clarity.

(d) The best RB model of WT alpha-actinin-2 in solvent-accessible surface representation superimposed on the crystal structure. For WT RB modeling, only ABD was allowed a variable position, whereas the EF hand helix–loop–helix motifs 3–4 were fixed in contact with the neck.

In all models, N-terminal residues missing from the crystal structure were modeled as dummy atoms. Arrows highlight the movement of ABO and EF hands 3–4 relative to the superimposed crystal structure.

COMMENT G2.20 MALLS COMBINED WITH CD, SMALL-ANGLE X-RAY SCATTERING AND NMR: CONVERSION OF PHOSPHOLAMBAN INTO A SOLUBLE PENTAMERIC HELICAL BUNDLE

In an exploration of membrane protein folding and solubility, surface, usually lipid-exposed, amino acid residues of the transmembrane domain of phospholamban (PLB) were mutated and replaced with charged and polar residues. Wild-type PLB forms a stable helical homopentamer within the sarcoplasmic reticulum membrane in eukaryotic cells. The experiments were designed to test whether the packing inside the membrane protein could maintain a similar fold even with the lipid-exposed surface redesigned for solubility in an aqueous environment. The CD spectra (see Chapter E4) indicated that the full-length soluble PLB is highly α-helical, similar to the wild-type protein. For the MALLS investigation, samples were subjected first to size exclusion chromatography. Peaks were detected as they eluted from the column with a UV detector at 280 nm (where the protein has a maximum absorbance), a light scattering detector at 690 nm, and a refractive index detector. The exact protein concentration was calculated from the change in the refractive index with respect to protein concentration. The weight-average molecular weight of the protein in each elution volume for different concentrations and pH values was then determined by the application of Eq. (G2.53). Small-angle X-ray scattering experiments on selected conditions confirmed the molecular mass of 120 kDa (exactly five times the protein monomer) found by MALLS, and determined the radius of gyration of the particle to be about 50 Å. However, NMR experiments (see Part J) suggested that the redesigned protein exhibits molten globule-like properties, indicating some alteration in native contacts at the core of the protein by the surface mutations. The experiments nevertheless established that the interior of a membrane protein contains at least some of the determinants necessary to dictate folding in an aqueous environment (Li *et al.*, 2001).

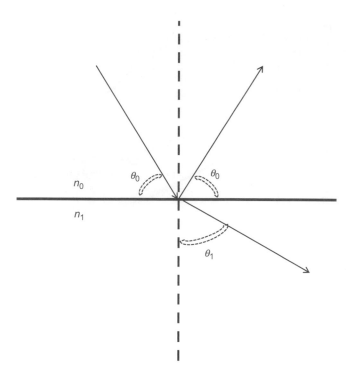

Fig. G2.22 Reflection and transmission at an interface between materials of different refractive index.

COMMENT G2.21 BIOLOGISTS BOX: DERIVING EQ. (G2.57)

$\cos^2\theta + \sin^2\theta = 1$

so that at small θ

$\cos^2\theta = 1 - \theta^2$

combining the above with equations G2.55, G2.56

for $n_1 = 1$:

$$\theta_c = \lambda\sqrt{\frac{\rho}{\pi}}$$

occurs at the vacuum/material interface, in contrast to light and X-rays, where it occurs in the denser medium. The critical angle, θ_c, ($\theta_0 = \theta_c$, $\theta_1 = 90°$, in Fig. G2.22) for total reflection at an interface between medium 1 and medium 2 is given by

$$\cos\theta_c = \frac{n_2}{n_1} \tag{G2.56}$$

Eq. (G2.57) is obtained as a good approximation by combining Equations (G2.55) and (G2.56) for small values of θ_c at an interface between air ($n = 1$) and a medium of refractive index n and scattering length density ρ (Comment G2.21).

$$\theta_c = \lambda\sqrt{\frac{\rho}{\pi}} \tag{G2.57}$$

Reflections and diffuse scattering in the off-specular directions provide information on inhomogeneities,

$n = 1 - \delta - i\beta$

β is an absorption coefficient, which is usually
  negligible for neutrons but not for X-rays

$\delta = \dfrac{\lambda^2\rho}{2\pi}$

λ is wavelength

ρ for neutrons = scattering length density

ρ for X-rays = $r_e n_a$, where r_e is the classical electron
  radius and n_a is atomic number density (G2.55)

Interestingly, in the neutron case, the refractive index of most materials is less than 1. Total reflection, therefore,

roughness, and fluctuations along the surface. The analysis is more complex than for specular reflections (Comment G2.22).

In a specular reflection experiment, the scattered intensity is measured as a function of the scattering vector component along z, $Q_z = 2k_z$ (see Fig. G2.19), where $Q = 4\pi \sin\theta/\lambda$; the analysis informs on the structure of the film in the direction perpendicular to the interface.

In the case where the scattering is weak compared to the incident wave, intensity and multiple scattering can be neglected, the reflectivity, $R(Q_z)$, is interpreted as the square of a one-dimensional function, $F(Q_z)$, which represents the Fourier transform of the scattering length density distribution along z, $\rho(z)$.

$$R(Q_z) = \frac{16\pi^2}{Q_z^2}|F(Q)|^2 \tag{G2.58}$$

$$F(Q) = \int \rho(z)\exp(iQz)dz \tag{G2.59}$$

$$\rho(z) = \int F(Q)\exp(-iQz)dQ \tag{G2.60}$$

Note the similarity of Eqs (G2.59) and (G2.60), with the diffraction equations in Chapter G1, Section G1.5.3.

The weak scattering approximation is valid for NR experiments on soft matter layers at the air–water interface, where the scattering length density of the water can be set to match that of air (zero contrast). It should be noted that the validity of the approximation has been extended to non-zero bulk contrast systems.

Equation (G2.58) is strictly valid only in the case of a smooth interface, and corrections are applied to account for surface roughness.

A typical reflectivity signal from a fluid film adsorbed on a smooth solid block in aqueous solution is shown in Fig. G2.23.

G2.8.2 Instrumental Set-up

An example of a state of the art experimental set-up for NR is given in Fig. G2.24. The horizontal sample geometry and vertical detector is especially adapted to the study of thin films at air–liquid, liquid–liquid or solid–liquid interfaces.

XRR experiments are performed on laboratory sources (e.g., www.ansto.gov.au/ResearchHub/Bragg/Facilities/Instruments/X-rayReflectometer) and also on high-brilliance, high-resolution scattering and diffraction instruments at synchrotron radiation facilities, such as ID10 at the European Synchrotron Radiation Facility in Grenoble (www.esrf.eu/UsersAndScience/Experiments/SoftMatter/ID10) or the SIRIUS beam line at SOLEIL near Paris (www.synchrotron-soleil.fr/Recherche/LignesLumiere/SIRIUS).

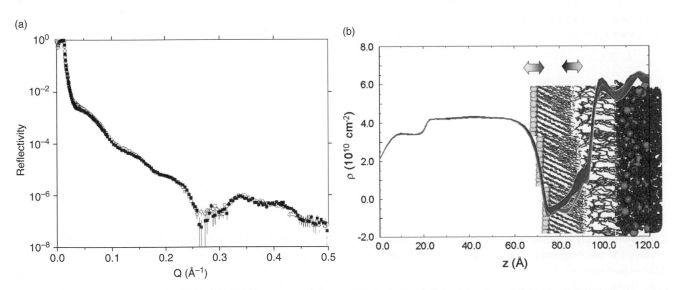

Fig. G2.23 (a) Neutron reflectivity from a lipid bilayer containing mellitin in D_2O solution. Sample no. 1 (open circles) was prepared five months earlier than sample no. 2 (black circles). Both samples were formed on the same gold-coated silicon substrate. **(b)** Molecular representation of the lipid bilayer from a molecular dynamics simulation, superimposed on the fitted neutron scattering length density (SLD) profiles derived from the reflectivity. (Adapted from Krueger *et al.*, 2001.)

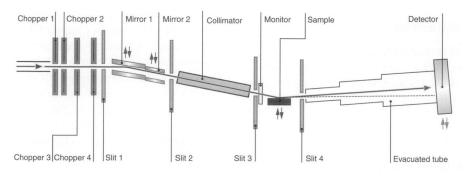

Fig. G2.24 FIGARO at the Institut Laue Langevin (www.ill.eu/instruments-support/instruments-groups/instruments/figaro/description/instrument-layout). The four "choppers" rotate independently to select neutron velocity and velocity spread, defining the wavelength and wavelength resolution. The following mirror sections deflect the beam upwards or downwards to choose the incoming angle. The collimator "cleans up" the beam and focuses it horizontally. The sample stage is equipped with active and passive anti-vibration systems, x-y-z translation axes, and a goniometer for solid samples. The two-dimensional detector is at a distance of about 3 m from the sample and can move up and down to detect the reflected and direct beams at all angles. FIGARO can also be adapted for grazing angle diffraction and small-angle scattering (GISANS) to look at large-scale structures perpendicular and along the surface of planar samples.

COMMENT G2.23 REFLECTOMETRY INSTRUMENTS MAINLY AVAILABLE TO USERS

An updated list of NR instruments in the world is kept at http://material.fysik.uu.se/Group_members/adrian/reflect.htm. See also the review of neutron instrumentation by Teixeira et al. (2008).

In the USA, there are NR instruments at:

- The National Center for Neutron Research (NCNR) of the National Institute of Science and Technology (NIST): www.ncnr.nist.gov/instruments
- The Spallation Neutron Source (SNS), Oak Ridge National Laboratory (ORNL): http://neutrons.ornl.gov/lr

In Europe:

- At the Institut Laue Langevin, Grenoble, France: www.ill.eu/instruments-support/instruments-groups
- At ISIS, Science and Technology Facilities Council, Oxfordshire, UK: www.isis.stfc.ac.uk/instruments/reflectometry2594.html
- At the FRM2 reactor near Munich, Germany: www.mlz-garching.de/refsans

In Australia:

- At the Bragg Institute: www.ansto.gov.au/ResearchHub/Bragg/Facilities/Instruments/Platypus/index.htm

---

Neutron reflectometry instruments are available to external users at various neutron sources around the world (Comment G2.23).

G2.8.3 Examples

Conformational Transition of a Membrane-Associated Protein Involved in the Progression of AIDS (Akgun et al., 2013)

Many proteins are targeted to lipid membranes to fulfill their function. Their precise position relative to a membrane is difficult to determine by usual structural methods and remains largely unknown. Akgun et al. (2013) used NR and XRR to measure the displacement of the HIV Nef protein core from lipid membranes, when it undergoes a functional modification. In this modification a myristate

group is inserted in the N-terminal of the protein. Nef is a soluble protein. When it is modified to myr-Nef, the myristate group is exposed, which promotes Nef binding to lipid headgroups on the membrane surface. Nef is one of several HIV-1 accessory proteins and an essential factor in AIDS progression. Absence of the modification dramatically reduces HIV-1 infectivity. NRR and XRR revealed that, upon modification, Nef undergoes a conformational change to an open conformation that positions the core domain 70 Å from the lipid membrane surface, ruling out previous speculation that the Nef core remains closely associated with the membrane.

Neutron reflectometry and X-ray reflectometry data were collected for a Langmuir lipid monolayer made up of deuterated dipalmitoyl phosphatidyl glycerol (dDPPG)

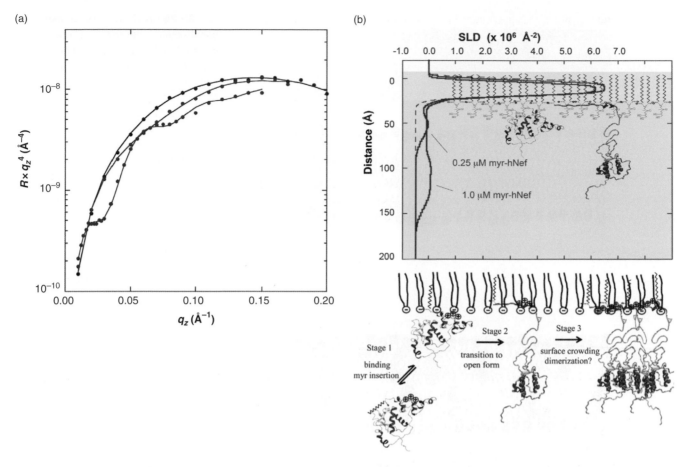

Fig. G2.25 (a) NR results for myr-Nef adsorbed to a dDPPG monolayer at 30 mN m^{-1} membrane pressure; the black line is the scattering length density (SLD) derived for the deuterated monolayer; the blue line is the SLD derived with myr-Nef adsorbed from solutions at 0.25 μM; and the red line is with myr-Nef adsorbed from solutions at 1.0 μM. The molecular models of dDPPG and of Nef are not drawn precisely to scale but were scaled to coincide approximately with the corresponding features in the SLD profiles. **(b)** Mechanism of myr-Nef binding to a dDPPG membrane. Myr-Nef is adsorbed through electrostatic attraction and myristate insertion at the first stage. At higher surface pressures (35 mN m^{-1}), there is no change in the conformation of myr-Nef after the first stage. At lower pressures (20 mN m^{-1}), with the insertion of the N-terminal arm into the membrane, the core domain is displaced 70 Å away from the membrane. Surface coverage of myr-Nef in open conformation increases very slowly as a function of time, perhaps involving membrane-driven dimerization. Stages 1 and 2 occur on timescales that are too fast to be detected by NR or XRR. Adapted from Akgun *et al.* (2013).

(for favorable neutron scattering density contrast) monolayer interacting with myr-Nef injected in the aqueous (H$_2$O) medium below it. The concentration of myr-Nef and lateral surface pressure on the monolayer were varied to provide structural details of membrane-bound Nef as a function of surface coverage (Fig. G2.25).

Ganglioside Forces the Redistribution of Cholesterol in a Biomimetic Membrane (Rondelli *et al.*, 2012)

Cholesterol and sphingolipids show cooperative effects on many biological processes, not limited to lipid-driven membrane organization. For example, they are thought to synergistically act as regulators of the function of transmembrane receptors either by direct interaction or by modulating the structure of their environment. Their

concentration at different cellular sites is subject to tight regulation. On the basis of the lipid composition of isolated lipid rafts, a strong enrichment in cholesterol is expected in the inner leaflet of biological membranes, while the outer leaflet ends up highly enriched in ganglioside and phosphatidylcholine lipids. Ganglioside–cholesterol asymmetric distribution in the cytofacial/exofacial leaflets of the lipid bilayer may be reflected in asymmetric fluidity and molecular partition. In this view, gangliosides and cholesterol could constitute a collective-pair acting as a structural unit across the membrane. By using the floating membrane technique (Fig. G2.26a) Rondelli *et al.* studied how cholesterol distributed between outer and inner leaflets in the presence and absence of ganglioside to verify the hypothesis (Fig. G2.26b).

(a)

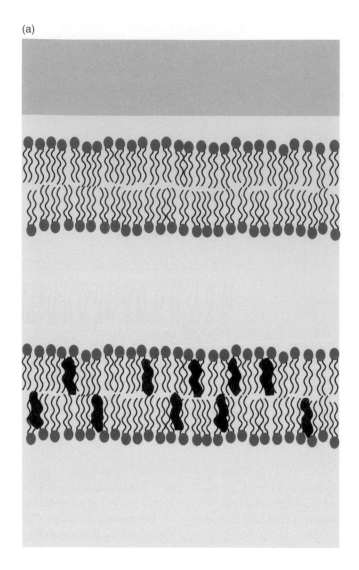

Fig. G2.26 (a) Scheme of sample construction, showing the cholesterol inserted in the floating lipid bilayer. **(b)** Cholesterol disposition in the inner (light gray) and outer (dark gray) leaflets of the floating membrane. Left panel: without GM1 galactoside, casted asymmetric disposition of cholesterol is destroyed by annealing. Central and right panels: GM1 forces redistribution of cholesterol in an optimal coupling ratio. Sample A: GM1 is inserted a posteriori into a pre-casted membrane where cholesterol is symmetrically distributed; sample B: GM1 participates from deposition to the outer layer of a membrane where cholesterol is initially hosted only in the inner layer. (Adapted from Rondelli *et al.* 2012)

(b)

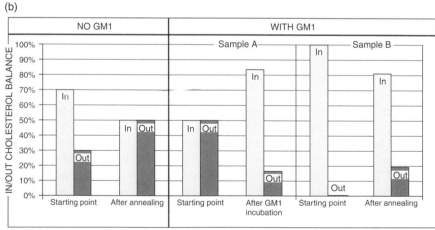

G2.9 CHECKLIST OF KEY IDEAS

- Information on particle shapes and interactions in solution can be obtained from SAS of X-rays and neutrons.

- The two assumptions required for the interpretation of the scattering curve in terms of a single-particle structure are: (1) the solution is monodisperse, i.e., all the particles in the solution are identical; and (2) the solution is infinitely dilute, i.e., there is no correlation between the positions or orientations of the particles in the solution.

- At small scattering vector (Q) values the scattering curve can be interpreted in the Guinier approximation, from which the radius of gyration of the particle and its molar mass are derived independently and in a model-independent manner.
- At large scattering vector values the intensity scattered by a particle with a well-defined border with solvent follows Porod's Q^{-4} law.
- The distance distribution function in the particle $p(r)$ is calculated from the Fourier transform of the scattered intensity $I(Q)$.
- Scattering from polydisperse solutions (mixtures of different particles) can be interpreted in the Guinier approximation in terms of the different radii of gyration and molar masses.
- A "structure factor" related to the organization of particles can be derived from the scattering curve of a concentrated solution.
- Scattering curves can be calculated analytically or numerically for particles of different shapes or atomic assemblies as in a crystal structure, for example, for comparison with experimental observations.
- The spherical particle is unique in that its internal structure expressed as the scattering density distribution as a function of radius, $\rho(r)$, directly corresponds to the rotational average; it is the only shape whose scattering amplitude, $F(Q)$, can be calculated directly from the square root of the scattering intensity, $I(Q)$; $\rho(r)$ is the Fourier transform of $F(Q)$.
- By using appropriate plots of the scattering curve, the mass per unit length and cross-sectional radius of gyration of rod-like particles (fiber structures) and the mass per unit area and transverse radius of gyration of flat sheet-like particles (membrane structures) can be determined.
- Model structures consistent with scattering curves can be determined by using different approaches,

including fitting with *spherical harmonics*, *Monte Carlo* and *simulated annealing* algorithms, and *genetics* algorithms.
- Contrast variation methods allow one to focus on different parts of a complex particle; contrast can be varied by changing the radiation (because X-ray and neutron scattering amplitudes are different) and/or by deuterium labeling of solvent and/or particle for neutron scattering.
- Different methodological approaches have been developed to plan and interpret contrast variation experiments in terms of the structures of different components within a particle.
- The forward-scattered intensities in neutron and X-ray SAS and light scattering are related to the thermodynamic and hydrodynamic parameters of macromolecule–solvent interactions in solution, and can be used to study these interactions.
- Application examples illustrate the power of SAS as a tool for the study of structural aspects of complex formation and interactions in solution, conformational changes in the reaction cycles of macromolecular machines, and the low-resolution structure of membrane proteins in lipid vesicles (by using contrast matched lipids).
- In the second decade of this century, SAS is combined with structural results from X-ray crystallography, size-exclusion chromatography multiangle laser light scattering (SEC-MALLS), ligand docking from molecular dynamics (MD) simulations, site-directed mutants, site-directed spin labeling EPR (electron spin resonance), fluorescence recovery after photobleaching (FRAP), NMR, and superresolution fluorescence microscopy with green fluorescence protein (GFP) labeling, in a strongly complementary way to solve important biological structures.
- Neutron (NR) and X-ray reflectometry (XRR) emerged as powerful tools to characterize the nanoscale structure of hard and soft thin films. Especially, but not only, in the case of soft structures in solution, such as biological membranes, neutrons have the advantage of H/D contrast variation, since, even though reflectometry and scattering are based on different phenomena, they both depend on scattering length density.

Suggestions for Further Reading

Glatter, O., and Kratky, O. (eds.) (1982). *Small Angle Scattering*. London: Academic Press.

Zaccai, G., and Jacrot, B. (1983). Small angle neutron scattering. *Ann. Rev. Biophys. Bioeng.*, **12**, 139–157.

Svergun, D. I. (2000). Advanced solution scattering data analysis methods and applications. *J. Appl. Crystallogr.*, **33**, 530–534.

Teixeira, S. C., Ankner, J., Bellissent-Funel, M. C., *et al.* (2008). New sources and instrumentation for neutrons in biology. *Chem Phys.*, **345**, 133–151.

Svergun, D. I., Koch. M. H. J., Timmins, P. A., and May, R. P. (2013). *Small Angle X-ray and Neutron Scattering from Solutions of Biological Macromolecules*. Oxford: Oxford University Press.

Gerelli, Y., Porcar, L., Lombardi, L., *et al.* (2013). Lipid exchange and flip-flop in solid supported bilayers. *LANGMUIR*, **29** (41), 12 762–12 769.

Penfold, J., and Thomas, R. K. (2014). Neutron reflectivity and small angle neutron scattering: An introduction and perspective on recent progress. *Curr. Opin. Colloid Interface Sci.*, **19**, 198–206.

X-RAY AND NEUTRON MACROMOLECULAR CRYSTALLOGRAPHY

G3

G3.1 HISTORICAL REVIEW

Nobel Prizes

In **1901**, the very first Nobel Prize in physics was awarded to **W. C. Röntgen** "in recognition of the extraordinary services he has rendered by the discovery of the remarkable rays subsequently named after him." **J. J. Thomson** obtained the Nobel Prize in physics in **1906** "in recognition of the great merits of his theoretical and experimental investigations on the conduction of electricity by gases." **M. von Laue** was awarded the physics Prize in **1914** "for his discovery of the diffraction of X-rays by crystals," and **W. H. Bragg** and **W. L. Bragg** in **1915**, "for their services in the analysis of crystal structure by means of X-rays." **J. Chadwick** was awarded the Nobel Prize in physics in **1935** "for the discovery of the neutron." **L. Pauling** was awarded the Nobel Prize in **1954** for his research into the nature of the chemical bond.

The birth of molecular biology was honored appropriately with the award of the **1962** Nobel Prize in physiology or medicine to **J. Watson**, **M. Wilkins**, and **F. Crick** "for their discoveries concerning the molecular structure of nucleic acids and its significance for information transfer in living material," and, in the same year, the award of the chemistry prize to **M. F. Perutz** and **J. C. Kendrew** "for their studies of the structures of globular proteins." **D. C. Hodgkin**, who had been recommended for the prize in the same year as Perutz and Kendrew, was awarded the **1964** prize in chemistry "for her determinations by X-ray techniques of the structures of important biochemical substances." The **1982** Nobel Prize in chemistry was awarded to **A. Klug** "for his development of crystallographic electron microscopy and his structural elucidation of biologically important nucleic acid–protein complexes," and the **1985** chemistry Prize was given to **H. A. Hauptman** and **J. Karle** "for their outstanding achievements in the development of direct methods for the determination of crystal structures." The first high-resolution X-ray crystallography structure of a membrane protein complex was rewarded by the **1988** chemistry Prize to **J. Deisenhofer**, **R. Huber**, and **H. Michel** "for the determination of the three-dimensional structure of a photosynthetic reaction center." The physics prize for **1992** was awarded to **G. Charpak** "for his invention and development of

particle detectors, in particular the multiwire proportional chamber"; these detectors also paved the way for the area detectors developed for X-ray and neutron crystallography, which replaced film. The **1994** Nobel Prize in physics was awarded to **B. N. Brockhouse** and **C. G. Shull** for the development of neutron scattering (spectroscopy and diffraction, respectively). The Nobel Prize for chemistry in **1997** was shared by **J. E. Walker** and **P. D. Boyer** "for their elucidation of the enzymatic mechanism underlying the synthesis of adenosine triphosphate (ATP)" and by **J. C. Skou** "for the first discovery of an ion-transporting enzyme, Na^+, K^+-ATPase." **P. Agre** and **R. MacKinnon** were awarded the Nobel Prize in chemistry in **2003** for discoveries concerning channels in cell membranes. In **2009** the Nobel Prize in chemistry was given to **V. Ramakrishnan**, **T. A. Steitz**, and **A. Yonath** for studies of the structure and function of the ribosome. **Robert J. Lefkowitz** and **B. K. Kobilka** shared the **2012** chemistry Nobel Prize for their studies of G-protein coupled receptors.

Nobel Prizes, however, although they are symbolic of the importance given to discoveries that profoundly mark a field of research, were awarded in fact to only a small number of the scientists whose contributions were absolutely fundamental. **J. D. Bernal** and **D. Crowfoot** observed the first X-ray diffraction pattern from a protein (pepsin) in **1934** by using a sample mounting procedure to keep the crystal "wet" that is still in use today. **D. M. Green**, **V. M. Ingram**, and **Perutz** opened the way to solving the phase problem with their publication on *isomorphous replacement* in **1954**; the rigorous application of the method was developed by **D. M. Blow** and **Crick** in a **1959** publication. **Blow**, with **M. Rossman**, also pioneered the application of *molecular replacement* in the **1960s**. *Anomalous dispersion*, another approach to solving the phase problem, was introduced in the study of lysozyme, the first enzyme to be solved by protein crystallography, by **C. C. F. Blake**, **D. F. Koenig**, **G. A. Mair**, **A. C. T. North**, **D. C. Philips**, and **V. R. Sarma** in **1965**.

In the **1970s**, the special properties of neutrons, especially with respect to locating hydrogen atoms in molecular structures, were successfully exploited in biological crystallography. Structures of the amino acids were solved by **M. S. Lehmann**, **T. F. Koetzle**, and

W. C. Hamilton; M. Ramanadham, R. Chidambaram, B. Schoenborn, and collaborators initiated the neutron crystallography study of myoglobin. Contrast variation methods were developed for neutron scattering (H. B. Stuhrmann); it was shown that, in nucleosomes, DNA is wrapped around the histone core (J. F. Pardon, D. L. Worcester, and their collaborators); the organization of nucleic acid and protein in spherical viruses was described (B. Jacrot and collaborators). The method of label triangulation was proposed independently by Engelman and Moore, and W. Hoppe. Engelman and Moore suggested using neutron scattering and specific deuterium labeling and applied the method to resolve the organization of proteins in the 30 S ribosomal subunit. K. Nierhaus and collaborators used label triangulation on the 50 S ribosomal subunit, and H. Heumann and his group applied the method to solve subunit distances and shapes in DNA-dependent RNA polymerase.

A. Brunger and coworkers in 1987 successfully solved the structure of *crambin* by using an NMR structure in a crystallographic molecular replacement approach. Brunger has also developed computational refinement methods to fit models to both experimental data and chemical restraints. With the advent of high-brilliance synchrotron sources, the use of cryo-protection techniques, developed by H. Hope and others for the freezing of crystals, became essential because they greatly reduce X-ray-induced radiation damage. Cryotechniques such as *flash cooling* also led to the development of *kinetic crystallography* for the study of "trapped" intermediates in an enzymatic reaction pathway, for example. Progress in protein crystallography in the last decades of the twentieth century was astounding, with huge complex structures with internal symmetry such as viruses (M. Rossman, D. Stuart, and their collaborators), and later cellular molecular machines without internal symmetry such as the ribosome (A. Yonath, T. Steitz, V. Ramakrishnan, and their collaborators) being crystallized and solved to high resolution.

The advent of new X-ray and neutron instrumentation is now allowing smaller and more fragile biological crystals to be studied.

G3.2 FROM CRYSTAL TO MODEL

The molecular structure of a protein is the basis of its function. A detailed understanding of this relationship is one of the fundamental aims of modern biology. Through the efforts of X-ray crystallography, 2000–3000 biological structures are determined yearly. This chapter details how it is possible to obtain an atomic structure from the X-ray diffraction pattern of a crystal (Comment G3.1). We describe: (1) how proteins and other biological molecules can be coaxed into forming crystals; (2) the steps involved in collecting diffraction data from these crystals; (3) the

---

COMMENT G3.1 BIOLOGIST'S BOX: CRYSTAL STRUCTURE AND RESOLUTION

A picture is worth a thousand words, but how reliable is the information in the structure of a protein obtained from crystallography? The *resolution* in ångström units to which a structure has been solved is a key value in order to judge this reliability. It usually appears in the title of the paper, or at least in the abstract – the lower the number, the "higher" the resolution.

In low-resolution structures, worse than 3.5 Å, it is difficult to see individual amino acid side-chains. On the other hand, secondary structure elements, especially α-helices, but sometimes even β-sheets, can be identified, as well as their organization in domains. At medium resolution (between 2 and 3 Å), most of the individual atoms should be visible and we start to see solvent molecules (water and ions). As the resolution approaches 2 Å, alternative conformations become clear if the same amino acid side-chain moves between different positions. At 1.6 Å resolution, the electron density of an aromatic ring is so well defined that we can see the "hole" in the middle. At 1.1 Å resolution, hydrogen atoms can be positioned in X-ray crystallography, despite the fact the X-rays have only been scattered by that atom's sole electron. In neutron crystallography, however, hydrogen scattering is similar to other atoms, and they can be positioned in a structure at much lower resolution.

---

evaluation of the data by using crystal symmetry; (4) different methods to determine the structure factor phases in order to calculate an electron density distribution; (5) how to build, refine, and assess reliability of a macromolecule structure model that fits the electron density.

G3.2.1 Reciprocal Lattice, Ewald Sphere, and Structure Factors

We recall from Chapter G1 that the structure of a macromolecule can be calculated from the Fourier transform of its diffraction waves. (Eq. (G1.11):

$$f(\mathbf{r}) = \int F(\mathbf{Q}) \exp(-i\mathbf{Q} \cdot r_j) dV_Q$$

In the case of a crystal, $F(\mathbf{Q})$ is replaced by $G(\mathbf{Q})$, the crystal diffraction, where from Eq. (G1.14),

$$G(\mathbf{Q}) = F(\mathbf{Q}) \times \sum_t \exp(it\mathbf{Q} \cdot \mathbf{a}) \times \sum_u \exp(iu\mathbf{Q} \cdot \mathbf{b}) \times \sum_v \exp(iv\mathbf{Q} \cdot \mathbf{c})$$

Following crystallographic convention, we replace $\mathbf{Q}$ by $\mathbf{S}$ where $\mathbf{Q} = 2\pi \mathbf{S}$. Recall that $S = 2 \sin \theta / \lambda$, and that peaks

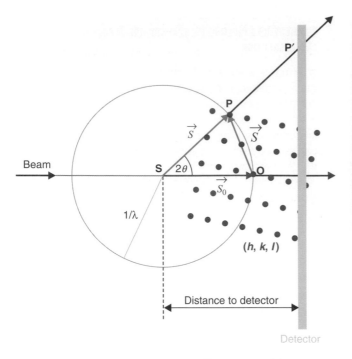

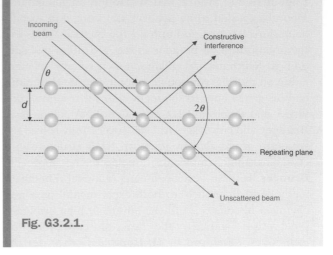

It is useful to picture a crystal as a set of periodic planes throughout space. Constructive interference only occurs in certain angular directions, given by Bragg's law:

$$n\lambda = 2d \sin\theta$$

where n is an integer, λ the wavelength, d the distance between the repeating planes in the crystal, and θ the incident angle of the radiation.

Fig. G3.2.1.

Fig. G3.1 The Ewald construction: diffraction conditions are satisfied at point P, and the diffraction spot appears on the detector at point P'. The angle OSP = 2θ.

only appear for values of **S** given by the *hkl* indices (Eq. (G1.15)),

$$\mathbf{S} \cdot \mathbf{a} = h \quad \mathbf{S} \cdot \mathbf{b} = k \quad \mathbf{S} \cdot \mathbf{c} = l$$

Similarly to the one-dimensional *reciprocal lattice* defined in Chapter G1, **S** defines a reciprocal lattice in three dimensions. The *Ewald construction* is a geometric representation of the conditions in Eq. (G1.15). The *Ewald sphere* of radius $1/\lambda$ is centered on the crystal. Each point of the reciprocal lattice has *hkl* coordinates that satisfy Eq. (G1.15). The origin ($h = 0$, $k = 0$, $l = 0$) is placed at the point where the incident beam intersects the Ewald sphere (Fig. G3.1).

The *hkl* lattice is rotated about the origin (O) by rotating the crystal. Diffraction is observed when a lattice point intersects the Ewald sphere (Comment G3.2).

We rewrite Eq. (G1.14) in terms of **S**

$$G(\mathbf{S}) = F(\mathbf{S}) \times \sum_{n} \exp\left(i2\pi n \mathbf{S} \cdot \mathbf{R}\right) \quad \text{(G3.1)}$$

$F(\mathbf{S})$ is the Fourier transform of the unit cell contents. In the case of a crystal this is called a *structure factor*. The unit cell has the shape of a parallelepiped, defined by three vectors **a**, **b**, and **c**, or more commonly by three lengths (a, b, c) and three angles (α, β, and γ). The cell contains atoms of scattering amplitude f_j at positions $\mathbf{r}_j$. We write

$$\mathbf{r}_j = x_j\mathbf{a} + y_j\mathbf{b} + z_j\mathbf{c} \quad \text{(G3.2)}$$

where x, y, z are fractional coordinates in the unit cell.

$$\begin{aligned} F(\mathbf{S}) &= \sum_{j} f_j \exp\left(i2\pi\mathbf{S} \cdot \mathbf{r}_j\right) \\ &= \sum_{j} f_j \left(\exp\left[i2\pi\mathbf{S} \cdot \left(x_j\mathbf{a} + y_j\mathbf{b} + z_j\mathbf{c}\right)\right]\right) \end{aligned} \quad \text{(G3.3)}$$

Since diffraction only occurs for the special values given by the Miller indices Eq. (G3.3) becomes

$$F(\mathbf{S}) = F(hkl) = \sum_{j} f_j \exp\left[i2\pi\left(hx_j + ky_j + lz_j\right)\right] \quad \text{(G3.4)}$$

The structure factor wave can be expressed as $F(hkl) = |F(hkl)| \exp[i\alpha(hkl)]$ (see Chapter A3) since only its absolute value can be measured from the square root of the intensity. $\alpha(hkl)$ is the relative phase of the wave.

G3.2.2 Space Group Symmetry

Recall that the proportion of X-rays scattered by matter is low and consequently the scattering from a single atom or molecule is too weak to be detected (see Chapter G1). However, in a crystal the molecules are periodically repeated throughout space. Low amplitudes scattered from sets of individual molecules add to form a significant scattered wave that can be observed. The symmetry of the crystal, therefore, is reflected in its diffraction pattern.

A nomenclature has been developed to describe a space group in which a letter and a set of numbers correspond to specific mathematical operations.

The letter refers to the type of crystal lattice. For example, P stands for primitive (lattice points only at the corners of the unit cell), F for face-centered (additional lattice points on the faces of the unit cell), and I for body-centered (an extra lattice point at the center of the unit cell).

The large numbers denote the symmetry around axes a, b, and c of the space group respectively, while the subscript numbers indicate if there is a screw axis; see, for example, the space group $P4_32_12$, discussed in Comment G3.4.

TABLE G3.1 SPACE GROUP NOMENCLATURE

Crystal system	Symbols of conventional unit cells	Unit cell parameters
Triclinic	P	$a \neq b \neq c$; $\alpha \neq \beta \neq \gamma$
Monoclinic	P, C	$a \neq b \neq c$; $\alpha = \gamma = 90$, $\beta > 90°$
Orthorhombic	P, C, I, F	$a \neq b \neq c$; $\alpha = \beta = \gamma = 90°$
Tetragonal	$P, I,$	$a = b \neq c$; $\alpha = \beta = \gamma = 90°$
Cubic	P, I, F	$a = b = c$; $\alpha = \beta = \gamma = 90°$
Hexagonal	P	$a = b \neq c$; $\alpha = \beta = 90°$, $\gamma = 120°$
Trigonal	R	$a = b = c$; $\alpha = \beta = \gamma \neq 90°$, $< 120°$

A crystal lattice can be described in terms of one of 230 symmetry relationships called *space groups* (Comment G3.3). The *unit cell* is the basic repeating motif, which contains one or more *asymmetric units*, depending on the space group (Table G3.1). The asymmetric unit may contain one or several molecules and even have its own internal symmetry, called *non-crystallographic symmetry* (e.g., two-fold symmetry in a protein dimer).

In protein crystals, certain space groups are not allowed due to the *chirality* or *handedness* of biological structures (see Chapter A2). Proteins are created from L-amino acids and secondary structure helices are preferentially right-handed in order to avoid steric hindrance in the side-chain conformation. Although there are 230 space groups, only 65 are possible for these chiral objects.

The symmetry results in certain space groups exhibiting missing reflections, called *systematic absences* (Table G3.2). Furthermore, *symmetry related* reflections should have

TABLE G3.2 SYMMETRY EFFECTS ON REFLECTION

Symmetry elements	Affected reflection	Systematic absences when
2-fold screw (2_A)	Along a	$h00$ $h = 2n + 1$ where
4-fold screw (4_2)	Along b	$0k0$ $k = 2n + 1$
6-fold screw (6_3)	Along c	$00l$ $l = 2n + 1$
3-fold screw ($3_1, 3_2$)	Along c	$00l$ $l = 3n + 1, 3n + 2$
6-fold screw ($6_2, 6_4$)		
4-fold screw ($4_1, 4_3$)		
	Along a	$h00$ $h = 4n + 1, 2,$ or 3
	Along b	$0k0$ $k = 4n + 1, 2,$ or 3
	Along c	$00l$ $l = 4n + 1, 2,$ or 3
C-centered (C)		$h + k = 2n + 1$
F-centered (F)		$h + k = 2n + 1$
		$k + l = 2n + 1$
		$h + l = 2n + 1$
Body-centered lattice (I)		$h + k + l = 2n + 1$

The effect of the crystal symmetry can be observed in space group $P4_322$. The four-fold screw axis (4_3) (along the c-axis) means that indices $00l$ are present only if $l = 4n$ (where n is an integer). The presence of two-fold screw axes (2_3) (along b) leads to systematic absences for odd reflections for the axial reflections $0k0$. Note that scaling and merging of diffraction intensities for $P4_32_12$ and $P4_12_12$ cannot resolve to which member of the possible pair of space groups the crystal form belongs.

For detailed information about a particular space group, crystallographers usually refer to the *International Tables for Crystallography*.

equivalent intensities in the same diffraction pattern, so that measured values can be averaged (Comment G3.4).

G3.2.3 Electron Density

In the case of X-ray crystallography it is convenient to define the atomic distribution in the unit cell by an electron density $\rho(x, y, z)$. Following the arguments developed in Chapters A3 and G1, the electron density can be calculated from the Fourier transform of the structure factors:

$$\rho(x,y,z) = \frac{1}{V} \sum_{h=-\infty}^{+\infty} \sum_{h=-\infty}^{+\infty} \sum_{l=-\infty}^{+\infty} |F(h,k,l)|$$
$$\exp\left[-2\pi i(hx + ky + lz) + i\alpha(h,k,l)\right] \quad \text{(G3.5)}$$

where V is the volume of the unit cell. Note that both the structure factor amplitude and its phase are needed to define electron density in the unit cell.

The theoretical electron density calculation of Eq. (G3.5) has to be modified in order to take into account experimental limitations. Protein crystals are imperfect and of limited size. Consequently, they produce diffraction patterns of limited resolution so that only a limited set of (hkl) reflections is measured. Furthermore, phase information is lost since only the intensities of the diffracted waves are recorded (see Chapter G1). The electron density calculated from the experimental data, $|F_{obs}(hkl)|$, is given by

$$\rho_{calc}(x,y,z) = \frac{1}{V} \sum_{h} \sum_{k} \sum_{l} |F_{obs}(h,k,l)|$$
$$\cos\left[-2\pi i(hx + ky + lz) + i\alpha_{estim}(h,k,l)\right]$$
$$\text{(G3.6)}$$

The sum is over the measured (h, k, l) indices only and the phases have to be estimated from other information. Several methods have been developed to solve the phase problem (see Section G3.6).

G3.2.4 Technical Challenges and the Crystallographic Model

Macromolecular crystallography involves overcoming a number of technical challenges. The structure factor phases must be estimated. The number of useful observations is low compared with the number of parameters required to define an atomic model of the macromolecule in the unit cell (Comment G3.5). Biological crystals also diffract weakly because they have a high solvent content. However, an X-ray structure provides a wealth of biological information, which amply justifies and rewards the invested effort.

COMMENT G3.5 ATOMIC COORDINATES AND TEMPERATURE FACTORS

In principle, each atom in the unit cell is defined by four parameters: the three spatial coordinates x, y, z of its average position, and a *temperature factor B*, which describes how the atom moves. A 10 kDa protein contains about 1000 atoms, which means there are 4000 parameters to be determined.

However, in a crystal structure determination, atomic coordinates are constrained by chemical information, e.g., the length of a carbon–carbon bond. In practice this reduces the number of parameters to be determined.

Structures of biological molecules are currently deposited in the Protein Data Bank (PDB) database. Each atom of the model is described by its position in the unit cell (x, y, z coordinates in ångströms), its temperature factor B and its occupancy (between 0 and 100%, according to the probability of finding it at that position). Experimental details and structure quality information are also contained in the PDB entry.

G3.3 CRYSTAL GROWTH: GENERAL PRINCIPLES INVOLVED IN THE TRANSFER OF A MACROMOLECULE FROM SOLUTION TO A CRYSTAL FORM

The initial step in crystallography is to transfer the soluble macromolecule to a solution in which it will form crystals (Comment G3.6). It is striking that such flexible, complex, highly hydrated molecules can arrange themselves in an ordered fashion. A protein crystal suitable for X-ray crystallography contains on average 10^{13}–10^{15} individual molecules.

Protein crystals are very fragile (when poked with a hair whisker they have the consistency of a soft French cheese). They are maintained by weak forces, of the level of a single hydrogen bond. Macromolecular crystallization is an entropy-driven process. The local increase in order gained by organizing the macromolecule on a crystal lattice is counterbalanced by the gain in freedom of other species in the solution. The crystallization process is dependent on the concentrations of different solutes, and specific physical chemical parameters, such as pH, which affect macromolecular surface charge. Crystallization trials aim to identify favorable conditions for crystal growth. The main parameters varied during these trials are the ionic conditions, the pH, and the concentration of so-called *precipitants*, which include certain salts, polymers, and organic solvents (Fig. G3.2).

G3.3.1 Purity and Homogeneity

It is imperative that the macromolecule solution is as pure and homogeneous (*monodisperse*) as possible. The presence

---

COMMENT G3.6 ORDER AND DISORDER IN PROTEIN CRYSTALS

Bernal realized that protein crystals would only diffract if they were not dried out. This implied that an "ordered" protein crystal would contain disordered solvent. The fraction of the crystal volume occupied by solvent has been observed to be between 20 and 80%, but is usually between 40 and 60% (Bernal and Crowfoot, 1934).

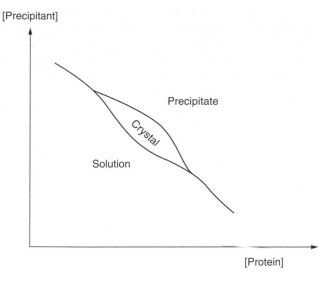

[Precipitant]

Precipitate

Crystal

Solution

[Protein]

Fig. G3.2 Phase diagram for the crystallization of a protein showing soluble crystal growth and amorphous (non-crystalline) precipitate regions as a function of precipitant and protein concentrations.

of glycosylation is often a hindrance to crystallization as the floppy nature of the saccharide chains may impede the creation of stable intramolecular contacts. In practice it is best to remove or shorten the sugar moieties before crystallization trials or to use recombinant protein that is not glycosylated. Large unstructured purification tags may also prevent crystallization.

G3.3.2 Crystallization Screens

Since each protein behaves differently, growing crystals is an empirical process. Multifactorial experiments are set up to find the optimum conditions for crystallization (Comment G3.7). The physical parameters usually varied in the first instance are the pH, the ion concentration, and the precipitating agent concentrations. There are three categories of precipitant: salts (e.g., ammonium sulfate), polymers (e.g., polyethylene glycol, PEG), and organic solvents (e.g., ethanol). Sample concentration and temperature (e.g., 37 °C, 20 °C, and 4 °C) can also be varied. The crystallization process can also be modified by altering the sample volume

---

(which changes the kinetics) or by small concentrations of additives (such as metal ions, detergents, urea, etc.).

Importantly, chemically modified or ligand-bound forms of the protein may not crystallize in the same conditions, since the molecule's surface characteristics, and consequently putative intermolecular contacts, are altered. Crystallization screens should therefore continue to be as exhaustive as possible.

G3.3.3 Crystallization Methods

Several methods have been developed to grow protein crystals by moving the protein into an environment (solution) that is appropriate for crystal nucleation and growth.

Vapor Diffusion

The vapor diffusion method is the preferred method for many crystallographers. Not only is it relatively straightforward to set up, but also the resultant crystals can be harvested with ease for X-ray data collection. A drop containing the protein is equilibrated against a larger reservoir of solution (the mother liquor). Volatile species, like water and certain ions and small solutes, diffuse between the two solutions until equilibrium is reached, i.e., the vapor pressure in the drop is equal to that in the reservoir (Comment G3.8). The hope (and goal) is that then the conditions in the drop are such that the protein will crystallize.

The drop is usually part reservoir solution and part protein-containing weak buffer. The typical protocol is to experiment with several types of reservoir solution in order to obtain the best crystallization conditions. As the drop equilibrates, the protein concentration within it usually increases and in the optimum case brings it within the crystallization phase. Drop volumes typically vary from 100 nl to 2 μl.

Dialysis

Crystals can also be obtained by dialyzing the protein solution against a crystallization solution. In this approach, the

---

COMMENT G3.7 CRYSTALLIZATION KITS

Several kits are available commercially for rapid and effective screening of potential crystallization conditions. A screen developed by Jancarik and Kim, for example, has already facilitated the determination of crystallization conditions for more than 500 proteins, peptides, oligonucleotides, and small molecules (Jancarik and Kim, 1991).

COMMENT G3.8 VAPOR DIFFUSION

In vapor diffusion, the drop size usually decreases during equilibration, increasing the constituents' concentrations. Although most molecules crystallize upon concentration, for some proteins the solubility decreases when they are diluted by water (through reverse diffusion). In such cases, the reservoir contains fewer solutes (especially salts) than the drop, which then grows in size as it equilibrates with the reservoir.

In each case, however, due to the changing conditions inside the drop, the protein either crystallizes or precipitates out (Jeruzalmi and Steitz, 1997; Richard *et al.*, 2000).

protein concentration is kept approximately constant, but its buffer is changed gradually.

Seeding

When the protein crystals are too small for X-ray crystallography, *seeding* offers the opportunity to increase the crystal size. It involves taking a crystal and adding it to a new drop containing soluble "fresh" protein. The crystal can then act as a nucleus from which a larger crystal can grow. Seeding can be done with all crystal sizes. Microcrystals can be seeded by a cat's hair whisker touching the crystals and then being drawn across a new drop. Larger crystals are often "etched" by passing through water, prior to adding them to a new drop. Crystals of one protein can also help initiate the growth of a crystal of a different but similar protein (Comment G3.9).

Membrane Proteins

Compared with soluble proteins, very few structures of integral membrane protein have been determined by X-ray crystallography (Comment G3.10). Several difficulties

---

COMMENT G3.9 THE MAJOR HISTOCOMPATIBILITY COMPLEX (MHC)

Major histocompatibility complex (MHC) protein molecules bind and display specific peptides to the immune system. An MHC molecule displaying one type of peptide has been used to grow crystals of the same MHC molecule presenting a different peptide (Bjorkman *et al.*, 1987).

COMMENT G3.10 EXAMPLES OF MEMBRANE PROTEIN CRYSTALLIZATION

Several difficulties are associated with working with integral membrane protein. In particular, since these proteins are expressed embedded in cellular lipid bilayers, they have to be solubilized (often with detergents) and purified prior to crystallization.

The crystallization step can then be attempted by incorporating in the usual trials either hydrophobic detergents or amphiphilic polymers, named amphipols, which bind to the transmembrane surface of the protein in a non-covalent manner. Another method, showing much promise, is to crystallize the protein from lipidic cubic phases. An integral membrane protein is confined to the three-dimensional network of the curved lipid bilayers, and protein crystals grow within the bulk cubic phase.

Interestingly, in several cases endogenous membrane lipids co-crystallized with the protein and were observed in the structure (Navarro *et al.*, 2002; Popot *et al.*, 2003).

---

are associated with the purification and crystallization of these proteins, because of their insolubility in the usual buffers of biochemistry. The usual approach is to extract the integral membrane protein in a solubilized form with detergents, and to proceed with crystallization trials as in the case of soluble proteins. The choice of detergent and its neutral or ionic character are crucial and depend on the specific characteristics of each protein. Amphiphilic polymer molecules, called *amphipols*, have also been developed for the solubilization of membrane proteins.

Another approach has been to rely upon lipid cubic phases for the crystallization of membrane proteins. These phases are formed by lipids, such as monoolein, under certain conditions of temperature and hydration (see Chapter A2). They present lipid–water interfaces of varying curvature. The membrane proteins are thought to diffuse to patches of lower curvature where they incorporate into a lamellar organization that associates to form highly ordered three-dimensional crystals.

G3.3.4 Identifying Crystals and Precipitates: Crystal Shapes and Sizes

Under certain conditions, salt crystals may appear. They are usually easy to identify as not being due to protein (Comment G3.11).

The presence of a clear drop indicates that either the drop has not finished equilibrating or that the sample concentration is too low. The presence of precipitate suggests that the sample concentration is too high. As a "rule of thumb," if more than 75% of the trials contain protein precipitant, the screen should be repeated with the starting sample concentration halved. However, in propitious circumstances, protein crystals may even grow in a drop containing precipitant.

Crystals have to be grown sufficiently large for X-ray data collection. The advent of synchrotrons has allowed full data sets to be collected from crystals approaching micron size, grown in 0.2 μl or smaller drops. Partial data sets can be collected from even smaller crystals (Comment G3.12).

COMMENT G3.11 SALT CRYSTALS

Certain buffers are not optimum for use in crystallography. In particular, phosphate, borate, and carbonate readily crystallize when they interact with divalent cations like Mg^{2+}, Zn^{2+}, and Ca^{2+}. These salt crystals can be identified prior to mounting and data collection, as they are harder, larger, often dichromatic, and more beautiful than "typical" protein crystals. Of course, some protein crystals also have every one of these characteristics....

COMMENT G3.12 X-RAY FREE ELECTRON LASER (XFEL)

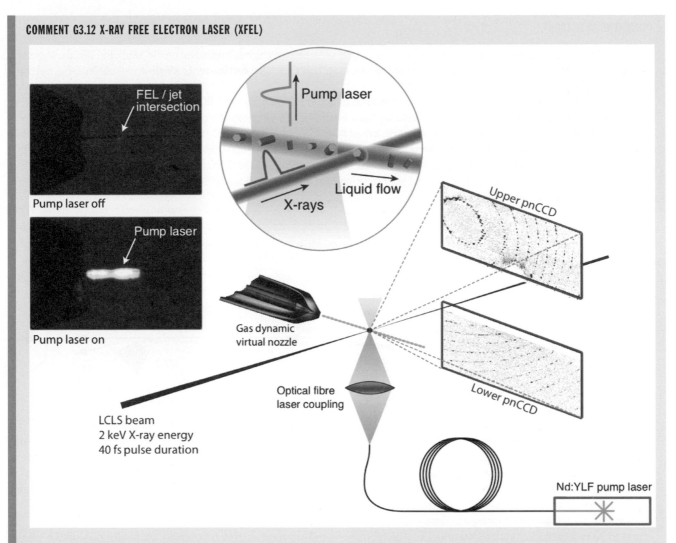

Fig. G3.12.1 General arrangement used for serial femtosecond nanocrystallography (SFX) at the Linac Coherent Light Source (LCLS) at SLAC near Stanford. Hydrated bioparticles are sprayed in single file, in vacuum, across the pulsed X-ray beam. The method of optical excitation of the particles is also shown, using a pump laser. The inset images show (top right) the geometric arrangement for a second, low-angle detector and (at left) experimental images of the particles producing a bright flash (top inset) as they are vaporized by the beam or (below) illuminated by the visible-light pump laser. For a $10\,\mu s$ delay between pump laser and X-ray pulse, the particles travel about $130\,\mu m$. Some arrangements allow on-demand triggering of particle injection using a piezo device. Reproduced with permission from Aquila *et al.* (2012). Copyright 2012 The Optical Society.

The peak brilliance ($\sim 10^{12}$ photons per pulse) and pulsed nature ($\sim 10\,fs$ duration) of a submicrometer focused beam make the XFEL a revolutionary new source of X-rays that is opening exciting new perspectives in the life sciences. It is making possible the imaging of protein nanocrystals, single particles such as viruses, *in vivo* crystallography following the discovery of protein crystals growing in cells, and snapshot wide-angle X-ray scattering from macromolecules in solution, as well as the study of protein structural dynamics on the femtosecond to picosecond timescale, e.g., by pump–probe experiments for time-resolved nano-crystallography, and time-resolved diffraction studies of non-cyclic reactions. A typical instrumental set-up on a XFEL source is shown in Fig. G3.12.1.

Neutze *et al.* (2000) combined molecular dynamics simulations with X-ray scattering calculations to argue that by rapidly collecting X-ray scattering data from a sample undergoing an X-ray damage-induced Coulomb explosion, it would be possible to recover interpretable diffraction data if the X-ray exposure were of shorter duration than the timescale needed for the sample to explode. Recent experiments showed that it is in fact possible to outrun radiation damage. The method requires a constantly refreshed supply of identical particles (see Fig. G3.12.1). So far, atomic resolution by this method could only be

Continued

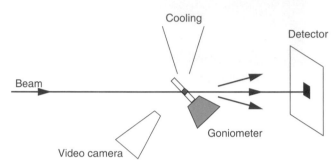

Fig. G3.3 Layout of an X-ray crystallography experiment. The beam (red) has been collimated and made monochromatic. The crystal (blue) in a capillary or a cryo-loop is mounted on a goniometer (green) and centered in the beam with the help of a video camera. The crystal is continuously cooled by a stream of cold dry nitrogen.

G3.3.5 Cryo-Crystallography and Cryo-Protectants

Synchrotron beam lines are preferred in X-ray crystallography because they permit the rapid collection of high-resolution diffraction data. The high intensity is problematic, however, as X-rays deposit energy into the crystal, causing heating and other radiation damage. Direct damage occurs when atoms in the crystal absorb X-ray photons, while indirect damage is due to diffusing reactive radicals produced by the radiation (Comment G3.13).

In order to reduce radiation damage, the crystal is kept at a low temperature (~100 K) such that the indirect damage is greatly reduced because of the lower diffusion rates, allowing sufficient time for data collection. The crystal has to be cooled very rapidly to cryo-temperature (*flash-cooled*) in order to avoid ice formation. Crystalline ice distorts both the lattice of the protein crystal and the structure of the protein itself. The key to the success of the method is the formation of a vitreous water phase. Low-freezing-point chemicals are often added to enable the solvent within the crystal to remain disordered, and favor the formation of an amorphous glass upon freezing. Commonly used cryo-protectants include glycerol, 2-methyl-2, 4-pentanediol (MPD), low molecular weight PEG, and oils. Further additions may not be necessary if the crystal growth solution itself is already cryo-protected.

It is always important to check that the cryo-protectant solution can be flash-cooled as an amorphous glass on its own. The protein crystal is then drawn through, or equilibrated against the cryo-protectant solution and subsequently flash-cooled.

G3.3.6 Crystal Mounting

Goniometer

The crystal is attached to a goniometer, which enables it to be rotated freely around an axis in order to examine a large volume of reciprocal space (Fig. G3.3). During data collection, the crystal is usually oscillated over a short angular range. The data corresponding to each angular position is called the *oscillation image*.

(a)

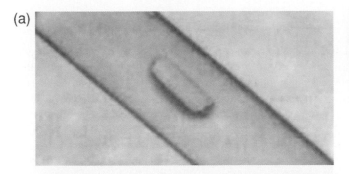

(b)

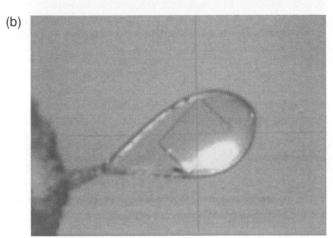

Fig. G3.4 Crystals mounted in: **(a)** a glass capillary and **(b)** in a cryo-loop.

Room Temperature Mounting in a Capillary Tube

In the early days of crystallography, before the advent of cryo-techniques, crystals were exposed to X-rays at room temperature in sealed glass capillary tubes (Fig. G3.4a). A drop of mother liquor is included in the tube near the crystal to prevent drying. Data collection is still conducted at room temperature if no suitable freezing conditions have been found.

Cryo-Loops

The most widely used method for freezing is to scoop the crystal with a small fiber loop (Fig. G3.4b). A thin film of solution spreads across the loop and supports the crystal. An advantage of the method is that it minimizes absorption due to glass. Rapid cooling is achieved by plunging the loop into a cryogen at low temperature (e.g., a stream of dry nitrogen at ~100 K). The crystal is then kept continuously frozen during data collection (Fig. G3.3).

G3.3.7 Labeling

Heavy Atom Derivatives

Binding heavy atoms to a protein can play a substantial role in obtaining structure factor phases (see Sections G3.6.4

TABLE G3.3 MOST COMMONLY CITED HEAVY-ATOM DERIVATIZING REAGENTS AS COMPILED FROM MACROMOLECULAR STRUCTURES FOR 1991–1994 (ADAPTED FROM ROULD, 1997)

Reagent	Reagent
K_2PtCl_4	$K_2Pt(CN)_4$
$KAu(CN)_2$	PIP
$Hg(CH_3COO)_2$	$Pb(CH_3COO)_2$
$Pt(NH_3)_2Cl_2$	K_2HgI
$UO_2(CH_3COO)_2$	Mersalyl
$HgCl_2$	p-Chloromercuribenzoate
$K_3UO_2F_3$	$CH_3Hg(CH_3COO)$
Ethyl mercurithiosalicylate	TAMM
$(K/Na)AuCl_4$	$SmCl_3$
$(Na/K)_3IrCl_2$	K_2OsO_4
$CH_3CH_2HgPO_4$	$(K/Na)_2OsCl$
K_2PtCl_6	UO_2SO_4
$UO_2(NO_3)_2$	Baker's dimercurial
$K_2Pt(NO_2)_4$	2-Chloromercuri-4-nitrophenol
$(CH_3)_3Pb(CH_3COO)$	$AgNO_3$
CH_3HgCl	CH_3CH_2HgCl
p-Chloromercuribenzene sulfonate	p-Hydroxymercuribenzoate

and G3.6.5). Heavy atom ions are usually soaked directly into the crystal in the hope that they will bind to the protein without destroying the crystal. Protein with bound heavy atoms is called a *heavy-atom derivative*. Table G3.3 lists a selection of commonly used derivatizing reagents.

Biosynthetic Labeling for Anomalous Dispersion

Sulfur, which is contained naturally in methionine and cysteine, and phosphorous, which is contained naturally in nucleic acid and lipids, can in principle be used for phasing by anomalous dispersion (see Section G3.6.5). Their absorption edges, however, are at X-ray wavelengths at which absorption is high for most elements, and weak anomalous dispersion signals make experiments difficult and sometimes impossible. Selenium produces a strong anomalous signal at the X-ray wavelengths more usual for data collection. It has become a standard technique to express proteins containing the modified amino acid selenomethionine to make use of the anomalous dispersion phasing method. Selenocysteine can also be used, as well as the incorporation of synthetic residues (like bromophenylalanine).

COMMENT G3.14 AUTO-INDEXING PROGRAMS

The auto-indexing routines determine the orientation and unit cell parameters of the crystal from the diffraction patterns and have been implemented in the computer programs XDS, Mosflm, and Denzo (Kabsch, 1988; Leslie, 1999; Otwinowski and Minor, 1997).

COMMENT G3.15 MOSAIC SPREAD AND EXTINCTION

A single crystal can be seen as made up of smaller microcrystalline domains that are slightly misaligned with respect to each other. The *mosaic spread* is an angular measure of the misalignment. The mosaic nature of crystals was proposed by Darwin in 1922. The mosaicity is the measure of the angular range over which a given reflection satisfies the diffraction condition. Protein crystals usually have very low mosaic spread, less than ~1°. Because of this, strong Bragg reflections are effectively diffracted by the first domains encountered so that the inner volume of the crystal is shielded from the X-ray beam. This effect is called *extinction*.

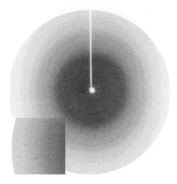

Fig. G3.5 Example of an X-ray diffraction pattern from a myoglobin crystal (Miele *et al.*, 2003).

G3.4 FROM INTENSITY DATA TO STRUCTURE FACTOR AMPLITUDES

Oscillation photography is the preferred method for the measurement of the diffraction patterns in protein crystallography. In classical crystallography, the axes of the crystal are carefully aligned to the beam in a series of preliminary exposures. This takes time, however, and, in the case of proteins, it is important to minimize exposure to X-rays because of radiation damage. The *American Method* developed by Rossman is the "shoot first, think later" technique, in which data are collected from the crystal in whatever orientation, then analyzed and interpreted (Comment G3.14). In fact, diffraction sets are now routinely automatically processed, integrated, and scaled.

Spots measured on the detector during data collection (Figs. G3.3, G3.5) have to be interpreted and converted to *hkl* intensity data, from which structure factor amplitudes can be calculated.

G3.4.1 Data Collection and Processing

The measured intensity has to be corrected for background and other effects. The Lorentz correction accounts for the different rates with which reciprocal lattice points cross the Ewald sphere. There are corrections also for X-ray polarization, extinction and absorption effects and radiation damage (Comment G3.15).

In classical crystallography, Wilson statistics allows an absolute scale *C* factor to be derived from a set of corrected diffraction intensities. The Wilson equation relates the mean intensity to the sum of squared scattering amplitudes and introduces an overall temperature factor *B*. It is a standard structure factor equation, assuming one atom per unit cell with a scattering intensity equal to the sum of atomic scattering intensities in the unit cell and a single isotropic temperature factor (see Chapters A3 and G1),

$$\langle I(h,k,I) \rangle = C \sum_n (f_n)^2 \exp\left(-2B\frac{\sin^2\theta}{\lambda^2}\right)$$

or

$$\ln\frac{\langle I(h,k,l)\rangle}{\sum\limits_n (f_n)^2} = \ln C - 2B\left(\frac{\sin^2\theta}{\lambda^2}\right) \tag{G3.7}$$

A plot of $\ln\left\langle I(h,k,l)/\sum_n(f_n)^2\right\rangle$ against $(\sin^2\theta)/\lambda^2$ is known as a Wilson plot. It should give a straight line from which the experimental values of *B* and *C* can be derived. This is not the case for protein crystals for which the Wilson plot looks like the one shown in Fig. G3.6.

The Wilson plot of a protein is characterized by a dip at around 6 Å resolution and a maximum at 4.5 Å resolution. This is due to the fact that because of the presence of α-helices and β-strands, interatomic distances are not distributed evenly. Recall that the protein crystal also contains solvent. At low resolution, the diffraction intensity is not due to the protein's density, but to the contrast in density between the protein and the surrounding solvent (see Chapter G1). As a result, the Wilson plot behaves linearly only at high resolution.

G3.4.2 Indexing Bragg Reflections

A diffraction spot is *indexed* by attribution of its *hkl* values. Initial estimates of the unit cell dimensions are made from the positions of the spots in the detector image and from

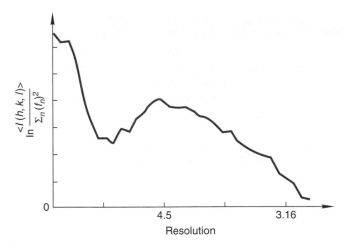

Fig. G3.6 Wilson plot for a protein crystal. The logarithm of average intensity is plotted against $[(\sin\theta)/\lambda]^2$. The corresponding resolution $(1/S)$ is given on the x-axis.

the physical parameters of the experiment (e.g., sample–detector distance). The parameters and indices are refined in a cyclic procedure.

In a theoretically perfect crystal, each diffraction spot's width is infinitesimally small, but due to mosaic spread, this is not the case (Comment G3.15), and the diffraction spot intensity may not be recorded fully. This is corrected for in the scaling step (see Section G3.4.3).

Indexing programs usually work on one oscillation image at a time. They convert raw diffraction data into a file, which contains hkl indices, the background, the corrected intensities of the spots on the image, and an estimate of their error.

G3.4.3 Scaling the Reflection Intensities

The images obtained from several orientations of the crystal then have to be incorporated into an overall data set. Numerous reflections are measured in more than one image and these partial data must be scaled and combined.

The scaling and merging of the data are important stages in the treatment of diffraction data as they allow a separate refinement of the orientation of each image, but with the same unit cell for the whole data set. The global refinement of the crystal parameters should produce precise unit cell values. The integrated data set therefore contains a list of measured hkl indices and their intensities.

G3.4.4 Twinning

Crystal growth anomalies may produce the phenomenon of *twinning*. This possibility must always be considered if the diffraction data cannot be solved. Twinning refers to a specific case of disorder in which there is partial or complete coincidence between the lattices of distinct crystal domains. Each domain diffracts X-rays independently

from each other so that their diffraction is not in phase; so that we have to add intensities rather than structure factors (see Chapters G1 and G2). In the presence of twinning, each measured reflection is the result of two or more overlapping Bragg spots, which have separate indices. Accordingly, since at least two of their axes must have equal length, tetragonal, trigonal, hexagonal, or cubic crystals are the most prone to such twinning.

Several procedures have been developed to *de-twin* diffraction data. In the case of two subcrystal twins (A and B), the measured overall intensity is given by

$$I_{overall} = \alpha I_A + (1-\alpha)I_B \qquad \text{(G3.8)}$$

where α is fraction of subcrystal A. Perfect twinning occurs when $\alpha = 0.5$, if not the twining is said to be partial.

In the case of *merohedral twinning* (from the Greek *meros*, part, and *hedron*, face), two or more lattices coincide exactly in three dimensions. This is not immediately obvious from the diffraction pattern. However, the presence of twinning can still be identified from the average intensity, $\langle I \rangle$, in each resolution range (Fig. G3.7):

$$\langle I^2 \rangle / \langle I^2 \rangle = 2 \,\text{untwinned}$$
$$\langle I^2 \rangle / \langle I^2 \rangle = 1.5 \,\text{twinned} \qquad \text{(G3.9)}$$

Similarly, for the structure factors:

$$\langle F^2 \rangle / \langle F^2 \rangle = 0.885 \,\text{untwinned}$$
$$\langle F^2 \rangle / \langle F^2 \rangle = 0.785 \,\text{twinned} \qquad \text{(G3.10)}$$

It is important to note that certain space groups, like H32, P422, and P4$_3$2$_1$2, cannot be twinned, while twinned crystals may display an apparent higher space group (for example perfectly twinned H3 will index like H32, and both P3$_1$21 and P6i will appear like P622).

G3.4.5 Radiation Damage

The issue of radiation damage is treated extensively in the chapter on electron microscopy. In fact, X-rays are much more damaging than electrons and the high intensities obtained from synchrotrons have led to the development of cryo-cooling procedures (see above), which limit indirect radiation damage (Comment G3.16). Very fast detectors (milliseconds per image) are now able to collect meaningful data from crystals prior to their destruction by X-rays.

G3.4.6 Determination of the Unit Cell Dimensions

Initial estimates of the unit cell size and orientation are obtained from data collected over a small oscillation of the crystal. Vectors are drawn between all the diffraction spots in the image. The two shortest non-linear vectors define the lattice of diffraction spots. These are the unit vectors from

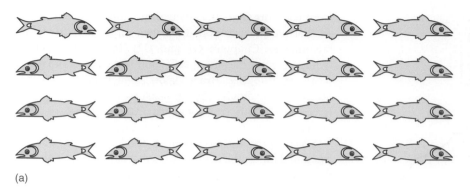

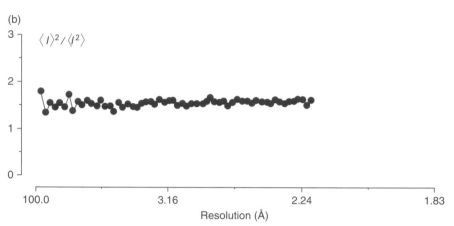

Fig. G3.7 (a) Example of twinning in a crystal. Several of the fish are oriented in the opposite direction with respect to the others. **(b)** Graph of $\langle I \rangle^2 / \langle I^2 \rangle$ plotted against resolution for a twinned crystal.

(a)

(b)

COMMENT G3.16 X-RAY RADIATION DAMAGE

Radiation damage is an inherent problem in X-ray crystallography. Experiments with 1 Å wavelength synchrotron radiation suggest that ten absorbed photons are sufficient to "kill" a 100 Å unit cell, and therefore correspondingly on average such a unit cell could contribute only one photon to total Bragg diffraction.

X-ray radiation can cause highly specific damage. For example, disulfide bridges can break, and, in the case of *Torpedo californica* acetylcholinesterase and hen egg white lysozyme, exposed carboxyls including those in these enzymes' active sites, appear more susceptible than other residues to radiation damage (Sliz *et al.*, 2003; Weik *et al.*, 2000).

---

which all other vectors can be obtained through combination and summation. Each of these two vectors defines the orientation and the length of one unit cell axis.

Diffraction spots in protein crystallography often cluster in patterns called *lunes*. A lune (from the Latin *luna*, moon) is the surface bounded by two intersecting arcs. The length and orientation of the third axis are obtained from the shape and distance between the lunes. Recall that the two-dimensional detector records in fact the angular position of the diffraction spots on the Ewald sphere. Based on these starting values, the diffraction pattern at other rotation angles is predicted and spurious spots that do not stem from the diffraction data, such as the *zingers* due to cosmic rays, are identified and omitted.

G3.4.7 Determination of the Space Group

The space group to which the crystal belongs can be identified from the systematic absences in the diffraction pattern. The presence of symmetry also implies that certain intensities should have similar values. It is therefore not necessary to rotate the crystal fully in order to collect a complete data set.

G3.4.8 Redundancy and Statistics

The signal-to-noise ratio (I/σ, where I is the intensity and σ the standard deviation) is a good criterion for assessing the resolution to which meaningful data have been collected. The minimum value of the signal-to-noise ratio is often set to 2 for the outer resolution shell of an individual image. This can be understood from the following statistical relationship:

$$\sigma(\langle I \rangle) = \frac{\sigma(I)}{\sqrt{N}} \tag{G3.11}$$

where N is the number of times each reflection is measured. We can assume that each individual intensity is similar to the average. If a reflection with $I/\sigma(I) \approx 2$ were measured four times during the data collection, the redundancy would yield $\langle I \rangle / \sigma \langle I \rangle \approx 4$.

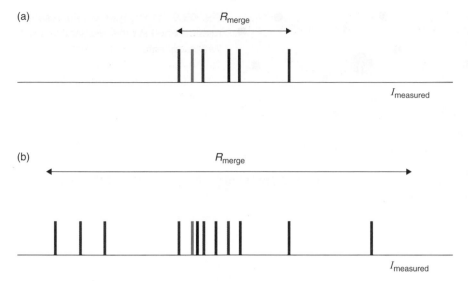

(a)

R_{merge}

$I_{measured}$

(b)

R_{merge}

$I_{measured}$

Fig. G3.8 Redundancy and R_{merge}.
(a) A reflection is measured four times (red bars) to give the values shown by the red bars. **(b)** The same reflection is measured a further six times to give a total of ten measurements. The estimated intensity value of the reflection (blue bar) is the average of the measured intensities (red bars). This average approaches the "real" intensity value (green bar) as more and more reflections are measured. However, the larger spread in the measured values (see **(b)**) results in a larger R_{merge} value.

It is accepted practice nevertheless to measure as much data as possible, as many times as possible. For example, synchrotron data are now often collected by fine slicing (<0.2 degree oscillation per image), so an individual reflection is measured multiple times over adjacent images.

Several algorithms assist in determining resolution of diffraction data (Comment G3.1). It is, however, important to note that images should still be checked by eye, as the high-resolution limit can vary from image to image. For example, a source of this anisotropy could be due to a needle-shaped crystal, as X-rays scatter from more unit cells in one direction compared to another.

Correlation coefficients (CC) can be calculated by randomly splitting the data into two data sets. A similar procedure for determining cryo-EM resolution is described in Part H. Correlation coefficients have the advantage of demonstrating if weak data have any information content.

The R_{merge} parameter has, however, traditionally provided an estimate of the precision of individual measurements. It is expressed according to the following formula:

$$R_{merge} = \frac{\sum\limits_{hkl}\sum\limits_{i=1}^{M} |I - \langle I \rangle|}{M \times \sum\limits_{hkl} I} \qquad (G3.12)$$

The parameter M takes into account the number of times a given reflection is measured., However, the R_{merge} term increases with the redundancy of the data, even though the more times a reflection is measured, the more accurate its value should become. R_{merge} is consequently not as informative about the quality of the data as are either the correlation calculation or the signal-to-noise ratio, and may be misleading (Fig. G3.8). The $R_{p.i.m.}$ term was therefore introduced to account for redundancy:

$$R_{p.i.m.} = \frac{\sum_{hkl}\sqrt{\frac{1}{n-1}}\sum_{j=1}^{n}|I_{hkl,j} - \langle I_{hkl}\rangle|}{\sum_{hkl}\sum_{j}I_{hkl,j}} \qquad (G3.13)$$

where n is the number times each reflection (hkl) was measured. This multiplicity-weighted R_{merge} gives a better estimate of data quality after merging multiple observations.

G3.4.9 Molecular Packing in the Unit Cell and the Patterson Function

The diffraction pattern intensities from which the unit cell dimensions and crystal symmetry can be determined contain information about the molecular packing within the unit cell.

Solvent Content

The fraction of the crystal volume occupied by solvent can be calculated from the crystal density and the partial specific volumes, respectively, of the macromolecule and the solvent. The partial specific volume of all proteins is close to $0.74\,cm^3\,g^{-1}$ or $1.23\,Å^3\,Da^{-1}$ (see Chapter A1). *Matthew's coefficient* is calculated as the ratio of protein volume to solvent volume in the crystal. The volume of the asymmetric unit is estimated from the unit cell dimensions and the space group. The solvent content of protein crystals has been observed to have values between 20 and 80%, but is usually between 40 to 60% (Matthew's coefficient between 1.5 and 0.66). If the molecule's molecular weight is known, this range is sufficiently restrictive to give a reliable estimate of the total number of molecules present per asymmetric unit.

Patterson Function

The Patterson map, which is directly calculated from the intensities in the diffraction pattern, can be used to deduce the relative orientation of the unit cell's components. The Patterson map is a vector map, with peaks at the positions of vectors between atoms in the unit cell (Fig. G3.9).

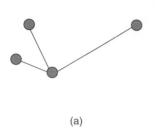
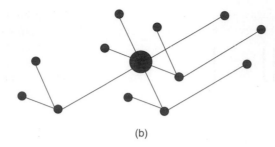

(a)   (b)

The Patterson function $P(uvw)$ does not rely upon phase information. It is the Fourier transform of the intensity of structure factors, with their phases set to zero:

$$P(uvw) = \frac{1}{V} \sum_h \sum_k \sum_l \exp\left[-2n(hu + kv + lw)\right]$$

(G3.14)

where u, v, w, are relative coordinates in the unit cell, and V is the cell's volume. This nomenclature is used instead of x, y, z in order to avoid confusion even though their dimensions are similar. This function can then be mathematically developed into the following,

$$P(uvw) = \iiint_{x\ y\ z} \rho(xyz)\rho(x+u, y+v, z+w)\partial_x\partial_y\partial_z$$ (G3.15)

The Patterson map's origin therefore contains the vector of each atom with itself, and the peaks throughout the map represent interatomic vectors of the crystal. Following upon this, the intermolecular vectors within the crystal can also be determined from the Patterson map.

These characteristics enable the calculation of relative orientation between similar subunits of the crystal. The Patterson map provides a starting point for many phasing methods (Section G3.6).

Non-Crystallographic Symmetry

The unit cell consists of a number of asymmetric units that are related through symmetry operators. *Non-crystallographic symmetry* (NCS) arises when there are two or more similar molecules (monomers) present in the asymmetric unit.

NCS can be determined from the diffraction intensity pattern. A *self-rotation function* searches for the direction and angle of rotation of the individual NCS operations, while a *cross-rotation function* searches for the relationship of a structure in one unit cell with similar structures in another cell (Fig. G3.10).

G3.5 FINDING A MODEL TO FIT THE DATA

G3.5.1 The Model

The ultimate aim of protein crystallography is to build a model of the crystal contents in terms of atomic positions and temperature factors. The steps involved are: (1) phasing the structure factors; (2) calculating an electron density distribution; (3) fitting an atomic model to the electron density; (4) refining the model with respect to the initial experimental observations and structural chemistry (and of course step; (5) is the deposition of the structure in the data bank and its publication). It is important to emphasize, however, that the experimental observations are the diffracted intensities. The reliability of the model is assessed with respect to: (1) the experimental observations and (2) the chemistry of the structure.

G3.5.2 Assessing Agreement Between the Model and the Data

The agreement between the calculated model and the observed data can be expressed in several ways. Structure factor amplitudes calculated from the model, $|F_{calc}|$, can be compared with the observations, $|F_{obs}|$. Recall Eq. (G3.6); the electron density is calculated from

$$\rho_{calc}(x, y, z) = \frac{1}{V} \sum_h \sum_k \sum_l |F_{obs}(h, k, l)|$$

$$\cos\left[-2\pi i(hx + ky + lz) + i\alpha_{estimate}(h, k, l)\right]$$

A standard linear CC between the observed and calculated structure factor amplitudes is determined from the following.

$$CC = \frac{\sum_{hkl} (|F_{obs}(h,k,l)| - \langle|F_{obs}(h,k,l)|\rangle) \times (|F_{calc}(h,k,l)| - \langle|F_{calc}(h,k,l)|\rangle)}{\left[\sum_{hkl}(|F_{obs}(h,k,l)| - \langle|F_{obs}(h,k,l)|\rangle)^2 \times \sum_{hkl} (|F_{calc}(h,k,l)| - \langle|F_{calc}(h,k,l)|\rangle)^2\right]^{1/2}}$$

(G3.16)

where $\langle|F_{obs}(h,k,l)|\rangle$ and $\langle|F_{calc}(h,k,l)|\rangle$ are the average structure factor amplitudes for the observed and calculated data, respectively. The values of the correlation coefficient range between 0 and 1. The larger value occurs when the calculated and observed structures factors are identical.

The R_{factor} is traditionally used to indicate the "correctness" of a model structure. It is defined by

$$R_{factor} = \frac{\sum_{hkl} \|F_{obs}(hkl)| - K|F_{calc}(hkl)\|}{\sum_{hkl} |F_{obs}(hkl)|}$$

(G3.17)

where K is a scaling factor.

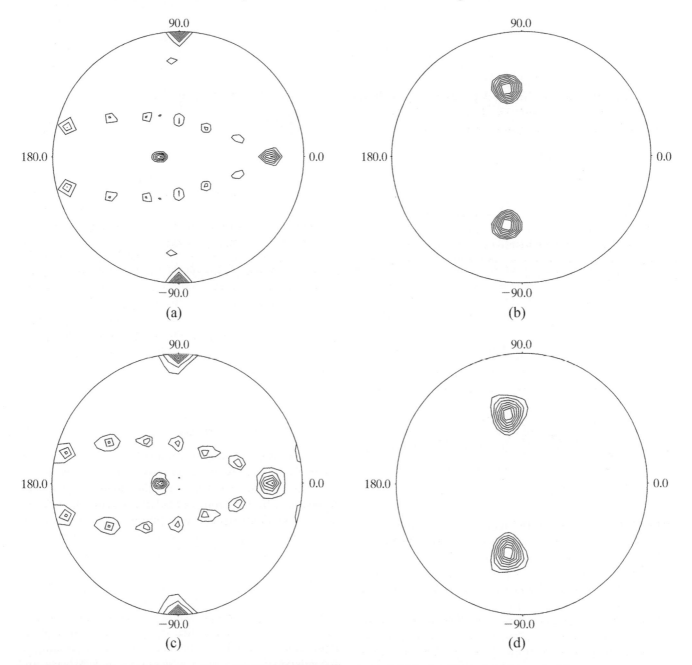

Fig. G3.10 Preliminary crystallographic analysis of the heptamer *Mycobacterium tuberculosis* chaperonin 10. The crystals belong to the monoclinic space group P2$_1$. Self-rotation functions were calculated using diffraction data with a resolution between 10 and 5.1 Å and a 30 Å integration radius, and are displayed in **(a)** and **(b)** as stereographic projections. The reciprocal space rotation is described in polar coordinates (ϕ, ω, κ).
In **(a)**, at $\kappa = 180°$, the peaks corresponding to the crystallographic 2$_1$-axis can be seen at the perimeter of the plot ($\omega = 90°$) and occur at φ values of 90 and −90°. Two strings of seven unique NCS two-fold axes are generated from each double heptamer in the unit cell. These appear as arcs above and below the equator and reflect the tilt in the plane of each heptamer with respect to the xz plane owing to their positioning about the 2$_1$-axis.
In **(b)**, at $\kappa = 51°$, the displacement of each peak from the perimeter at $\varphi = 90°$ and −90° represents the 35° tilt of the NCS seven-fold axis of each double heptamer related by 2$_1$ symmetry from the y-axis (Roberts *et al.*, 1999).

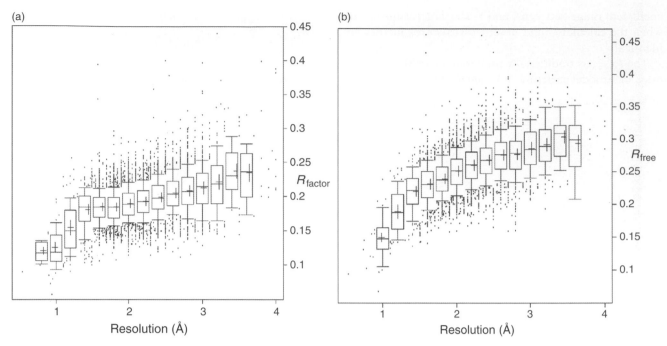

Fig. G3.11 (a) The distribution of R_{factor} values versus resolution.**(b)** The distribution of R_{free} values versus resolution. The data detail 10 888 macromolecular structures deposited in the PDB between 1991 and 2000. The "whiskers" indicate the tenth and ninetieth percentile of the data in the bin, whereas the upper and lower boundaries of the box indicate the twenty-fifth and seventy-fifth percentile. The horizontal line inside the box indicates the median (i.e., the fiftieth percentile), and the cross-hair inside the box indicates the average value (in both directions). Finally, in order to show the distribution of the "outliers," all individual data points outside the whiskers are shown (Kleywegt and Jones, 2002). (Figures reproduced with permission from Elsevier.)

A distribution of atoms placed at random in the unit cell would have an R_{factor} of 59% for *acentric* (non-centro-symmetric) reflections; the figure is 83% for *centric* reflections (the phase of a centro-symmetric reflection is 0 or π). During protein structure refinement (see below), the R_{factor} is ordinarily lowered to values in the 10–25% range (Fig. G3.11a).

It is possible to *overfit* the model, which results in an artificial lowering of the R_{factor}. The term overfitting refers to the case in which too many parameters are refined for the number of observations present. In order to guard against overfitting, an R_{free} term was introduced. It is calculated from structure factor amplitudes that are not included in the refinement (Fig. G3.11b). The R_{free} there-fore validates the extent to which the model explains the diffraction data.

$$R_{free} = \frac{\sum_{hkl \subset T} \|F_{obs}(hkl)| - K|F_{calc}(hkl)\|}{\sum_{hkl \subset T} |F_{obs}(hkl)|} \tag{G3.18}$$

The scaling factor K used is the same as for the R factor. The subset T contains the reflections set aside from the refinement. An estimated 500 reflections are necessary for the R_{free} to be statistically significant.

G3.5.3 Assessing Agreement Between the Model and Chemistry

Chemistry imposes certain rules on the conformation of molecules. Standard bond lengths and bond angles are typically obtained from crystallographic studies of small molecules. Information from high-resolution (subång-strom resolution) structures of proteins is also frequently incorporated when defining the stereochemical parameters of atomic models.

Due to steric clashes between the amino acid side-chains, the main chain of a protein is allowed only certain conform-ations. Recall that the conformation of the polypeptide chain is described by the φ and ψ rotation angles around the C_α carbon bonds and that there are areas for pairs of these values on a Ramachandran plot corresponding to secondary structures such as α-helices and β-sheets and other "acceptable" conformations (Fig. G3.12 and Chapter A2). The plot contains forbidden areas and if an amino acid residue in the model falls there, its conformation is obvi-ously wrong and has to be altered to a correct chemistry.

An analysis based on the notion that residues have pre-ferred positions in a protein structure can also be used to assess the plausibility of the model. For example, apolar residues are more likely to be in a hydrophobic environ-ment (Comment G3.17).

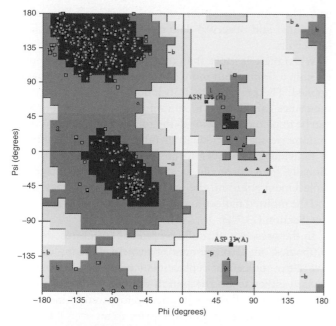

Fig. G3.12 The Ramachandran plot for protein CD1b (PDB accession code 1GZQ). The structure was solved at 2.26 Å resolution and has two residues (Asn128 and Asp33) in the generously allowed regions. Triangles represent glycines and squares the other residues in the protein structure (Gadola *et al.*, 2002).

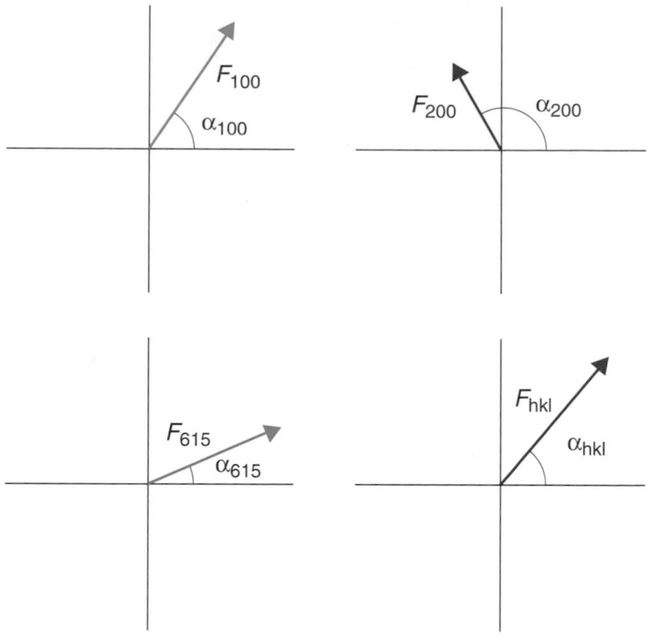

Fig. G3.13 Argand diagrams. Each structure factor can be represented on its own diagram. Examples are shown for reflections (100), (200), (615), and for a generic reflection (*hkl*).

COMMENT G3.17 VERIFY3D

The program VERIFY3D checks the plausibility of a model. The incorrect, but first published structures of Rubisco small subunit p21 ras and HIV protease would have been flagged by VERIFY3D (Bowie *et al.*, 1991; Luthy *et al.*, 1992).

COMMENT G3.18 MOLECULAR REPLACEMENT PROGRAMS

Several computational procedures to optimize the orientation superimposition have been developed and these procedures often have differing target functions. They include the programs Molrep, AMoRe, and phases inp script (Vagin and Teplyakov, 2000; Navaza and Saludjian; 1997; McCoy *et al.*, 2007).

G3.6 FROM THE DATA TO THE ELECTRON DENSITY DISTRIBUTION: INITIAL PHASE ESTIMATE

Phase information has to be obtained in order to calculate an electron density map. Several experimental procedures are used to phase observed structure factor amplitudes.

G3.6.1 Argand Diagram

Recall from Chapter G1 that each structure factor can be expressed as $\mathbf{F} = |\mathbf{F}|.\exp i\alpha$, which can be plotted on an Argand diagram, and where α is a phase angle (see Chapter A3). Each reflection has its own Argand diagram (Fig. G3.13).

The Argand diagram is a quick and easy way of visualizing structure factors and phases. A generic Argand diagram (for reflection (*hkl*)) will be extensively referred to in order to help clarify the different methods used to solve the phase problem.

G3.6.2 Molecular Replacement

A starting model, which is sufficiently similar to the molecule(s) in the crystal, can be used to provide an initial estimate for the phases (Comment G3.18).

Reasonable homology can be defined as a structure with at least 30% sequence identity, which typically implies an overall root mean square deviation between the coordinates of less than 1.5 Å. Sometimes, superposed X-ray structures of homologous proteins lead to better results than a single search model as these ensembles highlight conserved structural features.

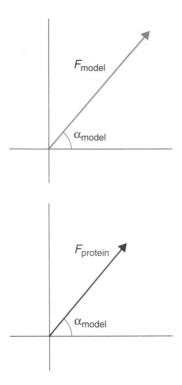

Fig. G3.14 Argand diagram for molecular replacement. The measured structure factor amplitude of the protein is given the phase of the one calculated from the model.

The aim of molecular replacement is to judiciously choose the position and orientation of the search model, and hence the phase origin, in order that the structure factors of the search model and of the unknown structure would be similar. The phases calculated for the search model are then transcribed to the crystal's structure factors (Fig. G3.14). In the case of one molecule per asymmetric unit, six parameters have to be determined in the molecule replacement approach. It is, however, simpler and less demanding computationally to separate the three orientation parameters of the model from the translation coordinates. For example, in real space, after the superposition of the centers of mass of two objects, a rotation function can be used to maximize the overlap between them in space. The translation function corresponds to the initial distance between the centers of mass.

The rotation step in molecular replacement is usually calculated from the overlap between the Patterson functions (see Section G3.4.9) calculated, respectively, from the observed data and from the model. The *rotation function* is defined as the overlap between the two Patterson functions. It is maximum when the two overlap optimally. The Patterson function of the model is preferably calculated from the molecule placed in a very large unit cell with no symmetry in order to avoid the contribution of intermolecular vectors.

A translation search is subsequently undertaken with the oriented model in the crystal space group and unit cell. The aim is to obtain the best overlap between the calculated and observed data now including intermolecular vectors.

When there is more than one molecule in the asymmetric unit, a similar (but more complicated) procedure can be applied. The presence of translational symmetry between molecules can be used to simplify the search, as this NCS can be implemented prior to the rotation function to create multimeric search models, thereby reducing the number of individual molecules to position in the unit cell. Additionally, if a molecule is missing from the molecular replacement solution, the crystal packing may have holes and the omitted molecule should appear as missing density (see Section G3.7.1).

In the presence of multicomponent structures, such as antibodies, slight movement of protein domains will not give clear molecular replacement solutions. It is hence advised either to do the search domain by domain or to attempt an overall rotation function followed by independent rotations for each domain, prior to an overall translation function.

G3.6.3 Direct Methods

Ab initio phasing methods were essential when the first molecular structures were determined and are still to be used for new protein structures for which there is no known suitable homologous structure to serve as a search model. Furthermore, molecular replacement suffers from *phase bias*; the resultant structure may resemble the search model, even though it will not refine properly.

Phases Derived from Patterson Maps

It is possible to interpret the phases directly from the Patterson map. If the individual Patterson peaks are fully resolved, the interatomic vectors can be determined accurately and are therefore sufficient to construct a model structure. In practice, however, only very small molecules are capable of providing such data.

Phase Relationships Between Structure Factors

A certain amount of phase information is contained in the reflection intensities. The diffracting crystal is a real object composed of atoms, whose electron density is positive everywhere in the unit cell. These assumptions of atomicity and positivity limit the number of possible phases but more importantly create phase relationships between the different reflections. For example, the phase of a reflection must be such that the respective real-space wave maxima overlap positions of high electron density in the crystal. The phases of three waves intersecting at a position of high electron density (for instance, an atom) therefore have a phase relationship, in order for their maxima to overlap in the crystal. In the case of the indices of three reflections which sum to zero, the *triplet relationship* between their respective phases φ is

$$\varphi_h - \varphi_k - \varphi_{h-k} \cong 0[2\pi] \tag{G3.19}$$

where $[2\pi]$ is the modulus. It is obvious then that the phase for reflection $\mathbf{h}$ can be determined if phases for reflections $\mathbf{k}$ and $\mathbf{h-k}$ are known, or

$$\varphi_h = \varphi_k + \varphi_{h-k} \tag{G3.20}$$

or, less succinctly,

$$\varphi_h = \varphi_{k1} + \varphi_{h-k1}, \quad \varphi_h = \varphi_{k2} + \varphi_{h-k2},$$
$$\varphi_h = \varphi_{k3} + \varphi_{h-k3} \quad \text{etc.} \tag{G3.21}$$

The *tangent formula* lets us refine the phase of reflections $\mathbf{h}$ if the phases of reflections $\mathbf{h}$ and $\mathbf{h-k}$ are known approximately,

$$\tan(\varphi_h) = \frac{\sum_k |E_k E_{h-k}| \cos(\varphi_k + \varphi_{h-k})}{\sum_k |E_k E_{h-k}| \sin(\varphi_k + \varphi_{h-k})} \tag{G3.22}$$

where E_h is the normalized structure factor:

$$E_h^2 = \frac{F_h^2}{\langle F_h^2 / \varepsilon \rangle} \tag{G3.23}$$

and ε is determined directly from the space group and takes the multiplicity of the reflection into account.

Direct methods are based on these relationships. However, few protein structures have been solved *ab initio* in such a way because direct methods require very precise and high-resolution data ($<1\,\text{Å}$). This methodology is nonetheless often used in conjunction with other phasing techniques. Provided there are a small number of starting phases, a complete data set can be iteratively deduced.

G3.6.4 Single and Multiple Isomorphous Replacement (SIR, MIR)

The addition of heavy atoms to a crystal changes the original diffraction pattern. If the protein is not distorted, i.e., it is *isomorphous*, the structure factors and phases due to the protein component are unchanged, but now the observed intensities also take into account the presence of the extra atoms. The influence of the heavy atoms can then be separated from that of the rest of the molecule as observed in an Argand diagram (Fig. G3.15).

If we recall the Patterson function (Eq. (G3.15))

$$P(uvw) = \int\limits_x \int\limits_y \int\limits_z \rho(xyz)\rho(x+u, y+v, z+w)\partial_x\partial_y\partial_z$$

a peak at position $\mathbf{r}$ in the Patterson map is proportional to the sum of the product of the electron densities of atoms separated by the vector $\mathbf{r}$. Since heavy atoms have greater density than the atoms usually found in proteins (hydrogen, carbon, oxygen, nitrogen), peaks at interatomic vectors between heavy atoms appear clearly in the scatter map obtained by subtracting the Patterson diagram of the native protein data set from that of the protein with heavy atoms.

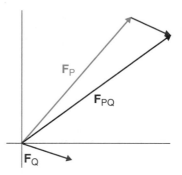

Fig. G3.15 Argand diagram for the presence of a heavy atom. F_P, F_Q, are respectively the structure factors of the protein and heavy atom alone. F_{PQ} is the structure factor of the protein structure containing the heavy atom.

For example, the extra scattering intensity of one mercury atom (containing 80 electrons) added to a protein would be proportional to $1 \times 80^2 = 6400$. If the protein contains 1000 atoms (equivalent to about 70 residues), the mean scattering intensity of the protein on its own is proportional to $1000 \times 7^2 = 49\,000$ (as there are on average seven electrons per protein atom). The single mercury atom added therefore represents 1.3% of the mean scattering intensity, which is observable. The effect of a derivative on the intensities is often quantified by

$$R_{\text{derivative}} = \frac{\sum\limits_{hkl} |I_{\text{derivative}} - I_{\text{native}}|}{\sum\limits_{hkl} I_{\text{native}}} \tag{G3.24}$$

where the *hkl* reflections can be limited to a small oscillation angle ($\sim 10°$).

Due to the small number of heavy atoms per asymmetric unit it is possible to estimate their coordinates from standard methods used in small-molecule crystallography, and therefore phase the structure $\mathbf{F_Q}$ corresponding to the heavy atoms alone. For each *hkl* value, the Argand vector of the derivative ($\mathbf{F_{PQ}}$) is the sum of the Argand vectors of the protein on its own ($\mathbf{F_P}$) and of the heavy atoms ($\mathbf{F_Q}$). As the phases for $\mathbf{F_Q}$ are known, only two possibilities exist where the structure factor amplitudes of the protein and its derivative are compatible (Fig. G3.16). In order to choose the correct phase angle, it is therefore preferable to use data from two or more different heavy atom crystals in order to produce a single solution for the phase of the protein crystal. Using statistical approaches, including error estimates for the intensities, it is also possible to estimate likely phases from a single isomorphous replacement derivative.

G3.6.5 Single and Multiple Anomalous Dispersion (SAD, MAD)

All atoms have the property that their electrons can be excited from a lower to a higher energy by incident photons of a given wavelength. A secondary photon is emitted as the electron returns to its shell. An anomalous signal refers to a

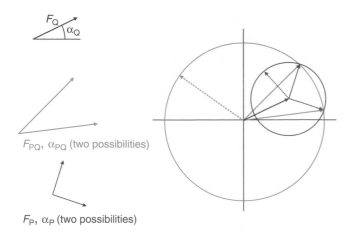

Fig. G3.16 Argand diagram for isomorphous replacement.

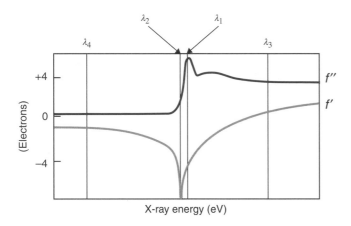

Fig. G3.17 Typical absorption edge chart of an atom.

Fig. G3.18 Anomalous scattering amplitudes.

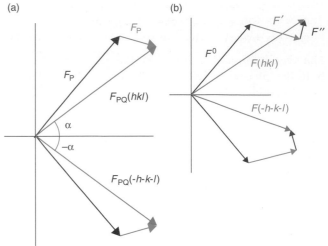

Fig. G3.19 (a) Friedel's law; (b) the effect of anomalous dispersion on **F**(hkl) and **F**($-h-k-l$).

phase change in scattering due to the absorption contribution, which results in a complex scattering amplitude written as,

$$f' + if' \tag{G3.25}$$

This effect depends on the wavelength (λ) of the incident X-ray and does not diminish with the diffraction angle (θ), because in practice the electron acts as point scatterer. The strength of the signal is expressed in electrons. The X-ray absorption spectra of atoms are well known and tabulated. The example in Fig. G3.17 displays its key characteristics.

The total scattering amplitude of the anomalous scatterer is written as

$$f = f^0(\theta) + f'(\lambda) + if''(\lambda) \tag{G3.26}$$

where $f^0(\theta)$ is the normal form factor (Fig. G3.18). The structure factor hkl due to the anomalous atom alone can now be written

$$\boldsymbol{F}(hkl) = \boldsymbol{F}^0(\theta) + \boldsymbol{F}'(\lambda) + \boldsymbol{F}''(\lambda) \tag{G3.27}$$

where $\boldsymbol{F}^0(\theta)$ and $\boldsymbol{F}'(\lambda)$ are parallel and perpendicular to $\boldsymbol{F}''\theta$) (Fig. G3.18).

In the case of normal scattering, *Friedel's law* states that the structure factors for (hkl) and ($-h-k-l$) (*Bijvoet pairs*) have equivalent amplitude but opposite phases (Fig. G3.19). The presence of an anomalous signal would add extra vectors $\boldsymbol{F}'$ and $\boldsymbol{F}''$ to the vector and result in the breaking of the symmetry in the Argand diagram (Fig. G3.19b).

An *anomalous Patterson map* is calculated from the intensity differences between Bijvoet pairs. The anomalous scattering is very weak, usually equivalent to the strength of a few electrons. In an ideal case, the ratio of the Bijvoet pairs' differences to normal diffraction of as low as 0.5% can be utilized for successful structure determination (Ramagopal *et al.*, 2003). Therefore, in order to reduce to a minimum the noise of the Patterson map, highly redundant data are a requisite for the precise determination of each reflection's structure factor amplitude. Vectors between anomalous scatters are nevertheless present and, when identified, provide phasing information.

During data collection, the X-ray wavelengths must be chosen with care. The first choice is to choose the wavelength that produces the largest anomalous signal (i.e., where $|f''|$ is largest). The wavelength is then varied to collect at maximal f'. Third and fourth data sets are usually collected at remote points of the absorption spectrum. As in the case of MIR, it is better to have as many anomalous derivatives as possible, each with a different anomalous scatterer.

The anomalous signal produced by heavy atoms, soaked into the crystal, is also often used in combination with MIR.

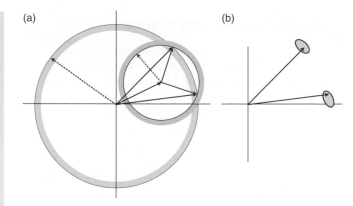

Fig. G3.20 Argand diagram with errors. If the measured structure factor amplitudes are not exactly known **(a)**, this results in a level of uncertainty for both the amplitude and phase of the protein's structure factors **(b)**.

The anomalous Patterson map is correlated with the difference Patterson map so as to unambiguously position the anomalous scatter in the crystal with less ambiguity. Ligands with a number of anomalous scatters bound together are hence particularly valuable. Since the distances between each atom in the ligand are known, they can be readily identified in the Patterson maps and consequently help phase the rest of the structure (Comment G3.19).

G3.7 FROM THE ELECTRON DENSITY TO THE ATOMIC MODEL: REFINEMENT OF THE MODEL – PHASE IMPROVEMENT

Because of uncertainties in the data, the structure factor Argand diagrams are composed of vectors whose amplitudes are not known exactly (Fig. G3.20). The ambiguity in each measured vector results in further error in the phase angle estimate. Several procedures have been developed to refine the phases in order to create a plausible structural model.

G3.7.1 Fitting to the Electron Density by Eye and by Hand

The interpretation of the electron density map of the crystal is done with the aid of interactive molecular graphics programs (Comment G3.20). In order to view the electron density accounted for by the model, it is usual to calculate as so-called 2F_o-F_c map,

$$\rho(x,y,z) = \frac{1}{V}\sum_{hkl}(2|F_{obs}| - |F_{calculated}|)$$
$$\exp[-2\pi i(hx + ky + lz) + i\alpha_{calculated}] \quad (G3.28)$$

The structure factor $F_{calculated}$ and phase angles $\alpha_{calculated}$ of the model are calculated from the model's theoretical electron density. In such a map, F_{obs} is said to be weighted twice with respect to $F_{calculated}$.

In order to visualize the location of the electron density not accounted for by the model, a 2F_{obs}-F_{calc}, or more often called F_o-F_c map, is calculated instead:

$$\Delta\rho(x,y,z) = \frac{1}{V}\sum_{hkl}(|F_{obs}| - |F_{calculated}|)$$
$$\exp[-2\pi i(hx + ky + lz) + i\alpha_{calculated}] \quad (G3.29)$$

Note that the calculated phase $\alpha_{calculated}$ is applied to both the F_{obs} and the F_{calc}. There are other weighting schemes for viewing the electron density and some crystallographers prefer to view their data in 3F_{obs}-2F_{calc}, or 2F_{obs}-F_{calc} type maps as they are considered less model-biased (Comment G3.21). These are therefore "native" maps (F_{obs} exp(i$\alpha_{calculated}$)) into which difference Fourier terms are introduced.

G3.7.2 Minimization of a Target Function (Maximum Likelihood)

Phases have a very strong influence on the appearance of the electron density. In molecular replacement, the phases of the search model are used with the diffraction data in order to calculate the starting electron density of the structure. The initial solution of the phase problem fits the information gleaned from the diffraction data, but the resultant macromolecular structure may not make sense physically, e.g., there may be steric clashes within the unit cell. The purpose of the refinement procedure is to optimize the agreement between an atomic model and both observed diffraction data and chemical restraints.

An entirely general refinement method would treat the chemical restraint information like the observed diffraction information. The squared residuals arising from the calculated and ideal chemical restraints and from the suitably weighted observed and calculated data can then be added together and minimized. The refinement can hence be conceptualized in terms of finding a minimum for the target function E as parameters are improved in cycles of refinement,

$$E = E_{\text{chem}} + W_{\text{X-ray}} . E_{\text{X-ray}} \tag{G3.30}$$

The term E_{chem} describes empirical information about the chemical interactions in the crystal. $W_{\text{X-ray}}$ is a weight that controls the relative contributions of the E_{chem} and $E_{\text{X-ray}}$ terms. The $E_{\text{X-ray}}$ term describes the difference between the observed and calculated diffraction data and can be understood as the experimental potential energy. It was initially calculated from

$$E_{\text{X-ray}} = \sum_{hkl} (|F_{\text{obs}}| - k |F_{\text{calculated}}|)^2 \tag{G3.31}$$

where k is a relative scale factor.

The $E_{\text{X-ray}}$ term has been subsequently modified into a *maximum likelihood* target function:

$$E_{\text{X-ray}} = \sum_{hkl} \frac{1}{\sigma^2_{\text{ML}^{cv}}} (|F_{\text{obs}}| - \langle |F_{\text{obs}}| \rangle^{cv})^2 \tag{G3.32}$$

where the expected value, $\langle |F_{\text{obs}}| \rangle^{cv}$, is derived from the calculated structure factors taking into account, approximately, the errors in the atomic model, and where the variance $\sigma^2_{\text{ML}^{cv}}$ embodies both data and model errors. The $|F_{\text{obs}}|^2$ intensities, for instance, should follow a Wilson distribution.

The model then becomes more and more complete during the course of the refinement. Missing side-chains, ions, and water molecules can be added as soon as the noise levels in the electron density maps decline sufficiently. Water molecules, for example, can be identified if

they are at the correct distance and orientation from donor and acceptor atoms in the model. Ions and small molecules are usually distinguished by their position with respect to the protein's surface and appear in the F_o-F_c maps, as missing density that the electron density of a water molecule in that location cannot fill (Comment G3.22).

G3.7.3 Crystallographic Refinement Restraints

Bulk Solvent Correction

As detailed in the previous discussion on the Wilson plot, the effect of the solvent on low-resolution diffraction data must be accounted for and a bulk correction must be applied.

Chemical Restraints

The E_{chem} energetic term describes the physical restrictions that the crystallographic model has to obey. In order for the refinement to proceed, it is therefore necessary to explicitly define the chemical characteristics of the model, such as the presence of covalent bonds and atomic charges.

E_{chem} is a function of the atom positions (bonds, angles, dihedrals, chiral volume, non-bonded contacts, etc.). It was initially designed for macromolecular simulations but subsequently modified for structure refinement (see Chapter I1). Weighting factors for certain of the stereochemical restraints had to be increased, for instance in order to keep aromatic rings planar. Similarly, the charges of certain residues have to be set to zero; the charges on arginine, lysine, and glutamic and aspartic acids are often screened by the bulk solvent. The covalent bond length distortions are also kept smaller than 0.02 Å and bond angle distortions to less than 3.0°.

The stereochemical constraints within molecules can be established from comparison with other molecules previously characterized by X-ray crystallography. Perusal of the Chemical Database Service, which contains structures of several hundred thousand small molecules, provides reference to the conformation geometries available to each molecule. As more macromolecules are determined, statistical analysis will further underpin information about a protein structure's general properties. This geometric information is then often codified into dictionaries for use in different refinement programs.

The alteration of the weight $W_{X\text{-ray}}$ in the target function E decides which term, the model's chemistry or the observed X-ray data, should have the greatest influence during the refinement. The optimal weight $W_{X\text{-ray}}$ can be estimated from the gradients between the two energy terms after an unrestrained molecular dynamics simulation and, as the X-ray refinement progresses, it must be adjusted as the estimated positions of the atoms approach their true value in the crystal structure.

Molecular Averaging

When more than one molecule is present inside the asymmetric unit cell, the molecules can be related by NCS. The electron densities of the related molecules are essentially equal, although differences may exist in the contact regions between neighboring molecules. The NCS operators can therefore be imposed to constrain or restrain the molecules to be similar during the refinement. Constraints confine the molecule's features to specific values, while restraints restrict each feature to a realistic range of possibilities.

G3.7.4 Refinement Procedures

Phase Extension

As discussed in Section G3.6.3, phase relationships exist between different structure factors. The phase extension enables phases for higher-resolution structure factors to be estimated from the phases of lower-resolution data.

Solvent Flattening

In solvent regions of the crystal, atoms are disordered so they are not represented by discrete positions in the electron density map. It is therefore possible to set an overall density in those regions. For example, a molecular envelope, or mask, can be created around protein and any density outside that mask is set to a uniform level. This procedure increases the interpretability of the electron density of the model.

Gradient Descent Refinement Methods

The refinement procedure tries to find the minimum for the target function E. In *rigid-body refinement*, the atoms of the model are kept static with respect to each other and the whole model is moved in order to optimize correlation with the experimental data. In the refinement of multicomponent complexes, protein domains can be treated as a rigid body.

Subsequently, the positions of individual atoms are varied in order to increase agreement between the observed X-ray data and the chemical parameters describing the model (Fig. G3.21).

B Factor Refinement

The scattering power of an atom is strongly dependent on its temperature factor B. In theory, this term reflects the atom's thermal motion, but it may also indicate local differences in molecular conformation. Recall that for atom j, its B factor is

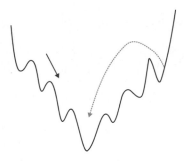

Fig. G3.21 Energy landscape showing the path followed in a gradient descent (red) and in simulated annealing (blue).

$$B_j = 8\pi^2 \langle u^2 \rangle$$

where $\langle u^2 \rangle$ is the atom's mean square displacement relative to the average position. The structure factor amplitude contribution of each atom is therefore modified:

$$f_j = f_{0,j} \exp\left[-B_j \left(\frac{\sin^2 \theta}{\lambda^2} \right) \right] \tag{G3.33}$$

where $f_{0,j}$ is the scattering factor for atom j calculated at 0 K.

Temperature factor refinement enables a better fit of the calculated structure factor to the observed data but introduces additional parameters into the refinement. It may therefore be necessary to limit the refinement to the overall B factor of the model in the case of low parameter/observation ratios. At very high resolution (1 Å), an anisotropic description of the B factor for each atom is possible, as an atom's thermal motions parallel and perpendicular to its bonds are different. At high resolution, (~2 Å), an isotropic B factor has to suffice. Nearer 3 Å resolution, it is suggested to have two B factors per residue; one for the main chain and the other for side-chain atoms, and at very low resolution, only one B factor per domain is appropriate.

Simulated Annealing

In gradient descent methods, the refinement may become trapped in one of several local minima. *Simulated annealing* can overcome such problems as it allows large areas of space to be sampled by the model's atoms because the refinement can proceed against the target energy gradient and therefore escape from local minima. It is therefore particularly useful after molecular replacement as it lowers the amount of bias toward the search model at the beginning of the refinement.

Simulated annealing is based on molecular dynamics simulation protocols (see Chapter I1). At its origin, this procedure involves solving the Langevin equation of motion, which describes the force acting on each atom in terms of the bond force, a frictional force due to solvent effects, and a random force due to temperature. Kinetic energy is initially added by increasing the temperature (to 1000 K, for example) in order to permit the model to

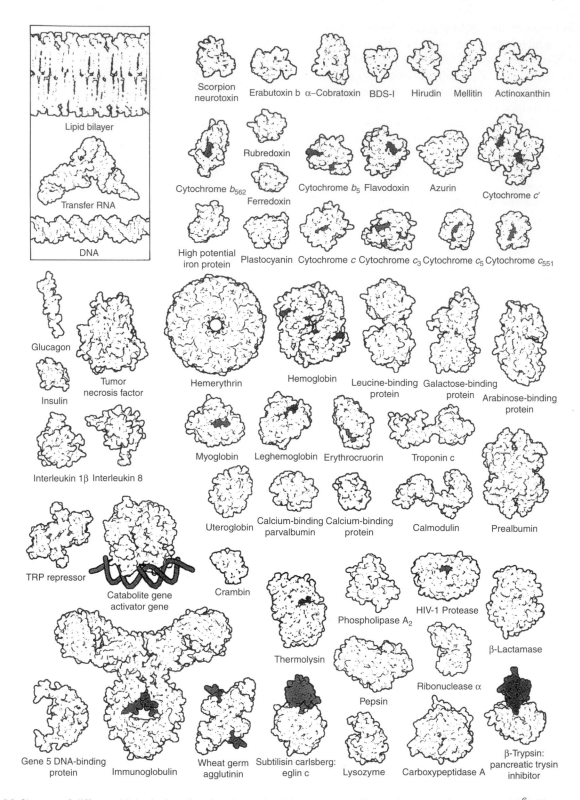

Fig. G3.22 Shapes of different biological molecules determined by X-ray crystallography (magnification 4×10^6). The scale is such that individual atoms have a diameter of 1–2 mm. The orientation of every molecule is chosen so that specific features of its shape are defined. (After Goodsell and Olson, 1993.)

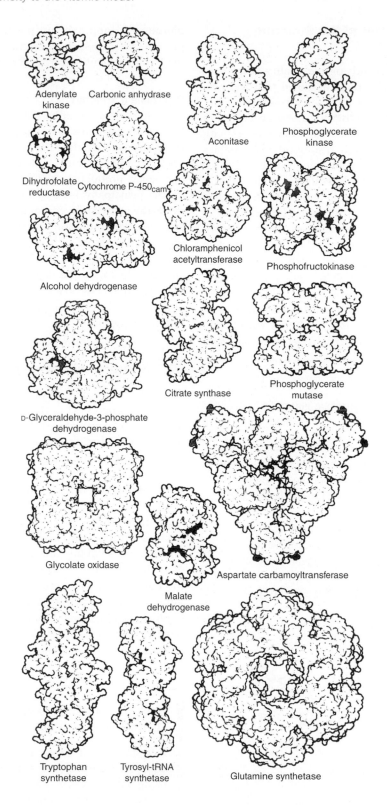

Adenylate kinase

Carbonic anhydrase

Aconitase

Phosphoglycerate kinase

Dihydrofolate reductase

Cytochrome P-450_{cam}

Chloramphenicol acetyltransferase

Phosphofructokinase

Alcohol dehydrogenase

D-Glyceraldehyde-3-phosphate dehydrogenase

Citrate synthase

Phosphoglycerate mutase

Glycolate oxidase

Aspartate carbamoyltransferase

Malate dehydrogenase

Tryptophan synthetase

Tyrosyl-tRNA synthetase

Glutamine synthetase

Fig. G3.22 (*cont.*)

sample many alternative conformations. The model is then slowly cooled until the atoms end up in a conformation more energetically compatible with the X-ray diffraction data. The simulated annealing temperature term has no physical meaning and is, in fact, unreasonable from the biological point of view. It can, however, be understood as the likelihood of overcoming the energetic barriers present in the target function E. The diagram in Fig. G3.22 shows an energy landscape and the simulated annealing procedure.

G3.7.5 Final Assessment of the Structure

Information and Resolution

The quality of a model depends on the quality of the data used in its creation. It is usually assumed that the positional error is roughly 10% of the quoted resolution.

High resolution enables one to place atoms accurately in the model (Fig. G3.22). It may also uncover residues with alternative conformations. High resolution also enables more complete models to be built since water and other molecules can be identified. As the resolution is lowered, it becomes more and more difficult to place first atoms, then individual side-chains, in the structure. The presence of NCS and independent phasing increases the amount of information. For example, although data only extend to 4 Å, the structure of the highly symmetric blue tongue virus is very informative. The 60-fold averaging procedure creates a model from which side-chains can be discerned. Similarly, the origins of anomalous diffraction can be identified with certainty even in a low-resolution map (Comment G3.23). In fact, in the initial maps derived from MAD phasing, the approximate positions of the residues in a protein are deduced from the locations of the selenomethionines identified from the anomalous signal.

Publishing a Structure

Scientific journals require that a model's coordinates and B factors are available prior to publication. Structures currently deposited in the PDB list further information such as relevant publications and details about the protein (sequence, production method, etc.). The crystallographic details principally include the crystallization conditions, the X-ray source, the computer programs used during the structure determination and refinement, and, importantly, the statistics associated with the data processing and the final model (Comment G3.24). Current trends are forcing structure factors to be deposited also as interested scientists can then check the quality of the electron density maps and decide on the validity of the published interpretation.

G3.7.6 Structural Genomics

Structure determination of a protein by X-ray crystallography has progressed to the point that several steps in the process can be automated. For example, robots have been designed for cloning and expression trials, and software can index diffraction intensities, and refine and build structural models which fit the X-ray data.

In conjunction with the short data-collection times associated with synchrotron radiation, it has become possible to study hundreds of proteins in parallel. This has several obvious advantages. Numerous protein constructs can be screened for solubility and crystallization. The elucidation of all the proteins of an organism (*structural genomics*) and the comparison of a protein's homologues expressed in different species (*comparative genomics*) offer very interesting biological insights. Finally, the determination of every possible protein fold provides crucial empirical knowledge for structure-from-sequence modeling approaches.

COMMENT G3.23 STRUCTURAL SEQUENCING GUIDED BY ANOMALOUS DIFFRACTION

In low-resolution maps, it is often impossible to directly identify the correct protein side-chains. The inclusion of an anomalous scatterer (like selenium) provides an anchor by which to position the rest of the amino acid sequence within the structure.

For example, a series of four selenomethionine mutants were generated in order to clearly identify the composition of the β–hinge domain bound to the core of AP-2 (Fig. G3.23.1). The resolution did not extend further than 3.07 Å. Maps are contoured at 4.5σ (a) or 3.5σ (b–d).

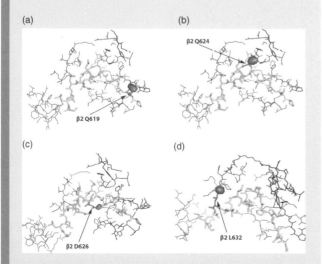

Fig. G3.23.1. (From Kelly *et al.* 2014)

COMMENT G3.24 BIOLOGIST'S BOX: CRYSTALLOGRAPHIC STATISTICS REPORTED IN PUBLICATIONS

A crystallographic paper usually has two tables. The first describes the diffraction data and usually includes: space group and unit cell, wavelength, resolution, number of unique reflections, completeness, redundancy, signal-to-noise ratio, R_{merge} and $R_{p.i.m}$. The second describes the refinement and details R_{factor}, R_{free}, the number of non-hydrogen atoms, the number of water molecules, the root mean square deviation (RMSD) from standard values in bond lengths and in bond angles, average B values, and Ramachandran plot statistics.

G3.8 KINETIC CRYSTALLOGRAPHY

All chemical and biological reactions involve atomic motion, embodied in dynamic structural changes. The aim of kinetic crystallography is to observe a number of structures, associated with reaction intermediates. The lifetime of these states varies between reactions; it can be extremely brief (in the femtosecond range), or last for hours.

G3.8.1 Trapping of Intermediate States

During the course of a reaction, a molecule goes through a series of intermediate states. In an enzymatic reaction these are the steps leading from substrates to products. The associated structures can be determined by physical or chemical trapping techniques, designed both to increase the lifetime of the molecules in a desired state and to trap a structurally homogeneous species.

An obvious trapping mechanism is to initiate a reaction (e.g., with a laser pulse), wait for the reaction to proceed sufficiently, and then flash-cool the crystal. It is often preferred, however, to use substrate analogues, which block the reaction from going any further. For example, hydrolysis of ATP to ADP, and subsequently to AMP, provides energy to drive many enzymatic reactions. The use of non-hydrolyzable forms of ATP and of ADP can inhibit the reaction at specific stages, from which structural mechanisms involved in the reaction can be identified.

G3.8.2 Laue Diffraction and Time-Resolved Crystallography

In contrast to trapping the crystal in specific states, time-resolved crystallography follows the evolution of the space-averaged structure in real time. The biochemical reaction is initiated in the crystal and followed. However, in order for this to succeed, the crystal must not become disordered during the reaction. As *time* is of the essence, data collection has to be as rapid as possible. The advent of high-brilliance synchrotrons allows collection of meaningful data over very short exposure times. A special diffraction setup called a *Weissenberg camera* is often used as it measures more data in one exposure than a flat detector. A Laue diffraction approach significantly increases the amount of data that can be collected concurrently. This method differs from the usual crystallographic studies as it uses polychromatic radiation. If we recall Comment G3.2, in order to satisfy Bragg's law, $n\lambda = 2d \sin\theta$, a crystal only diffracts at particular wavelengths for specific values of d and θ. The use of a broad spectrum of radiation consequently increases the number of crystal planes diffracting constructively for one crystal orientation. Figure G3.23 details the relevant Ewald construct. If wavelengths ranging from λ_{min} to λ_{max} fall on a crystal, all lattice points lying between the Ewald spheres of radii λ_{max}^{-1} and λ_{min}^{-1} are detected. A large number of reflections is therefore observed in a Laue experiment. In a typical experiment, for a crystal diffracting at 2 Å resolution and $\lambda_{max} - \lambda_{min} = 2$ Å, 10–60 times more reflections can be collected in one exposure than for a monochromatic experiment. Although spatial overlap of spots may become a problem, this usually does not affect more than 17% of spots in realistic experimental conditions.

In time-resolved Laue diffraction studies, the structure can be determined directly from the diffraction pattern, although it is much simpler to determine the starting and final conditions from static structures solved with monochromatic resolution. As the reaction proceeds, the

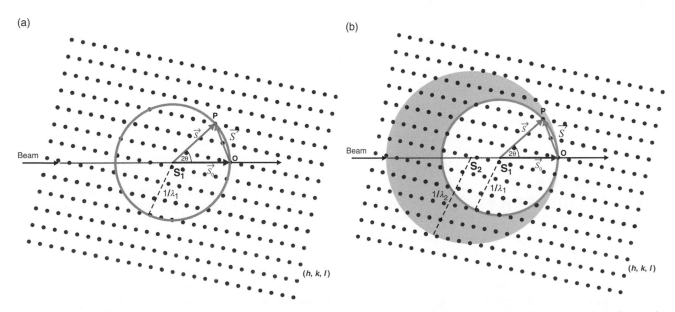

(a)

(b)

Beam

Beam

Fig. G3.23 Ewald constructs for: (**a**) monochromatic radiation (wavelength λ); and (**b**) polychromatic radiation (wavelengths ranging between λ_1 and λ_2).

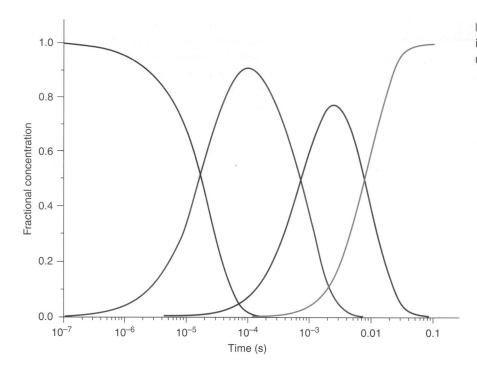

Fig. G3.24 Example of a simple irreversible sequential chemical kinetic mechanism with four intermediates.

time-dependent average structure in the crystal varies due to the time-dependent rise and fall of the concentrations of the intermediate states of the molecule (Fig. G3.24). The structures themselves do not vary with time, but since their concentrations do, the diffraction varies over time. If the crystal contains a set of transient structures $(1, 2, 3, \ldots, i)$ with fractional occupancies $(C_1, C_2, C_3, \ldots, i)$ and structure factors $\mathbf{F}_i(\mathbf{k})$, the total structure factor of the crystal is

$$\mathbf{F}(\mathbf{k}, t) = \sum_i C_i(t)\mathbf{F}_i(\mathbf{k}) \qquad \text{(G3.34)}$$

where $\sum_i C_i(t) = 1$ at all time t. The data can then be analyzed in order to identify the individual transient structures in the crystal.

A typical time-resolved experiment involves *kick-starting* the biochemical reaction in the crystal with a burst of laser light. A subsequent train of short X-ray pulses is then used for data collection with temporal resolution as short as 150 ps. The photolysis of the ligand, protein relaxation, and rebinding of CO to myoglobin, for example, has been determined to nanosecond time resolution (see Chapter A3).

G3.9 NEUTRON CRYSTALLOGRAPHY

Neutron diffraction is similar in theory and practice to X-ray diffraction. However, it provides a different information set due to the differences in atomic scattering and absorption between these two types of radiation (see Chapter G1). Significantly, neutron crystallography has the ability to locate hydrogen atoms in large molecules experimentally. X-rays, on the other hand, require high-resolution diffraction (<1 Å)

and hydrogen atoms must be often inferred from the stereochemistry of the heavier atoms to which they are bound. Since typically half the atoms in a protein are hydrogens, a protein's chemistry and physical structure can be precisely detailed by neutron crystallography; in particular in structures where water and protein hydrogens have unusual, but biochemically relevant, conformations.

Neutrons scatter off the atom nuclei of the structure, which act like point sources (see Chapter G1). In neutron diffraction from a crystal (Comment G3.26), we can rewrite Eq. (G3.4), and replace the scattering amplitude f_j of atom j of the structure by the nuclear coherent scattering amplitude b_j:

$$F(hkl) = \sum_j b_j \exp\left[i2\pi\left(hx_j + ky_j + lz_j\right)\right] \qquad \text{(G3.35)}$$

where atom j is at position (x_j, y_j, z_j) in the unit cell. The interaction of the neutron's magnetic moment with the crystal has not been included, as its effect is negligible for most biological material (Comment G3.25). We recall from Chapter G1 that the neutron scattering amplitude is usually a real number (either positive or negative) and does not vary much in magnitude between atom types. Hydrogen and larger atoms like carbon, nitrogen, and oxygen are consequently all discernible in 2 Å maps. The atomic structure can then be determined from the neutron diffraction in a similar manner to X-ray crystallography. However, phases are usually obtained from a previously determined X-ray model of the crystal.

The most serious limitation of neutron crystallography is due to the low neutron fluxes available, so a neutron structure is never determined before its X-ray counterpart. Large crystals (in the cubic millimeter range) and long

COMMENT G3.25 INCOHERENT NEUTRON SCATTERING FROM CRYSTALS AND DEUTERIUM LABELING

There are no interference effects due to the incoherent component of the scattering processes, although these contribute to background noise. In order to reduce the substantial incoherent scattering of neutrons by hydrogen atoms, crystals are therefore routinely soaked in D_2O buffer prior to data collection. The labile protons of the crystal are replaced by deuterium atoms, which have much lower incoherent scattering than hydrogen atoms. These atoms are consequently modeled as deuterium or hydrogen in the neutron structure, depending on whether or not they are accessible to solvent exchange. In order to make more efficient use of the neutron fluxes available, crystals are prepared with fully deuterated protein (obtained, for example, by recombinant expression in *E. coli*, grown on fully deuterated cultures), which significantly reduces the incoherent background and increases the coherent scattering signal (Kossiakoff, 1983).

COMMENT G3.26 MAGNETIC SCATTERING AMPLITUDE

The magnetic scattering amplitude is due to the interaction of the neutron spin with the spins of unpaired electrons. The magnetic scattering amplitude of certain metals, like Co, Ni, and Fe, can approach the magnitude of their nuclear coherent scattering amplitude.

collection times (weeks) are therefore required. Nonetheless, the development of Weissenberg camera-type detectors for neutrons and the use of Laue diffraction methods with neutrons of different energies has expanded the scope of neutron crystallography.

G3.10 CHECKLIST OF KEY IDEAS

- The molecular structure of a protein is the basis of its function. A detailed understanding of this relationship is one of the fundamental aims of modern biology.
- Phase information is lost during a diffraction experiment since only the diffraction intensities are recorded. The aim of crystallography is to solve this phase problem in order to create a model of the crystal structure.
- The *Ewald construction* is a geometric representation of the diffraction conditions.

- In protein crystals, certain space groups are not allowed due to the handedness of biological structures. The crystal symmetry results in certain space groups exhibiting missing reflections, called *systematic absences*.
- Structures of biological molecules are deposited in the PDB database. Each atom of the model is described by its position in the unit cell (x, y, z coordinates in ångstroms), its temperature factor B, and its occupancy (between 0 and 100%, according to the probability of finding it at that position).
- Protein crystals are very fragile. They are maintained by weak forces, of the level of a single hydrogen bond.
- Crystallization trials aim to identify favorable conditions for crystal growth. The main parameters varied are the ionic conditions, pH, and *precipitant* concentration.
- The vapor diffusion method is the method preferred by many crystallographers. In this method, a drop containing the protein is equilibrated against a larger reservoir of solution.
- In order to reduce radiation damage, a crystal is often flash-frozen.
- During data collection, the intensities measured on the detector have to be interpreted and converted to *hkl* intensity data, from which structure factor amplitudes can be calculated.
- Crystal growth anomalies may produce the phenomenon of *twinning*. Twinning refers to a specific case of disorder in which there is partial or complete coincidence between the lattices of distinct crystal domains.
- Initial estimates of the unit cell size and orientation can be obtained from diffraction data collected over a small oscillation of the crystal. The space group to which the crystal belongs can be identified from the systematic absences in the diffraction pattern.
- The signal-to-noise ratio and the $R_{p.i.m}$ value are good criteria for assessing the quality of the collected data. The R_{merge} value can be misleading if the data are highly redundant.
- The solvent content of protein crystals (denoted by *Matthew's coefficient*) has been observed to have values between 20 and 80%, but is usually between 40 and 60%. In many cases, this range is sufficiently restrictive to enable reliable estimation of the total number of molecules per asymmetric unit from the protein's molecular weight.
- *Non-crystallographic symmetry* arises when there are two or more similar molecules present in the asymmetric unit. It can be determined directly from the diffraction intensities.
- The ultimate aim of protein crystallography is to build a model of the crystal contents in terms of atomic positions and temperature factors.
- The R_{factor} is traditionally used to indicate the "correctness" of a model structure. In order to guard against overfitting, R_{free} is used. This is calculated from structure factor amplitudes that are not included in the refinement.

- The *Argand diagram* is a fast and easy way of visualizing structure factors and phases.
- In molecular replacement, a model that is similar to the molecule(s) in the crystal can be used to provide an initial estimate of the phases.
- *Ab initio* phasing methods were essential when the first protein structures were determined and must still be used for new protein structures for which there are no known suitable homologous structures to serve as search models.
- High-resolution diffraction data contain phase information.
- In SIR and MIR, the addition of heavy atoms to a crystal produces a different diffraction pattern, from which phase information can be obtained.
- In SAD and MAD, the presence of an anomalous signal in crystal diffraction can provide phase information.
- A model of the crystal is built that agrees with both the diffraction data and chemical knowledge of the structure. Several refinement procedures have been developed to refine the model against the data.

- Kinetic crystallography aims to interpret crystallographic data in terms of a small number of time-independent structures, each associated with a reaction intermediate. Intermediate states can be trapped or followed in real time.
- Neutron crystallography is particularly useful for the experimental determination of hydrogen atom positions in medium-resolution structures. The technique requires large crystals, however, because of the low neutron fluxes available.

Suggestions for Further Reading

Rossmann, M. G., and Arnold, E. (eds.) (2001). *International Tables for Crystallography. Volume F: Crystallography of Biological Macromolecules*. Dordrecht: Kluwer Academic.

Drenth, J. (2002). *Principles of Protein X-ray Crystallography*. 2nd edn. Berlin: Springer-Verlag.

ELECTRON DIFFRACTION

ELECTRON MICROSCOPY

H1

H1.1 HISTORICAL REVIEW

1820

J.-B. Biot and **F. Savart** collaborated on the theory of magnetism. Their eponymous law (formulated in **1820**) describes the motion of a charge in a magnetic field.

1872

Ernst Abbe formulated his wave theory of microscopic imaging. In **1878**, he was able to correlate resolution and wavelength of light mathematically. A few years later, through the company Zeiss, he commercialized the first ever lenses to have been designed based on sound optical theory and the laws of physics.

1873

In the **second half of the nineteenth century, J.C. Maxwell** established the laws of electromagnetism. His equations first appeared in fully developed form in his book *Electricity and Magnetism* (1873).

1897

J.J. Thomson discovered electrons by demonstrating cathode rays were actually units of electrical current made up of negatively charged particles of subatomic size. For these investigations, he won the Nobel Prize for physics in 1906.

1924

In his doctoral thesis, **de Broglie** theorized that matter has the properties of both particles and waves. He received the 1929 Nobel Prize for physics. The particle–wave duality was confirmed experimentally for the electron a few years later by **C.J. Davisson** and **L.H. Germer,** and by **G.P. Thomson**. Thomson and Davisson jointly shared the Nobel Prize in physics in 1937.

1926

H. Busch described a lens to focus electrons. **H. Ruska** subsequently built the first working electron microscope in **1930** (Nobel Prize in 1986). Ruska's **1931** paper contains for the first time the term "electron microscopy." The principle behind this type of microscope has not varied significantly since.

1932

Fritz Zernike invented the phase-contrast light microscope, which uses a quarter-wave plate. He was awarded the 1953 Nobel Prize for physics for this work, which opened up the study of transparent biological materials. The feasibility of a scanning electron microscope was demonstrated by **Knoll** in **1935**, and the first prototype was built by **M. Van Ardenne** three years later. By the **late 1930s**, electron micrographs of recognizable value to biologists were being published. In **1942**, **S.E. Luria** obtained the first high-quality electron micrograph of a bacteriophage.

1940s

The negative staining technique came into extensive use in the **1940s**, although it was only in **1955** that **C.E. Hall** recognized, and deliberately used, phosphotungstic acid as a negative stain. In **1959**, **S.J. Singer** used ferritin-coupled antibodies to detect cellular molecules.

1954

V.E. Cosslet and **M.E. Haine** suggested the use of a field emission gun (FEG) for electron microscopy. Due to technical difficulties associated with the extremely high vacuum required, a manageable FEG microscope was not available until one was built by **A.V. Crewe** in **1996**.

1968

A. Klug and **D.J. DeRosier** introduced three-dimensional image reconstruction techniques of electron micrographs and opened the way for the interpretation of images obtained with very low radiation doses and without the use of heavy metal stains. Aaron Klug was subsequently awarded the Nobel Prize in Chemistry in 1982 "for his development of crystallographic electron microscopy and his structural elucidation of biologically important nucleic acid–protein complexes."

1975

R. Henderson and **N. Unwin** solved the first structure of an unstained integral membrane protein (bacteriorhodopsin) by combining electron micrographs and electron diffraction. The structure was extended to atomic resolution by the same technique and published by **Henderson** and coworkers in 1990.

1984

J. Dubochet and coworkers developed methods to rapidly freeze biological specimens to achieve a near native frozen-hydrated state. The use of extremely low temperatures reduces the extent of radiation damage by as much as an order of magnitude more than at room temperature.

1990s

The single particle reconstructions of the ribosome by the groups of **J. Franck** and of **M. van Heel** breached nanometer resolution.

2010s

The advent of direct electron detectors and more powerful electron microscopes revolutionized data collection, enabling atomic resolution structures without the need for symmetry.

H1.2 INTRODUCTION TO BIOLOGICAL PROBLEMS

H1.2.1 The Electron Microscope Image

Transmission electron microscopy (or EM) offers structural information, which is complementary and competitive with crystallography, small-angle scattering (SAS) and nuclear magnetic resonance (NMR). Its resolution for biological specimens is usually of the order of 10 Å but has reached ~1 Å.

Electron microscopy is based on the wave-like behavior of electrons. It can be understood as operating in a similar manner to light microscopy, with electrons, instead of photons, focused by electromagnetic lenses (see Chapter F1). The features in the EM image are the result of the interaction of electrons with the electrostatic potential distribution of the atoms in the sample.

H1.2.2 Applications of EM

Advances in EM have opened up a wide variety of macromolecular assemblies to structural and dynamical analysis. The method enables the description of biological macromolecules at nanometer resolution and better. Furthermore, it can deal with a wide range of specimen geometries, ranging from asymmetric particles, through helical filaments (one-dimensional crystals), to two-dimensional crystals (and even three-dimensional crystals at low resolution).

EM is hampered primarily by radiation damage leading to a poor signal-to-noise ratio in the data, which limits both resolution and the size of the studied complex. Currently, it

is typical to obtain reconstructions with resolutions in the nanometer range, although two-dimensional crystals, and some well-behaved molecular complexes, have yielded atomic resolution reconstructions. In theory, it should be possible to determine the structures of biological molecules of mass down to 38 kDa. However, the current lower limit lies between 64 and 500 kDa depending on the shape of the unstained specimens.

H1.2.3 Techniques Covered

This chapter focuses on the methods for obtaining EM images of biological macromolecules.

Chapter H2 subsequently details how these images are processed in order to determine the three-dimensional structure of the specimen. It will cover helical reconstruction, two-dimensional crystallography, and single-particle reconstruction methods.

H1.3 PRINCIPLES OF ELECTRON DIFFRACTION AND IMAGING

H1.3.1 Properties of Electrons

Wave-like Properties of Electrons

Electrons, like photons, have the dual property of behaving both as a wave and as a particle. Electron microscopy is based on their wave-like behavior. The de Broglie relationship is thus applicable to electrons and relates the momentum and wavelength of the particle by Planck's constant h,

$$h = \lambda_e p \tag{H1.1}$$

where λ_e is the wavelength of the electron and p its momentum (Comment H1.1).

Effect of the Electron Charge

Electrons can also be considered as charged particles; consequently their movement is influenced by electromagnetic fields. A particle of charge q and mass m placed in an electric potential difference of Φ (in volts) accelerates until the particle's kinetic energy is balanced:

$$q\Phi = \frac{1}{2}\frac{p^2}{m} \tag{H1.2}$$

COMMENT H1.1 EFFECT OF AN ELECTRIC FIELD ON A CHARGED PARTICLE

A charge q placed in an electric field (**E**) experiences a force q**E**. The field can be expressed as the gradient of a potential (Φ) such that in one dimension we can write $E = d\Phi/dx$

where p is the equilibrium momentum. Since an electron has charge $(-e)$, its momentum can be written as

$$p = \sqrt{2m_e e \Phi} \tag{H1.3}$$

where the mass of an electron m_e is 9.1×10^{-28} g.

H1.3.2 Electromagnetic Lens

In electron microscopes, suitably shaped magnetic (or electrostatic) fields are used to control the direction of moving electrons. As in the case of a glass lens that focuses a light beam, these electromagnetic lenses focus electron beams. The electron microscope allows the study of a large range of length scales which overlaps with that of light microscopy (Fig. H1.1). The energy (or wavelength) of an electron is a feature specific to the accelerating voltage. It imposes limits on the electron microscope's resolution. The wavelength is given by Eqs. (H1.3) and (H1.1) as

$$\lambda_e = \frac{h}{\sqrt{2m_e e \Phi}} \approx 12\Phi^{-1/2} \tag{H1.4}$$

An electron microscope of 100 keV energy therefore produces electrons of 0.037 Å wavelength. Electron wavelengths are five orders of magnitude smaller than those of visible light.

Recall from Chapter G1 that resolution, wavelength, and maximum scattering angle are related. The angular aperture of a lens (2α), i.e., the angle between the most divergent rays that can pass through a lens to form the image of an object, therefore controls the lens' maximum resolution, d:

$$d = \frac{\lambda_e}{\sin \alpha} \tag{H1.5}$$

Since λ_e is very small, it is often convenient to assume that the EM data are not diffraction-limited. However, in the presence of spherical aberration, the resolution limit is lowered, as rays passing at larger angles through the lens are focused closer to the lens than the central ray. It is given in terms of geometrical optics by

$$d = \left(C_S \lambda_e^3\right)^{1/4} \tag{H1.6}$$

where C_S is the spherical aberration coefficient. With $C_S = 1$ mm, the maximum resolution of a 100 keV microscope is

5 Å. Note also that the resolution of an electron microscope depends on $\lambda_e^{3/4}$, and therefore on its voltage.

H1.3.3 The Image Recorded by an Electron Microscope

Signal-to-Noise Ratio

The principal problem in EM is the radiation damage to the imaged specimen, which limits the amount of information that can be collected. The *signal-to-noise ratio* of the image expresses the confidence with which a structure can be distinguished from the noise in the image. The signal is usually defined as resulting from elastic scattering of the electrons from the sample, and the noise is due to inelastic scattering events, but also to statistical fluctuations in the number of electrons detected in different parts of the image.

The *Rose model* describes the signal-to-noise ratio (S/σ_{Rose}) necessary for the detection at a certain contrast (C) of a uniform object of area A in a uniform background:

$$S/\sigma_{Rose} = C\sqrt{An_{backgrouund}} \tag{H1.7}$$

where the contrast is given by $|(n_{object} - n_{background})/n_{background}|$, with the mean number of electrons per area due to the background and to the object being respectively $n_{background}$ and n_{object}, and A the area of the object. Signal-to-noise ratios in the range of 5–7 are adequate to identify features in the image. $S/\sigma = 5$ represents an 83% confidence level, while $S/\sigma = 1$ represents only a 50% confidence level that a structural feature can be recognized from the noise.

Noise and Radiation Damage

The image produced by an electron microscope contains a lot of noise. As written above, this is due to the statistical fluctuations in the number of electrons between different parts of the image and to the inelastic scattering of electrons as they travel through the sample.

In *elastic scattering*, the incident electrons do not lose energy when scattered by the sample. In *inelastic scattering*, energy from the incident electrons is deposited into the sample. Electrons from the sample can be expelled with energies 5–10 times greater than valence bond energies. The resulting radiation damage severely limits the amount

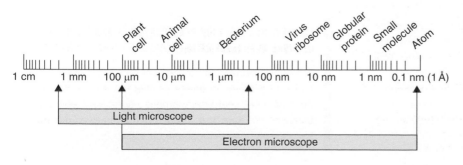

Fig. H1.1 Resolution ranges of optical and electron microscopes.

of structural information that can be obtained. It is, in fact, the main resolution-limiting feature.

In EM of biological objects, due to the energies of the incident electrons, the number of inelastic scattering events always outnumbers the number of scattering events by a factor of 3–4. Energy filters, although not routinely used in biological EM, can be used to eliminate certain electron energies in order to reduce the number of inelastic events. Current trends, however, favor an increase in accelerating voltage since this reduces inelastic scattering.

As with any scientific experiment, data collection protocols have to be adapted not only to the type of specimen studied, but also to the instrument used. In the case of rapid electron detectors, a series of images are often taken in quick succession. The first image is omitted as it often displays the drift induced by the electrons on encountering the sample. The subsequent images are then averaged with each other, until obvious electron-induced damage appears. These later images are consequently discarded.

H1.3.4 Transfer of Information from Sample to Image

Comparison with Optical Lenses

Magnetic lenses suffer from the same defects as optical lenses (Comment H1.2); in particular, *spherical aberration* and *astigmatism*. *Chromatic defects* happen when certain electron frequencies are affected, but not others.

Effect of the Microscope on the Path of the Electron

As the electron wave moves through the microscope and the sample to the image recording plane, its amplitude is modified at different points of its journey. By convention, the electron wave travels along the z-axis.

The incoming electron wave, τ_{in}, is modified as it travels through the sample. On leaving the sample, the scattered electrons form a wave τ_{out} that contains information of the projected specimen potential $\phi_p(x, y)$ along the z-axis:

$$\phi_p(x,y) \propto \int_{-t/2}^{t/2} \phi(x,y,z)\,dz \tag{H1.8}$$

where $\phi_p(x, y)$ is the Coulomb potential distribution within the object and t is the width of the object. For simplicity, constants have been cast aside and electron wavelength and mass have been corrected for relativistic effects (the electron velocity is close to the velocity of light). This wave can be depicted as a phase shift induced in the incoming wave, where

$$\tau_{out}(x,y) = \tau_{in} \exp\left[i\phi_p(x,y)\right] \tag{H1.9}$$

In EM of biological samples, the specimen can be viewed as a weak phase object, since very few electrons are scattered by the specimen ($\phi_p(x, y)$ is very small), and the electron beam is coherent. This results in a further approximation of $\tau_{out}(x, y)$ (see Chapter A3 and Comment H1.3):

$$\tau_{out}(x,y) = \tau_{in}\left[1 + i\phi_p(x,y)\right] \tag{H1.10a}$$

However, since some absorption occurs even for thin biological samples, an amplitude component should be included:

$$\tau_{out}(x,y) = \tau_{in}\left[1 + i\phi_p(x,y) + \mu(x,y)\right] \tag{H1.10b}$$

In the back focal plane of the microscope, the Fourier transform of $\tau_{out}(x, y)$, $T_{out}(u, v)$, is observed:

$$T_{out}(u,v) = \tau_{in}\left[\delta(u,v) + i\Phi_p(u,v) + M(u,v)\right] \times P(u,v) \tag{H1.11}$$

where $\delta(u, v)$ is a Dirac function, $\Phi_p(u, v)$ and $M(u, v)$ denote the Fourier transform of $\phi_p(x, y)$ and $\mu(x, y)$, respectively, and the $P(u, v)$ term is the transfer function of the microscope. $P(u, v)$, which is discussed in further detail in Chapter H2, arises essentially from the wave aberrations of the objective lenses. The extra phase shift due to the lens can be described by

$$T_{out}(u,v) = \tau_{in}\left[\delta(u,v) + i\Phi_p(u,v) + M(u,v)\right]$$
$$\times \exp\left(i\frac{2\pi}{\lambda}\gamma(u,v)\right) \tag{H1.12}$$

COMMENT H1.2 MAGNETIC LENSES

The magnetic lens influences the helical path of the electron. In old microscopes, the image rotated when the defocus was changed.

Light lenses are optically almost perfect. Magnetic lenses are not. Their ever-present spherical aberration results in variable phase shifts across each lens. Their astigmatism produces stretched images, which are clearly noticeable in the image's diffraction pattern; however, the effects of astigmatism can be corrected for prior to recording the image, and subsequently during image processing.

COMMENT H1.3 ELECTRON BEAM COHERENCE AND TRANSFER OF STRUCTURAL INFORMATION

It is very important to have a coherent beam of electrons as its waves keep in phase as they travel through space. An incoherent beam has no phase correlation and is therefore not able to transfer information to an electron micrograph from weak-phase objects.

assuming a rotationally symmetric astigmatism, with

$$\gamma(u,v) = \gamma(\theta) = \frac{-\Delta z \theta^2}{2} + \frac{C_S \theta^4}{4} \qquad \text{(H1.13)}$$

where θ is the scattering angle, Δz represents the objective lens defocus, and C_S the spherical aberration constant.

As in light microscopy, the (real-space) image on the electron micrograph is obtained by a further Fourier transform.

Contrast Transfer Function and Defocus

To understand the effect of the transfer function on the image, it is useful to recall that the image intensities can also be given by the product $\tau_{out}(x,y) \times \tau_{out}^*(x,y)$. The Fourier transform of image intensity $I(u,v)$ is then the convolution product:

$$I(u,v) = T_{out}(u,v) \otimes T_{out}(-u,-v)$$

Omitting the constant τ_{in} for simplicity, and since $\Phi_p(u,v)$ and $M(u,v)$ are both very small, the quadratic terms can be neglected, and

$$I(u,v) = \delta(u,v)$$
$$+ 2\left\{ \Phi_P \sin\left[\frac{2\pi}{\lambda}\gamma(u,v)\right] - M(u,v)\cos\left[\frac{2\pi}{\lambda}\gamma(u,v)\right] \right\} \qquad \text{(H1.14a)}$$

Experimentally, the amplitude component can be of the order of 35% for negatively stained samples and 7% for unstained samples in ice (see Sections H1.5.2 and H1.5.3). If we define $W(u,v)$ as the ratio of amplitude to phase component, Eq. (H1.14a) simplifies further to

$$I(u,v) = \delta(u,v)$$
$$+ 2\Phi_P\left\{ \sin\left[\frac{2\pi}{\lambda}\gamma(u,v)\right] - W(u,v)\cos\left[\frac{2\pi}{\lambda}\gamma(u,v)\right] \right\} \qquad \text{(H1.14b)}$$

or

$$I(u,v) = \delta(u,v) + \Phi_P \times \text{CTF} \qquad \text{(H1.15)}$$

where we have found it useful to define a *contrast transfer function* (CTF),

$$\text{CTF} = 2\left\{ \sin\left[\frac{2\pi}{\lambda}\gamma(u,v)\right] - W(u,v)\cos\left[\frac{2\pi}{\lambda}\gamma(u,v)\right] \right\} \qquad \text{(H1.16)}$$

where we recall

$$\gamma(u,v) = \gamma(\theta) = \frac{-\Delta z \theta^2}{2} + \frac{C_S \theta^4}{4}$$

For each microscope operating at a given high voltage, C_S and λ are fixed. The image defocus, Δz, is the experimental variable. In order to account for the gaps in the transfer of amplitude and phase information that occur because of the oscillating CTF, exposures are recorded over a broad defocus range (e.g., between 1000 Å and –5 μm defocus).

The Fourier transform of the intensities of each image therefore displays a series of ripples, where the observed amplitudes drop to zero at a certain spatial frequencies (denoted by θ) (Fig. H1.2). At each of these nodes, the phases of the image amplitudes reverse. Contrast transfer function correction procedures will be described in Chapter H2.

The Image as a Projection

If we recall the Fourier transform of the image intensities, and reintroduce τ_{in} (the incoming wave amplitudes),

$$I(u,v) = \{\delta(u,v) + \Phi_P \times \text{CTF}\}\tau_{in}^2 \qquad \text{(H1.17)}$$

The reverse Fourier transform gives the image intensities $(i(x,y))$,

$$i(x,y) = \{1 + \phi_P \otimes ctf\}\tau_{in}^2 = \tau_{in}^2 + \tau_{in}^2 \phi_P \otimes ctf \qquad \text{(H1.18)}$$

where *ctf* is the Fourier transform of the CTF.

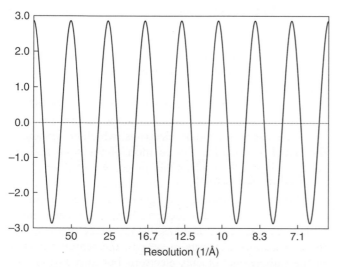
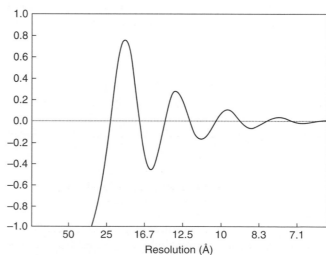

Fig. H1.2 Example of variation of average amplitude according to resolution.

In the presence of a perfect lens ($CTF = 1$) or, more realistically, a corrected CTF, this further reduces to

$$i(x,y) = \tau_{in}^2 + \tau_{in}^2 \phi_P = A + C \int_{-t/2}^{t/2} \phi(x,y,z)dz \qquad \text{(H1.19)}$$

using Eq. (H1.8) and where C is a constant. The constant A represents the intensities (τ_{in}^2) that would be recorded in the absence of sample. If we omit this background, the image intensities are a linear description of the sample's potential projection along the path of the electron beam, or expressed mathematically

$$i(x,y) \propto \int_{-t/2}^{t/2} \phi(x,y,z)dz \qquad \text{(H1.20)}$$

H1.4 ELECTRON MICROSCOPES

H1.4.1 Electron Beam Generation

Two main methods have been implemented in electron microscopes to generate electron beams (Fig. H1.3). A thermionic electron gun contains a loop of filament (the cathode), heated by an electric current. A negatively charged Wehnelt cylinder surrounds the loop, and a more positively charged anode then attracts the electrons. Due to the configuration of the gun, the electrons accelerate through the anode and on toward the sample. In the case of a tungsten thermionic gun, the filament has a diameter of ~100 µm and is heated to ~2700 K. The accelerating voltage applied to the Wehnelt cylinder is generally between –500 V and –50 000 V.

Field emission guns are now preferred as they produce much brighter electron beams. The increased spatial and temporal coherency in the electron beams produced by field emission guns appear to also give a real advantage in terms of achievable resolution. The tip of the pointed cathode has a very small radius and is perpendicular to the surface of a flat anode. A voltage of several thousand volts is applied between the anode and the cathode. Due to its shape, there is a very strong electric field gradient close to the cathode, forcing electrons to be emitted from the tip. These travel through a small hole in the anode and on toward the sample. The cathode tip is often a tungsten crystal with a diameter between 100 and 1000 Å.

H1.4.2 Transmission and Scanning Electron Microscopes

The transmission electron microscope (TEM) has a layout similar to the phase contrast light microscope (see Chapter F1 and Fig. H1.4). The image in the TEM is formed by the interference between the electrons that have passed

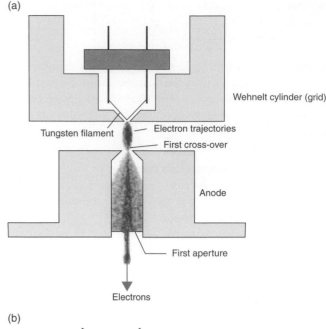

(a)

Wehnelt cylinder (grid)

Tungsten filament — Electron trajectories

— First cross-over

Anode

— First aperture

Electrons

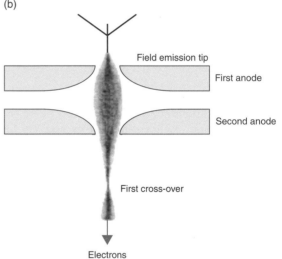

(b)

Field emission tip

First anode

Second anode

First cross-over

Electrons

Fig. H1.3 Schematic diagrams of **(a)** a thermionic electron gun and **(b)** a field emission gun.

through (and interacted with) the sample and with those electrons which were not affected by the sample, giving a phase contrast image.

In contrast, in the scanning electron microscope (SEM), a beam of electrons is scanned across the sample, resulting in the release of secondary electrons, which contribute to the signal. Correlating the position of the scan and the resulting signal then forms the image. Nevertheless, SEM provides a poor estimate of the sample's height so it is therefore not appropriate for generating high-resolution three-dimensional reconstructions. However, when coupled with X-ray detectors, SEM has the advantage that it can be used to identify elements present in the scanned portion of the sample, as not only electrons but also X-rays with characteristic elemental spectra are emitted.

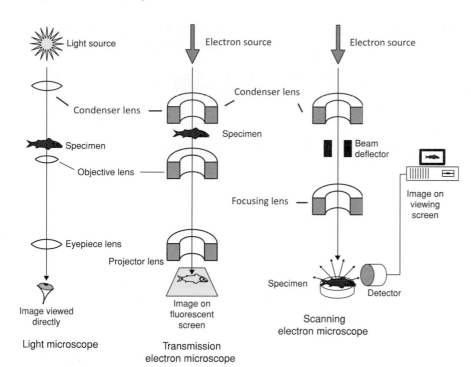

Fig. H1.4 Schematic diagrams of a light microscope, a TEM and an SEM.

H1.4.3 Electron Images

Features of an Electron Microscope Image

Electrons interact with a sample by being repelled by negative charges and attracted to positive charges. The electron microscope records the resulting interaction in an *electron micrograph*, which contains information regarding the electrostatic potential distribution within the sample. The *contrast* in the image is approximately proportional to the atomic number of the atoms imaged.

As we would expect, the information contained in the image decreases with increasing resolution. This fall-off is due to a number of factors, including beam-induced damage, sample movement (drift and temperature factors), loss of coherence of the electron beam at higher resolutions, etc.

Imaging Phase Objects

High-resolution transmission EM is closely analogous to optical phase contrast microscopy, in which ultra-thin low-contrast samples are imaged. Amplitude contrast and phase contrast are treated in Chapter F1. In optics, transparent objects do not absorb but just scatter photons. They can be conceptualized as pure phase objects. In order to visualize their details, the insertion of a quarter-wave plate in the light microscope transforms a phase difference induced by the imaged object into an amplitude difference. However, a practical quarter-wave plate for electrons does not exist. In order to obtain useful phase contrast from which structural details can be discerned, the samples are imaged at a slight defocus (Fig. H1.5).

Effect of Focus

Varying the focus of the electron microscope alters the features of the image (see Comment H1.2). At focus, the image has very low contrast. Amplitude differences are very weak. They result from the loss of electrons scattered at high angles and removed by the objective lens. In fact, a good indication that an image was recorded at focus is its featurelessness and smooth background. At underfocus, the image has enhanced contrast, with a white ring around the imaged object. The presence of a mass is observed as an increase in intensity. The phase shift is in the same direction as the intensity contrast and thus they both accentuate the contrast of the image. At overfocus, the imaged object appears white, with a dark rim around it. The phase shift and amplitude contrast are in opposite directions.

H1.4.4 Image Recording System

Optical Density

Electron microscope images have been typically documented on photographic films, which are extremely efficient recording media for electrons. Up to a certain electron dose (because the film will saturate), the recorded optical density is proportional to electron exposure. The intensity measured on the film must therefore be rescaled to reflect the actual dose received (Fig. H1.6). Electronic area charge-coupled device (CCD) detectors are also often used. Although they do not provide as high resolution as film, these cameras provide immediate quantitative feedback in a digital format (see below).

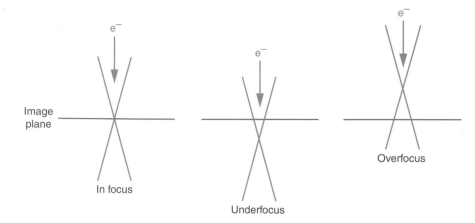

Fig. H1.5 Position of image plane: in focus, underfocus, and overfocus.

Although still prohibitively expensive for most EM labs, complementary metal-oxide (CMOS) cameras overcome the limitations of both film and CCD imaging systems. These cameras directly detect electrons with efficiencies better than photographic film. Their other advantage is that they can image at very high rates (some up to 2000 images per second), and basically record an EM video of the sample.

Pixelization

Although photographic film is able to capture high-resolution data better than most CCD detectors, the recorded data has to be digitized in order to be processed computationally. This process is never perfect so information is always lost. It should, however, be emphasized that the image can be sampled so that the information of interest is not lost. The *Shannon sampling theorem* explains that the sampling size should be at least half the final desired resolution. In practice, a finer sampling is used (1/3–1/4 resolution per pixel), as it reduces interpolation during the data processing.

The gray level of each pixel must also be quantified. Fourteen-bit CCD detectors have a very large dynamic range as each pixel can have 16 384 different gray values. Nonetheless, 256 gray levels are typically used in EM reconstructions, since the resulting file is easily manageable with currently available computer power.

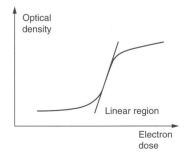

Fig. H1.6 Response of film to electron dose: optical density electron exposure relationship.

H1.5 TECHNIQUES IN SPECIMEN PREPARATION

H1.5.1 Specimen Support

During imaging, the sample has to be kept free from drift. It is usually deposited on a grid, which has numerous holes large enough for the electron beam to travel through but small enough for the sample to form a fine film over the surface (Comment H1.4). For cryo-EM, gold grids dramatically improved image quality as substrate motion during electron irradiation was nearly eliminated (Russo and Passmore, 2014).

A goniometer-mounted stage allows the grid, and therefore specimen, to be tilted with respect to the electron beam.

COMMENT H1.4 EM GRID

Grids have been made of many different materials and of differing meshiness. G400 grids, for instance, have 400 meshes per square inch. Figure H1.4.1 offers an overview of a grid.

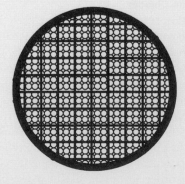

Fig. H1.4.1.

H1.5.2 Negative Staining

Principles of Negative Staining

Negative staining has been applied widely to the study of biological samples. It is a simple and rapid method for determining the overall morphology of a macromolecule and leads to high-contrast images whose best resolution, from three-dimensional reconstructions, is near 15 Å.

The main constituents of biological molecules (oxygen, carbon, and nitrogen) are not very electron dense and so do not interact strongly with the electron beam. The sample is therefore *negatively stained* by embedding it in an electrodense material, like an uranium or lead salt (Fig. H1.7, Comment H1.5). In the electron micrograph, the sample appears as a light region surrounded by a dark background originating from the stains. The particle itself is not observed but its surrounding stain is. The sample structure is hence inferred by the distribution of the heavy stain (and therefore has limited achievable resolution).

In the ideal case, the particle is not distorted by the stain; for example, negative staining can vividly display the symmetry of viruses (Fig. H1.7b). More often, however, the particle becomes distorted, leading to the higher-resolution data in the EM reconstruction being lost. The stain may also cause protein aggregation or dissociation.

Method for Negative Staining

The grids for negative staining are prepared by laying a thin film of plastic (usually formvar) over their dull surface and covering this with a further layer of carbon (the plastic could in fact be omitted) (see Comment H1.4). The amount of carbon that has been deposited can be estimated by eye from the darkness of the grid. The grids

---

COMMENT H1.5 POSITIVE STAIN

Positive staining is when the electrodense molecules bind only to specific sites of the sample.

---

are then glow discharged to render them hydrophilic. A small volume of sample is pipetted onto the carbon-coated side of the grid and left to spread out. The grid is then rapidly dipped several times in deionized water to wash off excess sample, and subsequently dipped for some time (seconds or longer if necessary) in an electro-dense solution containing the stain (e.g., uranyl acetate). Filter paper is then used to remove any excess fluid, and the grid air-dried prior to imaging.

H1.5.3 Freezing of the Sample

Reasons for Electron Cryo-Microscopy

In electron cryo-microscopy, the sample is rapidly cooled and observed at liquid nitrogen (and even liquid helium) temperatures. The method improves the preservation of the native structure and significantly reduces radiation damage. The sample can be subjected to more electrons before it is destroyed, or at least before high-resolution information is lost. Radiation damage is discussed in greater detail in Chapter G3.

Flash-Cooling

The aim of rapid cooling is to avoid ice crystal formation, which would damage the specimen. A small volume (about a microliter) of the sample is pipetted onto a mesh metal (often copper, but also gold) grid, with the excess sample fluid blotted away. This procedure can be carried out in a humidified atmosphere to avoid water evaporation from the sample and a consequent increase in salt concentration.

The flash-cooling is then usually performed by plunging the grid into liquid ethane or propane (temperature of 11 K). These liquids have high heat conductance and the water on the grid is transformed into vitreous ice (as it does not have sufficient time to form ice crystals) and thus remains in an amorphous transparent state. The procedure is described schematically in Fig. H1.8. The grid is then transferred to liquid nitrogen (110 K) for storage before loading into the electron microscope.

(a)

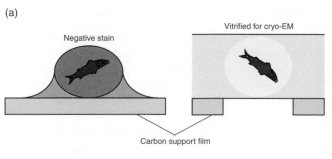

(b)

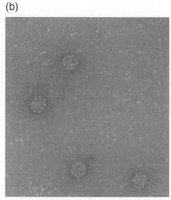

Fig. H1.7 (a) Diagram to show how the electron dense material covers the biological sample. **(b)** Example of a negatively stained virus.

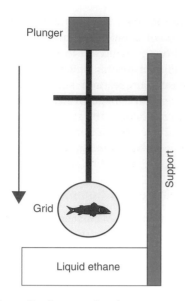

Plunger

Support

Grid

Liquid ethane

Fig. H1.8 Schematic diagram of a plunger.

H1.6 DATA COLLECTION

H1.6.1 Factors to Consider during Data Collection

The aim is to collect sufficient information from the electron images in order to achieve a reconstruction at the desired resolution of the sample structure. The amount of data must not be too large, however, as this would affect the speed and efficiency of the computer calculations.

Magnification

Electron microscopes provide a choice of magnification, from 1000 times to 200 000 times. The choice is dependent on a number of factors. These include the targeted resolution, the required number of imaged particles, the amount of electron radiation the sample can survive, and the resolution limits of the recording medium (film or CDD) and image digitizer. Under specified illumination conditions, the image intensity decreases with increasing magnification (M) and the dose per unit area decreases with $1/M^2$. Magnifications in the 40 000–60 000 range are often used for reconstructions of biological samples to 10–20 Å resolution.

Sample Concentration

Each micrograph should contain a sufficient number of particles from which to choose, but not be overcrowded. The balance is between the need to reduce the time required for digitization of the micrograph and particle aggregation. The optimum concentration depends on the sample studied. Protein concentrations in the region of 1 mg ml^{-1} are often appropriate. In the case of smaller molecular mass molecules (500 kDa or less), it is preferable to have a less concentrated sample (e.g., 0.1 mg ml^{-1}), as it is then easier to identify the molecules in the electron micrographs.

Electron Dose

A typical biological sample is very sensitive to the electron beam, and can only tolerate a dose of 10–15 electrons Å^{-2} before radiation-based structural damage renders the micrograph's data unusable. High-resolution data are generally more sensitive to electron beam damage than lower-resolution data.

Stained samples can often be imaged at higher electron doses because it is the stain envelope around the particle that is observed, rather than the particle itself. The imaging of unstained specimens at very low temperatures helps to reduce the radiation damage. Since inelastic events create highly charged and reactive radicals in the sample, low temperatures can limit the progress of the diffusion of these damaging molecules and ions through the sample.

H1.6.2 Data from Single Particles

The aim in collecting EM data from single particles is to produce micrographs that contain a sufficient number of well-spaced and well-behaved particles, as this will be crucial for particle identification, classification, and subsequent structural analysis (Chapter H2; Comment H1.6).

Data collection often uses the SEARCH/FOCUS/EXPOSURE mode. An initial low-dose/low-magnification SEARCH identifies areas of interest on the grid. Each area is then assigned a specific de-FOCUS, and a high-dose/high-magnification EXPOSURE is then recorded.

For example, pictures can initially be taken under low-dose conditions in order to identify the locations on the grids that should be imaged at high resolution. The image is then focused at high magnification (×230 000) at a site adjacent (5 μm) to the area of interest. The presence of drift is checked at even higher magnification (×490 000). An image of the area of interest is then taken at a magnification of 60 000 with a chosen unique defocus. This procedure is repeated at different places on the grid, sampling a range of defocus conditions.

H1.6.3 Imaging Crystals and Helical Molecules

The presence of symmetry in a sample facilitates its three-dimensional reconstruction. This is discussed in detail in

COMMENT H1.6 GLUTARALDEHYDE CROSS-LINKING

Although in EM it is possible to actively select which particle images to include in the reconstruction, sample polydisparity can still cause problems, especially if it is not possible to easily categorize the sample's different structural states.

By treating the sample with glutaraldehyde, the resultant chemical cross-links to the different proteins in the complex will stabilize a specific population of molecules. A specific conformation and/or oligomeric state can hence be imaged.

Chapter H2. Several biological molecules have inherent symmetry. For example, biological filaments, like microtubules and tubulin, assemble into helical-shaped objects (Fig. H1.9a), and many viruses have icosahedral symmetry. The power of using an ordered sample is that symmetry related parts of the image can be averaged, thus allowing the use of very low electron doses.

The integral membrane protein bacteriorhodopsin spontaneously assembles into two-dimensional crystals in the so-called purple membrane (see Comment A3.6). Other proteins can be attached to a lipid layer, where their movement is restricted to the lipid plane. Integral membrane proteins embed themselves directly into the lipids, while soluble protein can, for example, bind to specially functionalized lipids. Under the right biochemical conditions (lipid-to-protein ratio and temperature), they self-assemble in an ordered two-dimensional lattice (Fig. H1.9b). This crystallization process is very similar to three-dimensional crystallization, discussed in Chapter G3.

A similar procedure to single-particle imaging is implemented in the imaging of two-dimensional crystals and helical molecules. A rapid scan at low resolution over the grid identifies potential areas of interest. Crystals or helices do not, in fact, have to cover the whole field of view of the microscope, since each ordered region is processed individually before being combined during data processing. A Fourier transformation of the image is then computed to provide a diffraction pattern of the ordered sample, with

the added advantage over X-ray crystallography that the phase information associated with each reciprocal lattice point is also obtained. The extent of the diffraction pattern highlights the maximum resolution which could be theoretically achieved (see Chapter G1). The EM reconstruction of bacteriorhodopsin crystals, for instance, was determined to 3.5 Å resolution (see Chapter H2). The direct collection of diffraction patterns from two-dimensional crystals offers several advantages, in particular that specimen drift and defocus do not affect the data. In order to have sufficient signal-to-noise at high resolution, crystals must be large (at least several micrometers across).

H1.6.4 Tomography

In tomography (from the Greek *tomos*, slice) data are collected by imaging the same sample at different tilt angles. This technique is particularly advantageous when dealing with non-uniform samples in which the varying shapes make it impossible to average them for reconstruction. Nevertheless, the orientation of each image relative to the others of the tomographic series is known. Typically, the specimen is rotated about a fixed axis in 1° to 2° angular increments over ±70°. By using such an approach it has been possible to identify 26 S proteasomes and actin filaments in the cellular environment from their distinctive shapes (50–60 Å resolution) (Fig. H1.10). The cellular regions examined were between 300 and 600 nm in thickness.

In general, a tomography experiment aims to collect results at as many orientations as possible in order to produce an accurate reconstruction; however, the cumulative electron dose limits the number of exposures because of radiation damage. The resolution of a tomography experiment therefore is usually worse than the resolution

(a)   (b)

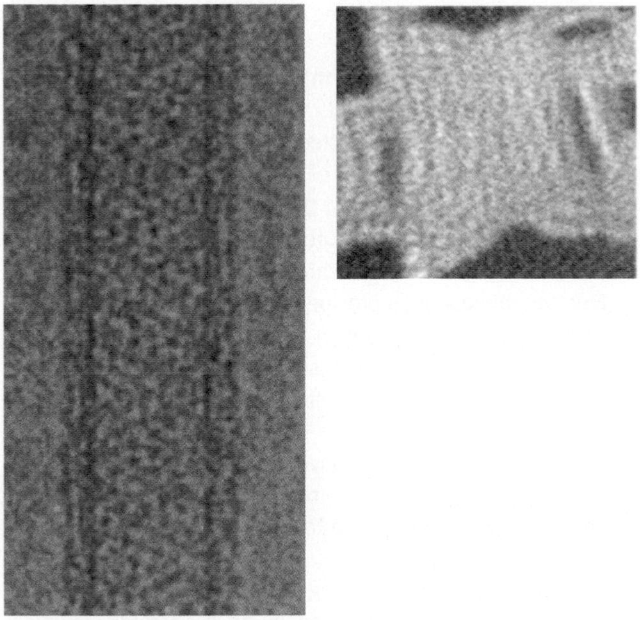

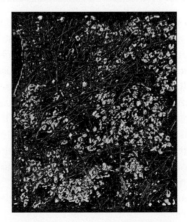

Fig. H1.9 (a) Helical fragment of Ca-ATPase, and **(b)** two-dimensional crystal of streptavidin. (Toyoshima *et al.*, 1993; Frey *et al.*, 1996) (Figures reproduced with permission from Macmillan Publishers Ltd. and National Academy of Sciences USA, respectively.)

Fig. H1.10 EM tomography of a frozen *Dictyostelium discoideum* cell. Colors were subjectively attributed to linear elements to mark in red the actin filaments, in green the other macromolecular complexes (mostly ribosomes), and in blue membranes (Medalia *et al.*, 2002). (Figure reproduced with permission from AAAS.)

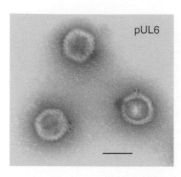

pUL6

Fig. H1.11 Electron micrograph of the herpes simplex virus (HSV-1) after staining with an antibody specific for the portal protein pUL6 followed by an anti-antibody conjugated to gold beads. Note that the gold clusters are found at a single capsid vertex. Bar, 100 nm. (Newcomb *et al.*, 2001.)

achieved from single-particle reconstruction. A current aspiration is to achieve a 20 Å resolution tomographic reconstruction for high-molecular-mass samples.

It is important to note that the field of view in a tomogram is typically under $3\,\mu m^2$, which represents less than 2% of a typical eukaryotic cell's surface. Electron microscopy imaging has therefore been combined with light and fluorescent microscopy (Part F). The latter instrument is first used to scan the biological sample for areas of interest; as for example cellular processes identified by using endogenously expressed fluorescent markers. These areas are studied subsequently in greater structural detail by EM tomography.

Freezing chambers have therefore been developed to prepare the same vitrified sample for analysis by both light/fluorescence microscopy and by EM.

H1.7 IMMUNOCHEMISTRY

An SEM can easily identify gold clusters due to the strong emission of secondary electrons and backscattered electrons. Labeling antibodies with gold has therefore provided a tool by which components in complex biological samples can be localized. The gold particles may be conjugated to primary antibodies for a one-step identification of antigens, but are more often employed as secondary antibody/protein labels (Fig. H1.11). The location of the antibody is therefore determined from the presence of the gold. Gold has the further advantage that it is available in different sizes (1–40 nm), so the binding pattern of different antibodies can be identified in a single micrograph.

H1.8 CHECKLIST OF KEY IDEAS

- EM is a powerful tool used in the study of biological specimens.
- EM reconstructions typically attain resolutions in the nanometer range, although two-dimensional crystals have yielded atomic-resolution reconstructions.
- An electron microscope is similar to a light microscope but instead of photons it uses electrons.
- In single-particle reconstruction, the data from differently oriented molecules can be combined to produce an accurate three-dimensional reconstruction of the molecule.
- In tomography, the same object is imaged at different orientations, and the images are then combined to give the three-dimensional structure of that object.
- EM is hampered primarily by the poor signal-to-noise ratio of the data, which places a lower limit on both resolution and the size of the studied complex.
- An electron has the dual property of behaving both as a wave and as a particle.
- In the same manner that a light lens focuses photons, electromagnetic lenses can focus electron beams.
- The EM map is the result of the interaction of the incident electrons with the electrostatic distribution of the electrons and nuclei in the sample.
- High-resolution EM is closely analogous to optical phase contrast microscopy in the analysis of weak phase objects.
- EM images can be recorded on photographic film, CCD detectors, and on CMOS cameras.
- The Shannon sampling theorem says that the sampling size should be at least half the final desired resolution.
- In negative staining, it is not the particle itself that is observed but the surrounding electrodense material in which it is embedded.
- The presence of symmetry in a sample eases its three-dimensional reconstruction.
- Certain biological macromolecules can be coaxed to form two-dimensional crystals.
- In electron cryo-microscopy, the sample is frozen rapidly for imaging. The low temperature reduces damage to the sample on exposure to electrons.
- Electron microscopes provide a choice of magnifications, from ×1000 to ×200 000.
- Biological samples are very sensitive to an electron beam. Typically they can only tolerate a dose of 10–15 electrons $Å^{-2}$.
- The technique of tomography is particularly advantageous when dealing with non-uniform samples in which the varying shapes make it impossible to average them for reconstruction.
- In immunochemistry, labeling antibodies with gold has provided a tool by which components in complex biological samples can be localized by EM.
- EM has been combined with light/fluorescent microscopy in order to study cellular processes in greater structural detail.

Suggestions for Further Reading

Asano, S., Engel, B. D., and Baumeister, W. (2016). In situ cryo-electron tomography: A post-reductionist approach to structural biology. *J. Mol. Biol.*, **428**, 332–343.

Henderson, R. (1995). The potential and limitations of neutrons, electrons and X-rays for atomic resolution microscopy of unstained biological molecules. *Q. Rev. Biophys.*, **28**, 171–193.

Lau, C. Y. L., and Rubinstein, J. L. (2013). Single particle electron microscopy *Meth. Mol. Biol.,* **955**, 401–426.

Edited by Jensen, J. G. (2010). Cryo-EM: Part A: Sample preparation and data collection. *Meth. Enzymol.*, **481**, 2–410.

THREE-DIMENSIONAL RECONSTRUCTION FROM TWO-DIMENSIONAL IMAGES

H2

H2.1 EM IN BIOLOGY

H2.1.1 Structural Biology with EM

Transmission electron microscopes (TEMs) can provide vivid details of biological macromolecules at the nanometer scale.

Recall from Chapter H1 that an electron image is a two-dimensional projection of the object along the electron beam. A three-dimensional reconstruction of the object can be obtained by combining data measured at different angular projections. In *single-particle reconstruction*, a population of molecules is imaged. If these are identical (monodisperse) and present in sufficient quantities, the data from differently oriented molecules can be combined to produce an accurate three-dimensional reconstruction of the structure. The presence of symmetry is taken advantage of in *helical reconstruction* and *two-dimensional crystallography*. In *tomography*, the same object is imaged at different orientations. The different images are then combined together to recreate its structure.

H2.1.2 Examples of Electron Cryo-Microscopy Reconstructions

In EM, the image is the result of the interaction of the incident electrons with the electrostatic distribution due to the atomic structure of the sample. The reconstruction of the structure results from the analysis of many images. Three-dimensional reconstructions at different resolutions depict the object at different levels of detail (Fig. H2.1).

At atomic resolution, it is a good approximation to understand this structural density as being equivalent to the electron density obtained by X-ray diffraction, but with a larger contribution from hydrogen atoms in the electron diffraction case. It should furthermore be possible to distinguish the charge state of amino acids.

At medium resolution (better than 10 Å), the scattering depends on protein-specific features such as secondary structures. α-helices produce a stronger signal at around 10 Å resolution and β-sheets at about 4.5 Å. Comparable features are also observed in X-ray diffraction experiments; the reader should refer to the chapters on X-ray

crystallography (Chapter G3) and solution scattering (Chapter G2) for more details.

At low resolution, the electron scattering is determined by the shape of the macromolecule. Reconstructions based on either negative staining or unstained samples therefore yield roughly the same information at 20 Å resolution.

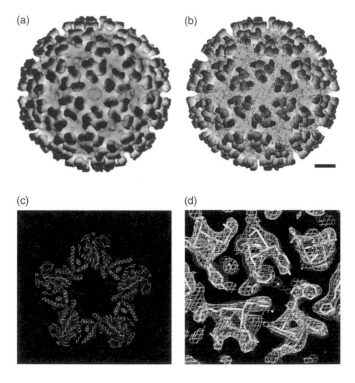

Fig. H2.1 EM reconstructions at different resolutions. **(a)**, **(b)** Semliki virus: reconstruction at: **(a)** 22 Å (Vénien-Bryan and Fuller, 1994) and **(b)** at 9 Å (Mancini et al., 2000). The color scheme reflects the radial distance from the center, increasing from blue to red, as shown in the inserted color scale. Scale bar, 100 Å. **(c)** Pyruvate kinase: density with superimposed coordinate model (Rosenthal and Henderson, 2003).

(d) Bacteriorhodopsin: weighted density maps for wild-type bacteriorhodopsin (3.5 Å, purple) and the D96G, F171C, F219L triple mutant (3.2 Å, yellow). The atomic model of the wild-type is included (Subramaniam and Henderson, 2000). (Figures reproduced with permission from Elsevier ((a), (b), (c)) and Macmillan Publishers Ltd. (d).)

H2.2 EM DATA PREPARATION

H2.2.1 Preliminary Analysis of the Image

Prior to any data processing, the quality of the electron micrograph must be checked. The preliminary analysis can be based on a direct examination of the micrograph by eye, as particles should be identifiable. The particles should be monodisperse and in sufficient number. Aggregated particles are not useful for reconstructions.

Optical diffraction is used to evaluate the putative resolution, the degree of underfocus, lack of astigmatism, and drift in the electron micrographs (see Chapter H1). Examples of these defects are given in Fig. H2.2. In two-dimensional crystallography, the contrast is often so low that optical diffraction is essential to check that the crystal has been imaged.

H2.2.2 Particle Selection

The particles are identified in the digitized image file of the micrographs and selected and stored in smaller files. Each file contains the image of just one particle and its surrounding background. All the subsequent data manipulation for the reconstruction will be performed on these files. For example, the data can be *band-pass filtered* to remove spurious high-/low-resolution information in order to improve the signal-to-noise ratio.

A reference image (or template) will facilitate the selection process if the shape of the particle is roughly known. However, a wrong template will influence what is being picked in the EM micrograph, and this will strongly bias the resulting reconstruction.

Initial images are typically taken at a large defocus, with relatively high-dose exposure (80 to 140 electrons Å^{-2}), so the particles can be easily selected as they stand out against the background. Although several automatic particle selection software are also available, it is important to note that they are not foolproof. Visual inspection of the selected data is therefore strongly advised to identify and remove any obvious artifacts.

Perfect lens   Astigmatism   Drift

Fig. H2.2 Diffraction patterns with different defects such as astigmatism and drift. Drift leads to the Thon ring fading in a direction perpendicular to the direction of the drift.

H2.2.3 Correction for the Contrast Transfer Function

Recall that a micrograph is taken at a specific defocus because of the contrast transfer function (CTF; see Chapter H1). Expressed as a Fourier transform (see Chapter A3), the CTF gives the average amplitudes of the spatial frequencies to oscillate with a resolution (Fig. H2.3a). The lack of information at certain resolution ranges can be overcome by combining data from micrographs with different CTFs. This enables uniform coverage of the resolution range to the highest possible resolution. The CTF can also be corrected by mathematically modifying the signal in certain resolution shells of the data (Fig. H2.3). The simplest correction to the CTF is *phase sloping* (Fig. H2.3b). Phases in the resolution ranges with negative amplitudes are flipped, i.e., 180° is added to the phase of structure factors with negative CTF. Hence the amplitudes are not modified.

The CTF can be also be corrected by either multiplying or dividing the data by the CTF itself (Fig. H2.3c). The phase gains the correct sign while the amplitudes at higher resolutions are increased. When dividing by the CTF, a problem arises at the CTF zeros, so a Weiner-like filter of the form $1/(CTF + s)$ is used, where s is a noise-suppression term. In the case of images of tilted specimens, the variation in defocus across the image can be accounted for by a *tilt transfer function*.

Finally, the decrease in amplitude with resolution can also be corrected for. Weighting schemes are described in further detail in Section H2.7.

H2.3 SINGLE-PARTICLE RECONSTRUCTION PROCEDURES

H2.3.1 Coordinate System

An EM reconstruction merges the data from images of differently oriented particles in order to recreate a three-dimensional reconstruction of the particle. The basis of the reconstruction is to identify the spatial relationship between the different images.

A particle is described in terms of its position in the plane of the image (x, y) and a rotation. The description of the z-axis is not possible as the particle is projected in that direction to produce the electron image. Depending on the reconstruction program, different angular conventions are used (Comment H2.1).

H2.3.2 Reconstruction from Projections

From the arguments detailed in Chapter H1, once the CTF has been corrected, we can assume the electron image, $i_{corrected}$, is a projection of the particle onto the xy plane, along the z-axis,

503

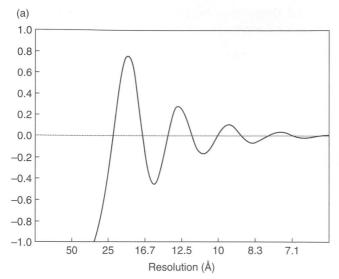

(a)

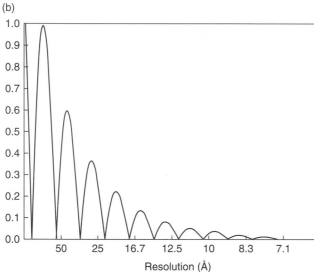

(b)

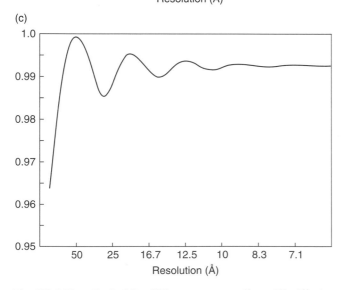

(c)

Fig. H2.3 The effect of the CTF on a cross-section of the Fourier transform of an image, and the effect of different correction procedures: **(a)** CTF with decay; **(b)** CTF correction by *phase sloping*; **(c)** Weiner correction of the CTF.

$$i_{\text{corrected}}(x,y) = \int_{-\frac{t}{2}}^{\frac{t}{2}} \phi(x,y,z)\mathrm{d}z \tag{H2.1}$$

where $i_{\text{corrected}}(x, y)$ is the two-dimensional CTF-corrected image and $\phi(x, y, z)$ is the three-dimensional particle potential.

The Fourier transform of an object $\phi(x, y, z)$ can be expressed as

$$\Phi(u,v,w) = \int_{-\infty}^{+\infty}\int_{-\infty}^{+\infty}\int_{-\infty}^{+\infty} \phi(x,y,z)$$
$$\exp\left[-i2\pi(ux + vy - wz)\right]\partial x \partial y \partial z \tag{H2.2}$$

If we restrict $\Phi(u, v, w)$ to the line defined by $w = 0$,

$$\Phi(u,v,0) = \int_{-\infty}^{+\infty}\int_{-\infty}^{+\infty}\int_{-\infty}^{+\infty} \phi(x,y,z) \exp\left[-i2\pi(ux + vy)\right]\partial x \partial y \partial z \tag{H2.2a}$$

and therefore the integral can be separated into two:

$$\Phi(u,v,0) = \int_{-\infty}^{+\infty}\int_{-\infty}^{+\infty}\left\{\int_{-\infty}^{+\infty} \phi(x,y,z)\partial z\right\} \exp\left[-i2\pi(ux + vy)\right]\partial x \partial y \tag{H2.2b}$$

The term in braces is the projection of the object along the z-axis, which, on referring back to Eq. (H2.1) is the corrected image, or

$$\Phi(u,v,0) = \int_{-\infty}^{+\infty}\int_{-\infty}^{+\infty} i_{\text{corrected}}(x,y) \exp\left[-i2\pi(ux + vy)\right]\partial x \partial y \tag{H2.3}$$

In words, the two-dimensional Fourier transform of the image is equivalent to a section of the three-dimensional Fourier transform of the particle along the direction in which the projection was taken.

Moreover, by the nature of the Fourier transform, if an object $\phi(x, y, z)$ is rotated by an angle with respect to one or more axes, the Fourier transform $\Phi(u, v, w)$ is correspondingly rotated by the same angle with respect to the corresponding axes. A sufficient number of differently oriented sections can therefore be merged to reconstruct the three-dimensional object by reverse Fourier transform-ation (Fig. H2.4). Even though the handedness of the par-ticle is still ambiguous as each image is a projection of the object, imaging *focal pairs* of the same particle solves this problem (see Section H2.7.2).

H2.3.3 Iteration Procedure: Reprojection Method

Due to the noise in each micrograph, the reconstruction procedures have to be iterated until the quality of the final reconstruction no longer improves. The three-dimensional model of the reconstructed particle is reprojected compu-tationally into different directions to create a new set of reference images with known orientations, against which the raw data set can be compared to determine the orien-tations more accurately. A new model is then generated and the procedure continued until the reconstruction solu-tion becomes stable.

H2.3.4 Common Lines Reconstruction Procedure

If the views of a particle are different, two different sections in Fourier space intersect along a common line. The corresponding planes can each rotate about the line. A third section would also display two other common lines, one for each of the other two sections. The common lines reconstruction procedure is based on the fact that the three planes are now locked in space.

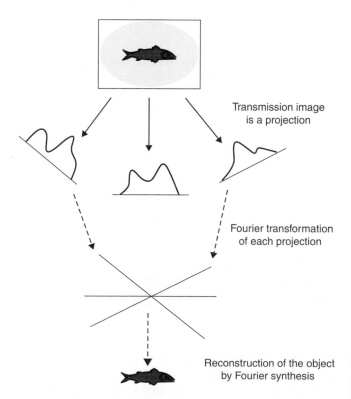

Transmission image is a projection

Fourier transformation of each projection

Reconstruction of the object by Fourier synthesis

Fig. H2.4 Concept of common lines: sections cut in space along a common line.

Theoretically images of three differently oriented par-ticles are therefore sufficient to orient each particle with respect to the other as soon as the common lines are identified. In many experimental cases, this is unfortu-nately not the case since the EM images are too noisy for the common lines in the Fourier transform to be identified accurately. Furthermore, because the common lines extend across Fourier space, they are resolution-dependent. A statistical analysis of the section pairs must therefore be implemented.

When symmetry is present in the particle, two types of common lines appear in the Fourier transform of its image. Application of the symmetry operators leads to the appear-ance of *self-common lines*. In icosahedral objects, a max-imum of 37 self-common lines can be observed due to the application of five-fold, three-fold, and two-fold symmetry and removal of non-unique lines. The *cross-common lines*, on the other hand, are due to the application of the par-ticle's symmetry between pairs of its images. In icosahedra, there are at most 60 pairs of cross-common lines.

H2.3.5 Polar Fourier Transform Reconstruction

Polar Fourier transform reconstruction relies on the use of a preliminary model (or template), which helps orient the images in order to create a more appropriate model of the imaged sample (Comment H2.2). The analysis starts by constructing a database, which contains differently projected views of a model representing the particle. For icosahedra, these views cover one-half of the asymmetric unit or 1/120 of the icosahedron (i.e., θ = 69–90° and ϕ = 0–32°). In each view, the Cartesian coordinate system is translated into a polar reference frame (r, γ). The radius r varies from 0 (at the chosen origin) to a radius just outside the particle edge, and the angle γ indicates the azimuthal direction of the projection. The particle image is also con-verted to polar coordinates.

In order to implement a polar Fourier transform recon-struction, each particle image must be initially centered and scaled, e.g., by cross-correlation with the circularly averaged projection of the model. The particle images and model projections are then Fourier transformed along the azimuthal axis to produce one-dimensional rotational spectra. The spectrum of every image is then multiplied

---

COMMENT H2.2 INITIAL MODEL FROM NEGATIVE STAINING

EM chiefly records the surface of stained particles. The resolution is often limited, but the staining can nevertheless highlight the particles' overall shape and variability. A model based on stained particles can therefore be used in the initial stages of the reconstruction process. In fact, just 5–10 particles are often sufficient.

by the spectrum of each model projection in order to produce a cross-correlation map. A high correlation indicates that the particle in the image is oriented similarly to the projected model. Images with the highest correlation are then used to create a better model. The process is then iterated until the correlation between the image and final model projections does not improve.

This method has the advantage of using all the data in the image. In contrast, the common-lines method makes use of only a small fraction of the Fourier data along the common line.

In a polar Fourier transform reconstruction, the choice of template obviously influences the reconstruction. For scientists' peace of mind, the final reconstruction can be confirmed by using different starting models, and the resulting reconstructions compared. A sufficiently large data set of images also permits reconstructions based on different subsets of the data. The comparison of the different reconstructions is the basis of the Fourier shell correlation (FSC) method, which is treated in greater detail below (see Section H2.6.2).

Polar Fourier transform reconstructions are not recommended for non-symmetrical objects since they require the chosen origin to be determined separately and not refined against the average model projections. The use of the center of mass of the image as an origin could be a way around this problem.

H2.3.6 Real-Space Reconstruction Procedure

Most of the data manipulation and analysis can also be performed in real space. Recall that the particle image is a two-dimensional projection of the particle. The image itself can be projected onto a single line. The rotation of the particle image over 360° creates a sinogram, the collection of all possible line projections (Fig. H2.5).

Sinograms can be used to find common lines between two projections of a particle. The largest correlation between two sinograms corresponds to a pair of shared-line projections. This method offers several advantages. It does not depend on defining the particle's image center. The presence of symmetry is also exhibited in the sinograms if the sinogram of the image is cross-correlated with itself.

H2.3.7 Focal and Tilt Pairs

Focal pairs can be recorded in order to facilitate image processing. Immediately afterward the first image, a second image is recorded at a large defocus. The resultant strong contrast in the low-resolution features of the second image can then provide estimates of the positions and orientations of the particle in the low-contrast first image.

In a similar fashion, *tilt pairs* of micrographs can be recorded in which the sample has been rotated for the second image. The hand of the particle can hence be

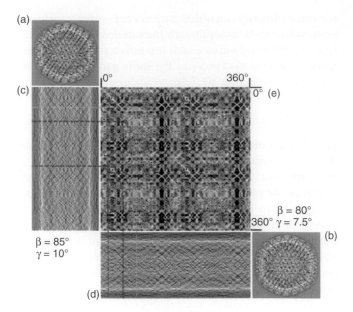

Fig. H2.5 A graphical overview of the relations between two projection images, their "sinograms," and the sinogram correlation function between two sinograms. The images (a) and (b) are class averages deduced from a large data set of herpes simplex virus type 1 (HSV1) electron cryo-microscopy images. Each line of the sinogram images (c), (d) is generated from the two-dimensional projection image by summing all one-dimensional lines of the two-dimensional images, from top to bottom, after rotation of the image over angles ranging from 0° to 360°. Each point of the sinogram correlation function contains the correlation coefficient of two lines of the two sinograms one is comparing (e). (Van Heel et al., 2000.)

identified in the initial steps of the reconstruction. As the rotation angle is known, the computational process can furthermore be optimized, e.g., by varying the resolution range, reconstruction radii, contrast, etc. in order to accurately predict the spatial relationship between the two images of the same particle. The optimized procedure can then be used to deduce efficiently the orientations between all the particles present in the first micrograph. Since it is of poorer quality due to radiation damage, the data in the second image are not used in the subsequent refinement.

H2.4 RECONSTRUCTION PROCEDURES FOR ONE- AND TWO-DIMENSIONAL CRYSTALS

H2.4.1 Helical Reconstruction

Many biological molecules have a helical organization. The best known is DNA, but key intracellular proteins such as actin and tubulin also assemble into structures with helical symmetry. The common helix can be visualized as the curve formed by the thread of an ordinary screw. Helical symmetry implies that the neighboring subunits in the helix are related by a displacement and a rotation along the screw axis. For example, seven repeating subunits could be spaced between two helical turns. An image of a helix therefore has the advantage of displaying the repeating subunit in a number of orientations. There is therefore no need to tilt the helix during data collection.

An image reconstruction of the helix makes use of its underlying geometrical relationships. The Fourier transform of a continuous helix is a series of lines fanning out as a cross (see Fig. 1d in the Introduction). The presence of a periodic displacement (of length a) between the different subunits creates a more complicated pattern due to the repetition of the X-like pattern every $1/a$ (Fig. H2.6). These short lines are called *layer lines*. Each layer-line pair corresponds to a set of helices. In a similar fashion to X-ray crystallography (Chapter G3), they have to be indexed in order to undertake the reconstruction. Since helical symmetry involves just two parameters, only two indices are needed for each layer plane. These indices (l and n) are derived from the helix's pitch ($l = p^{-1}$) and its rotational frequency n (a similar view of the helix appears after rotating the helix by $2\pi/n$ radians).

The advantage of the treatment in reciprocal space is that most of the image noise can be discarded by using only the layer-line amplitudes and phases in the reconstruction. Furthermore, several data sets are often combined in the helical reconstruction. Their size is usually a compromise between keeping the length of helical fragments as short as possible (to eliminate areas with distortions) and generating useful phase and amplitude data for alignment and averaging.

H2.4.2 Electron Crystallography

Data from thin two-dimensional crystals can be processed as in X-ray crystallography (Chapter G3). However, in electron crystallography the crystals are usually imaged in real space and the data are computer analyzed in Fourier space.

The Fourier transform of the crystal image displays a diffraction lattice from which structure factor intensities and phase information can be estimated. Noise can then be dramatically reduced by using only amplitudes and phases in the diffraction spots (see Fig. H2.7). Often, different orientations within a crystal can be patched together computationally by an *unbending procedure* to give the equivalent of one large crystal. This significantly increases the resolution that can be obtained from the image of a single crystal. Non-crystallographic symmetry, as in X-ray crystallography, can further increase the signal-to-noise ratio (see Chapter G3).

Recall that the electron image is the projection of the sample projections along the electron beam direction. Data from differently oriented crystals must therefore be integrated into a data set from which a three-dimensional structure can be calculated. For example, merged electron diffraction patterns from 402 crystals have been used to

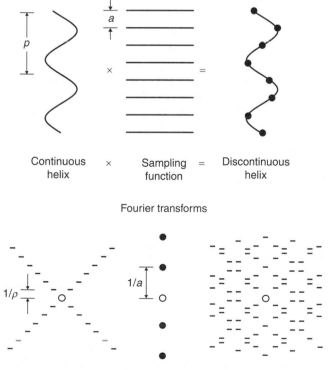

Fig. H2.6 Fourier transforms of a helix with pitch p and distance a between repeating subunits along its axis.

Fig. H2.7 Example of the electron diffraction pattern from a two-dimensional crystal of bacteriorhodopsin. (Subramaniam et al., 2002.) (Figure reproduced with permission from the Royal Society.)

COMMENT H2.3 MISSING DATA

The cone of missing information means that data are missing in one direction. The reconstruction of a circle would thus give an ellipse, elongated in that direction. It is therefore essential to check the effect of the missing data in any electron crystallographic reconstruction. For example, helices at low resolution appear as elongated densities, but these also often lie perpendicular to the membrane, which would be the direction of the missing information.

calculate a 3.2-Å structure of the protein bacteriorhodopsin. Since the 2-d crystal lies parallel to the grid, it is necessary to orient the grid at different angles to the electron beam. As angles above a certain orientation are not accessible experimentally due to the grid's thickness (for the bacteriorhodopsin specimen the maximum tilt was 62.5°), a *cone of missing data* occurs in Fourier space resulting in reconstructions with anisotropic resolution, which are therefore elongated perpendicularly to the crystal plane (Comment H2.3).

Data from the integral membrane protein bacteriorhodopsin displayed diffraction spots extending out to 2 Å, although only data from 2.8–3 Å were successfully phased for reconstruction. This resolution, however, was sufficient to allow molecular models to be built, integrating both stereochemistry and EM data, with the electron diffraction data (intensities transformed into amplitudes) and images (phases) providing the amplitude/phases of the structure factor components. A refinement procedure, similar to that used for X-ray crystallography (Chapter G3), was then used to match the crystal's structure factors and phases with an atomic model of the structure.

Finally, the structure can be solved by collecting diffraction data and solving the phases by *molecular replacement* (see Chapter G3), thus avoiding the need to record images of tilted crystals, as these, in particular, have different defocuses across the crystal plane.

H2.5 CLASSIFICATION PROCEDURES

H2.5.1 Statistical Analysis

Electron micrographs are unfortunately characterized by a low signal-to-noise ratio and *statistical analysis* is required. The signal-to-noise ratio is proportional to the square root of number of particles averaged.

Images that look similar can, however, be grouped together into *classes*. Each represents a specific orientation of the particle. The average image of a class, however, has an increased *signal-to-noise* ratio, which in turn facilitates the determination of the angular relationship between the different classes.

This classification procedure is now most often automated. Maximum-likelihood algorithms can simultaneously align and classify different images of the particle, by comparing an individual image with all the other data. Nonetheless, the overall number of classes generated is limited, since for signal-to-noise reasons, it is common practice to have at least 200 cryo-EM particle images per class.

H2.5.2 Multivariate Statistical Analysis

Multivariate statistical analysis is implemented in certain reconstruction algorithms. This *eigenvector analysis* of the image reduces the total number of data treated.

An initial step of the procedure is to reassign the data in the two-dimensional image to a different frame of reference, called *hyperspace*. Each pixel of the image corresponds to an axis within this hyperspace, and the pixel's gray level to the coordinate along that axis. For an $n \times n$ ($= m$) pixel image, the corresponding hyperspace has a total of m dimensions. Each image can therefore be reduced to a single point inside the hyperspace, with similar-looking images clustering together.

The multivariate statistical analysis aims to collapse the m-dimensional hyperspace into fewer dimensions and still represent the data truthfully. New axes in this hyperspace are therefore determined that enable a simplified, but truthful, representation of the data. The first axis is oriented according to the data's average. The second axis is oriented in the direction of the next largest variance. The third axis is oriented toward the next largest variance, and so on for each extra axis, until all the data's information has been used. Typically, the first ten axes contain most of the information about the data set, and are sufficient to summarize the data. Each axis in hyperspace can be monitored by eye as it is equivalent to an m-pixel image, called an *eigenimage*. Examples of eigenimages are shown in Fig. H2.8. Each picked particle is described by a linear combination of these new axes. Similar-looking particles can then be rapidly identified and clustered together based on these axes. It is optimum to perform this classification on previously aligned images.

H2.6 DETERMINATION OF THE RESOLUTION

H2.6.1 Number of Images Required for a Reconstruction

The number of views required to calculate a reconstruction up to a resolution d depends on the n-fold symmetry and diameter (D) of the particle. The minimum number of views required, N, is

(a)

(b)

(c)

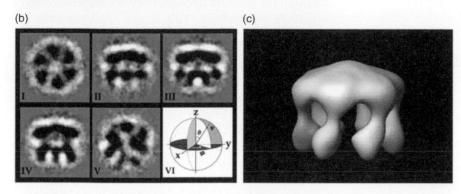

Fig. H2.8 (a) Typical eigenimages.
(b) Four characteristic projection views out of 50 class-sum images obtained by multivariate statistical analysis of the H$^+$-ATPase in the ligand-free state.
(c) The resultant three-dimensional map. (Rhee *et al.*, 2002.) (Figure reproduced with permission from Macmillan Publishers Ltd.)

$$N = \frac{\pi D}{nd} \tag{H2.4}$$

where d is the target resolution of the reconstruction. The number of images used in the reconstruction is, however, inherently much larger since Eq. (H2.4) is for the perfect case with noise-free images, and an evenly spaced distribution of particle orientations. Hence, more appropriately, the number of images required ($N_{required}$) can be estimated as

$$N_{required} = N_{proj} \left(\frac{\pi D}{nd} \right) \tag{H2.5}$$

where N_{proj} is the average number of images in one projection required to reach a certain signal-to-noise threshold. The calculation must also account for the loss of contrast in the high-resolution data due to the signal's amplitude decay.

For example, icosahedral viruses, due to their 60-fold symmetry, would require 60 times less views than for the asymmetric ribosome. The very conservative criterion of a signal-to-noise ratio greater than 3σ means 60 000 particles would be required for a 9 Å resolution reconstruction. A resolution criterion related to map interpretability brings this theoretical number down to 2200.

H2.6.2 Definition of Resolution

The resolution in EM is difficult to define, as there is often no independent way of monitoring it. In X-ray crystallography, the atomic resolution forces the reconstruction to obey certain chemical rules in order to be plausible. By analogy, due to the lower resolution of EM, particular care should be taken that visible structural features accord with the claimed resolution of the reconstruction. For example, at medium resolution, α-helices would appear as rod-shaped densities.

Shannon Sampling Theorem
The digitization of the image imposes a maximum resolution. The Shannon sampling theorem states that the measurement (or sampling) frequency must be at least twice the maximum frequency to be measured. The image must therefore be scanned at a step size which is at least twice as fine as the desired resolution: for example, if the pixel size is 4.8 Å, the maximum resolution of the reconstruction is 9.6 Å.

Phase Residual
A *phase residual* can be used to underscore the consistency of the phases corresponding to its common-line pairs. After a particle's orientation is estimated, the common lines are calculated. A phase residual of 0° average phase along the common lines means there is perfect agreement with the projection of the model, while a phase difference of 90° indicates complete lack of agreement.

Fourier Shell Correlation
Fourier shell correlation (FSC) is often used to qualify the final reconstruction. For this statistic to be most effective, the EM data should be randomly split into two at the very start. The two resultant independent reconstructions would then be compared.

The correlation of each reconstruction in Fourier space is calculated as

$$\text{FSC} = \frac{\sum F_1 F_2^*}{\sqrt{\sum |F_1|^2 \sum |F_2|^2}} \tag{H2.6}$$

where F_1 and F_2 are the structure factors of two maps calculated from each data set, and the sum runs over a resolution shell. The FSC plotted against resolution is shown in Fig. H2.9. The resolution can be understood as the spatial frequency at which the FSC drops below a certain threshold. The often-quoted value of 0.5 is quite

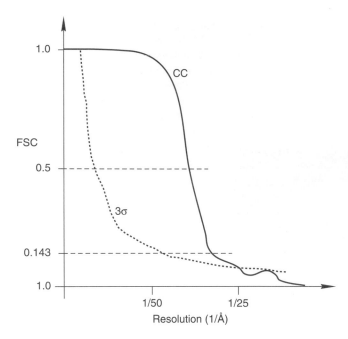

Fig. H2.9 FSC plotted against resolution for the reconstruction of the yeast RNA polymerase I. Different thresholds are shown (1, 0.5, and 0.143) as well as the 3σ level (dotted line). (Bischler *et al.*, 2002.)

pessimistic. However, at the resolution described by an FSC threshold of 0.143, the information contained in such maps should be equivalent to the information gained at that resolution in X-ray crystallography, and this value has been adopted by many research groups.

In addition, a resolution cut-off can be determined from the background noise (three standard deviations or 3σ) of the reconstruction (see Chapter A2). The resolution is found where the FSC is equal to

$$3\sigma \times \sqrt{n} \qquad \text{(H2.7)}$$

where *n* is the number of asymmetric units in the reconstruction. The icosahedral symmetry of viruses leads to a resolution with FSC threshold of approximately 0.5.

H2.6.3 Over-Fitting and Validation of the EM Reconstruction

Unlike X-ray crystallography and NMR (Chapter G3 and Part J, respectively), where a model can be validated by its chemistry, the lower resolution EM reconstructions compel the use of other validation criteria.

First, it is important to check there is sufficient coverage in particle orientations. This is not often the case for two-dimensional crystals, but it also happens for other macromolecules, as some may have preferred orientation on the EM grid. Reducing the number of unknown relative orientations (and therefore incompleteness) will reduce ill-posedness and thus lead to better reconstructions.

Second, it is imperative not to over-fit the model to the data; for example, including spurious data in the reconstruction in order to claim higher resolution, or biasing the reconstruction toward the starting template (see Section H2.2.2). In the former case, the resultant map lacks features expected for the claimed resolution; in the latter case, over-fitting can be noticed by the "hairy" aspect of the final model, as the low-resolution ghost of the search model is superposed with high-resolution features (Fig. H2.10).

The simplest way to avoid over-fitting is to randomly divide the data in two and impose a strict quarantine as the two half data sets are each used in independent reconstructions. A-priori knowledge can also help validate the reconstruction. For example, the reconstruction should have non-negativity and the solvent background can be assumed to be flat.

H2.7 MAP ENHANCEMENT

H2.7.1 Symmetry

The presence of symmetry in the particle is a boon to the reconstruction process. Since symmetry elements can be imposed during the particle's reconstruction, identifying and then applying a symmetry operator on the

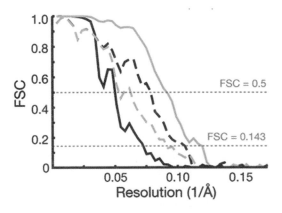

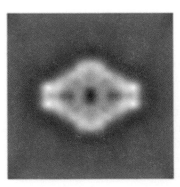

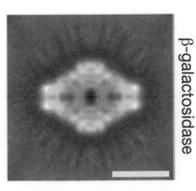

Fig. H2.10 Cross-section for the reconstruction of β-galactosidase based on two independently treated half data sets shows less noisy maps than a reconstruction based on a full dataset. Scale bars, 100 Å. (From Scheres and Chen, 2012)

reconstruction map drastically increases the signal-to-noise ratio of the data (Comment H2.4).

The task is relatively straightforward for crystalline samples as their translational symmetry can be deduced easily from the Fourier transforms of the images. In a crystal, the symmetry is determined from layout and systematic absences in the diffraction pattern. Furthermore, the extent of the crystallographic symmetry can be evaluated by analyzing the phase relationship between symmetry-related reflections in the image's computed Fourier transform, since the apparent symmetry breaks down at high resolution (probably due to low signal-to-noise ratios). The observed symmetry is then applied to the reconstruction.

H2.7.2 Weighting

The reconstructed density maps can be sharpened by imposing an amplitude profile on the final map in order to down-weight the low-resolution data and boost the higher-resolution features. The argument for weighting is that maps are phase-dominated. Provided the phases are reliable to a given resolution, the structure factor amplitudes can be scaled to correct for their fall-off with resolution. For example, the structure factors of the map could be multiplied by

$$\exp\left(-X/d^2\right) + \exp\left(Y/d^2\right) \tag{H2.8}$$

where the real and positive values X and Y and the resolution d are in ångströms. By varying the values of X and Y it is possible to highlight features in the map which are only noticeable at certain resolutions. In the case of a two-dimensional crystal, a negative temperature factor is applied, estimated from the Wilson plot of structure factor amplitude decay against resolution (see Chapter G3). It is typically -300 Å^2 to -400 Å^2 for a field emission gun microscope.

An *envelope decay function* can also be applied to correct for the image degradation. Some variant of the Wiener filter is generally used. The corrections can be estimated from either comparison with focal image pairs or on the basis of X-ray solution scattering data.

The decay in the EM map's scattering factors can be scaled to follow the decay that would be observed if the structure were determined by X-ray crystallography. The EM map is noise-weighted by incorporating information from the FSC. The map's structure factors are multiplied by

$$\sqrt{\frac{2\text{FSC}}{1 + \text{FSC}}} \exp\left(-B_{\text{restore}}/d^2\right) \tag{H2.9}$$

The exponential is a sharpening term to compensate for the decline in scattering factors with resolution (d) (B_{restore} reflects intrinsic image defects and in experiments is often approximately -1000 Å^2).

H2.7.3 Use of Other Structural Information

If the atomic structure of a component of the particle has been determined, it is possible to try to fit that model into the EM reconstruction.

The absolute magnification of the EM structure has to be adjusted accordingly, and due to experimental limitations the EM map's distance scale may vary by up to several percentage points. The relative values of the EM density distribution also have to be scaled to match the densities observed in the X-ray structure. Finally, in order to ease the scaling and fitting procedures, lower-resolution maps of the X-ray or NMR models can be generated.

The positioning of the atomic structures within the EM map permits the description of details of the EM reconstruction at a much greater resolution. In the case of a 20 Å EM map of a protein complex, amino acids can be modeled with 3–5 Å accuracy and thus detail the protein surfaces within the electron map (Comment H2.5). This exercise is akin to protein crystallography (see Chapter G3). Most crystals only diffract X-rays to between 2 and 3 Å, and it is only knowledge of the stereochemistry and the amino acid sequence that enables the description of the protein structure to atomic detail.

H2.8 APPLICATIONS AND EXAMPLES

H2.8.1 The Ribosome

The ribosome is a large RNA–protein complex involved in the synthesis of the protein chain from information encoded in an RNA chain. Single-particle methods of reconstruction of the ribosome have provided much understanding about the structures and dynamics involved in this translation process. Due to its large size (~2.5 MDa), lack of symmetry and the fact that RNA has twice the contrast and half the radiation sensitivity of proteins, this macromolecular structure has provided an ideal system on which to develop and test EM reconstruction procedures. The resultant EM reconstructions now often approach near-atomic resolution (for example, 30 000 particle images were used to generate a 4.5 Å map (Bai *et al.*, 2014)).

Importantly, as crystallization is not required, it has been possible to study ribosomes just after purification from their host cells and in near native (mild) conditions. For example, the 3.5 Å reconstruction of the large ribosomal subunit from human mitochondria revealed 48 proteins, 21 of which were specific to mitochondria (Brown *et al.*, 2014).

Several interesting ideas have emerged from the analysis of numerous EM reconstructions of functional ribosome complexes. There are quite significant structural differences between species. High-resolution reconstructions have identified drug-binding sites and the basis of their molecular action. The ribosome is also able to assume a large number of states, with the different domains being able to act either together or independently (Fig. H2.11).

H2.8.2 Icosahedral Viruses

The capsids of many viruses have icosahedral symmetry: a geometric construct defined by 12 pentagonal vertices and 20 triangular faces. The presence of such high symmetry has facilitated EM reconstruction of numerous viruses. Fewer than ten particles were used in the first ever EM reconstruction of viruses, while a 22 Å resolution reconstruction of Semliki forest virus was based on 50 viral particles. In order to reach 9 Å resolution, 5000 particle images were then used (Fig. H2.12a).

Radial density functions are often plotted in the analysis of the reconstructions of viruses (Fig. H2.12b). They show how the density varies throughout the virus. For instance, an outer peak would correspond to the viral capsid. A secondary peak, which represents an inner shell of density, could indicate the presence of packaged DNA or RNA. The radial density function can be used to scale reconstructions of other similar viruses.

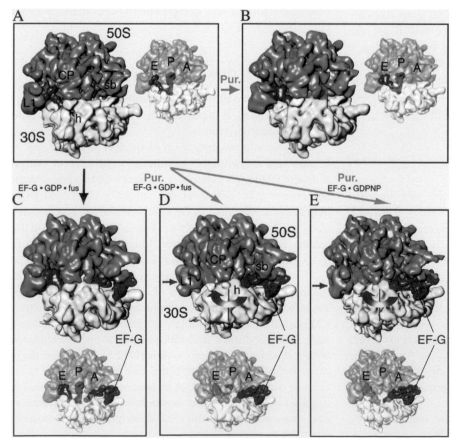

Fig. H2.11 Details of EM structures of complexes of ribosome 70 S and different ligands. The 50 S subunit is depicted in blue; the 30 S in yellow; EF-G in red; P-site tRNAs in green; E-site tRNAs in orange. (Valle *et al.*, 2003.) (Figure reproduced with permission from Elsevier.)

Epitopes can also be analyzed by EM. Particles can be incubated with the relevant ligands and imaged. The reconstructions then illustrate the interaction interface between the particles and their ligand (Fig. H2.13). In fact, the presence of bound ligands can often be rapidly appraised from raw micrographs. Native particles usually appear very smooth. In comparison, images of vitrified particle–ligand complexes appear fuzzy because of the presence of bound ligand.

Three-dimensional reconstructions have been made of antibody–particle Fab–particle and receptor–virus complexes (Fig. H2.13a). It is often noticed that only the particle and the ligand domains in immediate contact with the particles are well defined. The other domains of the ligand

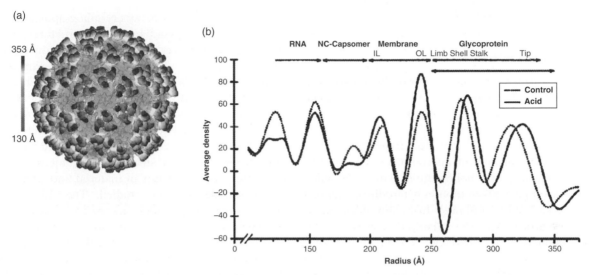

Fig. H2.12 (a) Example of the Semliki forest virus determined at 9 Å resolution. (b) Radial density function of the Semliki forest virus.

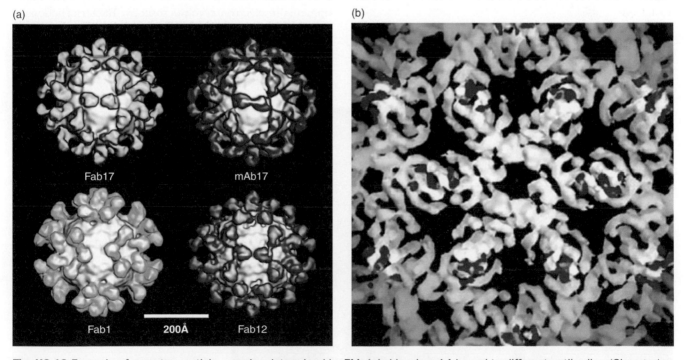

Fig. H2.13 Example of receptor–particle complex determined by EM: (a) rhinovirus-14 bound to different antibodies (Che *et al.*, 1998); (b) HBV complexed with a peptide: in red is the difference map between the reconstruction of the virus with peptide and that of the virus on its own (Böttcher *et al.*, 1998). (Figures reproduced with permission from the American Society for Microbiology and from Macmillan Publishers Ltd., respectively.)

are not seen due to their inherent mobility and flexibility, so they are averaged out during the reconstruction procedure.

The site of the ligand binding identifies the ligand's epitope. This footprint can be correlated with mutagenesis data to enable the positioning of key amino acids within the reconstruction. It has been shown that by reconstructing the same virus with and without a bound peptide it is possible to decipher where the peptide is bound to the viral capsid by omitting the peptide-viral map from the map with just the virus (Fig. H2.12b).

H2.8.3 Microtubules

Microtubules are essential components of the cytoskeleton of eukaryotic cells. In conjunction with motor proteins like dynein and kinesin, they engage in intracellular transport and cell division.

Reconstructions from electron micrographs detailed the structural organization of microtubules (Fig. H2.14). They are 25 nm-diameter hollow tubes with walls made from linear arrangements of tubulin heterodimers, named "protofilaments." When followed from protofilament to protofilament around the tube, tubulin describes a 12 nm pitch, left-handed, helical pathway around the microtubule. The number of protofilaments may vary from 9 to 16, resulting in microtubules with different configurations and symmetry. Some of them are truly helical. In the case of the 15-protofilament microtubules, the final map was produced from averaging helical data from 24 data sets, representing over 17 000 asymmetric units and layer lines to $1/18$ Å^{-1}.

Docking the atomic X-ray structure of the tubulin $\alpha\beta$-dimer into the lower-resolution EM maps provides an unambiguous orientation of the tubulin dimer within a protofilament. This pseudoatomic resolution microtubule model revealed that the tubulin heterodimer was stacked head-to-tail and also interacted laterally with tubulin from neighboring protofilaments. Reconstructions of microtubules bound to Ncd, a type of kinesin, subsequently revealed that the binding interface was extensive and primarily constituted of the β-tubulin monomer (Fig. H2.14).

H2.8.4 Integral Membrane Proteins

Through the efforts of X-ray crystallography and solution NMR, several thousand biological structures are determined yearly; however, less than a dozen concern integral membrane proteins. This is because of experimental difficulties, such as obtaining suitable crystals and collecting interpretable NMR spectra from lipid- or detergent-embedded proteins. Electron cryo-microscopy offers an alternative approach to determining the architecture of macromolecular assemblies. The method has several advantages over other structural techniques. It does not necessitate large amounts of material and even heterogeneous samples can be studied. The two-dimensional crystal, moreover, provides a more native-like environment for the membrane protein than three-dimensional crystals, as it is surrounded by lipid, with little or no detergent, and has the further advantage that it can be characterized by using essentially the same techniques that would be used for the protein in its native membrane (spectroscopy, ligand-binding assays, etc.).

Work on the structure of the integral membrane protein bacteriorhodopsin has been the basis of several key developments in electron cryo-microscopy. Bacteriorhodopsin, which is a 27 kDa light integral membrane protein found in *Halobacterium salinarum*, uses the energy of an absorbed photon to move a proton (H$^+$) across the cell membrane. Amazingly, this light-driven proton pump occurs naturally

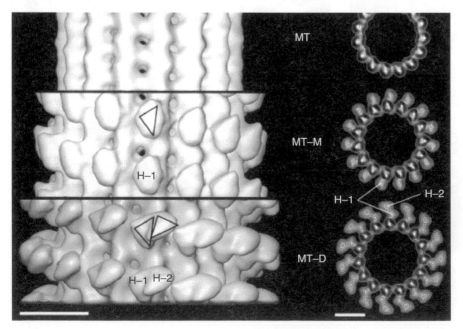

Fig. H2.14 Surface representations and end-on projections of the three-dimensional reconstruction of an undecorated microtubule (MT) in a complex with monomeric Ncd (MT-M), and in a complex with dimeric Ncd (MT-D). The side-view representations are oriented with the microtubule plus end at the top. The end projections represent the view from the plus end. Scale bars, 100 Å. (Sosa *et al.*, 1997.) (Figure reproduced with permission from Macmillan Publishers Ltd.)

(a)

(b)

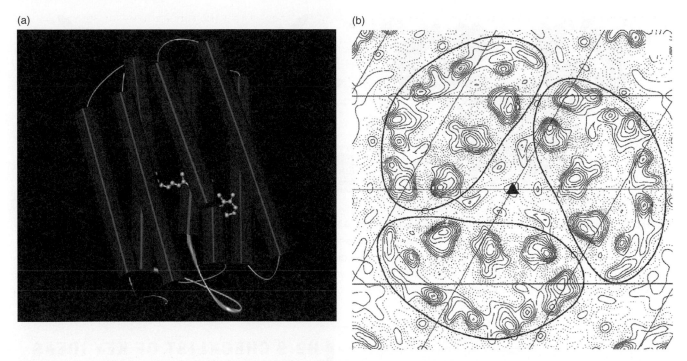

Fig. H2.15 The integral membrane protein bacteriorhodopsin and the two-dimensional projection maps identifying its α-helices. (a) Structure of bacteriorhodopsin: the retinal is shown as ball-and-stick. (b) Two-dimensional projection Fourier map of trimers of wild-type bacteriorhodopsin at a resolution of 3.5 Å, identifying locations of the seven transmembrane helices. The three-fold axis is indicated by the solid triangle. The solid lines around the densities indicate the boundary of each monomer. (Subramanian *et al.*, 2002; Subramanian and Henderson, 1999). (Figure reproduced with permission from the Royal Society and Elsevier, respectively).

as a two-dimensional crystal in the organism's inner membrane, so this system is particularly amenable to electron crystallographic analysis. The ordered lattice enabled collection of electron diffraction data (3.5 Å in-plane resolution), which, combined with phases from the images, enabled the determination of an atomic model of bacteriorhodopsin. The data were, in fact, adequate to resolve unambiguously the majority of the bulkier side-chains (Trp, Tyr, Phe, and Leu). The protein consists of seven helices with a lysine covalently bound via a protonated Schiff base linkage to a retinal molecule (Fig. H2.15a).

The flash-cooling of the sample offered the opportunity to trap the protein at different steps of the pumping mechanism, as the two-dimensional crystals could be frozen after different times following illumination. The EM data established that a single, large protein conformational change occurs within 1 ms of illumination, as the retinal isomerization from the all-*trans* to the 13-*cis* configuration initiates the release of a proton from the Schiff base to the outside of the cell. An intermediate protein structure follows that enables the base's reprotonation from the cytoplasmic side of the membrane. Retinal reisomerization brings the protein back into the resting state, and thus ready to start the pumping cycle again. Full three-dimensional reconstructions are not, in fact, necessary, as even from the two-dimensional projection maps it is possible to identify the location of the seven α-helices and any structural rearrangements (Fig. H2.15b).

In contrast to X-ray crystallography (Part G) and NMR (Part J), electron cryo-microscopy allows even partially heterogeneous samples to be studied. Single-particle reconstructions of integral membrane protein therefore offer the opportunity to determine the structure of integral membrane proteins. Independent high-resolution single particle EM reconstructions of the 2.2 MDa ryanodine receptor clarified the protein's architecture, highlighting how a calcium-binding EF-hand domain was positioned as a conformational switch for the allosteric gating of the ion channel. Intriguingly, the reconstructions also demonstrated that the core of the protein belongs to the six transmembrane ion channel superfamily (Zalk *et al.*, 2015; Efremov *et al.*, 2015) (Fig. H2.16).

The smaller size of many membrane proteins is still a limiting factor; however, negative staining may be relied upon to facilitate the identification and subsequently the determination of a low-resolution reconstruction. Jiang and coworkers, for example, analyzed the opening of a voltage gated ion channel (~200 kDa) at 10.5 Å resolution using negatively stained samples (Fig. H2.17).

In conclusion, the advent of better microscopes, sample preparation, and reconstruction procedures is bringing single-particle analysis of integral membrane protein into the realm of the feasible.

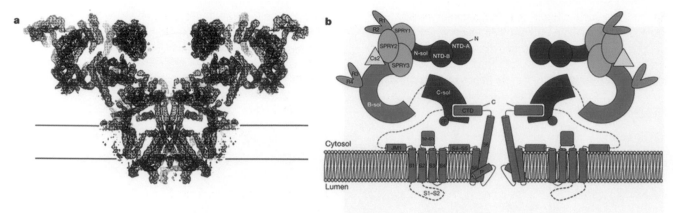

Fig. H2.16 The architecture of the ryanodine receptor RyR1 at 4.8 Å resolution. Slab through the EM map highlighting secondary structure features. The color-coded schematic representation of the domain structure of RyR1 derived from that EM map. Colors are green: bridge solenoid domain; red: core solenoid domain; blue: N-terminal domain; orange: transmembrane domain; cyan: SPRY domains; salmon: clamp region domains; purple: putative Ca^{2+}-binding domain; and yellow: bound calstabin. Dashed lines represent major disordered regions in the protein. (Adapted from Zalk *et al.*, 2015.)

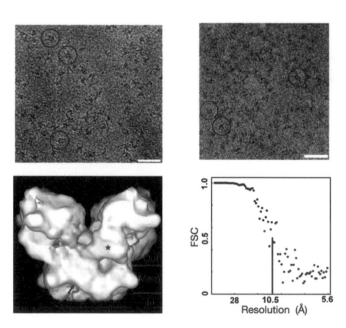

Fig. H2.17 Voltage gated opening of an ion channel KvAp. Top: Images of negatively stained (left) and cryo-negatively stained (right) KvAp–33H1 complexes, with selected particles marked with red circles. The scale bar is 50 nm. The Fab fragment 33H1 only binds to KvAp in its open conformation. Bottom left: three-dimensional reconstruction of the KvAP–33H1 complex at 10.5 Å resolution. The red arrow identifies a putative structural change on channel opening. The red asterisk marks the position of the variable immunoglobulin domain. Bottom right: FSC for the three-dimensional map of the KvAp–33H1 complex from two maps calculated using two halves of the data. The 0.5 threshold corresponds to 10.5 Å. (Jiang *et al.*, 2003.) (Figure reproduced with permission from Macmillan Publishers Ltd.)

H2.9 CHECKLIST OF KEY IDEAS

- Biological samples are very sensitive to the electron beam. Typically they can only tolerate a dose of 10–15 electrons $Å^{-2}$.
- The CTF is due to the imperfections of the electron microscope. Once it has been estimated, corrections can be applied.
- The common-lines reconstruction procedure is performed in Fourier space. The Fourier transforms of two different images of the same object intersect along a common line.
- Polar Fourier transform reconstruction is done with Fourier transforms of the EM images described by polar coordinates. The method relies on the use of a template that helps orient the images in order to create a more appropriate model of the imaged sample.
- The back projection method relies on matching EM images and that of a model which has been projected in several directions.
- Focal pairs can ease the initial reconstruction procedures, e.g., by identifying the hand of the object.
- Statistical analysis is used to help in the reconstruction procedures due to the poor signal-to-noise of the EM micrograph.
- In multivariate statistical analysis, the EM images are transformed to reduce the total number of data treated.
- The minimum number of particles needed for a reconstruction at a desired resolution depends on the size of the object, its symmetry, and the signal-to-noise ratio deemed acceptable.
- Resolution is not clearly defined in EM.
- The phase residual can be used to underscore the consistency of the phases of common-line pairs to a certain resolution.

- FSC is used to qualify the final reconstruction: It consists of a comparison between two reconstructions independently calculated from separate halves of the data set.
- An amplitude profile can be imposed on the final map in order to down-weight the low-resolution data and boost the higher-resolution features.
- An effective way to prevent over-fitting of the model to the EM data is to separate the data into two and undertake two independent reconstructions.
- The lower-resolution EM reconstruction can be combined with atomic models of its different constituents derived from X-ray crystallography to create a "quasi" atomic-resolution structure.

Suggestions for Further Reading

Several excellent reviews have been written on electron cryo-microscopy, including:

Baker, T. S., Olson, N. H., and Fuller, S. D. (2000). Adding the third dimension to virus life cycles: Three-dimensional reconstruction of icosahedral viruses from cryo-electron micrographs. *Microbiol. Mol. Biol. Rev.*, **63**, 862–922.

Van Heel, M., Gowen, B., Matadeen, R., *et al.* (2000). Single-particle electron cryo-microscopy: Towards atomic resolution. *Q. Rev. Biophys.*, **33**, 307–369.

Cheng, Y., Grigorieff, N., Penczek, P. A., and Walz, T. (2015) A primer to single particle cryo-electron microscopy. *Cell*, **161**, 438–449.

Few publications on three-dimensional reconstructions have been published, but those that have include:

Glaeser, K., Dowrling, K., Derosier, D., Baumeister, W., and Frank, J. (2004). *Electron Crystallography of Biological Macromolecules*. Oxford: Oxford University Press.

Frank, J. (2005). *Three-Dimensional Electron Microscopy of Macromolecular Assemblies*. New York: Oxford University Press.

Edited by Jensen, G. J. (2010). Cryo-EM: Part B. 3-D reconstruction. *Meth. Enzymol*, **482**, 2-410.

MOLECULAR DYNAMICS

ENERGY AND TIME CALCULATIONS   11

11.1 HISTORICAL REVIEW OF BIOLOGICAL APPLICATIONS

The first molecular dynamics simulation of a biological macromolecule, the small stable protein bovine pancreatic trypsin inhibitor (BPTI) for which an accurate X-ray crystallographic structure was available, was published by **J. A. McCammon**, **B. R. Gelin**, and **M. Karplus** in **1975**. Based on a simple approximation for the molecular mechanics potential, the simulation was performed for the macromolecule in vacuum, and limited to 9.2 ps by the computing power available at the time. The BPTI study was nevertheless effectively instrumental, together with the earlier hydrogen exchange experiments of **K. Linderstrom-Lang** and his collaborators **(1955)**, in establishing the view that proteins are not rigid bodies but dynamic entities whose internal motions must play a role in their biological activity.

In **2003**, **M. Karplus** wrote that molecular dynamics simulations of biological macromolecules developed from two lines of work: *individual particle trajectory calculations in chemical reactions* and *physical calculations to predict the dynamic behavior of large particle assemblies.*

The first line of work goes back to the two-body scattering problem for which analytical solutions are available. It is well known in physics, however, that no analytical solutions have been found for the three-body (or more) problem. A prototype trajectory calculation was attempted in **1936** by **J. A. Hirschfelder**, **H. Eyring**, and **B. Topley** for the simple chemical reaction involving the interconversion between hydrogen atoms and molecules, but the calculation could only be completed by using the computers that became available in the **1960s**. Classical chemical trajectory calculations have now evolved to incorporate quantum mechanical effects.

The second line of work has its origins in the work of **J. D. van der Waals** on particle–particle interactions and the statistical mechanics of **L. Boltzmann**. In the **1950s**, **B. J. Alder** and **T. E. Wainwright** published a study of liquids whose molecules behaved like "hard spheres." **A. Rahman** in **1964** published a molecular dynamics simulation of liquid argon, assuming a "soft" Lennard–Jones potential; the first simulation of liquid water, a much more complex liquid, was published in **1971**, by **F. H. Stillinger** and **A. Rahman**.

The two independent approaches were brought together in the simulation of protein molecular dynamics. In the **late 1960s**, good starting points for the calculations were provided, on the one hand, by the empirical energy functions that had been developed for bonded and non-bonded interactions in small organic molecules in the laboratories of **H. Scheraga** and **S. Lifson** and, on the other hand, by the precise three-dimensional structures that were being solved by X-ray crystallography.

The decades following the publication of the BPTI calculation saw a significant extension of the calculations to larger proteins and longer simulation times, as higher-resolution structures and more powerful computers became available. Calculation methods combining quantum and molecular mechanics (QM/MM) were developed. Macromolecular dynamics simulations are necessarily approximate due to the complexity and heterogeneity of the systems studied. Parallel developments in experimental approaches provided information on macromolecular internal motions and were essential in order to validate the simulation methods. These included the analysis of Debye–Waller factors in crystallography, fluorescence depolarization of tryptophan residues, NMR, inelastic neutron scattering, Mössbauer spectroscopy, and Fourier transform IR spectroscopy. Symbiotic relationships were established between molecular dynamics calculations and NMR and X-ray crystallography, starting with the incorporation of energy minimization in structural refinement.

In the last decade massive progress in computing power has made possible molecular dynamics simulations of systems of more than 100 000 atoms and timescales up to the microsecond. Molecular dynamics simulations are now being applied on the supramolecular and even cellular scale. In the study of molecular machines, like the ATP synthase (the enzyme responsible for most ATP synthesis in living organisms), macromolecular dynamics calculations are building the link between thermodynamics data and the crystallographic structure. Kinetic cryocrystallography, in which intermediate structures are "frozen" and examined, has become one of the powerful molecular dynamics experimental techniques.

The **2013 Nobel Prize in chemistry** was awarded to **Martin Karplus**, **Michael Levitt**, and **Arieh Warshel** for their work on developing computational methods to study

COMMENT I1.1 EPPUR SI MUOVE!

Galileo's possibly apocryphal remark when the inquisition forced him to deny his theory that the Earth rotated around the Sun, "and yet it moves," is the title of Jeremy Smith's and Benoit Roux's benchmark article in *Structure* about the award of the 2013 Nobel Prize in Chemistry to Martin Karplus, Michael Levitt, and Arieh Warshel. Smith and Roux (2013) trace the development of computational approaches to molecular dynamics from the 1960s, when all three future Nobelists were strongly influenced by Schneior Lifson at the Weizmann Institute. Lifson was developing ideas for using molecular mechanics' empirical functions to calculate the energies of large molecules. Smith and Roux conclude:

Perhaps, though, the singular ability of numerical simulation to furnish a firm energetic and thermodynamic foundation for the formation and functional use of three-dimensional structure meant that it was inevitable that this field would slowly but surely take a hold in molecular biophysics. At any rate there's no turning back. Computations are at the forefront of modern-day scientific planning, and simulation is now firmly established as the third pillar of science, linking experiment to theory for complex systems resisting the back of an envelope. For all their infuriating aspects, maybe accurate computer simulations are indeed the only way to unlock a deep understanding of how a biological system works.

complex chemical systems – work that has had enormous impact in structural biology (Comment I1.1)

I1.2 DYNAMICS, KINETICS, KINEMATICS, AND MOLECULAR STABILIZATION FORCES

The concepts of *dynamics*, *kinetics*, *kinematics*, and *molecular stabilization forces* were discussed in Chapter A3.

Our understanding of macromolecular structures is based on an understanding of the forces underlying them – the forces acting to maintain the atoms in their positions. And if, in turn, we gain an understanding of these forces, then we shall also understand how the atoms in the structure move about their mean locations. Expressed in the language of physics, our aim is the understanding of the *force field* or *potential energy function* around each atom in a structure. The forces that maintain biological macromolecular structure are known (see Section C2.4.9). They arise from hydrogen bonds, salt bridges, or screened electrostatic interactions, so-called hydrophobic interactions,

and van der Waals interactions. Covalent bonds other than the disulfide bond (that represents a special case because of its sensitivity to the reducing capacity of the environment), which are very strong with respect to temperature, are not usually broken when a macromolecule unfolds, so that in general they do not contribute to the stabilization energy of macromolecular tertiary or quaternary structure.

I1.3 LENGTH AND TIMESCALES IN MACROMOLECULAR DYNAMICS

The amplitudes of atomic motions in macromolecules at ambient temperature (300 K) range from 0.01 Å to >5 Å for time periods from 10^{-15} to 10^3 seconds (1 fs for electronic rearrangements to about 20 min for protein folding or local denaturation) (see Chapter A3).

I1.4 NORMAL MODE ANALYSIS

The *normal mode* approach to molecular dynamics has the advantage that it provides an analytical solution for the description of atomic motions and other dynamic parameters of a macromolecular system, such as thermodynamic averages of amplitudes, fluctuations, and correlations. The approach is based on the assumption that the atoms are coupled by simple harmonic potentials, and their vibration results from a superposition of fundamental modes (see Chapter A3).

The equations of motion of N harmonically coupled atoms in a macromolecule are given in Chapter A3. Their solutions form a set of $3N$ periodic functions. Six of these correspond to the macromolecular translational and rotational degrees of freedom, for which the atoms do not move relative to each other. The system, therefore, has $3N - 6$ normal modes. Atomic motions are a superposition of all the normal modes.

At a given temperature (mean energy) the mean square amplitude of a mode is inversely proportional to its effective force constant, so that the low-frequency modes dominate motion amplitudes (see the example calculation for BPTI in Fig. A3.14).

A normal mode calculation is expected to provide a reasonable description of vibrational protein dynamics at low temperature, below the dynamical transition above which atomic motions sample different conformational substates (see Chapter A3). Normal mode calculations are based on a number of assumptions, however, that are not always accurate when applied to proteins at ambient temperature. These are: (1) atoms move in simple harmonic potentials (quadratic functions of the spatial coordinates); (2) the macromolecular conformation is at an energy minimum (this assumption is not specific to normal mode calculations); (3) interactions with solvent can only be represented implicitly. In particular, solvent interactions and other anharmonic effects are far from negligible at room temperature. As we pointed out at the beginning of this section, however, the precise analytical nature of normal mode calculations of protein dynamics yields considerable advantages, and various approaches have been developed to make such calculations applicable even when the assumptions above are not strictly valid. The conformational energy landscape at ambient temperature is anharmonic, made up of the many local minima of the conformational substates (see Chapter A3). It may be approximated by an empirical potential calculated from molecular dynamics simulations (Section I1.5.3) in which solvent effects can be introduced in different ways (Section I1.5.4).

I1.5 MOLECULAR DYNAMICS SIMULATIONS

In a molecular dynamics simulation, the behavior of a system is determined as a function of time from an algorithm that includes a starting *structural model* and *force field*. Our knowledge of the determinants involved, however, is very far from complete, and the molecular dynamics simulation approach remains firmly anchored in experiment.

The quality of a molecular dynamics result may be sensitive to the quality of the starting structure (usually a high-resolution structure from X-ray crystallography) and on force field assumptions, which can be tested against experimental spectroscopic and diffraction data. In fact, the simulation procedure itself is often seen as an *experiment*, in which conditions are varied until good agreement is obtained with data from structural and spectroscopic methods.

Simulation tools fall into two categories: *classical* molecular mechanics approaches to study Newtonian and *stochastic* (statistically random) dynamics (in order to determine atomic motions, conformational changes, ligand binding, solvent effects, etc.) and *quantum* mechanics to study electronic distributions (in order to determine enzyme reaction mechanisms, etc.). *Ab initio* quantum mechanical calculations are limited by computer power. By using a semi-empirical quantum mechanical approach for a similar number of atoms, the time range is extended to about 1 ns. Molecular mechanics at the atom level deals with the energy of the system and the forces exerted on each of its atoms. Molecular mechanics calculations are relatively fast; the ANTON supercomputer now allows 1 ms calculation with 500 000 atoms; the TITAN supercomputer allows calculations involving 100 million atoms for 1 ms. The QM/MM approach combines a quantum mechanical calculation (around the atoms in an enzyme active site, for example) with a molecular mechanics calculation (for all the other atoms in the system).

I1.5.1 Force Field

A typical potential energy function in molecular mechanics is given by

$$V_{\text{total}} = V_{\text{cov}} + V_{\text{non-cov}} \tag{I1.1a}$$

$$V_{\text{total}} = V_{\text{bond}} + V_{\text{angle}} + V_{\text{dihedrals}} + V_{\text{elec}} + V_{\text{LJ}} \tag{I1.1b}$$

The first three terms on the right-hand side of Eq. (I1.1b) make up the covalent energy, V_{cov}; they correspond, respectively, to covalent bond stretching, angle, and dihedral angle energy. The last two terms refer to non-covalent bond interactions, $V_{\text{non-cov}}$; they are the electrostatic and Lennard–Jones potentials (which describe the van der Waals interaction) (Comment I1.2). Other energy terms that may be considered include a polarization term due to charge distribution distortion by the internal electric field in the macromolecule and the special case of the hydrogen bond, which can be represented by a combination of Van der Waals interaction and electrostatics.

The force on atom i along a coordinate vector $\mathbf{r}_i$ is given by the slope of the potential:

$$F_i = -\frac{\partial V_{\text{total}}}{\partial r_i} \tag{I1.2}$$

I1.5.2 Parameterization of the Force Field

The potentials in Eq. (I1.1) are described by phenomenological functions, expressed in terms of coordinates, coordinate deviations, and coefficients (Comment I1.2), which are optimized to reproduce experimental data and quantum calculations. The total energy function is itself empirical. It is expressed as a function of the relative position of each pair of atoms, so that for folded macromolecules it is important to have as good a starting structure as possible.

The parameters in the potential energy terms are obtained from experimental structural data (X-ray crystallography or NMR for bond lengths and angles, for example), spectroscopy (IR and Raman, for the force constants), and quantum calculations on model systems for charges and Lennard–Jones potential constants.

Molecular dynamics simulation programs propose different expressions for the force field and include parameter files for amino acids, proteins, nucleic acids, and many other organic molecules, especially enzyme substrates and inhibitors (Comment I1.3).

I1.5.3 Potential Energy Surface and Energy Minimization

We recall from Chapter A3 that the potential energy of a protein as a function of the positions of all its atoms (expressed in terms of a conformational coordinate) can be described as a rugged landscape with dips and rises (Fig. I1.1).

The *global minimum* is the lowest energy state. Depending on the available energy (proportional to the

COMMENT I1.2 COVALENT AND NON-COVALENT ENERGY

Covalent bond stretching:

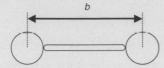

Fig. I1.2.1.

$$V_{bond} = \sum_{bonds} \frac{1}{2} k_b (b - b_0)^2$$

Covalent angle energy:

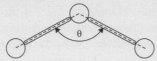

Fig. I1.2.2.

$$V_{angle} = \sum_{angles} \frac{1}{2} k_\theta (\theta - \theta_0)^2$$

Covalent dihedral angle energy:

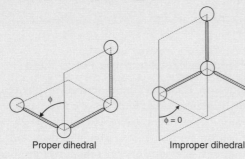

Proper dihedral   Improper dihedral

Fig. I1.2.3.

$$V_{dihedral} = \sum_{dihedrals} \frac{1}{2} k_\phi [1 + \cos(n\phi - \delta)]^2$$

Electrostatic energy:

Fig. I1.2.4.

$$V_{elec} = \frac{1}{4\pi\varepsilon_0\varepsilon} \sum_{ijpairs} \frac{q_i q_j}{r_{ij}^2}$$

Lennard–Jones potential:

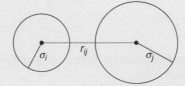

Fig. I1.2.5.

$$V_{LJ} = \sum_{ijpairs} 4\varepsilon_{ij} \left\{ \left(\frac{\sigma_{ij}}{r_{ij}} \right)^{12} - \left(\frac{\sigma_{ij}}{r_{ij}} \right)^6 \right\}$$

COMMENT I1.3 MD PROGRAMS

The CHARMM web page is www.charmm.org. The AMBER web page is www.amber.ucsf.edu.

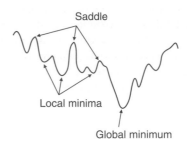

Fig. I1.1 Schematic potential energy surface as a function of conformational coordinate for a protein.

temperature), however, the molecule may be trapped in a *local minimum*, one of the conformational substates discussed in Chapter A3. Transitions between local minima pass through *saddle points*.

Before starting a molecular dynamics simulation, it is necessary to allow a structure to reach a locally stable energy state. At a minimum in the energy surface, the first derivative or gradient is zero and the second derivative is positive ($\partial V/\partial r = 0$ and $\partial^2 V/\partial r^2 > 0$), and the set of atomic coordinates that satisfy these equations must be found. This is a very complex problem to solve for proteins because of the large number of atoms and interactions involved, and various mathematical approaches have been developed to address it at different levels of approximation. It is, in particular, very difficult to explore large areas of conformational space by crossing saddle point barriers.

One approach used for energy minimization is based on *simulated annealing*. The temperature of the system is increased (e.g., to 800 K), then rapidly decreased (e.g., at a rate of 50 K ps^{-1}) to 300 K. Simulated annealing tends to bring the system into a relatively stable area of potential energy space, and increases the chances of finding a global minimum in conformational space.

I1.5.4 Modeling the Solvent

The importance of the role played by the solvent in protein dynamics was discussed in Chapter A3. In molecular dynamics simulations, the solvent can be accounted for by different approaches.

In the *explicit solvent* approach, the protein is positioned in a box of water molecules. The infinite character of the solvent is taken into account by introducing periodic boundary conditions in which a tractable number of water molecules is placed in an effective unit cell that is repeated a great number of times. Periodic boundary conditions have also been applied to simulate a protein crystal with a unit cell containing protein and water molecules (Comment I1.4). Solvent and protein mutually influence each other, and molecular dynamics simulations, such as the one described in Comment I1.4, also address the important point concerning how the hydration layer structure and dynamics are strongly influenced by interactions with the protein surface.

Where they are applicable, *implicit solvent* approaches can allow a substantial gain in computational time. Water molecules are not represented explicitly and solvent effects are included in the simulations in various ways, by attributing continuum dielectric constant values to the solvent and protein volumes, or by introducing a dielectric constant within the macromolecule that depends on the distance from the protein–solvent interface effectively screening the charges on the protein surface.

Solvent effects can also be described by using a friction coefficient and random force on the macromolecular atoms due to water molecule collisions. The general equation of motion (Eq. (I1.3)) used is due to Langevin, and describes what has been called *Langevin dynamics*:

$$m\frac{d^2r}{dt^2} = -\frac{\partial V}{\partial r} - m\beta\frac{dr}{dt} + R(t) \tag{I1.3}$$

where m is the mass of the atom under consideration, $\mathbf{r}$ is its positional vector, V is the potential energy function at that position, β is a friction coefficient, and $\mathbf{R}(t)$ is a random force as a function of time, t. Simulation of the random force implies simulating the macromolecule in a thermal water bath at constant temperature.

Note that the frictional force due to solvent viscosity reduces the particle energy, whereas the collisions due to the thermal agitation of the water molecules increase the energy.

I1.5.5 Typical Molecular Mechanics Simulation Protocol

The starting point for a molecular mechanics simulation of a macromolecule is the best possible available X-ray crystallography structure. A solvent model is then chosen and the potential energy function of the system is defined and minimized.

Each atom, i, is represented by a point at the position described by the vector, $\mathbf{r}_i$, that obeys Newtonian dynamics (or Langevin dynamics if this solvent model was adopted):

$$m_i\frac{d^2r_i}{dt^2} = -\frac{\partial V}{\partial r_i} \tag{I1.4}$$

where subscript i refers to atom i.

A velocity $\mathbf{v}_i$ associated with each atom is chosen in order to obtain the desired temperature

$$T = \frac{1}{3Nk_B}\sum_i m\langle v_i^2\rangle \tag{I1.5}$$

where N is the number of atoms.

The equations of motion are integrated (by various algorithms) to determine the *trajectory* of each atom, i.e., its position and velocity between time t and time $t + d_t$. The trajectories of all the atoms expressed as a chosen number of time-steps are then stored for further analysis.

A calculation time-step of 1 fs is usually used in protein dynamics simulations, so that a 1 ns simulation requires 10^6 force and energy calculations per atom.

I1.5.6 Analysis of Results

Temperature
The temperature of the simulated system is defined by Eq. (I1.5).

Radius of Gyration
The radius of gyration, R_G, is a useful parameter to define the overall conformation of the system (see Chapter G2). It can be calculated in two steps. The center-of-mass coordinates, $\mathbf{R}_C$, are obtained first from

$$\sum m_i(\mathbf{r}_i - \mathbf{R}_C) = 0 \tag{I1.6}$$

The radius of gyration is related to the moment of inertia and defined by

$$R_G^2 = \sum m_i(\mathbf{r}_i - \mathbf{R}_C)^2/M \tag{I1.7}$$

where M is the total mass of the system.

The radius of gyration can also be obtained from the system coordinates, without going through the center-of-mass calculation, from

$$R_G^2 = \sum\sum m_i m_j(\mathbf{r}_i - \mathbf{r}_j)^2/M^2 \tag{I1.8}$$

(d)

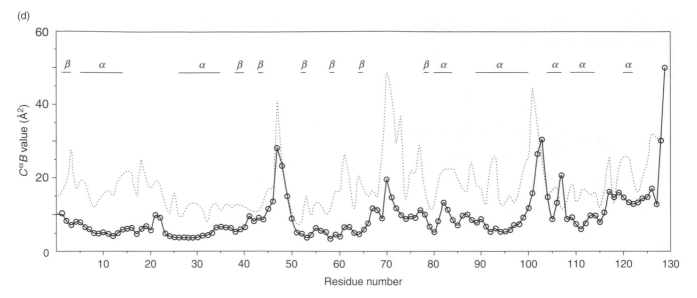

Fig. I1.2 Experimental *B*-factors (dotted line) and *B*-factors from the full set of normal modes calculated for 300 K in lysozyme (solid line), as a function of residue number. The *y*-axis corresponds to $\langle u^2 \rangle\, 8\pi^2/3$ (see Eq. (I1.10)). The secondary structure stretches in the sequence are shown also. (Levitt *et al.*, 1985.) (Figure reproduced with permission from Elsevier.)

Root Mean Square Deviation (RMSD)

The RMSD follows how the structure evolves during the simulation. It is defined by

$$\text{RMSD} = \sqrt{\frac{1}{N}\sum_{i=1}^{N}\left(\mathbf{r}_i - \mathbf{r}_i^{\text{ref}}\right)^2} \qquad (\text{I1.9})$$

where $\mathbf{r}_i^{\text{ref}}$ are the coordinate vectors of the reference structure (usually, the starting structure, used for the simulation). The RMSD plotted during the trajectory of a structure as it is allowed to equilibrate at a constant temperature may reach a plateau value in times of the order of 100 ps.

Mean Square Fluctuations and Debye–Waller Factors

Crystallographic *Debye–Waller factors* describe the fluctuations of atoms about their mean positions in a crystal structure (see Part G). Since crystallographic studies are time-averaged, fluctuation amplitudes include terms due to time-dependent thermal motions and terms due to deviations from atomic mean positions because of disorder in the crystal.

$$\langle u_i^2 \rangle_{\text{total}} = \frac{3B_i}{8\pi^2} = \langle u_i^2 \rangle_{\text{thermal}} + \langle u_i^2 \rangle_{\text{disorder}} \qquad (\text{I1.10})$$

where B_i is the crystallographic *B*-factor as it is usually defined. Where a structure does not "freeze" into alternative conformations, it may be possible to distinguish between the static and dynamic contributions to the Debye–Waller factor by collecting crystallographic data as a function of temperature. The disorder term is independent of temperature, while the square of the thermal amplitudes for harmonic vibrations varies linearly with absolute temperature (see Chapter A3).

Mean square fluctuations, calculated from a molecular dynamics or normal mode calculation, compare very favorably with fluctuations derived from Debye–Waller factors (Fig. I1.2), showing the larger mobility of loops outside secondary structure elements.

Diffusion Coefficients

When the atomic displacements are large during the time of the simulation, it is useful to express them in terms of a diffusion coefficient (see Part D). The diffusion coefficient of atom i is given by

$$D_i = \frac{1}{6t}\left\langle \left(\mathbf{r}_i(t) - \mathbf{r}_i(0)\right)^2 \right\rangle \qquad (\text{I1.11})$$

where t is time.

I1.6 APPLICATION EXAMPLES

I1.6.1 BPTI and Lysozyme

The bases for molecular dynamics simulations of proteins were set between 1975 and 1977. The first molecular dynamics simulation of a protein was published for BPTI, a small, 58-amino-acid residue protein of known structure, in 1977. BPTI was also the first protein for which normal mode calculations were performed in the early 1980s. As discussed above, normal mode calculations are

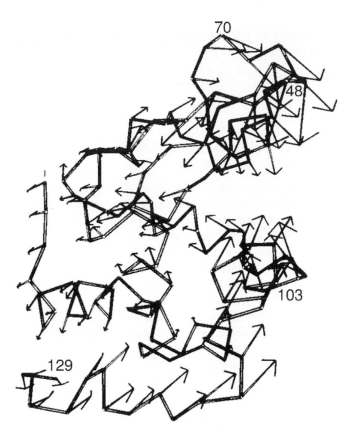

Fig. I1.3 Domain motion about the active site in the low-frequency normal mode of period 11.2 ps, in lysozyme. (Levitt *et al.*, 1985.) (Figure reproduced with permission from Elsevier.)

complementary to molecular dynamics simulations. The normal mode approach has the limitation that a quadratic approximation must be used for the free-energy potential, but the advantage is that it provides a complete description of atomic vibrations in the system, in the fundamental 0.1–10 ps time interval. The normal modes approach provides analytical expressions for thermodynamic averages of amplitudes, fluctuations, and correlations. Early normal mode calculations illustrated collective atomic motions in globular proteins, and showed how the lower-frequency modes dominated amplitudes (see Section I1.4). When calculated from the normal mode fluctuations, Debye–Waller factors were found to be in reasonable agreement with the experimental crystallographic values corresponding to thermal motions (Fig. I1.2).

Normal mode calculations revealed similar dynamics in the collective motions of small monodomain proteins, like BPTI, and in enzymes, like lysozyme and ribonuclease A, whose active sites are sandwiched between two domains. It is interesting to note, however, that in the larger molecules they also showed collective domain motions about the active site (Fig. I1.3).

I1.6.2 Protein Folding

The first microsecond molecular dynamics simulation of a protein was published in 1998. It described the thermodynamic folding of HP-36, a 36-residue subdomain.

In-vitro HP-36 folds spontaneously between 10 and 100 μs, so that only the very first steps of the process could be simulated. The 1 μs calculation on the 36 residues and about 3000 explicit water molecules occupied 256 processors for four months.

I1.6.3 Structure Refinement

The objective of structure refinement in high-resolution crystallography and NMR (see Parts G and J) is to reach the best possible agreement between a model structure and experimental data. This is obtained in practice by minimizing a function representing differences between observed parameters and corresponding parameters calculated from the model. The function can be expressed as a total energy to be minimized:

$$E = E_{chem} + w_{data}E_{data} \qquad (I1.12)$$

where E_{chem} represents the potential energy calculated for the model structure and E_{data} is related to the difference between measured and model calculated diffraction data; w_{data} is a factor to weight the relative contributions of the two terms (Comment I1.5).

I1.6.4 ATP Synthase: A Molecular Machine

Spectacular progress in macromolecular crystallography, yielding high-resolution structures of large protein complexes and molecular machines in their various states, has stimulated a parallel effort in molecular dynamics simulations.

The enzyme F_1F_0-ATPase is responsible for most of the ATP synthesis in living organisms (Comment I1.6). It is a large membrane-bound molecular machine made up of several protein subunits. The catalysis of ATP synthesis from ADP is driven by proton translocation across the membrane through the F_0 domain – the proton gradient having been created previously by respiration or photosynthesis. The catalytic domain, F_1, is made up of a nine-subunit $\alpha_3\beta_3\gamma\delta$ globular structure outside the membrane. The $\alpha_3\beta_3$ subunits are arranged in a ring around a central stalk made up of the γ subunit associated to δ, and whose foot makes extensive contact with a ring of c subunits in the F_0 domain. The central stalk in

F_1 and the c-ring in F_0 are believed to act similarly to the rotor in an electric motor and rotate together relative to the rest of the enzyme by using the proton translocation energy. The rotation modulates the binding affinities of the β subunits for substrate and product according to the *binding change* mechanism, in which the catalytic subunits cycle between three states: *open* (or *empty*), *loose*, and *tight*. During the cycle, $120°$ rotation of the stalk converts an open site (with low affinity for both substrate and product) to a loose site (which binds substrate); a further $120°$ rotation converts the loose site to the tight site (in which the reaction product is formed), and a further $120°$ rotation brings the subunit back to the open state to release the product. The three β states have been identified in crystal structures according to their binding properties for ADP and various derivatives. The F_1 domain retains its ability to hydrolyze ATP to ADP (enzyme catalysis can go in both forward and backward reaction directions). Hydrolysis of ATP leads to the rotation of the central stalk, strikingly visualized in a microscope by attaching an actin filament or a bead to its exposed foot (Comment I1.7; see also Chapter F4). Important hints on the mechanisms underlying the

rotation were obtained from comparisons of the different subunit structures observed in the crystallographic studies, and MD simulations were performed in order to analyze the dynamics of the changes and their sequence in time (Comment I1.8).

There will certainly be great progress in the coming years in molecular dynamics simulations of ATP synthase

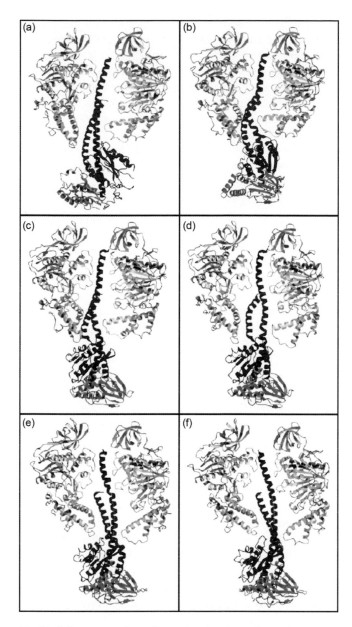

Fig. I1.4 Structures from the molecular dynamics trajectory: **(a)** is the initial structure; **(b)**, **(c)**, **(d)**, **(e)** are after 150, 300, 350, and 425 ps, respectively; **(f)** is the final structure. Subunit colors are red for α, yellow for β, purple for γ, green for δ, and light yellow for ε. (Ma *et al.*, 2002.) (Figure reproduced with permission from Elsevier.)

mechanisms in the wake of "better" crystal structures and improved calculation methods. The first results, however, have already provided fascinating insights into the workings of this molecular motor.

Structures during the molecular dynamics trajectory are shown in Fig. I1.4. The molecular dynamics simulations showed how the rotation of the γ subunit induces the opening and closing of the catalytic β subunits, with, of particular interest, the demonstration of the existence of an *ionic track* guiding their motion during the transition (Fig. I1.5). (See also Section F2.6.2.)

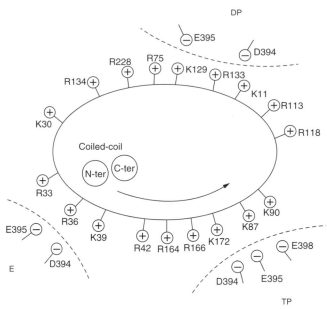

Fig. I1.5 Schematic view of the ionic track interactions between the γ subunit and β subunits (Ma *et al.*, 2002). (Figure reproduced with permission from Elsevier.)

I1.7 CHECKLIST OF KEY IDEAS

- Molecular dynamics involves the calculation of forces that underlie molecular structure and motions.
- The folded native structures of biological macromolecules are maintained by forces arising from hydrogen bonds, salt bridges, or screened electrostatic interactions, so-called hydrophobic interactions and van der Waals interactions.
- The *force field* is the *potential energy function* determined by the relative positions of all pairs of atoms in the macromolecule and solvent.
- The amplitudes of atomic motions in macromolecules at ambient temperature range from 0.01 Å to >5 Å for time periods from 10^{-15} to 10^3 s (1 fs for electronic rearrangements to about 20 min for protein folding or local denaturation).
- Our experimental understanding of protein dynamics is based on results from various spectroscopies, and crystallographic studies of time-averaged structures and transient intermediates.
- Normal mode calculations are based on harmonic (quadratic) potentials and provide analytical solutions for the atomic motions and thermodynamic averages of amplitudes, fluctuations, and correlations.
- A three-dimensional system of N atoms has $3N - 6$ normal modes.
- Low-frequency normal modes dominate vibration amplitudes.

- In a molecular dynamics simulation, the behavior of a system is determined as a function of time from an algorithm that includes a starting structural model and force field.
- Classical *molecular mechanics* simulations are used to study Newtonian or stochastic dynamics (atomic motions, conformational changes, ligand binding, solvent effects, etc.). By using supercomputers, molecular mechanics calculations can currently be performed for 1 ms duration with 500 000 atoms and up to 100 million atoms for 1 μs.
- *Quantum mechanics* is used to study electronic distributions (enzyme reaction mechanisms, etc.). Current limits are 100 atoms moving for 10 ps.
- The QM/MM approach combines QM for the study of a limited part of the structure (for example, the active site of an enzyme) and MM for the rest of the molecule.
- Force fields include covalent bond terms and terms referring to non-covalent bond interactions, such as electrostatic and Lennard–Jones potentials.
- The potential energy of a protein as a function of a conformational coordinate is represented by a rugged *energy landscape* with local minima (corresponding to the conformational substates), separated by saddle points, and a global energy minimum.
- In an *energy minimization* procedure, a structure is allowed to evolve toward a local or global minimum.
- *Simulated annealing* is a method that increases the chances of finding a global energy minimum by controlling the temperature during a molecular dynamics simulation. Currently, the method is not much used. Other advanced sampling techniques are preferred.
- Solvent and protein mutually influence each other during a molecular dynamics simulation.
- In the *explicit solvent* simulation approach, the protein is positioned in a box of water molecules and periodic boundary conditions are applied.
- In *implicit solvent* approaches, solvent effects are included in the simulations by attributing continuum dielectric constant values to the solvent and protein volumes.

- In *Langevin dynamics*, the solvent is taken into account by introducing a friction coefficient and random force on the macromolecular atoms due to water molecule collisions.
- Results of a molecular dynamics simulation include a *trajectory* describing the time evolution of conformational and fluctuation parameters, and values for the temperature and pressure of the system, its radius of gyration, atomic RMSD, and atomic diffusion coefficients.
- RMSDs from molecular dynamics simulations compare favorably with experimental Debye–Waller factor derived values from crystallography.
- Normal mode calculations in two-domain enzymes, like lysozyme and ribonuclease A, reveal collective domain motions about the active site.
- The first microsecond molecular dynamics simulation of a protein described the first steps in the thermodynamic folding of HP-36, a 36-residue subdomain, in the presence of 3000 explicit water molecules.
- Energy minimization by molecular dynamics simulation approaches is used in the refinement of high-resolution structures in crystallography and NMR.
- Targeted molecular dynamics simulations are being developed to study the mechanisms in large molecular machines such as ATP synthase or chaperones.

Suggestions for Further Reading

McCammon, J. A., and Harvey, S. C. (1987). *Dynamics of Proteins and Nucleic Acids.* Cambridge: Cambridge University Press.

Karplus, M. (2003). Molecular dynamics of biological macromolecules: A brief history and perspectives. *Biopolymers*, **68**, 350–358.

Field, M. (2005). *A Practical Introduction to the Simulation of Molecular Systems.* Cambridge: Cambridge University Press.

Smith, J. C., and Roux, B. (2013) Eppur si muove! The 2013 Nobel Prize in Chemistry. *Structure*, **21**, 2102–2105.

NEUTRON SPECTROSCOPY 12

12.1 HISTORICAL OVERVIEW AND INTRODUCTION TO BIOLOGICAL APPLICATIONS

We recall from Chapter A1 that biological events occur on an immensely extended range of timescales – from the femtosecond of electronic rearrangements in the first step of vision, to the 10^9 years of evolution. Thermal energy is expressed as atomic fluctuations on the picosecond–nanosecond timescale that constitutes the basis of molecular dynamics. These fluctuations are of particular interest in biophysics because they result from and reflect the forces that structure biological macromolecules and the atomic motions and molecular flexibility associated with biological activity.

Thermal energy propagates through solids in waves of atomic motion, such as the normal modes discussed in Chapters A3 and I1. We can estimate values for the frequencies, wavelengths, and amplitudes of thermal excitation waves from an order of magnitude calculation, e.g., by considering the movement of a mass similar to the mass of an atom moving in a simple harmonic potential of energy equal to Boltzmann's constant multiplied by 300 K (ambient temperature). It turns out that the frequencies are of the order of 10^{12} s^{-1}, while the wavelengths and amplitudes are on the ångström scale. We saw in Part E that the energy associated with thermal vibrations corresponds to the IR frequency range in optical spectroscopy. Similarly, *neutron spectroscopy* takes advantage of the fact that the energies of *thermal neutron* beams match vibrational energies in solids and liquids. This is not surprising because *thermal neutrons* are produced by equilibration at ambient temperature, so that, by definition, their energy is equal to the corresponding thermal energy. The kinetic energy of a neutron is given by the same equation as for a billiard ball: $E = \frac{1}{2}mv^2$. The quantum mechanical wavelength associated with a neutron beam is h/v, where h is Planck's constant (see Chapter A3). The mass of the neutron is very close to that of a proton and therefore to that of a hydrogen atom, or about one atomic mass unit ($1/N_A = 1.66 \times 10^{-24}$ g), where N_A is Avogadro's number. Together, neutrons and protons make up 99.9% of an atom's mass, the mass of the electron being several orders of magnitude smaller. Equating the neutron kinetic energy to thermal energy at ambient temperature, we can calculate the neutron velocity (about 4 km s^{-1}) and associated wavelength (about 1 Å).

We recall that thermal fluctuation amplitudes are of the order of 1 Å, so that not only do thermal neutron energies match thermal fluctuation energies, but also thermal neutron wavelengths match thermal fluctuation amplitudes. In contrast, the wavelength of IR radiation is of the order of 1000 Å. There is, therefore, an important difference between neutron and optical spectroscopy, because neutron experiments, as well as yielding information on excitation frequencies, also yield fluctuation amplitudes, i.e., they allow the direct determination of the *dispersion relations* of thermal excitations, the relations between wavelength and frequency of the corresponding waves. This unique physical property provided an impetus to develop neutron sources and spectrometers dedicated to the study of excitations in solids and liquids.

The neutron was discovered in **1932** by **J. Chadwick**. Neutrons are emitted spontaneously by certain radioactive nuclei and various elements undergo fission when bombarded by neutrons emitting additional neutrons. Because they are electrically neutral, neutron beams penetrate deeply into matter. These and other properties, as we have seen above, make the neutron a particularly useful probe for investigating structure and dynamics at the molecular level.

In **1935**, **J. R. Dunning** showed neutrons could be sorted according to their velocity by using "choppers," rotating discs with slits, thus setting the basis for the development of time-of-flight spectroscopy decades later.

Elsasser, H. v. Halban, and **P. Preiswerk**, and **D. P. Mitchell** and **P. N. Powers**, demonstrated diffraction of neutrons from a radium–beryllium source in **1936**, and thus their wave nature according to **de Broglie**'s relation, paving the way for the use of neutrons in crystallography.

Following the developments in nuclear science stimulated by the tragedy of World War II, neutron beams from pile reactors became available for diffraction experiments and crystallography. **C. Shull** performed the first neutron diffraction experiments to investigate material structures. The first inelastic neutron scattering experiments were performed by **B. Jacrot** and published in **1955**. He used the time-of-flight method to measure the energy distribution of neutrons scattered by a copper sample. **B. N. Brockhouse** invented the triple-axis spectrometer and measured vibrations in solids by neutron scattering. **Shull** and **Brockhouse** shared the Nobel Prize for physics in **1994**.

In the **late 1960s** and **early 1970s**, high-flux neutron beam reactors were dedicated to condensed matter physics and chemistry and biophysics. Neutron spectrometers were built with increasing energy resolution, based on time of flight, back-scattering, and later the spin echo technique. Experiments on protein dynamics became possible and contributed strongly to our present understanding (see Chapter A3). The high-flux reactor of the international Institut Laue Langevin (ILL) in Grenoble, France, delivered its first neutron beams in **1973**; the ILL remains the foremost neutron scattering center in the world due to the high flux of its neutron beams and its variety of instrumentation.

Since **2000**, neutron spectroscopy experiments on biological macromolecules have been performed at several neutron scattering centers around the world. We are now waiting for the second-generation neutron spallation sources that should become operational in the next decade and which should significantly extend the possibilities of the method for characterizing macromolecular dynamics.

I2.2 THEORY

The treatment of energy-resolved neutron scattering offered in this chapter uses many concepts introduced in Chapters A3 and G1.

I2.2.1 Momentum and Energy, Distance Traveled, and Time

We recall the expressions for neutron momentum and energy in terms of wavelength, velocity, and frequency in Comment I2.1.

COMMENT I2.1 NEUTRON MOMENTUM AND ENERGY (RECALLED FROM CHAPTER G1)

Momentum of a neutron:

$$\mathbf{p} = mv = \hbar k$$

Wavelength:

$$\lambda = \frac{h}{p} = \frac{2\pi}{k}$$

Energy:

$$E = \hbar\omega$$

Dispersion relation:

$$E = \frac{\hbar^2 k^2}{2m} = \frac{p^2}{2m}$$

Scattering neutrons from a sample and observing their energy and momentum changes is, in principle, a direct way to obtain information on the atomic motions in that sample. Because we have chosen their direction and wavelength or velocity carefully, the incident neutron particles have well-defined energy and momentum. Neutron spectroscopy is based on the observation of the momentum and energy of scattered neutrons, by measuring their scattering angle and wavelength or velocity, respectively. The momentum ($\mathbf{p}$) and energy ($\hbar\omega$) gained by the neutron from the sample in the scattering process are then calculated simply from conservation laws (Fig. I2.1):

$$\hbar Q = \Delta \mathrm{p} = \hbar(\mathbf{k}_1 - \mathbf{k}_0) \tag{I2.1a}$$

$$\hbar\omega = \hbar\omega_1 - \hbar\omega_0 \tag{I2.1b}$$

It is usual to drop Planck's constant in Eq. (I2.1) and write momentum and energy changes, respectively, as the scattering vector $\mathbf{Q}$ and angular frequency ω.

Note that incident and scattered wave vectors and scattering vector in Fig. I2.1 and Eq. (I2.1) refer to the same quantities as the ones already discussed in Chapter G1. In diffraction, however, only elastic scattering, for which the energy change is zero and $|\mathbf{k}_1| = |\mathbf{k}_0|$, is taken into account. In spectroscopy we deal with the more general case in which *energy as well as momentum* can be exchanged between the moving atoms and the neutrons.

In a similar way to scattered intensity $I(\mathbf{Q})$ in diffraction being expressed as a function of $\mathbf{Q}$, in spectroscopy we write the scattered intensity as $I(\mathbf{Q}, \omega)$. The information we seek on a system of moving atoms is the vector of each atomic displacement as a function of time (*how far each atom moves in a given direction in a given lapse of time*).

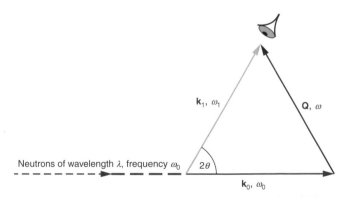

Fig. I2.1 A neutron beam of wavelength λ_0 and corresponding energy $\hbar\omega_0$ incident on a sample S. Scattered waves of wavelength λ_1 and energy $\hbar\omega_1$ are observed by a detector (the "eye") at an angle 2θ with respect to the incident beam. The incident and scattered wave vectors $\mathbf{k}_0$, $\mathbf{k}_1$ have amplitudes $2\pi/\lambda_0$, $2\pi/\lambda_1$, respectively. $\mathbf{Q} = \mathbf{k}_1 - \mathbf{k}_0$, is the scattering vector, and the energy increase of a neutron in the scattering process is $\hbar\omega = \hbar\omega_1 - \hbar\omega_0$.

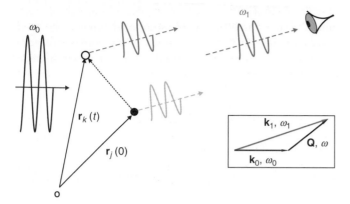

Fig. 12.2 Interference of waves scattered by moving atoms. The black atom is at the position given by the vector $\mathbf{r}_j$ at time zero; the white atom is at the position given by the vector $\mathbf{r}_k$ at time t. The incident beam of frequency ω_0 is in red. As the neutron wave passes through the sample it is scattered by both atoms. Waves (blue and green) scattered by the atoms interfere to yield the purple wave. Because of the energy exchanged between the neutron and moving atoms the frequency ω_1 of the purple wave is different from ω_0. The inset shows the diffraction diagram with the incident and scattered wave vectors and momentum transfer (see Fig. 12.1).

By analogy with scattered wave interference in diffraction methods discussed in Chapter G1, $I(\mathbf{Q}, \omega)$ is the result of the interference of waves scattered by atoms at different positions in space and at different times (Fig. I2.2). The neutron wave speed approximately matches the speed of the atomic motions, so that it can be diffracted by both atoms shown in the figure at different times.

The momentum transfer $\mathbf{Q}$ gives us information on the length scale of fluctuations, and the energy transfer ω gives us information on their timescale. In a similar way to the scattering vector $\mathbf{Q}$ being the reciprocal of a real-space vector $\mathbf{r}$ in the Fourier transform procedure (see Chapter G1), the frequency ω is the reciprocal of time t (see Chapter A3). As in the treatment of diffraction from a static group of atoms, the information in terms of $\mathbf{r}$ and t is calculated from the scattered intensity $I(\mathbf{Q}, \omega)$ by a succession of Fourier transforms.

12.2.2 The Dynamic Structure Factor, Intermediate Scattering Function and Correlation Function

The scattered intensity $I(\mathbf{Q}, \omega)$ is expressed in absolute units as a *double differential cross-section*:

$$\frac{d^2\sigma}{d\Omega d\omega} = \frac{I}{I_0} \tag{I2.2}$$

where I is the number of scattered neutrons per unit time in solid angle $d\Omega$, with an energy change $d\omega$, and I_0 is the incident flux (neutrons per unit area per unit time). Recall from Chapter G1 that the units of cross-section in

diffraction are *area* or $(length)^2$; the units of the double differential cross-section are $(length)^2/energy$.

The scattered intensity is related to the sample dynamic properties via the *dynamic structure factor* $S(\mathbf{Q}, \omega)$:

$$\frac{d^2\sigma}{d\Omega d\omega} = \frac{k_1}{k_0} N b^2 S(\mathbf{Q}, \omega) \tag{I2.3}$$

The equation is written for a system containing N atoms of the same type with scattering length b. Equation (I2.3) contains a further complication, when compared with simple diffraction, in the k_1/k_0 term. Because of the energy difference, the velocities of incident and scattered neutrons (proportional to k_0 and k_1, respectively) are different and the k_1/k_0 term accounts for the difference between the rates of the incident and scattered waves entering and exiting the sample, respectively.

Following the wave interference arguments of Section I2.2.1, the dynamic structure factor is the double Fourier transform of a *space-time correlation function* (defined below). The transformation is performed in two steps. First, integrating over time

$$S(\mathbf{Q}, \omega) = \frac{1}{2\pi} \int_{-\infty}^{+\infty} (I(\mathbf{Q}, t) \exp(-i\omega t) dt \tag{I2.4}$$

where $I(\mathbf{Q}, t)$ is called the *intermediate scattering function*. Second, integrating over space

$$I(\mathbf{Q}, t) = \int_{-\infty}^{+\infty} g(\mathbf{r}, t) \exp(-iQ \cdot \mathbf{r}) d\mathbf{r} \tag{I2.5}$$

where $g(\mathbf{r}, t)$ is an atomic correlation function in space and time (see Chapter D4):

$$g(\mathbf{r}, t) = \frac{1}{N} \sum_{j,k}^{N} \langle \delta(\mathbf{r} - |\mathbf{r}_j(0) - \mathbf{r}_k(t)|) \rangle \tag{I2.6}$$

where the sum is over all atoms and the angular brackets refer to the thermal average. The Dirac delta function in Eq. I2.6 is zero except when the position of atom j at time zero and the position of atom k at time t are separated by the vector $\mathbf{r}$ in the integration volume, where the function will be equal to 1. The correlation function, therefore, counts the fraction of atoms that are separated by the vector $\mathbf{r}$ after a lapse of time t. Recall that a similar formulation is used in the analysis of dynamic light scattering (Section D4.3).

Combining Eqs. (I2.5) and (I2.6), we find

$$I(\mathbf{Q}, t) = \frac{1}{N} \sum_{jk} \langle \exp[i\mathbf{Q} \cdot \mathbf{r_k}(t) \; \exp[-i\mathbf{Q} \cdot \mathbf{r_j}(0)] \rangle \tag{I2.7}$$

12.2.3 Coherent and Incoherent Cross-Sections

Coherent and incoherent scattering due to neutron spin was discussed in Chapter G1. We recall the relation between the total cross-section and the coherent and incoherent scattering lengths and cross-sections:

$$b_{inc}^2 = \langle b^2 \rangle - \langle b \rangle^2 \tag{I2.8}$$

where $\langle b^2 \rangle$ is b_{total}^2 and $\langle b \rangle^2$ is b_{coh}^2.

By using Eq. (I2.8), we separate the $b^2 S(\mathbf{Q}, \omega)$ term in Eq. (I2.3) into coherent and incoherent terms:

$$b_{total}^2 S(\mathbf{Q}, \omega) = b_{coh}^2 S_{coh}(\mathbf{Q}, \omega) + b_{inc}^2 S_{inc}(\mathbf{Q}, \omega) \tag{I2.9}$$

Equation (I2.9) introduces the coherent and incoherent dynamical structure factors.

Coherent scattering occurs when waves scattered by *different* atoms interfere, whereas incoherent scattering is scattering from a single atom. When the atom is motionless, it behaves like a point scatterer for neutrons so that the incoherent scattering is constant as a function of $\mathbf{Q}$.

A very interesting situation arises, however, when an atom is moving. The wave scattered by the atom at time zero interferes with the wave scattered by the *same atom* at time t, so that the displacement of the atom in the time lapse is reflected in a $\mathbf{Q}$ dependence of the scattered intensity. In the notation of Fig. I2.2, there is interference of waves from atom k at time zero and position $\mathbf{r}_k(0)$ and the same atom k at time t and position $\mathbf{r}_k(t)$. For incoherent scattering, therefore, the correlation function $g(\mathbf{r}, t)$ only contains *autocorrelation* terms for which $j = k$. The intermediate *incoherent* scattering function is given by

$$I_{inc}(\mathbf{Q}, t) = \frac{1}{N} \sum_k \langle \exp[i\mathbf{Q} \cdot \mathbf{r}_k(t)] \exp[-i\mathbf{Q} \cdot \mathbf{r}_k(0)] \rangle \tag{I2.10}$$

Similarly, the incoherent dynamic structure factor reports on the motions of individual atoms.

The coherent dynamic structure factor is sensitive to interference between waves scattered by different atoms according to their positions in space and as they move in time; it is, therefore, sensitive to *collective dynamics* in the sample, as, e.g., when atoms move coherently in a normal mode of vibration.

I2.3 APPLICATIONS

Most neutron spectroscopy applications to biological macromolecules are based on incoherent scattering for the practical reason that the incoherent cross-section of the hydrogen nucleus is more than an order of magnitude larger than the total scattering cross-sections of the other atoms. The information obtained, therefore, is on the motions of individual hydrogen atoms in the structures. It is nevertheless very useful information because hydrogen atoms are distributed homogeneously in biological macromolecules. In the picosecond–nanosecond timescale of the neutron experiments, hydrogen atoms reflect the motions of the larger groups (methyl groups, amino acid side-chains, etc.) to which they are bound, so they are a good gage of molecular dynamics. The data can also be used to test collective dynamics models through simulations of their effects on individual motions. Furthermore, since

the incoherent cross-section of deuterium, ^{2}H, is very much smaller than that of ^{1}H, isotope labeling can be used to reduce the contribution of parts of a complex structure in an approach similar to contrast variation in small-angle scattering (SAS) (see Chapter G2).

I2.3.1 Energy and Time Resolution

The *energy resolution* of an experiment is the smallest energy change ω_{min} that can be measured. The *energy range* is given by the maximum ω value. In the reciprocal time frame, the energy resolution corresponds to the *longest* lapse of time t_{max} over which the experiment is sensitive. Consider an atom moving over a distance equal to its own diameter in time t. If t is shorter than t_{max}, then there is interference between waves scattered by the atom in its two end positions and the displacement is observed. If t is longer than t_{max}, then the atom appears to be immobile.

The energy resolution of neutron spectrometers is usually given in electron volts (eV). A resolution of 1 µeV corresponds to a maximum time of about 1 ns, 10 µeV to 0.1 ns, etc. (see Section I2.4).

I2.3.2 Space-Time Window

The elastic peak in a diffraction or spectroscopy experiment is the peak of radiation scattered without energy change, i.e., for $\omega = 0$. In an ideally perfect spectrometer the elastic peak is an infinitely narrow Dirac delta function. In a real spectrometer the width of the elastic peak corresponds to the energy resolution (Fig. I2.3). Quasi-elastic and inelastic scattering, which are also shown in the figure, are discussed further below.

A neutron spectrometer opens *a window in space and time* defined by the minimum and maximum scattering

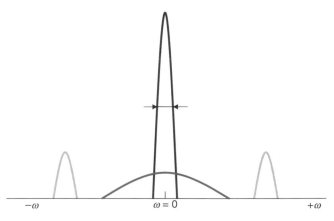

Fig. I2.3 The elastic peak (in red, the energy resolution is the full width at half height shown by the double arrow), quasi-elastic scattering (blue), and inelastic scattering (green) in a neutron spectroscopy experiment (see text). The intensity peaks are observed at a given value of $\mathbf{Q}$. Peak shapes are schematic, not realistic functions.

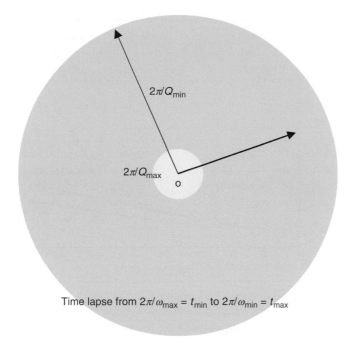

Time lapse from $2\pi/\omega_{max} = t_{min}$ to $2\pi/\omega_{min} = t_{max}$

Fig. I2.4 A window in space and time. The displacement **r** of an atom from the origin O in a time lapse between the minimum and maximum time given by the energy range and resolution, respectively, is observed if it lies in the light-blue area of the circle which is defined by the scattering vector range.

vector values, Q_{min} and Q_{max}, accessible, and by the energy resolution and range (Fig. I2.4).

Referring to Fig. I2.4, the motion of an atom is observed only if it lies in the light-blue part of the diagram. If the atom remains within the yellow circle it appears as an immobile point and its incoherent scattering is a constant as a function of **Q**. If its motion takes it outside the light-blue circle in the time lapse of the experiment, its incoherent scattering peaks close to **Q** = 0, too close to the direct beam to be observed.

We now recall the properties of Fourier transforms discussed in Chapter A3 to discuss the information content in $S(\mathbf{Q}, \omega)$. At **Q** = 0, the intensity is sensitive to all distances in the sample (in practice, up to a maximum distance given by the minimum Q value) but cannot distinguish between them. Displacements $2\pi/Q$ are resolved at a scattering vector value Q. Analogously, at ω = 0, the intensity is sensitive to events taking place in the total time lapse up to the maximum time given by the energy resolution (minimum energy transfer observable) but cannot distinguish between different times within that time frame. An event taking place during a time $2\pi/\omega$ is resolved at energy transfer ω.

The Space-Time Window as a Filter

Samples studied by neutron spectroscopy are usually complex, displaying motions on different time and length scales. A pure protein solution, for example, contains: (1) atoms that fluctuate while remaining firmly bound within

COMMENT I2.2 EXPERIMENTS IN H_2O SOLUTION AND THE SPACE-TIME WINDOW

A neutron spectrometer with an energy resolution of about 10 μeV, corresponding to a time lapse of about 100 ps, is not sensitive to water diffusion in a protein solution sample. A hydrogen atom bound to the protein fluctuates over about 1 Å, whereas the mean square displacement of a freely diffusing water molecule (calculated from its translational diffusion coefficient) is about 100 $Å^2$, well outside the space-time window of the experiment (Tehei *et al.*, 2001; Gabel, 2005).

the macromolecule; and (2) atoms in water molecules, which diffuse freely. Because neutron scattering is strongly dominated by the incoherent scattering of ^1H, experiments to study protein dynamics were performed in D_2O solutions, to minimize water scattering. This was not very satisfactory, however, because D_2O itself can modify dynamics, since the hydrogen bonds and deuterium bonds have different properties. Experiments on proteins in H_2O solution became possible when it was realized that the motion of freely diffusing water and hydrogen atoms bound to the protein are so different that they can be separated by choosing spectrometers with appropriate space-time windows (Comment I2.2).

I2.3.3 Q Dependence of the Elastic Intensity

We recall from Part G that the Debye–Waller factors in crystallography define the fluctuations in atomic positions due to thermal energy, from the scattered intensity dependence on **Q**. Briefly, the motion of each atom in the structure describes a cloud (also called the *thermal ellipsoid*), which is not time resolved in the elastic intensity. Scattering by the cloud has a **Q** dependence (a *form factor*) according to its shape. Similarly, the dependence of the elastic incoherent intensity on **Q**, even though it cannot resolve the motion of the individual atom as a function of time, contains information on the shape of the cloud traced out by the atom within the time window of the experiment (Fig. I2.5).

In the case of a sample structure that is not aligned in a particular direction, and for motion displacements that are well contained within the space-time window (*localized motion*), the atomic motion ellipsoids take up random orientations. The problem is strictly analogous to that of calculating $I(Q)$ in SAS of a particle in solution (Chapter G2). The particle in our case is the motion ellipsoid of an individual atom. We can then apply a Gaussian approximation to interpret $I(Q)$ similar to the Guinier approximation in SAS, at Q values for which $QR_G \sim 1$, where R_G is the radius of gyration of the motion ellipsoid,

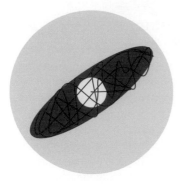

Fig. I2.5 Localized motion of an atom (black line) describing an ellipsoid (red) within the space-time window. A particularly complex diffusive pathway is shown in the example, but a more oscillatory path would be expected for an atom anchored within a macromolecule.

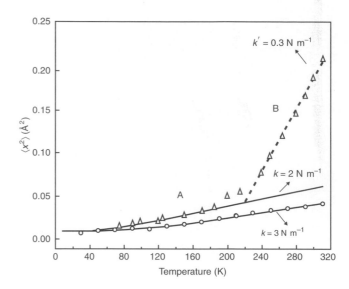

Fig. I2.6 Mean square fluctuations as a function of absolute temperature, in a myoglobin powder, hydrated by heavy water D_2O (Δ) and in a trehalose glass (o). The break in slope at about 180 K is called a dynamical transition. Effective force constants, k and k' were calculated for different parts of the curves as explained in the text. (From Zaccai, 2000, using data from Doster et al., 1989, and Cordone et al., 1999). (Figure reproduced with permission from *Science*.)

$$I(Q) = I(0) \exp\left(-\frac{1}{3} R_G^2 Q^2\right) \quad (12.11)$$

In the analysis of the incoherent neutron scattering intensity, however, it is conventional to use the mean square fluctuation $\langle u^2 \rangle$ notation instead of the radius of gyration notation, where $\langle u^2 \rangle = 2R_G^2$; note that the mean square fluctuation refers to the full amplitude of the motion, while the radius of gyration, R_G, refers to the displacement from the center of mass of the motion of the ellipsoid:

$$I_{\text{elastic}}(Q) = I(0) \exp\left(-\frac{1}{6} \langle u^2 \rangle Q^2\right) \quad (12.12)$$

The mean square fluctuation is calculated from the slope of the $\ln I_{\text{elastic}}(Q)$ versus Q^2 linear plot.

Mean Square Fluctuations and Effective Force Constants

In an *elastic temperature scan* the neutron incoherent elastic intensity $I_{\text{elastic}}(Q)$ is measured for different temperatures, and the mean square fluctuation plotted as a function of temperature (Fig. I2.6). Mean square fluctuations against absolute temperature T for myoglobin powders are shown in the figure; the elastic intensity data from which they were calculated were collected for Q values around 1 Å^{-1} with an energy resolution of about 10 μeV, i.e., the mean square fluctuations refer to about 1 Å amplitude motions taking place in 100 ps or less. They correspond to the thermal dynamics of the atoms in the protein.

The plot is discussed in Section A3.3.3. The temperature axis (T) is, in effect, an energy scale ($k_B T$, where k_B is Boltzmann's constant) for the motions. We recall that for a simple harmonic oscillator the square of the amplitude is proportional to the energy, with a proportionality constant related to the spring force constant. The straight-line dependence of $\langle u^2 \rangle$ versus T at low temperatures, therefore, indicates that the mean atomic motions within the protein in this temperature range can be described in terms

of a mean simple harmonic oscillator with a mean force constant, which can be calculated from the slope of the line:

$$\langle k \rangle = 2k_B \left(\frac{\mathrm{d}\langle u^2 \rangle}{\mathrm{d}T}\right)^{-1} \quad (12.13)$$

where the $\langle x^2 \rangle$ value plotted in Fig. I2.6 is equal to $\langle u^2 \rangle / 6$. Equation (I2.13) is an extremely useful relation because it allows quantification of the forces that maintain an atom within a macromolecular structure.

Above what has been called a dynamical transition at about 180 K, atomic motions are no longer harmonic, but an effective force constant $\langle k' \rangle$ can still be obtained from the slope of the line by applying a quasi-harmonic approximation (Comment I2.3). The force constants associated with different parts of the plot are shown in Fig. I2.6. There is a "softening" of an order of magnitude between the protein structure in the trehalose glass and above the dynamical transition in the hydrated state, the force constant decreasing from 3 to 0.3 N m^{-1}. It is interesting to note that values in the same range are obtained by direct force measurements in single-molecule manipulation experiments (see Part F).

The neutron elastic temperature scan uniquely allows a direct probe of the mean forces acting on atoms in a macroscopic sample. Such scans have been used effectively to explore stabilization forces in pure protein samples and even *in-situ* in live bacterial cells, illustrating the potential

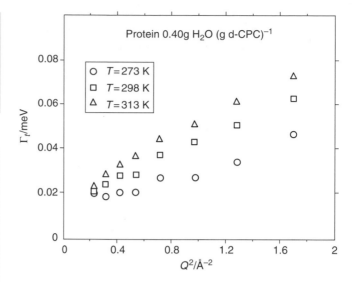

Fig. I2.7 Translational diffusion line width of the protein hydration water (H_2O) in a sample of fully deuterated C-phycocyanin (purified from algae grown in D_2O). The meV energy range of the spectrometer as well as the full deuterium labeling of protein hydrogen atoms ensured that the scattering signal was predominantly from the water. Data are shown for three temperatures. The sample was a powder containing 0.4 g of water per gram of protein. (From Bellissent-Funel *et al.*, 1996; reviewed by Gabel *et al.*, 2002.)

of the method for the characterization of the crowded internal cell environment (Comment I2.4).

I2.3.4 Quasi-Elastic Scattering and Diffusion

As we start to look outside the elastic peak, we gain information on the time constants of different dynamics processes. The *quasi-elastic scattering* is shown as the blue curve in Fig. I2.3. As its name suggests, it is centered on zero energy transfer and appears as a broadening of the elastic peak. Similarly to the case of dynamic light scattering (Chapter D4), quasi-elastic neutron scattering arises from a diffusing particle, with the energy transfer width of the curve inversely proportional to a relaxation time τ.

In the simplest case of linear diffusion obeying Fick's law (see Chapter D1), the intermediate scattering function calculated from the correlation function is written:

$$I(Q,t) = \exp\left(-\frac{t}{\tau}\right) = \exp\left(-DQ^2 t\right) \qquad (I2.14)$$

where D is the diffusion coefficient and $DQ^2 = 1/\tau$ (τ is the relaxation time). We recall that the Fourier transform of an exponential decay function is a Lorentzian function (Chapter A3), so that the dynamic structure factor of the quasi-elastic scattering is given by (from Eq. (I2.4)):

$$S(Q,\omega) = \frac{1}{\pi}\frac{DQ^2}{\left[DQ^2\right]^2 + \omega^2} \qquad (I2.15)$$

which describes a curve of full width at half maximum Γ equal to $2DQ^2$ in frequency units.

If the diffusion model is appropriate, plotting experimental $\Gamma(Q)$ versus Q^2 yields a straight line passing through

zero from the slope of which it is possible to calculate the translational diffusion coefficient. In practice, however, a simple diffusion model is more likely to be valid only in a certain range of Q. For example, in the case of confined motion, simple diffusion will be hampered at small Q values corresponding to the dimensions of the confinement volume. We can see this in Fig. I2.7, in which measured curve width values, assuming a single Lorentzian fit to the quasi-elastic scattering of protein hydration water, are plotted versus Q^2 for different temperatures. Clearly at small Q the data do not lie on straight lines passing through the origin. Diffusion coefficients could nevertheless be calculated from the slopes at high Q, quantifying the slowing down of protein hydration water with respect to bulk water, at three different temperatures.

Considerably more complex models than simple diffusion have been used in the analysis of quasi-elastic scattering data, with respect to the behavior of water in different biological environments, as well as to the slower motions of macromolecule-bound atoms, leading to quite a sophisticated understanding of the phenomena involved. These are discussed in Gabel *et al.* (2002).

I2.3.5 Inelastic Scattering and Vibrations

In Fig. I2.3, inelastic neutron scattering is shown in green. Similarly to the Stokes and anti-Stokes lines in light scattering (see Section E3.2), there are peaks for neutron energy gain (at positive values of ω) and energy loss (at negative values of ω). Although the peaks in the figure are shown as symmetrical, this is not the case experimentally, because the probability of scattering with neutron energy gain is higher.

As with the lines observed in light scattering, the neutron peaks arise from energy exchange with atomic vibrations in the sample. In contrast to light scattering, however, there are no selection rules for neutrons since the scattering is due to a direct interaction with the moving nuclei, and not with electric dipoles or other such effects. Also, because the wavelength of light is large compared with the fluctuation amplitudes, light scattering experiments are effectively limited to measurements at zero Q values. The wavelengths associated with neutrons of comparable energy are similar to the fluctuation amplitudes, and the whole Q range can be examined.

Coherent inelastic scattering arises from the interaction between the neutron and collective vibrational modes, which can be seen as waves of given frequency and momentum propagating through an ordered sample. These waves are called *phonons*, and the neutron–phonon scattering interaction obeys energy and momentum conservation rules. Inelastic neutron scattering has been extremely powerful for the determination of phonon *dispersion relations*, the functions relating the energy with the momentum of the vibration waves. Phonon propagation has been observed in collagen samples and in oriented membrane samples but, so far, suitably large protein crystals have not been available for phonon dispersion measurements by neutron scattering.

In the case of *incoherent inelastic scattering*, the peaks arise from the vibrations of individual atoms (here, too, dominated by the hydrogen atoms because of their higher scattering cross-section). Inelastic incoherent spectra have been measured for a wide variety of proteins under different conditions. The results have guided the development of molecular dynamics simulations (Chapter I1) and have been essential in understanding the forces acting in a protein structure and the important role of the hydration environment. It is interesting to note that because of the $Q{\cdot}r$ term in the intermediate scattering function (Eq. (I2.5)) the experimental intensity is dominated by the larger vibration amplitudes and, therefore, by the lower-frequency modes. Recall from Chapters I1 and A3 that the first few normal modes in a protein dominate the atomic displacement amplitudes. Since the effect of a vibration mode on biological activity is likely to increase with the extent of the atomic displacements, we find there is a good match between the sensitivity to amplitude of the neutron scattering measurements and biological relevance.

Neutron incoherent inelastic scattering spectra of wet and dry powder samples of the protein myoglobin at low and high temperature, measured with an energy resolution of 30 μeV, are shown in Fig. I2.8. The elastic peak below 0.1 meV, fitted by the full line, represents the experimental

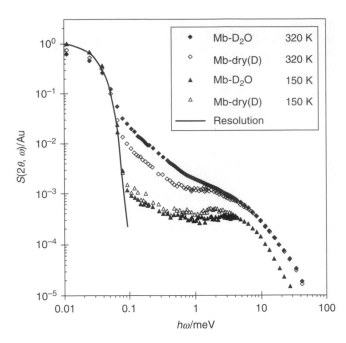

Fig. I2.8 The dynamical structure factor expressed in arbitrary units in terms of scattering angle 2θ instead of scattering vector measured for D_2O hydrated (0.35 g g^{-1}) and dried (from D_2O so that the exchangeable hydrogen atoms are deuterated) at 150 K and 320 K (Diehl *et al.*, 1997, reviewed by Gabel *et al.*, 2002).

energy resolution. The inelastic scattering is best seen at low temperature with a peak at 2 meV (corresponding to a period of a few picoseconds) for the hydrated sample, which moved to lower frequencies when the protein was dried. This low-frequency peak in the protein vibrations has been simulated approximately by molecular dynamics calculations. The fact that the hydrated sample at low temperature is stiffer than the dry one has been interpreted in terms of the hydrogen (deuterium) bond network surrounding the protein. At high temperature, the inelastic peak appears at about the same frequency values, and quasi-elastic scattering from diffusional rather than vibrational motions fills in the gap between 0.1 and 1 meV. The higher level of quasi-elastic scattering from the wet sample indicates the contribution of the solvent diffusion. Within the context of the physical model for protein dynamics (see Chapter A3), the quasi-elastic scattering from the protein at higher temperatures has been suggested to arise from the sampling of different conformational substates.

12.4 SAMPLES AND INSTRUMENTS

Relatively large samples (of the order of a few hundred milligrams) are required for neutron spectroscopy, because incident fluxes and scattering cross-sections are low. Samples for the triple-axis spectrometer (discussed below) must be crystalline or show preferred orientation as in the case of membranes or fibers. Ordered samples as well as disordered hydrated powder or solution samples can be used on the other instruments. Powders of hydrated proteins or other macromolecules do not contain free bulk water and are suitable for measurements over a wide temperature range, including below the freezing point of water. Experimental data and MD calculations have shown that provided there is sufficient hydration, the internal dynamics of proteins is similar in powder samples and in solution. Powder samples are also suitable for studies of the effects of different hydration levels on macromolecular dynamics. Care must be exercised with macromolecular solution samples in order to account for the contribution of the solvent to the signal, as well as for effects due to diffusion of the macromolecular particles themselves.

The dynamics of macromolecule and solvent or of different parts of a macromolecule can be measured separately by using H_2O/D_2O exchange and/or specific H-D labeling. Because of important applications in neutron scattering and NMR, specific deuteration methods are in full development for the labeling of amino acids within proteins or domains in complex structures. Neutron spectroscopy has been used successfully to study the dynamics of hydrogen labels in a fully deuterated natural membrane (see Fig. A3.29).

Below we present neutron spectrometers in order of decreasing energy resolution width (increasing timescale): ~1 meV (ps) for triple-axis spectrometry to ~10 μeV (100 ps) for time-of-flight, spectrometry to 1 μeV (1 ns) for back-scattering spectrometry to 10 ns for spin echo spectrometry, which measures dynamics directly in the time domain.

The Triple-Axis Spectrometer

B. Brockhouse was awarded the Nobel Prize in 1994 for the invention of the triple-axis spectrometer. By measuring scattered neutrons with accurate values of momentum and energy, the triple-axis spectrometer allowed fundamental discoveries in solid-state physics through the determination of phonon dispersion curves in a wide variety of materials. The principles underlying a triple-axis spectrometer are shown schematically in Fig. I2.9. The basic principle of the instrument is the selection of specific wavelengths by Bragg reflection off *monochromator* and *analyzer* crystals. On the first axis, a monochromator crystal M diffracts neutrons of a given wavelength (blue line) from the incident beam (red line). The "blue line" neutrons are scattered by the sample S, on the second axis, with an energy change (green line). The wavelength of the "green line" neutrons is measured by Bragg diffraction off an analyzer crystal A on the third axis and scattered toward the "eye" detector. By rotating M, S, and A around their respective axes, it is possible to scan the wavelengths, the diffraction angles of neutrons incident and scattered by the sample, as well as the orientation of the sample reciprocal lattice (see Chapter G1) and thus map the coherent $S(\mathbf{Q}, \omega)$ distribution in great detail.

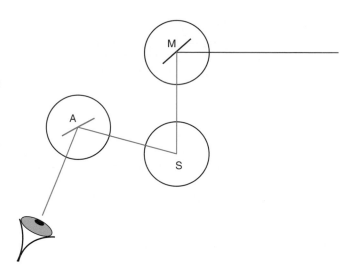

Fig. I2.9 Schematic diagram of a triple-axis spectrometer. M is the monochromator, S the sample, and A the analyzer. The "eye" is the detector. The beam from the neutron source is in red.

Because of the difficulty in producing suitable samples, the application of triple-axis spectrometry to biological samples is still in its infancy.

Time of Flight

In biophysical experiments, time-of-flight spectrometers are mainly used to measure quasi-elastic and inelastic incoherent scattering from proteins. A time-of-flight spectrometer is shown schematically in Fig. I2.10. Choppers are rotating discs with slots on their rims that select neutrons of a given velocity (wavelength). A pair of choppers selects neutrons of wavelength λ but also wavelengths $\lambda/2$, $\lambda/3$, etc. Several choppers in line are required to form a well-shaped pulse of neutrons with a specific velocity. The energy resolution of time-of-flight spectrometers is typically in the 10 μeV to 1 μeV range depending on the incident wavelength chosen, which also determines the Q range of the experiment.

Back-Scattering

Back-scattering spectrometers are based on the fact that Bragg's law is extremely wavelength selective for diffraction with scattering angle $2\theta = 180°$ ("back" scattering). Bragg's law (see Chapter G3) is written

$$2d \sin \theta = \lambda \qquad (\text{I2.16})$$

where d is the crystal spacing, 2θ is the scattering angle, and λ is the wavelength. By taking the derivative of the equation, we can estimate the wavelength spread dependence on scattering angle:

$$2d \cos \theta \cdot \Delta\theta = \Delta\lambda$$
$$\cot \theta \cdot \Delta\theta = \frac{\Delta\lambda}{\lambda} \qquad (\text{I2.17})$$

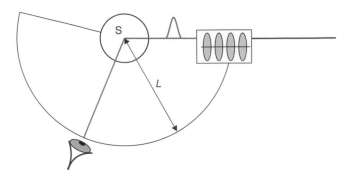

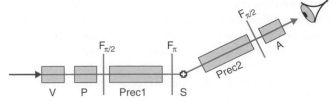

Fig. I2.12 Schematic diagram of a neutron spin echo spectrometer. V is a velocity selector; P a polarizer; $F_{\pi/2}$ and F_{π} are 90° and 180° spin flippers, respectively; Prec1 and Prec2 are paths in which the neutron spins are made to precess by a magnetic field; S is the sample; A an analyzer; and the "eye" is the detector. The beam from the neutron source is in red.

Fig. I2.10 Schematic diagram of a time-of-flight spectrometer. The neutron beam (red line) passes through a series of "choppers" (see text) that shape a neutron pulse (blue), which is incident on the sample S at a defined time. The scattered beam (green) is detected by a detector bank (eye) located at a distance L from the sample. L being known, the velocity of the scattered neutron is calculated from the time of flight between S and the "eye," which detects it. The scattering angle is given by the position of the detector on the circumference of the instrument.

wavelength is analyzed. In back-scattering spectrometers, the scattered beam wavelength is fixed by the choice of analyzer spacing and the incident beam wavelength is varied by scanning the monochromator. The Q value of the scattering is defined by the position of the analyzer crystal and the corresponding detector. Back-scattering spectrometers can attain lower than micro-eV resolution and are used to analyze elastic and quasi-elastic scattering from macromolecular samples.

Spin Echo

Neutron spin echo spectrometers are based on the same phenomenon as spin echo NMR (see Chapter J2). The phenomenon being very sensitive to small energy changes, the neutron spin echo technique achieves a higher energy resolution than back scattering. The measurements, however, are performed in the time domain (similarly to NMR) and the resolution is expressed accordingly (~10 ns) and not in energy units. A neutron spin echo spectrometer is shown schematically in Fig. I2.12. Neutrons are spin 1/2 particles (see Chapter G1) that can be aligned (*polarized*). A polarized neutron beam behaves like a classical magnetic moment and displays Larmor precession when exposed to a magnetic field perpendicular to it (see Section J1.2.2). In a neutron spin echo experiment, the neutron beam (red line in the figure) passes through a velocity selector V, which reduces the white beam wavelength spread, then through a polarizer P, which aligns the spins to produce a *polarized beam*; the spins are rotated by 90° or 180° in spin *flippers* F; magnetic fields along the paths Prec1 and Prec2 (the echo) set the spins precessing in opposite directions, so that in absence of interaction with the sample S, the beam reaches the analyzer A in the same state as when it emerged from P, i.e., the total precession angle comes back to zero. The precessing spins are very sensitive to small energy changes. An exchange of energy with the sample results in a total precession angle different from zero, and the analyzer direction has to be set accordingly to allow the neutron beam to pass to the detector. It can be shown that the method provides a direct measure of the intermediate scattering function, through the relation of $I(Q, t)$ with the

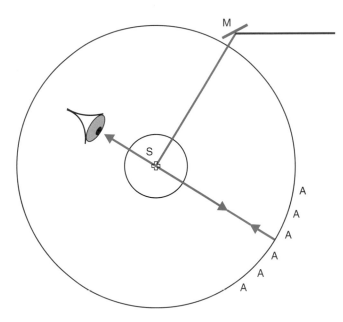

Fig. I2.11 Schematic diagram of a back-scattering spectrometer. The neutron beam (red line) is reflected by a monochromator M, which selects a wavelength (blue line) that is varied during the experiment. After scattering by the sample S, only neutrons of a specified wavelength (green lines) are back scattered by analyzer crystals A and detected by the detector (eye).

Note that $\cot \theta = 0$ for $\theta = 90°$ leading to zero wavelength spread (excellent wavelength resolution) in the back-scattering position regardless of angular spread. A back-scattering spectrometer is shown schematically in Fig. I2.11. Back-scattering spectrometers are "reverse geometry" spectrometers. In usual spectrometers, the incident beam wavelength is fixed and the scattered beam

analyzer direction. The Q value is set by the angle formed by the arm of the spectrometer downstream of the sample. An important advantage of the neutron spin echo method is that all neutrons are counted and beyond the velocity selector there is no selection of neutrons by their energy either before or after scattering by the sample.

The neutron spin echo timescale is suitable for probing diffusion coefficients of macromolecules in solution or within cellular environments, or slow domain movements in macromolecular complexes.

12.5 CHECKLIST OF KEY IDEAS

- The energies of *thermal neutron* beams match vibrational energies in solids and liquids (with periods in the picosecond to nanosecond timescale), and their wavelengths match thermal fluctuation amplitudes (in the ångström range), so that neutron experiments allow the direct determination of the *dispersion relations* of thermal excitations.
- Neutron spectroscopy is based on the observation of the momentum and energy of scattered neutrons, by measuring their scattering angle and wavelength or velocity, respectively.
- Momentum transfer **Q** gives us information on the length scale of fluctuations, and energy transfer ω gives us information on their timescale.
- Scattering intensity is analyzed in terms of an intermediate scattering function $I(\mathbf{Q}, t)$ and its Fourier transform, the dynamic structure factor $S(\mathbf{Q}, \omega)$; the intermediate scattering function is itself the Fourier transform of a space-time correlation function, relating the positions of the atoms in the sample as they move in time.
- Coherent scattering occurs when waves scattered by *different* atoms interfere, whereas incoherent scattering is scattering from a single atom; the incoherent dynamic structure factor reports on the motions of individual atoms; the coherent dynamic structure factor is sensitive to *collective dynamics* in the sample, as when atoms move coherently in a normal mode of vibration, for example.
- The incoherent cross-section of hydrogen nuclei strongly dominates the scattering signal; deuterium labeling enables one to modulate the scattering contribution of different parts of a complex sample; in the picosecond–nanosecond timescale hydrogen atoms reflect the motions of the larger groups (methyl groups, amino acid side-chains, etc.) to which they are bound, so that they are a good gage of molecular dynamics.
- The energy resolution of neutron spectrometers is usually given in electron volts (eV). A resolution of 1 μeV corresponds to a maximum time of about 1 ns, 10 μeV to 0.1 ns, etc.

- A neutron spectrometer opens *a window in space and time* defined by the minimum and maximum scattering vector values, Q_{min} and Q_{max}, accessible, and by the energy resolution and range; the space-time window is a filter that permits one to focus on specific motions in the sample.
- The elastic intensity provides information on localized motions occurring during the maximum time given by the instrumental resolution; mean square fluctuations and effective force constants can be derived from an analysis of the elastic intensity as a function of temperature; the quasi-elastic intensity informs on the correlation times of diffusional motions; the inelastic intensity informs on vibrational modes.
- Relatively large samples (of the order of a few hundred milligrams) are required for neutron spectroscopy, because incident fluxes and scattering cross-sections are low.
- By measuring scattered neutrons with accurate values of momentum and energy, the triple-axis spectrometer allowed fundamental discoveries in solid-state physics through the determination of phonon dispersion curves in a wide variety of materials.
- Time-of-flight spectrometers are used to measure quasi-elastic and inelastic incoherent scattering from macromolecules with an energy resolution in the 0.01–1 meV range; back-scattering spectrometers can attain lower than micro-eV resolution and are used to analyze elastic and quasi-elastic scattering from macromolecular samples.
- Neutron spin echo spectrometers, based on the neutron being a quantum spin 1/2 particle and its interactions with magnetic fields, achieve high energy resolution with a corresponding time domain ~10 ns; they provide a direct determination of the intermediate scattering function $I(\mathbf{Q}, t)$; the timescale is suitable for probing diffusion coefficients of macromolecules in solution or within cellular environments, or slow domain movements in macromolecular complexes.

Suggestions for Further Reading

Bée, M. (1988). *Quasielastic Neutron Scattering*. Bristol: Adam Hilger.

Smith, J. C. (1991). Protein dynamics: Comparison of simulations with inelastic neutron scattering experiments. *Q. Rev. Biophys.*, **24**, 227–291.

Gabel, F., Bicout, D., Lehnert, U., *et al.* (2002). Protein dynamics studied by neutron scattering. *Q. Rev. Biophys.*, **35**, 327–367.

Tehei, M., Franzetti, B., Madern, D., *et al.* (2004). Adaptation to extreme environments: Macromolecular thermal dynamics in psychrophile, mesophile and thermophile bacteria compared, *in-vivo*, by neutron scattering. *EMBO Rep.*, **5**, 66–70.

NUCLEAR MAGNETIC RESONANCE

J

DISTANCES AND ANGLES FROM FREQUENCIES

J1

J1.1 HISTORICAL REVIEW

1924

W. Pauli proposed the theoretical basis for NMR spectroscopy. He suggested that certain atomic nuclei have properties of spin and magnetic moment and, as a consequence, exposure to a magnetic field leads to splitting of their energy levels. **W. Gerlach** and **O. Stern** observed the splitting in atomic beam experiments, providing proof for the existence of nuclear magnetic moments.

1938

I. I. Rabi and colleagues first observed NMR by applying electromagnetic radiation in atomic beam experiments. Energy was absorbed at a sharply defined frequency, causing a small but measurable deflection of the beam. **Rabi** received the Nobel Prize for physics in **1944**.

1946

Research groups led by **F. Bloch** and **E. M. Purcell** reported the observation of proton NMR in liquid water and solid paraffin wax. **Bloch** and **Purcell** shared the **1953** Nobel Prize for physics.

1946

F. Bloch suggested a new method of excitation using a short radio-frequency pulse and in **1949 E. L. Hahn** showed that this did indeed produce a free precession signal. **Hahn** also established that pulse sequences could be used to generate additional information in the form of a spin echo. For many years, however, these methods were of little use to chemists because of the complexity of the signal obtained. In **1956, I. J. Lowe** and **R. E. Norberg** pointed out that the time-domain signal and the frequency-domain spectrum are related by Fourier transformation. The first high-resolution multichannel Fourier transform NMR spectrum was measured by **R. R. Ernst** and **W. A. Anderson**.

1950

W. G. Proctor and **F. C. Yu** observed two unexpected ^{14}N resonance frequencies for NH_4NO_3. At about the same time, **W. C. Dickinson** noticed similar effects for ^{19}F in several compounds. In **1951, J. T. Arnold** and colleagues introduced the term *chemical shift* following the observation of several resonance peaks for ^1H in ethanol, with the relative intensity in each peak corresponding to the relative number of protons in each chemical environment.

1951

H. S. Gutowsky and **D. W. McCall** suggested that interactions between spins of neighboring nuclei were responsible for multiple resonance lines. In **1951, N. F. Ramsey** and **E. M. Purcell** proposed the concept of indirect *spin–spin coupling* or *scalar coupling*. It was found that in certain cases spin coupling failed to produce the expected multiplets, leading to the development of the concept of *chemical exchange*.

Early 1950s

A. W. Overhauser explored the dynamic polarization of nuclei in metals where the electron spin resonance had been saturated. The effect he discovered was called the "Overhauser effect." The potential of nuclear Overhauser enhancement (NOE) signals for providing information on the conformation of molecules in solution was first demonstrated by **F. A. L. Anet** and **A. J. R. Bourn** in **1965**. In **1970, R. A. Bell** and **J. K. Saunders** reported a direct correlation between NOE and internuclear distances, and **R. E. Schrimer** and colleagues demonstrated that relative internuclear distances can be determined quantitatively from NOE measurements on a system containing three or more spins.

By the **mid-1950s** the basic physics of NMR and its potential value in chemistry had been elucidated, and commercial instruments were available. In **1956**, the observation frequency for ^1H NMR spectroscopy on the HR-30 Varian Spectrometer with a 0.7 T electromagnet was fixed by a crystal at 30 MHz. In order to improve sensitivity and increase chemical shift dispersion, commercial instrument development focused on increasing the magnetic field strength. The development of persistent superconducting solenoids (cryo-magnets) in the **early 1960s** constituted a major milestone for NMR applications. In the **late 1990s**, 18.8 T (corresponding to a resonance frequency of 800 MHz for ^1H NMR) spectrometers were installed in many NMR laboratories, with the first 900 MHz (21.1 T) instruments becoming available.

Late 1950s

M. Sauders, **A. Wishnia**, and **J. G. Kirkwood** reported the first NMR spectrum of a protein, and a small number of similar studies were reported in the following decade. Because of technical limitations in sensitivity and spectral resolution, however, these early NMR applications in structural biology did not bear directly on macromolecular three-dimensional structure. The fundamental theory of NMR was published in **1961** in a landmark book, *The Principles of Nuclear Magnetism*, by **A. Abragam**. Contributing to the development of solid state NMR, **E. R. Andrew** and **I. J. Lowe** showed that certain anisotropic interactions could be controlled by magic angle spinning (MAS).

1966

R. R Ernst proposed a Fourier transform method, which provided a major leap forward with respect to the amount of information accessible by NMR. The inherent advantages of greater sensitivity, high resolution, and the absence of line-shape distortions contributed to make Fourier transform spectroscopy the preferred experimental technique in the field.

1971

J. Jeener first suggested the idea of two-dimensional Fourier transform NMR (FT-NMR), based on the Fourier transformation of signals in two independent time domains to yield a plot with respect to two orthogonal frequency axes. In **1975**, **R. R. Ernst** and colleagues reported the first two-dimensional NMR (2D-NMR) ^{13}C spectrum of hexane. This was followed in **1976** by a seminal publication presenting a comprehensive theoretical treatment of 2D-NMR correlation spectroscopy (COSY). **Ernst** received the **1991** Nobel Prize for chemistry for his many contributions to NMR.[1]

1972

P. C. Lauterbur demonstrated the feasibility of macroscopic imaging by NMR. In the same year, **R. Damadian** used the method for investigations of the human body, in particular for cancer detection, paving the way for the non-invasive imaging of entire biological organisms. In **2003**, **Lauterbur** and **P. Mansfield** shared the Nobel Prize for medicine or physiology for contributions to magnetic resonance imaging (MRI).

1983

T. A. Cross and **S. J. Opella** showed that high-resolution structural constraints could be obtained from solid-state

NMR experiments, and the potential of the approach was rapidly established.

1985

K. J. Wüthrich and coworkers reported the complete three-dimensional structure of a protein, BPTI, in solution based on NOE distance constraints only. There has since been spectacular progress in the development and application of NMR methodology to protein structure determination in solution. Because of the growing number of peaks and larger peak widths with increasing molecular mass, the method was initially limited to macromolecules of a few tens of kilodaltons. Isotopic labeling extended the molecular mass range to 40–50 kDa. In **1998**, **Wüthrich** with coworkers discovered that in very high magnetic fields narrow resonance peaks can result from interference between dipole–dipole coupling and chemical shift anisotropy, and proposed the technique called transverse relaxation optimized spectroscopy (TROSY). TROSY experiments should make possible the determination of three-dimensional structures of proteins close to 100 kDa in molecular mass. **Wüthrich** was the **2002** Nobel laureate in chemistry.

Late 1980s

C. Griesinger, **R. R. Ernst**, **H. Oschkinat**, **D. Marion**, **A. Bax**, and **G. W. Vuister** introduced a third frequency dimension in NMR spectra (3D-NMR), and in the **early 1990s L. E. Kay** and coworkers and **G. M. Clore** and coworkers expanded the technique to 4D-NMR. By using isotopic enrichment, 3D- and 4D-NMR have become powerful experimental approaches, widely applicable in structural biology.

2000–2014

NMR has become one of the most powerful spectroscopic techniques in physics, chemistry, and biology. Powerful experimental methods have been devised for observing different NMR phenomena in detail. NMR in structural biology maintains all the typical signs of a young, emerging field of research, with fundamental contributions continuing to be made by many scientists, including **I. Campbell**, **P. Wright**, **S. Grzesiek**, **A. M. Groenborn**, and others. **P. Güntert** and collaborators proposed and developed fully automated protein structure analysis from NMR spectra. **M. Blackledge** and his collaborators applied NMR to characterize partially folded and intrinsically disordered proteins (IDPs). **C. Dobson**, **M. Blackledge**, **A. Bax**, and others further developed NMR to study dynamics. **R. G. Griffin** and **A. Lange** made major contributions to solid-state NMR. The field of NMR proved to be remarkable through the number of revolutionary innovations that have occurred since the first experimental observation of the phenomenon more than 55 years ago. The introduction of the

[1] J. Jeener originated the idea of two-dimensional NMR spectroscopy in 1971. Unfortunately, his first two-dimensional spectra were never published; the only reference to the work is in a set of lecture notes for a summer school.

second frequency dimension constituted a critical step for biological applications. A large variety of experimental schemes were developed to extend NMR applications to the characterization of complex molecules, such as small synthetic polymers, peptides, and sugars. Small proteins and oligonucleotides became accessible to study following the introduction of new procedures. The addition of a third frequency dimension (3D-NMR) was the next important development, with advances in genetic engineering enabling the overproduction of proteins and their labeling in microorganisms with NMR-stable isotopes. Isotope enrichment, in fact, increased NMR resolution sufficiently that it became possible to add a fourth frequency dimension to the spectra.

NMR occupies a very special place in the armory of physical techniques available to biologists. In the autumn of 2014, the PDB contained close to 11 000 NMR-derived structures out of a total of over 100 000. NMR also provides information on protein and nucleic acid dynamics in a time domain spanning from picoseconds to days. Its unique versatility has made NMR one of the most powerful tools in modern structural biology. In this, the importance of high technology and biotechnology cannot be overestimated. It includes not only the capability of producing strong homogeneous magnetic fields, but also the development of sophisticated isotope labeling, outstandingly precise mechanical engineering (for example, for the design and implementation of solid-state NMR sample holders, a fraction of a millimeter in diameter, spinning at the magic angle at rates up to 100 KHz), state-of-the-art electronics, and computing.

J1.2 FUNDAMENTAL CONCEPTS

Nuclear magnetic resonance is a field of spectroscopy based on the absorption of electromagnetic radiation in the radio-frequency region, 10 MHz–1 GHz. In contrast to UV, visible, and IR absorption spectroscopy, which involve outer-shell atomic electrons, NMR arises from the magnetic properties of atomic nuclei, which, when placed in an intense magnetic field, develop the energy states required for absorption to occur.

Figure J1.1 shows the electromagnetic spectrum from the radio-frequency region (frequency $v = 10^4$ Hz) through microwaves, IR, visible, UV, and X-rays to γ-rays ($v = 10^{22}$ Hz). NMR utilizes the low-frequency end of the spectrum. Other spectroscopic methods are concerned with larger energy-level splitting and hence higher frequencies.

In common with optical spectroscopy (Part E), quantum mechanics is essential for a full understanding of NMR in terms of absorption frequencies and nuclear energy states. A classical treatment, however, while limited, may yield a clearer physical picture of the process and of how it is measured.

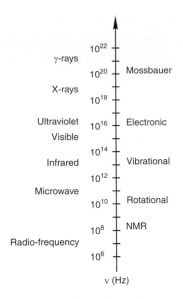

Fig. J1.1 The electromagnetic spectrum from γ-rays ($v = 10^{22}$ Hz) to radio-frequency (10^4 Hz) radiation.

Fig. J1.2 Schematic view of a spherical non-spinning nucleus.

J1.2.1 Quantum Mechanical Description

Magnetic Properties of Nuclei: Nuclear Spin

The concept of *spin* in quantum mechanics cannot be explained rigorously in classical terms. However, the properties of certain nuclei can be understood in terms of a model in which they behave as spherical bodies with the nuclear charge distributed uniformly over their surfaces. A non-spinning nucleus does not have a magnetic moment because there is no circulation of charge (Fig. J1.2). It is said to have a nuclear spin value equal to zero, and does not give an NMR signal. All nuclei with an even mass number and an even nuclear charge Z have a nuclear spin of zero (Table J1.1 and Comment J1.1). Two nuclei of considerable importance in biology, ^{12}C and ^{16}O, are of this type.

It is not as unfortunate as it might seem that the principal isotope of carbon has no NMR signal. Most NMR information is obtained from proton resonance peaks and if ^{12}C had a nuclear magnetic moment, the ^{1}H-NMR spectra of most biological macromolecules would be much more complicated than they are, and probably impossible to

TABLE J1.1 CHARGE Z, NUMBER OF NEUTRONS N AND NUCLEAR SPIN QUANTUM NUMBER I FOR NUCLEI OF PARTICULAR INTEREST IN BIOLOGY			
Nucleus	Z	N	I
^{1}H	1	0	1/2
^{2}H	1	1	1
^{12}C	6	6	0
^{13}C	6	7	1/2
^{14}N	7	7	1
^{15}N	7	8	1/2
^{16}O	8	8	0
^{17}O	8	9	5/2
^{19}F	9	10	1/2
^{31}P	15	16	1/2
^{32}S	16	16	0

COMMENT J1.1 SPIN QUANTUM NUMBER

The spin quantum number of a nucleus is determined by the number of unpaired protons and neutrons it contains. For example, ^{12}C has even numbers of protons and neutrons: Each proton pairs with a proton of opposite sign, as does each neutron, giving a net spin angular momentum of zero ($I = 0$). A nucleus with odd numbers of protons and neutrons (e.g., ^{14}N) generally has an integral non-zero quantum number, because the total number of unpaired nucleons is even, and each contributes 1/2 to the quantum number. The discussion can be extended to nuclei with even numbers of protons and odd numbers of neutrons, or vice versa, which usually have half-integral quantum numbers due to the odd number of unpaired nucleons.

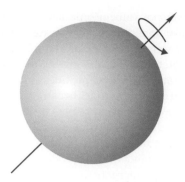

Fig. J1.3 Schematic view of a spinning spherical nucleus.

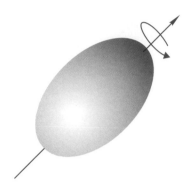

Fig. J1.4 Schematic view of an ellipsoidal (prolate) spinning nucleus.

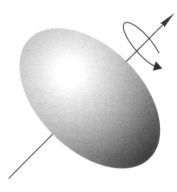

Fig. J1.5 Schematic view of an ellipsoidal (oblate) spinning nucleus.

interpret. Moreover, ^{13}C has a magnetic moment, so that this isotope can be used, either at its low natural abundance concentration or in labeling experiments, for the observation of carbon resonance peaks.

A number of nuclei of particular importance to structural biology (^{1}H, ^{13}C, ^{15}N, ^{19}F, and ^{31}P) have nuclear spin values of 1/2. In the model, they act as though they were *spinning* spherical bodies of uniform surface charge distribution (Fig. J1.3). A spinning nucleus has circulating charge, and this generates a magnetic field so that a nuclear magnetic moment results. The spherical charge distribution ascribed to nuclei with a spin of 1/2 means that a probing charge approaching them experiences the same electrostatic field regardless of the direction of

approach and, therefore, as with the spherical non-spinning nuclei, the electric quadropole moment is zero.

Most magnetic nuclei, however, act as though they were spinning bodies with non-spherical charge distributions and are assigned spin values of unity or larger integral multiples of 1/2. Such nuclei may be considered as approximate ellipsoids spinning about a principal axis (Figs. J1.4 and J1.5). A spinning prolate ellipsoid, with charge uniformly distributed over its surface, presents an anisotropic electrostatic field. The electrostatic work in bringing a unit charge to a given distance is different if the charge approaches along the spin axis or at some angle to it. By convention, the electric quadrupole moment of a nucleus ascribed the shape of a prolate ellipsoid is assigned a value greater than zero. Two nuclei of considerable importance in biology, ^{2}H and ^{14}N, are of this type. Nuclei that behave like charged, oblate ellipsoids also present an anisotropic electric field to a probing charge and by convention are assigned negative electric-quadrupole-moment values. Nuclei of this type include ^{17}O, ^{33}S, and ^{35}Cl.

Spin quantum numbers for nuclei of interest in biology are given in Table J1.1.

Magnetic Quantum Number

An important quantum mechanical property of a spinning nucleus is that the average value of the component of its magnetic moment vector along a defined direction takes up specific values described by a set of magnetic quantum numbers $m = 2I + 1$, in integral steps between $+I$ and $-I$

$$m = I, I - 1, I - 2, ..., -I + 1, -I \qquad (J1.1)$$

where I is the nuclear spin value. The spin angular momentum vector has magnitude $[I(I + 1)]^{1/2}\hbar$ and its z-component is m, where m is given by Eq. (J1.1) (Comment J1.2). Note that the z-axis (the axis of quantization) is arbitrary in the absence of an external magnetic field, so that the spin angular momentum has no preferred direction. The direction of an external magnetic field defines the spin z-axis. The angular momentum component of a spin $\pm 1/2$ nucleus (e.g., ^{1}H, ^{13}C) along the external field has two permitted directions, $I_z = \pm 1/2\hbar$, while a nucleus with $I = 1$ has three possible states, $I_z = 0, \pm\hbar$ (Fig. J1.6).

Nuclear Magnetization

A charged spinning nucleus creates a magnetic field that is analogous to the field produced when electricity flows through a coil of wire. The resulting magnetic moment μ (a vector quantity) is oriented along the axis of spin and is directly proportional to the angular momentum vector $\mathbf{I}$ with a proportionality constant, γ, known as the *gyromagnetic* ratio or, less commonly, the *magnetogyric* ratio:

COMMENT J1.2 BIOLOGIST'S BOX: SPIN ANGULAR MOMENTUM

Bohr's work on the spectrum of the hydrogen atom introduced the postulate that the angular momentum of a system was quantized, i.e., it could only take values which are integer multiples of $h/2\pi$, where h is Planck's constant. It was suggested later by Sommerfeld that the directions of orientation of the electronic angular momentum vector were restricted to certain orientations when the electron was in a closed orbit. In other words, the direction as well as magnitude of the angular momentum vector is quantized. Spin angular momentum $\mathbf{I}$ is a vector quantity, which has magnitude and orientation and should not be confused with the spin quantum number I.

Atom	$\gamma(10^7\,T^{-1}s^{-1})$
^{1}H	26.75
^{13}C	6.73
^{15}N	−2.71
^{19}F	25.18
^{31}P	10.84

TABLE J1.2 GYROMAGNETIC RATIOS OF SOME NMR NUCLEI. THE SI UNITS FOR THESE CONSTANTS ARE RAD T^{-1} S^{-1}

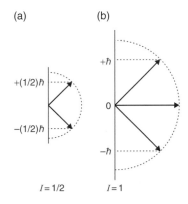

Fig. J1.6 Space quantization of **(a)** spin 1/2 and **(b)** spin 1 nuclei.

$$\mu = \gamma\mathbf{I} \qquad (J1.2)$$

Gyromagnetic ratios of some NMR nuclei are given in Table J1.2. The magnetic moment of a nucleus is parallel (or antiparallel for nuclei with negative γ) to the spin angular momentum, $\mathbf{I}$.

In the absence of an external magnetic field, all $2I + 1$ orientations of a spin-I nucleus have the same energy. This degeneracy is removed when a magnetic field is applied: The energy of a magnetic moment μ in a magnetic field $\mathbf{B}_0$ is minus the scalar product of the two vectors:

$$E = -\mu \cdot \mathbf{B}_0 \qquad (J1.3)$$

In the presence of a strong field, the quantization axis z is no longer arbitrary, but coincides with the field direction z. Therefore

$$E = -\mu_z B_0 \qquad (J1.4)$$

where μ_z is the z component of μ (the projection of μ onto $\mathbf{B}_0$) and B_0 is the strength of the field (the magnitude of $\mathbf{B}_0$) (Fig. J1.7). From Eqs. (J1.1) and (J1.2) $I_z = m$ and $\mu_z = \gamma I_z$; and so

$$E = \mu\hbar\gamma B_0 \qquad (J1.5)$$

Spin 1/2 nuclei give rise to only two states corresponding to $m = +1/2$ and $-1/2$ (Fig. J1.8). The energy spacing between them is given by

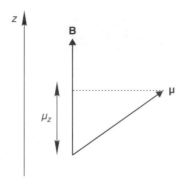

Fig. J1.7 The relationship between the magnetic field $\mathbf{B}_0$, the nuclear magnetic moment μ, and its component along the field direction, μ_z (the scalar product of $\mathbf{B}_0$ and μ).

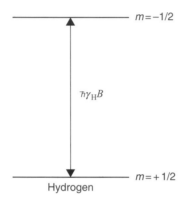

Fig. J1.8 Energy levels for hydrogen ($I = \frac{1}{2}$) nucleus in magnetic field $\mathbf{B}_0$.

$$\Delta E = \hbar \gamma B_0 \qquad (J1.6)$$

or, expressed in terms of frequency,

$$v = \gamma B_0 / 2\pi \, \text{Hz} \qquad (J1.7)$$

If the gyromagnetic ratio is positive (e.g., for ^{1}H and ^{13}C), then the +1/2 state lies lower in energy, and vice versa for negative γ values (e.g., ^{15}N).[2]

Distribution of Nuclei Between Magnetic Quantum States

In the absence of a magnetic field, there is no preference for one or other of the two possible states, for a spin 1/2 nucleus, so that in a large assemblage of such nuclei there are exactly equal numbers with m equal to +1/2 and m equal to –1/2.

When an external magnetic field is applied, positive γ nuclei tend to assume the magnetic quantum number +1/2, which represents alignment with the field; $m = +1/2$ represents a more favorable energy state than $m = -1/2$. The tendency of the nuclei to align with the field is opposed

---

[2] The electron has a spin of ½ and a magnetic moment that results in electron spin resonance in a magnetic field.

by thermal agitation. The equilibrium percentage of nuclei in each quantum state can be calculated as a Boltzmann distribution by using values for the nuclear moment, the external field strength and the temperature:

$$N_{\text{upper}}/N_{\text{lower}} = \exp\left(-\Delta E / kT\right) = \exp\left(-h v / kT\right) \qquad (J1.8)$$

where N_{upper} is the number of nuclei in the higher energy state ($m = -1/2$), N_{lower} is the number in the lower state ($m = +1/2$), and ΔE is the energy difference between the states. Since $h v$ is very much smaller than kT at temperatures normally used in an NMR experiment, there is a small excess of spins in the low state. This can be calculated to be 1 part in 10^9 at normal temperatures. Since this excess is proportional to the signal inducible in the probe, NMR is a very insensitive technique compared with other forms of spectroscopy, where the energy difference is very much larger (Comment J1.3). Substituting Eq. (J1.4) into Eq. (J1.6) gives

$$N_{\text{upper}}/N_{\text{lower}} = \exp\left(-\gamma h B_0 / 2\pi kT\right) \qquad (J1.9)$$

The energy-level diagram for a Boltzman distribution of $2n$ nuclei is shown in Fig. J1.9a. At equilibrium the individual magnetic moment vectors are distributed in two cones about the +z- and –z-axes (Fig. J1.9b). If γ is positive, slightly more than half the nuclei are aligned with the applied field. The excess spin population results in a net magnetization in the +z direction (Fig. J1.9c). Expanding the right-hand side of Eq. (J1.9) as a Maclaurin series, and truncating after the second term, we obtain the important result that

$$N_{\text{upper}}/N_{\text{lower}} = 1 - \gamma h B_0 / 2\pi kT \qquad (J1.10)$$

(a)   (b)   (c)

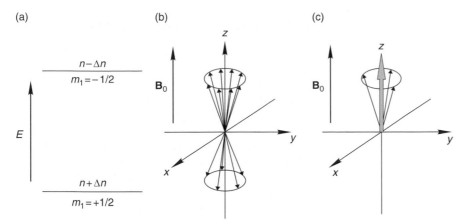

Fig. J1.9 (a) Energy-level diagram for a total of $2n$ nuclei (with positive γ), showing the excess population in the lower-energy state. **(b)** Schematic presentation showing a greater number of spins aligned with the magnetic field (lower-energy state). **(c)** The excess spin population aligned with the field results in a net magnetization in the $+z$ direction. The actual excess is much smaller than shown. (Adapted from King and Williams, 1989.)

The ratio of nuclei in the upper- and lower-energy states is linearly related to the magnetic field. Resonant absorption of radio-frequency energy corresponds to a transition between the lower- and upper-energy states. The number of transitions depends on the population ratio between the two states. Equation (J1.8) shows that the intensity of an NMR signal increases linearly with field strength, leading manufacturers to produce increasingly powerful magnets for NMR.

J1.2.2 Classical Mechanical Description

In order to understand the workings of an NMR experiment, and thus to predict the results, we require a formalism with which to "visualize" the evolution in time of a spin system. If we were to consider a single spin, a quantum mechanical formalism would be required. This is because atomic phenomena do not behave classically, i.e., they do not obey Newtonian mechanics. For example, if we were to attempt to measure the x or y components of the magnetization of a single proton, we would get one of two answers, $+1/2$ or $-1/2$ (in units of $\hbar$). However, if we repeat the measurement, the result would not always be the same. In other words, a single nucleus does not behave classically. The signal observed in an NMR experiment derives from a large ensemble of spins, so the detected signal (expectation value of the x or y component of the magnetization) behaves in a classical manner. In some respects this must be intuitively obvious, since the NMR sample (as opposed to a single nucleus) is a classical object and we therefore expect it to behave classically.

Now we consider a sample composed of many identical nuclei with spin quantum number $I = 1/2$. As we saw in Section J1.2.1, an angular momentum can be represented by a vector of length $[I(I+1)]^{1/2}$ units with a component of length m_i units along the z-axis. Since the uncertainty principle does not allow us to specify the x and y components of the angular momentum, all we know is that the vector lies somewhere on the cone around the z-axis (Fig. J1.10). In the absence of a magnetic field, the sample consists of equal numbers of spins of different energy

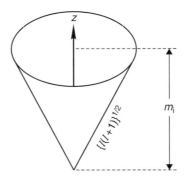

Fig. J1.10 The vector model of angular momentum for a single spin 1/2 nucleus. The angle around the z-axis is indeterminate.

with their vectors lying at random angles ϕ on the cones (Fig. J1.11). The angles ϕ are unpredictable, and at this stage we consider the spin vectors as stationary. The magnetization **M** of the sample, its net nuclear moment, is zero.

The Effect of the Static Field

Two changes occur in the magnetization when a magnetic field is present. The first is nuclear precession.

The behavior of a compass needle is a good starting point for the discussion of this phenomenon. If displaced from alignment with the field and then released, in the absence of friction the needle fluctuates back and forth indefinitely in a plane about the field axis (Fig. J1.12).

A quite different kind of motion occurs if the needle is spinning rapidly around its north–south axis. Because of the gyroscopic effect, the force applied by the field to the axis of rotation causes movement not in the plane of the force but perpendicular to this plane. The axis of the rotating particle moves in a circular orbit. It is said to *precess* around the magnetic field vector (Fig. J1.13). The precession angular velocity, ω_0, is proportional to the applied field strength B_0

$$\omega_0 = \gamma B_0 \,\text{rad/s} \tag{J1.11}$$

The angular velocity can be converted to the precession frequency, ν_0, by dividing by 2π

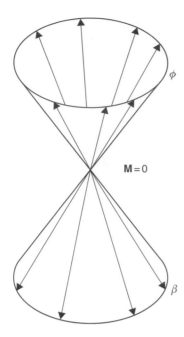

Fig. J1.11 The magnetization of the sample of spin 1/2 nuclei is the resultant of all their magnetic moments. In the absence of an externally applied field, there are equal numbers of spins with different energy at random angles around the z-axis (the field direction) and magnetization is zero.

Fig. J1.12 Magnetic compass needle.

$$\nu_0 = \gamma B_0/2\pi \; \text{Hz} \qquad \qquad (\text{J1.12})$$

Equation (J1.12) is called the Larmor equation. By considering the nucleus as a spinning magnet, the Larmor equation can be used to describe the fundamental phenomenon of NMR. A comparison of Eq. (J1.5) with Eq. (J1.12) shows that we can equate the Larmor frequency with the resonant frequency derived from quantum mechanical considerations. The proportionality constant, γ, therefore, corresponds to the gyromagnetic ratio of the nucleus in the quantum mechanical description. In the classical description, the spinning nucleus "resonates" with a precession (Larmor) frequency proportional to the applied field via its gyromagnetic ratio. It is clear, from the lack of any term in Eq. (J1.10) involving the angle of precession, that the energy of the system does not depend on the magnetic moment of the spinning nucleus, μ, but only on the

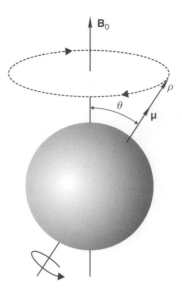

Fig. J1.13 According to classical mechanics, an individual magnetic moment μ precesses about the axis of the applied magnetic field B_0 under angle. This is called Larmor precession. It is analogous to the precession of a spinning gyroscope allowed to topple in the Earth's gravitational field.

TABLE J1.3 NMR FREQUENCIES (IN 14.1 T), AND NATURAL ABUNDANCE OF SELECTED NUCLEI		
Nucleus	ν (MHz)	Natural abundance
^{1}H	600	99.985
^{13}C	150.9	1.108
^{15}N	60.75	0.37
^{19}F	564.75	100.0
^{31}P	243.15	100.0

projection of μ onto the magnetic field B_0 axis (conventionally defined as the z-axis).

In a magnetic field strength of 14.1 T, which is typical for current NMR experiments, Eq. (J1.10) predicts a resonance frequency, $\nu = 6 \times 10^8$ Hz or 600 MHz for the ^{1}H nucleus (see Table J1.2 for the γ value). The frequency falls in the radio-frequency region of the electromagnetic spectrum and corresponds to a wavelength of 50 cm. Since the proton is by far the most popular NMR nucleus, NMR spectrometers are usually classified by their proton frequencies rather than by the strength of their magnetic fields.

Table J1.3 summarizes the NMR properties of several magnetic nuclei in an external magnetic field of 14.1 T. Note that of these nuclei ^{1}H has the largest gyromagnetic ratio.

The second change that occurs in an external magnetic field is that the population in the two spin states alters according to the Boltzmann distribution (Section J1.2.1). Despite a tiny imbalance of populations there is a net

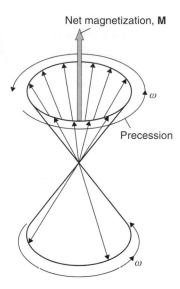

Net magnetization, **M**

Precession

ω

ω

Fig. J1.14 In the presence of an external magnetic field, the spins precess around their cones and there are also changes in the populations in the two spin states. As a result, there is a net magnetization along the z-axis.

magnetization that can be represented by a vector **M** pointing in the z-direction and with a length proportional to the population difference (Fig. J1.14).

Laboratory Frame and Rotating Frame

The classical formalism describes nuclear resonance in terms of the precession of magnetization vectors about the applied static magnetic field $\mathbf{B}_0$ with Larmor precession frequency. In order to visualize the effect of radio-frequency pulses, it is convenient to work with this vector model as it stands. One approach by which this can be achieved is to view the system in terms of the rotating frame of coordinates. In this representation, the normal (laboratory) Cartesian axes x, y, z are replaced by axes x', y', z', which are presumed to rotate at the Larmor frequency of the spins. This representation is known as the rotating frame of coordinates. In this frame the bulk magnetization vector appears stationary (Comment J1.4).

The concept of the rotating frame is enormously useful, since it simplifies the treatment of the otherwise complex gyration of the bulk magnetization vector. It is far easier to understand the workings of complex pulse sequences (see Chapter J2) if the bulk magnetization vector can be considered to be stationary at equilibrium.

The Effect of the Radio-Frequency Field

We now consider the effect of the magnetic component of the radio-frequency field in the xy plane. Suppose we choose the frequency of the oscillating field to be equal to the Larmor frequency of the spins. The nuclei now experience a steady $\mathbf{B}_1$ field (Fig. J1.15) because the rotating magnetic field is in step with the precessing spins. Under

the influence of this steady field, the magnetization vector begins to precess around its direction. If we apply the $\mathbf{B}_1$ field in a pulse of certain duration, the magnetization precesses into the xy plane (Fig. J1.16) at the Larmor frequency. The rotating magnetization induces a signal in the coil, which can be amplified and processed.

Relaxation Processes

There are two kinds of relaxation process in NMR. The first is related to the establishment of thermal equilibrium in an assemblage of nuclear magnets with different energy. In the absence of a magnetic field the two energy levels available to spin 1/2 nuclei are of equal energy and are hence equally populated. As we have discussed above, in the presence of a magnetic field the two energy levels and their populations are no longer equal. Provided the Larmor frequencies of nuclei are similar, they are in phase and capable of exchanging energy. As the system reverts to thermal equilibrium exponentially, the z component of magnetization approaches its equilibrium value M_0 with a time constant called the *longitudinal relaxation time* T_1. The constant T_1 reflects the efficiency of the coupling between a nuclear spin and its surroundings (lattice) and is also called the *spin–lattice relaxation time*. Spin–lattice relaxation is an energy effect. A shorter T_1 value means that coupling is more efficient and vice versa. Spin–lattice relaxation times lie between 10^{-3} and 10^2 s for liquids, and the range is even larger for solids.

The second kind of relaxation is illustrated in Fig. J1.17. Consider a group of nuclei, precessing in phase about a common magnetic field along the z-axis, like a tied-up bundle of sticks. They produce a resultant rotating magnetic vector with the component in the xy plane. If by any process the nuclei lose their phase coherence, there are as many positive as negative components in the xy plane and the resultant vector moves toward the z axis (Fig. J1.17b). The randomization, i.e., the decay of the y or x component of magnetization to zero, occurs exponentially with a time constant called the *transverse relaxation time*, T_2.

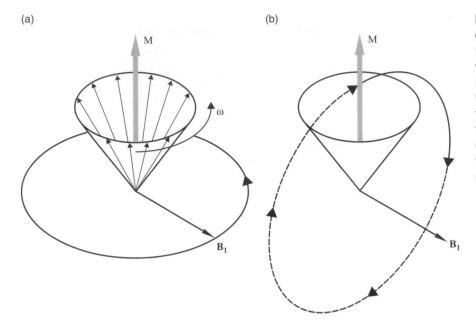

Fig. J1.15 (a) In a resonance experiment, a radio-frequency magnetic field **B**$_1$ is applied in the *xy* plane. **(b)** If we step into a frame rotating at the Larmor frequency, the radio-frequency field **B**$_1$ appears to be stationary if its frequency is the same as the Larmor frequency. When the two frequencies coincide, the magnetization vector of the sample begins to rotate around the direction of the **B**$_1$ field.

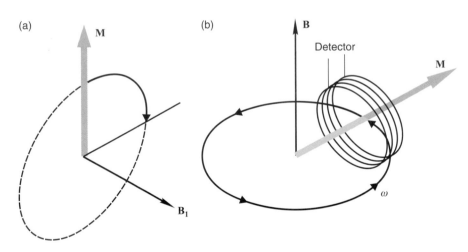

Fig. J1.16 (a) If the radio-frequency field is applied for a certain time, the magnetization vector is rotated into the *xy* plane. **(b)** To an external observer, the vector is rotating at the Larmor frequency, and can induce a signal in the receiver coil.

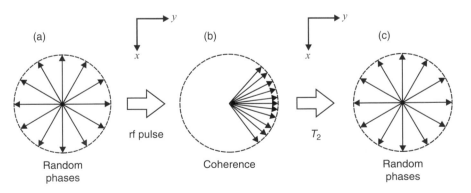

Fig. J1.17 The effect of a radio-frequency (rf) pulse on the magnetic moments of the individual spins in an NMR sample (looking down the *z*-axis). Starting from the equilibrium state with random phases **(a)**, a pulse along the *x*-axis in the rotating frame causes the spins to bunch together to some extent **(b)**, producing a net *y* magnetization in the sample. This phase correlation among the spins is known as coherence. After switching the field the randomization occurs with transverse relaxation time T_2 **(c)**.

The T_2 relaxation time is related to the width of spectral lines as

$$\Delta v_{1/2} = \frac{1}{\pi T_2} \qquad (J1.13)$$

where $\Delta v_{1/2}$ is the line width at half the maximum height. Typical values of T_2 in proton NMR are of the order of seconds corresponding to line widths of about 0.1 Hz.

So far, we have assumed that the equipment, and in particular the magnet, are perfect, and that the differences in Larmor frequencies arise solely from interactions within the sample. In practice, the magnet is not perfect, and the field is different at different locations in the sample, despite sample spinning. The inhomogeneity dominates the broadening resonance lines. It is usual to express the extent of inhomogeneous broadening in terms of an effective transverse relaxation time T_2^*, similar to Eq. (J1.13)

$$T_2^* = \frac{1}{\pi \Delta v_{1/2}} \qquad (J1.14)$$

where $\Delta v_{1/2}$ is the observed width at a half the maximum height. For example, if a line in a ^{1}H-NMR spectrum has a width of 5 Hz, then the effective transverse relaxation time is

$$T_2^* = \frac{1}{\pi \, 5\text{s}} = 64\,\text{ms} \qquad (J1.15)$$

Experimental NMR schemes for measuring T_1 and T_2 are described in Section J2.3.

J1.2.3 Nuclear Environment Effects on NMR

Nuclear magnetic resonance would not be a very useful technique if, during an experiment, every nucleus of the same species in a sample were subjected to exactly the same magnetic field defined by the spectrometer magnet. If this were the case, we could measure gyromagnetic ratios with great accuracy but not much else. The Larmor frequency of a given nucleus, however, is strongly affected by its chemical environment. As a consequence, NMR signals from molecules provide a wealth of spectral information that can serve to elucidate their chemical structure.

The spectra of ethanol, shown in Fig. J1.18, illustrate two types of environmental effect. The curves in Fig. J1.18a, obtained with a low-resolution instrument, show three resonance lines, whose surface areas, in the ratio 1:2:3, correspond to protons with different precession frequencies. It appears logical to attribute the peaks to the hydroxyl, methylene, and methyl protons, respectively (Comment J1.5). The shift in absorption frequency of a nucleus depending on the group to which it is bound is called the *chemical shift*. The higher-resolution spectrum in Fig. J1.18b reveals that two of the three proton peaks are further split into additional peaks. This secondary environmental effect is called *spin–spin splitting*.

Both the chemical shift and spin–spin splitting are very important in structural analysis. Experimentally, the two effects are easily distinguished. The peak separation resulting from a chemical shift is directly proportional to the field strength, while spin–spin splitting is, in general, independent of the strength of the external magnetic field.

Chemical Shift

Chemical shifts arise because the magnetic field $\mathbf{B}$ experienced by an atomic nucleus differs slightly from the external field $\mathbf{B}_0$: $\mathbf{B}$ is slightly smaller than $\mathbf{B}_0$ because of shielding by surrounding electrons. The external field induces the electrons to circulate within their atomic orbitals, much like an electric current passing through a coil of wire. This generates a small magnetic field $\mathbf{B}'$ in the opposite direction to $\mathbf{B}_0$ (Fig. J1.19). $\mathbf{B}'$ is proportional to $\mathbf{B}_0$ and

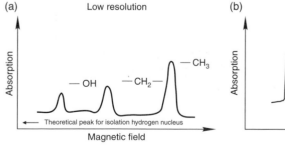
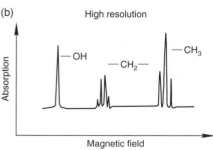

Fig. J1.18 60 MHz NMR spectra of ethanol at different resolutions. **(a)** Low resolution: The areas under peaks stand roughly in the ratio 1:2:3, as would be expected if each peak corresponded to the chemically different OH, CH_2, and CH_3 protons. **(b)** High resolution: The proton spectra show a considerably greater number of lines – the CH_2 resonance is split into four lines and the CH_3 resonance into three lines. (Adapted from Skoog, *et al.*, 1995.)

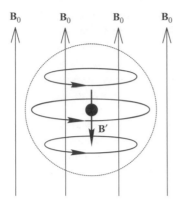

Fig. J1.19 An applied magnetic field $\mathbf{B}_0$ causes the electrons in an atom to circulate within their orbitals. This motion generates an extra field $\mathbf{B}'$ at the nucleus in opposition to $\mathbf{B}_0$.

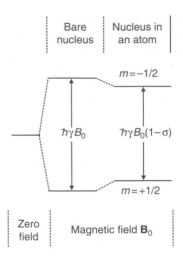

Fig. J1.20 Energy levels of a spin 1/2 nucleus.

COMMENT J1.6 MOLECULAR SCREENING

Molecular screening is not isotropic as it differs along various axes within the molecule; σ is therefore a tensor. However, in the gaseous and liquid phases, due to rapid molecular motion, a nucleus is subject only to an average value of σ. The individual elements of σ can be significant for samples in which isotropic motion is impossible, e.g., in liquid crystals and solids. The degree of anisotropy within chemical shifts is also important when discussing nuclear relaxation (Section J3.4).

COMMENT J1.7 REFERENCE SIGNALS

The reference signal is most conveniently obtained by adding a small amount of a suitable compound to the NMR sample. For ^1H and ^{13}C spectra this is usually tetramethylsilane $(CH_3)_4Si$, or TMS. This molecule is inert, soluble in most organic solvents, and gives a single, strong ^1H resonance from its 12 identical protons. Moreover, both ^1H and ^{13}C nuclei are strongly shielded (large δ values), so that the TMS resonance falls at the low-frequency end of the spectrum. Unfortunately, TMS is not water soluble and in aqueous media the sodium salt of 2,2-dimethyl-2-silapentane-5-sulfonic acid (DSS) is normally used as a reference. The methyl protons of this compound produce a peak at virtually the same place in the spectrum as TMS. However, the methylene protons of DSS give a series of small peaks that may interfere with the measurements. For this reason, most of the DSS now on the market contains deuterated methylene groups, which eliminate these undesirable peaks.

typically 10^4–10^5 times smaller. The field at the nucleus may be written

$$\mathbf{B} = \mathbf{B}_0 - \mathbf{B}' = \mathbf{B}_0(1 - \sigma) \qquad (J1.16)$$

where the proportionality constant σ is called the *shielding* or *screening* constant (Comment J1.6). The resonance condition (Eq. J1.5) becomes

$$\nu = \gamma \mathbf{B}_0(1 - \sigma)/2\pi \qquad (J1.17)$$

i.e., the resonance frequency of a nucleus within its atom is slightly lower than that of the same nucleus if it were bare, "stripped" of all its electrons (Fig. J1.20). The shielding constant is very sensitive to the chemical environment. For protons in a methyl group, it is larger than for methylene protons, and smaller than for the proton in a hydroxyl group. For an isolated hydrogen nucleus, the shielding constant is zero. In order to bring any of the protons in ethanol into resonance at a given excitation frequency ν, therefore, it is necessary to employ an external field $\mathbf{B}$ correspondingly greater than $\mathbf{B}_0$, the resonance value for the isolated proton (Eqs. (J1.13) and (J1.14)) Alternatively, if the applied field is held constant, the excitation frequency must be increased in order to bring about the resonance condition.

The shielding constant σ is an inconvenient measure of the chemical shift. Since absolute shifts are rarely needed and are difficult to determine, it is common practice to define the chemical shift in terms of the difference in resonance frequencies between that of the nucleus of interest (ν_x) and a reference nucleus (ν_{ref}) by means of a dimensionless parameter δ:

$$\delta = (\nu_x - \nu_{ref})/\nu_0 \qquad (J1.18)$$

where ν_0 is the frequency of the spectrometer. The frequency difference $\nu_x - \nu_{ref}$ is divided by ν_0 in order to define δ as a molecular property, independent of the external magnetic field. Values of δ are quoted in parts per million (or ppm) (Comment J1.7). A distinct advantage of the definition is that for a given peak, δ is the same regardless of the frequency of the instrument used to measured it; e.g., the same value of δ is obtained with a 60 and a 100 MHz spectrometer (Fig. J1.21). Conventionally, NMR peaks are plotted on a linear δ scale, with the field increasing from left to right. In the example in Fig. J1.21, the tetramethylsilane (TMS) reference peak defines the zero value for the δ scale and the value of δ increases from right to left.

The chemical shift of a nucleus depends on many factors, but the surrounding electron density is often the dominant one. A high electron density causes a large shielding effect, which means that the applied magnetic field must be increased to obtain resonance. The up-field shift leads to a lower value of δ. Conversely, a low electron density leads to a down-field shift and an increase in δ value. For example, the chemical shifts of the methyl protons (in italics below) in CH_3X relative to TMS become larger as X becomes a better electron-withdrawing group: $\delta(CH_3–CH_3) \approx 1$, $\delta(CH_3–C_6H_6) \approx 2$, $\delta(CH_3–OH) \approx 4$. If the proton is attached directly to an electro-negative atom such as in a carboxyl group, which has a very low electron density, chemical shifts can have a very high δ value, $\delta(COOH) \approx 10$. Figure J1.22 shows the approximate values of chemical shifts for various types of proton. Most proton peaks lie in the range $\delta = 1$ to $\delta = 13$.

For a given compound the appearance of the spectrum is governed by an intramolecular chemical shift difference, i.e., a difference in resonance frequencies for different nuclei of the same molecules. As an example, Fig. J1.23 shows the 60 MHz proton resonance spectrum of 2,2-dimethoxypropane: The methyl and methoxy protons give absorption at different positions in the spectrum. Figure J1.23 illustrates another of the important features of NMR spectra, namely that the intensity of absorption is strictly proportional to the concentration of the nuclei (in

the present case protons). This is of great importance for the structure determination. For instance, if the compound used in Fig. J1.23 was of unknown structure, the NMR spectrum would immediately show that it contained two types of proton and that there were equal numbers of each type. This use of NMR could be described as providing a relative proton count.

Figure J1.24 shows the one-dimensional ^{1}H NMR spectrum of BPTI, in which the resonance positions of different hydrogen atoms are given by their chemical shift δ in ppm. It is seen that chemical shift is determined primarily by the chemical structure. The resonance lines of all methyl groups appear on the extreme right at around 1 ppm; those of the labile amide protons are on the extreme left from about 7–11 ppm; in between, we observe the methylene groups at 2–3 ppm, the α-protons at 4–5 ppm, and the protons of the aromatic rings at 6–7.5 ppm.

All amino acid side-chains in an extended, flexible polypeptide chain are exposed to the same solvent environment so that multiple copies of a specified amino acid in the sequence have nearly identical ^{1}H chemical shifts (Fig. J1.25a). Therefore, the ^{1}H NMR lines of random coil polypeptides correspond closely to the sum of the resonance peaks of the constituent amino acid residues. Figure J1.26 shows the random-coil spectrum of denatured BPTI computed as such a sum. The increased complexity of the NMR spectrum of native BPTI (Fig. J1.24) results primarily from conformation-dependent ^{1}H chemical-shift dispersion, which is a consequence of a generalized solvent effect: interior peptide segments in globular proteins are shielded from the solvent and are surrounded by other peptide segments (Fig. J1.25b). When the interior of a globular protein is highly aperiodic, each amino acid residue is subjected to a unique microenvironment. In Fig. J1.26 these effects are shown in detail for the threonyl and tyrosyl residues in

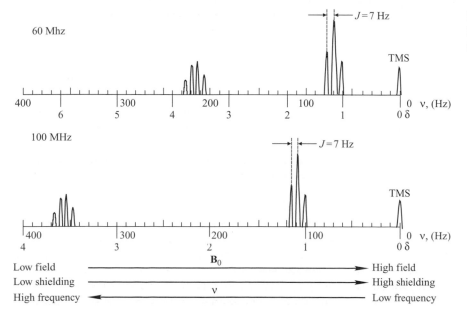

Fig. J1.21 Standard abscissa scales for NMR spectra. (Adapted from Skoog *et al.*, 1995.)

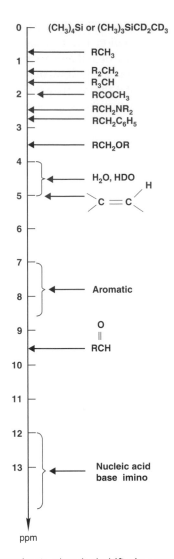

Fig. J1.22 Approximate chemical shifts in ppm relative to a reference for different types of proton. The smaller the value of δ, the greater the chemical shielding and the more up-field the proton signals occur. The reference signal is TMS for non-aqueous solutions or DSS for aqueous samples. (Adapted from Tinoco et al., 1998.)

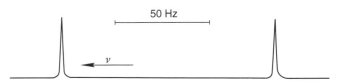

Fig. J1.23 Proton NMR spectrum of 2,2-dimethoxypropane, $(CH_3)_2C(OCH_3)_2$, in CCl_4 solution: The methyl and methoxy protons give absorption at different positions in the spectrum.

BPTI. In the unfolded polypeptide chain the methyl groups of the three threonyl residues give rise to a single line corresponding in intensity to nine protons, whereas in the folded BPTI the chemical shifts of the three methyl groups are dispersed so that three separate lines, corresponding to

three protons each, are observed. The NMR lines between 7.5 and 11 ppm in a globular protein correspond to amide protons that exchange slowly with solvent ^2H_2O, because of their location in the tertiary structure. They have no counterpart in the unfolded protein because the spectrum of Fig. J1.26 was calculated with the assumption that these protons had been completely exchanged with ^2H.

The detailed examination of the spectra of compounds containing double or triple bonds reveals that local effects are not sufficient to explain the position of certain proton peaks. For example, the δ values change in an irregular fashion for protons in the following hydrocarbons, arranged in order of increasing acidity of the groups to which they are bonded:

$$CH_3 - CH_3(\delta = 0.9), \ CH_2 - CH_2(\delta = 5.8),$$
$$HC \equiv CH(\delta = 2.9)$$

The effect of multiple bonds on the chemical shift can be explained by taking into account the anisotropic magnetic properties of these compounds. For example, the magnetic properties of crystalline aromatic compounds were found to differ appreciably, depending upon the orientation of the aromatic ring with respect to the applied field. The effect is called the *ring-current effect*. The anisotropy can be understood readily from the model shown in Fig. J1.27a. When the plane of the ring is perpendicular to the magnetic field, a ring current of π electrons is induced. The induced field is in the opposite direction to the applied field above and below the plane of the molecule, and in the same direction on the sides of the molecule as shown in Fig. J1.27a. Therefore, a proton near an aromatic group is shielded (decrease in δ) if it is above or below the center of the planar ring, and its δ is increased if it is at the outside of the planar ring. This effect is either absent or self-cancelling in other orientations of the ring. Ring effects are most useful for interpreting local protein conformations near phenylalanine or tyrosine residues, and nucleic acid conformations, in which the purines, adenine, and guanine have the largest ring-current effects. An analogous model is valid for carbonyl double bonds. In this case, we may imagine π electrons circulating in a plane along the bond axis where the molecule is oriented with the field as presented in Fig. J1.27b. Again, the secondary field produced acts upon the proton to reinforce the applied field.

Finally, it is necessary to point out that the chemical shift range is greater for certain nuclei other than ^1H. The chemical shift for ^{13}C in various functional groups typically lies in the range 0–220 ppm. For ^{19}F, the range of chemical shifts may be as large as 800 ppm, while for ^{31}P it is 300 ppm or more.

Spin–Spin Coupling

As may be seen in Fig. J1.28, the absorption bands for the methyl and methylene protons in ethanol consist of

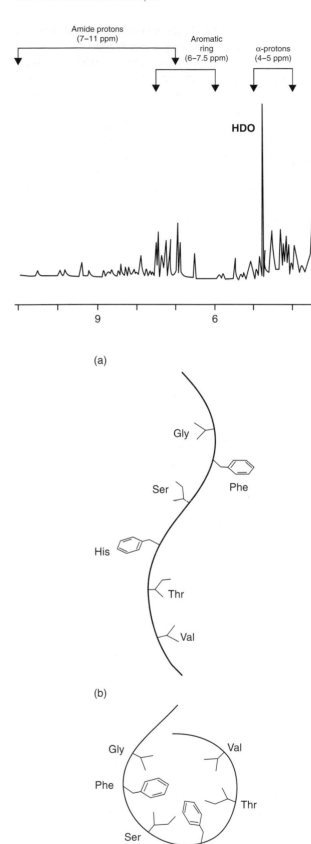

Amide protons
(7–11 ppm)

Aromatic
ring
(6–7.5 ppm)

α-protons
(4–5 ppm)

Methylene
protons
(2–3 ppm)

Methyl protons
(~1 ppm)

HDO

δ, (ppm)

Fig. J1.24 One-dimensional ^{1}H NMR spectrum of a freshly prepared D$_2$O solution of the protein, BPTI (protein concentration = 5 mM, pD = 4.5, T = 318 K, ^{1}H NMR frequency = 360 MHz). HDO identifies the solvent water resonance. (Adapted from Wuthrich, 1995.)

(a)

Gly
Ser
His
Phe
Thr
Val

(b)

Gly
Phe
Ser
His
Val
Thr

Fig. J1.25 Schematic presentation of amino acid side-chains in **(a)** an extended, random-coil conformation; and **(b)** a folded, globular conformation. (Adapted from Wuthrich, 1995.)

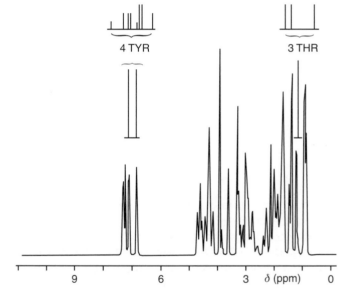

4 TYR

3 THR

δ (ppm)

Fig. J1.26 Computed random-coil ^{1}H NMR spectrum of BPTI in ^{2}H$_2$O solution, where all the labile protons in N–H and O–H groups are replaced by ^{2}H. All resonance lines in these spectra have been assigned to distinct H atoms of BPTI. Stick diagrams indicate the positions and intensities of the γCH$_3$ resonances, the threonyl residues in the sequence positions 11, 32, and 54, and the aromatic protons of the tyrosyl residues 10, 21, and 35. (Adapted from Wuthrich, 1995.)

several narrow peaks that can be easily separated with a high-resolution spectrometer (Comment J1.8). Careful examination of these peaks shows that the spacing of the three components of the methyl band is identical to that of the four peaks of the methylene band. This spacing in hertz is called the *coupling constant* for the interaction and is given the symbol *J*. Moreover, the peak areas within a multiplet are in an integer ratio to one another. The ratio

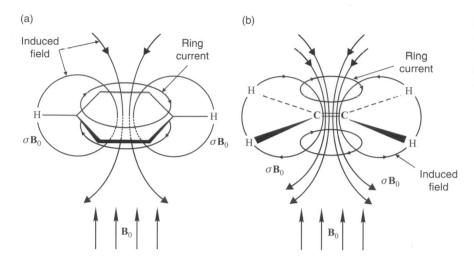

Fig. J1.27 Approximate model of the ring-current effect in NMR. "Deshielding" of **(a)** aromatic and **(b)** ethylene protons. (After Skoog *et al.*, 1995.)

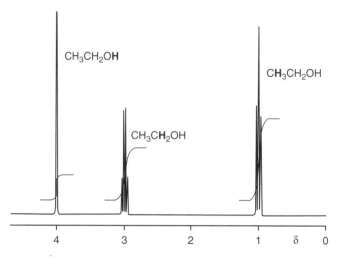

Fig. J1.28 The high-resolution ^{1}H NMR spectrum of ethanol showing the splitting produced by spin–spin coupling. The bold letters denote the protons giving rise to the resonance peak. The step-like curve is the integrated peak. The CH$_3$ protons form one group of nuclei with $\delta \approx 1$. The two CH$_2$ protons are in a different part of the molecule. They experience a different local magnetic field, and resonate at $\delta \approx 3$. Finally, the OH proton is in another environment, and has a chemical shift of $\delta \approx 4$.

COMMENT J1.8 SPECTRUM OF ETHANOL

The methylene protons and the OH proton are separated by only three bonds, so coupling should increase the multiplicity of both OH and methylene peaks. Indeed, the spectrum of highly purified ethanol shows additional splitting of OH (the triplet) and CH$_2$ (the eight methylene peaks). However, if we add a trace of acid or base to a pure sample, the spectrum reverts to the form shown in Fig. J1.28. Both acids and bases, and also impurities in ethanol, catalyze the exchange of OH protons. It is thus plausible to associate the decoupling observed in the presence of these catalysts to an exchange process. If exchange is rapid, each OH group has several protons associated with it during any brief period; within this interval, all of the OH protons experience the effects of the three spin arrangements of the methylene protons. Thus, the magnetic effects on the ethanol proton are averaged, and a single sharp peak is observed.

of areas for the methyl triplet is 1:2:1, whereas it is 1:3:3:1 for the methylene quartet.

The results of detailed theoretical calculations are consistent with the concept that coupling take place via interaction between the nuclei and the bound electrons rather than through free space. Let us first consider the effect of the methylene protons in ethanol on the resonance of the methyl protons. Reference to Fig. J1.29 shows that the two methylene protons may have any one of four possible magnetic quantum number combinations.

The spins of the two methylene protons are paired and aligned against or toward the external field in two of the combinations. There are two other combinations in which

the spins oppose one another. The magnetic effect that is transmitted to the methyl protons on the adjacent carbon atom is determined by the instantaneous spin combinations in the methylene group. If the spins are paired and opposed to the external field, the effective applied field on the methyl protons is slightly decreased, and a somewhat higher field is needed to bring them into resonance, resulting in an up-field shift. Spin pairs that are aligned with the field result in a down-field shift. Neither of the opposed spin combinations has an effect on the resonance of methyl protons. There results a splitting of the methyl resonance into three peaks, with the unperturbed resonance in the middle. The area under the middle peak is twice that of either of the other two, since two spin combinations are involved.

Let us now consider the effect of the three methyl protons upon the methylene peak (Fig. J1.29). We have eight possible spin combinations. Among these, however,

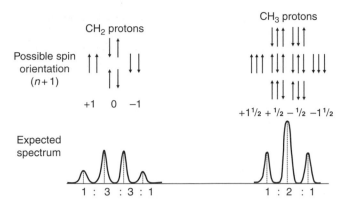

Fig. J1.29 Possible nuclear spin orientations of ethyl and methyl group protons and expected spin–spin splitting patterns.

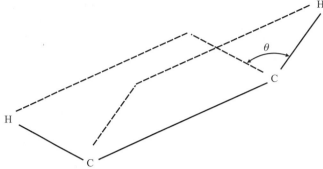

Fig. J1.30 Definition of the dihedral angle θ in the Karplus equation.

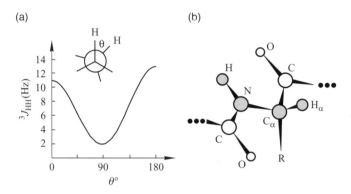

Fig. J1.31 (a) Typical dependence of a three-bond H–C–C–H coupling constant on the dihedral angle θ, calculated using Eq. (J1.19). (b) Part of the backbone of a peptide chain, showing the H–N–C_α–H dihedral angle. R is an amino acid side-chain. (After Hore, 1995.)

there are two groups containing three combinations that have equivalent magnetic effects. The methylene peak is thus split into four peaks having areas in the ratio 1:3:3:1.

The example of adjacent methyl and methylene groups in ethanol suggests the general rule that the number of peaks in a split band in a first-order spectrum is equal to $n + 1$, where n is the number of magnetically equivalent protons.

The spin–spin splitting expressed as a frequency J is independent of the applied magnetic field (unlike the chemical shift). J values for protons range from 0 to about 20 Hz. If we are measuring proton NMR in a hydrocarbon or carbohydrate, there are no effects from the carbon or oxygen nuclei, because they have no magnetic moment (except from the very small amounts of ^{13}C and ^{17}O). Naturally occurring ^{14}N has a spin of 1 and tends to broaden neighboring proton lines rather than split them. Consequently, in practice we need to consider only proton–proton splittings.

Three-Bond Couplings
The most useful spin–spin couplings are those involving nuclei separated by three bonds, e.g., $^3J_{HH}$ in H–C–C–H fragments. The value of this coupling is given by the Karplus equation:

$$J(\theta) = A\cos^2\theta + B\cos\theta + C \tag{J1.19}$$

where θ is the dihedral angle between protons (Fig. J1.30). The values A, B, C depend upon the precise system under investigation, and in particular are sensitive to the electronegativity of the substituents on the carbon. Typical values are $A = 2$ Hz, $B = -1$ Hz, and $C = 10$ Hz, which give a θ variation of the type shown in Fig. J1.31a. Due to its empirical nature, care should be taken in the generalization of any parameterization to an unknown system.

The Karplus equation finds valuable applications in studies of protein structure. For example, the couplings between the amide (NH) and C_α protons in a polypeptide chain provide information on the conformation of the protein

backbone (Fig. J1.31b). In particular, two major elements of secondary structure in proteins – helices and sheets – have characteristic H–N–C_α–H dihedral angles; $\approx 120°$ and $\approx 180°$, respectively. Thus, amide–C_α proton–proton coupling constants smaller than 6 Hz usually indicate a helix structure, while couplings larger than 7 Hz generally arise from sections of the protein with a β-sheet structure.

Proton–proton coupling constants are generally very small (<1 Hz) when the nuclei are separated by more than three bonds.

Weakly and Strongly Coupled Spins
There are two frequently encountered spin systems (sets of coupled nuclei), which give rise to distinctive multiplets (doublets, triplets, quartets, etc.). Figure J1.32 shows spectral simulations for two coupled spins for a range of Δv values between $16J$ and zero. For the first system it is assumed that all pairs of spins are *weakly coupled*, i.e., that the difference in resonance Larmor frequency of the two nuclei Δv greatly exceeds their mutual coupling ($\Delta v/J \gg 1$). Each spin in such a spin system is labeled with a letter of the Roman alphabet. According to convention, these letters

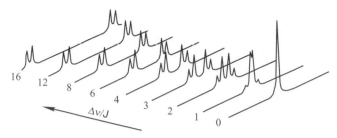

Fig. J1.32 Spectral simulations of a pair of spin 1/2 nuclei for a range of Δv values between $16J$ and zero. As the difference in resonance frequencies is reduced, keeping J fixed, the inner component of each doublet steadily increases in amplitude, while the outer components become weaker. The extreme case is that of equivalent nuclei, for which $\Delta v = 0$. This case demonstrates that equivalent nuclei do not split each other. (After Hore, 1995.)

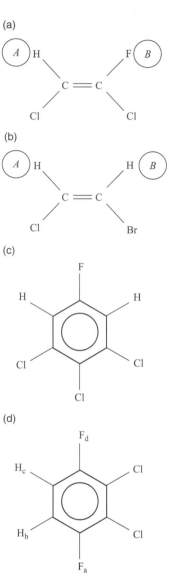

Fig. J1.33 Classification of molecules according to their spin system: **(a)** *cis* 1,2-dichlorofluoroethylene is an example of *AX* proton spin system; **(b)** chlorbromethylene is an example of *AB* proton spin system; **(c)** 1-fluoro-3,4,5-trichlorobenzene: two protons are magnetically equivalent; **(d)** 1,4-difluoro-2,3-dichlorobenzene: two protons are magnetically non-equivalent.

are chosen to be well spaced in the alphabet to indicate that the spin system is loosely coupled. Such a two-spin system is referred to as an *AX* system. This always applies if the nuclei are of different isotopes or elements. Thus, *cis*-1,2-dichlorofluoroethylene (Fig. J1.33a) is said to possess an *AX* spin system (it is irrelevant whether *A* refers to H and *X* to F or vice versa; the chlorines may be ignored).

In the second system all pairs of spins are *strongly coupled*, i.e., the difference in resonance frequency is close to their mutual coupling ($\Delta v/J \sim 1$). This system is referred to as an *AB* system. The letters are chosen to be close in alphabet to indicate that the system of interest is strongly coupled. Chlorbromethylene (Fig. J1.33b) is an example of such a system.

A group of n equivalent spins with identical chemical shifts and identical spin–spin couplings to all other spins in the system are denoted A_n. In 1-fluoro-3,4,5-trichlorobenzene (Fig. J1.33c), two protons are magnetically equivalent by symmetry. In the case of 1,4-difluoro-2,3-dichlorobenzene (Fig. J1.33d) protons are chemically equivalent, but magnetically non-equivalent, because the (H_b, F_d) relationship is not the same as the (H_c, F_d) one – they are *meta* and *ortho*, respectively.

In the case of three spins, we can describe an *AMX* spin system which represents three loosely coupled spins. Similarly, we can define an *ABX* spin system to indicate that protons *A* and *B* are weakly coupled to *X*. Multiplet patterns expected for some of the simple spin systems discussed below are collected together in Fig. J1.34.

The concept of a spin system can help to show how a multiplet structure arises. Moreover, in some simple cases we can illustrate how the multiplet pattern can be used to determine or verify the structures of molecules, without prior knowledge of the magnitudes of the chemical shifts or coupling constants involved.

The simplest case is the *AX* system. In this case interaction of nucleus *A* with *X* causes the *A* resonance to split into two equally intense lines centered at the chemical shift

of *A* (a doublet), with spacing equal to the *AX* coupling constant, J_{AX}.

A step up from the previous case is the *AMX* spin system, which consists of three nuclei with different chemical shifts and three distinct coupling constants: J_{AM}, J_{AX}, J_{MX}. Four lines are expected because there are four non-degenerate arrangements of the *M* and *X* spins ($M{\uparrow}X{\uparrow}$, $M{\uparrow}X{\downarrow}$, $M{\downarrow}X{\uparrow}$, $M{\downarrow}X{\downarrow}$). These peaks are displaced from the chemical shift of *A* by simple combination of the couplings to spin *A* (J_{AM} and J_{AX}). The *A* multiplet should therefore be a doublet of doublets, as shown in Fig. J1.35. The two equivalent spin 1/2 nuclei (AX_2) are a special case of the

Fig. J1.34 Multiplet patterns for the A nucleus in various spin systems. M and X are spin 1/2 systems. Weak coupling is assumed throughout. Spectra are drawn for $|J_{AX}| > |J_{AM}|$.

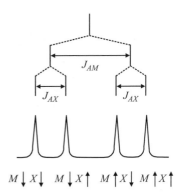

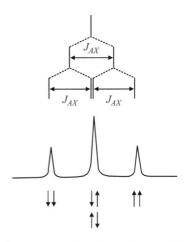

$M \downarrow X \downarrow \quad M \downarrow X \uparrow \quad M \uparrow X \downarrow \quad M \uparrow X \uparrow$

Fig. J1.35 NMR spectrum of nucleus A in an AMX spin system. The four components of the A multiplet, a doublet of doublets, arise from the four combinations of M and X spins, indicated $\uparrow (m = +1/2)$ and $\downarrow \uparrow (m = -1/2)$.

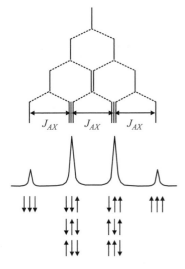

$\downarrow\downarrow\downarrow \quad \downarrow\downarrow\uparrow \quad \downarrow\uparrow\uparrow \quad \uparrow\uparrow\uparrow$
$\downarrow\uparrow\downarrow \quad \uparrow\downarrow\uparrow$
$\uparrow\downarrow\downarrow \quad \uparrow\uparrow\downarrow$

Fig. J1.37 NMR spectrum of nucleus A in an AX_3 spin system. The quartet arises from the eight combinations of the three X spins, as indicated.

COMMENT J1.9 RESONANCE LINES FOR AX, AX_2, AND AX_3 SPECTRA

The results for AX, AX_2, and AX_3 can be generalized. For n equivalent X (spin 1/2) nuclei, the A resonance is split into $n + 1$ equally spaced lines, with the relative intensity given by the coefficient in the binomial expansion of $(1 + x)^n$.

$\downarrow\downarrow \quad \downarrow\uparrow \quad \uparrow\uparrow$
$\uparrow\downarrow$

Fig. J1.36 NMR spectrum of nucleus A in an AX_2 spin system. The triplet arises from the four combinations of the two X spins as indicated.

AMX spin system, with $J_{AM} = J_{AX}$. As seen from Fig. J1.36, the two central lines of the doublet of doublets coincide to give a triplet centered at the chemical shift of A, with a line spacing equal to the coupling constant, and relative intensities 1:2:1.

The multiplet of A in an AX_3 spin system (three identical AX coupling constants) is a four-line quartet (Fig. J1.37). The quartet arises from the eight combinations of the three X spins (Comment J1.9).

The high-resolution spectrum of ethanol discussed before (Fig. J1.28) can now be understood. The ethyl protons make up an A_3X_2 spin system: The triplet arises

because each of the CH_3 protons couples equally with the two equivalent CH_2 protons interacting identically with each of the CH_3 protons. The rapid internal rotation around the C–C bond averages out the chemical shift differences associated with the different conformations of the molecule, and renders the three methyl protons magnetically equivalent, and similarly the two methylene protons. The absence of splittings from coupling between the CH_2 group and the OH proton in the ethanol spectrum (Fig. J1.28) were discussed before (Comment J1.10).

The concept of a spin system can in many cases considerably simplify calculations in macromolecules. For example, important features of many NMR spectra can be calculated by considering only those spins that are scalar coupled. In the case of proteins, each amino acid residue can often be considered essentially an isolated spin system. Likewise, the bases in nucleic acids and monosaccharide

COMMENT J1.10 FIRST-ORDER SPECTRA

The interpretation of spin–spin splitting patterns is relatively straightforward for first-order spectra. First-order spectra are those in which the chemical shift between interacting groups of nuclei is large with respect to their coupling constant J. Rigorous first-order behavior requires that J be smaller than 0.05δ. The ethanol spectrum shown in Fig. J1.28 is an example of a pure first-order spectrum, in which J is 7 Hz for both the methyl and methylene peaks, while the separation between the centers of the two multiplets is about 140 Hz.

Interpretation of second-order NMR spectra is relatively complex and will not be dealt with in this book. Note, however, that because δ increases with increases of the magnetic field while J does not, spectra obtained with a spectrometer having a high magnetic field are much more readily interpreted than those produced by a spectrometer with a weaker magnet (Skoog *et al.*, 1995)

residues of oligosaccharides can often be considered in isolation. Complete listings for the aliphatic spin systems and for aromatic spin systems can be found in the literature.

J1.3 CHECKLIST OF KEY IDEAS

- Certain atomic nuclei can be considered as having "spin"; the term spin implies that each nucleus can be considered as a rotating electrical charge and, consequently, along with its electrical properties, it also possesses an angular magnetic momentum.
- A number of nuclei of particular importance in structural biology may be assigned the nuclear spin value I of $1/2$ (^1H, ^{13}C, ^{15}N, ^{19}F, and ^{31}P). The nuclei of ^{12}C and ^{16}O have an I value equal to zero.
- A nucleus of spin I has $2I+1$ energy levels, equally spaced with separation $\Delta E = \mu B_0/I$, where B_0 is the applied magnetic field and μ is the nuclear magnetic moment.
- Nuclear magnetic moment μ is given by $\mu = \gamma hI/2\pi$, where γ is the gyromagnetic ratio, a constant for a given nucleus, and h is Planck's constant; if the gyromagnetic ratio is positive (e.g., ^1H and ^{13}C), then the $+1/2$ state lies lower in energy, but the opposite is true for a nucleus with negative γ (e.g., ^{15}N).

- Nuclei with large γ and high natural abundance are favorable for use in practice; hence the popularity of ^1H as an NMR nucleus.
- In the absence of a magnetic field the two energy levels available to spin 1/2 nuclei are of equal energy and are hence equally populated; in the presence of a magnetic field the energy levels are no longer equal in energy.
- The torque exerted on a magnetic moment by a magnetic field inclined at any angle relative to the moment causes the nuclear magnetic moment to precess about the direction of the field with a frequency given by the Larmor equation $v_0 = \gamma B_0/2\pi$; such a movement is analogous to the motion of a gyroscope.
- In a single-pulse experiment, the flip angle of magnetization α is given by $\alpha = \gamma B_1 t_p$, where t_p is the duration of the pulse. At a constant amplitude of B_1, the pulse length can be varied so as to produce, e.g., a 45°, 90°, or 180° pulse.
- A typical magnetic field strength B_0 used today for NMR is about 14 T; for hydrogen nuclei, the Larmor equation predicts a resonance frequency of $v = 6 \times 10^8$ Hz or 600 MHz; for ^{13}C the frequency is 150 MHz.
- In a standard one-dimensional NMR spectrum, the relative intensities of different resonance lines reflect the number of nuclei manifested by these lines.
- NMR spectroscopy differs fundamentally from optical spectroscopy: In optical spectroscopy, an excited molecule returns to equilibrium by spontaneous emission almost instantaneously, whereas in NMR spectroscopy the probability of spontaneous emission is negligible.
- The nuclear spin system returns to equilibrium with its surroundings (the "lattice") by a relaxation process characterized by a time T_1, the *spin–lattice* relaxation time, or *longitudinal* relaxation time.
- The nuclear spin system returns to internal equilibrium by a relaxation process characterized by a time T_2, the *spin–spin* relaxation time or the *transverse* relaxation time.
- The *chemical shift* δ defines the location of the NMR line along the radio-frequency axis; it is a characteristic measure of nuclear–electron interaction and exquisitely sensitive to local geometry.
- The *chemical shift* is proportional to the applied magnetic field B_0; it is commonly indicated in parts per million (ppm) relative to a reference compound.
- The *spin–spin splitting*, J, is a measure of the interactions (through-bond) of two or more neighboring nuclei, where the interaction is transmitted by the intervening electrons.
- The spin–spin splitting constant J does not depend on the applied magnetic field and is customarily quoted in hertz.

Suggestions for Further Reading

Historical Review

Tinoco, I. Jr., Sauer, K., and Wang, J. C. (1998). *Physical Chemistry: Principles and Applications in Biological Science*. Upper Saddle River, NJ: Prentice Hall.

Jeener, J. and Alewaeters, G. (2016). "Pulse pair technique in high resolution NMR": a reprint of the historical 1971 lecture notes on two-dimensional spectroscopy. *Prog. Nucl. Magn. Reson. Spectrosc.*, **94–95**, 75–80.

Fundamental Concepts

Harris, R. (1983). *Nuclear Magnetic Resonance Spectroscopy*. Marshfield, MA: Pitman.

King, R. W., and Williams, K. R. (1989). The Fourier transform in chemistry: Part 1: Nuclear magnetic resonance – introduction. *J. Chem. Education*, **66**, A213–A219.

Skoog, D. A., Holler, F. J., and Nieman, T. A. (1995). *Principle of Instrumental Analysis*, Philadelphia, PA: Saunders College Publishing.

Hore, P. J. (1995). *Nuclear Magnetic Resonance*. Oxford: Oxford University Press.

Lian, Lu-Yun and Roberts, G. (eds.) (2011). *Protein NMR Spectroscopy: Practical Techniques and Applications*. Chichester: John Wiley and Sons.

EXPERIMENTAL TECHNIQUES

J2

J2.1 FOURIER TRANSFORM NMR SPECTROSCOPY

J2.1.1 Principles

Experiments in the early years of nuclear magnetic resonance (NMR) spectroscopy (1945–1970) used so-called continuous wave methods, in which the sample was irradiated with a weak, fixed amplitude, radio-frequency field (Fig. J2.1a). Spectra were obtained either by keeping the electromagnetic frequency fixed, while slowly sweeping the magnetic field strength, or vice versa, so as to bring spins with different chemical shifts sequentially into resonance. The 1970s were dominated by the revolutionary development of pulse Fourier spectroscopy (Fig. J2.1b), which paved the way for modern NMR and an unprecedented expansion of its applications. The starting point was the design of a multichannel spectrometer, which allowed the simultaneous measurement of many points of a frequency spectrum. It was soon recognized, however, that the instrumental effort became exorbitant as the number of channels increased.

Traditional continuous wave spectrometers have now been almost completely replaced by pulse Fourier instruments. The inherent advantages of greater sensitivity, high resolution, and the absence of line-shape distortions contributed to make Fourier spectroscopy the preferred experimental technique in NMR.

In pulse Fourier instruments, data are invariably collected in the time domain; i.e., they are stored in the computer memory as a function of time. However, spectroscopists are interested in the frequency-domain response of a spin system since the energy differences between spin states possess characteristic resonance lines at specific frequencies. In fact, the time domain and the frequency domain are inextricably linked, and we can convert between the two using a procedure known as Fourier transformation (Comment J2.1).

The Fourier transform relates the time-domain data $f(t)$ with the frequency-domain data $f(v)$ by the following equation:

$$f(v) = \int_{-\infty}^{+\infty} f(t) \exp(ivt)dt \qquad (J2.1)$$

The existence of two related domains allows us to define *Fourier pairs*. Several of these are of particular importance in NMR. For example, the Fourier transform of a time-domain decaying exponential is a Lorentzian line (see Chapter A3) at zero frequency (Fig. J2.2a). This is identical to the well-known relationship between the free induction decay and the NMR spectrum (see Section J2.2).

If the time-domain signal is an exponentially decaying sinusoidal or cosinusoidal oscillation, then again the frequency-domain signal is a Lorentzian line, but offset from zero frequency by the frequency of oscillation of the sinusoidal or cosinusoidal waveform (Fig. J2.2b). Recall from Chapter A3 that the Lorentzian is the Fourier transform of an exponential decay function.

A third and equally important example of Fourier pairs is the Fourier transform of a radio-frequency pulse. A basic result of the time–frequency Fourier transform relation is that a short pulse in time can be considered as a

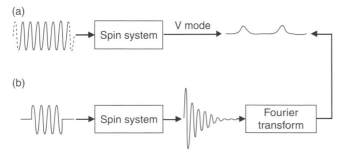

Fig. J2.1 The two basic methods of obtaining an NMR spectrum: **(a)** by applying a continuous excitation and varying energy; **(b)** by applying a pulse of energy and Fourier transforming the result.

multifrequency source, which could allow the simultaneous excitation of different resonance frequencies in an NMR experiment. In order to clarify this point, we examine the magnetization of a sample under the influence of static and radio-frequency fields. In the frame of reference rotating at the angular frequency, ω, of the radio-frequency

field, the nuclear magnetization M_j of nuclei of resonant frequency ω_i, precesses about an effective field given by

$$|B_{\mathrm{eff}}| = (1/\gamma)\left[(\nu_i - \nu)^2 + (\gamma B_1)^2\right]^{1/2} \tag{J2.2}$$

If B_1 is chosen large enough so that

$$\gamma B_1 \gg 2\pi\Delta \tag{J2.3}$$

where Δ (in hertz) is the entire range of chemical shifts in the sample, measured with respect to the radio-frequency, then for any ν_i within the spectrum the term $(\nu_i - \nu)$ can be neglected, and

$$B_{\mathrm{eff}} \approx B_1 \tag{J2.4}$$

The magnetization vector precesses about B_1, which is along the x'-axis, (Section J1.2), for *all* nuclei with Larmor frequencies in the range Δ. The width of the time pulse required to cover this frequency range should therefore be $\ll 1/\Delta$.

We call Δ' the frequency range to be examined, as indicated in Fig. J2.3. Because of the heterodyne nature of the detection scheme normally employed, it is not the actual resonance frequencies, ν_i, that are important, but the differences between them and the applied radio-frequency field, $(\nu_i - \nu)$. If ν is chosen within the range Δ', as indicated by the vertical dashed line in Fig. J2.3, then some frequency differences are positive and some are negative. During data acquisition, however, the detector, which measures in the time domain, cannot distinguish positive and negative frequencies (lines 8 and 12 of Fig. J2.3, for example, will appear to be very close together). Since positive and negative frequencies do differ in phase, two phase detectors are used to unravel the spectrum.

Fourier analysis is also essential for the understanding of the effect of the pulse itself. A rectangular pulse in time of monochromatic radiation of frequency, ν_0 can be described in the frequency domain as a band about ν_0 (Fig. J2.4). The monochromatic frequency, ν_0, is produced by a pulse generator and is called the spectrometer frequency. Following the rules of Fourier transformation, as the pulse length decreases, the width of the frequency band

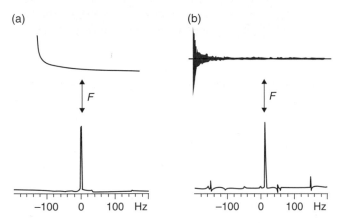

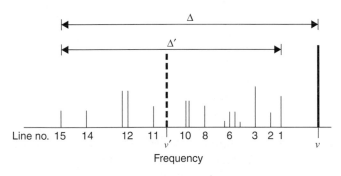

Fig. J2.2 Two "Fourier pairs": **(a)** a decaying exponential gives a Lorentzian line at zero frequency after Fourier transformation; **(b)** an exponentially decaying cosinusoid gives a Lorentzian line offset from zero frequency by an amount equal to the frequency of the oscillation of the cosinusoid.

Fig. J2.3 Schematic spectrum at constant B_0, showing resonance lines covering a range of frequencies Δ'. Reference frequency values are shown within the range (ν') and outside the range (ν).

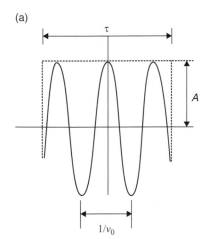

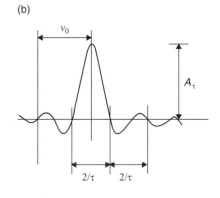

Fig. J2.4 A time pulse of monochromatic radiation of frequency ν_0 with a rectangular envelope, **(a)**, that can be described in the frequency domain as a band centered on ν_0, **(b)**.

must increase (Fig. J2.4). For a typical pulse length of 10 μs, the flat central portion of the frequency band, where the amplitude is within 1% of the peak value, is about 16 kHz wide. For a 7.05 T field ($v = 300$ MHz for protons, 75 MHz for ^{13}C), this region easily spans the range of chemical shifts of the commonly observed nuclei (15 ppm or 4.5 kHz for protons, 200 ppm or 15 kHz for ^{13}C) (see Section J1.2.3).

The waveform in Fig. J2.5a illustrates a typical pulse train, pulse width, and time interval between pulses. The expanded view of one of the pulses is actually a packet of radio-frequency radiation (10^2–10^3 MHz). The width of the pulse, τ, is usually less than 10 μs. The interval, T, between pulses is typically one to several seconds. During T, a time domain radio-frequency signal, called the free induction decay (FID), is emitted by the excited nuclei as they relax. Free induction decay is detected with a radio receiver coil, perpendicular to the static magnetic field.

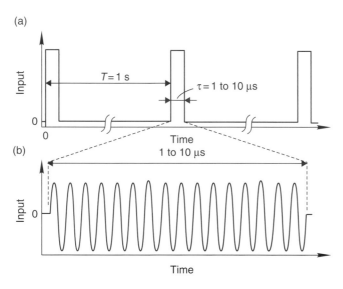

Fig. J2.5 Typical input signal for pulsed NMR: **(a)** pulse sequence; **(b)** expanded view of radio-frequency pulse, typically at a frequency of several hundred megahertz. The time axis is not drawn to scale. It is assumed further that the length of the pulse is short relative to T_1 and T_2, so that no relaxation occurs during the pulse time. (Adapted from Skoog et al., 1995.)

J2.1.2 The Fourier Transform NMR Spectrometer

A schematic layout of a typical NMR spectrometer designed for the liquid state is shown in Fig. J2.6. The sensitivity and resolution of the spectrometer depend critically upon the strength and quality of the magnet, which is thus the key component of the instrument. It is advantageous to operate at the highest possible field strength. In addition, the field must be highly homogeneous and reproducible. These very stringent specifications are only met by superconducting solenoids. At the time of writing, the highest magnetic field available for NMR is 21 T, corresponding to a proton frequency of 900 MHz (Comment J2.2).

The radio-frequency coil acts as both a transmitter and a detector of the resonance frequency. The measured signal processed by the computer is a low-frequency line resulting from the difference between the transmitted and detected frequencies.

The sample is placed in the center of the cylindrical magnet to ensure that all the magnetic nuclei experience the same average field. Although a superconducting magnet operates at liquid helium temperature (4 K), the sample itself is normally at room temperature.

In order to perturb the spin system with radio-frequency energy, the spectrometer contains sophisticated pulse programmer and transmitter units, which allow the application of complex pulse sequences to the sample of interest (Section J2.5). The radio-frequency source is a very stable crystal oscillator unit which runs continuously, and all frequencies in the spectrometer are derived from it. The continuous-wave signal derived from the source is

COMMENT J2.2 MAGNETIC FIELD STRENGTH

A 21 T field is more than 400 000 times stronger than the Earth's magnetic field. Obtaining a homogeneous field of this magnitude poses several technical challenges. The wire for the superconducting solenoid is made from an alloy of niobium and tin, in a ratio of 3:1, that is able to provide the homogeneity and stability required.

Fig. J2.6 A schematic presentation of a typical NMR spectrometer.

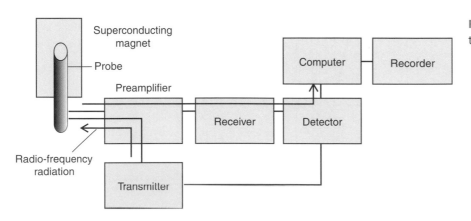

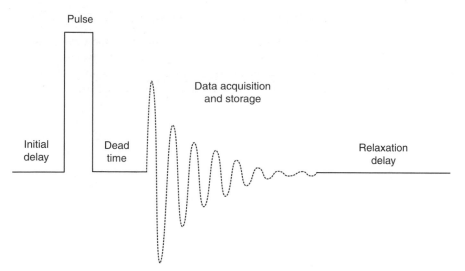

Pulse

Data acquisition
and storage

Initial
delay

Dead
time

Relaxation
delay

controlled in amplitude and phase by the gate unit. Current pulse programmers have better than 1° phase control.

The pulse/acquisition/delay sequence is repeated N times until signal-to-noise ratio is satisfactory. The detection of weak signals such as ^{13}C NMR at natural isotopic abundance (1%) is now routine.

J2.2 SINGLE-PULSE EXPERIMENTS

J2.2.1 Data Acquisition and Processing

The benefits of the Fourier transform method have been enhanced by the introduction of techniques using various radio-frequency pulse sequences. In this section we describe single-pulse NMR experiments to obtain proton or ^{13}C spectra. Figure J2.7 shows a typical sequence of events. After an *initial delay*, the pulse is applied to the sample with the radio-frequency coil. The receiver (detector) is turned on, and after a very short *dead time* the response of the spin system is measured, digitized, and added to the computer memory. The acquisition time should be long enough for the response of the spin system to have decayed to a negligible level (a few times T_2). A further delay allows the sample spins to equilibrate fully (T_1 relaxation). We recall that in the rotating frame of reference the sample is in an external field $\mathbf{B}_0$ along the z-axis and the radio-frequency pulse creates a non-oscillating field $\mathbf{B}_1$ along the x'-axis. In the absence of the radio-frequency pulse, i.e., in the fixed external magnetic field, sample magnetization is small and along $\mathbf{B}_0$, reflecting the imbalance of low- and high-energy spins described by the Boltzmann distribution (Fig. J1.9). When the pulse frequency, v_0, corresponds exactly to the Larmor frequency, the nuclear magnetization vector "tips" out of the z direction (Fig. J1.15). The pulse is along x', and the tip angle is toward y', in accordance with the gyroscopic effect (Comment J2.3). An insight into why NMR is a "resonance" technique can be

COMMENT J2.3 OPTIMUM COMBINATION OF TIP ANGLE AND T_1 RELAXATION TIME

In order to avoid signal saturation, the sum of the acquisition time and the relaxation delay should be several times T_1, but the resulting long experimental times may make unreasonable demands on the available instrumental resources. For best sensitivity, there is an optimum combination of the tip angle, α, the T_1 relaxation time, and the time between pulses, t_{rep}, given by Ernst's equation

$$\cos(\alpha) = \exp(-t_{rep}/T_1)$$

The length of a typical 90° pulse is about 10^{-5} s. For the proton this requires a $\mathbf{B}_1$ field of about 6×10^{-4} T. Note that this is much smaller than the external field, $\mathbf{B}_0$.

gained from Fig. J2.8. Resonance is a condition in which energy is transferred in such a way that a small periodic perturbation produces a large change in some parameter of the system being perturbed. NMR is a resonance technique because the small periodic perturbation $\mathbf{B}_1$ produces a large change in the orientation of the sample magnetization vector $\mathbf{M}$. In most experiments, $\mathbf{B}_1$ is a few orders of magnitude smaller than $\mathbf{B}_0$. The tip angle, α, is given (in radians) by

$$\alpha = \gamma B_1 t_p \tag{J2.5}$$

The radio-frequency pulse is usually designated by the value of α that it produces and the axis to which it is applied. The term $90°_x$ or $(\pi/2)_x$ refers to a pulse directed along the $+x'$-axis. At the end of a $90°_x$ pulse the magnetization vector points along the minus y' direction. Application of the $\mathbf{B}_1$ field for twice as long ($\theta = \pi$) results in inversion of $\mathbf{M}$. Such a pulse is called a 180° or π pulse. Single pulses are

(a)

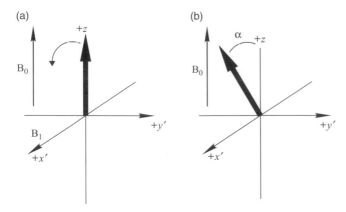

(b)

Fig. J2.8 The effect of a short radio-frequency pulse along x' viewed in the rotating frame. Initially the magnetization vector is parallel to the external field along the z-axis. **(a)** A pulse creating a field along x' exerts a force on the magnetization vector tipping it toward minus y'. **(b)** After the $\mathbf{B}_1$ field has been removed, the magnetization vector maintains its tip angle α with the z-axis.

COMMENT J2.4 SELECTIVE AND NON-SELECTIVE PULSES

A non-selective pulse is a radio-frequency pulse with a wide frequency bandwidth (short, high-power pulse) that excites all nuclei of a given type (e.g., all protons in the sample). A selective pulse is a radio-frequency pulse with a narrow frequency bandwidth (long, low-power pulse) that excites nuclei in a limited chemical shift range.

A crucial feature of pulsed NMR is the ability to excite nuclei with different chemical shifts uniformly and simultaneously. For example, a typical range of ^1H resonance frequencies is 4 kHz (10 ppm × 400 MHz); a 90° pulse of strength $\gamma B_1/2\pi \gg$ 4 kHz therefore rotates the magnetization vectors of all protons irrespective of their resonance frequencies.

---

used both as a method for perturbing the spin system from equilibrium and as a means for detecting the magnetization (Comment J2.3).

The pulse produces a band of frequencies of almost equal amplitude ($\mathbf{B}_1$ value) (Fig. J2.3), so that the nuclei throughout the entire chemical shift range are all effectively tipped by the same angle (Eq. J2.5). As far as the excitation pulse is concerned, each chemically shifted nucleus may be considered to belong to a frame rotating at the corresponding Larmor frequency. However, in order to understand what happens after excitation, it is necessary to assign the frame to a single frequency, which is chosen to be the spectrometer frequency, v_0, the value at the center of the frequency band (Comment J2.4).

J2.2.2 Free Induction Decay

Figure J2.9 shows the behavior of the magnetization vector following excitation by a 90_x° pulse. The $\mathbf{B}_0$ field is still present, and the nuclei continue to precess about it. Focusing first on the nuclei whose Larmor frequency corresponds exactly to the frequency of the rotating frame, the magnetization vector remains directed along the $-y'$-axis, in the absence of relaxation effects (Fig. J2.9a). However, T_1 and T_2^* processes both act to reduce the $-y'$ component of magnetization (Section J1.2.2). $\mathbf{B}_0$ inhomogeneities, the principal source of T_2^* effects, cause groups of nuclei in different parts of the sample to experience slightly different local $\mathbf{B}_0$ fields. Figure J2.9b shows only two of the many such microscopic sample sections (called *isochromats* because the field is the same within the section). One group is shown precessing slightly ahead of the frame, and the other group is lagging behind. At the same time, T_1 relaxation processes cause a gradual return of the magnetization toward the z-axis (Fig. J2.9c), further decreasing the component along $-y'$. Because field inhomogeneities usually cause T_2^* to be less than T_1, an intermediate situation often occurs as in Fig. J2.9d, where the $-y'$ component of the magnetization decays to zero before the spin population can achieve Boltzmann equilibrium. At some later time (after at least five times T_1) the magnetization has again returned to its equilibrium value (Fig. J2.9e). Returning now to the laboratory frame, the $-y'$ component corresponds to a magnetization vector rotating in the xy plane. As the $-y'$ component decreases, the oscillating voltage from the coil decays, and it reaches zero when the condition of the spins corresponds to the situation depicted in Fig. J2.9d. The record of the receiver voltage in the time domain is called the FID, because the nuclei are allowed to precess "freely" in the absence of the $\mathbf{B}_1$ field. The FID is the sum of the individual oscillating voltages from the various nuclei in the sample, each with a characteristic offset frequency (i.e., chemical shift and spin–spin couplings), amplitude, and T_2. It contains all the information necessary to obtain an NMR spectrum. The FID is the Fourier partner of the NMR frequency spectrum.

The FID of ethanol is shown in Fig. J2.10. The frequency-domain spectrum obtained by Fourier transformation is shown, in Fig. J1.18b. The FID curve in Fig. J2.10 is very complex because it is composed of eight components (minimum), each with a characteristic frequency.

J2.3 MULTIPLE-PULSE EXPERIMENTS

In a multiple-pulse experiment a specific radio-frequency pulse sequence is applied to the sample before the FID is measured. Such experiments have greatly enriched the potential of NMR applications because they provide information that is difficult or impossible to obtain by the

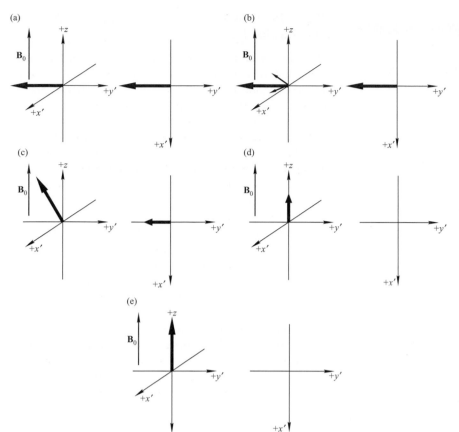

Fig. J2.9 Behavior of nuclei after a 90°_x pulse. The frame is rotating at the spectrometer frequency, f_s, which is assumed to be equal to the Larmor frequency. **(a)** Immediately after the pulse the tipped magnetization vector is fixed in the rotating frame. **(b)** T_2^* effects (mainly $\mathbf{B}_0$ inhomogeneity) cause divergent local Larmor frequencies. Only two divergent vectors are shown here. **(c)** At the same time, T_1 relaxation is taking the magnetization back toward the z-axis. **(d)** T_2^* effects have resulted in complete loss of magnetization in the $x'y'$ plane, but the nuclei have not returned to equilibrium. **(e)** Complete restoration of equilibrium z-axis magnetization. (Adapted from King and Williams, 1989a.)

Time

Fig. J2.10 FID of a sample of ethanol. (King and Williams, 1989b.)

single-pulse technique. A pulse sequence is defined by the amplitude and width of each pulse and the time delay in between (Comment J2.5).

There is an enormous variety of multiple-pulse experiments. They fall into three main groups presented in Table J2.1.

The first group (group 1 in Table J2.1) includes pulse sequences for *relaxation time* measurements. A simple two-pulse sequence is the basis of the *inversion-recovery* method for the measurement of the *spin–lattice longitudinal relaxation* time T_1. The spin echo method for measurements of the *spin–spin transverse relaxation characteristic time T_2* is based on a different pulse sequence, which was originally designed to refocus effects of $\mathbf{B}_0$ inhomogeneity.

The second group (group 2 in Table J2.1) includes heteronuclear polarization transfer techniques for sensitivity

COMMENT J2.5 HSQC, MULTIPLE-PULSE SEQUENCES AND QUANTUM MECHANICS

Heteronuclear single quantum coherence (HSQC) is used frequently in NMR spectroscopy of proteins. The resulting spectrum is two-dimensional, with one axis for ^1H and the other for a *heteronucleus* (a nucleus other than a proton, usually ^{15}N or ^{13}C). The spectrum contains a peak for each proton–heteronucleus pair being considered. The HSQC provides for highly sensitive 2D-NMR experiments. The basic scheme of this experiment involves the transfer of magnetization on the proton to the second nucleus, ^{15}N or ^{13}C, via INEPT (see below). After a time delay (t_1), the magnetization is transferred back to the proton via a retro-INEPT step and the signal is then recorded. In HSQC, a series of experiments is recorded where the time delay t_1 is incremented. The ^1H signal is detected in the directly measured dimension in each experiment, while the chemical shift of ^{15}N or ^{13}C is recorded in the indirect dimension, formed from the series of experiments.

In many cases the explanation of how a pulse sequence acts requires quantum mechanical analysis at a level of complexity that is beyond the scope of this book. The interested reader is referred to the various texts and publications on this subject such as Derome (1987), Ernst *et al.* (1987), and Brey (1988).

enhancement. These techniques greatly contribute to the success of NMR for the observation of rare and low-γ nuclei. The basic trick involves "borrowing" polarization from rich spin nuclei of high γ-value (see Section J2.3.3). The group includes many very sophisticated pulse sequences with special time delays and sensitive polarization transfer simultaneously for all spins in the sample. They are represented in Table J2.1 by the insensitive nuclei enhanced by polarization transfer (INEPT) method, which is conceptually one of the simplest.

Almost all heteronuclear multidimensional NMR experiments use INEPT to transfer magnetization between different spin species, but, unfortunately, the efficiency of INEPT deteriorates with increasing rotational correlation time (molecular tumbling). In contrast, transfer polarization by the cross-relaxation induced polarization technique (CRIPT) is independent of rotational correlation time. The combination of INEPT and CRIPT leads to another highly efficient transfer protocol (CRINEPT) for solution NMR on very high molecular mass samples (up to 100 kDa).

The third group in Table J2.1 uses the nuclear Overhauser effect (NOE) based on polarization transfer via space dipolar coupling.

In this section we discuss the simplest multiple pulse schemes (inversion recovery and spin echo for relaxation measurements and the INEPT polarization technique). Nuclear Overhauser effect will be discussed in Section J2.4 and CRINEPT in Section J2.5.

J2.3.1 The Inversion Recovery Method to Measure Spin–Lattice Relaxation Time T_1

The method is shown schematically in Fig. J2.11. The nuclei are first subjected to a radio-frequency pulse of sufficient width to cause **M** to rotate through 180° to the z-axis. Following the pulse there is a delay period, τ, whose length is chosen appropriately. The delay period, τ, is a key component of most multiple-pulse methods. Typically, eight or ten values of τ are used, ranging from a small value to four or five times T_1, in an inversion-recovery experiment (Comment J2.6). During the τ interval the system returns to equilibrium by the spin–lattice relaxation process. Spin–spin relaxation, which involves magnetization components only in the xy plane, is not involved, because **M** lies along the z-axis. As the individual magnetic moments gradually return to their favored orientations along the +z-axis, the net magnetization vector becomes shorter in the –z direction. Depending on the length of the delay, **M** passes through zero and eventually recovers its full original magnitude along the +z-axis.

In order to determine the extent of relaxation, **M** must be converted into observable magnetization in the xy plane. This is done by applying a second pulse of half the length of the first to the +x-axis; the 90°_x pulse rotates **M** to lie along the y-axis (Fig. J2.12). If **M** is still negative before the 90° pulse, then the magnetization is rotated to the +y direction, but when relaxation is complete, the magnetization after the 90° pulse lies along the –y-axis. The intensity I of the peak resulting from the Fourier transform of the FID varies with τ from a maximum negative value for $\tau = 0$ to a maximum positive value for $\tau = \infty$ (in practice, four or five times T_1). The intensities are related exponentially to T_1 by the equation:

$$I = I_\infty[1 - 2\exp(-\tau/T_1)] \tag{J2.6}$$

A typical plot of peak intensity as a function of τ is shown in Fig. J2.13. The peak intensity goes through zero when $\tau/T_1 = \ln(2)$, i.e., when $\tau = 0.693T_1$.

J2.3.2 The Spin-Echo Effect to Measure T_2

We recall that magnetization in the xy plane may be lost by processes that do not affect the z component. The rate constant for this relaxation is the reciprocal of the time constant of the FID, T_2^*, the effective spin–spin relaxation time. T_2^* may be broken down into a $\mathbf{B}_0$ inhomogeneity component (the major cause of line broadening in NMR) and the spin–spin relaxation time, T_2. The value of T_2 is determined by the same factors that are responsible for T_1, plus other processes such as spin and chemical exchange.

A spin echo is a magnetic analogue of an audio echo: a pulse of magnetization is formed, is allowed to spread, is reflected, and is then detected as another pulse a short time later.

TABLE J2.1 EXPERIMENTAL MULTIPLE-PULSE EXPERIMENTS

Multiple-pulse scheme	Physical phenomenon	Pulse sequence	Applications
Group 1			
Inversion recovery	Relaxation process	180°_x 90°_x Relaxation Delay, τ, Variable Delay, Acquisition	Measurements of longitudinal relaxation time T_1
Spin echo	Relaxation process	90°_x 180°_x 180°_x τ_1 τ_2 τ_3 τ_4	Measurements of transversal relaxation time T_2^*
Group 2			
Insensitive nuclei enhanced by polarization transfer (INEPT)	Polarization transfer via scalar couplings	^{1}H: 90°_x τ 180°_x τ 90°_y; ^{13}C: 180°_x τ 90°_x	Enhancement sensitivity of NMR experiments on rare and low gyromagnetic ratio nuclei
Cross-relaxation insensitive nuclei enhanced polarization transfer (CRINEPT)	Highly efficient polarization transfer for ^{15}N, ^{1}H and other nuclei	(c) I: $-x$, τ, $-x/y$, $x/-y$, $-x$, ϕ_1, Ψ_1, t_1; S	Key element of multidimensional NMR (building block to create magnetization and coherence where required)
Group 3			
Nuclear Overhauser phenomenon (NOE)	Polarization transfer via space dipolar coupling	S, I	Key constraint for three-dimensional structure determination (distance measurements)

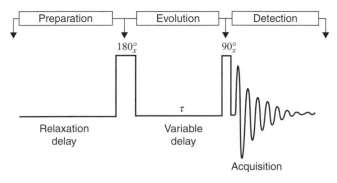

Fig. J2.11 Preparation — Evolution — Detection

180°_x 90°_x

Relaxation delay, τ, Variable delay, Acquisition

Fig. J2.11 Schematic pulse sequences for relaxation time T_1 measurements by *inversion recovery*.

COMMENT J2.6 COMPACT REPRESENTATION OF A PULSE SEQUENCE

Although the schematic pulse sequence shown in Fig. J2.11 is a useful pictorial device, it may be represented more compactly as:

$$\text{equilibration delay} - 180^\circ_x - \tau - 90^\circ_x - \text{acquisition}$$

The sequence may be divided into three periods: preparation (equilibrium delay and the 180°_x pulse), evolution (the time delay), and detection (the 90°_x pulse and the data acquisition). These periods are easy to pick out in the simple sequences and are important when analyzing the more involved pulse sequences present.

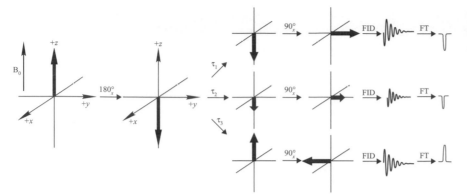

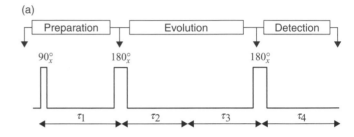

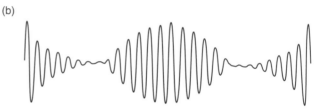

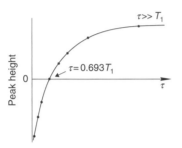

Fig. J2.13 Plot of transformed peak intensity as a function of τ. (After Williams and King, 1990a.)

(a)

Preparation	Evolution	Detection
90°_x	180°_x	180°_x

τ_1　τ_2　τ_3　τ_4

(b)

Fig. J2.14 (a) Pulse sequence in a spin-echo experiment. **(b)** NMR signal showing "echoes" at the end of τ_2, τ_4. The NMR signal is shown throughout the entire sequence, but in practice the receiver is activated only at the time of appearance of the echo (i.e., after τ_2, τ_4, etc.).

The *spin-echo* pulse sequence shown in Fig. J2.14 allows the independent measurement of the $\mathbf{B}_0$ inhomogeneity component (Comment J2.7). The initial 90°_x pulse turns the total magnetization vector to the $-y$-axis. If the axes are rotating at the exact Larmor frequency of the nuclei, the effect of T_2 is a gradual shortening of $\mathbf{M}$ in the $-y$ direction. However, because $\mathbf{B}_0$ is not perfectly homogeneous, some of the sample isochromats (Section J2.1) precess faster than the rotating frame and move in a counterclockwise direction; the fastest group is labeled 1 in Fig. J2.15, and the slightly slower one is labeled 2. Similarly, isochromats 3 and 4 precess slower than the frame and rotate clockwise. As a result of this disparity, the spin isochromat vectors fan out in the xy plane during the delay τ_1, after the 90° pulse. At the end of the first τ delay a 180°_x pulse flips the spins about the x-axis. The isochromats are still precessing in the same directions at the same rates, so, for example, isochromat 1 that moved fastest counterclockwise now has farthest to go to reach the $+y$-axis. After a second τ period of equal length all the isochromats refocus along the $+y$-axis to give a maximum xy magnetization. If the NMR signal is monitored during this $-90^{\circ}_x - \tau - 180^{\circ}_x - \tau-$ sequence, it dies away during the first τ interval (τ_2, τ_4 in Fig. J2.15), then it recovers, to another maximum, the spin echo, when the isochromats are refocused after τ_2. Since the magnetization is now along the $+y$-axis, the echo signal is inverted with respect

---

COMMENT J2.7 ABBREVIATED FORM OF THE ECHO SEQUENCE

In customary abbreviated form, the echo sequence is
equilibration delay $-90^{\circ}_x - \tau - 180^{\circ}_x - \tau-$ acquisition
Again the three periods corresponding to preparation (equilibrium delay and 90° pulse), evolution (the $\tau - 180^{\circ}_x - \tau$ segment), and detection are evident in the pulse sequence.

---

to the first signal observed. After the period shown as τ_3 in Fig. J2.15, the FID again decays to a minimum because of the same $\mathbf{B}_0$ inhomogeneity effects as before. A second $180^{\circ}_x - \tau$ sequence results in an echo at the end of τ_4 along the $-y$-axis. The amplitude is reduced by T_2 relaxation and also by any residual spin–lattice effects in the system. The time constant for the decrease in the magnitude of the echo

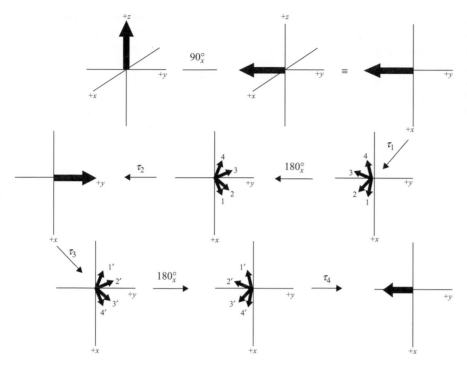

Fig. J2.15 Vector description of a spin-echo experiment. A spin echo is a magnetic analogue of an audio echo: A pulse of magnetization is formed, is allowed to spread, is reflected, and is then detected as another pulse a short time later (After Williams and King, 1990a.)

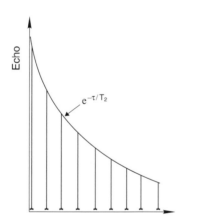

Fig. J2.16 The exponential decay of the spin echoes gives a transverse relaxation time T_2^*.

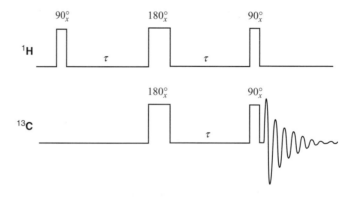

Fig. J2.17 Pulse sequence for INEPT experiments.

is the "true" T_2, since the effects of $\mathbf{B}_0$ inhomogeneity were eliminated by the refocusing process (Fig. J2.16).

With small modifications, the spin-echo sequence can provide some very useful structural information. The simplest example is the attached proton test, which allows the determination of whether each carbon in a molecule has an odd or an even number of protons attached to it.

J2.3.3 Polarization Transfer

Another important group of multiple-pulse experiments includes the *polarization transfer* methods for sensitivity enhancement. A nucleus such as ^{13}C or ^{15}N suffers from poor sensitivity because of its low natural abundance

(Table J1.3) and also because of its small gyromagnetic ratio, γ, which, for a given $\mathbf{B}_0$, determines the energy difference between the two spin states (Eq. (J1.6)). Since the gyromagnetic ratio of the proton is very nearly four times that of ^{13}C, the population difference is in the same proportion (Table J1.2). If some of the polarization of the proton can be transferred by some means to a less sensitive nucleus, then the signal of the latter will be enhanced. One pathway for polarization transfer is the NOE, which operates *via the same through-space dipolar interactions* that are mostly responsible for spin–lattice relaxation in the $^{13}C-^1H$ system. In contrast, polarization transfer pulse sequences were developed to achieve the desired result solely *via spin coupling effects*.

Conceptually one of the simplest experiments of this type uses INEPT, which is shown in Fig. J2.17 (Comment J2.8). The initial 90_x° preparation pulse non-selectively rotates the vectors for all proton resonances to the $-y$-axis (Comment J2.9). During the first τ interval, vectors rotate in the xy plane as a result of chemical shift and spin–spin coupling. However, only the precession due to spin coupling must be considered, because the $\tau-180^\circ-\tau$ segment refocuses dispersion of the spin vectors arising from the chemical shift difference. One essential feature of the experiment is the length of the τ delay, which is made equal to $1/(4J_{C\text{-}H})$. This means that at the end of the first τ interval the two vectors have rotated 90° apart from one another, shown as + and −45° from the $-y$-axis in Fig. J2.18. The protons are then subjected to a 180°_x pulse, which ensures that chemical shift effects are refocused. However, because of the second part of the sequence, the usual spin echo at the $+y$-axis is not observed. Simultaneously with the proton 180°_x pulse, a 180°_x pulse applied to the carbons causes the labels of the spin states of the protons to be preserved. Protons that were formerly bonded to carbons in an α spin state now find themselves coupled to ^{13}Cs with β spins, and vice versa. This means that the rotation direction of the vectors is also switched; the counterclockwise-rotating vector becomes the clockwise-rotating one, etc. Instead of meeting on the point of the y-axis at the end of the second $1/(4J_{C\text{-}H})$ interval, the vectors arrive 180° out of phase along + and $-x$-axes. The 90°_y pulse on the protons rotates the antiphase vectors to the z-axis, and the polarization transfer to the coupled carbons is produced. The immediate 90°_x pulse rotates the carbon magnetization to the $-y$-axis, and the FID is obtained in the usual manner.

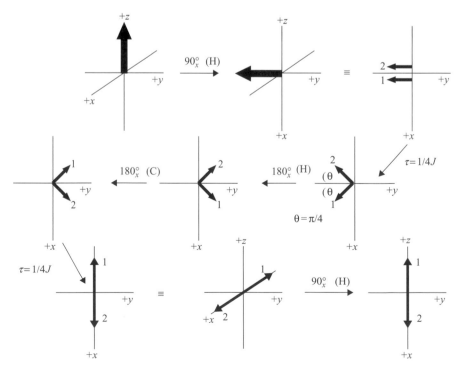

Fig. J2.18 Vector description of the spin system in INEPT experiments.

J2.4 NUCLEAR OVERHAUSER ENHANCEMENT

Nuclear Overhauser enhancement is illustrated for a two-spin system in Fig. J2.19. A saturating radio-frequency field is applied to the high-γ spins S. The resulting population redistribution leads to a polarization enhancement of the I spins, provided the relaxation processes are favorable. This transfer of polarization is called NOE.

In one-dimensional NMR, NOE can be observed in a number of ways, all of which involve the application of a selective radio-frequency pulse at the position of one of the resonance peaks in a system containing two or more. Quantitatively, the NOE is expressed in terms of the relative increase in signal intensity

$$NOE = (I - I_0)/I_0 \tag{J2.7}$$

where I and I_0 are the intensities with and without the enhancement due to the Overhauser effect.

The principle of NOE is relatively simple (Fig. J2.20). Consider a system of two protons in a molecule, the

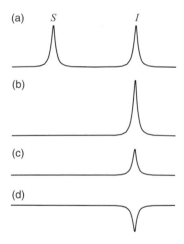

(a) S I

(b)

(c)

(d)

Fig. J2.19 Schematic presentation of NOE. A molecule contains two inequivalent spins, I and S, with no scalar coupling, so that the NMR spectrum consists of a singlet at each of the chemical shifts. (a) Conventional spectrum of two neighboring spins S and I. (b)–(d) Possible spectra resulting from saturation of S resonance: the I peak gets stronger (b), weaker (c), or inverts (d) depending on the conditions.

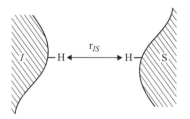

Fig. J2.20 The basis of NOE experiments. r_{IS} is the distance between the protons I and S.

resonance peak of one of which is saturated as described above. NOE is a consequence of the modulation of dipole–dipole coupling between the two spins by the molecular Brownian motion. The equation describing NOE has the general form

$$NOE \propto f(r_{IS})f(\tau_c) \tag{J2.8}$$

where $f(r_{IS})$ is a function of the distance r_{IS} between the protons and $f(\tau_c)$ is a function of the molecular rotational correlation time τ_c (Section D2.3).

First, we discuss the $f(r_{IS})$ term. This term is a cross-relaxation effect due to the dipole–dipole interaction

$$\left\langle \frac{1}{r^3} \times \frac{1}{r^3} \right\rangle = \left\langle \frac{1}{r^6} \right\rangle \tag{J2.9}$$

A relationship between the observed intramolecular NOE and the sixth power of the internuclear distance was confirmed from studies of single proton–proton interactions and of the interaction between a proton and the protons of a methyl group in a series of molecules related to alkaloids,

$$f_I(S) = 1/\left(2 + Ar_{IS}^6\right) \tag{J2.10}$$

We recall that r_{IS} is the separation between proton I (whose resonance peak is observed) and proton S (whose resonance peak has been saturated), and A is a constant.

It follows from Eq. (J2.10) that NOE is applicable only for very close neighbors, typically, nuclei closer to each other than 5 Å. The short range of the effect limits its applications, but provides great specificity for the assignment of proton NMR peaks (Section J2.5).

The maximum value of NOE for the heteronuclear case is given by

$$NOE_{max} = 0.5\gamma_S/\gamma_I + 1 \tag{J2.11}$$

where γ_S/γ_I is the ratio of gyromagnetic ratios of the saturated and observed nuclei. The maximum NOE is 4.0 if protons are irradiated while ^{13}C nuclei are observed. For nuclei with negative gyromagnetic ratios such as ^{15}N (Table J1.2), irradiation of protons causes a negative NOE, leading to partial or complete loss of the signal.

Spin diffusion is the cause of one of the fundamental difficulties that arise when trying to correlate NOE intensities with intramolecular distances. In a network of like-spins contained in a macromolecule tumbling with a correlation time τ_c such that $\omega\tau_c > 1$, spin diffusion by two or several subsequent cross-relaxation steps can greatly influence the observed NOE. In the simple example of three spins shown in Fig. J2.21, a two-step pathway for cross-relaxation, spin 1 to spin 2 followed by spin 2 to spin 3, may under certain experimental conditions be more efficient than direct cross-relaxation between spin 1 and spin 3. The NOE on spin 3 is then no longer a faithful manifestation of the internuclear distance $r_{1,3}$. In the spatial structure of proteins and nucleic acids, the geometric arrangement of hydrogen atoms usually

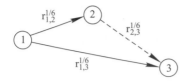

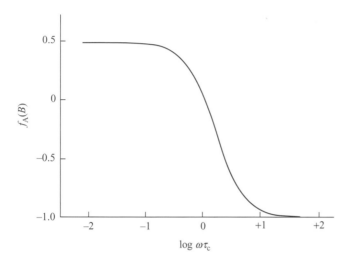

Fig. J2.21 Direct cross-relaxation between spins 1 and 2, and 1 and 3 (solid arrows) and spin diffusion pathway from spin 1 via spin 2 to spin 3 (broken arrow).

Fig. J2.22 Dependence of NOE on the product of the Larmor frequency ω_0, and the rotational correlation time τ_c. calculated according to Eq. (J2.8) on a double logarithmic scale.

allows for a variety of spin diffusion pathways in addition to direct cross-relaxation between distinct groups of protons. Next, we discuss the rotational correlation time dependence of NOE, i.e., $f(\tau_c)$ in Eq. (J2.8). Theory shows that for two dipolar protons

$$\text{NOE} = \frac{5 + \omega^2 \tau_c^2 - 4\omega^4 \tau_c^4}{10 + 23\omega^2 \tau_c^2 + 4\omega^4 \tau_c^4} \qquad (J2.12)$$

Equation (J2.12) shows that the NOE phenomenon is intimately related to spin relaxation. Analogously to the spin relaxation times T_1 and T_2, the NOE varies as a function of the product of the Larmor frequency ω_0, and τ_c (Fig. J2.22). Rotational correlation times are of the order of 10^{-10}–10^{-12} s for small molecules; in the range of field strengths used in NMR spectrometers, Larmor frequencies are in the range 3.6×10^8–3.6×10^9 rad s^{-1}. Thus, $\omega_0 \tau_c$ is about 0.1–0.01. The correlation times, however, become longer with increasing molecular mass and solvent viscosity.

We see from Fig. J2.22 that as long as $\omega \tau_c < 0.1$, the term in $\omega^2 \tau_c^2$ contributes very little and the NOE is close to +0.5. When $\omega \tau_c = 10$ or more, NOE approaches a limit of –1, which corresponds to the disappearance of the I signal. The change from positive to negative NOE occurs at $\omega \tau_c = 1.118$.

An example of changing the sign of NOE in the same spin system by changing the Larmor frequency is shown in Fig. J2.23. The 90 and 250 MHz spectra of valinomycin in

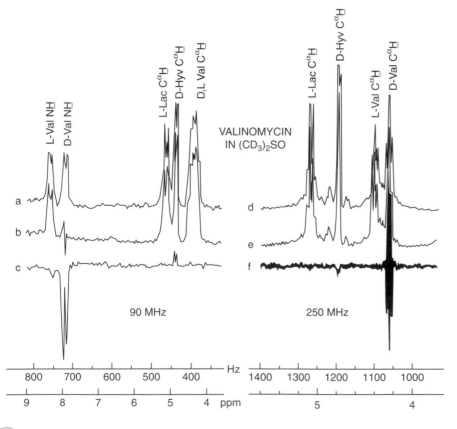

VALINOMYCIN IN (CD₃)₂SO

Fig. J2.23 Demonstration of positive NOE at 90 MHz and negative NOE at 250 MHz for the D-hydroxyvaline α-proton in valinomycin in deutero, dimethylsulfoxide solution: **(a)**, **(d)** normal spectra; **(b)**, **(e)** spectra with irradiation of the D-Val NH resonance; **(c)**, **(f)** difference spectra. (After Pitner et al., 1976.)

dimethylsulfoxide solution are plotted with and without irradiation on the D-Val-NH resonance (*S*). The effect on the hydroxyvaline α-proton (*I*) is positive at 90 MHz, and negative at 250 MHz.

The four curves in Fig. J2.24 describe the maximum NOEs for four nuclear spins (^{13}C, ^{31}P, ^{15}N, ^{1}H) interacting with ^{1}H, if the pre-irradiation is on ^{1}H and the relaxation of the observed spin is entirely by dipole–dipole coupling with the pre-irradiation proton. For ^{1}H–^{13}C and ^{1}H–^{31}P the NOE is positive over the entire τ_c range and becomes very small for long τ_c. For ^{15}N–^{1}H NOE is negative throughout because of the negative value of γ (Table J2.1).

J2.5 TWO-DIMENSIONAL NMR

The term one-dimensional NMR refers to experiments in which the transformed signal is presented as a function of a single frequency. By analogy, in a two-dimensional NMR spectrum the coordinate axes correspond to two frequency domains.

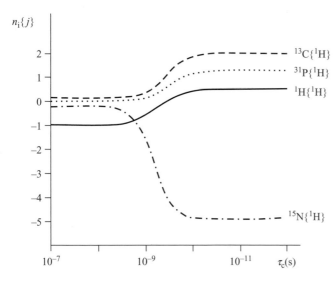

Fig. J2.24 Plots of the maximum NOE (Eq. (J2.12)) versus log τ_c for ^{1}H, ^{13}C, ^{15}N, and ^{31}P interacting with ^{1}H: $B_0 = 11.74$ T, pre-irradiation on ^{1}H; it is assumed that the relaxation is entirely by dipole–dipole coupling with pre-irradiated proton. (Wutrich, 1986.)

The general scheme of two-dimensional NMR spectroscopy is demonstrated in Fig. J2.25. There are four successive time periods: *preparation*, *evolution*, *mixing*, and *detection*. The preparation period usually consists of a delay time, during which thermal equilibrium is attained. Following this period, the spin system is prepared to an initial out-of-equilibrium state. Normally this is done by a 90° radio-frequency pulse as shown in Fig. J2.7. Coherence means that an ensemble of like spins in the sample have the same phase in the $x'y'$ plane (see Fig. J1.17).

Unlike one-dimensional NMR, two-dimensional NMR is characterized by the introduction of the *evolution* and *mixing* periods. The former defines a time variable t_1. Signals from the evolution and detection time periods, t_1 and t_2, respectively, are mixed in the time period, τ_m. The data acquisition of NMR signals is performed only during the detection period t_2. By varying the delay time t_1 systematically, while keeping other experimental conditions unaltered, we obtain a matrix type of data set $s(t_1, t_2)$. Double Fourier transformation yields a two-dimensional spectrum frequency $S(\omega_1, \omega_2)$.

The general scheme of two-dimensional NMR is very flexible (Fig. J2.25), and can be adapted to the type of sample, and to the different parameters to be measured. Different classifications have been proposed for two-dimensional NMR experiments. Wuthrich proposed a classification into two groups (Comment J2.10). Below we describe the classification in three groups, according to Ernst:

(1) Experiments designed to *correlate* transitions of coupled spins by transferring transverse magnetization or multiple-quantum coherence from one transition to another in the course of a suitably designed mixing process. This type of experiment is called correlation spectroscopy known under the acronym (COSY).

(2) Experiments designed to *separate* different interactions (e.g., chemical shifts and spin–spin couplings) in orthogonal frequency dimensions, with the purpose of resolving one-dimensional spectra by spreading overlapping resonances in a second dimension. These experiments require conditions such that the spectra in the evolution and detection periods contain different information. The method is called homonuclear or heteronuclear two-dimensional *J*-resolved spectroscopy or spin echo spectroscopy.

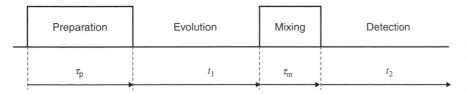

Fig. J2.25 Basic scheme for two-dimensional time-domain spectroscopy, with four distinct intervals leading to a time domain signal $s(\tau_p, t_1, \tau_m, t_2)$. In standard experiments τ_p and τ_m are fixed values and the time-domain signal is a function of only two parameters, t_1 and t_2.

(3) Nuclear Overhauser enhanced spectroscopy (NOESY), which is concerned with the study of *dynamic processes* such as chemical exchange, cross-relaxation, or transient Overhauser effects.

After Ernst proposed this classification, the transverse relaxation-optimized spectroscopy method was developed. Transverse relaxation-optimized spectroscopy suppresses transverse nuclear spin relaxation, which is the direct cause of the deterioration of NMR spectra of large molecular structures. The combination of TROSY and CRINEPT (Section J2.5.3) allows the collection of high-resolution spectra from structures with molecular masses >100 kDa, significantly extending the range of macromolecular systems that can be studied by NMR in solution.

COMMENT J2.10 ON THE CLASSIFICATION OF TWO-DIMENSIONAL NMR EXPERIMENTS

According to Wuthrich, it is useful to classify two-dimensional NMR experiments into two groups:

(1) experiments for delineating *through-bond*, scalar spin–spin connectivity, such as *correlated NMR spectroscopy*; and (2) experiments for delineating through-space, dipolar spin–spin connectivity, such as *nuclear Overhauser spectroscopy*.

J2.5.1 Correlation Spectroscopy

The COSY technique is one of the simplest, and yet most useful, of the various two-dimensional NMR techniques. In fact, COSY was the first two-dimensional NMR experiment to be described. The simplest COSY experiment is shown in Fig. J2.26. The pulse sequence of the type $90°$–kt_1–$90°$– data acquisition, where $k = 0, 1, 2, \ldots, 2^n$. The experiment is repeated for each k value. FID signals are measured and plotted as $s(t_1, t_2)$, where t_1 and t_2 are independent time variables corresponding to the two sampling times. Digital values of t_1 and t_2 are taken in order to apply a fast Fourier transformation algorithm to obtain two-dimensional spectra $S(\omega_1, \omega_2)$ in frequency space.

Before describing the theoretical and practical aspects of this important technique in detail, it is worthwhile demonstrating in a purely classical manner how a two-dimensional spectrum can result from this sequence. Consider two protons A and B, which are scalar coupled. A simple pulse and collect sequence, followed by Fourier transformation, generates an NMR spectrum as shown diagrammatically in Fig. J2.27. We use the terms ω_A, ω_B, J_{AB} in obvious notation to denote the Larmor frequencies of spins A and B and their scalar (J) coupling. In essence, spin A has precessed with a Larmor frequency of ω_A during t_1, and spin B has precessed with a Larmor frequency of ω_B during t_1, although in fact two frequencies are involved for each spin ($\omega_A \pm 1/2 J_{AB}$ and $\omega_B \pm 1/2 J_{AB}$) due to the spin–spin coupling. Now consider what would happen if we were to use the pulse sequence of Fig. J2.28. Let us postulate that we can record data during both

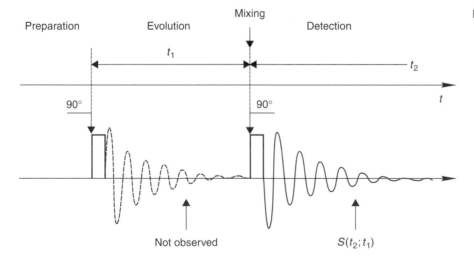

Preparation · Evolution · Mixing · Detection

t_1 · t_2 · t

$90°$ · $90°$

Not observed · $S(t_2; t_1)$

Fig. J2.26 The COSY pulse sequence.

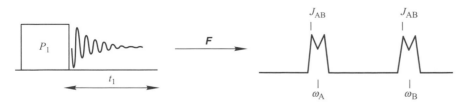

P_1 · t_1 · $\xrightarrow{F}$ · J_{AB} · J_{AB} · ω_A · ω_B

Fig. J2.27 Fourier transformation of the FID resulting from the application of a radio-frequency pulse to a two-spin system results in the NMR spectrum.

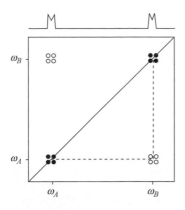

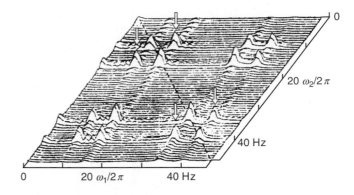

Fig. J2.28 Schematic illustration of the contour plot of an ^1H–^1H COSY spectrum of two coupled spins. The filled circles represent diagonal peaks, whereas the open circles represent off-diagonal peaks or cross-peaks. The dashed line illustrates how cross-peaks correlate diagonal peaks derived from scalar coupled spins. (Adapted from Homans *et al.*, 1989.)

Fig. J2.29 The two-dimensional spectrum of 2,3-dibromothiophene (an *AX* spin system), obtained by applying the pulse sequence in Fig. J2.26.

> **COMMENT J2.12 TWO-DIMENSIONAL NMR AND TWO-DIMENSIONAL ELECTROPHORESIS**
>
> The separation of interactions by two-dimensional spectroscopy can be compared with two-dimensional electrophoresis.

> **COMMENT J2.11 2,3-DIBROMOTHIOPHENE**
>
>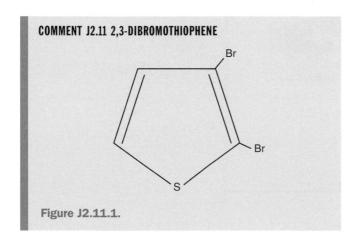
>
> **Figure J2.11.1.**

the t_1 and t_2 intervals. After 90° pulse and the delay t_1, the situation is analogous to that of Fig. J2.27, i.e., each proton resonates with its characteristic Larmor frequency during t_1. After the second 90° pulse, the situation becomes more complicated. The second pulse may appear to have no effect on the spins, i.e., they continue to resonate at ω_A and ω_B during t_2, just as they did in t_1. Alternatively, a portion of the magnetization associated with spin A during t_1 may transfer to spin B during t_2. This result derives from the quantum mechanical process known as *coherence transfer*. In an analogous manner, a proportion of the magnetization which precessed at ω_B during t_1 now precesses at ω_A during t_2. Now, since we are observing the NMR spectrum with respect of two time periods, it follows that we must employ a two-dimensional Fourier transform to observe the frequency components

described above. This means that we may display the spectrum in two orthogonal dimensions on a plane. Such a display is shown in Fig. J2.28 for the two-spin case described above.

The real spectrum obtained for 2,3-dibromothiophene (Comment J2.11) with a pulse sequence like the one in Fig. J2.26 is shown in Fig. J2.29. Two types of two-dimensional peaks, spread with non-vanishing intensities, can be seen. Eight peaks around the diagonal line (the dashed straight line in Fig. J2.29) on the spectra result from the evolution of the magnetization which has transition frequencies originating from spin flips of the same spins in both transitions. The other eight peaks off the diagonal line correspond to the spin flips belonging to different coupled spins in the two transitions. The latter eight peaks are very important in this two-dimensional NMR method since the appearance of these peaks just indicates that there is scalar coupling between two nuclei.

These cross-peaks are utilized in protein NMR to elucidate the coupling of various proton species even in a complicated spectrum where many resonance lines overlap.

In a second example, Fig. J2.30a illustrates the one-dimensional NMR and the COSY spectra for a small 57-amino-acid residue protein, the protease inhibitor K. The two-dimensional spectrum can be plotted in different ways (Comment J2.12). A *stacked plot* (Fig. J2.30b) conveniently shows the peak heights in the third dimension, while a *contour* plot (Fig. J2.30c) is more convenient to show peak positions in the two-dimensional plane. In the stacked plot of the ^1H COSY spectrum (Fig. J2.30b), the complete one-dimensional NMR spectrum (Fig. J2.30a) can be recognized on the diagonal from the upper right to the lower left. For example, the highest field methyl resonance at -0.9 ppm has coordinates ($\omega_1 = -0.9$ ppm, $\omega_2 = -0.9$ ppm),

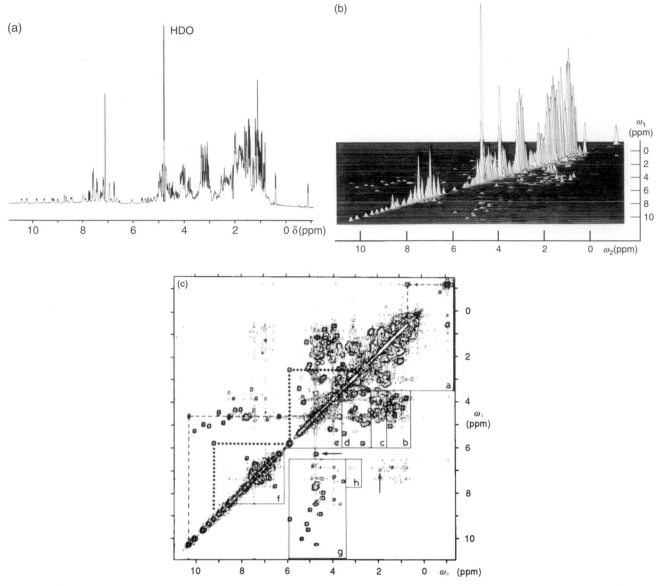

Fig. J2.30 ^{1}H COSY spectrum of a D$_2$O solution of inhibitor K (0.01 M, pD 3.4, 25 °C, 360 MH): **(a)** the one-dimensional ^{1}H NMR spectrum; **(b)** stacked plot representation of the COSY spectrum; **(c)** contour plot of the COSY spectrum. (After Wuthrich, 1986.)

and the lowest field amide-proton line at 10.3 ppm is at ($\omega_1 = 10.3$ ppm, $\omega_2 = 10.3$ ppm). The arrangement of the cross-peaks is best seen in the contour plot (Fig. J2.30c). The spectrum is *symmetrical* with respect to the diagonal. Using the empirical rule that COSY cross-peaks are with few exceptions observed only between protons separated by three or less covalent bonds in the amino acid structure allows one to outline the regions a–h (Fig. J2.30c, below the diagonal) between special protons in amino acids.

J2.5.2 Nuclear Overhauser Enhanced Spectroscopy

The NOESY pulse sequence is similar to that for proton spin systems in COSY, with an additional third 90° pulse (Fig. J2.31). The resulting sequence is: 90°–t_1–90°–τ_m–90°–t_2–acquisition.

As in COSY, the equilibrium delay and initial 90° pulse prepare the nuclei by turning z-axis magnetization to the *xy* plane. During the variable t_1 evolution period the nuclei are frequency-labeled by their chemical shift values. The second 90° pulse rotates part of the magnetization back to the z-axis. During the τ_m interval (which is usually kept constant), the longitudinal magnetization is allowed to relax. This produces a mixing of the magnetization of nuclei that are related by the dipolar relaxation mechanism, thus correlating their chemical shifts. The third 90° pulse is needed to rotate the vectors back to the *xy* plane, where they can be detected during the t_2 acquisition period.

The general features of a two-dimensional NOESY spectrum are outlined in the lower part of Fig. J2.31. Magnetization components which do not exchange with other

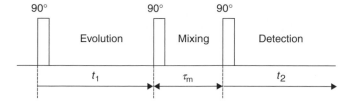

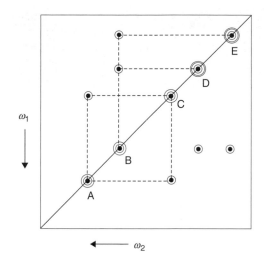

Fig. J2.31 The top trace shows the pulse sequence for two-dimensional NOESY. The bottom trace shows a contour plot of a schematic two-dimensional NOE spectrum. For the five resonance lines A–E, spin proximity is manifested by the NOE cross-peaks between A and C, B and D, B and E. (Adapted from Kumar *et al.*, 1980.)

components during the mixing time τ_m maintain their frequencies during t_1 and t_2. Hence the corresponding peaks in the two-dimensional spectrum lie on the diagonal. An exchange of magnetization between two components due to dipolar coupling during the mixing period is manifested by cross-peaks. Peak A is dipole–dipole coupled with peak C, and peak B with D and E. The two-dimensional NOESY spectrum of BPTI is plotted in Fig. J2.32. It shows, for example, that the backbone amide proton of Glu 31 is connected with the α-proton of Cys 30, and the proximity of the 3.5 ring protons of Tyr 23 and the α- and methyl protons of Ala 25. Fundamentally, the two-dimensional NOESY spectrum manifests all possible proton combinations in the protein. A cross-peak indicates that the proton pair is separated by less than about 5 Å, while the absence of a peak is indicative of longer distances. A NOESY spectrum contributes in an important way to macromolecular structure resolution by NMR (Section J3.2).

J2.5.3 Transverse Relaxation-Optimized Spectroscopy

During the past 20 years, the highest magnetic field available for NMR has increased in steps corresponding to

proton resonance frequencies of 400, 500, 600, 750, 800, and now 900 MHz. Each of these steps benefited NMR in structural biology through improved intrinsic sensitivity and peak separation. For commonly used heteronuclear experiments, however, the advantages of higher magnetic fields are offset partly by field-dependent line broadening due to increased transverse relaxation rates. TROSY was developed to overcome this limitation. The TROSY pulse sequence for two-dimensional ^{15}N–^{1}H correlation spectroscopy is shown in Fig. J2.33.

In this technique, one first creates proton magnetization. This is then transferred to ^{15}N via a polarization transfer element. After "frequency labelling" of the ^{15}N magnetization during the evolution period, t_1, magnetization is transferred back to ^{1}H via a reverse polarization transfer element and then detected on ^{1}H during the acquisition period, t_2. Transverse relaxation (Section J2.5.3) occurs throughout multidimensional NMR experiments (i.e., during polarization transfer periods as well as during frequency labeling periods). The rate of transverse nuclear spin relaxation increases with molecular mass and has a dominant impact on the upper size limit for macromolecular structures that can be studied by NMR in solution.

In conventional heteronuclear NMR experiments, magnetization is transferred between the different types of nuclei via scalar spin–spin couplings in so-called INEPT transfers (Section J2.3). The time periods T required for the INEPT transfers can be comparable to the mean duration of the evolution period (Fig. J2.33). CRINEPT overcomes this limitation by combining INEPT with cross-correlated relaxation-induced polarization transfer (CRIPT), which becomes a highly efficient transfer mechanism for molecular sizes above 200 kDa in aqueous solution at ambient temperature (Section J2.6). Furthermore, in contrast to INEPT, TROSY is active during the CRINEPT transfer periods.

Technically, the TROSY approach has the following basis. In heteronuclear two-spin systems, such as ^{15}N–^{1}H and aromatic ^{13}C–^{1}H, the NMR signal of each nucleus is split into two components by the scalar spin–spin coupling. Therefore, in two-dimensional correlation experiments a four-line fine structure is observed (Fig. J2.34b) (see also the example in Fig. J2.35). With the advent of modern multidimensional NMR, this four-line pattern has routinely been collapsed into a single, centrally located line using broad-band decoupling (Fig. J2.34a) during the evolution and acquisition periods to obtain a simplified spectrum and improved sensitivity. However, at high magnetic fields, chemical shift anisotropy (CSA) of ^{1}H, ^{15}N, and ^{13}C nuclei can be a significant source of transverse relaxation, in addition to the omnipresent relaxation as a result of dipole–dipole coupling. TROSY exploits constructive interference between dipole–dipole coupling and CSA relaxation, and actually uses CSA relaxation at higher fields to cancel field-independent dipolar relaxation. Using the TROSY technique, the multiplet structure is not

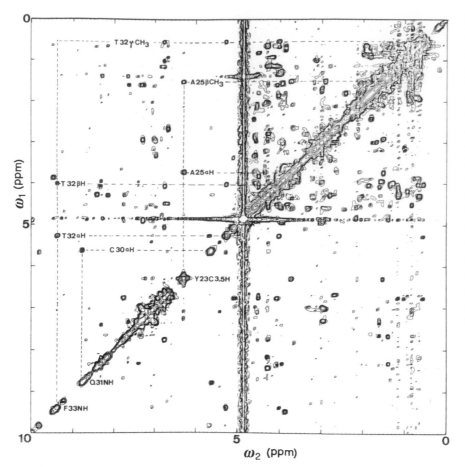

Fig. J2.32 Contour plot of a proton two-dimensional NOESY spectrum at 360 MHz for BPTI. The protein concentration was 0.02 M, solvent ^2H_2O, pH 3.8, $T = 18\,°C$. The mixing time was 100 ms. The total accumulation time was 18 h. Cross-relaxation connectivities for selected amino acid residues are indicated by the broken lines (see text). Connected peaks are identified by the one-letter symbol for amino acids (A = alanine, T = threonine, C = cysteine, Q = glutamine, F = phenylalanine, X = tyrosine), the position in the amino acid sequence, and the type of protons observed. Bands of intense signals extend from the water line at 4.85 ppm parallel to both the ω_1 and ω_2 axes. (After Kumar et al., 1980.)

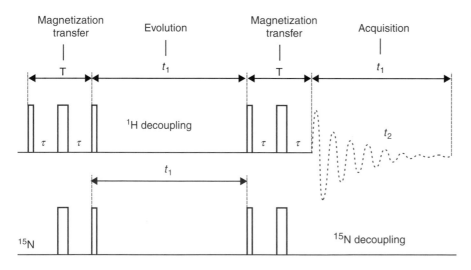

Fig. J2.33 The basic pulse sequence of two-dimensional $^{15}N–^1H$ correlation spectroscopy experiments.

decoupled and only the narrowest, most slowly relaxing line of each multiplet is retained (Fig. J2.34c).

A comparison of the $^{15}N–^1H$ correlation spectra of a protein with molecular mass of 45 kDa recorded by using COSY and TROSY, respectively, is presented in Fig. J2.35. The figure demonstrates the essential improvement in quality obtained by the TROSY technique.

Finally, we would like to point out that the implementation of [$^{15}N–^1H$] TROSY in triple resonance experiments results in a several-fold improved sensitivity for $^2H/^{13}C/^{15}N$-labeled proteins and approximately two-fold sensitivity gain for $^1H/^{13}C/^{15}N$-labeled proteins. By applying TROSY, the spectra have been obtained of proteins of molecular mass close to 100 kDa.

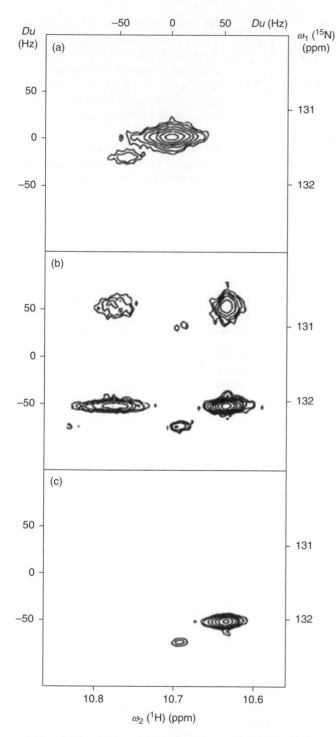

Fig. J2.34 Contour plots of a tryptophan indole $^{15}N–^{1}H$ cross-peak from different types of $^{15}N–^{1}H$ correlation spectra: **(a)** conventional broad-band decoupled [$^{15}N–^{1}H$]-COSY spectrum; **(b)** the same as **(a)**, without decoupling the evolution and detection periods (see Fig. J2.33). **(c)** [$^{15}N–^{1}H$] TROSY spectrum consisting of the sharp component. (After Wider and Wuthrich, 1999.)

J2.6 MULTI-DIMENSIONAL, HOMO- AND HETERONUCLEAR NMR

By analogy with two-dimensional NMR, the coordinate axes in three- and four-dimensional NMR spectra correspond to three and four frequency domains, respectively. Figure J2.36 summarizes the relationship between two-, three-, and four-dimensional NMR pulse sequences.

We recall that two-dimensional experiments all comprise the same basic scheme, namely a preparation pulse (P) followed by an evolution period ($E(t_1)$), during which the spins are labeled according to their chemical shifts; a mixing period (M), during which the spins are correlated with each other; and finally a detection period ($D(t_2)$).

A three-dimensional pulse scheme simply combines two two-dimensional pulse sequences, leaving out the detection period of the first and the preparation pulse of the second (Fig. J2.36). The two evolution periods, t_1 and t_2, are incremented independently, and are followed by a detection period t_3. Further extension to a fourth dimension is easily implemented by combining three two-dimensional schemes, leaving out the preparation pulses of the second and third experiments and the detection period of the first and second (Fig. J2.36).

The first three-dimensional experiments performed on proteins were of the ^{1}H homonuclear variety, in which a scalar correlation pulse scheme was combined with a NOESY sequence. Despite the elegance of the method, the applicability of ^{1}H homonuclear three-dimensional experiments is limited to molecular masses less than or equal to about 10 kDa, line widths becoming too wide for larger proteins. A more useful approach in three- and four-dimensional NMR employs uniformly (>95%) labeled ^{15}N and/or ^{13}C proteins (see Section J2.8.1). The larger heteronuclear couplings can be resolved more efficiently to allow the study of molecular masses up to about 25 kDa.

As an illustration, Fig. J2.37 shows a three-dimensional ^{15}N-correlated [^{1}H, ^{1}H] NOESY spectrum. The third axis corresponds to the NMR frequencies of the ^{15}N. As a result, the same number of NMR peaks that would have been observed in a two-dimensional [^{1}H, ^{1}H] NOESY spectrum are now distributed among a number (typically 64 or 128) of ^{1}H, ^{1}H ones (Fig. J2.38). In addition to the ^{1}H NMR data, the approach provides ^{13}C and ^{15}N NMR information, which may be used to support the structure determination and to provide supplementary data on dynamic features of the molecule studied. Nonetheless, the key purpose of heteronuclear NMR experiments is to obtain the maximum possible number of individually assigned $^{1}H–^{1}H$ NOE peaks in the system studied.

(a)

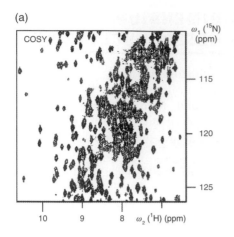

(b)

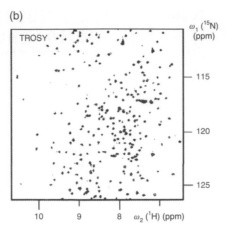

Fig. J2.35 A comparison of the ^{15}N–^1H correlation spectra of a protein of molecular mass 45 kDa recorded using: **(a)** a conventional procedure (COSY); and **(b)** TROSY. Both spectra were measured at a proton resonance frequency of 750 MHz, using a 0.8 mM sample of uniformly ^{15}N- and ^2H-labeled gyrase-45 from *Staphylococcus aureus* in water at 25 °C and pH 8.6. (Adapted from Wider and Wuthrich, 1999.)

2D NMR: $P_a \rightarrow E_a(t_1) \rightarrow M_a \rightarrow D_a(t_2)$

3D NMR: $P_a \rightarrow E_a(t_1) \rightarrow M_a \rightarrow E_b(t_2) \rightarrow M_b \rightarrow D_b(t_3)$

4D NMR: $P_a \rightarrow E_a(t_1) \rightarrow M_a \rightarrow E_b(t_2) \rightarrow M_b \rightarrow E_c(t_3) \rightarrow M_c \rightarrow D_c(t_4)$

Fig. J2.36 Relationship between the pulse sequences for recording two-, three-, and four-dimensional NMR spectra. Abbreviations are: P, preparation; E, evolution; M, mixing; and D, detection. (Adapted from Clore and Gronenborn, 1991.)

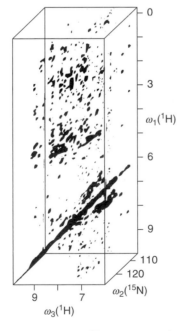

Fig. J2.37 Three-dimensional ^{15}N-correlated [^1H, ^1H] NOESY spectrum of oxidized *E. coli* glutaredoxin (protein concentration = 1.5 mM; solvent = 90% H_2O/10% D_2O; pH = 7.0; T = 301 K; ^1H frequency = 500 MHz; mixing time = 100 ms). The protein was uniformly labeled with ^{15}N to the extent of ≥95%. (After Sodano *et al.*, 1991.)

J2.7 STERICALLY INDUCED ALIGNMENT

Toward the end of the 1990s, a new class of experiment was reported that dramatically changed the range of applicability of NMR structural methods. It is based on anisotropic magnetic interactions that were not normally observable in high-resolution liquid-state NMR spectra. The experiments provide structural long-range constraints that are *orientational*, rather than *distance*-based. They rely on the measurement of residual dipolar couplings, and, in some cases, CSA, under particular conditions. Powerful applications include structural studies of multidomain proteins, macromolecular complexes, and nucleic acids. The global orientational information contained in dipolar couplings ideally complements the distance information contained in NOE peaks. Moreover, dipolar coupling provides a tool for evaluating the quality of NMR structures in an objective manner.

J2.7.1 Residual Dipolar Couplings

Dipolar couplings are potentially large interactions caused by the magnetic flux lines of the nucleus affecting the static magnetic field at the site of another nucleus (Fig. J2.39). Only the component parallel to the external much stronger magnetic field leads to the interaction. Components orthogonal to $\mathbf{B}_0$ have a negligible effect on the total vector sum of the external and dipolar field. The z component of the dipolar field of nucleus I changes the resonance frequency of nucleus S by an amount that depends on the internuclear distance r and on the angle θ between the internuclear vectors I–S and the direction of the magnetic field $\mathbf{B}_0$ (Fig. J2.40).

A dipolar interaction D_{IS} between nuclei I and S is described by

$$D_{IS} = \text{const} \frac{\gamma_I \gamma_S h \mu_0}{4\pi^2 r_{IS}^3} \left\langle \frac{(3\cos 2\theta - 1)}{2} \right\rangle \qquad (J2.13)$$

(a)

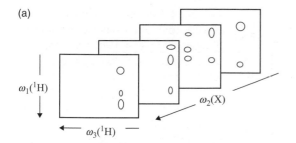

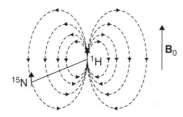

(b)

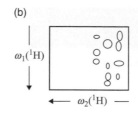

Fig. J2.38 Scheme illustrating the improved peak separation in a three-dimensional heteronuclear-resolved [^{1}H, ^{1}H] NMR experiment **(a)**, when compared with the corresponding two-dimensional [^{1}H, ^{1}H] NMR experiment **(b)**. The two spectra contain the same number of peaks. In the three-dimensional spectrum these are distributed among multiple $\omega_1(^1H)\omega_3(^1H)$ planes that are separated along the heteronuclear chemical-shift axis, $\omega_2(X)$. (Otting and Wuthrich, 1990.)

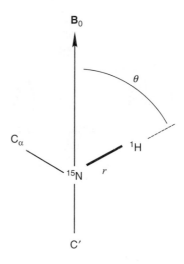

Fig. J2.39 Magnetic dipole–dipole coupling, illustrated for a ^{15}N–^1H spin pair. ^{15}N and ^1H magnetic moments are aligned parallel (or antiparallel) to the static magnetic field $\mathbf{B}_0$. The total magnetic field in the $\mathbf{B}_0$ direction at the ^{15}N position can increase or decrease relative to $\mathbf{B}_0$, depending on the orientation of the ^{15}N–^1H vector and the spin state of the proton (parallel or antiparallel to $\mathbf{B}_0$). (Bax, 2003.)

Fig. J2.40 Dipolar coupled ^{15}N–^1H spin pair in an amide bond. The bond length, r, is assumed fixed and the primary variable is the angle, θ, between the magnetic field, $\mathbf{B}_0$, and the internuclear vector. (Hansen et al., 1998.)

where the γ_i are the gyromagnetic ratios, r_{IS} the internuclear distance and the symbol $\langle \rangle$ denotes a time average. Equation (J2.13) shows that the dipolar interaction, D_{IS}, provides direct information on the angle θ, i.e., on the orientation of the internuclear vector, and that it scales with the inverse of the cubed internuclear distance.

In an isotropic solution, rotational Brownian diffusion rapidly averages out the internuclear dipolar interaction of Eq. (J2.13) to zero. As a result, a solution NMR spectrum shows narrow resonance lines, which can be assigned to individual nuclei in the protein, but which no longer contain valuable orientational information.

The properties of scalar and dipolar couplings are summarized in Table J2.2. Most biological macromolecules display some degree of anisotropic magnetic susceptibility that causes them to align in a strong magnetic field and leads to anisotropic Brownian motion (tumbling). For example, additional structural parameters were derived from the ^1H–^{15}N dipolar coupling measured in a magnetically oriented paramagnetic protein cyanmethemoglobin. However, the degree of alignment in the systems was relatively low and the largest dipolar coupling was less than 5 Hz.

Tunable degrees of alignment can be achieved in a magnetic field in the presence of an orienting agent such as planar phospholipid bicelles, rod-shaped viruses and phages, purple membrane, and polymeric strained gels. Methods for creating weakly aligned states also include direct magnetic alignment in electrical fields.

The large magnetic susceptibility of lipid bicelles was used to achieve high degrees of alignment (Comment J2.13). At 30–40 °C, bicelles orient with the normals to the bilayer surfaces perpendicular to the magnetic field. Figure J2.41 illustrates protein orientation in the presence of a dilute solution of phospholipid bicelles. Proteins and peptides become oriented by the presence of the oriented discoidal surfaces. The protein tumbles rapidly, but anisotropically, in the large aqueous interbicelle space.

The best tunable degree of macromolecular alignment has been achieved in a filamentous phage medium (Comment J2.14). The intrinsic structural properties of these phages make them an ideal system for inducing alignment of nucleic acids and proteins. These properties

TABLE J2.2 COMPARATIVE PROPERTIES OF SCALAR AND DIPOLAR COUPLINGS

Scalar couplings	Dipolar couplings
Through bonds	Through space
Strength $\leq 100\,Hz$	Strength depends on distance and nucleus (e.g., for N–H pair ~20 kHz)
Isotropic	$3\cos^2\theta - 1$
Small for >3 bonds	$1/r^3$
Produce splittings	Splittings when motion restricted

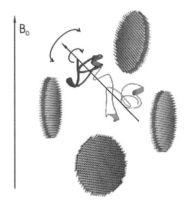

Fig. J2.41 Induced protein orientation by dilute phospholipid bicelles. The protein tumbles rapidly, but anisotropically, in large aqueous interbicelle spaces. (Prestegard, 1998.)

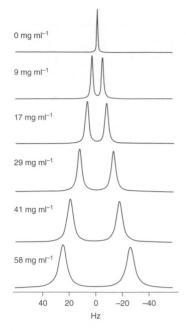

Fig. J2.42 1D ^{2}H spectra of a 10 mM Tris, pH 8.0, 90% H_2O/10% D_2O sample containing 0, 9,17, 29, 41, and 58 mg ml^{-1} Pf1 phage. The vertical scales of individual spectra were adjusted so that all peaks have the same vertical heights. (After Hansen *et al.*, 1998.)

COMMENT J2.13 BICELLES

Bicelles are fragments of lipid bilayers, a well-accepted building block of biological membranes. They were found to form liquid crystal arrays when prepared at 20–30 weight% lipid to aqueous buffer using mixtures of long-chain phospholipids such as dimyristoylphosphatidylcholine (DMPC) and lipids with detergent-like properties such as dihexanoylphosphatidylcholine (DHPC). The bilayer fragments appear to be small discoidal particles a few hundred ångstroms in diameter, and have become known as bicelles (Sanders *et al.* 1994).

COMMENT J2.14 BACTERIOPHAGE Pf1

The bacteriophage Pf1 consists of a 7349-nucleotide single-stranded circular DNA genome that is packaged at a ~1:1 nucleotide/coat protein ratio into a ~60 Å diameter by ~20 000 Å long particle that has a negatively charged surface. The coat protein forms an α-helical structure that runs roughly parallel to the long axis of the phage. The coat proteins form a repeating network of carbonyl groups that are believed to be the source of the phage's large anisotropic magnetic susceptibility, with the long axis of the phage aligning parallel to the magnetic field. A solution of oriented bacteriophage particles forms a liquid crystalline organization.

---

are: (1) rod-like phages are already fully aligned in 7 T magnetic fields (300 MHz proton frequency) so that ultra-high magnetic fields are not required for creating aligned systems; (2) the degree of alignment of the dissolved macromolecule can be tuned easily, in order to modulate the size of the dipolar couplings, by simply changing the phage concentration; (3) Pf1 phages are extremely stable under usual protein buffer conditions and readily produced, with high purity and in large quantities; (4) for all nucleic acids and proteins studied so far, there appears to be no effect on the rotational correlation time of the macromolecule, which means that standard high-resolution spectra are still obtained; at the concentrations used, the large phage particles lead to high macroscopic viscosity, but do not affect the microscopic tumbling rates of the individual macromolecules (see also Section J2.7.2).

The magnetic alignment of Pf1 phage solutions has been monitored by one-dimensional ^{2}H NMR. Figure J2.42 shows the ^{2}H NMR spectra of a 90% H_2O/10% D_2O solution as a function of phage concentration. The splitting of the HOD signal arises from the large deuterium quadrupole moment, which appears because it is not isotropically averaged for water molecules bound to the aligned phage particles. The observed splitting varies approximately linearly with phage concentration (up to 60 mg ml^{-1} of phage). Figure J2.42 also shows that the deuterium quadrupole splitting of the water resonance is a good indicator of the purity and aggregation state of the phage.

The observation of dipolar coupling in a weakly aligned protein is illustrated in Fig. J2.43. ^{15}N-labeled calmodulin

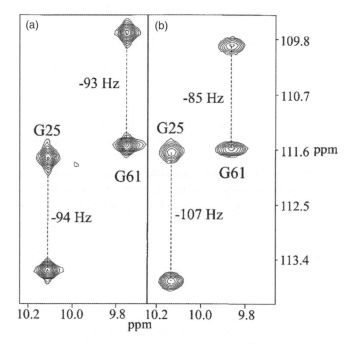

Fig. J2.43 A portion of the two-dimensional (^{15}N–^{1}H) NMR spectra collected on bovine apo-calmodulin: **(a)** in absence of phage and **(b)** in a 25 mg ml^{-1} phage solution. This region of the spectrum shows the effect of phage on the ^{1}H–^{15}N couplings for the G25 and G61 amides. (Hansen et al., 1998.)

was dissolved in 25 mg ml^{-1} Pf1 phage solution. A comparison of parts of the amide regions of the two-dimensional (^{15}N, ^{1}H) NMR spectra collected in the presence and absence of phage shows that the ^{1}H–^{15}N amide couplings for residues G25 and G61 are -13 ± 2 and $+8 \pm 2$ Hz, respectively.

The Pf1 phage approach appears to align RNA, DNA, or protein systems by a steric mechanism, with no evidence of binding between macromolecules and phages. There are no significant changes in chemical shifts or other NMR parameters that would suggest structural changes in the macromolecule in the presence of phage. Alignment appears to be induced by collisions of the asymmetric macromolecule with the magnetically aligned phage.

Figure J2.44a schematically illustrates the steric interaction between Pf1 phage and the DNA duplex d(GGCAAAAACGG)/d(**CCGTTTTTGCC**). The bold letters represent nucleotides uniformly labeled with ^{13}C and ^{15}N. The residual dipolar couplings are shown in Fig. J2.44b.

J2.7.2 Chemical Shift Anisotropy (CSA)

Residual dipolar coupling is not the only anisotropic spin interaction that can provide useful structural information. Chemical shifts are also anisotropic. Chemical shifts arise because nuclei in various molecular functional groups resonate at different frequencies depending on shielding by the local electronic environment. Electronic environments are seldom isotropic, and hence shielding is different for different orientations of the functional groups. An angular dependence of the chemical shift occurs that is analogous to the one seen for residual dipole interaction.

An illustration of anisotropic chemical shift effects is given for ^{13}C of the peptide carbonyl group in Fig. J2.44; if the peptide bond has a preferred orientation with the field perpendicular to its plane, the carbonyl carbon would resonate at a higher field (or lower frequency) than in the isotropic case.

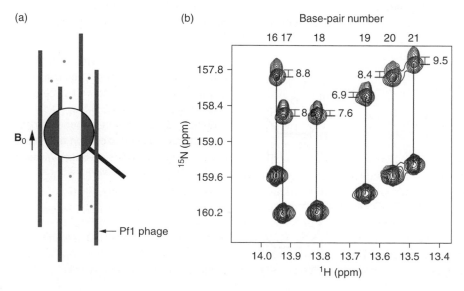

Fig. J2.44 Acquisition of residual dipolar couplings. **(a)** Sketch showing the steric interaction between Pf1 phage and the DNA, resulting in a slightly anisotropic environment and thus preventing the dipolar couplings from averaging to zero. B_0 is a static magnetic field. **(b)** Overlaid [^{15}N, ^{1}H] NMR spectra of the DNA duplex (d(GGCAAAAACGG)/ d(**CCGTTTTTGCC**). The spectra were recorded in the absence (black) and presence (red) of Pf1 phage. The bold letters represent nucleotides that are uniformly labeled with ^{13}C and ^{15}N. (After MacDonald and Lu, 2002.)

NMR spectra of a two-domain fragment of a carbohydrate-binding protein from barley, which has been ordered in a 5% 3:1 DMPC/DHPC bicelle dispersion, are shown in Fig. J2.45. Comparing isotropic (Fig. J2.45a) and

oriented (Fig. J2.45b) spectra, Lys 53 shows a negative 5 Hz residual dipolar contribution and Lys 95 shows a positive 2 Hz contribution. These are indicative of an average N–H vector orientation perpendicular and parallel to the magnetic field, respectively.

The results above show that residual dipolar coupling and CSA in oriented macromolecules no longer average to zero, and are readily measured. The resulting parameters provide valuable information on long-range order for structure determination.

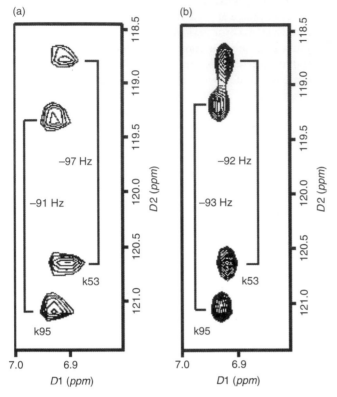

Fig. J2.45 Segments from a proton coupled, nitrogen decoupled, ^{15}N–^{1}H NMR spectra of a 0.4 mM solution of a barley lectin fragment in a 5% DMPC/DHPC 3:1 bicelle (doped with a positively charged amphiphile): **(a)** an isotropic spectrum at 25 °C; **(b)** an oriented spectrum at 35 °C. The bicelle preparations have a convenient property in that an isotropic phase containing the same lipids, presumably in more symmetric micelles, is obtained by lowering the temperature slightly. Both increases and decreases in couplings are observed. (After Prestegard, 1998.)

J2.8 ISOTOPE LABELING OF PROTEINS AND NUCLEIC ACIDS

J2.8.1 Labeling Strategies for Proteins

Proteins used for NMR studies can be labeled by using a number of different protocols to produce molecules with different patterns of ^{2}H, ^{13}C, and ^{15}N incorporation.

Uniform or random labeling strategies result in ^{2}H incorporation throughout a protein in a roughly site-independent manner. Amino-acid-specific labels have been used since the earliest ^{1}H-based NMR studies of proteins. Specifically protonated amino acids were introduced in otherwise fully deuterated molecules by growing microorganisms on minimal media with high levels of D$_2$O, supplemented with fully or partially protonated amino acids. Most of the 20 natural amino acids have been successfully incorporated in this manner, either individually or in different residue. Amino-acid-specific auxotrophic strains have been employed in cases where it is necessary to use smaller amounts of the potentially expensive labeled compounds or to limit undesirable metabolic spreading of the label. Similar approaches have been used in the context of specific ^{15}N and ^{13}C labeling.

A high level of isotopic labeling at specific sites within residues can also be achieved. Uniformly ^{15}N, ^{13}C, and ^{2}H-labeled proteins have been produced, in which methyl groups in alanine, valine, leucine, and isoleucine were protonated.

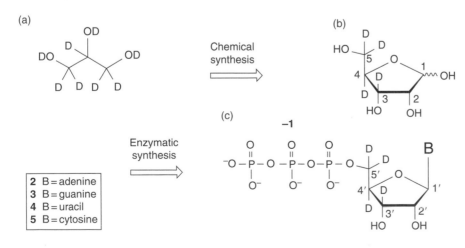

Fig. J2.46 Strategy for synthesis of deuterated nucleotides. Chemical synthesis is used to convert glycerol-d_8 **(a)** into D-ribose-3,4,5,5-d_4 (D-d$_4$-ribose, **–1**) **(b)**, and then four enzymatic reactions are used to convert that ribose into the four d_4-NTPs **(2–5)** **(c)**. (Adapted from Tolbert and Williamson, 1996.)

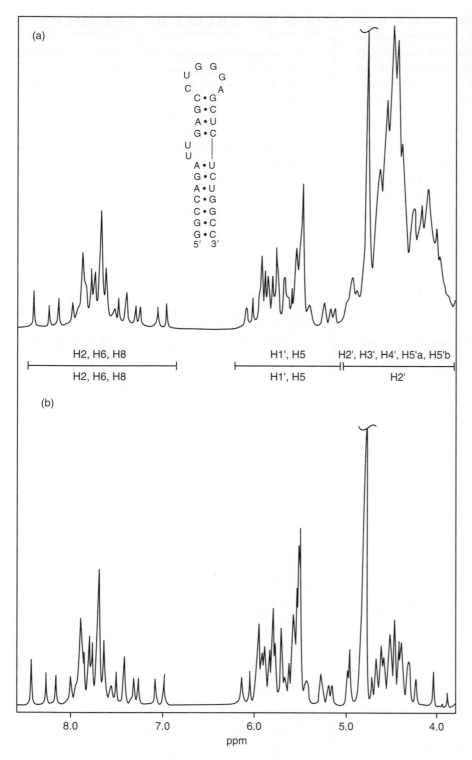

(a)

(b)

H2, H6, H8

H2, H6, H8

H1', H5

H1', H5

H2', H3', H4', H5'a, H5'b

H2'

8.0   7.0   6.0   5.0   4.0

ppm

Fig. J2.47 500 MHz NOESY spectrum of HIV-2 TAR RNA: **(a)** unlabeled (inset is the secondary structure of HIV-2 TAR RNA); **(b)** d_4-labeled TAR RNA. (After Tolbert and Williamson, 1996.)

An approach that promises to further extend studies to high-molecular-mass, single polypeptide chains is one in which a segmentally isotope-labeled protein is produced through the ligation of labeled and unlabeled polypeptides. The ligation of two protein fragments under very mild reaction conditions has been shown to be possible through a self-catalytic protein-splicing process involving naturally occurring domains called inteins. The method has been used to selectively ^{15}N-, ^{13}C-label various segments of maltose-binding protein, as well as the C-terminal domain of the RNA polymerase α subunit.

The majority of isotopically labeled proteins studied by NMR have been obtained by heterologous protein expression in *E. coli*. Although bacterial expression remains the most economical method for producing labeled proteins, it is not always possible to obtain samples in this manner, particularly if there are problems with protein folding or bacterial toxicity. In addition, many eukaryotic proteins undergo post-translational modifications that do not occur in bacterial expression systems. Alternative methods of protein expression that can address these limitations were developed, using the yeast *Pichia pastoris*, Chinese hamster ovary cells, and cell-free expression systems.

J2.8.2 Labeling Strategies for RNA

The two significant problems encountered in the application of NMR to larger RNA molecules are spectral crowding and large line widths in the ribose region. Uniform ^{13}C and ^{15}N labeling has been applied to RNA structure determination to great advantage. However, introduction of these isotopes results in broader line widths for protons, due to the strong dipolar interaction with directly bonded ^{13}C and ^{15}N nuclei. This effect is particularly problematic for larger RNA molecules.

Deuteration offers the advantage of spectral simplification without causing broader line widths. Since large RNAs are most readily prepared by transcription with T7 RNA polymerase, the main problem is how to prepare deuterated nucleoside triphosphates (NTPs). One method is to harvest total cellular RNA from bacteria that have been grown on deuterated media, enzymatically digest the RNA to nucleoside monophosphates (NMPs), and finally enzymatically convert the NMPs to NTPs. Uniform ^{13}C, ^{15}N, and ^{13}C/^{15}N labeling of NTPs has also been accomplished by this method, by growing *E. coli* or *Methylophilus methylotrophus* on ^{13}C- and/or ^{15}N-labeled media. Chemical approaches and approaches combining chemical and biochemical steps have also been developed to produce isotopically labeled NTP (Fig. J2.46). D,L-ribose-3′,4′,5′,5′-d_4 is first synthesized from glycerol-d_8 by chemical methods. Four multi-enzyme reactions are used that mimic purine salvage and pyrimidine biosynthetic pathways to convert D-d_4-ribose (1) into d_4-ATP (2), d_4-GTP (3), and d-$_4$UTP (4). Using this strategy, a 30-nucleotide RNA derived from the HIV-2 TAR RNA has been prepared with protonated and deuterated NTP at specific positions.

Figure J2.47 illustrates the dramatic effect of deuteration on the NMR spectra of RNA. The spectral crowding in the ribose region of the deuterated RNA is greatly reduced, and individual H2′ (see Fig. J2.47b) resonance peaks are clearly observed between 4 and 5 ppm.

J2.9 CHECKLIST OF KEY IDEAS

- FID describes the response of the sample to the pulse. FID is converted to a frequency spectrum by a Fourier transform: An *amplitude versus time* signal is transformed into an *amplitude versus frequency* signal.
- NOE is observed experimentally as the fractional change in the intensity of one NMR line when another resonance is irradiated in a double irradiation experiment.
- NOE is due to dipolar interactions (through space) between different nuclei and is correlated with the inverse *sixth power* of the internuclear distance; as a consequence, NOE is usually observed only for protons pairs separated by ≤5–6 Å.
- Polarization transfer uses a suitable pulse combination to impart the greater equilibrium polarization of protons to a coupled nucleus with a smaller gyromagnetic ratio (e.g., ^{13}C or ^{15}N); polarization transfer is an alternative to NOE for sensitivity enhancement.
- The term one-dimensional NMR refers to experiments in which the transformed signal is presented as a function of a single frequency. By analogy, in two-, three-, and four-dimensional NMR spectra the coordinate axes correspond to two, three, and four frequency domains.
- COSY is a multidimensional method in which peaks appear at the coordinates of two nuclei related by a mutual interaction (*J*-coupling, NOE, or chemical exchange).
- COSY identifies pairs of protons that are coupled to each other via scalar spin–spin coupling. NOESY identifies pairs of protons that are related by the NOE.
- TROSY makes use of transverse relaxation optimization to record high-quality solution NMR spectra of large molecular structures.
- The main source of geometric *short-range* information contained in the experimental NMR restraints is provided by the NOE; additional structural refinements include the use of three-bond coupling constants and secondary ^{13}C and ^{1}H shifts.
- The main source of geometric *long-range* information is measurements of residual dipolar couplings and, in some cases, chemical shift anisotropy in a weakly aligned medium.

Suggestions for Further Reading

Fourier Transform NMR Spectroscopy
King, R. W., and Williams, K. R. (1989). The Fourier transform in chemistry, Part 2: Nuclear magnetic resonance – the single pulse experiment. *J. Chem. Education*, **66**, A243–A248.

Williams, K. R., and King, R. W. (1990). The Fourier transform in chemistry: NMR. Part 3. Multiple-pulse experiments. *J. Chem. Education*, **67**, A93–A99.

Williams, K. R., and King, R. W. (1990). The Fourier transform in chemistry: NMR. Part 4. Two-dimensional methods. *J. Chem. Education*, **67**, A125–137.

Single-Pulse Experiments

Brey, W. S. (1988) (ed.). *Pulse Methods in 1D and 2D Liquid-Phase NMR*. San Diego, CA: Academic Press.

Skoog, D. A., Holler, F. J., and Nieman, T. A. (1995). *Principles of Instrumental Analysis.* Philadelphia, PA: Saunders College Publishing.

Multiple-Pulse Experiment

Wuthrich, K. (1986). *NMR of Proteins and Nucleic Acids*. New York: Wiley-Interscience.

Nuclear Overhauser Effect

Kumar, A., Ernst, R. R., and Wuthrich, K. (1980). A two-dimensional nuclear Overhauser enhancement (2D NOE) experiment for the elucidation of complete proton–proton cross-relaxation networks in biological macromolecules. *Biochem. Biophys. Res. Commun.*, **95**, 1–6.

Two-Dimensional NMR

Derome, A. E. (1987). *Modern NMR Techniques for Chemistry Research*. New York: Pergamon.

Ernst, R. R., Bodenhausen, G., and Wokaun, A. (1987). *Principles of Nuclear Magnetic Resonance in One and Two Dimensions*. Oxford: Oxford University Press.

Multi-Dimensional, Homo- and Heteronuclear NMR

Clore, G. M., and Gronenborn, A. M. (1991). Two-, three-, and four-dimensional NMR methods for obtaining more precise three-dimensional structure of proteins in solution. *Ann. Rev. Biophys. Chem.*, **20**, 29–63.

Sterically Induced Alignment

Sanders, C. R., Hare, B. J., Howard, K. P., and Prestegard, J. H. (1994). Magnetically oriented phospholipid micelles as a tool for the study of membrane-associated molecules. *Prog. Nucl. Magn. Res. Spectr.*, **26**, 421–444.

Prestergard, J. H. (1998). New techniques in structural NMR: Anisotropic interactions. *Nature Struct. Biol.*, **5**, 517–522.

Hansen, M. R., Mueller, L., and Pardi, A. (1998). Tunable alignment of macromolecules by filamentous phage yields dipolar coupling interaction. *Nature Struct. Biol.*, **5**, 1065–1074.

Bax, A. (2003). Weak alignment offers new NMR opportunities to study protein structure and dynamics. *Protein Sci.*, **12**, 1–16.

Isotope Labeling of Proteins and Nucleic Acids

Tolbert, T. J., and Williamson, J. R. (1996). Preparation of specifically deuterated RNA for NMR studies using a combination of chemical and enzymatic synthesis. *JACS*, **116**, 7929–7940.

Gardner, K. H., and Kay, L. E. (1998). The use of ^{2}H, ^{13}C, ^{15}N multidimensional NMR to study the structure and dynamics of protein. *Ann. Rev. Biophys. Biomol. Struct.*, **27**, 357–406.

Goto, N. K., and Kay, L. E. (2000). New developments in isotope labeling strategies for protein solution NMR spectroscopy. *Curr. Opin. Struct. Biol.*, **10**, 585–592.

J3.1 STRUCTURE CALCULATION STRATEGIES FROM NMR DATA

Unlike X-ray crystallography, which generates only one set of coordinates, an ensemble of sometimes more than 50 models is produced from nuclear magnetic resonance (NMR) data analysis. This situation is inherent in the method, and arises from the interpretation of the data and also from the dynamic nature of the protein molecule itself. Because of this, the general consensus is that NMR-derived models should be judged by different criteria than those used in the assessment of the more "rigid" structures from X-ray crystallography, and it has been proposed that for the solution structures a type of statistical description might eventually have to be used (Comment J3.1).

Irrespective of the algorithm used, any structure determination by NMR seeks to find the global minimum region of a target function E_{tot} given by

$$E_{tot} = E_{cov} + E_{vdw} + E_{NMR} \qquad (J3.1)$$

where "E_{cov}," "E_{vdw}," and "E_{NMR}" are terms representing the covalent geometry (bonds, angles, planarity, and chirality), the non-bonded contacts, and the experimental NMR restraints, respectively. The uncertainties associated with the first two terms are relatively small, and the major determinant of accuracy resides in the number and quality of the experimental NMR restraints that enter into the third term, E_{NMR}. Comparable minimization algorithms are also used in X-ray crystallography (see Chapter G3).

The main source of geometric information contained in the experimental NMR restraints is provided by the nuclear Overhauser effect (NOE). The NOE is proportional to the inverse sixth power of the distance between the protons, so its intensity falls off very rapidly with increasing distance between proton pairs (see Section J2.4). Despite the short-range nature of the observed interactions, the short approximate interproton distance restraints derived from NOE measurements can be highly conformationally restrictive, particularly when they involve residues that are far apart in the sequence but close together in space.

The accuracy of NMR structures is also affected by errors in the interproton distance restraints, arising from three sources: misassignments, errors in distance estimates, and incomplete sets of NOE restraints. The best check on the correctness is provided by verifying that all short interproton distances (<3.5 Å) in a structure are in fact observed in the NOE spectra.

Although the interproton distance restraints derived from NOEs provide the mainstay of NMR structure determination, direct refinement against other experimental NMR restraints is both feasible and desirable. These additional experimental restraints that also provide *short-range* structural information are: specifically three-bond coupling constants, secondary ^{13}C chemical shifts, and ^{1}H chemical shifts.

NMR structure determination used to rely exclusively on these restraints whose information a priori is entirely local and restricted to atoms close in space. To date there are two new approaches in which restraints that define *long-range* order a priori can supply invaluable information. The first is based on the dependence of heteronuclear (^{15}N or ^{13}C) longitudinal (T_1) and transverse (T_2) relaxations times, specially T_1/T_2 ratios, on rotational diffusion anisotropy, and the second relies on residual dipolar couplings (RDCs) in oriented macromolecules (Section J2.7). The two methods provide restraints that are related in a simple geometric manner to the orientation of one-bond internuclear vectors (e.g., N–H and C–H) relative to external tensor. In the case of T_1/T_2 ratios, the tensor is the diffusion tensor. In the case of RDCs, the tensor may be the *magnetic susceptibility tensor* for molecules aligned in magnetic field, or the *molecular alignment tensor* for molecules aligned by anisotropic media such as liquid crystals. The two methods permit the relative positioning of structural elements that do not have many short interproton distance contacts between them. Examples of such systems are discussed in Section J3.2.

COMMENT J3.1 QUALITY OF A SOLUTION STRUCTURE

A commonly used criterion for the "quality" of an NMR solution structure has been the average root-mean-square deviation (RMSD) for the coordinates of different conformations in the ensemble of structures, which are compatible with the set of experimental data. The term *resolution* does not really apply to NMR structures, but it may be useful, figuratively, to denote the precision with which a solution structure can be specified.

As mentioned above, the addition of dipolar coupling terms to the NMR energy function during simulated annealing results in a highly rippled surface with innumerable sharp local minima. In the absence of long-range distance information from NOEs, it is usually very difficult to find a good starting model close to the true structure. One simple method for obtaining such a starting model simply breaks the protein of interest into overlapping fragments of 7–10 residues in length. Then, the entire Protein Data Bank (PDB) is searched for fragments that are compatible with the experimental dipolar couplings and chemical shifts. Figure J3.1 illustrates the assembly of the backbone of the RecA-inactivating protein DinI by the molecular fragment replacement approach. The approach yields a backbone structure that differs by less than 2 Å from the previously determined NMR structure.

Table J3.1 summarizes all experimental NMR constraints used today for macromolecular structure determination.

J3.2 THREE-DIMENSIONAL STRUCTURE OF BIOLOGICAL MACROMOLECULES

J3.2.1 Proteins

Solution NMR structural studies are based on high-quality spectra recorded with good sensitivity and resolution. We

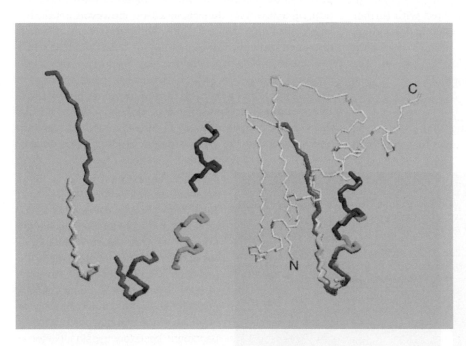

Fig. J3.1 Assembly of a protein backbone by the molecular fragment replacement approach, using seven-residue fragment lengths. The assembly is illustrated for the backbone of the RecA-inactivating protein DinI. The partially overlapping fragments are strung together, while maintaining their correct orientation. (After Bax, 2003.)

TABLE J3.1 EXPERIMENTAL NMR CONSTRAINTS USED IN MACROMOLECULAR STRUCTURE DETERMINATION

Conformational constraints	Main geometrical parameter	Range order
Interproton distances from NOE measurements (Section J2.4)	Interproton distances	Short, <5 Å
Three-bond coupling constants from J correlated spectroscopy (Section J1.2.3)	Torsion angle θ	Short
^{13}C secondary chemical shifts (Section J1.2.3)	Torsion angles θ and ψ	Short
^1H chemical shifts (Section J1.2.3)	Distance and orientation of aromatic ring to the proton of interest	Short
Heteronuclear relaxation time measurements (Section J1.2.2)	Orientation of the internuclear vectors relative to molecular diffusion tensor	Long
Residual dipolar coupling (Section J2.7)	Orientation of the internuclear vectors relative to molecular magnetic susceptibility tensor (alignment tensor)	Long

recall that with increasing molecular mass, these basic requirements are difficult to achieve, because of line broadening and overlap. Advances have been obtained with both novel NMR techniques and new biochemical approaches. In particular, the application of TROSY and CRINEPT (Section J2.5), in combination with suitable isotope-labeling schemes (Section J2.8), allowed a seven-fold extension of the protein size limit by several-fold. A few examples of NMR studies of proteins follow, classified according to molecular mass range.

Small Proteins (6–10 kDa)

In principle, the solution structures of small compact proteins (100 amino acids or less) can be obtained from constraints imposed by NOE signal information alone. In practice, these are commonly supplemented with further constraints based on *J* coupling and chemical bonding. Figure J3.2 shows the first *de novo* atomic resolution NMR structure determination of a globular protein, the bull seminal protease inhibitor (Comment J3.2).

Proteins of Medium Size (10–60 kDa)

For proteins of molecular mass greater than 10 kDa, the standard sequential assignment procedure may not yield an unambiguous result because of very extensive overlap in the spectra. The overlap problem can be alleviated significantly by employing isotope labeling.

In a double-labeling technique, one type of amino acid is labeled with ^{13}C in the carbonyl position and a second type is labeled with ^{15}N. Dipeptide fragments, in which the ^{13}C label precedes the ^{15}N label, can be identified easily via the *J* coupling peak splitting in the ^{1}H–^{15}N correlation spectrum.

The development of heteronuclear triple-resonance NMR experiments (three-dimensional NMR), in conjunction with essentially complete ^{13}C and ^{15}N enrichment, dramatically increased the scope of the NMR method, and structure determinations of proteins of 150–200 amino acid residues became a matter of routine. With the exception of a few very favorable cases, however, studies of larger proteins (beyond 200–300 amino acid residues) remained difficult. The sensitivity of heteronuclear triple-resonance NMR experiments was improved by partial random deuteration, which decreased peak overlap by decreasing the number of proton resonances observed. The optimum deuteration level depends on molecular mass. For example, a level of ~50% deuteration gave the best results for the SH3 domain protein from chicken

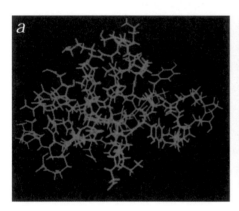 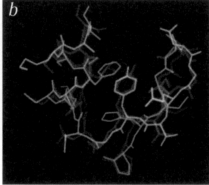

Fig. J3.2 The first protein structure determined by NMR. **(a)** A representation of the NMR structure of the proteinase inhibitor IIA from bull seminal plasma (BUSI IIA). **(b)** Superposition of the core region (residues 23–42) in the NMR structure of BUSI IIA (green) on the corresponding polypeptide segment in the X-ray crystal structure of the homologous porcine pancreatic secretory trypsin inhibitor (PSTI) (blue). (After Wuthrich, 2001.)

COMMENT J3.2 THE FIRST STEPS OF NMR PROTEIN STRUCTURE DETERMINATION

K. Wuthrich, in the article "The way to NMR structure of proteins," wrote:

The completion of the first protein NMR structure brought new, unexpected challenges. When I presented the structure of BUSI [Fig. J3.2a] in some lectures in the spring of 1984, the reaction was one of disbelief, and because of the close coincidence [Fig. J3.2b] with results from an independent crystallographic study of the homologous protein PSTI (porcine pancreatic secretory trypsin inhibitor) it was suggested that our structure must have been modeled after this crystal structure. In the discussion following a seminar in Munich on May 14, 1984, Robert Huber (Nobel Prize in Chemistry, 1988) proposed that we settle the matter by independently solving a new protein structure by X-ray crystallography and by NMR. For this purpose, each one of us received an sample supply of the α-amylase inhibitor tendamistat. Virtually identical three-dimensional structures of tendamistat were obtained in our laboratory by NMR in solution and in R. Huber's laboratory by X-ray diffraction in single crystals.

(Wuthrich, 2001)

A new era of NMR had dawned.

brain alpha spectrin $M = 7.2$ kDa. For proteins with a molecular mass of about 30 kDa, a higher level of deuteration is required. The solution structure of the 259-residue, 30 kDa, N-terminal domain of enzyme I of the *Escherichia coli* phosphoenolpyruvate (EIN) was solved at a deuteration level of 90% (for the non-exchangeable protons). Figure J3.3 shows ribbon diagrams of EIN in solution (red) and in the crystal (blue).

Structural studies of proteins of more than 30 kDa require higher levels of deuterium enrichment. For example, assignment of over 90% of the backbone amide ^{1}H, ^{15}N, ^{13}C$^{\alpha}$, and ^{13}C$^{\beta}$ resonance peaks of a 64 kDa ternary complex consisting of two tandem dimers of trp-repressor was achieved on samples with 95% deuteration, by using site-specific isotope labeling (Section J2.8) and special pulse sequences (Section J2.5).

The history of protein structure determination is treated by Günter (2011) who, with his collaborators, proposed and developed methods for fully automated structure determination of proteins in solution (FLYA) from NMR spectra.

Large and Very Large Macromolecular Complexes

Special techniques (TROSY and CRINEPT) were developed for NMR studies of large and very large macromolecular complexes (Section J2.5). Their resolving power was tested on isotope-labeled heptameric co-chaperonin GroES (72 kDa), free in solution, in complex with the homotetra-decameric chaperonin GroEL (800 kDa), or with the single-ring GroEL variant SR1 (400 kDa) (Fig. J3.4; Comment J3.3). Most amino acids of GroES showed the same resonance peaks, whether free in solution or in complex with chaperonin; however, residues 17–32 showed large chemical shift changes upon binding. These amino acids belong to a mobile loop region that forms contacts with GroEL.

(a)

(b)

Fig. J3.4 (a) GroES (72 kDa) is an oligomeric protein with seven identical units, one of which is shown in gold. **(b)** A molecular model of the complex of GroES and SR1 (472 kDa). The blue colored SR1 is not isotope-labeled and therefore not visible in this NMR experiment. (After Flaux *et al.*, 2002.)

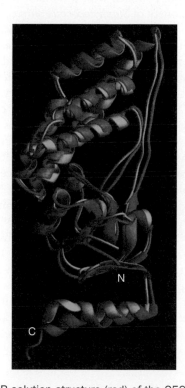

Fig. J3.3 NMR solution structure (red) of the 259-residue, 30 kDa, N-terminal domain of enzyme I of *Escherichia coli* (EIN). The structure determination was based on a total of 4251 experimental constraints, and the precision of the coordinates for the final 50 simulated structures is 0.79 ± 0.18 Å for the backbone atoms and 1.06 ± 0.15 Å for all atoms. A comparison with the X-ray structure of EIN (blue) indicates that there are no significant differences between the solution and crystal structures within the errors of the coordinates. (After Garret *et al.*, 1997.)

COMMENT J3.3 PHYSICIST'S BOX: MOLECULAR CHAPERONES (SEE ALSO COMMENT G2.19)

Molecular chaperones are required for the correct folding, transport, and degradation of many proteins in vivo. In the *Escherichia coli* chaperonin system, the double ring-shaped homotetradecamer GroEL (800 kDa) interacts with the dome-shaped homoheptamer GroES (72 kDa; Fig. J3.4) in the presence of ATP to form a chamber that is the site of productive polypeptide folding (Xu *et al.*, 1997). The homo-oligomeric nature and high symmetry of GroEL, SRI, GroES, and their complexes can facilitate NMR spectral analysis because the number of resonances corresponds to the size of an individual subunit, here 58 kDa for GroEL and 10 kDa for GroES, even though the large functional assembly is investigated.

Protein–Protein Interactions

An elegant NMR strategy for the determination of the interface residues in a large protein–protein complex has been proposed. The approach requires an unlabeled target protein (containing ^{1}H and ^{14}N spins), and a uniformly ^{2}H-,^{15}N-labeled ligand protein, in which only the amide proton positions are occupied with ^{1}H. The relative populations of ^{15}N–^{1}H and ^{15}N–^{2}H groups in the ligand protein are further adjusted by variation of the solvent ^{1}H$_2$O–^{2}H$_2$O ratio. Figure J3.5 illustrates the resulting situation. The target protein (T) contains a dense population of aliphatic ^{1}H spins, which resonate at chemical shifts, δ, from 0 to 3 ppm in the ^{1}H NMR spectrum (green circles), and aromatic and amide ^{1}H spins (pink circles), which resonate between 6 and 10 ppm and are hence well separated from the aliphatics (Fig. J3.6). The ligand protein (L) contains ^{15}N–^{1}H groups (red circles), but no other hydrogen atom positions are occupied with ^{1}H spins. The complex is immersed in solvent ^{1}H$_2$O (green circles), which has a ^{1}H resonance line near 5 ppm. (Fig. J3.6). If only the aliphatic proton resonance lines in a large protein with natural isotope abundance (T in Fig. J3.5) are irradiated with a radio-frequency field, then its aromatic and amide protons are instantaneously saturated due to spin diffusion. Spin diffusion is highly effective only for closely spaced ^{1}H spins. Therefore, although the uniformly

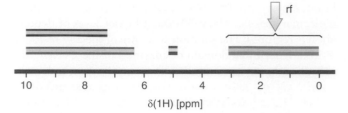

Fig. J3.6 Distribution of NMR signals along the ^{1}H chemical shift axis. The color code is the same as in Fig. J3.5: green, aliphatic ^{12}C–^{1}H; pink, aromatic ^{12}C–^{1}H and ^{14}N–^{1}H; red, ^{15}N–^{1}H; purple, ^{1}H$_2$O. The green arrow indicates the radio-frequency irradiation on the spectral region of the aliphatic protons. (After Wuthrich, 2000.)

^{2}H-, ^{15}N-labeled protein is not directly affected by the radio-frequency field (Fig. J3.6), saturation can be transferred from the unlabeled target molecule to the labeled protein through close ^{1}H–^{1}H contacts across the protein–protein interface in the complex (Fig. J3.5).

This strategy was applied to studies of a 64 kDa immunoglobulin complex with a domain of protein A, a cell wall component of *Staphylococcus aureus* that specifically binds to the Fc portion of immunoglobulin G (IgG). It was shown that helices I (Gln 10–His 19) and II (Glu 25–Asp 37) are primarily responsible for the binding of FB to the Fc fragment. The contact residues thus identified overlap with those determined by the X-ray crystallography study (Fig. J3.7a). Interestingly, the values of the NMR intensity ratios of the cross-peaks for helicies I and II are smaller for every third or fourth residue (Gln 11, Tyr 15, and Leu 18 in helix 1; Asn 29, Ile 32, and Lys 36 in helix II), suggesting that one side of each of the helices is responsible for binding the Fc fragment (Fig. J3.7b).

In identifying the interface of a large protein–protein complex, the method proposed is obviously superior to traditional NMR methods that rely on chemical shift perturbation and H–D exchange. This is because the method described here extracts more direct information on through-space interactions between the two molecules in a complex. The method should be generally applicable to large protein complexes and, with appropriate modifications, to other complexes, such as protein–DNA and protein–lipid complexes.

J3.2.2 Nucleic Acids

There has been significant progress in NMR structure determination of nucleic acids, despite the fact that, in principle, it is much more difficult than for proteins (Comment J3.4). The low proton density (0.35 protons per atom versus 0.52 protons per atom for proteins), and the absence of a globular fold result in insufficient NOE constraints (Fig. J3.8). Even the accurate determination of local geometry is difficult because of the low redundancy in the

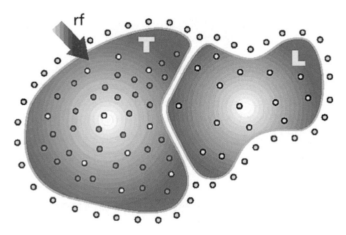

Fig. J3.5 Isotope labeling scheme used for identification of intermolecular contacts between two proteins, T and L, by NMR saturation transfer. T denotes the unlabeled target protein, which contains proton spins in aliphatic ^{12}C–^{1}H groups (green circles) and in aromatic ^{12}C–^{1}H, as well as ^{14}N–^{1}H groups (pink circles), which can be saturated but are not observed in the presently discussed experiment. L denotes the uniformly ^{2}H-, ^{15}N-labeled ligand protein, which contains NMR-observable ^{15}N–^{1}H groups (red circles), but is otherwise devoid of ^{1}H spins. The complex is surrounded by solvent water, of which only a first hydration layer is shown (purple circles). The green arrow represents the radio-frequency field, which is applied to saturate the aliphatic ^{12}C–^{1}H protons (green circles). (After Wuthrich, 2000.)

(a) (b)

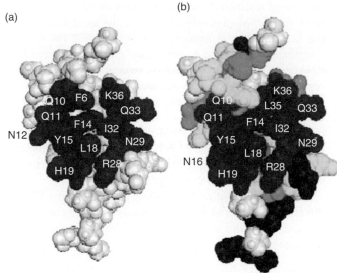

(a) (b)

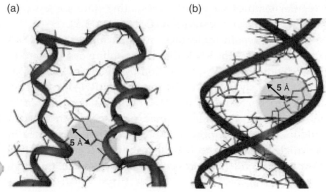

Fig. J3.7 Comparison of the binding sites of the FB–Fc complex. **(a)** The X-ray crystallography study: Residues with accessible surface areas that are covered upon binding of the Fc fragment (Phe 6, Gln 10, Gln 11, Asn 12, Phe 14, Tyr 15, Leu 18, His 19, Arg 28, Asn 29, Ile 32, Gln 33, and Lys 36) are colored red. **(b)** The cross-saturation NMR experiment: Residues Gln 10, Gln 11, Phe 14, Tyr 15, Glu 16, Leu 18, His 19, Arg 28, Asn 29, Ile 32, Gln 33, Leu 35, and Lys 36 are given in a different color. The color reflects different intensity ratio of the cross peaks. (Adapted from Takahashi *et al.*, 2000.)

Fig. J3.8 A comparison of NOE contacts in proteins and nucleic acids. **(a)** Proteins have NOE contacts that are >5 Å along their primary sequences, but <5 Å apart in their tertiary structure. This allows long-range NOE constraints to be acquired. **(b)** Nucleic acids lack these long-range NOE constraints between nucleotides distant in their primary structure; this makes their total structure less defined in the absence of residual dipolar couplings. (Adapted from MacDonald and Lu, 2002.)

COMMENT J3.4 X-RAY AND NMR STRUCTURES IN THE PDB

Although in absolute figures the number of deposited protein structures is still much larger, the proportion of DNA and RNA structures in the PDB is increasing steadily. The statistics show a substantial improvement in multidimensional NMR spectroscopy and in the capabilities of advanced isotope-labeling techniques for both RNA and DNA. The absolute values for NMR-determined structures are as follows: in 1997, 137 nucleic acid structures and 784 protein structures were deposited; in 2001, 419 nucleic acid structures and 1775 protein structures were deposited. The numbers for nucleic acids also include the structures of protein–DNA/RNA and drug–DNA/RNA complexes (Zidek *et al.*, 2001). In autumn 2014, the PDB contained a total of 2726 nucleic acid structures, of which 1093 were determined by NMR and 1559 by X-rays.

(a) (b)

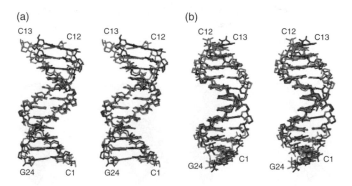

Fig. J3.9 Two stereoviews of structures of d(CGCGAAT-TCGCG)$_2$. The NMR structure is shown in blue and the X-ray structure in red. **(b)** The view in **(b)** is rotated by 90° around the helix axis relative to **(a)** and highlights the deviation from cylindrical symmetry, caused by the very different widths of the minor and major grooves. A nearly complete set of one-bond ^1H–^{13}C dipolar couplings, supplemented by ^1H–^{15}N and ^1H–^1H dipolar couplings, was obtained in the bicelle-based liquid crystal medium. (After Tjandra *et al.*, 2000.)

labeling (Section J2.8) and molecular alignment in ordered media, such as solutions of disc-shaped phospholipid bicelles and filamentous bacteriophage (Section J2.7).

Figure J3.9 demonstrates the solution structure of the DNA dodecamer d(CGCGAATTCGCG)$_2$ in an aqueous liquid crystalline medium containing 5% w/v bicelles. The structure, determined by using dipolar couplings, together with torsion angles derived from *J* couplings and interproton distances obtained from NOE, shows a well-defined B-form helix. Overall agreement with the corresponding X-ray structures is reasonable, although several of the

measured NOE. It could be argued, however, that the need for accurate solution structures is even greater for nucleic acids than for proteins because of their flexible structures, whose conformations can be significantly affected by crystallization. Nucleic acid NMR has benefited from the development of methodology based on selective isotopic

irregular features seen in the crystalline state are less pronounced or absent in solution (Comment J3.5).

In multidomain molecules, such as higher-order DNA structures, tRNA, or RNA, dipolar coupling can be used to determine the relative orientation of the various domains. Figure J3.10 shows the NMR model for tRNA^{Val} superimposed on the X-ray structure. The global orientation of helical domains was determined precisely from a set of residual dipolar couplings. There is a slight difference in the angle between the helical arms for the two models (Comment J3.6). These results demonstrated that a small

number of residual dipolar couplings can be used to determine the global structure of tRNA if the local structure of the helical stems is known. To test how well the dipolar couplings can be used to determine global structure of an RNA with a known secondary structure, the four stem regions in tRNA^{Val} were replaced with A-form helices. A conformational search was performed to find the orientations of the anticodon and acceptor arms that gives the best fit to the measured dipolar couplings. The resulting structure is given in Fig. J3.10b. It has an interarm angle of $101 \pm 2°$, in excellent agreement with the angle determined with the helices from the X-ray structure.

Finally, we point out that NMR has provided a unique direct observation of hydrogen bonds in nucleic acid base pairs, through the scalar coupling of imino proton donors and acceptor nitrogens.

COMMENT J3.5 X-RAY AND NMR DATA FOR DNA DODECAMER

The original X-ray structure exhibited an 18° bend in its helix axes over 12 base pairs, whereas the NMR dipolar structure has significantly less curvature: 7° over 12 base pairs. Most parameters, such as sugar pucker, propeller twist, roll, and the near absence of helical bending, are well determined by the NMR data. In contrast, other parameters such as the base-pair opening and the $^{31}P-^{31}P$ distance across the minor groove, are less accurately defined by the NMR.

J3.2.3 Carbohydrates

The structural characterization of carbohydrates is especially challenging because of their conformational flexibility. There are two kinds of internal motions in oligosaccharides. The first kind consists of modest variations of the glycosidic dihedral angles, of the order of $\pm 15°$. Oligosaccharides exhibiting this type of internal motion are called "rigid" in the sense that experimental NOE data may agree with those calculated for a model having a single conformation. The second kind features larger variations of glycosidic dihedral angles, between distinct energy minima (Fig. J3.11). Oligosaccharides exhibiting this type of internal motion are called "flexible" in the sense that experimental NOE data cannot be reconciled with any single conformation. In this case, molecules can be considered as an ensemble of conformers.

Advances in high-resolution NMR allow direct measurements of dipolar coupling by dissolving the molecules in oriented media such as liquid crystals composed of bicelles or phage (Section J2.7). Methyl 3,6-di-O-(α-D-mannopyranosyl)-α-D-mannopyranoside (trimannoside, Fig. J3.12) has been used to illustrate the application of residual dipolar coupling to the conformational analysis of oligosaccharides. The residual dipolar couplings were calculated as the difference of the measured couplings in aligned media (bicelle and phage samples at 38 °C) and isotropic media (bicelle sample at 20 °C).

Figure J3.13a shows the average structure of trimannoside ring I and ring III determined from bicelle media, and Fig. J3.13b illustrates the average structure of trimannoside ring I and ring III determined from phage media. The structure is oriented to show the similarity of conformations about the linkage connecting ring I and ring III. Figure J3.14a shows the average structure of trimannoside ring II and ring III determined from bicelle media and Fig. J3.14b illustrates the average structure of trimannoside ring II and ring III determined from phage media. The

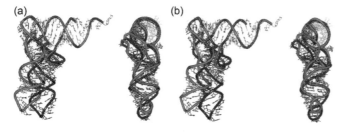

(a)   (b)

Fig. J3.10 Comparison of the model structure of tRNA^{Val} (red) with the global structure determined from residual dipolar couplings: **(a)** global structure resulting from domain orientation using the X-ray structure-based model for the helical arms (green): **(b)** global structure resulting from domain orientation using an A-form model for the helical arms (green). The tRNA was partially oriented by addition of 10 mg ml^{-1} of filamentous phage Pf1. (Mollova *et al.*, 2000.)

COMMENT J3.6 X-RAY AND NMR DATA FOR tRNA

The interarm angles are 86° for the tRNA^{Phe} crystal structure and $99 \pm 2°$ for the tRNA^{Val} structure determined from the dipolar couplings. There is significant variation for the interarm angles in three-dimensional X-ray structures of different tRNAs, ranging from 76° to 96°, and the angle determined here for tRNA^{Val} in solution is near the upper end of the range that been observed by X-ray (Westhof *et al.*, 1988; Basavappa and Sigler, 1991).

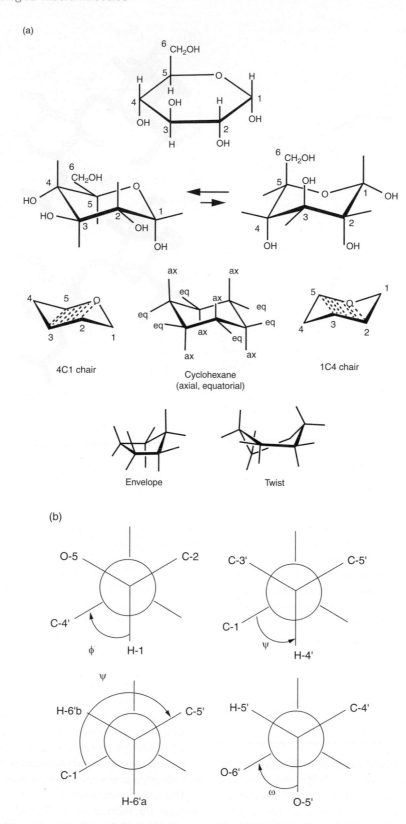

Fig. J3.11 (a) So-called Haworth projection of glucose with conventional labelling of sugar atoms shown above "chair" representations of two conformations; **(b)** diagram showing conventional angles and motion directions.

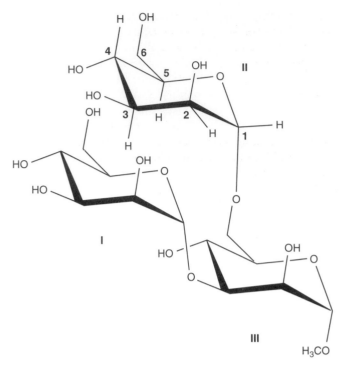

Fig. J3.12 Structure of trimannoside, methyl 3,6-di-O-(α-D-mannopyranosyl)-α-D-mannopyranoside. Rings are labeled with roman numerals. This trimannoside forms a common core in N-linked oligosaccharides. Three mannoses are connected together through two glycosides linkage α-(1,3) and α-(1,6). (Adapted from Tian *et al.*, 2001.)

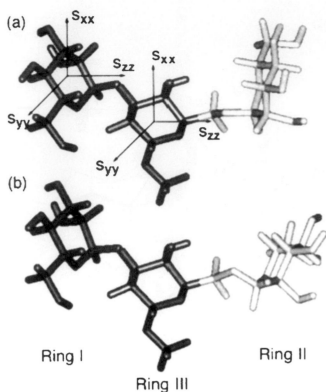

Fig. J3.13 (a) The average structure of trimannoside ring I and ring III assembled from bicelle media. **(b)** The same from phage media. The structure is oriented to show the similarity of conformations about the linkage connecting ring I and ring III. (After Tian *et al.*, 2001.)

structure is aligned to superimpose ring III with that in Fig. J3.14a to show the similarity of conformations about the linkage connecting ring II and ring III.

Comparison of data presented in Figs. J3.13 and J3.14 indicates that internal motion between rings I and II (at the α-(1,3) linkage) is limited, while significant motion is suggested between rings II and III (at the α-(1,6) linkage). The results clearly demonstrate that trimannoside exists as a set of conformers.

J3.3 DYNAMICS OF BIOLOGICAL MACROMOLECULES

Molecular dynamics on a broad range of timescales can be monitored using various types of NMR experiments. Nuclear spin relaxation is sensitive to motions occurring in shorter times than the molecular correlation time (10–20 ns for a protein in solution; see Section D3.5.3). Chemical shift exchange between two sites occurs in microseconds to milliseconds. Line broadening and spectral resolution effects take longer, of the order of seconds. These features make NMR a unique and powerful tool for the study of macromolecular dynamics.

J3.3.1 Protein Dynamics from Relaxation Measurements

We assume that the internal atomic motions in a macromolecule are independent of its overall geometric motion arising from tumbling or translational diffusion. In other words, the effective internal correlation time, τ_e, is much shorter than τ_{corr}, the rotational correlation time (see Section D2.3). The three commonly measured NMR relaxation rates, the spin–lattice relaxation rate, $R_1 = 1/T_1$, the spin–spin relaxation rate, $R_2 = 1/T_2$, and the heteronuclear NOE are all sensitive to internal motion on the subnanosecond timescale. In a model-free approach, one obtains two independent parameters: (1) the order parameter S^2, which is a measure of spatial restrictions on the motion; and (2) τ_e, the effective internal correlation time. The analysis does not invoke a specific model for internal motions.

The order parameter S^2 measures the magnitude of the angular fluctuation of a chemical bond vector such as the N–H, C–H, C–C bonds in a protein and thus reflects the flexibility of the polypeptide chain at these sites (Fig. J3.15). S^2 is interpreted as describing the amount of spatial freedom the vector **μ** has as a result of internal motion. If the bond vector **μ** diffuses in a cone as depicted in the inset, S^2 decreases rapidly as the cone angle θ_c

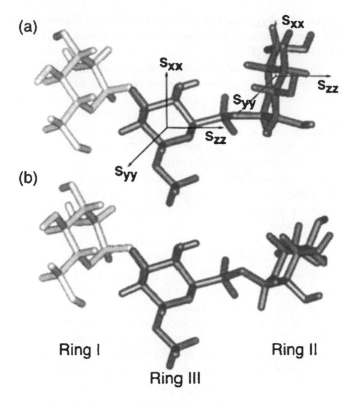

(a)

S_{xx}

S_{xx}

S_{zz}

S_{yy}

S_{zz}

S_{yy}

(b)

Ring I   Ring II

Ring III

Fig. J3.14 (a) The average structure of trimannoside ring II and ring III determined from bicelle media. **(b)** The same from phage media. The structure is aligned to superimpose ring III with that in part **(a)**. Note that there is a variation in the apparent orientation of ring II. (After Tian *et al.*, 2001.)

increases from $0°$ to $75°$. If the motion is completely free (i.e., all orientations of μ are equally probable), then $S = 0$. On the other hand, if the motion is completely restricted, then $S = 1$. Since ^{15}N labeling of proteins became commonly available for proteins with efficient expression systems (Section J2.8), studies of mobility based on the model-free approach have become routine.

The model-free approach is based on a very simple form for the total autocorrelation function that describes the overall and internal motions when it is assumed that they are independent:

$$G(\tau) = \frac{1}{5} G_0(\tau) G_i(\tau) \qquad (J3.2)$$

where $G_0(\tau)$ is the correlation function for overall motion (Section D3.5.2):

$$G_0(\tau) = \exp(-\tau/\tau_{corr}) \qquad (J3.3)$$

and $G_i(\tau)$ is the simplest form for internal motions:

$$G_i(\tau) = S^2 + (1 - S^2) \exp(-\tau/\tau_i) \qquad (J3.4)$$

Internal correlation times τ_e, τ_i, and τ_{corr} are related as

$$\frac{1}{\tau_i} = \frac{1}{\tau_{corr}} + \frac{1}{\tau_e} \qquad (J3.5)$$

After Fourier transformation, the resulting spectral density $J(\omega)$ is

$$J(\omega) = \frac{2}{5} \left\{ \frac{S^2 \tau_{corr}}{1 + (\omega \tau_{corr})^2} + \frac{(1 - S^2) \tau_i}{1 + (\omega \tau_i)^2} \right\} \qquad (J3.6)$$

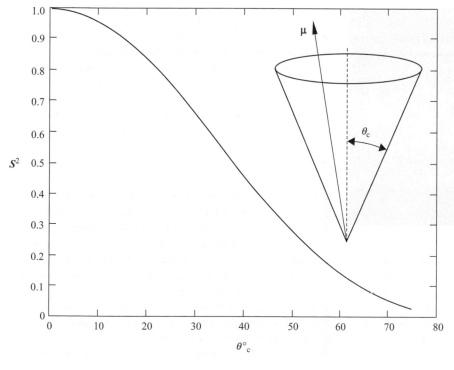

μ

θ_c

S^2

$\theta°_c$

Fig. J3.15 Relationship between internal motion and the model-free parameter S^2. The parameter S^2 provides information about the angular amplitude of the internal motions of bond vectors (N–H, C–H, C–C) at numerous sites in the protein sequences. If the bond vector μ diffuses in a cone as depicted in the inset, S^2 decreases rapidly as the cone angle θ_c increases from $0°$ to $75°$, and remains small for $75° < \theta_c < 180°$. When the bond is rigid $S^2 = 1$, and when internal motion is isotropic $S^2 = 0$. (After Ishima and Torchia, 2000.)

In the limit that S^2 is unity $J(\omega)$ reduces to the familiar Lorentzian distribution for the rotational diffusion of rigid tumbling molecules (see Sections A3.2.3 and D4.3.1). The spectral density function of Eq. J3.6 contains essentially only two fitting parameters, S^2 and τ_e. The overall correlation time τ_{corr} is assumed to be the same for all nuclei of a protein. If τ_{corr} is known, two measurements of relaxation parameters, such as T_1 and T_2 are sufficient to determine these two parameters. The measurement of a third parameter is a check on the validity of the model. In the original experiments the values of S^2 and τ_e were obtained from a pair of relaxation parameters (NOE and T_1) at two fields.

As an example, extensive studies of the relaxation properties were carried out for *Bacillus subtilis* glucose permease IIA domain. The order parameters are represented mapped onto the protein structure in Fig. J3.16. The amide nitrogen positions are color-coded according to their calculated order parameters, S^2. Atoms are colored on a continuous scale from red ($S^2 \leq 0.4$) to indigo ($S^2 = 0.9$). All regions of higher mobility were located in three areas: at the termini, surface loops, or regions of irregular secondary structure.

In the simplest analysis, overall motion is assumed to be isotropic and this contribution is described by a single exponential. In the case of rotational diffusion anisotropy, the time correlation is a multiexponential function whose relative weighting depends on the orientation of the different relaxation interactions with respect to the principal components of the diffusion tensor. Thus, the determination of the rotational diffusion tensor is a prerequisite in the analysis of internal dynamics using heteronuclear relaxation. The determination of the overall rotational diffusion

tensor can also provide important information concerning the overall form of globular proteins (Comment J3.7; see also Section D1.7.2).

The rotational diffusion tensor is specified by three principal components – the diffusion constants for rotation about the X, Y, Z principal axes – and by the orientation of the principal axes relative to that of the protein. The principal axes, X, Y, Z, are fixed in the protein, and defined as the coordinate axes in which the diffusion tensor assumes a simple, diagonal form.

The rotational diffusion can be characterized accurately using the relaxation rate ratio (R_1/R_2), a parameter which becomes independent of internal dynamics in the fast-motion limit while remaining highly sensitive to overall diffusion. As an example, the analysis of the rotational tensor of the chitin binding of the anti-fungal protein from *Streptomyces tendae* (AFP1) has been performed by measuring ^{15}N relaxation parameters. Relaxation rates R_1 and R_2, as well as ^{15}N–^{1}H heteronuclear NOE values, were obtained for a total 65 backbone amides (78% of all possible). AFP1 exhibits a large degree of asymmetry in its overall shape. In fact, the relative magnitudes of the principal components of inertial tensor **I** are calculated as 1:0.94:0.61 for the average regularized structure. The orientation of the principal components of **I** with respect to the molecular frame is shown in Fig. J3.17. The figure also shows the orientation of the diffusion tensor **D** obtained for the three non-isotropic solutions and its relation to the inertial tensor **I**. The principal component of the prolate solution, $D_{\parallel}^{PRO}$, or the direction about which diffusion is maximal (green in Fig. J3.17) is coincident with the direction of $\mathbf{I}_{yy}$ and it is perpendicular to the direction

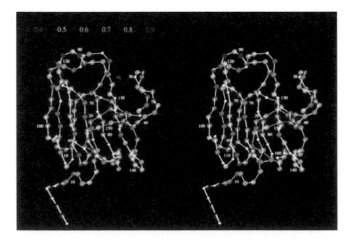

Fig. J3.16 Schematic representation of the solution structure of the *Bacillus subtilis* IIA domain showing the amide nitrogen positions color-coded according to their calculated order parameters, S^2. Atoms are colored on a continuous scale from red ($S^2 \leq 0.4$) to indigo ($S^2 = 0.9$). Backbone nitrogens for which order parameters were not calculated are shown as small white spheres. Every tenth backbone nitrogen is numbered. (After Stone *et al.*, 1992.)

COMMENT J3.7 HYDRONMR PROGRAM

Hydrodynamic calculations have been used to interpret the ^{15}N NMR relaxation data of globular proteins using the program HYDRONMR. HYDRONMR uses a shell model (Section D2.2.2), which is constructed from the atomic representation of the protein contained in the PDB file. This procedure avoids the bead overlap problems, but requires computer time. HYDRONMR contains a single adjustable parameter, the atomic element radius (the radius of the bead in the terminology of Section D2.2.2). Comparison shows that there is good agreement between experimental relaxation values (R_1/R_2) and those calculated using HYDRONMR and the three-dimensional structure using an atomic element radius that ranges from 2.8 Å to 3.8 Å, with the distribution centered at 3.3 Å. Deviations from the usual range toward larger values are associated with aggregation of proteins. Deviations to lower values may be related to large-scale motions or inappropriate model structures (Bernado *et al.*, 2002).

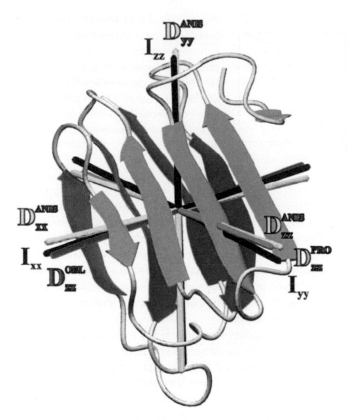

Fig. J3.17 Representation of the inertial and diffusion tensors of AFP1. The inertial tensor **I**, with its principal components labeled I_{xx}, I_{yy}, and I_{zz} (where $I_{xx}:I_{yy}:I_{zz} = 1.00:0.94:0.61$) is shown in black. The principal axes of the two axially symmetric diffusion tensors, corresponding to the oblate or prolate ellipsoid model, are drawn in red and green, respectively. The axes for the fully anisotropic solution are colored yellow. (After Campos-Olivas et al., 2001.)

of the maximal inertia I_{xx}. Coincidence of $D_{\parallel}^{PRO}$ (fastest diffusion) with I_{zz} (minimal inertia) would be expected if the anisotropy were simply determined by molecular asymmetry. However, for the oblate solution, the direction of the slowest diffusion, $D_{\parallel}^{OBL}$ (red in Fig. J3.17), is coincident with the direction of maximal inertia, as expected if molecular shape determines the rotational anisotropy. Therefore, the oblate solution, representing the best fit to relaxation data, is in perfect agreement with the asymmetry of the molecule and should be considered as the most appropriate description of the real rotational tumbling behavior of AFP1 in solution.

Protein function, protein–ligand, and protein–protein interactions are often affected by motions on much slower timescales, ranging from microseconds to seconds. These interactions often contribute to the rate of transverse spin relaxation, R_2, via the chemical exchange mechanism. In the "model-free" formalism the evidence of this class of motions emerges via the introduction of the "exchange" term R_{ex}, which is the exchange contribution to R_2. The exchange term R_{ex} may or may not be present (Comment J3.8).

COMMENT J3.8 TWO-SITE CHEMICAL EXCHANGE IN THE SPIN ECHO METHOD

The exchange term R_{ex} in the following form approximates the effect of two-site chemical exchange in modified spin-echo method for measuring nuclear relaxation times

$$R_{ex} = \frac{p_1 p_2 (\Delta\omega)^2}{k_{ex}} \times \left[1 - \frac{2}{k_{ex}\tau_{cp}} \tanh\left(\frac{k_{ex}\tau_{cp}}{2}\right)\right]$$

where τ_{cp} is the time between 180° pulses, $\Delta\omega$ is the difference in chemical shift of the nuclei between two conformational states; $k_{ex} = k_{-1}/p_1 = k_1/p_2$; $1 \geq p_1 \geq 0.5$, and $p_2 = (1 - p_1)$ are the populations of the conformational states; k_1 is the forward exchange rate constant; and k_{-1} is the reverse exchange rate constant (Luz and Meiboom 1963).

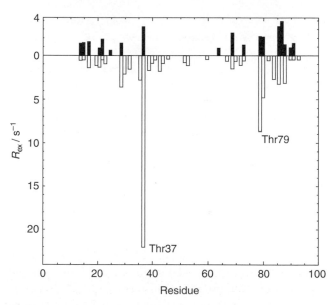

Fig. J3.18 Exchange contribution R_{ex} along the polypeptide chain for rat CD2d1 at 10 mM phosphate buffer, pH = 6.0, $T = 25$ °C, a proton resonance frequency of 500 MHz and protein concentrations of 0.1 mM (black bars) and 1.2 mM (white bars). A small number of residues – most notably Thr 37 and Thr 79 – show very large deviation on the R_{ex} panel. (After Pfuhl et al., 1999.)

As an example, the NMR exchange broadening arising from specific low-affinity protein self-association of the N-terminal domain of the rat T-cell adhesion protein (CD2d1) has been performed by measuring of ^{15}N nuclear relaxation at wide intervals of protein concentration. Figure J3.18 shows exchange contribution R_{ex} along the polypeptide chain for two protein concentrations, of 0.1 mM (black bars) and 1.2 mM (white bars). The magnitudes of the R_{ex} terms for Thr 37 ($22\,s^{-1}$) and Thr 79 ($8\,s^{-1}$) are exceptionally large and very strongly dependent on the

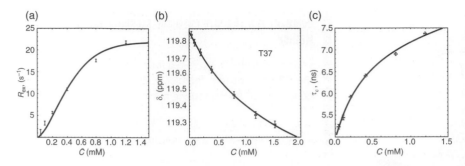

Fig. J3.19 (a) Variation of ^{15}N line broadening as a function of protein concentration for Thr 37 at 10 mM phosphate buffer, pH 6.0 and $T = 25\,°C$. **(b)** The same dependence for variation of ^{15}N chemical shifts δ in ppm. **(c)** The same dependence for the apparent rotational correlation time τ_c. (After Pfuhl et al., 1999.)

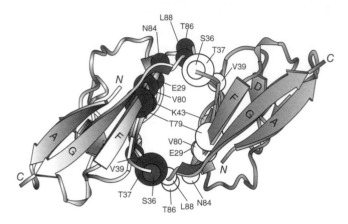

Fig. J3.20 CD2d1 dimer interface as seen in the rat CD2 crystal structure. For clarity the two monomers are shown in different shades of gray. Residues with $R_{ex} > 1.5\,s^{-1}$ are marked by a sphere centered on the position of the corresponding backbone nitrogen. The radius of the sphere is approximately proportional to the magnitude of R_{ex}. (After Pfuhl et al., 1999.)

the dynamics of local structural fluctuations. Exchange between amide protons in proteins and bulk solvent can be measured in a variety of ways. Slow exchange lifetimes (from minutes to days) are determined by following the loss of amide proton signal intensity of a protein dissolved in D$_2$O, and provide information about the relative solvent accessibility of various components of protein secondary and tertiary structure.

Faster exchange lifetimes (5–500 ms) can be monitored by following the exchange of amide proton magnetization with that of water protons. At high pH the base-catalyzed proton exchange reaction is very fast ($\sim 10^5\,s^{-1}$ at pH 10), and these magnetization exchange measurements directly measure the timescale of rate-limiting protein conformational openings that permit solvent access to the exchanging amide sites. Such measurements have been performed on a highly thermostable protein, rubredoxin from *Pyrococcus furiosus*. At 28 °C, solvent access to all amide sites in the protein occurs on the millisecond timescale and solvent exchange protection factors for all sites are similar to those observed for mesophilic homologues. This observation suggests that the flexibility of thermophilic rubredoxin is similar to that of the mesophilic homologues, and brings into question the general hypothesis that the high stability of thermophilic protein comes from the rigidity of its native structure.

J3.3.3 Dynamics of Slow Events: Translational Diffusion

Traditionally, diffusion coefficients of biological macromolecules are determined by the numerous physical methods presented in Table D1.2. Although effective, these techniques are often carried out at micromolar concentrations that are three orders of magnitude less than the millimolar concentrations required for NMR. This discrepancy is particularly important for the concentration dependence of diffusion processes. In contrast, diffusion measurements of the NMR sample report under conditions identical to those used for structure determination. This provides knowledge of the oligomeric state of macromolecules prior to structure calculations using NMR data.

Diffusion coefficients of molecules can be measured using a spin echo experiment in pulsed field gradient

protein concentration. Figure J3.19 shows the variation of experimental parameters as a function of protein concentration for one residue, Thr 37. The value of R_{ex} drops to very small values in more dilute samples (Fig. J3.19a). The broadened resonance also displays considerable concentration-dependent chemical shift variation (Fig. J3.19b). These findings, together with the observation of concentration-dependent values of the rotational correlation time τ_c (Fig. J3.19c) allow low-affinity self-association of CD2d1 to be identified as a major factor in the explanation of its NMR behavior.

Concentration-dependent line-width effects are observed for the resonances of a specific limited region of the protein surface, which is coincident with the major intermolecular contact in crystals of the complete extracellular portion of the CD2 protein (Fig. J3.20).

These observations clearly demonstrate that very weak protein–protein interactions can be detected by NMR.

J3.3.2 Protein Dynamics from Amide Proton Exchange

Measurements of amide proton exchange rates in proteins can provide useful information about global stability and

(PFG) NMR (Comment J3.9). The method relies on two gradient pulses surrounding the 180° pulse in the spin echo: The first dephases the transverse magnetization in a spatially dependent manner along the z-axis and the second gradient then rephases the magnetization. If the molecules move along the z-axis during the time between the two gradients, the magnetization does not refocus completely. Therefore the attenuation of its resonance is large, if the molecule diffuses rapidly; the attenuation is relatively small if the molecule diffuses slowly. The echo measured amplitude $A(t)$ is differentially attenuated in each spectrum due to translational diffusion, which is related to D_t by

$$A(t) = A(0) \exp \left[-(\gamma \delta)^2 D_t (\Delta - \delta/3) G^2 \right] \qquad \text{(J3.7)}$$

where γ is the ^{1}H gyromagnetic ratio of a proton, δ is the PFG duration time (s), G is the gradient strength (G cm^{-1}), and Δ is the time between PFG pulses (s). Equation (J3.7) shows that correct measurements of D_t require at least one order of magnitude signal decay to accurately fit the data to Eq. (J3.7). The limit on measurable D_t values, therefore, depends on the area of the gradient pulses (δG) and the time allowed for diffusion (Δ). Larger values of D_t ($\geq 10^{-5}$ cm^2 s^{-1}) can be reliably measured using this type of NMR experiment. However, as one attempts to measure D_t for proteins (where D_t is about ~10^{-6} cm^2 s^{-1}), the time necessary to observe diffusion, Δ, becomes long relative to the transverse relaxation time $(T_2^* \sim 12 - 40$ ms), and the signal decays from spin–spin interactions, independent of diffusion.

The stimulated spin echo experiment (90°–τ (PFG)–90°–T–90°–τ(PFG)–Acq) was designed to avoid T_2 relaxation effects by storing the magnetization along the z–axis during time T, so that relaxation depends primarily on T_1, which is usually much longer than T_2 for proteins. Longer diffusion times (Δ) can then be used for larger proteins without significant signal loss from relaxation.

Stimulated spin echo NMR experiments have been used (see Section J2.2) to determine the diffusion coefficient of lysozyme (Comment J3.10). The diffusion coefficient was obtained by fitting the signal attenuation to Eq. (J3.7), which yields a value of 10.8×10^{-7} cm^2 s^{-1}. This result is in good agreement with data obtained by other methods (see Table D1.2).

The stimulated spin echo NMR technique has been used for the characterization of protein unfolding. Figure J3.21a and c show stack plots of the signal attenuation of the aromatic signal from His^{15} taken from the spectrum of lysozyme in different concentrations of urea (0 M and 8 M, respectively). Figure J3.21b shows the attenuation of the dioxane signal used as the reference molecule in the same solution. The observed diffusion coefficients of both dioxane and lysozyme decreased with increasing urea concentration as a result of the increased viscosity, but the ratio D_{diox}/D_{lys} increased, as shown in Fig. J3.22. The effective hydrodynamic radius increases by 38% on unfolding in urea, a result in a good agreement with a study by small-angle X-ray scattering. Unfortunately, the small diffusion coefficients ($D < 10 \times 10^{-7}$ cm^2 s^{-1}) are difficult to measure by PFG NMR methods using stimulated echoes described above. This limitation results from the rapid longitudinal relaxation of the nuclei (usually protons) that carry the information about the localization of the molecules during the diffusion interval. The duration of this interval can be increased by about one order of magnitude by storing the information in the form of longitudinal magnetization of heteronuclei, such as ^{15}N, that have much longer spin–lattice relaxation times than protons. This novel method, called "heteronuclear stimulated echoes," allows one to gain about one order of magnitude in measurement of D, so that the diffusion coefficient of

---

COMMENT J3.9 SPIN ECHO IN PULSED FIELD NMR

The original experimental scheme for measuring the diffusion coefficient using spin-echo experiment in pulsed field gradient NMR was (Stejskal and Tanner, 1965):

$$(90° - \tau(PFG) - 180° - \tau(PFG) - \text{Acquisition})$$

COMMENT J3.10 STIMULATED SPIN ECHO IN PULSED FIELD NMR

The principal experimental scheme for a stimulated spin echo in pulsed field NMR is:

$$(90° - \tau(PFG) - 90° - T - 90° - \tau(PFG) - \text{Acquisition})$$

In practice the scheme is much more complex and requires special sequences to provide sufficient water suppression and amplitude distortions, which complicate accurate measurements of diffusion coefficients (Altieri et al., 1995).

(a) (b) (c)

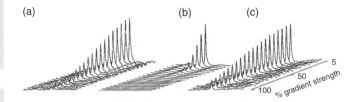

Fig. J3.21 Stimulated spin echo spectra obtained from a 1.4 mM solution of hen egg white lysozyme in D$_2$O at pH 2.0 and 20 °C. A small amount of 1,4-dioxane was added as the radius standard. The diffusion gradients were varied between 5% and 100% of their maximum strength (60 G cm^{-1}).
(a) Attenuation of the aromatic signal from His^{15} taken from the spectrum of lysozyme in 0 M of urea. **(b)** Attenuation of the dioxane signal used as a standard reference molecule.
(c) Attenuation of the aromatic signal from His^{15} taken from the spectrum of lysozyme in 8 M of urea. (After Jones et al., 1997.)

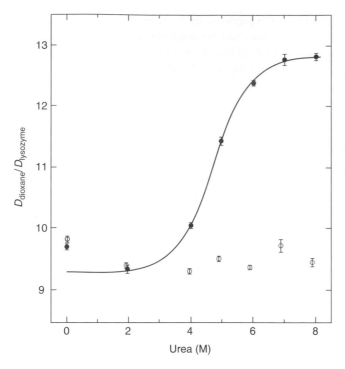

Fig. J3.22 Diffusion coefficients for 1.4 mM of hen egg white lysozyme as a function of urea concentration. The filled circles and the solid line show data at pH 2.0, while open circles show data at pH 5.5. Diffusion coefficients are presented as a ratio of D_{diox}/D_{lys}, in which the hydrodynamic radius of dioxane was taken as a standard (1.7 Å). It is assumed that for protein and reference molecules in the same solution, $R_H^{prot} = \frac{D_{ref}}{D_{prot}} \times R_H^{ref}$ (After Jones *et al.*, 1997.)

molecules with molecular masses in excess of 100 kDa should be readily measured.

Finally, we point out that stimulated spin echo NMR has been used to measure translational diffusion coefficients of the three short DNA duplexes (12, 14, and 24 base pairs). The results were in good agreement with data obtained by the DLS method (see Table D1.3).

J3.4 SOLID-STATE NMR

In 1983 the first paper that demonstrated that solid-state NMR could provide high-resolution structural constraints was published. The potential of this approach was rapidly established and it evolved through the development of new types of constraints and advances in NMR technology.

In solid-state NMR studies, the sites of interest are immobile on the timescales defined by the relevant nuclear spin interactions, ranging from about 10^3 Hz for chemical shift interactions to 10^6 Hz for nuclear quadrupole interactions. Suitable immobile sites are present in supramolecular structures such as large nucleoprotein or membrane–protein complexes in solution, and in sedimented protein

microcrystals. In single crystals any arbitrary orientation of the sample can be studied. In uniaxially oriented samples, the direction of orientation must be parallel to that of applied magnetic field.

Obtaining the information contained in solid-state NMR spectra is more difficult than analyzing liquid NMR spectra. It is necessary to apply more complicated pulse sequences, use special sample spinning techniques, and apply higher radio-frequency energies (Comment J3.11). All of the experiments benefit from using a high magnetic field.

J3.4.1 Solution and Solid-State NMR: Comparative Analysis

From an NMR spectroscopist's viewpoint, the main difference between solids and liquids is the mobility of the molecules in the sample. The main restriction for solution NMR, that samples must be tumbling isotropically, does not apply to solid-state NMR: For solid-state NMR, the more rigid the sample the better. This reflects the distinction between solid-state NMR (observation of anisotropic systems) and solution NMR (observation of isotropic systems). Consequently, the "correlation time problem" discussed as a molecular mass limit for solution NMR (Section J2.4) does not apply to solid-state NMR. The molecular mass range that can be studied by solid-state NMR is nearly without bounds, hence there have been studies of silk fibroin, the 16 MDa filamentous virus fd, and colicin E1 P190 in planar lipid bilayers.

The NMR spectra of solids actually contain more information than is available by liquid NMR, but the information is hidden under broad, overlapping peaks with poor resolution. Note that line broadening has different origins in solution and solid-state NMR. To eliminate or greatly reduce the spectral line-broadening caused by dipolar interactions and CSA (see Section J3.4.2), most solid-state NMR experiments rely on magic angle spinning (MAS) The sample is tilted to the "magic angle" of 54.7° relative to the external magnetic field (Fig. J3.23). In MAS, the sample is

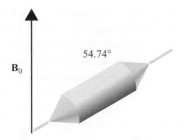

Fig. J3.23 The magic angle in solid-state NMR.

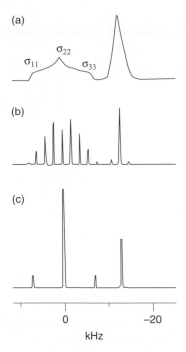

and physically rotated at high speeds in the magnetic field in order to average the nuclear spin interactions. Both the chemical shift and the dipolar coupling contribution have terms that contain factors of $(3\cos^2\theta - 1)$, and at the magic angle these terms vanish (see also Section J2.7.1). For example, the CSA, σ_{izz}, can conveniently be expressed in terms of the angle of rotation of the sample with respect to the applied field:

$$\sigma_{izz} = (3\cos^2\theta - 1)(\text{other terms}) + (3/2\sin^2\theta)\sigma_i \quad (J3.8)$$

When θ is chosen to be $54.7°$, $3\cos^2\theta - 1 = 0$, $\sin^2\theta = 2/3$, and $\sigma_{izz} = \sigma_i$, the isotropic chemical shift, i.e., that which is observed in NMR of liquids. This results in high-resolution ^{13}C NMR spectra of solids.

Figure J3.24 illustrates the distinctive ^{13}C line-shapes for the carboxyl and α-carbon resonances of free glycine. These line-shapes arise from the random orientation of glycine molecules in a polycrystalline sample and are representative of the broad NMR resonances mentioned above. A single molecule or crystallite in the sample contributes a narrow Lorentzian component to the overall line-shape at a frequency that is dependent on its orientation with respect to the external magnetic field. In Fig. J3.24a, the line-shapes represent the sum of all of the molecular orientations in the sample and result solely from anisotropy in the chemical-shift interaction because dipolar interactions between ^{13}C and ^1H spins have been eliminated by proton decoupling. In Fig. J3.24b, c, the CSA has been averaged by spinning the sample at the magic angle.

J3.4.2 Solving Three-Dimensional Structures in Solid-State NMR

Solid-state NMR has the inherent ability to detect single atomic sites and, hence, the potential to yield high-resolution data. The method, however, has been limited by low signal-to-noise ratios. There is hope, however, for significant improvement. The dynamic nuclear polarization (DNP) method powerfully enhances NMR sensitivity, and is in the process of development for biophysical applications (Comment J3.12).

Fig. J3.24 Solid-state ^{13}C NMR spectra of glycine illustrating: **(a)** the broad NMR resonances in static samples; **(b)** the effect of magic spinning at 2.0 kHz; and **(c)** the effect of magic spinning at 7.2 kHz. The principal values of the chemical-shift tensor are shown for the ^{13}C carboxyl resonance and correspond to the down-field inflection point (σ_{11}), the maximum (σ_{22}), and the up-field inflection point (σ_{33}). MAS collapses the broad line-shapes into sharp center-bands at the isotropic chemical shifts (σ_{iso}) and the rotational side-bands spaced at the spinning frequency. The frequency scale is centered on the carboxyl center-band. (After Smith and Peersen, 1992.)

COMMENT J3.12 DYNAMIC NUCLEAR POLARIZATION

Overhauser first proposed DNP in NMR in 1953. In an NMR experiment, radio-frequency radiation is used to excite polarized nuclear spins in a magnetic field. Only a low level of polarization is achieved, however, leading to weak signal-to-noise ratios. Spin polarization of electrons in paramagnetic compounds, such as stable free radicals, is several orders of magnitude larger than that of nuclei. DNP consists of transferring spin polarization from such electrons to sample nuclei. The method consists of irradiating the electron paramagnetic resonance (EPR) spectrum with microwaves to excite electron–nucleus transitions, leading to a transfer of polarization. The combination of DNP and MAS can yield high-resolution spectra of solids. Major developments that contributed to the usefulness of DNP in solid-state NMR include gyrotrons, which provide microwave power over a broad frequency range, innovative polarizing agents, and cryogenic MAS probes to work at liquid nitrogen temperature, since DNP functions optimally at low temperature. (Griffin, 2010.)

Orientational constraints are observed in samples that have unique orientation with respect to the magnetic field axis of the NMR spectrometers. By obtaining numerous constraints, all with respect to the same axis, three-dimensional structure can be achieved.

Distance constraints that define the relative separation of two atoms can be obtained through observation of residual dipolar interactions. Both homonuclear and heteronuclear interactions can be observed with considerable precision. As is routine in solution NMR, three-dimensional structures can be solved by distance constraints.

Torsional constraints are defined through the observation of a relative orientation of spin interaction tensors in adjacent sites within the protein. Such constraints lead directly to the definition of the structural torsion angles, the ultimate goal for defining three-dimensional structure.

pertinent features of the protein backbone structure are illustrated with the small 46-residue protein crambin (Fig. J3.25). The backbone of a protein consists of linked planes with quite regular geometry. The α-carbons of adjacent amino acids in a polypeptide chain are joined by the amide C–N bonds. The six atoms that form this peptide linkage lie roughly in a plane, and the relative orientation of adjacent planes is defined by the ϕ and ψ torsion angles (Fig. J3.26). The secondary structure of the peptide backbone can be determined by sequentially establishing the orientation of each peptide plane relative to the common axis, the z-axis of the external magnetic field. Measurements of both the dipolar and chemical-shift interactions are necessary to limit the number of possible orientations and define the peptide structure.

Dipole Interactions, CSA, and Quadrupole Interactions

The dipole–dipole interactions, CSA, and quadrupole interactions in polypeptide backbone sites are highly anisotropic. The dipole–dipole interaction tensor between two

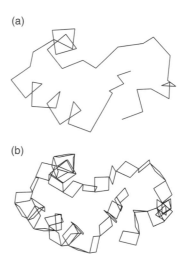

Fig. J3.25 Representation of the peptide backbone structure of crambin: (a) α-carbon atoms connected by vectors; (b) peptide planes outlined. (Adapted from Opella and Stewart, 1989.)

There are many approaches for gaining structural insights into proteins from solid-state NMR. The main ones have orientational, distance, and torsional constraints (Comment J3.13).

Orientation-Dependent Approaches

The assembly of three-dimensional structures is similar to that in solution NMR where distances and torsional constraints are involved (Section J3.1). However, orientational constraints are unique in this regard. Orientational constraints are absolute and independent, unlike the torsional and distance ones. Moreover, because these constraints are very precise even when there are relatively few constraints, high-resolution structure is achievable.

One solid-state NMR method focuses on the determination of torsion angles in the polypeptide backbone of protein structures solely from orientation constraints. The

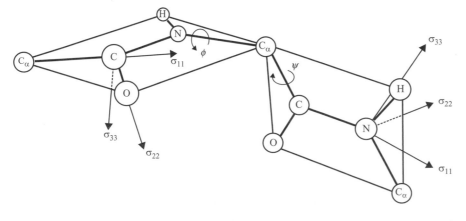

Fig. J3.26 Illustration of the peptide planes of a dipeptide and the approximate orientations of the amide ^{15}N and carbonyl ^{13}C chemical-shift tensors with respect to the molecular frame. The rotation axes of the torsion angles ϕ and ψ are defined with the arrows. The angles ϕ and ψ define the orientation of the planar peptide linkages and are determined by measuring the N–C and N–H bond orientations, as well as the ^{15}N and ^{13}C chemical-shift tensors. (After Smith and Peersen, 1992.)

spins depends on the distance between the two spins as well as on the orientation of the internuclear vector with respect to the direction of the applied magnetic field; therefore this spin interaction provides spatial as well as angular information. In contrast, the CSA and quadrupole interactions, which reflect local electronic properties, provide only orientational information.

Additional constraints for peptide-plane orientations are derived from the amide ^{15}N and carbonyl ^{13}C chemical-shift tensors. The orientation dependence of the chemical-shift interaction has the form:

$$\sigma = \sigma_{11}\cos^2\alpha\sin^2\beta + \sigma_{22}\sin^2\alpha\cos^2\beta + \sigma_{33}\cos^2\beta \quad (J3.9)$$

where σ is observed chemical shift, σ_{11}, σ_{22}, and σ_{33} are the principal components of the chemical-shift tensor and α and β are the Euler angles relating the principal axis system of the chemical-shift tensor to the laboratory frame.

The quadrupole interaction of the spin $S = 1$, ^{14}N amide site is useful for determining angles in solid-state NMR studies and has several important advantages over other observable interactions. ^{14}N is over 99% naturally abundant, thus no isotopic labeling is required and sensitivity is high; the large quadrupole interaction provides high spectral resolution. The ^{14}N quadrupole tensor has been well characterized for model peptides. Like the chemical-shift tensor, it is non-axially symmetric and thus there is a dependence on two angles, α and β, which describe the orientation of the principal axes of the tensor with respect to the magnetic field vector. Because ^{14}N is an $S = 1$ nucleus, there are two fundamental $\Delta m = 1$ transitions and the spectra from oriented samples are doublets centered at the ^{14}N Larmor frequency. The observed splitting between the two frequencies gives angular information described by

$$\Delta\nu_Q = (3/4)\varsigma\left[3\cos^2\beta - 1 + \kappa\sin^2\beta\cos 2\alpha\right] \quad (J3.10)$$

where ς is the quadrupole coupling constant and κ is the asymmetry parameter.

^{2}H quadrupole interactions can also be useful for determining the peptide backbone structure. For a single deuterium bonded to carbon, the quadrupole tensor is axially symmetric with the largest principal component along the C–^{2}H bond. Thus, the observed quadrupole splitting depends on a single angle. By specifically replacing C_α protons with deuterons, it is possible to measure the ^{2}H quadrupole splitting. The angular dependence of the splitting is described by

$$\Delta\nu_Q = (3/4)\varsigma\left[3\cos^2\beta - 1\right] \quad (J3.11)$$

where ς is the quadrupole coupling constant, which is approximately 180 kHz, and β is the angle between the C–^{2}H bond and the magnetic field. This angle is the same as would be found from ^{1}H–^{13}C$_\alpha$ dipole–dipole splitting and both measurements have the same angular

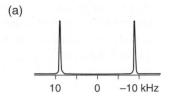

(a)

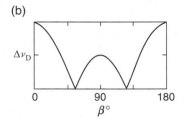

(b)

Fig. J3.27 (a) Calculated spectrum for heteronuclear dipole–dipole splitting of 17.8 kHz. **(b)** Plot of the dipole–dipole splitting, $\Delta\nu_D$, versus the angle β. (After Opella and Stewart, 1989.)

dependence. Figure J3.27 shows a calculated spectrum that is a doublet with a splitting, $\Delta\nu_D$, corresponding to a heteronuclear ^{15}N–^{1}H dipole–dipole coupling of 17.8 kHz.

Distance-Dependent Approaches

Long-distance restraints are arguably the most powerful for NMR structure determination. The second class of methods used to determine protein structure by solid-state NMR relies on an accurate determination of weak dipolar couplings between nuclear spins. A set of distances determined from dipolar couplings serve to constrain the structure, analogous to the use of ^{1}H...^{1}H distance constraints from NOE data in solution NMR (Section J2.4). The difficulty in these measurements is that the weak couplings are virtually impossible to observe in the broad NMR spectra, while they average to zero in magic spinning spectra. Two special methods, rotational echo double-resonance (REDOR) and rotational resonance (RR), have been developed to overcome the difficulty.

Recent illustrative applications of solid-state NMR to important biological problems are described in the following subsections.

J3.4.3 Atomic Structure of the Injection Needle Used by Pathogenic Bacteria (Loquet et al., 2012)

In the type III secretion system (T3SS) of bacteria that cause serious diseases, including typhoid fever and bubonic plague, effector proteins are injected into the human cell through a hollow needle-like protein filament. Lange and collaborators produced an atomic model of the *Salmonella typhimurium* T3SS needle by a combination of wild-type needle production, solid-state NMR, electron microscopy, and Rosetta molecular modeling software.

Secondary structure assignment was performed by chemical shift analysis of ^{13}C, with positive and negative secondary chemical shifts indicating α-helices and β-sheets, respectively (Fig. J3.28a). Sparse ^{13}C labeling of the molecule was achieved by using [1–^{13}C] or [2–^{13}C] glucose (Glc) to produce the recombinant protein. Sparse labeling reduces magnetization transfer between close atoms to significantly improve the NMR spectra so that cross-peaks could be readily assigned to long-distance restraints (Fig. J3.28b). The architecture of the needle is in Fig. J3.29, showing the intra- and inter-subunit restraints determined by solid-state NMR.

The complete atomic model derived from the solid-state NMR data combined with electron microscopy and Rosetta molecular modeling is shown in Fig. J3.30.

J3.4.4 Rapid Proton-Detected NMR Assignment for Proteins with Fast MAS (Barbet-Massin et al., 2014)

In this collaboration between nine laboratories from France, Germany, the USA, Latvia, and Italy, Pintacuda and coworkers established a method that greatly facilitates the sequence-specific resonance assignment of solid-state NMR spectra. Sequence-specific resonance assignment of MAS spectra is an essential step in the determination of distances, torsion angles, and molecular structures. The method enables the use, in solid-state NMR, of computational structure-solving protocols already developed for solution NMR. In the paper, the method was applied to five challenging "solid-state" biological systems: two microcrystalline proteins, a sedimented virus capsid, a predominantly α-helical, and a predominantly β-sheet membrane protein.

Standard MAS NMR assignment procedures were based on double- and triple-resonance spectra, using correlations between ^{13}C and ^{15}N signals. Experiments required large amounts ($\sim$1 mg kDa^{-1}) of $^{13}C/^{15}N$-labeled samples, long data acquisition times, and manual analysis of the spectra.

Thousands of structures have been solved by solution NMR and very efficient objective analysis protocols have been developed, based on triple resonance pulse schemes correlating backbone and side-chain ^{1}H, ^{13}C, and ^{15}N resonances for sequential assignment. The proton (^{1}H) is the optimal choice for detection in solution. In solids, however, strong homonuclear ^{1}H dipolar couplings dramatically lower the resolution. Barbet-Massin et al. overcame this obstacle by

(1) dilution of the proton content (similarly to sparse labeling discussed in the previous section);
(2) fast MAS;
(3) a high magnetic field.

Proton dilution in the proteins was controlled by expression in suitable deuterated media and H–D solvent exchange. Magic angle spinning at a rate of 60 kHz was applied to the sample, in the strong magnetic field from spectrometers operating at 800 or 1000 MHz ^{1}H-Larmor frequencies.

Using appropriate pulse sequences, the ^{15}N–^{1}H correlation spectra recorded at 1 GHz and 60 kHz MAS are in Fig. J3.31.

Note how the peaks are well separated. The 2D spectra served as modules to design 3D procedures correlating ^{1}H and ^{15}N with neighboring ^{13}C, as well as more complex experiments. The resulting spectra were of sufficient quality to allow the application of modern computational protocols for efficient and robust NMR analysis.

The method is of particular interest for the NMR study of membrane proteins. Membrane proteins are not only fundamental for a wide range of vital biological functions and pathologies, they also make up the largest class of drug targets. But their structural biology is much more difficult to elucidate than that of soluble proteins.

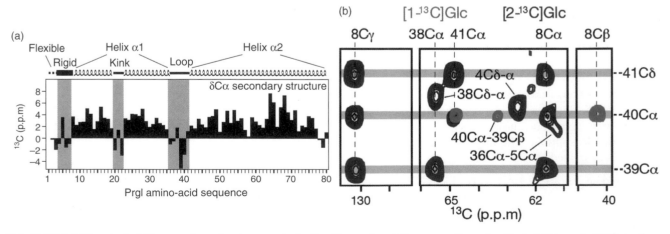

Fig. J3.28 (a) Chemical shift secondary structure analysis. Positive and negative secondary chemical shifts are indicative of α-helical and β-sheet structure, respectively. (b) Solid-state NMR distance restraints. ^{13}C–^{13}C spectra of [1–^{13}C] Glc-labeled (in green), [2-^{13}C] Glc-labeled (in magenta) needles. (From Loquet et al., 2012.)

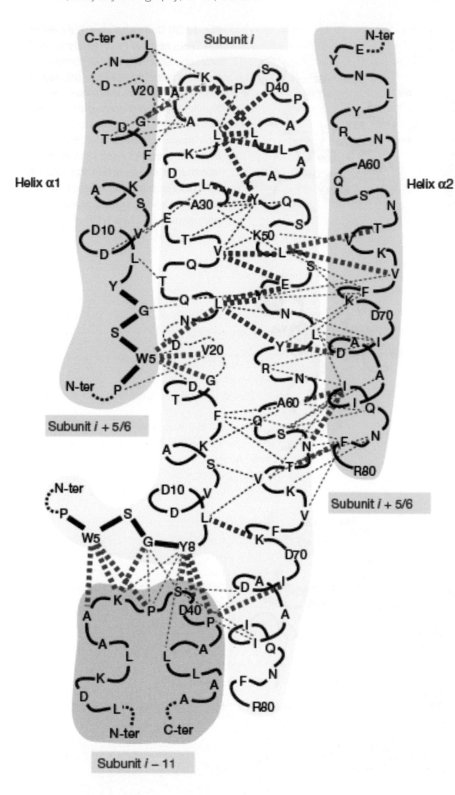

Fig. J3.29 Distance restraints. Architecture of the T3SS needle assembly. Dashed lines represent the intra-subunit (in red) and inter-subunit (in blue) solid-state NMR distance restraints. Bold lines indicate the detection of more than three distance restraints. Lateral interactions between subunits i and i15or i16 are not distinguishable at this stage of the analysis, but only after first rounds of Rosetta calculations. N-ter, N terminus; C-ter, C terminus. (From Loquet et al., 2012.)

J3.5 NMR, X-RAY CRYSTALLOGRAPHY, SMALL-ANGLE X-RAY AND NEUTRON SCATTERING (SAXS AND SANS)

In autumn 2014, there were more than 100 000 structures deposited in the PDB, of which about 10% had been solved by NMR. One reason that NMR has not played a larger role in structure determination is its limitation on protein size. Resolution depends inversely on resonance line widths, and the widths depend in turn on the effective molecular mass. Limits imposed by resolution have been pushed back over the years by the use of isotope enrichment (^{13}C, ^{15}N, and ^{2}H), by the extension of spectra to three and four

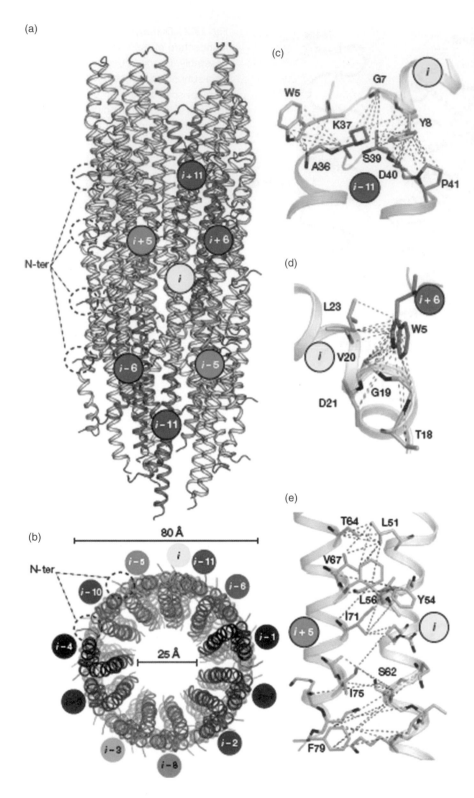

Fig. J3.30 Complete atomic model of the T3SS needle. (**a, b**) Ribbon representation showing different subunits: side perspective (**a**); top view (**b**). The N-terminal domains (N-ter) are highlighted by red dashed circles. (**c–e**) Selections of the subunit–subunit interfaces. (**c**) The axial interface between subunits i and i211. (**d**) The lateral interface between subunits i and i16. (**e**) The lateral interface between subunits i and i15. Blue dashed lines represent solid-state NMR restraints (Loquet *et al.*, 2012.)

dimensions, by measuring residual coupling in oriented samples, and more recently by TROSY techniques in combination with high magnetic fields. However, complete structure determination of monomeric proteins still has not pushed beyond a molecular mass limit of 50 kDa, and determinations not requiring deuteration of the protein still seem to be limited to 25–30 kDa. While this is a severe limitation, it is mitigated by the fact that the average domain size of encoded proteins appears to be about 100 amino acids (Comment J3.14).

Since X-ray diffraction in crystals and NMR in solution can both be used independently to determine the complete three-dimensional structure of proteins, application of the two methods to the same proteins provides a basis for

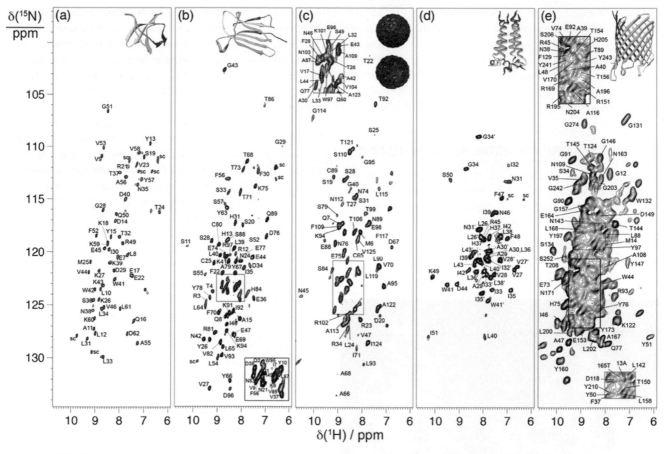

Fig. J3.31 ^{15}N–^{1}H correlation spectra recorded on a 1 GHz spectrometer under 60 kHz MAS for [U-HN,^{2}H,^{13}C,^{15}N]-labeled (a) microcrystalline SH3, (b) microcrystalline β2m, and (c) sedimented nucleocapsids of AP205, (d) M2 channel, and (e) OmpG. (From Barbet-Massin, 2014).

COMMENT J3.14 TIME FOR DATA COLLECTION

More severe limitations in the application of NMR are that the time required for data acquisition and analysis is long and sample preparation requires the use of isotopically labeled media (^{15}N-, ^{13}C-, and ^{2}H-labeled proteins). The weeks of acquisition and subsequent months-long periods required for assignment and structure determination are still a major obstacle for NMR.

The promise of reducing actual data collection time has come from the introduction of NMR cryoprobes in which the receiver coil and associated electronic components are cooled to a very low temperature (Montelione et al., 2000). Sensitivity improvements of a factor of ~3 can be achieved for protein samples. This saves a factor of 9 in time when substantial signal averaging is required. For example, a protein of 180 amino acids required 1.5 days for backbone assignments (Medek et al., 2000).

studies may coincide closely with the natural physiological environment of the protein, or they may be varied over a wide range for studies of structural transitions with pH, temperature, or ionic strength. Extensive similarities between corresponding crystal and solution structures as well as major differences in the conformational features of the two states have now been well documented, and are briefly surveyed in this section.

J3.5.1 Structure of Macromolecules in Crystal and in Solution: Comparative Studies

A large number of protein structures have been determined by both methods, and comparison of these structures allows a further important assessment of the "reality" of the models. In general, such comparisons provide reassuring confirmation of the similarity of protein structures in the crystalline and solution states, in addition to providing a method for identifying errors in one or other of the techniques (Comment J3.15).

Provided the structures have been correctly determined in both methods, the comparisons show that the fold and conformations both of interior residues and of

meaningful comparisons of the corresponding structures in single crystals and in non-crystalline states. This is highly relevant, since the solution conditions for NMR hydrogen-bonded secondary structures are very similar. However, it has generally been concluded that protein surfaces can have different structures and dynamic properties in crystals and in solution, and there are several examples in which the differences appear to be real and significant.

One example is human recombinant interleukin-4. The protein has a 129-residue four-helix bundle, for which four

---

COMMENT J3.15 MOLECULAR INTERNAL MOTION AND FLEXIBILITY

The occurrence of internal motion and flexible parts in molecules always presents a challenge in NMR-based structure determination. It is often difficult to distinguish whether the multiple conformers observed in NMR structures reflect real motion, or results simply from insufficient restraints (see Section J3.1).

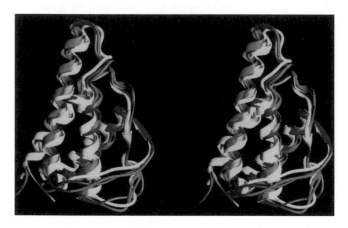

Fig. J3.32 Stereo diagram showing ribbon traces of the superimposed backbone of the four independently determined human interleukin-4 structures. Two different NMR structures (green and yellow) and two different X-ray structures (blue and red) are shown. (After Smith et al., 1994.)

independent structures, two by NMR techniques and two by X-ray crystallography, have been compared (Fig. J3.32). The largest differences between the four structures were found in the exposed surface loop regions, which were inadequately defined in all four structures. In the X-ray structures, the diffuse electron density made chain tracing difficult (see Chapter G3).

A second example is the structure of estrogen receptor DNA-binding domain (84 residues). The domain is monomeric in solution, but two molecules bind cooperatively to specific DNA sequences. The NMR-derived structure is compared with the X-ray crystal structure in Fig. J3.33. Although the two structures are very similar over the regions of ordered residues (C_a RMS deviation = 1.07 Å), the NMR-derived structure of the monomer shows that the 15 internal residues (Cys43–Cys59) disordered in solution make contact with both DNA and the corresponding region of the monomer. The results suggest that these residues become ordered during DNA binding, forming the dimer interface and thus contributing to the cooperative interaction between monomers.

A third example is calmodulin. Calmodulin has two globular domains that bind calcium (see also Section F2.6.8). In the protein crystal, a helix connects the two domains, yielding a dumb-bell structure (Fig. J3.34a). In fact, the connecting helix appears to be an artifact of crystallization. When calmodulin is in solution, part of the helical rod melts into a flexible linker, which enables calmodulin to wrap itself around its target (Fig. J3.34b).

J3.5.2 NMR in Structural Genomics and for Intrinsically Disordered Proteins

The majority of eukaryotic genes do not encode for aqueous single-domain proteins, but for multidomain, membrane, and "unstructured" proteins. NMR spectroscopy is unique

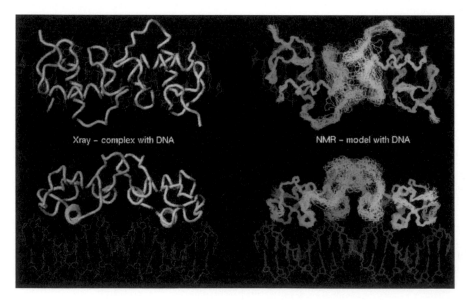

Fig. J3.33 Orthogonal views (top and bottom) of the X-ray structure of estrogen receptor DNA–binding domain–DNA complex: (left) X-ray structure and (right) three-dimensional NMR structure modeled as a dimer with DNA. Two monomers are shown, for clarity, in blue and yellow. (After Schwabe et al., 1993).

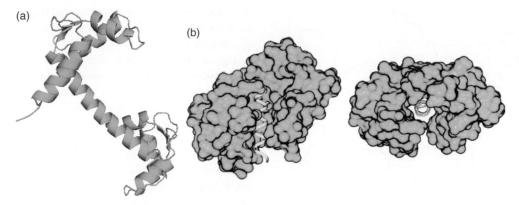

Fig. J3.34 (a) Calmodulin structure in crystal (PDB file 1osa). **(b)** Calmodulin (blue) bound to the target helix from calmodulin-dependent protein kinase II (yellow) is shown from two orientations: looking from the side of the target helix (left) and looking down the target helix (right).

COMMENT J3.16 PHYSICIST'S BOX: INTRINSICALLY DISORDERED PROTEINS AND DISEASE

Intrinsically disordered proteins or intrinsically disordered regions intervene in a wide range of fundamental cellular processes, including molecular recognition, signaling, replication, transcription, and cell cycle control. Intrinsically disordered proteins are studied intensively because they constitute the majority of proteins (~80% compared to 40% in the global human proteome) involved in severe pathologies such as neurodegenerative disease (see example in Fig. J3.36) and cancer (Blackledge, 2011). The roles of IDP in fundamental processes and disease converge in the measles virus. The genome of the virus is in a capsid formed by multiple copies of a nucleoprotein complex. The intrinsically disordered C-terminal domain of the complex (N_{TAIL}) is essential for transcription and replication of the virus during infection in a human cell by binding to viral polymerase complex through MoRE, a molecular recognition element (see Figs. J3.38, J3.39) (Jensen *et al.*, 2011).

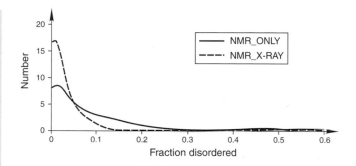

Fig. J3.35 The influence of unstructured regions in proteins on the choice of structure determination approach. The subset NMR_ONLY means that only the NMR method was used for structure determination, the subset NMR_X-RAY means that NMR and X-ray methods were used for structure determinations. (After Prestegard *et al.*, 2001.)

NMR-based structures persist to a much greater percentage of disorder. Such observations clearly support the suggestion that NMR-based structure determination is more applicable to proteins with disordered regions, possibly because these proteins may be difficult to crystallize.

Structure determinations have revealed the existence of native, folded proteins that contain long, flexible coils attached to well-structured globular domains (Comment J3.16). A striking example is the prion protein (Fig. J3.36), with a globular domain containing α-helical and β-sheet secondary structure, and an N-terminal domain of nearly equal size that forms a highly mobile extended coil. The length of this extended coil exceeds the diameter of the globular domain by almost ten-fold. A similar structure consisting of a globular domain and a flexibly extended coil has been observed for a yeast heat-shock transcription factor and many other proteins.

among the techniques of structural biology in its ability to observe and characterize such polypeptide chains in solution. Following important and at the time controversial research in the 1990s, it became evident that over 40% of the human proteome represents proteins that are not folded in their functional form. These are the intrinsically disordered proteins (IDPs) or regions (IDRs) within a protein (Comment J3.16).

Figure J3.35 shows the influence of unstructured regions in proteins on the choice of structure determination method. For proteins with a low percentage of disorder, related X-ray structures are abundant, but for proteins with a higher percentage of disorder (more than 10%), the number of related X-ray structures drops dramatically.

NMR makes a valuable contribution to the study of IDP through its simple ability to identify and characterize these

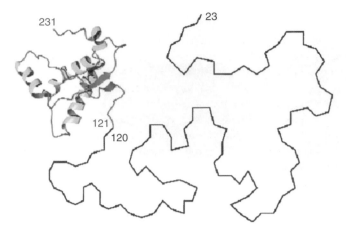

Fig. J3.36 NMR structure of the recombinant murine prion protein. In the intact protein the segment with residues 126–226 forms a globular domain, whereas the segment with residues 23–126 forms an extended coil. (After Riek *et al.*, 1997.)

COMMENT J3.17 DETECTION OF DISORDER SEGMENTS IN NMR

At pH 7, amide protons of unstructured regions of polypeptide chains undergo exchange with protons of water on timescales of tenths of seconds or less. In structured regions, amide protons are either buried in the hydrophobic interior of the protein or involved in hydrogen bonding. These amides exchange much more slowly (minutes to hours). A simple magnetization transfer experiment that uses magnetization associated with the protons of water as they rapidly exchange sites to provide a detectable protein signal selectively shows amides in disordered regions.

disordered protein regions. One such approach is based on a simple test for rapidly exchanging amide protons (Comment J3.17).

Figure J3.37a shows the amide region of a one-dimensional spectrum of BFP in the presence of Ca^{2+} (see also Section F3.6). The few signals in the 8.1–8.9 ppm region are typical of a well-folded protein with a few unstructured or surface-exposed amides. The spectrum suggests that at least in the presence of Ca^{2+} production of quality crystals may be possible. The intensity of the spectrum in Fig. J3.37b, in the same region, is abnormally high and indicative of partial unfolding of the backbone. The data suggest that the protein is difficult to crystallize in the absence of Ca^{2+}, but that it may maintain folded regions that could be characterized by an NMR study.

The description of an unfolded protein domain in terms of a single set of atomic coordinates is clearly inappropriate. A dynamic ensemble of interconverting structures, which may include transient ordered tertiary structures that may be important for physiological function, would be the best description of an IDP domain. The conformational space sampled is expected to be very large, requiring a combination of complementary experimental techniques to map it in the appropriate length and timescales. Circular dichroïsm, Raman, and infrared spectroscopy have reported on local structures in IDP, while long-range interactions affecting overall dimensions have been probed by size-exclusion chromatography, DLS, SAXS, and SANS (see below). Nuclear magnetic resonance can report on site-specific, local, and long-range conformational behavior on timescales covering many orders of magnitude. Intrinsically disordered protein studies by NMR include interpretation of chemical shifts, scalar couplings, NOE, paramagnetic relaxation

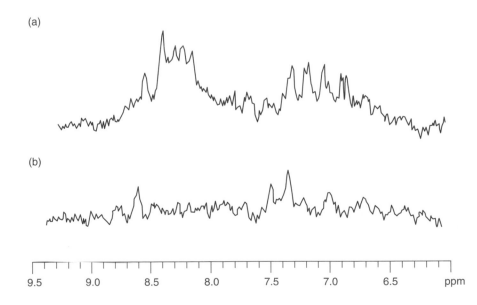

Fig. J3.37 One-dimensional NMR amide proton exchange spectra of BFP at 0.5 mM: (**a**) with Ca^{2+} and (**b**) without Ca^{2+}. The spectra were each acquired in about 15 min from 0.5 mM protein samples using a standard 600 MHz spectrometer. Such experiments can be made quite efficient, and even automated, using NMR flow probes and micromanipulator robots. A special NMR technique that eliminates artifacts due to transfers from α-protons underlying the water resonance was used. (Adapted from Prestegard *et al.*, 2001.)

enhancements, and RDCs. Residual dipolar couplings (see also Sections J2.7 and J3.1), in particular, have been extraordinarily powerful in describing the conformational behavior of IDP.

The characterization of the *functional* conformational behavior of N_{TAIL}, the disordered C-terminus "tail" measles virus nucleoprotein (Fig. J3.38), provides a beautiful example of an NMR application to IDP. Chemical shifts and RDC, measured in a weakly ordering alignment medium, were combined to reveal that while most of N_{TAIL} behaves like an intrinsically disordered chain, the MoRE (molecular recognition element) exists in a rapidly interconverting conformational equilibrium between an unfolded form and conformers containing one of four discrete α-helical elements situated around the interaction site. Small-angle scattering and electron microscopy complemented the NMR results to describe the conformational behavior and mechanistic role of N_{TAIL} *in situ*, in the intact virus (Fig. J3.39).

J3.5.3 NMR Combined with SAXS and SANS

Nuclear magnetic resonance, small-angle X-ray and neutron scattering have been shown to be powerfully complementary for the study of complexes in solution.

Small-angle X-ray provides molecular mass and shape information, while SANS, with its additional advantages of D-labeling and H_2O/D_2O contrast variation, provides information on individual components and distances between their centers of mass within a complex (Chapters G1 and G2).

Nuclear magnetic resonance RDC analysis (Sections J2.7, J3.1) and paramagnetic relaxation enhancement (PRE; Comment J3.12) data provide the mutual orientation of the interacting domains in the complex. In PRE, a tag carrying a free electron is coupled with a single cysteine engineered, for example, in one of the domains, and the relaxation enhancements elicited on the methyl groups of other domains by the unpaired electron in the

(a)

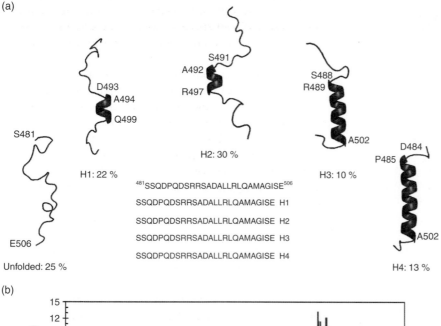

(b)

(c)

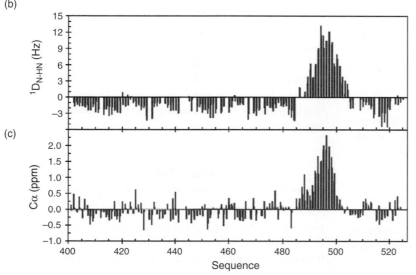

Fig. J3.38 Ensemble description of the MoRE of N_{TAIL}. **(a)** N_{TAIL} preferentially adopts a dynamic equilibrium between a completely unfolded state and different partially helical conformations, each represented by a single cartoon structure for clarity. All helices are stabilized by N-capping interactions through aspartic acids or serines (blue residues). The location of the helices within the MoRE is shown in the primary sequence. **(b)** Comparison of experimental (blue) and back-calculated (red) $^1D_{N-HN}$ RDC(*) from the model of N_{TAIL} shown in **(a)**. **(c)** Comparison of experimental (blue) and back-calculated (red) Cα secondary chemical shifts from the model of N_{TAIL} shown in **(a)**. (*) $^1D_{N-HN}$ is the dipolar interaction for a N-HN pair (Section J2.7.1). (From Jensen et al., 2011).

(a)

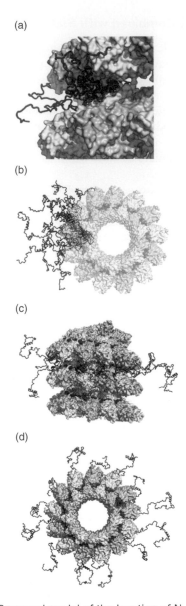

(b)

(c)

(d)

Fig. J3.39. Proposed model of the location of N_{TAIL} in intact nucleocapsids from NMR, EM, and SAS. The three-dimensional coordinates of the respiratory syncytial virus (RSV, belonging to the same family as measles virus, MeV). Nucleoprotein(N)–RNA subunit docked into the EM density map of MeV N–RNA were used (34). Amino-acid-specific conformational sampling allows the chain to escape from the interstitial space of the capsid helix. **(a)** Representation of the conformational sampling of N_{TAIL} from a single N protein in the capsid. Different copies of N_{TAIL} (red) are shown to indicate the available volume sampling of the chain. The first 50 amino acids of N_{TAIL} are shown. **(b)** Representation of the conformational sampling of N_{TAIL} from a single N protein in the capsid, shown along the axis of the nucleocapsid. **(c, d)** Representation of the 13 N_{TAIL} conformers from a single turn of the nucleocapsid. In the interests of clarity, **(b–d)** deliberately show more conformers outside the capsid, and fewer bound to the surface, than are probable at any one time. The position of the RNA is shown in blue. From Jensen *et al.* (2011).

label are translated into distances, which can be used to determine the mutual orientation of the interacting protein domains. NMR can also elucidate interaction surfaces between components via methyl (^{13}C, ^{1}H) labeling of specific residues and monitoring chemical shift perturbations. The examples below illustrate these points.

Post-transcriptional modifications of RNA are essential to the cell life cycle (see Comment B2.10). The box C/D ribonucleoprotein (RNP, RNA–protein complex) enzyme that specifically methylates ribosomal RNA (rRNA) uses a multitude of guide RNAs as templates for the recognition of rRNA target sites. Two methylation guide sequences are combined on each guide RNA, the significance of which remains unclear. Lapinaite *et al.* (2013) used a powerful combination of NMR spectroscopy and SANS to solve the structure of the 390 kDa archaeal RNP enzyme bound to substrate RNA (Fig. J3.40). It was shown that the two methylation-guide sequences are located in different environments in the complex (Fig. J3.41), and that the methylation of physiological substrates targeted by the same guide RNA occurs sequentially. The structure provides a means for differential control of methylation levels at the two sites and at the same time offers an unexpected regulatory mechanism for rRNA folding.

In another example of a study combining different methods, NMR/SANS/SAXS and crystallography achieved a characterization of the ternary ribonucleoprotein complex involved in establishing genetic equality. Genetic equality between males and females is established by chromosome-wide dosage-compensation mechanisms. In the fruit fly *Drosophila melanogaster*, the dosage-compensation complex promotes two-fold hypertranscription of the single male X-chromosome and is silenced in females. Hennig *et al.* (2014) reported on the 2.8 Å crystal structure, NMR/SAXS/SANS solution structure of the ternary Sxl–Unr–msl2 ribonucleoprotein involved complex in the dosage compensation mechanism. The results suggested that repression of dosage compensation, necessary for female viability, is triggered by specific, cooperative molecular interactions that lock a ribonucleoprotein switch to regulate translation. The solution structure validation by NMR/SAXS/SANS is illustrated in Fig. J3.42. The authors proposed that the structure serves as a paradigm for how a combination of general and widespread RNA-binding domains expands the code for specific single-stranded RNA recognition in the regulation of gene expression.

J3.6 IN-CELL NMR

NMR structural studies could, in principle, be performed in close to physiological solutions. In order to obtain good data, however, buffers are chosen to favor solubility with, furthermore, D_2O replacing H_2O to avoid solvent proton

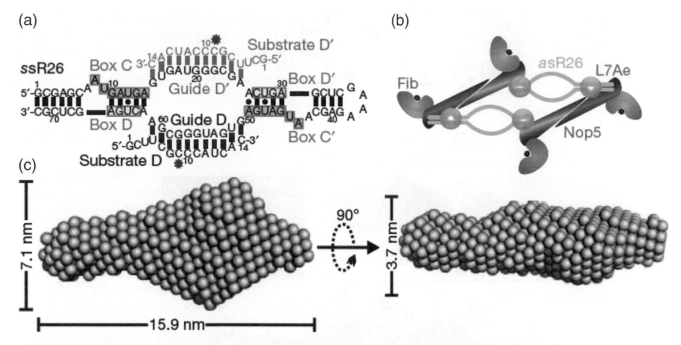

Fig. J3.40 SSR26RNA in the *P. furiosus* box C/D complex. (a) ssR26RNA used in the sRNP for NMR and SAS experiments. Substrates D (red) and D' (salmon) have the same sequence. A star marks the 2'-O-methylation site. (b) Scheme of the sRNP explaining the concept of contrast matching in SANS. The data are collected in experimental conditions where the scattering intensity of the [^{1}H]proteins is masked by matching with the solvent (42%/58% D_2O/H_2O), whereas the [^{2}H]RNA scattering dominates the curve. Fib, fibrillarin. (c) *Ab initio* modeling of the ssR26 RNA in the context of the sRNP from SANS data collected as described in (b). The length of 15.9 nm is considerably larger than the 11 nm expected for the mono-RNP model 16 and accommodates two ssR26 molecules. (From Lapinaite et al., 2013).

resonances obscuring on the NMR spectra of the macromolecule. These buffers are quite far from the natural environment, and it is of considerable interest to be able to characterize a protein structure directly within the cell. In-cell NMR is a powerful method to permit this. Results compared favorably with traditional in-vitro experiments, but also demonstrated that differences between the in-vitro and the in-vivo states of macromolecules can be detected and characterized. In-cell NMR was established in the beginning of the 2000s with first experiments on isotopically labeled macromolecules in bacterial cells. Since then it has been extended to many cellular systems, including mammalian cells.

The principle of in-cell NMR is based on the sensitivity of chemical shifts in the environment. Post-translational modifications, conformational changes, and binding events are detected through shifts in resonance peaks in 1D and 2D NMR spectra. A limitation of in-cell NMR is its requirement of large amounts of material in order to obtain good signal-to-noise ratio – typically concentrations of up to 250 μm are required, several orders of magnitude higher than the natural concentration of a macromolecule in the cell (~1 μm). Because of this, rather than investigating a specific macromolecule, in-cell NMR is most often used to monitor enzymatic, "nonspecific" interactions or bulk

properties, such as post-translational modifications or enzymatic degradation. In this type of interaction, a low cellular concentration of enzyme leads to the modification of a much higher concentration of effects. Interactions studied include the effects of molecular crowding on macromolecular conformation, e.g., to address the question as to whether or not IDPs (see Comment J3.16) adopt a more compact state within the intracellular environment. The CyberCell database provides an important store of such experimental information on the bacterial cell (Comment J3.18).

The observation of a particular macromolecule within the cell is done by specific labeling with ^{13}C and ^{15}N in experiments such as 2D [^{15}N, ^{1}H] HSQC or 2D [^{13}C, ^{1}H] HSQC, which select only hydrogen atoms directly bound either to the labeled N or C nuclei (see Comment J2.5). Two-dimensional [^{15}N, ^{1}H]-HSQC, in particular, measures the chemical shifts of the amide hydrogen and nitrogen atoms, and provides information about the conformational state of the protein backbone. Biological macromolecules can be labeled endogenously by overexpression in the cell to be studied, or exogenously by chemical synthesis or cellular expression, purification, and subsequent microinjection or transfection into the cell to be studied (Fig. J3.43).

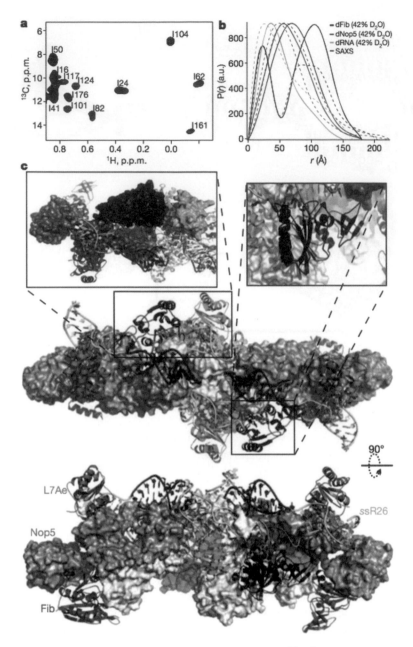

Fig. J3.41 Catalytic structure of the box C/D sRNP. **(a)** Isoleucine region of the ^{13}C-^{1}H correlations of the fibrillarin ILV-methyl groups in the apo-enzyme (blue) and after addition of substrate RNA (pink). Methyl groups of I24, L58, I62, I82, I117, I176, V196, and L210 split in two sets. **(b)** Pair-wise distance distribution P(r), calculated from the scattering curves of the box C/D complex in the absence (solid lines) and in the presence (dashed lines) of substrate RNA and assembled from all non-labeled components (SAXS), [^{2}H]RNA and [^{1}H]proteins (dRNA, SANS), [^{2}H]Nop5, [^{1}H]others (dNop5, SANS), [^{2}H]fibrillarin, [^{1}H]others (dFib, SANS). a. u., arbitrary units. **(c)** Structure of the holo-box C/D sRNP (color code as in **(b)**). Two fibrillarin copies (light blue) are in the "off" position, on the opposite side from the corresponding guide–substrate D duplexes (firebrick). Two fibrillarin copies (dark blue) contact the guide–substrate D9 duplexes (salmon, right insert) and are able to perform methylation. The fibrillarin is directed to the methylation site by packing with L7Ae and Nop5-CTD (left insert). (From Lapinaite *et al.*, 2013.)

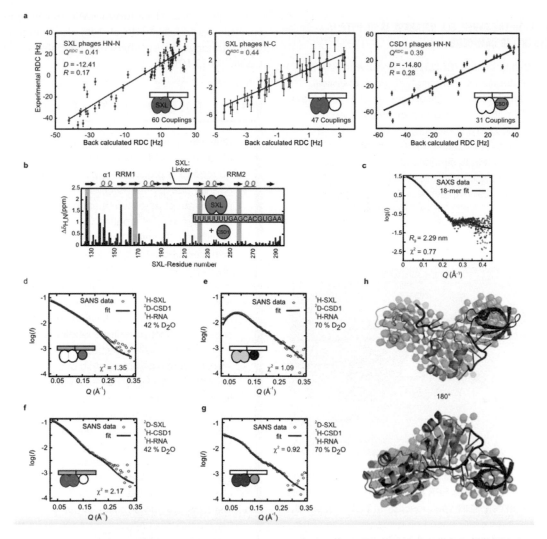

Fig. J3.42 Structure validation in solution. (**a**) Residual dipolar couplings (RDCs) validate the domain orientations for the Sxl RRMs and Unr CSD1 seen in the crystal structure in solution. Correlations between experimental and back-calculated RDC values are shown. The type of dipolar coupling, the observed protein within the complex, and the number of couplings are indicated. All measurements were performed in Pf1 phages. The RDC quality factors QRDC 55 are shown in red and indicate a good agreement between the crystal structure and the domain arrangements in solution. Error bars indicate measurement uncertainties depending on the spectral resolution. (**b**) NMR chemical shift perturbation (combined ^1H and ^{15}N chemical shift difference DdH-N) of the Sxl–RNA complex upon titration with CSD1. Major shifts occur only on residues located in RRM1, consistent with the crystal structure. (**c**) Small-angle X-ray scattering (SAXS) analysis of the ternary complex showing back-calculated scattering densities (red line) and experimental scattering data (blue dots). A x^2 of 0.77 confirms that the structure of the complex in solution is the same as in the crystal structure. (**d**) Small-angle neutron scattering (SANS) of the ternary dRBD3–CSD1–msl2-mRNA complex, where CSD1 is perdeuterated and dRBD3/msl2–mRNA is protonated, at 42% D_2O buffer concentration. In this composition, scattering of dRBD3 matches the scattering density of the buffer and therefore does not contribute to the signal (indicated with a white dRBD3 component in the schematic complex used throughout this study). The back-calculated scattering curve (red line) is fitted against the experimental data (blue dots), and the resulting χ^2 is shown as inset. (**e**) as (**d**) but using 70% D_2O in the buffer, where RNA matches the buffer contrast, and where the perdeuterated component (CSD1) has a positive contrast (indicated with a dark blue color), and dRBD3 a negative contrast (light green). (**f**) as (**d**), but here dRBD3 is perdeuterated and CSD1 is protonated. Thus, CSD1 matches the contrast of the buffer and does not contribute to the signal, whereas dRBD3 exhibits a positive contrast. (**g**) as (**e**) but inversely labeled, thus, CSD1 having a negative contrast to the buffer and dRBD3a positive. These data enabled us to localize each component to be localized within the overall *ab initio* bead models as shown in (**h**), obtained with the program MONSA50 (two views are shown; green, Sxl; blue, CSD1; magenta, RNA). (From Hennig *et al.*, 2014).

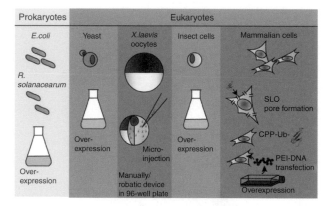

Fig. J3.43 Summary of established methods for the preparation of samples for in-cell NMR or EPR experiments in prokaryotes and eukaryotes. In-cell NMR spectroscopy in bacteria, yeast, and insect cells is generally based on overexpression of the target protein. In contrast, for in-cell NMR experiments with *Xenopus laevis* oocytes, samples are prepared by the microinjection of macromolecules. The expansion of in-cell NMR spectroscopy to mammalian cells resulted in a variety of procedures, including pore formation with streptolysin O (SLO), the use of cell-penetrating peptides (CPPs), and simple overexpression preceded by polyethylenimine (PEI)-assisted transfection of a DNA plasmid encoding the protein of interest. (From Hänsel *et al.*, 2014).

Complementary to solution NMR, in-cell solid-state NMR spectroscopy has been developed, also combined with electron paramagnetic spectroscopy (see also Comment J3.12; Comment J3.19).

J3.7 CHECKLIST OF KEY IDEAS

- Unlike X-ray crystallography, which generates only one set of coordinates, an ensemble of sometimes more than 50 models is produced from NMR data analysis.
- A range of experimental NMR restraints are applied in protein structure determination, including information from ^{1}H, ^{13}C, ^{15}N labeling.
- Alignment of biological macromolecules can be achieved in high magnetic fields by including orienting agents such as planar phospholipid bicelles, rod-shaped viruses and purple membrane phages, polymeric strained gels, and so on.
- Nucleic acid NMR is more difficult than protein NMR and has benefited from the development of methodology based on selective isotope labeling and molecular alignment in ordered media.
- The structural characterization of carbohydrates is especially challenging because of their conformational flexibility. Advances in high-resolution NMR allowed direct measurements of dipolar coupling for carbohydrates by dissolving the molecules in oriented media.
- Molecular dynamics on a very broad range of timescales can be monitored using various types of NMR experiments: relaxation measurements, amide proton exchange, spin echo methods.
- The NMR spectra of solids actually contain more information than is available from liquid NMR, but the information is hidden under broad, overlapping peaks with poor resolution. To eliminate the spectral line-broadening, most solid-state NMR experiments rely on using the magic angle (54° 44′ relative to the external magnetic field).
- Provided the structures have been correctly determined using NMR and X-ray crystallography methods, comparisons show that the fold and conformations of both interior residues and hydrogen-bonded secondary structures are very similar, but it has been concluded that protein surfaces can have different structures and dynamic properties in crystals and in solution.
- NMR is able to produce three-dimensional structures for partially disordered proteins in solution.
- NMR, SAXS, and SANS are powerfully complementary for the study of complexes in solution.
- In-cell NMR has been developed to characterize a protein structure directly within the cell. The principle of the method is based on the sensitivity of chemical shifts on the environment. Post-translational modifications, conformational changes, and binding events are detected through shifts in resonance peaks in 1D and 2D NMR spectra.

Suggestions for Further Reading

Protein NMR Spectroscopy

Lu-Yun Lian, and Roberts, G. (eds.) (2011). *Protein NMR Spectroscopy: Practical Techniques and Applications*. Chichester: John Wiley and Sons.

Three-Dimensional Structure of Biological Macromolecules

Wuthrich, K. (1995). NMR: This other method for protein and nucleic acid structure determination. *Acta Cryst.*, **D51**, 249–270.

Garret, D. S., Seok, Y.-J., Liao, D. L., *et al.* (1997). Solution structure of the 30 kDa N-terminal domain of enzyme I of the *Escherichia coli* phosphoenolpuruvate: Sugar phosphotransferase system by multidimensional NMR. *Biochemistry*, **36**, 2517–2530.

Wider, G., and Wuthrich, K. (1999). NMR spectroscopy of large molecules and multimolecular assemblies in solution. *Curr. Opin. Struct. Biol.*, **9**, 594–601.

Wuthrich, K. (2000). Protein recognition by NMR. *Nat. Struct. Biol.*, **7**, 188–189.

Takahashi, H., Nakanishi, T., Kami, K., Arata, Y., and Shimada, I. (2000). A novel NMR method for determining the interfaces of large protein–protein complexes. *Nat. Struct. Biol.*, **7**, 220–223.

Tjandra, N., Tate, S., Ono, A., Kainosho, M., and Bax, A. (2000). The NMR structure of a DNA dodecamer in an aqueous dilute liquid crystalline phase. *JACS*, **122**, 6190–6200.

Mollova, E. T., Hansen, M. R., and Pardi, A. (2000). Global structure of RNA determined with residual dipolar coupling. *JACS*, **122**, 11 561–11 562.

Tian, F., Al-Hashimi, H. M., Craighead, J. L., and Prestegard, J. H. (2001). Conformational analysis of a flexible oligosaccharide using residual dipolar coupling. *JACS*, **123**, 485–492.

Zidek, L., Stefl, R., and Sklenar, V. (2001). NMR methodology for the study of nucleic acids. *Curr. Opin. Struct. Biol.*, **11**, 275–281.

Flaux, J., Bertelsen, E. B., Horwich, A. L., and Wuthrich, K. (2002). NMR analysis of a 900K GroEL–GroES complex. *Nature*, **418**, 207–211.

Dynamics of Biological Macromolecules

Stejskal, E. O., and Tanner, J. E. (1965). Spin diffusion measurements: Spin echoes in the presence of a time dependent field gradient. *J. Chem. Phys.*, **42**, 288–292.

Jones, J. A., Wilkins, D. K., Smith, L. J., and Dobson, C. M. (1997). Characterization of protein unfolding by NMR diffusion measurements. *J. Biomolecular NMR*, **10**, 199–203.

Ishima, R., and Torchia, D. (2000). Protein dynamics from NMR. *Nat. Str. Biol.*, **7**, 740–743.

Solid-State NMR

Opella, S. J., and Stewart, P. L. (1989). Solid-state nuclear magnetic resonance structural studies of proteins. *Meth. Enzymol.*, **176**, 242–275.

Smith, S. O., and Peersen, O. B. (1992). Solid-state NMR approaches for studying membrane protein structure. *Ann. Rev. Biophys. Biomol. Str.*, **21**, 25–47.

Marassi, F. M., Ma, C., Gratkowski, H., *et al.* (1999). Correlation of the structural and functional domains in the membrane protein Vpu from HIV-1. *Proc. Natl. Acad. Sci (USA)*, **96**, 14 336–14 341.

Griffin, R. G. (2010) Clear signals from surfaces. *Nature*, **468**, 381–382.

Loquet, A., Sgourakis, N. G., Gupta, R., *et al.* (2012). Atomic model of the type III secretion system needle. *Nature*, **486**, 276–279.

Barbet-Massin, E., Pell, A. J., Retel, J. S., *et al.* (2014). Rapid proton-detected NMR assignment for proteins with fast magic angle spinning. *J. Am. Chem. Soc.*, **136**, 12 489–12 497.

NMR and X-ray Crystallography

Schwabe, J. W. R., Chapman, L., Finch, J. T., Rhodes, D., and Neuhaus, D. (1993). DNA recognition by the oestrogen receptor: From solution to the crystal. *Structure*, **1**, 187–204.

Smith, L. J., Redfield, C., Smith, R., *et al.* (1994). Comparison of four independently determined structures of human recombinant interleukin-4. *Struct. Biol.*, **1**, 301–310.

Prestegard, J. H., Valafar, H., Glushka, J., and Tian, F. (2001). Nuclear magnetic resonance in the era of structural genomics. *Biochemistry*, **40**, 8677–8685.

Intrinsically Disordered Proteins

Blackledge, M. (2011). Studying partially folded and intrinsically disordered proteins using NMR residual dipolar couplings. In *Protein NMR Spectroscopy: Practical Techniques and Applications*, eds., Lu-Yun Lian and Gordon Roberts. Chichester: John Wiley and Sons.

Jensen, M. R., Communie, G., Ribeiro, Jr E. A., *et al.* (2011). Intrinsic disorder in measles virus nucleocapsids. *PNAS*, **108**, 1839–1844.

NMR Combined with SAXS and SANS

Lapinaite, A., Simon, B., Skjaerven, L., *et al.* (2013). The structure of the box C/D enzyme reveals regulation of RNA methylation. *Nature*, **502**, 519–523.

Hennig, J., Militti, C., Popowicz, G. M., *et al.* (2014). Structural basis for the assembly of the Sxl–Unr translation regulatory complex. *Nature*, **515**, 287–290.

In-Cell NMR

Hänsel, R., Luh, L. M., Corbeski, I., Trantirek, L., and Dotsch, V. (2014). In-Cell NMR and EPR spectroscopy of biomacromolecules. *Angewandte Chemie*, **53**, 10 300–10 314.

MEDICAL IMAGING

RADIOLOGY AND POSITRON EMISSION TOMOGRAPHY

K1

K1.1 HISTORICAL REVIEW

Ivan Pulyui at the University of Vienna in **1877**, and **Fernando Sanford** at Stanford University in **1891**, reported phenomena of photographic paper impressed by the radiation produced by Crookes tubes. The credit for appreciating the importance of the discovery of X-rays for medical imaging, however, goes to **Wilhelm Roentgen** who, in **1895**, appreciated that bones and dense foreign bodies like jewelry strongly absorb X-rays. The X-ray image taken by Roentgen of his wife's hand is shown in Fig. K1.1; the bones and her ring are obvious while soft tissue appears transparent. Soon after, the first successful chest X-ray was taken by **Francis Henry Williams** in **1896**, and **Emil Grubbe** in Chicago pioneered X-ray radiotherapy to treat cancer. In the following decades, X-rays were used extensively to visualize bone fractures and foreign bodies such as bullets and shrapnel in war wounded.

Radiology progressed rapidly from the **middle of the twentieth century** and it is now possible to image both bone and soft tissue. By introducing a variety of **contrast media** via intravenous injection, or by swallowing or enema into the digestive tract, it is possible to visualize blood flow and movement in the gastric tract. Combined with the development of **image intensifiers**, **high performance X-ray sources** and **detection devices**, and **computer-assisted tomography** (**CT** or **CAT**) radiography can now provide density information to reveal internal body structures in exquisite detail, with a spatial resolution down to 1 mm.

The mathematical basis of tomography was established in **1917**, when **Johann Radon** demonstrated that three-dimensional images could be reconstructed from an infinite number of two-dimensional projections. In **1921**, **André Bocage** developed focal-plane tomography. And in **1963**, **Allan Cormack** published a theoretical analysis and the experimental results using a computer to reconstruct cross-sectional images by **back-projection**, paving the way for the first clinical CT scan, which was performed in **1971** with an instrument developed by **Godfrey Hounsfield** and his team in London (Fig. K1.2).

Cormack and **Hounsfield** were awarded the Nobel Prize in physiology or medicine in **1979**.

In positron emission tomography (PET), the fate of specific radioactive compounds introduced into the body is followed and imaged (Comment K1.1). The first PET scanner was built by **James Robertson** and his team in **1961** and the first commercial instrument became available in **1975**.

In **1999**, **David Townsend** and **Ron Nutt** developed PET/CT, which simultaneously combines functional information from PET and high-resolution images from X-ray CAT. The main benefit of such a combined scanner is that it is easy to correlate points in the scans. For example, if a metastasis is detected by the PET scan, radiation therapy can be planned, targeted to the relevant location on the CT scan.

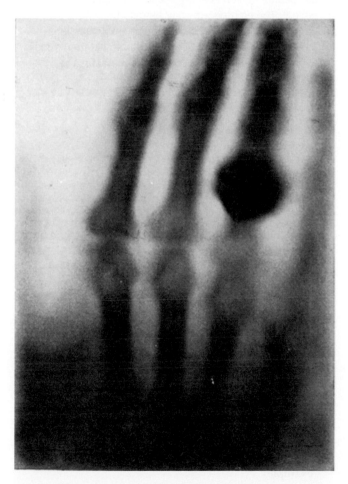

Fig. K1.1 The X-ray photograph of Anna Roentgen's hand taken by her husband (Public Domain via Wikimedia Commons).

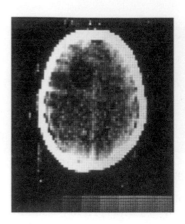

Fig. K1.2 The first clinical CT scan (from www.impactscan.org).

K1.2 HEALTH PHYSICS

Ionizing radiation rips off electrons from atoms to cause severe chemical modifications. In direct action, the radiation creates single- or double-strand breaks. In indirect action, the radiation chemically may ionize water molecules, creating OH ions, which may then interact with the tissue. X-rays and γ-rays used in radiology and PET constitute ionizing radiation. They pose a health hazard for medical staff running the procedure, who must take necessary steps to avoid exposure. The patient's exposure to the radiation cannot be avoided. Nevertheless, dose awareness and instrumental developments that reduce exposure contribute importantly to minimize risk.

K1.2.1 Simple Rules to Protect the Medical Practitioner

(1) Even though radiation propagates in straight lines from the source and neither the hands nor any part of the body should enter the ray path, diffuse radiation from scattering of the source beam by material it crosses, including air, makes a significant contribution to doses absorbed by medical staff.

(2) The intensity of radiation is attenuated according to an inverse square law. At a distance twice as far from the source, for example, the intensity is four times weaker. Thus, distance to the source or the patient is a powerful tool for protecting staff. Handle radioactive isotopes with tweezers or tongs if practicable, stand away from radioactive patients when possible.

(3) Time spent in the proximity of radioactive materials should be minimized. Carry out as many tasks as possible away from the radioactive materials.

(4) Metals and heavy atoms stop X-rays. Lead shields, lead-lined aprons, neck protectors, gloves, and glasses containing lead will shield the practitioner from diffuse radiation scattering. Lead aprons exist in different thicknesses (lead equivalences). Different thicknesses

COMMENT K1.1 BIOLOGIST'S BOX: POSITRONS AND PET

The positron is a particle of the same mass as the electron, but carrying a positive charge. It was the first of the **antiparticles** to be predicted on theoretical physics grounds by **P. Dirac**, who named it the antielectron in a paper published in **1931**. It was also the first antiparticle to be discovered experimentally in **1932** through its track pattern in cloud chamber photographs of cosmic rays (Fig. K1.1.1), by **Carl Anderson**. Positrons are stable in a vacuum, but when a low energy positron meets a low energy electron, they annihilate with the production of two or more gamma (γ) rays.

Positrons are emitted in beta plus (β^+) decay of proton-rich light radioactive nuclei, such as ^{11}C (six protons, five neutrons in the nucleus), ^{15}O (eight protons, seven neutrons), ^{13}N (seven protons, six neutrons), ^{18}F (nine protons, nine neutrons). Note that these elements are found or can be easily incorporated in biological material. In PET, compounds labeled with positron emitting isotopes are imaged via the emitted γ-ray pair after being introduced specifically into the body.

In FDG-PET, the radioactive positron-emitting compound ^{18}F-fluorodeoxyglucose is injected into the patient. The emitted γ-ray pair permits to localize the tracer isotope in the body. Since tumors uptake large amounts of glucose, PET can be used to determine their location and whether the cancer has metastasized.

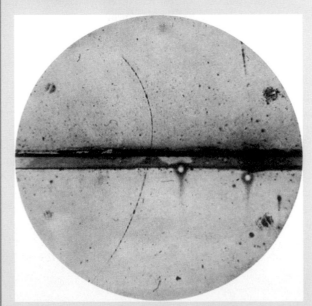

Fig. K1.1.1 Cloud chamber photograph of the first positron ever observed. (By Carl D. Anderson (1905–1991) [Public domain], via Wikimedia Commons).

are appropriate for different tasks (X-ray vs CT, for example).

(5) Dosimeters should always be worn to measure integrated exposure to ionizing radiation and also to

monitor current exposure and warn in case of inappropriate practice or instrument defect.

While health workers may be exposed to ionizing radiation, the key principle of radiation protection is ALARA – keeping their personal doses As Low As Reasonably Achievable.

K1.2.2 Ionizing Radiation Dose

Absorbed dose D corresponds to the energy absorbed per mass. The unit of absorbed dose is the **Gray (Gy)**.

1 Gy = 1 joule of energy absorbed in one kilogram of mass

Different types of ionizing radiation have specific effects on biological tissue beyond the energy absorbed. These are expressed in terms of equivalent dose (takes into account the radiation) and effective dose (takes into account the tissue). The measurement unit for equivalent and effective dose is the **Sievert** (Sv).

1 Sv = 1 joules absorbed per kilogram multiplied by weighting factors depending on radiation type and tissue involved

The Gray and Sievert represent enormous amounts of energy and their 1/1000 values (mGy and mSv) are more relevant for the purposes of medical imaging.

Equivalent dose H_T is the dose D_T in Grays absorbed by a particular tissue T multiplied by a dimensionless weighting factor W_R according to radiation type.

$$H_T \text{ (Sv)} = W_R \, D_T \text{ (Gy)}$$

W_R, the relative biological effectiveness of radiation, is 1 for photons (X-rays and γ-rays). Values for other radiation types are given in Comment K1.2. W_R are determined and reappraised by the ICRP (International Commission on Radiological Protection).

Effective dose E to the entire organism accounts for tissue specificity as well as radiation type. Where the whole body is irradiated uniformly, effective dose is equal to equivalent dose. Where there is partial or non-uniform irradiation, weighting factors W_T takes into account susceptibility to genetic mutation of the organs irradiated. The effective dose for the whole body is the sum of effective doses on the organs. Note that the sum of W_T for all the organs is 1, in accordance with effective dose being equal to equivalent dose for uniform irradiation.

$$E = \sum W_T \, H_T$$

Comment K1.2 includes values of W_R for different radiation types and the table of W_T for different organs.

In the USA, legal radiation dose limits are set by the Environmental Protection Agency (EPA), Nuclear Regulatory Commission (NRC), Department of Energy (DOE), and state agencies. For example, the NRC annual limit for a member of the general public is 1 mSv.

COMMENT K1.2 VALUES FOR RADIATION TYPES

Equivalent Dose Weighting Factors (W_R)

photons (X-rays, γ-rays) = 1
electrons (β radiation) = 1
α-particles, solar particles, heavy ions = 20
neutrons = 5–20 (depending on their energy)
protons = 2

Tissue Weighting Factors (*)

Gonads = 0.08
Red bone marrow = 0.12
Colon = 0.12
Lung = 0.12
Stomach = 0.12
Breasts = 0.12
Bladder = 0.04
Liver = 0.04
Esophagus = 0.04
Thyroid = 0.04
Skin = 0.01
Bone surface = 0.01
Salivary glands = 0.01
Brain = 0.01
Remainder of body = 0.12
Total = 1.00

(*) Values may vary slightly according to regulating agency.

K1.2.3 Dosimetry

Three dosimetric measurements are to be distinguished in **conventional radiology**.

The **ambient dose** (in air) is an important control of instrumental safety. **Kerma** (kinetic energy released per unit mass) is defined as the sum of the initial kinetic energies of all the charged particles liberated by uncharged ionizing radiation such as X-ray photons. Air kerma values are fundamental for the practical calibration of radiation protection instruments for photon measurements.

The **dose at the skin surface** integrates scattered radiation and informs on the exposure due to the radiological examination.

The effective **dose at the organ** is calculated from determined coefficients based on a model.

In a **CT scan**, the **dose index** is equal to the mean dose measured at the center of a series of slices. In practice, the useful parameter is the product **dose index multiplied by length explored**, which is correlated to the effective dose in the explored volume.

Effective doses received in radiological exams are given in Comment K1.3.

K1.2.4 Practice to Reduce Effective Dose in the Patient

All instrumentation used must be properly maintained under high-quality control and medical staff are required to follow regulatory procedures to minimize the effective doses received by patients during radiological exams (Comment K1.4).

For diagnostic X-rays, suitable values of voltage and current must be selected in order to keep patient doses as low as possible. The optimal values depend on the body part (head, abdomen, extremity) and patient size (e.g., small child vs. obese adult).

For CT scans, the same considerations apply and in addition an appropriate slice thickness must be selected. The fine detail obtained from thin slices of 1 mm may be appropriate to look for small structures in a head CT; thicker slices of 5 mm may be appropriate for an abdominal CT.

CT scans can give quite high effective doses (Comment K1.3) and should be avoided if imaging using non-ionizing radiation such as MRI or ultrasonography can be applied, if the diagnostic significance of images were equivalent in terms of contrast and spatial resolution. The risk of low-level radiation is still an open question and particular precautions should be taken with respect to the eyes, thyroid, gonads, and bone marrow.

It should be noted that some rare patients are hypersensitive to radiation. This is due to a repair deficiency of DNA double-strand breaks. Particular care should be applied in dealing with such patients in radiography and radiation therapy.

K1.3 INTERACTION OF X-RAYS AND γ-RAYS WITH MATTER

Radiation interaction with matter is treated in Chapter G1, in which absorption is discussed as a nuisance for scattering, diffraction, and crystallographic studies. Quoting from Section G1.2.2:

X-ray absorption is due to photons exciting electrons to higher energy levels. It is, therefore, energy- (and

wavelength-) dependent; absorption increases with increasing wavelength. In practice, X-ray absorption leads to non-negligible radiation damage in the sample that should be corrected for in diffraction experiments. Absorption is severe for wavelengths above 2.5 Å, where even air in the beam path absorbs significantly.

In contrast, differential X-ray absorption by matter forms the basis of medical radiology. X-rays and γ-rays are found toward the high-energy end of the electromagnetic spectrum (Comment K1.5).

X-rays interact with matter via three processes: Compton scattering; the photoelectric effect; and pair production. Of these, only the photoelectric effect contributes usefully to imaging. For this reason, the energy of X-rays should be as low as possible. However, with very low X-ray energies, the photons required for imaging get absorbed before they can exit the body. In practice, the optimal X-ray energy is as low as possible for the selected body part (70 kV may be adequate for a head; 120 kV may be required for a large abdomen).

Electromagnetic waves are produced by accelerating charges in different energy scales. Gamma rays are produced by high-energy events inside the atomic nucleus, while X-rays arise from tightly bound inner shell (closest to the nucleus) electrons.

The *penetration* of X-rays depends on their photon energy. Higher energies are attenuated less and penetrate more into matter.

The concept of *cross-section* was defined in Section G1.3. It represents a measure of the probability that radiation will interact with matter in a certain manner. The *X-ray scattering cross-section* of an atom, for example, measures the probability of X-ray scattering by that atom. Similarly, the *absorption cross-section* measures the probability of absorption. Cross-section is expressed in units of area (e.g., cm^2) (Fig. K1.3).

$$I_{abs} = \sigma_{abs} I_0 \qquad (K1.1)$$

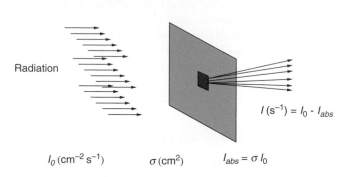

Fig. **K1.3** Cross-section, expressed as an area.

$\sim Z^3 / E^3$ (K1.2)

It increases strongly, the heavier the element, and decreases with increasing energy (nevertheless above minimum binding energy). It is the main mechanism behind X-ray absorption in bone, which contains calcium ($Z = 20$), dominating that in soft tissue, composed of water and organic compounds containing elements with $Z < 10$. The energy term in the above probability can be used to amplify absorption differences between various tissue types. The absorption difference between bone and tissue will be reduced at higher X-ray energies – an effect that is applied in *dual energy radiography* in order to adjust image contrast between tissue types. It must be noted, however, that at lower energies patient effective dose will be increased because of the higher probability of the photoelectric effect absorbing all the incident energy.

(2) **The Compton effect** is due to partial absorption of the energy of a photon, which is scattered with energy loss by ejecting an electron of low binding energy from an atom. In high atomic number materials, the effect involves outer shell electrons. In soft tissue, composed essentially of low atomic number material, all electrons are involved because their binding energy is less than 1 keV. The ejected electron is called the *recoil electron*. From the physics point of view, the Compton effect is the main interaction with matter of high energy X-rays and intermediate energy γ-rays. At radiological X-ray energies, because it gives rise to radiation being scattered in all directions, the Compton effect contributes significantly to the diffuse background in the image. Compton scattered X-rays emerge in all directions from the irradiated area, posing also a health risk for medical staff in the vicinity if they do not take appropriate protective measures.

(3) **Pair production** constitutes a third process by which a photon loses energy, but considering the energy involved it is not relevant for radiology. A minimum photon energy of about one million eV (1 MeV) is required for this process, in which a γ-ray photon is fully converted into an electron/positron pair.

In Eq. (K1.1), I_{abs} is the absorbed intensity in photons s^{-1}, I_0 is the incident flux in photons $cK^1 sec^{-1}$, and σ_{abs} is the absorption cross-section in cm^2.

K1.3.1 Processes of X-ray and γ-ray Absorption in Matter

Absorption of electromagnetic radiation depends on elemental composition and density of the matter and on the energy of the radiation.

The binding energy of an orbital electron is the energy required to remove it from its atom. When radiation of photon energy larger than the minimum binding energy is incident on the atom, part or all of its energy may be transformed through two possible mechanisms:

(1) **The photoelectric effect**: The energy of a photon is fully converted for the ejection of an inner electron. The ejected electron is called the *photoelectron*. If, for example, the electron is ejected from the K-shell, another electron can fill the vacancy with the emission of a K-fluorescent X-ray photon.

The photoelectric effect is the main process by which contrast is obtained in radiology. The probability of it occurring for a photon of energy E in an element of atomic number Z is approximately proportional to Z^3 and inversely proportional to E^3:

In order to calculate the *attenuation* of the incoming beam, *coherent scattering* should be added to the *absorption* due to the *photoelectric* and *Compton effects*. Coherent scattering corresponds to the elastic (i.e., without energy change) scattering of the radiation by the atomic electrons. It is the useful part in crystallography (Chapter G1); at the X-ray energies of medical radiology, however, *coherent scattering* represents a very small fraction of the attenuation.

K1.3.2 Mass Attenuation Coefficient

Attenuation of a beam of radiation of incident intensity I_0 as it crosses a material of mass per unit area x $(\mathrm{g\,cm^{-2}})$ and density ρ $(\mathrm{g\,cm^{-3}})$ follows an exponential law:

$$I = I_0[\exp(-\mu/\rho)x] \qquad (\text{K1.3})$$

where I is the intensity of the beam after attenuation and (μ/ρ) is called the mass attenuation coefficient (in units of $\mathrm{cm^2\,g^{-1}}$).

Plots of mass attenuation coefficients as a function of photon energy are shown in Fig. K1.4 for cortical bone, brain gray matter, and lead. The usual photon energy range for medical X-rays is ~40–60 keV. It should be noted that the scales for bone and brain are similar, whereas the scale for lead is a factor of 10 bigger. Furthermore, at ~60 KeV the difference in attenuation between bone and soft tissue is less pronounced than at the lower energy, an effect that is applied in dual energy radiography. The sharp peak for lead at ~90 keV is an absorption edge where the photon energy exceeds that of an inner shell electron. At this energy, the peak in absorption corresponds to that required to eject a K-shell electron.

K1.4 PRODUCTION OF X-RAYS

There are two general ways to produce X-rays for medical imaging: from electrons hitting a metal target and synchrotron radiation. In conventional radiology and usual hospital CT scan facilities, X-ray radiation is produced in an X-ray tube in which a beam of electrons is accelerated by a voltage to hit a metal target. Beam lines for specialized medical applications have been built at several synchrotron radiation facilities (Comment K1.6)

The *penetration* of X-rays depends on the accelerating voltage of the electrons that produce them.

The energy spectrum produced depends on the nature of the metal target. The spectrum from a tungsten target is shown in Fig. K1.5. The electrons are slowed down in the metal, releasing the Bremsstrahlung (braking radiation) body of the spectrum. Note the distinction between N–kV and N–keV; N–kV is the voltage required to accelerate the electrons to an energy of N–keV.

Bremsstrahlung (dashed lines in Fig. K1.5) covers an energy range from zero to that of the initial electron energy; 40–90 kV accelerating voltage produces lower-energy photons and less penetration; 100–140 kV produces higher-energy photons and higher penetration. The lower-energy part of the spectrum loses intensity by "filtration" effects where the X-rays cross the glass wall etc. as they emerge from the tube. Bremsstrahlung is used for radiation therapy.

The peaks at specific energy values in the spectrum are due to electrons ejecting inner shell electrons from the metal atoms. *Characteristic X-rays* are emitted when the vacancies are filled. The K lines in the tungsten spectrum of Fig. K1.5 correspond to X-rays emitted when K shell vacancies are filled.

K1.5 DETECTION OF X-RAYS

X-rays were discovered when it was noticed that an invisible radiation blackened photographic film; photographic film is still in use today to reveal X-ray images, though increasingly in first-world countries electronic film is used rather than photographic film. Improving the efficiency of film, however, was important in order to obtain the best

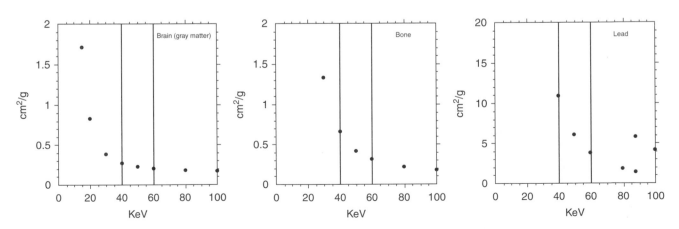

Fig. K1.4 Mass attenuation coefficients as a function of photon energy. Medical X-rays are in the range ~15–100 keV, spanning the mean energies used from head imaging to mammography.

image with the minimum amount of effective dose for the patient. This was achieved, for example, by a two-step detection process in which a sensitive fluorescent screen that emits light when hit by X-rays is put before the film. In recent instruments, position-sensitive electronic detectors replace film.

COMMENT K1.6 MEDICAL BEAM-LINES AT THE EUROPEAN SYNCHROTRON RADIATION FACILITY (WWW.ESRF.EU)

Synchrotron-based medical applications involve research in both diagnostic imaging and irradiation for therapy. In an *angiography* application, coronary arteries are visualized by *dual-energy* images acquired during the passage of an iodine contrast agent following intravenous infusion. Dynamic tomographic and projection imaging of xenon concentrations in the airways are used to map regional lung ventilation in asthma and emphysema studies. Analyzer-based imaging (ABI) is based on the use of highly monochromatic radiation, passing through an analyzer crystal in order to obtain edge enhancement at interfaces between regions of different refractive index. ABI is especially promising for radiation-sensitive, soft-tissue imaging in mammography. The method offers a new kind of contrast for soft-tissue imaging based on the fact that each tissue type has its own small-angle scattering (see Chapter G2) characteristics. Synchrotron radiation CT also permits very precise imaging in cerebral perfusion studies. Direct measurement of contrast agent concentration allows the determination of cerebral blood volume or flow and permeability coefficient – parameters essential for the analysis of the physiopathology of brain diseases.

X-rays constitute ionizing radiation, which produces charges as it passes through matter. The amount of charge (exposure) produced by the passage of X-rays in an ionization chamber is proportional to the energy absorbed. Measurements are extremely sensitive and following exposure/dose calibration can detect absorbed energies as low as 10–100 nGy (10^{-9} Gy), 1000 times smaller than the reading from a medical X-ray instrument (10–100 mGy).

X-ray detection in a CT scanner is performed either in an ionization chamber containing xenon gas or in ceramic scintillators, in which X-ray photons produce flashes of light that are detected by a photodiode and converted into electric signals. Xenon gas at high pressure is chosen because of its large interaction cross-section for X-rays to produce charged ions in a small volume. The gas detector is made up of a single chamber subdivided by electrodes, while the solid-state detector is made up of an array of individual scintillators. Depending on scan geometry, the detectors are in rows in a ring around the patient, or may rotate with the tube.

K1.6 PRINCIPLES OF THE CT SCAN

The **scan principle** is shown in Fig. K1.6. In fourth-generation scanners, the X-ray tube rotates around the patient at ~1 s per 360°, and the intensity distribution of the attenuated fan beam is recorded in a fixed ring of ~1000 detector elements. In a **helical scan**, the patient table moves concomitantly through the gantry. The **pitch** is defined as the ratio between the **displacement of the table for one rotation** and the **collimation width** that determines slice thickness (t in Fig. K1.6). This represents major progress with respect to previous scanners that, because of instrumental limitations such as electrical cable connections and cooling of the X-ray tubes, had to use an **irradiate–step the table–irradiate cycle** with patients *holding their breath* at each step. The continuous scan also makes possible real-time CT by injection of contrast media for a vascular examination and **CT fluoroscopy**.

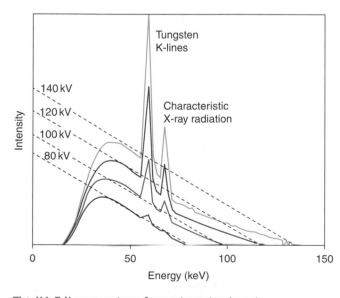

Fig. **K1.5** X-ray spectrum from a tungsten target.

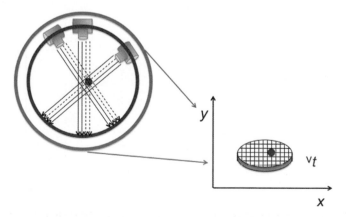

Fig. **K1.6** CT scan principle

Image formation. The X-ray beams are more or less attenuated by passage through the body, according to the local mass absorption coefficient, to form "shadows" in the intensity distribution. In Fig. K1.6, the dashed lines represent X-ray beams that have been attenuated by passage through the object in the center. The **collimation** of the fan beam determines t in an incremental scan. In helical mode the value of t is also influenced by the speed of the table and **interpolation calculation** applied to reduce artifacts due to table motion. There is a trade-off between collimation width (smaller width gives higher resolution)

and length of scan, and an appropriate compromise should be chosen in each clinical case: Better resolution will reduce artifacts due to peristaltic movement, cardiac contractility, and respiration, but shorter scan rates will favor shorter periods for patients to hold their breath; finer t values will favor better image reconstruction; faster scans with poorer collimation will favor better real-time vascular imaging.

Image reconstruction calculations to convert the 2D density plots of the irradiated slices into a 3D image proceed in a similar way to those in electron microscopy described in Chapter H2, by applying **Fourier transform** methods. Fan projection data have to be **interpolated** and **re-binned** in order to obtain parallel projections from adjacent detectors. **Back projection** reverses the measurement process to reconstruct a 3D image from 2D projections. **Filtered back projection** reduces the blurring inherent in the procedure by selectively amplifying certain frequencies in the Fourier transform of the image. Sharp filters amplify higher frequencies to increase image resolution, while smooth filters lead to a fall-off in amplification for higher frequencies and are more adapted to view soft tissue.

Effective image density is expressed on the **Hounsfield scale** (Comment K1.7; Fig. K1.7) – a linear gray scale on which **air** corresponds to **–1000** (black: the object is transparent to the X-ray beam), **water** to **0** (gray), and **dense bone** to **+1000** (white: the object is opaque to the X-ray beam). The image scale can be usefully calibrated, however, on a shorter range of the scale as shown in Fig. K1.7; for example, with black at –150 and white at +200 to distinguish various types of soft tissue.

COMMENT K1.7 HOUNSFIELD SCALE (HU)

Bone	~300 to 1000
Liver	~40 to 60
White matter	~20 to 30
Gray matter	~37 to 45
Blood	~40
Muscle	~10 to 40
Kidney	~30
Cerebrospinal fluid	~15
Water	0
Fat	–50 to –100
Air	–1000

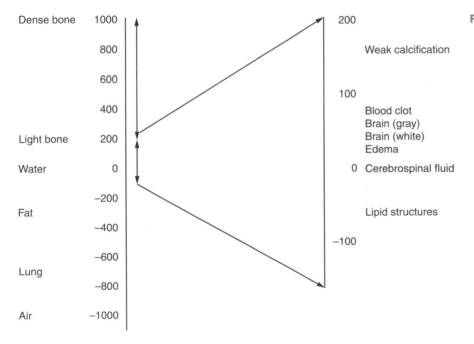

Fig. K1.7 Hounsfield scale.

K1.7 POSITRON EMISSION TOMOGRAPHY

In PET, a biologically active compound with a positron-emitting radioactive label (Comment K1.1) is injected intravenously into the patient; the γ-ray pair arising from positron-electron annihilation is observed and permits the localization of the compound. In contrast to radiology, which produces images of density distribution, PET informs on biological function (**functional imaging**) by following the injected compound (Comment K1.8).

^{18}F labeled fluorodeoxyglucose is an analogue of glucose, which is rapidly taken up by tissue. Positron emission tomography detects its concentration to inform on metabolic activity. Its use in the exploration of cancer metastasis accounts for most PET examinations.

K1.7.1 Principles of PET

Positron emission and annihilation is shown in Fig. K1.8.

Positron-electron annihilation emits a pair of γ-rays each of energy 511 keV, which are detected by a complete circle of thousands of detectors around the patient. The γ-ray trajectories from an annihilation event are at 180° from each other. The position-sensitive detection of γ-ray pairs around the ring forms **lines of reconstruction** that locate the positron-emitting compound, without the requirement for collimation (Fig. K1.9a).

The γ-ray detectors are made up of small, solid-state scintillator crystals coupled to photomultipliers, using materials and designs that are in constant development to increase performance in spatial resolution, response time, and efficiency. The **annihilation coincidence detection**

principle is applied. A γ-ray striking a detector opens a time window in which, if a second γ-ray is detected in a detector across from the first one, the pair is accepted as true coincidence. The time window corresponds to the **coincidence**

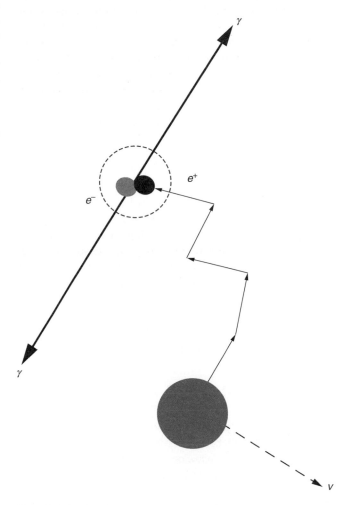

Fig. K1.8 Positron emission and annihilation

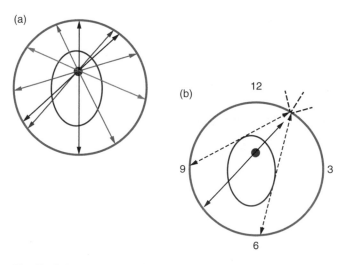

Fig. K1.9 Schematics of PET.

COMMENT K1.8 PET MOLECULES

^{18}F-fluorodeoxyglucose (FDG)	Metabolism, tumor imaging
^{18}F-sodium fluoride	Bone imaging
^{18}F-fluorocholine	Prostate tumor imaging
^{15}O-water	Blood flow imaging (brain, heart)
^{11}C-L-methyl-methionine	Brain tumor imaging, parathyroid imaging
^{18}F-sodium fluoride	Bone imaging
^{18}F-fluorocholine	Prostate tumor imaging
^{68}Ga-dotatate	Neuroendocrine tumor imaging
^{13}N-ammonia	Blood flow imaging (heart)

resolving time of the system. The example of a γ-ray striking a detector at a position corresponding to one o'clock in the detector ring is shown in Fig. K1.9b. The ellipse in the center of the detector ring represents the part of the body that is being examined and the smaller circle the position of the positron-emitting compound. The dotted double arrows indicate the extremities of the directions for the parallel and antiparallel "real event" γ-rays originating in the patient. The electronics will, therefore, "look" to see if a γ-ray is detected on the ring between six and nine o'clock within the coincidence resolving time. **Time-of-flight (TOF)** methods have been introduced in PET to increase the time resolution of the detection. This allows a better discrimination between real events and accidental coincidences due to scattered γ-rays, for example.

Modern **PET/CT scanners** combine PET and X-ray CAT to provide complementary imaging and functional information. Positron emission tomography imaging resolution is limited by inherent factors such as the path length of the positron before annihilation. The higher resolution of CT imaging provides a clearer picture of the tissue examined and also contributes to correcting the PET data for attenuation of the γ-rays within the body. **Image reconstruction** in PET proceeds by using filtered back-projection, similarly to CT (see Section K1.5).

Functional PET to study brain activation is discussed with MRI at the end of Chapter K3.

K1.8 CHECKLIST OF KEY IDEAS

- X-rays and γ-rays constitute ionizing radiation that poses a severe health hazard. Medical staff must take steps to avoid exposure. Following the ALARA principle, staff may protect themselves by maximizing practical distance to the source, minimizing the time spent with the source, and by wearing appropriate protective clothing.
- Dose awareness and instrumental developments reduce exposure to minimize risk to the patient.
- Differential X-ray absorption by matter forms the basis of medical radiology. The photoelectric and Compton effects are the main mechanisms by which radiation is absorbed.
- The mass attenuation coefficient (in units of $cm^2 g^{-1}$) depends on the elemental composition of the material and radiation energy.

- Synchrotron facilities produce extremely intense X-ray beams for medical applications. In conventional radiology and usual hospital CT scan facilities, however, X-ray radiation is produced in sealed *X-ray tubes* in which electrons are accelerated by a voltage to hit a metal target.
- X-rays are detected in ionization chambers containing xenon gas or ceramic scintillators, in which X-ray photons produce flashes of light that are detected by a photodiode and converted into electric signals.
- In fourth-generation scanners, the X-ray tube rotates around the patient at ~1 s per 360°, and the intensity distribution of the attenuated fan beam is recorded in a fixed ring of ~1000 detector elements.
- Moving the patient table concomitantly with the rotating tube leads to a continuous scan that makes possible real-time CT by injection of contrast media for a vascular examination and CT fluoroscopy.
- Back-projection reverses the measurement process to reconstruct a 3D image from the 2D density projections of irradiated slices. Filtered back-projection reduces the blurring inherent in the procedure by selectively amplifying certain frequencies in the Fourier transform of the image.
- Effective image density is expressed in the Hounsfield scale – a linear gray scale on which air corresponds to – 1000 (black: transparent to the X-ray beam), water to 0 (gray), and dense bone to 300–1000 (white: opaque to the X-ray beam).
- In PET, a biologically active compound with a positron-emitting radioactive label is injected intravenously into the patient; the γ-ray pair arising from positron-electron annihilation is observed and permits the localization of the compound.
- PET informs on biological function (functional imaging) by following the injected compound.
- Modern PET/CT scanners combine PET and X-ray CAT to provide complementary imaging and functional information.

Suggestion for Further Reading

Barrie Smith, N., and Webb, A. (2010). *Introduction to Medical Imaging: Physics, Engineering and Clinical Applications*. Cambridge: Cambridge University Press.

ULTRASOUND IMAGING

K2

K2.1 HISTORICAL REVIEW

Ultrasound refers to sound waves of frequencies higher than the hearing capacity of the human ear (between 20 Hz and 20 kHz). Frequencies used in medicine are usually greater than 2.5 MHz. Issuing from military and industrial applications for SONAR (sound navigation and ranging) and metal flaw detection, ultrasonic energy to "view" inside the human body was developed in parallel in hospitals of many countries around the world, including the **USA, Europe, Japan, China**, and **Australia.** The brief history below summarizes the main developments in the USA and Europe.

Ultrasound emerged in the **1940s** as a diagnostic, therapeutic, and then surgical tool using high intensities to break down tissue. The first diagnostic application is accepted to be due to **Karl Theodore Dussik** and collaborators in Austria. The work described in a paper published in **1942** showed images claimed to be of cerebral ventricles and included a discussion that low-intensity ultrasound could be used to locate brain tumors. Dussik and his collaborators had used a transmission method and the published images were later criticized as being artifacts due to reflections of sound waves within the skull. The transmission method of ultrasound imaging was subsequently dropped.

In the USA, **George Ludwig** applied ultrasound to systematic imaging of various tissues and organs from animals and measured the sound impedance of different types of gallstone, muscle, and fat, in the body. He was a naval officer and his work was classified and only released for publication in **1949**. At about the same time, **John Wild**, a surgeon, used the one-dimensional (A-mode) ultrasound investigation method to assess the thickness of bowel tissue under different surgical conditions. Then, with **John Reid**, a recent graduate in electrical engineering, they built a B-mode (two-dimensional) instrument able to visualize tumors by scanning breast lumps.

In **1953** in Sweden, **Carl Helmut Hertz**, a graduate student in physics (son of Gustav Ludwig Hertz in honor of whom the unit of frequency, in s^{-1}, was named) collaborated with cardiologist **Inge Edler** to make the first successful ultrasound measurement of heart activity, and generated the first echo-encephalogram (ultrasound brain scan). C. H. Hertz was already familiar with the application of ultrasound to non-destructive materials testing, and they

made the measurements on an instrument borrowed from a ship construction company.

The work of obstetrician **Ian Donald** in Scotland led to the first diagnostic applications of ultrasonography. Again, by using borrowed industrial testing equipment, Donald and collaborators started by measuring the ultrasonic properties of various anatomical specimens. In **1958**, Donald, obstetrician **John MacVicar**, and physicist **Tom Brown**, after developing the equipment to diagnose pathological states in volunteer patients, published what many consider as the seminal paper on diagnostic medical ultrasonography (Fig. K2.1). Obstetric applications, including measurements to assess fetal size and growth, were developed by Donald and **James Willocks** and later refined by **Stuart Campbell**. And from the late **1960s**, ultrasonography became the method of choice to study pregnancy from start to birth and to diagnose its many complications.

In **1962**, **Joseph Holmes**, **William Wright**, and **Ralph Meyerdirk** developed the first compound contact B-mode scanner (which yields two-dimensional images). In the **late 1960s**, **Gene Strandness** and his team developed Doppler ultrasound as a diagnostic tool for vascular disease.

K2.2 ULTRASOUND WAVES

Sound waves of frequencies higher than 20 kHz, the upper limit that can be heard by the human ear, are called ultrasound. In contrast to electromagnetic radiation that can propagate in a vacuum, sound waves propagate in a material medium, inducing cyclical local modifications in density, pressure, and temperature. They are **longitudinal waves**, i.e., the modifications are along the direction of wave propagation – not perpendicular to it as would be the case for transverse waves (Fig. K2.2a). Ultrasound frequencies used in medical imaging are in the range ~1.5–7.5 MHz, with higher frequency images more suitable for more superficial tissues.

The speed of propagation of a wave, c (ms^{-1}), its frequency, f (s^{-1}), and wavelength, λ (m) are logically related by:

$$c = f\lambda \qquad (K2.1)$$

A typical value of c in tissue is ~1500 m s^{-1}; for a frequency of 1.5 MHz the wavelength is 1.0 mm.

. ARTICLES   THE LANCET

INVESTIGATION OF ABDOMINAL MASSES BY PULSED ULTRASOUND

IAN DONALD
M.B.E., B.A. Cape Town, M.D. Lond., F.R.F.P.S., F.R.C.O.G.
REGIUS PROFESSOR OF MIDWIFERY IN THE UNIVERSITY OF GLASGOW

J. MacVICAR
M.B. Glasg., M.R.C.O.G.
GYNAECOLOGICAL REGISTRAR, WESTERN INFIRMARY, GLASGOW

T. G. BROWN
OF MESSRS. KELVIN HUGHES LTD.

VIBRATIONS whose frequency exceeds 20,000 per second are beyond the range of hearing and therefore termed " ultrasonic ". One of the properties of ultrasound is that it can be propagated as a beam. When such a beam crosses an interface between two substances of differing specific acoustic impedance (which is defined as the product of the density of the material and the velocity of the sound wave in it), five things happen:

(1) Some of the energy is reflected at the interface, the amplitude of the reflected waves being proportional to the difference of the two acoustic impedances divided by their sum (Rayleigh's law). Therefore the greater the difference in specific acoustic impedance between two adjacent materials the higher will be the percentage of energy reflected. This fact makes a liquid-gas interface almost impenetrable to ultrasound and is important in relation to gas-filled intestine within the abdominal cavity.

(2) Much of the energy which is not reflected is transmitted into the second medium but is somewhat attenuated.

(3) Some refraction may occur, particularly when the ultrasonic beam is not at right-angles to the plane of the interface.

(4) Some of the energy may be absorbed and produce heat. The ability to absorb ultrasound varies with different tissues —e.g., that of bone is considerable.

(5) Cavitation may be produced if considerable energies are present at the lower ultrasonic frequencies. This phenomenon, whose mechanism is not yet fully understood, can develop when the negative sound pressure exceeds the ambient hydro-static pressure, giving rise to small temporary voids in the material. Cavitation becomes increasingly difficult to produce as the frequency of the ultrasound is raised, and usually develops only when the ultrasonic energy is applied continuously or in

Fig. K2.1 First page of the paper considered to be one of the most important in medical imaging: **Donald**, I., **MacVicar**, J., and **Brown**, T. G. (**1958**) Investigation of abdominal masses by pulsed ultrasound. *Lancet*, **1**, 1188–1195.

The amplitude A of a wave corresponds to the maximum displacement from equilibrium of the particles composing the medium induced during the wave cycle (Fig. K2.2). The energy of the wave is proportional to A^2 (see Chapter A3 for a more detailed treatment of waves).

The speed of sound c, in a medium, depends on the average local density ρ (in units of $g\,m^{-3}$ in Table K2.1), and compressibility κ ($m\,g^{-1}\,s^2$) (Comment K2.1) according to:

$$c = \sqrt{\frac{1}{\kappa\rho}} \qquad (K2.2)$$

The **acoustic impedance** Z, the square root of the density/compressibility ratio, is a physical property of the medium.

$$Z = \sqrt{\frac{\rho}{\kappa}} \qquad (K2.3)$$

Combining Eqs (K2.2) and (K2.3):

$$Z = \rho c \qquad (K2.4)$$

where c is the speed of sound in the medium (Comment K2.2).

Formally, Z is defined as the ratio of the pressure in the sound wave at a given point to the particle velocity at that point. Its units are ($g\,cm^{-2}\,s^{-1}$). As is seen in Comment K2.2 and below, it is a useful parameter to describe the behavior of an acoustic wave as it traverses different media.

Acoustic impedance values are also useful for describing the pressure amplitude and intensity ratios between the reflected and transmitted signals and the incident intensity (p_r/p_i, p_t/p_i, I_r/I_i, I_t/I_i, respectively).

The strongest reflected signal is obtained when the incident beam is at right angles to the interface (incident angle equal to zero in Fig. K2.3). The reflected beam is **back-scattered**. In this case the ratios are given by:

$$R_P = \frac{p_r}{p_i} = \frac{(Z_2 - Z_1)}{(Z_1 + Z_2)}$$
$$T_P = \frac{p_t}{p_i} = \frac{2Z_2}{(Z_1 + Z_2)} \qquad (K2.5)$$

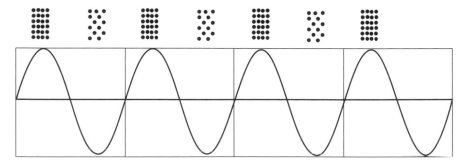

Fig. K2.2 Longitudinal sound waves are pressure waves in air.

TABLE K2.1 ACOUSTIC PARAMETERS AND THEIR UNITS FOR DIFFERENT TISSUE TYPES (ADAPTED FROM SMITH AND WEBB, 2010)

Medium	$Z\,(10^{-5}\,\mathrm{g\,cm^{-2}\,s^{-1}})$	$c\,(\mathrm{m\,s^{-1}})$	$\rho\,(\mathrm{g\,m^{-3}})$	$\kappa\,(10^{-11}\,\mathrm{cm\,g^{-1}\,s^2})$
Air	0.00043	330	1.3	70 000
Fat	1.38	1450	925	5.0
Brain	1.58	1540	1025	4.2
Blood	1.59	1570	1060	4.0
Kidney	1.62	1560	1040	4.0
Liver	1.65	1570	1050	3.9
Muscle	1.7	1590	1075	3.7
Bone	7.8	4000	1908	0.3

COMMENT K2.1 BIOLOGIST'S BOX: COMPRESSIBILITY

The compressibility of a medium is defined as the relative volume change when pressure is applied.

$$\kappa = \frac{1}{V}\left(\frac{\partial V}{\partial p}\right)$$

In thermodynamics, the value of κ depends on whether the change is observed at constant temperature or constant entropy. In the case of a solid medium, these two values are very close to each other. The units of compressibility are inverse pressure. Qualitatively, the compressibility of a medium is inversely proportional to the forces that maintain its average structure; the softer the medium, the higher its compressibility.

$$R_I = \frac{I_r}{I_i} = \frac{(Z_2 - Z_1)^2}{(Z_1 + Z_2)^2}$$

$$T_I = \frac{I_t}{I_i} = \frac{4Z_1 Z_2}{(Z_1 + Z_2)^2}$$

(K2.6)

where the reflection and transmission pressure coefficients (R_p, T_p) and reflection and transmission intensity coefficients (R_I, T_I) are related by:

$$T_p - R_p = 1$$
$$T_I + R_I = 1$$

(K2.7)

The intensity of a wave is the square of its amplitude. The amplitude of a sound wave corresponds to the maximum pressure in the cycle, so that

$$R_I = |R_p|^2$$

The reflected signal intensity is proportionally stronger with the square of the difference between the impedance values (Eq. (K2.6)). The extreme case is when one of the Z values is zero. But then the T ratio will be zero; the wave will not penetrate to deeper structures! The opposite

COMMENT K2.2 REFRACTIVE INDEX AND ACOUSTIC IMPEDANCE

The refractive index n of a medium is a measure of how much the speed of light in a vacuum c_{LV} is decreased when it propagates in the medium:

$$v = c_{LV}/n$$

The change of direction (refraction) of a light wave as it crosses between media of different refractive index (n_1, n_2) is governed by Snell's Law:

$$n_1 \sin\theta_i = n_2 \sin\theta_t$$

which can be written

$$\frac{1}{v_1}\sin\theta_i = \frac{1}{v_2}\sin\theta_t$$

where θ_i and θ_t are the angles of incidence and transmission (refraction) of the wave, respectively (Fig. K2.3).

A similar equation holds for an acoustic wave.

$$\frac{1}{c_1}\sin\theta_i = \frac{1}{c_2}\sin\theta_t$$

where c_1 and c_2 are the speed of sound, respectively, in medium 1 and medium 2.

Applying Eq. (K2.4)

$$\frac{\rho_1}{Z_1}\sin\theta_i = \frac{\rho_2}{Z_2}\sin\theta_t$$

The refractive index analogy for sound is the ratio of density to acoustic impedance.

extreme is when two tissue types have equal impedances, $Z_1 = Z_2$. But then there will be no acoustic interface between them and no reflected beam ($R_I = 0$).

The image is formed by the reflected waves from acoustic boundaries between different tissue types. It is, nevertheless, important to have good transmitted intensity in

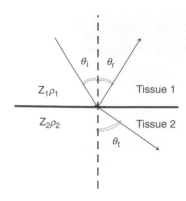

Fig. K2.3 Snell's law diagram

COMMENT K2.3 WATTS, DECIBELS AND BEAM ATTENUATION

The intensity of the ultrasound beam is given in Watts cm^{-2}, units of energy per area.

The ratio between the attenuated intensity and the incident intensity can be expressed in decibels; the decibel is a logarithmic expression of an intensity ratio.

The attenuation coefficient in units of length^{-1} is defined by Eq. (K2.8). The conversion to decibel units is given by:

$$\mu(dB\,cm^{-1}) = 10 \times \log_{10}(e)\mu(cm^{-1})$$

$$\log_{10}(e) = 0.4343 \text{ so that } \mu(dB\,cm^{-1}) = 4.343\,\mu(cm^{-1}).$$

A useful rule-of-thumb for the interpretation of Eq. (K2.8) is that each 3 dB reduction corresponds to a reduction of a factor of 2 in intensity.

---

order for the sound waves to penetrate to the deeper tissue layers (Section K2.4).

Intensity and intensity ratio are measured in units of Watts cm^{-2} and decibels, respectively (Comment K2.3). Values of acoustic impedance and speed of sound in different tissue types are shown in Table K2.1, together with corresponding density and compressibility.

According to Eq. (K2.2), sound waves travel faster in more rigid (less compressible) and/or the less dense media of propagation. In bone, even though it is about twice as dense, sound travels almost three times faster than in other tissues (Table K2.1) because of its very low compressibility (a factor of 13 lower than for blood).

The values for soft tissue in Table K2.1 are very similar. In practice, this indicates that for ultrasound waves only a small fraction of the wave intensity will be reflected at an acoustic interface between two tissue types. At a fat–muscle interface, for example, $Z_1 \sim 1.38$ and $Z_2 \sim 1.7$ in the units of Table K2.1. Substituting in Eq. (K2.5), $R_I = 1\%$; $T_I = 99.9\%$; most of the beam is transmitted to deeper tissue. The imaging information is

contained in the 1% of the incident intensity that is reflected (see Section K2.4).

For a wave that crosses a boundary between soft tissue ($Z_1 \sim 1.6$) and bone ($Z_2 \sim 7.8$), $R_I = 43.5\%$; $T_I = 56.5\%$; the reflected signal is strong but less than 60% of the beam is transmitted to deeper tissue.

K2.3 HEALTH PHYSICS, ABSORPTION, AND ATTENUATION OF ULTRASOUND WAVES IN BIOLOGICAL TISSUE

K2.3.1 Effects of Ultrasound

Biological effects of sound waves have only been observed in animals for extremely high incident intensities. During medical examinations, the absorbed energy is small and no ill effects on human health have been observed. It is nevertheless recommended not to use high intensities in prenatal examinations.

Thermal effects. Negligible heat is produced in tissue by ultrasound waves during usual imaging procedures. More heat is produced in Doppler mode (Section K2.4.2), in which it is recommended to use the minimum power for a useful signal and to limit exposure time. Local heating effects at higher intensities are used in ultrasound heat therapy to relieve pain.

Mechanical effects of ultrasound include **cavitation** (the production of microscopic bubbles in the tissue), which has been applied in the development of contrast agents for ultrasound imaging. High-intensity ultrasound has therapeutic applications; for example, to break up deposits such as gall or kidney stones or to ablate tumors or specific tissue in **focused ultrasound surgery (FUS)**.

K2.3.2 Attenuation

Ultrasound beam intensity is reduced as it penetrates into tissue through successive reflections at acoustic boundaries, and attenuation by **scattering** and **absorption** or energy loss through production of heat. Note that a relatively new technique, HIFU (high-intensity focused ultrasound), makes use of this heat to thermally ablate tumors.

As in the case of radiation (Eq. (K1.3)), the attenuation of a sound wave follows an exponential law. At a depth z (cm), the incident intensity I_0 is reduced to I_z

$$I_z = I_0 \exp(-\mu z) \tag{K2.8}$$

where μ (cm^{-1}) is the linear attenuation coefficient.

It turns out that the attenuation coefficient in tissue is approximately linear with frequency with a value of **~1 dB MHz^{-1} cm^{-1}** for soft tissue. The decibel unit (dB) and its relation to μ are defined in Comment K2.3. In-depth exploration becomes more difficult the higher the

(a)   (b)

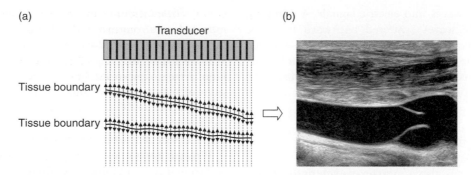

Fig. K2.4 Principles of ultrasound imaging. (a) At boundaries between tissues, a small fraction of the energy of pressure waves from a transducer is scattered back to the transducer, where it is detected. The depth of the boundary is detected from the speed of sound in the tissue. (b) The image formed by electronic steering of the beam. The intensity in each pixel is proportional to the back-scattered signal. (From Smith and Webb, 2010, with permission.)

frequency, and ultrasound imaging instruments apply corrections to compensate for the attenuated signal as a function of depth.

As in the general case for the diffraction of waves by different-sized objects (Chapter G1), sound waves are scattered by structures of smaller size than the wavelength. In soft tissue ($c \sim 1500\,\mathrm{ms^{-1}}$) the wavelength is 1 mm for a frequency of 1.5 MHz (Eq. (K2.1)). Much smaller particles, such as red blood cells, will act as point objects and scatter isotropically. As for the general radiation case, there will also be interference between waves scattered by neighboring particles. Back-scattering will be constructive for particles that are close together. Analogously to the formation of diffraction fringes, interference will be constructive or destructive as a function of angle for particles that are further apart, resulting in a complicated intensity pattern called **speckle**, which contributes to image noise.

K2.4 PRINCIPLES OF ULTRASOUND IMAGING

The basic principle of ultrasound imaging is illustrated in Fig. K2.4. Ultrasound waves are sent through the tissue under examination. The back-scattered reflected waves (the "echoes") at acoustic boundaries are detected as a function of time, so that the depth of the boundary can be calculated from the knowledge of the speed of sound through the tissue. The image is formed by the successive lines of reflected intensity as the beam sweeps the sample. Image intensity is proportional to the intensity of the echo.

K2.4.1 Imaging Mode

The instrumental setup for ultrasound imaging is illustrated in Fig. K2.5. A frequency generator produces short periodic voltage pulses that are amplified and transmitted

to a transducer, which converts them into a sound wave that is transmitted into the body. The transducer both transmits the high-power pressure wave pulses and detects the low-intensity echo signals. It is, therefore, important to keep them apart. This is done by a transmit/receive switch, which opens sequentially to allow passage either from the frequency generator to the transducer or back from the transducer to the imaging line. The signals from back-scattered pressure waves are analyzed according to the time they reached the transducer and processed electronically to yield a real-time image display on a monitor.

The **transducer** contains a piezoelectric element that produces the incident pressure waves when it receives an electric signal, and in the inverse process transforms

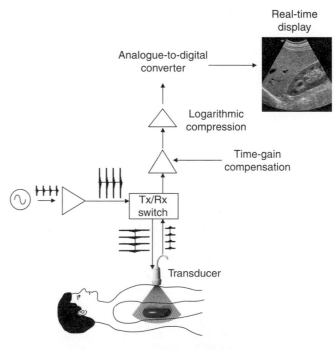

Fig. K2.5 Ultrasound imaging setup. (From Smith and Webb, 2010, with permission.).

the echo pressure waves into electric signals. A **damping layer** absorbs the waves scattered backward from the piezoelectric element. A **matching layer** at the external face increases the efficiency of the setup significantly by providing acoustic coupling between the element and the patient, through a zone of intermediate acoustic impedance.

$$Z_{matching layer} = \sqrt{Z_{piezoelectric} Z_{skin}} \qquad (K2.9)$$

The matching layer is also a **quarter-wavelength plate** – its thickness equal to one-quarter the ultrasound wavelength to maximize energy transmission in both directions.

Transducers have **large bandwidths** around a specified central frequency – e.g., ±2 MHz for a central frequency of 3 MHz. This means that a single transducer can be used for many applications.

In single-element instruments, the transducer is mounted on a rotating or oscillating system to allow sectorial scanning (Fig. K2.6a).

Transducers in recent instruments, however, contain **large arrays** of piezoelectric elements that enable the acquisition of two-dimensional images while the transducer is held in a fixed position (Fig. K2.6b). The array can be linear (for a rectangular image) or convex (for a sectorial image), with each element generating an image line.

Phased arrays are rectangular probes with which a sectorial scan is obtained by applying voltage pulses to each element at slightly different times (out of phase). The sum of waves produces an effective wave front that can be focused and steered within the body by appropriate phasing. Phased array probes have the advantage of being able to scan a large area with a relatively small probe – e.g., for exploration via an intercostal pathway.

Lateral resolution, focusing, and axial resolution (Fig. K2.7). **Lateral resolution** describes the minimum

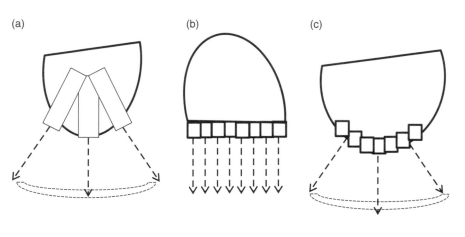

(a) (b) (c)

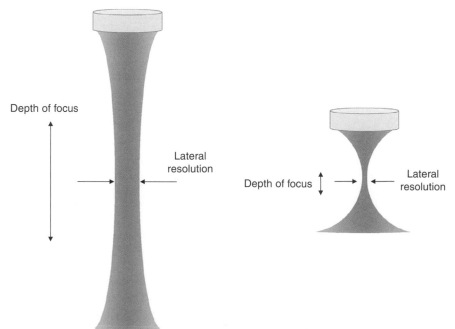

Fig. **K2.6** Transducers (schematic). (**a**) Single-element rotating or oscillating system to allow sectorial scanning. (**b**) Fixed large array of piezoelectric elements (planar). (**c**) Fixed large array (curved).

Fig. **K2.7** Lateral resolution, focusing and axial resolution. (From Smith and Webb, 2010, with permission.)

Depth of focus

Lateral resolution

Depth of focus

Lateral resolution

distance for objects at the same depth to be distinguished as separate. Lateral resolution is determined by the width of the ultrasound beam as it diverges from the transducer. **Beam focusing** reduces beam width and concentrates ultrasound energy in a certain zone, the focal point; the finer the focused beam, however, the shorter the depth of focus. In transducer array instruments, lateral resolution will also depend on the number of beams and their spacing. Focusing is obtained by using a curved plastic lens (similarly to the optical case) or by electronic means in phased arrays (see above).

Axial resolution describes the minimum distance for two acoustic boundaries along the beam propagation line to be distinguished as separate. Recall that depth is determined from the time it takes the echo wave to reach back to the transducer. Axial resolution will, therefore, depend on the vibration frequency of the probe and quality of the damping layer, since the probe should have stopped vibrating after detecting an echo before being able to detect a subsequent one. In principle, axial resolution cannot be better than one-half the wavelength of the sound wave.

Contrast resolution describes the capability to distinguish between structures whose acoustic impedance is close. Current transducers are sensitive to $150\,dB$ and can detect echoes of weak intensity. Microbubbles of gas in a lipid or protein shell of diameter $2–10\,\mu m$ are used as **contrast agents**. The highly compressible gas has an enhanced acoustic response compared to surrounding tissue or blood that reflects back not only the main frequency component (f_o) of the incident ultrasound beam, but also higher harmonics ($2f_o$, $3f_o$, etc.). The presence of strong harmonics makes it possible to distinguish the echoes from microbubbles from those from other tissue boundaries. The transducer bandwidth should be sufficiently broad to emit the (f_o) incident beam and to detect the higher harmonic echoes.

A-mode, M-Mode, and B-Mode Scanning

B-mode or **brightness scanning** is the most common ultrasound procedure. It produces a two-dimensional sectorial or rectangular (depending on transducer geometry) image of the tissue cross-section examined. The image is built up of echo lines from the transducer array or by mechanically scanning a single transducer plotted as a function of depth (obtained from the time between incident pulse emission and echo reception). Brightness is proportional to the intensity of the echo.

A one-dimensional image is obtained from an **amplitude (A-) mode scan**. The line-image is a plot of the back-scattered amplitude as a function of time, which provides the depth from which the echo originated. A major application of A-mode scanning is to measure corneal thickness by applying a small probe directly on the center of the previously anesthetized eyeball.

In an **M-mode** or **motion scan**, a series of A-mode lines, representing different depths, is displayed as a function of time, with the brightness of the lines proportional to the amplitude of the echo. A straight line represents a motionless boundary. Pulses will appear on the line when an acoustic boundary moves perpendicular to the ultrasound beam direction (the surface of the heart, for example). M-mode scans are used in echography of the heart and the fetal heart.

Correcting for Beam Geometry and Other Artifacts

Lobes develop on either side of the divergent main ultrasound beam from interference effects between wavelets as they leave the transducer. The side beams will reflect back off acoustic barriers, outside the region studied, to create artifacts in the image called **clutter**. Clutter will also arise from other undesired interference effects, e.g., from multiple reflections. Electronic processing of the signals can reduce this. Recall from Section A3.3.2 that signals in the time and frequency domains, respectively, are related by Fourier transformation. Multiplication in one domain corresponds to convolution in the other. A time window introduced in the time domain to eliminate echoes from side lobes will correspond to a convolution of all echo frequencies by the Fourier transform of the time window and will act as an appropriate filter. Clutter, mainly due to the main frequency reflections, will also be reduced by **harmonic imaging** (see above), in which the second harmonic of the echo is analyzed. In the absence of contrast agents, the second harmonic is quite weak but the analysis profits from much better lateral resolution and reduced clutter.

The depth dependence of attenuation is discussed in Section K2.3.2. It leads to a wide dynamic range in the received signal, with, for example, strong echoes from boundaries with large differences in acoustic impedance close to the surface and much weaker ones from boundaries between two types of soft tissue deeper down. This effect is corrected by instrument electronics in a process called **time-gain compensation**. Echo signals are not amplified by the same factor but according to the time received. Early echoes from shallow structures are amplified by a smaller factor than those received later from deeper structures.

Speckle (see Section K2.3.2), which produces a grainy appearance, is the main artifact in ultrasound imaging. It is reduced by **compound scanning**, an approach in which the ultrasound image is acquired from multiple angles. Speckle is due to interference effects, which are angle dependent. When the "views" from various angles are combined to form the final image, "true" features will appear in most of them, while the speckle pattern will depend on angle and be reduced by averaging out. Compound scanning is also useful in image boundaries of irregular curvature that are parallel to the beam and in reducing image artifacts due to **acoustic enhancement** when the beam crosses a volume of exceptionally low attenuation (such as water-containing cysts), or the opposite effect, **shadowing**, in which structures are "hidden" behind a strongly attenuating medium (such as bone or a solid tumor).

K2.4.2 Doppler Mode

A classic illustration of the Doppler effect is the rising frequency of sound from a fast-approaching train that goes through a station without stopping and the falling frequency after the train goes past the platform and recedes into the distance. It is as if the wave is pushed "compressed" by the approaching speed and expanded as the source moves away. The same principle applies to sound reflected off a moving object. Blood flow in Doppler mode ultrasound is measured by reflecting ultrasound waves off red blood cells. The effective velocity v of the red blood cells is obtained from the frequency shift Δf of the echo with respect to the incident frequency. The Doppler effect is highest when the incident beam is parallel to the direction of motion and null when it is perpendicular to the direction of motion. The shift is given by Eq. (K2.10), in which the cosine term determines the component of the incident beam that is parallel to the motion.

$$\Delta f = f_R - f_0 = \frac{2f_0 v \cos\theta}{c} \qquad (K2.10)$$

In the case of blood flowing in a direction roughly parallel to the transducer, the signal will be very low, taking up small positive and negative values around $\Delta f = 0$ (cos 90° = 0). In **power Doppler** mode, the intensity of the signal is integrated, with both positive and negative amplitudes contributing as the squares. The advantages of the method are its higher sensitivity and that artifacts due to high flow rates are eliminated. A disadvantage is that information on the directional of flow is lost.

A Doppler mode measurement is illustrated schematically in Fig. K2.8.

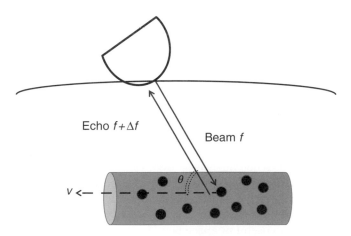

Fig.K2.8 Ultrasound Doppler measurement. The beam of frequency f is in red, the echo of frequency f+Δf, from the particle moving with velocity v, is in blue.

In **continuous mode**, the transducer array is divided into two sections, one that emits a continuous incident beam and one that receives the Doppler shifted echo. The measurement area is the overlap of the areas delimited by the two beams, with no depth resolution. The measured speed is the average speed of blood flow in the measurement area.

In **pulsed mode**, the transducer alternates between emitting the ultrasound beam and receiving the echo. An important advantage of pulsed wave Doppler is good spatial resolution that enables focusing on a specific blood vessel, by combining with a B-mode image (duplex image). Results are plotted with frequency on the y-axis, time on the x-axis, and brightness proportional to the density of red blood cells contributing to the particular frequency. There is a range of velocities in a blood vessel that changes significantly over the cardiac cycle. Fourier transform analysis splits the signal into its component frequencies to inform on the velocities present. By using fast Fourier transform methods, the analysis is performed and plotted on the screen in real time in addition to the duplex image (triplex image). In **color flow** imaging, the Doppler information is overlaid on the B-mode image. Blue and red represent, respectively, flow toward or away from the transducer, with color intensity proportional to flow velocity. Frequency dispersion can also be coded in color, usually in green or yellow.

Similarly to B-mode imaging, the incident frequency must be adapted to the exam. Low frequencies (2–5 MHz) are used for deep vascular exploration (e.g., in the abdomen), while higher frequencies (7.5–10 MHz) are required to examine superficial blood vessels. Bandwidth settings should also be adapted to the velocity range. Pulses beyond the high end of the range will create artifacts due to *aliasing* (an *alias* pulse that appears at a "wrong" frequency within the range). Pulses corresponding to slow velocities will be missed if outside the low end of the range. Finally, the θ angle between the incident beam and the vessel (Eq. (K2.10)) should be kept below about 40° for good signal sensitivity (cos 40° = 0.766).

K2.5 CHECKLIST OF KEY IDEAS

- Ultrasound refers to sound waves of frequencies higher than the hearing capacity of the human ear (between 20 Hz and 20 kHz). Frequencies used in medicine are usually greater than 2.5 MHz.
- Sound waves are longitudinal pressure waves.
- A typical value for the speed of sound in tissue is ~1500 m s⁻¹; for a frequency of 1.5 MHz, the wavelength is 1.0 mm.
- The compressibility of a medium is the relative volume change when pressure is applied.

- The acoustic impedance is the square root of the density/compressibility ratio. It is a physical property of the medium.
- Sound propagation across boundaries follows Snell's law. The refractive index analogy for sound is the ratio of density to acoustic impedance.
- The strongest reflected signal is obtained when the incident beam is at right angles to the interface and the reflected beam is back-scattered. It is proportional to the square of the difference between the acoustic impedance values.
- The image is formed by the reflected waves from acoustic boundaries between different tissue types.
- Sound waves travel faster in more rigid (less compressible) and/or less dense media of propagation.
- The absorbed energy is small during medical examinations and no ill effects on human health have been observed. It is nevertheless recommended not to use high intensities in prenatal examinations.
- Thermal effects are negligible, except in the case of high-intensity focused ultrasound, a technique that is used for the thermal ablation of tumors.
- Mechanical effects include cavitation, which has been applied in the development of contrast agents for ultrasound imaging. High-intensity ultrasound has therapeutic applications, for example to break up deposits such as gall or kidney stones or to ablate tumors or specific tissue in focused ultrasound surgery (FUS).
- The attenuation coefficient in tissue is approximately linear with frequency, with a value of $\sim 1\,\mathrm{dB\,MHz^{-1}\,cm^{-1}}$ for soft tissue. In-depth exploration becomes more difficult at higher frequency and ultrasound imaging instruments apply corrections to compensate for the attenuated signal as a function of depth.
- Sound waves are diffracted by structures of smaller size than the wavelength (1 mm in soft tissue for $\sim 1500\,\mathrm{ms^{-1}}$ speed of sound and a frequency of 1.5 MHz). Much smaller particles such as red blood cells will act as point objects and scatter isotropically.
- In imaging mode, echoes at acoustic boundaries are detected as a function of time. The depth of the boundary is calculated from the knowledge of the speed of sound through the tissue. The image is formed by the successive lines of reflected intensity, as the beam sweeps the sample. Image intensity is proportional to the intensity of the echo.
- The transducer contains a piezoelectric element that produces the incident pressure waves when it receives an electric signal, and in the inverse process transforms the echo pressure waves into electric signals.
- Lateral resolution describes the minimum distance for objects at the same depth to be distinguished as separate. It is determined by the width of the ultrasound beam as it diverges out from the transducer.
- Beam focusing reduces beam width and concentrates ultrasound energy at a focal point; the finer the focused beam, however, the shorter the depth of focus.
- Axial resolution describes the minimum distance for two acoustic boundaries along the beam propagation line to be distinguished as separate.
- Contrast resolution describes the capability to distinguish between structures whose acoustic impedance is close.
- B-mode, or brightness scanning, is the most common ultrasound procedure. It produces a two-dimensional sectorial or rectangular (depending on transducer geometry) image of the tissue cross-section examined.
- An amplitude or A-mode scan produces a one-dimensional line-plot of the back-scattered amplitude as a function of time, which provides the depth from which the echo originated.
- In an M-mode or motion scan, a series of A-mode lines representing different depths is displayed as a function of time, with the brightness of the lines proportional to the amplitude of the echo.
- Instruments include electronic signal processing to correct for beam geometry and other image artifacts.
- Compound scanning, an approach in which the ultrasound image is acquired from multiple angles, is used to correct for speckle.
- In Doppler mode, blood flow is measured by reflecting ultrasound waves off red blood cells. The effective velocity of the red blood cells is obtained from the frequency shift of the echo with respect to the incident frequency.
- In continuous Doppler mode, the transducer array is divided into two sections, one that emits a continuous incident beam and one that receives the Doppler shifted echo. There is no depth resolution. The measured speed is the average speed of blood flow in the measurement area.
- In pulsed Doppler mode, the transducer alternates between emitting the ultrasound beam and receiving the echo. An important advantage of pulsed wave Doppler is good spatial resolution that enables focusing on a specific blood vessel, by combining with a B-mode image.

Suggestion for Further Reading

Barrie Smith, N., and Webb, A. (2010). *Introduction to Medical Imaging: Physics, Engineering and Clinical Applications*. Cambridge: Cambridge University Press.

MAGNETIC RESONANCE IMAGING

K3

K3.1 INTRODUCTION TO MRI AND HISTORICAL REVIEW

Magnetic resonance imaging (MRI) was introduced commercially in 1980, a few years after the introduction of the PET scanner. It is the youngest of the current powerful medical imaging methods. At first sight, it is not obvious how an imaging concept could be fulfilled by using the physics of NMR. . When we consider attenuation of radiation by human tissue, we note that Nature provided three windows through which we can look inside the human body: X-rays of wavelength <20 Å; radio-frequency (rf) waves >1 m wavelength; and acoustic radiation of wavelength >1 mm.

The X-ray window has been exploited and has completely revolutionized medical diagnosis (see Chapter K1). The low-frequency ultrasonic window led to the development of ultrasound imaging (see Chapter K2). The rf window, however, was not exploited until the early 1970s. This is not astonishing, considering the achievable resolution, which is usually limited by the wavelength of the applied radiation through the uncertainty relation. The maximum rf useful for imaging is about 100 MHz, leading to a resolution of 3 m, which is not sufficient even for imaging elephants.

The crucial idea in MRI is the application of a magnetic field *gradient* to disperse the nuclear magnetic resonance frequencies of the various volume elements in the body. Nuclear magnetic resonance theory (Section J1.2.3) shows that nuclei will precess with identical Larmor frequencies (ω) if they all experience the same effective field B_0. In a high-resolution MRI experiment, we detect the sensitivity of each nucleus to small changes in B_0 dictated by the microenvironment. Since these changes are very small, it is necessary to employ a magnetic field that is spatially homogeneous to a very high degree across the sample volume. A linear magnetic field gradient imposes corresponding linear shifts in the Larmor frequencies of nuclei across the sample volume since each experiences an effective field that is different from B_0 (Comment K3.1).

As treated in more detail in the introduction to Chapter J1, it is accepted that nuclear magnetic resonance (NMR) spectroscopy was born with the independent work of **Felix Bloch** and **Edward Purcell** in **1946**. Bloch and Purcell were awarded the Nobel Prize in physics in **1952**. In **1949**, **Erwin Lewis Hahn** discovered the **spin echo** principle, which plays an important role in MRI. **Raymond Damadian**, in **1971**, discovered that normal and cancerous tissue displayed different NMR response. Two years later, in **1973**, **Paul Lauterbur** obtained the first MRI images (of capillary tubes and a green pepper) by applying the **back-projection** mathematical method, similar to the approach in CT (see Section K1.1; Fig. K3.1). **William Moore** and **Waldo Hinshaw** obtained the first human images from MRI in **1976** and in the same year **Peter Mansfield** developed the **echo planar** technique that allows very fast imaging. In a further important development for MRI, **Richard Ernst** introduced Fourier transform methods to encode NMR signals in terms of their phase and frequency. Ernst was awarded the Nobel Prize in chemistry in **1991**; in **2003**, Lauterbur and Mansfield were awarded jointly the Nobel Prize in physiology or medicine.

Of all the imaging methods, MRI is the one that most requires a good knowledge of physical principles, essential not only for the understanding and execution of a

---

COMMENT K3.1 MAGNETIC FIELD GRADIENT AND LARMOR FREQUENCY

Consider the Larmor equation (Eq. (J1.12)):

$$v_0 = \gamma B_0/2\pi \, \text{Hz}$$

For example, a magnetic field gradient of $1 \times 10^{-5}\,\text{T cm}^{-1}$, applied along the bore of an MRI magnet, will result in a resonance frequency range of $(2.68 \times 10^8$ radians $\text{s}^{-1}\,\text{T}^{-1})(1 \times 10^{-5}\,\text{T cm}^{-1})/(2\omega$ radians$) = 425\,\text{Hz cm}^{-1}$. In other words, protons 1 cm apart along the field gradient in the subject have resonance frequencies that differ by 425 Hz. Thus, by changing the center frequency of the probe pulse in increments of 425 Hz, it is possible to probe successive 1 cm positions in the direction of the magnetic field gradient. Each consecutive rf pulse produces a free induction decay (FID; see Chapter J2) signal that encodes the concentration of protons at each 1 cm position along the direction of the field gradient.

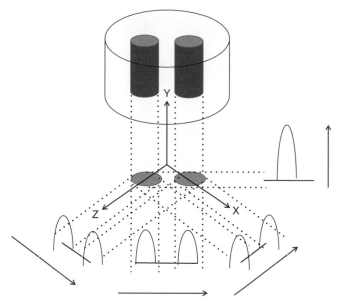

Fig. K3.1 The first picture of an object taken by NMR. The test object consisted of two thin-walled capillaries containing H_2O (blue cylinders), inside a larger capillary containing D_2O (red outline cylinder). The relative intensities in the image were due to the different NMR signals from H and D. The figure shows the relationship between the three-dimensional object, its two-dimensional projection down the y-axis (pink) and four one-dimensional projections separated by 45° in the xz plane. The NMR signal in the presence of a field gradient represents a one-dimensional projection of the H_2O content. One method of creating a two-dimensional image is to combine several projections by rotating the field gradient around the object (blue arrows). Redrawn from Lauterbur (1973).

procedure but also and especially for the interpretation of the results (see Chapter J1).

K3.2 PRINCIPLES OF NUCLEAR MAGNETIC RESONANCE

The fundamentals of NMR are described in Chapters J1 and J2. Here, we recall the salient points for medical imaging applications (Fig. K3.2). Body tissue is rich in hydrogen atoms through its water and lipid content. The behavior of the proton (nucleus of the H atom) in a magnetic field is at the origin of the MR image.

(a) Because of a nuclear property called spin 1/2, the proton has a magnetic moment that in the absence of an external field does not take up a specific orientation, so that the average magnetic moment of a proton population is zero. In a strong external magnetic field, B_0, a preferred direction, is created along the z-axis and the magnetic moments of a proton population will align parallel or antiparallel to the field. In a quantum mechanical analysis, the parallel and antiparallel orientations represent, respectively, a low and high level of well-defined energy, which depends on the nature of the nucleus (the proton in this case) and the strength of the external field. Only transitions between these two states are allowed. They can be stimulated by an external energy source of the appropriate magnitude (the **resonance energy** of the nucleus). A slightly higher number of protons will be in the low-energy configuration than in the higher one, resulting in a small net magnetic moment for the system, **M**, in the direction of B_0 (Comment K3.2).

(b) A pulse of resonance frequency radiation will create a magnetic field, B_1, in the xy plane, perpendicular to B_0 that is along the z-axis. The energy absorbed will cause transitions between them and disturb the equilibrium distribution between the two quantum states. In the classical mechanics model, the nucleus is represented as a spinning top precessing around the z-axis with a frequency (called the Larmor frequency) equal to the resonance frequency. The B_1 field will exert a force that, as in a gyroscope, will be perpendicular to both B_1 and B_0, tilting **M** to form an angle with B_0. The magnitude of the angle will depend on the length of the pulse. A so-called 90° pulse, for example, will be long enough to generate a tilt of **M** by 90° into the xy plane. The pulse will not only reduce the z-component of **M** to zero (equal populations of the high and low energy levels, in the quantum mechanical interpretation) but also bring the spins into phase, creating components of magnetic moment rotating around the z-axis in the xy plane to give an oscillating magnetic field transverse to the B_0 field.

(c) After the pulse (B_1 switched off) **M** will **relax** back to its equilibrium configuration in B_0, along z. The relaxation time T_1 corresponds to the recovery of the z-component. The relaxation time T_2 corresponds to the disappearance of components in the xy plane, as spin phases are again randomized. T_1 is called the *longitudinal* or *spin–lattice* relaxation time because recovery of the equilibrium distribution between the two energy levels occurs through energy coupling with magnetic moments due to atomic electrons. T_2 is called the *transverse* or *spin–spin* relaxation time because the randomizing of spin phases occurs via interactions between spins, creating fluctuations in magnetization. T_2 also contains a less interesting contribution from local small fluctuations in B_0, which, in practice, is not perfectly homogeneous. Since the time to reach equilibrium is essentially infinite, the relaxation time is defined as the time at which 63% of the equilibrium configuration has been recovered in both T_1 and T_2 cases. The basis of MRI is that protons in different tissue and different environments display different T_1 and T_2 relaxation times.

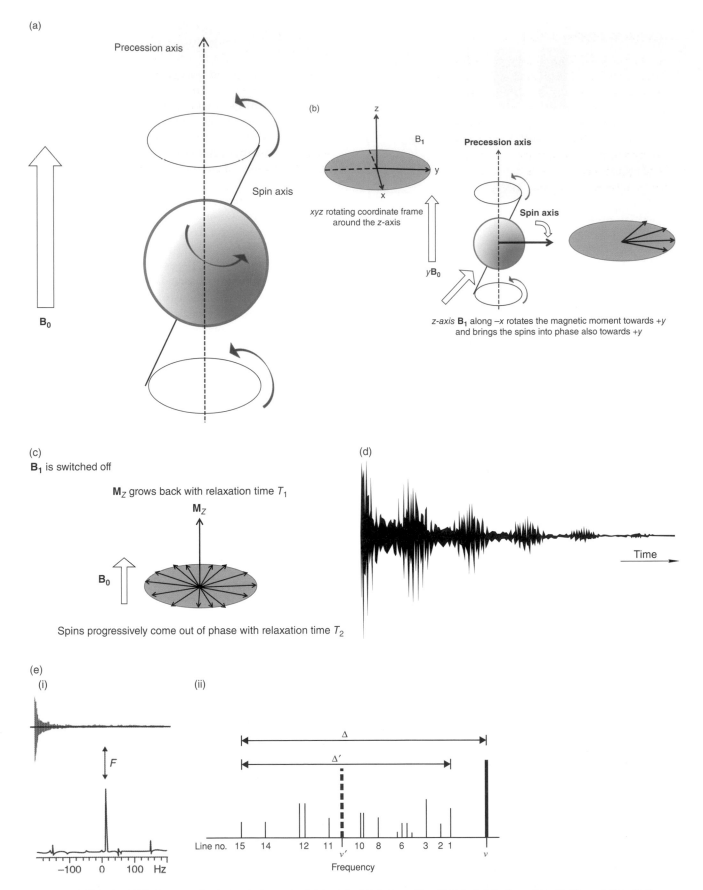

(a)

Precession axis

Spin axis

$\mathbf{B}_0$

(b)

z

B_1

x

y

xyz rotating coordinate frame
around the *z*-axis

Precession axis

Spin axis

$y\mathbf{B}_0$

z-axis $\mathbf{B}_1$ along –*x* rotates the magnetic moment towards +*y*
and brings the spins into phase also towards +*y*

(c)
$\mathbf{B}_1$ is switched off

$\mathbf{M}_Z$ grows back with relaxation time T_1

$\mathbf{M}_Z$

$\mathbf{B}_0$

Spins progressively come out of phase with relaxation time T_2

(d)

Time

(e)

(i)

F

−100   0   100   Hz

(ii)

Δ

Δ′

Line no.   15   14   12   11   10   8   6   3 2 1

v'

Frequency

v

Fig. K3.2 NMR for MRI. See text for legends to parts **(a)** to **(e)**. **(d)** is from Fig. J2.10; **(e)** is from Fig. J2.2b and Fig. J2.3.

The **resonance energy** between the two levels for a spin 1/2 nucleus is proportional to the external field magnitude, B_0, through the **gyromagnetic ratio**, γ, a constant depending on the nucleus; h is Planck's constant. Resonance energy can be expressed as a frequency, ν, called the **Larmor frequency**, by analogy between nuclear spin and the precession of a spinning top around an axis in the classical mechanical model of NMR (see Chapter J1):

$$\Delta E = \frac{h}{2\pi}\gamma B_0$$
$$\Delta E = h\nu$$
$$\nu = \frac{\gamma B_0}{2\pi}$$

The gyromagnetic ration of the proton is 26.75×10^7 radians (Tesla)$^{-1}$ s^{-1}. For an external field of 3 T, used in MRI, the Larmor frequency will be equal to ~100 MHz, at the high end of the rf range (see Fig. J1.1).

de Bazelaire *et al.* (2004) measured T_1 and T_2 relaxation times of normal human abdominal and pelvic tissues and lumbar vertebral bone marrow at 3.0 T.

Relaxation times (mean ± SD) at 3.0 T are reported for kidney cortex (T_1, 1,142 ms ± 154; T_2, 76 ms ± 7), kidney medulla (T_1, 1,545 ms ± 142; T_2, 81 ms ± 8), liver (T_1, 809 ms ± 71; T_2, 34 ms ± 4), spleen (T_1, 1,328 ms ± 31; T_2, 61 ms ± 9), pancreas (T_1, 725 ms ± 71; T_2, 43 ms ± 7), paravertebral muscle (T_1, 898 ms ± 33; T_2, 29 ms ± 4), bone marrow in L4 vertebra (T_1, 586 ms ± 73; T_2, 49 ms ± 4), subcutaneous fat (T_1, 382 ms ± 13; T_2, 68 ms ± 4), prostate (T_1, 1,597 ms ± 42; T_2, 74 ms ± 9), myometrium (T_1, 1,514 ms ± 156; T_2, 79 ms ± 10), endometrium (T_1, 1,453 ms ± 123; T_2, 59 ms ± 1), and cervix (T_1, 1,616 ms ± 61; T_2, 83 ms ± 7). On average, T_1 relaxation times were 21% longer ($p < 0.05$) for kidney cortex, liver, and spleen and T_2 relaxation times were 8% shorter ($p < 0.05$) for liver, spleen, and fat at 3.0 T; however, the fractional change in T_1 and T_2 relaxation times varied greatly with the organ. At 1.5 T, no significant differences ($p > 0.05$) in T_1 relaxation time between the results of this study and the results of other studies for liver, kidney, spleen, and muscle tissue were found.

In conclusion, T_1 relaxation times are generally higher and T_2 relaxation times are generally lower at 3.0 T than at 1.5 T, but the magnitude of change varies greatly in different tissues.

(d) The NMR signal is the oscillating magnetization in the transverse direction as the system relaxes after receiving a pulse of resonance energy. It is detected as a voltage induced in a Faraday coil as a function of time. This is the FID. The FID can contain different contributions (e.g., from protons in different environments) decaying with different T_2 values. Its Fourier transform is easier to interpret.

(e) Recall from Part J that Fourier transformation converts the time domain into the frequency domain. The Fourier transform of the FID will, therefore, identify the waves of different time periods making up the FID as peaks in the frequency domain with widths that are proportional to $1/T_2$ (Fig. K3.2e (i)). The frequency lines when the B_0 field is constant (Fig. K3.2e (ii)) does not contain any information on the location where the signal originated. In MRI, the information is obtained by imposing field gradients. The origin of a signal is then calculated from its frequency and the magnitude of the field at that location. T_1 and T_2 are obtained from the FID by using specific resonant radiation pulse sequences. The relaxation times are measured in milliseconds and although there is no direct relationship between them, T_1 is always greater than T_2 (Comment K3.3).

K3.3 PRINCIPLES OF MRI

A photograph of an MRI spectrometer with a schematic view of the spectrometer is shown in Fig. K3.3.

In the instrument "tunnel," the patient is lying down in the strong magnetic field, $\mathbf{B_0}$, of the superconducting magnet. *Gradient coils* create gradients in $\mathbf{B_0}$ along all three dimensions. Recall that the NMR frequency is proportional to the field intensity. The field gradients, therefore, organize space so that a specific resonance frequency corresponds to a specific *xyz* coordinate. The *RF (radio-frequency) coil* transmits resonant frequency pulses and records the FID.

By suitable choice of field gradients, the image can be obtained in any two-dimensional plane or three-dimensional volume. It is a map of H nuclei (protons) in the chosen area, with intensity at a given point depending on the proton number density and properties of the environment (e.g., H nuclei in water molecules and lipid molecules can be distinguished by different NMR signals). H nuclei in different tissue types will differ by their T_1 and T_2 relaxation times. Pulse sequences are chosen to emphasize these differences in order to maximize contrast in the image.

MRI instrumentation is expensive compared to CT or ultrasound and image acquisition is slower. MRI has significant advantages, however, in that it does not use ionizing radiation and provides anatomical images with excellent contrast and close to 1 mm resolution. Its

(a)
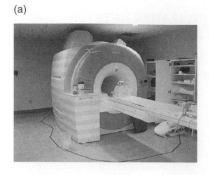

(b)
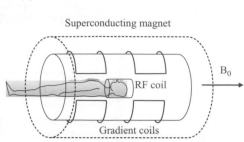

Superconducting magnet

RF coil

B_0

Gradient coils

Fig. K3.3 (a) Photograph of an MRI instrument; **(b)** Schematic view of an MRI system (only one of three gradient coils is shown for clarity); **(c)** MRI images can be acquired in any direction; shown are coronal, sagittal, and axial images, respectively, of the brain. From Smith and Webb (2011) with permission.

(c)
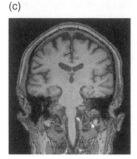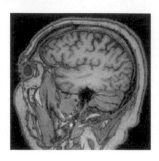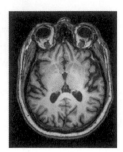

sensitivity to blood flow, blood perfusion, and water diffusion makes available a powerful tool for functional angiography and studies of localized brain activation.

K3.3.1 Health Physics

The high magnetic fields (in the order of 100 000 times the Earth's field) used in MRI have been shown not to constitute a health hazard. Before an MRI examination, however, it must be ensured that the patient does not have implants or surgical clips made of magnetic metals. Titanium, which is often used in artificial joints, is not a magnetic metal. Red tattoos that contain iron may produce an MRI response that will leave the patient with severe burn wounds. Fortunately, most red tattoos contain pigments that do not constitute an MRI risk.

The rf electric field associated with the rf magnetic field produces electric currents in conductive tissue. The corresponding power deposition is quantified via the local and average specific absorption rate (SAR) values in watts per kilogram of tissue.

$$SAR = \frac{\sigma}{2\rho} |E|^2 \tag{K3.1}$$

where ρ, σ, are tissue density and conductivity, respectively, and E is electric field.

The SAR is proportional to the $\mathbf{B}_1$ field multiplied by the time for which it is applied. Certain NMR sequences can result in considerable power deposition in the patient and there are strict regulatory safeguards on SAR values. Commercial MRI instruments have built-in hardware and software to estimate SAR for each sequence run in order to adjust parameters to respect safety limits.

K3.3.2 T_1 and T_2 Image Formation, Signal Localization, and Pulse Sequences

The NMR signal from protons in different tissue and different environments occurs with different T_1 and T_2 relaxation times. The values for various tissue types are given in Comment K3.3.

Slice selection. Recall that in CT scans, the selected slice lies in the plane of the detector circle, transverse to the patient axis (it is an axial slice). In contrast, in MRI, a slice in any of the three – axial, coronal, or sagittal – orientations can be defined by suitably chosen field gradients. In combination with the field gradients, the application of an appropriate pulse sequence followed by Fourier transform analysis of the corresponding FID yields T_1 and T_2 relaxation times for the different proton populations as a function of their xyz coordinates. The computer converts this information to form an image in shades of gray.

An axial slice (e.g., in an xy plane) in the volume to be examined is defined by applying a frequency-selective rf pulse at the same time as a field gradient along the z-axis. The nuclear resonance frequency, being proportional to the external magnetic field, will be proportional to the position of the nuclei along z. The frequency and bandwidth of the pulse are, therefore, selected in order to stimulate only the protons in the chosen slice. The slice will be thinner for steeper field gradients or smaller rf bandwidth. The application of successive gradient/rf pulse couples in the three spatial directions will select axial, sagittal, or coronal slices, respectively perpendicular to the z, x, or y axes (Fig. K3.4).

Coordinates in the plane of the slice (phase and frequency encoding). In the case of a slice perpendicular to the z-axis, the x and y coordinates of the NMR signal are encoded in its phase and frequency, respectively. Phase encoding is obtained by applying a gradient in the y direction for a given period to induce a spatially dependent

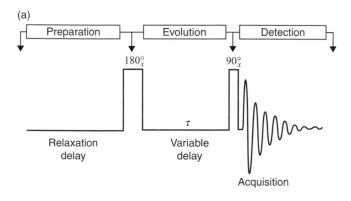

Selected slice

Magnetic field gradient

Tissue resonance frequency

rf pulse

Fig. K3.4 Slice selection and phase and frequency encoding.

phase shift along that direction. The x value is encoded by applying a frequency-encoding gradient along x.

Each voxel (volume element) of the image is, therefore, characterized by a thickness and z coordinate (defined by the slice) and x and y coordinates defined, respectively, by a specific frequency and a specific phase.

Pulse sequences. As presented in more detail in Section J2.3, NMR is a very powerful method because of the richness of information that can be obtained from the FID, by applying a wide variety of resonant frequency pulse sequences. Inversion recovery and spin echo pulse sequences are the simplest to measure T_1, and T_2 relaxation times, respectively (Fig. K3.5).

In the **inversion recovery pulse sequence** (Fig. K3.5a), the nuclei are first subjected to a 180° pulse (i.e., a radio-frequency pulse of sufficient width to cause **M** to rotate from up to down the z-axis). A variable delay period, τ, follows the pulse. Typically, ten values of τ are used, ranging from a small value to four or five times T_1. During the delay period the system returns to equilibrium by the spin–lattice relaxation process. As the individual magnetic moments gradually return to their initial orientations along $+z$, the net magnetization vector becomes shorter in

(a)

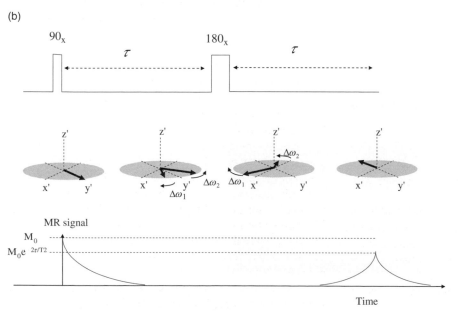

| Preparation | Evolution | Detection |

180°_x 90°_x

Relaxation delay

Variable delay

τ

Acquisition

(b)

90_x 180_x

τ τ

z' z' z' z'

$\Delta\omega_2$

x' y' x' y' $\Delta\omega_2$ $\Delta\omega_1$ x' y' x' y'

$\Delta\omega_1$

MR signal

M_0

$M_0 e^{2\tau/T2}$

Time

Fig. K3.5 (a) Schematic pulse sequence for relaxation time T_1 measurement by inversion recovery (from Fig. J2.11); **(b)** Spin echo pulse sequence (From Smith and Webb (2011), with permission).

the –z direction. Depending on the length of the delay, **M** passes through zero and eventually recovers its full original magnitude along the +z-axis. Since the FID measures magnetization fluctuations in the xy plane, in order to determine the extent of relaxation, **M** must be converted into observable magnetization in the xy plane. Applying a 90° pulse along x rotates **M** to lie along the y-axis. If **M** is still negative before the 90° pulse, then the magnetization is rotated to the +y direction, but, when relaxation is complete, the magnetization after the 90° pulse lies along the –y-axis. The intensity I of the frequency peak resulting from the Fourier transform of the FID varies with τ from a maximum negative value for $\tau = 0$ to a maximum positive value for $\tau = \infty$ (in practice, four or five times T_1). The intensities are related exponentially to T_1 by the equation:

$$I = I_\infty[1 - 2\exp(-\tau/T_1)] \tag{K3.2}$$

In **spin echo**, an initial 90° pulse along x turns the total **M** to the –y-axis. The effect of T_2 is a gradual shortening of **M** in the –y direction. However, because **B**$_0$ is not perfectly homogeneous, spin phases will fan out in the xy plane during the delay τ that follows the pulse, causing the FID to die away from a maximum. After the end of the first τ delay, a 180° pulse flips the spins about the x-axis (Fig. K3.5b). This will refocus their phases after a period equal to τ and the NMR signal recovers, to another maximum, the spin echo. Since the magnetization is now along the +y-axis, the echo signal is inverted with respect to the first signal observed. The sequence is repeated. The time constant for the decrease in the magnitude of the echo is the "true" T_2, since effects due to **B**$_0$ inhomogeneities were eliminated by the refocusing process.

A careful choice of repetition time (TR) and spin echo time (TE = 2τ) leads to images that are either T_1 or T_2 weighted and permits to discriminate between tissue types and environments.

TR interferes with the recovery of longitudinal magnetization. For example, a tissue with a long T_1 will not

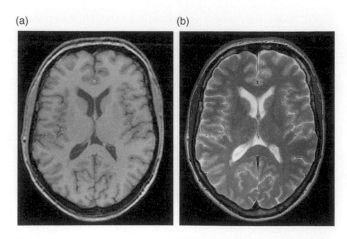

Fig. K3.6 Brain matter contrast. **(a)** T_1 weighted; **(b)** T_2 weighted axial slice through the brain. In **(a)**, the cerebrospinal fluid filling the ventricles is much darker than the white/gray matter, whereas in **(b)** it is brighter.

have time to recover if TR is too short, and its NMR signal will be weaker. Contrast between tissues of different T_1 values is higher at low repetition times but the signals are weaker.

The choice of TE will influence T_2 weighting in the image. A long TE increases contrast due to T_2 differences between tissue types, but decreases signal-to-noise ratio.

Relative image contrast for brain matter is shown in Fig. K3.6 as a function of different TR and TE intervals, respectively.

Acquisition times in MRI can be quite long, with a full clinical examination taking about an hour, consisting of several kinds of scan. The **gradient echo** pulse sequence allows very rapid image acquisition (~10 s), but has the disadvantage that the measured T_2 includes a non-negligible contribution from **B**$_0$ inhomogeneities, so that in contrast to the spin echo approach, the "true" T_2 values associated with different tissues are not determined.

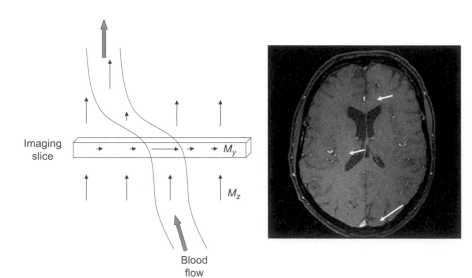

Fig. K3.7 TOF NMR angiographic technique. Blood flowing through the imaging slice has full M_z magnetization before the rf pulse. A 90° pulse creates maximum M_y magnetization. In contrast, stationary tissue experiences every rf pulse during the imaging sequence and, since TR ≪ 3 T, is "saturated," i.e., has an M_z magnetization below its thermal equilibrium value. The 90° pulse, therefore, creates a much lower M_y magnetization in the tissue than in the blood. Vessels with flow perpendicular to the image slice will appear as bright spots in the MRI image. (From Smith and Webb, 2011.)

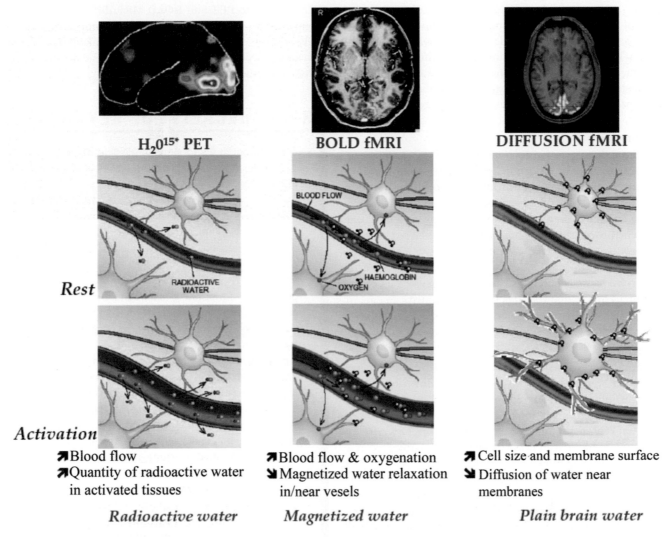

H₂0¹⁵* PET **BOLD fMRI** **DIFFUSION fMRI**

Rest

Activation

↗ Blood flow
↗ Quantity of radioactive water
  in activated tissues

Radioactive water

↗ Blood flow & oxygenation
↘ Magnetized water relaxation
  in/near vesels

Magnetized water

↗ Cell size and membrane surface
↘ Diffusion of water near
  membranes

Plain brain water

Fig. K3.8 Functional PET and MRI to measure brain water. General principles of PET and MRI functional neuroimaging methods. Left: brain activation mapping with PET. Radioactive water (H_2O^{15}) is used as a tracer to detect neuronal activation-induced increases in blood flow. Middle: brain activation mapping with BOLD fMRI. Water magnetization in and around small vessels is modulated by the flow of red blood cells containing paramagnetic deoxyhemoglobin. Right: brain activation mapping with diffusion fMRI. The reduction in water diffusion, which occurs during activation, is thought to originate from a membrane-bound water layer which expands during activation-induced cell swelling. While with PET and BOLD fMRI water is only an indirect means to detect changes in blood flow through the imaging scanner, the changes in water properties seen with diffusion fMRI seem to be an intrinsic part of the activation process. (From Le Bihan (2007), with permission).

Efficient signal treatment approaches have been developed by expressing the NMR signal, including the encoding steps, in terms of spatial frequencies in so-called **k-space formalism**, mathematically similar to scattering vectors in diffraction (see Chapter G1). The image in an xy plane is then obtained from the two-dimensional Fourier transform of the k-space data.

Contrast Agents

Contrast agents are used to modify the magnetic parameters of the tissue to which they bind, for cases in which the natural contrast-to-noise ratio is insufficient (e.g., for the detection of very small lesions). At variance with other imaging techniques, MRI contrast agents are not visualized directly, but through their action on the T_1 and T_2 relaxation of the adjoining tissue. MRI contrast agents modify the magnetization of protons in their close environment.

Positive contrast agents are based on specific chemical chelates of the gadolinium ion. The Gd^{3+} ion is paramagnetic and significantly shortens the T_1 relaxation time of a water molecule bound in its vicinity.

Negative contrast agents reduce the NMR signal from tissue in which they accumulate. They are superparamagnetic and cause large inhomogeneities in the local magnetic field that reduce T_2 relaxation times with a resulting reduction in signal intensity.

Contrast agents are specific for the visualization of certain types of lesion in a given tissue. For example, ultra-small superparamagnetic iron oxide particles will enter only healthy liver cells and not accumulate in tumors. They are introduced intravenously during the MRI exam and should be considered as potentially toxic drugs, and so proper precautions must be taken with respect to dosage etc.

Image Artifacts

Synchronizing the repetition rate TR with the respiration or cardiac rhythm can reduce image artifacts due to motion. Truncation errors in phase encoding will result in striations in the image at an interface representing a large transition in signal. Interfaces between environments of very different magnetic properties (air–water, bone–air, or close to metal) will produce signal loss. Chemical shift (see Chapter J2) differences at an interface may also introduce image artifacts in frequency encoding. There is also the artifact of the edge of an image that folds back to form a mirror image, in cases where object size is larger than the image field.

Angiography

The Fourier transform of the NMR signal from a moving object will appear as extended along the direction of the encoding phase. Considered in early MRI as a source of noise, the effect is now applied in angiography. The time-of-flight (TOF) technique is based on the shorter *effective* T_1 relaxation of blood as it flows through an imaging slice in a direction perpendicular to it (Fig. K3.7).

A strongly T_1 weighted pulse sequence is used to differentiate between water protons moving in blood and the stationary tissue. The technique does not require the use of contrast agents, but they can be used to increase the signal difference between flowing blood and stationary tissue.

K3.4 BRAIN WATER, FUNCTIONAL MRI, AND PET

Based on very different physical approaches, functional positron emission tomography (PET) and functional (fMRI) rely on the principle that neuronal activation and blood flow are coupled through metabolism. They both make brain activation imaging possible through measurements involving water molecules.

Brain activation in PET is mapped via the observation of radioactive water (H_2O^{15}) (see Section K1.7) as a tracer to detect neuronal activation-induced increases in blood flow.

Blood oxygen level dependent (BOLD) MRI is based on the modulation of water magnetization through the paramagnetic deoxyhemoglobin molecules in red blood cells.

Blood brain activation can also be mapped directly by diffusion fMRI because there appears to be a slowing down of water diffusion in activated cells.

The different protocols are illustrated Fig. K3.8.

In contrast to PET and BOLD MRI, diffusion fMRI can reveal changes in the local behavior of water during brain activation, which could be more directly linked to neuronal activation mechanisms and ultimately lead to improved spatial and time resolution.

K3.5 CHECKLIST OF KEY IDEAS

- The behavior in a strong magnetic field, $\mathbf{B_0}$, of H spin 1/2 nuclei in water and lipid molecules is at the origin of MR imaging.
- The resonance energy between the two levels for a spin 1/2 nucleus is proportional to the external field magnitude, B_0, through the gyromagnetic ratio, γ, a constant depending on the nucleus. Resonance energy can be expressed as a frequency, v, called the Larmor frequency, by analogy between nuclear spin and the precession of a spinning top around an axis in the classical mechanical model of NMR. In the classical description there will be a small net magnetic moment for the system, $\mathbf{M}$, in the direction of $\mathbf{B_0}$.
- A pulse of resonance frequency radiation will create a magnetic field, $\mathbf{B_1}$, in the xy plane, perpendicular to $\mathbf{B_0}$ that is along the z-axis. The energy absorbed will cause transitions between them and disturb the equilibrium distribution between the two energy levels. The $\mathbf{B_1}$ field will exert a force perpendicular to both $\mathbf{B_1}$ and $\mathbf{B_0}$, tilting $\mathbf{M}$ to form an angle with $\mathbf{B_0}$. The magnitude of the angle will depend on the length of the pulse.
- A 90° pulse will be long enough to generate a tilt of $\mathbf{M}$ by 90° into the xy plane. The pulse will reduce the z-component of $\mathbf{M}$ to zero and bring the spins into phase, creating components of magnetic moment rotating around the z-axis in the xy plane to give an oscillating magnetic field transverse to the $\mathbf{B_0}$ field.
- When $\mathbf{B_1}$ is switched off, $\mathbf{M}$ will relax back to its equilibrium configuration in $\mathbf{B_0}$, along z, with a relaxation time T_1. The relaxation time T_2 corresponds to the disappearance of components in the xy plane, as spin phases are again randomized. T_1 is called the *longitudinal* or *spin–lattice* relaxation time. T_2 is called the *transverse* or *spin–spin* relaxation time.
- The NMR signal is the oscillating magnetization in the transverse direction as the system relaxes. It is detected as a voltage induced in a Faraday coil as a function of time. This is called the free induction decay (FID).
- T_1 and T_2 are obtained from the FID by using specific resonant radiation pulse sequences. The relaxation times are measured in milliseconds and although there is no direct relationship between them, T_1 is always greater than T_2.
- During an MRI examination, the patient in the instrument "tunnel" is lying down in the strong magnetic field, $\mathbf{B_0}$, produced by a superconducting magnet. Gradient coils create gradients in $\mathbf{B_0}$ along all three dimensions. NMR

frequency is proportional to the field intensity. The field gradients, therefore, organize space so that a specific resonance frequency corresponds to a specific *xyz* coordinate. An rf (radio-frequency) coil transmits resonant frequency pulses and records the FID.

- A slice in any of the three – axial, coronal, or sagittal – orientations is defined by suitably chosen field gradients.

- The application of an appropriate pulse sequence followed by Fourier transform analysis of the corresponding FID yields T_1 and T_2 relaxation times for the different proton populations as a function of their *xyz* coordinates. The computer converts this information to form an image in shades of gray.

- "Inversion recovery" and "spin echo" are commonly used pulse sequences, leading to images that are either T_1 or T_2 weighted, which permits discrimination between tissue types.

- Acquisition times in MRI can be quite long, with a full clinical examination taking about an hour, consisting of several kinds of scan.

- Contrast agents modify the magnetic parameters of the tissue to which they bind, for cases in which the natural contrast-to-noise ratio is insufficient.

- In angiography, the time-of-flight (TOF) technique is based on the shorter *effective* T_1 relaxation of blood as it flows through an imaging slice in a direction perpendicular to it. A strongly T_1 weighted pulse sequence will differentiate between water protons moving in blood and in stationary tissue.

- Based on very different physical approaches, functional PET and functional MRI rely on the principle that neuronal activation and blood flow are coupled through metabolism. They both make brain activation imaging possible through measurements involving water molecules.

Suggestions for Further Reading

Le Bihan, D. (2007) The "wet mind": Water and functional neuroimaging. *Phys. Med. Biol.*, **52**, R57–90.

Barrie Smith, N. and Webb, A. (2011). *Introduction to Medical Imaging: Physics, Engineering and Clinical Applications*. Cambridge: Cambridge University Press.

www.sprawls.org/resources. Online resources for learning and teaching physics and technology of medical imaging.

REFERENCES

Agalarov, S. C., and Williamson, J. R. (2000). A hierarchy of RNA subdomains in assembly of the central domain of the 30S ribosomal subunit. *RNA*, **6**, 402–408.

Agalarov, S. C., Sheleznyakova, E. N., Selivanova, O. M., *et al.* (1998). In vitro assembly of a ribonucleoprotein particle corresponding to the platform domain of the 30S ribosomal subunit. *Proc. Natl. Acad. Sci. USA*, **95**, 999–1003.

Agalarov, S. C., Selivanova, O. M., Zheleznyakova, E., *et al.* (1999). Independent *in vitro* assembly of all three major morphological parts of the 30S ribosomal subunit of *Thermus thermophilus*. *Eur. J. Biochem.*, **266**, 533–537.

Agnew, C., Borodina, E., Zaccai, N.R., *et al.* (2011). Correlation of in situ mechanosensitive responses of the *Moraxella catarrhalis* adhesin UspA1 with fibronectin and receptor CEACAM1 binding. *Proc. Nat. Acad. Sci. USA*, **108**, 15 174–15 178.

Aia, J. (2013). Glycosaminoglycan glycomics using mass spectrometry. *Mol. Cell Proteomics*, **12**, 885–892.

Akgun, B., Satija, S., Nanda, H., *et al.* (2013). Conformational transition of membrane-associated terminally acylated HIV-1 Nef. *Structure*, **21**, 1822–1833.

Allemand, J.-F., Bensimon, D., Lavery, R., and Croquette, V. (1998). Stretched and overwound DNA forms a Pouling-like structure with exposed base. *Proc. Natl. Acad. Sci. USA*, **95**, 14 152–14 157.

Allison, S. A. (2001). Boundary element modelling of biomolecular transport. *Biophys. Chem.*, **93**, 197–213.

Altieri, A. S., Hinton, D., and Byrd, R. A. (1995). Association of biomolecular system via pulse field gradient NMR self-diffusion measurements. *JACS*, **117**, 7566–7567.

Altose, M. D., Zheng, Y., Dong, J., Palfey, B. A., and Carey, P. R. (2001). Comparing protein–ligand interactions in solution and single crystals by Raman spectroscopy. *Proc. Natl. Acad. Sci. USA*, **98**, 3006–3011.

Ando, T. (2012). High-speed atomic force microscopy coming of age. *Nanotechnology*, **23**, 062001.

Ando, T., Uchihashi, T., and Scheuring, S. (2014). Filming biomolecular processes by high-speed atomic force microscopy. *Chemical Rev.*, **114**, 3120–3188.

Aquila, A., Hunter, M.S., Doak, R.B., *et al.* (2012). Time-resolved protein nanocrystallography using an X-ray free-electron laser. *Opt. Express*, **20**, 2706–2716.

Babu, M., Vlasblom, J., Pu, S., *et al.* (2012). Interaction landscape of membrane–protein complexes in *Saccharomyces cerevisiae*. *Nature*, **489**, 585–598.

Bacia, K., and Schwille, P. (2003). A dynamic view of cellular processes by in vivo fluorescence auto- and cross-correlation spectroscopy. *Methods*, **29**, 74–85.

Bai, X. C., Fernandez, I.S., McMullan, G., and Scheres, S. H. (2013). Ribosome structures to near-atomic resolution from thirty thousand cryo-EM particles. *Elife*, **2**, e00461.

Bailey, B., Farkas, D. L., Taylor, D. L., and Lanni, F. (1993). Enhancement of axial resolution in fluorescence microscopy by standing-wave excitation. *Nature*, **366**, 44–48.

Ban, N., Nissen, P., Hansen, J., *et al.* (1999). Placement of protein and RNA structures into a 5 Å -resolution map of the 50S ribosomal subunit. *Nature*, **400** (6747), 841–847.

Banachowicz, E., Gapinski, J., and Patkowski, A. (2000). Solution structure of biopolymers: A new method of constructing a bead model. *Biophys. J.*, **78**, 70–78.

Bandecar, J. (1992). Amide modes and protein conformation. *Biochim. Biophys. Acta*, **1120**, 123–143.

Barbet-Massin, E., Pell, A.J., Retel, J.S., *et al.* (2014). Rapid proton-detected NMR assignment for proteins with fast magic angle spinning. *J. Am. Chem. Soc.*, **136**, 12 489–12 497.

Barnes, W. L., Dereux, A., and Ebbesen, T. W. (2003). Surface plasmon subwavelength optics. *Nature*, **424** (6950), 824–830.

Barron, L. D., Hecht, L., Blanch, E. W., and Bell, A. F. (2000). Solution structure and dynamics of biomolecules from Raman optical activity. *Prog. Biophys. Mol. Biol.*, **73**, 1–49.

Basavappa, R., and Sigler, P. B. (1991). The 3 A crystal structure of yeast initiator tRNA: Functional implications in initiator/ elongator discrimination. *EMBO J.*, **10**, 3105–3111.

Bastiaens, P. I. H., and Pepperkok, R. (2000). Observing proteins in their natural habitat: The living cell. *TIBS*, **25**, 631–636.

Baumann, C. G., Bloomfield, V. A., Smith, S. B., *et al.* (2000). Stretching of single collapsed DNA molecules. *Biophys. J.*, **78**, 1965–1978.

Bax, A. (2003). Weak alignment offers new NMR opportunities to study protein structure and dynamics. *Protein Sci.*, **12**, 1–16.

Belke, J., and Ristau, O. (1997). Analysis of interacting biopolymer systems by analytical centrifugation. *Eur. Biophys. J.*, **25**, 325–332.

Bellissent-Funel, M. C., Zanotti, J. M., and Chen, S. H. (1996). Slow dynamics of water molecules on the surface of globular proteins. *Faraday Discuss.*, **103**, 281–294.

Belov, M. E., Gorshkov, M. V., Udeseth, H. R., Anderson, G. A., and Smith, R. D. (2000). Zeptomole-sensitivity electrospray ionization: Fourier transform ion cyclotron resonance mass spectrometry proteins. *Anal. Chem.*, **72**, 2271–2279.

Benner, W. H. (1997). A gated electrostatic ion trap to repetitiously measure the charge and *m/z* of large electrospray ions. *Anal. Chem.*, **69**, 4162–4168.

Bennett, M. J., Choe, S., and Eisenberg, D. (1994). Refined structure of monomeric diphtheria toxin at 2.3 Å resolution. *Protein Sci.*, **3** (9), 1464–1475.

Bennink, M. L., Leuba, S. H., Leno, G. H., *et al.* (2001). Unfolding individual nucleosomes by stretching single chromatin fibers with optical tweezers. *Nature Str. Biol*, **8**, 606–610.

Berg, H. (1983). *Random Walks in Biology*. Princeton, NJ: Princeton University Press.

Bergethon, P. R. (1995). *The Physical Basis of Biochemistry: The Foundation of Molecular Biophysics*. New York: Springer.

Bernado, P., Garcia de la Torre, J., and Pons, M. (2002). Interpretation of 15N NMR relaxation data for globular proteins using hydrodynamic calculations with HYDRONMR. *J. Biomol. NMR*, **23**, 139–150.

Bernal, J. D., and Crowfoot, D. (1934). X-ray photographs of crystalline pepsin. *Nature*, **134**, 794–795.

Biemann, K. (1992). Mass spectrometry of peptides and proteins. *Annu. Rev. Biochem.*, **61**, 977–1010.

Bischler, N., Brino, L., Carles, C., *et al.* (2002). Localization of the yeast RNA polymerase I-specific subunits. *Embo. J.*, **21** (15), 4136–4144.

Bjorkman, P. J., Saper, M. A., Samraoui, B., *et al.* (1987). Structure of the human class I histocompatibility antigen, HLA-A2. *Nature*, **329** (6139), 506–512.

Blackledge, M. (2011). Studying partially folded and intrinsically disordered proteins using NMR residual dipolar couplings. In *Protein NMR Spectroscopy: Practical Techniques and Applications*, eds., Lu-Yun Lian and Gordon Roberts. Chichester: John Wiley and Sons.

Block, S. M., Blair, D. F., and Berg, H. C. (1989). Compliance of bacterial flagella measured wih optical tweezers. *Nature*, **338**, 514–518.

Boeri Erba, E. (2014). Investigating macromolecular complexes using top-down mass spectrometry. *Proteomics*, **14**, 1259–1270.

Bon, C., Lehmann, M. S., and Wilkinson, C. (1999). Quasi Laue neutron-diffraction study of the water arrangement in crystals of triclinic hen egg-white lysozyme. *Acta Crystallogr. D*, **55**, 978–987.

Bon, C., Dianoux, A. J., Ferrand, M., and Lehmann, M. (2002). A model for water motion in crystals of lysozyme based on an incoherent quasielastic neutron-scattering study. *Biophys. J.*, **83** (3), 1578–1588.

Booth, D. R., Sunde, M., Bellotti, V., *et al.* (1997). Instability, unfolding and aggregation of human lysozyme variants underlying amyloid fibrillogenesis. *Nature*, **385**, 787–793.

Bottcher, B., Tsuji, N., Takahashi, H., *et al.* (1998). Peptides that block hepatitis B virus assembly: Analysis by cryomicroscopy, mutagenesis and transfection. *Embo. J.*, **17** (23), 6839–6845.

Bowie, J. U., Luthy, R., and Eisenberg, D. (1991). A method to identify protein sequences that fold into a known three-dimensional structure. *Science*, **253** (5016), 164–170.

Braiman, M. S., and Rothschild, K. J. (1988). Fourier transform infrared techniques for probing membrane protein structure. *Ann. Rev. Biophys. Biophysical Chem.*, **17**, 541–570.

Brändén, C.-I., and Tooze, J. (1999). *Introduction to Protein Structure*. New York: Garland Pub.

Brant, D. A. (1999). Novel approaches to the analysis of polysaccharide structure. *Curr. Opin. Struct. Biol.*, **9**, 556–562.

Brey, W. S. (ed.) (1988). *Pulse Methods in 1D and 2D Liquid-Phase NMR*. San Diego, CA: Academic Press.

Brocca, S., Testa, L., Sobott, F., *et al.* (2011). Compaction properties of an intrinsically disordered protein: Sic1 and its kinase-inhibitor domain, *Biophys. J.*, **100**, 2243–2252.

Brodbelt, J. S. (2014) Photodissociation mass spectrometry: New tools for characterization of biological molecules. *Chem. Soc. Rev.*, **43**, 27–57.

Brower-Toland, B. R., Smith, C. L., Yeh, R. S., *et al.* (2002). Mechanical disruption of individual nucleosomes reveals a reversible multistage release of DNA. *Proc. Natl. Acad. Sci. USA*, **99**, 1960–1966.

Brown, A., Amunts, A., Bai, X. C., *et al.* (2014). Structure of the large ribosomal subunit from human mitochondria. *Science*. **346** (6210), 718–722.

Brudler, R., Rammelsberg, R., Woo, T. T., *et al.* (2001). Structure of the I1 early intermediate of photoactive yellow protein by FTIR spectroscopy. *Nat. Struct. Biol.*, **8** (3), 265–270.

Brune, D., and Kim, S. (1993). Predicting protein diffusion coefficients. *J. Am. Chem. Soc.*, **90**, 3835–3839.

Brunger, A. T., and Adams, P. D. (2002). Molecular dynamics applied to X-ray structure refinement. *Acc. Chem. Res.*, **35** (6), 404–412.

Buchanan, M. V., and Hettich, R. L. (1993). Fourier transform mass spectrometry of high mass molecules. *Anal. Chem.*, **65**, 245A–259A.

Buldt, G., Gally, H. U., Seelig, J., and Zaccai, G. (1979). Neutron diffraction studies on phosphatidylcholine model membranes. I. Head group conformation. *J Mol. Biol.*, **134**, 673–691.

Bustamante, C., Erie, D. A., and Keller, D. (1994). Biochemical and structural applications of scanning force microscopy. *Curr. Opin. Struct. Biol.*, **4**, 750–760.

Byron, O. (1997). Construction of hydrodynamic bead models from high resolution x-ray crystallographic or nuclear magnetic resonance data. *Bioph. J.*, **72**, 406–415.

Caffrey, M. (2000). A lipid's eye view of membrane protein crystallization in mesophases. *Curr. Opin. Struct. Biol.*, **10** (4), 486–497.

Cai, S., and Singh, B. R. (1999). Identification of beta-turn and random coil amide III infrared bands for secondary structure estimation of proteins. *Biophys. Chem.*, **80** (1), 7–20.

Callender, R., and Deng, H. (1994). Nonresonance Raman difference spectroscopy: A general probe of protein structure, ligand binding, enzymatic catalysis, and the structures of other biomacromolecules. *Annu. Rev. Biophys. Biomol. Struct.*, **23**, 215–245.

Campos-Olivas, R., Horr, I., Bormann, C., Jung, G., and Gronenborn, A. M. (2001). Solution structure, backbone dynamics and chitin binding of the anti-fungal protein from *Streptomyces tendae* TU901. *J. Mol. Biol.*, **308**, 765–782.

Canet, D., Doering, K., Dobson, C. M., and Dupont, Y. (2001). High-sensitivity fluorescence anisotropy detection of protein-folding events: Application to α-lactalbumin. *Biophys. J.*, **80**, 1996–2003.

Cantor, C., and Schimmel, P. (1980). *Biophysical Chemistry: Part II. Technique for the Study of Biological Structure and Function*. San Francisco, CA: W. H. Freeman and Company.

Capel, M. S., Kjeldgaard, M., Engelman, D. M., and Moore, P. B. (1988). Positions of S2, S13, S16, S17, S19 and S21 in the 30 S ribosomal subunit of *Escherichia coli*. *J. Mol. Biol.*, **200**, 65–87.

Caprioli, R. M., and Suter, M. J.-F. (1995). Mass spectrometry. In *Introduction to Biophysical Methods for Protein and Nucleic Acid Research*. San Diego, CA: Academic Press.

Carr, S. A., and Burlingame, A. L. (1996). The meaning and usage of the terms monoisotopic mass, average mass, mass resolution, and mass accuracy for measurements of biomolecules. In *Mass Spectrometry in the Biological Sciences*, eds. A. L. Burlingame and S. A. Carr, pp. 546–552. Totowa, NJ: Humana Press.

Carra, J. H., Murphy, E. C., Privalov, P. L. (1996). Thermodynamic effects of mutations on the denaturation of T4 lysozyme. *Biophys. J.*, **71** (4), 1994–2001.

Carrion-Vazquez, M., Overhauser, A. F., Fisher, T., *et al.* (2000). Mechanical design of proteins studied by single-molecule force spectroscopy and protein engineering. *Prog. Biophys. Mol. Biol.*, **74**, 63–91.

Cate, J. H., Yusupov, M. M., Yusupova, G., *et al.* (1999). X-ray crystal structures of 70S ribosome functional complexes. *Science*, **285** (5436), 2095–2104.

Chacon, P., Diaz, J. F., Moran, F., and Andreu, J.M. (2000). Reconstruction of protein form with X-ray solution scattering and a genetic algorithm. *J. Mol. Biol.*, **299**(5), 1289–1302.

Chang, C. T., Wu, C. S., and Yang, J. T. (1978). Circular dichroic analysis of protein conformation: Inclusion of the beta-turns. *Anal. Biochem.*, **91**, 13–31.

Che, Z., N. Olson, H., Leippe, D., *et al.* (1998). Antibody-mediated neutralization of human rhinovirus 14 explored by means of cryoelectron microscopy and X-ray crystallography of virus–Fab complexes. *J. Virol.*, **72** (6), 4610–4622.

Checovich, W. J., Bolger, R. E., and Burke, T. (1995). Fluorescence polarization: A new tool for cell and molecular biology. *Nature*, **375**, 254–256.

Cheow, L. F., Viswanathan, R., Chin, C. S., *et al.* (2014). Multiplexed analysis of protein–ligand interactions by fluorescence anisotropy in a microfluidic platform. *Anal. Chem.*, **86**, 9901–9908.

Cherry, R. J., and Schneider, G. (1976). A spectroscopic technique for measuring slow rotational diffusion of macromolecules. 2: Determination of rotational correlation times of protein in solution. *Biochemistry*, **15**, 3657–3661.

Chervenka, C., 1969. *A Manual of Methods for the Analytical Ultracentrifuge*. Palo Alto, CA: Spinco Division, Beckman Instruments.

Clore, G. M., and Gronenborn, A. M. (1991). Two-, three-, and four-dimensional NMR methods for obtaining more precise three-dimensional structure of proteins in solution. *Ann. Rev. Biophys. Chem.*, **20**, 29–63.

Cluzel, P., Lebrun, A., Heller, C., *et al.* (1996). DNA: An extensible molecule. *Science*, **271**, 792–794.

Collins, K. D. (1997). Charge density-dependent strength of hydration and biological structure. *Biophys. J.*, **72** (1), 65–76.

Cordone, L., Ferrand, M., Vitrano, E., *et al.* (1999). Harmonic behavior of trehalose-coated carbon-monoxy-myoglobin at high temperature. *Biophys. J.*, **76**, 1043–1047.

Crick, F. H. (1968). The origin of the genetic code. *J. Mol. Biol.*, **38**, 367–379.

Dasgupta, S., and Spiro, T. G. (1986). Resonance Raman characterization of the 7-ns photoproduct of (carbonmonoxy) hemoglobin: Implications for hemoglobin dynamics. *Biochemistry*, **25**, 5941–5948.

Davidson, I. W., and Secrest, W. L. (1972). Determination of chromium in biological materials by atomic absorption spectrometry using a graphite furnace atomizer. *Anal. Chem.*, **44** (13), 1808–1813.

de Bazelaire, C. M., Duhamel, G. D., Rofsky, N. M., and Alsop, D. C. (2004). MR imaging relaxation times of abdominal and pelvic tissues measured in vivo at 3.0 T: Preliminary results. *Radiology*, **230**, 652–659.

de la Torre, G. (2001) Hydration from hydrodynamics: General consideration and applications to bead modelling to globular proteins. *Biophys. Chem.*, **93**, 159–170.

Dekker, N. H., Rybenkov, V. V., Duguet, M., *et al.* (2002). The mechanism of type IA topoisomerases. *Proc. Natl. Acad. Sci. USA*, **99**, 12 126–12 131.

Denninger, A. R., Deme, B., Cristiglio, V., *et al.* (2014). Neutron scattering from myelin revisited: Bilayer asymmetry and water-exchange kinetics. *Acta Crystallogr. D Biol. Crystallogr*, **70**, 3198–3211.

Derome, A. E. (1987). *Modern NMR Techniques for Chemistry Research*. New York: Pergamon.

Dessen, P., Blanquet, S., Zaccai, G., and Jacrot, B. (1978). Antico-operative binding of initiator transfer RNAMet to methionyl-transfer RNA synthetase from *Escherichia coli*: Neutron scattering studies. *J. Mol. Biol.*, **126**, 293–313.

Dickerson, R., and Geiss, I. (1969). *The Structure and Action of Proteins*. Menlo Park, CA: Benjamin Cummings.

Diehl, M., Doster, W., Prety, W., and Schober, H. (1997). Water-coupled low-frequency modes of myoglobin and lysozyme observed by inelastic neutron scattering. *Biophys. J.*, **73**, 2726–32.

Dobo, A., and Kaltashov, I. A. (2001). Detection of multiple protein conformational ensembles in solution via deconvolution of charge-state distribution in ESI MS. *Anal. Chem.*, **73**, 4763–4773.

Dolgikh, D. A., Gilmanshin, R. I., Brazhnikov, E. V., *et al.* (1981). Alpha-lactalbumin: Compact state with fluctuating tertiary structure? *FEBS Lett.*, **136**, 311–315.

Dong, J., Wan, Z., Popov, M., Carey, P. R., and Weiss, M. A. (2003). Insulin assembly damps conformational fluctuations: Raman analysis of amide I linewidths in native states and fibrils. *JMB*, **330**, 431–442.

Doster, W., Cusack, S., and Petry, W. (1989). Dynamical transition of myoglobin revealed by inelastic neutron scattering. *Nature*, **337**, 754–756.

Dubin, S. B., Clark, N. A., and Benedek, G. B. (1971). Measurement of the rotational diffusion coefficient of lysozyme by depolarised light scattering: Configuration of lysozyme in solution. *J. Chem. Phys.*, **54**, 5158–5164.

Dunkerk, A. K., Williams, R. W., and Peticolas, W. (1979). Ultraviolet and LASER Raman investigation of the buried tyrosines in fd phage. *J. Biol. Chem.*, **254**, 6446.

Durchschlag, H. (1965). Specific volumes of biological macromolecules and some other molecules of biological interest. In *Proteins, Amino Acids and Peptides as Ions and Dipolar Ions*, eds, E. J. Cohn, and J. T. Edsall. New York: Hafner Publ. Co.

Dutta, R. K., Hammons, K., Willibey, B., and Haney, M. A. (1991). Analysis of protein denaturation by high-performance continuous differential viscometry. *J. Cromatog.*, **536**, 113–121.

Dykxhoorn, D. M., Novina, C. D., and Sharp, P. A. (2003). Killing the messenger: Short RNAs that silence gene expression. *Nat. Rev. Mol. Cell Biol.*, **4**, 457–467.

Eastman, J. E., Taguchi, A. K., Lin, S., *et al.* (2000). Characterization of a *Rhodobacter capsulatus* reaction center mutant that enhances the distinction between spectral forms of the initial electron donor. *Biochemistry*, **39**, 14 787–14 798.

Efremov, R. G., Leitner, A., Aebersold, R., and Raunser, S. (2015). Architecture and conformational switch mechanism of the ryanodine receptor. *Nature*, **517** (7532), 39–43.

Eghiaian, F., Rico, F., Colom, A., Casuso, I., and Scheuring, S. (2014). High-speed atomic force microscopy: Imaging and force spectroscopy. *FEBS Lett.*, **588**, 3631–3638.

Eigen, M., and Rigler, R. (1994). Sorting single molecules: Applications to diagnostic and evolutionary biotechnology. *Proc. Natl. Acad. Sci. USA*, **91**, 5740–5747.

Eimer, W., and Pecora, R. (1991). Rotational and translational diffusion of short rodlike molecules in solution: Oligonucleotides. *J. Chem. Phys.*, **94**, 2324 2329.

Eisenberg, H. (1981). Forward scattering of light, X-rays and neutrons. *Q. Rev. Biophys.*, **14**, 141–172.

Elöve, G. A., Chaffotte, A. F., Roder, H., and Goldberg, M. E. (1992). Early steps in cytochrome c folding probed by time-resolved circular dichroism and fluorescence spectroscopy. *Biochemistry*, **31**, 6876–6883.

Emsley, P., Lohkamp, B., Scott, W., and Cowtan, K. (2010). Features and development of Coot. *Acta Cryst.*, **D66**, 486–501.

Engel, A., Lyubchenko, Y., and Muller, D. (1999). Atomic force microscopy: A powerful tool to observe biomolecules at work. *Trends Cell Biol.*, **9**, 77–80.

Enriquez-Algeciras, M., and Bhattacharya, S. K. (2013). Lipidomic mass spectrometry and its application in neuroscience. *World J. Biol. Chem.*, **4** (4), 102–110.

Ernst, R. R., Bodenhausen, G., and Wokaun, A. (1987). *Principles of Nuclear Magnetic Resonance in One and Two Dimensions*. Oxford: Oxford University Press.

Essevaz-Roulet, B., Bockelman, U., and Heslot, F. (1997). Mechanical separation of the complementary strands of DNA. *Proc. Natl. Acad. Sci. USA*, **94**, 11 935–11 940.

Fabris, D. (2011). MS analysis of nuclcic acids in the post-genomic era. *Anal. Chem.*, **83**, 5810–5816.

Fan, E., Merritt, E. A., Verlinde, C. L., and Hol, W. G. (2000). AB(5) toxins: Structures and inhibitor design. *Curr. Opin. Struct. Biol.*, **10**, 680–686.

Fasman, G. D. (ed.) (1996). *Circular Dichroism and the Conformational Analysis of Biomolecules*. New York: Plenum Press.

Ferrer, M. L., Duchowicz, R., Carrasco, B., Garcia de la Torre, J., and Acuna, A. U. (2001). The conformation of serum albumin in solution: A combined phosphorescence depolarization–hydrodynamic modelling study. *Biophys. J.*, **80**, 2422–2430.

Feynman, R. P., Leighton, R. B., and Sands, M. (1963). *The Feynman Lectures on Physics*. Reading, MA: Addison-Wesley Pub. Co.

Fisher, T. E., Marszalek, P. E., and Fernandez, J. M. (2000). Stretching single molecules into novel conformations using the atomic force microscopy. *Nature Stuctr. Biol.*, **7**, 719–724.

Flaux, J., Bertelsen, E. B., Horwich, A. L., and Wuthrich, K. (2002). NMR analysis of a 900K GroEL–GroES complex. *Nature*, **418**, 207–211.

Franklin, S. E., and Gosling, R. G. (1953). Molecular configuration in sodium thymonucleate. *Nature*, **171**, 740–741.

Franks, F., Gent, M., and Johnson, H. (1963). Solubility of benzene in water. *J. Chem. Soc.*, **8**, 2716–2723.

Franzen, S., and Boxer, S. G. (1997). On the origin of heme absorption band shifts and associated protein structural relaxation in myoglobin following flash photolysis. *J. Biol. Chem.*, **272**, 9655–9660.

Frauenfelder, H., Parak, F., and Young, R. D. (1988). Conformational substates in proteins. *Ann. Rev. Biophys. Chem.*, **17**, 451–479.

Freire, E. (1995). Thermal denaturation methods in the study of protein folding. *Methods Enzymol.*, **259**, 144–168.

Frey, W., Schief, W. R., Jr., Pack, D., *et al.* (1996). Two-dimensional protein crystallization via metal-ion coordination by naturally occurring surface histidines. *Proc. Natl. Acad. Sci. USA*, **93**, 4937–4941.

Frohn, J. T., Knapp, H. F., and Stemmer, A. (2000). True optical resolution beyond the Rayleigh limit achieved by standing wave illumination. *Proc. Natl. Acad. Sci. USA*, **97**, 7232–7236.

Fuller, S. D., and Argos, P. (1987). Is *Sindbis* a simple picornavirus with an envelope? *Embo. J.*, **6**, 1099–1105.

Fuller, W., Forsyth, T., and Mahendrasingam, A. (2004). Water–DNA interactions as studied by X-ray and neutron fibre diffraction. *Phil. Trans. R. Soc. Lond. B Biol. Sci.*, **359**, 1237–1247; discussion 1247–1248.

Gabel, F. (2005). Protein dynamics in solution and powder measured by incoherent elastic neutron scattering: The influence of Q-range and energy resolution. *Eur. Biophys. J.*, **34**, 1–12.

Gabel, F., Bicout, D., Lehnert, U., *et al.* (2002). Protein dynamics studied by neutron scattering. *Q. Rev. Biophys.*, **35**, 327–367.

Gadola, S. D., Zaccai, N. R., Harlos, K., *et al.* (2002). Structure of human CD1b with bound ligands at 2.3 Å : A maze for alkyl chains. *Nat. Immunol.*, **3**, 721–726.

Ganem, B., Li, Y.-T., and Henion, J. D. (1991). Observation of noncovalent enzyme. Substrate and enzyme product complexes by ion spray mass spectrometry. *J. Am. Chem. Soc.*, **113**, 7818–7819.

Garces-Chavez, V., McGloin, D., Melville, H., Sibbett, W., and Dholakia, K. (2002). Simultaneous micromanipulation in multiple planes using a self-reconstruction light beam. *Nature*, **419**, 145–147.

Garcia de la Torre, J. (2001). Hydration from hydrodynamics: General consideration and applications of bead modelling to globular proteins. *Biophys. Chem.*, **93**, 159–170.

Garret, D. S., Seok, Y.-J., Liao, D.-I., *et al.* (1997). Solution structure of the 30 kDa N-terminal domain of enzyme I of the *Escherichia coli* phosphoenolpuruvate:sugar phosphotransferase system by multidimensional NMR. *Biochemistry*, **36**, 2517–2530.

Gelles, J., and Landick, R. (1998). RNA polymerase as a molecular motor, *Cell*, **93**, 13–16.

Giegé, R., Lorber, B., Ebel, J., *et al.* (1982). Formation of a catalytically active complex between tRNAAsp and aspartyl-tRNA synthetase from yeast in high concentrations of ammonium sulphate. *Biochimie*, **64**, 357–362.

Giessing, A. M., and Kirpekar, F. (2012). Mass spectrometry in the biology of RNA and its modifications. *J. Proteomics*, **75**, 3434–3449.

Gimzewski, J. K., and Joachim, C. (1999). Nanoscale science of single molecules using molecular probes. *Science*, **283**, 1683–1688.

Gluehmann, M., Zarivach, R., Bashan, A., *et al.* (2001). Ribosomal crystallography: From poorly diffracting microcrystals to high-resolution structures. *Methods*, **25**, 292–302.

Go, N., Noguti, T., and Nishikawa, T. (1983). Dynamics of a small globular protein in terms of low-frequency vibrational modes. *Proc. Natl. Acad. Sci. USA*, **80**, 3696–3700.

Goldberg, D. E. (1989). *Genetic Algorithms in Search, Optimization, and Machine Learning*. Reading, MA: Addison-Wesley Pub. Co.

Gomez, J., Hilser, V. J., Xie, D., and Freire, E. (1995). The heat capacity of proteins. *Proteins*, **22**, 404–412.

Goodsell, D. S., and Olson, A. J. (1993). Soluble proteins: Size, shape and function. *TIBS*, **18**, 65–68.

Gordon, D. B. (2000). Mass spectrometric technique. In *Principles and Techniques of Practical Biochemistry*, 5th edn., eds. K. Wilson and J. Walker. Ch. 11. Cambridge: Cambridge University Press.

Greis, K. D., Hayes, B. K., Comer, F., *et al.* (1996). Selective detection and site-analysis of O-GlcNAc-modified glycopeptides by beta-elimination and tandem electrospray mass spectrometry. *Anal. Biochem.*, **234**, 38–49.

Grier, D. (2003). A revolution in optical manipulation. *Nature*, **424**, 810–816.

Griffin, R. G. (2010). Clear signals from surfaces. *Nature*, **468**, 381–382.

Griko, Y. V., Freire, E., Privalo, G., *et al.* (1995). The unfolding thermodynamics of c-type lysozymes: A calorimetric study of the heat denaturation of equine lysozyme. *J. Mol. Biol.*, **252**, 447–459.

Gross, S. (2003). Application of optical traps in vivo. *Methods in Enzymol.*, **361**, 162–174.

Grotjahn, L., Frank, R., and Blocker, H. (1982). Ultrafast sequencing of oligodeoxyribonucleotides by FAB-mass spectrometry. *Nucl. Acids Res.*, **10**, 4671–4677.

Güntert, P. (2011). Calculation of structures from NMR restraints. In *Protein NMR Spectroscopy: Practical Techniques and Applications*, eds. Lu-Yun Lian and Gordon Roberts. Chichester: John Wiley and Sons.

Gutsche, I., Holzinger, J., Rauh, N., Baumeister, W., and May, R. P. (2001). ATP-induced structural change of the thermosome is temperature-dependent. *J. Struct. Biol.*, **135**, 139–146.

Hafner, J. H., Cheung, C.-L., Wooley, A. T., and Lieber, C. M. (2001). Structural and functional imaging with carbon nanotube AFM probes. *Progr. Biophys. Mol. Biol.*, **77**, 73–110.

Hamm, P., Lim, M., and Hochstrasser, R. M. (1999). Structure of the amide I band of peptides measured by femtosecond non-linear-infrared spectroscopy. *Proc. Natl. Acad. Sci. USA*, **96**, 6123–6128.

Han, W., Lindsay, S. M., Dlakic, M., and Harrington, R. E. (1997). Kinked DNA. *Nature*, **386**, 563.

Hänsel, R., Luh, L. M., Corbeski, I., Trantirek, L., and Dotsch, V. (2014). In-cell NMR and EPR spectroscopy of biomacromolecules. *Angewandte Chemie*, **53**, 10 300–10 314.

Hansen, J. C., Lebowitz, J., and Demeler, B. (1994). Analytical ultracentrifugation of complex macromolecular systems. *Biochemistry*, **33**, 13 155–13 163.

Hansen, M. R., Mueller, L., and Pardi, A. (1998). Tunable alignment of macromolecules by filamentous phage yields dipolar coupling interaction. *Nat. Struct. Biol.*, **5**, 1065–1074.

Harpaz, Y., Gerstein, M., and Chothia, C. (1994). Volume changes on protein folding. *Structure*, **2**, 641–649.

Harris, R. (1983). *Nuclear Magnetic Resonance Spectroscopy*. London: Pitman.

Haupts, U., Tittor, J., Bamberg, E., and Oesterhelt, D. (1997). General concept for ion translocation by halobacterial retinal proteins: The isomerization/switch/transfer (IST) model. *Biochemistry*, **36**, 2–7.

Haupts, U., Tittor, J., and Oesterhelt, D. (1999). Closing in on bacteriorhodopsin: Progress in understanding the molecule. *Ann. Rev. Biophys. Biomol. Struct.*, **28**, 367–399.

Haustein, E., and Schwille, P. (2003). Ultrasensitive investigations of biological systems by fluorescence correlation spectroscopy. *Methods*, **29**, 153–166.

Hazlett, T. L., Moore, K. J. M., Lowe, P. N., Jameson, D. M., and Eccleston, J. F. (1993). Solution of p21 ras proteins bound with fluorescent nucleotides: A time-resolved fluorescence study. *Biochemistry*, **32**, 13 575–13 583.

Heberle, J., and Gensch, T. (2001). When FT-IR spectroscopy meets X-ray crystallography. *Nat. Struct. Biol.*, **8**, 195–197.

Hellweg, T., Eimer, W., Krahn, E., Schneider, K., and Muller, A. (1997). Hydrodynamic properties of nitrogenase: The MoFe protein from *Azotobacter vinelandii* studied by dynamic light-scattering and hydrodynamic modelling. *Biochim. Biophys. Acta*, **1337**, 311–318.

Hennig, J., Militti, C., Popowicz, G. M., *et al.* (2014) Structural basis for the assembly of the Sxl-Unr translation regulatory complex. *Nature*, doi:10.1038/nature13693.

Hensley, P. (1996). Defining the structure and stability of macromolecular assemblies in solution: The re-emergence of analytical ultracentrifugation as a practical tool. *Structure*, **4**, 367–373.

Hillisch, A., Lorenz, M., and Diekmann, S. (2001). Recent advances in FRET: Distance determination in protein–DNA complexes. *Curr. Opin. Struct. Biol.*, **11**, 201–207.

Hirao, I., and Ellington, A. D. (1995). Re-creating the RNA world. *Curr. Biol.*, **5**, 1017–1022.

Holzwarth, G., Doty, P., and Gratzer, W. B. (1962). Optical activity of polypeptides in far ultraviolet. *J. Am. Chem. Soc.*, **84**, 3194–3196.

Homans, S. W., Edge, C. J., Ferguson, M. A., and Dwek, R. A. (1989). Solution structure of the glycosylphosphatidylinositol membrane anchor glycan of *Trypanosoma bruccei* variant surface glycoprotein. *Biochemistry*, **28**, 2881–2887.

Hore, P. J. (1995). *Nuclear Magnetic Resonance*. Oxford: Oxford University Press.

Horwitz, J., Strickland, E. H., and Billups, C. (1970). Analysis of the vibrational structure in the near-ultraviolet circular dichroism and absorption spectra of tyrosine derivatives and ribonuclease-A at 77 K. *J. Am. Chem. Soc.*, **92**, 2119–2129.

Hu, C.-M., and Zwanzig, R. (1974). Rotational friction coefficients for spheroids with the slipping boundary conditions. *J. Chem. Phys.*, **60**, 4354–4357.

Huang, T.-Y., Liu, J., and McLuckey, S. A. (2010) *J. Am. Soc. Mass Spectrom.*, **21**, 890–898.

Hunt, J. F., McCrea, P. D., Zaccai, G., et al. (1997). Assessment of the aggregation state of integral membrane proteins in reconstituted phospholipid vesicles using small angle neutron scattering. *J. Mol. Biol.*, **273**, 1004–1019.

Hutchens, J. O. (1970). *Handbook of Chemistry and Selected Data for Molecular Biology*, ed. H. A. Sober, Cleveland, OH: Chemical Rubber Co.

Huygens, C. (1690). *Treatise on Light*, New York: Dover (1962 edition of the English translation, first published by Macmillan and Co. in 1912).

Ishijima, A., Kojima, H., Funatsu, T., et al. (1998). Simultaneous observation of individual ATPase and mechanical events by a single myosin molecule during interaction with actin. *Cell*, **92**, 161–171.

Ishima, R., and Torchia, D. (2000). Protein dynamics from NMR. *Nat. Struct. Biol.*, **7**, 740–743.

Jacrot, B. (1976). The study of biological structures by neutron scattering from solution. *Rep. Prog. Phys.*, **39**, 911–953.

Jacrot, B., and Zaccai, G. (1981). Determination of molecular weight by neutron scattering. *Biopolymers*, **20**, 2414–2426.

Jacrot, B., Chauvin, C., and Witz, J. (1977). Comparative neutron small-angle scattering study of small spherical RNA viruses. *Nature*, **266** (5601), 417–421.

Jaenicke, R. (2000). Do ultrastable proteins from hyperthermophiles have high or low conformational rigidity? *Proc. Natl. Acad. Sci. USA*, **97**, 2962–2964.

Jancarik, J., and Kim, S.-H. (1991). Sparse matrix sampling: A screening method for crystallization of proteins. *J. Appl. Cryst.*, **24**, 409–411.

Jeener, J. and Alewaeters, G. (2016). "Pulse pair technique in high resolution NMR": a reprint of the historical 1971 lecture notes on two-dimensional spectroscopy. *Prog. Nucl. Magn. Reson. Spectrosc.*, **94–95**, 75–80.

Jensen, M. R., Communie, G., Ribeiro, Jr E. A., et al. (2011). Intrinsic disorder in measles virus nucleocapsids. *Proc. Natl. Acad. Sci. USA*, **108**, 1839–1844.

Jeruzalmi, D. and Steitz, T. A. (1997). Use of organic cosmotropic solutes to crystallize flexible proteins: Application to T7 RNA polymerase and its complex with the inhibitor T7 lysozyme. *J. Mol. Biol.*, **274**, 748–756.

Jiang, Y., Ruta, V., Chen, J., et al. (2003). The principle of gating charge movement in a voltage-dependent K+ channel. *Nature*, **423**, 42–48.

Johnson, K. H., and Gray, D. M. (1992). Analysis of an RNA pseudoknot structure by CD spectroscopy. *J. Biomol. Struct. Dyn.*, **9**, 733–745.

Johnson, W. C. (1985). *Circular Dichroism and Its Empirical Application to Biopolymers*. Chichester: John Wiley and Sons.

Jolly, D., and Eisenberg, H. (1976). Photon correlation spectroscopy, total intensity light with laser radiation, and hydrodynamic studies of a well fractionated DNA sample. *Biopolymers*, **15**, 61–95.

Jones, J. A., Wilkins, D. K., Smith, L. J., and Dobson, C. M. (1997). Characterization of protein unfolding by NMR diffusion measurements. *J. Biomolecular NMR*, **10**, 199–203.

Kabsch, W. (1988). Evaluation of single-crystal X-ray diffraction data from a position-sensitive detector. *J. Appl. Cryst.*, **21**, 916–924.

Kailemia, M. J., Ruhaak, L. R., Lebrilla, C. B., and Amster, I. J. (2013). Oligosaccharide analysis by mass spectrometry: A review of recent developments. *Anal. Chem.*, **86**, 196–212.

Kam, Z., and Rigler, R. (1982). Cross-correlation laser scattering. *Biophys. J.*, **39**, 7–13.

Kassas, S., Thomson, N. H., Smith, B. L., et al. (1997). *Esherichia coli* RNA polymerase activity observed using atomic force microscopy. *Biochemistry*, **36**, 461–468.

Keiderling, T. A. (2002). Protein and peptide secondary structure and conformational determination with vibrational circular dichroism. *Curr. Opin. Chem. Biol.*, **6**, 682–688.

Kelly, B. T., Graham, S. C., Liska, N., et al. (2014). Clathrin adaptors: AP2 controls clathrin polymerization with a membrane-activated switch. *Science*, **25** (6195), 459–463.

Kincaid, J. R. (1995). Structure and dynamics of transient species using time-resolved resonance Raman spectroscopy. *Methods Enzymol.*, **246**, 460–501.

King, R. W., and Williams, K. R. (1989a). The Fourier transform in chemistry. Part 1. Nuclear magnetic resonance: Introduction. *J. Chem. Education*, **66**, A213–A219.

King, R. W., and Williams, K. R. (1989b). The Fourier transform in chemistry, Part 2. Nuclear magnetic resonance: The single pulse experiment. *J. Chem. Education*, **66**, A243–A248.

Kinosita, K., Jr. (1998). Linear and rotary molecular motors. *Adv. Exp. Med. Biol.*, **453**, 5–13; discussion 13–14.

Kinosita, K., Jr., Yasuda, R., Noji, H., Ishiwata, S., and Yoshida, M. (1998). F1-ATPase: A rotary motor made of a single molecule. *Cell*, **93**, 21–24.

Klar, T. A., Jacobs, S., Dyba, M., Egner, A., and Hell, S. W. (2000). Fluorescence microscopy with diffraction resolution barrier broken by stimulated emission. *Proc. Natl. Acad. Sci. USA*, **97**, 8206–8210.

Klein, G., Satre, M., Zaccai, G., and Vignais, P. V. (1982). Spontaneous aggregation of the mitochondrial natural ATPase inhibitor in salt solutions as demonstrated by gel filtration and neutron scattering: Application to the concomitant purification of the ATPase inhibitor and F1-ATPase. *Biochim. Biophys. Acta*, **681**, 226–232.

Kleywegt, G. J., and Jones, T. A. (2002). Homo crystallographicus: Quo vadis? *Structure (Camb)*, **10**, 465–472.

Koch, M. H., and Stuhrmann, H. B. (1979). Neutron-scattering studies of ribosomes. *Methods Enzymol.*, **59**, 670–706.

Konermann, L., and Douglas, D. J. (1998a). Equilibrum unfolding of proteins monitored by electrospray ionization mass spectrometry: Distinguishing two-state from multi-state transition. *Rapid Commin. Mass Spectr.*, **12**, 435–442.

Konermann, L., and Douglas, D. J. (1998b). Unfolding of proteins monitored by electrospray ionization mass spectrometry: A comparison of positive and negative ion modes. *J. Am. Soc. Mass Spectrom.*, **9**, 1248–1254.

Konijnenberg, A., Butterer, A., and Sobott, F. (2013) Native ion mobility-mass spectrometry and related methods in structural biology. *Biochima. Biophys. Acta.*, **1834**, 1239–1256.

Koppel, D. E. (1979). Fluorescence redistribution after photobleaching. *Biophys. J.*, **28**, 281–291.

Korgel, B. A., Van Zanten, J. H., and Monbouquette, H. G. (1998). Vesicle size distributions measured by flow field-flow fractionation coupled with multiangle light scattering. *Biophys. J.*, **74**, 3264–3272.

Kossiakoff, A. A. (1983). Neutron protein crystallography: Advances in methods and applications. *Ann. Rev. Biophys. Bioeng.*, **12**, 159–182.

Kroes, S. J., Canters, G. W., Giardi, G., van Hoek, A., and Visser, A. J. W. G. (1998). Time-resolved fluorescence study of azurin variants: Conformational heterogeneity and tryptophan mobility. *Biophys. J.*, **75**, 2441–2450.

Krueger, S., Meuse, C. W., Majkrzak, C. F., *et al.* (2001). Investigation of hybrid bilayer membranes with neutron reflectometry: Probing the interactions of melittin. *Langmuir*, **17**, 511–521.

Kumar, A., Ernst, R. R., and Wuthrich, K. (1980). A two-dimensional nuclear Overhauser enhancement (2D NOE) experiment for the elucidation of complete proton–proton cross-relaxation networks in biological macromolecules. *Biochem. Biophys. Res. Commun.*, **95**, 1–6.

Kuntz, I. D., Jr., and Kauzmann, W. (1974). Hydration of proteins and polypeptides. In *Advances in Protein Chemistry*, eds., C. B. Anfinsen, J. T. Edsall, and F. M. Richards, **vol. 28**. New York: Academic Press.

Lakowicz, J. R. (ed.) (1999). *Principles of Fluorescence Spectroscopy*, 2nd edn. New York: Kluwer Academic/Plenum.

Lakowicz, J. R., Gryczynski, I., Piszczek, G., *et al.* (2000). Microsecond dynamics of biological macromolecules. *Methods Enzymol.*, **323**, 473–509.

Langan, P., Nishiyama, Y., and Chanzy, H. (1999). A revised structure and hydrogen bonding system in cellulose II from a neutron fibre diffraction analysis. *J. Am. Chem. Soc.*, **121**, 9940–9946.

Langley, K. H. (1992). Developments in electrophoretic laser light scattering and some biochemical applications. In *Laser Scattering in Biochemistry*, eds., S. E. Harding, D. B. Sattelle, and V. A. Bloomfield. Cambridge: Royal Society of Chemistry.

Langowski, J., Kremer, W., and Kapp, U. (1992). Dynamic light scattering for study of solution conformation and dynamics of superhelical DNA. *Methods Enzymol.*, **211**, 431–448.

Lapinaite, A., Simon, B., Skjaerven, L., *et al.* (2013). The structure of the box C/D enzyme reveals regulation of RNA methylation. *Nature*, **502**, 519–523.

Lauterbur, P. (1973). Image formation by induced local interactions: Examples employing nuclear magnetic resonance. *Nature*, **242**, 190–191.

Lay, J. O. (2001). MALDI-TOF mass spectrometry of bacteria. *Mass Spectr. Rev.*, **20**, 172–194.

Leavitt, S., and Freire, E. (2001). Direct measurement of protein binding energetics by isothermal titration calorimetry. *Curr. Opin. Struct. Biol.*, **11**, 560–566.

Le Bihan, D. (2007). The "wet mind": Water and functional neuroimaging. *Phys. Med. Biol.*, **52**, R57-90.

Lee, R. C., Feinbaum, R. L., and Ambros, V. (1993). The *C. elegans* heterochronic gene lin-4 encodes small RNAs with antisense complementarity to lin-14. *Cell*, **75**, 843–854.

Lee, S. H., Shin, J. Y., Lee, A., and Bustamante, C. (2012). Counting single photoactivatable fluorescent molecules by photoactivated localization microscopy (PALM). *Proc. Natl. Acad. Sci. USA*, **109**, 17 436–17 441.

Lehnert, U. (2002). Hydration dependence of local thermal motions in the Purple Membrane explored by neutron scattering and isotopic labeling. PhD thesis. Université Joseph Fourier, Grenoble.

Lemasters, J. J., Chacon, E., Zahrebelski, G., Reece, J. M., and Nieminen, A.-L. (1993). Laser scanning confocal microscopy of living cells. In *Optical Microscopy: Emerging Methods and Application*, eds. B. Herman and J. J. Lemasters. San Diego, CA: Academic Press.

Leone, M., Cupane, A., Militello, V., and Cordone, L. (1994). Thermal broadening of the Soret band in heme complexes and in heme-proteins: Role of iron dynamics. *Eur. Biophys. J.*, **23**, 349–352.

Leslie, A. G. (1999). Integration of macromolecular diffraction data. *Acta Crystallogr. D Biol. Crystallogr.*, **55**, (Pt 10), 1696–1702.

Levitt, M., Sander, C., and Stern, P. (1985). Protein normal-mode dynamics: Trypsin inhibitor, crambin, ribonuclease and lysozyme. *J. Mol. Biol.*, **181**, 423–447.

Lewis, A., Lieberman, K., Ben-Ami, N., *et al.* (1995). New design and imaging concepts in NSOM. *Ultramicroscopy*, **61**, 215–220.

Lewis, A., Radko, A., Ben-Ami, N., Palanker, D., and Lieberman, K. (1999). Near-field scanning optical microscopy in cell biology. *TIBS*, **9**, 70–73.

Li, H., Cocco, M. J., Steitz, T., and Engelman, D. M. (2001). Conversion of phospholamban into a soluble pentameric helical bundle. *Biochemistry*, **40**, 6636–6645.

Li, M., Yang, L., Bai, Y., and Liu, H. (2014). Analytical methods in lipidomics and their applications. *Anal. Chem*, **86**, 161–175.

Li, Y., Hunter, R. L., and Mciver, R. T., Jr. (1994). High-resolution mass spectrometer for protein chemistry. *Nature*, **370**, 393–395.

Liphardt, J., Onoa, B., Smith, S. B., Tinoco, I. Jr., and Bustamante, C. (2001). Reversible unfolding of single RNA molecules by mechanical forces. *Science*, **292**, 733–737.

Liu, Y., Huttenhain, R., Collins, B., and Aebersold, R. (2013). Mass spectrometric protein maps for biomarker discovery and clinical research. *Expert Rev. Mol. Diag.*, **13**, 811–825.

Loquet, A., Sgourakis, N. G., Gupta, R., et al. (2012). Atomic model of the type III secretion system needle. *Nature*, **486**, 276–279.

Luthy, R., Bowie, J. U., and Eisenberg, D. (1992). Assessment of protein models with three-dimensional profiles. *Nature*, **356**, 83–85.

Luz, Z., and Meiboom, S. (1963). Nuclear magnetic resonance study of the protolysis of trimethylammonium ion in aqueous solution: Order of the reaction with respect to solvent. *J. Chem. Phys.*, **39**, 366–370.

Luzzati, V., and Tardieu, A. (1980). Recent developments in solution X-ray scattering. *Annu. Rev. Biophys. Bioeng.*, **9**, 1–29.

Ma, J., Flynn, T. C., Cui, Q., et al. (2002). A dynamic analysis of the rotation mechanism for conformational change in F(1)-ATPase. *Structure (Camb)*, **10**, 921–931.

MacDonald, D., and Lu, P. (2002). Residual dipolar couplings in nucleic acid structure determination. *Curr. Opinion Str. Biol.*, **12**, 337–343.

MacGregor, I. K., Anderson, A. L., and Laue, T. M. (2004). Fluorescence detection for the XLI analytical ultracentrifuge. *Biophys. Chem.*, **108**, 165–185.

Machtle, W. (1999). High-resolution, submicron particle size distribution analysis using gravitational-sweep sedimentation. *Biophys. J.*, **76**, 1080–1091.

Madern, D., Ebel, C., and Zaccai, G. (2000). Halophilic adaptation of enzymes. *Extremophiles*, **4**, 91–98.

Makarov, A., Denisov, A., Kholomeev, A., et al. (2006). Performance evaluation of a hybrid linear ion trap/orbitrap mass spectrometer. *Anal. Chem.*, **78**, 2113–2120

Makhatadze, G. I., and Privalov, P. L. (1990). Heat capacity of proteins: I. Partial molar heat capacity of individual amino acid residues in aqueous solution: Hydration effect. *J. Mol. Biol.*, **213**, 375–384.

Makhatadze, G. I., and Privalov, P. L. (1996). On the entropy of protein folding. *Protein Sci.*, **5**, 507–510.

Manavalan, P., and Johnson, W. C., Jr. (1987). Variable selection method improves the prediction of protein secondary structure from circular dichroism spectra. *Anal. Biochem.*, **167**, 76–85.

Mancini, E. J., Clarke, M., Gowen, B. E., Rutten, T., and Fuller, S. D. (2000). Cryo-electron microscopy reveals the functional organization of an enveloped virus, Semliki forest virus. *Mol. Cell.*, **5**, 255–266.

Mandelkow, E., and Mandelkow, E. M. (2002). Kinesin motor and disease. *TICB*, **12**, 585–591.

Manor, D., Weng, G., Deng, H., Cosloy, S., and Chen, C. X. (1991). An isotope edited classical Raman difference spectroscopic study of the interactions of guanine nucleotides with elongation factor Tu and H-*ras* p21. *Biochemistry*, **30**, 10 914–10 920.

Marassi, F. M., Ma, C., Gratkowski, H., et al. (1999). Correlation of the structural and functional domains in the membrane protein Vpu from HIV-1. *Proc. Natl. Acad. Sci. USA*, **96**, 14 336–14 341.

Mattei, B., Borch, J., and Roepstorff, P. (2004). Biomolecular interaction analysis and MS. *Anal. Chem.*, **76**, 18A–26A.

McCoy, A. J., Grosse-Kunstleve, R. W., Adams, P. D., et al. (2007). Phase crystallographic software. *J. Appl. Crystallogy*, **40**, 658–674.

Medalia, O., Weber, I., Frangakis, A., et al. (2002). Macromolecular architecture in eukaryotic cells visualized by cryoelectron tomography. *Science*, **298**, 1209–1213.

Medek, A., Olejniczak, E. T., Meadows, R. P., and Fesik, S. W. (2000). An approach for high-throughput structure determination of proteins by NMR spectroscopy. *J. Biomol. NMR*, **18**, 229–238.

Medina, P. P., Nolde, M., and Slack, F. J. (2010). OncomiR addiction in an in vivo model of microRNA-21-induced pre-B-cell lymphoma. *Nature*, **467**, 86–90.

Mehta, A. D., Rief, M., Spudich, J. A., Smith, D. A., and Simmons, R. M. (1999). Single-molecule biomechanics with optical methods. *Science*, **283**, 1689–1695.

Michielsen, S., and Pecora, R. (1981). Solution dimensions of the gramicidin dimer by dynamic light scattering. *Biochemistry*, **20**, 6994–6997.

Miele, A. E., Federici, L., Sciara, G., et al. (2003). Analysis of the effect of microgravity on protein crystal quality: The case of a myoglobin triple mutant. *Acta Crystallogr. D Biol. Crystallogr.*, **59**, (Pt 6), 982–988.

Millero, F. (1969). The partial molal volumes of ions in seawater. *Limnol. Oceanogr.*, **14**, 376–385.

Mittelbach, P., and Porod, G. (1962). Zur Röntgenkleinwinkelstreuung die Berechning der Steukurven von Dreiachsigen Ellipsoiden. *Acta Physica Austriaca*, **15**, 122–147.

Mollova, E. T., Hansen, M. R., and Pardi, A. (2000). Global structure of RNA determined with residual dipolar coupling, *JACS*, **122**, 11 561–11 562.

Montelione, G. T., Zheng, D., Huang, Y. J., Gundalus, K. S., and Szyperski, T. (2000). Protein NMR spectroscopy in structural genomics. *Nat. Struct. Biol.*, **7**, 982–985.

Moore, J. L., Caprioli, R. M., and Skaar, E. P. (2014). Advanced mass spectrometry technologies for the study of microbial pathogenesis. *Curr. Opin. Microbiol.*, **19**, 45–51.

Moscowitz, A. (1962). Theoretical aspects of optical activity. Part one: Small molecules. *Adv. Chem. Phys.*, **4**, 67–112.

Mou, J., Csajkovsky, D. M., Zhang, Y., and Shao, Z. (1995). High-resolution atomic force microscopy of DNA: The pitch of the double helix. *FEBS Lett.*, **371**, 279–282.

Muders, V., Kerruth, S., Lorenz-Fonfria, V. A., et al. (2014). Resonance Raman and FTIR spectroscopic characterization of the closed and open states of channelrhodopsin-1. *FEBS Lett.*, **588**, 2301–2306.

Muller, D. J., Janovjak, H., Lehto, T., Kuerschner, L., and Anderson, K. (2002). Observing structure, function and assembly of single proteins by AFM. *Progress Biophys. Mol. Biol.*, **79**, 1–43.

Myszka, D. G., Sweet, R. W., Hensley, P., *et al.* (2000). Energetics of the HIV gp120-CD4 binding reaction. *Proc. Natl. Acad. Sci. USA*, **97**, 9026–9031.

Nagorni, M., and Hell, S. W. (1998). 4Pi-Confocal microscopy provides three-dimensional images of the microtubule network with 100- to 150-nm resolution. *J. Struct. Biol.*, **123**, 236–247.

Navarro, J., Landau, E. M., and Fahmy, K. (2002). Receptor-dependent G-protein activation in lipidic cubic phase. *Biopolymers*, **67**, 167–177.

Navaza, J., and Saludjian, P. (1997). AMoRe: An automated molecular replacement program package. *Methods Enzymol.*, **276**, 581–594.

Neutze, R., Wouts, R., van der Spoel, D., Weckert, E., and Hajdu, J. (2000). Potential for biomolecular imaging with femtosecond X-ray pulses. *Nature*, **406**, 752–757.

Newcomb, W. W., Juhas, R. M., Thomsen, D. R., *et al.* (2001). The UL6 gene product forms the portal for entry of DNA into the herpes simplex virus capsid. *J. Virol.*, **75**, 10 923–10 932.

Nie, S., and Zare, R. N. (1997). Optical detection of single molecule. *Annu. Rev. Biophys. Biomol. Struct.*, **26**, 567–596.

Nielsen, M. L., Bennet, K. L., Larsen, B., Monniate, M., and Mann, M. (2002). Peptide end sequencing by orthogonal MALDI tandem mass spectroscopy. *J. Proteome Res.*, **1**, 63–71.

Nishiyama, Y., Langan, P., and Chanzy, H. (2002). Crystal structure and hydrogen-bonding system in cellulose Ibeta from synchrotron X-ray and neutron fiber diffraction. *J. Am. Chem. Soc.*, **124**, 9074–9082.

Noji, H., Yasuda, R., Yoshida, M., *et al.* (1997). Direct observation of the rotation of F1-ATPase. *Nature*, **386**, 299–302.

Oberg, K. A., Ruysschaert, J. M., and Goormaghtigh, E. (2004). The optimization of protein secondary structure determination with infrared and circular dichroism spectra. *Eur. J. Biochem.*, **271**, 2937–2948.

Opella, S. J., and Stewart, P. L. (1989). Solid-state nuclear magnetic resonance structural studies of proteins. *Methods Enzymol.*, **176**, 242–275.

Orgel, L. E. (1968). Evolution of the genetic apparatus. *J. Mol. Biol.*, **38** (3), 381–393.

Ormo, M., Cubbit, A. B., Kallio, K., Gross, L. A., and Tsien, R. Y. (1996). Crystal structure of the *Aequorea victoria* green fluorescent protein. *Science*, **273**, 1392–1395.

Oster, G., and Wang, H. (2003). Rotary protein motor. *Trends Cell Biol.*, **13**, 114–121.

Otting, G., and Wuthrich, K. (1990) Heteronuclear filters in two-dimensional [1H, 1H]-NMR spectroscopy: Combined use with isotope labelling for studies of macromolecular conformation and intermolecular interactions. *Q. Rev. Biophys.*, **23**, 39–56.

Otwinowski, Z., and Minor, W. (1997). Processing of X-ray diffraction data in oscillation mode. *Methods Enzymol.*, **276**, 307–326.

Pecora, R. (1968). Spectral distribution of light scattered by monodisperse rigid rods. *J. Chem. Phys.*, **48**, 4126–4128.

Perrakis, A., Sixma, T. K., Wilson, K. S., and Lamzin, V. S. (1997). wARP: improvement and extension of crystallographic phases by weighted averaging of multiple-refined dummy atomic models. *Acta Cryst. D.*, **53**, 448–455.

Peticolas, W. L. (1995). Raman spectroscopy of DNA and proteins. *Methods Enzymol.*, **246**, 389–416.

Peticolas, W. L., and Evertsz, E. (1992). Conformation of DNA in vitro and in vivo from laser Raman scattering. *Methods Enzymol.*, **211**, 335–352.

Pfuhl, M., Chen, H. A., Kristinsen, S., and Driscool, P. C. (1999). NMR exchange broadening arising from specific low affinity protein self-association: Analysis of nitrogen-15 relaxation for rat CD2 domain 1. *J. Biomolecular NMR*, **14**, 307–320.

Picco, L. M., Bozec, L., Ulcinas, A., *et al.* (2007). Breaking the speed limit with atomic force microscopy. *Nanotechnology*, **18**, 44 030–44 033.

Piston, D. W. (1999). Imaging living cells and tissues by two-photon excitation microscopy. *Trends Cell Biol.*, **9**, 66–69.

Pitner, T. P., Walter, R., and Glickson, J. D. (1976). Mechanism of the intramolecular 1H nuclear Overhauser effect in peptides and depsipeptides. *Biochem. Biophys. Res. Commun.*, **70**, 746–751.

Pohl, F. M., and Jovin, T. M. (1972). Salt-induced co-operative conformational change of a synthetic DNA: Equilibrium and kinetic studies with poly(dG-dC). *J. Mol. Biol.*, **67**, 375–396.

Popot, J. L., Berry, E. A., Charvolin, D., *et al.* (2003). Amphipols: Polymeric surfactants for membrane biology research. *Cell Mol. Life Sci.*, **60**, 1559–1574.

Prestegard, J. H. (1998). New techniques in structural NMR: Anisotropic interactions. *Nature Struct. Biol.*, **5**, 517–522.

Prestegard, J. H., Valafar, H., Glushka, J., and Tian, F. (2001). Nuclear magnetic resonance in the era of structural genomics. *Biochemistry*, **40**, 8677–8685.

Price, N. C., Dwek, R. A., Ratcliffe, G., and Wormald, M. (2001). *Principles and Problems in Physical Chemistry for Biochemists*. Oxford: Oxford University Press.

Price, P. B. (2000). A habitat for psychrophiles in deep Antarctic ice. *Proc. Natl. Acad. Sci. USA*, **97**, 1247–1251.

Privalov, G. P., and Privalov, P. L. (2000). Problems and prospects in microcalorimetry of biological macromolecules. *Methods Enzymol.*, **323**, 31–62.

Privalov, G. P., Kavina, V., Freire, E., and Privalov, P. L. (1995). Precise scanning calorimeter for studying thermal properties of biological macromolecules in dilute solution. *Anal. Biochem.*, **232**, 79–85.

Privalov, P. L. (1980). Scanning microcalorimeters for studying macromolecules. *Pure & Appl. Chem.*, **52**, 479–497.

Privalov, P. L. (1982). Stability of proteins: Proteins which do not present a single cooperative system. *Adv. Protein Chem.*, **35**, 1–104.

Privalov, P. L., and Khechinashvili, N. N. (1974). A thermodynamic approach to the problem of stabilization of globular protein structure: A calorimetric study. *J. Mol. Biol.*, **86**, 665–684.

Privalov, P. L., and Makhatadze, G. I. (1990). Heat capacity of proteins. II. Partial molar heat capacity of the unfolded polypeptide chain of proteins: Protein unfolding effects. *J. Mol. Biol.*, **213**, 385–391.

Privalov, P. L., and Makhatadze, G. I. (1992). Contribution of hydration and non-covalent interactions to the heat capacity effect on protein unfolding. *J. Mol. Biol.*, **224**, 715–723.

Purcell, E. M. (1977). Life at low Reynolds number. *Am. J. Phys.*, **45**, 311.

Ramagopal, U. A., Dauter, M., and Dauter, Z. (2003). Phasing on anomalous signal of sulfurs: What is the limit? *Acta Crystallogr. D Biol. Crystallogr.*, **59**, 1020–1027.

Ramakrishnan, V., and Moore, P. B. (2001). Atomic structures at last: The ribosome in 2000. *Curr. Opinion in Str. Biol.*, **11**, 144–154.

Rayment, I. (1996). Kinesin and myosin: Molecular motor with similar engine. *Structure*, **4**, 501–504.

Reinhart, B. J., Slack, F. J., Basson, M., *et al.* (2000). The 21-nucleotide let-7 RNA regulates developmental timing in *Caenorhabditis elegans*. *Nature*, **403**, 901–906.

Reviakine, I. I., Bergsma-Schutter, W., and Brisson, A. (1998). Growth of protein 2-D crystals on supported planar lipid bilayers imaged in situ by AFM. *J Struct. Biol.*, **121**, 356–361.

Rhee, K. H., Scarborough, G. A., and Henderson, R. (2002). Domain movements of plasma membrane H(+)-ATPase: 3D structures of two states by electron cryo-microscopy. *Embo. J.*, **21**, 3582–3589.

Ribeiro Ede, A., Jr., Pinotsis, N., Ghisleni, A., *et al.* (2014) The structure and regulation of human muscle alpha-actinin. *Cell*, **159**, 1447–1460.

Richard, S. B., Madern, D., Garcin, E., and Zaccai, G. (2000). Halophilic adaptation: Novel solvent protein interactions observed in the 2.9 and 2.6 A resolution structures of the wild type and a mutant of malate dehydrogenase from *Haloarcula marismortui*. *Biochemistry*, **39**, 992–1000.

Rief, M., Oesterhelt, T. F., Heymann, B., and Gaub, H. E. (1997). Single molecule force spectroscopy on polysaccharides by atomic force microscopy. *Science*, **275**, 1295–1297.

Rief, M., Fernandez, J. M., and Gaub, H. E. (1998). Elastically coupled two-level system as a model for biopolymer extensibility, *Phys. Rev. Letters*, **81**, 4764–4767.

Rief, M., Pascual, J. Saraste, M., and Gaub, H. E. (1999). Single molecule force spectroscopy of spectrin repeats: Low unfolding forces in helix bundles. *J. Mol. Biol.*, **286**, 553–561.

Riek, R., Hornemann, S., Wider, G., Glockshuber, R., and Wuthrich, K. (1997). NMR characterization of the full-length recombinant murine prion protein, mPrP(23–231). *FEBS Lett.*, **413**, 282–288.

Roberts, M. M., Coker, A. R., Fossati, G., *et al.* (1999). Crystallization, X-ray diffraction and preliminary structure analysis of *Mycobacterium tuberculosis* chaperonin 10. *Acta Crystallogr. D Biol. Crystallogr.*, **55** (Pt 4), 910–914.

Rondelli, V., Fragneto, G., Motta, S., *et al.* (2012). Ganglioside GM1 forces the redistribution of cholesterol in a biomimetic membrane. *Biochimica et Biophysica Acta*, **1818**, 2860–2867.

Rosenheck, K., and Doty, P. (1961). The far ultraviolet absorption spectra of polypeptide and protein solutions and their dependence on conformation. *Proc. Natl. Acad. Sci. USA*, **47**, 1775–1785.

Rosenthal, P. B., and Henderson, R. (2003). Optimal determination of particle orientation, absolute hand, and contrast loss in single-particle electron cryomicroscopy. *J. Mol. Biol.*, **333**, 721–745.

Rostom, A., and Robinson, C. V. (1999). Disassembly of intact multiprotein complexes in the gas phase. *Curr. Opin. Struct. Biol.*, **9**, 135–141.

Rostom, A., Fucini, P., Benjamin, D., *et al.* (2000). Detection and selective dissociation of intact ribosomes in a mass spectrometer. *Proc. Natl. Acad. Sci. USA*, **97**, 5185–5190.

Rould, M. A. (1997). Screening for heavy-atom derivatives and obtaining accurate isomorphous differences. *Methods Enzymol.*, **276**, 461–472.

Rubtsov, I. V., Wang, J., and Hochstrasser, R. M. (2003). Dual-frequency 2D-IR spectroscopy heterodyned photon echo of the peptide bond. *Proc. Natl. Acad. Sci. USA*, **100**, 5601–5606.

Salmeen, I., Rimai, L., Liebes, L., Rich, M. A., and McCormick, J. J. (1975). Hydrodynamic diameters of RNA tumor viruses: Studies by laser beat frequency light scattering spectroscopy of avian myeblastosos and Rauscher murine leukemia viruses. *Biochemistry*, **14**, 134–141.

Salmena, L., Poliseno, L., Tay, Y., Kats, L., and Pandolfi, P. P. (2011). A ceRNA hypothesis: The Rosetta Stone of a hidden RNA language? *Cell*, **146**, 353–358.

Salza, R., Peysselon, F., Chautard, E. *et al.* (2014). Extended interaction network of procollagen C-proteinase enhancer-1 in the extracellular matrix. *Biochemistry*, **457**, 137–149.

Sanders, C. R., Hare, B. J., Howard, K. P., and Prestegard, J. H. (1994). Magnetically oriented phospholipid micelles as a tool for the study of membrane-associated molecules. *Prog. NMR Spectros.*, **26**, 5.

Schachman, H. K. (1989). Analytical ultracentrifugation reborn. *Nature*, **341**, 259–260.

Scheres, S., and Chen, S. (2012). Prevention of overfitting in cryo-EM structure determination. *Nat. Meth.*, **9** (9), 853–854.

Schmidt, B., and Reisner, D. (1992). A fluorescence detection system for analytical ultracentrifuge and its application to proteins, nucleic acids, viroid and viruses. In *Analytical Ultracentrifugation in Biochemistry and Polymer Science*, eds., S. E. Harding, A. J. Rowe, and J. C. Horton. Cambridge: Royal Society of Chemistry.

Scholtan, W., and Lange, H. (1972). Bestimmung der Teilchegrobenverteilung von Latices mit der Ultracentrifuge. *Kolloid-Z., u. Z. Polimere.*, **250**, 782–796.

Schouten, S., Hopmans, E. C., Pancost, R. D., and Damste, J. S. (2000). Widespread occurrence of structurally diverse tetraether membrane lipids: Evidence for the ubiquitous presence of low-temperature relatives of hyperthermophiles. *Proc. Natl Acad. Sci. USA*, **97**, 14 421–14 426.

Schrader, M., Bahlmann, K., Giese, G., and Hell, S. W. (1998). 4Pi-confocal imaging in fixed biological specimens. *Biophys. J.*, **75**, 1659–1668.

Schuck, P. (2004). A model for sedimentation in inhomogeneous media: I. Dynamic density gradients from sedimenting co-solutes. *Biophys. Chem.*, **108**, 187–200.

Schultz, D. A. (2003). Plasmon resonant particles for biological detection. *Curr. Opin. Biotechnol.*, **14**, 13–22.

Schwabe, J. W. R., Chapman, L., Finch, J. T., Rhodes, D., and Neuhaus, D. (1993). DNA recognition by the oestrogen receptor: From solution to the crystal. *Structure*, **1**, 187–204.

Schwille, P., Meyer-Almes, F. J., and Rigler, R. (1997). Dual-colour fluorescence cross-correlation spectroscopy for multicomponent diffusional analysis in solution. *Biophys. J.*, **72**, 1878–1886.

Seelig, J. (2004). Thermodynamics of lipid–peptide interactions. *Biochim. Biophys. Acta.*, **1666**, 40–50.

Seils, J., and Dorfmuller, T. (1991). Internal dynamics of linear and superhelical DNA as studied by photon correlation spectroscopy. *Biopolymers*, **31**, 813–825.

Serdyuk, I. N., Grenader, A. K., and Zaccaï, G. (1979). Study of the internal structure of *Escherichia coli* ribosomes by neutron and x-ray scattering. *J. Mol. Biol.*, **135**, 691–707.

Serdyuk, I. N., Pavlov, M., Rublevskaya, I. N., Zaccaï, G., and Leberman, R. (1994). The triple isotopic substitution method in small angle neutron scattering: Application to the study of the ternary complex EF-Tu.GTP.aminoacyl-tRNA. *Biophys. Chem.*, **53**, 123–130.

Serdyuk, I., Ulitin, A., Kolesnikov, I., *et al.* (1999). Structure of a beheaded 30S ribosomal subunit from *Thermus thermophilus*. *J. Mol. Biol.*, **292**, 633–639.

Sheetz, M. P., Turne, Y. S., Qian, H., and Elson, E. L. (1989). Nanometre level analysis demonstrates that lipid flow does not drive membrane glycoprotein movements. *Nature*, **340**, 284–285.

Shiku, H., and Dunn, R. C. (1999). Near field scanning optical microscopy. *Anal. Chem.*, **71**, 23A–29A.

Shingyoji, C., Higuchi, H., Yoshimura, M., Katayama, E., and Yanagida, T. (1998). Dynein arms are oscillating force generators. *Nature*, **393**, 711–714.

Sigler, P. B., Xu, Z., Rye, H. S., *et al.* (1998). Structure and function in GroEL-mediated protein folding. *Ann. Rev. Biochem.*, **67**, 581–608.

Simpson, A. A., Tao, Y., Leiman, P., *et al.* (2000). Structure of the bacteriophage ϕ29 DNA packaging motor. *Nature*, **408**, 745–750.

Skoog, D. A., Holler, F. J., and Nieman, T. A. (1995). *Principle of Instrumental Analysis*. Philadelphia, PA: Saunders College Publishing.

Sliz, P., Harrison, S. C., and Rosenbaum, G. (2003). How does radiation damage in protein crystals depend on X-ray dose? *Structure (Camb)*, **11**, 13–19.

Smith, D. E., Tans, S. J., Smith, S., *et al.* (2001). The bacteriophage ϕ 29 portal motor can package DNA against a large internal force. *Nature*, **413**, 748–752.

Smith, J. C., and Roux, B. (2013). Eppur si muove! The 2013 Nobel Prize in chemistry. *Structure*, **21**, 2102–2105.

Smith, L. J., Redfield, C., Smith, R. A., *et al.* (1994). Comparison of four independently determined structure of human recombinant interleikin-4. *Struct. Biol.*, **1**, 301–310.

Smith, M. H. (1970). Molecular weight of proteins and some other materials including sedimentation diffusion and frictional coefficients and partial specific volumes. In *Handbook of Biochemistry.:Selected Data for Molecular Biology*, ed., H. A. Sober, pp. C3–C47. Cleveland, OH: Chemical Rubber Co.

Smith, N., and Webb, A. (2011). *Introduction to Medical Imaging: Physics, Engineering and Clinical Applications*, Cambridge: Cambridge University Press.

Smith, R. D., Bruce, J. E., Wu, Q., *et al.* (1996). The role of Fourier transform ion cyclotron resonance mass spectrometry in biological research: New development and applications. In *Mass Spectrometry in the Biological Science*, eds., A. L. Burlingame and S. A. Carr, pp. 25–68. Totowa, NJ: Humana Press.

Smith, S. B., Cui, Y., and Bustamante, C. (1996). Overstretching B-DNA: The elastic response of individual double-stranded and single-stranded DNA molecules. *Science*, **271**, 795–799.

Smith, S. O., and Peersen, O. B. (1992). Solid-state NMR approaches for studying membrane protein structure. *Ann. Rev. Biophys. Biomol. Str.*, **21**, 25–47.

Snatzke, G. (1994). Circular dichroism: An introduction. In *Circular Dichroism: Principles and Applications.*, eds., K. Nakanishi, N. Berova, and R. W. Woody. New York, VCH Publishers.

Sober, H. A. (ed.) (1970). *Handbook of Biochemistry*, 2nd edn. Cleveland: CRC Press.

Sodano, P., Chary, K. V., Björnberg, O., *et al.* (1991). Nuclear magnetic resonance studies of recombinant *Escherichia coli* glutaredoxin: Sequence-specific assignments and secondary structure determination of the oxidised form. *Eur. J. Biochem*, **200**, 369–377.

Sorlie, S. S., and Pecora, R. (1988). A dynamic light scattering study of a 2311 base pair DNA restriction fragment. *Macromolecules*, **21**, 1437–1441.

Sosa, H., Dias, D. P., Hoenger, A., *et al.* (1997). A model for the microtubule-Ncd motor protein complex obtained by cryo-electron microscopy and image analysis. *Cell*, **90**, 217–224.

Spiro, T. G., and Chernuszevich, R. S. (1995). Resonance Raman spectroscopy of metalloprotein. *Methods Enzymol.*, **246**, 416–459.

Spolar, R. S., and Record, M. T., Jr. (1994). Coupling of local folding to site-specific binding of proteins to DNA. *Science*, **263**, 777–784.

Sportsman, J. R. (2003). Fluorescence anisotropy in pharmacologic screening. *Methods Enzymol.*, **361**, 505–529.

Spronk, C. A. E. M., Linge, J. P., Hilbers, C. W., and Vuister, G. W. (2002). Improving the quality of protein structures derived by NMR spectroscopy. *J. Biomolecular NMR*, **22**, 281–289.

Spudlish, J. A. (1994). How molecular motors works. *Nature*, **372**, 515–518.

Srajer, V., Ren, Z., Teng, T., *et al.* (2001). Protein conformational relaxation and ligand migration in myoglobin: A nanosecond to millisecond molecular movie from time-resolved Laue X-ray diffraction. *Biochemistry*, **40**, 13 802–13 815.

Sreerama, N., and Woody, R. W. (2003). Structural composition of betaI- and betaII-proteins. *Protein Sci.*, **12**, 384–388.

Stafford, W. F., and Braswell, E. H. (2004). Sedimentation velocity, multi-speed method for analyzing polydisperse solutions. *Biophys. Chem.*, **108**, 273–279.

Steely, H. T., Jr., Gray, D. M., and Lang, D. (1986a). Study of the circular dichroism of bacteriophage φ6 and φ6 nucleocapsid. *Biopolymers*, **25**, 171–188.

Steely, H. T., Jr., Gray, D. M., Lang, D., and Maestre, M. F. (1986b). Circular dichroism of double-stranded RNA in the presence of salt and ethanol. *Biopolymers*, **25**, 91–117.

Stejskal, E. O., and Tanner, J. E. (1965). Spin diffusion measurements: Spin echoes in the presence of a time dependent field gradient. *J. Chem. Phys.*, **42**, 288–292.

Stoeckli, M., Chaurand, P., Hallahan, D. E., and Caprioli, R. (2001). Imaging mass spectrometry: A new technology for the analysis of protein expression in mammalian tissues. *Nat. Medicine*, **7**, 493–496.

Stone, M. J., Fairbrother, W. J., Palmer, A., *et al.* (1992). Backbone dynamics of the *Bacillus subtilis* glucose permease IIA domain determined from 15N NMR relaxation measurements. *Biochemistry*, **31**, 4394–4406.

Strathmann, F. G., and Hoofnagle, A. N. (2011). Current and future applications of mass spectrometry to the clinical laboratory. *Am. J. Clin .Pathol.*, **136**, 609–616.

Strick, T., Allemand, J.-F., Bensimon, D., Bensimon, A., and Croquettet, V. (1996). The elasticity of a single supercoiled DNA molecule. *Science*, **271**, 1835–1837.

Strick, T., Allemand, J.-F., Croquete, V., and Bensimon, D. (2000). Twisting and stretching single DNA molecules. *Progr. Biophys. Mol. Biol.*, **74**, 115–140.

Stroebel, D., Choquet, Y., Popot, J.L., and Picot, D. (2003). An atypical haem in the cytochrome b(6)f complex. *Nature*, **426**, 413–418.

Stryer, L. (1968). Fluorescence spectroscopy of proteins. *Science*, **162**, 526–533.

Stryer, L., and Hougland, R. P. (1967). Energy transfer: A spectroscopic ruler. *Proc. Natl. Acad. Sci. USA*, **58**, 719–726.

Subramaniam, S., and Henderson, R. (1999). Electron crystallography of bacteriorhodopsin with millisecond time resolution. *J. Struct. Biol.*, **128**, 19–25.

Subramaniam, S., and Henderson, R. (2000). Molecular mechanism of vectorial proton translocation by bacteriorhodopsin. *Nature*, **10**, 653–657.

Subramaniam, S., Hirai, T., and Henderson, R. (2002). From structure to mechanism: Electron crystallographic studies of bacteriorhodopsin. *Phil. Trans. A Math. Phys. Eng. Sci.*, **360**, 859–874.

Surewicz, W. K., Mantsch, H. H., and Chapman, D. (1993). Determination of protein secondary structure by Fourier transform infrared spectroscopy: A critical assessment. *Biochemistry*, **32**, 389–394.

Susi, H., and Byler, D. M. (1986). Resolution-enhanced Fourier transform infrared spectroscopy of enzymes. *Methods Enzymol.*, **130**, 290–311.

Suzuki, Y., Yasunaga, T., Ohkura, R., Wakabayashi, T., and Sutoh, K. (1998). Swing of the lever arm of a myosin motor at the isomerization and phosphate-release steps. *Nature*, **396**, 380–383.

Svergun, D. I. (1999). Restoring low resolution structure of biological macromolecules from solution scattering using simulated annealing. *Biophys. J.*, **76**, 2879–2886.

Svergun, D. I. (2000). Advanced solution scattering data analysis methods and their applications. *J. Appl. Crystallog.*, **33**, 530–534.

Svergun, D. I., Barberato, C., and Koch, M. (1995). CRYSOL: A program to evalate X-ray solution scattering of biological macromolecules from atomic coordinates. *J. Appl. Crystallogr.*, **28**, 768–773.

Svergun, D. I., Volkov, V. V., Kozin, M. B., and Stuhrmann, H. B. (1996). New developments in direct shape determination from small-angle scattering 2: Uniqueness. *Acta. Crystallog. A*, **52**, 419–426.

Svergun, D. I., Richard, S., Koch, M., *et al.* (1998). Protein hydration in solution: Experimental observation by x-ray and neutron scattering. *Proc. Natl. Acad. Sci. USA*, **95**, 2267–2272.

Svergun, D. I., Malfois, M., Koch, M. H., Wigneshweraraj, S. R., and Buck, M. (2000). Low resolution structure of the sigma54 transcription factor revealed by X-ray solution scattering. *J. Biol. Chem.*, **275**, 4210–4214.

Szubiakowski, J. P. (2014). Identifiability analysis of rotational diffusion tensor and electronic transition moments measured in time-resolved fluorescence depolarization experiment. *J. Chem. Phys.*, **140**, http://dx.doi.org/10.1063/1.4881257.

Taillandier, E., and Liquier, J. (1992). Infrared spectroscopy of DNA. *Methods Enzymol.*, **211**, 307–335.

Takahashi, H., Nakanishi, T., Kami, K., Arata, Y., and Shimada, I. (2000). A novel NMR method for determining the interfaces of large protein–protein complexes. *Natur. Struct. Biol.*, **7**, 220–223.

Tamm, L. K. (1993). Total internal reflectance fluorescence microscopy in optical microscopy. In *Emerging Methods and Applications*, eds., B. Herman and J. Lemasters. San Diego, CA: Academic Press.

Tanford, C., Kawahara, K., and Lapanje, S. (1967). Proteins as random coils: I. Intrinsic viscosities and sedimentation coefficients in concentrated guanidine hydrochloride. *J. Am. Chem. Soc.*, **89**, 729–736.

Tarek, M., and Tobias, D. J. (2002). Single-particle and collective dynamics of protein hydration water: A molecular dynamics study. *Phys. Rev. Lett.*, **89**, 275501.

Tcien, R. Y., and Miyawaki, A. (1998). Seeing the machinary of live cells. *Science*, **280**, 1954–1955.

Tehei, M., Madern, D., Pfister, C., and Zaccai, G. (2001). Fast dynamics of halophilic malate dehydrogenase and BSA measured by neutron scattering under various solvent conditions influencing protein stability. *Proc. Natl. Acad. Sci. USA*, **98**, 14 356–14 361.

Tehei, M., Franzetti, B., Madern, D., *et al.* (2004). Adaptation to extreme environments: Macromolecular dynamics in bacteria compared in vivo by neutron scattering. *EMBO Rep.*, **5**, 66–70.

Teixeira, S. C., Ankner, J., Bellissent-Funel, M. C., *et al.* (2008). New sources and instrumentation for neutrons in biology. *Chem .Phys.*, **345**, 133–151.

Thalhammer, S., Stark, R. W., Müller, S., Weinberg, J., and Heck, W. M. (1997). The atomic force microscope as a new microdissecting tool for the generation of genetic probes. *J. Struct. Biol.*, **119**, 232–237.

Thomas, G. J., and Tsuboi, M. (1993). Raman spectroscopy of nucleic acids and their complexes. *Adv. Biophys. Chem.*, **3**, 1–69.

Tian, F., Al-Hashimi, H. M., Craighead, J. L., and Prestegard, J. H. (2001). Conformational analysis of a flexible oligosaccharide using residual dipolar coupling, *JACS*, **123**, 485–492.

Tinoco, I., Jr., Sauer, K., and Wang, J. C. (1998). *Physical Chemistry: Principles and Applications in Biological Science.* Englewood Cliffs, NJ: Prentice Hall.

Tirado, M. M., Martinez, C. L., and Garcia de la Torre, J. (1984). Comparison of theories for the translational and rotational diffusion coefficients of rod-like macromolecules: Application to short DNA fragments. *J. Chem. Phys.*, **81**, 2047–2051.

Tjandra, N., Tate, S., Ono, A., Kainosho, M., and Bax, A. (2000). The NMR structure of a DNA dodecamer in an aqueous dilute liquid crystalline phase. *JACS*, **122**, 6190–6200.

Tolbert, T. J., and Williamson, J. R. (1996). Preparation of specifically deuterated RNA for NMR studies using a combination of chemical and enzymatic synthesis. *JACS*, **118**, 7929–7940.

Toyoshima, C., Sasabe, H., and Stokes, D. L. (1993). Three-dimensional cryo-electron microscopy of the calcium ion pump in the sarcoplasmic reticulum membrane. *Nature*, **362**, 469–471.

Ulitin, A. B., Agalarov, S. C., and Serdyuk, I. N. (1997). Preparation of a "beheaded" derivative of the 30S ribosomal subunit. *Biochimie*, **79**, 523–526.

Vagin, A., and Teplyakov, A. (2000). An approach to multi-copy search in molecular replacement. *Acta Crystallogr. D Biol. Crystallogr.*, **56**, 1622–1624.

Vale, R. D. (1996). Switches, latches, and amplifier: Common themes of proteins and molecular motors. *J. Cell Biol.*, **135**, 291–302.

Valle, M., Zavialov, A., Sengupta, J., *et al.* (2003). Locking and unlocking of ribosomal motions. *Cell*, **114**, 123–134.

van der Groot, F. G., Gonzàlez-Mānas, J. M., Lakey, J. H., and Pattus, F. (1991). A "molten globule" membrane-insertion intermediate of the pore-forming domain of colicin A. *Nature*, **354**, 408–410.

van Heel, M., Gowen, B., Matadeen, R., *et al.* (2000). Single-particle electron cryo-microscopy: Towards atomic resolution. *Q. Rev. Biophys.*, **33**, 307–369.

Váró, G., and Lanyi, J. K. (1991). Effects of the crystalline structure of purple membrane on the kinetics and energetics of the bacteriorhodopsin photocycle. *Biochemistry*, **30**, 7165–7171.

Velazquez-Campoy, A., Kiso, Y., and Freire, E. (2001). The binding energetics of first- and second-generation HIV-1 protease inhibitors: Implications for drug design. *Arch. Biochem. Biophys.*, **390**, 169–175.

Velazquez-Campoy, A., Leavitt, S. A., and Freire, E. (2004). Characterization of protein–protein interactions by isothermal titration calorimetry. *Methods Mol. Biol.*, **261**, 35–54.

Venyaminov, S. Y. and Vassilenko, K. S. (1994). Determination of protein tertiary structure class from circular dichroism spectra. *Anal. Biochem.*, **222**, 176–184.

Vénien-Bryan, C., and Fuller, S. D. (1994). The organization of the spike complex of Semliki forest virus. *J. Mol. Biol.*, **236**, 572–583.

Walther, D., Cohen, F. E., and Doniak, S. (2000). Reconstruction of low-resolution three dimensional density maps from one dimensional small-angle X-ray solution scattering data for biomolecules. *J. Appl. Crystallogr.*, **33**, 350–363.

Wang, K., Forbes, J. G., and Jin, A. J. (2001). Single molecule measurements of titin elasticity. *Progr. Biophys. Mol. Biol.*, **77**, 1–44.

Wang, M. (1999). Manipulation of single molecules in biology. *Curr. Opin. Biotech.*, **10**, 81–86.

Wang, M. D., Schnitzer, M. J., Yin, H., *et al.* (1998). Force and velosity measured for single molecules of RNA polymerase. *Science*, **282**, 902–907.

Watrous, J. D., Phelan, V. V., Hsu, C. C., *et al.* (2013). Microbial metabolic exchange in 3D. *ISME J.*, **7**, 770–780.

Watson, J. D., and Crick, F. H. (1953). Molecular structure of nucleic acids: A structure for deoxyribose nucleic acid. *Nature*, **171**, 737–738.

Webb, R. H. (1996). Confocal optical microscopy. *Rep. Progr. Phys.*, **59**, 427–471.

Weik, M., Ravelli, R. B., Kryger, G., *et al.* (2000). Specific chemical and structural damage to proteins produced by synchrotron radiation. *Proc. Natl. Acad. Sci. USA*, **97**, 623–628.

Weiskopf, A. S., Vouros, P., and Harvey, D. (1997). Characterization of oligosaccharide composition and structure by quadropole ion trap mass spectrometry. *Rapid Commun. Mass Spectr.*, **11**, 1493–1504.

Weiss, S. (1999). Fluorescence spectroscopy of single biomolecules. *Science*, **283**, 1676–1683.

Weiss, S. (2000). Shattering the diffraction limit of light: A revolution in fluorescence microscopy? *Proc. Nat. Acad. Sci. USA*, **97**, 8747–8749.

Weissman, M., Schindler, H., and Feher, G. (1976). Determination of molecular weights by fluctuation spectroscopy: Application to DNA. *Proc. Nat. Acad. Sci. USA*, **73**, 2776–2780.

Westhof, E., Dumas, P., and Moras, D. (1988). Restrained refinement of two crystalline forms of yeast aspartic acid and phenylalanine transfer RNA crystals. *Acta Cryst.*, **A44**, 122–123.

Wetlaufer, D. B. (1962). Ultraviolet spectra of proteins and nucleic acids. *Adv. Prot. Chem.*, **17**, 303–390.

Wider, G., and Wuthrich, K. (1999). NMR spectroscopy of large molecules and multimolecular assemblies in solution. *Curr. Opin. Struct. Biol.*, **9**, 594–601.

Willcox, B. E., Gao, G. F., Wyer, J., *et al.* (1999). TCR binding to peptide-MHC stabilizes a flexible recognition interface. *Immunity*, **10**, 357–365.

Williams, K. R., and King, R. W. (1990a). The Fourier transform in chemistry: NMR. Part 3. Multiple-pulse experiments. *J. Chem. Educ.*, **67**, A93–A99.

Williams, K. R., and King, R. W. (1990b). The Fourier transform in chemistry: NMR. Part 4. Two-dimensional methods. *J. Chem. Educ.*, **67**, A125–A137.

Wimberly, B. T., Brodersen, D. E., Clemons, W. M., *et al.* (2000). Structure of the 30S ribosomal subunit. *Nature*, **407**, 327–339.

Winston, R. L., and Fitzgerald, M. C. (1997). Mass spectrometry as a readout of protein structure and function. *Mass Spectr. Rev.*, **16**, 165–179.

Woese, C. R. (1967). *The Genetic Code: The Molecular Basis for Genetic Expression*. New York: Harper & Row.

Wuthrich, K. (1986). *NMR of Proteins and Nucleic Acids*, New York: Wiley-Interscience.

Wuthrich, K. (1995). NMR: This other method for protein and nucleic acid structure determination. *Acta Cryst.*, **D51**, 249–270.

Wuthrich, K. (2000). Protein recognition by NMR, *Nat. Struct. Biol.*, **7**, 188–189.

Wuthrich, K. (2001). The way to NMR structures of protein. *Nat. Struct. Biol.*, **8**, 923–925.

Wyer, J. R., Willcox, B. E., Geo, G. F., *et al.* (1999). T cell receptor and coreceptor CD8 alphaalpha bind peptide-MHC independently and with distinct kinetics. *Immunity*, **10**, 219–225.

Xu, Z., Horwich, A. L., and Sigler, P. B. (1997). The crystal structure of the asymmetric GroEL-GroES-(ADP)7 chaperonin complex. *Nature*, **388**, 741–750.

Yan, Y., Winograd, E., Viel, A., *et al.* (1993). Crystal structure of the repetitive segments of spectrin. *Science*, **262**, 2027–2030.

Yanagida, T., Kitamura, K., Tanaka, H., Iwane, A. H., and Esaki, S. (2000). Single molecule analysis of the actomyosin motor. *Curr. Opin. Cell. Biol.*, **12**, 20–25.

Yguerabide, J., Epstein, H. F., and Stryer, L. (1970). Segmental flexibility in an antibody molecule. *J. Mol. Biol.*, **51**, 573–590.

Yonath, A., Miissig, J., Tesche, B., *et al.* (1980). Crystallization of the large ribosomal subunit from *B. stearothermophilus*. *Biochem. Int.*, **1**, 428.

Yoshida, K., Yoshimoto, M., Sasaki, K., *et al.* (1998). Fabrication of new substrate for atomic force microscopic observation of DNA molecules from an ultrasmooth sapphire plate. *Biophys. J.*, **74**, 1654–1657.

Zaccai, G. (2000). How soft is a protein? A protein dynamics force constant measured by neutron scattering. *Science*, **288**, 1604–1607.

Zaccai, G., Buldt, G., Seelig, A., and Seelig, J. (1979a). Neutron diffraction studies on phosphatidylcholine model membranes: II. Chain conformation and segmental disorder. *J. Mol. Biol.*, **134**, 693–706.

Zaccai, G., Morin, P., Jacrot, B., *et al.* (1979b). Interactions of yeast valyl-tRNA synthetase with RNAs and conformational changes of the enzyme. *J. Mol. Biol.*, **129**, 483–500.

Zalk, R., Clarke, O. B., des Georges, A., *et al.* (2015). Structure of a mammalian ryanodine receptor. *Nature*, **517** (7532), 44–49.

Zanni, M., and Hochstrasser, R. (2001). Two-dimensional infrared spectroscopy: A promising new method for the time resolution of structure. *Curr. Opin. Struct. Biol.*, **11**, 516–562.

Zhang, H., Cui, W., and Gross, M. L. (2013). Native electrospray ionization and electron-capture dissociation for comparison of protein structure in solution and the gas phase. *Int. J. Mass. Spectrom.*, 354–355.

Zhang, Q., and Wakelam, M. J. O. (2014). Lipidomics in the analysis of malignancy. *Adv. Biol. Regul.*, **54**, 93–98.

Zheng, R., Zheng, X., Dong, J., and Carey, P. R. (2004). Proteins can convert to β-sheet in single crystals. *Protein Sci.*, **13**, 1288–1294.

Zhou, H.-X. (1995). Calculation of translational friction and intrinsic viscosity: I. General formulation for arbitrarily shaped particles. *Biophys. J.*, **69**, 2286–2297.

Zhou, H.-X. (2001). A unified picture of protein hydration: Prediction of hydrodynamic properties from known structures. *Biophys. Chem.*, **93**, 171–179.

Zidek, L., Stefl, R., and Sklenar, V. (2001). NMR methodology for the study of nucleic acids. *Curr. Opin. Struct. Biol.*, **11**, 275–281.

Zipper, P., and Durchshlag, H. (2000). Prediction of hydrodynamic and small angle scattering parameters from crystal and electron microscopic structure. *J. Appl. Cryst.*, **33**, 788–792.

Zlatanova, J., Lindsay, S. M., and Leuba, A. H. (2000). Single molecule force spectroscopy in biology using the atomic force microscope. *Progr. Biophys. Mol. Biol.*, **74**, 37–61.

Zubarev, R. A., and Makarov, A. (2013). Orbitrap mass spectrometry. *Anal. Chem.*, **85**, 5288–5296.

AUTHOR INDEX

Luria, S. E., 489
Luzzati, V., 441

MacKinnon, R., 456
MacVicar, J., 639
Magde, D., 162, 230
Mair, G. A., 456
Mamyrin, B., 71
Manavalan, P., 315
Mandelshtam, L., 161, 229, 288
Mansfield, P., 546, 648
Marion, D., 546
Marshall, A., 71
Maxwell, J. C., 115, 161, 489
McCall, D. W., 545
McCammon, J. A., 43, 521
Mellors, R. C., 253
Meselson, M., 184, 207
Meyerdirk, R., 639
Michel, H., 456
Miles, H. T., 253
Miles, M. J., 336, 348
Millar, D. P., 216
Minsky, M., 326
Mitchell, D. P., 407, 528, 531
Miyazawa, T., 253, 280
Moerner, W. E., 385–386
Moffit, W., 307
Moore, P. B., 457
Moore, W., 648
Müller, D. J., 336

Nernst, W., 115
Newton, I., 1, 161, 173, 253
Nicholls, G., 325
Nichols, J. B., 184
Nierhaus, K., 457
Noji, H., 335, 528
Norberg, R. E., 545
North, A. C. T., 456
Nutt, R., 629

Ohtani, H., 335
Opella, S. J., 546, 610–611
Orgel, L. E., 26
Orrit, M., 385
Oschkinat, H., 546
Ostwald, W., 13
Overhauser, A. W., 545, 609

Parak, F., 44
Pardon, J. F., 457
Pasteur, L., 25, 306
Pastor, R., 162

Pauli, W., 71, 217, 253, 545
Pauling, L., 307, 369, 456
Pecora, R., 229–230, 238
Pedersen, K. O., 184
Perrin, F., 215
Perrin, J., 2, 184, 215
Perutz, M., xvii, 1, 25, 60, 408, 456
Petran, M., 326
Philips, D. C., 456
Philpot, J. St. L., 184
Placzek, G., 229
Planck, M., 253
Pohl, D., 325
Poisson, S.-D., 19
Porod, G., 426
Powers, P. N., 407, 531
Preiswerk, P., 407, 531
Priestley, J., 25
Prigogine, I., 116
Pringsheim, P., 215
Privalov, P. L., 116, 126, 130, 134
Proctor, W. G., 545
Ptitsyn, O. B., 317
Pulyui, I., 629
Purcell, E. M., 161, 163, 545, 648
Putman, C. A. J., 335
Pythagoras, 43

Quate, C. F., 335–336

Rabi, I. I., 545
Radmacher, M., 335
Radon, J., 629
Raether, H. Z., 149
Rahman, A., 521
Ramakrishnan, V., 456–457
Raman, C. V., 288
Ramanadham, M., 457
Ramsey, N. F., 545
Randall, M., 13, 19
Raoult, F.-M., 13
Reid, J., 639
Reynolds, O., 161
Rief, M., 335, 380
Rigler, R., 162, 230, 385
Ritter, J. W., 253
Roberts, F., 325
Robertson, J., 629
Rohrer, H., 9, 335–336
Röntgen, W. C., 2, 407, 456, 629
Rossman, M. G., 31, 456–457, 466
Rotman, B., 384
Rowe, A. J., 161

Ruska, H., 489
Rutherford, E., 253

Sanford, F., 629
Sanger, F., 25
Sarma, V. R., 456
Sauders, M., 546
Saunders, J. K., 545
Savart, F., 489
Schachman, H. K., 184–185
Scheraga, H., 521
Schoenborn, B., 457
Schrimer, R. E., 545
Schrödinger, E., 2, 9, 253
Schuck, P., 193, 196, 205
Schwille, P., 230
Senebier, J., 25
Sheetz, M. P., 173
Shivashankar, G. V., 335
Shull, C., 408, 456, 531
Simha, R., 161
Simmons, R. M., 358
Singer, S. J., 489
Singh, B. R., 254
Singh, S., 384
Skou, J. C., 456, 528
Smoluchowski, M., 229, 441
So, P., 384
Spudlish, J. A., 335
Stahl, F. W., 184, 207
Staudinger, H., 2
Steinwedel, H., 71
Steitz, T. A., 456–457
Stern, O., 545
Stillinger, F. H., 521
Stokes, A. R., 1
Stokes, G. G., 2, 161, 166–167
Strandness, G., 639
Stryer, L., 215, 384, 396
Stuart, D. I., 457
Stuhrmann, H. B., 457
Sturtevant, J. M., 116, 126
Sumner, J. B., 25
Svedberg, T., 2, 162, 184, 195
Svenson, H., 184
Svergun, D. I., 435
Svoboda, K., 335
Synge, E., 325

Talbot, H. F., 325
Tanaka, K., 71
Tao, T., 215
Tardieu, A., 441, 443
Teller, D., 162
Thomson, G. P., 489

SUBJECT INDEX